Handbook of
Thin Film Materials

Handbook of Thin Film Materials

Volume 3
Ferroelectric and Dielectric Thin Films

Edited by

Hari Singh Nalwa, M.Sc., Ph.D.
Stanford Scientific Corporation
Los Angeles, California, USA

*Formerly at
Hitachi Research Laboratory
Hitachi Ltd., Ibaraki, Japan*

ACADEMIC PRESS

A Division of Harcourt, Inc.

San Diego   San Francisco   New York   Boston   London   Sydney   Tokyo

ACADEMIC PRESS
A division of Harcourt, Inc.
525 B Street, Suite 1900, San Diego, CA 92101-4495, USA
http://www.academicpress.com

Academic Press
Harcourt Place, 32 Jamestown Road, London, NW1 7BY, UK
http://www.academicpress.com

Library of Congress Catalog Card Number: 00-2001090614
International Standard Book Number, Set: 0-12-512908-4
International Standard Book Number, Volume 3: 0-12-512911-4

Printed in the United States of America
01 02 03 04 05 06 07 MB 9 8 7 6 5 4 3 2 1

100266 6983

To my children
Surya, Ravina, and Eric

Preface

Thin film materials are the key elements of continued technological advances made in the fields of electronic, photonic, and magnetic devices. The processing of materials into thin-films allows easy integration into various types of devices. The thin film materials discussed in this handbook include semiconductors, superconductors, ferroelectrics, nanostructured materials, magnetic materials, etc. Thin film materials have already been used in semiconductor devices, wireless communication, telecommunications, integrated circuits, solar cells, light-emitting diodes, liquid crystal displays, magneto-optic memories, audio and video systems, compact discs, electro-optic coatings, memories, multilayer capacitors, flat-panel displays, smart windows, computer chips, magneto-optic disks, lithography, microelectromechanical systems (MEMS) and multifunctional protective coatings, as well as other emerging cutting edge technologies. The vast variety of thin film materials, their deposition, processing and fabrication techniques, spectroscopic characterization, optical characterization probes, physical properties, and structure-property relationships compiled in this handbook are the key features of such devices and basis of thin film technology.

Many of these thin film applications have been covered in the five volumes of the *Handbook of Thin Film Devices* edited by M. H. Francombe (Academic Press, 2000). The *Handbook of Thin Film Materials* is complementary to that handbook on devices. The publication of these two handbooks, selectively focused on thin film materials and devices, covers almost every conceivable topic on thin films in the fields of science and engineering.

This is the first handbook ever published on thin film materials. The 5-volume set summarizes the advances in thin film materials made over past decades. This handbook is a unique source of the in-depth knowledge of deposition, processing, spectroscopy, physical properties, and structure–property relationship of thin film materials. This handbook contains 65 state-of-the-art review chapters written by more than 125 world-leading experts from 22 countries. The most renowned scientists write over 16,000 bibliographic citations and thousands of figures, tables, photographs, chemical structures, and equations. It has been divided into 5 parts based on thematic topics:

Volume 1: Deposition and Processing of Thin Films
Volume 2: Characterization and Spectroscopy of Thin Films
Volume 3: Ferroelectric and Dielectric Thin Films
Volume 4: Semiconductor and Superconductor Thin Films
Volume 5: Nanomaterials and Magnetic Thin Films

Volume 1 has 14 chapters on different aspects of thin film deposition and processing techniques. Thin films and coatings are deposited with chemical vapor deposition (CVD), physical vapor deposition (PVD), plasma and ion beam techniques for developing materials for electronics, optics, microelectronic packaging, surface science, catalytic, and biomedical technological applications. The various chapters include: methods of deposition of hydrogenated amorphous silicon for device applications, atomic layer deposition, laser applications in transparent conducting oxide thin film processing, cold plasma processing in surface science and technology, electrochemical formation of thin films of binary III–V compounds, nucleation, growth and crystallization of thin films, ion implant doping and isolation of GaN and related materials, plasma etching of GaN and related materials, residual stresses in physically vapor deposited thin films, Langmuir–Blodgett films of biological molecules, structure formation during electrocrystallization of metal films, epitaxial thin films of intermetallic compounds, pulsed laser deposition of thin films: expectations and reality and b″-alumina single-crystal films. This volume is a good reference source of information for those individuals who are interested in the thin film deposition and processing techniques.

Volume 2 has 15 chapters focused on the spectroscopic characterization of thin films. The characterization of thin films using spectroscopic, optical, mechanical, X-ray, and electron microscopy techniques. The various topics in this volume include: classification of cluster morphologies, the band structure and orientations of molecular adsorbates on surfaces by angle-resolved electron spectroscopies, electronic states in GaAs-AlAs short-period superlattices: energy levels and symmetry, ion beam characterization in superlattices, *in situ* real time spectroscopic ellipsometry studies: carbon-based materials and metallic TiNx thin films growth, *in situ* Faraday-modulated fast-nulling single-wavelength ellipsometry of the growth of semiconductor, dielectric and metal thin films, photocurrent spectroscopy of thin passive films, low frequency noise spectroscopy for characterization of polycrystalline semiconducting thin films and polysilicon thin film transistors, electron energy loss spectroscopy for surface study, theory of low-energy electron diffraction and photoelectron spectroscopy from ultra-thin films, *in situ* synchrotron structural studies of the growth of oxides and metals, operator formalism in polarization nonlinear optics and spectroscopy of polarization inhomogeneous media, secondary ion mass spectrometry (SIMS) and its application to thin films characterization, and a solid state approach to Langmuir monolayers, their phases, phase transitions and design.

Volume 3 focuses on dielectric and ferroelectric thin films which have applications in microelectronics packaging, ferroelectric random access memories (FeRAMs), microelectromechanical systems (MEMS), metal–ferroelectric–semiconductor field-effect transistors (MFSFETs), broad band wireless communication, etc. For example, the ferroelectric materials such as barium strontium titanate discussed in this handbook have applications in a number of tunable circuits. On the other hand, high-permittivity thin film materials are used in capacitors and for integration with MEMS devices. Volume 5 of the *Handbook of Thin Film Devices* summarizes applications of ferroelectrics thin films in industrial devices. The 12 chapters on ferroelectrics thin films in this volume are complimentary to Volume 5 as they are the key components of such ferroelectrics devices. The various topics include electrical properties of high dielectric constant and ferroelectrics thin films for very large scale integration (VLSI) integrated circuits, high permittivity (Ba, Sr)TiO$_3$ thin films, ultrathin gate dielectric films for Si-based microelectronic devices, piezoelectric thin films: processing and properties, fabrication and characterization of ferroelectric oxide thin films, ferroelectric thin films of modified lead titanate, point defects in thin insulating films of lithium fluoride for optical microsystems, polarization switching of ferroelecric crystals, high temperature superconductor and ferroelectrics thin films for microwave applications, twinning in ferroelectrics thin films: theory and structural analysis, and ferroelectrics polymers Langmuir–Blodgett films.

Volume 4 has 13 chapters dealing with semiconductor and superconductor thin film materials. Volumes 1, 2, and 3 of the *Handbook of Thin Film Devices* summarize applications of semiconductor and superconductors thin films in various types of electronic, photonic and electro-optics devices such as infrared detectors, quantum well infrared photodetectors (QWIPs), semiconductor lasers, quantum cascade lasers, light emitting diodes, liquid crystal and plasma displays, solar cells, field effect transistors, integrated circuits, microwave devices, SQUID magnetometers, etc. The semiconductor and superconductor thin film materials discussed in this volume are the key components of such above mentioned devices fabricated by many industries around the world. Therefore this volume is in coordination to Volumes 1, 2, and 3 of the *Handbook of Thin Film Devices*. The various topics in this volume include; electrochemical passivation of Si and SiGe surfaces, optical properties of highly excited (Al, In)GaN epilayers and heterostructures, electical conduction properties of thin films of cadmium compounds, carbon containing heteroepitaxial silicon and silicon/germanium thin films on Si(001), germanium thin films on silicon for detection of near-infrared light, physical properties of amorphous gallium arsenide, amorphous carbon thin films, high-T_c superconducting thin films, electronic and optical properties of strained semiconductor films of group V and III-V materials, growth, structure and properties of plasma-deposited amorphous hydrogenated carbon–nitrogen films, conductive metal oxide thin films, and optical properties of dielectric and semiconductor thin films.

Volume 5 has 12 chapters on different aspects of nanostructured materials and magnetic thin films. Volume 5 of the *Handbook of Thin Film Devices* summarizes device applications of magnetic thin films in permanent magnets, magneto-optical recording, microwave, magnetic MEMS, etc. Volume 5 of this handbook on magnetic thin film materials is complimentary to Volume 5 as they are the key components of above-mentioned magnetic devices. The various topics covered in this volume are; nanoimprinting techniques, the energy gap of clusters, nanoparticles and quantum dots, spin waves in thin films, multi-layers and superlattices, quantum well interference in double quantum wells, electro-optical and transport properties of quasi-two-dimensional nanostrutured materials, magnetism of nanoscale composite films, thin magnetic films, magnetotransport effects in semiconductors, thin films for high density magnetic recording, nuclear resonance in magnetic thin films, and multilayers, and magnetic characterization of superconducting thin films.

I hope these volumes will be very useful for the libraries in universities and industrial institutions, governments and independent institutes, upper-level undergraduate and graduate students, individual research groups and scientists working in the field of thin films technology, materials science, solid-state physics, electrical and electronics engineering, spectroscopy, superconductivity, optical engineering, device engineering nanotechnology, and information technology, everyone who is involved in science and engineering of thin film materials.

I appreciate splendid cooperation of many distinguished experts who devoted their valuable time and effort to write excellent state-of-the-art review chapters for this handbook. Finally, I have great appreciation to my wife Dr. Beena Singh Nalwa for her wonderful cooperation and patience in enduring this work, great support of my parents Sri Kadam Singh and Srimati Sukh Devi and love of my children, Surya, Ravina and Eric in this exciting project.

Hari Singh Nalwa
Los Angeles, CA, USA

Contents

Chapter 1. THE ELECTRICAL PROPERTIES OF HIGH-DIELECTRIC-CONSTANT AND FERROELECTRIC THIN FILMS FOR VERY LARGE SCALE INTEGRATION CIRCUITS

Joseph Ya-min Lee, Benjamin Chihming Lai

Chapter 2. HIGH-PERMITTIVITY $(Ba, Sr)TiO_3$ THIN FILMS

M. Nayak, S. Ezhilvalavan, T. Y. Tseng

Chapter 3.   ULTRATHIN GATE DIELECTRIC FILMS FOR Si-BASED MICROELECTRONIC DEVICES

C. Krug, I. J. R. Baumvol

Chapter 4.   PIEZOELECTRIC THIN FILMS: PROCESSING AND PROPERTIES

Floriana Craciun, Patrizio Verardi, Maria Dinescu

**Chapter 5.   FABRICATION AND CHARACTERIZATION OF
FERROELECTRIC OXIDE THIN FILMS**

Jong-Gul Yoon, Tae Kwon Song

Chapter 6.   FERROELECTRIC THIN FILMS OF MODIFIED LEAD TITANATE

J. Mendiola, M. L. Calzada

**Chapter 7.   POINT DEFECTS IN THIN INSULATING FILMS OF LITHIUM
FLUORIDE FOR OPTICAL MICROSYSTEMS**

Rosa Maria Montereali

Chapter 8. POLARIZATION SWITCHING OF FERROELECTRIC CRYSTALS

Lung-Han Peng

Chapter 9. HIGH-TEMPERATURE SUPERCONDUCTOR AND FERROELECTRIC THIN FILMS FOR MICROWAVE APPLICATIONS

Félix A. Miranda, Joseph D. Warner, Guru Subramanyam

Chapter 12. OPTICAL PROPERTIES OF DIELECTRIC AND SEMICONDUCTOR THIN FILMS

I. Chambouleyron, J. M. Martínez

About the Editor

Dr. Hari Singh Nalwa is the Managing Director of the Stanford Scientific Corporation in Los Angeles, California. Previously, he was Head of Department and R&D Manager at the Ciba Specialty Chemicals Corporation in Los Angeles (1999–2000) and a staff scientist at the Hitachi Research Laboratory, Hitachi Ltd., Japan (1990–1999). He has authored over 150 scientific articles in journals and books. He has 18 patents, either issued or applied for, on electronic and photonic materials and devices based on them.

He has published 43 books including *Ferroelectric Polymers* (Marcel Dekker, 1995), *Nonlinear Optics of Organic Molecules and Polymers* (CRC Press, 1997), *Organic Electroluminescent Materials and Devices* (Gordon & Breach, 1997), *Handbook of Organic Conductive Molecules and Polymers*, Vols. 1–4 (John Wiley & Sons, 1997), *Handbook of Low and High Dielectric Constant Materials and Their Applications*, Vols. 1–2 (Academic Press, 1999), *Handbook of Nanostructured Materials and Nanotechnology*, Vols. 1–5 (Academic Press, 2000), *Handbook of Advanced Electronic and Photonic Materials and Devices*, Vols. 1–10 (Academic Press, 2001), *Advanced Functional Molecules and Polymers*, Vols. 1–4 (Gordon & Breach, 2001), *Photodetectors and Fiber Optics* (Academic Press, 2001), *Silicon-Based Materials and Devices*, Vols. 1–2 (Academic Press, 2001), *Supramolecular Photosensitive and Electroactive Materials* (Academic Press, 2001), *Nanostructured Materials and Nanotechnology*–Condensed Edition (Academic Press, 2001), and *Handbook of Thin Film Materials*, Vols. 1–5 (Academic Press, 2002). The *Handbook of Nanostructured Materials and Nanotechnology* edited by him received the 1999 Award of Excellence in Engineering Handbooks from the Association of American Publishers.

Dr. Nalwa is the founder and Editor-in-Chief of the *Journal of Nanoscience and Nanotechnology* (2001–). He also was the founder and Editor-in-Chief of the *Journal of Porphyrins and Phthalocyanines* published by John Wiley & Sons (1997–2000) and serves or has served on the editorial boards of *Journal of Macromolecular Science-Physics* (1994–), *Applied Organometallic Chemistry* (1993–1999), *International Journal of Photoenergy* (1998–) and *Photonics Science News* (1995–). He has been a referee for many international journals including *Journal of American Chemical Society, Journal of Physical Chemistry, Applied Physics Letters, Journal of Applied Physics, Chemistry of Materials, Journal of Materials Science, Coordination Chemistry Reviews, Applied Organometallic Chemistry, Journal of Porphyrins and Phthalocyanines, Journal of Macromolecular Science-Physics, Applied Physics, Materials Research Bulletin*, and *Optical Communications*.

Dr. Nalwa helped organize the First International Symposium on the Crystal Growth of Organic Materials (Tokyo, 1989) and the Second International Symposium on Phthalocyanines (Edinburgh, 1998) under the auspices of the Royal Society of Chemistry. He also proposed a conference on porphyrins and phthalocyanies to the scientific community that, in part, was intended to promote public awareness of the *Journal of Porphyrins and Phthalocyanines*, which he founded in 1996. As a member of the organizing committee, he helped effectuate the First International Conference on Porphyrins and Phthalocyanines, which was held in Dijon, France in 2000. Currently he is on the organizing committee of the BioMEMS and Smart Nanostructures, (December 17–19, 2001, Adelaide, Australia) and the World Congress on Biomimetics and Artificial Muscles (December 9–11, 2002, Albuquerque, USA).

Dr. Nalwa has been cited in the *Dictionary of International Biography, Who's Who in Science and Engineering, Who's Who in America,* and *Who's Who in the World.* He is a member of the American Chemical Society (ACS), the American Physical Society (APS), the Materials Research Society (MRS), the Electrochemical Society and the American Association for the Advancement of Science (AAAS). He has been awarded a number of prestigious fellowships including a National Merit Scholarship, an Indian Space Research Organization (ISRO) Fellowship, a Council of Scientific and Industrial Research (CSIR) Senior fellowship, a NEC fellowship, and Japanese Government Science & Technology Agency (STA) Fellowship. He was an Honorary Visiting Professor at the Indian Institute of Technology in New Delhi.

Dr. Nalwa received a B.Sc. degree in biosciences from Meerut University in 1974, a M.Sc. degree in organic chemistry from University of Roorkee in 1977, and a Ph.D. degree in polymer science from Indian Institute of Technology in New Delhi in 1983. His thesis research focused on the electrical properties of macromolecules. Since then, his research activities and professional career have been devoted to studies of electronic and photonic organic and polymeric materials. His endeavors include molecular design, chemical synthesis, spectroscopic characterization, structure-property relationships, and evaluation of novel high performance materials for electronic and photonic applications. He was a guest scientist at Hahn-Meitner Institute in Berlin, Germany (1983) and research associate at University of Southern California in Los Angeles (1984–1987) and State University of New York at Buffalo (1987–1988). In 1988 he moved to the Tokyo University of Agriculture and Technology, Japan as a lecturer (1988–1990), where he taught and conducted research on electronic and photonic materials. His research activities include studies of ferroelectric polymers, nonlinear optical materials for integrated optics, low and high dielectric constant materials for microelectronics packaging, electrically conducting polymers, electroluminescent materials, nanocrystalline and nanostructured materials, photocuring polymers, polymer electrets, organic semiconductors, Langmuir-Blodgett films, high temperature-resistant polymer composites, water-soluble polymers, rapid modeling, and stereolithography.

List of Contributors

Numbers in parenthesis indicate the pages on which the author's contribution begins.

S. PAMIR ALPAY (517)
Department of Metallurgy and Materials Engineering, University of Connecticut,
Storrs, Connecticut, USA

I. J. R. BAUMVOL (169)
Instituto de Física, Universidade Federal do Rio Grande do Sul, Porto Alegre,
RS 91509-900, Brazil

M. L. CALZADA (369)
Instituto Ciencia de Materiales de Madrid (CSIC), Cantoblanco, 28049 Madrid, Spain

I. CHAMBOULEYRON (593)
Institute of Physics "Gleb Wataghin," State University of Campinas—UNICAMP,
13073-970 Campinas, São Paulo, Brazil

FLORIANA CRACIUN (231)
Istituto di Acustica "O.M. Corbino," Consiglio Nazionale delle Ricerche, 00133 Rome, Italy

MARIA DINESCU (231)
National Institute for Lasers, Plasma and Radiaton Physics, Institute of Atomic Physics,
Bucharest, Romania

STEPHEN DUCHARME (545)
Department of Physics and Astronomy, Center for Materials Research and Analysis,
University of Nebraska, Lincoln, Nebraska, USA

S. EZHILVALAVAN (99)
Department of Electronics Engineering and Institute of Electronics,
National Chiao-Tung University, Hsinchu, Taiwan, Republic of China

V. M. FRIDKIN (545)
Shubnikov Institute of Crystallography, Russian Academy of Sciences,
Moscow 117333, Russia

C. KRUG (169)
Instituto de Física, Universidade Federal do Rio Grande do Sul, Porto Alegre,
RS 91509-900, Brazil

BENJAMIN CHIHMING LAI (1)
Department of Electrical Engineering and Institute of Electronics, Tsing-Hua University,
Hsinchu, Taiwan, Republic of China

JOSEPH YA-MIN LEE (1)
Department of Electrical Engineering and Institute of Electronics, Tsing-Hua University,
Hsinchu, Taiwan, Republic of China

J. M. Martínez (593)
Institute of Mathematics and Computer Science, State University of Campinas—UNICAMP, 13083-970 Campinas, São Paulo, Brazil

J. Mendiola (369)
Instituto Ciencia de Materiales de Madrid (CSIC), Cantoblanco, 28049 Madrid, Spain

Félix A. Miranda (481)
NASA Glenn Research Center, Communication Technology Division, Cleveland, Ohio, USA

Rosa Maria Montereali (399)
ENEA C.R. Frascati, Applied Physics Division, 00044 Frascati (RM), Italy

M. Nayak (99)
Department of Electronics Engineering and Institute of Electronics, National Chiao-Tung University, Hsinchu, Taiwan, Republic of China

S. P. Palto (545)
Shubnikov Institute of Crystallography, Russian Academy of Sciences, Moscow 117333, Russia

Lung-Han Peng (433)
Department of Electrical Engineering, National Taiwan University, Taipei, Taiwan, Republic of China

Tae Kwon Song (309)
Department of Ceramic Science and Engineering, Changwon National University, Changwon, Kyungnam 641-773, Korea

Guru Subramanyam (481)
Department of Electrical and Computer Engineering, University of Dayton, Dayton, Ohio, USA

T. Y. Tseng (99)
Department of Electronics Engineering and Institute of Electronics, National Chiao-Tung University, Hsinchu, Taiwan, Republic of China

Patrizio Verardi (231)
Istituto di Acustica "O.M. Corbino," Consiglio Nazionale delle Ricerche, 00133 Rome, Italy

Joseph D. Warner (481)
NASA Glenn Research Center, Communication Technology Division, Cleveland, Ohio, USA

Jong-Gul Yoon (309)
Department of Physics, University of Suwon, Hwaseung, Kyung-gi-do 445-743, Korea

Handbook of Thin Film Materials

Edited by H.S. Nalwa

Volume 1. DEPOSITION AND PROCESSING OF THIN FILMS

Volume 2. CHARACTERIZATION AND SPECTROSCOPY OF THIN FILMS

Volume 3. FERROELECTRIC AND DIELECTRIC THIN FILMS

Volume 4. SEMICONDUCTOR AND SUPERCONDUCTING THIN FILMS

Volume 5. NANOMATERIALS AND MAGNETIC THIN FILMS

Chapter 1

THE ELECTRICAL PROPERTIES OF HIGH-DIELECTRIC-CONSTANT AND FERROELECTRIC THIN FILMS FOR VERY LARGE SCALE INTEGRATION CIRCUITS

Joseph Ya-min Lee, Benjamin Chihming Lai

Department of Electrical Engineering and Institute of Electronics, Tsing-Hua University, Hsinchu, Taiwan, Republic of China

Contents

Handbook of Thin Film Materials, edited by H.S. Nalwa
Volume 3: Ferroelectric and Dielectric Thin Films
Copyright © 2002 by Academic Press
All rights of reproduction in any form reserved.

ISBN 0-12-512911-4/$35.00

1. INTRODUCTION

1.1. Scaling of Very Large Scale Integration Circuits

Dielectric films serve important functions for semiconductor devices and integrated circuits. The gate dielectric is an important ingredient of the metal–oxide–semiconductor field effect transistor (MOSFET). The capacitor dielectric is one of the two components of a dynamic random-access memory (DRAM) cell. Dielectric regions on the surfaces of a water provide the necessary isolation between active devices. The interlayer dielectric separates different metallic layers on the surface of integrated circuits. These dielectric films are indispensable for the successful operation of solid-state semiconductor devices. As the device dimension (represented by the minimum feature size) of very large scale integration (VLSI) integrated circuits continues to scale down, several problems associated with the dielectric films become apparent [1, 2]: (1) When the SiO$_2$ gate oxide thickness becomes very thin (down to about 2.5 nm in deep submicrometer devices), the phenomenon of carrier direct tunneling becomes significant. (2) The area of the DRAM unit cell scales much faster than the charges stored in the capacitor of the DRAM cell. This will cause significant problems for the fabrication of a DRAM storage capacitor.

The unit cell of a DRAM chip consists of a transistor and a storage capacitor. The charge stored in the capacitor Q is given by

$$Q = \varepsilon_r \varepsilon_0 A / t$$

where Q is the charge stored in the capacitor, A is the area of the capacitor, ε_r is the relative dielectric constant of the ca-pacitor dielectric, ε_0 is the permittivity of the vacuum, and t is the thickness of the dielectric film. In the earlier generations of DRAM chips, silicon dioxide was always used as the capacitor dielectric. Because silicon dioxide has a relatively small dielectric constant of 3.9, the stored charge is limited. Starting with the 256-Kb DRAM generation, Si$_3$N$_4$–SiO$_2$ sandwich dielectric layers began to be used [1, 2]. Beginning with 1-Mb DRAM, the planar structure of the capacitor was no longer applicable. To increase the stored charge density, the approach taken in the 4- to 64-Mb DRAM generations was a combination of the following methods: (1) reducing the thickness of the capacitor dielectric and (2) increasing the capacitor area A by using either stacked or trenched capacitor structures. Table I shows the trend of DRAM scaling. It can be seen that, beginning with the 256-Mb DRAM generation, capacitors with high-dielectric-constant films will have to be employed [12]. The stacked capacitor approach suffers from the problem of depth focus during VLSI processing. The trench capacitor approach suffers from process complexity and leakage current. The most attractive alternative for future DRAM chips is to increase the relative dielectric constant of the capacitor dielectric. The application of silicon nitride Si$_3$N$_4$, which has a dielectric constant of 7, will increase the stored charge density to some extent. Nitride–oxide (NO) and oxide–nitride–oxide (ONO) films have been used since the generation of 256-Kb DRAM. However, the gain from silicon nitride is relatively small and will not be sufficient for future DRAMs. The application of higher dielectric constant dielectric films in future memory devices will be necessary.

Table I.   Trend of DRAM Scaling

Memory capacity (bit)	1 Mb	4 Mb	16 Mb	64 Mb	256 Mb	1 Gb	4 Gb	16 Gb	Scaling per generation
Design rule (μm) [3–5]	1.2	0.8	0.5	0.18–0.35	0.18	0.15	0.13	0.1	0.7
Lithography, light source [5]	Hg G-line 436 nm	Hg G-line 436 nm	Hg I-line 365 nm	Kr–F+ phase shift mask 248 nm	Kr–F+ phase shift mask 248 nm	Ar–F+ phase shift mask 193 nm	Ar–F+ phase shift mask 193 nm	Ar–F or F$_2$+ phase shift mask 157 nm	—
Chip size (mm^2) [4–6]	60	90	150	100–200	130–300	450	480	526	1.7
Cell size (μm^2) [5–9]	30	10	3.6	1.3	0.5	0.125–0.28	0.065–0.12	0.031	0.33
Cell structures [3, 4]	Planar, STC, trench	STC, trench	STC, trench	Rugged STC, trench	HSGSTC + cylindrical, planar,	HSGSTC + cylindrical trench	HSGSTC + cylindrical, trench	HSGSTC + cylindrical trench	—
Gate oxide thickness (nm) [4, 10]	23	12–15	10–12	7–9	5–6	3.5	2.5	—	0.7
Row address strobe T_{RAS} (ns) [4–6]	100	70	60	45	35	30	25	—	0.75
Dielectric materials [8]	ON, ONO	ONO, NO	ON, ONO	ON, ONO	ON, Ta$_2$O$_5$, BST	Ta$_2$O$_5$, BST	BST	BST, high ε	—
Cell capacitance (fF) [4, 6]	50	40	30	30	25	25	25	—	—
Power supply (V) [4, 5]	5.0	5.0	5.0/3.3	3.3	2.5/3.3	1.8/2.0	1.2/1.5	1.2/1.5	0.7
ISSCC reported year [4, 8, 11]	1984–1986	1986–1987, 1991	1987–1990	1991–1992	1993–1996	1995	1996	—	2–3
Production year [4, 5, 11]	1987	1990	1993	1996	1999	2002	2005	2008	3

STC, stacked capacitor; HSG, hemispherical grain; ON, oxide–nitride; ONO, oxide–nitride–oxide.

There are two kinds of high-dielectric-constant materials that have been actively pursued for this application. The first group consists of the "simple" high-dielectric-constant materials with an atomic composition similar to that of SiO$_2$. These dielectric materials are compounds of two elements, a single metal and oxygen. These oxides include Ta$_2$O$_5$, TiO$_2$, Al$_2$O$_3$, Y$_2$O$_3$, Gd$_2$O$_3$, HfO$_2$, and ZrO$_2$. They have a dielectric constant ranging from 8 to about 160. Table II shows the electrical prop-erties of some of the simple high-dielectric-constant materials. However, because the energy bandgaps of these single metal oxides are usually less than that of SiO$_2$ (9 eV), the dielectric strength and the leakage current properties are therefore usually inferior compared with those of SiO$_2$. Of all the simple high-dielectric-constant materials, the most important and intensively studied is Ta$_2$O$_5$. The electrical properties of Ta$_2$O$_5$ will be discussed in Section 2. The application of the other

Table II.   Physical Parameters of "Simple" High-Dielectric-Constant Films

Dielectrics	SiO_2	Si_3N_4	Al_2O_3	ZrO_2	HfO_2	Ta_2O_5	TiO_2
Relative dielectric constant	3.9	7.5	8	20–22	25–40	25–60	20–160
Energy bandgap (eV)	9.0	5.1	7.0–8.0	5.2–7.8	5.68	3.9–4.4	3.0–3.5
Breakdown field (MV/cm)	>10	>7	6–8	3–4	~4	5–6	2–3

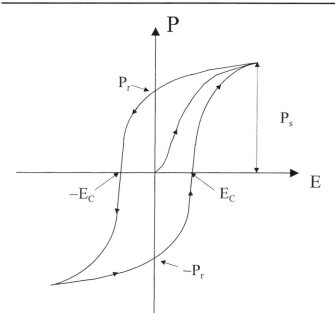

Fig. 1. A schematic polarization–electric field (P–E) curve of ferroelectric materials.

simple high-dielectric-constant materials will be presented in Sections 3–5.

The second group of high-dielectric-constant materials consists of the ferroelectric thin films [13–19]. A ferroelectric material exhibits an electric dipole moment even in the absence of an external electric field. The electric polarization of the ferroelectric material can be reversed by an applied electric field. The ferroelectric materials are noted for their nonlinear polarization–electric field (P–E) curve, which shows hysteresis characteristics. A schematic polarization–electric field (P–E) curve is shown in Fig. 1, where P_s is spontaneous polarization, P_r is remanent polarization, and E_c is the coercive field. The dielectric constants of ferroelectric materials are very large in bulk form, often exceeding 1000, and still on the order of several hundreds in the form of thin films. The ferroelectric materials actively pursued for VLSI applications are mostly of the perovskite crystal structure. The perovskite structure has the atomic composition of ABO_3, where A represents a cation with a larger ionic radius, B represents a cation with a smaller radius, and O is oxygen [13, 16]. The ferroelectric materials will show ferroelectric properties below a critical temperature, the Curie temperature T_C, and show paraelectric properties similar to

those of the conventional dielectrics above the Curie temperature. (This is mostly, but not always the case. Some ferroelectric materials have more than one transition temperature. For example, the first discovered ferroelectric material, Rochelle salt, has two Curie temperatures. The ferroelectric state is confined to the region between approximately −18°C and +24°C [13].) The ferroelectric materials that show high application potential are lead zirconate titanate (PZT), barium strontium titanate (BST), and strontium bismuth tantalate (SBT). In the following sections, we will refer to the single metal oxides simply as the "high-dielectric-constant materials" and designate the perovskite-structured materials as the "ferroelectric materials," including those that show either ferroelectric or paraelectric properties at room temperature.

The ferroelectric materials have two potential applications in VLSI memory devices [15–22]. As we mentioned earlier, the ferroelectric materials have very large dielectric constant in bulk form and still show quite large dielectric constant on the order of several hundred in thin-film form. As a result, these ferroelectric films can be used as the capacitor dielectric for DRAM devices.

There is another unique application of these ferroelectric materials in memory devices. Only those materials that show ferroelectric properties in the operation temperature range are applicable. Due to their hysteresis property in the polarization–electric field (P–E) curve, the ferroelectric materials have two stable states at zero electric field. This provides a natural memory element for nonvolatile memories. This approach was taken early in the history of semiconductor device development [23–25]. However, this approach was not very successful due to the difficulty of material preparation and the required operation voltage. Interest in this approach has been revitalized due to the improvement of ferroelectric thin-film preparation and by the development of ferroelectric random-access memory (FRAM) [15–19].

Several nonvolatile memory technologies are available, including read-only memory (ROM), metal–nitride–oxide–semiconductor (MNOS) memory, erasable programmable read-only memory (EPROM), electrically erasable programmable read-only memory (EEPROM), and flash memory [20]. However, there is no single technology that can satisfy all the requirements. Because ferroelectric materials exhibit at least two stable states that can be switched back and forth, it should be feasible to build a ferroelectric memory device that can store information in digital form [21]. The applications of ferroelectric materials in DRAM and FRAM are discussed in Sections 6–10. The application of PZT is discussed in Section 6, BST in Section 7, and SBT in Section 8. Ferroelectric materials other than PZT, BST, and SBT are discussed in Section 9. Applications utilizing the metal–ferroelectric–insulator–semiconductor (MFIS) structure are discussed in Section 10.

In addition to these applications, high-dielectric-constant materials can potentially be used as the gate dielectrics for future deep submicrometer MOSFETs. Because the transistor current is directly proportional to the capacitance of the gate dielectric, there is a significant advantage of this approach in future VLSI devices. MOSFET transistors with high-dielectric-constant gate dielectrics have already been fabricated using Ta_2O_5, TiO_2, Al_2O_3, HfO_2, and ZrO_2. The application of these high-dielectric-constant materials as gate dielectrics will demand even tighter requirements on their electrical and material properties.

Because the interface between the ferroelectric materials and the semiconductor surface is still inferior compared with that of Si–SiO$_2$, various buffer layers have been employed. Metal–ferroelectric–insulator–semiconductor-type structures have been studied and metal–ferroelectric–insulator–semiconductor field effect transistors (MFISFETs) fabricated. Development in this direction is discussed in Section 10.

1.2. Applications of High-Dielectric-Constant Films

As we discussed previously, the "simple" high-dielectric-constant materials, of which Ta_2O_5 is the most important example, have two important potential applications in future VLSI devices. They can be used as the capacitor dielectric for DRAMs and they can also be used as the gate dielectric for deep submicrometer MOSFETs. The application of high-dielectric-constant films for DRAM will simplify the memory cell structure, because planar geometry can again be used. However, the tradeoff is that these materials are new to the integrated circuit industry and a great deal of effort has to be done to integrate them in a full complementary metal-oxide-semiconductor (CMOS) process.

The properties of these high-dielectric-constant materials are listed in Table II. To be useful for DRAM applications, the dielectric constant should be significantly larger than that of SiO_2. The leakage current should be acceptable to memory device applications, usually less than 10^{-7} A/cm^2. The dielectric strength should be preferably larger than 5 MV/cm. These electrical specifications lead to some requirements on the physical properties of these high-dielectric-constant materials, such as the energy bandgap and the electron affinity. For gate dielectric applications, the interface between the high-dielectric-constant materials and silicon is critical. The interface state charges and the charges in the high-dielectric-constant materials will decide the stability of the MOSFETs built with these gate dielectrics.

1.3. Applications of Ferroelectric Films

The second group of high-dielectric-constant materials is the "ferroelectric" materials. The ferroelectric materials are those with a spontaneous polarization that is reversible by the application of an electric field. If the material is paraelectric at room temperature, it behaves very much like a normal dielectric material. The Curie temperature depends on the composition of the ferroelectric material.

Ferroelectric materials can be classified as either displacive type or order–disorder type, depending on the origin of the dipole moment in the material. The ferroelectric materials most important for the application of microelectronics are of the perovskite crystal structure. These are oxide ceramics having the

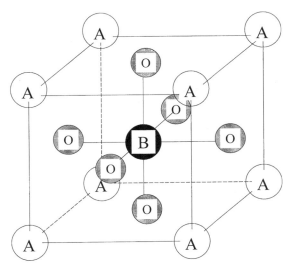

Fig. 2. Cubic ABO$_3$ perovskite unit cell.

general chemical formula of ABO_3, where O is oxygen, A represents a cation with a larger ionic radius, and B represents a cation with a smaller ionic radius. Figure 2 shows a cubic ABO_3 (e.g., A is Ba and B is Ti in $BaTiO_3$) perovskite-type unit cell. Most of the ferroelectrics with perovskite-type structure are compounds with either $A^{2+}B^{4+}O^{2-}$-, or $A^{1+}B^{5+}O_3^{2-}$-type formula. In the perovskite family, there are many compounds with the formula $A^{3+}B^{3+}O^{2-}$, but among them no ferroelectrics have been found [16].

Although some of the perovskite materials have a low Curie temperature and are paraelectric at room temperature, it is more convenient for our purposes to group all the materials with the perovskite crystal structure under the category as "ferroelectric".

These materials [13–16] include barium titanate (BT, $BaTiO_3$), lead titanate (PT, $PbTiO_3$), strontium titanate (ST, $SrTiO_3$), lead zirconate titanate [PZT, $Pb(Zr, Ti)O_3$], barium strontium titanate [BST, $(Ba, Sr)TiO_3$], lead lanthanum titanate [PLT, $(Pb, La)TiO_3$], lanthanum-doped lead zirconate titanate [PLZT, $(Pb, La)(Zr, Ti)O_3$], potassium niobate (KN, $KNbO_3$), and lead magnesium niobate [PMN, $Pb(Mg, Nb)O_3$]. Other ferroelectric materials include lithium niobate ($LiNbO_3$), lithium tantalate ($LiTaO_3$), strontium barium tantalate (SBT, $SrBi_2Ta_2O_9$), and lead iron niobate [$Pb(Fe, Nb)O_3$]. The most intensively studied ferroelectric films are PZT and SBT, and the most studied paraelectric film is BST. As a result, we will choose these three materials, PZT, BST, and SBT, as the main focus of our discussion.

The important physical properties of these ferroelectric materials are listed in Table III [12–44]. As we discussed earlier, ferroelectric thin films can be used as the capacitor dielectrics in DRAMs because of their high dielectric constants. For example, the dielectric constant of lead zirconate titanate (PZT) is on the order of 700–1500 [21, 30] in bulk and several hundreds in thin-film form. The switched charge of a ferroelectric capacitor is greater than 20 μC/cm^2, an order of magnitude

Table III. Electrical Properties of Ferroelectric Materials

	E_g (eV)	χ (eV)	P_s (μC/cm^2)	P_r (μC/cm^2)	E_c (kV/cm)	ε_r	T_C (°C)
PZT	3.4 ± 0.1 [18]	3.5 ± 0.2 [18]	32–45 [15][a]	21–35 [15][a]	35–150 [15][a]	>1000	260–410 [15][a]
	3.5 [46]	1.75 [46]				(70 nm) [12]	180–350 [16]
	3.41 [45]	2.6 [45]					
BST	3.2 ± 0.1 [18]	4.1 ± 0.2 [18]	Paraelectric	Paraelectric	Paraelectric	800	−50 ($x = 0.5$)
						(60 nm)	[17]
						300 [27]	
SBT	4.0 ± 0.1 [18]	3.4 ± 0.2 [18]	5.8 [14]	8.8 [47]	35 [37]	250 [37]	335 [14]
					29.3 [47]		
BT	3.9 [39]	2.52 [39]	7 [15]	5 [15]	15 [15]	∼200–300	115 [15]
			26 [13, 14]		100 [30]	[27, 40–42]	120 [13, 16]
			13 [30]			1000 [30]	135 [14]
ST	3.1 [43]	4.1 [43]	Paraelectric	Paraelectric	Paraelectric	230 (53 nm)	−233 [16]
						200 [27]	−163 [14]
PT			48 [15]	35 [15]	160 [15]		515–535 [15]
			27 [30]		200 [30]		490 [14, 16, 38]
PLT			22–45 [15][a]	20–38 [15][a]	34–50 [15][a]	1400	−25 (28% La)
						(500 nm)	[38]
						∼80–690	140 (20% La)
						(La from 0 to 25%)	
						[44]	
						1000–3000	
						[1, 27]	

[a] Depending on concentration.

higher than the 1.7 μC/cm^2 typical of current DRAM capacitors. Because the polarization vs. electric field (P–E) curve of ferroelectric materials is nonlinear and shows a hysteresis loop, the application of ferroelectric capacitors in DRAMs is limited to the unipolar region; that is, the polarity of the applied voltage will not be reversed during the operation of the DRAM chip. The DRAM operation of a ferroelectric capacitor is shown in Fig. 3 [31–33]. The bits "0" and "1" correspond to the points of remanent polarization and saturated polarization, respectively.

The second application of ferroelectric films in memory devices is in the area of nonvolatile memories. There are two ways to build a nonvolatile memory element using a ferroelectric thin film. One approach is to replace the gate insulator of a field effect transistor with a ferroelectric layer. The threshold voltage of such a device will have two distinct values, depending on whether the dipoles in the ferroelectric material are oriented up or down. However, this device is very difficult to build reliably [22]. The second implementation is to use a ferroelectric capacitor [18, 22, 34]. A schematic operation of a ferroelectric nonvolatile memory is shown in Fig. 4. Two types of ferroelectric memory cells have been used: (1) one-transistor/one-capacitor design (1T/1C) (1 cell per bit) and (2) two-transistor/two-capacitor design (2T/2C) (2 cells per bit).

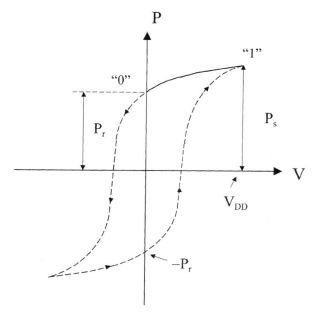

Fig. 3. DRAM operation with ferroelectric capacitor.

There are some significant advantages of ferroelectric memory compared with other nonvolatile memories. To program (write) a state into a FRAM cell, the electric field needs to

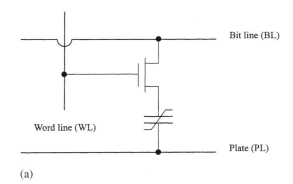

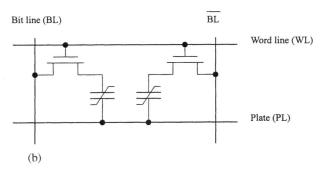

Fig. 4. FRAM memory cells: (a) 1T/1C cell, (b) 2T/2C cell.

be applied for less than 100 ns in order to polarize the non-volatile elements. In a standard EEPROM, it takes a millisecond or more for sufficient charges to travel through the insulating oxide to charge up the gate element. In addition, the high-voltage-generation circuitry takes some time to stabilize before it can cause this transfer to take place. These differences allow a FRAM memory cycle time of 500 ns worst case compared to 10 ms for an EEPROM.

In an EEPROM, the charge tunneling across the oxide degrades its characteristics, causing catastrophic breakdown or excessive trapped charges. For these reasons, EEPROM devices are guaranteed for only 10,000 to 100,000 write cycles. FRAM memories do not suffer from these same limitations, and so can provide more than 10 billion (10^{10}) cycles, although both read and write operations must adhere to these limits [34].

Other authors have made similar comparisons. For example, Hori [26] mentioned that, as compared with nonvolatile flash memories, FRAMs have the following advantages: (1) The write operation is much faster (typically, <100 ns) due to the short time required for polarization reversal. (2) The power consumption is much smaller, because the write operation does not involve such a large amount of current flow as in flash memories. (3) The radiation hardness is generally better, which may arise from its operation principle of storing charges by discrete dipoles.

However, FRAMs also suffer from several drawbacks [26]: (1) The endurance characteristics are unsatisfactory. This is usually called fatigue. (2) The retention characteristics could also be a problem. (3) There are also problems associated with imprint.

To be applicable to semiconductor devices, a ferroelectric material must satisfy several demanding requirements. These requirements are as follows: (1) The Curie temperature must be above the operation temperature of the semiconductor device. This means that the Curie temperature should be above 125°C or so. (2) The ferroelectric material must have adequate ferroelectric hysteresis properties. (3) To serve as storage element, the ferroelectric material must have sufficient charge density, that is, above about 5 μC/cm^2. (4) The ferroelectric material must be able to operate in a voltage range compatible with integrated-circuit applications. This means that the ferroelectric film must be switchable with a relatively low voltage, such as 3–5 V, or even lower voltages for future VLSI technologies. This means that the ferroelectric films must have sufficiently low coercive field and adequate dielectric breakdown field. (5) The ferroelectric material should be able to be deposited as thin films compatible with integrated-circuit fabrication: for example, it should have good enough step coverage to be used in an integrated circuit. (6) The ferroelectric film must be able to operate in the VLSI circuit in a reliable manner, with proper performance in such reliability requirements as retention and endurance [15, 22, 27, 28].

The ferroelectric materials that satisfy all of the preceding requirements and have been chosen by semiconductor companies pursuing ferroelectric memories are PZT and SBT. Commercial companies such as McDonnell Douglas, Raytheon, Ramtron, and National Semiconductor are all using PZT [22, 29]. Because most ferroelectrics have a polarization that corresponds to about 100 times the switched charge of a Si DRAM of the same area, it is not necessary to choose a ferroelectric material with a large spontaneous polarization: 0.1–1.0 μC/cm^2 is sufficient [35].

2. HIGH-DIELECTRIC-CONSTANT FILMS: TANTALUM OXIDE (Ta$_2$O$_5$)

Tantalum oxide is a high-dielectric-constant ($\varepsilon_r = 26$–50) material with reasonably low leakage current (1×10^{-8} A/cm^2 at 1 V) and high breakdown strength (5–6 MV/cm). It has been used for applications in semiconductor devices such as the capacitor dielectric of DRAM, gate dielectric of MOSFET and thin-film transistor (TFT), lightly doped drain spacer, and ion-sensitive field effect transistor (ISFET) sensor. Although the dielectric constant of Ta$_2$O$_5$ is lower than those of perovskite-structure films such as SrTiO$_3$, (Ba$_x$Sr$_{1-x}$)TiO$_3$, and Pb$_x$Zr$_{1-x}$TiO$_3$, Ta$_2$O$_5$ has been used as the capacitor dielectric for 256-Mb or 1-Gb DRAM due to the following advantages: (1) The dielectric constant is only slightly depend on the applied voltage across the capacitor. In contrast, the dielectric constant of ferroelectric films depends strongly on the applied voltage. (2) Ta$_2$O$_5$ has lower leakage current and higher dielectric breakdown strength than those of other high-dielectric-constant films such as TiO$_2$ or (Ba$_x$Sr$_{1-x}$)TiO$_3$. (3) Ta$_2$O$_5$ can be deposited by (LPCVD), which is suitable for mass production. Ferroelectric films are mostly deposited by

Table IV.   Prototype DRAM Chips with Ta_2O_5 Storage Capacitor

Reference/ company	DRAM capacity	Process technology	Cell area (μm^2)	Chip size (mm^2)	Capacitor structure	Supply voltage (external/ internal)
Hamada et al. [51]/ NEC	4 Mb (for merged DRAM/logic application)	0.18 μm, CMOS	0.45		Capacitor under bit line, HSG polysilicon bottom electrode, 24 fF/cell	1.8 V
Kim et al. [50]/ Samsung	4 Gb	0.15 μm, 1-W/2-Al, CMOS	0.18	13.32 (16-Mb experimental chip)	HSG polysilicon bottom electrode, 25 fF/cell	
Lee et al. [49]/ Samsung	1 Gb	0.16 μm, 2-well, 3-metal, CMOS	0.334	569.7	OCS cell	2.5 V/2.0 V
Takai et al. [52]/ NEC	1 Gb	0.18 μm, 3-well, CMOS	0.32	547		2.5 V/2.1 V
Yoo et al. [48]/ Samsung	1 Gb	0.16 μm, 2-well, 3-poly/1-TiSi$_2$/2-W/2-Al, CMOS	0.334	652.1	OCS cell, 25 fF/cell	2.0 V/1.8 V
Yoon et al. [53]/ Samsung	1 Gb	0.14 μm, 3-well, 3-metal, CMOS	0.181	349	OCS cell	2.5 V/2.0 V

OCS, one-cylindrical-stacked bottom electrode.

sputtering due to chemical composition problems. (4) The stoichiometry of Ta_2O_5 can be easily controlled compared with that of perovskite films.

2.1. Dynamic Random-Access Memory Applications

The study of Ta_2O_5 films for DRAM storage capacitor applications started around 1980. Prototypes for 1-Gb DRAM have already been successfully fabricated in research laboratories [48–53]. In the last 10 years, suitable Ta_2O_5 capacitor dielectrics for 256-Mb and 1 Gb-DRAM application have been developed. However, BST is more suitable for 1-Gb DRAM than Ta_2O_5 because the dielectric constant of BST is higher than that Ta_2O_5.

Table IV lists the literature results of Ta_2O_5 storage capacitors for 1- or 4-Gb prototype DRAMs. The capacitance was maintained at about 25 fF per cell for both 1- and 4-Gb DRAMs. The surface area of the polysilicon bottom electrode was increased to keep the same capacitance. A cylindrical-shaped bottom electrode was used by Samsung [48–50, 53]. A hemispherical grain (HSG) polysilicon bottom electrode was used by NEC [51, 52]. The cell capacitance can be doubled this way. Because the cell size of 4-Gb DRAM is about half that of 1-Gb DRAM, the dielectric constant of the capacitor dielectric or the capacitor area of the 4-Gb DRAM capacitor should be double that of the 1-Gb DRAM capacitor. Therefore, the dielec-

tric constant of Ta_2O_5 should be further improved. This can be made by depositing Ta_2O_5 on a metal bottom electrode such as Ru. The dielectric constant of Ta_2O_5 can be as high as 90–110 when Ta_2O_5 is crystallized at (001) preferred orientation [54].

2.2. Metal–Oxide–Semiconductor Field Effect Transistor Gate Dielectric Applications

When the thickness of a MOSFET gate dielectric is scaled down below 1.5 nm, the gate leakage current level is unacceptable due to direct tunneling current [55]. A metal gate instead of polysilicon was used to prevent the poly-depletion effect and the boron penetration in p-channel MOSFETs [55]. MOSFETs with $L_{eff} = 0.08$ μm using TiN metal gates were successfully fabricated [55]. The 1999 International Technology Roadmas for Semiconductors (ITRS) pointed out that when the gate oxide thickness is thinner than 2.0 nm, the advantage of gate oxide scaling would be offset by the poly-depletion effect. Therefore, a metal gate is required when the MOSFET channel length is scaled down below $L_{eff} < 0.10$ μm.

The deposition of the Ta_2O_5 gate dielectric and the gate metal after source–drain formation is not consistent with the present self-aligned gate process. This is because the crystallization of Ta_2O_5 and the diffusion of silicon atoms into Ta_2O_5 will result in high leakage current. The TiN gate electrode cannot withstand the temperature of source–drain dopant activation

Table V. Electrical Characteristics of MOSFETs with Ta_2O_5 Gate Dielectric

Reference	Gate dielectric	Ta_2O_5 dielectric constant	Equivalent SiO_2 thickness	MOSFET structure	Electrical parameters
Kim and Kim [56]	61-nm PECVD Ta_2O_5/native SiO_2 (3.5 nm)	18	13.2 nm	Polysilicon gate, n-channel with $L = 4\ \mu m$ and 100 μm	$S_t = 68$–74 mV/decade, $\mu_e = 500$–600 cm^2/V-s when $V_D = 50$ mV
Autran et al. [57]	61-nm ECR-PECVD Ta_2O_5	25	9.5 nm	Ta_2O_5 was deposited after S/D diffusion, Al metal gate p-channel with $L = 3$–10 μm	$Q_{ot} = 4 \times 10^{10}$ cm^{-2}, $\langle D_{it} \rangle = 3 \times 10^{10}$ eV^{-1}cm^{-2} and $\langle \sigma \rangle = 10^{-16}$ cm^2 from charge pumping method
Chatterjee et al. [55]	5.5–6.5-nm LPCVD Ta_2O_5	10.7–11.8	2.0–2.15 nm	Replacement metal TiN gate process, p-channel with $L = 0.1$ μm, n-channel with $L = 0.06\ \mu m$	$\langle D_{it} \rangle = 2.8 \times 10^{11}$ eV^{-1}cm^{-2} from charge pumping method, $\mu_c \cong 170$ cm^2/V-s and $\mu_p \cong 70$ cm^2/V-s when $V_D = 50$ mV
Kizilyalli et al. [58]	SiO_2 (0.5–1.5 nm)/5 nm MOCVD Ta_2O_5/SiO_2 (0.5–1.5 nm)	12–13	2.0–3.5 nm	Polysilicon gate, p-channel with $L = 2.0$ μm, n-channel with $L = 0.35\ \mu m$	$D_{it} = 3.2 \times 10^{11}$ eV^{-1}cm^{-2}
Park et al. [59]	6-nm LPCVD Ta_2O_5/native SiO_2 (0.8–1.0 nm)	14	1.8 nm	TiN metal gate n-channel with $L = 30$ μm	$\mu_e \cong 150$ cm^2/V-s when $V_D = 50$ mV
Yu et al. [60]	18.6-nm PECVD Ta_2O_5	24	3.1 nm	Ta_2O_5 was deposited after S/D diffusion, Al metal gate n-channel with $L = 3\ \mu m$	$\mu_e \cong 333$ cm^2/V-s when $V_D = 50$ mV, $D_{it} = 7.8 \times 10^{12}$ cm^{-2} eV^{-1}, surface recombination velocity $S_0 = 780$ cm/s and minority carrier lifetime $\tau = 3 \times 10^{-6}$ s

annealing. Therefore, a replacement gate process was proposed to solve these problems [55]. The conventional MOSFET was fabricated by the self-aligned gate process. After removing the polysilicon and the SiO_2 layer of the MOSFET, the Ta_2O_5 gate dielectric and TiN gate electrode were deposited on the silicon substrate to form a MOSFET with a Ta_2O_5 gate dielectric.

Table V shows the published results of MOSFETs with Ta_2O_5 gate dielectrics. The disadvantages of a Ta_2O_5 film as the MOSFET gate dielectric are as follows: (1) The interface trapped charge density Q_{it} at the Ta_2O_5/silicon interface is about 10^{11}–10^{12} eV^{-1}cm^{-2}, which is 1–2 orders of magnitude higher than that of the SiO_2/silicon interface. Therefore, a thin SiO_2 buffer layer is required to reduce the interface trapped charge density. (2) The oxide trapped charge density Q_{ot} of Ta_2O_5 film is much higher than that of SiO_2 film, which will cause the same instability problem for MOSFET threshold voltage. Although the oxide trapped charge of Ta_2O_5 film can be minimized by annealing, the flatband voltage shift is

still too large. (3) Silicon atoms will diffuse into Ta_2O_5 film after high-temperature annealing, which will result in high leakage current and lower dielectric breakdown voltage. Although low-temperature annealing such as O_2-plasma annealing can reduce this problem, the hole oxide trapped density will increase after O_2-plasma annealing [61]. (4) Because the TiN gate electrode cannot withstand source–drain dopant activation annealing [62], the source and drain areas must be fabricated before deposition of the gate dielectric and the gate electrode. Therefore, the conventional self-aligned gate MOSFET process is not suitable for a MOSFET with a Ta_2O_5 gate dielectric. Although this problem can be avoided by using the replacement gate process [55], the complex process will affect the yield of integrated circuits. The situation concerning the hot carrier effect and other reliability problems of MOSFETs with Ta_2O_5 gate dielectrics are unclear from the literature. It was reported that the electron barrier height at Ta_2O_5/silicon interface was not higher than 0.85 eV [63]. This is much smaller than the

3.2 eV of the SiO₂/silicon interface. More research on the reliability issues of MOSFETs with Ta₂O₅ film gate dielectrics is need in the future.

2.3. Physical Properties of Ta₂O₅ Thin Films

2.3.1. Energy Bandgap

As shown in Table VI, there are two reported ranges of Ta₂O₅ energy bandgap, one between 4.0 and 4.4 eV [64–71] and the other at 5.3 eV [72–74]. The former is more widely accepted. The energy bandgap E_g of Ta₂O₅ was often found by plotting the absorption coefficient α vs. the wavelength of incident light and letting α be extrapolated to 0 [68, 70, 75]. Murawala et al. measured the transmittance spectrum of Ta₂O₅ film deposited by plasma-enhanced chemical vapor deposition (PECVD) using optical transmission spectroscopy [74]. An energy bandgap of $E_g = 5.28$ eV for Ta₂O₅ film was obtained by plotting the transmission intensity vs. the wavelength of incident light and letting the transmission intensity be extrapolated to 0. The energy bandgap of Ta₂O₅ film was overestimated because the transmission intensity data were not transferred to the absorption coefficient α.

2.3.2. Electron Affinity

Electron affinity is an important parameter for constructing the energy band diagram of the metal–insulator–semiconductor structure. If the electron affinity of the dielectric film is high, the barrier height between the metal contact and the dielectric will be low, which will result in higher leakage current. The energy band diagram is also important in analyzing MOSFET hot carrier effects. High-dielectric-constant materials often have larger electron affinity values. Flannery and Pollock [76] studied the Ta₂O₅ capacitor using Ta and Au as the bottom and top electrodes. The barrier heights are 1.1 and 1.6 eV at the Ta₂O₅/Ta interface and the Au/Ta₂O₅ interface, respectively. The work

functions of Au and Ta are 4.7 and 4.25 eV. Therefore, an electron affinity of Ta₂O₅ film of about 3.1 eV can be calculated from the analysis of thermionic emission conduction. Lai and Lee deposited Ta₂O₅ film on a W bottom electrode [63]. A Ta₂O₅/W barrier of 1.32 eV was obtained from Fowler–Nordheim tunneling by assuming an effective electron mass of $m^* = 0.5m_0$. Because the work function of W is 4.55 eV, an electron affinity of 3.2 eV was obtained. Robertson and Chen assumed an electron affinity of 3.3 eV of Ta₂O₅ in studying the Schottky barrier heights of metal oxides [77]. The electron affinity of Ta₂O₅ thus lies between 3.1 and 3.3 eV. Figure 5 shows the energy band diagram of a metal/Ta₂O₅/silicon capacitor. If the electron affinity of 3.2 eV of Ta₂O₅ is used,

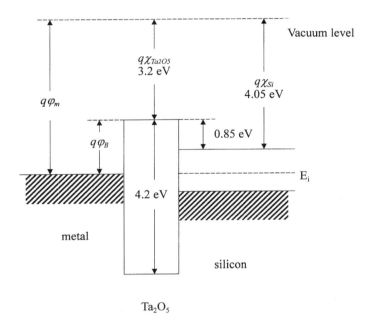

Fig. 5. Energy band diagram of metal/Ta₂O₅/silicon capacitor at flatband condition.

Table VI. Ta₂O₅ Energy Bandgap Values and Characterization Methods

Reference	Device structure	Experiment method	Energy bandgap
Kaplan et al. [67]	480-nm LPCVD Ta₂O₅ on fused silica	Optical absorption	4.38 eV
Kimura et al. [68]	2-μm-thick Ta₂O₅ on quartz with furnace N₂ annealing for 1 h at 600 or 700°C	Optical absorption	4.1–4.2 eV
Kukli et al. [64]	Ta₂O₅ on glass	Optical transmittance spectrum	4.2 eV for amorphous Ta₂O₅ 3.9–4.5 eV for crystallized Ta₂O₅
Metikos-Hukovic and Ceraj-Ceric [72]	Ta₂O₅ on Ta electrode	Photocurrent spectrum	4.0 eV
Murawala et al. [74]	PECVD Ta₂O₅ on glass	Optical transmittance spectrum	5.28 eV
Quarto et al.[70]	Ta₂O₅ on Ta electrode	Photocurrent spectrum	4.3 eV

the barrier height at the Ta_2O_5/silicon interface will be about 0.85 eV. This value is much small than that of 3.2 eV at the SiO_2/silicon interface.

2.3.3. Dielectric Constant

2.3.3.1. Thickness Dependence

As shown in Table VII, the dielectric constant of Ta_2O_5 increases with film thickness. However, the dielectric constant of Ta_2O_5 is affected by the formation of native SiO_2 at the interface of the Ta_2O_5/silicon substrate [78, 80–83]. The thickest interfacial SiO_2 was reported by Nagahiro and Raj using electron cyclotron resonance (ECR)-PECVD [81]. Ta_2O_5 films 40–80 nm thick were deposited on an n-type silicon substrate. A native SiO_2 of about 10–16 nm thick was formed at the Ta_2O_5/silicon interface. Therefore, ECR-PECVD-deposited Ta_2O_5 films are not suitable for DRAM and MOSFET gate dielectric applications. The decrease in the overall dielectric constant is due to the existence of the transition layer at the Ta_2O_5/silicon interface rather than the size effect. Seki et al. reported that when a 110-nm-thick Ta_2O_5 was etched back to 50 nm and the dielectric constant was measured, the dielectric constant was 8.5, which was the same as 50-nm as-deposited Ta_2O_5 [82]. The dielectric constant of Ta_2O_5 film is uniform throughout the film. A thin low-dielectric-constant layer is formed at the Ta_2O_5/silicon interface and the thickness of the transition layer is independent of the total thickness of the capacitor.

2.3.3.2. Annealing Condition and Bottom Electrode Dependence

As shown in Table VIII, the reported phases of Ta_2O_5 include hexagonal δ-Ta_2O_5 [64, 84–91] and tetragonal and orthorhombic β-Ta_2O_5 [81, 91–96] and α-Ta_2O_5 [92]. Aoyama et al. annealed Ta_2O_5 films from 620 to 700°C [72]. The temperature increment was 20°C. The X-ray diffraction (XRD) result shows that Ta_2O_5 films are amorphous below 640°C and become crystallized after 660°C. The dielectric constant of Ta_2O_5 films is dependent on the annealing conditions. Park et al. reported that the dielectric constant of Ta_2O_5 films increases when the annealing temperature increases from 700 to 750°C [99]. However, increasing the annealing temperature will reduce the dielectric constant. The increase in dielectric constant is due to the crystallization of Ta_2O_5 film at 750°C. The dielectric constant decreases at higher annealing temperature because thin SiO_2 is formed at the Ta_2O_5/silicon interface. The dielectric constant of Ta_2O_5 film increases significantly when Ta_2O_5 crystallizes with some preferred orientation. Kishiro et al. deposited Ta_2O_5 films on Pt, Ru, and Si_3N_4/silicon electrodes using LPCVD [93]. The films were then annealed by rapid thermal annealing at 600–800°C for 1 min in N_2 ambience. Ta_2O_5 film was crystallized as β-Ta_2O_5 after 700°C annealing. The preferred orientation for Ta_2O_5 films deposited on a Pt electrode is (110) and the preferred orientation for Ta_2O_5 films deposited on a Ru electrode is (001). However, there is no preferred orientation for Ta_2O_5 films deposited on a Si_3N_4/silicon electrode. The dielectric constant of Ta_2O_5 films annealed at 750°C with preferred orientation is as high as 60 and for films without preferred orientation only 30. Lin et al. deposited 30-nm Ta_2O_5 films on a Ru bottom electrode [54]. They found that the Ru bottom electrode has a (002) preferred orientation and

Table VII. Thickness-Dependent Dielectric Constants of Ta_2O_5 Films

Reference	Capacitor structure	Annealing condition	Dielectric constant
Hitchens et al. [78]	7.5–40-nm LPCVD Ta_2O_5/silicon	Furnace O_2 annealing at 800°C	8–23, increasing with dielectric thickness
Joshi and Cole [79]	17–260-nm MOSD Ta_2O_5/Pt	Furnace O_2 annealing at 700°C	42–50 when dielectric thickness <36 nm 51 when thickness >51 nm
Mikhelashvili et al. [80]	7.5–130-nm e-beam evaporated Ta_2O_5/silicon	Furnace O_2 annealing at 750°C for 60 min	12–31 for as-deposited films 8–28 for annealed films
Nagahiro and Raj [81]	40–80-nm ECR-PECVD Ta_2O_5/silicon	None	8–12, the low dielectric constant due to a thin SiO_2 about 10–16 nm formed at the Ta_2O_5/silicon interface
Seki et al. [82]	17–110-nm sputtered Ta_2O_5/silicon	Postmetallization H_2 annealing at 450°C for 30 min	Increasing from 3.9 to 12.5 after annealing
Zaima et al. [83]	18–96-nm LPCVD Ta_2O_5/silicon	Furnace O_2 annealing at 700°C for 30 min	5–25 for as-deposited films 10–35 for annealed Ta_2O_5

Table VIII. Dependence of Ta$_2$O$_5$ Dielectric Constant on the Annealing Condition and Electrode Material

Reference	Capacitor structure	Annealing condition	Crystallization temperature and the phase
Aoyama et al. [72]	LPCVD Ta$_2$O$_5$/silicon	620–700°C in furnace	Ta$_2$O$_5$ crystallized at 660°C
Aoyama et al. [97]	LPCVD Ta$_2$O$_5$/Ru	1. RTA N$_2$ annealing at 700°C for 1 min 2. O$_2$-plasma annealing at 400°C for 5 min 3. RTA N$_2$+ O$_2$ − plasma annealing	RTA N$_2$ annealing at 700°C can increase the dielectric constant of Ta$_2$O$_5$ due to crystallization
Joshi and Cole [79]	17–260-nm MOSD Ta$_2$O$_5$/Pt	Furnace O$_2$ annealing at 600–750°C for 60 min	β-Ta$_2$O$_5$ after 650°C annealing
Kim et al. [90]	ECR-PECVD Ta$_2$O$_5$/silicon	Furnace O$_2$ annealing at 700–850°C for 30 min	δ-Ta$_2$O$_5$ after 750°C annealing
Kishiro et al. [93]	LPCVD Ta$_2$O$_5$ on Pt, Ru, or Si$_3$N$_4$	RTA N$_2$ at 700–800°C annealing for 1 min	β-Ta$_2$O$_5$ after 700°C annealing Preferred orientation: (110) for Ta$_2$O$_5$/Pt (001) for Ta$_2$O$_5$/Ru
Kamiyama et al. [96]	LPCVD Ta$_2$O$_5$/polysilicon	RTA at 600–950°C for 1 min in O$_2$ ambience	β-Ta$_2$O$_5$ after 700°C annealing t_{eq} = 3.0 nm at 600°C and 4.1 nm at 950°C
	LPCVD Ta$_2$O$_5$/900°C NH$_3$ annealed polysilicon		t_{eq} = 2.7 nm for as-deposited film 2.5 nm at 600°C and 3.9 nm at 950°C
	LPCVD Ta$_2$O$_5$/1100°C NH$_3$ annealed polysilicon		t_{eq} = 2.8 nm for as-deposited film 2.5 nm at 700°C and 3.6 nm at 950°C
Kamiyama et al. [98]	LPCVD Ta$_2$O$_5$/NH$_3$ annealed polysilicon	O$_2$-plasma annealing at 100–400°C	t_{eq} = 2.6 nm independent of O$_2$-plasma annealing temperature
Lin et al. [54]	LPCVD Ta$_2$O$_5$/Ru	N$_2$O-plasma at 350°C for 3 min + RTA O$_2$ annealing at 800°C for 1 min	Ru preferred orientation at (002) Ta$_2$O$_5$ preferred orientation at (001)
Park et al. [99]	1. LPCVD Ta$_2$O$_5$/silicon 2. Thermal oxidation Ta$_2$O$_5$ (at 500°C)/Ta	Furnace N$_2$ or O$_2$ annealing at 700–950°C for 20 min	SiO$_2$ formed at the Ta$_2$O$_5$/silicon interface when >750°C

the Ta$_2$O$_5$ also has a preferred (001) orientation. The dielectric constant of Ta$_2$O$_5$ can be as high as 90–110. The high dielectric constant of Ta$_2$O$_5$ comes from the crystallized structure with a preferred (001) orientation of Ta$_2$O$_5$ film. Therefore, crystallization with a preferred orientation is important to obtain high-dielectric-constant Ta$_2$O$_5$ films.

In summary, the crystallization of Ta$_2$O$_5$ film will induce a higher leakage current density. If the Ta$_2$O$_5$ film is deposited on a silicon or polysilicon substrate, silicon atoms will diffuse into the grain boundary of the Ta$_2$O$_5$ film and further increase the leakage current during high-temperature annealing. The advantage of crystallized Ta$_2$O$_5$ film is that the dielectric constant will increase significantly especially for Ta$_2$O$_5$ films with a preferred crystallization orientation. The dielectric constant of Ta$_2$O$_5$ films deposited on a Ru electrode is high (90–110) and with reasonably low leakage current 5×10^{-8} A/cm^2 at

1.0 V [54]. Therefore, it is suitable for future 1-Gb DRAM storage capacitor application.

2.3.4. Metal–Insulator Barrier Height

The leakage current mechanism of the dielectric film depends on the interfacial potential barrier height when the conduction mechanism is electrode limited. When the influence of the interface state is not significant, the interface barrier height is determined by the difference between metal work function Φ_m and the electron affinity χ_s of the semiconductor [77].

Although most studies reported that the current mechanism of Ta_2O_5 is bulk limited [100–103], some authors also reported electrode-limited conduction [63, 94, 104, 105]. If there is a thin SiO_2 layer formed at the Ta_2O_5/silicon interface, electron transport through the Ta_2O_5/SiO_2 double layer will be limited by the thin SiO_2 layer. The conduction mechanism through the Ta_2O_5/SiO_2 double layer is Fowler–Nordheim tunneling with a barrier height of 3.2 eV [100, 101, 106–111]. Matsuhashi and Nishikawa used Mo, W, Ti, Ta, and their metal nitrides as the top electrode of metal/Ta_2O_5/polysilicon capacitors [112]. The leakage current of the capacitor decreases when the work function of the top electrode increases. The leakage current is independent of the top electrode material if the leakage current densities are plotted as a function of $V_G - \Phi_{ms}$ when electrons are injected from the polysilicon bottom electrode. Therefore, they concluded that the current mechanism of a metal/Ta_2O_5/polysilicon capacitor is electrode limited for both voltage polarities. Miki et al. used CVD TiN as the top electrode of a Ta_2O_5/Si_3N_4/polysilicon capacitor [107]. When electrons are injected from TiN into Ta_2O_5, the conduction mechanism is find to be Schottky emission. They also found that the TiN/Ta_2O_5 barrier height depends on the deposition rate of the TiN metal. The barrier height decreases when the TiN deposition rate increases. Therefore, the barrier height not only depends on the work function of the metal but also on the process conditions.

2.3.5. Trap Type and Trap Level

As shown in Table IX, photocurrent spectroscopy, photoluminescence spectra [66], activation energy [72, 102, 112–114], and zero-bias thermally stimulated current (ZBTSC) measurements [65, 73, 115, 116] have been used to study the trap properties in Ta_2O_5 films. Temperature-dependent leakage current measurements are commonly used to study the relationship between the leakage current and the traps. The activation energy E_a is obtained from an Arrhenius plot at different applied voltages. The reason for the formation of traps in Ta_2O_5 film is not totally clear at this time. Alers et al. reported that the traps come from hydrogen atoms present in the film after the CVD process or from Ta^+ defects in the oxide [66]. They claimed that the defects did not come from the carbon impurities because the carbon concentration in the Ta_2O_5 film had little influence on the photoluminescence spectra. Aoyama et al. reported that the traps of the amorphous Ta_2O_5 film come from hydrocarbon

contamination and oxygen vacancy [72]. The traps of polycrystallized Ta_2O_5 come from silicon atoms diffusing along the grain boundary and penetrating into Ta_2O_5 grains. Ezhilvalavan and Tseng also reported that the traps of polycrystallized Ta_2O_5 come from silicon atoms diffusing into Ta_2O_5 and resulting in the formation of oxygen vacancy [102]. Lau et al. studied the trap characteristics of Ta_2O_5 films by ZBTSC [73, 65, 116]. The dominant defects in Ta_2O_5 film (defect B) are due to the interaction between Ta_2O_5 and the silicon substrate during postannealing. Because Ta_2O_5 deposited on the p-silicon has a higher density of defect B than that of deposited on n-silicon, defect B was an acceptor state about 0.3 eV above the valence band of the Ta_2O_5 film. Another reason for the formation of defects in Ta_2O_5 may come from the oxygen vacancies [116].

Donors [92, 117, 118], shallow neutral trap [115, 118], and acceptor center [65, 73, 102, 106] have been found in Ta_2O_5 films. As listed in Table X, most papers reported that since Poole–Frenkel emission is the leakage current mechanism of Ta_2O_5 films and electrons are ejected from positive Coulombic centers in the films. Therefore, traps in Ta_2O_5 film are donor centers [115, 117, 118]. Simmons reported that the traps in Ta_2O_5 film are donor centers because the leakage current is due to Poole–Frenkel emission [118]. He also found that if there are shallow neutral traps E_t above the donor centers E_d, an anomalous Poole–Frenkel effect will be observed when the Fermi level lies between the E_t and E_d trap states. The anomalous Poole–Frenkel effect is often confused with Schottky emission because both of their expressions have the same exponential term $\exp(q\beta E^{1/2}/kT)$, where $\beta = (q/\pi\varepsilon\varepsilon_0)^{1/2}$. Lee et al. reported that oxygen vacancy acts as a donor level and becomes charged with one or two electronic charges at the Ta_2O_5/silicon interface [92]. O_2-plasma annealing reduces hydrocarbon contamination and oxygen vacancies in the Ta_2O_5 film. Therefore, leakage current was reduced after O_2-plasma annealing. Ezhilvalavan and Tsang reported that because the valence of Ta is 5 and the valence of silicon is 4, traps come from silicon contamination diffusion from the silicon substrate [102]. Acceptor centers are created in Ta_2O_5 films because Ta atoms are replaced by silicon. When Ta_2O_5 is deposited on the crystallized β-Ta electrode, the traps at Ta_2O_5 grain boundaries and at the interface are donor states. Lau et al. measured the traps of Ta_2O_5 by using ZBTSC [65, 73]. They found that Ta_2O_5 films deposited on a p-silicon substrate have a higher concentration of defect B. Defect B is an acceptor state positioned 0.3 eV above the valence band of the Ta_2O_5. Defect B comes from the interaction between the Ta_2O_5 film and the silicon substrate during postannealing because the concentration of defect B increases at higher annealing temperature. The leakage current of Ta_2O_5 film can be reduced by using the two-step annealing process proposed by Shinriki and Nataka [106]. Ta_2O_5 film was initially annealed in either UV-O_2 or UV-O_3 ambience at 300°C and then subsequently annealed by dry O_2 in the furnace. The oxygen vacancy acts as an acceptor with an electron in the conduction band of the Ta_2O_5 to keep the charge neutrality. The oxygen vacancy can be charged by either one or two

Table IX.   Trap Energy Levels of Ta$_2$O$_5$ Films

Reference	Experiment method	Trap depth (eV)	Comments
Alers et al. [66]	Photoluminescence spectra	2.2 eV below the conduction band E_c of Ta$_2$O$_5$	Traps come from hydrogen or Ta$^+$ defect in Ta$_2$O$_5$
Aoyama et al. [72]	Current–temperature measurement with top electrode biased at $V_G = -2$ V	$E_a = 0.71$ eV for amorphous Ta$_2$O$_5$ $E_a = 0.32$ eV for polycrystallized Ta$_2$O$_5$	Traps come from hydrocarbon contamination and oxygen vacancies
Devine et al. [115]	ZBTSC method with illumination; the top Al electrode was biased at 3.0 V	$E_a = 1.2$ eV; trap level is 1.2 eV below the E_c of Ta$_2$O$_5$	Shallow neutral trap is 1.0 eV below the E_c of Ta$_2$O$_5$
Ezhilvalavan and Tseng [102]	Current–temperature measurement with top electrode biased at $V_G = 2$ V	1. $E_a = 0.36$ eV for amorphous Ta$_2$O$_5$ $E_a = 0.22$–0.31 eV for polycrystallized Ta$_2$O$_5$ 2. 0.30 eV from Poole–Frenkel emission, 0.36 eV from Schottky emission	Silicon atoms diffuse into Ta$_2$O$_5$, resulting in the formation of oxygen vacancies
Lau et al. [65, 73]	ZBTSC method illuminated by ultraviolet light; 98.6-nm Ta$_2$O$_5$ film on either n$^+$- or p$^+$-silicon	O$_2$ annealed Ta$_2$O$_5$: defect B, 0.3 eV; defect C 0.6 eV N$_2$O annealed Ta$_2$O$_5$: defect B 0.3 eV; defect C, 0.6 eV; defect D, 0.8 eV	Leakage current is limited by the shallowest defects in the Ta$_2$O$_5$ (defect B); N$_2$O annealed film has lower leakage current than O$_2$ annealed film due to smaller defect B concentration
Lau et al. [116]	ZBTSC method, 8-nm Ta$_2$O$_5$ film deposited on n$^+$-silicon	N$_2$O annealed Ta$_2$O$_5$: defect A, 0.2 eV; defect B, 0.3 eV O$_2$ annealed Ta$_2$O$_5$: defect A, 0.2 eV; defect B, 0.3 eV; defect B$'$, 0.4 eV	Leakage current is limited by the shallowest defect in the Ta$_2$O$_5$ (defect A); defect A comes from the oxygen vacancies in the Ta$_2$O$_5$ film
Lo et al. [113]	Time to breakdown measurement with top TiN electrode biased at -3.7 V	$E_a = 0.2$ eV for Ta$_2$O$_5$ on rugged polysilicon $E_a = 0.7$ eV for Ta$_2$O$_5$ on smooth polysilicon	

Table X.   Trap Type of Ta$_2$O$_5$ Films

Reference	Trap type	Comments
Choi and Ling [117]	Poole–Frenkel emission donor trap	$N_d = 1.7 \times 10^{14}$–5×10^{17} cm^{-3}
Devine et al. [115]	Poole–Frenkel emission donor trap	Trap level is 1.0 eV below E_c of Ta$_2$O$_5$
Ezhilvalavan and Tseng [102]	Traps at Ta$_2$O$_5$ grain boundary and interface defects are donor states	Acceptor traps, caused by silicon diffusing into Ta$_2$O$_5$
Lau et al. [65, 73]	Acceptor state due to the interaction of Ta$_2$O$_5$ and silicon during annealing	Trap level is 0.3 eV above E_V of Ta$_2$O$_5$
Lee et al. [92]	Oxygen vacancies act as donor trap	O$_2$-plasma annealing can minimize the oxygen vacancies
Shinriki and Nataka [106]	Oxygen vacancies act as acceptor traps	O$_2$-plasma annealing can minimize the oxygen vacancies
Simmons [118]	Poole–Frenkel emission donor trap	Shallow neutral trap will result in anomalous Poole–Frenkel emission

electrons. The excited oxygen ions will diffuse into the Ta$_2$O$_5$ film and accept the electrons in the conduction band of Ta$_2$O$_5$ during UV-O$_2$ or UV-O$_3$ annealing. Therefore, the leakage current density decreases because of the decreases in the oxygen vacancies.

2.4. Fabrication Processes of Ta$_2$O$_5$ Thin Films

2.4.1. Deposition Methods

Many deposition methods, such as low-pressure CVD, plasma-enhanced CVD, metal–organic CVD, ECR-CVD, photo-CVD, magnetron sputtering, thermal oxidation, atomic layer deposition, and anodic oxidation, have been used to deposit Ta_2O_5 films. Most often, Ta_2O_5 is deposited by low-pressure CVD (LPCVD) for DRAM capacitor and MOSFET gate dielectric applications. Ta_2O_5 film deposition by LPCVD is performed at 300–450°C in an O_2 ambience. The $Ta(OC_2H_5)_5$ precursor is vaporized in the source tank and nitrogen is used as the carrier gas. The deposition pressure is maintained at 0.1–0.4 mTorr. The uniformity of the Ta_2O_5 film can be maintained best by using LPCVD as compared with other deposition methods. As a result, LPCVD is most suitable for mass production. The main drawback of LPCVD is hydrocarbon contamination and the formation of a large number of oxygen vacancies due to the incomplete decomposition of the $Ta(OC_2H_5)_5$ precursor. The high leakage current density of as-deposited Ta_2O_5 films is unacceptable for DRAM applications. Postdeposition oxygen annealing is required to reduce the leakage current of Ta_2O_5 films. Ta_2O_5 deposited by plasma deposition systems such as PECVD or ECR-PECVD has fewer oxygen vacancies as compared to Ta_2O_5 deposited by LPCVD. However, the silicon substrate will be oxidized by high-energy oxygen plasma so that a thin SiO_2 layer about 1.0–2.0 nm will be formed at the interface of Ta_2O_5 and silicon. The native oxide will increase the equivalent SiO_2 thickness and reduce the capacitance of the Ta_2O_5 capacitor. Furthermore, the native oxide will limit the scaling capability of the Ta_2O_5 gate dielectric, which is not acceptable for future application in sub-0.1-μm channel length MOSFETs. The other deposition methods such as radiofrequency (rf) sputtering, thermal oxidation, and atomic layer deposition are not suitable for depositing very thin Ta_2O_5 films due to uniformity problems.

The quality of the Ta_2O_5 film is determined not only by the deposition method but also by the precursor material. The precursors used to deposit Ta_2O_5 films include $Ta(OCH_3)_5$, $Ta(OC_2H_5)_5$, $TaCl_5$, and TaF_5. The fluoride precursors $TaCl_5$ and TaF_5 require water vapor during Ta_2O_5 deposition, whereas the metal–organic precursors $Ta(OCH_3)_5$ and $Ta(OC_2H_5)_5$ react with oxygen gas to form Ta_2O_5 film during film deposition. The fluoride precursors are not as acceptable as the metal–organic precursors due to their lower thermal decomposition ability. Therefore, the metal–organic precursors $Ta(OCH_3)_5$ and $Ta(OC_2H_5)_5$ are more suitable for mass production. Ta_2O_5 is often deposited by LPCVD using $Ta(OC_2H_5)_5$ as the precursor. $Ta(OC_2H_5)_5$ can be fully vaporized when the source tank temperature is above 150°C [74].

2.4.2. Annealing Conditions

Maintaining the leakage current density of Ta_2O_5 films at an acceptable level is a very important issue for both DRAM capacitor and MOSFET gate dielectric applications. High leakage current will result in the loss of data in the storage capacitor and reduce the DRAM refresh cycle time. However, Ta_2O_5 films deposited by LPCVD using $Ta(OC_2H_5)_5$ as the precursor have the following problems: (1) Ta_2O_5 films suffer from hydrocarbon contamination [72, 84, 106], which results in high electron and

hole oxide trap densities, (2) The incompletely oxidized tantalum metals result in high oxygen vacancy density [56, 72, 84, 85]. However, the results of Shimizu et al. are different [119]. They deposited Ta_2O_5 films by LPCVD and magnetron sputtering. The Ta_2O_5 films were annealed under three conditions. The Rutherford backscattering (RBS) result showed that the Ta/O or C/O composition ratio had no direct relation with the leakage current density of Ta_2O_5 films.

As shown in Table XI, the reported annealing methods can be classified into three groups: (1) high-temperature annealing, such as rapid thermal annealing (RTA) and furnace annealing at 700–900°C in N_2, O_2, or N_2O ambience; (2) low-temperature annealing, such as O_2-plasma annealing and UV-O_3 irradiation; and (3) two-step annealing, which is a mixed annealing method that includes a high-temperature annealing step and a low-temperature annealing step. The advantages of O_2-plasma annealing and UV-O_3 irradiation include the following: (1) The generated high-energy oxygen ions can penetrate the Ta_2O_5 film to repair the oxygen vacancies and to react with carbon to form CO_2. Therefore, hydrocarbon contamination can be minimized by these low-temperature annealing methods [72, 92, 98, 123]. (2) The purpose of low substrate temperature annealing is to avoid Ta_2O_5 crystallization, which will induce high leakage current [72, 86, 93, 124]. (3) The high leakage current due to silicon diffusion can also be avoided by using a low-temperature process [65, 72, 73, 99, 125]. (4) Low-temperature annealing can prevent the growth of native oxide at the interface of the polysilicon bottom electrode and Ta_2O_5. The advantages of UV-O_3+RTA O_2 two-step annealing are as follows: (1) Oxygen vacancies and hydrocarbon contamination in the Ta_2O_5 film are minimized by UV-O_3 irradiation. (2) The defect density of Ta_2O_5 film is then reduced by the following 800°C rapid thermal annealing [106]. The advantages of RTA O_2 annealing+O_2-plasma annealing include the following: (1) Ta_2O_5 film will be crystallized after the rapid thermal annealing step, which will give a higher dielectric constant value. (2) Hydrocarbon contamination in the Ta_2O_5 film can be reduced by oxygen plasma annealing. The reported dielectric constant can be as high as 57, which is higher than that of amorphous films due to Ta_2O_5 crystallization [93]. The use of N_2O as the annealing gas with either rapid thermal annealing or furnace annealing replacing oxygen can further reduce the leakage current level of the Ta_2O_5 film [65, 120, 126] because N_2O annealing can more effectively reduce hydrocarbon contamination. However, it will introduce a 2- to 4-nm-thick SiO_2 at the Ta_2O_5/silicon interface. The effective dielectric constant of the overall Ta_2O_5/SiO_2 stack will be greatly reduced after N_2O annealing. Therefore, furnace N_2O annealing is not suitable for Ta_2O_5 gate dielectric application because the SiO_2 layer will limit the scaling of the MOSFET gate dielectric.

2.4.3. Bottom Electrode

The dielectric constant of Ta_2O_5 is not quite high enough for 256-Mb and 1-Gb DRAM applications if a simple planar bottom electrode is used. Some way to increase the capacitor

Table XI. Annealing Methods of Ta_2O_5 Films

	Annealing method	Annealing step	Advantages and disadvantages	Leakage current density at 1.0 V
High-temperature annealing	O_2 or N_2O rapid thermal annealing O_2 or N_2O furnace annealing	700–900°C for 30–60 s 700–900°C for 30–60 min	Dielectric constant was reduced by the formation of SiO_2 at Ta_2O_5/silicon interface	Smaller than 1×10^{-8} A/cm^2 [120] Smaller than 1×10^{-9} A/cm^2 [120]
Low-temperature annealing	UV-O_2 or UV-O_3 annealing O_2-plasma annealing	300°C for 5–10 min 400–500°C, rf frequency 50 kHz or 13.56 MHz	1. Repairs oxygen vacancies and removes hydrocarbon contamination 2. Low-temperature annealing can avoid silicon atoms diffusing into Ta_2O_5 film	1×10^{-5}–1×10^{-8} A/cm^2 [121] 1×10^{-8} A/cm^2 [98, 122]
Two-step annealing	UV-O_3 annealing+ RTA O_2 annealing at	UV-O_3 at 300°C RTA O_2 annealing 800°C	1. Repairs oxygen vacancies and removes hydrocarbon contamination by UV annealing 2. Reduces defect density by RTA annealing	1×10^{-8} A/cm^2 [106]
	RTA O_2 annealing+ O_2-plasma annealing	RTA O_2 annealing at 550°C+500°C O_2-plasma annealing RTA O_2 annealing at 750°C+570°C O_2 plasma annealing	1. The dielectric constant was increased due to the crystallization of Ta_2O_5 after RTA annealing 2. Oxygen vacancies and hydrocarbon contamination were removed by O_2 plasma annealing	4×10^{-8} A/cm^2 [92] 1×10^{-6} A/cm^2 [93]

area is necessary. The reduction of Ta_2O_5 dielectric thickness is not desirable because the decrease in dielectric thickness will worsen the dielectric reliability. Table XII shows the suitable electrode materials for Ta_2O_5 as the DRAM capacitor dielectric. Rugged or hemispherical grained (HSG) polysilicon bottom electrodes are most widely used [3, 56, 62, 85, 107, 113, 123, 130, 131]. The bottom electrode is etched into become a cylindrical shape. The area of the bottom electrode is thus increased. To avoid the formation of SiO_2 on the polysilicon bottom electrode, nitridation of the polysilicon electrode is required at 800–950°C in NH_3 ambience for 30–60 s. A thin Si_3N_4 layer about 1–2 nm is formed on the polysilicon electrode after the nitridation process. The nitridation process can also be performed in NO ambience at 800°C by rapid thermal annealing [132]. Some authors reported the deposition of 2–3 nm of Si_3N_4 on the polysilicon to suppress oxygen diffusion toward the bottom electrode interface [61, 87, 93]. To fulfill the charge density requirement of 25 fF/cell for 1-Gb DRAM, the height of the cylindrical-shaped bottom electrode should be about 0.4–0.7 μm. This will cause difficulty during the lithography and etching steps. Therefore, approaches such as using

a metal bottom electrode to reduce native SiO_2 or increasing the dielectric constant of the Ta_2O_5 film by high-temperature annealing are required for future 1-Gb DRAM application.

The use of a metal bottom electrode has received much attention. Materials such as W, WSi_x, Pt, TiN, and Ru have been extensively studied. The material selected for the bottom electrode of a DRAM storage capacitor should satisfy either of the following conditions [129]: (1) It must not react with Ta_2O_5, or (2) it must react with Ta_2O_5 to form a high-dielectric-constant insulator, or (3) it must react with Ta_2O_5 to form a conductor. The disadvantage of W and TiN is that their metal oxides WO_3 [63, 127] and TiO_2 [127, 128] will be formed on the bottom electrode interface after high-temperature annealing in oxygen ambience. SiO_2 will also be formed on a tungsten silicide bottom electrode after oxygen annealing [121, 122]. These unwanted oxides will reduce the capacitance of the storage capacitor and lower the effective dielectric constant. The use of a noble metal such as Pt as the bottom electrode can avoid the formation of interfacial oxide. However, the higher leakage current about 1×10^{-7} A/cm^2 at 1.0 V [93, 104], and the metal etching process [93] are the main problems for this application.

Table XII.   Bottom Electrode Materials for Ta_2O_5 Storage Capacitors in DRAM

Reference	Bottom electrode	Advantages and disadvantages	Leakage current density at 1.0 V (A/cm^2)
Aoyama et al. [72] Kamiyama et al. [96, 98] Kim and Kim [85] Lo et al. [113] Ohji et al. [123] Shinriki and Nataka [106] Sun and Chen [120, 126]	Polysilicon with 1. NH_3 nitridation annealing 2. Hemispherical grain (HSG) rugged surface 3. Cylindrical shape	SiO_2 will form at the Ta_2O_5/polysilicon interface, which reduces the effective dielectric constant of Ta_2O_5	Smaller than 1×10^{-8} A/cm^2
Kamiyama et al. [122] Lai and Lee [63] Kim et al. [127] Matsui et al. [121]	W, WSi_x	1. The capacitance of Ta_2O_5/W capacitor is twice that of Ta_2O_5/polysilicon capacitor 2. WO_3 will form at Ta_2O_5/W interface, which reduces the effective dielectric constant of Ta_2O_5	1×10^{-8} A/cm^2
Kim et al. [127] Moon et al. [128] Alers et al. [66]	TiN	TO_2 will form at Ta_2O_5/TiN interface, which reduces the effective dielectric constant of Ta_2O_5	1×10^{-8} A/cm^2
Kishiro et al. [93] Mikhelashvili and Eisenstein [104] Chiu et al. [94] Joshi and Cole [79] Lee et al. [92]	Pt	1. No interface oxide formed at Ta_2O_5/Pt interface 2. Pt is difficult to etch 3. Silicon under the Pt electrode can penetrate through the columnar grain of Pt into Ta_2O_5	1×10^{-7} A/cm^2
Hashimoto et al. [129] Matsuhashi and Nishikawa [112]	Mo	Silicon under Mo electrode can penetrate through the columnar grain of Mo into Ta_2O_5 and result in high leakage current	Higher than 1×10^{-8} A/cm^2
Aoyama et al. [72] Kishiro et al. [93] Moon et al. [128]	Ru	RuO_2 is a conductor; the effective dielectric constant will not be affected by the formation of RuO_2 at Ta_2O_5/Ru interface	1×10^{-8} A/cm^2

Furthermore, the grains of Pt film are of columnar grain structure. Silicon atoms under the Pt electrode will penetrate through the Pt and go into the Ta_2O_5 film, hence increasing the leakage current of the Ta_2O_5 capacitor. The same problem is also observed when Mo is used as the capacitor bottom electrode [129]. The use of Ru as the metal bottom electrode has attracted much attention. The metal oxide of the Ru electrode still acts as a conductor so that the dielectric constant of the Ta_2O_5 capacitor does not decrease. The dielectric constant of Ta_2O_5 film deposited on a Ru electrode can be increased from 26 to 57 [74] after high-temperature annealing because Ta_2O_5 film will become polycrystallized when the annealing temperature is higher than 660°C. It was reported that there is no reaction between the Ru bottom electrode and the Ta_2O_5 film [93]. Therefore, no interfacial oxide is formed at the Ta_2O_5/Ru interface.

2.4.4. Top Electrode

The leakage current density of the storage capacitor for 1-Gb DRAM should be smaller than 1×10^{-8} A/cm^2 at a half power supply voltage of about 1.0 V. The thermal stability of the capacitor top electrode is an important concern. A top electrode with poor thermal stability will increase capacitor leakage current density after a back-end high-temperature process. The

selected top electrode materials including molybdenum, tungsten, titanium, tantalum, and their metal nitrides MoN, WN, TiN and TaN [112]. The thermal stability and the leakage current level of TiN and WoN are better than those of the other metals. The TiN top electrode is compatible with postmetallization at about 400°C and WoN with an 800°C annealing process.

The TiN top electrode and phosphorous-doped polysilicon bottom electrode are used for the Ta_2O_5 capacitor. TiN is also used as the gate metal for Ta_2O_5 gate dielectric applications [55, 59, 130, 132, 133]. However, TiN cannot withstand the high-temperature boron phosphosilicate glass (BPSG) reflow process because of the formation of TiO_2 between the Ta_2O_5 and the top TiN electrode [62, 131]. Cracks will be generated in Ta_2O_5 films due to the difference in the thermal expansion coefficients of Ta_2O_5 and TiO_2. This problem can be minimized by deposition of a 160-nm polysilicon between the BPSG and the TiN layer. After the deposition of the polysilicon layers, the TiN electrode can withstand 850°C, 30-min annealing without damaging the Ta_2O_5 film. The other method to avoid cracks in the Ta_2O_5 film is to use a low-temperature annealing process [134]. SiO_2 was deposited on the TiN top electrode using either PECVD or the spin-on-glass method. Ta_2O_5 capacitors were then annealed at 500°C under a N_2 ambience for 1 h after the metal contact processes.

2.5. Leakage Current Mechanisms of Ta₂O₅ Thin Films

Both bulk-limited current mechanisms and electrode-limited current mechanisms have been observed in Ta₂O₅ films. The bulk-limited current mechanisms include Poole–Frenkel emission, ohmic conduction, hopping conduction, and space-charge-limited current. The electrode-limited current mechanisms include Schottky emission and Fowler–Nordheim tunneling. Poole–Frenkel emission is often observed when Ta₂O₅ film is biased at high electric field or under high temperature. At low electric field, ohmic conduction or hopping conduction is the dominant current mechanism [135]. Schottky emission was also observed at high electric field and high temperature [100]. The current mechanism was reported to be Schottky emission at low electric field and Poole–Frenkel emission at high electric field [94, 102, 105]. When a thin SiO₂ layer is formed at the Ta₂O₅/silicon interface, the leakage current mechanism of the Ta₂O₅/SiO₂ double layer is often Fowler–Nordheim tunneling [108, 111].

2.5.1. Poole–Frenkel Emission

The interaction between the positively charged trap and the electrons gives rise to the Coulombic barrier [118]. As shown in Fig. 6, when the Coulombic potential barrier is lowered by the applied electric field, the electron will be ejected from the traps to the conduction band of the dielectric by field-assisted thermal ionization. The conductivity of the dielectric film will increase due to this effect. This is Poole–Frenkel emission. The traps in Ta₂O₅ film are often denoted as donor centers with energy level E_d [117, 118, 136]. The general expression of current density due to Poole–Frenkel emission is

$$J = q\mu N_C E \exp\left(\frac{-E_d}{kT} + \frac{q\beta_{\text{PF}}E^{1/2}}{kT}\right)$$

where $\beta_{\text{PF}} = (e/\pi\varepsilon_0\varepsilon_i)^{1/2}$. As shown in Table XIII, Poole–Frenkel emission is often observed at high electric field and high temperature [100–103]. Banerjee et al. deposited 10.4

to 30.4-nm Ta₂O₅ films on either an n⁺-silicon or a Ta substrate by sputtering [100]. At relatively higher temperatures (50–100°C), the leakage current mechanism is Poole–Frenkel emission. However, Schottky emission also occurs in this temperature range. To distinguish these two mechanisms, Banerjee et al. deposited Ta₂O₅ film on a p⁺-silicon substrate so that there was a work function difference of about 1.1 eV between the top Al electrode and the p⁺-silicon substrate. Because the observed leakage current is independent of the voltage polarity, the leakage current mechanism should be bulk-limited Poole–Frenkel emission. For Ta₂O₅ films deposited on a Ta substrate, the leakage current mechanism is Poole–Frenkel emission in the high-temperature range (50–100°C). Chaneliere et al. studied the leakage current mechanisms of six different stacked Ta₂O₅ structures [101], namely, a-Ta₂O₅^PECVD/Si₃N₄^plasma, a-Ta₂O₅^PECVD/Si₃N₄^LPCVD, a-Ta₂O₅^PECVD/SiO₂^dry oxidation, a-Ta₂O₅^PECVD/SiO₂^deposition, δ-Ta₂O₅^PECVD/SiO₂^annealing, and δ-Ta₂O₅^LPCVD/SiO₂^annealing. Most of these stacked Ta₂O₅ structures show Poole–Frenkel emission leakage current at medium and high electric field. The leakage current at high electric field for δ-Ta₂O₅^LPCVD/SiO₂^annealing is not Poole–Frenkel emission, however, due to the hydrocarbon contamination in the LPCVD Ta₂O₅ film and high inhomogeneity and cracks at the surface of the Ta₂O₅ film. Ezhilvalavan and Tseng reported that the leakage current of Ta₂O₅ is ohmic conduction at low electric field (<100 kV/cm), Schottky emission at medium electric field (100–350 kV/cm), and Poole–Frenkel emission at high electric field (>350 kV/cm) [102]. They used the static dielectric constant rather than the high-frequency dielectric constant in their analysis of the leakage current mechanism.

Angle and Talley reported that the conduction mechanism of Ta₂O₅ film is ohmic at low applied electric field and Poole–Frenkel emission at high electric field [135]. If there are other defect states such as trap states E_t or acceptor centers E_a in the films, the number of emitted free electrons from the donor centers N_d will decrease. Therefore, the Poole–Frenkel emission is dependent on the concentration of N_t and N_d. When $N_t < N_d$, the conduction mechanism is called normal Poole–Frenkel emission:

$$J = q\mu N_C \frac{N_d - N_t}{N_t} E \exp\left(\frac{-E_d}{2kT} + \frac{q\beta_{\text{PF}}E^{1/2}}{kT}\right) \qquad N_t < N_d$$

When $N_t \cong N_d$, the conduction mechanism is called modified Poole–Frenkel conduction or the anomalous Poole–Frenkel effect:

$$J = q\mu N_C \left(\frac{N_d}{N_t}\right)^{1/2} E \exp\left(\frac{-(E_d + E_t)}{2kT} + \frac{q\beta_{\text{PF}}E^{1/2}}{2kT}\right)$$
$$N_t \cong N_d$$

The exponential term of modified Poole–Frenkel conduction is the same as that of Schottky emission. Therefore, modified Poole–Frenkel conduction is sometimes confused with Schottky emission. Chaneliere et al. deposited Ta₂O₅ film on a silicon substrate using LPCVD [87]. Modified Poole–Frenkel conduction was used to describe the current transport of the Ta₂O₅

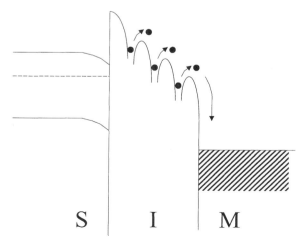

S   I   M

Fig. 6.   Schematic energy band diagram of Poole–Frenkel emission.

Table XIII. Poole–Frenkel Emission in Ta_2O_5 Films

Reference	Capacitor structure	Current mechanism	Comments
Angle and Talley [135]	50–100-nm anodic oxidation or thermal oxidation Ta_2O_5/silicon	1. Ohmic at low field 2. Poole–Frenkel emission at high field	1. Normal Poole–Frenkel emission when $N_t < N_d$ 2. Anomalous Poole–Frenkel emission when $N_t \cong N_d$
Banerjee et al. [100]	10.4–30.4-nm sputtered Ta_2O_5/silicon	1. -50–$25°C$, tunneling 2. 50–$100°C$, Poole–Frenkel emission	Flower–Nordheim tunneling with $\Phi_b = 3.07$ eV for Ta_2O_5 annealed at $900°C$ in O_2 ambience for 30 min
	10.4–30.4-nm sputtered Ta_2O_5/Ta	1. -50–$25°C$, hopping 2. 50–$100°C$, Poole–Frenkel emission	
Chaneliere et al. [87]	LPCVD as-deposited Ta_2O_5/silicon	Modified Poole–Frenkel emission ($r = 2$)	$J = CE \exp\left(\dfrac{-E_d}{kT} + \dfrac{q\beta_{PF}E^{1/2}}{rkT}\right)$ where $1 \le r \le 2$ With high oxide trap density
	Ta_2O_5 annealed at $800°C$ in O_2 ambience for 30 min	Poole–Frenkel emission ($r = 1.4$)	With lowest oxide trap density
Chaneliere et al. [101]	a-$Ta_2O_5^{PECVD}$/$Si_3N_4^{plasma}$ a-$Ta_2O_5^{PECVD}$/$Si_3N_4^{LPCVD}$	Poole–Frenkel emission 1. Hopping at low field 2. Poole–Frenkel emission at medium and high field	
	a-$Ta_2O_5^{PECVD}$/$SiO_2^{dry\ oxidation}$	Poole–Frenkel emission at medium and high field	
	a-$Ta_2O_5^{PECVD}$/$SiO_2^{deposition}$	Poole–Frenkel emission at medium and high field	
	δ-$Ta_2O_5^{PECVD}$/$SiO_2^{annealing}$ δ-$Ta_2O_5^{LPCVD}$/$SiO_2^{annealing}$	Poole–Frenkel emission at medium E field	
Choi and Ling [117]	Simulation, $n < N_t \ll N_d$	Poole–Frenkel emission	n, the free electron density N_t, electron trap density or acceptor center density, $N_t = 8 \times 10^9$–5×10^{10} cm^{-3}
	$N_t < n \ll N_d$	Modified Poole–Frenkel emission	N_d, donor center, $N_d = 1.7 \times 10^{14}$–5×10^{17} cm^{-3} with $E_d = 1.0$–1.5 eV
Devine et al. [115]	85-nm ECR-PECVD Ta_2O_5/silicon	Modified Poole–Frenke emission	$E_t = 1.2$ eV below the E_c of Ta_2O_5; a neutral trap is at 1.0 eV below the E_c of Ta_2O_5
Ezhilvalavan and Tseng [102]	1000-nm sputtered Ta_2O_5/Pt with furnace O_2 annealing at 500–$800°C$ for 30 min	1. Ohmic at low field (<100 kV/cm) 2. Schottky at 100–350 kV/cm 3. Poole–Frenkel emission when $E > 350$ kV/cm	Static dielectric constant was used during analysis
Houssa et al. [103]	6–10-nm MOCVD Ta_2O_5/silicon furnace O_2 annealing at $600°C$ for 30 min	1. Fowler–Nordheim tunneling at low field 2. Poole–Frenkel emission at high field	Refractive index n calculated from Poole–Frenkel emission: As-deposited Ta_2O_5, $n = 2.26$; postannealed Ta_2O_5, $n = 2.44$
Joshi and Cole [79]	17–260-nm MOSD Ta_2O_5/Pt with furnace O_2 annealing at 500–$750°C$ for 60 min	1. Poole–Frenkel emission for amorphous Ta_2O_5 2. Modified Poole–Frenkel emission for crystallized β-Ta_2O_5	β-Ta_2O_5 when annealing temperature higher than $650°C$

Table XIII.   Continued

Reference	Capacitor structure	Current mechanism	Comments
Kaplan et al. [67]	100-nm LPCVD Ta_2O_5/silicon	1. Poole–Frenkel emission at low and medium field 2. SCLC at high field with $I \propto V^{2.4}$	
Miki et al. [107]	10–20-nm MOCVD Ta_2O_5/polysilicon with 800°C annealing	1. Fowler–Nordheim tunneling when $V_G > 0$ V 2. Schottky emission when $V_G < 0$ V at low temperature 3. Poole–Frenkel emission for thicker Ta_2O_5 when $V_G < 0$ V	Leakage currents were measured at −50–120°C
Metikos-Hukovic and Ceraj-Ceric [75]	Anodic oxidation 1–80-nm Ta_2O_5/Ta	1. Modified Poole–Frenkel emission for 20–30-nm-thick films 2. Films thicker than 30 nm, ohmic at low applied voltage, SCLC at high voltage	Trap barrier height $\Phi = (E_t + E_d)/2$; neutral traps E_t formed in Ta_2O_5 film due to water species diffusing into Ta_2O_5 film during anode oxidation
Wu et al. [71]	Sputtered 100-nm Ta_2O_5 on metal	Poole–Frenkel emission	Trap barrier height Φ_b is proportional to $1/\varepsilon$
Zhang et al. [105]	ALD Ta_2O_5 (10 nm)/HfO_2/Si_3N_4/Si ALD Ta_2O_5 (10 nm)/ZrO_2/Si_3N_4/Si	1. Schottky emission when $E < 0.4$ MV/cm 2. Poole–Frenkel emission when $E > 0.4$ MV/cm	Static dielectric constant was used during analysis

film. The expression of Poole–Frenkel emission can be rewritten as

$$J = CE \exp\left(\frac{-E_d}{kT} + \frac{q\beta_{PF}E^{1/2}}{rkT}\right)$$

where $1 \leq r \leq 2$. When $r = 1$, it is the normal Poole–Frenkel effect. When $r = 2$, it represents modified Poole–Frenkel conduction. At high electric field (0.4–1.6 MV/cm), the leakage current mechanism for as-deposited Ta_2O_5 is modified Poole–Frenkel emission ($r = 2$). To satisfy the $\varepsilon_i = n^2$ relationship, r found to be 1.4 for 800°C O_2 annealed Ta_2O_5. Because modified Poole–Frenkel emission is often observed when Ta_2O_5 contains a high trap concentration, therefore as-deposited Ta_2O_5 has a higher density of oxide traps than 800°C O_2 annealed Ta_2O_5. Choi and Ling considered Ta_2O_5 films containing not only donor centers E_d but also electron traps E_t or acceptor centers. They found that the number of electrons ejected from the donor centers to the conduction band of Ta_2O_5 will decrease due to theses energy states [117]. Poole–Frenkel emission leakage current will decrease due to this effect. The conductivity of Ta_2O_5 depends on the donor center density N_d, the electron density n, and the trap density N_t in the film. When $n < N_t \ll N_d$, the leakage current mechanism is normal Poole–Frenkel emission:

$$\sigma = q\mu \frac{N_d - N_t}{2N_t} N_C \exp\left(\frac{-E_d}{kT} + \frac{q\beta_{PF}E^{1/2}}{kT}\right)$$

When $N_t < n \ll N_d$, the concution mechanism is modified Poole–Frenkel emission:

$$\sigma = q\mu \left(\frac{N_d N_C}{2}\right)^{1/2} \exp\left(\frac{-E_d}{2kT} + \frac{q\beta_{PF}E^{1/2}}{2kT}\right)$$

The density of state of Ta_2O_5 in the conduction band and the mobility of electrons are assumed to be $N_C = 10^{18}$ cm^{-3} and $\mu = 0.1$–1.0 cm^2 V^{-1} s^{-1}, respectively. The calculated donor center concentration ranges from 1.7×10^{14} cm^{-3} to 5×10^{17} cm^{-3}. The trap energy level E_d is 1.0–1.5 eV. The trap concentration N_t is about 8×10^9–5×10^{10} cm^{-3}. These parameters are dependent on the film deposition and annealing process conditions.

2.5.2. Hopping Conduction

The leakage current mechanism for amorphous Ta_2O_5 at low electric field is hopping conduction [75, 100, 111]. Hopping conduction in Ta_2O_5 films is listed in Table XIV. Hopping conduction is due to the thermal excitation of electron traps or electrons hopping from one trap level to another in the film. Figure 7 shows a schematic drawing of the energy band diagram of hopping conduction. The expression for hopping conduction is as follows [111]:

$$J = qan\nu \exp\left(-\frac{U}{kT} + \frac{qaE}{kT}\right)$$

where a is the mean hopping distance, n is the electron concentration in the conduction band of Ta_2O_5, ν is the frequency of thermal vibration of electrons at trap sites, and U is the energy of the trap level. Ohmic conduction is often observed at

Table XIV.   Hopping Conduction in Ta_2O_5 Films

Reference	Capacitor structure	Current mechanism	Extracted parameters
Banerjee et al. [100]	10.4–30.4-nm sputtered Ta_2O_5/Ta	1. Hopping when $V_g = 0.1$– 0.4 V with $E_a = 0.20$ eV 2. SCLC when $V_G = 0.5$– 1.0 V	Leakage currents were measured at -50–$25°C$
Metikos-Hukovic and Ceraj-Ceric [75]	Anodic oxidation 1–80-nm Ta_2O_5/Ta	When Ta_2O_5 thicker than 30 nm, ohmic at low applied voltage, SCLC at high voltage	
Zaima et al. [111]	20–40-nm as-deposited LPCVD Ta_2O_5/silicon	Hopping at low field	
	Ta_2O_5 annealed at 700– 900°C in O_2 ambience for 30 min	1. Hopping for 700 and 900°C annealed Ta_2O_5 2. Tunneling for 800°C annealed Ta_2O_5 at low temperature	Mean hopping distance for 900°C annealed Ta_2O_5 film is $a = 4.8$ nm

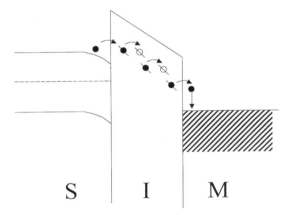

Fig. 7.   Schematic energy band diagram of hopping conduction.

low voltage regions with the space-charge-limited current at medium and higher voltages [75, 100]. Banerjee et al. deposited 10.4 to 30.4-nm Ta_2O_5 films on a Ta substrate using sputtering [100]. When the top electrode is biased at 0.1–0.4 V, the conduction mechanism is hopping with an activation energy E_a of 0.20 eV. When the top electrode is biased at 0.5–1.0 V, the activation energy E_a is between 0.49 and 0.89 eV and the I–V relationship is $I \propto V^2$. Thus, the current mechanism is space-charge-limited conduction at this voltage range. Zaima et al. reported that the conduction mechanism for as-deposited Ta_2O_5 film at low electric field is ohmic because as-deposited Ta_2O_5 film is though to have so many defects that trapped electrons are thermally excited from one trap level to another [111]. The calculated mean hopping distance is 4.8 nm, which only depends slightly on the annealing temperature. The leakage current of 800°C O_2 annealed Ta_2O_5 film is lower than that of 700°C O_2 annealed film by 4 orders of magnitude. This is due to the fact that the density of defects with a shallow energy level is reduced by high-temperature annealing.

2.5.3. Space-Charge-Limited Current

Space-charge-limited current (SCLC) is one of the bulk-limited current mechanisms. Ohmic conduction is often observed at low electric field and the SCLC at medium and high electric field. Therefore, the I–V relationship is $J \propto V^1$ at low voltage and $J \propto V^2$ at higher voltage. $J \propto V^2$ when the injected free carrier densitiy is higher than the intrinsic thermally generated carrier density. Assuming a single trap level, the SCLC can be expressed as

$$J = \frac{9\theta\mu\varepsilon_0\varepsilon_r}{8d^3}V^2$$

where μ is the carrier mobility in the absence of traps, d is the film thickness, θ is the ratio of free to trapped charges:

$$\theta = \frac{N_C}{N_t}\exp\left(\frac{-E_t}{kT}\right)$$

and $\mu_{eff} = \mu\theta$ is the effective mobility. When electrons fill the trap levels, the current density will increase sharply by a factor of $1/\theta$. This is the trap-filled-limited (TFL) condition. Finally, the leakage current returns to the $J \propto V^2$ relationship. The Ta_2O_5 film has a very slow mobility μ of about 10^{-6} cm^2 V^{-1} s^{-1} or even smaller [100]. Metikos-Hukovic and Ceraj-Ceric reported that when the thickness of the Ta_2O_5 film is thinner than 30 nm, the leakage current mechanism is ohmic conduction at low applied voltage [75]. The J–V relationship at high electric field followed $J \propto E^2$. The leakage current mechanism is space-charge-limited current. The calculated effective mobility μ_{eff} is 10^{-9} cm^2 V^{-1} s^{-1}.

2.5.4. Schottky Emission

Schottky emission is one of the electrode-limited current mechanisms. Figure 8 shows the schematic drawing of the energy band diagram of Schottky emission. Schottky emission can be represented as

$$J = A^* T^2 \exp\left(\frac{-q\Phi_B}{kT} + \frac{q\beta_{SK}E^{1/2}}{kT}\right)$$

where $\beta_{SK} = (q/4\pi\varepsilon\varepsilon_0)^{1/2}$ is due to image force barrier lowering. When the expression of Schottky emission is compared to that of normal Poole–Frenkel emission ($r = 1$), $\beta_{SK} = \beta_{PF}/2$ in the exponential term. If the expression of Schottky emission is compared to that of modified Poole–Frenkel emission ($r = 2$), theses two mechanisms have the same β factor in the barrier-lowering term. Bulk-limited mechanisms are independent of voltage polarity and show symmetric current

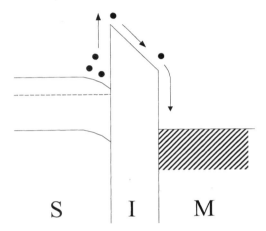

Fig. 8.   Schematic energy band diagram of Schottky emission.

characteristics. Schottky emission can be distinguished from modified Poole–Frenkel emission by the dependence of the leakage current on the voltage polarity. Some authors reported the observation of Schottky emission at low electric field with Poole–Frenkel emission at high electric field [94, 105]. A static dielectric constant was used in their analysis. Thinner Ta_2O_5 films tend to show Schottky emission, whereas thicker Ta_2O_5 films show Poole–Frenkel emission [107]. Schottky emission in Ta_2O_5 films is listed in Table XV. Lai and Lee deposited 14-nm Ta_2O_5 film on a tungsten electrode using PECVD [63]. When the bias on the top Au electrode is smaller than 2.5 V, the leakage current is due to Schottky emission with a high-frequency dielectric constant of 3.6. This high-frequency dielectric constant is in agreement with the refractive index of 1.9–2.0 measured by ellipsometry. The potential barrier between the W and Ta_2O_5 interface is found to be 1.0 eV. Matsuhashi and Nishikawa selected Mo, W, Ti, Ta, and their metal nitrides as the top electrodes of Ta_2O_5/n^+-polysilicon capacitors [112]. When electrons are injected from the top electrode, the leakage current decreases as the work function of the top electrode increases. When electrons are injected from the polysilicon bottom electrode, the leakage current is independent of the top electrode material when the current density is plotted as function of the voltage drop on the Ta_2O_5 film ($V_G - \Phi_{ms}$). This indicates that current mechanism of a metal/Ta_2O_5/polysilicon capacitor is electrode limited for both voltage polarities. However, according to Matsuhashi and Nishikawa the current mechanism is not Schottky emission because the extrapolated dielectric constant of 0.3–0.8 is smaller than the static dielectric constant of 21.

Table XV.   Schottky Emission in Ta_2O_5 Films

Reference	Capacitor structure	Measurement conditions	Comments
Chiu et al. [94]	52.5-nm PECVD Ta_2O_5/Pt	1. Schottky emission when $E < 2$ MV/cm 2. Poole–Frenkel emission when $E > 2$ MV/cm	Static dielectric constant was used during analysis
Lai and Lee [63]	14.4-nm PECVD Ta_2O_5/W	1. Schottky emission when $V_G < 2.5$ V 2. Fowler–Nordheim tunneling when $V_G > 2.5$ V	1. High-frequency dielectric constant $\varepsilon = 3.6$ 2. Ta_2O_5/W barrier height is 1.0 eV
Matsuhashi and Nishikawa [112]	LPCVD 10–50-nm Ta_2O_5/polysilicon Top electrode materials: Mo, W, Ti, Ta, and their metal nitrides	Electrode-limited mechanism	High-frequency dielectric constant obtained from Schottky emission mechanism was 0.3–0.8
Miki et al. [107]	10–20-nm MOCVD Ta_2O_5/rugged polysilicon	1. Schottky emission when $V_G < 0$ V at low temperature 2. Poole–Frenkel emission for thicker Ta_2O_5 when $V_G < 0$ V	Leakage currents were measured under -50–$120°C$
Zhang et al. [105]	ALD Ta_2O_5 (10 nm)/HfO_2/Si_3N_4/Si ALD Ta_2O_5 (10 nm)/ZrO_2/Si_3N_4/Si	1. Schottky emission when $E < 0.4$ MV/cm 2. Poole–Frenkel emssion when $E > 0.4$ MV/cm	Static dielectric constant was used during analysis

2.5.5. Fowler–Nordheim Tunneling

When the energy of the incident electrons is smaller than the potential barrier Φ_B, classical physics predicts that the electron will be reflected. However, when the barrier is thin enough (<10 nm), quantum mechanics shows that the electron wave function will penetrate through the potential barrier. Therefore, the probability of electrons existing at the other side of the potential barrier is not 0. This phenomenon is called tunneling. Figures 9 and 10 show the energy band diagrams for Fowler–Nordheim tunneling and direct tunneling, respectively. Fowler–Nordheim tunneling occurs when the applied electric field on the film is large enough so that the electron wave function can penetrate through the triangular potential barrier into the conduction band of the dielectric. Fowler–Nordheim tunneling current can be represented as

$$J = \frac{q^3 E^2}{8\pi h \Phi_B} \exp\left(\frac{-8\pi (2m*)^{1/2}}{3qhE}\right)$$

As listed in Table XVI, Fowler–Nordheim tunneling was often observed when a thin Ta_2O_5 film was deposited on a silicon substrate [100, 108, 110]. However, the extracted barrier height interface obtained from the Fowler–Nordheim tunneling

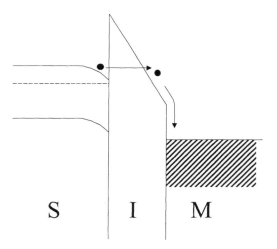

Fig. 9. Schematic energy band diagram of Fowler–Nordheim tunneling.

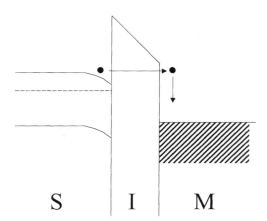

Fig. 10. Schematic energy band diagram of direct tunneling.

mechanism between Ta_2O_5 and silicon is not the true barrier height. The extracted barrier height of about 3.07 eV is the barrier height of native SiO_2 and silicon [100, 108]. Banerjee et al. pointed out that because the polaron mobility of Ta_2O_5 film is very small, the polaron effective mass m^* should be large [100]. Therefore, the electron potential barrier height at the Ta_2O_5/silicon interface should be smaller than half of the bandgap of Ta_2O_5. Lai and Lee found that the potential barrier height at the Ta_2O_5/W interface is 1.32 eV using the Fowler–Nordheim tunneling equation and assuming an effective mass of $m^* = 0.5m_0$ [63]. There is a thin $WO_{2.9}$ layer at the Ta_2O_5/W interface because the tungsten bottom electrode was oxidized by oxygen plasma during the deposition of the Ta_2O_5 film. Because the energy bandgap of $WO_{2.9}$ is only 3.0 eV, this 1.32-eV barrier is considered as the barrier height between the tungsten electrode and Ta_2O_5. The barrier height of 1.0 eV obtained from Schottky emission is lower than that obtained from Fowler–Nordheim tunneling. This may be due to the fact that $WO_{2.9}$ acts like a stepping stone for electron transport from the W electrode to the Ta_2O_5 conduction band. Zaima et al. reported that the leakage current mechanism for Ta_2O_5 film with an annealing temperature higher than 850°C is Fowler–Nordheim tunneling [111]. The potential barrier heights for 850 and 900°C annealed Ta_2O_5 films are 1.76 and 1.69 eV, respectively, assuming an effective mass of $m^* = 0.42m_0$.

2.6. Dielectric Charges of Ta_2O_5 Thin Films

The main problem for the application of Ta_2O_5 as the MOSFET gate dielectric is the instability of the C–V characteristics. The flatband voltage shift of Ta_2O_5 is very large, on the order of a few tenths of a volt. This will cause instability in the threshold voltage. Figure 11 shows four kinds of charges in the metal–dielectric–silicon structure. The mechanism of oxide trap formation in Ta_2O_5 film is not clear. Most literature results reported that the traps in LPCVD Ta_2O_5 come from hydrocarbon contamination [72, 84, 92, 106]. The oxygen vacancy is also reported to be responsible for the oxide trap formation in Ta_2O_5 film [72, 84, 85, 92, 95, 137]. However, Rutherford backscattering (RBS) results showed that the Ta/O or C/O composition ratio has no direct relationship with the leakage current density of Ta_2O_5 films [119]. Other authors reported that the traps come from the diffusion of silicon atoms into Ta_2O_5 film [65, 72, 95, 116]. These silicon atoms formed SiO_2 and SiO in the Ta_2O_5 grains and on the grain boundaries [72]. Because high-temperature annealing will result in the diffusion of silicon atoms and increase the number of oxygen vacancies [102], low-temperature (300–550°C) annealing such as O_2-plasma annealing is required for future DRAM and MOSFET gate dielectric applications.

2.6.1. Fixed Oxide Charge

Angle and Talley deposited 50- to 100-nm Ta_2O_5 films on a silicon substrate by using anodic oxidation or thermal oxidation [135]. A fixed oxide charge density of $Q_f = -8 \times$

Table XVI.   Fowler–Nordheim Tunneling in Ta_2O_5 Films

Reference	Capacitor structure	Current mechanism	Extracted parameters
Banerjee et al. [100]	10.4–30.4-nm sputtered Ta_2O_5/silicon with 900°C furnace O_2 annealing for 30 min	Fowler–Nordheim tunneling with $\Phi_B = 3.07$ eV and $m^* = 0.39m_0$	Thin SiO_2 is formed at Ta_2O_5/silicon interface
Lai and Lee [63]	14-nm PECVD Ta_2O_5/W	1. Schottky emission when $V_G < 2.5$ V 2. Fowler–Nordheim tunneling when $V_G > 2.5$ V	Ta_2O_5/W barrier height obtained from Fowler–Nordheim tunneling is 1.32 eV with $m^* = 0.5m_0$
Lo et al. [108]	60-nm LPCVD Ta_2O_5/SiO_2 (3.4 nm)/silicon	1. Poole–Frenkel emission when $|E_{ox}| < 4$ MV/cm 2. Fowler–Nordheim tunneling when $|E_G| > 0$ V	When $V_G > 0$, the barrier height obtained from Fowler–Nordheim tunneling is 3.0 eV, which is the barrier height of the SiO_2/silicon interface
	SiO_2 (2 nm)/60-nm LPCVD Ta_2O_5/SiO_2 (3.4 nm)	1. Poole–Frenkel emission when $|E_{ox}| < 4$ MV/cm 2. Fowler–Nordheim tunneling when $|E_{ox}| < 4$ MV/cm 3. Fowler–Nordheim tunneling when $V_G > 0$ V	
	Si_3N_4 (3 nm)/60-nm LPCVD Ta_2O_5/SiO_2 (3.4 nm)/ silicon	1. Poole–Frenkel emission when $V_G < 0$ V 2. Fowler–Nordheim tunneling when $V_G > 0$ V	
Miki et al. [107]	10–20-nm MOCVD Ta_2O_5/rugged polysilicon with 800°C annealing in O_2 ambience	Fowler–Nordheim tunneling when $V_G > 0$ V because the leakage currents is independent of the temperature	Leakage currents were measured under −50–120°C
Nishioka et al. [110]	6–40-nm sputtered Ta_2O_5/silicon	1. Bulk-limited mechanism when Ta_2O_5 thicker than 10 nm 2. Fowler–Nordheim tunneling when Ta_2O_5 thinner than 10 nm	
Shinriki and Nataka [106]	7–15-nm LPCVD Ta_2O_5/polysilicon with two-step annealing	1. Bulk-limited mechanism for Ta_2O_5 $t_{eq} > 4$ nm and $V_G > 0$ V 2. Fowler–Nordheim tunneling or direct tunneling when Ta_2O_5 $t_{eq} < 4$ nm and $V_G > 0$ V	Two-step annealing: UV-O_3 annealing at 300°C+ dry O_2 annealing at 800°C for 30 min
	7–15-nm LPCVD Ta_2O_5/polysilicon with UV-O_3 annealing	1. Bulk-limited mechanism for Ta_2O_5 thicker than 10 nm and $V_G > 0$ V 2. Direct tunneling for Ta_2O_5 thinner than 10 nm and $V_G > 0$ V	
Zaima et al. [111]	20–40-nm LPCVD Ta_2O_5/silicon with 800°C annealing in O_2 ambience for 30 min	Tunneling current	
	20–40-nm LPCVD Ta_2O_5/silicon with 850°C annealing in O_2 ambience for 30 min	Fowler–Nordheim tunneling with barrier height of 1.76 eV and $m^* = 0.42m_0$	
	20–40-nm LPCVD Ta_2O_5/silicon with 900°C annealing in O_2 ambience for 30 min	Fowler–Nordheim tunneling with barrier height of 69 eV and $m^* = 0.42m_0$	

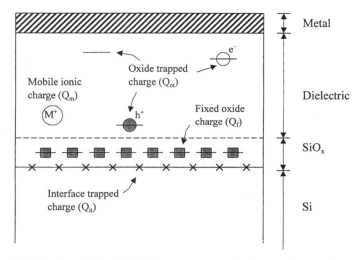

Fig. 11. Four kinds of dielectric charges and their locations in the metal–dielectric–silicon structure.

10^{11} cm^{-2} was calculated from the flatband voltage. The negative fixed oxide charge density of the Ta$_2$O$_5$/silicon interface is different from the positive fixed oxide charge density of the SiO$_2$/silicon structure. The reason for the negative fixed oxide charge density of Ta$_2$O$_5$ film is not clear. Doping Ta$_2$O$_5$ with nitrogen was found to reduce the Q_f by about 30%. The average fixed oxide charge density of Ta$_2$O$_5$ films without nitrogen doping is $Q_f = -5.2 \times 10^{11}$ cm^{-2}, compared with $Q_f = -7.8 \times 10^{11}$ cm^{-2} in nitrogen-doped films. Nitrogen doping had the effect of providing more uniform V_{FB} and lower Q_f.

2.6.2. Oxide Trapped Charge

As shown in Table XVII, flatband voltage shifts due to electron trapping [99, 115, 138] or hole trapping [33, 71, 75, 82, 87, 108, 124, 139] were both observed for Ta$_2$O$_5$ films. Devine et al. deposited 85-nm Ta$_2$O$_5$ film on p-type silicon using ECR-PECVD [115]. The C–V curves were measured at 1 khz at a sweep rate of 0.2 V/s from −3 V to +3 V. A positive flatband voltage shift due to electron trapping by oxide trapped charges Q_{ot} was observed in Ta$_2$O$_5$. The activation energy E_a measured by ZBTSC for the trap is 1.2 eV. Devine et al. explained this result by an electron trap of 1.2 eV below the conduction band of Ta$_2$O$_5$. Postmetallization annealing at 200°C can remove the shallow energy levels in Ta$_2$O$_5$. Hwu et al. reported a clockwise C–V hysteresis no matter whether the start voltage of the C–V measurement is from accumulation or inversion [138]. The direction of C–V hysteresis is also independent of the return voltage or the hold time at the return voltage. This indicates that the traps in Ta$_2$O$_5$ tend to capture electrons. The increase in leakage current of Ta$_2$O$_5$ was related to electron trapping in Ta$_2$O$_5$ film. Park et al. stressed Ta$_2$O$_5$ capacitors at −4 V for 1 min and measured the flatband voltage shift [99]. When the annealing temperature is higher than 750°C, the flatband voltage shift ΔV_{FB} of the stressed film increases much

faster than that of the unstressed film. This is due to negative charge trapping in the Ta$_2$O$_5$ bulk, at the Ta$_2$O$_5$/silicon and Ta$_2$O$_5$/SiO$_2$ interfaces. Grahn et al. reported that when the C–V curve was swept from inversion to accumulation and back, the C–V hysteresis was counterclockwise with a flatband voltage shift ΔV_{FB} of −0.1–0.2 V [124]. The negative value of ΔV_{FB} shows that the Ta$_2$O$_5$ film contains net positive trapped charge. The flatband voltage V_{FB} changes its sign from positive to negative when the thermal oxidation temperature of the Ta$_2$O$_5$ film is higher than the crystallization temperature. Seki et al. studied the influence of postmetallization annealing on the flatband voltage shift and oxide trapped charge Q_{ot} [82]. The flatband voltage shift ΔV_{FB} decreases linearly with an increase in annealing temperature. The Q_{ot} after 450°C annealing is -1.4×10^{11} cm^{-2}. The flatband voltage shift ΔV_{FB} is 0.03 V, which is independent of the postmetallization annealing temperature. The Q_{ot} for 17 to 110-nm-thick Ta$_2$O$_5$ capacitor is about -3.0×10^{11}–-3.8×10^{11} cm^{-2}, which is independent of Ta$_2$O$_5$ thickness. If the oxide trapped charge is uniformly distributed in the Ta$_2$O$_5$ film, thicker Ta$_2$O$_5$ film will have higher concentration of Q_{ot}. Therefore, Q_{ot} is not uniformly distributed across the Ta$_2$O$_5$ film.

In fact, the flatband voltage shift depends on the C–V sweep direction and the stress conditions. Oehrlein reported that the C–V hysteresis loop of a Ta$_2$O$_5$/p-silicon capacitor after 15-V 30-s stress has larger flatband voltage than the unstressed film [140]. This is due to the negative charges injected into the Ta$_2$O$_5$ film or generated within the film. When the stress voltage becomes more negative, the hysteresis loop will shift to more negative value due to positive charge injected into the Ta$_2$O$_5$ film or generated within the film. The flatband voltage value does not depend on the positive starting voltage (inversion voltage) but depends on the negative starting voltage (accumulation voltage). This indicates that the hole-trapping effect in accumulation is more significant than electron trapping in inversion. Kato et al. studied the relationship between the start voltage of the C–V sweep and the flatband voltage shift [61]. The C–V curves were measured with different starting voltages from inversion (4–10 V) to accumulation (−10 V) and back. For a Ta$_2$O$_5$/Si$_3$N$_4$/p-silicon capacitor annealed by 800°C RTA, the flatband voltage becomes more negative when the starting of voltage decreases. This is because the electrons trapped in the Ta$_2$O$_5$ film move back to the depletion layer of silicon when the capacitor is biased under accumulation. However, there is no flatband voltage shift for a Ta$_2$O$_5$ capacitor after O$_3$ annealing+800°C RTA annealing biased under these conditions. The effective trapped charge densities $Q_{eff} = -C_{ox}\Delta V_{FB}$ for 10-, 20-, and 31-nm-thick Ta$_2$O$_5$ film were studied. The calculated Q_{eff} is about 1.9×10^9–3.1×10^9 cm^{-2}. The effective trapped charge density is linearly dependent on the thickness of the Ta$_2$O$_5$ film. The electron trap centers are distributed uniformly throughout the Ta$_2$O$_5$ layer. The C–V curves were also measured using different starting voltage from accumulation (−4–−10 V) to inversion (10 V) and back. The flatband voltage for a capacitor after O$_3$ annealing+800°C RTA annealing decreases with

Table XVII. Oxide Trapped Charges in Ta_2O_5 Films

Reference	Capacitor structure	Measurement conditions	Extracted parameters
Devine et al. [115]	85-nm ECR-PECVD Ta_2O_5/p-silicon	1. $C-V$ sweep from -3 V to $+3$ V 2. -3 V stress with illumination for 30 s 3. Second $C-V$ sweep from -3 V to $+3$ V	1. Flatband voltage shift $\Delta V_{FB} = 1.7$ V 2. Electrons were trapped by oxide trapped charge Q_{ot} in Ta_2O_5 3. $E_a = 1.2$ eV was measured from ZBTSC
	85-nm ECR-PECVD Ta_2O_5/p-silicon with post-metallization annealing at 150°C for 10 min	1. $C-V$ sweep from -3 V to $+3$ V 2. Postmetallization annealing 3. Second $C-V$ sweep from -3 V to $+3V$	Flatband voltage shift $\Delta V_{FB} = 1.0$ V
	85-nm ECR-PECVD Ta_2O_5/p-silicon with post-metallization annealing at 200°C for 10 min	1. $C-V$ sweep from -3 V to $+3$ V 2. Postmetallization annealing 3. Second $C-V$ sweep from -3 V to $+3V$	1. Flatband voltage shift $\Delta V_{FB} = 0$ V 2. Shallow trap energy levels were removed by 200°C postmetallization annealing
Grahn et al. [124]	1. 96-nm e-beam-deposited Ta/p-silicon 2. O_2-plasma oxidation to form Ta_2O_5 at 170–750°C	$C-V$ hysteresis was measured from inversion to accumulation and back	1. Flatband voltage V_{FB} changes from positive to negative when Ta_2O_5 becomes crystallized 2. $C-V$ hysteresis is negative with $\Delta V_{FB} = -0.1-$ -0.2 V, indicating a net positive trapped charge in Ta_2O_5
Hwu et al. [138]	1. 62-nm e-beam-deposited Ta/p-silicon 2. Thermal oxidation to form Ta_2O_5 at 500°C	$C-V$ measured from inversion (accumulation) to accumulation (inversion) and back with different return voltages or different hold times for the return voltage	1. $C-V$ hysteresis is clockwise, independent of start and return voltages and hold time of the return voltage 2. Traps in Ta_2O_5 tend to trap electrons
Kato et al. [61]	10–31-nm LPCVD Ta_2O_5/3-nm Si_3N_4/p-silicon with 300°C O_3 annealing 5 min+O_2 RTA 800°C annealing for 30 s	$C-V$ sweep from inversion (5 V) to accumulation (-8 V) and back	1. $C-V$ hysteresis is counterclockwise 2. Some $C-V$ curve stretch-out was observed
		$C-V$ sweep from inversion (4–10 V) to accumulation (-10 V) and back	No flatband voltage shift
		$C-V$ sweep from accumulation ($-4--10$ V) to inversion (10 V) and back	Flatband voltage becomes more negative when the absolute value of start voltage increases
	10–31-nm LPCVD Ta_2O_5/3-nm Si_3N_4/p-silicon with O_2 RTA 800°C annealing for 30 s	$C-V$ sweep from inversion (5 V) to accumulation (-8 V) and back	1. ΔV_{FB} is smaller than that of O_3 annealed+O_2 RTA 800°C annealing Ta_2O_5 film 2. No $C-V$ curve stretch-out
		$C-V$ sweep from inversion (4–10 V) to accumulation (-10 V) and back	Flatband voltage becomes more negative when the start voltage decreases
		$C-V$ sweep from accumulation ($-4--10$ V) to inversion (10 V) and back	No flatband voltage shift

Table XVII.   (Continued)

Reference	Capacitor structure	Measurement conditions	Extracted parameters
Lai et al. [139]	20-nm PECVD Ta_2O_5/p-silicon	$C-V$ curve hysteresis measurement: 1. $C-V$ curve measured from -3 V to $+3$ V 2. Capacitors were stressed at $+3$ V or -3 V 3. $C-V$ curve measured from -3 V to $+3$ V again	1. ΔV_{FB} is negative and is independent of stress voltage; oxide traps in Ta_2O_5 tend to trap holes 2. Hole trap density is independent of Ta_2O_5 thickness
	14-nm PECVD Ta_2O_5/p-silicon	Tunneling front model measurement: 1. Capacitors were stressed at $+3$ V or -3 V for 100 s 2. Capacitors were shorted to measure the discharging transient current	1. Longer stress time is required to generate electron traps for 14-nm Ta_2O_5 film 2. Electron trap density for 14-nm-thick Ta_2O_5 is smaller than that of 20-nm-thick Ta_2O_5
Lo et al. [108]	60-nm LPCVD Ta_2O_5/SiO_2 (3.4 nm)/silicon	Constant current stresses at 8×10^{-6} A/cm^2 for both $+V_G$ and $-V_G$ polarity	1. Little positive-charge buildup under $-V_G$ stress 2. Initial negative-charge buildup followed by a positive-charge buildup under $+V_G$ constant current stress
	SiO_2 (2 nm)/60-nm LPCVD Ta_2O_5/SiO_2 (3.4 nm) Si_3N_4 (3 nm)/60-nm LPCVD Ta_2O_5/SiO_2 (3.4 nm)/silicon		1. Significant positive-charge buildup was observed under $-V_G$ stress 2. Only positive-charge trapping was observed under $+V_G$ constant current stress
Oehrlein [140]	1. E-beam-evaporated Ta/p-silicon 2. Furnace O_2 annealing at 430–675°C to form 75-nm-thick Ta_2O_5	$C-V$ hysteresis curve was swept from 4 V to -4 V and back 2 times	1. $C-V$ curve hysteresis is counterclockwise when temperature <490°C; $C-V$ hysteresis loop moves to positive direction for the second time measurement 2. $C-V$ curve hysteresis is clockwise when temperature >520°C; $C-V$ hysteresis loop moves to positive direction for the second time measurement
Park et al. [99]	1. Sputtered Ta/p-silicon 2. Furnace O_2 annealing at 500°C for 1 h to form 19-nm Ta_2O_5 3. Furnace N_2 or O_2 annealing at 700–950°C for 20 min	Ta_2O_5 capacitors were stressed at -4 V for 1 min to measure the flatband voltage shift	1. ΔV_{FB} increases significantly after 750°C annealing due to negative-charge trapping in Ta_2O_5 2. The increase in ΔV_{FB} is due to the increase in thickness of SiO_2 for annealing temperature at 750–950°C
Park and Im [125]	1. Sputtered Ta/p-silicon 2. Furnace O_2 annealing at 450–700°C for 1 h to form 65–212-nm-thick Ta_2O_5	Flatband voltages were measured at 100 kHz	1. V_{FB} becomes more negative with the increase in annealing temperature 2. V_{FB} increases with increasing Ta_2O_5 thickness

Table XVII.   (Continued)

Reference	Capacitor structure	Measurement conditions	Extracted parameters
		1. Constant voltage stress was performed at 15 V for 30 s 2. C–V hysteresis curve was swept from 4 V to −4 V and back	1. C–V hysteresis loop moves to positive direction as compared to unstressed sample 2. Ta_2O_5 tends to trap negative charge
		C–V curves were measurement at 1 MHz	1. V_{FB} decreases with increasing annealing temperature or increasing annealing time 2. V_{FB} increases with increasing Ta_2O_5 thickness 3. V_{FB} value depends on the negative start voltage (accumulation voltage) 4. V_{FB} measured under illumination is higher than that measured without illumination
Seki et al. [82]	17–110-nm sputtered Ta_2O_5/silicon Postmetallization H_2 annealing at 300 or 450°C for 30 min	C–V curves of Ta_2O_5 capacitor	1. V_{FB} decreases with increasing annealing temperature 2. Q_{ot} for 17–110-nm-thick Ta_2O_5 after 450°C postmetallization H_2 annealing ranges from -3.0×10^{11} to -3.8×10^{11} cm^{-2}; Q_{ot} is not uniformly distributed across the Ta_2O_5 film

increasing absolute value of the sweep starting voltage. This means that the trapped hole density increases with the increasing absolute value of accumulation voltage. However, there is no flatband voltage shift for an 800°C RTA annealed Ta_2O_5 capacitor under these biased conditions. Kato et al. concluded that O_3 annealed + 800°C RTA annealed Ta_2O_5 films have fewer electron traps but more hole traps than 800°C RTA annealed Ta_2O_5 films. This is related to the change in stoichiometry of Ta_2O_5 from oxygen deficient to oxygen abundant or to the change in the Ta–O bonding length. Lai et al. deposited Ta_2O_5 film on a silicon substrate using PECVD. The C–V curves of a 20-nm Ta_2O_5 capacitor were measured at 1 MHz from inversion (3 V) to accumulation (−3 V) [139]. The samples were stressed at 3 V or −3 V before C–V measurement. The flatband voltage shift ΔV_{FB} of Ta_2O_5 capacitors without stress is −0.22 V. The flatband voltage shifts for Ta_2O_5 capacitor with 3 V and −3 V stress are $\Delta V_{FB} = -0.138$ and −0.483 V, respectively. All of these capacitors show negative flatband voltage shift. This indicates that the oxide trapped charges in Ta_2O_5 tend to trap holes. As shown in Fig. 12, when electrons release their potential energy in the silicon substrate, impact ionization will occur. Holes with high energy will be injected into the Ta_2O_5 film either by jumping across the Ta_2O_5/silicon

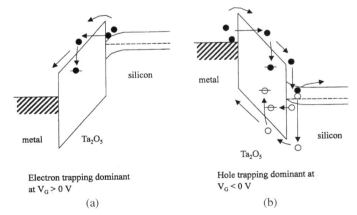

Electron trapping dominant at $V_G > 0$ V

(a)

Hole trapping dominant at $V_G < 0$ V

(b)

Fig. 12.   Schematic energy band diagram of metal/Ta_2O_5/silicon capacitor under (a) $V_G > 0$ V and (b) $V_G < 0V$ conditions.

hole barrier height or through tunneling. The generated holes will be trapped in the Ta_2O_5 film. The spatial distribution of the oxide trapped electron and hole charge densities can then be obtained by measuring the discharging transient current calculated using the tunneling front model. The electron and hole discharging transient currents were measured after 3 V or −3 V stress for 100 s, respectively. The electron trap levels and hole trap levels are assumed at the middle of the energy bandgap of Ta_2O_5 (2.1 eV). An effective mass $m^* = 0.5m_0$ and a characteristic time $t_0 = 10^{-13}$ s are used to calculate the oxide trapped charge density in Ta_2O_5. The calculated oxide trapped charge density shows that the hole trap density is independent of Ta_2O_5 thickness. The electron trap density for 20-nm-thick Ta_2O_5 is higher than that of 14-nm-thick Ta_2O_5. Because the flatband voltage shift is determined by net the sum of electron and hole trapped charges in the Ta_2O_5 film, the 20-nm-thick Ta_2O_5 has smaller flatband voltage shift than that of 14-nm-thick Ta_2O_5.

The origin of hole generation may come from impact ionization in the silicon substrate [139]. It is also possible that impact ionization occurs in Ta_2O_5 films under high-voltage stress because the energy bandgap of Ta_2O_5 is only about 4 eV. Lo et al. studied the flatband voltage shift of Ta_2O_5 stack layers [108]. Constant current stresses at 8×10^{-6} A/cm^2 were performed on three Ta_2O_5 stack layers to study the oxide trapped charge buildup. For $Si_3N_4/Ta_2O_5/SiO_2$/silicon and $SiO_2/Ta_2O_5/SiO_2$/silicon capacitors, significant positive-charge buildup was observed under $-V_G$ stress. This is due to the small energy bandgap of Ta_2O_5 of about 4 eV. Therefore, impact ionization can occur in Ta_2O_5 relatively easily. The impact-ionization-generated holes at the Ta_2O_5/SiO_2 interface will drift toward the polysilicon top electrode. If there is a thin SiO_2 or Si_3N_4 layer as the hole injection barrier between the polysilicon electrode and Ta_2O_5 film, the generated hole may not drift out of Ta_2O_5, which could result in more positive-charge trapping. The positive-charge trapping is more serious for the $Si_3N_4/Ta_2O_5/SiO_2$/silicon capacitor than for the $SiO_2/Ta_2O_5/SiO_2$/silicon capacitor because the energy bandgap of Si_3N_4 is also small (about 5 eV). The impact-ionization-generated hole also accumulated at the Si_3N_4/Ta_2O_5 interface because Si_3N_4 acts as a hole injection barrier between the polysilicon electrode and the Ta_2O_5 film. The Ta_2O_5/SiO_2/silicon capacitor under $+V_G$ constant current stress show an initial negative-charge buildup followed by a positive-charge buildup. This is due to initial negative-charge buildup by trapped electrons followed by trapped holes, which then recombined with electrons. The trapped holes come from either ionization occurring in the Ta_2O_5 film or hole injection from the top polysilicon electrode.

Oehrlein reported that the direction of the $C-V$ hysteresis curve changes from counterclockwise to clockwise when the thermal oxidation temperature of Ta_2O_5 films higher than 520°C [140]. The flatband voltage of Ta_2O_5 capacitor decreases with increasing thermal oxidation temperature and thermal oxidation time. The flatband voltage also decreases with decreasing Ta_2O_5 thickness. The decrease in the flatband voltage is due to the reduced oxygen deficiency in the Ta_2O_5 films. The flatband voltage shift ΔV_{FB} of Ta_2O_5 capacitors was measured after constant voltage stress at $0--17$ V for 20 s. The flatband voltage shift first increases with the absolute value of stress voltage. The maximum ΔV_{FB} shift occurred at -7-V stress (1 MV/cm). ΔV_{FB} then decreases significantly with further increasing stress voltage. The ΔV_{FB} of Ta_2O_5 capacitors is higher under illumination than without illumination. The increase in ΔV_{FB} under illumination is due to electron–hole pairs generated in the silicon depletion layer, which results in electrons being captured by the interface trap states. The flatband voltage instability is not due to ionic processes or dielectric polarization because ionic processes or dielectric polarization will lead to a negative ΔV_{FB} under positive stress voltage and give positive ΔV_{FB} under negative stress voltage.

2.6.3. Interface Trapped Charge

As shown Table XVIII, the interface trapped charge density can be characterized using the charge pumping method [55, 136], the quasi-static $C-V$ method [108], the Terman method [82, 83, 99, 137], and the conductance method [139]. The reported results show that the interface trapped charge for a Ta_2O_5/silicon capacitor are donor states [99, 137]. Zhang et al. studied the interface states at the Ta_2O_5/n-silicon interface by using deep-level transient spectroscopy (DLTS) [137]. The interface states at the Ta_2O_5/silicon interface are donor states because the DLTS signal comes from the majority carrier. The trap state is 0.4 eV below the conduction band of silicon. The same result was also found using the Terman method. Zhang et al. explained that the trapped states are due to Ta atoms diffusing into the silicon substrate during high-temperature annealing. The difference in thermal expansion coefficients between Ta_2O_5 and silicon may also cause the interface trapped charge density to increase. Park et al. reported that there is stress in the Ta_2O_5 film due to volume expansion by a factor of 2.3 during thermal oxidation [99]. The measured stress for as-deposited Ta is 2.03×10^{10} dyne/cm^2 in compression and 4.4×10^9 dyne/cm^2 for Ta_2O_5 films grown at 450°C. The stress for Ta_2O_5 grow at 650°C is about 2×10^8 dyne/cm^2. The compressive stress decreases linearly with increasing thermal oxidation temperature of Ta_2O_5. Therefore, both the interface state density and the compressive stress decrease with increasing thermal oxidation temperature of Ta_2O_5.

The charge pumping method can give an average interface trapped charge density by measuring the substrate current of MOSFETs. Autran et al. studied the interface trapped charge at Ta_2O_5/silicon interface using the charge pumping method [136]. The calculated mean interface state density is $\langle D_{it} \rangle = 3 \times 10^{10}$ eV^{-1} cm^{-2}. The average trap capture cross-sectional area is $\langle \sigma \rangle = 10^{-16}$ cm^2. The calculated interface state density and trap cross-sectional area are close to those of the SiO_2/silicon structure. This may be due to the existence of a thin SiO_2 layer of about 1 nm at the Ta_2O_5/silicon interface. Chatterjee et al. fabricated MOSFETs using a Ta_2O_5 gate dielectric. The channel length is smaller than 0.1 μm [55]. The Ta_2O_5 films deposited using LPCVD have a thickness of 5.5–6.5 nm. The interface state density measured from the charge pumping method is 2.8×10^{11} eV^{-1} cm^{-2}.

The high interface trapped charge density at the Ta_2O_5/silicon interface can be minimized by annealing under O_2 or

Table XVIII.   Interface Trapped Charges of Ta_2O_5/Silicon Interface

Reference	Capacitor structure	Measurement conditions	Extracted parameters
Autran et al. [57]	61-nm ECR-PECVD Ta_2O_5 as the gate dielectric of p-MOSFET Channel length, 3 μm	Charge pumping method with 100-kHz gate voltage pulse	1. Mean interface state density $\langle D_{it} \rangle = 3 \times 10^{10}$ $eV^{-1}cm^{-2}$ 2. Average trap capture cross-sectional area $\langle \sigma \rangle = 10^{-16}$ cm^2
Chatterjee et al. [55]	5.5–6.5-nm LPCVD Ta_2O_5 as the gate dielectric of MOSFET Channel length, $\sim$0.1 μm	Charge pumping method	Mean interface state density $\langle D_{it} \rangle = 2.8 \times 10^{11}$ $eV^{-1}cm^{-2}$
Lo et al. [108]	60-nm LPCVD Ta_2O_5/SiO_2 (3.4 nm)/silicon SiO_2 (2 nm)/60-nm LPCVD Ta_2O_5/SiO_2 (3.4 nm) Si_3N_4 (3 nm)/60-nm LPCVD Ta_2O_5/SiO_2 (3.4 nm)/silicon	High-frequency and quasi-static C–V curves were measured after the constant current stresses of 8×10^{-6} A/cm^2 at both $+V_G$ and $-V_G$ polarity for 180 or 1400 s	ΔD_{it} is the smallest among these capacitors for $+V_G$ stress ΔD_{it} is the smallest among these capacitors for $-V_G$ stress 1. ΔD_{it} for $+V_G$ constant current stress is higher than that of $-V_G$ constant current stress 2. ΔD_{it} is the largest among these three capacitors for both $+V_G$ and $-V_G$ stress
Park and Im [125]	1. Sputtered Ta/p-silicon 2. Furnace O_2 annealing at 450–700°C for 1 h to form 65–212-nm-thick Ta_2O_5	1. D_{it} was calculated from the 100-kHz C–V curves 2. Compressive stresses at the Ta/silicon interface and at the Ta_2O_5/silicon interface were measured	Both D_{it} and the compressive stress decrease when the formation temperature of Ta_2O_5 film increases
Seki et al. [82]	1. 17–110-nm sputtered Ta_2O_5/silicon 2. Postmetallization H_2 annealing at 300 or 450°C for 30 min	C–V curves of Ta_2O_5 capacitor were measured	1. ΔV_{FB} between 1 kHz and 1 MHz is about 8 V for as-deposited Ta_2O_5 and ΔV_{FB} < 0.04 V for H_2-annealed Ta_2O_5 2. $D_{it} = 1 \times 10^{12}$ eV^{-1} cm^{-2} for as-deposited Ta_2O_5, $D_{it} = 1.5 \times 10^{11}$ eV^{-1} cm^{-2} for H_2-annealed Ta_2O_5
Zhang et al. [137]	260-nm laser-pulse-deposited Ta_2O_5/n-silicon with 700–900°C annealing	Study the interface state properties by DLTS and Terman method	1. Interface states are donor states 2. Trap level is 0.4 eV below the E_c of Ta_2O_5
Zaima et al. [83]	35-nm LPCVD Ta_2O_5/silicon with furnace O_2, at 800–900°C annealing for 30 min	D_{it} was calculated from the high-frequency C–V curves	1. $D_{it} \sim 10^{12}$ eV^{-1} cm^{-2} for 800 and 850°C annealed film 2. $D_{it} < 10^{11}$ eV^{-1} cm^{-2} for 900°C annealed film due to the formation of SiO_2 at Ta_2O_5/silicon interface

H_2 ambience [82, 83, 139]. Seki et al. reported that the flatband voltage becomes more negative when the frequency of the C–V measurement increases [82]. The flatband voltage difference between 1-kHz and 1-MHz measurement is about 8 V for as-deposited Ta_2O_5 film. This is due to the existence of high-density interface states with various response time constants. The flatband voltage difference of Ta_2O_5 capaci-

tors after H_2 annealed is smaller than 0.04 V. The interface state density D_{it} for as-deposited Ta_2O_5 film at midgap measured from a 1-MHz C–V curve using the Terman method is about 1×10^{12} eV^{-1} cm^{-2}. The interface state density D_{it} for 450°C annealed Ta_2O_5 film in H_2 at the midgap is about 1.5×10^{11} eV^{-1} cm^{-2}. The existence of interfacial native SiO_2 can also reduce the interface trapped charge density.

A SiO$_2$ transition layer, however, will decrease the dielectric constant.

Zaima et al. reported that the D_{it} for 800 or 850°C annealed Ta$_2$O$_5$ films is about 10^{12} eV^{-1} cm^{-2} [83]. However, the distortion of the C–V curve for 850°C annealed Ta$_2$O$_5$ film is smaller than that of 800°C annealed Ta$_2$O$_5$ film. This is due to the formation of Si–O bonds at the Ta$_2$O$_5$/silicon interface when the annealing temperature is higher than 850°C. The midgap D_{it} for Ta$_2$O$_5$ films annealed at 900°C is smaller than 1×10^{11} eV^{-1} cm^{-2} because of the existence of a thin SiO$_2$ layer at the Ta$_2$O$_5$/silicon interface. Lai et al. deposited 14- and 20-nm Ta$_2$O$_5$ films on a silicon substrate using PECVD [139]. The interface trapped charge density of Au/Ta$_2$O$_5$/p-Si capacitors were measured using the conductance method. The interface trapped charge density D_{it} at midgap is about 1×10^{13} eV^{-1} cm^{-2} for as-deposited Ta$_2$O$_5$, whereas $D_{it} = 2 \times 10^{11}$ eV^{-1} cm^{-2} for samples RTA annealed at 400°C for 1 min. The interface trapped charge density increases when the annealing temperature is higher than 400°C or the annealing time is longer than 1 min.

Lo et al. studied the generation of interface states in stack films after constant current stress [108]. Three stacked capacitors, namely, Si$_3$N$_4$/Ta$_2$O$_5$/SiO$_2$/silicon, SiO$_2$/Ta$_2$O$_5$/SiO$_2$/silicon, and Ta$_2$O$_5$/SiO$_2$/silicon, were used. A constant current stress was applied at a current level of 8×10^{-6} A/cm^2. Both $+V_G$ and $-V_G$ stress conditions were used. When the top electrode is biased under the $-V_G$ condition, the electric field at the anode for the Si$_3$N$_4$/Ta$_2$O$_5$/SiO$_2$/silicon capacitor is the highest among these three capacitors because electrons will be accumulated at the Si$_3$N$_4$/Ta$_2$O$_5$ interface due to the larger bandgap of Si$_3$N$_4$ compared with Ta$_2$O$_5$. Therefore, the generated interface state density ΔD_{it} of the Si$_3$N$_4$/Ta$_2$O$_5$/SiO$_2$/silicon capacitor is the largest among these three capacitors under $-V_G$ stress. Another possible reason is that the high concentration of hydrogen in Si$_3$N$_4$ results in a high concentration of defects generated after stress. ΔD_{it} due to constant current stress under $+V_G$ is higher than that of constant current stress under $-V_G$. Lo et al. explained this effect by assuming that the interface states caused by damage creation and migration (under $+V_G$) are probably much more significant than those created by direct electron bombardment (under $-V_G$).

2.6.4. Mobile Ionic Charge

Ta$_2$O$_5$ films are deposited by LPCVD from the reaction of a Ta(OC$_2$H$_5$)$_5$ liquid source with oxygen. This is different from the thermally grown SiO$_2$. Metal impurities may be incorporated into Ta$_2$O$_5$ film during deposition. Therefore, the instability of the Ta$_2$O$_5$ C–V curve may come from the drift of mobile ionic charges. Chaneliere et al. deposited Ta$_2$O$_5$ films on a silicon substrate using LPCVD [87]. The Ta$_2$O$_5$ films were then annealed at 800°C under N$_2$ ambience for 30 min. The C–V hysteresis of amorphous Ta$_2$O$_5$ film is counterclockwise, whereas the C–V hysteresis of polycrystallized δ-Ta$_2$O$_5$ film is clockwise. The flatband voltage shift of amorphous Ta$_2$O$_5$ ($\Delta V_{FB} = -0.153$ V) is smaller than that of δ-Ta$_2$O$_5$ ($\Delta V_{FB} =$

0.432 V). The counterclockwise C–V hysteresis is due to surface state trapping. The clockwise C–V hysteresis may be due to metal contamination in the Ta$_2$O$_5$ film. The secondary-ion mass spectroscopy (SIMS) result shows that Na$^+$ ion contamination in the amorphous Ta$_2$O$_5$ is higher than that in δ-Ta$_2$O$_5$. Kaplan et al. deposited 100-nm Ta$_2$O$_5$ film on n-type silicon using LPCVD [67]. The mobile ionic charge density Q_m was studied using the bias–temperature test. The capacitor was initially biased at 10 V and placed at 165°C for 10 min to force the metal impurity to move toward the Ta$_2$O$_5$/silicon interface. The capacitor was then biased at -5 V for 10 min at the same temperature to force the metal impurity to move toward the Al/Ta$_2$O$_5$ interface. No Q_m was found after this measurement.

2.7. Dielectric Reliability of Ta$_2$O$_5$ Thin Films

The channel length of state-of-the-art MOSFETs is about 0.13 μm. According to the 1999 ITRS, when the effective channel length L_{eff} is smaller than 0.18 μm, the electric field on the gate oxide will be higher than 5 MV/cm. Constant electric field scaling has been used to avoid MOSFET degradation. However, the threshold voltage of the MOSFET cannot be scaled down with device size, because a lower threshold voltage will result in high subthreshold current and high standby power dissipation. As a result, the supply voltage V_{DD} is scaled slower than the device size and the electric field on the gate oxide increases as scaling continues. For DRAM applications, capacitor thickness is scaled to keep the required capacitance per cell. Hence, the electric field on the storage capacitor increases when the device size is scaled down. The reliability of the DRAM capacitor dielectric is therefore an important issue.

2.7.1. Breakdown Strength

For dielectric film with smaller energy bandgaps, electron–hole pairs will be generated relatively easily by either thermal excitation or impact ionization, which, in turn, will result in larger leakage current. The dielectric breakdown strength will also decrease accordingly. Generally, the dielectric breakdown strength and dielectric constant are related to the energy bandgap. The dielectric breakdown strength decreases and the dielectric constant increases with smaller energy bandgap. As a result, high-dielectric-constant films often have smaller dielectric breakdown strength [81]. Ta$_2$O$_5$ film is used as the DRAM capacitor dielectric because it has a reasonably high dielectric breakdown strength of 5–6 MV/cm as well as a high dielectric constant of 25 even at a thickness of 10 nm.

As shown in Table XIX, the dielectric breakdown of Ta$_2$O$_5$ is affected by defects in the film such as pinholes. Reducing defect density will increase the breakdown strength of Ta$_2$O$_5$ film. Byeon and Tzeng compared the breakdown voltages of sputtered Ta$_2$O$_5$ film, Ta$_2$O$_5$ film grown by anodic oxidation, and sputtered Ta$_2$O$_5$ film with anode reoxidation processing [141]. The dielectric breakdown strength of sputtered Ta$_2$O$_5$ film with anodic oxidation is higher than those of sputtered Ta$_2$O$_5$ film and Ta$_2$O$_5$ film grown by anodic oxidation. The breakdown

Table XIX.　　Dielectric Breakdown of Ta_2O_5 Films

Reference	Capacitor structure	Breakdown field E_{BD}	Comments
Byeon and Tzeng [141]	120-nm sputtered Ta_2O_5/Ta	1–3 MV/cm	Ta_2O_5 breakdown due to defects in the film such as pinholes
	120-nm anode oxidation Ta_2O_5/Ta	3–4.5 MV/cm	
	120-nm sputtered Ta_2O_5/Ta with anode oxidation	6–7 MV/cm	Pinholes were sealed after anode oxidation
Hwu and Jeng [142]	Sputtered Ta/n-silicon and thermal oxidation at 500°C to form Ta_2O_5	3.0–3.6 MV/cm	Breakdown voltage was defined as the physical breakdown of Ta_2O_5
Kim et al. [86]	16-nm ECR-PECVD Ta_2O_5/p-silicon with 650–850°C O_2 annealing for 30 min	1. As-deposited Ta_2O_5, 4.3 MV/cm 2. 750°C annealed Ta_2O_5 film 3.3 MV/cm 3. 850°C annealed Ta_2O_5 film, 4.1 MV/cm	1. Breakdown voltage was defined as the physical breakdown of Ta_2O_5 2. Increase in breakdown strength for 850°C annealed film is due to formation of 10-nm SiO_2 at Ta_2O_5/silicon interface
Kim et al. [89]	16-nm ECR-PECVD Ta_2O_5/p-silicon with different substrate temperatures	The breakdown strengths for Ta_2O_5 deposited at 145 and 205°C are 3.2 and 4.5 MV/cm, which are related to the SiO_2 thickness	Native SiO_2 thicknesses at 145, 165, and 205°C substrate temperatures are 2.5, 2.9, and 3.8 nm
Kim et al. [90]	14-nm ECR-PECVD Ta_2O_5/n-silicon with 700–850°C O_2 furnace annealing for 30 min	1. As-deposited Ta_2O_5, 0.6 MV/cm 2. 800°C O_2-annealed Ta_2O_5 2.3 MV/cm 3. Breakdown before measurement for 850°C annealed film	Breakdown voltage of Ta_2O_5 capacitor was defined at the leakage current level of 1×10^{-6} A/cm^2
	14-nm ECR-PECVD Ta_2O_5/n-silicon with 700–850°C N_2 furnace annealing for 30 min	1. Smaller than 1.2 MV/cm for 700–850°C annealed film 2. Breakdown before measurement for 850°C annealed film	
Nagahiro and Raj [81]	25–2000-nm ECR-PECVD Ta_2O_5/n-silicon	6.8 MV/cm for 44.6-nm-thick Ta_2O_5 and 1.9 MV/cm for 99-nm-thick Ta_2O_5; E_{BD} decreases with increasing Ta_2O_5 thickness	1. Breakdown voltage of Ta_2O_5 capacitor was defined at the leakage current level of 1×10^{-6} A/cm^2 2. High particle density in Ta_2O_5 will result in low breakdown voltage
Nishioka et al. [110]	27-nm sputtered Ta_2O_5/2.4-nm SiO_2 on silicon or W	5 MV/cm	Electrons accumulate at the Ta_2O_5/SiO_2 interface and cause the reduction of electric field in SiO_2 before catastrophic breakdown of SiO_2
Shinriki et al. [143]	15-nm LPCVD Ta_2O_5/polysilicon with UV-O_3 annealing 15-nm LPCVD Ta_2O_5/polysilicon with UV-O_3 annealing at 300°C+ furnace O_2 annealing at 800°C	Smaller than 0.5 MV/cm and the defect density is higher than 5 cm^{-2} 6–6.5 MV/cm and the defect density is smaller than 0.04 cm^{-2}	Breakdown voltage of Ta_2O_5 capacitor was defined at the leakage current level of 1×10^{-6} A/cm^2

Table XIX. (Continued)

Reference	Capacitor structure	Breakdown field E_{BD}	Comments
Zaima et al. [83]	24–46-nm LPCVD Ta_2O_5/silicon with furnace O_2 at 850°C annealing for 15–60 min	1. A thin transition layer about 3–4 nm formed at Ta_2O_5/silicon interface 2. E_{BD} of $\varepsilon_r = 10$ and 38 films, 4.5 and 0.8 MV/cm, respectively	1. Breakdown voltage of Ta_2O_5 capacitor was defined at the leakage current level of 1×10^{-7} A/cm^2 2. E_{BD} increases and ε_r decreases with increasing transition layer thickness

strength of sputtered Ta_2O_5 film and Ta_2O_5 film grown by anodic oxidation is related to the defects in the films and shows a widespread distribution. The weak spots are sealed by anodic reoxidation for sputtered Ta_2O_5 films. The distribution of breakdown voltage is thus more concentrated. Nagahiro and Raj found that the breakdown strength decreases when the Ta_2O_5 thickness increases [81]. They also found that the breakdown strength of Ta_2O_5 film is related to the particle density. The breakdown of Ta_2O_5 films is thus dominated by the presence of stray particles. Shinriki et al. deposited 15-nm Ta_2O_5 film on a polysilicon bottom electrode using LPCVD [143]. The Ta_2O_5 film was then annealed by two-step annealing. The breakdown strength of Ta_2O_5 film annealed by UV-O_3 annealing is smaller than 0.5 MV/cm and the defect density is larger than 5 cm^{-2}. For Ta_2O_5 films annealed by two-step annealing, the breakdown strength is about 6–6.5 MV/cm and the defect density is less than 0.04 cm^{-2}. Shinriki et al. pointed out that there are two kinds of defects in Ta_2O_5 film. One comprises weak spots where the Ta_2O_5 film is locally thin. These weak spots were eliminated during dry O_2 annealing. The other type of defect is carbon introduced into Ta_2O_5 through thermal decomposition of $Ta(OC_2H_5)_5$ during film deposition. The carbon concentration in the Ta_2O_5 film will decrease when the temperature of dry O_2 annealing is increased. The breakdown strength of Ta_2O_5 film after two-step annealing increases because the defect density decreases after annealing.

When Ta_2O_5 film is deposited on silicon or polysilicon, the dielectric breakdown strength will be affected by the inerfacial native SiO_2. The interfacial SiO_2 will increase the breakdown strength of the Ta_2O_5/SiO_2 stack film. Kim et al. found that the breakdown strength increases with the deposition substrate temperature [88]. A thin layer of SiO_2 is formed on the silicon substrate within 1 min after the beginning of the deposition of Ta_2O_5 film by ECR-PECVD. The increase in breakdown strength with substrate temperature is related to the increase in SiO_2 thickness and the improvement in Ta_2O_5 quality. Zaima et al. found that the breakdown strength decreases as the dielectric constant increases [83]. The breakdown strengths of Ta_2O_5 capacitors with dielectric constant of $\varepsilon_r = 10$ and 38 are $E_{BD} = 4.5$ and 0.8 MV/cm, respectively. Zaima et al. explained this result by assuming that there is a thin transition layer of about 3–4 nm at the Ta_2O_5/silicon interface. The transition layer is a compound formed by Ta, Si, and O atoms.

The breakdown voltage increases and the dielectric constant decreases as the thickness of the transition layer increases.

Because the electric displacement should be continuous across the Ta_2O_5/SiO_2 stack layer, most of the voltage drop will lie across the interfacial native SiO_2 because the dielectric constant of SiO_2 is much smaller than that of Ta_2O_5. Therefore, the thin SiO_2 layer will be the first layer to show breakdown phenomena and thus trigger the dielectric breakdown of the stacked layer [86]. However, breakdown of the thin SiO_2 layer was not observed [144]. This may be due to electron accumulation at the Ta_2O_5/SiO_2 interface, which causes a reduction in the electric field. Nishioka et al. compared the leakage current densities of W/Ta_2O_5 (27 nm)/W and Al/Ta_2O_5 (27 nm)/SiO_2 (2.4 nm)/silicon capacitors [144]. The density of trapped electrons at the Ta_2O_5/SiO_2 interface calculated by the equation $\varepsilon_{Ta_2O_5} E_{Ta_2O_5} = \varepsilon_{SiO_2} E_{SiO_2} + Q/\varepsilon_0$ is 9.3×10^{-6} C/cm^2. Electron accumulation at the Ta_2O_5/SiO_2 interface and reduction in the electric field in SiO_2 occur before the catastrophic breakdown of SiO_2. These effects limit the voltage drop across the SiO_2 film at the interface to less than 1 V.

2.7.2. Time-Dependent Dielectric Breakdown

For Ta_2O_5 to be applicable to DRAM, the lifetime of the Ta_2O_5 dielectric should be longer than 10 years at half of the supply voltage $V_{DD}/2$ or about 1.0 V. As generally known, the dielectric breakdown of SiO_2 is due to holes generated during impact ionization and trapped close to the cathode, which cause an increase in the electric field. If the dielectric lifetime is inversely proportional to the charge fluence through the SiO_2 film, the time to breakdown t_{BD} will be inversely proportional to the leakage current density of SiO_2, that is, $t_{BD} \propto J^{-1} \propto \exp(1/E)$. The lifetime of thin SiO_2 at $V_{DO}/2$ can be calculated from a $\log(t_{BD})$ vs. $1/E$ plot. However, almost all of the reported results used a $\log(t_{BD})$ vs. V_G plot to calculate the lifetime of Ta_2O_5 capacitors [3, 62, 96, 103, 113, 120, 130, 133, 143]. The leakage current mechanism was not discussed when using $t_{BD} \propto \exp(E)$. This assumption is compatible with a leakage current mechanism of $J \propto e^{\alpha E}$. The only leakage current mechanism that satisfies this relationship is hopping conduction, which occurs when the applied voltage is small. However, high electric field stress is used to extrapolate the lifetime at low electric field operation. The commonly found current mechanisms in the literature are Poole–Frenkel emission and Schottky emission. If either Poole–Frenkel emission or Schottky emission is used to predict the lifetime of Ta_2O_5 film, a relationship like $t_{BD} \propto J^{-1} \propto \exp -(\beta E)^{1/2}$ should be used. Kamiyama and Sacki used a $\log(t_{BD})$ vs. $1/E$ plot to predict the lifetime of Ta_2O_5 film [145]. This indicates that Fowler–Nordheim tunneling was assumed as the leakage current mechanism of the Ta_2O_5 film.

The lifetime of the capacitor dielectric in a DRAM chip is not just affected by the t_{BD} of the Ta_2O_5 film. It is also affected by the working temperature of the DRAM chip, the total capacitor area on the chip, and the surface roughness of bottom electrode. A bottom electrode with a rough surface will have higher leakage current density due to increasing surface area and increasing electric field caused by surface asperity. Asano et al. deposited 15-nm Ta_2O_5 film on a rugged polysilicon electrode [130]. The effects of capacitor area ($A = 10^{-3}$ or 10^{-2} cm^2), stress voltage (4.25–6.0 V), measurement temperature (25°C, 90°C, or 125°C), and surface roughness of the polysilicon electrode on the time-dependent dielectric breakdown (TDDB) of Ta_2O_5 films were studied. For Ta_2O_5 films deposited on a rugged polysilicon electrode, the activation energy measured from the t_{BD} vs. temperature plot is 0.3 eV. The time to breakdown t_{BD} decreases by a factor of 10 when the temperature increases from 25°C to 90°C. The t_{BD} also decreases by a factor of 10 when the capacitor area increases from 10^{-3} to 10^{-2} cm^2. The t_{BD} of Ta_2O_5 film deposited on smooth polysilicon is smaller than that deposited on rugged polysilicon by 1.5 decades. The extrapolated lifetime of Ta_2O_5 films deposited on a smooth polysilicon electrode with an area A of 10^{-3} cm^2 is 3×10^{19} s at 1.0 V (25°C). Assuming that the total area of Ta_2O_5 capacitors deposited on a rugged polysilicon electrode in a 256-Mb DRAM chip is 10 cm^2 and the working temperature of the DRAM chip is 90°C, the lifetime of Ta_2O_5 films should decrease by 6.5 decades compared with that of a single capacitor. A mean time to failure (MTTF) of 9×10^{12} s of Ta_2O_5 capacitors in the 256-Mb DRAM can be obtained by including the effects of capacitor area, the roughness of the electrode, and the working temperature of the DRAM chip.

Degraeve et al. studied the TDDB effect by constant current stress of a Pt/Ta_2O_5 (6 nm)/SiO_2 (1.4 nm)/silicon capacitor [146]. The dielectric breakdown of Ta_2O_5/SiO_2 stack at high electric field is determined by the breakdown of SiO_2. The leakage current of Ta_2O_5 (6 nm)/SiO_2 (1.4 nm) is less than that of 1.4-nm SiO_2 when the applied voltage is lower than 2.5 V. The leakage currents are the same when the applied voltage is larger than 2.5 V due to electron direct tunneling through the thin SiO_2 layer into the Ta_2O_5 film. The Ta_2O_5 film acts as a conducting layer when the applied voltage is high. Therefore, the SiO_2 interfacial layer degrades as if the Ta_2O_5 layer were not present when the applied voltage is high and the breakdown of the capacitor is determined by the breakdown of the thin SiO_2 layer. The possible leakage current mechanism of the Ta_2O_5/SiO_2 capacitor at low electric field is either Fowler–Nordheim tunneling or a more complicated mixture of

tunneling and Poole–Frenkel conduction. Ta_2O_5 merely acts as a voltage distributor. Therefore, predicting the lifetime at low electric field using the extrapolated lifetime of the Ta_2O_5/SiO_2 capacitor based on the $\log(t_{BD})$ vs. E plot at high electric field will result in a wrong t_{BD} value. This is because different leakage current mechanisms are at work in high and low electric fields.

3. HIGH-DIELECTRIC-CONSTANT FILMS: SILICON NITRIDE (SI$_3$N$_4$)

Si_3N_4 has been widely used in semiconductor devices due to its high dielectric constant ($\varepsilon_r = 7.5$), high breakdown field (>8 MV/cm), radiation hardness, and an extremely good barrier to the diffusion of water and sodium. The energy bandgap of Si_3N_4 is 5.1 eV. Because direct tunneling current limits the MOSFET gate dielectric to further scaling down below 1.5 nm, an alternative material with a high dielectric constant to replace SiO_2 as the MOSFET gate dielectric is required. Si_3N_4 is a candidate material to replace SiO_2 as the MOSFET gate dielectric due to its good hot carrier immunity and resistance to boron penetration. The leakage current density of Si_3N_4 film is 3 orders of magnitude smaller than that of SiO_2 film with the same equivalent SiO_2 thickness due to the thicker physical thickness of Si_3N_4 film [147].

3.1. Fabrication Processes of Si$_3$N$_4$ Thin Films

3.1.1. Deposition Methods

LPCVD and PECVD are the most common Si_3N_4 deposition method in the semiconductor process. Si_3N_4 is often deposited by reacting SiH_4 [148–151] or SiH_2Cl_2 [152–154] with NH_3 [149, 150] or N_2 [151] at temperatures between 650 and 850°C. Because the trap density increases due to hydrogen atoms incorporated into the Si_3N_4 film, a N_2 annealing is required to reduce the hydrogen concentration and increase the nitrogen concentration in the film [155, 156]. Ultrathin Si_3N_4 can be formed on a silicon substrate by nitridation of silicon by rapid thermal processing (RTP) at 700–800°C for 10 s in NH_3 ambience followed by an *in situ* N_2O oxidation at 800–900°C for 15–30 s [157]. High-quality Si_3N_4 films deposited using jet vapor deposition (JVD) have attracted much attention [158]. A nozzle forms a supersonic jet of inert carrier gas that transports vapor to stationary or moving substrates. Highly diluted SiH_4 from the inner nozzle and N_2 from the outer nozzle flow into a helium microwave discharge sustained near the outer nozzle exit. Reactive Si species and N atoms generated in the discharge are carried by the supersonic-speed helium jet toward the substrate where they form silicon nitride [158]. Because the depositing species have the same supersonic speed as that of the helium carrier gas, the kinetic energies of these species could be on the order of 1 eV. Because of the separation of the constituent depositing species, and their short transit times, there is very little chance for gas-phase nucleation. The high impact energies of the depositing species also

Table XX. Annealing Methods of Si_3N_4 Films

Reference	Annealing method	Annealing condition	Advantages
Ando et al. [150] Kobayashi et al. [153]	*In situ* native oxide cleannig anneal	RTA in H_2 ambience at 850–950°C for 60 s at a reduced pressure of 50–100 Torr before Si_3N_4 deposition	*In situ* H_2 cleaning is very useful for eliminating the bottom oxide at the Si_3N_4/Si interface
Eriguchi et al. [152] Kim et al. [149] Song et al. [157, 162]	RTA NH_3+RTA N_2O two step annealing	1. RTA in NH_3 ambience at 950–1000°C for 30 s 2. RTA in N_2O ambience at 850–900°C for 30 s	Improves Si–N stoichiometry and TDDB reliability
Kim et al. [149]	RTP NH_3 annealing	950–1000°C for 30 s in pure NH_3	NH_3 anneal can make the as-deposited Si-rich Si_3N_4 film more stoichiometric and increase its oxidation resistance significantly by incorporating more N into the films
Wu and Lucovsky [151]	RTP He annealing	RTA in He ambience for 30 s	Reduces the hydrogen concentration in Si_3N_4

contribute to the improved film quality [159]. JVD Si_3N_4 has strong resistance to boron penetration, lower leakage current for the same equivalent thickness compared to SiO_2, very low interface and low bulk oxide trap generation, high resistance to hot carrier damage, and hardly any stress-induced leakage current (SILC) after hundreds of Coulombs/square centimeter of constant current stress [158–161]. Sekine et al. reported that high-quality ultrathin Si_3N_4 was deposited using a radial-line slot antenna (RLSA) high-density plasma system in the ambience of Ar/NH_3 or Ar/H_2/N_2 mixing gas at 400°C [147]. The leakage current density of RLSA Si_3N_4 is smaller than that of JVD Si_3N_4 by 1 order of magnitude for the same equivalent SiO_2 thickness. The breakdown strength of RLSA Si_3N_4 is 15 MV/cm. After a constant voltage stress of 100 C/cm^2, there is no discernible SILC at all.

3.1.2. Annealing Conditions

As shown in Table XX, Si_3N_4 is often annealed by rapid thermal annealing with various kinds of gases. The annealing gases include N_2, NH_3, H_2, and He. Annealing in N_2 ambience is required for LPCVD Si_3N_4 to reduce the hydrogen concentration in the film [155, 156, 159]. Annealing in N_2 ambience for JVD Si_3N_4 is also required to reduce both the surface and the bulk hydrogen concentration [161]. Postdeposition annealing in He ambience for JVD Si_3N_4 for 30 s can reduce the hydrogen concentration [163]. Two-step annealing using RTA NH_3+RTA N_2O [149, 152, 157, 162] is reported to improve the Si–N stoichiometry and the TDDB reliability. NH_3 annealing can convert the as-deposited, Si-rich Si_3N_4 film to be more stoichiometric and can also increase its oxidation resistance significantly by incorporating more N atoms into the films [149]. *In situ* H_2 cleaning is very useful for eliminating the native oxide at the Si_3N_4/Si interface [149, 153]. *In situ*

cleaning of the native oxide on the silicon wafer is carried out by H_2 gas at a pressure of 50–100 Torr and at a substrate temperature between 850 and 950°C for 60 s. The X-ray spectroscopy (XPS) depth profile shows that the oxygen level at the nitride/n$^+$-polysilicon interface is reduced with *in situ* H_2 cleaning [153].

3.2. Leakage Current Mechanisms of Si_3N_4 Thin Films

The leakage current mechanism of LPCVD Si_3N_4 is reported to be mostly due to Poole–Frenkel emission [154, 156, 158, 164–167]. Sze studied the leakage current mechanism of 30 to 300-nm-thick CVD Si_3N_4 [167]. At high electric fields and high temperatures (>50°C), Poole–Frenkel emission dominates the current conduction band. At low temperatures (<70°C) and high fields, the current is tunneling current mainly due to field ionization of trapped electrons into the conduction band. At low electric fields and moderate temperatures, the current is due to hopping of thermally excited electrons from one isolated state to another. The leakage current mechanisms of Au/Si_3N_4/n$^+$-silicon and Au/Si_3N_4/Mo capacitors were studied [167]. It was found that, at room temperature and at a given electric field, the I–V curves are essentially independent of the film thickness, the device area, the electrode materials, and the polarity of the electrodes. The symmetry of the I–V characteristic is evidence that current transport in Si_3N_4 films is bulk controlled. Recently, high-quality Si_3N_4 films were obtained due to improvement of the semiconductor fabrication process and scaling of the dielectric thickness. The leakage current mechanism of high-quality Si_3N_4 was found to be Fowler–Nordheim tunneling current [147, 159]. The leakage current of the Si_3N_4/SiO_2 double layer was found to be trap-assisted tunneling [149].

Table XXI.   Poole–Frenkel Emission in Si$_3$N$_4$ Films

Reference	Device structure	Current mechanism	Comments
Lim and Ling [166]	SiO$_2$/LPCVD Si$_3$N$_4$ (5.0 nm)/SiO$_2$/silicon	Poole–Frenkel emission	High-frequency dielectric constant $\varepsilon = 3.6$, which is close to the reported value of 4 in the optical range
Sze [167]	30–300 nm CVD Si$_3$N$_4$/n$^+$-silicon or Si$_3$N$_4$/Mo	1. Poole–Frenkel emission at high field and high temperature (>50°C) 2. Tunneling current at low temperature (<70°C) and high field 3. Hopping at low field and moderate temperature	The dynamic frequency dielectric constant ε_d obtained from the slopes of I vs. $E^{1/2}$ plots is found to be 5.5 ± 1
Tanaka et al. [154]	SiO$_2$/LPCVD Si$_3$N$_4$ (9.0 nm)/SiO$_2$/silicon	1. Fowler–Nordheim tunneling current or direct tunneling current in SiO$_2$ layer 2. Poole–Frenkel emission via traps in Si$_3$N$_4$ layer, depending on the film thickness	
Wang et al. [158]	JVD 15-nm Si$_3$N$_4$/silicon	Poole–Frenkel emission	High-frequency dielectric constant $\varepsilon = 7.5$
Yoon et al. [156]	Reoxidation SiO$_2$ (2.5 nm)/Si$_3$N$_4$ (8 nm)/silicon *In situ* rapid thermal reoxidation in N$_2$O or O$_2$ ambience at 1000°C for 30 s	Poole–Frenkel emission	Hole injection from the top polysilicon is dominant for $+V_G$ bias due to Poole–Frenkel emission

3.2.1. Poole–Frenkel Emission

As shown in Table XXI, the leakage current mechanism of Si$_3$N$_4$ is often reported as Poole–Frenkel emission. Sze studied the leakage current mechanism of 30- to 300-nm-thick CVD Si$_3$N$_4$. The leakage current of Si$_3$N$_4$ films can be classified into three kinds of currents [167]:

$$J = J_1 + J_2 + J_3$$

where

$$J_1 = C_1 E \exp\left[-q\left(\Phi_1 + \sqrt{qE/\pi\varepsilon_0\varepsilon_d}\right)/kT\right]$$
$$J_2 = C_2 E^2 \exp(-E_2/E)$$
$$J_3 = C_3 E \exp(-q\Phi_3/kT)$$

J_1 is due to the internal Schottky effect or the Poole–Frenkel effect. C_1 is a function of the density of the trapped centers. J_2 is due to field ionization of the trapped electrons into the conduction band, presumably from the same centers as for J_1. This is a tunneling process that is essentially independent of temperature. J_3 is due to the hopping of thermally excited electrons from one isolated state to another. This process gives an ohmic I–V characteristic, exponentially dependent on temperature with a thermal activation energy $q\Phi_3$. Sze also studied the leakage current mechanism of Au/Si$_3$N$_4$/n$^+$-silicon and

Au/Si$_3$N$_4$/Mo capacitors [167]. At high electric field and high temperature J_1 dominates the current conduction. The dynamic frequency dielectric constant ε_d obtained from the slopes of I vs. $E^{1/2}$ plots is found to be 5.5 ± 1. Because the measured dielectric constant in the visible light range ($\sim$0.5 μm) is about 4 and the static dielectric constant of Si$_3$N$_4$ film is about 7, the dynamic dielectric constant (wavelength for 20 μm) should be between 4 and 7. Therefore, it is believed that the Poole–Frenkel effect is observed in the Si$_3$N$_4$ films. Lim and Ling studied the leakage current mechanism of a SiO$_2$/LPCVD Si$_3$N$_4$ (5.0 nm)/SiO$_2$ stack layer [166]. A larger current passes through the interpoly ONO film under negative gate bias than under positive gate bias. This is attributed to the thicker top oxide, which blocks hole injection from the anode under positive gate bias more effectively than the thinner bottom oxide does under negative bias. The leakage current mechanism is Poole–Frenkel emission. The calculated high-frequency dielectric constant from the slope of the Poole–Frenkel plot is around 3.6, which is close to the reported value of 4 in the optical range.

3.2.2. Tunneling Current

Kim et al. reported that because there is a high trap density in LPCVD Si$_3$N$_4$, the leakage current of ultrathin (1.5–2 nm)

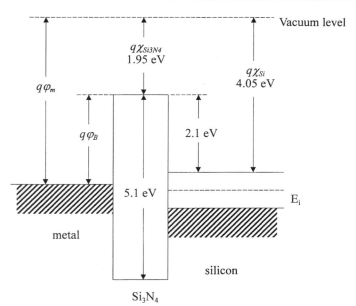

Fig. 13. Energy band diagram of metal/Si$_3$N$_4$/silicon capacitor at flatband condition.

LPCVD Si$_3$N$_4$ is trap-assisted tunneling current [149]. The as-deposited Si$_3$N$_4$ films tend to be silicon rich. NH$_3$ anneal can reduce the trap density in Si$_3$N$_4$ films and therefore reduce the trap-assisted tunneling current considerably. High-quality Si$_3$N$_4$ films were obtained due to improvements in the semiconductor fabrication process and the scaling of dielectric thickness. Both Ma [159] and Sekine et al. [147] reported that the leakage current mechanism in Si$_3$N$_4$ is due to Fowler–Nordheim tunneling. The electron barrier height at the Si$_3$N$_4$/silicon interface is 2.1 eV as calculated by assuming an effective mass m^* of $0.5m_0$. The energy band diagram of a metal/Si$_3$N$_4$/silicon capacitor is shown in Fig. 13. The electron affinity of 1.95 eV of Si$_3$N$_4$ is obtained by calculating the difference between the barrier height at the Si$_3$N$_4$/silicon interface and the electron affinity of silicon.

3.3. C–V Characteristics of Si$_3$N$_4$ Thin Films

3.3.1. Fixed Oxide Charge

The high fixed oxide charge density and high interface trapped charge density are the main disadvantages of using Si$_3$N$_4$ as the MOSFET gate dielectric. These are due to nitrogen bonds located at the SiO$_2$/silicon interface [157, 168]. For Si$_3$N$_4$/SiO$_2$ stack capacitors, a slight negative shift of the C–V curve relative to the control oxide was observed, indicating a positive fixed oxide charge buildup. A positive oxide charge buildup has been found on nitrided oxides and is attributed to the incorporation of nitrogen at the silicon/SiO$_2$ interface [168].

3.3.2. Oxide Trapped Charge

Thicker Si$_3$N$_4$ film show larger oxide trapped charge density than thinner Si$_3$N$_4$. This gives the thicker films a larger flat-band voltage shift after stress. Kim et al. reported that C–V hysteresis in thicker film can be reduced by NH$_3$ postannealing [149]. However, the effect of NH$_3$ postannealing on the hysteresis curve becomes much less as the film thickness is reduced. The NH$_3$ anneal also increases TDDB lifetime and reduces charge trapping significantly under constant voltage stress due to reduced trap density and leakage current. Wang et al. [158] reported that the interface of JVD Si$_3$N$_4$ is very resistant to hot carrier damage, as indicated by the very small increase in D_{it} and Q_{ot} after Fowler–Nordheim (F–N) injection. The C–V shift after F–N hot electron injection at an injection current level of 1 μA/cm^2 for 1000 s is about 0.2 V. By using the constant current injection technique and monitoring the voltage shift required to maintain this constant current, the effective density of electron traps was estimated to be approximately 6–7 $\times$ 10^{11} cm^{-2}, which is comparable to the values for thermal SiO$_2$, but much lower than other deposited Si$_3$N$_4$. Yoon et al. reported that RT-N$_2$O annealing reduces the excess silicon atoms and hydrogen-related species such as Si–H bonds in the Si$_3$N$_4$ films, resulting in Si$_3$N$_4$/SiO$_2$ layers with less electron trapping [156]. Less electron trapping in the Si$_3$N$_4$/SiO$_2$ stacked films with RT-N$_2$O annealing leads to less field buildup within the Si$_3$N$_4$/SiO$_2$ stacked films, resulting in prolonged t_{BD}.

As shown in Table XXII, Si$_3$N$_4$ film with thickness smaller than 3.0 nm shows negligible flatband voltage shift after constant current stress. Khare et al. reported that the C–V curve shift of ultrathin JVD Si$_3$N$_4$ (<2.0 nm) after 0.1-A/cm^2 constant current stress for either polarity is very small [160], indicating little trap generation and/or carrier trapping under these conditions. There is hardly any SILC after hundreds of constant current stress. There is also hardly any boron penetration through a 3.5-nm JVD Si$_3$N$_4$ film after 1050°C annealing for 30 min [169]. Song et al. deposited Si$_3$N$_4$ films (<2 nm) on a silicon substrate by NH$_3$ nitridation of the silicon substrate performed in RTP at 700–800°C for 10 s [157], followed by in situ N$_2$O oxidation at 800–950°C for 15–30 s. A negligible amount of hysteresis ($\Delta V_{FB} \sim$ a few millivolts) is observed due to lesser bulk traps after N$_2$O oxidation. Identical V_{FB} values with different C_{ox} (65–85 pF) imply a negligible amount of fixed oxide charges in these films with different Si$_3$N$_4$ thicknesses. Angle-resolved XPS data of NH$_3$ + N$_2$O films indicated that an ultrathin (<1.0 nm) pure SiO$_2$ layer was formed between Si$_3$N$_4$ and silicon after N$_2$O reoxidation, suggesting N atoms are removed from the SiO$_2$/silicon interface. NH$_3$ + N$_2$O-annealed Si$_3$N$_4$ samples show no boron penetration up to 1050°C, 30 s of dopant activation annealing due to the large amount of nitrogen incorporation from thermally grown nitride. Sekine et al. reported that the C–V curve hysteresis of RLSA-deposited Si$_3$N$_4$ and the leakage current can be improved by terminating the dangling bonds in the Si$_3$N$_4$ film with hydrogen [147]. There is no perceptible hysteresis attributed to charge traps in the Si$_3$N$_4$ film and the excellent match between high-frequency and quasi-static C–V curve indicates low density of interface trap. After a constant current stress of 100 C/cm^2, there is no discernible SILC at all and the threshold shift is only 11 mV after stress. After 10 C/cm^2 of

Table XXII.   Oxide Trapped Charges in Si_3N_4 Films

Reference	Device structure	Measurement condition	Comments
Khare et al. [160]	JVD Si_3N_4 (<2.0 nm)/ silicon with 800°C annealing in N_2 ambience for 10 min	$C-V$ curve shift after 0.1 A/cm^2 for either polarity is very small, indicating little trap generation and/or carrier trapping under these conditions	$I-V$ curves after hundreds of C/cm^2 constant current stress, there is hardly any SILC
Kim et al. [149]	1.5–2-nm Si_3N_4/SiO_2 (0.6 nm)/silicon with RTP NH_3 anneal (950–1000°C, 30 s) and RTP N_2O post-deposition anneal (850–900°C, 30 s)	NH_3 anneal increases TDDB lifetime and reduces charge trapping significantly under constant voltage stress due to reduced trap densities and lower leakage current	
Sekine et al. [147]	RLSA Si_3N_4 (t_{eq} = 2.1–4.5 nm)/silicon	1. After constant current stress of 100 C/cm^2, there is no discernible SILC and the threshold shift is only 11 mV after stress 2. After 10 C/cm^2 of constant current stress, gate voltage shift of $I-t$ measurement is only 18 mV	There is no perceptible hysteresis attributed to charge trap in Si_3N_4 film and excellent match between high-frequency and quasi-static $C-V$ curves indicates low density of interface traps
Song et al. [157]	Si_3N_4 films (<2 nm)/silicon Si_3N_4 is formed by using NH_3 nitridation of silicon substrate by RTP at 700–800°C for 10 s, followed by *in situ* N_2O oxidation at 800–950°C for 15–30 s	1. Negligible amount of hysteresis ($\Delta V_{FB} \sim$ few mV) is observed due to eliminated bulk traps by N_2O oxidation 2. Identical V_{FB} values with different C_{ox} (65–85 pF) imply negligible amount of fixed oxide charge in these films with different Si_3N_4 thicknesses	
Wang et al. [158]	15-nm JVD Si_3N_4/silicon	1. $C-V$ shift after $F-N$ hot electron injection at an injection current level of 1 μA/cm^2 for 1000 s is about 0.2 V 2. The effective density of electron traps is 6–7$\times10^{11}$ cm^{-2}	
Yoon et al. [156]	8-nm RPCVD Si_3N_4/polysilicon with *in situ* rapid thermal reoxidation in N_2O or O_2 ambient at 1000°C for 30 s	Less electron trapping of the Si_3N_4/SiO_2 stacked films with RT N_2O leads to less field buildup within the Si_3N_4/SiO_2 stacked films resulting in prolonged t_{BD}	RT N_2O annealing probably reduces the excess silicon atoms and hydrogen-related species such as Si–H bonds in the Si_3N_4 films, resulting in Si_3N_4/SiO_2 layer with less electron trapping

constant current stress, the gate voltage shift is only 18 mV. These results show that there is very little trap generation in Si_3N_4 films grown by RLSA high-density plasma.

3.3.3. Interface Trapped Charge

Ultrathin Si_3N_4 (<3.0 nm) is suitable for replacing SiO_2 as a MOSFET gate dielectric due to its high resistance to hot carrier damage and boron penetration. The interface trapped charge density of ultrathin Si_3N_4 is less than 1×10^{11} cm^{-2} eV^{-1} [158], which is slightly higher than that of the SiO_2/silicon interface. Iwai et al. deposited 3- to 6-nm Si_3N_4 films on SiO_2 (3 or 5 nm)/silicon substrate as the MOSFET gate dielectric. When the Si_3N_4 film is thick, the shift in threshold voltage is significant [170]. The charge pumping current I_{CP} increases with increasing stacked nitride thickness. The I_{CP} for 6-nm Si_3N_4 film is about twice that of 3-nm Si_3N_4 film. When the Si_3N_4 thickness was decreased to 3–4 nm, the threshold voltage shift was less of a problem, but the generation of interface states after stress became more serious. The n- and p-MOSFETs after 1000 s of hot carrier stress show that the threshold voltage shift was extremely large because the Si_3N_4 layer easily traps electrons. This means that the Si_3N_4 layer needs to be less than 4 nm, if possible as thin as 3 nm for MOSFET gate dielectric application. Wang et al. deposited 15-nm Si_3N_4 films on a silicon substrate using JVD [158]. The interface trapped charge density D_{it} is less than 5×10^{10} $cm^{-2} eV^{-1}$. The JVD Si_3N_4/silicon interface is very resistant to hot carrier damage, as indicated by the very small increase in D_{it} and Q_{ot} after F–N injection. The improved Si_3N_4 properties may be partly attributed to the lack of hydrogen in the film, which is shown by comparing the infrared (IR) transmission spectra for the JVD Si_3N_4 films and the remote-plasma CVD (RPCVD) Si_3N_4 films grown in $NH_3 + SiH_4$ ambience. Khare et al. deposited $t_{eq} = 3.8$ nm of JVD Si_3N_4 film a on silicon substrate as the MOSFET gate dielectric [171]. MOSFETs with Si_3N_4 or oxynitride as the gate dielectric contain higher densities of border traps [171] with long trapping time constants. In the low gate field region, the trapping of carriers into the border traps causes a reduction in the number of carriers in the channel, resulting in reduced transconductance. In the high gate field region, the electronic screening effect gives rise to a smoothing of the electronic interface, leading to an increased transconductance [159]. The charge pumping current I_{CP} shows that the interface state density D_{it} for a MOSFET ($L = 100$ nm) with JVD Si_3N_4 gate dielectric is only higher than that of a MOSFET with SiO_2 gate dielectric by a factor of 2 [172]. The ΔI_{CP} and also the ΔD_{it} after 1000 s $I_{sub,max}$ hot carrier stress for a MOSFET with JVD Si_3N_4 gate dielectric is about half that of a MOSFET with SiO_2 gate dielectric.

3.4. Metal–Oxide–Semiconductor Field Effect Transistor Gate Dielectric Applications of Si_3N_4 Thin Films

Table XXIII shows the reported results of MOSFETs with a Si_3N_4 gate dielectric. Song et al. deposited Si_3N_4 film (<2.0 nm) as the MOSFET gate dielectric using rapid-thermal CVD [162]. The interfacial SiO_2 passivation layer (0.6–0.7 nm) was grown in NO at 800°C. This passivation layer provides excellent interface properties and suppresses interface state generation during hot carrier stress. The subthreshold swing values of 10-μm channel length n-MOSFET and p-MOSFET devices were approximately 66 mV/decade and 74 mV/decade, respectively. The comparable subthreshold swing values for MOSFET

Table XXIII. Electrical Characteristics of MOSFETs with Si_3N_4 Gate Dielectric

Reference	Gate dielectric	Equivalent SiO_2 thickness	MOSFET structure	Electrical parameters
Mahapatra et al. [172]	JVD Si_3N_4	3.1 nm	Channel length $L = 0.1$ μm for n-channel MOSFET	1. ΔI_{cp} (hence ΔN_{it}) is lower in JVD Si_3N_4 MOSFET than in SiO_2 MOSFET 2. Peak g_m for JVD Si_3N_4 gate dielectric is about 70% of SiO_2 gate dielectric
Song et al. [157, 162]	RPCVD Si_3N_4 (<2.0 nm) with interface passivation layer (0.6–0.7 nm) grown in NO at 800°C	2.2 nm	Two kinds of channel length $L = 0.9$ μm and $L = 10$ μm for both n-channel and p-channel MOSFETs	1. $S_t = 66$ mV/decade for n-MOSFET and 74 mV/decade for p-MOSFET 2. Lesser g_m degradation compared with SiO_2 attributed to the stronger Si–N bonds due to NO interface passivation layer 3. C–V hysteresis $\Delta V_{FB} \sim 5$ mV, indicating very few oxide trapped charges
Tseng et al. [155, 161]	JVD Si_3N_4 (5.0 nm)	3.1 nm	Channel length $L = 0.4$ μm both n-channel and p-channel MOSFET	1. $S_t = 81.7$ mV/decade 2. JVD Si_3N_4 has a strong resistance to fluorine-enhanced boron penetration 3. g_m degradation is better than SiO_2 MOSFET
Wu and Lucovsky [151, 163]	Si_3N_4 (1.5 nm)/N_2O oxidation SiO_2 (0.4–0.7 nm)	1.6 nm	Channel length $L = 0.8$ μm for p-channel MOSFET	1. Peak $\mu_p = 80$ cm^2/V-s 2. p-MOSFET shows only ~5% mobility degradation compared to SiO_2 MOSFET 3. $S_t = 78$ mV/decade for $L = 0.7$ μm p-MOSFET

with Si$_3$N$_4$ gate dielectric and SiO$_2$ gate oxide suggest that CVD Si$_3$N$_4$ with a NO passivation layer has comparable interface state density with controlled SiO$_2$. Less transconductance degradation was observed for Si$_3$N$_4$ n-MOSFETs. This was attributed to the stronger Si–N bonds as a result of the NO interface passivation layer. The p-MOSFET with SiO$_2$ gate oxide shows turnaround phenomenon in transconductance degradation due to electron trapping and interface state generation. The p-MOSFET with Si$_3$N$_4$ gate dielectric shows strong immunity to hot carrier stress due to its extremely low electron trap density as well as better interface properties. Wu and Lucovsky deposited an ultrathin nitride–oxide ($\sim$1.5/0.7 nm) double layer as the MOSFET gate dielectric [163]. The interfacial SiO$_2$ passivation layer was formed by N$_2$O plasma oxidation. The C–V curves shows that there is no measurable boron penetration effect. The p-channel MOSFET with Si$_3$N$_4$/SiO$_2$ double-layer gate dielectric and a t_{eq} of 1.6 nm shows only approximately 5% mobility degradation when compared with the 2.6-nm SiO$_2$ MOSFET. The 0.7-μm p-MOSFET shows a good on–off characteristic with subthreshold swing of 78 mV/decade. No hard breakdown was found for −3.6-V stress after 155,700-C/cm^2 carrier injection. The MOSFET still functions but shows a 10% transconductance degradation and a 15% threshold voltage shift. Ma and his associates studied high-quality Si$_3$N$_4$ films deposited by jet vapor deposition [155, 159, 171, 172]. They deposited 5.0-nm JVD Si$_3$N$_4$ film as the MOSFET gate dielectric with a t_{eq} of 3.1 nm [161]. The JVD Si$_3$N$_4$ film shows much less C–V shift with a boron-doped p$^+$ gate, revealing the efficiency of boron penetration reduction. Furthermore, the negligible C–V difference between B- and BF$_2$-implanted p$^+$ gate capacitors using JVD Si$_3$N$_4$ suggests that JVD Si$_3$N$_4$ has a strong resistance to fluorine-enhanced boron penetration. The current drive for p-MOSFETs is relatively low, which may be due to the high $V_{th,p}$ caused by fixed charge and interface states. This problem can be addressed by in situ N$_2$O-plasma pretreatment in the JVD Si$_3$N$_4$ deposition process. The subthreshold swing for MOSFETs with a channel length of 0.35 μm at $V_{DS} = 0.1$ V is about 81.7 mV/decade. Mahapatra et al. fabricated a MOSFET with JVD Si$_3$N$_4$ film as the gate dielectric [172]. The channel length is 100 nm. The peak transconductance of the MOSFET using a JVD Si$_3$N$_4$ gate dielectric is about 70% compared to that using a SiO$_2$ gate dielectric. Although the transconductance of JVD Si$_3$N$_4$ devices at low gate biases is lower than that of SiO$_2$ gate dielectric devices, the transconductance of JVD Si$_3$N$_4$ devices is higher than that of the SiO$_2$ MOSFET at higher gate biases. The saturation transconductance is consistently higher for JVD Si$_3$N$_4$ MOSFETs. The less than 5% degradation in subthreshold slope of JVD Si$_3$N$_4$ MOSFETs compared to SiO$_2$ MOSFETs is essentially due to the slightly higher interface states in JVD Si$_3$N$_4$ film. It is evident that ΔI_{CP} and hence ΔN_{it} generation is lower in JVD Si$_3$N$_4$ MOSFETs. The fact that JVD Si$_3$N$_4$ shows less degradation indicates improved robustness against hot carrier stressing in spite of the lower energy barrier (2.1 eV) compared to silicon dioxide.

3.5. Dielectric Reliability

Because the dielectric constant of Si$_3$N$_4$ is only twice that of SiO$_2$, the MOSFET with a Si$_3$N$_4$ gate dielectric is not sufficient for further scaling. An alternative dielectric with dielectric constant higher than Si$_3$N$_4$ is required. The 2000 ITRS reported that the electric field on the gate dielectric might be higher than 5 MV/cm when the MOSFET channel length L_{eff} is smaller than 0.18 μm. It was reported that the equivalent breakdown strength of Si$_3$N$_4$ film is 16–22 MV/cm [147, 159, 160, 162, 165, 172]. The reliability of using Si$_3$N$_4$ as a deep submicrometer MOSFET gate dielectric is thus acceptable.

4. HIGH-DIELECTRIC-CONSTANT FILMS: TITANIUM OXIDE (TiO$_2$)

TiO$_2$ has been used in semiconductor applications such as sensors, antireflection coatings, photocatalysts [173], and MOSFET gate dielectrics due to its high dielectric constant ($\varepsilon_r = 20$–160), high refractive index (2.7), good optical transmittance, and high dielectric breakdown strength (1–3 MV/cm). There are three kinds of TiO$_2$ crystallization phases, namely, anatase, rutile, and brookite [174, 175]. Brookite is an unstable rhombic structure and has not been observed in thin films [176]. The dielectric constant of rutile is 86–89 perpendicular to the c axis and 160–170 parallel to the c axis. The dielectric constant of anatase is 36 [174].

4.1. Physical Properties of TiO$_2$ Thin Films

4.1.1. Energy Bandgap

The energy bandgap of TiO$_2$ film is about 3.0–3.5 eV, depending on the crystallization phase of TiO$_2$. It was reported that the energy bandgap of amorphous TiO$_2$ is 3.5 eV; anatase phase TiO$_2$, 3.2 eV; and rutile phase TiO$_2$, 3.0 eV [176–178]. Fuyuki and Matsunami deposited TiO$_2$ using LPCVD [184]. TiO$_2$ is amorphous when the substrate temperature is 200°C. The TiO$_2$ phase becomes anatase when the substrate temperature is 400°C. The bandgap can be deduced from the dependence of the absorption coefficient α on the wavelength of incident light in the near-ultraviolet region. The bandgap of amorphous film is 3.44 eV and increases up to 3.98 eV for films deposited at 400°C. Wiggins et al. deposited TiO$_2$ on fused silica using sputtering [185]. The energy bandgap of TiO$_2$ was obtained from the steepest sloped section of the absorption edge curve extrapolated to zero transmittance. The bandgap is about 3.65 eV ($\lambda_g \sim 340$ nm) for as-deposited amorphous TiO$_2$. The TiO$_2$ phase becomes anatase when annealed at 700–800°C and the bandgap is unchanged. However, the phase of TiO$_2$ film becomes rutile when annealed at 1150°C. The obtained bandgap is 3.35 eV ($\lambda_g \sim 370$ nm).

4.1.2. Electron Affinity

If the electron affinity of a dielectric layer is large, the potential barrier height at the dielectric–metal electrode interface

Table XXIV. Metal/TiO$_2$ Barrier Heights

Reference	Device structure	Measurement conditions	Comments
He et al. [178]	100-nm LPCVD TiO$_2$/p-silicon	1. Schottky emission at medium applied voltage 2. TiO$_2$/silicon barrier height: 1.05 eV for electron, 0.9 eV for hole	
Kim et al. [179, 180] Campbell et al.[177] Yan et al.[183]	19-nm MOCVD TiO$_2$/p-silicon	1. When $V_G < 0$ V: as-deposited TiO$_2$, $\Phi_B = 0.5$ eV 750°C O$_2$-annealed TiO$_2$, $\Phi_B = 1.0$ eV 2. When $V_G > 0$ V: at 100–160°C, $\Phi_B = 0.94$–1.0 eV	1. The leakage current mechanism is Schottky emission 2. TiO$_2$/silicon barrier height was obtained from J/T^2 vs. $1/T$ plot
Sun and Chen [187]	18-nm LPCVD TiO$_2$/n-silicon	1. Electrode limited-current mechanism 2. Leakage current decreases with increase in top electrode work function for as-deposited TiO$_2$ film	The top electrode materials and their work functions: Mo (4.25 eV), W (4.64 eV), Ta (4.75 eV) and the metal nitrides MoN (4.95 eV), WN (5.0 eV), TaN (5.33 eV), TiN (5.41 eV)
Szydio and Poirier [173]	Au/TiO$_2$ Schottky diode using the rutile single crystal	1. Au/n-TiO$_2$ Schottky barrier height = 0.87–0.94 eV 2. TiO$_2$ electron affinity $q\chi = 4.0$ eV	Assuming an effective mass of $m^* = 9m_0$, Richardson constant $A^* = 1100$ A cm^{-2} K^{-2}

will be low. This will give a high leakage current density. Szydio and Poirier studied the Au/TiO$_2$ Schottky diode using the rutile single crystal. TiO$_2$ slices are reduced under a vacuum of 10^{-6} Torr at 800°C for about 5 h [173]. Point defects such as oxygen vacancies will act as donor centers and change the insulating rutile into an n-type semiconductor. Assuming an effective mass m^* of $9m_0$ for the carriers in TiO$_2$ and a Richardson constant A^* of approximately 1100 A cm^{-2} K^{-2}, the calculated Au/n-TiO$_2$ Schottky barrier height is about 0.87–0.94 eV. Because the work function of Au is 4.9 eV, an electron affinity of 4.0 eV for TiO$_2$ is obtained [173].

4.1.3. Metal–Insulator Barrier Height

The leakage current is limited by the barrier height at the dielectric–electrode interface when the current mechanism is electrode limited. As shown in Table XXIV, the current mechanisms of TiO$_2$ reported most often are electrode limited [173, 177–183, 186]. Kim et al. [179–182], Campbell et al. [177], and Yan et al. [183] found that when the top Pt electrode was biased positive, the leakage current mechanism is Schottky emission. The calculated potential barrier height at the TiO$_2$/silicon interface for as-deposited TiO$_2$ is 0.5 eV. For TiO$_2$ annealed at 750°C in O$_2$ ambience, the TiO$_2$/silicon barrier height is 1.0 eV. The increase in the barrier height with annealing may be due to the crystallization of the as-deposited films or it may represent a change in the band alignment, perhaps due to a change in the charge states or strain at the TiO$_2$/silicon interface. Sun and Chen studied the influence of postannealing on the top electrode of TiO$_2$ capacitors [187]. For as-deposited TiO$_2$ film, the

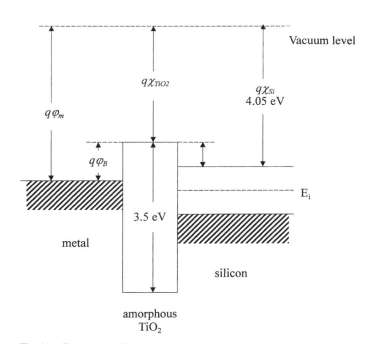

Fig. 14. Energy band diagram of metal/TiO$_2$/silicon capacitor at flatband condition.

leakage current decreases with increasing top electrode work function when the top electrode is biased negative. Because the leakage current depends on the potential barrier height at the interface of the top electrode and TiO$_2$, the leakage current mechanism is electrode limited. Figure 14 shows the energy band diagram of a metal/TiO$_2$/silicon capacitor. When the TiO$_2$

Table XXV.   Trap Type and Trap Energy Level of TiO$_2$ Films

Reference	Device structure	Measurement conditions	Comments
Hada et al. [189]	150–200-nm sputtered TiO$_2$/Ti	1. $V_G = 1.2$ V, the current increases gradually with time constant τ_i 2. $V_G < -1.2$ V, current goes to a negative value and gradually returns to 0 with time constant τ_r 3. Activation energy for τ_i, $E_a = 0.45$–0.5 eV; for τ_r, $E_a = 0.07$ eV	1. When $V_G > 0$ V, $E_a = 0.52$ eV 2. H$^+$ sites at interstitials in TiO$_2$ structure and acts as electron trap or donor levels in the forbidden band gap of the oxide
Lalevic and Taylor [190]	2–11-μm-thick TiO$_2$ on electrode Ti by dry O$_2$ thermal oxidation	Poole–Frenkel emission at high field	Trap ionization energy $E_i = 0.145$ eV
Lee et al. [191]	10–30-nm sputtered TiO$_2$/silicon 10–30-nm sputtered TiO$_2$/Ir bottom electrode	Leakage currents were measured at ± 2 V at different temperatures	Leakage current is independent of temperature $E_a = 0.78$ eV
Won et al. [188]	700–1900-nm MOCVD TiO$_2$/silicon	High-frequency C–V curves were measured	V_{FB} increases with increasing oxygen content

electron affinity of 4.0 eV proposed by Szydio and Poirier is used [173], there is no barrier at the TiO$_2$/silicon interface. This is in conflict with the barrier height obtained from Schottky emission current. Further study on the energy band diagram of the TiO$_2$/silicon structure is required in the future.

4.1.4. Trap Type and Trap Level

The trap type and trap energy level of TiO$_2$ films is listed in Table XXV. When the TiO$_2$ is oxygen deficient, the oxygen vacancies will act as donor centers. This will make TiO$_2$ an n-type semiconductor. The conductivity will also increase [173, 183]. Won et al. reported that the flatband voltage increases with increasing oxygen content [188]. The increase in oxygen content causes an increase in the defect density such as excess oxygen atoms at the interface, an increase in acceptor traps, and a shift the flatband voltage to more positive values. Hada et al. measured the transient current characteristic of Au/TiO$_2$/Ti capacitors [189]. The measurement was carried out in an atmosphere of relatively high humidity. The transient current increases gradually and then becomes saturated with a time constant τ_i with the top Au electrode biased at 1.2 V. When the bias is -1.2 V, the transient current goes to a negative value and gradually returns to 0 with a time constant τ_r. The recovery current is attributed to the migration of hydrogen ions in the oxide film. Hence, the activation energy of the time constant τ_r is considered to be an activation energy for hydrogen ions to diffuse through the TiO$_2$ film. There is a possibility that the H$^+$ at the interstitial site in TiO$_2$ act as an electronic trap or donor level in the forbidden bandgap of TiO$_2$. It is likely that the recovery is due to H$^+$ moving back to the Au/TiO$_2$ contact and recombining with the surface

trap center. Lalevic and Taylor reported that the leakage current mechanism of a Cr/TiO$_2$/Ti capacitor at high electric field is Poole–Frenkel emission with a trap ionization energy E_i of 0.145 eV [190]. Lee et al. reported that the activation energy E_a of the leakage current of a Pt/TiO$_2$/Ir metal–insulator–metal (MIM) capacitor at either 2 V or -2 V bias is approximately 0.78 eV [191].

4.1.5. Dielectric Constant

The dielectric constant of TiO$_2$ is dependent on the film thickness as shown in Table XXVI. Takeuchi et al. deposited 0.27- to 2.8-μm-thick TiO$_2$ film on a brass strip substrate [197]. Two kinds of TiO$_2$ powders, anatase and rutile, were used as the starting material of the sputtering target. The X-ray diffraction result shows that when the TiO$_2$ thickness is less than 1.0 μm, TiO$_2$ is amorphous and the dielectric constant increases with increasing TiO$_2$ thickness. The crystallization phase of TiO$_2$ is rutile when the TiO$_2$ thickness is greater than 1.0 μm. The dielectric constant of rutile TiO$_2$ is 100, which is not very thickness dependent. The variation of dielectric constant mentioned previously is independent of the crystallization phase of the TiO$_2$ sputtering target.

The dielectric constant of TiO$_2$ is also depends on the crystallization phase. Abe and Fukuda showed that when TiO$_2$ film was formed by O$_2$-plasma oxidation at a substrate temperature of 180°C [192], TiO$_2$ is amorphous with $\varepsilon_r = 20$. The TiO$_2$ film becomes rutile and the dielectric constant increases with increasing substrate temperature when the substrate temperature is higher than 280°C. When the substrate temperature is higher than 390°C, the dielectric constant is 120. Fuyuki and Matsunami reported that TiO$_2$ film deposited by LPCVD

Table XXVI. Crystallization Phases and Dielectric Constants of TiO_2 Films

Reference	Device structure	Crystallization phase	Dielectric constant
Abe and Fukuda [192]	Sputtered Ti on SiO_2/silicon or Pt/SiO_2/silicon substrate and then annealed by ECR O_2-plasma to form 3.5–26-nm TiO_2	1. Amorphous when the substrate temperature is 180°C 2. Rutile when the substrate temperature is 280°C	1. $\varepsilon_r = 20$ for amorphous TiO_2 2. $\varepsilon_r = 120$ when the substrate temperature is higher than 380°C
Fukushima and Yamada [194]	Reactive ionized-cluster beam TiO_2/silicon	When the acceleration voltage V_a was low, a weak rutile and a strong anatase phase were observed	A decrease in the dielectric constant is observed by either increasing V_a or increasing electron current for ionization I_e
Fuyuki and Matsunami [184]	78-nm LPCVD TiO_2/silicon	Amorphous when the substrate temperature is 200°C	$\varepsilon_r = 20$–86, depending on the water concentration in the precursor
Fuyuki et al. [193]	LPCVD TiO_2; the precursor is a mixture of wet oxygen and TPT		ε_r increases from 16 to 32 as the water vapor concentration increases from 0 to 300 ppm
Kim et al. [182]	19-nm MOCVD TiO_2/silicon	Anatase without any rutile phase after 750°C dry O_2 annealing	$\varepsilon_r = 30$
Lee et al. [191]	100-nm LPCVD TiO_2/silicon	1. Anatase when substrate temperature ranges from 300 to 700°C 2. Rutile when substrate temperature higher than 800°C	
Lee et al. [195]	Excimer-laser-deposited TiN/silicon; 230-nm-thick TiO_2 was formed by 780°C dry O_2 annealing	Rutile	$\varepsilon_r = 49$
Lee et al. [176]	20–200-nm ECR-CVD TiO_2/n/silicon	1. Anatase when substrate temperature is 400–600°C 2. Rutile when substrate temperature is 900°C	$\varepsilon_r = 60$ for rutile phase
Shin et al. [196]	SAM TiO_2; silicon wafer was immersed in $TiCl_4$ liquid to form 63-nm TiO_2 at 80°C	Anatase	1. As-deposited film $\varepsilon_r = 24$–57, depending on the thickness of interface SAM layer and SiO_2 layer 2. After 400°C annealing in air for 2 h, $\varepsilon_r = 67$–97
Wiggins et al. [185]	73–830-nm sputtered TiO_2/silicon	1. Both anatase and rutile were observed at 350–750°C 2. Anatase disappears when TiO_2 annealed at 750–850°C 3. Rutile when annealing temperature higher than 850°C	

is amorphous when the deposition substrate temperature is 200°C [184]. The dielectric constant ranges from 20 to 86. A small amount of water added to the source has an important effect on the dielectric properties of the deposited films, which may be the cause of the scatter in the value of the dielectric constant. Fuyuki and Matsunami then deposited TiO_2 film by using a tetra-isopropyl-titanate (TPT) precursor and added

some wet oxygen to it [193]. The deposition substrate temperature was kept at 200°C. When the water vapor concentration P_w increases from 0 to 300 ppm, the dielectric constant increases from 16 to 32. The reason that the dielectric constant increases with P_w is not clear. The experimental results of increased refractive index and decreased etching rate with increasing P_w suggested that there is some variation of the configuration of

Ti–O bond, which may cause the dielectric constant to increase.

Fukushima and Yamada deposited TiO_2 film on a silicon substrate using a reactive ionized-cluster beam [194]. A small rutile peak is observed in the X-ray diffraction result when the acceleration voltage V_a is 3 kV and the electron current for ionization I_e is 400 mA. A decrease in the dielectric constant is observed when either V_a or I_e increases. When V_a is low, a weak rutile phase and a strong anatase phase are observed. When $V_a = 2$ kV, only the rutile (100) peak appears. Further increase of V_a to 4 kV produces the anatase peak again. This indicates that the crystalline structure was affected by the ion kinetic energy. Wiggins et al. reported that when the annealing temperature is at 350–750°C, TiO_2 film contains both anatase and rutile phases [185]. However, the anatase peaks decrease and the rutile peaks increase when the annealing temperature is at 750–850°C. TiO_2 becomes rutile when the annealing temperature is higher than 850°C. Shin et al. used the self-assembled organic monolayer (SAM) method to deposit TiO_2 film on a silicon substrate [196]. Silicon wafer was immersed in $TiCl_4$ liquid to form 63-nm-thick TiO_2 at 80°C. Two thin interfacial layers are formed at the TiO_2/silicon interface, the SAM layer and the SiO_2 layer. The crystallization phase of TiO_2 is anatase. The dielectric constant of this TiO_2 film is about 24–57, depending on the thickness of the SAM layer and the SiO_2 layer. The dielectric constant is about 67–97 when TiO_2 film is annealed at 400°C for 2 h. This is because the capacitance is only determined by the thicknesses of the TiO_2 layer and the SiO_2 layer as the SAM layer is pyrolyzed at 400°C.

4.2. Fabrication Processes of TiO_2 Thin Films

4.2.1. Deposition Methods

The deposition methods of TiO_2 films include LPCVD [187, 193, 198], plasma-enhanced CVD [199], metal–organic CVD [177–183, 188], ECR-PECVD [171, 192, 200] sputtering [185, 186, 189, 197], thermal oxidation [175, 201, 202], excimer laser ablation [195], and jet vapor deposition [203, 204]. The precursors for TiO_2 LPCVD, PECVD, and MOCVD systems include tetra-isopropyl-titanate [TPT, $Ti(OC_3H_7)_4$] [187, 193], TET [$Ti(OC_2H_5)_4$] [188], nitrato-titanium [NT, $Ti(NO_3)_4$] [198], $TiCl_4$ [199], and titanium tetrakis-isopropoxide (TTIP) [177–179, 183]. The vaporization temperature of these precursors is about 60–70°C. The TiO_2 films were deposited at a substrate temperature of 200–400°C in oxygen ambience. In the past, TiO_2 films were grown by oxidation of titanium or TiN films [174, 175, 190, 192, 195, 200, 205–207]. The titanium or TiN films were deposited on silicon by electron-beam (e-beam) evaporation, sputtering, excimer laser ablation, or physical vapor deposition. The oxidation methods to form TiO_2 include RTA at 400–900°C in dry O_2 ambience, ECR-PECVD O_2-plasma [192, 200], and inductive coupled O_2-plasma [175].

Burns et al. compared the electrical properties of TiO_2 films deposited by different kinds of processes [175]. They found that e-beam-evaporated TiO_2 or sputtered TiO_2 in Ar ambience will cause a reduction in the TiO_2 during deposition, which gives TiO_2 films of low resistivity and low dielectric constant. TiO_2 films deposited by MOCVD have high conductivity (about 10^6 Ω-cm) and the film is either amorphous or of anatase phase, hence giving a low dielectric constant. The thermal oxidation of titanium to TiO_2 requires an annealing step at 500–600°C for 20–30 h. Although RTA can reduce the annealing time to 40 s, it requires a high temperature of about 900°C. The oxidation of titanium to TiO_2 can also be performed using O_2-plasma annealing. The leakage current density of O_2-plasma-oxidized TiO_2 can be as small as 1×10^{-8} A/cm^2 at 1.0 V.

4.2.2. Annealing Conditions

The leakage current density of TiO_2 film can be reduced by annealing. The annealing gases include O_2, N_2, and N_2O. The common annealing process for TiO_2 is furnace annealing in dry O_2 at 600–800°C. The crystallization phase of TiO_2 annealed at 300–700°C is a mixture of anatase and rutile. When the annealing temperature is higher than 800°C, the crystallization phase is high-dielectric-constant phase rutile [173, 178, 185]. Therefore, an annealing temperature higher than 800°C is necessary to obtain the high-dielectric-constant value [176]. Annealing under N_2 ambience cannot reduce the leakage current level effectively [191].

Sun and Chen annealed TiO_2 films at 800°C using either furnace annealing for 30 min or RTA for 60 s [187]. The annealing gas was either O_2 or N_2O. The leakage current of furnace-annealed TiO_2 is less than that of RTA-annealed TiO_2. The leakage current of TiO_2 annealed under N_2O ambience is less than that annealed under O_2 ambience. The N_2O furnace-annealed TiO_2 film has the lowest leakage current density, which is about 1×10^{-8} A/cm^2 at 1.0 V. This is due to the reduction in oxygen vacancies and carbon concentration in TiO_2 by the reactive atomic species generated by the N_2O dissociation during the furnace thermal cycle. However, the reduction in leakage current is related to the decrease in TiO_2 dielectric constant. The N_2O furnace-annealed TiO_2 film has the lowest dielectric constant value. This is due to the existence of a thin SiO_2 layer formed at the TiO_2/silicon interface [187].

4.2.3. Bottom Electrode Materials

The reported bottom electrode materials for TiO_2 as a DRAM storage capacitor include n^+-polysilicon [187, 199], Pt [192] and Ti [192, 200]. Abe and Fukuda deposited Ti on SiO_2/silicon or Pt/SiO_2/silicon substrates using sputtering [192]. Ti was then oxidized to form TiO_2 using O_2-plasma generated by ECR. The top electrode of the TiO_2 capacitor was Au. The leakage current density of TiO_2 with a Pt bottom electrode is 1×10^{-8} A/cm^2 at 1.25 V. However, the leakage current density for TiO_2 with a Ti bottom electrode at 1.25 V is 1×10^{-2} A/cm^2. Abe and Fukuda explained this result by the low work function of the Ti bottom electrode. Because the work function of Pt is 5.65 eV and is much higher than 4.33 eV for Ti, the barrier height at the TiO_2/Pt interface is higher than that at the TiO_2/Ti interface.

Another possible reason is that an oxygen-deficient TiO_{2-x} layer with low resistivity may remain near the TiO_2/Ti interface, resulting in a higher leakage current.

4.2.4. Top Electrode Materials

Sun and Chen studied the effect of postannealing on the top electrode of a TiO_2 capacitor [187]. The top electrodes studied include Mo, W, Ta, and the metal nitrides MoN, WN, TaN, and TiN. The work functions of Ta, Mo, W, TiN, WN, MoN, and TaN are 4.25, 4.64, 4.75, 4.95, 5.00, 5.33, and 5.41, respectively. For as-deposited TiO_2 film, the leakage current decreases with increasing top electrode work function when the top electrode is biased negatively. The leakage current density for TiO_2 annealed at 400°C under N_2 ambience for 30 min ranges from 1×10^{-6} to 1×10^{-7} A/cm². Because the work functions of these top electrode metals obtained from the $C-V$ flatband voltage are 4.5–4.8 eV after N_2 annealing, the leakage current densities of these capacitors with different top electrode materials after 400°C annealing are independent of the top electrode material. The leakage current of TiO_2 with a Ta top electrode increases significantly due to the existence of metallic Ta as an interstitial impurity in TiO_2. For TiO_2 films with Mo and MoN top electrodes, the leakage current increases significantly after 800°C annealing in N_2 ambience due to Mo diffusion into TiO_2. The leakage current of TiO_2 with a WN top electrode after 800°C annealing is the smallest among these top electrode materials. The leakage current is 3×10^{-6} A/cm² at 1.0 V. The Auger electron spectroscopy (AES) depth profile of the WN/TiO_2/Si structure shows that no appreciable W diffusion in TiO_2 film.

4.3. Leakage Current Mechanisms of TiO_2 Thin Films

As shown in Table XXVII, most literature results show that the leakage current mechanism of TiO_2 film at high electric field and at high temperature is due to Schottky emission. At low electric field, the leakage current mechanism is due to either ohmic or hopping conduction. However, there are reports claiming that the leakage current mechanism at high electric field and room temperature is Poole–Frenkel emission [190, 196]. Schottky emission at low electric field and Poole–Frenkel emission at high electric field have also been reported [201]. When TiO_2 film is deposited on a silicon or polysilicon substrate and annealed at high temperature in O_2 ambience, Fowler–Nordheim tunneling is observed because a thin SiO_2 layer is formed at the TiO_2/silicon or TiO_2/polysilicon interface [178].

4.3.1. Poole–Frenkel Emission

Lalevic and Taylor used dry O_2 thermal oxidation to form a 2- to 11-μm-thick TiO_2 film on Ti at 800°C [190]. The top electrode was Cr. The leakage current mechanism at high electric field is Poole–Frenkel emission. The expression for Poole–Frenkel emission with temperature dependence is as follows:

$$J = AT^n \exp(-E_i/kT) \sinh \beta V^{1/2}$$

where n is a constant between 2 and 4. The calculated result shows that $n = 2$ and the trap ionization energy $E_i = 0.145$ eV. Shin et al. deposited a thin TiO_2 layer on a silicon wafer using the self-assembled organic monolayer (SAM) method [196]. The silicon wafer was immersed in $TiCl_4$ liquid to form a 63-nm-thick TiO_2 at 80°C. A thin SAM layer with thickness of 2.5 nm and a thin SiO_2 layer with a thickness of 2.5 nm formed at the TiO_2/silicon interface after deposition. When the capacitor is biased in the accumulation region, the leakage current density of TiO_2 after 400°C annealing is higher than that of as-deposited TiO_2 by 2 orders of magnitude. This is due to the fact that grain boundaries of polycrystalline TiO_2 films probably become leakage current paths, as occurs in other polycrystalline oxide layers. Also, a new interface between the TiO_2 and SiO_2 layers will be formed because the SAM layer is decomposed after annealing. When the applied electric field is higher than 0.3 MV/cm, the leakage current mechanism is Poole–Frenkel emission with a calculated dielectric constant of 30.

4.3.2. Schottky Emission

The leakage current mechanism in TiO_2 film reported most often is Schottky emission [177–182, 187, 198, 199, 201]. Brown and Grannemann reported that when the Al/as-deposited TiO_2/n-silicon capacitor is biased at $V_G > 0$ V, the leakage current follows a $J \propto V^1$ relationship due to electron hopping [201]. The leakage current increases sharply and then follows ohmic conduction, again at a very high applied voltage. A plot of $\log(I)$ vs. V_G for the non-ohmic portion of the $I-V$ curve yields two straight lines, indicating that there are at least two different conduction mechanisms. The nonlinear part of the leakage current is Poole–Frenkel current at low voltage and Schottky emission current at high applied voltages and high temperatures. He et al. deposited 100-nm-thick TiO_2 film on a silicon substrate using LPCVD [178]. The leakage current at low bias is electron hopping between trap sites. The leakage current follows Schottky emission behavior as the bias voltage or the temperature increases. The electron and hole barrier heights for TiO_2/n-silicon and TiO_2/p-silicon are 1.05 and 0.9 eV, respectively. Kim et al. [179–182], Campbell et al. [177], and Yan et al. [183] reported that the Schottky barrier for as-deposited TiO_2 is 0.5 eV and for 750°C O_2 annealed TiO_2 1.0 eV. The increase in barrier height with O_2 annealing is due to a change in the band alignment, perhaps due to a change in the charge states or strain at TiO_2/silicon interface. Abe and Fukuda reported that the leakage current of TiO_2 deposited on a Pt bottom electrode is 1×10^{-8} A/cm² at 1.25 V, which is smaller than that of TiO_2 deposited on a Ti bottom electrode by 6 orders of magnitude [192]. This is due to the larger work function of Pt (5.65 eV) than that of Ti (4.33 eV).

4.3.3. Fowler–Nordheim Tunneling

Jeon et al. reported that for a TiO_2/Si_3N_4/silicon structure annealed at 900°C O_2 ambience, the leakage current is inde-

Table XXVII.   Schottky Emission in TiO_2 Films

Reference	Device structure	Measurement condition	Barrier height
Abe and Fukuda [192]	Sputtered Ti/SiO$_2$/ silicon or Pt/SiO$_2$/silicon and then oxidized in ECR O$_2$-plasma to form 3.5–26-nm thick TiO$_2$	Schottky emission	Leakage current for TiO$_2$ with Pt bottom electrode is smaller than that with Ti electrode by 6 orders of magnitude at 1.25 V due to the higher work function of Pt (5.65 eV) compared with that of Ti (4.33 eV)
Brown and Grannemann [201]	50–200-nm e-beam-evaporated TiO$_2$/n-silicon	When $V_G > 0$ V: 1. $J \propto V^1$ at low voltage due to electron hopping 2. Poole–Frenkel current at medium applied voltage 3. Schottky emission current at high applied voltage	
He et al. [198]	10–11-nm MOCVD TiO$_2$/silicon with 700°C N$_2$ annealing	When $V_G < 0$ V, the current can be well fit by either thermionic emission or Poole–Frenkel emission	1. Φ_B extracted from J vs. $1/T$ plot decreases with increasing bias 2. Φ_B becomes temperature independent at larger bias
He et al. [178]	100-nm LPCVD TiO$_2$/silicon	1. Hopping at low voltage 2. Thermionic emission at medium applied voltage 3. Fowler–Nordheim tunneling at high voltage	Electron barrier height of 1.05 eV and hole barrier height 0.9 eV were obtained from thermionic emission current
Kim et al. [179, 180] Campbell et al. [177] Yan et al. [183]	19-nm MOCVD TiO$_2$/p-silicon	1. When $V_G < 0$, Schottky emission 2. When $V_G > 0$ V and current was measured under 100–160°C Schottky emission	when $V_G < 0$ V: 1. As-deposited TiO$_2$, $\Phi_B = 0.5$ eV 2. 750°C O$_2$-annealed TiO$_2$, $\Phi_B = 1.0$ eV When $V_G > 0$ V, $\Phi_B = 0.94$–1.0 eV
Sun and Chen [187]	18-nm LPCVD/n-silicon	1. Electrode-limited mechanism for as-deposited TiO$_2$ 2. The leakage current was limited by the barrier height at the top electrode/TiO$_2$ interface	The studied top electrodes include Mo, W, Ta, MoN, WN, TaN, and TiN

pendent of the measurement temperature when the capacitor is biased for both positive and negative polarities at either $+2$ V or -2 V [186]. Hence, the conduction current mechanism is tunneling. The direct tunneling current is limited by the interfacial barrier layer. Kim et al. [179–182], Campbell et al. [177], and Yan et al. [183] reported that for 19-nm-thick TiO$_2$ deposited on p-type silicon, the leakage current mechanism of the metal–insulator–semiconductor (MIS) capacitor biased at $V_G > 0$ V and in the temperature range of 100–160°C is Schottky emission. The calculated barrier height at the TiO$_2$/silicon interface is about 0.94–1.0 eV. When the measurement temperature is less than 100°C or higher than 160°C, the leakage current density is independent of the temperature. Hence, the leakage current mechanism is due to tunneling. Guo et al. used jet vapor deposition (JVD) to deposit TiO$_2$ (7 nm) and Si$_3$N$_4$

(1.0–1.5 nm) on a silicon substrate [203]. The JVD Si$_3$N$_4$ acts as a buffer layer because it is an excellent diffusion barrier and has a good electrical interface with silicon. The leakage current of the TiO$_2$/Si$_3$N$_4$/silicon capacitor shows very weak temperature dependence over a wide temperature range (27–200°C), suggesting tunneling as the dominant conduction mechanism.

4.3.4. Hopping Conduction

He et al. [178] reported that the leakage current of 100-nm-thick TiO$_2$ deposited by LPCVD is due to electron hopping between trap sites at low voltage. This current is strongly dependent on film deposition and annealing conditions. When the applied voltage or the measurement temperature increases, the leakage current follows Schottky emission behavior. Lalevic and Taylor used dry O_2 thermal oxidation to form 2- to 11-μm-thick TiO_2 on a Ti electrode [190]. Alternating-current (ac) conductivity of $Cr/TiO_2/Ti$ was measured at a frequency range of 10^2–2×10^6 Hz. The leakage current mechanism is noncorrelated hopping. The expression of conductivity for noncorrelated hopping is

$$\sigma(\omega) = \frac{1}{96} \pi^3 N^2(E_F) kTaq^2 r_\omega^4 \omega$$

where $N(E_F)$ is the probability density of localized states at the Fermi level, r_ω is the separation between a pair of localized states, and a is the lattice constant. The transition from $\sigma(\omega) \propto \omega$ dependence to $\sigma(\omega) \propto \omega^2$ dependence occurred at a frequency of 10^6 Hz. The quadratic frequency range 2×10^6–10^8 Hz corresponds to Debye-type ac conductivity and loss resulting from thermally activated hopping from one localized site to another. Thus,

$$\sigma(\omega) \propto \frac{\omega^2 \tau}{1 + \omega^2 \tau^2}$$

where τ is the mean time for a transition from one site to another. For $\omega \ll \tau$, this becomes $\sigma(\omega) \propto \omega^2$, which corresponds to the observed experimental results.

4.3.5. Space-Charge-Limited Current

Space-charge-limited current (SCLC) was often observed at medium applied voltage. Brown and Grannemann reported that for Al/as-deposited TiO_2/n-silicon biased at $V_g > 0$ V, the leakage current follows $I \propto V^1$ at low voltage due to electron hopping, followed by a non-ohmic current region and then a second ohmic region at high voltages [201]. For TiO_2 film annealed at 1000°C under dry O_2 ambience for 10 min, the $I \propto V^1$ region is followed by an $I \propto V^2$ region at low voltage. Taylor and Lalevic reported that the I–V relationship for the $Cr/TiO_2/Ti$ capacitor corresponds, in general, to Lampert's double-carrier SCLC model [206, 207]. In this model, ohmic conduction is observed at low voltage. At higher voltages, the density of injected holes and electrons is greater than that of thermal carriers. The injected hole encounters a recombination barrier and a region of single-carrier SCLC current is produced where $I \propto V^2$. At a threshold voltage V_{th}, the hole transit time across the insulator is on the order of the low-level hole lifetime, and a negative-resistance region is caused by a hole lifetime that increases with the injection level. Beyond this negative resistance region, both injected electrons and holes contribute to the current, producing a double-injection SCLC current with $I \propto V^2$.

4.4. C–V Characteristics of TiO2 Thin Films

Instability of the C–V characteristics is the main problem for the application of TiO_2 as a MOSFET gate dielectric. The reason for trap formation in TiO_2 film is not clear. Most papers attribute the origin of traps to oxygen vacancies. If the concentration of oxygen vacancies is sufficiently high, TiO_2 will change from an insulator to an n-type semiconductor [173].

4.4.1. Fixed Oxide Charge

Won et al. reported that the flatband voltage becomes more positive as TiO_2 film thickness increases [188]. Acceptor traps due to excess oxygen will result in negative fixed oxide charges. The concentration of defects such as acceptor traps increases with increasing TiO_2 thickness. The flatband voltage increases with increasing oxygen content. Higher deposition temperature causes more negative flatband voltage. This may be explained by reduced excess oxygen in the TiO_2 films. Yan et al. compared TiO_2 films deposited using a TTIP precursor only and using TTIP with the addition of water [183]. The films deposited with water have a higher interface state density of about 5×10^{12} cm^{-2} eV^{-1}. A lateral shift of the C–V curve of about 0.65 V was observed. The calculated fixed charge density is approximately 10^{12} cm^{-2}.

4.4.2. Oxide Trapped Charge

As shown in Table XXVIII, the flatband voltage shift due to oxide charge trapping in TiO_2 is much larger than that of SiO_2. Jeon et al. reported that the C–V hysteresis loop of TiO_2/Si_3N_4/silicon stack layers is counterclockwise, which suggest that the C–V hysteresis is caused by positive charge trapping rather than ferroelectricity [186]. Positive charges generated by oxygen vacancies are the main cause of hysteresis and flatband voltage shift. Lee et al. reported that there is a flatband voltage difference of about -1.0 V between the 600 and the 1000°C annealed TiO_2 films due to the positive oxide charge generated during the 1000°C annealing in Ar [176]. Loss of oxygen in the TiO_2 lattice creates oxygen vacancies, presumably generating the oxide charges. Kim et al. deposited 19-nm-thick TiO_2 film on a p-type silicon using MOCVD [179–182]. Both the 100-kHz and the 1-MHz C–V curves displayed a pronounced voltage-dependent accumulation capacitance. This might be caused by hole trapping in TiO_2 close to native SiO_2 at the TiO_2/silicon interface. The leakage current increased by a factor of 10^4 at -2.0 V after 100-C/cm^2 stress. High-frequency C–V measurement indicates that there is little charge trapping during electrical stress. By measuring the flatband voltage shift $\Delta V_{fb}(t)$, the change in the bulk trap density ΔN_{ot} under constant current injection stress is given by

$$\frac{d}{dt} \Delta V_{fb}(t) = \frac{\sigma J N_{eff}}{C_{ox}} \exp\left(\frac{-\sigma J t}{q}\right)$$

where N_{eff} is the effective trap density per unit area and σ is the trap cross section. By assuming an N_{eff} of 10^{14} cm^{-2}, the trap cross section σ was estimated to be roughly 10^{-23} cm^2. The increase in leakage current upon electrical stress suggests that uncharged, near-interface states may be created in the TiO_2 film near the SiO_2 interface layer that allows a tunneling current component at low bias. Alternatively, the increase in current may be due to the creation of neutral traps near the Pt gate electrode. Guo et al. used JVD to deposit TiO_2 (7 nm)/Si_3N_4 (1.0–1.5 nm) stack layers on a silicon substrate [203]. The TiO_2 stack layer was stressed by constant current injection

Table XXVIII.   Oxide Rapped Charges in TiO$_2$ Films

Reference	Device structure	Measurement condition	Comments
Guo et al. [203]	JVD TiO$_2$(7 nm)/Si$_3$N$_4$ (1.0–1.5 nm)/silicon	Constant current injection (0.1 A/cm^2) for both polarities after 1000 C/cm^2 of stressing charge fluence	Gate voltage shift $\Delta V_G < 20$ mV, indicating a low bulk trap density and low trap generation rate and high charge to breakdown values
He et al. [198]	11-nm MOCVD TiO$_2$/silicon with 700°C N$_2$ annealing	About -80 mV of hysteresis is observed for sweeping from 1.0 V to -2.0 V and back	TiO$_2$ tends to trap holes
Jeon et al. [186]	10–35-nm sputtered TiO$_2$/2-nm JVD Si$_3$N$_4$/silicon with 600–900°C dry O$_2$ annealing	C–V curve hysteresis loops were measured	Counterclockwise C–V hysteresis loops suggest that hysteresis is caused by positive-charge trapping rather than ferroelectricity
Kim et al. [179–181]	19-nm MOCVD TiO$_2$/p-silicon	Leakage current increased by a factor of 10^4 at -2.0 V after 100 C/cm^2 stress	Assuming on N_{eff} on the order of 10^{14} cm^{-2}, the trap cross section σ was estimated to be roughly 10^{-23} cm^2
Lee et al. [176]	20–200-nm ECR-PECVD TiO$_2$/n-silicon	$\Delta V_{\text{FB}} = -1.0$ V between the 600 and the 1000°C annealed TiO$_2$ films	Loss of oxygen in the TiO$_2$ lattice creates vacancies at the oxygen site, presumably generating the oxide charges

(0.1 A/cm^2) of both polarities. After 1000 C/cm^2 of stressing charge fluence, the gate voltage shift is less than 20 mV, indicating a low bulk trap density, a low trap generation rate, and a high charge to breakdown value. There is hardly any stress-induced leakage current (SILC) after ± 12 MV/cm of stress. The small shift in the C–V curve after high electric field stress of ± 12 MV/cm is consistent with V_G vs. stress time results. The TiO$_2$/Si$_3$N$_4$ stack deposited by JVD was proposed as an alternative gate dielectric for future VLSI applications.

4.4.3. Interface Trapped Charge

The interface trapped charge density for a TiO$_2$/silicon structure is about 1–2 orders of magnitude larger than that of a SiO$_2$/silicon structure as shown in Table XXIX. Yan et al. compared TiO$_2$ films deposited by MOCVD using a TTIP precursor only and TiO$_2$ films deposited using TTIP with the addition of water [183]. The midgap interface state density of the first type of samples was in the mid-10^{11}-cm^{-2}-eV^{-1} range. A conductance measurement showed a peak in the G_p/ω vs. frequency plot at 50 kHz when the gate was biased at -0.5 V. The peak height corresponded to a midgap interface state density of 1.7×10^{12} cm^{-2} eV^{-1}. TiO$_2$ films deposited with the addition of water had a much larger interface state density, with midgap levels in the mid-10^{12}-cm^{-2}-eV^{-1} range measured using the Terman method. Campbell et al. deposited 19-nm-thick TiO$_2$ on a p-type silicon substrate [177]. The midgap interface state density D_{it} was about 3.0×10^{10}–1.0×10^{11} cm^{-2} eV^{-1} obtained from high- and low-frequency C–V measurements. The distribution of the interface state density in the silicon bandgap

followed a V shape rather than the common U shape seen in MOS capacitors. A conductance measurement for a 65-nm-thick TiO$_2$ sample showed a peak in the G_p/ω vs. frequency plot at 50 kHz when the gate was biased at 0.05 V. The interface state density of 1.0×10^{12} cm^{-2} eV^{-1} is in agreement with high- and low-frequency C–V analysis. Fuyuki and Matsunami deposited a 78-nm-thick TiO$_2$ film on a silicon substrate using LPCVD with a substrate temperature of 200–400°C [184]. The interface state density decreases with increasing substrate temperature. The minimum interface trapped charge density at the midgap was 2.0×10^{11} cm^{-2} eV^{-1}. The interface state density distribution in the silicon bandgap showed a profile similar to that of a thermally grown SiO$_2$/silicon system.

4.4.4. Mobile Ionic Charge

Lee et al. deposited TiN film on silicon using an excimer laser. The TiO$_2$ film was formed by dry O$_2$ annealing at 780°C [195]. The final thickness of the TiO$_2$ film was 230 nm. The hysteresis loop of TiO$_2$ was measured by first sweeping the C–V curve from 0 V to 2.0 V. The C–V measurement was then swept from 2.0 V to -2.0 V and back to 2.0 V. The C–V characteristic of the TiO$_2$ film shows a large hysteresis about 1.5 V, which suggests the existence of ionic drift in the oxide.

4.5. Metal–Oxide–Semiconductor Field Effect Transistor Gate Dielectric Applications of TiO$_2$ Thin Films

The application of TiO$_2$ film as the dielectric of a DRAM capacitor is not as promising as Ta$_2$O$_5$. This is due to the fact that

Table XXIX.   Interface Trapped Charges of the TiO_2/Silicon Interface

Reference	Device structure	Measurement condition	Comments
Brown and Grannemann [201]	50–200-nm e-beam evaporated TiO_2/silicon with 1000°C dry O_2 annealing	Lowest value of D_{it} was 10^{11} $cm^{-2}\,eV^{-1}$ with a dielectric constant of 5.8	D_{it} is characteristic of the SiO_2/silicon interface due to the low dielectric constant
Campbell et al. [177]	19-nm MOCVD TiO_2/p-silicon	1. Midgap $D_{it} = 3.0 \times 10^{10}$– $1.0 \times 10^{11}\,cm^{-2}\,eV^{-1}$ 2. $D_{it} = 1.0 \times 10^{12}\,cm^{-2}\,eV^{-1}$ obtained from the conductance method	Distribution of interface trap state in the silicon bandgap followed a V shape rather than the common U shape
Fukushima and Yamada [194]	100–450-nm reactive ionized-cluster beam TiO_2/silicon with 600°C dry O_2 annealing for 30 min	D_{it} density was calculated by the Terman method; the minimum D_{it} densities of TiO_2/n-silicon and TiO_2/p-silicon at the midgap were about $1.5 \times 10^{11}\,cm^{-2}\,eV^{-1}$ and $2 \times 10^{11}\,cm^{-2}\,eV^{-1}$, respectively	
Fuyuki and Matsunami [184]	78-nm LPCVD TiO_2/silicon with 200–400°C substrate temperature	Minimum D_{it} density at the midgap was $2.0 \times 10^{11}\,cm^{-2}\,eV^{-1}$	Interface state density distribution in the silicon bandgap is similar to thermally grown SiO_2/silicon
Yan et al. [183]	10–100-nm MOCVD TiO_2/p-silicon, TiO_2 films were deposited using a TTIP precursor with the addition of water	1. Midgap D_{it} for TTIP samples was in the mid-10^{11}-$cm^{-2}\,eV^{-1}$ range, $D_{it} = 1.7 \times 10^{12}\,cm^{-2}\,eV^{-1}$ calculated from the conductance method 2. Midgap D_{it} for TTIP + H_2O samples was in the mid-10^{12}-cm^{-2}-eV^{-1} range	

the leakage current of TiO_2 is higher than that of Ta_2O_5 by at least 1 order of magnitude and the dielectric strength of TiO_2 is smaller than that of Ta_2O_5. Furthermore, the dielectric constant of TiO_2 film depends on both the crystallization phase and the film thickness. The dielectric constant of amorphous TiO_2 film is about 20 [193, 200], which is smaller than that of Ta_2O_5.

Table XXX lists the published results of MOSFETs with TiO_2 gate dielectric. Campbell et al. fabricated MOSFETs with TiO_2 gate dielectric using three different kinds of gate electrode materials, n^+-doped polysilicon, aluminum, and platinum [177]. The gate leakage current for the MOSFET with TiO_2 gate dielectric was lowest with the platinum gate electrode. Polysilicon was found to dramatically increase the leakage current, particularly after subsequent thermal cycles. The electron mobility of the MOSFET with the TiO_2 gate dielectric is only one-third to two-thirds of conventional MOSFETs with SiO_2 gate dielectric at the same equivalent SiO_2 thickness [177, 184]. Guo et al. used JVD to deposit TiO_2 (7 nm)/Si_3N_4 (1.0–1.5 nm) stack layer as the MOSFET gate dielectric [203]. The gate electrode of this MOSFET is aluminum. Because the quality of the JVD Si_3N_4/silicon interface is comparable to that of thermal SiO_2/silicon, the electron mobility of MOSFETs fabricated with JVD TiO_2/Si_3N_4 gate dielectric is comparable to that of MOSFETs with SiO_2 gate dielectric. The hot carrier

problem for MOSFETs with TiO_2 gate dielectric is also different from that of MOSFETs with SiO_2 gate dielectric. Kim et al. measured the hot carrier stress of a 1.25-μm channel length TiO_2 (18 nm) gate dielectric MOSFET at $V_D = 3.5$ V and $V_G = 2$ V [182]. The threshold voltage decreases and the transconductance increases rather than decreases after hot carrier stress. The reasons may be due to improved device properties after trap site filling.

4.6. Dielectric Reliability of TiO_2 Thin Films

The dielectric breakdown strength depends on the energy bandgap. Because the energy bandgap of TiO_2 (3.0–3.5 eV) is smaller than those of Ta_2O_5 ($E_g = 4.2$ eV) and SiO_2 ($E_g = 9.0$ eV), TiO_2 has a low breakdown strength of 1–3 MV/cm [175, 177, 191, 196]. Although the leakage current and breakdown strength properties of TiO_2 are worse than those of Ta_2O_5, the high dielectric constant of TiO_2 still makes it useful as a potential MOSFET gate dielectric. TiO_2 film coupled with a JVD Si_3N_4 buffer layer is a candidate for future sub-100-nm channel length MOSFETs due to its good reliability behavior. Guo et al. used JVD to deposit a TiO_2 (7 nm)/Si_3N_4 (1.0–1.5 nm) stack layer on silicon [204]. The equivalent SiO_2 thickness of the TiO_2/Si_3N_4 stack layer is 2.0 nm. Constant cur-

Table XXX. Electrical Characteristics of MOSFETs with TiO$_2$ Gate Dielectric

Reference	Gate dielectric	TiO$_2$ dielectric constant	Equivalent SiO$_2$ thickness	MOSFET structure	Electrical parameters
Campbell et al. [177]	19-nm LPCVD TiO$_2$	30	2.5 nm	Pt gate with MOSFET channel length $L = 100 \ \mu$m	1. $\mu_e = 160$ cm^2/V-s, about 1/3 of that in an MIS using SiO$_2$ as gate dielectric 2. $D_{it} = 3 \times 10^{10} - 1 \times 10^{11}$ eV^{-1} cm^{-2} by high- and low-frequency methods
Fuyuki and Matsunami [184]	78-nm LPCVD TiO$_2$	42	7.2 nm	S/D contacts were made with Schottky contacts using Mg	$\mu_e = 200$–300 cm^2/V-s, about 2/3 of that in an MIS using SiO$_2$ as gate dielectric
Hobbs et al. [205]	3.5-nm TiO$_2$/3.5-nm native SiO$_2$	6–7	4.1 nm for n-MOS 4.3 nm for p-MOS	TiN gate with MOSFET channel length $L = 0.1$–20 μm	$S_t = 81$–83 mV/decade when $L = 0.25 \ \mu$m
Kim et al. [180]	19-nm MOCVD TiO$_2$	30	2.5 nm	Pt gate with MOSFET channel length $L = 1.5 \ \mu$m	1. $\mu_e = 160$ cm^2/V-s, about 1/3 of that in an MIS using SiO$_2$ as gate dielectric 2. $D_{it} \cong 1 \times 10^{11}$ eV^{-1} cm^{-2} 3. g_m was improved after $V_D = 3.5$ V and $V_G = 2$ V hot carrier stress for 4000 s

rent injection (0.1 A/cm^2) at both polarities with a stress charge fluence of 1000 C/cm^2 was performed. The gate voltage shift is less than 20 mV, which indicates a low bulk trap density, a low trap generation rate, and a high charge to breakdown value. There is hardly any stress-induced leakage current after ± 12 MV/cm of stress. The small shift on the C–V curves after high electric field stress of ± 12 MV/cm is consistent with the V–t results. The dielectric breakdown field shows a tight distribution around 17–18 MV/cm. Guo et al. used sputtered TiO$_2$ (12 nm) as the MOS capacitor dielectric and JVD Si$_3$N$_4$ (1.5 nm) as the buffer layer [203]. The equivalent SiO$_2$ thickness of the TiO$_2$/Si$_3$N$_4$ stack layer is 1.55 nm. Constant current injection (5 A/cm^2) after stressing charge fluence of 10^4 C/cm^2 was performed. The gate voltage shift is 12 mV with no hard breakdown was observed. A tight dielectric breakdown field distribution centered at 20 MV/cm for the TiO$_2$/Si$_3$N$_4$ capacitor was obtained for the TiO$_2$/Si$_3$N$_4$ stack layer with an equivalent SiO$_2$ thickness of 1.5–1.8 nm. The projected cumulative failure rate (50% failure on a t_{BD} vs. E_{eff} plot) shows that the TiO$_2$/Si$_3$N$_4$ stack layer can be operated at an effective electric field E_{eff} of 10 MV/cm for 10 years. The TiO$_2$/Si$_3$N$_4$ stack layer deposited by JVD shows excellent interface quality, low leakage currents, high reliability, and comparable transistor performance compared to

thermal oxide. These results suggest that the TiO$_2$/Si$_3$N$_4$ stack could be an alternative gate dielectric for future DRAM devices.

5. OTHER HIGH-DIELECTRIC-CONSTANT FILMS: Al$_2$O$_3$, Y$_2$O$_3$, HfO$_2$, ZrO$_2$, AND Gd$_2$O$_3$

5.1. Aluminum Oxide (Al$_2$O$_3$)

Al$_2$O$_3$ has been used as the DRAM storage capacitor dielectric due to the fact that its dielectric constant is higher than that of Si$_3$N$_4$ and SiO$_2$ [208, 209]. Table XXXI shows the published results of DRAM chips with Al$_2$O$_3$ as the capacitor dielectric. The dielectric constant of Al$_2$O$_3$ film is 9 and the breakdown strength is about 6–8 MV/cm, which is comparable to that of Si$_3$N$_4$ film. The energy bandgap of Al$_2$O$_3$ is 8.7 eV [211]. Although the dielectric constant of Al$_2$O$_3$ is much smaller than that of Ta$_2$O$_5$, the leakage current level of 1×10^{-9} A/cm^2 at 1.0 V is smaller than Ta$_2$O$_5$. Furthermore, Al$_2$O$_3$ film is more suitable than Ta$_2$O$_5$ film for the embedded DRAM application because postannealing is not required. The overall thermal process for Al$_2$O$_3$ capacitor formation can be accomplished below 600°C so that the short channel effect will not be further aggravated [208].

Table XXXI. Interface Trapped Charges of the Al_2O_3/Silicon Interface

Reference	Device structure	Measurement condition	Comments
Chin et al. [210]	4.8-nm Al_2O_3 was grown by the thermal oxidation of Al film in an MBE chamber under 450–500°C in O_2 ambience	D_{it} with a midgap value of 1×10^{11} eV^{-1} cm^{-2} was obtained from Al_2O_3 capacitor	ΔV_G less than 0.04 V after 0.1 mA/cm^2 (~5.4 V) constant current stress for 1000 s
Kolodzey et al. [211]	Al_2O_3 films (101–435 nm) were grown by the thermal oxidation of AlN film under 800–1100°C in O_2 ambience for 1–3 h	1. $D_{it} < 1 \times 10^{11}$ eV^{-1} cm^{-2} 2. Samples oxidized at 1000°C and above have Q_{ot} near 10^{11} cm^{-2}	D_{it} was obtained from the high-frequency and quasi-static C–V measurements
Manchanda et al. [212]	Using Zr or silicon-doped (0.1–5 wt%) Al target to deposit 3–5-nm Al_2O_3 films	Addition of dopants reduces the midgap D_{it} to less than 5×10^{10} eV^{-1} cm^{-2}	D_{it} was measured from the high-frequency C–V, quasi-static C–V and conductance methods

5.1.1. Deposition Methods

Kim and Ling [213] and Lim et al. [214] deposited 6-nm Al_2O_3 on polysilicon using atomic layer deposition at 350°C. The Al_2O_3 capacitor shows no degradation of the leakage current level after 830°C high-temperature BPSG reflow annealing for 30 min. The leakage current level is 1×10^{-9} A/cm^2 at 1.0 V, which is suitable for future 256-Mb DRAM storage capacitors. The application of Al_2O_3 film as the MOSFET gate dielectric was also reported. Al_2O_3 gate dielectrics formed by using sputtered AlN film on a silicon substrate and followed by thermal oxidation at temperatures ranging from 800 to 1000°C for 1–2 h were also studied [211, 215]. Chin et al. reported on Al deposited on a silicon substrate using molecular beam epitaxy (MBE) followed by thermal oxidation at a temperature of 400–500°C in the MBE chamber to form Al_2O_3 gate dielectrics for MOSFETs [210]. Manchanda et al. reported using Zr or silicon-doped Al_2O_3 targets to deposit Al_2O_3 film as the MOSFET gate dielectric [212]. The addition of dopants can reduce the midgap interface trap density to less than 5×10^{10} eV^{-1} cm^{-2} at the dielectric–silicon interface without inserting SiO_2 between the silicon and the Al_2O_3 film.

5.1.2. Metal–Insulator Barrier Height

Ludeke et al. deposited 8-nm Al_2O_3 on a silicon substrate by sputtering [216]. They studied the electronic and transport properties of Al_2O_3 measured on a local, nanometer-sized scale by using the tip of a scanning tunneling microscope (STM) as a hot electron injector. The collector current I_C was measured as a function of voltage V_T applied on the tungsten tip. The linear dependence of V_{th} on $V_{ox}^{1/2}$ implies that the barrier lowering is due to the image force effect. The threshold voltage V_{th} is defined as the tip bias at which I_C becomes measurable. Ludeke et al. found that the Al_2O_3/W barrier height is about 3.90 eV with a high-frequency dielectric constant ε of 1.89. The barrier height at the Al_2O_3/p-silicon interface is 2.78 ± 0.06 eV. Figure 15 shows the energy band diagram of the

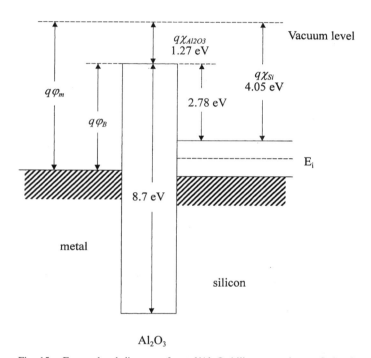

Fig. 15. Energy band diagram of metal/Al_2O_3/silicon capacitor at flatband condition.

metal/Al_2O_3/silicon capacitor. If a barrier height of 2.78 eV at the Al_2O_3/silicon interface is used, an electron affinity of about 1.27 eV of Al_2O_3 is obtained.

5.1.3. Leakage Current Mechanisms of Al_2O_3 Thin Films

The possible leakage current mechanism reported in the literature include Poole–Frenkel emission [211], phonon-assisted tunneling [211], and Fowler–Nordheim tunneling [212, 214, 217]. Kolodezy et al. deposited AlN on a silicon substrate using sputtering. The TiO_2 film was formed on the silicon substrate by annealing the AlN films at 800–1100°C in dry O_2 ambience

Table XXXII.   Prototype DRAM Chips with Al$_2$O$_3$ Storage Capacitors

Reference/ company	DRAM capacity	Process technology	Cell area (μm^2)	Chip size (mm^2)	Capacitor structure	Supply voltage (external/ internal)
Ha et al. [209]/ Samsung	For embedded DRAM application	0.15 μm, 1-W/2-Al, CMOS	0.182		OCS with HSG electrode (OCS cell), 25 fF/cell	
Kim et al. [208]/ Samsung	1 Gb	0.13 μm, CMOS	0.138	267	PAOCS cell	2.5 V/2.0 V

PAOCS, Polyamorphous one-cylindrical-stacked bottom electrode; OCS, one-cylindrical-stacked bottom electrode; HSG, hemispherical grain.

for 1–2 h [211]. The leakage current mechanism of the Al$_2$O$_3$ films is Poole–Frenkel emission and the current density is described by the following equation as given in Section 2.5.1:

$$J = q\mu N_C E \exp\left(\frac{-E_d}{kT} + \frac{q\beta_{PF}E^{1/2}}{kT} \right)$$

The trap energy obtained from the temperature dependence of the Poole–Frenkel emission from 25 to 100°C was studied. The terms in the numerator of the exponent in Poole–Frenkel emission can be considered as a field-dependent effective activation energy, $E_{act} = q(E_{ox})^{1/2} - q\Phi_t$. The trap energy of 1.6 eV was obtained from the Arrhenius plot of effective activation energy. Kim et al. deposited 6-nm-thick Al$_2$O$_3$ films on n$^+$-polysilicon using atomic layer deposition [217]. The temperature dependence of the leakage characteristics of the Al$_2$O$_3$ capacitor is negligible at 20°C–85°C. Therefore, the leakage current mechanism is considered as tunneling current.

5.1.4. C–V Characteristics of Al$_2$O$_3$ Thin Films

As shown in Table XXXII, the interface trapped charge density of the Al$_2$O$_3$/silicon interface is about 1×10^{11} eV^{-1} cm^{-2}, which is comparable to that of the Si$_3$N$_4$/silicon interface. Kolodzey et al. reported that the oxide trapped charge density Q_{ot} is about 10^{11} cm^{-2} for Al$_2$O$_3$ oxidized at 1000°C or above [211, 215]. Lower oxidation temperature yielded higher values of flatband voltage and higher Q_{ot}, which were attributed to the presence of unoxidized AlN. Chin et al. reported that the interface state at the midgap for MBE-deposited Al$_2$O$_3$ film is 1×10^{11} eV^{-1} cm^{-2} [210]. The gate voltage changes less than 0.04 V after 0.1 mA/cm^2 (~5.4 V) constant current stress for 1000 s. Manchanda et al. used a Zr or silicon-doped (0.1–5 wt%) Al target to deposit 3- to 5-nm Al$_2$O$_3$ films a on silicon substrate for MOSFET gate dielectric application [212]. The selection of dopants in the Al target is based on the enthalpy of formation of the dopant oxide, which should be smaller than the enthalpy of the metal oxide being deposited. The bond strength of the dopant oxide should be larger than the bond strength of the metal oxide. The enthalpies of formation of Al$_2$O$_3$, SiO$_2$, and ZrO$_2$ films are -390, -217, and -266 kCal/mol, respec-

tively. The bond strengths of Al$_2$O$_3$, SiO$_2$, and ZrO$_2$ films are 122, 190, and 181 kCal/mol, respectively. The interface state densities were measured by the high- and low-frequency C–V methods and the conductance method. The addition of dopants can reduce the midgap interface trap density to less than 5×10^{10} eV^{-1} cm^{-2} without inserting SiO$_2$ between the silicon and the Al$_2$O$_3$ film.

5.1.5. Metal–Oxide–Semiconductor Field Effect Transistor Gate Dielectric Applications of Al$_2$O$_3$ Thin Films

Chin et al. fabricated a MOSFET with Al$_2$O$_3$ as the gate dielectric using MBE [210]. The gate length and the gate dielectric thickness were 10 μm and 8.0 nm, respectively. The leakage current is approximately 7 orders of magnitude lower than the equivalent 2.1-nm thermal SiO$_2$. The dielectric constant was about 9.0, which is higher than that of Si$_3$N$_4$. The electron mobility is comparable to the conventional MOSFET with a SiO$_2$ gate oxide. Jin et al. deposited 50-nm Al$_2$O$_3$ on a SiGe substrate by reactive sputtering as the thin-film transistor gate dielectric [218]. The subthreshold slope and electron field effective mobility are 0.44 V/decade and 47 cm^2/V-s, respectively. The subthreshold slope is better than those obtained from heavily hydrogenated devices with SiO$_2$ gate insulators, with or without a silicon buffer. Part of this improvement resulted from the higher relative dielectric constant of Al$_2$O$_3$, which is about 2.5 times that of SiO$_2$. A more important reason is the improved interface between the Al$_2$O$_3$ dielectric and the SiGe channel layer.

5.2. Yttrium Oxide (Y$_2$O$_3$)

The dielectric constant ε_r of Y$_2$O$_3$ is about 20, which is higher than that of SiO$_2$ and is close to that of Ta$_2$O$_5$ ($\varepsilon_r = 26$). The energy bandgap and the breakdown field of Y$_2$O$_3$ film are 5.6 eV [219] and 3–6 MV/cm [220–222], respectively. Because the electrical properties of Y$_2$O$_3$ film are close to those of Ta$_2$O$_5$ film, the application of Y$_2$O$_3$ film for DRAM storage capacitor dielectric and MOSFET gate dielectric was also studied extensively. Because the lattice constant of Y$_2$O$_3$

($a = 1.006$ nm) is roughly twice that of silicon ($2a = 1.086$ nm) [219–223], MOSFETs with Y_2O_3 as the gate dielectric are better than those using other high-dielectric-constant films. The reported deposition methods of Y_2O_3 include sputtering [220, 222], yttrium evaporation followed by thermal oxidation at 500°C [221], and KrF pulse laser ablation deposition [219, 223].

5.2.1. Annealing Conditions

Hunter et al. deposited 50- to 80-nm-thick Y_2O_3 film on a silicon substrate using a KrF excimer laser [223]. The effect of postdeposition annealing on the thickness of the transition layer at the Y_2O_3/silicon interface was studied. Y_2O_3 samples were annealed in Ar ambience for either 30 min or 60 min. Some Y_2O_3 samples were annealed in O_2 ambience for 5, 15, and 30 min at 900°C. High-resolution transmission electron microscopy (HRTEM) images of the annealed samples show two distinct amorphous layers at the interface, namely, the SiO_2 layer and the Y_2O_{3-x} layer. SiO_2 layers about 3.3 nm thick were found to remain roughly the same under all annealing conditions, while the thickness of the Y_2O_{3-x} (3.5–5.9 nm) layers was found to decrease with increasing annealing time in O_2 ambience. The slow diffusion of O_2 from the surface into Y_2O_3 explains the presence of amorphous Y_2O_{3-x} at the Y_2O_3/silicon interface even after annealing in O_2 up to 30 min. This also accounts for the lack of increased SiO_2 thickness during annealing in O_2. Yip and Shih [222] found that both the C–V shift of Y_2O_3 films at different frequencies and the frequency-dependent accumulation capacitance could be eliminated by a postdeposition heat treatment at 600°C in O_2 ambience.

5.2.2. Metal–Insulator Barrier Height

Rozhkov et al. studied the photoelectrical properties of metal/Y_2O_3/silicon capacitors with different top electrode metals [221]. The top electrodes studied include Al, Ag, and Ni. The dependence of the photocurrent on the photoenergy was observed to follow a square law above the threshold of the photocurrent. The work functions of Al, Ag, and Ni are 4.25, 4.25, and 4.75 eV, respectively. The Al/Y_2O_3, Ag/Y_2O_3, and Ni/Y_2O_3 barrier heights obtained from the photocurrent spectrum were 3.3–3.4 eV, 3.3–3.35 eV, and 3.7–3.8 eV, respectively. Therefore, an electron affinity of about 0.9 eV of Y_2O_3 can be obtained from the difference between the metal work function and the metal/Y_2O_3 barrier height. Figure 16 shows the energy band diagram of the metal/Y_2O_3/silicon capacitor. The leakage current mechanism for Y_2O_3 film is Poole–Frenkel emission [221].

5.2.3. Interface Trapped Charge

Manchanda and Gurvitch deposited yttrium metal on silicon or a SiO_2 (4–7 nm)/silicon substrate using sputtering followed by 700°C annealing for 1 h to form Y_2O_3 film [220]. The

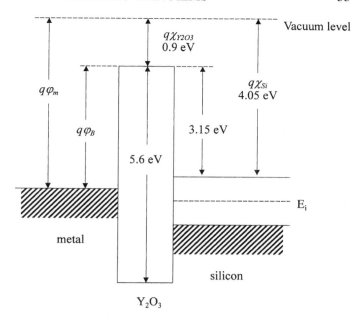

Fig. 16. Energy band diagram of metal/Y_2O_5/silicon capacitor at flatband condition.

thickness of the Y_2O_3 film was 26 nm. The interface state densities were measured from high-frequency and quasi-static C–V characteristics. The midgap interface state densities for the metal/Y_2O_3/SiO_2/silicon capacitor and the metal/Y_2O_3/silicon capacitor are 2×10^{11} and 1×10^{11} eV^{-1} cm^{-2}, respectively.

5.3. Zirconium Oxide (ZrO_2)

High-dielectric-constant films are required as the MOSFET gate dielectric to suppress the direct tunneling gate leakage current when the SiO_2 gate dielectric is scaled below 1.5 nm. However, a transition layer with low dielectric constant about 1.0–2.0 nm thick is usually formed at the interface for high-dielectric-constant films such as Ta_2O_5, TiO_2, or Y_2O_3. Another problem is the interface quality of these high-dielectric-constant films is not as good as thermally grown SiO_2. A thin passivation SiO_2 layer about 1.0 nm thick is required to obtain good interface quality. Therefore, the scaling of the MOSFET gate dielectric is limited by the low-dielectric-constant transition layer or the thin passivation layer. ZrO_2 is attractive because it is thermodynamically stable in contact with silicon [224]. The energy bandgap and the breakdown field of ZrO_2 are 5.16–7.8 eV [225] and 2.2 MV/cm [226], respectively.

5.3.1. Trap Energy Level

Houssa et al. deposited ZrO_2 (7.4 nm) or Ta_2O_5 (5.7 nm) films on a silicon substrate using atomic layer CVD (ALCVD) [227]. There is a thin SiO_2 layer with a thickness of about 1.2 nm at the Ta_2O_5/silicon or the ZrO_2/silicon interface. A ZrO_2/silicon barrier height of 2 eV and a ZrO_2 bandgap of 5.4 eV were found from the electron photoinjection experiment. Figure 17 shows the energy band diagram of the metal/ZrO_2/silicon capacitor.

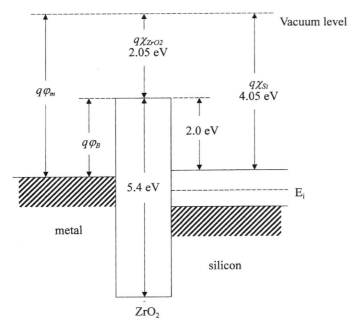

Fig. 17. Energy band diagram of metal/ZrO₂/silicon capacitor at flatband condition.

The electrical parameters shown in the energy band diagram of the metal/ZrO₂/silicon capacitor are very close to those of the metal/Si₃N₄/silicon capacitor because the energy bandgap of ZrO₂ and the ZrO₂/silicon barrier height are similar to those of Si₃N₄/silicon structure. By using the trap-assisted tunneling model, the trap level energy is found to be about 0.7 and 0.8 eV below the conduction band of Ta₂O₅ and ZrO₂, respectively. The trap densities of Ta₂O₅ and ZrO₂ films are 6×10^{17} and 3×10^{18} cm^{-3}, respectively.

5.3.2. Fabrication Processes of ZrO₂ Thin Films

5.3.2.1. Deposition Methods

ZrO₂ ($\varepsilon_r = 25$) can deposit on a silicon substrate without any passivation layer with a low interface trapped charge density of about 1×10^{11} eV^{-1} cm^{-2} [226, 228]. Because the electrical properties ZrO₂ film are close to those of Ta₂O₅, ZrO₂ is suitable for future ultrathin ($t_{eq} \sim 1.0$ nm) MOSFET gate dielectric applications. Although there is a thin silicate layer formed at the dielectric–silicon interface, the dielectric constant of the silicate layers ($\varepsilon_r = 9$–10) is still higher than that of SiO₂. The most often used deposition method for ZrO₂ is sputtering [224–226, 228–232].

5.3.2.2. Annealing Conditions

ZrO₂ tends to crystallize at low temperatures, resulting in high leakage current paths along the grain boundaries and nonuniform dielectric constant [232]. The crystallization temperature of ZrO₂ is 400°C [232]. ZrO₂ doped with Al (~25%) can raise the crystallization temperature to 800°C in oxygen ambience [230]. ZrO₂ doped with Si will be stable in direct contact

with Si. ZrSi$_x$O$_y$ (2–8 at% Zr) remains amorphous in direct contact with Si after 1050°C for 20 s in N₂ ambience [232]. No SiO₂ layer formed at the ZrSi$_x$O$_y$/Si interface after such high-temperature annealing. Qi et al. observed that the interface layer increased from 1.5 to 2.5 nm after 500–650°C annealing in O₂ ambience [225]. The interface layer is not pure SiO₂. It is a Zr silicate, which can be converted into a more stoichiometric SiO₂ under higher annealing temperature.

5.3.3. Leakage Current Mechanisms of ZrO₂ Thin Films

The leakage current mechanism of ZrO₂ is tunneling [224, 225, 227]. Houssa et al. reported the leakage current of a ZrO₂ (7.4 nm)/SiO$_x$ (1.2 nm) stack layer at low voltage is trap-assisted tunneling current [227], which can be approximately expressed as

$$J_G \propto N_t \exp\left[(qV_{ox} - \Phi_1 + \Phi_2 + \Phi_t)/kT \right]$$

where N_t is the trap density, V_{ox} is the voltage across the SiO$_x$ film, $\Phi_1 = 3.2$ eV is the electron barrier height between the silicon substrate and the SiO$_x$ layer, $\Phi_2 = 1.2$ eV is the difference between the Si/SiO$_x$ and Si/ZrO₂ barrier heights, and Φ_t is the trap level energy below the conduction band of ZrO₂. The trap level energy Φ_t and the trap density calculated from the trap-assisted tunneling current model are 0.8 eV and 6×10^{17} cm^{-3}, respectively. Qi et al. reported the leakage current of 2.4- to 3.2-nm-thick ZrO₂ is almost independent of temperature in the range of 10–200°C, which suggests that the leakage current is tunneling-like [225].

5.3.4. C–V Characteristics of ZrO₂ Thin Films

5.3.4.1. Fixed Oxide Charge

Houssa et al. deposited 7.4-nm ZrO₂ film on an n-type silicon substrate using ALCVD [227]. There is about 1.2-nm SiO$_x$ film formed at the ZrO₂/silicon interface. The flatband voltage V_{FB} of the Au/ZrO₂/SiO$_x$/silicon capacitor is about 1.2 V. Considering the work function difference Φ_{ms} between the n-type silicon substrate and the Au electrode, the flatband voltage should be about 0.5 V. Consequently, negative fixed charges are present in the ZrO₂/SiO$_x$ gate dielectric stack. The effective density of these negative charges is estimated to be about 5.2×10^{12} cm^{-2}.

5.3.4.2. Oxide Trapped Charge

As shown in Table XXXIII, the flatband voltage shift of the ZrSi$_x$O$_y$/silicon capacitor can be less than 10 mV, which is much smaller than that of the Ta₂O₅/silicon capacitor. This suggests that ZrSi$_x$O$_y$ is suitable for replacing SiO₂ as the MOSFET gate dielectric. Wilk et al. deposited zirconium silicate (ZrSi$_x$O$_y$) films on a silicon substrate for MOSFET gate dielectric application [231, 232]. Because ZrO₂ is ionic conductor, O ions can diffuse through the oxide and leave vacancies behind. Furthermore, ZrO₂ tends to crystallize at low temperatures, leading to polycrystalline films, which results in

Table XXXIII. Oxide Trapped Charges in ZrO_2 Films

Reference	Device structure	Measurement condition	Comments
Houssa et al. [227]	7.4-nm ALCVD ZrO_2/1.2-nm SiO_x/n-silicon	1. $C–V$ hysteresis ΔV_{FB} on the order of 200 mV 2. Effective density of trapped electrons is found to be about 1.5×10^{12} cm^{-2}	V_{FB} shift is due to the trapping of electrons in the ZrO_2/SiO_x stack during the first voltage sweep
Qi et al. [225]	2.4–3.2-nm sputtered ZrO_2/silicon followed by 1 h furnace annealing in N_2 or O_2 ambience	Charge trapping rate is small, ΔV_G is 21.5 mV for gate injection and 7.7 mV for substrate injection after 100-C/cm^2 stress fluence	No significant SILC was observed up to 100 C/cm^2 at a stress current of 50 mA/cm^2 under both stressing polarities
Wilk et al. [231, 232]	5-nm Zr silicate ($ZrSi_xO_y$ with ~2–8 at% Zr) on silicon substrate by sputtering	$C–V$ hysteresis of $t_{eq} = 2.08$ nm Zr silicate films less than 10 mV	Very small ΔV_{FB} indicates small amount of oxide trapped charges in the films

high leakage paths along grain boundaries. The hetero-interface formed between the ZrO_2 and silicon may also lead to a significant degradation of MOSFET channel electron mobility. Silicate films with approximately 2–8 at% Zr in direct contact with silicon exhibit excellent electrical properties and high thermal stability. The silicate gate dielectric can be viewed as only another modification of SiO_2, in this case by "doping" with Zr, to further improve device performance while remaining compatible with the rest of the CMOS processes. The $C–V$ hysteresis of $t_{eq} = 2.08$ nm in Zr-silicate films is less than 10 mV, which indicates that only a small amount of oxide trapped charges is present in the films. Qi et al. reported that no significant SILC was observed for 2.4-nm-thick ZrO_2 film up to 100 C/cm^2 at a stress current of 50 mA/cm^2 under both stressing polarities [225]. This indicates that no traps or defects are generated during the constant current stress. The charge trapping rate is small. ΔV_G is 21.5 mV for gate injection and 7.7 mV for substrate injection after 100-C/cm^2 stress fluence. Houssa et al. reported that a flatband shift on the order of 200 mV is observed after double bias sweep, that is, when going from accumulation to inversion and back to accumulation [227]. The shift of V_{FB} is due to the trapping of electrons

in the ZrO_2/SiO_x stack during the first voltage sweep. From the value of ΔV_{FB}, the effective density of trapped electrons is found to be about 1.5×10^{12} cm^{-2}.

5.3.4.3. Interface Trapped Charge

As shown in Table XXXIV, the interface trapped charge density of the ZrO_2/silicon interface is about 1×10^{11} eV^{-1} cm^{-2}, which is comparable to that of the Si_3N_4/silicon interface. Nagi et al. deposited 3- to 8-nm ZrO_2 films on an n-type SiGe substrate by reactive sputtering, followed by a 5-min anneal at 550°C in a N_2 ambience [226]. Although ZrO_2 is thermally stable on silicon and no passivation layer grown is required, interfacial layers are believed to be formed between ZrO_2 and SiGe, due to the oxygen plasma used in ZrO_2 deposition. The interface layer thickness is found to be about 0.9 nm. The $C–V$ hysteresis has a thickness dependence and becomes negligible for 1.65-nm films. Small hysteresis of thin films could be attributed to the fact that less sputtering time for thin films introduces less mobile charges and large capacitance of thin films further reduces the shift of threshold voltages. The interface state density was found to be about 2×10^{11} eV^{-1} cm^{-2}.

Table XXXIV. Interface Trapped Charges of the ZrO_2/silicon Interface

Reference	Device structure	Measurement condition	Comments
Nagi et al. [226]	3–8-nm sputtered ZrO_2/n-type SiGe followed by a 5-min anneal at 550°C in a N_2 ambience	Interface state density was found to be about 2×10^{11} eV^{-1} cm^{-2}	$C–V$ hysteresis has thickness dependence and becomes negligible for 1.65-nm film
Qi et al. [225]	2.4–3.2-nm sputtered ZrO_2/silicon followed by a furnace annealing in N_2 or O_2 ambience	1. $C–V$ hysteresis from -2.0 V to 1.0 V was about 50 mV 2. D_{it} calculated from the 1-MHz $C–V$ curves was less than 1×10^{11} eV^{-1} cm^{-2}	

5.3.5. Metal–Oxide–Semiconductor Field Effect Transistor Gate Dielectric Applications of ZrO₂ Thin Films

Ma et al. fabricated a MOSFET with 3.9-nm-thick ZrO_2 film as the gate dielectric [230]. The 3.9-nm-thick ZrO_2 corresponds to an equivalent SiO_2 thickness of 1.1 nm. ZrO_2 films were deposited by sputtering a Zr target in a mixture of oxygen and argon ambience at room temperature. The subthreshold swing of the 0.6-μm channel length p-MOSFET with ZrO_2 gate dielectric is 72 mV/decade. The hole mobility of the MOSFET with ZrO_2 gate dielectric is about 15% lower than that of the SiO_2 gate oxide predicted by the universal mobility curves.

5.4. Hafnium Oxide (HfO₂)

HfO_2 is attractive for MOSFET gate dielectric application because it is thermodynamically stable in contact with silicon. Hf forms the most stable oxide with the highest heat of formation ($H_f = 271$ kCal/mol) among the IV-A group of elements of the periodic table (Ti, Zr, Hf). Hf can reduce the negative SiO_2 layer to form HfO_2 [224]. HfO_2 is also very resistant to impurity diffusion and intermixing at the interface because of its high density (9.68 g/cm²). HfO_2 has high dielectric constant (~30), high heat of formation (271 kCal/mol), relatively large energy bandgap (5.68 eV), and high breakdown strength (4 MV/cm) [224].

5.4.1. Fabrication processes of HfO₂ Thin Films

5.4.1.1. Deposition Methods

HfO_2 ($\varepsilon_r = 28$–30) can be deposited on a silicon substrate without any passivation layer with low interface trapped charge density of about 1×10^{11} eV^{-1} cm^{-2} [226, 228]. Because the electrical properties of HfO_2 are close to those of Ta_2O_5 film, HfO_2 is suitable for future ultrathin ($t_{eq} \sim 1.0$ nm) MOSFET gate dielectric applications. Although there is a thin silicate layer formed at the dielectric–silicon interface, the dielectric constant of these silicate layers ($\varepsilon_r = 9$–10) is still higher than that of SiO_2. The deposition method for HfO_2 is sputtering [231, 232].

5.4.1.2. Annealing Conditions

HfO_2 tends to crystallize at low temperatures, resulting in high leakage current paths along the grain boundaries and nonuniform dielectric constant [232]. The crystallization temperature of HfO_2 is 700°C [224]. HfO_2 doped with Si will be stable in direct contact with Si. $HfSi_xO_y$ (2–8 at% Hf) remains amorphous in direct contact with Si after 1050°C for 20 s in N_2 ambience [232]. No SiO_2 layer is formed at the $HfSi_xO_y$/Si interface after such high-temperature annealing. In contrast, a thin SiO_2 layer is formed at the HfO_2/Si interface after 600°C annealing in O_2 ambience, resulting in an increase in the equivalent SiO_2 thickness [224, 228]. Lee et al. observed that the equivalent SiO_2 thickness of HfO_2 films increases with increasing annealing temperature in the range of 600–900°C in N_2 ambience [228]. The composition of the interface layer is believed to be Hf silicate because the estimated dielectric constant of the interface layer is higher than that of SiO_2.

5.4.2. Leakage Current Mechanisms of HfO₂ Thin Films

The leakage current mechanism of HfO_2 films is tunneling [224]. The leakage current of HfO_2 is decided by the combined layers of the silicate interface layer and the HfO_2 layer [225]. As the physical thickness of the HfO_2 layer is decreased, leakage current shows more tunneling-like behavior governed by the silicate layer. Therefore, the temperature dependence of leakage current is weaker for thinner HfO_2 films.

5.4.3. C–V Characteristics of HfO₂ Thin Films

5.4.3.1. Oxide Trapped Charge

The flatband voltage shift of a $HfSi_xO_y$/silicon capacitor can be less than 10 mV, which is much smaller than that of a Ta_2O_5/silicon capacitor. This suggests that $HfSi_xO_y$ is suitable for replacing SiO_2 as the MOSFET gate dielectric. Wilk et al. deposited hafnium silicate ($HfSi_xO_y$) on a silicon substrate for MOSFET gate dielectric application [231, 232]. Because HfO_2 is an ionic conductor, O ions can diffuse through the oxides and leave vacancies behind. Furthermore, HfO_2 tend to crystallize at low temperatures, leading to polycrystalline films, which results in high leakage paths along grain boundaries. The hetero-interface formed between HfO_2 and silicon may also lead to significant degradation of the MOSFET channel electron mobility. Silicate films with approximately 2–8 at% Hf in direct contact with silicon exhibit excellent electrical properties and high thermal stability. Silicate gate dielectric can be viewed as only another modification of SiO_2, in this case by "doping" with Hf, to further improve device performance while remaining compatible with the rest of the CMOS process. The C–V hysteresis of $t_{eq} = 1.78$ nm in Hf silicate is less than 10 mV, which indicates that only a small amount of oxide trapped charges is present in the films.

5.4.3.2. Interface Trapped Charge

As shown in Table XXXV, the interface trapped charge density of the HfO_2/Si interface is about 1×10^{11} eV^{-1} cm^{-2}, which is comparable to that of the Si_3N_4/silicon interface. Lee et al. deposited $t_{eq} = 1.0$–2.5 nm of HfO_2 film on a silicon substrate by sputtering for MOSFET gate dielectric application [228]. The hysteresis of the C–V curves for as-deposited film is about 200 mV, but could be reduced to a negligible level by postmetallization annealing without any increase in equivalent SiO_2 thickness. The midgap interface state density obtained from the high-frequency C–V curve using the Terman method is about 7×10^{10} eV^{-1} cm^{-2}. Kang et al. reported that electron trapping was observed for low constant current stress and hole trapping for high constant current stress [229]. The hole trapping was generated by energetic electrons as the stress current increased.

Table XXXV. Interface Trapped Charges of the HfO_2/Silicon Interface

Reference	Device structure	Measurement condition	Comments
Kang et al. [229]	4.5–13.5-nm sputtered HfO_2/silicon followed by 500°C annealing in N_2 ambience for 5 min	$D_{it} = 1 \times 10^{11}$ eV^{-1} cm^{-2} was extracted from the 1-MHz $C–V$ curve	The amount of charge trapping (ΔV_G) is negligible after 1–50 mA/cm^2 of constant current stress
Lee et al. [228]	$t_{eq} = 1.0$–2.5-nm sputtered HfO_2/silicon	Midgap $D_{it} = 7 \times 10^{10}$ eV^{-1} cm^{-2} obtained from the high-frequency $C–V$ curve	Hysteresis of $C–V$ curves for as-deposited film is about 200 mV
Wilk et al. [231, 232]	5-nm Hf silicate ($HfSi_xO_y$ with ~2–8 at% Hf) on silicon substrate by sputtering	1. $D_{it} \sim 1$–5×10^{11} eV^{-1} cm^{-2} obtained from the 100-kHz $C–V$ curves 2. The hysteresis obtained from 100-kHz $C–V$ curves was less than 10 mV	

Wilk et al. deposited 5-nm Hf silicate ($HfSi_xO_y$ with ~2–8 at% Hf) on a silicon substrate by sputtering [231, 232]. The equivalent SiO_2 thickness t_{eq} for Hf silicate was 1.78 nm. $HfSi_xO_y$ films that have a composition close to $HfSiO_4$ are expected to exhibit a higher dielectric constant. Obtaining $HfSiO_4$ stoichiometry is not necessary because it is desirable to maintain an amorphous structure. Any film composition that is somewhat silicon rich will be suitable for preventing the formation of crystalline HfO_2 precipitates and associated structural and electrical defects during any postprocessing steps. The TEM result shows that these silicate films remain amorphous and stable up to at least 1050°C in direct contact with the silicon substrate without any interface reaction. The interface state density obtained by comparison of the 100-kHz $C–V$ curves to an ideal one was $D_{it} \sim 1$–5×10^{11} eV^{-1} cm^{-2}. The hysteresis obtained from 100-kHz $C–V$ curves was less than 10 mV.

5.5. Gadolinium Oxide (Gd_2O_3)

The development of a III-V compound semiconductor MOSFET using Ga_2O_3 (Gd_2O_3) as the gate dielectric has attracted much attention [233–237]. The dielectric constant of Ga_2O_3 is 14.2 [233]. The energy bandgap and the breakdown field of Ga_2O_3 are 4.4 eV [234] and 3.5 MV/cm [233, 234], respectively. The Ga_2O_3 (Gd_2O_3) gate dielectric was deposited using e-beam evaporation with a $Ga_5Gd_3O_{12}$ single crystal as the evaporation source. In the past, the high interface trapped density was the main problem for developing a III-V compound semiconductor MOSFET. The interface trap density of the III-V compound semiconductor MOSFET using Ga_2O_3 (Gd_2O_3) as the gate dielectric can be as low as $D_{it} = 5 \times 10^{10}$ eV^{-1} cm^{-2} [233]. Therefore, very high speed VLSI circuits using III-V compound semiconductor MOSFETs are possible in the future.

Ren et al. demonstrated the first In-channel enhancement-mode $In_{0.53}Ga_{0.47}As$ MOSFET on an InP semi-insulating substrate using a Ga_2O_3 (Gd_2O_3) gate dielectric [234]. Ga_2O_3 (Gd_2O_3) was electron beam deposited from a high-purity single-crystal $Ga_5Gd_3O_{12}$ source. The midgap interface state density and the interface recombination velocity are 5×10^{10} eV^{-1} cm^{-2} and 9000 cm/s, respectively. A 0.75-μm channel length device with 40-nm Ga_2O_3 (Gd_2O_3) exhibits an extrinsic transconductance of 190 mS/mm, which is an order of magnitude improvement over the enhancement-mode InGaAs MOSFET. The effective mobility is 470 cm^2/V-s, calculated from the drain conductance g_D vs. V_G curve at a small drain voltage (~0 V). The current gain cutoff frequency f_t and the maximum frequency of oscillation f_{max} for the 0.75×100-μm^2 gate dimension device at $V_G = 3$ V and $V_D = 2$ V are 7 and 10 GHz, respectively. The current gain cutoff frequency f_t and the maximum frequency of oscillation f_{max} for the 1×200-μm^2 gate dimension depletion-mode GaAs MOSFET with 24-nm Ga_2O_3 (Gd_2O_3) gate dielectric at $V_G = 1$ V and $V_D = 2$ V are 14 and 36 GHz, respectively [234]. The maximum transconductance of 210 mS/mm was obtained with a threshold voltage of −0.5 V for 1×100-μm^2 gate dimension devices. The transconductance of depletion-mode devices is higher than that of enhancement-mode devices because the current density is determined by channel doping. For enhancement-mode devices, the drain current is induced by bending the conduction band and creating an inversion channel. Therefore, the transconductance is dominated by the quality of the surface state density and the smoothness of the oxide–semiconductor interface [234].

6. FERROELECTRIC FILMS: LEAD ZIRCONATE TITANATE (PZT)

Lead zirconate titanate (PZT) is a solid solution of lead titanate ($PbTiO_3$) and lead zirconate ($PbZrO_3$). It has a chemical composition of $Pb(Zr_xTi_{1-x})O_3$. PZT is the most studied ferroelectric film for nonvolatile memory device applications. The requirements for memory device applications are discussed in Section 1; PZT satisfies nearly all of these requirements. PZT has a transition temperature well above room temperature, a relatively large remanent polarization of about 35 μC/cm^2, and a comparatively small coercive field of 35 kV/cm. It also has a high charge density.

PZT is easy to prepare using conventional film deposition techniques, for example, magnetron sputtering. PZT is also the ferroelectric material used most often by the commercial semiconductor companies, including Ramtron, McDonnell Douglas, Raytheon, among others [29]. It is generally felt that PZT and strontium barium tantalate (SBT) are the most suitable materials for nonvolatile memory application and barium strontium titanate (BST) is the most suitable for DRAM capacitors [18].

6.1. Physical Properties of PZT Thin Films

PZT is a mixture of lead titanate and lead zirconate. Lead titanate (PbTiO$_3$) is ferroelectric with a Curie temperature of 490°C and a spontaneous polarization larger than 50 μC/cm^2 at room temperature [14]. Lead zirconate (PbZrO$_3$) is antiferroelectric with a Curie temperature of 230°C [13, 14]. The mixture, lead zirconate titanate Pb(Zr$_{1-x}$Ti$_x$)O$_3$, is tetragonal when $x = 0$–0.53 and rhombohedral when $x > 0.53$ [15, 16, 238]. Most applications of PZT use a concentration close to the morphotropic phase boundary with a Zr/Ti concentration ratio of 53/47. At compositions near the morphotropic phase boundary, both the coupling coefficient and the relative permittivity reach their peak values, a feature that is often exploited in commercial compositions [14, 239].

Pb(Zr$_x$Ti$_{1-x}$)O$_3$ films with $x = 0$–0.53 show a Curie temperature of 410°C (at $x = 0.45$), a spontaneous polarization P_s of 45 μC/cm^2, a remanent polarization P_r of 32 μC/cm^2, and a coercive field E_c of 150 kV/cm. Pb(Zr$_x$Ti$_{1-x}$)O$_3$ films with a concentration $x > 0.53$ show a Curie temperature of 260°C (at $x = 0.9$), a spontaneous polarization of 42 μC/cm^2, a remanent polarization of 35 μC/cm^2, and a coercive field of 35 kV/cm [15]. Other reported values of PZT thin films include

a storage density of 15 μC/cm^2 (at 125 kV/cm) for films prepared by the sol–gel method with a thickness of 200–600 nm (at $x = 0.5$) [31]. Sayer et al. [30] listed a remanent polarization of 36 μC/cm^2, a coercive field greater 20 kV/cm, and a dielectric constant of 700–1200 for PZT. Kidoh and Ogawa [240] reported a remanent polarization of 20 μC/cm^2, a coercive field of 130 kV/cm, and a dielectric constant of 350 using excimer laser ablation. Kim et al. [241] reported the electrical properties of PZT prepared by MOCVD. Krupanidhi et al. [242] reported the properties of PZT prepared by rf magnetron sputtering. The spontaneous polarization and the coercive field are 20.75 μC/cm^2 and 10 kV/cm, respectively. Moazzami et al. [32] reported a remanent polarization of 15 μC/cm^2 and a coercive field of 25 kV/cm for films of 400 nm thickness at a voltage of 5 V for films prepared by the sol–gel method. Dey and Zuleeg [243] reported on 100–2000-nm-thick PZT films with a remanent polarization of 18–20 μC/cm^2 and a coercive field of 35–100 kV/cm. Sanchez et al. [244] reported their results on PZT prepared by the sol–gel method. Sudhama et al. [33] mentioned that their magnetron-sputtered 100-nm-thick film shows 12 μC/cm^2. Udayakumar et al. [245] showed a remanent polarization of 36 μC/cm^2, a coercive field of 30 kV/cm, a dielectric constant of 1300, and a Curie temperature of 366°C for PZT with a composition of PbZr$_{0.52}$Ti$_{0.48}$O$_3$ close to the morphotropic phase boundary. Kundu and Lee [246] prepared a series of PZT thin-film capacitors with film thickness ranging from 70 to 680 nm. Normal ferroelectric characteristics are observed for PZT films with thickness above 130 nm. The spontaneous polarization ranges from 29 to 40 μC/cm^2. The remanent polarization ranges from 10 to 19 μC/cm^2. The coercive field ranges from 72 to 142 kV/cm. The dielectric constant is about 350. These physical parameters of PZT thin films are summarized in Table XXXVI.

The electrical properties of ferroelectric films are often measured in capacitor form. The capacitor is constructed using a metal–ferroelectric film–metal structure. The energy bandgap

Table XXXVI.   Dielectric and Ferroelectric Properties of PZT Thin Films

Reference	Structure	Film thickness (nm)	Remanent polarization (μC/cm^2)	Coercive field (kV/cm)	Dielectric constant	Fatigue cycle
Carrano et al. [31]	Pt/PZT/Pt	200–600	12.2		577 (350 nm)	10^{13} (unipolar)
Dey and Zuleeg [243]		100–2000	18–20	35–100	1000–1250	10^{10}–10^{12}
Kidoh and Ogawa [240]		1300	20	130	350	
Sayer et al. [30]			36	>20	700–1200	
Kundu and Lee [246]	Au/PZT/Pt	130–680	10–19	72–142	332–350	
Moazzami et al. [32]	Pt/PZT/Pt	400	15	25		
Krupanidhi et al. [242]	Top: Au or Al	~2000	20.75	10	850–950	
	Bottom: Pt, conducting glass					
Sudhama et al. [247]	Pt/PZT/Pt	100	12			10^{10} (unipolar)
Udayakumar et al. [245]	Au/PZT/Pt	~300–500	36	30	1300	

and the electron affinity of the ferroelectric material and the work function of the metal are therefore important parameters in determining the electrical properties of the metal–ferroelectric–metal system. Metallic oxides sometimes are also used as the electrodes for the capacitors.

The energy bandgap of PZT is listed as around 3–4 eV in Lines and Glass's book on ferroelectrics [14]. A value of about 3.0 eV was used by Sudhama et al. [33]. The energy bandgap of PZT is listed as 3.4 ± 0.1 eV in Scott's book on ferroelectric memories [18]. PZT is known to have a p-type conductivity [16]. The p-type conductivity is caused by PbO vapor evaporation from the PZT sample during the sintering process, which creates Pb vacancies in the sintered ceramic sample. A Pb vacancy can play the role as center of negative electric charges with an effective valence of -2. One Pb vacancy, just like an acceptor impurity in a semiconductor, offers an acceptor level and two holes. Thus, undoped PZT usually presents a p-type conductivity [16].

6.2. Electrical Characteristics of PZT Thin Films

6.2.1. Electrical Conduction

One of the most important electrical properties of ferroelectric films is their electrical conductivity. However, electrical conduction is a complex subject. Many factors influence the electrical conductivity of ferroelectric films. The measurement of true leakage current in ferroelectric films is also more complicated than conventional insulator films because the ferroelectric film has polarization current. The measurement of true leakage current requires a waiting time that is long compared to the relaxation time but short compared to the onset of time-dependent dielectric breakdown (TDDB) [18].

There is some dispute on the major carriers that constitute the conduction current in PZT. Some reports indicate that the conduction in PZT is dominated by electrons [18, 248, 249]. The dependence of leakage current on the cathode work function, together with the independence of leakage current on the anode work function, establishes that in BST and PZT the carriers are electrons only (no holes), injected from the cathode electrode [250]. There are other reports that postulate that the injected carriers are holes [45, 251, 252].

A number of authors discussed electrical conduction in PZT thin films.

1. Moazzami et al. [32] reported that the leakage current of PZT exhibits ohmic behavior at low field and exponential behavior at moderately high field. A low field resistivity of 3.5×10^{10} Ω-cm and an activation energy of 0.35 eV are obtained. At higher field, Moazzami et al. ascribe the leakage current to ionic or electronic hopping, assuming single-particle transport.
2. Sudhama et al. [33] described how the current–voltage characteristics of a Pt/PZT/Pt capacitor structure fit a two-carrier injection metal–semiconductor–metal model incorporating blocking contacts. By considering the PZT layer as a large bandgap semiconductor, the $I–V$ characteristics are

described by a work-function-driven double-Schottky-barrier-limited conduction mechanism, observed previously in metal–silicon–metal structures along with a space-charge-limited current (SCLC) mechanism for high injection levels. A double-injection metal–semiconductor–metal model is used to qualitatively explain the $I–V$ characteristics of Pt/PZT/Pt capacitors. It is believed that in the low-voltage regime hole conduction dominates in ferroelectric PZT, followed by a steep increase in the minority electron current after the ferroelectric layer is fully depleted. The Pt/PZT Schottky barrier is shown to have a barrier height of 0.6 eV.

3. Scott and co-workers [253–255] considered space-charge-limited current (SCLC). The SCLC model for the cases of single- and double-carrier injection was invoked for the interpretation of conduction in ferroelectric thin films.
4. Hu and Krupanidhi [256] described a unified grain boundary model. A double-depletion-layer barrier model is used to describe the origin of high resistivity of the grain boundaries. It is suggested that the barrier height varies significantly with the applied field due to nonlinear ferroelectric polarization, and the barrier is overcome by tunneling at sufficiently high fields. In some other cases, the resistivity of the grain boundaries is comparable to that of the grains, and therefore the intrinsically heterogeneous films degenerate into quasi-homogeneous media, to which SCLC theory is applicable. A unified grain boundary model thus reconciles different types of conduction mechanisms in the ultrafine-grained ferroelectric films.
5. Hwang et al. [257] used different electrodes and measured the current for two different polarities to distinguish the Schottky and Poole–Frenkel emission mechanisms. They found that the leakage current mechanism of PZT film is not a barrier-limited process such as Schottky emission, but a bulk-limited process.
6. Bernacki [46] proposed a model based on Schottky barrier formation due to PZT/platinum work function differences and barrier height modulation due to remanent polarization. This model predicts a conduction current dependence on both remanent polarization state and electrode material. This model also suggests that the conduction current is limited by electrode/PZT interface effects rather than the bulk conductivity of the PZT.
7. Mihara and Watanabe [258, 259] found two regions in the $I–V$ characteristics. Below about 800 kV/cm, the $I–V$ characteristics had ohmic-like current with strong decay time dependence. Above about 1 MV/cm, the current had strong electric field dependence. The conduction current mechanism at high field was considered to be a mixture of Schottky emission and Poole–Frenkel emission, taking the dynamic dielectric constant into account. The barrier height and trapped level were estimated to be 0.58 and 0.5 eV, respectively.
8. A "ferroelectric Schottky diode" model was proposed by Blom et al. [260]. A bistable conduction characteristic is observed in PbTiO$_3$. The observations are explained by a model in which the depletion width of the ferroelectric Schottky diode is determined by the polarization dependence of the internal elec-

tric field at the metal–ferroelectric interface. The model was considered to be unlikely by Stolichnov and Tagantsev [251] because the effect of direct-current (dc) pretreatment (i.e., poling) can be erased by heating to a temperature much lower than is needed to depolarize the PZT material. The negation of the effect by illumination is also evidence against a ferroelectric phenomenon.

9. Wouters et al. [45] analyzed the possible conduction processes in Pt/PZT/Pt capacitors based on the electronic properties of PZT and the Pt/PZT interface. PZT behaves as an insulator, resulting in carrier-blocking Pt/PZT contacts, except when mobile ionic defects can provide the space charge to prevent full depletion.

10. Stolichnov and Tagantsev [251] proposed a space-charge-influenced injection model based on hole injection. There are two regimes of carrier injection. A critical field is associated with the crossover from the ohmic to the Schottky regime. The time dependence of the measured current is attributed to a time drift of parameters of the conductive system of the films rather than to the dielectric relaxation current. It was stated by the authors that the scenario proposed is not the only one possible. The models based on Poole–Frenkel conduction or space-charge-limited conduction can also be applied to the interpretation of the measured data.

11. Yoo and Desu [261] described the leakage current by Schottky emission.

12. Chen et al. [262] and Kundu and Lee [246] studied the mechanisms of leakage current of PZT films. Chen et al. found that the leakage current in PZT consists of electronic conduction current and polarization current. For PZT films with a thickness of 500 nm, the polarization current is dominant in the electric field lower than 100 kV/cm, whereas the electronic conduction current is dominant for an electric field higher than 100 kV/cm. The electronic conduction current is due to ohmic conduction in the electric field lower than 40 kV/cm and due to Poole–Frenkel emission in the electric field higher than 40 kV/cm. Kundu and Lee [246] studied PZT films with different thicknesses and showed that different conduction mechanisms are involved for PZT films with different thicknesses. At field lower than 50 kV/cm, ohmic conduction still holds for films of all thicknesses. For films thicker than 280 nm, a grain-boundary-limited conduction (GBLC) is found to be operative in the field range of 50–150 kV/cm and Poole–Frenkel emission is operative in the field range of 150–200 kV/cm. For films thinner than 200 nm, the conduction is ohmic below 200 kV/cm and is dominated by trap-controlled space-charge-limited current (SCLC) above 200–250 kV/cm. In the latter case, the distribution of traps is such that the grain boundary does not result in any significant potential barrier. As a result, SCLC rather than GBLC is observed.

13. Waser [263] considered the role of grain boundaries in the conduction and breakdown of perovskite-type titanates.

14. Scott [18] discussed the leakage current and energy band diagrams of the Pt/SBT and Pt/PZT systems. The PZT energy band is shown to bend downward in contact with the metal and the conduction is n type. It was pointed out that even in PZT

specimens in which the Fermi level is nearly midgap (compensated), the conduction can still be predominately n type. This is because the electron mobility is about twice that for holes, so given an equal number of thermally excited electrons and holes, the electron current will be larger.

It was stated that the conduction band structure in the ferroelectric semiconductor near the metal electrode interface is rather complicated (due in part to the fact that the defect density is a few times 10^{20} cm^{-3} near the interface, compared with a few times 10^{18} cm^{-3} in the film interior). This produces a band structure in applied fields, which is neither simple band tilting nor simple Schottky structure.

It was stated that a general consensus exists that PZT and BST both behave as fully depleted devices at the intended voltages (2–5 V) and that, although nominally p type, they manifest only electron conduction [250].

15. Waser [264] reported evidence for an ionic conduction mechanism involving oxygen vacancies in SrTiO$_3$ which, like PZT, is a perovskite titanate.

16. Nagaraj et al. [265] fabricated PZT capacitors with (La, Sr)CoO$_3$ top and bottom electrodes. The leakage current at low fields (less than 10 kV/cm) shows ohmic behavior with a slope of nearly 1 and is nonlinear at higher voltages and temperatures. Bulk-limited field-enhanced thermal ionization of trapped carriers (Poole–Frenkel emission) is the controlling mechanism at higher fields and temperatures. The activation energies are in the range of 0.5–0.6 eV. These energies are compatible with Ti^{4+} ion acting as the Poole–Frenkel centers.

The status of electrical conduction in PZT thin films is still undetermined. The literature results on the electrical conduction properties of PZT are summarized in Table XXXVII.

According to metal–semiconductor contact theory, the work function and the electron affinity of the semiconductor will determine the electrical characteristics of the contact, if the influence of the interface states can be neglected. For ionic semiconductors, the barrier height generally depends strongly on the metal and a correlation has been found between interface behavior and electronegativity [266]. The classical band-bending model should work for PZT, BST, or SBT on metal electrodes [18, 19]. The electrical characteristics of the metal–semiconductor contact can be divided into four different types, according to whether the semiconductor is n type or p type and whether the work function of the metal $q\phi_m$ is larger or smaller than the work function of the semiconductor $q\phi_s$. This situation is shown in Figs. 18–21.

Figure 18 shows the case for an n-type semiconductor with $q\phi_m > q\phi_s$. The figure on the left-hand side shows the band diagrams before joining and the figure on the right-hand side shows the band diagram after joining. Because the Fermi level of the semiconductor is higher than that of the metal, electrons will be transferred from the semiconductor to the metal after joining. The surface region of the semiconductor will be in depletion. A Schottky barrier will be formed and the barrier height for electrons $q\phi_{Bn}$ is $q\phi_{Bn} = q(\phi_m - \chi)$, where χ is the electron affinity of the semiconductor.

Table XXXVII. Electrical Conduction and Leakage Current of PZT Thin Films

Reference	Device structure	Film thickness (nm)	Leakage current level (A/cm^2)	Current mechanism
Bernacki [46]				Polarization dependent
Chen et al. [262]	Au/PZT/Pt	500	10^{-7} at 100 kV/cm	Ohmic below 40 kV/cm, Poole–Frenkel above 40 kV/cm
Hu and Krupanidhi [256]	Au/PZT/Pt	700	10^{-9} at 100 kV/cm	GBLC and SCLC
Hwang et al. [257]	Al/PZT/Pt	200	10^{-5} at 150 kV/cm	Poole–Frenkel
Kundu and Lee [246]	Au/PZT/Pt	70–680	10^{-8} at 100 kV/cm (70 nm)	Ohmic below 50 kV/cm, Poole–Frenkel, GBLC, and SCLC
Mihara and Watanabe [258]	Pt/PZT/Pt	280	10^{-6} at 500 kV/cm	Ohmic below 800 kV/cm, mixture of Schottky and Poole–Frenkel above 1 MV/cm
Moazzami et al. [32]	Pt/PZT/Pt	400	10^{-5} at 375 kV/cm, 10^{-4} at 1.3 MV/cm	Ohmic at low field, exponential at higher field (ionic or electronic hopping)
Scott et al. [254]	Pt electrode	216	5×10^{-4}	SCLC
Stolichnov [251]	Pt/PZT/Pt	220–350	10^{-7} at 100 kV/cm	Space-charge-influenced injection
Sudhama et al. [33]	Pt/PZT/Pt	150	2.2×10^{-7} at 100 kV/cm, 7.5×10^{-5} at 200 kV/cm	Double-injection metal–semiconductor–metal model
Wouters et al. [45]				
Yoo and Desu [261]	Pt/PZT/Pt and RuO$_x$/PZT/RuO$_x$		10^{-5} at 200 kV/cm	Ohmic at low field, SCLC, Schottky emission, tunneling

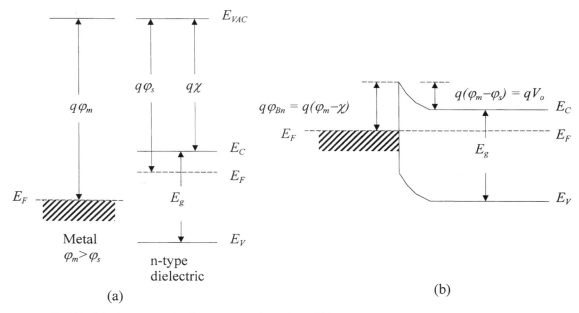

Fig. 18. Metal–n-type semiconductor contact: $\phi_m > \phi_s$, rectifying contact: (a) before contact, (b) after contact.

Figure 19 shows the case for an n-type semiconductor but with $q\phi_m < q\phi_s$. The figure on the left-hand side shows the band diagrams before joining and the figure on the right-hand side shows the band diagram after joining. Because the Fermi level of the semiconductor is lower than that of the metal, electrons will be transferred from the metal to the semiconductor after joining. The surface region of the semiconductor will be in accumulation and the electrical characteristic of the interface will be an ohmic contact. The barrier to electron flow between the metal and the semiconductor will be small. The current is then determined by the bulk resistance of the semiconductor. Such a contact is termed an ohmic contact.

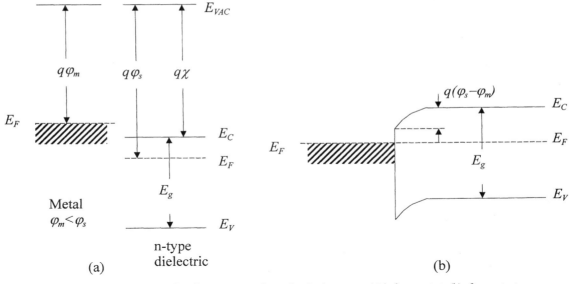

Fig. 19.   Metal–n-type semiconductor contact: $\phi_m < \phi_s$, ohmic contact: (a) before contact, (b) after contact.

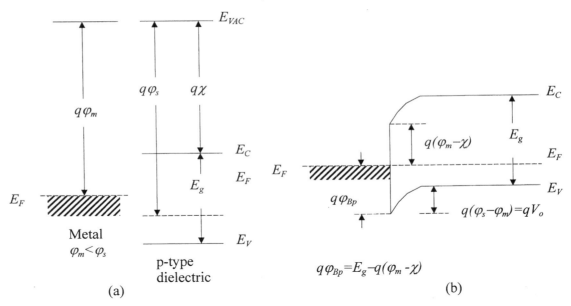

Fig. 20.   Metal–p-type semiconductor contact: $\phi_m < \phi_s$, rectifying contact, (a) before contact, (b) after contact.

Figure 20 shows the case for a p-type semiconductor with $q\phi_m < q\phi_s$. The figure on the left-hand side shows the band diagrams before joining and the figure on the right-hand side shows the band diagram after joining. Because the Fermi level of the metal is higher than that of the semiconductor, electrons will be transferred from the metal to the semiconductor after joining. The surface region of the semiconductor will be in depletion. A Schottky barrier will be formed. Because the current in a p-type semiconductor is carried by holes, one needs to look for a barrier for holes and the barrier height for holes $q\phi_{Bn}$ is equal to $q\phi_{Bp} = E_g - q(\phi_m - \chi)$, where E_g is the energy bandgap and χ is the electron affinity of the semiconductor.

Figure 21 shows the case for a p-type semiconductor but with $q\phi_m > q\phi_s$. The figure on the left-hand side shows the band diagrams before joining and the figure on the right-hand side shows the band diagram after joining. Because the Fermi level of the semiconductor is higher than that of the metal, electrons will be transferred from the semiconductor to the metal after joining. The surface region of the semiconductor will have less electrons and more holes and therefore will be in accumulation. The barrier to hole flow between the metal and the semiconductor will be small. The current is then determined by the bulk resistance of the semiconductor. This contact will be an ohmic contact.

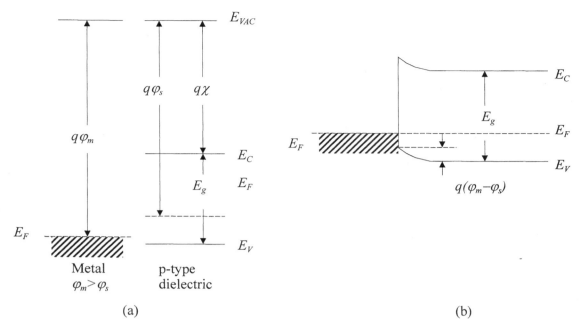

Fig. 21. Metal–p-type semiconductor contact: $\phi_m > \phi_s$, ohmic contact, (a) before contact, (b) after contact.

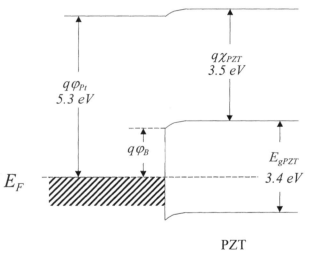

Fig. 22. Energy band diagram of the Pt/PZT system (partial depletion).

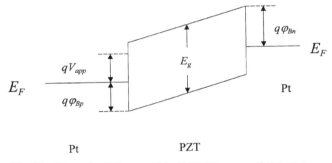

Fig. 23. Energy band diagram of the Pt/PZT/Pt system (full depletion).

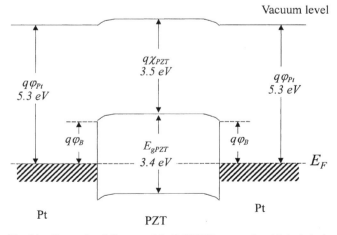

Fig. 24. Energy band diagram of the Pt/PZT/Pt system (partial depletion).

The Pt/PZT contact can be analyzed as follows. The work function of Pt is about 5.3. The energy bandgap and the electron affinity of PZT are about 3.4 and 3.5 eV, respectively [18]. If the Fermi level of PZT is at the middle of the energy bandgap, $q\phi_m$ will be very close to $q\phi_s$. If PZT is p type and the Fermi level is at the lower half of the PZT bandgap, the energy band diagram after contact will resemble Fig. 22. If the concentration of carriers is low and the PZT is fully depleted under bias, the energy band of PZT will tilt as shown in Fig. 23.

A symmetrical Pt/PZT/Pt capacitor structure, assuming the situation of Fig. 22 applies to both sides, will resemble Fig. 24.

A symmetrical Pt/PZT/Pt capacitor structure, assuming that the injected carriers are holes, is shown in Fig. 25. The top figure is the situation without an external applied voltage

and the bottom figure is with an external applied voltage of V_{app}.

If the PZT is assumed to be n type, the Fermi level of PZT will be in the top half of the energy bandgap and the situation

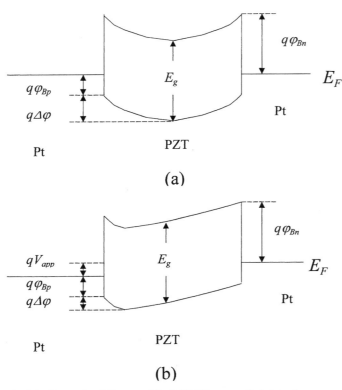

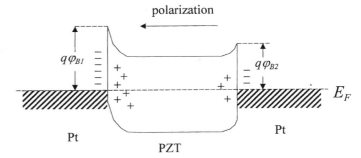

Fig. 27. Energy band diagram of the Pt/PZT/Pt system, assuming polarization dependence.

6.2.2. Energy Band Diagram

As can be seen from the preceding discussion, in order to properly discuss the electrical conduction of PZT films, it is very important to understand the energy band structures of the metal–ferroelectric system or the metallic oxide electrode–ferroelectric system. Although many people have discussed the leakage current of PZT films, relatively few references discussed the energy band diagram of the PZT capacitor system. The reported results are briefly discussed next.

1. The energy band diagrams of Pt/PZT and Pt/BST systems are shown in [18, 19]. Both of these diagrams show p-type ferroelectric films and the ferroelectric energy bands bend downward in contact with the metal electrode Pt. (This is actually called "bending upward" in the original reference.)

2. Bernacki [46] claimed that the PZT film behaves electrically as if it were fully depleted, and hence, like an insulator. The Fermi level is midway between the valence band and the conduction band. This changes the model of the PZT from a p-type semiconductor to an intrinsic semiconductor, where carriers of both polarities within the thermal energy of the band edges participate in the conduction process, their relative contribution being determined by their density of state and mobility. With this modification, an energy band diagram of platinum and unpolarized polycrystalline PZT is drawn. With the Fermi level in the middle of the PZT bandgap, the PZT electron affinity is taken as 1.75 eV. Note that this electron affinity value is much smaller than the value 3.5 eV listed in [18]. The work functions of Pt and undoped PZT are taken as 5.5 and 3.5 eV, respectively. The energy bandgap of PZT is taken as approximately 3.5 eV. Due to the difference in work functions, electrons flow from the PZT into the metal, leaving behind a positive space-charge region of ionized donor sites within the PZT. This situation is analogous to the depletion region concept in semiconductor physics. This flow of charge continues until the Fermi levels reach equipotential values in the platinum and the PZT, resulting in a bending of the energy bands within the PZT. The energy band bends upward in contact with the metal.

Now consider that the ferroelectric PZT is polarized with a negative bias on the metal electrode, inducing a positive remanent polarization on the surface of the ferroelectric PZT. This polarization is equivalent to a bound positive sheet

Fig. 25. Energy band diagram of the Pt/PZT/Pt system (hole injection, assuming $\phi_{Pt} > \phi_{PZT}$ and positive space charge at the PZT surface region) (a) without bias voltage, (b) with applied voltage.

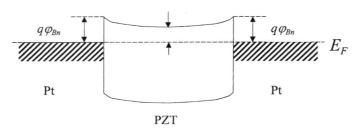

Fig. 26. Energy band diagram of the Pt/PZT/Pt system, assuming n-type PZT and $\phi_{Pt} > \phi_{PZT}$.

will be most likely similar to the situation of Fig. 18. A Schottky barrier for electrons will be formed at the metal/PZT interface. The situation for a Pt/PZT/Pt capacitor structure will resemble Fig. 26. There will be a Schottky barrier on both sides.

If the polarization of the ferroelectric material such as PZT has an effect on the conductivity as proposed by Bernacki [46], the situation will resemble Fig. 27, where the polarization is pointing toward the left and there will be more bound positive sheet charges on the left-hand side of the ferroelectric film. This will decrease the electron affinity at the PZT surface and increase the Schottky barrier height as shown in Fig. 27.

Table XXXVIII.   Energy Band Diagram Parameters of the Metal/PZT System

Reference	PZT energy bandgap (eV)	PZT electron affinity (eV)	Metal work function (eV)	Electron barrier height (eV)	Hole barrier height (eV)	Conduction type and band bending
Bernacki [46]	3.5	1.75	5.5 (Pt)	(3.75)		Bending upward
Lee et al. [268]						Bending upward
Lines and Glass [14]	3–4					
Scott [18]	3.4 ± 0.1	3.5 ± 0.2	5.3 (Pt)	1.8	1.6	Bending downward
Stolichnov and Tagantsev [251]						Hole injection, bending upward
Sudhama et al. [33]	~3.0			0.6 (Pt) 0.1 (In)	2.4 (Pt) 2.9 (In)	Bending downward
Wouters et al. [45]	3.41	2.6	5.65 (Pt)	3.05	0.36	Fully depleted

charge on the surface of the PZT and hence will add to the positive charge density within the PZT depletion region. This will decrease the electron affinity at the PZT surface and hence increase the Schottky barrier height. Therefore, the reverse leakage current of this Schottky diode will decrease. On the opposite side of the capacitor, a negative sheet of bound charge will be created on the surface of the PZT, which will have the effect of increasing the reverse leakage current of that Schottky diode. Thus, the barrier heights on either side of the capacitor are now unequal, as shown in the energy band diagram in Fig. 27, causing the electron component of the leakage current through the capacitor to be unequal, greater in one direction than the other [46]. This model differs from models in which the Schottky barrier height does not depend on the remanent polarization state of the PZT.

3. Hu and Krupanidhi [256] discussed the energy band in their grain boundary model.

4. Larson et al. [267] depicted the energy band, but did not indicate the direction of bending.

5. Lee et al. [268] proposed an energy band diagram of the Pt/PZT/Pt system. n-type PZT was assumed. There is a surface depletion region and a blocking layer at the electrode interface. The band bending upward at the interface is similar to Fig. 26.

6. Stolichnov and Tagantsev [251] discussed the energy diagram. They postulated that the injected carriers are holes. The energy band of PZT bends upward as shown in Fig. 25.

7. Sudhama et al. [33] discussed the energy band diagram of the Pt/PZT/In capacitor system. p-type PZT was assumed. The energy band of PZT bends downward to form a Schottky barrier for holes. The energy band diagram is drawn under the assumptions that (a) PZT is a p-type semiconductor, (b) the work function (on PZT) of In is about 0.5 eV less than that of Pt, and (c) the doping profile is flat across the thickness of the device.

8. Yoshida et al. [269] in their work on nanoscale conduction modulation showed an energy band diagram for the Au/PZT/SrRuO$_3$ system. p-type PZT was assumed. The PZT energy band bends downward at the electrode contacts.

9. Nagaraj et al. [265] described the energy band diagram of the LSCO/PZT system. The p-type PZT has a work function of 4.4 eV and LSCO a work function of 4.2 eV. Holes flow from PZT to LSCO, and a depletion layer is formed at the interface that rectifies the holes. The same argument would hold true for a Pt electrode (work function 5.3 eV) with PZT (when considering electron exchange) and as such one would not expect a Schottky barrier.

The literature results on the energy band diagrams of metal/PZT systems are summarized in Table XXXVIII. As we can see here, there is really no clear consensus on the energy band diagram of the metal/PZT/metal system.

6.2.3. Trap Charges

Chen et al. [270] studied the correlation between trap density and leakage current of PZT capacitors by examining the current–time (I–t) and the current–voltage (I–V) characteristics. The increase in leakage current after dc electrical field stress is correlated with the number of charges trapped inside the films. The spatial density distribution of trapped charges is calculated by analyzing the decay of discharging current after the application of dc stress. The discharging current density J is well fitted by a $1/t$ relationship, where t is the discharging time. This behavior is explained using the tunneling front model [271, 272] in treating the SiO$_2$/Si system.

The discharging current measured after removing the stress voltage is plotted using a log–log scale, where t is the discharging time. The log J vs. log t curves are almost straight lines with a slope of -1. The $J \propto 1/t$ relationship can be explained by using a tunneling front model, which was developed to calculate the trap density and distribution inside SiO_2 thin films after high dc stressing. This model [271, 272] assumes that the stress-induced trapped holes in a SiO_2/Si system will discharge from the trap states to the valence band of silicon by a tunneling process (or, equivalently, electrons tunnel from silicon to recombine with the trapped holes). This model is applied to describe the discharging process in PZT films [270].

The tunneling front moving through the PZT films with a tunneling distance $x(t)$ from an electrode can be modeled as [271, 272]

$$x(t) = \frac{1}{2\beta} \ln\left(\frac{t}{t_0}\right)$$

where t_0 is a characteristic time and β is a tunneling parameter that can be calculated as

$$\beta = \left(\frac{2m_t^*}{\hbar^2} E_b\right)^{1/2}$$

where m_t^* is the carrier effective mass and E_b is the potential barrier height for the discharging holes. β is calculated to be 1.14×10^{-1} nm^{-1} using $m_t^* = 10^{-3}m_0$ and $E_b = 0.5$ eV. The parameter m_t^* is extracted by assuming a two-carrier injection metal–semiconductor–metal model for metal/PZT/metal capacitors and E_b is assumed to be a single-hole trap level located at 0.5 eV above the PZT valence band edge. The characteristic time t_0 is related to the discharging process and assumed to be 10^{-13} s (the same value as used in SiO_2). The traps located closer than $x(t)$ are recombined by time t and those beyond $x(t)$ still occupied by holes. The current associated with the tunneling front model can be calculated as

$$I(t) = qnvA = qN\big(x(t)\big)\frac{dx}{dt}A = qNx(t)A \times (2\beta t)^{-1}$$
$$\cong qNA(2\beta t)^{-1}$$

where $I(t)$ is the time-dependent current due to the discharge of trapped charges, q is the electron charge, n is the carrier density substituted by the trapped charge density, v is the velocity of the tunneling front, which can be calculated by dx/dt, and A is the capacitor area. If the trapped charge density is assumed not to change very rapidly across the film, the $\log(I)$ vs. $\log(t)$ plot is a straight line with a slope of -1. The measured result agreed with this relationship very well.

The trap distribution within the film can be calculated from

$$N\big(x(t)\big) = \frac{2\beta I(t)}{qA} t$$

The calculated spatial distribution of trapped charges for different stress fields is as follows. The trapped charge density increases with the stress voltage and ranges from 2.5×10^{18} to 5.0×10^{18} cm^{-3} for dc stress voltage from 117 to 333 kV/cm. The calculated trap density is of the same order of magnitude as the acceptor doping concentration N_A.

The result shows that dc electrical field stress of PZT induces trapped positive charges in the thin films. This phenomenon will influence the carrier transport conduction in the thin films

and consequently change the $I–V$ characteristics. The trapped charges near both the top and the bottom electrodes will discharge by a tunneling process when the sample is connected to a short circuit. This discharging behavior causes the time-dependent current to decay nonexponentially when the applied electrical field is removed. This time-dependent discharging current can be analyzed with a tunneling front model. The logarithmic plot of discharging current vs. discharging time shows a straight line with a slope equal to the theoretical value of -1. The tunneling front model of the discharging current for the SiO_2/Si system is hence applied to ferroelectric thin films such as PZT. The spatial distribution of trapped charges therefore can be deduced based on hole tunneling from the trap states to the metal electrodes in the PZT.

6.2.4. Voltage Stress

Chen and Lee [273] studied the degradation of capacitance of PZT ferroelectric thin-film capacitors under a dc stress electric field. Two stress effects are observed in the PZT capacitance–voltage ($C–V$) characteristics. The first is that the capacitance is reduced, and the second is a voltage shift of the $C–V$ curve. These effects are found to depend on the stress electric field and the injected charge fluence. A correlation between the stress effects and the electron trapping inside the films is established. The injected charge fluence can be calculated from the leakage current. The trapping efficiency, electron capture cross section, and trap density are obtained from the measured injected charge influence and the $C–V$ voltage shift.

The degradation of capacitance and the voltage shift of the $C–V$ curves are correlated with electron trapping inside the film. Electron trapping at the domain walls can cause domain wall pinning, which, in turn, leads to capacitance reduction. The voltage shift of the $C–V$ curves is caused by the buildup of trapped electron charges. The electron trapping efficiency depends on the applied electric field and the injected charge fluence. The calculated value of the electron capture cross section is 1.89×10^{-19} cm^2 and the neutral trap density is 1.20×10^{13} cm^{-2}. The trapping efficiency is on the order of 10^{-6}–10^{-7}.

6.2.5. Transient Current

Chen and Lee [274] studied the temperature dependence of the transient current of PZT thin-film capacitors by examining the current density vs. time ($J–t$) characteristics in the temperature range from room temperature to 140°C. The transient current consists of two components, the conduction current and the polarization current. The conduction current is related to traps inside the films. The trapped holes cause an increase in the local electric field and hence the conduction current. The polarization current decreases and saturates as time increases. The dependence of the transient current on temperature therefore relies on the relative magnitudes of the two components. Depending on the relative magnitudes of these two current components, the transient current characteristics can be separated into two different types, the degradation type and the polarization type. The transient current increases with time in the degradation-type region and decreases exponentially with time in the polarization-type region. The increase in the degradation-type transient current can be explained by the accumulation of trapped holes. The effect of the carrier trapping becomes less significant as the temperature increases. At room temperature, the transient current of PZT is dominated by trap-related conduction current at field higher than 83 kV/cm and by polarization current at field lower than 83 kV/cm. This transition voltage or transition electric field separating the two regions increases linearly with temperature. The increase in temperature enhances the emission rate of captured holes and the conduction current component decreases. The discharging current in the PZT capacitor follows the tunneling front model in the temperature range below 100°C, whereas a different model is required above 100°C.

6.2.6. Size Effect

The ferroelectric properties are a function of film thickness and grain size. The decrease in the size of an object below a critical limit leads to a change in its properties, which is often called the size effect. Batra and Silverman [275] first estimated the critical thickness for ferroelectric triglycine sulfate to be about 400 nm. These films are unstable against depolarization field. Tilley and Zeks [276] developed a detailed theory. Tilley [277] made numerical calculations including Thomas–Fermi screening in the metal electrode. Tilley calculated a minimum ferroelectric film thickness of 90 nm for strong ferroelectric material. Experimentally, PZT film with a thickness of 25 nm has been made [19].

Desu et al. [278] studied the ferroelectric properties and switching characteristics of sputtered PZT thin films as a function of film thickness (30–300 nm) and grain size. A Au/PZT/Pt structure was used. The theoretical models accounting for the dimensional effects in the ferroelectric layers predict the following: lowering of P_s, E_c, dielectric constant, and T_c and changes in the lattice spectrum with decreasing film thickness. The critical size is around 30 nm. The experimental result demonstrates definite switching properties for films having a thickness around the value of the predicted critical size. This can be attributed to the fact that films with thickness around 30 nm show grain size around 200 nm, which is much larger than the predicted critical size. Thus, the grain size, when compared to film thickness, is important in accounting for the size effects on the properties of ferroelectric films.

The electrical properties of PZT as a function of thickness using the Au/PZT/Pt/Ti capacitor structure were studied by Kundu and Lee [246]. The PZT thickness ranges from 70 to 680 nm. The dielectric constant remains pretty much constant around 340 except for the 70-nm film. The coercive field decreases from 142 kV/cm for 130-nm to 72 kV/cm for 680-nm film. The remanent polarization is around 15 μC/cm^2 except for the 70-nm film, which may have lower film quality. The conduction mechanism is found to vary with film thickness.

Cho et al. [279] studied the thickness and annealing temperature dependence of PZT films deposited by the sol–gel method. The film thickness is in the range of 120–360 nm. The coercive field calculated from the C–V curves decreases with increasing thickness. As the film thickness increases, the remanent polarization increases up to about 25 μC/cm^2, whereas the coercive field decreases down to 47 kV/cm for 360-nm-thick films annealed at 600°C.

6.3. Very Large Scale Integration Applications of PZT Thin Films

Ferroelectric films can be used in DRAM and nonvolatile memories. Most studies have emphasized PZT for nonvolatile memory elements and BST for DRAM capacitors [250].

Table XXXIX lists the reports on nonvolatile memory applications. The ferroelectric material used in most applications is PZT. Some use SBT and PLZT.

Philofsky [34] reviewed the status of ferroelectric random-access memory (FRAM). Two key aspects of FRAM technology are considered to make FRAM superior to those other nonvolatile memories manufactured with EEPROM technology. First, FRAM employs a polarization technique instead of a charge tunneling mechanism. Second, it permits all internal operations to utilize 5 V instead of the 12–15 V requried by conventional EEPROM technologies.

Kraus et al. [284] reported the fabrication of 1-Mb ferroelectric random-access memory (FRAM) with 15.8-μm^2 cell size and 60-ns read/write time. The FRAM incorporates a one-transistor/one-capacitor (1T/1C) architecture. The memory cell size is 3.95 μm × 4.00 μm. The cell capacitor area is 3.025 μm^2. The die size is 7.49 mm × 5.67 mm. PZT ferroelectric capacitors utilizing platinum electrodes are formed during a 0.5-μm planarized CMOS process using tungsten plugs.

Jones [283] discussed the application of ferroelectric nonvolatile memories for embedded applications. Ferroelectric nonvolatile memories offer low-voltage operation and fast write speeds, making them attractive for embedded applications. The bit cell used a two-transistor/two-capacitor (2T/2C) architecture, and the basic design feature size was 1.2 μm. The capacitors were fabricated using spin-coated SBT and sputter-deposited Pt electrodes. A test chip that embedded FRAM into an 8-bit microcontroller was fabricated. An 8-Kb FRAM was designed to replace 4 Kb of EEPROM and 4 KB of SRAM.

Takasu et al. [285] also discussed ferroelectric embedded devices. PZT capacitors with IrO$_2$ electrodes were used. Test chips with 0.5-μm design rule ferroelectric memory cells were fabricated.

6.4. Reliability Issues of PZT Thin Films

The problem of reliability is important for VLSI application. The reliability issues for ferroelectric memory device applications include fatigue, aging, resistance degradation, retention, and imprint.

Table XXXIX.   Nonvolatile Memory Applications

Reference	Company	Chip	Ferroelectric film	Technology	Status
Aoki et al. [280]	Fujitsu	1 Gb	SBT		Research
Bondurant [22]	Ramtron	1 Kb, 2 Kb, 4 Kb, 8 Kb, 16 Kb	PZT	1.5-μm CMOS	Product
Bondurant and Gnadinger [21]	Ramtron	256 bit	PZT	3.5-μm CMOS	Test chip
Bondurant and Gnadinger [21]	Ramtron	512 × 8 bit	PZT	1.5-μm CMOS	Product
Bondurant and Gnadinger [21]	Krysalis	2048 × 8 bit	PLZT	2.0-μm CMOS	Product
Evans and Suizu [282]	Radiant	64 Kb	PZT	Static FRAM	Research
Evans and Womack [281]	Krysalis	512 bit		3.0-μm CMOS	Research
Geideman [29]	McDonnell Douglas	64 Kb	PZT	GaAs	
Jones [283]	Motorola	8 Kb in micro-controller	SBT	1.2μm 2T/2C	Test chip
Kraus et al. [284]	Ramtron, Fujitsu	1 Mb	PZT	0.5-μm CMOS 1T/1C	Test chip
Philofsky [34]	Ramtron	4 Kb, 8 Kb, 16 Kb, 64 Kb	PZT	1.0-μm CMOS	Product
Takasu et al. [285]	Rohm	Embedded	PZT	0.5 μm	Test chip

6.4.1. Fatigue

Fatigue in ferroelectrics is defined as the decrease in switchable polarization with electric field cycling [19, 286]. The loss of switchable polarization is due to pinning of domain walls, which inhibits switching of the domains [286]. A variety of mechanisms for domain wall pinning have been proposed, including electron charge pinning [287–289], pinning by oxygen vacancies [35, 290–292], as well as pinning by extended defects [293].

Experiments with photoillumination, which can excite electron–hole pairs, show that the switchable polarization of the fatigued samples can be restored. These results indicate that the pinning of domains by trapped charges at internal domain boundaries. In optical restoration, the photoexcited carriers recombine with the trapped charge and allow the dc bias to reorient the locked domains [286].

The involvement of oxygen vacancies is suggested by experiments that show improved fatigue behavior due to donor doping [254, 286]. Scott et al. [253, 254] showed that the oxygen concentration near the electrodes decreases after fatigue, indicating an increase in the concentration of oxygen vacancies near the interface. Accumulation of oxygen vacancies could promote fatigue by stabilizing near surface charge traps or by creating an n-type layer in the near-electrode region, which would increase the electron injection rate and thus the trapping rate.

Experiment also shows that there is some change in the defect distribution during fatigue testing. Fatigued samples restored using saturating bias or bandgap light fatigue faster when retested. This result shows that even though the switchable polarization can be recovered through an electronic process, this process does not restore the material to its original state [289].

Al-Shareef et al. [296] and Dimos et al. [297] showed that nominally fatigue-free LSCO/PZT/LSCO and Pt/PZT/Pt capacitors can both be made to exhibit significant polarization fatigue when illuminated with UV light during fatigue testing. In addition, capacitors that have been fatigued under illumination can be fully rejuvenated by applying a dc saturating bias with light or by electric field cycling without light. These results suggest that fatigue is really a competition between domain wall pinning and unpinning and that fatigue-free behavior occurs when the pinning rate does not exceed the unpinning rate. The role of light is solely to generate free carriers in the film.

Experiments by Du and Chen [298] showed that electron injection is a necessary condition for polarization degradation, whereas hole injection is not. Injected electrons can decrease the valence of oxygen vacancy and increase its mobility. This mechanism promotes fatigue either by oxygen vacancy segregation at the PZT/electrode interface or by pinning domain boundaries.

The application of metallic oxide electrodes, such as (La, Sr)CoO_3, RuO_2, and IrO_2, has greatly improved the fatigue performance. This clearly shows the critical role of the ferroelectric–electrode interface in determining the properties of fatigue. It has been suggested that the oxide electrodes inhibit fatigue primarily by acting as a sink for oxygen vacancies, so that they do not accumulate at the electrode–ferroelectric interface [299].

6.4.2. Aging

Ferroelectric aging is generally defined as a spontaneous change in the $P-V$ hysteresis behavior with time. This spontaneous change is typically ascribed to a gradual stabilization of the domain configuration, which can manifest itself in a number of ways, such as a shift or constriction in the $P-V$ response, a steady decrease in the dielectric constant, or a change in the piezoelectric or electrooptic response [286, 300–304]. Aging occurs only in the ferroelectric state.

For thin-film devices, the voltage shift is probably the most important manifestation of ferroelectric aging. The voltage shift can be attributed to the introduction of a space-charge field due to charge trapping [286]. The sign of the induced space-charge field is such that it compensates the internal depolarization field. It has also been shown that trapped electrons, rather than holes, are primarily responsible for the space-charge field [305].

This space-charge field is due to a combination of trapped electrons and accumulated oxygen vacancies, where the driving force for inducing the space-charge field is the net polarization that is uncompensated by charge on the electrodes [305, 306].

As for the reason for aging [307], Mason [308] suggested that the aging effect is a reduction of the effective polarization produced by the domain-wall motion. The rearrangement of ferroelectric domains with time is considered to be the cause of aging phenomena [309]. Other mechanisms have been proposed for aging effects. Bradt and Ansell [310] assumed that the aging phenomenon results from the relief of the residual stress due to domain switching arising from electric and elastic fields.

6.4.3. Resistance Degradation

Resistance degradation is a deterioration in the insulating properties of a dielectric (ferroelectric or paraelectric) under dc bias and elevated temperature [286]. Resistance degradation is also sometimes referred to as time-dependent dielectric breakdown (TDDB) [311]. Resistance degradation can occur at temperatures and fields much lower than the critical onset values for thermal or dielectric breakdown. Resistance degradation is a general dielectric phenomenon and not related to the ferroelectric nature of the material.

TDDB in dielectric films such as silicon dioxide is usually explained by the mechanism of impact ionization [312]. The electron–hole pairs generated during the impact ionization process will be subjected to traps in the films. This will cause the gradual accumulation of trapped charges in the thin films. The trapping of holes will be more severe due to the lower mobility of the holes. The accumulated holes near the cathode will change the conduction band energy and increase the local electric field. This will eventually lead to TDDB.

The time-dependent dielectric breakdown in ferroelectric films was studied by a number of authors [31, 32, 313, 314]. The time to breakdown (t_{BD}) is measured by applying a relatively high voltage such as 20–30 V until dielectric breakdown. It is usually assumed that the t_{BD} is a function of the total number of carriers passing through the films. Therefore, the time to breakdown t_{BD} is inversely proportional to the leakage current density J, that is,

$$t_{BD} \propto 1/J$$

Carrano et al. [31] reported a linear dependence of $\log(t_{BD})$ on the applied electrical stress field E. The projected lifetime is relatively low. If the lifetime is extrapolated by plotting $\log(t_{BD})$ vs. $1/E$ as is normally done for silicon dioxide, the extrapolated lifetime would be much longer. Moazzami et al. [32] found that the PZT leakage current shows ohmic behavior at low field and exponential behavior at moderately high field. The high field characteristic is attributed to either ionic or electronic hopping conduction. A current density equation in the form of $J \propto \sinh(BE)$ is assumed, where B is a constant. If exponential dependence is assumed, the extrapolated lifetime is shorter than month. If an inverse-field relationship is used, the extrapolated lifetime is longer than 100 years. The divergence of the lifetimes in the two extrapolation methods is very obvious.

Chen et al. [313] reported a different method for extrapolating the lifetime. PZT capacitors with gold top electrode and platinum bottom electrode were fabricated. Applied voltages ranging from 2 to 30 V were used in the TDDB experiment. There is no significant difference in leakage current when the polarity is reversed. Therefore, the leakage current is due to bulk-limited mechanisms. It was observed that there are two clearly divisible regions in the current vs. voltage relationship. In the low field region (less than 100 kV/cm), the $\log(J)$ vs. $\log(E)$ curve has a slope of 0.88, which shows that the current in the low field region is ohmic-like. In the high field region (above 100 kV/cm), the $\log(J)$ vs. $\log(E)$ curve has a slope of 9.6. These curves show that the leakage current depends on the applied voltage in a power law relationship. A $J \propto E^m$ relationship can be explained by space-charge-limited current with deep traps.

If an impact ionization model for the leakage current and a Fowler–Nordheim-type current $J \propto \exp(-\alpha/E)$ are assumed, the lifetime t_{BD} should be extrapolated by a $\log(t_{BD})$ vs. $1/E$ plot. The extrapolated time is very long, on the order of 10^{17} s at 5 V. If a $J \propto \exp(BE)$-type electron hopping conduction mechanism is assumed, the lifetime should be extrapolated by a $\log(t_{BD})$ vs. E plot. The extrapolated lifetime is shorter than 3 days. The discrepancy is due to the fact that neither leakage current mechanism is adequate for PZT. The leakage current in PZT thin films is ohmic-like at low field and space-charge limited at high field. Because the PZT leakage current in PZT is dominated by different mechanisms at different voltage ranges, a general relationship $J \propto E^m$ with separate values of m at low and high fields can best describe the leakage current in PZT. The extrapolated lifetime can be obtained by assuming the general expression $J \propto E^m$. Therefore,

$$t_{BD} \propto 1/J \propto E^{-m}$$

The lifetime can be extrapolated by plotting $\log(t_{BD})$ vs. $\log(E)$. The slope measured from this curve should be $-m$ and m is found to be 9.1. This is in good agreement with the slope of 9.6 obtained from the leakage current measurement. Because the lifetime extrapolated in this way uses the breakdown time data from the high field region, the slope should be modified in the low field region. A lower m (slope = $-m$ = -0.88) is used for the low field to extrapolate the actual lifetime for operation at lower voltages (5 V or lower). The extrapolated lifetime should then be on the order of 3×10^7 s. This new TDDB lifetime extrapolation method gives a PZT lifetime that lies between the divergent lifetimes obtained from the two previous methods.

A similar method has also been used to extrapolate the lifetime of barium strontium titanate (BST) [314].

6.4.4. Imprint

The development of a voltage shift in a thin film left in a given polarization state for an extended period of time is the basis of the phenomenon of imprint [286, 315]. Imprint manifests itself as a bit error reading that occurs during combined thermal and electrical cycling of a ferroelectric thin-film memory. It has been shown to be caused by trapping of electrons in near interfacial traps within the ferroelectric film, a process that is dependent on the magnitude of the spontaneous polarization [305]. The asymmetrically trapped charge causes a shift along the voltage axis of the dielectric hysteresis loop analogous to internal bias field effects in bulk ceramics. If the voltage shift caused by imprint approaches the magnitude of the coercive voltage, considerable charge may be switched by disturb pulses from adjacent cells, which can result in bit error readings. A criterion that has been used for nonvolatile memory design is that the imprint voltage divided by the coercive voltage should be less than 0.5 under any operational conditions. Bias voltage, increased temperature, and ultraviolet excitation accelerate imprint. The use of donor doping and appropriate electrode technology has minimized imprint effects in PZT thin-film capacitors [316, 317].

Research [315, 318–323] has shown that the imprint phenomenon is caused by trapping of electronic charge carriers near the interface between the ferroelectric film and the electrode. The voltage shift arises from an asymmetric distribution of trapped charge [319]. Either thermal or optical processes can induce the voltage shift of the $P–V$ characteristics of ferroelectric thin films. For PZT films, the thermally induced voltage shifts are greater than those obtained optically. This is attributed to the role of oxygen-related defect dipoles throughout the film [319]. Dimos et al. [305] proposed that optically induced voltage shifts are engendered by trapping charge carriers at near interfacial sites to compensate for the depolarizing field. An uncompensated depolarizing field arises if the dipole charge does not ideally terminate at the electrode. The thermally induced shifts are generally larger than those obtained optically. Pike et al. [306] in studying a similar phenomenon in PLZT, attributed this phenomenon to the role of a defect-dipole component in the polarization. This component is due to a volumetric distribution of aligned defect-dipole complexes in the material. This difference in the thermal and the optical shifts is attributed to the role of orientable defect-dipole complexes involving oxygen vacancies. Processes that tend to reduce the oxygen vacancy concentration, such as substituting some of the Ti(Zr) sites with donor dopants, lead to smaller thermally induced voltage shifts.

For PZT material, a linear relationship has been observed between the remanent polarization and the magnitude of voltage offsets in the hysteresis curve. It has been proposed that the increased polarization lowers the electrostatic potential well for the trapping of electrons, thereby leading to greater voltage shifts. It has also been observed that the remanent polarization and defect occupancy are temperature dependent, which collectively impact the observed voltage offsets measured at elevated temperature [320]. A similar phenomenon has also been observed in SBT [318].

6.4.5. Retention

Retention is the ability of a ferroelectric capacitor cell to retain its stored charge and hence its 1 or 0 of stored information. This is a shelf life or aging test in which the cell is not cycled with bipolar applied voltages. A retention test therefore consists of storing a bit for a time t, then addressing it to test its polarity. This is a destructive test and therefore a reset voltage must be applied after the "read" operation to restore the cell to its original stored polarity [18].

A typical logic state retention test scheme would be to write the capacitor with a pulse of one polarity and subsequently read it with two pulses of the opposite polarity, thereby simulating a logic "0" state and a logic "1" read. The delay between the write and read pulses is the retention time.

Yin et al. [324] described the retention test as follows: the capacitor was written with a triangular pulse of 5 V and subsequently read with two triangular pulses of −5 and 5 V, respectively. The pulse width was 2 ms. The time delay between the write pulse and the first read pulse is the retention time.

The retention properties of ferroelectric films have been studied for a number of materials, including PZT by Jo et al. [325] and Zomorrodian et al. [326]; SBT by Zhang et al. [47, 327]; $SrBi_2(Ta, Nb)_2O_9$ (SBTN) by Shimada et al. [328]; and $Pb(Ta_{0.05}Zr_{0.48}Ti_{0.47})O_3^*$ (PTZT) using $La_{0.25}Sr_{0.75}CoO_3$ (LSCO) top and bottom electrodes by Yin et al. [324]. The application of LSCO electrodes on lead-based thin films has been studied by Aggarwal et al. [329].

Zhang et al. [47] found that SBT has quite good retention properties. Zomorrodian et al. [326] used the pyroelectric current and phase detection method and the conventional double-pulse method to measure the polarization retention in $Au/PZT/YBa_2Cu_3O_{7-x}$ (YBCO) capacitors. The retention properties were measured at the elevated temperatures of 150 and 200°C. A polarization degradation of less than 5.5% was detected for PZT at 200°C for more than 24 h.

On the various degradation issues of ferroelectric capacitors, solutions to the problem of fatigue in ferroelectric films have been provided by both layered perovskite thin films and PZT thin films with oxide electrodes. It has been demonstrated that fatigue is a direct result of charge trapping leading to domain pinning and that oxygen vacancies can play an important role in the process. Voltage shifts, which lead to the phenomenon of imprint, are a result of near interfacial charge trapping. Once again, oxygen vacancies can play an important role, primarily as the mobile component of defect dipoles. Resistance degradation, which is important for ferroelectrics and nonferroelectrics, is due to the migration and subsequent pileup of oxygen vacancies at the cathode. Substantial progress has been made in mitigating these reliability-limiting processes through donor doping and optimizing the perovskite composition and electrode material [330].

7. PARAELECTRIC FILMS: BARIUM STRONTIUM TITANATE (BST)

7.1. Physical Properties of BST Thin Films

Barium strontium titanate $(Ba_xSr_{1-x})TiO_3$ is a substitutional solid solution of barium titanate $(BaTiO_3)$ and strontium titanate $(SrTiO_3)$. $BaTiO_3$ has a cubic perovskite structure above 120°C. Below 120°C, it transforms successively to three ferroelectric phases: first to tetragonal, then to orthorhombic at about 5°C, and, finally, to trigonal below −90°C [14]. $SrTiO_3$ is a paraelectric material at room temperature. The addition of Sr will lower the Curie temperature of the $(Ba, Sr)TiO_3$ solid solution. Extrapolation would lead to an expected Curie temperature of pure $SrTiO_3$ around 40 K. However, this extrapolation is not very reliable. The actual Curie temperature may be higher [13]. The Curie temperature of $SrTiO_3$ is listed as about −163°C [14].

$Ba_{0.5}Sr_{0.5}TiO_3$ has a Curie temperature of about −50°C and $Ba_{0.75}Sr_{0.25}TiO_3$ a Curie temperature of about +40°C [331]. For DRAM applications, the Ba/Sr ratio should be adjusted to reduce the Curie temperature below the DRAM operating temperature. Most research has been performed with a Ba/Sr ratio of 50/50.

BST was chosen for DRAM capacitor application because it has the following properties [331]:

1. BST has a high dielectric constant at the high-frequency zone for writing to DRAMs and ensures a sufficient amount of accumulated electric charge.
2. It is in the paraelectric phase without spontaneous polarization; hence, there will be no fatigue problem.
3. Because it has insulation properties, the electric charge leaks during the refreshing cycle while the DRAM is operating will be small.
4. It does not contain ions harmful to the operation of the silicon transistor, such as Na, K, and Li.
5. Film deposition is relatively easy.

7.2. Dielectric Properties of BST Thin Films

The dielectric constant of bulk BST is 1000 or more and that for thin film is in the low value range of 200–800 [331]. The deposition temperature of BST thin films has a significant influence on the dielectric constant of BST films.

Table XL lists the dielectric properties of BST thin films reported in the literature. The dielectric constants of most BST thin films are on the order of 200–400. Most of the reports have BST thin-film thickness ranging from 15 to 200 nm. Platinum is the most often used metal electrode. Ru, RuO_2, $SuRuO_3$, and other oxide electrodes are also used.

Baumert et al. [332] characterized the dielectric properties of sputtered BST and $SrTiO_3$ films. Specific capacitance values of 96 $fF/\mu m^2$ for BST deposited at 600°C and 26 $fF/\mu m^2$ for $SrTiO_3$ are obtained. The BST film thickness is on the order of 60–120 nm.

The dielectric constant of BST decreases with film thickness [331]. Lee et al. [354] observed that the dielectric constant increases from 348 to 758 when the grain size increases from 32 to 82 nm in BST films deposited at 600°C. The grain size dependence of the dielectric constant has been explained by assuming a low-dielectric-constant layer with a fixed thickness on the crystal grain interface. However, there is also a Curie temperature change of several degrees. Present theory cannot quite explain both of these phenomena at the same time [331].

7.3. Electrical Characteristics of BST Thin Films

7.3.1. Electrical Conduction and Leakage Current

The most important electrical property of BST thin films is the leakage current. Many researchers have studied the leakage current of BST thin films. The reported results on leakage current and conduction mechanisms in BST thin films are summarized in Table XLI. Because ferroelectric thin-film capacitors are fabricated with top and bottom electrodes, the contact properties between metal electrodes such as Pt and perovskite oxides such as PZT [368, 369] and BST [370, 371] have been studied by various authors. The ideal potential barrier height at a metal/n-type semiconductor is $q(\phi_m - \chi)$, neglecting the image force effect as well as contributions by the surface states, where $q\phi_m$ is the metal work function and $q\chi$ is the electron affinity of the ferroelectric material. The reported results on the conduction mechanism and contact properties are briefly discussed next.

1. Dietz et al. [335] characterized the leakage current as a function of temperature and applied voltage. The experimental data can be interpreted by a thermionic emission model. Schottky emission accounts for non-ohmic behavior at higher fields. A systematic decrease in barrier height is found with increasing applied field. This is likely due to an additional lowering of the cathode barrier by a distribution of deep acceptor states in the bandgap of the ceramic. The resulting decrease in the barrier is stronger than the one caused by the image forces of the injected electrons.

2. Hsu et al. [355] separated the polarization current and the electronic leakage current by monitoring the discharging current when the applied voltage is turned off. Electronic current comes from electric field–enhanced Schottky emission at the electrode–dielectric interface and dominates the current flow at high electric field. At low electric field, polarization current prevails. The voltage and time dependence of the polarization current can be modeled by a distribution of Debye-type relaxations.

3. Hwang et al. [339, 356–359, 370] reported on the electrical properties of BST thin-film capacitors in a series of articles. The Pt/BST/Pt capacitor shows Schottky emission behavior with interface potential barrier heights of about 1.5–1.6 eV. The barrier height is largely determined by the interface electron trap states of the BST. The IrO_2/BST interface shows an ohmic contact nature due to the elimination of the surface trap states as the result of the formation of strong chemical

Table XL.   Physical Properties of BST Thin Films

Reference	Concentration $Ba_x Sr_{1-x} TiO_3$ (%)	Capacitor structure	Film thickness (nm)	Dielectric constant	Loss tangent (tan δ)	Capacitance per unit area (fF/μm^2)
Baumert et al. [332]	0.5	Pt/BST/Pt	60–120	239–538		45–96
Chu and Lin [333]	0.5	Pt/BST/Pt		170–250		
		Pt/BST/LaNiO₃				
Chu and Lin [334]		Pt/BST/BaRuO₃	100	300		
Dietz et al. [335]	0.7	Pt/BST/Pt	200	300		
Gao et al. [336]		Pt/BST/Ir	50–60	200–240		
Horikawa et al. [337]	0.65	Pt/BST/Pt	500	190–700		
Hou et al. [338]	0.5	Au/Ti/BST/SrRuO₃/ ZrO₂	100–200	360	0.01	
Hwang et al. [339]	0.5	Pt/BST/Pt	15–50	225–325	<0.01	
Izuha et al. [340]	0.5	Pt/BST/SRO	20–60	146–206		
Jia et al. [341]	0.5	Ag/BST/SRO	250	∼500	<0.01	
Kawahara et al. [342]		Ru, RuO₂, Pt electrodes	25	190	0.018	
Koyama et al. [343]	0.5	TiN/BST/Pt/Ta	70-200	300		40
Kuroiwa et al. [344]	0.5	Pt/BST/Pt	50	400		
Kuroiwa et al. [344]	0.75	Pt/BST/Pt	50	320		
Lee et al. [345]	0.5	Pt/BST/Pt	100	725		
Nagaraj et al. [346]		LSCO/BST/LSCO	75	350		35.1
Peng and Krupanidhi [347]		Au/BST/Pt	400	563		
Roy and Krupanidhi [348]	0.5	BST/Pt		330	0.02	
Shimada et al. [349]	0.7	Pt/BST/Pt	185	∼250		
Shin et al. [350]		Pt/BST/Pt	30–150	∼300–500		
		Pt/BST/IrO₂				
Tsai et al. [351]	0.5	Pt/BST/Pt	80	∼200–550		
		Pt/BST/Ru				
		Pt/BST/Ir				
Wu and Wu [352]	0.4	Pt/BST/LaNiO₃	80	160–320		
Yamamichi et al. [353]	0.4	Al/TiN/BST/RuO₂/ Ru/TiN/TiSi$_x$	34	200		

bonds between the IrO₂ and BST, which results in a Poole–Frenkel emission conduction mechanism. The Pt/BST/IrO₂ capacitor shows Schottky emission behavior and a positive temperature coefficient of resistivity (PTCR) effect, depending on the bias polarity. This may be due to the space-charge region within the BST at the Pt/BST contact causing this effect.

The theories that explain the electrical conduction properties are based on the metal–semiconductor contact theory, including band bending, surface trap state contribution, and carrier injection. The direction of band bending is not only dependent on the work function difference between the two materials but also strongly depends on the interface electron trap states and their distribution within the bandgap. The BST thin films have a large number of conduction electrons ($\sim10^{20}$ cm^{-3}) as a result of the large number of oxygen vacancies. The measured values of the barrier height (1.5–1.6 eV) are much larger than the ideal value, suggesting that the barrier height is determined not by the work functions of PT and BST, but by the surface states.

The IrO₂/BST contact is ohmic in nature. The IrO₂/BST/IrO₂ capacitor shows ohmic conduction behavior in the low-voltage region and Poole–Frenkel emission behavior in the high-voltage region with an electron trapping energy of 1.23 eV. The much stronger chemical bonds between IrO₂ and BST eliminate the surface electron trap states and, thus, the bulk properties of the two materials are major determinants of the contact properties. The small work function for IrO₂ appears to produce ohmic contact properties.

A partial depletion model with a very thin (about 1 nm) layer devoid of a space charge at the interface with the Pt electrode has been proposed to explain the $V^{1/2}$-dependent variation of $\ln(J_0)$, as well as the decreasing dielectric constant with decreasing film thickness. J_0 is the reverse saturation current of the Schottky emission equation. This denuded layer is assumed to originate from the partial cancellation of the space charge in the BST film due to the very large density of negative interface charge, which is generated by the trapping of electrons at the

Table XLI.   Electrical Conduction and Leakage Current of BST Thin Films

Reference	Device structure	BST film thickness (nm)	Leakage current level (A/cm^2)	Current mechanism	Barrier height (eV)
Dietz et al. [335]	Pt/BST/Pt	24–162	10^{-8} at 100kV/cm	Thermionic emission	1.17–1.02, decreasing with applied field
Hsu et al. [355]	Pt/BST/Pt	73		Possibly Schottky	
Huang et al. [314]	Au/BST/Pt	330	$\sim 3 \times 10^{-8}$ at 100kV/cm		
Hwang et al. [339]	Pt/BST/Pt	15–50	$\sim 4 \times 10^{-8}$ at 750 kV/cm	n-type conductivity, Schottky barrier	
Hwang et al. [356]	Pt/BST/IrO$_2$	40	10^{-6} at 500 kV/cm	Schottky at Pt/BST, ohmic at BST/IrO2	1.63 at Pt/BST
Hwang et al. [357]	IrO$_2$/BST/IrO$_2$ Pt/BST/Pt Pt/BST/IrO$_2$	20–150	$\sim 10^{-8}$ at 350 kV/cm	Schottky for Pt/BST, ohmic and Poole–Frenkel for IrO$_2$/BST	1.5–1.6 for Pt/BST
Hwang et al. [358]	Pt/BST/Pt IrO$_2$/BST/IrO$_2$	20–150		Schottky above 120°C, tunnel below 120°C	1.6–1.7 at Pt/BST
Hwang et al. [359]	Pt/BST/Pt	40		Schottky emission	0.81–0.83 for N$_2$-annealed BST
Joo et al. [360]	Pt/BST/Pt	50	10^{-7} at 3 V		
Joshi et al. [361]	Pt/BST/Pt	150	10^{-8} at 3V	Modified Schottky	1.3–1.5
Kim and Lee [362]	Pt/BST/RuO$_x$	200	$\sim 4 \times 10^{-9}$ at 200 kV/cm		
Koyama et al. [343]	TiN/BST/Pt/Ta	70–200	10^{-7} at 300 kV/cm		
Lin and Lee [363]	Au/BST/Pt	130	$< 10^{-8}$ at 100 kV/cm and 7×10^{-6} at 1 MV/cm	300–373 K, Schottky, higher than 403 K, Poole–Frenkel	0.88
Lee et al. [366]	Pt/BST/Pt	100	10^{-7} at 100kV/cm		
Maruno et al. [364]	Pt/BST/Pt	50	2×10^{-8} at 400 kV/cm	Schottky emission thermionic field emission	
Nagaraj et al. [346]	LSCO/BST/LSCO	75	10^{-7} at 133 kV/cm	Poole–Frenkel	
Peng and Krupanidhi [347]	Au/BST/Pt	400		SCLC	
Shimada et al. [349]	Pt/BST/Pt	185	10^{-8} at 162 kV/cm	Schottky at high field ($>$500 kV/cm) and Poole–Frenkel at higher temperature (423 K)	0.8
Shin et al. [350]	Pt/BST/Pt Pt/BST/IrO$_2$	30–150	10^{-8} at 300 kV/cm 10^{-5} at 900 kV/cm	Combined Schottky–tunneling model	1.32
Tsai et al. [351]	Pt/BST/Pt Pt/BST/Ru Pt/BST/Ir	80	2.2×10^{-8} at 200 kV/cm 4.4×10^{-7} at 200 kV/cm 1.9×10^{-8} at 200 kV/cm		
Wang and Tseng [365]	Pt/BST/Pt	80–90		Ohmic at low voltage, Schottky or Poole–Frenkel >6 V	0.46 for Schottky, 0.51 for Poole–Frenkel
Yamamichi et al. [353]	Al/TiN/BST/ RuO$_2$/Ru/TiN/ TiSi$_x$	34	10^{-6} at 1 V (about 294 kV/cm)		
Yamamichi et al. [354]	Au/Ti/BST/Pt	50–160	10^{-7} at 500 kV/cm (160 nm)		
Zafar et al. [367]	Pt/BST/Pt	40, 100		Ionic space-charge-limited current	

interface trap sites. This denuded layer has the smallest, constant dielectric constant owing to the constant and largest static field in the layer. There is a much stronger chemical interaction between IrO_2 and BST. Therefore, the denuded layer of space charge cannot be formed at the IrO_2/BST interface due to the lack of trapping interface.

4. Lin and Lee [363] studied the temperature dependence of the current–voltage characteristics of Au/BST/Pt capacitors. The conduction mechanism is dominated by Schottky emission from 300 to 373 K and dominated by Poole–Frenkel emission above 403 K. Because the conduction current mechanism changes from Schottky to Poole–Frenkel emission in this temperature range, the usual method of plotting $\log(J/T^2)$ vs. $1/T$ based on Schottky emission to extrapolate the barrier height is not really correct in this temperature range. Lin and Lee [363] calculated the Schottky barrier height from the interception of the fitting line in the Schottky plot with $\log(J/T^2)$ plotted vs. $E^{1/2}$. The Au/BST and Pt/BST Schottky barrier heights are 0.88 and 0.72 eV, respectively. The electron trap energy level in BST is extracted from Poole–Frenkel emission to be 0.83 eV.

5. Maruno et al. [364] modeled the leakage characteristics of Pt/BST/Pt thin-film capacitors. The net conductivity in BST thin films is assumed to be n type according to work done in [372, 373]. It is observed that the leakage current increases asymmetrically for negative and positive bias voltage with increasing annealing temperature. A model of the leakage current is proposed based on the assumption that oxygen vacancies are generated by annealing at the interfaces of the dielectric film adjacent to the Pt electrodes. The oxygen vacancies act as donors. A model of leakage characteristics is proposed on the basis of the Schottky emission mechanism. This model predicts that at the Pt/BST interfaces there exists a high density of oxygen vacancies, about 10^{20} cm^{-3}. The oxygen-deficient layers induce a large built-in electric field within the BST. This electric field results in enhanced leakage current through reduction of the Schottky barrier height.

6. Nagaraj et al. [346] studied the leakage current in LSCO/BST/LSCO capacitors. The conduction mechanism is not interface limited but predominantly bulk limited. The extracted dielectric constant from Poole–Frenkel emission agrees better with the reported value for the optical dielectric constant. This suggests that Poole–Frenkel emission is the dominant current mechanism in these structures.

7. Peng and Krupanidhi [347] found that the leakage current of BST films doped with high donor concentration showed a bulk space-charge-limited conduction with discrete shallow traps embedded in a trap-distributed background at high fields. Donor doping could significantly improve the time-dependent dielectric breakdown behavior of BST thin films, most likely due to the lower oxygen vacancy concentration resulting from donor doping.

8. Shimada et al. [349, 374] studied the temperature-dependent [349] and time-dependent [374] leakage current characteristics of Pt/BST/Pt capacitors. The leakage current at low field (<50 kV/cm) indicates ohmic conduction. At high field (>500 kV/cm), the Schottky mechanism plays a domi-

nant role, whereas Poole–Frenkel emission begins to contribute as the temperature is elevated. The time-dependent increase in leakage current is ascribed to a change in the conduction mechanism from the interface-controlled Schottky type to the bulk-related space-charge-limited type due to the accumulation of oxygen vacancies near the cathode as a result of interface barrier lowering and the migration of distributed oxygen vacancies across the film.

9. Shin et al. [350] analyzed the electrical conduction of BST thin films based on a fully depleted film and a combined Schottky–tunneling conduction model. The fact that the capacitance values in $C-V$ increase with increasing film thickness indicates that the films are fully depleted. The leakage current density under an electric field smaller than 120 kV/cm is controlled by a Schottky conduction mechanism, and over that field strength, themionic field emission is dominant.

10. Wang and Tseng [365] discussed the electrical conduction of Pt/BST/Pt capacitors. The capacitors show ohmic behavior at low voltage (<1 V) and Schottky emission or Poole–Frenkel emission at high voltage (>6 V). The barrier height and trap level are estimated to be 0.46 and 0.51 eV, respectively.

11. Joshi et al. [361] found that above a threshold voltage of 3–5 V, the leakage current density exhibits a modified Schottky emission, which states that $\log J$ is proportional to $V^{1/4}$. The barrier height is approximately 1.3–1.5 eV.

12. Scott et al. [375] showed that the dc leakage currents of BST films at voltage below breakdown are dominated by tunneling injection and image forces at low voltages (<3 V), Schottky behavior at intermediate voltages (between 3 and 6 V or between 150 and 300 kV/cm in electric field), and Fowler–Nordheim tunneling at high voltages.

It can be seen from the preceeding discussion and Table XLI that most research results show that the conduction mechanism of metal/BST systems is due to Schottky emission at room temperature. Poole–Frenkel emission is the most likely conduction mechanism at higher temperatures. As for the barrier height values, there is great variance in the literature as can be seen from Table XLI. There is also some variation in the work function and electron affinity values. It has come to the authors' attention that Hwang et al. [356, 357] used an electron affinity of 3.3 eV for BST and concluded that the calculated Schottky barrier height of 2.1 eV from $q(\phi_m - \chi)$ is much smaller than their measured value of 1.5–1.6 eV. The value of 3.3 eV is quoted from Dietz et al. [335]. However, Dietz et al. only mentioned that the electron affinities for $SrTiO_3$ and $BaTiO_3$ are 4.1 and 2.5 eV, respectively, and the electron affinity for BST solid solution is not known. The observed results suggest that it is closer to the $SrTiO_3$ value than to that of $BaTiO_3$. An average value of 3.3 eV was apparently used in [357]. If the electron affinity of BST is 4.1 eV as used in [18] and assuming a work function of 5.3 eV for Pt, the calculated barrier height would be 1.2 eV, which actually is smaller than the barrier height of 1.5–1.6 eV found in [357]. There is also the question of a possible change in conduction mechanism in this temperature

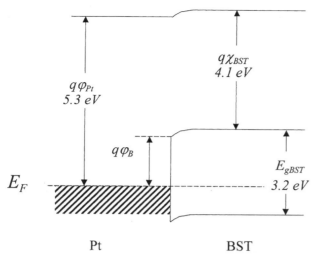

Fig. 28. Energy band diagram of the Pt/BST system.

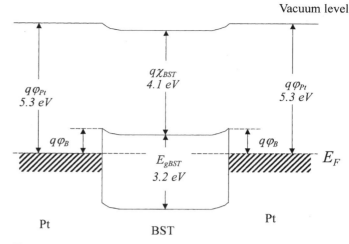

Fig. 30. Energy band diagram of the Pt/BST/Pt system (BST energy band bending upward).

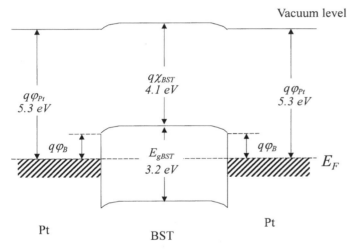

Fig. 29. Energy band diagram of the Pt/BST/Pt system (BST energy band bending downward).

range from Schottky emission to Poole–Frenkel emission reported by Lin and Lee [363]. Therefore, Hwang's conclusion in [357] that the barrier height of Pt/BST is determined not by the work functions but by the surface states may not be valid.

For the Pt/BST/Pt system, the work function of Pt is about 5.3 eV. The energy bandgap and the electron affinity of BST are about 3.2 and 4.1 eV, respectively [18]. If the Fermi level of BST is close to the middle of the energy bandgap, the work functions of Pt and BST are again very close. This situation may resemble Fig. 20 and the energy band diagram will be similar to Fig. 28. A possible Pt/BST/Pt energy band diagram is shown in Fig. 29. However, if BST is n type due to a high concentration of oxygen vacancies [364], the energy band diagram will resemble Fig. 30.

7.3.2. Energy Band Diagram

An energy band diagram is very helpful in understanding the electrical properties of a metal/BST or metal/BST/metal system. The reported parameters associated with such an energy band diagram are listed in Table XLII. BST is reported to have an energy bandgap of 3.2 eV and an electron affinity of 4.1 eV [18].

For ionic semiconductors, the barrier height generally depends strongly on the metal [266]. For oxide perovskite materials, the sensitivity of band bending to the metal electrode work function is very high. Thus, a classical band-bending model should work for PZT, BST, or SBT on metal electrodes such as Pt [19] due to the ionic character of the ferroelectrics and the high value of electronegativity difference. If $q\phi_m$ represents the work function of the metal electrode and $q\phi_s$ represents the work function of the ferroelectric, the nature of the metal–ferroelectric contact will depend on the relative size of the two work functions. It has been reported that BST, PZT, and SBT are all typically p-type semiconductors in bulk, if undoped. This is because there is a greater abundance of impurities such as Al, Na, and Mg in the starting growth materials than there are high-valence impurities, and because impurities are almost always substitutional in perovskites [18, 19]. However, the net conductivity in BST thin films is known to be n type [364, 372, 373]. A Schottky barrier is normally observed for Pt/BST contact. The energy band diagram is drawn with the BST energy band bending upward at the contact to form an energy barrier for the electrons. The calculated barrier height would be 1.2 eV from $q\phi_{Bn} = q(\phi_m - \chi) = 5.3\ \text{eV} - 4.1\ \text{eV}$. The reported values are listed in Tables XLI and XLII.

Some of the literature results on metal/BST energy band diagrams are briefly discussed next.

1. Scott [19] showed energy band diagrams of the Pt/PZT and Pt/BST systems with the ferroelectric energy band bending downward from inside the ferroelectric film toward the metal contact. The Pt/PZT and Pt/SBT energy band diagrams shown

Table XLII.   Energy Band Diagram Parameters of the Metal/BST System

Reference	BST energy bandgap (eV)	BST electron affinity (eV)	Metal work function (eV)	Electron barrier height (eV)	Comments
Hwang et al. [339]		1.7	5.5 (Pt)		
Hwang et al. [358]					Bending upward
Hwang et al. [357]	~3.5	3.3	5.5 (Pt)	2.1 (ideal) 1.5–1.6 (exp.)	Bending upward (blocking)
Hwang et al. [357]	~3.5	3.3	2–3 (IrO$_2$)		Bending downward (ohmic)
Joo et al. [376]					Bending downward
Lin and Lee [363]	3.2	4.1	5.3 (Pt)	0.88	
Lin and Lee [363]	3.2	4.1	4.7 (Au)	0.72	
Scott [19]	3.2 ± 0.1	3.9 ± 0.2	5.3 (Pt)	1.4 ± 0.2	Bending downward
Scott [18]	3.2 ± 0.1	4.1 ± 0.2	5.3 (Pt)		
Tsai et al. [351]	3.5	4.1	5.6 (Pt) 4.8 (Ru)		Bending upward

in [18] also have the energy band of the ferroelectric bending downward toward the metal–ferroelectric interface. The energy band diagram for Pt/BST shows the energy band of BST bending upward toward the Pt electrodes. Modification of the BST/Pt depletion width via acceptor or donor doping can be used to reduce BST leakage current [18].

2. Hwang et al. [357] showed the energy band diagrams of the Pt/BST/Pt and IrO$_2$/BST/IrO$_2$ systems. The Pt/BST/Pt system has the BST energy band bending upward and the IrO$_2$/BST/IrO$_2$ system bending downward toward the electrode–ferroelectric interface. The BST film close to the interface is in depletion for the Pt/BST/Pt system, whereas the BST film is in accumulation for the IrO$_2$/BST/IrO$_2$ case.

3. Tsai et al. [351] assumed an n-type BST film and the BST energy band bending upward at both the Pt/BST and the Ru/BST interfaces. The BST film tends to show an n-type conductivity because the oxygen vacancies act as donor dopants. The higher concentration of oxygen vacancies accumulated at the interface tends to show an n$^+$-type conductivity.

4. Maruno et al. [364] assumed that the net conductivity in BST thin film is n type according to other works [372, 373]. Because the work function of Pt is larger than that of BST, the Schottky barrier height is expected to be larger than 1 eV. The energy barrier formation therefore results in downward band bending within the BST film near the contacts. At equilibrium without an applied bias, it is assumed that the depletion region width is larger than the film thickness, which means that the film is wholly depleted. The annealing process in vacuum might induce heavily charged depletion layer due to oxygen vacancies at the top of the Pt/BST interface. The density of the oxygen vacancies is on the order of 10^{20} cm^{-3}. The oxygen-deficient layers induce a large built-in electric field within BST. This electric field results in enhanced leakage currents through reduction of the Schottky barrier height due to image force increase.

5. Shimada et al. [374] showed an energy band diagram of a metal/BST system with the energy band of BST bending upward at the interface.

7.3.3. Size Effect

Paek et al. [377] found that the electrical properties degraded markedly as the thickness of the films decreased. Electrical properties, such as leakage current, and the dielectric constant are closely related to the surface morphology, in particular, the grain size of the films. The existence of an interfacial layer between the BST film and the Pt bottom electrode was confirmed by HRTEM.

7.4. Very Large Scale Integration Circuit Applications of BST Thin Films

The literature results for BST VLSI applications are summarized in Table XLIII. Significant effort by industry is under way to incorporate the BST capacitor technology in future DRAMs. 256-Mb and 1-Gb test chips with BST capacitors have already been fabricated. Recent results show the equivalent SiO$_2$ thickness is on the order of 0.37–0.6 nm.

Table XLIII.   VLSI Applications of BST Thin Films

Reference	Company	Capacitor structure	Process technology	Capacitance per unit area	Equivalent SiO$_2$ thickness (nm)
Eimoil et al. [378]	Mitsubishi		0.25 μm, 256 Mb		0.47
Fujii et al. [379]	Matsushita, Symetrix	Planar		32 fF/μm^2	1.3
Hayashi et al. [380]	Matsushita		1 Gb	95 fF/μm^2	0.37
Kim et al. [381]	Samsung	Pt/TiN/BST/Pt/Ru	0.15 μm		
Lee et al. [382]	Samsung	Pt/BST/Pt		72 fF/cell, 256 Mb 25 fF/cell, 1 Gb	
Lee et al. [383]	Samsung	Pt/BST/Pt			0.51 (25-nm BST)
Nishioka et al. [384]	Mitsubishi	Ru/BST/Ru	0.14 μm		0.56 (25-nm BST)
Nitayama et al. [385]	Toshiba				
Ono et al. [386]	Mitsubishi	Ru/BST/Ru	0.14 μm		
Yuuki et al. [387]	Mitsubishi	Ru/BST/Ru	0.14 μm, 1 Gb	30 fF/cell	0.60

7.5. Reliability Issues of BST Thin Films

7.5.1. Resistance Degradation

1. Shimada et al. [374] found that the time-dependent increase in leakage current in Pt/BST/Pt capacitors can be ascribed to a change in the conduction mechanism from the interface-controlled Schottky type to the bulk-related space-charge-limited type. This is due to the accumulation of oxygen vacancies near the cathode as a result of interface barrier lowering and the migration of distributed oxygen vacancies across the film.

2. Zafar et al. [388] discussed resistance degradation of the Pt/BST/Pt structure. Under a constant applied voltage, the current density is observed to increase with time until it reaches a maximum value. The barrier height at the BST/Pt (cathode) interface is observed to decrease after prolonged electrical stressing. The resistance degradation effect is reversible, particularly at elevated temperatures. The measurement shows that the effective barrier height decreases by 0.17 eV after prolonged stressing. A quantitative model for resistance degradation is proposed. It is speculated that positively charged oxygen vacancies drift under the applied voltage and accumulate at the cathode, thereby causing the barrier height to decrease with time. The time dependence of the barrier height decrease is given by a stretched exponential equation.

3. Numata et al. [389] discussed resistance degradation of BST and SrTiO$_3$. The structures are Au/BST/Pt and Au/ST/Pt. The relationship between the temperature T and the breakdown time t_{BD} (called characteristic time in the paper) is well fitted by the Arrhenius equation, $t_{BD} = \alpha \exp(E_{act}/kT)$, where α is the overall coefficient, E_{act} the activation energy, k the Boltzmann constant, and T the absolute temperature. The breakdown time is defined as the time when the current density is increased by 1 order of magnitude. A good fit is obtained with an E_{act} of approximately 0.9 eV for both the SrTiO$_3$ and the BST thin films. No degradation is found for negative bias. A possible ex-

planation is the intrinsic inhomogeneous distribution of oxygen vacancies in the films.

4. Horikawa et al. [390] studied the degradation of Pt/BST/Pt capacitors under dc and bipolar stress conditions. The BST thickness was 30 nm. The leakage properties under a small bipolar stress are degraded more than those under dc stress of the same field strength. It is proposed that both degradations are caused by the deterioration of the Schottky barrier, and under bipolar stress conditions, the dielectric relaxation current in the dielectric film probably enhances the degradation. The breakdown time was found to be approximated by $t_B \approx \alpha \exp(\beta/E)$ for both dc and bipolar measurements. The value of the exponential β in the bipolar stress measurement was about half of the dc measurement.

5. Peng and Krupanidhi [311] discussed the models of time-dependent dielectric breakdown of dielectric films. These models include the following: (a) The grainboundary model suggests that a dc-field-induced deterioration of the grain boundaries leads to the local dielectric breakdown process or field-assisted emission of trapped charge carriers (Poole–Frenkel effect) because resistance at the grain boundary is higher than that at the grain interior. (b) The reduction model assumes that the positively charged oxygen vacancies with relatively high mobility electromigrate toward the cathode under dc electric field. The oxygen vacancies then pile up at the front of the cathode and are compensated by injected electrons from the cathode. At the anode, on the other hand, an electrode reaction leads to the generation of oxygen gas and electrons, leaving oxygen vacancies behind. In all, the ceramics suffer a chemical reduction that leads to the growth of an n-conducting cathodic region toward the anode and thus increases the electronic conductivity. (c) The grain boundary barrier height model modifies the reduction model and suggests that the preceding reactions can occur at grain boundaries. The net result is that the space-charge accumulation at the grain boundaries reduces the grain boundary barrier height and increases the leakage cur-

rent. (d) The demixing model suggests that the concentration depolarization of the oxygen vacancies between the anode and the cathode leads to the strong increase in local conductivity near the electrode (significant increase in the hole and electron concentration in the anodic and cathodic regions, respectively). The simplest way to improve TDDB is to reduce the concentration of oxygen vacancies in the materials by donor doping.

6. Huang et al. [314] studied the time-dependent dielectric breakdown of BST thin films. A method similar to that used for PZT [313] was adopted. Because different conduction mechanisms are involved at different field ranges, a single conduction mechanism cannot be assumed to extrapolate the breakdown time. The dependence of the current density J on the applied electric field E can be most conveniently expressed in a power law relationship, a method that takes into consideration the different functional dependence of current density on electric field in the high and the low field regions. The extracted time-dependent dielectric breakdown lifetime is on the order of 10^{12} s at 3.3 V (corresponding to a field of 100 kV/cm), which falls between the lifetimes obtained by the two previous methods.

7. Yamamichi et al. [366] studied the time-dependent dielectric breakdown and stress-induced leakage current (SILC) of BST thin films. Au/Ti is used as the top electrode and Pd is used as the bottom electrode. The thickness of BST is varied from 50 to 160 nm. Both time to breakdown (t_{BD}) vs. electric field (E) and t_{BD} vs. $1/E$ plots show straight lines, independent of the film thickness, and predict lifetimes longer than 10 years at $+1$ V for 50-nm BST films with a SiO_2 equivalent thickness of 0.70 nm. SILC is observed at $+1$ V after electrical stress of BST films. However, 10-year reliable operation is still predicted in spite of the charge loss by SILC. A lower (Ba + Sr)/Ti ratio is found to be beneficial for low leakage, low SILC, long t_{BD}, and, therefore, greater long-term reliability.

8. FERROELECTRIC FILMS: STRONTIUM BISMUTH TANTALATE (SBT)

Strontium bismuth tantalate (SBT) has a composition of $SrBi_2Ta_2O_9$. SBT is one member of a large class of multilayer interstitial compounds having the general chemical formula of $(Bi_2O_2)^{2+}(A_{x-1}B_xO_{3x+1})^{2-}$, where A and B represent ions of appropriate size and valence, $x = 2, 3, 4$, and 5. For $x = 2$, the crystal structure consists of two perovskite layers, infinite in two dimensions, alternating with a layer of Bi_2O_2, along the c axis [19, 391].

As can be seen from Table III, the Curie temperature of SBT is 335°C. SBT therefore shows ferroelectric properties at room temperature. The dielectric constant is reported to be about 140–330 [392–395] for thin films.

The application of SBT is advantageous for the following reasons [19]:

1. It exhibits negligible fatigue.

2. It has relatively low leakage current. The leakage current level is typically 100 times less than that of PZT of the same thickness and electrodes.

3. It can be prepared in extremely thin films without loss of bulk characteristics. This suggests that the damage layer near the electrode interface is very thin in SBT [19].

When compared with PZT, SBT has better endurance under repeated read/erase/rewrite operation (fatigue), better retention (shelf life), and the ability to maintain its characteristics when fabricated in very thin (<100 nm) films [250].

Auciello compared the advantages and disadvantages of PZT and SBT [396]:

The advantages of SBT are as follows:

1. SBT capacitors with negligible or no fatigue, long polarization retention, negligible imprint, and low leakage current can be fabricated using a simpler Pt electrode technology than the complex oxide electrode appproach required to produce PZT-based capacitors with comparable properties.

2. SBT layers appear to maintain good electrical properties even when they are very thin (<100 nm).

The disadvantages include the folowing:

1. SBT-based capacitors have lower polarization than highly oriented PZT-based capacitors. This may be an issue when developing high-density memories for which capacitors will shrink to less than 0.5-μm dimensions. However, SBT capacitors have similar polarization to the polycrystalline PZT-based ones, which are simpler to fabricate and less costly than highly oriented PZT capacitors, and therefore are the most probable choice for a first generation of FRAMs.

2. The synthesis of SBT layers currently requires higher processing temperatures than those needed to produce PZT films. However, work under way in various laboratories may contribute to lowering the processing temperature, although the same comment applies to the PZT-based technology [396].

8.1. Physical Properties of SBT Thin Films

SBT has an energy bandgap of 4.0 eV [18, 397] and an electron affinity of 3.4 eV [18]. The Fermi level is placed at 2.1 eV above the valence band.

For a stoichiometric SBT film, the Curie temperature is approximately 260°C, which is lower than that for the ceramic. SBT film with excess Bi composition has a T_C of about 300°C, and for Sr-deficient SBT higher than 340°C [398].

The process temperature of SBT is about 150°C higher than that of PZT. This will be a significant disadvantage to commercialize SBT memories [18]. Table XLIV summarizes the physical properties of SBT thin films.

Table XLIV.   Physical Properties of SBT Thin Films

Reference	Structure	Film thickness (nm)	Remanent polarization ($\mu C/cm^2$)	Coercive field (kV/cm)	Dielectric constant	Fatigue cycle
Chu et al. [399]	Pt/SBT/Pt/Si	200	5–6	55		10^{11}
Dat et al. [400]	Pt/SBT/Pt/ Ti/SiO$_2$/Si			25		10^{11}
Hayashi et al. [393]	Au/SBT/Pt/ SiO$_2$/Si	400	4.5	60	180	10^9
Hu et al. [401]	Au/SBT/Pt/ Ti/SiO$_2$/Si	300	$2P_r = 19.8$	$2E_c = 116$		
Ishikawa and Funakubo [402]	Pt/SBT/ SRO/STO	200	11.4	80	140	
Joshi et al. [394]	Pt/SBT/Pt	250	8.6	23	330	10^{10}
Kang et al. [403]	Au/SBT/Pt	1000	9.2	60		
Kwon et al. [404]	Pt/SBT/Pt	200	$2P_r = 13.3$	60		
Mihara et al. [392]	Pt/SBT/Pt	240	$2P_r = 20$	35	250	2×10^{11}
Park et al. [395]	Pt/SBT/Pt/ Ti/SiO$_2$/Si	200	$2P_r = 9.1$	$2E_c = 85$	210	10^{10}
Ravichandran et al. [405]	Au/SBT/Pt	400	5	$\sim$50		
Rastogi et al. [406]	Pt/SBT/Pt	160	6.6	51		
Seong et al. [407]	Pt/SBT/Pt	200	15	50		10^{11}
Yang et al. [408]	Pt/SBT/Pt	300	$2P_r = 18.5$	$2E_c = 150$		10^{10}
Zhang et al. [47]	Pt/SBT/Pt	640	8.8	29.3		

8.2. Electrical Characteristics of SBT Thin Films

8.2.1. Electrical Conduction

SBT thin films show relatively low leakage current with simple Pt electrode [396]. The leakage current and conduction mechanism in SBT films are listed in Table XLV. The leakage current is generally on the order of 10^{-6}–10^{-7} A/cm^2 at an applied field of 100–300 kV/cm.

Watanabe et al. reported on the dependence of leakage current of Pt/SBT/Pt capacitors on the applied voltage and film composition [413]. Their results showed that Schottky emission dominated the leakage current at voltages above the ohmic conduction regime, whereas space-charge-limited current (SCLC) appeared to dominate the leakage current in high-conductivity SBT thin films including bismuth-excess samples. Negative differential resistivity was observed in high-conductivity SBT thin films. The theory on space-charge-limited current [414, 415] was successfully applied to the temperature and voltage dependences of leakage currents in Bi-excess SBT. It was also observed that even 0.5% excess of Bi causes qualitative changes in leakage current and contact properties. Because excess Bi is often provided to compensate for bismuth evaporation during high-temperature processing, this is a serious problem [413].

Seong et al. [411] found out that in plasma-enhanced metal–organic chemical vapor deposition of SBT, a BTO (Bi$_4$Ti$_3$O$_{12}$ or Bi$_2$Ti$_4$O$_{11}$) phase is formed at the interface between SBT films and Pt/Ti bottom electrodes. The BTO phase decreases the leakage current density. The leakage current of the SBT films is controlled by Schottky emission. The Schottky barrier heights of SBT films deposited on Pt/Ti/SiO$_2$/Si and Pt/SiO$_2$/Si are about 1.2 and 0.8 eV, respectively. Because Schottky emission is controlled by the interface between the electrode and the SBT film, the BTO layer formed at the interface increases the Schottky barrier height and thus decreases the leakage current density of the SBT fims [411].

Mihara et al. [392] mentioned that the leakage current of SBT did not depend strongly on either thickness or temperature. The variation of leakage current with capacitor area shows that the current density decreases with area. The leakage current result shows that the area component of leakage current is much smaller than the edge component.

In the SBT family of layered-structure perovskites, there is some indication of double injection. This is indicated by the apparent negative differential conductivity and current–voltage hysteresis, and hole injection from the anode probably is initiated above a certain threshold voltage [250, 253].

Scott et al. [416, 417] reported that SBT films with columnar grains extending from cathode to anode are more likely to exhibit space-charge-limited currents along the grain boundaries, whereas films with grains in the form of globular polyhedra are more likely to manifest Schottky or Poole–Frenkel behavior.

For a Pt/SBT system, the energy bandgap and electron affinity of SBT are about 4.0 and 3.4 eV [18]. The work functions of Pt and SBT will be very close. The energy band diagram could be similar to Fig. 31.

Table XLV.   Electrical Conduction and Leakage Current of SBT Thin Films

Reference	Device structure	SBT film thickness (nm)	Leakage current level (A/cm^2)	Current mechanism	Schottky barrier height (eV)
Hu et al. [401]	Au/SBT/Pt				
Joshi et al. [394]	Pt/SBT/Pt/Si	250	10^{-8} at 150 kV/cm		
Kanehara et al. [409]	Pt/SBT/Pt	700	$\sim 10^{-7}$		
	IrO$_2$/SBT/IrO$_2$				
Kim et al. [410]	Pt/SBT/CeO$_2$/Si	250	10^{-7} at 500 kV/cm		
Park et al. [395]	Pt/SBT/Pt/Ti/ SiO$_2$/Si	200	7×10^{-7} at 150 kV/cm		
Seong et al. [411]	Pt/SBT/Pt	250	10^{-7} at 300 kV/cm	Schottky emission	1.2 for Pt/SBT/Pt/Ti 0.8 for Pt/SBT/Pt
Seong et al. [407]	Pt/SBT/Pt	200	5×10^{-8} at 300 kV/cm		
Taylor et al. [412]		140–200	2×10^{-6} at 3V		
Yang et al. [408]	Pt/SBT/Pt	300	8.6×10^{-8} at 100 kV/cm		
Watanabe et al. [413]	Pt/SBT/Pt	180	10^{-6} at $\sim$300 kV/cm	Ohmic at low field, Schottky or SCLC at high field	0.8 for Schottky emission

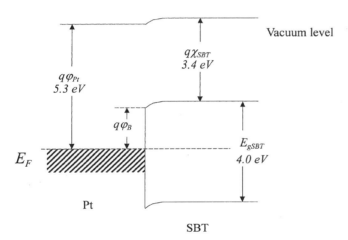

Fig. 31.   Energy band diagram of the Pt/SBT system.

8.2.2. Energy Band Diagram

The energy band diagram of a metal–ferroelectric system is important in understanding its electrical properties, especially in considering the conduction mechanism. The energy bandgap of SBT is 4.0 eV and the electron affinity is 3.4 eV. If we assume the Fermi level of SBT is at the middle of the energy bandgap, the work function of SBT will be about 5.4 eV. This value will be very close to that of the Pt metal electrode. A possible energy band diagram of the Pt/SBT system is shown in Fig. 31. This assumes that the Pt/SBT contact is determined by the work function and not by the interface states. The Pt/SBT barrier height of 1.8 eV is calculated by subtracting the electron affinity from the Pt work function.

The energy bond diagram parameters of the metal/SBT system is summarized in Table XLVI.

8.2.3. Dielectric and Ferroelectric Properties

The dielectric and ferroelectric properties of SBT thin films are listed in Table XLIV. The remanent polarization of SBT films is on the order of 10 μC/cm^2. The coercive field is on the order of 50–80 kV/cm. The dielectric constant is on the order of 200 in thin-film form. SBT capacitors show practically no polarization fatigue up to 10^{10}–10^{11} switching cycles.

Platinum electrodes are used in most of the fabricated capacitor structures. Complicated oxide electrodes are not necessary for SBT capacitors.

8.3. Very Large Scale Integration Circuit Applications of SBT Thin Films

Aoki et al. [280] reported a one-transistor (1T) and two-ferroelectric-capacitor (2C) scalable gain cell for ferroelectric random-access memory (FRAM) applications. Eshita et al. [419] reported a fully functional 64-Kb embedded SBT ferroelectric random-access memory chip using a low-temperature deposition technique. The crystallization temperature is lowered from 800 to 700°C.

The VLSI application of SBT thin films is summarized in Table XLVII.

8.4. Reliability Issues of SBT Thin Films

8.4.1. Fatigue

The major advantage of SBT is that it has a negligible fatigue effect up to 10^{10}–10^{11} switching cycles. Al-Shareef et al. [296] developed a theory that the domain walls of SBT have a depinning energy much less than that of PZT. This, in turn, arises from the lack of volatile Pb ions in the SrBi compounds, which

Table XLVI.   Energy Band Diagram Parameters of the Metal/SBT System

Reference	SBT energy bandgap (eV)	SBT electron affinity (eV)	Metal work function (eV)	Barrier height (eV)
Scott [18]	4.0 ± 0.1	3.4 ± 0.2	5.3 (Pt)	
Hartman and Lamb [397]	4.0			
Robertson et al. [418]	5.1			
Scott [19]	4.1 ± 0.1	3.4 ± 0.2		1.8 ± 0.2 (at 0 V)
				0.9 ± 0.2 (at 3 V)
Seong et al. [411]				0.8 for Pt/SBT/Pt
				1.2 for Pt/SBT/Pt/Ti
Watanabe et al. [413]				0.8 for Schottky emission

Table XLVII.   VLSI Applications of SBT Thin Films

Reference	Company	Device structure	Process technology and application
Aoki et al. [280]	Fujitsu	Pt/IrO$_2$/SBT/ Pt/Ti/SiO$_2$/Si	1T/2C FRAM
Eshita et al. [419]	Fujitsu	Pt/SBT/Pt	0.5-μm CMOS, 64-Kb FRAM

results in a smaller oxygen vacancy concentration. The weak domain wall pinning in SBT is postulated to result from two factors. One is the smaller magnitude of ferroelectric polarization, which results in weaker trapping of individual charges and/or lower trapped charge density at domain wall boundaries. The other is relatively low oxygen vacancy concentration in the perovskite sublattice.

However, optical illumination combined with a bias voltage can result in significant suppression of the switchable polarization. A similar effect was also observed in PZT. However, electric field cycling of the optically fatigued SBT capacitors results in nearly complete recovery of the suppressed polarization. In contrast, electric field cycling of optically fatigued PZT capacitors does not result in any polarization recovery. Their results suggest that optical fatigue in both SBT and PZT capacitors results from pinning of domain walls due to trapping of photogenerated carriers at domain boundaries, whereas recovery exhibited by SBT thin films indicates that the domain walls are more weakly pinned in SBT than in PZT thin films. Consequently, the fatigue-free behavior of SBT thin films during electric cycling can be viewed as a competition between domain wall pinning due to charge trapping and domain wall unpinning due to the cycling field [296].

For SBT thin films, the unpinning of the domain walls by the cycling field must therefore occur at least as rapidly as domain wall pinning by the trapped electronic charge carriers, resulting in no net loss in the switchable polarization. In PZT, little unpinning of the domain wall occurs during electric field cycling because the domain walls are strongly pinned by the electronic charge carriers.

The relatively weak domain wall pinning exhibited by SBT could be due to either (1) weak trapping of individual charges or (2) a lower concentration of trapped charge at the domain boundary. In the case of weak trapping, rejuvenation by field-assisted detrapping requires the applied field be sufficient to detrap carriers from shallow defect states. The trap depth for the charge is expected to depend on both the type of defects present and the depth of the electrostatic well at the domain boundary, which is determined by the magnitude of the ferroelectric polarization. Because SBT has lower polarization than PZT, the contribution of ferroelectric polarization to the trap depth should be smaller in SBT. In the other case, the density of trapped charge at the domain wall is determined by the availability of trapping sites (i.e., defects) and the value of the remanent polarization, P_r. The remanent polarization sets the upper limit on the trapped charge density, because the charge is trapped to compensate for the polarization discontinuity at the domain wall. The relatively small remanent polarizations of these materials may therefore lead to relatively small trapped charge densities and thus to weaker domain wall pinning than exhibited in PZT thin films.

The relatively weak domain wall pinning in SBT may result from other factors besides the magnitude of ferroelectric polarization. For instance, Pb loss in PZT leads to the formation of compensating point defects, such as oxygen vacancies. Oxygen vacancies in PZT tend to stabilize the trapped electronic charge carriers during electrical fatigue. In SBT, the absence of a volatile cation on the A site likely results in smaller oxygen vacancy concentration in the perovskite sublattice, which may also contribute to the weaker domain wall pinning in SBT [296].

Oxygen vacancies acting as trapping sites for electrons injected into the ferroelectric layer, during polarization switching, are being considered as major contributors to polarization fatigue in ferroelectric capacitors. Auciello et al. [420], in a study on SBT film growth and surface structural characterization, showed the following (1) Atomic oxygen originating from a multicomponent SBT target during the sputtering process is incorporated into the growing film more efficiently than molecular oxygen. (2) The SBT surface appears to terminate in an incomplete $(Bi_2O_2)^{2+}$ layer with a top surface of oxygen atoms. The smaller fatigue effect in Pt/SBT/Pt capacitors is thus due to an oxygen-rich layer at the Pt/SBT interface suppressing oxygen vacancies.

It was reported that the Bi_2O_2 layer should play an important role in preventing fatigue failures. The Bi_2O_2 layers have net electrical charges, and their positioning in the lattice is self-regulated to compensate for space charge near electrodes [421]. Kang et al. [403] studied the fatigue behavior of $SrBi_2Ta_2O_9$ and Bi_3TiTaO_9. The only structural differences between SBT and BTT are the double perovskite layers between the Bi_2O_2 layers. Because volatile elements such as Bi and Pb easily generate oxygen vacancies, the Bi ions in the perovskite layers of BTT would result in weak oxygen stability. Therefore, the poor fatigue characteristics of the BTT film should be attributed to the weak oxygen stability in the perovskite layer. The self-regulating adjustment of the Bi_2O_2 layers in preventing the fatigue problems is not enough [403].

Mihara et al. [392] reported that the SBT thin film on a platinum electrode has fatigue-free characteristics up to 2×10^{11} cycles without requiring any complicated electrode system such as conducting oxide. Moreover, SBT thin film has many advantages (e.g., high signal/noise ratio of 8 at 1.2 V, low-voltage operation at as low as 1 V, long data retention, little surface effect, superior imprint properties, and low leakage current). The authors considered that these advantages are due to (1) less space charge and (2) the inherent domain motion [392].

Park et al. [422], in studying SBT and $Bi_4Ti_3O_{12}$, also found that the difference in stability of the metal–oxygen octahedra should be related to different fatigue behaviors of the SBT and the BTO films.

Robertson et al. [418] showed some evidence that the Bi–O layers largely control the electronic response, whereas the ferroelectric response originates mainly from the perovskite Sr–Ta–O block. Bi- and Ta-centered traps are calculated to be shallow, which may account in part for its excellent fatigue properties.

Zhang et al. [423] found that the polarization of SBT thin films annealed in air shows more degradation than those annealed in oxygen, which indicates that the oxygen vacancy also plays an important role in the fatigue behavior of SBT thin films. They also discussed the effect of switching voltage and frequency on the fatigue properties [47, 423]. The fatigue endurance of SBT thin films at a higher frequency and a higher applied voltage could be better than that at a lower frequency and lower voltage.

8.4.2. Imprint

Imprint is defined as the tendency of one polarization state to become more stable than the opposite state. The tendency to-

ward imprint failure in a ferroelectric capacitor is manifested by a shift in the hysteresis curve along the voltage axis. Al-Shareef et al. [318] reported that the voltage shifts in the hysteresis response of SBT thin-film capacitors can be induced using both thermal and optical methods. The voltage shifts in the hysteresis curve of SBT are caused by trapping of electronic charge carriers near the film–electrode interfaces, similar to that reported for the PZT system. In addition, a correlation is established between the magnitude and sign of remanent polarization and the thermally induced voltage shift. Unlike the PZT system, the thermally induced voltage shifts in SBT are smaller than those optically induced. One possible explanation is that the contribution of defect-dipole complexes to the voltage shifts in SBT is negligible. It is suggested that the smaller contribution of defect-dipole complexes to the voltage shift in SBT may be related to the smaller oxygen vacancy concentration in the perovskite sublattice of SBT as compared to that of PZT [318].

8.4.3. Retention

Zhang et al. [327] showed that the polarization of SBT ferroelectric thin films synthesized by metal–organic decomposition decay increases with increasing write/read voltage within the first second. This could be attributed to the depolarization field, which increases with increasing retained polarization. They also found that the polarization loss is insignificant with different write/read voltages over a range of 1–30,000 s. The experimental result also indicates that there is a weak pinning of domain walls existing in SBT, which plays an important role for SBT thin films over a range of 1–30,000 s with a low write/read voltage.

8.4.4. Other Degradation Problems

Im et al. [424] studied the degradation of SBT characteristics due to hydrogen annealing. They found that the hydrogen-induced degradation in the electrical properties of SBT capacitors is mainly due to the loss of Bi in the near surface region of the SBT layer. They inferred that the compositional changes of SBT films during hydrogen annealing result in substantial alternation of the SBT film structure near the surface, which can be recovered by oxygen annealing.

Zafar et al. [425], in studying the effect of hydrogen annealing of SBT, showed that hydrogen induces both structural and compositional changes in the ferroelectric film. Hydrogen reacts with the bismuth oxide to form bismuth and the reduced bismuth diffuses out of the SBT film, causing the electrode to peel.

9. OTHER FERROELECTRIC AND PARAELECTRIC FILMS

In Section 1.3, we discussed the various requirements for ferroelectric materials to be useful in memory devices. Parker and Tasch [1] compared the physical properties of some of the ferroelectric materials. In this section, ferroelectric materials other than PZT, BST, and SBT are briefly discussed.

Table XLVIII. Physical Properties of BaTiO$_3$ Thin Films

Reference	Structure	Film thickness (nm)	Remanent polarization (μC/cm^2)	Coercive field (kV/cm)	Dielectric constant	Ferroelectricity	Leakage current or resistivity
Chang and Anderson [426]	Au/BT/Si	301					1.8 × 10^{-9} A/cm^2 at 130 kV/cm
Francombe [15]			5	15		(T_C = 115°C)	
Hayashi et al. [427]	Au/BT/Pt	250	4	90	230		
		580	8	30	1000		
Jia et al. [40]	Pd/BT/Pd/Si	300–500			330		
Kim et al. [428]	Ag/BT/Si	200			200		
Li and Lu [39]	Al/BT/Si (MIS)	61			14–16		
Shi et al. [41]	Pd/BT/Pd/Si				16 (amorphous)		1.3 × 10^{10} Ω-cm (amorphous)
					330 (poly)		3.7 × 10^4 Ω-cm (poly)
Shi et al. [429]	Pd/BT/Pd/Si						2.5 × 10^{-9} A/cm^2 (amorphous) 7.7 × 10^{-7} A/cm^2 (poly) at 4 V
Van Buskirk et al. [430]	Pt/BT/Pt				30 (amorphous) 300 (poly)	Absent Slight	10^8 Ω-cm (amorphous) 5 × 10^8 Ω-cm (poly)
Yeh et al. [42]	Au/BT/Si	680			200		1.57 × 10^{-6} A/cm^2 at 2.5 V

9.1. Barium Titanate (BaTiO$_3$)

The physical properties of barium titanate thin films are listed in Table XLVIII. Bondurant [22] discussed that a ferroelectric material needs a minimum of 5 μC/cm^2 signal charge to be useful in memory devices. For barium titanate, the storage charge densities of 0.3–5 μC/cm^2 are at the low end of the ULSI DRAM requirement range [1]. Amorphous BaTiO$_3$ may have lower charge densities than the paraelectric PLT, PLZT, or Pb(Mg, Nb)O$_3$ (PMN). However, easier processing and lower leakage currents may still make it a very good choice [1].

Barium titanate films deposited by rf magnetron sputtering were studied by Jia et al. [40] and Shi et al. [41, 429]. Films deposited at low substrate temperature were amorphous, whereas those deposited at high substrate temperature were polycrystalline. Amorphous films had a low conductivity, a high breakdown voltage, and a small dielectric constant of around 16. Polycrystalline films yielded a high dielectric constant up to 330, but also led to large conductivity. Plasma-enhanced metal–organic chemical vapor deposition of BT was studied by Van Buskirk et al. [430]. Barium titanate films were also deposited using laser ablation by Yeh et al. [42]. The dielectric constant and dielectric strength of the films were measured to be 200 and 1 MV/cm, respectively. The storage charge density and the leakage current density were measured to be 0.66 μC/cm^2 and 1.57 μA/cm^2, all measured at 2.5 V. The estimated write time is 0.1 ps for a 1-μm^2 storage area. Hayashi et al. [427] prepared barium titanate films by the sol–gel method. With increasing film thickness, the dielectric constant and remanent polarization increase and the coercive field decreases. Barium titanate thin, film with a thickness of 580 nm exhibits a dielectric constant of 1000, remanent polarization of 8 μC/cm^2 and coercive field of 30 kV/cm. Barium titanate thin films deposited on silicon were studied by Chang and Anderson [426] and Kim et al. [428] for possible application as high-dielectric-constant gate dielectrics in ferroelectric memory devices.

9.2. Strontium Titanate (SrTiO$_3$)

SrTiO$_3$ is a paraelectric material at room temperature. The physical properties of strontium titanate thin films are listed in Table XLIX. The thin films have been prepared by rf magnetron sputtering, sol–gel, ECR sputter, laser ablation, and metal–organic solution methods. The electrical properties are measured using metal–insulator–metal and metal–insulator–semiconductor structures. The measured dielectric constants are on the order of 200.

Dietz et al. [43] fabricated metal/SrTiO$_3$/metal structures. The conduction mechanism is based on totally depleted back-to-back Schottky barriers at the contact interfaces. The main

Table XLIX.   Physical Properties of Strontium Titanate ($SrTiO_3$) Thin Films

Reference	Deposition method	Structure	Dielectric constant	Conduction	Charge density
Abe and Komatsu [431]	rf magnetron sputtering	Ni/ST/Pt/MgO	100–300, depending on thickness	Grain boundary suppresses leakage current	
Cho et al. [432]	rf magnetron sputtering	ST on silicon			
Dietz et al. [43]	sol–gel	Various metal electrodes/ /ST/Pt/Si		Thermionic emission of electrons	
Fukuda et al. [433]	ECR sputter	Au/ST/Pt/Ti/SiO$_2$/Si		Ohmic at low field (<1 MV/cm), thermionic emission at high field	
Hirano et al. [434]	ArF excimer laser deposition	Ag/ST/ST, MgO, or YBCO	140 at room temp.		
Joshi and Krupanidhi [435]	metall–organic solution	Au/ST/Pt/Si	~200	Ohmic at low field, SCL at high field	Charge density 36.7 fC/cm^2 at 200 kV/cm
Lee and Koinuma [436]	Pulsed laser deposition	Al/ST/TiN/Si	270		
Lee et al. [437]	rf sputtering	Ag/ST/Si	300	4×10^{-5} A/cm^2 at 5 V	Charge storage density 2.5 μC/μm^2 at 100 kV/cm
Nakahira et al. [438]	MOCVD	Ru/ST/Ru	100	10 fA/cell at 0.8 V	
Nam et al. [439]	rf magnetron sputtering	Ag/ST/Si	75		
Rao and Krupanidhi [440]	KrF excimer laser deposition	Au/ST/Pt/Ti/SiO$_2$/Si	190–225	Ohmic at low field, SCL at high field	Charge density 40 fC/cm^2

conduction mechanism is thermionic emission of electrons from the cathode into the strontium titanate film. The current–voltage characteristics are influenced by the Schottky effect at this barrier. It is also thought possible that the leakage at very high fields, where the Schottky barrier can easily be surmounted, becomes space charge limited.

Fukuda et al. [433] found that the leakage current characteristics show an ohmic-like conduction for electric field lower than 1 MV/cm and, at higher fields, are limited by Schottky emission. The ohmic-like leakage current characteristics show strong dependence on the measurement conditions, such as the values of voltage step and delay time. This effect is attributed to the absorption current due to the dielectric relaxation phenomena.

Joshi et al. [435] found that the conduction process of the metal–strontium titanate–metal capacitors is bulk limited. The I–V characteristics are ohmic at low fields and space charge limited at high fields. The electronic transport in the films of different thickness is in agreement with the space-charge-limited

conduction model in the presence of a single shallow trapping level. For a 500-nm-thick strontium titanate film, a unit area capacitance of 3.5 fF/μm^2 and a storage charge density of 36.7 fC/μm^2 are obtained at an applied electric field of 200 kV/cm.

Lee and Koinuma [436] studied ST films grown on silicon by pulsed laser deposition using TiN as a buffer layer. The dielectric constant depends on the crystallinity and the surface roughness of the films. The highest dielectric constant is 270.

Lee et al. [437] deposited ST films on silicon by rf sputtering. The ST films exhibit a dielectric constant of 300 and leakage current of 4×10^{-5} A/cm^2 at a bias of 5 V.

Nakahira et al. [438] developed a low-temperature ST capacitor process for embedded DRAM. ST films deposited at 475°C were crystallized without additonal annealing. It is known that ST shows lower dielectric constant than BST. However, the crystallization temperature of ST is lower than that of BST. ST has an advantage for composition control, which affects dielectric constant and leakage current properties due to less metal constituents. A Ru/ST/Ru capacitor structure was used. The dielectric constant is 100. The leakage current of the concave structure capacitor is less than 1 fA/cell at 0.8 V.

9.3. Lead Titanate (PbTiO$_3$)

Lead titanate (PbTiO$_3$) has a Curie point at 490°C and is ferroelectric at room temperature [13]. The structure of lead titanate, at room temperature, involves a tetragonal distortion from the perovskite lattice, and the compound is isomorphous with tetragonal BaTiO$_3$ [13]. The physical properties of lead titanate thin films are listed in Table L.

PbTiO$_3$ thin films have a tetragonal structure, with a Curie temperature of 515–535°C, a spontaneous polarization of 48 μC/cm^2, a remanent polarization of 35 μC/cm^2, and a coercive field of 160 kV/cm [15].

9.4. Lead Lanthanum Titanate [(Pb, La)TiO$_3$]

PLT has a tetragonal structure (when the La concentration $x = 0$–0.15), a Curie temperature of 525°C–18°C × (La %), a spontaneous polarization of 22–45 μC/cm^2, a remanent polarization of 20–38 μC/cm^2, and a coercive field of 50–34 kV/cm in this concentration range [15]. The electrical properties of PLT thin films are shown in Table LI.

Lee et al. [444] deposited paraelectric (Pb$_{0.72}$La$_{0.28}$)TiO$_3$ thin films on platinum-coated silicon substrates by the sol–gel method. Two distinct groups of metals, M_T (Ni, Cr, and Ti, all transition metals) and M_N (Pt, Au, and Ag, all noble metals), form ohmic and Schottky contact, respectively. When a semi-insulating PLT thin film of thickness d with Schottky barriers is subjected to a dc voltage stress V, the electric field in the film is no longer equal to V/d because 95% of the voltage drop occurs across the reverse-biased Schottky

Table L. Physical Properties of Lead Titanate (PbTiO$_3$) Thin Films

Reference	Deposition method	Structure	Remanent polarization (μC/cm^2)	Coercive field (kV/cm)
Ching-Prado et al. [441]	sol–gel			
Francombe [15]			35	160
Imai et al. [442]	ArF and YAG laser ablation	On Si or on Pt/Si		
Kushida et al. [443]	rf magnetron sputtering	On SrTiO$_3$		

Table LI. Physical Properties of Lead Lanthanum Titanate (PLT) Thin Films

Reference	Deposition method	Structure	Dielectric properties	Remanent polarization (μC/cm^2)	Coercive field (kV/cm)	Conduction	Charge storage density
Dey and Lee [38]	sol–gel	Au/PLT/Pt/Ti/ SiO$_2$/Si	Paraelectric, dielectric constant 850–1400			0.5 μA/cm^2 at 200 kV/cm	15.8 μC/cm^2
Francombe [15]				20–38 (La = 0–15%)	50–34 (La = 0–15%)		
Lee et al. [444]	sol–gel	Transition metals and noble metals/PLT/Pt/ Ti/SiO$_2$/Si				Schottky emission, thermionic emission, and Fowler–Nordheim	
Schwartz et al. [44]	Spin casting	PLT on Pt bottom electrode	80–690 decreasing with increasing La content	1.8–2.9 decreasing with increasing La content	19.8 (PLT— 20/100)		
Wang et al. [445]	sol–gel	PLT on Pt	210 for 15% La	16.5 for 14% La	60 for 14% La		

barrier. Thus, the conventional Schottky emission and Fowler–Nordheim tunneling equations are modified to account for the voltage dependence of the interfacial permittivity. Schottky emission, thermionic tunneling, and Fowler–Nordheim tunneling mechanisms are predominant in the voltage ranges of $2 < V < 7$, $7 < V < 16$, and $V > 16$, respectively.

The electrical properties of PLT thin films for potential application in memory devices were also characterized by Dey and Lee [38] and Schwartz et al. [44]. The measured results are listed in Table LI.

9.5. Lead Lanthanum Zirconate Titanate (PLZT)

PLZT has a composition of $(Pb_{1-x}La_x)(Zr_yTi_{1-y})O_3$. The electrical properties of PLZT thin films are listed in Table LII.

Kumar et al. [448] reported the leakage current characteristics of Al/PLZT/Pt capacitors. The leakage current shows strong dependence on the processing temperature of the bottom electrode. A drop-in leakage current by 5 orders of magnitude has been observed in a capacitor with a platinum electrode deposited at room temperature as compared with platinum deposited at a higher substrate temperature of 500°C. The study suggests the possibility of controlling leakage current in PLZT capacitors through the use of platinum electrodes having good interface coherence with the PLZT film.

Tominaga et al. [452] studied the switching and fatigue properties of PLZT. The PLZT films were deposited by metal–organic chemical vapor deposition (MOCVD). The composition of the PLZT film was Zr/Ti = 45/55 and La = 0–11 at%.

The film thickness was 300–400 nm. The measurement results can be summarized as follows: (1) The relative dielectric constant increases as the La content increases up to about 5 at% and, beyond that, it becomes constant at 600. (2) The remanent polarization and coercive field decrease from 22 to 1 $\mu C/cm^2$ and from 90 to 20 kV/cm, respectively, as the La content increases. (3) The switching time and leakage current densities decrease with increasing La content. (4) The fatigue is relaxed as the La content increases. This suggests that fatigue of PLZT with reversal polarization does not occur up to about 10^{13} cycles at an access voltage of 3 V.

Tseng et al. [453] deposited PLZT films on LaNiO$_3$ layers by pulsed laser ablation. Ferroelectric and dielectric properties are optimized when the (100) LNO/Pt double layers are used as the bottom electrodes. The remanent polarization and coercive field are 14.9 $\mu C/cm^2$ and 3.5 kV/cm, respectively. The dielectric constant is 950.

Stolichnov et al. [454] studied the polarization switching and fatigue endurance of PLZT thin films with Pt and SrRuO$_3$ electrodes. The structures were Pt/PLZT/Pt and Pt/SRO/PLZT/Pt. It was found that the asymmetrical Pt/SRO/PLZT/Pt structure exhibit good fatigue performance in combination with low leakage similar to that on an identically processed Pt/PLZT/Pt capacitor. The experimental results suggest that the high polarization endurance of the SRO/PLZT/Pt capacitor is governed by only one interface, whereas the second one does not play a significant role as long as the driving voltage is above the critical field.

Table LII.   Physical Properties of PLZT Thin Films

Reference	Deposition method	Composition/ structure	Dielectric properties	Comments
Ibuki et al. [446]	rf sputtering	7/65/35	$\varepsilon_r = 5000$ at 1 kHz	ESCA study
Khamankar et al. [447]	sol–gel			Annealing study
Kumar et al. [448]	Electron beam evaporation	7/65/35 Al/PLZT/Pt/SiN/Si		Leakage current, STM study
Liu et al. [449]	Pulsed laser deposition	Au/PLZT/SrRuO$_3$/ Ru/Pt/Ti/Si	$\varepsilon_r = 1204$ $P_r = 25.6\ \mu C/cm^2$, $E_c = 47.1$ kV/cm	
Lu et al. [450]	rf sputtering	PLZT/Pt/Ti/SiO$_2$/Si	$P_r = 220$ mC/m^2 $E_c = 6.5$ MV/m	
Peterson et al. [451]	Laser ablation	7/0/100 and 28/0/100 (PLT)		
Tominaga et al. [452]	MOCVD	0–11/45/55 PLZT/Pt/SiO$_2$/Si	$\varepsilon_r = 400$–600 $P_r = 22$–1 $\mu C/cm^2$ $E_c = 90$–20 kV/cm with increasing La content of 0~11 at.%	No fatigue up to 1×10^{13} cycles at 3 V
Tseng et al. [453]	Laser ablation	PLZT/LNO/Si, PLZT/LNO/Pt/Si	$\varepsilon_r = 950$ $P_r = 14.9\ \mu C/cm^2$ $E_c = 3.5$ kV/cm	

Stolichnov et al. [455] studied the dielectric breakdown in PLZT thin film with Pt and $SrRuO_3$ (SRO) electrodes. It was observed that the constant current breakdown of PLZT films for the current range of 4×10^{-6}–5×10^{-4} A/cm^2 is controlled by the charge flown through the capacitor rather than the voltage applied. Only the charge flown through the grains determines the breakdown performance for this regime. Stolichnov et al. suggested that the breakdown onset is related to the formation of the critical concentration of defects in the bandgap. It was shown that the charge to breakdown of the Pt/PLZT/Pt capacitors can be substantially reduced by the polarization fatigue. This effect is related to defect generation due to the charge carriers injected into the film during fatigue bipolar voltage cycling. The strong increase in the charge to breakdown for the Pt/SRO/PLZT/Pt capacitor as compared with that of Pt/PLZT/Pt can be explained by high grain boundary conduction.

Yoon et al. [456] studied the relaxation current and the leakage current characteristics of PLZT thin films with various Ir-based top electrodes. The PLZT capacitors with Pt or IrO_2 top electrodes in contact with PLZT films show the strong time dependence of true leakage current, a result consistent with the space-charge-influenced injection model. On the other hand, the true leakage current of capacitors with Ir or IrO_2/Ir top electrodes is independent of time, contradicting space-charge injection model [251].

10. METAL–FERROELECTRIC–INSULATOR–SEMICONDUCTOR STRUCTURES

10.1. Memory Applications

Other than the applications we discussed in previous sections, ferroelectric films can also be used as the gate dielectric in metal–ferroelectric–semiconductor field effect transistors. Moll and Tarui [457] demonstrated the conductivity modulation in a device using a semiconducting CdS film on a ferroelectric TGS (triglycine sulfate) crystal. A CdS thin-film transistor was fabricated on a TGS single crystal. Wu [24, 458] fabricated metal–ferroelectric–semiconductor (MFS) transistors using bismuth titanate films on Si. The direction of the conduction change after polarization of the ferroelectric material, however, was opposite to the expected direction. The origin of this reversal was found to be the injection of electrons and holes from the semiconductor into the ferroelectric material [459]. Sugibuchi et al. [25] showed that this injection could be stopped by stacking a bismuth titanate film on a thermal oxide. Higuma et al. [460] reported that an MFS transistor using PZT and PLZT thin films deposited on single-crystal GaAs has fairly good memory hysteresis of the I_D–V_G characteristics. Efforts along this direction have also been carried out by others [461–463]. The early work was reviewed by Tarui et al. [459].

However, these devices have problems due to the interface reactions between the ferroelectric materials and the silicon substrate, making it difficult to obtain a good ferroelectric–silicon interface in the MFSFET gate [461]. To obtain a good

interface as the gate oxide for MFSFETs and to obtain oriented ferroelectric films on a silicon substrate, a buffer insulator layer between the silicon and the ferroelectric layer is very desirable [24, 25, 464, 465]. Therefore, a metal–ferroelectric–insulator–semiconductor (MFIS) structure has been proposed. The MFISFET-type memory is advantageous, because it is possible to read out memorized states nondestructively and a large remanent polarization is not required [466]. The nondestructive readout (NDRO) FRAM is discussed in [467, 468]. The literature results on MFIS capacitor and MFISFET are summarized in Tables LIII and LIV, respectively.

The fabrication technology of ferroelectric memories was discussed by Nakamura et al. [489]. Two types of ferroelectric memories have been proposed. One is one-transistor/one-capacitor (1T/1C) memory; the other is FET memory. 1T/1C memory has a structure similar to dynamic random-access memory (DRAM). The fabrication of the 1T/1C is not difficult, because the ferroelectric layer is separated from the CMOS layer by a thick interlayer dielectric. However, this type of memory has destructive readout (DRO) operation. In a read operation, the data are destroyed and then written again. The DRO has a disadvantage in access speed and number of endurance cycles. On the other hand, a FET-type ferroelectric memory has nondestructive readout (NDRO). It is possible for this type of memory to show high-speed operation and high reliability. This memory also has the potential for memory that has only 1 FET per cell. However, a ferroelectric FET is difficult to fabricate.

1T/1C Type of Ferroelectric Memory. This type of memory has one ferroelectric capacitor and one selective FET per cell. When the read bias is applied to the ferroelectric capacitor selected by the FET, the data are selected by the difference in the charge between the switching mode and the nonswitching mode. Damage to the CMOS layer during ferroelectric processing can be suppressed because the ferroelectric capacitor layer is separated from the CMOS layer by a thick interlayer dielectric.

FET type of Ferroelectric Memory. The ferroelectric memory FET has potential as an NDRO nonvolatile memory with a single transistor per cell as in a flash memory. This type of memory is expected to be the next generation of ferroelectric memory. In a read operation, the current between the source and drain is sensed. The ferroelectric polarization does not reverse and the data are not destroyed in the read operation. However, fabrication of a conventional MFSFET is very difficult because of deposition of the ferroelectric film directly on silicon.

10.2. Metal–Ferroelectric–Insulator–Semiconductor Capacitors

Various ferroelectric and insulator thin films have been used in the metal–ferroelectric–insulator–semiconductor (MFIS) approach. The most often used ferroelectric materials are PZT and SBT. The most often used insulator materials include SiO_2, CeO_2, and TiO_2. The literature results on MFIS capacitors are

Table LIII. Electrical Properties of MFIS Capacitors

Reference	Ferroelectric/thickness (nm)	Insulator/thickness (nm)	Memory window (V)	Wafer type/$C-V$ orientation[a]	Transistor
Byun et al. [469]	PbTiO$_3$ (150–180)	TiO$_2$ (100)		p/cw and ccw	
Choi et al. [470]	SBT (200)	YMnO$_3$ (25)	0.3–1.5	p/cw	
Hirai et al. [461]	PbTiO$_3$ (60)	CeO$_2$ (15)	~1	n/ccw	
Hirai et al. [464]	PbTiO$_3$ (81.3)	CeO$_2$ (18)	2.4	n/ccw	
Hirai et al. [471]	SBT (330)	CeO$_2$ (~20)	~1.1	n/ccw	Yes (NMOS)
Kanashima and Okuyama [466]	SBT, PZT	SiO$_2$, MgO, CeO$_2$, SrTiO$_3$	5	n/ccw	
Kashihara et al. [472]	PZT (150)	SrTiO$_3$ (35)	1–2	n/cw and p/ccw	
Kijima et al. [492]	Bi$_4$Ti$_3$O$_{12}$ (100–400)	Bi$_2$SiO$_5$ (30)	0.8–2.9	p/cw	
Kim and Shin [465]	SBT (160)	CeO$_2$ (25)/SiO$_2$ (5)	0.9–2.5	p/cw	
Kim et al. [474]	SBT (200–220)	CeO$_2$ (10–20) Y$_2$O$_3$ (10–20) Zr$_2$O$_3$ (10–20)	0.3–1.25 (from I_D–V_G)	n/ccw	Yes (NMOS)
Lee et al. [475]	SBT (180)	Si$_3$N$_4$ (5)/SiO$_2$ (5) or Al$_2$O$_3$ (10)	0.75–1.2 (for Si$_3$N$_4$/SiO$_2$)	p/cw	
Nagashima et al. [476]	SBT(220)	CeO$_2$(25)	0.5	n/ccw	
Noda et al. [477]	Sr$_{0.7}$Bi$_{2.8-2.0}$Ta$_2$O$_9$ (400)	SiO$_2$ (27)	2–3.6	n/ccw	
Okuyama et al. [478]	SBT (900) Bi$_4$Ti$_3$O$_{12}$ (450)	SiO$_2$ (28)	3.5 1.7	n/ccw	
Park et al. [479]	PZT (40–220)	Y$_2$O$_3$ (30–35)	2.6	n/ccw	
Sze and Lee [480]	PZT (700)	Ta$_2$O$_5$ (74)	13 at a ±15-V sweep	p/ccw	
Tarui et al. [459]		CeO$_2$, Ce$_x$Zr$_{1-x}$O$_2$		n/ccw	
Tokumitsu et al. [481]	PLZT (400)	STO (70)	0.6 0.9 (from FET)	n/ccw	Yes (PMOS)
Tokumitsu et al. [482]	PLZT (400)	STO (70)			Yes (PMOS)
Xiong and Sakai [483]	SBT (350–400)	SiO$_2$ (10), MgO buffered SiO$_2$	0.3–1.2	n/ccw	
Yu et al. [484]	PZT (250–400)	SiO$_2$ (40)	~1	n/ccw	Yes (PMOS) (retention problem)

[a] p, p-type wafer; n, n-type wafer; cw, clockwise; ccw, counterclockwise.

summarized in Table LIII. Some of the important results are briefly discussed next.

1. Byun et al. [469] fabricated Pt/PbTiO$_3$/TiO$_2$/Si structures. The ferroelectric characteristics are largely affected by the interface quality of TiO$_2$/Si, which depends on the growth temperature of the TiO$_2$ film; the TiO$_2$ film should be thick enough to prevent the diffusion of Pb during the deposition of PbTiO$_3$ film. The memory window of an MFIS capacitor is known to be proportional to the coercive field of the ferroelectric film [490].

2. Choi et al. [470] used a Pt/SBT/YMnO$_3$/Si structure. They found that there is no reaction between SBT and Si in the MFIS structure annealed at 900°C. The YMnO$_3$ buffer layer plays an important role in alleviating the interdiffusion between elements of SBT and Si. The leakage current of an MFIS structure annealed at 850°C is about 2×10^{-7} A/cm^2 at 200 kV/cm. To operate an MFIS structure in the low-voltage range, the capacitance of the buffer layer should be high compared to the capacitance of the ferroelectric. YMnO$_3$ is recommended as a dielectric layer because of its high dielectric constant (about 30 for bulk ceramics) and good thermal stability and because yttrium in YMnO$_3$ easily deoxidizes the SiO$_2$ formed on Si.

3. Hirai et al. [461, 464] studied Al/PbTiO$_3$/CeO$_2$/Si structures. The density of states at the CeO$_2$/Si interface is estimated from the $C-V$ measurement to be 8×10^{11}/cm^{2}eV.

Table LIV.   Electrical Properties of MFISFET Devices

Reference	Ferroelectric layer (thickness, nm)	Insulator layer (thickness, nm)	Channel	On–off ratio	Mobility $(cm^2/V\text{-}s)$	Transconductance (mS/mm)
Basit et al. [485]	PZT (300–2700)	MgO (7–70)/ SiO_2 (10–100)	NMOS			
Hirai et al. [471]	SBT (330)	CeO_2 (~20)	NMOS			
Kim [486]	$LiNbO_3$ (90)	None	NMOS	More than 4 orders of magnitude	600	0.16
Kim et al. [474]	SBT (200–220)	CeO_2 (10–20)	NMOS			
Lee et al. [487]	SBT (480)	Y_2O_3 (16)	NMOS			
Sugibuchi et al. [25]	$Bi_4Ti_3O_{12}$ (1000)	SiO_2 (50)	PMOS			
Tokumitsu et al. [481]	PLZT (400)	STO (70)	PMOS			1.4×10^{-3}
Tokumitsu et al. [482]	PLZT (400)	STO (70)	PMOS			
Tokumitsu et al. [488]	Pt/SBT (400)/Pt	$SrTa_2O_6$ (30)/ SiON(2)	PMOS (MFMISFET)			
Wu [24]	$Bi_4Ti_3O_{12}$	None	NMOS			
Yu et al. [484]	PZT (250–400)	SiO_2 (40)	PMOS (retention problem)			

The retention time of the MFIS structure is estimated to be 10^5 s.

4. Kanashima et al. [466] analyzed the $C–V$ characteristics of an MFIS structure. They found that the film thicknesses of the ferroelectric and insulator layers and the coercive field affect the memory window width markedly and have to be optimized in order to obtain a large memory window width. On the other hand, the $C–V$ characteristics and memory window width are not influenced by remanent polarization because the effective remanent polarization is small in the MFIS structure.

5. Noda et al. [477] used Al/SBT/SiO$_2$/Si structures. Regarding MFIS applications, memory window, fatigue, and retention are key factors for realizing a practical memory FET. The memory window is supposed to be around twice the value of the coercive voltage. An optimum range of memory window should be decided from the viewpoints of operation margin, operation voltage, and reliability. Requirements for the size of the memory window seem to depend on individual circuit operation. It is felt that memory windows approximately 3 V are adequate.

6. Okuyama et al. [478] analyzed the voltage dependence of the high-frequency capacitance of an MFIS structure by relating the potential profile to the dielectric hysteresis of the ferroelectric thin film. About one hundredth of the dielectric polarization of ferroelectric ceramic PZT is enough to control the silicon surface potential for ferroelectric gate FET memory, and a large coercive force is required to obtain enough voltage window.

7. Park et al. [479] used Al/PZT/Y$_2$O$_3$ structures. Because the dielectric constant of PZT is much larger than that of the buffer layer, the voltage applied to the gate electrode is not effectively transported to the PZT film and thus the operation

voltage of these devices is rather high. To solve this problem, the dielectric constant of the buffer layer should be sufficiently high or the layer thickness should be sufficiently thin, so that the capacitance of the buffer layer is comparable to that of the PZT film.

8. Xiong and Sakai [483] used SBT/SiO$_2$/Si and MgO-buffered SiO$_2$/Si structures. Ferroelectric nonvolatile memories have the potential to replace current nonvolatile memories. Ferroelectric gate field effect transistors have advantages compared with FRAM using passive capacitor, because nondestructive readout is possible and MFSFET obeys good scaling-down rule. However, MFS structures have not yet been successfully applied in practice because of charge injection between the ferroelectric and the semiconductor. Because ferroelectric thin-film gates were directly deposited on Si semiconductor substrates, the very high density of surface traps on such ferroelectric/Si interfaces results in large injected charge density, which precludes ferroelectric hysteresis in the electrical response of such MFS structures. The effect of charge injection can be minimized by inserting an insulating buffer layer (i.e., metal–ferroelectric–insulator–semiconductor structure). In this work, it was shown that the storage data were maintained without serious degradation over 7 days. The MgO buffer layer inserted between SBT and SiO$_2$/Si played important roles.

10.3. Metal–Ferroelectric–Semiconductor Field Effect Transistors

Because the deposition of ferroelectric thin film directly on silicon is difficult to make, there are comparatively fewer results in this area as compared with the MFISFET approach.

Some of the work on metal–ferroelectric–semiconductor field effect transistors (MFSFET) is briefly discussed next.

1. Wu [24, 458] fabricated MFS transistors using bismuth titanate films on Si. The ferroelectric field effect was demonstrated on bulk silicon using a thin ferroelectric film of bismuth titanate ($Bi_4Ti_3O_{12}$). An MFSFET device was fabricated. This device utilizes the remanent polarization of a ferroelectric thin film to control the surface conductivity of a bulk semiconductor substrate and perform a memory function. The C–V characteristics of the metal–ferroelectric–semiconductor structure were employed to study the memory behavior. The bismuth titanate has a polarization slightly higher than 4 $\mu C/cm^2$. Injection of carriers, both electrons and holes, from the semiconductor into the ferroelectric was observed. This was found from the shifts of flatband voltage on the measured C–V curves. The charges injected into the ferroelectric were presumably attracted by the remanent polarization and bound to ferroelectric domains after the external field was removed.

2. Rost et al. [491] fabricated an MFSFET using lithium niobate ($LiNbO_3$) as the gate dielectric. The channel conductance was strongly affected by the application of voltage pulses between the gate and the substrate. This behavior was found to be consistent with the influence of the polarization charge of the $LiNbO_3$ layer on the carriers in the channel.

3. Kim [486] fabricated n-channel MFSFETs using a $LiNbO_3$/Si structure and demonstrated nonvolatile memory operations of the MFSFETs. The I_D–V_G characteristic of the MFSFETs showed a hysteresis loop due to the ferroelectric nature of the $LiNbO_3$ films. The drain current of the "on" state was more than 4 orders of magnitude larger than the "off" state current at the same "read" gate voltage of 0.5 V, which means the memory operation of the MFSFET. The estimated field effect electron mobility and transconductance on a linear region of the fabricated FET were 600 cm^2/V-s and 0.16 mS/mm, respectively.

4. Yoon et al. [492] fabricated MFSFETs using SBT as the ferroelectric layer. These devices were applied to adaptive-learning neuron integrated circuits.

10.4. Metal–Ferroelectric–Insulator–Semiconductor Field Effect Transistors

Metal–ferroelectric–insulator–semiconductor field effect transistors (MFISFETs) have been reported in the literature. Some of work in this area is briefly discussed next.

1. Sugibuchi et al. [25] fabricated a ferroelectric field effect memory device using a ferroelectric $Bi_4Ti_3O_{12}$/SiO_2/Si structure, where the SiO_2 serves to prevent charge injection from silicon into the ferroelectric film. The electron injection process would degrade the retention of memoried states. The ferroelectric layer has a thickness of 1 μm. The remanent polarization of the film is large and sufficient to modulate the surface potential of silicon. A p-channel FET was fabricated on an n-type silicon wafer. The FET can be switched by a voltage of 15 V.

2. Hirai et al. [471] fabricated an n-channel MFISFET using an Al/SBT/CeO_2/Si structure. The I_D–V_G characteristics show threshold hysteresis of about 1.1 V. The memory effect of the MFISFET device was also shown in the I_D–V_D characteristics at read voltage after applying write voltage.

3. Kim et al. [474] fabricated an n-channel MFISFET using SBT as the ferroelectric layer and using ZrO_2, CeO_2, and Y_2O_3 insulator layers. Pt is used as the top electrode. Ideally, the nondestructive readout (NDRO) MFISFET has good input–output isolation, large nonlinearity and noise margin, good current fanout, and high tolerance of signal distortion for digital device applications.

4. Tokumitsu et al. [481, 482] fabricated a p-channel MFISFET using an Al/PLZT/STO/Si structure. The drain current can be significantly changed by the "write" voltage, which was applied before the measurement, even at the same "read" gate voltage. This demonstrates the memory operation of the MFISFET device.

5. Tokumitsu et al. [488] later fabricated a p-channel metal–ferroelectric–metal–insulator–semiconductor (MFMIS) field effect transistor using a Pt/SBT/Pt/$SrTa_2O_6$/SiON/Si structure. If one uses PZT as the ferroelectric layer and SiO_2 as the insulator layer, the excellent interface properties of SiO_2/Si are available. However, it becomes difficult to apply sufficient voltage to the PZT layer because the dielectric constant of PZT is much larger than that of SiO_2. Hence, when SiO_2 is used for the insulator layer in the MFIS and MFMIS structures, the SiO_2 layer must be very thin. Transistor operation of MOSFETs has been reported even when the gate SiO_2 thickness is less than 1.5 nm. However, if such a thin SiO_2 is used in ferroelectric-gate FETs using MFIS and MFMIS structures, the data retention time will be short, because of the large gate leakage current due to direct tunneling. Thererfore, a high-dielectric-constant material must be used as the insulator layer. It should be noted that most of available high-dielectric-contant materials such as TiO_2 and Ta_2O_5 are oxides. Hence, several nanometers of SiO_2 transition layer can be easily formed at the interface during the formation of the high-dieletric-constant material. In MFMIS structures, the applied voltage is divided between the MFM capacitor and the MIS diode according to the capacitances of each layer. By reducing the area of the MFM capacitor formed on the MIS diode, one can equivalently decrease the remanent polarization and dielectric constant, which make it possible to apply sufficient voltage to the ferroelectric layer even at low gate voltage. In this work, by reducing the MFM capacitor area, a large memory window of 3.0 V can be obtained.

6. Yu et al. [484] fabricated a p-channel MFISFET using a Au/PZT/SiO_2/Si structure. Retention was observed to be a problem in the operation of this MFISFET device.

The literature results on MFISFETs are summarized in Table LIV.

11. CONCLUSION

Further scaling of DRAM integrated circuits requires high-dielectric-constant or ferroelectric thin films for storage capacitor dielectrics. This is a significant departure from the more evolutionary improvement in the last several generations of DRAM chips. Among the high-dielectric-constant films, Ta_2O_5 can be used for 256-Mb and 1-Gb applications. For further application after the 1-Gb generation, barium strontium titanate (BST) is the material with the greatest potential. Ferroelectric materials can also be used if some of the reliability problems can be solved. Ferroelectric random-access memory (FRAM) is a nonvolatile memory technology with high potential. Among the ferroelectric materials, PZT and SBT are the most studied and have the greatest potential. Low-density ferroelectric memory chips are already in production. The future of the FRAM devices depends on the reliability of these ferroelectric films in terms of fatigue, imprint, and retention. The research of high-dielectric-constant and ferroelectric thin films is one of the research areas in VLSI technology that is more revolutionary than evolutionary in nature.

Acknowledgments

We thank the National Science Council of the Republic of China for financial support over the years in the technical area of dielectric films for VLSI application. We also thank all the previous and present students of the Microelectronic Devices and Materials Laboratory, Department of Electrical Engineering, Tsing-Hua University, for their research efforts.

REFERENCES

1. L. H. Parker and A. F. Tasch, *IEEE Circuits Devices Magazine* 17 (January 1990).
2. A. F. Tasch and L. H. Parker, *Proc. IEEE* 77, 374 (1989).
3. P. C. Fazan, V. K. Mathews, N. Sandler, G. Q. Lo, and D. L. Kwong, *IEDM Tech. Digest* 263 (1992).
4. B. Prince, "Semiconductor Memories," 2nd ed., Wiley, Chichester, 1991.
5. International Technology Roadmap for Semiconductors, 1999.
6. K. Itoh, K. Sasaki, and Y. Nakagome, *Proc. IEEE* 83, 524 (1995).
7. J. M. C. Stork, *Proc. IEEE* 83, 607 (1995).
8. H. Yamaguchi, T. Iizuka, H. Koga, K. Takemura, S. Sone, H. Yabuta, S. Yamamichi, P. Lesaicherre, M. Suzuki, Y. Kojima, N. Kasai, T. Sakuma, Y. Kato, Y. Miyasaka, M. Yoshida, and S. Nishimoto, *IEDM* 675 (1996).
9. K. Sunouchi, H. Kawaguchiya, S. Matsuda, H. Nomura, T. Shine, K. Murooka, S. Sugihara, S. Mitsui, K. Kondo, and Y. Kikuchi, *IEDM* 601 (1996).
10. B. Davari, R. H. Dennard, and G. G. Shahidi, *Proc. IEEE* 83, 595 (1995).
11. Nikkei Microdevices 28 (March 1995).
12. P. C. Fazan, *Integ. Ferroelec.* 4, 247 (1994).
13. F. Jona and G. Shirane, "Ferroelectric Crystals." Dover, New York, 1993.
14. M. E. Lines and A. M. Glass, "Principles and Applications of Ferroelectrics and Related Materials." Clarendon, Oxford, 1977.
15. M. H. Francombe, "Ferroelectric Films for Integrated Electronics," Physics of Thin Films, Vol. 17, p. 225. Academic Press, USA, 1993.
16. Y. Xu, "Ferroelectric Materials and Their Applications." Elsevier Science, Amsterdam, 1991.
17. R. Ramesh, Ed., "Thin Film Ferroelectric Materials and Devices." Kluwer Academic, Boston, 1997.
18. J. F. Scott, "Ferroelectric Memories." Springer-Verlag, Berlin, 2000.
19. J. F. Scott, in "The Physics of Ferroelectric Ceramic Thin Films for Memory Applications" OPA (Overseas Publishers Association) (G. W. Taylor and A. S. Bhalla, Eds.), Ferroelectric Review, Vol. 1, p. 1. 1998.
20. C. Hu, Ed., "Semiconductor Nonvolatile Memories." IEEE Press, New York, 1991.
21. D. W. Bondurant and F. P. Gnadinger, *IEEE Spectrum* 30 (July 1989).
22. D. W. Bondurant, *Ferroelectrics* 112, 273 (1990).
23. J. L. Moll and Y. Tarui, *IEEE Trans. Electron Devices* 10, 338 (1963).
24. S. Y. Wu, *IEEE Trans. Electron Devices* 21, 499 (1974).
25. K. Sugibuchi, Y. Kurogi, and N. Endo, *J. Appl. Phys.* 46, 2877 (1975).
26. T. Hori, "Gate Dielectrics and MOS ULSIs, Principles, Technologies, and Applications." Springer-Verlag, Berlin, 1997.
27. A. Ishitani, P. Lesaicherre, S. Kamiyama, K. Ando, and H. Watanabe, *IEICE Trans. Electron.* E76-C, 1564 (1993).
28. S. Sinharoy, H. Buhay, D. R. Lampe, and M. H. Francombe, *J. Vac. Sci. Technol., A* 10, 1554 (1992).
29. W. A. Geideman, *IEEE Trans. Ultrasonics, Ferroelectrics and Frequency Control* 38, 704 (1991).
30. M. Sayer, Z. Wu, C. V. R. Vasant Kumar, D. T. Amm, and E. M. Griswold, *Can. J. Phys.* 70, 1159 (1992).
31. J. Carrano, C. Sudhama, V. Chikarmane, J. Lee, A. Tasch, W. Shepherd, and N. Abt, *IEEE Trans. Ultrasonics, Ferroelectrics and Frequency Control* 38, 690 (1991).
32. R. Moazzami, C. Hu, and W. H. Shepherd, *IEEE Trans. Electron Devices* 39, 2044 (1992).
33. C. Sudhama, A. C. Campbell, P. D. Maniar, R. E. Jones, R. Moazzami, C. J. Mogab, and J. C. Lee, *J. Appl. Phys.* 75, 1014 (1994).
34. E. M. Philofsky, "1996 International Nonvolatile Memory Technology Conference," 1996, p. 99.
35. J. F. Scott and C. A. Paz de Araujo, *Science* 246, 1400 (1989).
36. R. E. Jones, P. D. Maniar, R. Moazzami, P. Zurcher, J. Z. Witowski, Y. T. Lii, P. Chu, and S. J. Gillespie, *Thin Solid Films* 270, 584 (1995).
37. T. Mihara, H. Yoshimori, H. Watanabe, and C. A. Pas de Araujo, *Jpn. J. Appl. Phys.* 34, 5233 (1995).
38. S. K. Dey and J. J. Lee, *IEEE Trans. Electron Devices* 39, 1607 (1992).
39. P. Li and T. M. Lu, *Phys. Rev.* 43, 14261 (1991).
40. Q. X. Jia, Z. Q. Shi, and W. A. Anderson, *Thin Solid Films* 209, 230 (1992).
41. Z. Q. Shi, Q. X. Jia, and W. A. Anderson, *J. Vac. Sci. Technol., A* 11, 1411 (1993).
42. M. H. Yeh, Y. C. Liu, K. S. Liu, I. N. Lin, and J. Y. Lee, *J. Appl. Phys.* 74, 2143 (1993).
43. G. W. Dietz, W. Antpohler, M. Klee, and R. Waser, *J. Appl. Phys.* 78, 6113 (1995).
44. R. W. Schwartz, B. A. Tuttle, D. H. Doughty, C. E. Land, D. C. Goodnow, C. L. Hernandez, T. J. Zender, and S. L. Martinez, *IEEE Trans. Ultrasonics, Ferroelectrics and Frequency Control* 38, 677 (1991).
45. D. J. Wouters, G. J. Willems, and H. E. Maes, *Microelectron Eng.* 29, 249 (1995).
46. S. E. Bernacki, *Mater. Res. Soc. Symp. Proc.* 243, 135 (1992).
47. Z. G. Zhang, J. S. Liu, Y. N. Wang, J. S. Zhu, J. L. Liu, D. Su, and H. M. Shen, *J. Appl. Phys.* 85, 1746 (1999).
48. J. H. Yoo, C. H. Kim, K. C. Lee, K. H. Kyung, S. M. Yoo, J. H. Lee, M. H. Son, J. M. Han, B. M. Kang, E. Haq, S. B. Lee, J. H. Sim, J. H. Kim, B. S. Moon, K. Y. Kim, J. G. Park, K. P. Lee, K. Y. Lee, K. N. Kim, S. I. Cho, J. W. Park, and H. K. Lim, *IEEE ISSCC* 378 (1996).
49. K. C. Lee, H. Yoon, S. B. Lee, J. H. Lee, B. S. Moon, K. Y. Kim, C. H. Kim, and S. I. Cho, *Symp. VLSI Circuit Digest Tech.* 103 (1997).
50. K. N. Kim, H. S. Jeong, G. T. Jeong, C. H. Cho, W. S. Yang, J. H. Sim, K. H. Lee, G. H. Koh, D. W. Ha, J. S. Bae, J. G. Lee, B. J. Park, and J. G. Lee, *Symp. VLSI Tech. Digest* 16 (1998).
51. M. Hamada, K. Inoue, R. Kubota, M. Takeuchi, M. Sakao, H. Abiko, H. Kawamoto, H. Yamaguchi, H. Kitamura, S. Onishi, K. Mikagi, K. Urabe,

T. Taguwa, T. Yamamoto, N. Nagai, I. Shirakawa, and S. Kishi, *IEDM Tech. Digest* 45 (1999).

52. Y. Takai, M. Fujita, K. Nagata, S. Isa, S. Nakazawa, A. Hirobe, H. Ohkubo, M. Sakao, S. Horiba, T. Fukase, Y. Takashi, M. Matsuo, M. Komuro, T. Uchida, T. Sakoh, K. Saino, S. Uchiyama, Y. Takada, J. Sekine, N. Kakanishi, T. Oikawa, M. Igeta, H. Tanabe, H. Miyamoto, T. Hashimoto, H. Yamaguchi, K. Koyama, Y. Kobayashi, and T. Okuda, *IEEE ISSCC* 418 (1999).

53. H. Yoon, G. W. Cha, C. S. Yoo, N. J. Kim, K. Y. Kim, C. H. Lee, K. N. Lim, K. C. Lee, J. Y. Jeon, T. S. Jung, H. S. Jeong, T. Y. Jeong, K. N. Kim, and S. I. Cho, *IEEE ISSCC* 412 (1999).

54. J. Lin, N. Masaaki, A. Tsukune, and M. Yamada, *Appl. Phys. Lett.* 76, 2370 (1999).

55. A. Chatterjee, R. A. Chapman, K. Joyner, M. Otobe, S. Hattangady, M. Bevan, G. A. Brown, H. Yang, Q. He, D. Rogers, S. J. Fang, R. Kraft, A. L. P. Rotodaro, M. Terry, K. Brennan, S. W. Aur, J. C. Hu, H. L. Tsai, P. Jones, G. Wilk, M. Aoki, M. Rodder, and I. C. Chen, *IEDM Tech. Digest* 777 (1998).

56. S. O. Kim and H. J. Kim, *J. Vac. Sci. Technol., B* 12, 3006 (1994).

57. J. L. Autran, R. Devine, C. Chaneliere, and B. Balland, *IEEE Electron Device Lett.* EDL-18, 447 (1997).

58. I. C. Kizilyalli, R. Y. S. Huang, and P. K. Roy, *IEEE Electron Device Lett.* EDL-19, 423 (1998).

59. D. Park, Y. C. King, Q. Lu, T. J. King, C. Hu, A. Kalnitsky, S. P. Tay, and C. C. Cheng, *IEEE Electron Device Lett.* EDL-19, 441 (1998).

60. J. C. Yu, B. C. Lai, and J. Y. Lee, *IEEE Electron Device Lett.* EDL-21, 537 (2000).

61. H. Kato, K. S. Seol, T. Toyoda, and Y. Ohki, "International Symposium on Electrical Insulating Materials," 1998, p. 131.

62. K. W. Kwon, C. S. Kang, S. O. Park, H. K. Kang, and S. T. Ahn, *IEEE Trans. Electron Devices* ED-43, 919 (1996).

63. B. C. Lai and J. Y. Lee, *J. Electrochem. Soc.* 146, 266 (1999).

64. K. Kukli, J. Aarik, A. Aidla, O. Kohan, T. Uustare, and V. Sammelselg, *Thin Solid Films* 260, 135 (1995).

65. W. S. Lau, K. K. Khaw, P. W. Qian, N. P. Sandler, and P. K. Chu, *J. Appl. Phys.* 79, 8841 (1996).

66. G. B. Alers, R. M. Fleming, Y. H. Wong, B. Dennis, A. Pinczuk, G. Redinbo, R. Urdahl, E. Ong, and Z. Hasan, *Appl. Phys. Lett.* 72, 1308 (1998).

67. E. Kaplan, M. Balog, and D. Frohman-Bentchkowsky, *J. Electrochem. Soc.* 123, 1571 (1976).

68. S. Kimura, Y. Nishioka, A. Shintani, and K. Mukai, *J. Electrochem. Soc.* 130, 2414 (1983).

69. G. Q. Lo, D. L. Kwong, and S. Lee, *Appl. Phys. Lett.* 62, 973 (1993).

70. F. D. Quarto, C. Gentile, S. Piazza, and C. Sunseri, *Corrosion Sci.* 35, 801 (1993).

71. X. M. Wu, S. R. Soss, E. J. Rymaszewski, and T. M. Lu, *Mater. Chem. Phys.* 38, 297 (1994).

72. T. Aoyama, S. Sadai, Y. Okayama, M. Fujisaki, K. Imai, and T. Arikado, *J. Electrochem. Soc.* 143, 977 (1996).

73. W. S. Lau, K. K. Khaw, P. W. Qian, P. Sandler, and P. K. Chu, *Jpn. J. Appl. Phys.* 35, 2599 (1996).

74. P. A. Murawala, M. Sawai, T. Tatsuta, O. Tsuji, S. Fujita, and S. Fujita, *Jpn. J. Appl. Phys.* 32, 368 (1993).

75. M. Metikos-Hukovic and M. Ceraj-Ceric, *Thin Solid Films* 145, 39 (1986).

76. W. E. Flannery and S. R. Pollack, *J. Appl. Phys.* 37, 4417 (1966).

77. J. Robertson and C. W. Chen, *Appl. Phys. Lett.* 74, 1168 (1999).

78. W. R. Hitchens, W. C. Krusell, and D. M. Dobkin, *J. Electrochem. Soc.* 140, 2615 (1993).

79. P. C. Joshi and M. W. Cole, *J. Appl. Phys.* 86, 871 (1999).

80. V. Mikhaelashvili, Y. Betzer, I. Prudnikov, M. Orenstein, D. Ritter, and G. Eisenstein, *J. Appl. Phys.* 84, 6747 (1998).

81. A. Nagahiro and R. Raj, *J. Am. Ceram. Soc.* 78, 1585 (1995).

82. S. Seki, T. Unagami, and B. Tsujiyama, *J. Electrochem. Soc.* 131, 2622 (1984).

83. S. Zaima, T. Furuta, and Y. Yasuda, *J. Electrochem. Soc.* 137, 1297 (1990).

84. B. K. Moon, C. Isobe, and J. Aoyama, *J. Appl. Phys.* 85, 1731 (1999).

85. S. O. Kim and H. J. Kim, *Thin Solid Films* 253, 435 (1994).

86. I. Kim, S. D. Ahn, B. W. Cho, S. T. Ahn, J. Y. Lee, J. S. Chun, and W. J. Lee, *Jpn. J. Appl. Phys.* 33, 6691 (1994).

87. C. Chaneliere, S. Four, J. L. Autran, R. A. B. Devine, and N. P. Sandler, *J. Appl. Phys.* 83, 4823 (1998).

88. C. Chaneliere, S. Four, J. L. Autran, and R. A. B. Devine, *Electrochem. Solid-State Lett.* 2, 291 (1999).

89. I. Kim, J. S. Kim, O. S. Kwon, S. T. Ahn, J. S. Chun, and W. J. Lee, *J. Electron. Mater.* 24, 1435 (1995).

90. I. Kim, J. S. Kim, B. W. Cho, S. D. Ahn, J. S. Chun, and W. J. Lee, *J. Mater. Res.* 10, 2864 (1995).

91. S. O. Kim, J. S. Byun, and H. J. Kim, *Thin Solid Films* 206, 102 (1991).

92. C. J. Lee, L. T. Huang, S. Ezhilvalavan, and T. Y. Tseng, *Electrochem. Solid-State Lett.* 2, 135 (1999).

93. K. Kishiro, N. Inoue, S. C. Chen, and M. Yoshimaru, *Jpn. J. Appl. Phys.* 37, 1336 (1998).

94. F. C. Chiu, J. J. Wang, J. Y. Lee, and S. C. Wu, *J. Appl. Phys.* 81, 6911 (1997).

95. S. Ezhilvalavan and T. Y. Tseng, *Appl. Phys. Lett.* 74, 2477 (1999).

96. S. Kamiyama, P. Y. Lesaicherre, H. Suzuki, A. Sakai, I. Nishiyama, and A. Ishitani, *J. Electrochem. Soc.* 140, 1617 (1993).

97. T. Aoyama, S. Yamazaki, and K. Imai, *J. Electrochem. Soc.* 145, 2961 (1998).

98. S. Kamiyama, H. Suzuki, H. Watanabe, A. Sakai, H. Kimura, and J. Mizuki, *J. Electrochem. Soc.* 141, 1246 (1994).

99. S. W. Park, Y. K. Baek, J. Y. Lee, C. O. Park, and H. B. Im, *J. Electron. Mater.* 21, 635 (1992).

100. S. Banerjee, B. Shen, I. Chen, J. Bohlman, G. Brown, and R. Doering, *J. Appl. Phys.* 65, 1140 (1989).

101. C. Chaneliere, J. L. Autran, and R. A. B. Devine, *J. Appl. Phys.* 86, 480 (1999).

102. S. Ezhilvalavan and T. Y. Tseng, *J. Appl. Phys.* 83, 4797 (1998).

103. M. Houssa, R. Degraeve, P. W. Mertens, M. M. Heys, J. S. Jeon, A. Halliyal, and B. Ogle, *J. Appl. Phys.* 86, 6462 (1999).

104. V. Mikhaelashvili and G. Eisenstein, *Appl. Phys. Lett.* 75, 2863 (1999).

105. H. Zhang, R. Solanki, B. Roberds, G. Bai, and I. Banerjee, *J. Appl. Phys.* 87, 1921 (2000).

106. H. Shinriki and M. Nataka, *IEEE Trans. Electron Devices* ED-38, 455 (1991).

107. H. Miki, M. Kunitomo, R. Furukawa, T. Tamaru, H. Goto, S. Iijima, Y. Ohji, H. Yamamoto, J. Kuroda, T. Kisu, and I. Asano, *VLSI Symp. Tech. Digest* 99 (1999).

108. G. Q. Lo, D. L. Kwong, and S. Lee, *Appl. Phys. Lett.* 60, 3286 (1992).

109. C. A. Mead, *Phys. Rev.* 128, 2088 (1962).

110. Y. Nishioka, H. Shinriki, and K. Mukai, *J. Appl. Phys.* 61, 2335 (1987).

111. S. Zaima, T. Furuta, Y. Koide, and Y. Yasuda, *J. Electrochem. Soc.* 137, 2876 (1990).

112. H. Matsuhashi and S. Nishikawa, *Jpn. J. Appl. Phys.* 33, 1293 (1994).

113. G. Q. Lo, D. L. Hwong, P. C. Pazan, V. K. Mathews, and N. Sandler, *IEEE Electron Device Lett.* EDL-14, 216 (1993).

114. J. Y. Zhang, B. Lim, and I. W. Boyd, *Appl. Phys. Lett.* 73, 2292 (1998).

115. R. A. B. Devine, L. Vallier, J. L. Autran, P. Paillet, and J. L. Leray, *Appl. Phys. Lett.* 68, 1775 (1996).

116. W. S. Lau, L. Zhong, A. Lee, C. H. See, and T. C. Chong, *Appl. Phys. Lett.* 71, 500 (1997).

117. W. K. Choi and C. H. Ling, *J. Appl. Phys.* 75, 3987 (1994).

118. J. G. Simmons, *Phys. Rev.* 155, 657 (1967).

119. K. Shimizu, M. Katayama, H. Funaki, E. Arai, M. Nakata, Y. Ohji, and R. Imura, *J. Appl. Phys.* 74, 375 (1993).

120. S. C. Sun and T. F. Chen, *IEDM Tech. Digest* 687 (1996).

121. M. Matsui, S. Oka, K. Yamagishi, K. Kuroiwa, and Y. Tarui, *Jpn. J. Appl. Phys.* 27, 506 (1988).

122. S. Kamiyama, H. Suzuki, H. Watanabe, A. Sakai, M. Oshida, T. Tatsumi, T. Tanigawa, N. Kasai, and A. Ishitani, *IEDM Tech. Digest* 93 (1993).

123. Y. Ohji, Y. Matsui, T. Itoga, M. Hirayama, Y. Sugawara, K. Torii, H. Miki, M. Nataka, I. Asano, S. Iijima, and Y. Kawamoto, *IEDM Tech. Digest* 112 (1995).

124. J. V. Grahn, P. E. Hellberg, and E. Olsson, *J. Appl. Phys.* 84, 1632 (1998).

125. S. W. Park and H. B. Im, *Thin Solid Films* 207, 258 (1992).

126. S. C. Sun and T. F. Chen, *IEEE Electron Device Lett.* EDL-17, 355 (1996).

127. I. Kim, J. S. Chun, and W. J. Lee, *Material Chemistry and Physics* 44, 288 (1996).

128. B. K. Moon, J. Aoyama, and K. Katori, *Appl. Phys. Lett.* 74, 824 (1999).

129. C. Hashimoto, H. Oikawa, and N. Honma, *IEEE Trans. Electron Devices* ED-36, 14 (1989).

130. I. Asano, M. Kunitomo, S. Yamamoto, R. Furukawa, Y. Sugawara, T. Uemura, J. Kuroda, M. Kanai, M. Nakata, T. Tamaru, Y. Nakamura, T. Kawagoe, S. Yamada, K. Kawakita, H. Kawamura, M. Nakamura, M. Morino, T. Kisu, S. Iijima, Y. Ohji, T. Sekiguchi, and T. Tadaki, *IEDM Tech. Digest* 755 (1998).

131. K. W. Kwon, S. O. Park, C. S. Kang, Y. N. Kim, S. T. Ahn, and M. Y. Lee, *IEDM Tech. Digest* 53 (1993).

132. H. F. Luan, S. J. Lee, C. H. Lee, S. C. Song, Y. L. Mao, Y. Senzaki, D. Roberts, and D. L. Kwong, *IEDM Tech. Digest* (1999).

133. H. F. Luan, B. Z. Wu, L. G. Kang, B. Y. Kim, R. Vrtis, D. Roberts, and D. L. Kwong, *IEDM Tech. Digest* 609 (1998).

134. Y. Takaishi, M. Sakao, S. Kamiyama, H. Suzuki, and H. Watanabe, *IEDM Tech. Digest* 839 (1994).

135. R. L. Angle and H. E. Talley, *IEEE Trans. Electron Devices* ED-25, 1277 (1978).

136. J. L. Autran, P. Paillet, J. L. Leray, and R. A. B. Devine, *Sens. Actuators, A* 51, 5 (1995).

137. S. K. Zhang, L. Ke, F. Lu, Q. Z. Qin, and X. Wang, *J. Appl. Phys.* 84, 335 (1998).

138. J. G. Hwu, M. J. Jeng, W. S. Wang, and Y. K. Tu, *J. Appl. Phys.* 62, 4277 (1987).

139. B. C. Lai, N. H. Kung, and J. Y. Lee, *J. Appl. Phys.* 85, 4087 (1999).

140. G. S. Oehrlein, *Thin Solid Films* 156, 207 (1988).

141. S. G. Byeon and Y. Tzeng, *IEEE Trans. Electron Devices* ED-37, 972 (1990).

142. J. G. Hwu and M. J. Jeng, *J. Electrochem. Soc.* 135, 2080 (1988).

143. H. Shinriki, T. Kisu, S. Kimura, Y. Nishioka, Y. Kawamoto, and K. Mukai, *IEEE Trans. Electron Devices* ED-37, 1939 (1990).

144. Y. Nishioka, S. Kimura, H. Shinriki, and K. Mukai, *J. Electrochem. Soc.* 134, 410 (1987).

145. S. Kamiyama and T. Saeki, *IEDM Tech. Digest* 827 (1991).

146. R. Degraeve, B. Kaczer, M. Houssa, G. Groeseneken, M. Heyns, J. S. Joen, and A. Halliyal, *IEDM Tech. Digest* 327 (1999).

147. K. Sekine, Y. Saito, M. Hirayama, and T. Ohmi, *Symp. VLSI Tech. Digest* 115 (1999).

148. H. C. Cheng, H. W. Liu, H. P. Su, and G. Hong, *IEEE Electron Device Lett.* EDL-16, 509 (1995).

149. B. Y. Kim, H. F. Luan, and D. L. Kwong, *IEDM Tech. Digest* 463 (1997).

150. K. Ando, A. Yokozawa, and A. Ishitani, *Symp. VLSI Tech. Digest* 47 (1993).

151. Y. Wu and G. Lucovsky, *IEEE Electron Device Lett.* EDL-19, 367 (1998).

152. K. Eriguchi, Y. Harada, and M. Niwa, *IEDM Tech. Digest* 323 (1999).

153. K. Kobayashi, Y. Inaba, T. Ogata, T. Katayama, H. Watanabe, Y. Matsui, and M. Hirayama, *J. Electrochem. Soc.* 143, 1459 (1996).

154. H. Tanaka, H. Uchida, T. Ajioka, and N. Hirashita, *IEEE Trans. Electron Devices* ED-40, 2231 (1993).

155. H. H. Tseng, P. G. Y. Tsui, P. J. Tobin, J. Mogab, M. Khare, X. M. Wang, T. P. Ma, R. Hegde, C. Hobbs, J. Veteran, M. Hartig, G. Kenig, V. Wang, R. Blumenthal, R. Cotton, V. Kaushik, T. Tamagawa, B. L. Halpern, G. J. Cui, and J. J. Schmitt, *IEDM Tech. Digest* 149 (1997).

156. G. W. Yoon, G. Q. Lo, J. Kim, L. K. Han, and D. L. Kwong, *IEEE Electron Device Lett.* EDL-15, 266 (1994).

157. S. C. Song, H. F. Luan, C. H. Lee, A. Y. Mao, S. J. Lee, J. Gelpey, S. Marcus, and D. L. Kwong, *Symp. VLSI Tech. Digest* 137 (1999).

158. D. Wang, T. P. Ma, J. W. Golz, B. L. Halpern, and J. J. Schmitt, *IEEE Electron Devices Lett.* EDL-13, 482 (1992).

159. T. P. Ma, *IEEE Trans. Electron Devices* ED-45, 680 (1998).

160. M. Khare, X. W. Wang, and T. P. Ma, *Symp. VLSI Tech. Digest* 218 (1998).

161. H. H. Tseng, "Fifth International Conference on SSIC Technology," 1998, p. 279.

162. S. C. Song, H. F. Luan, Y. Y. Chen, M. Gardner, J. Fulford, M. Allen, and D. L. Kwong, *IEDM Tech. Digest* 373 (1998).

163. Y. Wu and G. Lucovsky, *IEEE Electron Device Lett.* EDL-21, 116 (2000).

164. V. A. Gritsenko, J. B. Xu, and S. Lin, "Hong Kong Electron Device Meeting," 1998, p. 40.

165. G. Q. Lo, S. Ito, D. L. Kwong, V. K. Mathews, and P. C. Fazan, *IEEE Electron Device Lett.* EDL-13, 372 (1992).

166. K. S. Lim and C. H. Ling, *ICSE* 42 (1998).

167. S. M. Sze, *J. Appl. Phys.* 38, 2951 (1967).

168. W. Ting, J. H. Ahn, and D. L. Kwong, *Electron. Lett.* 27, 1046 (1991).

169. M. Khare, X. Guo, X. W. Wang, and T. P. Ma, *Symp. VLSI Tech. Digest* 51 (1997).

170. H. Iwai, S Momose, T. Morimoto, Y. Ozawa, and K. Yamabe, *IEDM Tech. Digest* 235 (1990).

171. M. Khare, X. W. Wang, and T. P. Ma, *IEEE Electron Device Lett.* EDL-20, 57 (1999).

172. S. Mahapatra, V. R. Rao, K. N. ManjuliaRani, C. D. Parikh, J. Vasi, B. Cheng, M. Khare, and J. C. S. Woo, *Symp. VLSI Tech. Digest* 79 (1999).

173. N. Szydio and R. Poirier, *J. Appl. Phys.* 51, 3310 (1980).

174. G. P. Burns, *J. Appl. Phys.* 65, 2095 (1989).

175. G. P. Burns, I. S. Baldwin, M. P. Hastings, and J. G. Wilkes, *J. Appl. Phys.* 66, 2320 (1989).

176. Y. H. Lee, K. K. Chan, and M. J. Brady, *J. Vac. Sci. Technol., A* 13, 569 (1995).

177. S. A. Campbell, D. C. Cilmer, X. C. Wang, M. T. Hsieh, H. S. Kim, W. L. Gladfelter, and J. Yan, *IEEE Trans. Electron Devices* ED-44, 104 (1997).

178. B. He, N. Hoilien, R. Smith, T. Ma, C. Taylor, I. St. Omer, S. A. Campbell, W. L. Gladfelter, M. Gribelyuk, and D. Buchanan, "University/Government/Industry Microelectronics Symposium and Proceedings 13th Biennial," 1999, p. 33.

179. H. S. Kim, D. C. Gilmer, S. A. Campbell, and D. L. Polla, *Appl. Phys. Lett.* 69, 3860 (1996).

180. H. S. Kim, S. A. Campbell, and D. C. Gilmer, "35th Proceedings of the Reliability Physics Symposium," 1997, p. 90.

181. H. S. Kim, S. A. Campbell, and D. C. Gilmer, *IEEE Electron Device Lett.* EDL-18, 465 (1997).

182. H. S. Kim, S. A. Campbell, D. C. Gilmer, and D. M. Kim, "Proceedings of the Fifth ICPADM," 1997, p. 1039.

183. J. Yan, D. C. Gilmer, S. A. Campbell, W. L. Gladfelter, and P. G. Schmid, *J. Vac. Sci. Technol., B* 14, 1706 (1996).

184. T. Fuyuki and H. Matsunami, *J. Appl. Phys.* 25, 1288 (1986).

185. M. D. Wiggins, M. C. Nelson, and C. R. Aita, *J. Vac. Sci. Technol., A* 14, 771 (1996).

186. Y. Jeon, B. H. Lee, K. Zawadzki, W. J. Qi, A. Lucas, R. Nieh, and J. C. Lee, *IEDM Tech. Digest* 797 (1998).

187. S. C. Sun and T. F. Chen, *Jpn. Appl. Phys.* 36, 1346 (1997).

188. T. K. Won, S. G. Yoon, and H. G. Kim, *J. Electrochem. Soc.* 139, 3284 (1992).

189. T. Hada, S. Hayakawa, and K. Wasa, *Jpn. J. Appl. Phys.* 9, 1078 (1970).

190. B. Lalevic and G. Taylor, *J. Appl. Phys.* 46, 3208 (1975).

191. B. H. Lee, Y. Jeon, K. Zawadzki, W. J. Qi, and J. Lee, *Appl. Phys. Lett.* 74, 3143 (1999).

192. Y. Abe and T. Fukuda, *Jpn. J. Appl. Phys.* 33, L1248 (1994).

193. T. Fuyuki, T. Kobayashi, and H. Matsunami, *J. Electrochem. Soc.* 135, 248 (1988).

194. K. Fukushima and I. Yamada, *J. Appl. Phys.* 65, 619 (1989).

195. M. B. Lee, M. Kawasaki, M. Yoshimoto, B. K. Moon, H. Ishiwara, and H. Koinuma, *Jpn. J. Appl. Phys.* 34, 808 (1995).

196. H. Shin, M. R. De Guire, and A. H. Heuer, *J. Appl. Phys.* 83, 3311 (1998).

197. M. Takeuchi, T. Itoh, and H. Nagasaki, *Thin Solid Films* 51, 1978 (1978).

198. B. He, T. Ma, S. A. Campbell, and W. L. Gladfelter, *IEDM Tech. Digest* 1038 (1998).

199. T. Kamada, M. Kitagawa, M. Shibuya, and T. Hirao, *Jpn. J. Appl. Phys.* 30, 3594 (1991).

200. Y. Abe and T. Fukuda, *Jpn. J. Appl. Phys.* 32, L1167 (1993).

201. W. D. Brown and W. W. Grannemann, *Thin Solid Films* 51, 119 (1978).

202. W. D. Brown and W. W. Grannemann, *Solid-State Electron.* 21, 837 (1978).

203. X. Guo, T. P. Ma, T. Tamagawa, and B. L. Halpern, *IEDM Tech. Digest* 377 (1998).

204. X. Guo, X. Wang, Z. Luo, T. P. Ma, and T. Tamagawa, *IEDM Tech. Digest* 1038 (1999).

205. C. Hobbs, R. Hegde, B. Maiti, H. Tseng, D. Gilmer, P. Tobin, O. Adetutu, F. Huang, D. Weddington, R. Nagabushnam, D. O'Meara, K. Reid, L. La, L. Grove, and M. Rossow, *Symp. VLSI Tech. Digest* 133 (1999).

206. G. Taylor and B. Lalevic, *Solid-State Electron.* 19, 669 (1976).

207. G. Taylor and B. Lalevic, *J. Appl. Phys.* 48, 4410 (1977).

208. K. N. Kim, T. Y. Chung, H. S. Jeong, J. T. Moon, Y. W. Park, G. T. Jeong, K. H. Lee, G. H. Koh, D. W. Shin, Y. S. Hwang, D. W. Kwak, H. S. Uh, D. W. Ha, J. W. Lee, S. H. Shin, M. H. Lee, Y. S. Chun, J. K. Lee, B. J. Park, J. H. Oh, J. G. Lee, and S. H. Lee, *Symp. VLSI Tech. Digest* 10 (2000).

209. D. Ha, D. Shin, G. H. Koh, J. Lee, S. Lee, Y. S. Ahn, H. Jeong, T. Chung, and K. Kim, *IEEE Trans. Electron Devices* ED-47, 1499 (2000).

210. A. Chin, C. C. Laio, C. H. Lu, W. J. Chen, and C. Tsai, *Symp. VLSI Tech. Digest* 135 (1999).

211. J. Kolodzey, E. A. Chowdhury, T. N. Adam, G. Qui, I. Rau, J. O. Olowolafe, J. S. Suehle, and Y. Chen, *IEEE Trans. Electron Devices* ED-47, 121 (2000).

212. L. Manchanda, W. H. Lee, J. E. Bower, F. H. Baumann, W. L. Brown, C. J. Case, R. C. Keller, Y. O. Kim, E. J. Laskowski, M. D. Morris, R. L. Opila, P. J. Silverman, T. W. Sorsch, and G. R. Weber, *IEDM Tech. Digest* 605 (1998).

213. K. S. Lim and C. H. Ling, *ICSE Proc.* 42 (1998).

214. J. S. Lim, Y. K. Kim, S. J. Choi, J. H. Lee, Y. S. Kim, B. T. Lee, H. S. Park, Y. W. Park, and S. I. Lee, "International Conference on VLSI CAD," 1999, p. 506.

215. J. Kolodzey, E. A. Chowdhury, G. Qui, J. Olowolage, C. P. Swann, K. M. Unruh, J. Suehle, R. G. Wilson, and J. M. Zavada, *J. Appl. Phys.* 77, 3802 (1997).

216. R. Ludeke, M. T. Cuberes, and E. Cartier, *Appl. Phys. Lett.* 76, 2886 (2000).

217. Y. K. Kim, S. M. Lee, I. S. Park, C. S. Park, S. I. Lee, and M. Y. Lee, *Symp. VLSI Tech. Digest* 52 (1998).

218. Z. Jin, H. S. Kwok, and M. Wong, *IEEE Electron Device Lett.* EDL-19, 502 (1998).

219. S. Zhang and R. Xiao, *J. Appl. Phys.* 83, 3842 (1998).

220. L. Manchanda and M. Gurvitch, *IEEE Electron Device Lett.* EDL-9, 180 (1988).

221. V. A. Rozhkov, V. P. Goncharov, and A. Y. Trusova, "IEEE Fifth International Conference on Conduction and Breakdown in Solid Dielectrics," 1995, p. 552.

222. L. S. Yip and I. Shih, *Electron. Lett.* 24, 1287 (1998).

223. M. E. Hunter, M. J. Reed, N. A. El-Masry, J. C. Roberts, and S. M. Bedair, *Appl. Phys. Lett.* 76, 1935 (2000).

224. B. H. Lee, L. Kang, W. J. Qi, R. Nieh, Y. Joen, K. Onishi, and J. C. Lee, *IEDM Tech. Digest* 133 (1999).

225. W. J. Qi, R. Nieh, B. H. Lee, L. Kang, Y. Jeon, K. Onishi, T. Ngai, S. Banerjee, and J. C. Lee, *IEDM Tech. Digest* 145 (1999).

226. T. Nagi, W. J. Qi, R. Sharma, J. Fretwell, X. Chen, J. C. Lee, and S. Banerjee, *Appl. Phys. Lett.* 76, 502 (2000).

227. M. Houssa, M. Tuominen, M. Naili, V. Afans'ev, A. Stesmans, S. Haukka, and M. M. Heyns, *J. Appl. Phys.* 87, 8615 (2000).

228. B. H. Lee, L. Kang, R. Nieh, W. J. Qi, and J. C. Lee, *Appl. Phys. Lett.* 76, 1926 (2000).

229. L. Kang, B. H. Lee, W. J. Qi, Y. Jeon, R. Nieh, S. Gopalan, K. Onishi, and J. C. Lee, *IEEE Electron Device Lett.* EDL-21, 181 (2000).

230. Y. Ma, Y. Ono, L. Stecker, D. R. Evans, and S. T. Hsu, *IEDM Tech. Digest* 149 (1999).

231. G. D. Wilk and R. M. Wallace, *Appl. Phys. Lett.* 74, 2854 (1999).

232. G. D. Wilk, R. M. Wallace, and J. M. Anthony, *J. Appl. Phys.* 87, 484 (2000).

233. M. Hong, "Proceedings of the Fifth International Conference on SSIC Technology," 1998, p. 319.

234. F. Ren, M. Hong, J. M. Kuo, W. S. Hobson, J. R. Lothian, H. S. Tsai, J. Lin, P. Mannaerts, J. Kwo, S. N. G. Chu, Y. K. Chen, and A. Y. Cho, "19th GaAs IC Symposium," 1997, p. 18.

235. S. J. Kim, J. W. Park, M. Hong, and J. P. Mannaerts, *IEE Proc. Circuit Devices Syst.* 145, 162 (1998).

236. F. Ren, J. M. Kuo, M. Hong, W. S. Hobson, J. R. Lothian, J. Lin, H. S. Tsai, J. P. Mannaerts, J. Kwo, S. N. G. Chu, Y. K. Chen, and A. Y. Cho, *IEEE Electron Device Lett.* EDL-19, 309 (1998).

237. Y. C. Wang, M. Hong, J. M. Kuo, J. P. Mannaerts, H. S. Tsai, J. Kwo, J. J. Krajewski, Y. K. Chen, and A. Y. Cho, *Electron. Lett.* 35, 667 (1999).

238. B. Jaffe, W. R. Cook, and H. Jaffe, "Piezoelectric Ceramics." Academic Press, London, 1971.

239. A. J. Moulson and J. M. Herbert, "Electroceramics." Chapman & Hall, London, 1990.

240. H. Kidoh and T. Ogawa, *Appl. Phys. Lett.* 58, 2910 (1991).

241. Y. M. Kim, W. J. Lee, and H. G. Kim, *Thin Solid Films* 279, 140 (1996).

242. S. B. Krupanidhi, N. Maffei, M. Sayer, and K. El-Assai, *J. Appl. Phys.* 54, 6601 (1983)

243. S. K. Dey and R. Zuleeg, *Ferroelectrics* 108, 37 (1990)

244. L. E. Sanchez, S. Y. Wu, and I. K. Naik, *Appl. Phys. Lett.* 56, 2399 (1990)

245. K. R. Udayakumar, P. J. Schuele, J. Chen, S. B. Krupanidhi, and L. E. Cross, *J. Appl. Phys.* 77, 3981 (1995).

246. T. K. Kundu and J. Y. Lee, *J. Electrochem. Soc.* 147, 326 (2000).

247. C. Sudhama, J. Kim, R. Khamankar, V. Chikarmane, and J. C. Lee, *J. Electron Mater.* 23, 1261 (1994).

248. J. F. Scott, B. M. Melnick, L. D. McMillan, and C. A. Paz de Araujo, *Integ. Ferroelec.* 3, 129 (1993).

249. J. F. Scott, M. Azuma, C. A. Paz de Araujo, L. D. McMillan, M. C. Scott, and T. Roberts, *Integ. Ferroelec.* 4, 61 (1994).

250. J. F. Scott, in "Thin Film Ferroelectric Materials and Devices" (R. Ramesh, Ed.), p. 115. Kluwer Academic, Boston, 1997.

251. I. Stolichnov and A. Tagantsev, *J. Appl. Phys.* 84, 3216 (1998).

252. I. Stolichnov, A. K. Tagantsev, E. L. Colla, and N. Setter, *Appl. Phys. Lett.* 73, 1361 (1998).

253. J. F. Scott, B. M. Melnick, J. D. Cuchiaro, R. Zuleeg, C. A. Araujo, L. D. McMillan, and M. C. Scott, *Integ. Ferroelec.* 4, 85 (1994).

254. J. F. Scott, C. A. Araujo, B. M. Meadows, L. D. McMillan, and R. Zuleeg, *J. Appl. Phys.* 70, 382 (1991).

255. B. M. Melnick, J. F. Scott, C. A. Paz de Araujo, and L. D. McMillan, *Ferroelectrics* 135, 163 (1992).

256. H. Hu and S. B. Krupanidhi, *J. Mater. Res.* 9, 1484 (1994).

257. Y. S. Hwang, S. H. Paek, and J. P. Mah, *Journal of Materials Science Letters* 15, 1039 (1996).

258. T. Mihara and H. Watanabe, *Jpn. J. Appl. Phys.* 34, 5664 (1995).

259. T. Mihara and H. Watanabe, *Jpn. J. Appl. Phys.* 34, 5674 (1995).

260. P. W. M. Blom, R. M. Wolf, J. F. M. Cillessen, and M. P. C. M. Krijn, *Phys. Rev. Lett.* 73, 2107 (1994).

261. I. K. Yoo and S. B. Desu, "Proceedings of the Eighth IEEE International Symposium on the Applications of Ferroelectrics," 1992, p. 225.

262. H. M. Chen, S. W. Tsaur, and J. Y. Lee, *Jpn. J. Appl. Phys.* 37, 4056 (1998).

263. R. Waser, *Ferroelectrics* 133, 109 (1992).

264. R. Waser, *J. Am. Ceram. Soc.* 73, 1645 (1990).

265. B. Nagaraj, S. Aggarwal, T. K. Song, T. Sawhney, and R. Ramesh, *Phys. Rev. B* 59, 16022 (1999).

266. S. M. Sze, "Physics of Semiconductor Devices," p. 276. Wiley, New York, 1981.

267. P. K. Larson, G. J. M. Dormans, D. J. Taylor, and P. J. van Veldhoven, *J. Appl. Phys.* 76, 2405 (1994).

268. J. J. Lee, C. L. Thio, and S. B. Desu, *J. Appl. Phys.* 78, 5073 (1995).

269. C. Yoshida, A. Yoshida, and H. Tamura, *Appl. Phys. Lett.* 75, 1449 (1999).

270. H. M. Chen, J. M. Lan, J. L. Chen, and J. Y. Lee, *Appl. Phys. Lett.* 69, 1713 (1996).

271. J. M. Benedetto, H. E. Boesch, F. B. McLean, and J. P. Mize, *IEEE Trans. Nucl. Sci.* NS-32, 3916 (1985).

272. D. J. Dumin, J. R. Maddux, R. S. Scott, and R. Subramonium, *IEEE Trans. Electron Devices* ED-41, 1570 (1994).

273. H. M. Chen and J. Y. Lee, *Appl. Phys. Lett.* 73, 309 (1998).

274. H. M. Chen and J. Y. Lee, *J. Appl. Phys.* 82, 3478 (1997).

275. I. P. Batra and B. D. Silverman, *Solid State Commun.* 11, 291 (1972).

276. D. R. Tilley and B. Zeks, *Solid State Commun.* 49, 823 (1984).

277. D. R. Tilley, in "Ferroelectric Crystals" (N. Setter and E. L. Colla, Eds.), p. 163. Birkhäuser, Berne, 1993.

278. S. B. Desu, C. H. Peng, L. Kammerdiner, and P. J. Schuele, *Mater. Res. Soc. Symp. Proc.* 200, 319 (1990).

279. C. R. Cho, W. J. Lee, B. G. Yu, and B. W. Kim, *J. Appl. Phys.* 86, 2700 (1999).

280. M. Aoki, M. Mushiga, A. Itoh, T. Eshita, and Y. Arimoto, "Symposium on VLSI Technology," 1999, p. 145.

281. J. T. Evans and R. Womack, *IEEE J. Solid State Circuit* 23, 1171 (1991).

282. J. T. Evans and R. I. Suizu, "1998 International Nonvolatile Memory Technology Conference," 1998, p. 26.

283. R. E. Jones, "IEEE 1998 Custom Integrated Circuits Conference," 1998, p. 431.

284. W. Kraus, L. Lehman, D. Wilson, T. Yamazaki, C. Ohno, E. Nagai, H. Yamazaki, and H. Suzuki, "IEEE 1998 Custom Integrated Circuits Conference," 1998, p. 242.

285. H. Takasu, T. Nakamura, and A. Kamisawa, *Integ. Ferroelec.* 21, 41 (1998).

286. D. Dimos, W. L. Warren, and H. N. Al-Shareef, in "Thin Film Ferroelectric Materials and Devices" (R. Ramesh, Ed.), p. 199. Kluwer Academic, Boston, 1997.

287. C. J. Brennan, R. D. Parrella, and D. E. Larson, *Ferroelectrics* 151, 33 (1994).

288. W. L. Warren, D. Dimos, B. A. Tuttle, R. D. Nasby, and G. E. Pike, *Appl. Phys. Lett.* 65, 1018 (1994).

289. W. L. Warren, D. Dimos, B. A. Tuttle, G. E. Pike, R. W. Schwartz, P. J. Clews, and D. C. McIntrye, *J. Appl. Phys.* 77, 6695 (1995).

290. S. B. Desu and I. K. Yoo, *Integ. Ferroelec.* 3, 365 (1993).

291. I. K. Yoo, S. B. Desu, and J. Xing, *Mater. Res. Soc. Symp. Proc.* 310, 165 (1993).

292. W. Y. Pan, C. F. Yue, and B. A. Tuttle, *Ceram. Trans.* 25, 385 (1992).

293. Z. Wu and M. Sayer, "Proceedings of the American Ceramic Society," 1993.

294. H. Watanabe, T. Mihara, H. Yoshimori, and C. A. Paz de Araujo, "Proceedings of the Fourth International Symposium on Integrate Ferroelectrics," 1992, p. 346.

295. W. I. Lee, J. K. Lee, I. Cheung, I. K. Yoo, and S. B. Desu, *Mater. Res. Soc. Symp. Proc.* 361, 421 (1995).

296. H. N. Al-Shareef, D. Dimos, T. J. Boyle, W. L. Warren, and B. A. Tuttle, *Appl. Phys. Lett.* 68, 690 (1996).

297. D. Dimos, H. N. Al-Shareef, W. L. Warren, and B. A. Tuttle, *J. Appl. Phys.* 80, 1682 (1996).

298. X. Du and I. W. Chen, *Appl. Phys. Lett.* 72, 1923 (1998).

299. H. N. Al-Shareef, B. A. Tuttle, W. L. Warren, T. J. Headley, D. Dimos, J. A. Voigt, and R. D. Nasby, *J. Appl. Phys.* 79, 1013 (1996).

300. U. Robels, L. Schneider-Stormann, and G. Arlt, *Ferroelectrics* 168, 301 (1995).

301. P. V. Lambeck and G. H. Jonker, *Ferroelectrics* 22, 729 (1978).

302. G. Arlt and H. Neumann, *Ferroelectrics* 87, 109 (1988).

303. W. A. Schulze and K. Ogino, *Ferroelectrics* 87, 361 (1988).

304. H. J. Hagemann, *J. Phys. C: Solid State Phys.* 11, 3333 (1978).

305. D. Dimos, W. L. Warren, M. B. Sinclair, B. A. Tuttle, and R. W. Schwartz, *J. Appl. Phys.* 76, 4305 (1994).

306. G. E. Pike, W. L. Warren, D. Dimos, B. A. Tuttle, R. Ramesh, J. Lee, V. G. Keramidas, and J. T. Evans, *Appl. Phys. Lett.* 66, 484 (1995).

307. S. Y. Chen and V. C. Lee, *J. Appl. Phys.* 87, 3050 (2000).

308. W. P. Mason, *J. Acoust. Soc. Am.* 27, 73 (1955).

309. R. Herbiet, H. Tenbrock, and G. Arlt, *Ferroelectrics* 76, 319 (1987).

310. R. C. Bradt and G. S. Ansell, *J. Am. Ceram. Soc.* 52, 192 (1969).

311. C. J. Peng and S. B. Krupanidhi, *J. Mater. Res.* 10, 708 (1995).

312. J. C. Lee, L. C. Chin, and C. Hu, *IEEE Trans. Electron Devices* 35, 2268 (1988).

313. J. L. Chen, H. M. Chen, and J. Y. Lee, *Appl. Phys. Lett.* 69, 4011 (1996).

314. S. C. Huang, H. M. Chen, S. C. Wu, and J. Y. Lee, *J. Appl. Phys.* 84, 5155 (1998).

315. J. Lee, R. Ramesh, V. G. Keramidas, W. L. Warren, G. E. Pike, and J. T. Evans, *Appl. Phys. Lett.* 66, 1337 (1995).

316. B. A. Tuttle, in "Thin Film Ferroelectric Materials and Devices" (R. Ramesh, Ed.), p. 145. Kluwer Academic, Boston, 1997.

317. B. A. Tuttle, H. N. Al-Shareef, W. L. Warren, M. V. Raymond, T. J. Headley, J. A. Voigt, J. Evans, and R. Ramesh, *Microelectron Eng.* 29, 223 (1995).

318. H. N. Al-Shareef, D. Dimos, W. L. Warren, and B. A. Tuttle, *J. Appl. Phys.* 80, 4573 (1996).

319. W. L. Warren, D. Dimos, G. E. Pike, B. A. Tuttle, M. V. Raymond, R. Ramesh, and J. T. Evans, *Appl. Phys. Lett.* 67, 866 (1995).

320. W. L. Warren, H. N. Al-Shareef, D. Dimos, B. A. Tuttle, and G. E. Pike, *Appl. Phys. Lett.* 68, 1681 (1996).

321. S. Sadashivan, S. Aggarwal, T. K. Song, R. Ramesh, J. T. Evans, B. A. Tuttle, W. L. Warren, and D. Dimos, *J. Appl. Phys.* 83, 2165 (1998).

322. J. Lee, C. H. Choi, B. H. Park, T. W. Noh, and J. K. Lee, *Appl. Phys. Lett.* 72, 3380 (1998).

323. J. Lee and R. Ramesh, *Appl. Phys. Lett.* 68, 484 (1996).

324. J. Yin, T. Zhu, Z. G. Liu, and T. Yu, *Appl. Phys. Lett.* 75, 3698 (1999).

325. W. Jo, D. C. Kim, and J. W. Hong, *Appl. Phys. Lett.* 76, 390 (2000).

326. A. R. Zomorrodian, H. Lin, N. J. Wu, T. Q. Huang, D. Liu, and A. Ignatiev, *Appl. Phys. Lett.* 69, 1789 (1996).

327. Z. G. Zhang, Y. N. Wang, J. S. Zhu, F. Yan, X. M. Lu, H. M. Shen, and J. S. Liu, *Appl. Phys. Lett.* 73, 3674 (1998).

328. Y. Shimada, K. Nakao, A. Inoue, M. Azuma, Y. Uemoto, E. Fujii, and T. Otsuki, *Appl. Phys. Lett.* 71, 2538 (1997).

329. S. Aggarwal, T. K. Song, A. M. Dhote, A. S. Prakash, R. Ramesh, N. Velasquez, L. Boyer, and J. T. Evans, *J. Appl. Phys.* 83, 1617 (1998).

330. W. L. Warren, D. Dimos, and R. M. Waser, *Mater. Res. Soc. Bull.* 40 (1996).

331. N. Mikami, in "Thin Film Ferroelectric Materials and Devices" (R. Ramesh, Ed.), p. 43. Kluwer Academic, Boston, 1997.

332. B. A. Baumert, L. H. Chang, A. T. Matsuda, T. L. Tsai, C. J. Tracy, R. B. Gregory, P. L. Fejes, N. G. Cave, W. Chen, D. J. Taylor, T. Otsuki, E. Fujii, S. Hayashi, and K. Suu, *J. Appl. Phys.* 82, 2558 (1997).

333. C. M. Chu and P. Lin, *Appl. Phys. Lett.* 70, 249 (1997).

334. C. M. Chu and P. Lin, *Appl. Phys. Lett.* 72, 1241 (1998).

335. G. W. Dietz, M. Schumacher, R. Waser, S. K. Streiffer, C. Basceri, and A. I. Kingon, *J. Appl. Phys.* 82, 2359 (1997).

336. Y. Gao, S. He, P. Alluri, M. Engelhard, A. S. Lea, J. Finder, B. Melnick, and R. L. Hance, *J. Appl. Phys.* 87, 124 (2000).

337. T. Horikawa, N. Mikami, T. Makita, J. Tanimura, M. Kataoka, K. Sato, and M. Nunoshita, *Jpn. J. Appl. Phys.* 32, 4126 (1993).

338. S. Y. Hou, J. Kwo, R. K. Watts, J. Y. Cheng, and D. K. Fork, *Appl. Phys. Lett.* 67, 1387 (1995).

339. C. S. Hwang, S. O. Park, H. J. Cho, C. S. Kang, H. K. Kang, S. I. Lee, and M. Y. Lee, *Appl. Phys. Lett.* 67, 2819 (1995).

340. M. Izuha, K. Abe, M. Koike, S. Takeno, and N. Fukushima, *Appl. Phys. Lett.* 70, 1405 (1997).

341. Q. X. Jia, X. D. Xu, S. R. Foltyn, and P. Tiwari, *Appl. Phys. Lett.* 66, 2197 (1995).

342. T. Kawahara, M. Yamamuka, A. Yuuki, and K. Ono, *Jpn. J. Appl. Phys.* 35, 4880 (1996).

343. K. Koyama, T. Sakuma, S. Yamamichi, H. Watanabe, H. Aoki, S. Ohya, Y. Miyasaka, and T. Kikkawa, *IEDM* 823 (1991).

344. T. Kuroiwa, Y. Tsunemine, T. Horikawa, T. Makita, J. Tanimura, N. Mikami, and K. Sato, *Jpn. J. Appl. Phys.* 33, 5187 (1994).

345. W. J. Lee, I. K. Park, G. E. Jang, and H. G. Kim, *Jpn. J. Appl. Phys.* 34, 196 (1995).

346. B. Nagaraj, T. Sawhney, S. Perusse, S. Aggarwal, V. S. Kaushik, S. Zafar, R. E. Jones, J. H. Lee, V. Balu, and J. Lee, *Appl. Phys. Lett.* 74, 3194 (1999).

347. C. J. Peng and S. B. Krupanidhi, *J. Mater. Res.* 10, 708 (1995).

348. D. Roy and S. B. Krupanidhi, *Appl. Phys. Lett.* 10, 1056 (1993).

349. Y. Shimada, A. Inoue, T. Nasu, K. Arita, Y. Nagano, A. Matsuda, Y. Uemoto, E. Fujii, M. Azuma, Y. Oishi, S. Hayashi, and T. Otsuki, *Jpn. J. Appl. Phys.* 35, 140 (1996).

350. J. C. Shin, J. Park, C. S. Hwang, and H. J. Kim, *J. Appl. Phys.* 86, 506 (1999).

351. M. S. Tsai, S. C. Sun, and T. Y. Tseng, *IEEE Trans. Electron Devices* 46, 1829 (1999).

352. C. M. Wu and T. B. Wu, *Jpn. J. Appl. Phys.* 36, 1164 (1997).

353. S. Yamamichi, P. Y. Lesaicherre, H. Yamaguchi, K. Takemura, S. Sone, H. Yabuta, K. Sato, T. Tamura, K. Nakajima, S. Ohnishi, K. Tokashiki, Y. Hayashi, Y. Kato, Y. Miyasaka, M. Yoshida, and H. Ono, *IEEE Trans. Electron Devices* 44, 1076 (1997).

354. W. J. Lee, H. G. Kim, and S. G. Yoon, *J. Appl. Phys.* 80, 5891 (1996).

355. W. Y. Hsu, J. D. Luttmer, R. Tsu, S. Summerfelt, M. Bedekar, T. Tokumoto, and J. Nulman, *Appl. Phys. Lett.* 66, 2975 (1995).

356. C. S. Hwang, B. T. Lee, H. J. Cho, K. H. Lee, C. S. Kang, H. Hideki, S. I. Lee, and M. Y. Lee, *Appl. Phys. Lett.* 71, 371 (1997).

357. C. S. Hwang, B. T. Lee, C. S. Kang, J. W. Kim, K. H. Lee, H. J. Cho, H. Horii, W. D. Kim, S. I. Lee, Y. B. Roh, and M. Y. Lee, *J. Appl. Phys.* 83, 3703 (1998).

358. C. S. Hwang, B. T. Lee, C. S. Kang, K. H. Lee, H. J. Cho, H. Hideki, W. D. Kim, S. I. Lee, and M. Y. Lee, *J. Appl. Phys.* 85, 287 (1999).

359. C. S. Hwang and S. H. Joo, *J. Appl. Phys.* 85, 2431 (1999).

360. J. H. Joo, J. M. Seon, Y. C. Jeon, K. Y. Oh, J. S. Roh, and J. J. Kim, *Appl. Phys. Lett.* 70, 3053 (1997).

361. V. Joshi, C. P. Dacruz, J. D. Cuchiaro, and C. A. Araujo, *Integ. Ferroelec.* 14, 133 (1997).

362. Y. T. Kim and C. W. Lee, *Jpn. J. Appl. Phys.* 35, 6153 (1996).

363. Y. B. Lin and J. Y. Lee, *J. Appl. Phys.* 87, 1841 (2000).

364. S. Maruno, T. Kuroiwa, N. Mikami, K. Sato, S. Ohmura, M. Kaida, T. Yasue, and T. Koshikawa, *Appl. Phys. Lett.* 73, 954 (1998).

365. Y. P. Wang and T. Y. Tseng, *J. Appl. Phys.* 81, 6762 (1997).

366. S. Yamamichi, A. Yamamichi, D. Park, T. J. King, and C. Hu, *IEEE Trans. Electron Devices* 46, 342 (1999).

367. S. Zafar, R. E. Jones, B. Jiang, B. White, P. Chu, D. Taylor, and S. Gillespie, *Appl. Phys. Lett.* 73, 175 (1998).

368. S. Dey, J. J. Lee, and P. Alluri, *Jpn. J. Appl. Phys., Part 1* 34, 3142 (1995).

369. S. Dey, P. Alluri, J. J. Lee, and R. Zuleeg, *Integ. Ferroelec.* 7, 715 (1995).

370. C. S. Hwang, B. T. Lee, S. O. Park, J. W. Kim, H. J. Cho, C. S. Kang, H. Horii, S. I. Lee, and M. Y. Lee, *Integ. Ferroelec.* 13, 157 (1996).

371. J. F. Scott, *Integ. Ferroelec.* 9, 1 (1995).

372. N. H. Chan, R. K. Sharma, and D. M. Smyth, *J. Am. Ceram. Soc.* 64, 556 (1981).

373. M. Copel, P. R. Duncombe, D. A. Neumayer, T. M. Shaw, and R. M. Tromp, *Appl. Phys. Lett.* 70, 3227 (1997).

374. Y. Shimada, A. Inoue, T. Nasu, Y. Nagano, A. Matsuda, K. Arita, Y. Uemoto, E. Fujii, and T. Otsuki, *Jpn. J. Appl. Phys.* 35, 4919 (1996).

375. J. F. Scott, M. Azuma, C. A. Paz de Araujo, L. D. McMillan, M. C. Scott, and T. Roberts, *Integ. Ferroelec.* 4, 61 (1994).

376. J. H. Joo, Y. C. Jeon, J. M. Seon, K. Y. Oh, J. S. Roh, and J. J. Kim, *Jpn. J. Appl. Phys.* 36, 4382 (1997).

377. S. H. Paek, J. Won, K. S. Lee, J. S. Choi, and C. S. Park, *Jpn. J. Appl. Phys.* 35, 5757 (1996).

378. T. Eimori, Y. Ohno, H. Kimura, J. Matsufusa, S. Kishimura, A. Yoshida, H. Sumitani, T. Maruyama, Y. Hayashide, K. Morizumi, T. Katayama, M. Asakura, T. Horikawa, T. Shibano, H. Itoh, K. Sato, K. Namba, T. Nishimura, S. Satoh, and H. Miyoshi, *IEDM* 631 (1993).

379. E. Fujii, Y. Uemoto, S. Hayashi, T. Nasu, Y. Shimada, A. Matsuda, M. Kibe, M. Azuma, T. Otsuki, G. Kano, M. Scott, L. D. McMillan, and C. A. Paz de Araujo, *IEDM* 267 (1992).

380. S. Hayashi, M. Huffman, M. Azuma, Y. Shimada, T. Otsuki, G. Kano, L. D. McMillan, and C. A. Paz de Araujo, "Symposium on VLSI Technology," 1994, p. 153.

381. K. N. Kim, D. H. Kwak, Y. S. Hwang, G. T. Jeong, T. Y. Chung, B. J. Park, Y. S. Chun, J. H. Oh, C. Y. Yoo, and B. S. Joo, "Symposium on VLSI Technology," 1999, p. 33.

382. B. K. Lee, K. H. Lee, C. S. Hwang, W. D. Kim, H. Horii, H. W. Kim, H. J. Cho, C. S. Kang, J. H. Chung, S. I. Lee, and M. Y. Lee, *IEDM* 249 (1997).

383. B. K. Lee, C. Y. Yoo, H. J. Lim, C. S. Kang, H. B. Park, W. D. Kim, S. H. Ju, H. Horii, K. H. Lee, H. W. Kim, S. I. Lee, and M. Y. Lee, *IEDM* 815 (1998).

384. Y. Nishioka, K. Shiozawa, T. Oishi, K. Kanamoto, Y. Tokuda, H. Sumitani, S. Aya, H. Yabe, K. Itoga, T. Hifumi, K. Marumoto, T. Kuroiwa, T. Kawahara, K. Nishikawa, T. Oomori, T. Fujino, S. Yamamoto, S. Uzawa, M. Kimata, M. Nunoshita, and H. Abe, *IEDM* 903 (1995).

385. A. Nitayama, Y. Kohyama, and K. Hieda, *IEDM* 355 (1998).

386. K. Ono, T. Horikawa, T. Shibano, N. Mikami, T. Kuroiwa, T. Kawahara, S. Matsuno, F. Uchikawa, S. Satoh, and H. Abe, *IEDM* 803 (1998).

387. A. Yuuki, M. Yamamuka, T. Makita, T. Horikawa, T. Shibano, N. Hirano, H. Maeda, N. Mikami, K. Ono, H. Ogata, and H. Abe, *IEDM* 115 (1995).

388. S. Zafar, B. Hradsky, D. Gentile, P. Chu, R. E. Jones, and S. Gillespie, *J. Appl. Phys.* 86, 3890 (1999).

389. K. Numata, Y. Fukuda, K. Aoki, and A. Nishimura, *Jpn. J. Appl. Phys.* 34, 5245 (1995).

390. T. Horikawa, T. Kawahara, M. Yamamuka, and K. Ono, "IEEE Reliability Physics Symposium," 1997, p. 82.

391. O. Auciello, A. Krauss, and J. Im, in "Thin Film Ferroelectric Materials and Devices" (R. Ramesh, Ed.), p. 91. Kluwer Academic, Boston, 1997.

392. T. Mihara, H. Yoshimori, H. Watanabe, and C. A. Paz de Araujo, *Jpn. J. Appl. Phys.* 34, 5233 (1995).

393. T. Hayashi, H. Takahashi, and T. Hara, *Jpn. J. Appl. Phys.* 35, 4952 (1996).

394. P. C. Joshi, S. O. Ryu, X. Zhang, and S. B. Desu, *Appl. Phys. Lett.* 70, 1080 (1997).

395. S. S. Park, C. H. Yang, S. G. Yoon, J. H. Ahn, and H. G. Kim, *J. Electrochem. Soc.* 144, 2855 (1997).

396. O. Auciello, *Integ. Ferroelec.* 15, 211 (1997).

397. A. J. Hartmann and R. N. Lamb, *Integ. Ferroelec.* 18, 101 (1997).

398. K. Takemura, T. Noguchi, T. Hase, and Y. Miyasaka, *Appl. Phys. Lett.* 73, 1649 (1998).

399. P. Y. Chu, R. E. Jones, P. Zurcher, D. J. Taylor, B. Jiang, S. J. Gillespie, Y. T. Lii, M. Kottke, P. Fejes, and W. Chen, *J. Mater. Res.* 11, 1065 (1996).

400. R. Dat, J. K. Lee, O. Auciello, and A. I. Kingon, *Appl. Phys. Lett.* 67, 572 (1995).

401. G. D. Hu, I. H. Wilson, J. B. Xu, W. Y. Cheung, S. P. Wong, and H. K. Wong, *Appl. Phys. Lett.* 74, 1221 (1999).

402. K. Ishikawa and H. Funakubo, *Appl. Phys. Lett.* 75, 1970 (1999).

403. B. S. Kang, B. H. Park, S. D. Bu, S. H. Kang, and T. W. Noh, *Appl. Phys. Lett.* 75, 2644 (1999).

404. O. S. Kwon, C. S. Hwang, and S. K. Hong, *Appl. Phys. Lett.* 75, 558 (1999).

405. D. Ravichandran, K. Yamakawa, R. Roy, A. S. Bhalla, S. Trolier-McKinstry, R. Guo, and L. E. Cross, "Proceedings of the 10th IEEE International Symposium on Applications of Ferroelectrics," 1996, p. 601.

406. A. C. Rastogi, S. Tirumala, and S. B. Desu, *Appl. Phys. Lett.* 74, 3492 (1999).

407. N. J. Seong, S. G. Yoon, and S. S. Lee, *Appl. Phys. Lett.* 71, 81 (1997).

408. C. H. Yang, S. S. Park, and S. G. Yoon, *Integ. Ferroelec.* 18, 377 (1997).

409. T. Kanehara, I. Koiwa, Y. Okada, K. Ashikaga, H. Katoh, and K. Kaifu, *IEDM* 601 (1997).

410. Y. T. Kim, D. S. Shin, Y. K. Park, and I. H. Choi, *J. Appl. Phys.* 86, 3387 (1999).

411. N. J. Seong, C. H. Yang, W. C. Shin, and S. G. Yoon, *Appl. Phys. Lett.* 72, 1374 (1998).

412. D. J. Taylor, R. E. Jones, P. Y. Chu, P. Zurcher, B. White, S. Zafar, and S. J. Gillespie, "Proceedings of the 10th IEEE International Symposium on Applications of Ferroelectrics," 1996, p. 55.

413. K. Watanabe, A. J. Hartmann, R. N. Lamb, and J. F. Scott, *J. Appl. Phys.* 84, 2170 (1998).

414. A. Rose, *Phys. Rev.* 97, 1538 (1955).

415. M. A. Lampert and P. Mark, "Current Injection in Solids." Academic Press, New York, 1970.

416. J. F. Scott, B. M. Melnick, L. D. McMillan, and C. A. Paz de Araujo, *Integ. Ferroelec.* 3, 129 (1993).

417. J. F. Scott, M. Azuma, C. A. Paz de Araujo, L. D. McMillan, M. C. Scott, and T. Roberts, *Integ. Ferroelec.* 4, 61 (1994).

418. J. Robertson, C. W. Chen, W. L. Warren, and C. D. Gutleben, *Appl. Phys. Lett.* 69, 1704 (1996).

419. T. Eshita, K. Nakamura, M. Mushiga, A. Itoh, S. Miyagaki, H. Yamawaki, M. Aoki, S. Kishii, and Y. Arimoto, "Symposium on VLSI Technology," 1999, p. 139.

420. O. Auciello, A. R. Krauss, J. Im, D. M. Gruen, E. A. Irene, R. P. H. Chang, and G. E. McGuire, *Appl. Phys. Lett.* 69, 2671 (1996).

421. C. A. Paz de Araujo, J. D. Cuchiaro, L. D. McMillan, M. C. Scott, and J. F. Scott, *Nature* 374, 627 (1995).

422. B. H. Park, S. J. Hyun, S. D. Bu, T. W. Noh, J. Lee, H. D. Kim, T. H. Kim, and W. Jo, *Appl. Phys. Lett.* 74, 1907 (1999).

423. Z. G. Zhang, J. S. Liu, Y. N. Wang, J. S. Zhu, F. Yan, X. B. Chen, and H. M. Shen, *Appl. Phys. Lett.* 73, 788 (1998).

424. J. Im, O. Auciello, A. R. Krauss, D. M. Gruen, R. P. H. Chang, S. H. Kim, and A. I. Kingon, *Appl. Phys. Lett.* 74, 1162 (1999).

425. S. Zafar, V. Kaushik, P. Laberge, P. Chu, R. E. Jones, R. L. Hance, P. Zurcher, B. E. White, D. B. Melnick, and S. Gillespie, *J. Appl. Phys.* 82, 4469 (1997).

426. L. H. Chang and W. A. Anderson, *Appl. Surf. Sci.* 92, 52 (1996).

427. T. Hayashi, N. Oji, and H. Maiwa, *Jpn. J. Appl. Phys.* 33, 5277 (1994).

428. T. W. Kim, Y. S. Yoon, S. S. Yom, and C. O. Kim, *Appl. Surf. Sci.* 90, 75 (1995).

429. Z. Q. Shi, Q. X. Jia, and W. A. Anderson, *J. Vac. Sci. Technol., A* 10, 733 (1992).

430. P. C. Van Buskirk, R. Gardiner, P. S. Kirlin, and S. Krupanidhi, *J. Vac. Sci. Technol., A* 10, 1578 (1992).

431. K. Abe and S. Komatsu, *Jpn. J. Appl. Phys.* 32, 4186 (1993).

432. N. H. Cho, S. H. Nam, and H. G. Kim, *J. Vac. Sci. Technol., A* 10, 87 (1992).

433. Y. Fukuda, K. Aoki, K. Numata, and A. Nishimura, *Jpn. J. Appl. Phys.* 33, 5255 (1994).

434. T. Hirano, T. Fujii, K. Fujino, K. Sakuta, and T. Kobayashi, *Jpn. J. Appl. Phys.* 31, L511 (1992).

435. P. C. Joshi and S. B. Krupanidhi, *J. Appl. Phys.* 73, 7627 (1993).

436. M. B. Lee and H. Koinuma, *J. Appl. Phys.* 81, 2358 (1997).

437. W. S. Lee, N. O. Kim, J. K. Kim, and S. Y. Kim, "Proceedings of the Sixth International Conference on the Properties and Applications of Diel," 2000.

438. J. Nakahira, M. Kiyotoshi, S. Yamazaki, M. Nakabayashi, S. Niwa, K. Tsunoda, J. Lin, A. Shimada, M. Izuha, T. Aoyama, H. Tomita, K. Eguchi, and K. Hieda, "Symposium on VLSI Technology," 2000, p. 104.

439. S. H. Nam, N. H. Cho, and H. G. Kim, *J. Phys. D: Appl. Phys.* 25, 727 (1992).

440. G. M. Rao and S. B. Krupanidhi, *J. Appl. Phys.* 75, 2604 (1994).

441. E. Ching-Prado, A. Reynes-Figueroa, R. S. Katiyar, S. B. Majumder, and D. C. Agrawal, *J. Appl. Phys.* 78, 1920 (1995).

442. T. Imai, M. Okuyama, and Y. Hamakawa, *Jpn. J. Appl. Phys.* 30, 2163 (1991).

443. K. Kushida and H. Takeuchi, *IEEE Trans. Ultrasonics, Ferroelectrics and Frequency Control* 38, 656 (1991).

444. J. J. Lee, P. Alluri, and S. K. Dey, *Appl. Phys. Lett.* 65, 2027 (1994).

445. P. Wang, M. Liu, Y. Rao, Y. Zeng, and N. Ji, *Sens. Actuators, A* 49, 187 (1995).

446. S. Ibuki, T. Nakagawa, M. Okuyama, and Y. Hamakawa, *Jpn. J. Appl. Phys.* 29, 532 (1990).

447. R. Khamankar, J. Kim, B. Jiang, C. Sudhama, P. Maniar, R. Moazzami, R. Jones, and J. Lee, *IEDM* 337 (1994).

448. P. S. A. Kumar, B. Panda, S. K. Ray, B. K. Mathur, D. Bhattacharya, and K. L. Chopra, *Appl. Phys. Lett.* 68, 1344 (1996).

449. K. S. Liu, T. F. Tseng, and I. N. Lin, *Appl. Phys. Lett.* 72, 1182 (1998).

450. D. X. Lu, E. Y. B. Pun, E. M. W. Wong, P. S. Chung, and Z. Y. Lee, *IEEE Trans. Ultrasonics, Ferroelectrics and Frequency Control* 44, 675 (1997).

451. G. A. Petersen, L. C. Zou, W. M. Van Buren, L. L. Boyer, and J. R. McNeil, *Mater. Res. Soc. Symp. Proc.* 200, 127 (1990).

452. K. Tominaga, A. Shirayanagi, T. Takagi, and M. Okada, *Jpn. J. Appl. Phys.* 32, 4082 (1993).

453. T. F. Tseng, C. C. Yang, K. S. Liu, J. M. Wu, T. B. Wu, and I. N. Lin, *Jpn. J. Appl. Phys.* 35, 4743 (1996).

454. I. Stolichnov, A. Tagantsev, N. Setter, J. S. Cross, and M. Tsukada, *Appl. Phys. Lett.* 74, 3552 (1999).

455. I. Stolichnov, A. Tagantsev, N. Setter, S. Okhonin, P. Fazan, J. S. Cross, and M. Tsukada, *J. Appl. Phys.* 87, 1925 (2000).

456. S. G. Yoon, A. I. Kingon, and S. H. Kim, *J. Appl. Phys.* 88, 6690 (2000).

457. J. L. Moll and Y. Tarui, *IEEE Trans. Electron Devices* 10, 338 (1963).

458. S. Y. Wu, *Ferroelectrics* 11, 379 (1976).

459. Y. Tarui, T. Hirai, K. Teramoto, H. Koike, and K. Nagashima, *Appl. Surf. Sci.* 113/114, 656 (1997).

460. Y. Higuma, Y. Matsui, M. Okuyama, T. Nakagawa, and Y. Hamakawa, *Jpn. J. Appl. Phys.* 17, 209 (1978).

461. T. Hirai, K. Teramoto, T. Nishi, T. Goto, and Y. Tarui, *Jpn. J. Appl. Phys.* 33, 5219 (1994).

462. Y. Matsui, M. Okuyama, M. Noda, and Y. Hamakawa, *Appl. Phys. A* 28, 161 (1982).

463. D. R. Lampe, D. A. Adams, M. Austin, M. Polinsky, J. Dzimianski, S. Sinhalov, H. Buhay, P. Braband, and Y. M. Liu, *Ferroelectrics* 133, 61 (1992).

464. T. Hirai, K. Teramoto, K. Nagashima, H. Koike, and Y. Tarui, *Jpn. J. Appl. Phys.* 34, 4163 (1995).

465. Y. T. Kim and D. S. Shin, *Appl. Phys. Lett.* 71, 3507 (1997).

466. T. Kanashima and M. Okuyama, *Jpn. J. Appl. Phys.* 38, 2044 (1999).

467. B. M. Melnick, J. Gregory, and C. A. Paz de Araujo, *Integ. Ferroelec.* 11, 145 (1995).

468. T. Nakamura, Y. Nakao, A. Kamisawa, and H. Takasu, *Ferroelectrics* 11, 161 (1995).

469. C. Byun, Y. I. Kim, W. J. Lee, and B. W. Lee, *Jpn. J. Appl. Phys.* 36, 5588 (1997).

470. K. J. Choi, W. C. Shin, J. H. Yang, and S. G. Yoon, *Appl. Phys. Lett.* 75, 722 (1999).

471. T. Hirai, Y. Fujisaki, K. Nagashima, H. Koike, and Y. Tarui, *Jpn. J. Appl. Phys.* 36, 5908 (1997).

472. K. Kashihara, T. Okudaira, H. Itoh, T. Higaki, and H. Abe, "Symposium on VLSI Technology," 1993, p. 49.

473. T. Kijima and H. Matsunaga, *Jpn. J. Appl. Phys.* 37, 5171 (1998).

474. Y. T. Kim, C. W. Lee, D. S. Shin, and H. N. Lee, *IEEE ISAF Proc.* 35 (1998).

475. W. J. Lee, C. H. Shin, C. R. Cho, J. S. Lyu, B. W. Kim, B. G. Yu, and K. I. Cho, *Jpn. J. Appl. Phys.* 38, 2039 (1999).

476. K. Nagashima, T. Hirai, H. Koike, Y. Fujisaki, and Y. Tarui, *Jpn. J. Appl. Phys.* 35, L1680 (1996).

477. M. Noda, Y. Matsumuro, H. Sugiyama, and M. Okuyama, *Jpn. J. Appl. Phys.* 38, 2275 (1999).

478. M. Okuyama, W. Wu, Y. Oishi, and T. Kanashima, *Appl. Surf. Sci.* 117/118, 406 (1997).

479. B. E. Park, S. Shouriki, E. Tokumitsu, and H. Ishiwara, *Jpn. J. Appl. Phys.* 37, 5145 (1998).

480. S. Y. Sze and J. Y. Lee, "11th International Semiconducting and Insulating Materials Conference," 2000, p. ii.

481. E. Tokumitsu, R. Nakamura, and H. Ishiwara, *IEEE ISAF* 107 (1996).

482. E. Tokumitsu, R. Nakamura, and H. Ishiwara, *IEEE Electron Device Lett.* 18, 160 (1997).

483. S. B. Xiong and S. Sakai, *Appl. Phys. Lett.* 75, 1613 (1999).

484. J. Yu, Z. Hong, W. Zhou, G. Cao, J. Xie, X. Li, S. Li, and Z. Li, *Appl. Phys. Lett.* 70, 490 (1997).

485. N. A. Basit, H. K. Kim, and J. Blaschere, *Appl. Phys. Lett.* 73, 3941 (1998).

486. K. H. Kim, *IEEE Electron Device Lett.* 19, 204 (1998).

487. H. N. Lee, M. H. Lim, Y. T. Kim, T. S. Kalkur, and S. H. Choh, *Jpn. J. Appl. Phys.* 37, 1107 (1998).

488. E. Tokumitsu, G. Fujii, and H. Ishiwara, *Appl. Phys. Lett.* 75, 575 (1999).

489. T. Nakamura, Y. Fujimori, N. Izumi, and A. Kamisawa, *Jpn. J. Appl. Phys.* 37, 1325 (1998).

490. S. L. Miller and P. J. McWhorter, *J. Appl. Phys.* 72, 5999 (1992).

491. T. A. Rost, H. Lin, and T. A. Rabson, *Appl. Phys. Lett.* 59, 3654 (1991).

492. S. M. Yoon, E. Tokumitsu, and H. Ishiwara, *IEEE Electron Device Lett.* 20, 526 (1999).

Chapter 2

HIGH-PERMITTIVITY (Ba, Sr)TiO$_3$ THIN FILMS

M. Nayak, S. Ezhilvalavan, T. Y. Tseng

Department of Electronics Engineering and Institute of Electronics, National Chiao-Tung University, Hsinchu, Taiwan, Republic of China

Contents

Handbook of Thin Film Materials, edited by H.S. Nalwa
Volume 3: Ferroelectric and Dielectric Thin Films
Copyright © 2002 by Academic Press
All rights of reproduction in any form reserved.

ISBN 0-12-512911-4/$35.00

1. INTRODUCTION

1.1. General Background

High-permittivity thin films have gained tremendous interest these days because of their vast applications, including non-volatile random access memory (NV-RAM), high-density dynamic random access memory (DRAM), electro-optic coplanar waveguides, microwave tunable devices, sensors, and actuators. Although these films have many applications, their real potential has yet to be explored. Hence high-permittivity films have garnered much interest in academia, research laboratories, and in the electronics industry alike. Because of the maturity of thin film growth and integration techniques, integrated ferro-electrics has become a flourishing field. Integrated ferroelectric memories, integrated pyroelectric sensor arrays, and integrated electro-optic wave guiding circuits have become the talk of industry.

Many perovskite thin films have been extensively investigated because of their interesting dielectric properties. Perovskite compounds such as $(Pb,La)TiO_3$ (PLT), $(Pb,Zr)TiO_3$ (PZT), $PbTiO_3$, $BaTiO_3$, $SrTiO_3$, and $Ba_{1-x}Sr_xTiO_3$ have been studied for various application purposes. However, the volatility and reactivity of the lead compounds causes processing difficulties, and the environmental concerns surrounding lead poisoning limits the use of lead-based compounds. The perovskite oxide compounds $BaTiO_3$ and $SrTiO_3$, which form a continuous solid solution with a high dielectric constant at room temperature, are of typical interest. $BaTiO_3$ is ferroelectric at room temperature, whereas $SrTiO_3$ is paraelectric at room temperature. The Curie temperature (T_c) of solid solution $Ba_{1-x}Sr_xTiO_3$ varies linearly with composition. Hence, T_c can be tailored by changing the composition, so that the solid solution is in the required state (paraelectric or ferroelectric) in the device's operating temperature range. For example, a pyroelectric sensor requires that the material be in the ferroelectric state to attain a maximum change in the polarization charge when the temperature is changed and that in the case unipolar non-switching devices such as DRAMs, dielectric material in the paraelectric state is highly preferred to avoid fatigue due to the domain switching in ferroelectric materials. Barium titanate is one of the most extensively used and currently most actively investigated ferroelectrics, because of its high dielectric constant and its composition-dependent Curie temperature, which can be modified by substitution and doping. Owing to its excellent dielectric, pyroelectric, thermoelectric, and optoelectronic properties, $BaTiO_3$ has been widely used in electronic industry in the applications such as capacitors, transducers, sensors, and actuators. In recent years, with substitution of strontium, the $(Ba,Sr)TiO_3$ (BST) system has been one of the top candidates to replace SiO_2 for fabricating the new generation of extremely high-density DRAMs. Moreover, these perovskite compounds have better thermal and chemical stability than their lead-based counterparts.

Successful fabrication of a device based on thin film material requires the development of sophisticated synthesis and processing techniques, understanding of the dependence of properties on the process parameters, and optimization of processing conditions to get novel properties and in-depth knowledge of structure property relationship. High-permittivity thin film processing methods include sputter deposition, laser ablation, sol-gel, chemical vapor deposition, metal-oxide chemical vapor deposition and metal-organic deposition. All of these film processing methods have experienced substantial growth in recent years, resulting in production of high-quality ceramic thin films that have enabled materials integration strategies to produce devices with unique, desirable properties.

This chapter reviews recent developments in $(Ba,Sr)TiO_3$ (BST) high-permittivity films for various applications, including DRAM, pyroelectric sensors, gas detection sensors, microwave voltage tunable devices, and optical properties of the film. The major portion of the review is dedicated to the DRAM applications because of the large number of studies reported in this field. Recently, investigations into other applications, such as microwave voltage tunable devices, sensors, and coplanar waveguides, are picking up momentum. We first describe the different methods of BST thin film material processing, since they are the ultimate factors determining the oxide properties. We then look into the main physical and electrical properties, optical properties, and finally some possible applications highlighting some well-established experimental results. We review specific examples from the recent literature to exemplify how the technique has been used to date in solid-state technology. We also review the general theories of electrical conduction mechanisms and the various methods of leakage current reduction, to check the limits of their applicability to ultra-large-scale integrated circuits (ULSI).

1.2. High-Permittivity Films for ULSI DRAM Capacitor Applications

DRAMs have been advanced by focusing mainly on how to make memory cells small to realize high-density DRAMs. The continuous "shrink technology" up to Gb density exposes many challenges. The most critical challenges in Gb-density DRAMs are yield loss due to large die size, and small feature size, standby current failure caused by large chip size, and small data retention times owing to reduced charge packet in the memory cell. Narrowing the bandwidth mismatch between fast processors and slower memories and achieving low-power consumption together with the aforementioned challenges drive DRAM technologies toward smaller cell sizes, faster memory cell operation, less power consumption, and longer data retention time. In addition, a tight control of increasingly complicated wafer processing requires that DRAM process technology be simpler and less sensitive to processing variation. Thus, DRAM technology in the Gb era should solve the challenges imposed by the shrink technology system application requirement and manufacturing technology [1–5].

One of the most critical challenges facing Gb-density DRAMs is memory cell capacitance. Memory cell capacitance is the crucial parameter that determines sensing signal voltage, sensing speed, data retention time, and endurance against the soft error event. It is generally accepted that the minimum cell capacitance is more than 25 fF/cell regardless of density. However, lower supply voltage and increased junction leakage current due to high doping density drive memory cell capacitance toward higher value more than 25 fF/cell in the Gb-density DRAMs. Up to now, strategies for increasing memory cell capacitance have focused on increasing memory cell capacitor area and decreasing the dielectric thickness.

For the memory cell capacitor, which is the most important technology in the Gb era, a high-dielectric constant capacitor seems to be the only solution [6–8]. In the recent years, thin film perovskite materials with high dielectric constant, such as PZT, $SrTiO_3$, and BST [9–29], have been investigated as dielectric materials for future DRAMs. The best-suited dielectric material would have a low leakage current and a high dielectric constant, and would also be in the paraelectric phase to avoid fatigue from ferroelectric domain switching. $SrTiO_3$ has a smaller dielectric constant than BST, and PZT is in the ferroelectric phase at room temperature. Thus BST is very appealing for DRAM capacitors.

BST thin films are being widely investigated as alternative dielectrics for ULSI DRAM storage capacitors because of their (a) high dielectric constant ($\varepsilon_r > 200$), (b) low leakage current, (c) low temperature coefficient of electrical properties, (d) small dielectric loss, (e) lack of fatigue or aging problems, (f) high compatibility with device processes, (g) linear relation of electric field and polarization, and (h) low Curie temperature [11–13, 30–35]. However, whether or not BST thin film can be successfully applied depends largely on more a thorough understanding of the material's properties. Accordingly, the deposition techniques and electrical properties of BST films have received increasing interest [36–38]. According to those investigations, the electrical and dielectric properties and reliability of BST films depend heavily on the deposition process, postannealing process, composition, base electrodes, microstructure, film thickness, surface roughness, oxygen content, and film homogeneity.

1.2.1. Trends in the Development of ULSI DRAM Capacitors

The capacitor materials currently used in DRAMs are either silicon dioxide (SiO_2) or a silicon oxide/nitride composite layer (ONO) with a relative dielectric constant of 6. Use of SiO_2 or ONO allows the memory cell to be fabricated as a metal-oxide-semiconductor (MOS) device. As the number of memory cells increase to gigabits, the available area for the capacitor decreases rapidly (approximately 0.4 μm^2 for a 256-Mb device and 0.2 μm^2 for a 1-Gb device) to maintain acceptable die sizes. Table I indicates that the capacitance-per-unit area should be increased to achieve higher DRAM densities [39]. For maintaining sufficient memory cell storage capacitance, manufacturers abandoned the idea of flat integrated circuits, and three-dimensional cell structures consequently have been incorporated by the use of deep trenches and stacked layers to offer more surface area. So far, these structures with ONO

Table I. The Road Map of DRAM Technology [39]

	Minimum feature size (μm)	C/area (fF/μm^2)	Capacitor area (μm^2)	Operating voltage (V)	Year* 1 million devices
16 Mb	0.60	25	1.10	3.3	1992
64 Mb	0.35	30	0.70	3.3	1995
256 Mb	0.25	55	0.35	2.2	1998
1 Gb	0.18	100	0.20	1.6	2001
4 Gb	0.15	140	0.10	1.1	2004

*Year in which 1 million devices were/are projected to be produced.

storage dielectrics can adhere to the requirements of 256-Mb DRAMs. The capacitor areas will be close to 0.2 μm^2 and 0.1 μm^2 for future 1-Gb and 4-Gb DRAMs, respectively, and there is a requirement that the capacitance per unit area be increased as shown in Table I. The ONO dielectrics are not used in these products, because the capacitor area cannot be maintained constant in a cell that can still be manufactured, and also because the ONO dielectric thickness has reached a lower limit set by electron tunneling through the dielectric [40, 41]. Consequently, an increasing effort has been made to find an alternative dielectric with substantially higher permittivity.

The first step in the direction of high-dielectric constant materials is to consider some single-metal oxide materials such as Ta_2O_5, TiO_2, and others. Table II illustrates the dielectric constant and critical capacitance (defined as the maximum capacitance-per-unit area that can be achieved for a film that satisfies DRAM leakage requirements) [42] for various single-metal oxides that present dielectric constant values in the range of 10–100. As can be seen, the highest capacitance values can be obtained for Ta_2O_5 films. These films are also compatible with MOS fabrication facilities and can be easily deposited by the chemical vapor deposition (CVD) technique required to form complex three-dimensional features. Integration issues relative to the choice of electrodes—in particular, a top CVD-TiN electrode is needed and postdeposition annealings have to be limited to obtain the best film properties before such film can be used in mass production [43, 44]. The reported storage capacitances for Ta_2O_5 are around 10–20 fF/μm^2 [45]. If Ta_2O_5 is used at all, it apparently will be appropriate for only one DRAM generation [45–48].

On the other hand, ferroelectric materials are considered the ideal DRAM dielectrics for the Gb era, since they exhibit dielectric constants in the 200–2000 range. These values are much lower when thin films are considered, however. For DRAM applications, ferroelectric films that are in the paraelectric phase of the DRAM operating temperature range should be considered to benefit from the full stored charge during the read operations. Table III lists some of the most promising ferroelectric material candidates for DRAM applications. Among these $(Ba_{1-x}Sr_x)TiO_3$ films have been investigated as the most promising capacitor material in future DRAM applications.

Table II.   Medium Dielectric Constant Materials

Dielectric	ε (thick films)	C_{crit} (fF/μm^2)	Growth	Reference
Ta$_2$O$_5$	25	13.8 (20.4)	MOCVD	[55, 56]
	50		Sputtering	[45–48]
TiO$_2$	30–40	9.3	MOCVD	[57]
ZrO$_2$	14–28	9.9	MOCVD	[58]
Nb$_2$O$_5$	30–100	–	–	[42]
Y$_2$O$_3$	17	4.7	Sputtering	[59]
Si$_3$N$_4$ (comparison)	7	7–8.6 (120)	MOCVD	[42]

The values in parentheses are given for the case of a HSG-rugged Si capacitor.

Reprinted with permission from [42], © 1994, Taylor & Francis, Ltd.

Table III.   High Dielectric Constant Materials for DRAM Applications

Dielectric	ε	t (nm)	C_{crit} (fF/μm^2)	Reference
SrTiO$_3$	230	53	55.0	[14]
(Ba$_{1-x}$Sr$_x$)TiO$_3$ (BST)	320	70	40.5	[60]
	800	60	118.1	–
Ba(Ti$_{0.8}$Sn$_{0.2}$)TiO$_3$ (BTS)	210	100	–	[61]
(Ba,Pb)(ZrTi)O$_3$ (BPZT)	200	200	8.9	[62]
(Pb$_{1-x}$La$_x$)TiO$_3$ (PLT)	1400	500	24.8	[63]
(Pb,La)(Zr,Ti)O$_3$ (PLZT)	1474	150	87.0	[64]
(PbZr$_{1-x}$)Ti$_x$O$_3$ (PLT)	>1000	70	>70.0	[65]

Reprinted with permission from [42], © 1994, Taylor & Francis, Ltd.

BST films have the advantages of a low leakage current, a para-electric room temperature with a high dielectric constant, and a large dielectric breakdown strength [48–50]. A high-dielectric BST capacitor is basically formed in a metal–insulator–metal (MIM) structure. In this MIM structure, the storage polysilicon electrode is replaced with metal electrode. By using a proper metal electrode that has a strong resistance to native oxide, the native oxide on storage electrode can be completely removed. The Pt storage node electrode is a good choice for metal to have strong resistance to oxidation. Therefore, a MIM cell capacitor using a BST dielectric seems to be the ultimate solution for the Gb era. Figure 1 compares the leakage capacitance characteristics of BST, Ta$_2$O$_5$, and ONO dielectrics to illustrate the considerable improvement that can be obtained with such materials. Although the capacitance improvements are clear, the integration issues raised by the introduction of these new materials are not simple.

Although the BST dielectric capacitor can provide the sufficient cell capacitance for the Gb era, many key issues regarding the BST capacitor need to be solved. These issues include the barrier height between the metal electrodes and the BST dielectric, the thickness-dependent dielectric constant, the crystallization temperature after BST film deposition, the barrier layers between the storage electrode and polysilicon plug, and its resistance to oxidation during crystallization temperature

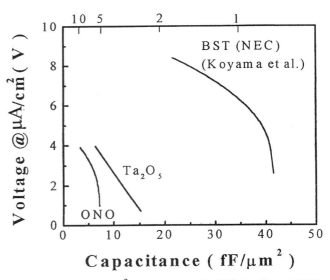

Fig. 1.   Voltage at 1 μA/cm^2 vs. capacitance for ONO, Ta$_2$O$_5$, and BST capacitors. Reprinted with permission from [42], © 1994, Taylor & Francis, Ltd.

and electrode formation. Current conduction of the BST capacitor is known to be governed by the Schottky emission current [51]. Since the Schottky conduction current is strongly dependent on the barrier height between the metal electrode and the BST capacitor, a metal electrode with a higher barrier height is needed. It is very important to control the surface properties of the metal electrode and the interface properties between the metal electrode and the BST dielectric to maintain good barrier properties. The dielectric constant of BST is known to be dependent on the thickness of the BST film [52]. As BST film thickness decreases, the dielectric constant decreases. This can be explained by the lower dielectric constant resulting in a depletion region between metal and the BST dielectric. The deposited BST dielectric requires high-temperature annealing at (about 750°C) in O$_2$ atmosphere to achieve crystallinity. Crystalline structures of BST film are found to have a higher dielectric constant and lower leakage current density [53]. During high-temperature annealing in O$_2$ ambient, a considerable amount of oxygen penetrates the Pt storage electrode, oxidizing the polysilicon at the interface between the plugged polysilicon and the BST film. Therefore, a barrier layer between the Pt electrode and plugged polysilicon is needed to block the oxygen penetration. The barrier layer should have resistance to oxygen penetration at a high annealing temperature. Unfortunately, the TiN/Ti barrier commonly used in current metallization schemes is not suitable, because of the loss of barrier property around 500°C. Another thing is the electrode formation because it determines the surface area of capacitor. A more vertical etching profile of the storage node and a larger capacitor area can be achieved. So far, the Pt electrode has been found to have superior leakage current characteristics as well as the highest capacitance [45, 48]. But Pt is very difficult to etch vertically [54]. Vertical etching of Ru or RuO$_2$ electrodes has been found to be easy. However, a Ru- or RuO$_2$-based BST capacitor suffers from high leakage current and low dielectric constant. Therefore, the aforementioned issues need to be fully understood before BST films are subjected to mass production. In the following sections, we address these issues in detail by making use of the extensive research work carried out on BST films by various groups.

1.2.2. Capacitor Structure and Design Rule

Since the concept of the DRAMs was patented by Dennard in 1968 [66] and the first commercial product was introduced by Intel in 1970, the development of DRAMs has progressed at a rapid pace, with a new generation introduced every three years. The advantages of lower cost per bit, higher device density, and flexibility of use (i.e., both read and write operations are possible) have made DRAMs the most used form of semiconductor memory to date, and the vehicle that helps drive a large part of the manufacturing infrastructure for the microelectronic industries.

Over the past three decades, the development of DRAMs has involved the struggle to increase the number of bits on a single silicon wafer, because the more bits on a single wafer, the lower the average cost per bit. For each generation, the bit density increases that of the previous generation by fourfold. The increase correlates directly with the development and improvement of the dynamic cell. Historically, the evolution of the dynamic memory cell began with the four-transistor cell; and later the one-transistor cell consisting of a single access transistor and a storage capacitor has become the basic cell due to the process, the lowest component count, the smallest chip size, and hence the highest device density compared to other dynamic cells. Figure 2 shows the schematic diagram of the one-transistor cell, which could have a bit of information by storing a charge on a storage capacitor. Later in the 16-kb generation, the double-polysilicon structure [67] was implemented into the one-transistor cell and become the standard DRAM cell architecture, known as the "planar" DRAM cell. The use of this basic cell construction lasts to 1 Mb. During this period, innovation focused on scaling the basic cell to smaller geometries, thereby producing a much higher device density. Because

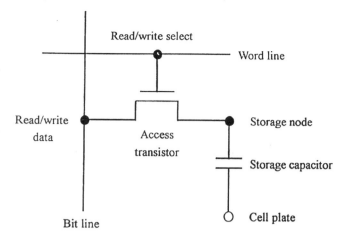

Fig. 2. Schematic diagram of a one-transistor cell in DRAMs [66].

a storage capacitor occupies a rather large area (about 30%) in a cell, shrinkage of the storage capacitor is inevitable. This reduced capacitor area leads to decreased capacitance and makes it difficult to maintain a sufficient storage charge to present the soft error effect induced by alpha particle incidence and environmental afects [68]. As a result, the dielectric film on a capacitor has to be simultaneously reduced to maintain a high enough charge to avoid soft error rate (SER). However, continuously thinning the dielectric film will result in the degradation of the long-term reliability due to the higher electrical field as the dielectric film becomes biased [69, 70]. Thus, the size requirement of the 1-Mb DRAM determines the need to develop novel cell structures for the next generations and beyond.

The advent of the 4-Mb DRAMs necessitated the introduction of three-dimensional memory cell structures, notably the stacked cell [71–74], the trench capacitor cell [75–80], and their variants [81, 82]. For the trench capacitor cell, in which the storage capacitor is buried in the surface of the chip, the main benefit is that the capacitance of the cell can be increased by increasing the depth of the trench without increasing the surface area of silicon occupied by the cell. In general, two types of trench cell are used in the 4-Mb DRAM generation: the traditional trench cell [originally called the corrugated capacitor cell (CCC)] [75] and the substrate plate trench cell (SPT) [78]. In the CCC cell, the electrode is located inside the trench, as the common capacitor plate and stores charge outside the trench. The cell cannot be used for 64 Mb or beyond, because of limitations on scaling to smaller dimensions. One problem is that the spacing between the cells must be limited to prevent punchthrough caused by the leakage of current between adjacent trenches. Another problem is SER susceptibility because of a large charge collection area outside the trench walls. Accordingly, the SPT cell structure was developed to overcome the disadvantages of the CCC cell by storing charges inside rather outside the trench and using the substrate as the common capacitor plate. The trench-to-trench punchthrough is eliminated, because the substrate region between trenches is at a common potential. The SER immunity is also very good, because the junction area for charge collection is small and the cell is con-

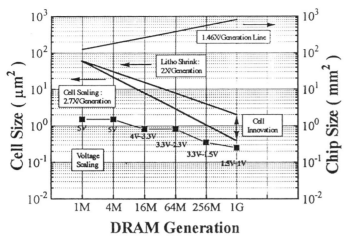

DRAM Generation

Fig. 3.   Scaling issues of DRAMs [96].

structed well inside, which blocks the collection of charges by the capacitor storage node. Recent study has shown that the cell is scalable to 1-Gb dimensions, and a capacitance of above 35 fF can be achieved through properly controlling the deep trench etching and scaling the SiO_2/Si_3N_4 (ONO) dielectric thickness to the range of 4–5 nm [83].

Meanwhile, the stack capacitor cell forms the storage capacitor between two dielectrically separated sheets of polysilicon that rise above the surface of the chip. Because the storage capacitor is stacked up and over the word line, the effective capacitance of a stacked cell is increased over that of a planar cell, not only because of the increased surface area, but also because of the curvature and sidewall field effect [74]. However, a simple stacked capacitor cannot provide sufficient capacitance, and thus the introduction of special high-permittivity materials as the dielectric for a storage capacitor is necessary. Figure 3 shows the DRAM trends for cell size, chip size, lithography, and the operation voltage from 1 Mb to 1 Gb [84–86]. Although the number of bits per chip has increased fourfold in every generation, the chips have increased in size by only about 40% per generation. This leads to a cell size reduction factor of 3. Furthermore, lithography scaling gives a reduction in area of only 2 as each linear dimension is reduced by a factor of 0.7. The remainder must be contributed from the innovations in cell structure. Thus several novel storage capacitor structures have been presented, including the multifine structure, which involves stacking more layers [87]. The capacitor-over-bit line (COB) cell [88], which places the bit line under the storage capacitor, has more freedom to increase the height of the capacitor for increasing the cell capacitance. Also, rough polysilicon and hemispherical silicon (HSG) have been introduced into simple stacked capacitors or cylindrical capacitors [89–95].

Along with shrinking the capacitor area, the operating voltage of the storage dielectric must be reduced to below 5 V for 16-Mb DRAMs and beyond, to increase the device's reliability [96]. Lower-power operation has been demonstrated to be indispensable in maintaining the favorable retention characteristics of high-density DRAMs [97]. However, this reduction has

made it difficult to keep the required amount of stored charge to maintain immunity against the soft error effect [68]. Keeping the required amount of stored charge requires increasing the capacitance of the dielectric. The limited surface area available for a stacked cell and a limitation of total capacitor height for reducing overall chip topography makes scaling of the capacitor dielectric a major concern for stacked technologies [98, 99]. The conventional ONO dielectric films deposited on polysilicon reach their physical limits as the effective SiO_2 thickness of the dielectric film is less than 4.0 nm [100]. This is because of the significant increase in direct tunneling current due to thinning of the dielectric. Therefore, the introduction of new capacitor dielectrics such as BST that have a high permittivity, immunity against SER and a low leakage current density for favorable memory operation appears to be indispensable for 1 Gb and beyond [101–108].

DRAM, as the simplest but most highly integrated kind of memory devices, took full advantage of the technical and economical merits of ULSI, resulting in cost benefits expressed by the "π rule" [109–111]. This was possible because there was enough area margin on the surface of the chip to allow more devices to be squeezed onto the chip. But ULSI has difficulties with storage capacitance, which must be kept to a certain value while the device dimensions are reduced. This difficulty is explained by the simple scaling rule predicting that capacitance will decrease in inverse proportion to dimension. The difficulty has promoted the adoption of complicated structures (e.g., the trench or stack), whose fabrication requires more process steps.

A new rule describing the relation between cost and the degree of integration thus became necessary. The analysis of DRAM sizes up to 4 Mb led to proposition of "The Bi rule" [110, 111]. The Bi rule accounts for increased chip size, process steps, and manufacturing cost by using a larger diameter wafer. As the DRAM integration level reaches 64 or 256 Mb, the overall investment for equipment will exceed that predicted by "The Bi rule." Keeping the Bi rule at higher levels of integration requires decreasing equipment cost, process steps, and so forth. Because the Bi rule implies that the bit cost is reduced by one-half at each generation, it divides the fruit of technical development equally between the user and the manufacturer [112].

1.2.3. BST Thin Films

$BaTiO_3$ is a ferroelectric perovskite that has been well studied in bulk ceramic form where the measured permittivities are well into the thousands. Use of the $BaTiO_3$–$SrTiO_3$ solid solution allows the Curie temperature (the ferroelectric–paraelectric transition temperature, T_c) of $BaTiO_3$ to be shifted from 120°C to around room temperature for $Ba_{1-x}Sr_xTiO_3$ films. When Sr is added to $BaTiO_3$, the linear drop of T_c is about 3.4°C per mol%. Therefore, 30 mol% Sr ($x = 0.3$) would bring the T_c down to room temperature. The effects of several isovalent substitutions on the transition temperatures (Curie temperature) of ceramic $BaTiO_3$ are shown in Figure 4 [6, 32, 60, 114–116].

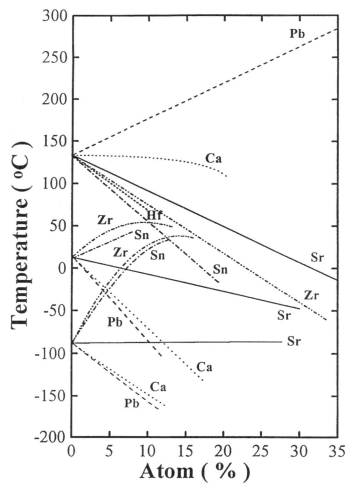

Fig. 4. Effect of several isovalent substitutions on the transition temperature of ceramic BaTiO$_3$ [115].

cise microscopic control of stoichiometry is essential in obtaining uniform single-phase films.

The basic parameters for applying capacitor thin films on DRAMs are dielectric constant, leakage current density, and reliability. The targets for ideal Gb-era DRAM dielectrics include [40] (a) SiO$_2$ equivalent thickness <0.2 nm for Gb, (b) leakage current density $<1 \times 10^{-7}$ A/cm^2 at 1.6 V, (c) lifetime of 10 years at 85°C and 1.6 V, (d) stability of 10^{15} cycles at >100 MHz, and (e) general compatibility to semiconductor processing.

2. THIN FILM DEPOSITION

Deposition process are generally divided into two categories: physical vapor deposition (PVD) and chemical vapor deposition (CVD). CVD is of the greatest interest, because PVD processes (e.g., evaporation and sputtering) do not generally produce films of the same quality as CVD processes. Commonly used techniques for depositing dielectric thin films include low-pressure chemical vapor deposition (LPCVD), metal-organic chemical vapor deposition (MOCVD), sputtering, pulse laser ablation, and sol-gel methods. Each technique has merits and drawbacks. For example, MOCVD can be used for large-scale production, but elevated growing temperature is required for cracking the metal-organic (MO) source. Pulse laser ablation is suitable for low-temperature epitaxial growth but can process samples only on a limited scale. Thin films are used in a host of different applications in ULSI fabrication and can be prepared using various techniques. Regardless of the method by which they are formed, the process must be economical and the resultant films must exhibit the following characteristics: (a) good thickness uniformity, (b) high purity and density, (c) controlled composition stoichiometries, (d) high degree of structural perfection, (e) good electrical properties, (f) excellent adhesion, and (g) good step coverage. The following sections introduce these deposition techniques [120].

2.1. Chemical Vapor Deposition (CVD)

CVD is defined as the formation of a nonvolatile solid film on a substrate by the reaction of vapor phase chemicals (reactants) that contain the required constituents. The reaction gases are introduced into a reaction chamber and are decomposed and reacted at a heated surface to form the film. A wide variety of thin films used in ULSI fabrication have been prepared by CVD [121, 122]. CVD processes are often selected over competing deposition techniques because they offer the following advantages: (a) High-purity deposits can be achieved, (b) a great variety of chemical compositions can be deposited, (c) some films cannot be deposited with adequate film properties by any other method, and (d) good economy and process control are possible for many films.

Ba$_{1-x}$Sr$_x$TiO$_3$ films not only are paraelectric at the DRAM operating temperature range (0–70°C ambient and 0–100°C on chips) [40], but also achieve maximum permittivity around the operating temperature. On the other hand, the BST components have lower volatilities than the Pb-based ferroelectric materials, thereby making BST films relatively easier to introduce into fabrication facilities [40, 41].

BST films are polycrystalline. Their properties depend heavily on composition, stoichiometry, microstructure (grain size and size distribution), film thickness, characteristics of the electrode, and homogeneity of the film. The BST thin film growth method significantly affects the composition, stoichiometry, crystallinity, and grain size of the film and, consequently, its dielectric properties. A variety of techniques, including rf-sputtering [37, 52, 53, 60, 64], laser ablation [50], metal-organic deposition (MOD) [117–119], chemical vapor deposition (CVD) [73–76], and sol-gel processing [80], have been used to deposit BST thin films. These methods are highly competitive, each having advantages and disadvantages in terms of homogeneity, processing temperature, and processing costs. Because of the multicomponent nature of BST materials, pre-

2.2. Atmospheric Pressure Chemical Vapor Deposition Reactors (APCVDs)

APCVD were the first to be used by the microelectronics industry [123]. Operation at atmospheric pressures kept reactor design simple and allowed high film deposition rates. However, APCVD is susceptible to gas phase reactions, and the films typically exhibit poor step coverage.

2.3. Low-Pressure Chemical Vapor Deposition (LPCVD)

The growth of thin films by LPCVD has become one of the most important methods of film formation in semiconductor fabrication. In some cases, LPCVD is able to overcome the uniformity, step coverage, and particulate contamination limitations of early APCVD systems [124–130]. By operating at medium vacuum (0.25–2.0 torr) and higher temperatures (550–600°C), LPCVD reactors typically deposit films in the reaction rate-limited regime. Low-pressure operation also decreases gas phase reactions, making LPCVD films less subject to particulate contamination. Two main disadvantages of LPCVD are the relatively low deposition rates and relatively high operating temperatures. Attempting to increase deposition rates by increasing the reactant partial pressures tends to initiate gas phase reactions.

2.4. Metal-Organic Chemical Vapor Deposition (MOCVD)

MOCVD is a specialized area of CVD that uses metal-organic compounds as precursors, usually in combination with hydrides or other reactants. The thermodynamic and other kinetic principles of CVD and its general chemistry also apply to MOCVD. Major concerns associated with the MOCVD technique are the choice of precursor sources and the substrate temperature required to crack the precursors. For depositing dielectric thin films, the ideal precursor for MOCVD should have the following features: (a) high vapor pressure and low vaporization temperature, (b) low cracking temperature, (c) large temperature difference between vaporization and cracking, and (d) no contamination from the organic constituent.

A wide variety of materials can be deposited by MOCVD as either single-crystal, polycrystalline, or amorphous films [120]. The most important application is in the deposition of the group III–V semiconductor compounds, including gallium arsenide (GaAs), indium arsenide (InAs), indium phosphide (InP), and gallium aluminum phosphide (GaAlP), particularly for epitaxial deposition. Most MOCVD reactions occur in the temperature range of 600–1000°C and at pressures varying from 1 torr to atmospheric. The equipment and chemicals used in MOCVD are expensive, and the production cost is high. For these reasons, MOCVD is most often considered when very high quality is required. It has recently been investigated for other applications in the area of very high temperature oxidation protection up to 2200°C.

2.5. Rapid Thermal Low-Pressure Metal-Organic Chemical Vapor Deposition (RT-LPMOCVD)

RT-LPMOCVD was explored as a way to carry out a number of critical steps in the processing of many electronic devices in a vacuum environment [131]. With shorter time duration cycles for wafers, the reaction temperature is a key characteristic in rapid thermal processing (RTP). Thus the incorporation of impurities into wafers is minimized. In addition, the unique time–temperature schedule provided by the RTP cycle allows for more highly controlled processes than can be obtained by conventional furnace sintering.

The RT-LPMOCVD technique appears to be very promising in the fabrication of thin, high-quality semiconductors and dielectric and metallic films with abrupt interfaces. The technique combines rapid thermal annealing and chemical deposition processes. The reactions are carried out by switching the radiant energy sources (halogen–tungsten lamps) on and off, to initiate and terminate reactions that occur in the presence of continuously flowing gases. Alternatively, the radiant sources can be left to operate continuously, while switching the gas flow. In the latter case, the RT-LPMOCVD reactor behaves as a standard LPMOCVD reactor.

2.6. Plasma-Enhanced Metal-Organic Chemical Vapor Deposition (PE-MOCVD)

MOCVD of BST thin films at temperatures of 600-700°C uses various precursor systems. Although such films have acceptable electrical properties, development of ULSI circuit requires low deposition temperatures and conformal step coverage. Although the latter requirement is achievable by thermal MOCVD, the former must be met by nonthermal modes of enhancement—for example, by plasma or photons. Some advantages of low-temperature plasma-enhanced MOCVD (PE-MOCVD) are as follows: (a) interfaces or doping profiles and previously deposited layers are not perturbed; (b) solid-state diffusion and defect formation are minimized; (c) compatibility with low-melting point substrates/films is maintained; and (d) complex and metastable compounds may be synthesized.

The plasma is generated by the application of an rf field to a low-pressure gas, thereby creating few electrons within the discharge region. The fact that the radicals formed in the plasma discharge are highly reactive presents some options as well as some problems to the process engineer. PE-MOCVD films in general are not stoichiometric, because the deposition reactions are so varied and complicated. Moreover, byproducts and incident species are incorporated into the resultant films (especially hydrogen, nitrogen, and oxygen) along with the desired products. Excessive incorporation of these contaminants may lead to outgassing and concomitant bubbling, cracking, or peeling during later thermal cycling and to threshold shifts in metal oxide semiconductor (MOS) circuits.

A plasma process requires control and optimization of several parameters besides those of the CVD process, including rf power density, frequency, and duty cycle. The deposition process is dependent in a very complex and independent way on these parameters, as well as on the usual parameters of gas composition, flow rate, temperature, and pressure. Furthermore, as with CVD, the PE-MOCVD method is surface reaction limited, and thus adequate substrate temperature control is necessary to ensure uniform film thickness [127, 128].

2.7. Photon-Induced Chemical Vapor Deposition (PHCVD)

A final CVD method, which may fill the need for an very-low-temperature deposition process without the film composition problems of PECVD, is photon-induced CVD (PHCVD) [123]. PHCVD uses high-energy, high-intensity photons to either heat the substrate surface or dissociate and excite reactant species in the gas phase. In substrate surface heating, the reactant gases are transparent to the photons, and thus the potential for gas phase reaction is completely eliminated. In reactant gas excitation, the energy of the photons can be chosen to provide efficient transfer of energy to either the reactant molecules themselves or to a catalytic intermediary, such as mercury vapor. This technique enables deposition at extremely low substrate temperatures. PHCVD films show good step coverage, but many suffer from low density and molecular contamination because of the low deposition temperature.

There are two classes of PHCVD reactors, depending on energy source ultraviolet (UV) lamp and laser. UV lamp reactors generally use mercury vapor for energy transfer between photons and the reactant gases [129]. UV radiation at 2537 Å is effectively absorbed by mercury atoms, which then transfer energy to the reactant species. Deposition rates for UV-PHCVD reactors are typically much slower than those for other low-temperature techniques. Laser PHCVD reactors offer the advantages of frequency tunability and a high intensity light source [130]. Tunability is useful in that specific photon energies targeted for particular dissociation reactions can be dialed in, enabling greater control of the deposition reaction. The energy from the high-intensity laser also increases reaction rates. Laser PHCVD opens the possibility of CVD writing. Nevertheless, the deposition rates of current PHCVD processes are still too low to allow their adoption in microelectronic production applications.

2.8. Electron Cyclotron Resonance Plasma Chemical Vapor Deposition (ECR-CVD)

ECR-CVD reactors use electron cyclotron resonance to produce plasma through the proper combination of an electric field and a magnetic field [120, 121]. Cyclotron resonance is achieved when the frequency of the alternating electric field matches the natural frequency of the electrons orbiting the lines of force of the magnetic field. This occurs, for instance, with a frequency of 2.45 GHz (a standard frequency for microwave) and a magnetic field of 875 Gauss. ECR plasma can be easily generated at much lower pressures of 10^{-5} to 10^{-3} torr. The microwave

ECR plasma typically has an ion source of low energy (<50 eV) and high flux density ($>10^{10}$ cm^{-3}) and is quite attractive in thin film processing because of the presence of electrodes and its ability to create high densities of charged and excited species at low pressures (about 10^{-4} torr).

An ECR plasma has two advantages. First, possibility of damaging the substrate by high intensity ion bombardment is minimized. In an rf plasma reactor, the ion energy may reach 100 eV, which can easily damage devices having submicron line with features and made from the more fragile compound semiconductor materials, such as gallium arsenide (III–V and II–VI compounds). Second, it operates at lower temperature than an rf plasma, which minimizes the risk of damaging heat-sensitive substrate. Also minimized is the possibility of forming hillocks, which are the result of the recrystallization of the conductor metals, particularly aluminum.

Limitations of ECR system include the need for very low pressure (10^{-3} to 10^{-5} torr) as opposed to 0.1 to 1 torr for rf plasma systems, as well as the need for a high-intensity magnetic field. This requires more costly equipment. In addition, the magnetic field makes the processing more difficult to control.

2.9. Liquid Source Chemical Vapor Deposition

Although each application has unique material requirements, in most cases the deposition process must have good conformability, composition control, and throughput, while minimizing the thermal budget. The absence of volatile and, in some cases, stable precursors has driven the development of the "liquid delivery technique" in which reactant gas composition is set by volumetric metering of liquids followed by flash vaporization [107]. The liquid delivery technique relies on the flash vaporization of liquid solutions and overcomes the limitations of bubbling. Neat liquids, as well as liquid solutions comprised of solids dissolved in organic or inorganic media, can be used with this technique. The liquid source reagent solutions are maintained at room temperature, and the composition of the inlet to the CVD chamber is controlled by one of several methods. In the preferred method, reactant gas composition is controlled through real-time volumetric mixing of the individual source reagent solutions. The liquid mixture is then flash-vaporized to generate a homogeneous gas at the inlet to the CVD tool. This method ensures process reproducibility as variations in delivery rate give rise to variations in the overall deposition rate but do not impact film composition.

The precursors for CVD of perovskite dielectrics are predominantly metal alkyls, cyclopentadienyls, alkoxides, and β-diketonates [132–134]. Metal alkyls have the formula MR_n, where n is the valency of the metal and R is an aliphatic, olefinic, or aromatic moeity. The compounds are volatile and have been used for the MOCVD of metals, semiconductors, nitrides, and oxides. In addition, the metal alkyls are extremely sensitive to oxygen and water and are very toxic.

Metal π-complexes include complexes such as metal cyclopentadienyls, $M(C_5H_5)_n$, where n is the valency of the metal. Lanthanides (e.g., lanthanum) can have large coordination numbers (typically 6 to 9 in organolanthanide compounds) because of their large size and the large number of orbitals available for bonding. The π-complexing ligands (e.g., as cyclopentadienyls) donate electrons to the central metal atoms. This electron donation, coupled with steric hindrance, promotes volatility of the complexes by satisfying the metals electronically and coordinatively. In essence, intermolecular associations with adjacent ligands are prevented. The vapor pressures of the lanthanide cyclopentadienyls, $M(C_5H_5)_3$, increase as the metals vary from lanthanum to lutetium across the lanthanide series. This is a direct result of the decreasing size of the central metal atom. Two advantages of these compounds over the trivalent β-diketonates are (1) there is less than one-half as much carbon and (2) volatilities are higher. But an external source of oxygen must be used, because there is no oxygen in the complex. Moreover, these compounds are extremely air-sensitive and difficult to handle.

Metal alkoxides are the predominant precursors in the MOCVD of titanium and zirconium oxides. The alkoxides of the alkaline-earth group metals are generally nonvolatile and are not useful as CVD precursors. Recently, liquid barium alkoxides with the formula $Ba[O(CH_2CH_2O)_nR]_2$, where $n = 2$ or 3 and $R = CH_3$ or C_2H_5, have been synthesized. When $n = 2$, the coordination number of barium is 6. When $n = 3$, the coordination number increases to 8. These molecules are very stable with respect to hydrolysis, but decompose in the gas phase. The alkoxides of the lanthanides are somewhat volatile. As in the π-complexes, the volatility increases with atomic number, with lanthanum the most difficult to volatilize. By using bulky alkoxide groups, the oligomeric character can be reduced and the volatility increased.

The β-diketonates are quite useful ligands, especially in metals that do not form any other type of volatile compounds, for example, the barium derivative barium bis(2,2,6,6-tetramethyl-3,5-heptanedionate) $[Ba(O_2C_{11}H_{19})_2$ or $Ba(thd)_2]$. A major problem associated with these solid compounds is that the volatilities are often low, and fairly high temperatures ($\sim 200°C$) must be used to cause sublimation. The fluoride derivative—for example, barium bis-heptafluorodimethylocta-dione $[Ba(O_2C_{10}H_{10}F_7)_2]$—can be vaporized at 30–50°C lower than the parent β-diketonates. The substitution of fluorine for hydrogen increases volatility but can lead to the deposition of undesired metal fluorides. Even though β-diketonates are not as reactive to water as alkoxides, they often form hydrates and are thus hygroscopic. Heating the hydrates at elevated temperatures causes ligand loss, leading to the formation of nonvolatile β-diketonate hydroxooligomers. Recently there have been a number of advances in metal β-diketonate chemistry. The first is the use of donor–acceptor bonding between neutral ligands and the Lewis acid metal acceptor. By using complexes such as $Ba(thd)_2 \cdot nD$, where D is an adduct such as ammonia, methanol, tetraglyme, trien/tetraen, or ethers, the molecular complexity can be reduced. As stated previously, by increasing the coordination number of the metal atoms, reduced intermolecular interactions can give rise to increased volatility. The second ad-

vance has been the synthesis of the first metal β-diketonate, a liquid bis-2,2dimethyl-8-methoxyoctane-3,5-dionato barium. This novel ligand is basically the thd ligand with an n-propyl methoxide group attached at the fifth carbon. Because the coordination number of the barium atom is 6, high volatility is achieved.

Other mixed ligand compounds used for CVD of $BaTiO_3$ are alkoxy β-diketonates, such as titanium bis(isopropoxide) bis(2,2,6-tertramethyl-3,5-heptanedionate) $[Ti(^iOC_3H_7)_2 (O_2C_{11}H_{19})_2]$. This compound was synthesized to avoid the gas phase reaction between $Ti(^iOC_3H_7)_4$ and $Ba(thd)_2$, which gave rise to nonvolatile barium alkoxy β-diketonate. From the previous discussions, it is obvious that to obtain reproducible, high-permittivity dielectric thin films at low temperatures, precursors must be handled under moisture-free conditions. Sometimes these precursors may require modifications or development, especially for the alkaline earths, lanthanides, and heavy metals (Pb and Bi). It is desirable to use component precursors that have similar volatilities (vapor pressures in excess of 50 mtorr at room temperature would be ideal) and that have adequate mass transport characteristics at the lowest possible temperatures. Precise control of stoichiometry may be difficult in the event that transport properties vary (due to decomposition) during the course of a CVD run.

The traditional delivery technique used for solids–liquids with low vapor pressures are inert or active gas-bubbler and direct sublimation methods [132–134]. These require well-controlled, high-temperature (which is precursor-dependent and can be as high as 230°C) delivery lines to avoid condensation of the precursors, but the complexity and the maintenance of delivery systems increase as the length (i.e., the source-to-reactor distance) and number of the delivery lines increase. The conventional resistively heated lines pose problems, because hot spots are very difficult to avoid. Using liquid–gas heating in coaxial tubes and ovens improves this method, but implementation for each precursor line with valves and other bulky fittings leads to difficulties. Moreover, some precursors are unstable at elevated temperatures over long periods. To circumvent these problems, the liquid-source injection delivery technique is particularly suitable for multicomponent systems where individual precursors (with or without adducts) are initially diluted in a common solvent (e.g., tetrahydrofuran) at room temperature [132–134]. Multicomponent solutions with selected flow rates are introduced into a vaporizer with a precision micropump. The vaporizer is kept near the reactor chamber to prevent premature decomposition of the vapor. Table IV lists advantages of and questions about traditional and liquid-source delivery methods.

2.10. Sputtering

The sputtering system used for depositing oxides can be divided into rf magnetron sputtering and ion beam sputtering (IBD). The operating principles of rf sputtering and IBD have been described in earlier work [135]. In recent years, IBD has gradually become a popular technique for the deposition of oxide thin films. In the IBS system, one or several Kaufman-type ion guns

Table IV.   Comparison of Conventional Delivery and Liquid-Source Injection Methods [132]

Bubbler/direct sublimation	Liquid source injection
Advantages	*Advantages*
Abundant data on many systems	Flexibility
Potential control of stoichiometry	ΔP-independent flow control
Multiple injection for thin-film uniformity on large wafers	Precursor stability due to low temperature
At low pressures, ability to deliver precursors independently	Potentially scaleable for simple systems
Questions	*Questions*
Potential decomposition when premixed at high pressures	Stability of multicomponent solutes in common solvents
Temperature control of source and lines	Window between vaporization and decomposition temperatures
Achievement of high molar flow rates	Clogging in vaporizer
Time-dependent precursor flow rate:	Gas-phase reactions at high temperatures
Δ depth of the source	Time-dependent incorporation efficiency of various components:
	amount and type of solvent;
	ΔM of source; Δ molar flow rate

ΔP, pressure differential; ΔM, time-dependent molarity.

are used to generate a broad inert gas ion beam with an accelerating voltage of 100–1500 eV. As the ion beam impinges on the target surface, the oxide materials are sputtered to deposit on the heated substrate.

2.11. Pulsed Laser Ablation

The fundamental principle of pulsed laser ablation deposition involves the interaction between the laser beam and a solid surface. This interaction involves steps: (1) absorption of photon energy by the target and heat conduction, (2) surface melting of oxide target, and (3) evaporation and ionization of the oxide target. It has been demonstrated that ferroelectrics or an oxide conductor (e.g., PZT, $SrBaTiO_3$, $SrRuO_3$, BST, or ITO) can be deposited or epitaxially grown by pulsed laser deposition [50, 136–138].

2.12. Sol-Gel Method

Probably the simplest method of thin film coating is the sol-gel method, which exploits the hydrolysis and polycondensation of relevant molecular precursors. The most striking feature of the sol-gel method is that, before gelation, the fluid solution is ideal for preparing thin films by such processes as dipping, spinning, or spraying [139]. The method involves preparation of a solution of the elements of the desired compound in an organic solvent using inorganic or organometallic compounds, such as alkoxides of appropriate solubility. This is followed by polymerizing the solution to form a gel, drying this gel to displace the organic solvent at appropriate temperature, and finally baking at a higher temperature to decompose the residual organic components and form a final inorganic oxide [139, 140]. The gelation can be done in the solution stage or after coating on the substrate by a brief exposure to the moist air. Owing to the intimacy of mixing of the constituents and to the extreme reactivity of the dried gel, the crystallization stage may occur

at temperatures several hundred degrees below those required for traditional mixed-oxide processing. These reduced temperatures could enable the direct integration of ceramic components with semiconductor devices and other substrate materials as well as the fabrication of unique material combinations with unusual phase assemblages and novel properties [141]. Compared to conventional film formation methods such as CVD, PLD, or sputtering, the sol-gel film formation method requires considerably less equipment and offers many advantages, including:

- deposition at room temperature, followed by postdeposition processing also at low temperature,
- easier compositional control and better uniformity of the films,
- better homogeneity because of the solution state mixing and also the ability to easily introduce dopants into the solution, which are used for the microstructure and property modification of the thin films,
- economical compared to other techniques and simple,
- the ability to coat a large area and geometrically complex substrates.

Metal alkoxides have varying affinity toward water, making the multicomponent homogeneous solution preparation difficult because of the premature and selective gelation of one of the components. This can be avoided by adding chelating organic ligands into the solution to control the hydrolysis rates of highly reactive alkoxides. These chelated solutions are more stable in air and easy to handle during processing [140]. Film thickness, uniformity, and morphology are highly dependent on solution concentration, viscosity, and prebaking and baking temperatures. This in turn has an effect on the thin film properties.

BST thin films of different composition have been prepared by many authors by using the sol-gel method [142–160] with different precursors and solvents. The most commonly used precursor as the Ba- and Sr-sources are Ba- and Sr-acetates [$Ba(CH_3COO)_2$ and $Sr(CH_3COO)_2$] and alkoxides. The commonly used solvent is acetic acid or 2-methoxy ethanol, depending on the precursor. The titanium source is a titanium alkoxide, such as titanium isopropoxide [$Ti(OC_3H_7)_4$] or titanium tetra-n-butoxide [$Ti(OC_4H_9)_4$]. The precursor solution content and formulation determines the baking temperature and also affects the crystallization temperature. Tahan et al. [142] prepared BST thin films using the acetic acid-based precursor solution by dissolving Ba-, Sr-, and Ti-isopropoxides in heated (90°C) acetic acid in the required molar ratio. Ethylene glycol was added to the solution as a chelating agent to prevent selective hydrolysis and condensation of titanium as titanium hydroxide. The solution was heated to promote a condensation reaction between the acetic acid and the ethylene glycol. The films were spin coated at 7500 rpm for 90 seconds, followed by solvent evaporation and organic pyrolysis at 250°C. Tahan and collegues observed that the film crystallization temperature depended on the acetic acid : ethylene glycol ratio in the precursor solution and that without ethylene glycol, the precursor solution precipitated. The crystallization of the films formed by the precursor solution with an acetic acid/ethylene glycol ratio of 3 : 1 started at 500°C as characterized by the XRD patterns. However, thin films prepared without addition of ethylene glycol required higher crystallization temperatures. A modified method was tried by Wang et al. [143], wherein the crystallization temperature was in the range of 600–650°C. They prepared BST thin films using Ba- and Sr-acetates and titanium isopropoxide precursors, with glacial acetic acid and 2-methoxy ethanol ($CH_3OCH_2CH_2OH$) as the solvents. Ba- and Sr-acetates were dissolved in heated acetic acid in the proportion of 2 g/ml. The solution was refluxed at about 120°C for 3 hours and cooled to room temperature. The required amount of titanium isopropoxide was dissolved in excess 2-methoxy ethanol, then refluxed for 3 hours at about 120°C and cooled to room temperature. Two solutions were mixed and agitated in an ultrasonic bath at room temperature until a uniform and clear solution was formed. This solution was spin coated onto the substrate at 5000 rpm for 15 seconds. After each coating, the samples were preheated to 300°C for 10 minutes to evaporate the residual solvents. The XRD and TG/DTA studies [143] established that the perovskite phase was formed only after annealing in the temperature range of 600–650°C. Jang et al. [144] prepared a BST precursor solution by reacting BaO, $SrCl_2$, and titanium isopropoxide with 2-methoxy ethanol. Acetylacetonate was used as the chelating agent to prevent rapid hydrolysis rate of the Ti-alkoxide. The crystallization temperature of the thin film was reported to be around 580°C. Researchers have modified their methods by changing either the starting materials or the solvent to try to obtain a crystalline BST thin film at a very low temperature compatible with the Si technology. The BST films prepared using the Ba- and Sr-isopropoxides and Ti-isopropoxide as starting materials with 2-methoxy ethanol as the solvent and acetyl acetonate as the chelating agent reportedly crystallize at 530–550°C [145]. Recently, Hayashi et al. [146] reported on a

BST thin film preparation using $Ba(OH)_2 \cdot nH_2O$ and $Sr(OH)_2$ as the alkaline-earth metal sources, Ti-isopropoxide as the Ti source, and methanol as the solvent. These hydroxides are the cheapest sources of the Ba and Sr. The perovskite phase started forming at a temperature above 650°C, making it less interesting. However, modification of the precursor solution methods may bring a decrease in the perovskite phase formation temperature, as previously observed in the case of films prepared from acetate precursors.

Burhanuddin et al. [147] studied a chemical solution deposition (CSD) method in which, after spin coating, the films were immediately pyrolized without formation of gel. (The gelation stage was avoided.) They used Ba- and Sr-hydroxides and Ti-isopropoxides as the source materials and 2-methoxy ethanol as the solvent. They claimed to have obtained high-quality polycrystalline films of BST at 550°C. Another important observation of their investigation is that as the barium-content in the BST film increased, the crystallization temperature of the film also increased [147]. The BST films obtained by spin coating using the alkoxide-isopropanol precursor solution showed a completely crystalline film around 700°C after annealing for 1 hour [148]. It was found that the density and microstructure of the samples were greatly influenced by the firing procedure. Dense samples were obtained only when they were fired to the maximum temperature after each deposition and before another layer was spun on top of the previous one. Also, the drying and baking processes for pyrolizing organics have some effect on crystallization of the films. The investigation showed that samples that did not go through drying and baking step had poor crystallinity. This is because of the trapping of the organics in the film and their unsuccessful burning, which may inhibit complete crystallization of the films [148].

Pontes et al. [149, 150] prepared BST thin films from polymeric precursor solutions. In this process, desired metal cations were chelated in a solution using a hydroxycarboxylic acid (e.g., citric acid) as the chelating agent. This solution was then mixed with a polyhydroxy alcohol (e.g., ethylene glycol) and heated to promote a polyesterification reaction in the solution. Metal ions were chelated by the carboxylic groups and remained homogeneously distributed in the polymeric network. The starting materials for this method were barium carbonates ($BaCO_3$), strontium carbonate ($SrCO_3$), and titanium isopropoxide, which were dissolved in aqueous citric acid solution at appropriate pH to form citrate solution. Ethylene glycol and citric acid were used as the polymerization/complexation agents for the process. Ammonium hydroxide was used to adjust the pH and prevent the precipitation of barium citrate. Titanium citrate was formed by dissolving titanium isopropoxide in an aqueous solution of citric acid at 60–70°C. After the homogenization of this solution, a stoichiometric amount of $SrCO_3$ was added to the solution, which was stirred slowly until a clear solution was obtained. Later, $BaCO_3$ was added slowly. To achieve complete solubility, ammonium hydroxide was added drop by drop until the pH was around 7–8. After this solution was homogenized, ethylene glycol was added to promote mixed citrate polymerization by a polyesterification reaction.

With continuous heating at 80–90°C, the solution became more viscous without any visible phase separation. After appropriate dilution, this solution was used for the film preparation by spin coating. The authors obtained·a polycrystalline BST thin film at 600°C [149, 150].

Hoffmann et al. [159] studied the decomposition behavior, phase formation kinetics, and influence of the chemical solution deposition parameters on film morphology and orientation. The films prepared from CSD solutions based on low-temperature decomposing carboxylate compounds (long-chain compounds of Sr and Ba) and Ti-alkoxide crystallized into the perovskite phase at about 450–550°C. In contrast, the films prepared from the Sr- and Ba-acetate (high-temperature decomposing precursors) and Ti-alkoxide precursors had a higher crystallization temperature of 650°C. The higher temperature is because crystallization into the perovskite phase proceeds through an intermediate phase with a high carbonate content, which is stable at a temperature of 550–650°C. Also, the films produced from these two methods differed in morphology. The films prepared from CSD solution based on "low-temperature" decomposing carboxylates exhibited a randoms oriented grainy morphology, whereas the films prepared from high-temperature decomposing precursors exhibited morphologies and grain orientations that depend on the film decomposition conditions, such as heat treatment and solution concentration [159].

The main objective of the present research on the sol-gel preparative route remains to find new precursors and precursor solution formulations so that well-crystallized, crack-free BST thin films can be obtained at lower temperatures. Also, to find a method to avoid repetitive coating–drying–baking process. This thermal cycling process causes the formation of hillocks on the bottom electrode, resulting in inferior electrical properties compared to films obtained by other methods.

2.13. Rapid Thermal Annealing

In recent years, rapid thermal annealing (RTA) has become more important in ULSI applications, such as thin dielectric deposition, polysilicon growth, shallow junction formation, silicidation, and annealing [161, 162]. The great advantages of the technique are a rather short processing time and relative process simplicity as compared to the conventional furnace apparatus. A short processing time helps reduce the time–temperature product such that the physical or chemical processes are completed while unwanted processes (e.g., dopant diffusion penetration, interface reactions, decomposition) are effectively controlled. Rapid thermal N_2O annealing (RTN_2O) has been applied to reduce the leakage current in the BST films prepared by CVD [40].

Table V compares the best properties of BST films prepared by various methods. Among the various techniques described earlier, the most common techniques that have been frequently used to deposit BST thin films are dc and rf sputtering, CVD, PECVD, MOCVD, LSCVD, ECR-CVD, laser ablation, and the sol-gel method. To be suitable for application to storage capacitors, dielectric films must have a very small leakage current

to maintain the favorable retention characteristics, and, in the case of topography for three-dimensional memory cells, have the capability of excellent step coverage. Fortunately, dc or rf sputtering deposition has the advantage of depositing BST films at low temperatures, which is very desirable for applications in which the processing temperature or thermal budget is a major concern. A major difficulty when using the sputtering deposition technique is choosing the process conditions to obtain stoichiometric BST films at the highest deposition rate. Several researchers have investigated the sputtering deposition process and have proposed different criteria of the deposition condition for preparing stoichiometric BST films.

Although as-deposited sputtered films have low leakage currents in the amorphous phase, the high-temperature treatments necessary for standard DRAM processes lead to the crystallization of these films and hence a drastic increase in the leakage current [40, 48]. This obviously limited their application in DRAMs in terms of the cells' refresh characteristics. The ECR-CVD process allows lower processing temperature ($\leq$500°C) for BST films, and thereby keeps the leakage current at a lower level ($<10^{-6}$ A/cm^2) and also allows the use of various multiple electrode structures, such as $RuO_2/Ru/TiN/TiSi_x$, which are not otherwise stable at temperatures above 750°C [102]. The as-deposited CVD BST films have rather leaky current characteristics due to oxygen deficiency and impurity contamination existing in films, this can be significantly reduced to acceptable levels for applications by annealing techniques. Reports indicate that RTA processed BST thin films in O_2 or N_2O ambients showed better electrical characteristics [40, 48]. CVD BST films can provide better step coverage ability and good thickness uniformity across the wafer. Thus, based on consideration of the electrical characteristics and step coverage, CVD BST is more suitable for application to mass production.

3. PHYSICAL AND ELECTRICAL PROPERTIES OF BST THIN FILMS

3.1. Factors that Influence BST Thin Film Properties

Parameters that have been identified to affect the properties of BST film capacitors are processing methods, annealing conditions, microstructure, interface structure, electrode materials, and their correlation. These factors are discussed briefly as follows:

3.1.1. Processing Methods

A variety of deposition techniques, including rf sputtering and IBS, laser ablation, CVD, MOD, and sol-gel have been successfully used to synthesize BST films. The various techniques used to fabricate BST films are designed to produce the films' specific microstructure and dielectric properties. In addition, it is also essential to obtain the lowest possible process temperature to comply with silicon technology and to minimize postdeposition thermal treatments under low oxygen partial pressure to maintain the resistance of the films. Meanwhile, maintaining precise microscopic control of the stoichiometry, achieving

Table V.   Comparison of Electrical Data from BST Samples Prepared by Various Deposition Techniques

Deposition technique	Composition	Film thickness (nm)	Dielectric constant	Leakage current density (A/cm^2) at 100 V/cm	Dielectric strength (MV/cm) at 10^{-6} A/cm^2	Capacitor structure (top/BST/bottom)	Reference
rf sputtering	Ba$_{0.75}$Sr$_{0.25}$TiO$_3$	80	320	1×10^{-8}	0.5	Pt/BST/Pt	[36]
rf sputtering	Ba$_{0.75}$Sr$_{0.25}$TiO$_3$	60	400	1×10^{-7}	0.5	Pt/BST/Pt	[149]
rf sputtering	Ba$_{0.65}$Sr$_{0.35}$TiO$_3$	100	400	5×10^{-9}	0.35	Pt/BST/Pt	[150]
rf sputtering	Ba$_{0.5}$Sr$_{0.5}$TiO$_3$	100	470	3×10^{-9}	0.5	TiN/BST/Pt	[160]
rf sputtering	Ba$_{0.5}$Sr$_{0.5}$TiO$_3$	100	600	3×10^{-7}	0.3	Pt/BST/Pt	[114]
rf sputtering	Ba$_{0.5}$Sr$_{0.5}$TiO$_3$	100	250	3×10^{-7}	0.7	Pt/BST/LNO	[151]
rf sputtering	Ba$_{0.5}$Sr$_{0.5}$TiO$_3$	60	200	1×10^{-8}	1	Pt/BST/SRO	[152]
rf sputtering	Ba$_{0.5}$Sr$_{0.5}$TiO$_3$	36	338	2×10^{-8}	0.69	Pt/BST/Ir Pt/BST/IrO$_2$	[153]
rf sputtering	Ba$_{0.5}$Sr$_{0.5}$TiO$_3$	100	375	8×10^{-9}	1.6	Pt/BST/Pt	[38]
rf sputtering	Ba$_{0.5}$Sr$_{0.5}$TiO$_3$	100	230	8×10^{-8}	0.5	BRO/BST/BRO	[154]
rf sputtering	Ba$_{0.5}$Sr$_{0.5}$TiO$_3$	20	274	8×10^{-9}	1	SRO/BST/SRO	[155]
rf sputtering	Ba$_{0.5}$Sr$_{0.5}$TiO$_3$	75	350	1×10^{-7}	–	LSCO/BST/LSCO	[113]
rf sputtering	Ba$_{0.5}$Sr$_{0.5}$TiO$_3$	120	600	10^{-9}	–	Pt/BST/TiO$_2$/Pt	[266]
rf sputtering	Ba$_{0.5}$Sr$_{0.5}$TiO$_3$	100	573	10^{-7}	–	Pt/BST/RuO$_2$	[247]
ECR sputtering	Ba$_{0.55}$Sr$_{0.45}$TiO$_3$	200	320	2×10^{-7}	0.09	Pt/BST/Pt	[156]
Excimer laser ablation	Ba$_{0.5}$Sr$_{0.5}$TiO$_3$	200	375	5×10^{-7}	0.15	Pt/BST/Pt	[157, 158]
Excimer laser ablation	Ba$_{0.5}$Sr$_{0.5}$TiO$_3$	500	467	10^{-7}	–	Au/BST/Pt	[143]
MOCVD	Ba$_{0.7}$Sr$_{0.3}$TiO$_3$	40	450	2×10^{-8}	0.7	Pt/BST/Pt	[159]
MOCVD	Ba$_{0.7}$Sr$_{0.3}$TiO$_3$	40	210	1×10^{-8}	0.37	Ir/BST/Pt	[160]
ECR-MOCVD	Ba$_{0.4}$Sr$_{0.6}$TiO$_3$	100	600	7×10^{-7}	0.15	Al/TiN/BST/RuO$_2$	[161]
ECR-PCVD	Ba$_{0.5}$Sr$_{0.5}$TiO$_3$	27	140	1×10^{-8}	0.44	Pt/BST/Pt	[162]
LS-CVD	Ba$_{0.5}$Sr$_{0.5}$TiO$_3$	200	300	3×10^{-7}	0.15	Pt/BST/Pt	[163, 164]
LS-CVD	Ba$_{0.5}$Sr$_{0.5}$TiO$_3$	50	200	1×10^{-8}	0.35	Pt/BST/Pt	[145]
LS-CVD	(Ba,Sr)$_{1+x}$TiO$_{3+x}$, $x = \pm 0.2$	30	260	1×10^{-7}	–	Pt/BST/Pt	[180]
CVD	Ba$_{0.5}$Sr$_{0.5}$TiO$_3$	100	400	8×10^{-8}	0.7	Pt/BST/Pt	[165]
MOD	Ba$_{0.7}$Sr$_{0.3}$TiO$_3$	140	420	1×10^{-9}	0.43	Pt/BST/Pt	[117]
MOD	Ba$_{0.7}$Sr$_{0.3}$TiO$_3$	300	563	10^{-6}	–	Au/BST/Pt	[197]
MOD	(Ba$_{0.7}$Sr$_{0.3}$)(Ti$_{0.95}$Nb$_{0.05}$)O$_3$	300	250	2×10^{-6}	–	Au/BST/Pt	[197]

large area deposition and achieving good step coverage are also relevant tasks [40, 48].

The deposition methods noted above are highly competitive. Each method has its own merits and limitations related to the deposition mechanics and film properties. For instance, the properties of rf-sputtered BST films can satisfy the requirements for use in a 256-Mb DRAM capacitor. The rf sputtering method using a multicomponent oxide target not only satisfies this requirement, but also is an appropriate method for producing BST films. NEC, Mitsubishi, and Samsung have made great studies in rf magnetron-sputtered (Ba$_{0.5}$Sr$_{0.5}$)TiO$_3$ films to a practical stacked DRAM capacitor [60]. They sputter-deposited thin BST films with an equivalent SiO$_2$ thickness of 8 Å over Pt/Ta electrodes, subsequently attaining an unit area capacitance of 40 fF/μm^2 and leakage current of $<10^{-7}$ A/cm^2. They also observed the dependence of the dielectric constant on film thickness; BST's dielectric constant decreased with reduced film thickness. The dielectric constant of the 70-nm film exceeds 300, and that of the 200-nm film exceeds 600—values that are much larger than those of SrTiO$_3$. Mitsubishi deposited thin (Ba$_{0.75}$Sr$_{0.25}$)TiO$_3$ films by rf sputtering at substrate temperatures of 480–750°C [163]. The 30-nm films deposited at 660°C on a Pt/SiO$_2$/Si substrate have a dielectric constant of 250, corresponding to an equivalent SiO$_2$ thickness of 0.47 nm, and a leakage current density about 1×10^{-8} A/cm^2, which partially satisfied the requirements for use in a 256-Mb DRAM capacitor. Samsung researchers have studied rf magnetron-sputtered (Ba$_{0.5}$Sr$_{0.5}$)TiO$_3$ films with thicknesses of 15–50 nm at 640–660°C on 6-inch Pt/SiO$_2$/Si substrates and postannealed at 550–850°C in O$_2$ or N$_2$ [53]. The 20-nm film, with an SiO$_2$ equivalent thickness of 0.24 nm, has a leakage current of 4×10^{-8} A/cm^2 and unit area capacitance of 145 fF/μm^2, which is the highest storage capacitance reported to date for BST films. They also contended that N_2 annealing of the BST thin film after the top electrode deposition is critical for obtaining a low leakage current, because n-type BST film is required to form a high interfacial potential energy barrier. However, their dielectric constants are insufficiently large for application to a Gb-era DRAM with a planar-type storage capacitor.

The PLD method has been used to successfully synthesize $(Ba_{0.5}Sr_{0.5})TiO_3$ thin films [50, 164]. Although this method has the ability to grow crystalline films at low substrate temperatures with a good control of stoichiometry, the reported leakage current density values are very high. The MOD technique provides advantages of reproducible coating thickness and composition and low deposition cost. Fujii et al. [117] used this technique to prepare BST films over a Pt/Ti/Si substrate based on alcohol-based precursor liquid. They obtained the dependence of the lattice constant and the dielectric constant of the fabricated BST film on the Sr composition. These results present maximum dielectric constants for films close to the composition of $Ba_{0.7}Sr_{0.3}TiO_3$. They also used this BST film, which allows incorporation of the planar-type single-stack structure into the ULSI DRAM storage capacitor and achieve equivalent SiO_2 thickness of 1.3 nm and leakage current density of 2×10^{-9} A/cm^2 at 3.3 V.

The advantages of CVD include a high deposition rate, uniform deposition over large areas, and satisfactory step coverage [125, 165]. However, CVD of BST film is restricted by low vapor pressure of source materials and deterioration during storage. Researchers at Mitsubishi have developed an alternative method of precursor transportation, in which suitable precursors are dissolved in organic liquids and the liquid is injected into a CVD reactor [166]. The liquid-source delivery methods have produced BST thin films on 6-inch Pt/SiO$_2$/Si substrates using Ba(DPM)$_2$, Sr(DPM)$_2$, and TiO(DPM)$_2$ (DPM = dipivaloylmethanato, $C_{11}H_{19}O_2$) dissolved in tetrahydrofuran and achieved a reproducibility of $\pm3\%$ for (Ba + Sr)/Ti, coverage of 72%, a dielectric constant of 230, an equivalent SiO_2 thickness of 7.8 Å, and leakage current density of 6.7×10^{-6} A/cm^2 at 1.65 V.

Mitsubishi also constructed a DRAM cell with dimensions appropriate for a 1-Gb device on the basis of a Ru/BST/Ru stacked capacitor. The BST films, with an equivalent SiO_2 thickness of 0.5 nm and excellent step coverage, were deposited at 420°C by a two-step process of LSCVD. NEC used ECR-CVD to develop a Gb storage capacitor. BST films with a thickness of 61 nm were deposited on Pt/TaO$_x$/Si substrates at 450°C and treated with RTA at 700°C for 1 minute. These films had a dielectric constant of 220 and leakage current density of 3×10^{-7} A/cm^2 at 1 V [167]. They also developed ECR-MOCVD BST-based stacked capacitors with RuO$_2$/Ru/TiN/TiSi$_x$ storage nodes for Gb DRAM generations [168]. Sol-gel processing involves hydrolysis and condensation of organo-metallic precursors, with the resulting sol coated onto the substrates and dried to a solid film (gel). The resulting gel film is then decomposed and densified by heat treatment to produce a crystallized film. Tahan et al. [142] sol-gel deposited

400-nm-thick $Ba_{0.8}Sr_{0.2}TiO_3$ films on Pt/Ti/SiO$_2$/Si substrates to obtain a dielectric constant of 400 and a leakage current density of 0.17 μA/cm^2 at 3 V. Although BST films with adequate dielectric properties can be deposited using various techniques, on the basis of the DRAM capacitor cell structures proposed by the NEC, Mitsubishi, Samsung, and U.S. DRAM consortium, CVD BST films are required in future Gb-era DRAMs to increase storage capacitance. Integration issues, including selection and stability of electrode and barrier layer materials, step coverage of dielectric films, high temperature endurance, and etching methods, must be solved before high-permittivity film-based DRAMs are commercialized.

Thus, the step coverage of BST films deposited using the CVD technique has attracted increasing attention for use in the side area of bottom electrodes in DRAM capacitors to increase storage capacitance. NEC researchers have applied ECR plasma MOCVD BST films to a practical stacked Gb DRAM capacitor. Applying this stacked capacitor technology can achieve a sufficient cell capacitance of 25 fF for 1-Gb DRAMs in a capacitor area of 0.225 μm^2 with only 0.3-μm-high storage nodes [102].

Electrical properties of the BST films prepared using various deposition techniques are compared in Table V. For the leakage current densities, an extremely large variation in results is found for capacitors using various electrodes, compositions, and thicknesses of BST films. This table also includes values of the dielectric constant for $(Ba_{0.5}Sr_{0.5})TiO_3$ between 140 and 600. Different studies vary markedly with respect to dielectric breakdown strength. Such a variation might be ascribed to intrinsic film properties (e.g., microstructure and stoichiometry), as well as to electrode and interface properties [40, 48].

According to a previous investigation, oxygen vacancies in BST films play a prominent role in leakage current [186]. In general, BST films sensitive to oxygen deficiencies are prepared or annealed in oxygen ambience to reduce the concentration of oxygen vacancies and improve the dielectric properties of the films. Table VI lists the previous publications of BST thin films prepared at various O$_2$ to Ar flow ratios during rf sputtering [13, 32, 35, 52, 114, 169, 187–199]. As was recently demonstrated, the dielectric properties depend on a gas ratio of O$_2$/(Ar + O$_2$) (OMR) for rf-sputtered BST films [38]. In that investigation, the dielectric constant increased with an increase of OMR and reached a maximum value at 50% OMR (Fig. 5). The leakage current density, although decreasing with an increasing oxygen flow, had a minimum value at 40% OMR. The film deposited at 450°C and 50% OMR had a dielectric constant of 375 and a leakage current density of 7.35×10^{-9} A/cm^2 at an electrical field of 100 kV/cm with a delay time of 30 seconds. The BST films can exhibit large dielectric constants due to polarization of electric dipoles. It has been reported that the dielectric constant of the films was influenced by oxygen stoichiometry [114], composition [190], grain size [53, 169, 170], grain boundary [194], and crystallinity (dipole density, polarization) [52, 170, 187]. High oxygen incorporation in the films seems to play an important role in promoting the polarization of electric dipoles. A related study that fabricated Pt/BST/Pt capacitors us-

Table VI.   Published O_2 to Ar Flow Ratio During rf Sputtering Deposition
of BST/Metal or Si Thin Films

Number	O_2 : Ar	Dielectric constant	Maximum deposition temperature (°C)	Reference
1	1 : 1	720	650	[114]
2	1 : 4	400	460	[167]
3	1 : 4	600	600	[170]
4	1 : 4	–	–	[171]
5	1 : 4	1300	600	[174]
6	5 : 22.5	–	650	[35]
7	1 : 9	540	650	[114]
8	1 : 9	700	700	[13]
9	1 : 9	400	600	[169]
10	1 : 9	550	750	[150]
11	1 : 9	30	400	[172]
12	1 : 9	500	750	[52]
13	1 : 9	420	700	[175]
14	1 : 9	–	550	[178]
15	2 : 8	375	640	[176]
16	2 : 8	–	650	[177]
17	Mixed	350	500	[168]
18	Mixed	–	550	[32]
19	Mixed	440	560	[179]
20	Mixed	550	360	[173]
21	Mixed	375	450	[38]

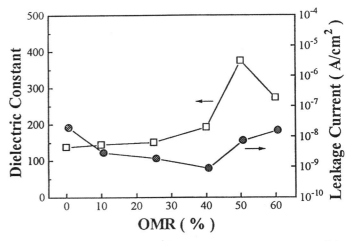

Fig. 5. Effect of OMR on the dielectric constant and leakage current of the $Ba_{0.5}Sr_{0.5}TiO_3$ films deposited at 450°C. Reprinted with permission from [38], © 1997, American Institute of Physics.

O_2 content of the sputtering gas. It was believed that this resulted from the sputtering yield of O^{2-} being lower than that of Ar^+. Under all deposition conditions of the study, perovskite-structure BST films were obtained and crystallinity was improved by annealing. However, no change in crystal orientation was observed. The dielectric constant of the BST films increased with increasing O_2 content of the sputtering gas and improved by annealing. In addition, the leakage current characteristics varied with varying O_2 content of the sputtering gas, and showed a decreasing trend with the increase of the O_2/Ar ratio. Using atomic force microscopy (AFM) images, the authors verified that the leakage current characteristics of the BST films were strongly related to the films' surface roughness.

An appropriate annealing procedure was found to produce improved crystallinity of BST films with little morphology degradation. Postannealing for as-deposited film is in general a promising method for improving the crystalline structure and consequently, increasing the dielectric constant of the film. However, the leakage current depends heavily on the postannealing temperature and the atmosphere after top electrode fabrication. Park et al. [196] examined how postannealing affects the electrical properties of 20-nm-thick sputtered BST films. According to their results, the dielectric constant increased without significantly increasing the leakage current with annealing at 750°C. Later, Horikawa et al. [200] developed a postannealing process for 30-nm-thick CVD-deposited BST films. Their results indicated that direct annealing of BST capacitors roughened the surface morphology of the upper Pt electrodes of the BST capacitors. However, the postannealing of capacitors with silicon dioxide passivation changed the surface morphology of Pt and BST only slightly and did not significantly deteriorate the leakage current.

Deposition temperature, a major parameter in the deposition process, determines the decomposition rate of the precursors and has a strong influence on the crystallinity and structure of the deposited films. Depositing good-quality BST thin films requires a rather high process temperature owing to BST's high

ing a sputtering technique and postannealed them under a N_2 or H_2 atmosphere indicated abnormally higher leakage current when the negative bias was applied to the top electrode [198]. In addition, the enhanced leakage currents were effectively reduced by annealing under an O_2 atmosphere. These results can be accounted for by compensating for the oxygen vacancy in the BST films by introducing oxygen through the top Pt electrodes, with the grain boundaries of the columnar structure acting as a diffusion path for the oxygen.

BST thin films prepared on (111) Pt/Ti/SiO_2/Si and (100) Si substrates by PLA have been reported as a function of the target composition and the oxygen pressure [190]. Surface morphology of films prepared at high oxygen pressure was rough compared to that of films prepared at low oxygen pressure. The dielectric constant of those films was found to be lower than that of the films prepared at low oxygen pressure. The authors suggested that excessively high oxygen pressure during PLA deposition deteriorated the crystal structure and the dielectric property of the BST films and that optimum oxygen pressure depends on the composition of BST films.

Lee et al. [199] deposited BST thin films on Pt/SiO_2/Si substrates with various O_2/Ar ratios by rf sputtering and investigated the crystallinity, microstructure, and electrical properties of the films. The deposition rate decreased with increasing crystallization temperature. Besides influencing the BST material, a high process temperature also affects the interfaces with electrodes, which controls the overall electrical properties of the capacitor, particularly when BST film thickness is very small [53].

3.1.2. Film Composition

Film composition has a pronounced effect on the dielectric constant [179, 182, 200]. A film with a composition of $Ba_{0.5}Sr_{0.5}TiO_3$ has the highest dielectric constant at room temperature. Several researchers have conferred that a film's maximum dielectric constant at room temperature is when the $(Ba + Sr)/Ti$ ratio is $1:1$. According to their results, the dielectric constant decreases when films are either titanium rich or titanium poor [166]. It has been found that in $(Ba_xSr_{1-x})Ti_{1+y}O_{3+z}$, the factor y, corresponding to the $(Ba + Sr)/Ti$ ratio, strongly affects most film properties at a given x and deposition temperature [134, 166, 201–205] and thus is one of the primary parameters used to control film performance. For example, BST films with $x = 0.7$ have a maximum resistance degradation lifetime at approximately $y = 0.083$, although the maximum value of the dielectric constant is found at $y = 0$ [134]. Reasonable film behavior is generally achieved up to $y = 0.15$, which greatly exceeds the solubility of excess Ti in bulk BST of approximately $y \leq 0.001$ [204]. Given this large stoichiometry, a necessary step in understanding the composition dependence of film properties is to determine the locations within the microstructure at which the excess titanium is accommodated in BST thin films. Stemmer et al. [205] reported measurements of the microstructural accommodation of nonstoichiometry in BST thin films grown by LS-CVD. Their observations indicate partial accommodation of excess titanium in the grain interiors of polycrystalline BST films, either concurrent with or followed by accommodation at the grain boundaries. At extreme titanium excess, an amorphous phase (possibly TiO_x) was found between grains. The increased grain boundary area in these nanocrystalline films compared to that in much larger grained bulk ceramics, in combination with the nonequilibrated microstructure of the films due to lower processing temperatures, were said to the reason why Ti contents well beyond the bulk solid solubility limit are tolerated by their BST film structure.

It has been reported that the addition of dopants seriously influences the electrical properties of BST thin film capacitors [206, 207]. The effects of Al and Nb doping on the leakage current behavior of $Ba_{0.5}Sr_{0.5}TiO_3$ thin films deposited by rf magnetron sputtering were reported by In et al. [207]. Al and Nb were known to replace Ti sites of the BST perovskite. BST thin films deposited at room temperature and annealed subsequently in air showed improved electrical properties. In particular, the leakage current density of the Al-doped BST thin film was measured at around 10^{-8} A/cm^2 at 125 KV/cm, which was much lower than that of the undoped or Nb-doped thin films. The grain boundary Schottky barrier formed in the BST film interior seems to be the reason why Al doping is particularly beneficial for the suppression of the leakage current in the postannealed films. It is well known that grain boundaries improve the insulation resistance in undoped or acceptor-doped titanates [208]. This is due to inherent grain boundary donor-type interface states, such as immobile $V_O^{\cdot\cdot}$ and hole [209–212]. The formation of such negative space charge regions should reduce local conductivity at grain boundaries, which may explain why Al-doped BST film has a lower leakage current than the undoped film. For the Nb (donor)-doped film, Nb ions first compensate the intrinsic acceptor ions, and the remaining portion seems to form the positive space charge region adjacent to the negatively charged grain boundary region similar to the n-doped $BaTiO_3$ ceramics [213, 214]. In reality, surface states originating from surface lattice defects and the interfacial segregation of acceptor or donor dopants greatly influences the barrier height [196, 215]. Studies on the type and concentration of dopants segregated in the grain boundaries or electrode interfacial layer should be pursued to elucidate the doping effects.

Copel et al. [216] investigated the effects of Mn impurities on $Ba_{0.7}Sr_{0.3}TiO_3$ films using X-ray photo emission spectroscopy. Mn acts as an electron acceptor, compensating for the charge density found in nominally undoped films. This causes a greatly increased depletion width in acceptor-doped films. The decreased leakage current in the acceptor-doped films was attributed to the increased barrier to thermionic emission of electrons from Pt contacts into the dielectric. Doping in the films lowered the dielectric constant. This lowering effect results from the incorporation of aliovalent ions, which hinders film crystallization due to the requirement for higher solution energies to form compensating point defects [217]. In addition, the composition of surfaces/interfaces also largely determines the properties of the films and the characteristics of the devices based on the films. These results suggest that segregation of acceptor or donor dopants at the grain boundaries in the film's interior heavily influences the barrier height, which could determine the leakage behavior in BST thin films.

3.1.3. Crystalline Structure

Crystalline BST films are usually obtained at relatively high substrate temperatures. During film growth, however, interlayers and specific grain structures are developed that cause serious problems of low dielectric constant and the leakage current [218–220]. An alternative approach is to grow amorphous BST films at low temperature and the crystallize them in a postannealing process. Improved dielectric constants and leakage currents has been reported on postannealed amorphous BST films [192, 221]. The crystallization of amorphous $Ba_xSr_{1-x}TiO_3$ thin film grown on a single-crystal MgO (001) substrate by rf sputtering was studied by Noh et al. [222] in a synchrotron X-ray scattering experiment. Their study shows that a metastable intermediate phase nucleated at around 500–600°C at the interface plays a crucial role in the crystallization process. In a 550-Å-thick film, the crystallization to perovskite

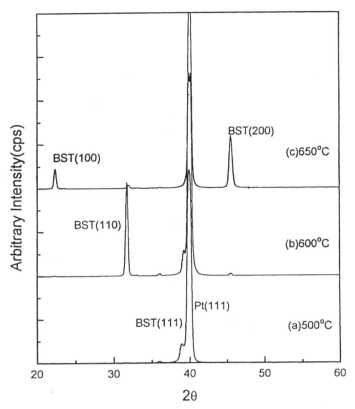

Fig. 6. XRD patterns of $Ba_{0.5}Sr_{0.5}TiO_3$ thin films deposited at various substrate temperatures [114].

Table VII.   Dielectric Constant of BST/Barrier Layer/Si Capacitor Structure
[223]

Barrier layer	Thickness (nm)	Dielectric constant
Deposited on Au/Mo/sapphire		170
Directly deposited on Si		98
Pd	50–150	<120
Pt	50–150	<94
TiSi$_2$	500	83
TaSi$_2$	500	71
RuSi	500	181
Pd/Ti	Pd: 50–150; Ti: 20	<86
Pt/Ti	Pd: 50–150; Ti: 2–50	>140
Pt/Ta	Pd: 50–150; Ti: 10–20	>140

However, the cause of the difference in crystal structures is not yet clearly understood. It is speculated that these differences are strongly associated with the crystallinity and the orientation of the Pt electrode. Furthermore, the change of the preferred orientation from the (110) to the (100) direction above 600°C is considered to be related to the surface energy. In the perovskite structure, the (100) plane is a closely packed oxygen plane that has the lowest surface energy. Even if there is large lattice mismatch, the (100) preferred orientation is strongly developed at high deposition temperatures. In general, the films deposited on the Si surface require sufficiently higher substrate temperatures to form the crystalline phase than are required on the Pt surface. These results, given in Table VII, reveal that the Pt surface can enhance nucleation of the BST film more effectively than the Si surface [223]. BST films on Pt/Ti and Pt/Ta have higher dielectric constant than that of BST films directly on Si, TiSi$_2$, and TaSi$_2$.

3.1.4. Microstructure

The dielectric property of polycrystalline BST films is affected not only by the composition and crystalline structure of the phase, but also by the microstructure. The dielectric film in the next-generation DRAM capacitor should have an equivalent SiO$_2$ thickness (t_{eq}) of less than 1 nm. When the BST film with a dielectric constant of about 300 is applied for the DRAM, the actual film thickness must be less than 130 nm to obtain t_{eq} less than 1 nm. The effect of grain size on dielectric properties is important for application to the DRAM capacitor, because the BST film has such a small thickness [13]. Notable size effects of the dielectric constant, including thickness dependence and grain size dependence, have been reported in BST films. Miyasaka and Matsubara [11] reported that the highest dielectric constant of a 500-nm-thick polycrystalline $(Ba_{0.5}Sr_{0.5})TiO_3$ film was 900, whereas that of bulk ceramics is known to be more than 5000. They also reported that thinner films (80 nm thick) showed a smaller dielectric constant, of about 400. Horikawa et al. [13] investigated the correlation between the dielectric constant and broadness of an X-ray

phase was occurred at around 700°C, whereas a 5500-Å-thick film became crystalline at 550°C. The thickness dependence of the crystallization was attributed to the observed intermediate phase nucleated at near 600°C at the interface.

In thin films, a high annealing temperature was required because of the energy barrier between the perovskite phase and the intermediate phase. In thick films, the perovskite phase was nucleated directly from the amorphous phase in the bulk of the film concurrent with nucleation of the intermediate phase at the interface. A transmission electron microscopy (TEM) study by Paek et al. [219] also reported an intermediate phase near the interface.

X-ray diffraction patterns of as-grown BST films deposited at various substrate temperatures are shown in Figure 6 [114]. A cubic perovskite structure of the BST films was typically obtained under all conditions. As can be seen in Figure 6, the films' crystal orientation and crystallinity were strongly dependent on the deposition temperature. Films deposited below 600°C had a cubic perovskite structure and showed the polycrystalline state. The crystallinity of the films increase with increasing deposition temperature. Films deposited at 600°C and 650°C had textured structures with (110) and (100) orientations, respectively. At 650°C, highly (100)-oriented BST films were obtained. Many researchers have reported obtaining BST thin films in the polycrystalline state on the polycrystalline Pt electrode at a substrate temperature of around 500–650°C.

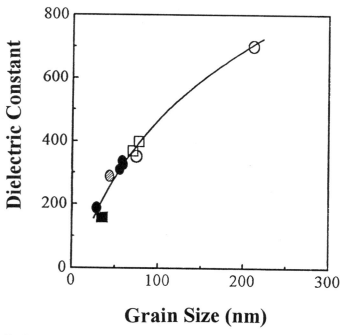

Fig. 7. Dependence of the dielectric constant on the grain size from XRD for the $Ba_{0.65}Sr_{0.35}TiO_3$ films deposited at substrate temperatures 500°C (■), 550°C (●), 600°C (△), 650°C (□), and 700°C (○) [13].

Kawahara et al. [225] observed that protrusions of BST crystallites appeared on BST film surfaces prepared by liquid source CVD (at 420°C). They felt that these protrusions (consisting of the cubic BST perovskite phase) appeared because the BST films deposited in the first ~150-Å layer were not sufficiently crystallized; that is, the density of nuclei in that layer was small. As the (Ba + Sr)/Ti ratio increases, the density of protrusions increases at a rate similar to that of the BST (110) peak intensity. The protrusions were successfully restrained by two-step deposition where BST films consisted of a buffer layer and a main layer, with the buffer layer a CVD BST film about 60 Å thick annealed in ambient N_2 before the second deposition. By this two-step deposition on Pt electrodes, a BST film of $t_{eq} = 0.56$ nm was achieved at a total film thickness of 230 Å. The annealing of the buffer layer effectively increases the density of nuclei and makes the BST film surfaces smooth. Recently, some research groups have shown that buffer layers, such as $PbTiO_3$ and $SrTiO_3$, play a very important role in the growth of high-quality perovskite structure thin films at low temperatures [226–228].

Miyasaka and Matsubara [11] reported the interesting fact that the dielectric constant of the BST thin film becomes maximum at a Ba/Sr atomic ratio of 1/1, although the dielectric constant is maximum at a Ba/Sr ratio of about 3/1 in the bulk of BST [229]. Arlt et al. [230] reported increased dielectric constant and increased temperature variation with increased grain size in bulk $BaTiO_3$ ceramics with grain size less than 1 μm. They explained that these phenomena were caused by the grain size effects. The same phenomena were observed in BST films. Thus the grain size effect certainly influences the dielectric constant and its temperature dependence in BST films. Generally, BST thin films show marked electrical degradation with decreasing thickness [60, 231]. As the thickness of BST thin films decreases, the leakage current increases and the dielectric constant decreases markedly.

It is generally accepted that the grain size of BST thin films deposited by rf-magnetron sputtering decreases as the film thickness decreases [232]. If the grain size is large, then the dielectric constant approaches that of the bulk. As the grain size decreases, the proportion of grain boundaries in the films increases. Grain boundaries act as leakage current paths. Therefore, the small grain size in the 50-nm-thick film results in a higher leakage current density and a lower dielectric constant than in thicker films. BST polycrystalline films have been prepared using rf sputtering by Miyasaka and Matsubara [11] and Horikawa et al. [13]. According to their experimental results, the dielectric constant of the polycrystalline film was much smaller than that of the bulk. Because the grain boundaries are considered a main factor in the decrease in the dielectric constant in the polycrystalline films, epitaxially grown BST film is expected to have a dielectric constant close to that of the bulk.

Mukhortov et al. [233] reported heteroepitaxial growth of BST films on MgO (100) single crystals. Abe et al. [234–236] studied $(Ba_{0.24}Sr_{0.76})TiO_3$ thin film epitaxially on a Pt/MgO (100) substrate, where the Pt film was also epitaxially grown on the MgO as a bottom electrode. The reason for

diffraction in $(Ba_{0.65}Sr_{0.35})TiO_3$ thin films. According to their result, the polycrystalline film with a grain size of 45 nm had a dielectric constant smaller than 200, whereas the film with a grain size of 220 nm had a dielectric constant larger than 700 (Fig. 7). Horikawa et al. [13] studied the effect of grain size on the dielectric properties of $(Ba_{0.65}Sr_{0.35})TiO_3$ films deposited at substrate temperatures of 500–700°C. The dielectric constant of these films ranged from 190 to 700 at room temperature. This value changed with grain size rather than with film thickness. In the film deposited at 550°C, a film structure with granular grains was observed. The grain size was 10–50 nm. Granular grains with a size of 10–30 nm were seen in the film at 600°C. On the other hand, columnar grains with width of 20–100 nm were found in the film deposited at 700°C. The authors concluded that the reduction of dielectric constant in the polycrystalline films was probably dominated by the grain size in the thickness direction [13]. Similar size effects have also been observed in $SrTiO_3$ thin films. However, these effects seem to be much weaker in $SrTiO_3$ films compared to those observed in BST films.

The effects of plasma bombardment on the initial growth of BST films and their properties were studied by Tsai et al. [224]. The films grown outside the plasma region exhibited better crystallinity, higher dielectric constants, higher electrical conductivity, and rougher surfaces than those grown inside the plasma region. However, plasma bombardment did not affect initial growth of the films on $Pt/SiO_2/Si$ or MgO substrates, as explored by AFM. The films grown on $Pt/SiO_2/Si$ showed island growth characteristics, whereas those grown on MgO substrates revealed layer-by-layer growth characteristics.

choosing the Sr-rich composition was explained as because of the closeness of the lattice constant ($a = 3.93$ Å) of BST to that of Pt ($a = 3.923$ Å) was expected to bring about easy epitaxial growth on Pt. However, from the surface morphology observations using scanning electron microscopy (SEM) and reflected high-energy electron diffraction (RHEED), they noticed "stitch"-like projections, which were assumed to have probably resulted from the lattice constant inequality between $(Ba_{0.24}Sr_{0.76})TiO_3$ and Pt.

Recently, Yoon et al. [237] successfully prepared BST thin films that were epitaxially grown on Pt/MgO and YBCO/MgO substrates by means of a laser ablation technique. The thickness of their BST films was about 200 nm. However, anomalous elongation of the lattice constant was not reported. In the reactive evaporation method, the evaporated atoms have small kinetic energy, so the enhanced tetragonality may disappear in thick films because accidental dislocation is easily introduced during the deposition. On the other hand, in a sputtering method, the epitaxially grown film maintains the epitaxial effect of the tetragonality to a much thicker region, because sputtered atoms have a large kinetic energy [238, 239]. The sputtering method is likely to have merit in epitaxial growth of BST films with a small probability of misfit dislocation.

Lee et al. [240] studied the microstructure dependence of the electrical properties of BST thin films deposited on Pt/SiO$_2$/Si using cross-sectional TEM and diffraction analysis. Accordingly, BST film has a columnar structure that grows from the Pt to the BST surface. Also, the different layers were dense and smooth. Generally, when BST films grow with random orientation, diffraction patterns show rings. However, spot patterns were observed by Lee et al. [240] for their BST films, indicating that their film growth has (110) preferred orientation. Their grain size measurement by the line intercept method on the surface of the SEM micrograph demonstrated that as the film thickness increased, grain size increased slightly. They observed that the dielectric constant increased from 348 to 758 when the grain size increased from 32 to 82 nm in the 600°C deposited BST films. They suggested that the abrupt decrease of dielectric constant in the thin film (below 75 nm) was due to another factor besides the low dielectric layer that formed during the initial deposition stage. They speculated that it is strongly associated with the grain size of BST thin films. A related study indicated that the low leakage current density of a very thin BST capacitor with a SrRuO$_3$ bottom electrode might be attributed to the local epitaxial BST film [172].

In the case of $(Ba_{0.75}Sr_{0.25})TiO_3$ and $(Ba_{0.5}Sr_{0.5})TiO_3$ films, film structure changed from granular to columnar with increasing substrate temperature and was columnar for the film deposited at 750°C [52]. The dielectric constant correlated closely with grain size in the direction parallel to film thickness for both films. The grain size in the direction perpendicular to film thickness increases with deposition temperature, as does the grain size in the direction parallel to film thickness. Therefore, it is thought that the dielectric constant is greatly affected by film crystallinity, grain size, and the ratio of Ba/Sr composition interactively.

The properties of interfaces between electrodes and BST depend not only on the electrode material, but also on the processing, such as deposition conditions and postannealing. The existence of an interfacial layer between the BST film and the Pt bottom electrode was confirmed by high-resolution transmission electron microscopy (HRTEM) [219]. The interfacial layer appeared to have a different crystallinity than both the BST thin film and the Pt bottom electrode, resulting in variation of the interfacial states between BST and Pt. As the thickness of the BST films decreased from 300 nm to 50 nm, the thickness of the interfacial layer increased from 9.5 nm to 11 nm. The dielectric constant of the interfacial layer, calculated from its measured overall capacitance and thickness and confirmed by HRTEM, was about 30. This low–dielectric constant interfacial layer has been shown to affect the electrical degradation of BST thin films with decreasing thickness. The role of the interface becomes increasingly dominant in the overall electrical conduction process when film thickness is typically less than 100 nm.

3.1.5. Surface Morphology

The bottom electrode materials greatly affect the electrical characteristic of BST thin films through resultant formation of the surface morphologies. Increasing the $O_2/(O_2 + Ar)$ mixing ratio (OMR) during the film deposition process increases the root mean square (RMS) surface roughness of BST films [38]. The RMS surface roughnesses are 1.67, 3.199, 4.179, and 3.782 nm for 0%, 25%, 50%, and 60% OMR BST films, respectively, and the RMS value decreases for BST films deposited above 60% OMR. The diffusion energy of sputtered atoms is probably reduced when OMR is increased, and lateral movement on the surface also may be reduced, because the collisions between sputtered atoms and oxygen atoms increase and mean free path may also may be shorter. Therefore, the surface roughness of the films prepared at below 50% OMR would be expected to increase. But above 50% OMR, the O$_2$ resputtering rate may be stronger than the Ar sputtering rate, and hence the surface roughness would decrease at 60% OMR. The film deposited at 450°C and 50% OMR exhibited good surface morphology and had a dielectric constant of 375, a tangent loss of 0.074 at 100 kHz, and a leakage current density of 7.35×10^{-9} A/cm^2 at 100 kV/cm with a delay time of 30 seconds. Lee et al. [199] envisaged similar surface morphology studies for as-deposited and annealed BST films deposited in 0% and 50% OMR. Their observations indicate that the surface roughness of the as-deposited film decreased with increasing O$_2$/Ar ratio and that the annealing resulted in a sharp increase in film roughness. The dielectric constants of the film increased with increasing O$_2$ content, whereas the leakage current density decreased. From the AFM analysis, the authors verified that the leakage current characteristics of the BST films are strongly related to the surface roughness of the films.

In addition, the RMS value showed greater variation for BST films deposited at different bottom electrode materials [49, 241]. Table VIII indicates greater RMS surface roughness of

Table VIII. Properties of BST Deposited on the Various Bottom Electrodes[a]

Properties	Bottom electrodes				
	Pt	Ir	IrO_2/Ir	Ru	RuO_2/Ru
Dielectric constant	219	309	234	548	322
	(503)	(593)	(501)	(325)	(433)
Leakage current (10^{-8} A/cm^2) at 100 kV/cm	2.2	4.9	2.5	39.4	3.5
	(2.5)	(2.1)	(3.3)	(2.1)	(2.4)
Tangent loss	0.014	0.046	0.016	0.32	0.017
	(0.015)	(0.019)	(0.02)	(0.019)	(0.012)
Work function (eV)	5.6	5.35	–	4.8	–
Breakdown field (MV/cm)	3.84	3.68	3.49	1.94	1.84
Fatigue endurance (cycling number)	$>10^{11}$	$>10^{11}$	$>10^{11}$	$>10^{11}$	$>10^{11}$
Surface roughness (nm)	1.9	1.27	2.25	4.40	4.12
H_2 damage endurance (dielectric constant variation)	−16%	−12%	−13%	−29%	−14%
Stability in O_2 ambient	up to 700°C	up to 700°C	up to 700°C	up to 500°C	up to 700°C

[a] Values in parentheses represent O_2 annealing for 20 minutes at 700°.

Reprinted with permission from [49], © 1999, American Ceramic Society.

BST deposited on Ru and RuO_2/Ru compared to the Pt, Ir, and IrO_2/Ir, which is attributed to the higher RMS roughness of the Ru and RuO_2/Ru bottom electrode itself. If the as-grown film tends to be an amorphous layer with significant surface mobility of adatoms, then the film usually has a smooth surface. The RMS surface roughness may also be affected by the numbers of bottom stack layers, it generally increases with an increasing number of stack layers. Tsai and Tseng [49] observed that BST films deposited on Pt, Ir, and IrO_2/Ir with small grain size and smooth surface roughness showed a higher breakdown field, whereas the BST on Ru or RuO_2/Ru with larger grain size and greater surface roughness exhibited lower breakdown fields.

3.1.6. Film Thickness

The effect of film thickness on the microstructure and the associated property variation of BST thin film has been discussed in earlier sections. The details on the role of film thickness on the BST film's electrical and dielectric parameters are briefly presented here. The thickness dependence of the dielectric constant of rf-sputtered BST films varies with the substrate temperature in connection with the grain size effect [170]. Among the explanations tentatively proposed to give insight into the field and thickness dependence of permittivity in ferroelectric and paraelectric thin films, the most common approach has been invocation of the Schottky barrier model [242]. In this model, the variation in apparent capacitance with bias is explained via a voltage-dependent interfacial depletion layer capacitance in series with the capacitance of the bulk of the film, whose permittivity is taken to be bias independent. This description can be used to account for the observed capacitance-voltage behavior, but only by judicious choice of a variety of parameters that are necessary for the analysis. However, discrepancies associated with this model appear when dealing with films of interest for DRAMs. Although true leakage currents are very

well accounted for the existence of Schottky barriers at the film–electrode interface when nonreactive contacts such as Pt are used, reasonable assumptions for the charge carrier density lead one to conclude that the films are fully depleted for the thickness range of most interest for DRAMs. This occurs because the charge carrier concentrations are lower than required to compensate for the difference between the electrode work function and the film Fermi level, so that the total width of the depletion layers resulting from the Schottky barriers exceeds the film thickness. The electric field is then approximately uniform in uniform microstructural regions of the sample rather than being concentrated in the depletion layer. This view is strongly supported by the fact that the true leakage current does not strongly depend on film thickness, except when due to lowering in the field-induced barrier height. With an uniform electric field in the film, the leakage current under the same external field is essentially independent of the film thickness over a wide thickness range (20 nm to at least 200 nm) [242, 243].

Film thickness has been established to impact primarily the zero bias permittivity through a thickness dependence of the first-order coefficient of the Landau–Ginzburg–Devonshire approach [242, 243]. The dependence of the inverse of the zero bias capacitance density of BST thin film to its thickness is often attributed to the presence of a constant-valued capacitance density, C_i/A, represented by the nonzero intercept, in series with the thickness-dependent capacitance density of the bulk of the film [242–244]. The constant capacitance is usually thought to represent some type of interfacial layer between the dielectric and one or both of the electrodes and might arise from surface contamination of the BST, nucleation or reaction layers at the film–electrode interfaces, or changes in the defect chemistry at the dielectric–electrode interfaces. The apparent capacitance density at the zero field may then be expressed as

$$\frac{A}{C_{app}} = \frac{A}{C_i} + \frac{A}{C_B} = \frac{t_i}{\varepsilon_i \varepsilon_0} + \frac{t - t_i}{\varepsilon_B \varepsilon_0} \quad (1)$$

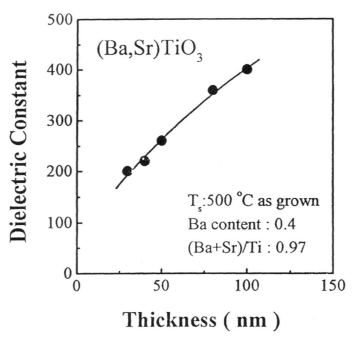

Fig. 8. Thickness dependence of the dielectric constant for $Ba_{0.4}Sr_{0.6}TiO_3$ films deposited on RuO_2 electrodes at 500°C by ECR-MOCVD. Reprinted with permission from [102], © 1997, IEEE.

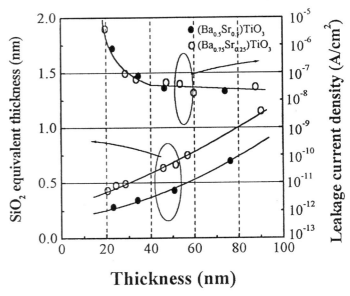

Fig. 9. Thickness dependences of leakage current density at an applied voltage of 1.65 V and SiO_2 equivalent thickness. $Ba_{0.5}Sr_{0.5}TiO_3$ and $Ba_{0.75}Sr_{0.25}TiO_3$ films were deposited at the substrate temperature of 660°C [52].

where A is area, C_{app} is the apparent capacitance, C_i is the interfacial capacitance, C_B is the bulk film capacitance, ε_B is the film bulk permittivity, ε_i is the interfacial layer permittivity, ε_0 is the permittivity of the free space, t is the total film thickness, and t_i is the interfacial layer thickness. Therefore, the nonlinear ferroelectric response is a long-range cooperative phenomenon, and the true permittivity may quite well change with film thickness.

Noh et al. [222] suggested that the observed thickness dependence of the crystallization in BST might be related to the film substrate interfacial behavior during crystallization. In thin films, the amorphous phase first transforms to the intermediate metastable phase. Since the energy barrier from the metastable phase to the perovskite phase is probably higher than that from the amorphous to the perovskite phase, the crystallization temperature in the thin film is higher. In thick films, the amorphous phase transforms directly to the perovskite phase at relatively low annealing temperature. The thickness dependence of dielectric constant for as-grown BST films deposited on RuO_2 at 500°C was studied by Yamamichi et al. [102]. Although the ε_r decreases with decreasing film thickness, the value larger than 400 was obtained for 100-nm BST without postannealing (Fig. 8). This crystallization in the as-grown states is one of the most important advantages of ECR-MOCVD, resulting in process step reduction for capacitor fabrication.

Film thickness dependences of the leakage current density at the applied voltage of 1.65 V and the SiO_2 equivalent thickness (t_{eq}) are shown in Figure 9, as reported by Kuroiwa et al. [52]. The t_{eq} is given by $(\varepsilon_{SiO_2}/\varepsilon_{BST}) \times t$, where t is the thickness of the BST film and ε_{SiO_2} and ε_{BST} are relative dielectric constants for SiO_2 and BST, respectively. The leakage current density

increased considerably for films less than 30 nm thick, but no marked difference was found between the $(Ba_{0.5}Sr_{0.5})TiO_3$ and $(Ba_{0.75}Sr_{0.25})TiO_3$ films. It is thought that leakage current depends on film thickness rather than on the difference in the target compositions. The t_{eq} is 0.35 nm for 30-nm-thick $(Ba_{0.5}Sr_{0.5})TiO_3$ film, in comparison with 0.47 nm for 30-nm-thick $(Ba_{0.75}Sr_{0.25})TiO_3$ film, and the leakage current density is less than 1×10^{-7} A/cm² in both films. Similar studies by Horikawa et al. [163] showed that the dielectric constant remains about 320 in the thickness range of 50 to 90 nm and the t_{eq} decreases linearly with film thickness. At a thickness of less than 50 nm, the dielectric constant is no longer constant, it drops to 25 in the 30-nm-thick film, and the minimum t_{eq} value of 0.47 nm is obtained in this film.

Paek et al. [219] observed that the leakage current of BST increases exponentially with decreasing film thickness. A leakage current of 9.8×10^{-8} A/cm² was obtained when their BST film thickness was 300 nm. However, this increased rapidly to 1.06×10^{-7} A/cm² for the 50-nm-thick BST film. The increased proportion of grain boundaries in 50-nm-thick BST film was said to be the main cause of the abrupt increase in the leakage current. On the other hand, Hwang et al. [53] noted that oxide equivalent thickness does not decrease in proportion with the decreasing BST film thickness, but rather decreases more slowly due to the decreased dielectric constant of the film when the films become thinner. They also found that the dielectric constant of the BST thin film sandwiched between two Pt electrodes decreases as the thickness decreases, because a low dielectric constant layer (which is a static space charge layer) exists at both interfaces with the electrodes due to the difference between the work function of Pt (5.5 eV) and electron affinity of the BST (about 1.7 eV). The abrupt decrease in the dielectric constant of the thinner BST films (below 75 nm) is attributed to another factor besides a low dielectric layer formed during the initial deposition stages. Many authors speculate that it is strongly associated with the grain size of BST thin films [60, 240].

3.1.7. Electrode Materials

The growth in density and complexity of integrated circuits has made flexibility in available metallization options very desirable. Metallization used for interconnects, contacts, diffusion barriers, and gate electrodes requires low resistivity and high thermal stability of the conducting material. It would be preferable in terms of processing simplicity to have one metal that is applicable to all these functions. However, because of more specific requirements, such as low contact resistance in contact applications [245] and a proper work function in gate metallization [246], the present metallization schemes cannot satisfy all of the above-mentioned functions. The choices are further reduced by the required compatibility with present fabrication techniques. Therefore, there is a need for novel materials that can fulfill one or more metallization requirements in today's and tomorrow's microelectronic devices.

The metallic oxides of transition metals may present a very attractive metallization option in a variety of very large scale integrated (VLSI) applications [247]. Dioxides of Ru, Ir, Os, Rh, V, Cr, Re, and Nb have bulk metallic resistivities ranging from 30 to 100 $\mu\Omega$ cm, with IrO_2 being the best conductor in this group. Other transition metal oxides merit attention, such as ReO_3, which has a resistivity of 10 $\mu\Omega$ cm (lower than that of the widely used $TiSi_2$). The heats of formation of transition metal oxides are comparable to those of transition metal nitrides [248], which emphasizes their normal stability. RuO_2 was reported to have a low contact resistance on Ti metal [249], comparable to that of pure ruthenium and gold. CrO_2, although fairly resistive, was used successfully as a diffusion barrier in the late 1970s [250] and recently, films of RuO_2 prepared by CVD were also reported to exhibit good diffusion barrier properties [251].

In the application of BST films as a dielectric, electrode materials must meet certain requirements, including high metallic conductivity, sufficient resistance against oxidation, good adhesion to BST, and interfacial smoothness, to achieve large capacitance and low leakage current. In addition, such factors as grain size distribution, crystalline orientation, interface, and surface structures significantly influence the performance of electrode materials. The electrical properties of storage electrodes and the nature of electrode/polysilicone plug contacts, as well as the electrical properties of dielectrics, dominate the overall capacitor performance. Oxidation and interdiffusion of materials constituting the electrode structures during film deposition can increase the resistance of electrodes and interconnections or cause the formation of a low dielectric layer. A resistance increase limits the speed of the charge–discharge response because of an increase in the capacitance–resistance (C–R) time constant, and the formation of a low dielectric layer

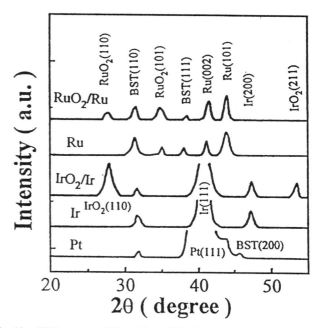

Fig. 10. XRD patterns of $Ba_{0.47}Sr_{0.53}TiO_3$ films deposited at 500°C on Pt, Ir, IrO_2/Ir, Ru, and RuO_2/Ru bottom electrodes. Reprinted with permission from [49], © 1999, American Ceramic Society.

results in a reduction in the effective capacitance. The electrode materials used in BST film capacitors can be classified into two general groups. The first group consists of noble metals, such as Pt, Ir, and Ru [94, 252, 253], whereas the second group involves conducting oxides, such as RuO_2, IrO_2, $BaRuO_3$, $Yba_2Cu_3O_7$, $SrRuO_3$, and $(La,Sr)CoO_3$ [34, 181, 254–256]. The metal electrodes normally have a lower leakage current density than the oxide electrodes, implying that the electrical conduction mechanisms are closely related to the BST–electrode interfaces. The greater leakage current in oxide electrodes appears to be related to the lack of a potential barrier at the BST–oxide electrode interface. High work function metals, such as platinum (~5.6 eV) or iridium (~5.3 eV) films, are primarily used as the electrode material. Pt appears to be the material of choice for use as electrodes for BST capacitors, given its excellent electrical properties. However, in practical application of Pt bottom electrodes, there are still a few drawbacks, including the formation of hillocks at higher temperatures, an amiability for oxygen diffusion, poor adhesion with Si, and difficulty in patterning [54].

The dielectric constant and leakage current density of BST films deposited by rf sputtering on the various bottom electrode materials (Pt, Ir, IrO_2/Ir, Ru, and RuO_2/Ru) before and after annealing in O_2 and N_2 ambient were investigated by Tsai and Tseng [49, 241, 257]. Improvement in crystallinity of BST films deposited on various bottom electrodes was observed with annealing. Figure 10 illustrates XRD patterns of 80-nm-thick BST films deposited on Pt, Ir, IrO_2/Ir, Ru, and RuO_2/Ru bottom electrode materials at 500°C. The dielectric constant of BST thin films deposited on various bottom electrode materials also increased with increasing annealing temperature (Fig. 11). The effect of bottom electrodes is summarized in Table VIII. The refractive index, dielectric constant, loss tangent, and leakage

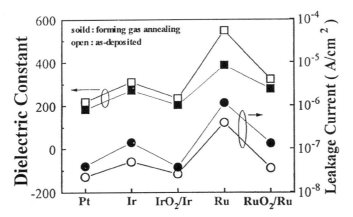

Fig. 11. Dielectric constant and leakage current of $Ba_{0.47}Sr_{0.53}TiO_3$ films deposited on Pt, Ir, IrO_2/Ir, Ru, and RuO_2/Ru bottom electrodes and annealed in forming gas at 400°C for 30 minutes. Reprinted with permission from [49], © 1999, American Ceramic Society.

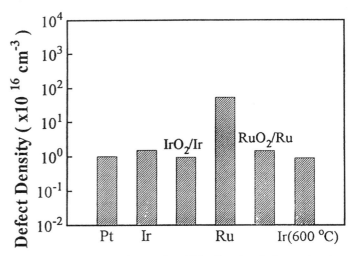

Fig. 12. Defect density of $Ba_{0.47}Sr_{0.53}TiO_3$ films deposited on various bottom electrodes. Reprinted with permission from [49], © 1999, American Ceramic Society.

current of the films were strongly dependent on the annealing conditions. BST thin film deposited on an Ir bottom electrode at 500°C, after 700°C annealing in O_2 for 20 minutes, had a dielectric constant of 593, a loss tangent of 0.019 at 100 kHz, a leakage current of 2.1×10^{-8} A/cm^2 at an electric field of 100 kV/cm with a delay time of 30 seconds, and a charge storage density of 53 fC/μm^2 at an applied field of 150 kV/cm. Based on the dielectric constant, leakage current, and reliability, they suggested that the optimum electrode material for the bottom electrode with annealing is Ir and that the Ru electrode is unstable, because interdiffusion of Ru and Ti occurs at the interface between BST and Ru after annealing. Figure 12 reveals that the defect density of BST films depends on the choice of electrode material. The defect density of BST is larger on Ru than on other bottom electrodes. From the perspective of dielectric constant, leakage, and reliability, the optimum material for the bottom electrode with annealing is Ir [49, 241, 257]. Based on a model of leakage characteristics for BST films deposited on Pt on the basis of Schottky emission mechanism, Maruno et al. [258] predicted that the predominant defect density in BST films is the oxygen vacancies, estimated at around 10^{20} cm^{-3}. This value is very close to the one reported by Tseng [48].

Conducting oxides, such as RuO_2 (rutile type), are known to be easily etched in the fabrication of Gb-density DRAMs [259–262] and acts as a good diffusion barrier against oxygen. It has also been reported that oxide electrodes greatly mitigate the fatigue problems encountered in ferroelectric memory capacitors [253, 263–266]. However, one serious issue related to the oxide electrode is its large leakage current. Lesaicherre et al. [259] reported a leakage current density of several μA/cm^2 at an applied voltage of 1.5 V from their ECR-MOCVD BST thin films deposited on a RuO_2 electrode [259]. The large leakage current appears to be related to the absence of a potential barrier at the BST–RuO_2 interface. The phenomena of reduction and reoxidation of RuO_2 during a BST deposition procedure appears to be the cause of the high leakage current and the large property variation of BST thin films deposited on RuO_2 electrodes.

Therefore, it is very important to set up the process condition to inhibit the reaction of $RuO_2 \Leftrightarrow Ru + O_2$, especially in the case of a high vacuum process, such as sputtering, PECVD, or ECR-CVD [267–269].

Surface roughening of RuO_2, due to reduction and reoxidation directly produces in the rough surface of BST thin films. The variation of surface morphology of BST thin films with heating conditions results in the variation of electrical properties of BST thin films deposited on RuO_2. By controlling the oxygen partial pressure through oxygen flow heating, it is possible to achieve BST capacitors with a smoother surface and lower leakage current (about two orders less) than BST thin films prepared by vacuum heating [267].

The bottom electrode is required to make contact with a polysilicone plug or Si substrate as a DRAM capacitor. However, when RuO_2 film is directly deposited on Si, a SiO_2 layer is formed between the RuO_2 and the Si [103, 219]. The SiO_2 layer induces high contact resistance at the RuO_2–Si interface, resulting in low capacitance in the DRAM capacitor. However, Ru films deposited directly on Si easily form Ru_2Si_3 by thermal treatment [270]. In this case, a SiO_2 layer is also formed between the high dielectric film and the Ru_2Si_3, leading to decreased capacitance. Aoyama et al. [271] found that Ru films deposited on Si in an ambient of 10% O_2 by dc sputtering do not form Ru_2Si_3, even after RTA at 700°C. Silicification of Ru is suppressed by the amorphous layer formed between Ru and Si that is composed of Ru, Si, and O.

$SrRuO_3$ is known to be a conductive oxide with a pseudocubic perovskite structure. It has metallic conduction with low resistivity ($\rho < 1$ mΩ cm), and has a pseudocubic lattice parameter of 0.393 nm, which provides a suitable base for heteroepitaxial growth of BST films [272, 273]. Jia et al. [274] and Abe et al. [34] described the heteroepitaxial growth of BST films on $SrRuO_3$/$LaAlO_3$ and $SrRuO_3$ electrodes, respectively, and verified good electrical properties. According to Abe et al. [224–226], their film deposited on the $SrRuO_3$ electrodes demonstrated a dielectric constant of 740 ($t = 42$ nm) and a leakage current density $<10^{-8}$ A/cm^2 (at 5 V). In addition, Hou et al. [275] made a Ba$_{0.5}$Sr$_{0.5}$TiO$_3$ (100–200 nm)/SrRuO$_3$/YSZ capacitor on an Si substrate using 90° off-axis sputtering, using Au/Ti as the top electrode.

Pt-based structures with Si diffusion barrier layers (e.g., Pt/TiN and Pt/Ta) are also used as storage electrodes because of their stability. However, difficulty in fine patterning of thick Pt will restrict its use for Gb DRAM capacitors, in which use of the side wall area of thick electrodes is necessary to obtain a sufficient storage charge density [102, 259]. Yamamichi et al. [102] and Lesaicherre et al. [259] proposed RuO$_2$/TiN storage nodes for a Gb DRAM capacitor. A thick RuO$_2$ layer can be easily patterned into a 0.15-μm line-and-space structure by O$_2$-C$_{12}$ plasma [102, 259, 276, 277]. In addition, the sputtered BST–RuO$_2$ interface was shown to be stable, and no hillocks were observed on the RuO$_2$ surface after BST deposition at 650°C. Consequently, high-dielectric BST thin film capacitors with low leakage current were obtained [278].

Grill et al. [279] and Yoshikawa et al. [280] reported structural changes in RuO$_2$-based stacked structures on Si during annealing and film deposition. Even a slight degradation of the electrode–barrier or barrier–contact–plug interface affects the electrical properties of the capacitors used for high-density DRAM application [281]. For multilayered electrode structures, such as RuO$_2$/TiN, it is difficult to evaluate the electrode resistance adequately with a conventional contact chain or a Kelvin pattern when a high-resistance region exists at the electrode–barrier interface. The electrical properties of a thick RuO$_2$/TiN-based storage electrode with polysilicone contact plugs for BST films have been studied by Takemura et al. [281]. Resistance of the storage electrodes including contact plugs can be evaluated from the dispersion observed in capacitance frequency measurements. A Ru layer inserted at the RuO$_2$–TiN interface, a TiN–TiSi$_2$–Si junction, and RTA annealing of the TiN layer in N$_2$ ambient are effective ways to reduce the resistance of RuO$_2$/TiN-based electrodes [281]. The barrier layer is required to prevent the electrode from reacting with the polysilicone plug. However, oxidation of the barrier during BST deposition and postannealing imposes a serious difficulty for the integration, because the barrier layer must remain conductive after the whole integration process [278, 282].

Continued efforts in Pt etching have generated a much improved storage node shape [54]. An integrated BST capacitor for 256 Mb, with Pt electrodes and TiSiN as a diffusion barrier and covered by SiO$_2$ spacers, was recently fabricated by Samsung [283]. The excellent diffusion barrier and oxidation-resistant properties of TiSiN, which further protect it from oxidization by the SiO$_2$ spacers, make postannealing up to 650°C possible. A capacitance of 72 fF/cell and a leakage current density of 1.0 fA/cell at ±1.0 V were obtained from a capacitor with a projected area of 0.3×0.8 μm^2 (0.58 μm pitch) and 200 nm with 256-M density. The capacitance corresponds to a value of 25 fF/cell of a DRAM with 0.30-μm pitch, which is the expected cell size of a 1-Gb DRAM.

An appropriately placed oxidation-resistant barrier and adhesion layers enhance the thermal and physical stability of the bottom electrode structure. Khamankar et al. [284] successfully demonstrated novel BST storage capacitor node technology using Pt electrodes for Gb DRAMs. Promising results were obtained in separate Pt etch experiments using TiAlN as a hard mask. A 40-nm TiAlN hard mask was used to etch 300-nm Pt. The absence of thick photo resist during the Pt etch and the high Pt/hard-mask etch selectively led to the formation of fence-free bottom electrodes with high sidewall angles ($\sim$70°). No elaborate postetch cleanup, regarded as a major issue with the Pt etch process [285], was required. A capacitance of 17 fF/cell, with a leakage current density of 1.2×10^{-7} A/cm^2 at 1 V, was obtained for a capacitor array with 0.5-μm features for 100-nm Pt bottom electrodes. The BST with Pt thickness of 250 nm showed a capacitance of $\sim$33 fF/cell and tan $\delta \sim 0.009$. These results demonstrate the promise of integrating Pt as an electrode material with BST as the capacitor dielectric for Gb DRAMs.

4. CONDUCTION MECHANISMS

The study of carrier transport in BST films is important from both fundamental and the practical standpoints. Although the conduction mechanisms in BST films have been studied very broadly [287, 288], the subject is still controversial and very often confusing. More recently, several conduction mechanisms have been reported in the literature to describe the nature of electrical conductivity in BST thin films, and controversy still exists regarding the major leakage mechanisms as suggested by different researchers depending on electrode materials, processing conditions, and the type of storage node structures used for the BST capacitors [51, 53, 289–293]. Many experimental results show that thin dielectric films subjected to an external high electric field displayed a liner relationship in the $\log_{10}(I/V)$ vs. $V^{1/2}$ plots, where I is the current passing through the film and V is the voltage applied across the film. This dependence is attributed to either the field-enhanced Schottky (SE) mechanism [288] or the Poole–Frenkel (PF) mechanism [288, 294, 295]. The former is a Schottky emission process across the interface between a semiconductor (metal) and an insulating film as a result of barrier lowering due to the applied field and the image force. The latter is associated with the field-enhanced thermal excitation of charge carriers from traps, sometimes called the "internal Schottky effect." These two transport mechanisms are very similar, except that in the PF mechanism, barrier lowering is twice as large as in the SE mechanism due to the fact that the positively charged trap in PF mechanism is immobile and the interaction between the electron and the charged trap is twice as large as the image force in the SE mechanism. This phenomenon leads to doubling of the slope, which is given by $(q^3/\pi\varepsilon)^{1/2}$, in the $\log_{10}(I/V)$ vs. $V^{1/2}$ plot. Here q is the electrical charge and ε is the dielectric constant of the film.

Current transportation in BST films at high fields is normally attributed to either the SE mechanism or the PF mechanism.

The leakage current, J_{SE}, governed by a SE mechanism is expressed as

$$J_{SE} = A^* T^2 \exp\left\{-\frac{q[\varphi_B - (qE/(4\pi\varepsilon_d\varepsilon_0))^{1/2}]}{kT}\right\} \quad (2)$$

where A^* is a constant, φ_B is the potential barrier height on the surface, ε_d is the dynamic dielectric constant in ferroelectric material in the infrared region, q is the unit charge, k is Boltzmann's constant, T is temperature, and E is the external electric field [296]. The barrier height (φ_B) of the SE mechanism depends on many parameters, including the work function of the metal, the electron affinity of the semiconductor, barrier lowering by image force, and the surface trap state [296, 297]. The temperature and electric field dependence are governed mainly by φ_B and ε_d, respectively. The SE model can be modified to a PF-type conduction model if the conduction is governed by carriers that are thermally emitted from the trapped centers under a strong electric field [298]. The PF current, J_{PF}, is expressed as

$$J_{PF} = BE \exp\left\{-\frac{q[\varphi_t - (qE/\pi\varepsilon_d\varepsilon_0)^{1/2}]}{kT}\right\} \quad (3)$$

where B is a constant and φ_t represents the depth of the trapped level [296]. Hence J_{SE} and J_{PF} are very similar; usually they are very hard to distinguish.

A natural way to distinguish these two mechanisms is as follows. Plotting the current–voltage (I–V) curve in the form of $\log_{10}(I/V)$ vs. $V^{1/2}$, which should be a straight line, allows the slope β to be measured. The slope can also be calculated using the dielectric constant measured from an independent measurement. For the two different mechanisms, the slopes are given by the following relations:

$$\beta_{PF} = \left(\frac{q^3}{\pi\varepsilon}\right)^{1/2} \quad \text{for the PF mechanism} \quad (4)$$

$$\beta_{SE} = \left(\frac{q^3}{4\pi\varepsilon}\right)^{1/2} \quad \text{for the SE mechanism} \quad (5)$$

By comparing the slopes determined from an independent measurement of the dielectric constant with the slope measured from I–V characteristics, the mechanism can be determined in principle. But thus far, the factor of two differences in the slope of the SE and PF mechanisms has not been demonstrated unambiguously in an experiment for the case of BST films. This is because in most experiments, the calculated and measured slopes are not very close. Most studies were done on materials of low dielectric constant [46]. Thus the difference caused by the factor of 2 in the two mechanisms is comparable to the experimental uncertainty, making it difficult to draw an unambiguous conclusion. A related issue, which has caused much confusion, is choosing the value of ε to use in describing the conduction mechanisms. Use of both the static [299] and the optical [288, 299] dielectric constants has been proposed. Both suggestions are supported by experimental data reported in literature [301–304]. An effective dynamic dielectric constant [296–298] has also been suggested to describe

some experiments. The key issue is how fast a carrier is assumed to move in the dielectric materials. A fast carrier induces only the electronic component of the polarization, and the dielectric response should be optical. On the other hand, if the carrier is slow (as in, e.g., the polaron model [299]), then the static dielectric constant should be used.

Jonscher and Hill [305] pointed out that the conduction mechanisms in insulating films is intrinsically related to the density of the charged defects in the films. If a film contains a very high density of positively charged traps, then a PF mechanism should dominate; otherwise, an SE mechanism is possible. Also, a pure SE conduction mechanism can in principle exist only in contact with no significant intrinsic barrier, namely a neutral contact. Otherwise, the dependence of $\log_{10}(I/V)$ on V will have a different form [288]. Here the intrinsic barrier refers to the barrier formed either due to the difference in the Fermi levels or due to the existence of surface states and a high charged trap density in the film and at the interface.

As the thickness of the oxide films decreases to a few tens of nm due to device requirements, the electrical behavior of the films deviates considerably from that of the bulk materials. There appear to be two reasons for this discrepancy. The first is due to the physical dimension of the material (i.e., very small thickness), which makes the electrical properties dependent on the interface. The second is due to the fact that the techniques used in film fabrication are quite different from those used for bulk material. The first reason may be regarded as intrinsic, whereas the second is rather extrinsic; this extrinsic effect makes the interpretation of the electrical properties of the electrode–BST interface quite controversial. It is not unusual to assume that different fabrication methods produce different films (in terms of defect and carrier concentrations) and different interface properties.

Many papers have reported on the leakage properties of BST thin films with high work function metals such as Pt [14, 25, 32, 196, 306–308]. Figure 13 shows a typical J–V behavior as reported by Fukuda et al. [27]. Two distinct regions are observed; in the low voltage region, the current density is almost proportional to the applied voltage, whereas in the high voltage region, it is proportional to the $V^{1/2}$. Some researchers attribute the former region is to dielectric relaxation and the latter to Schottky emission from the cathode [14, 25, 32, 196, 306–308]. Recently Fukuda et al. [307–309] reported that postannealing in ambient oxygen is very effective in reducing the currents from both mechanisms. Since the diffusion coefficient is proportional to the oxygen vacancy density in the BST [310], the increase in the Pt/BST Schottky barrier height by postannealing is thought to be caused by the decrease in oxygen vacancy density at the Pt–BST interface. In other words, the Fermi level of the BST thin film may be pinned at the interface because of oxygen vacancies in the film [309].

It is well known that oxygen vacancies in BST films play a prominent role in leakage current of films [38, 186, 293]. Tsai and Tseng [293] made an attempt to correlate the electrical leakage mechanism and possible concentration variation of oxygen vacancies in the BST films deposited at various OMRs.

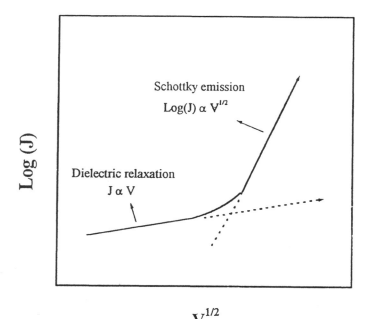

Fig. 13. Typical J–V behavior of $Ba_{0.5}Sr_{0.5}TiO_3$ capacitors [289].

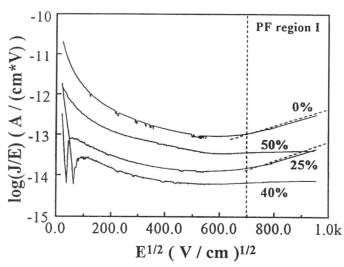

Fig. 14. The $\log(J/E)$ vs. $E^{1/2}$ plot showing the PF region for various OMR $Ba_{0.5}Sr_{0.5}TiO_3$ films [273].

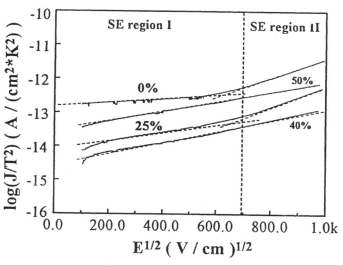

Fig. 15. The $\log(J/T^2)$ vs. $E^{1/2}$ plot showing the SE region for various OMR $Ba_{0.5}Sr_{0.5}TiO_3$ films [273].

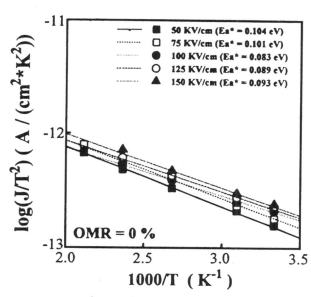

Fig. 16. The $\log(J/T^2)$ vs. T^{-1} for the $Ba_{0.5}Sr_{0.5}TiO_3$ capacitors under various applied electric fields [273].

The $\log(J/E)$ vs. $E^{1/2}$ curves for BST thin films deposited at 0%, 25%, 40%, and 50% OMR shown in Figure 14 indicate that the slopes for all of the films below the transition electric field of 490 kV/cm have negative values, but the slopes for the 0% and 25% OMR films beyond the transition electric field change to positive values and show the linear relation. Similar behaviors are seen in the $\log(J/T^2)$ vs. $E^{1/2}$ curves for the BST thin films deposited at 0%, 25%, 40%, and 50% OMR, as shown in Figure 15. For the 0% and 25% OMR BST films, two step slopes are observed with breaks at 490 kV/cm; for the 40% and 50% OMR films, only the single slope can be observed.

On the basis of Eqs. (2) and (3), the activation energies E_a^* listed in Table IX can be obtained from the slopes of the linear regions of the plots $\log(J_{SE}/T^2)$ vs. T^{-1} (Fig. 16) and $\log(J_{PF}/E)$ vs. T^{-1} (Fig. 17), respectively, in the BST capacitors sputtered at 0%, 25%, 40%, and 50% OMR at the various fields indicated (with positive bias applied on the top electrode). Figures 16 and 17 are typical plots for BST capacitors deposited at 0% OMR. The activation energy increased with increasing OMR at 50 kV/cm, but it had the maximum value at 40% OMR. Also, activation energy decreased with an increasing electric field. An extrapolation of the activation energy E_a^* vs. the square root of the electric field curve gives φ_B for $\ln(J_{SE}/T^2)$ vs. T^{-1} and φ_t for for $\ln(J_{PF}/E)$ vs. T^{-1}. The values of [φ_B (eV), φ_t (eV)] between 0%, 25%, 40%, and 50% OMR films and bottom electrodes are (0.16, 0.23), (0.20, 0.27), (0.24, 0.31), and (0.14, 0.21), respectively, as

Table IX.   Activation Energy (E_a^*) Obtained from $\ln(J/T^2)$ vs. T^{-1} and $\ln(J/E)$ vs. T^{-1} Plots of the 0%, 25%, 40%, and 50% OMR BST Capacitors at Various Fields as Indicated [273]

| | | Activation energy (E_a^*) (eV) | | | | |
Plot	OMR (%)	50 kV/cm	75 kV/cm	100 kV/cm	125 kV/cm	150 kV/cm
$\ln(J/T^2)$ vs. T^{-1}	0	0.104	0.101	0.083	0.089	0.093
	25	0.116	0.116	0.081	0.074	0.068
	40	0.136	0.081	0.078	0.076	0.039
	50	0.092	0.045	0.039	0.039	0.039
$\ln(J/E)$ vs. T^{-1}	0	0.169	0.164	0.144	0.150	0.154
	25	0.180	0.180	0.178	0.135	0.129
	40	0.200	0.142	0.137	0.137	0.097
	50	0.156	0.106	0.100	0.100	0.099

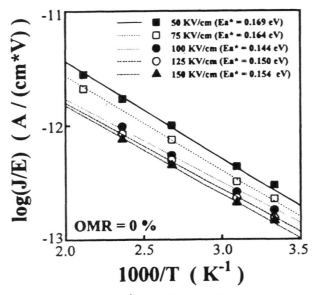

Fig. 17.   The $\log(J/E)$ vs. T^{-1} for the $Ba_{0.5}Sr_{0.5}TiO_3$ capacitors under various applied electric fields [273].

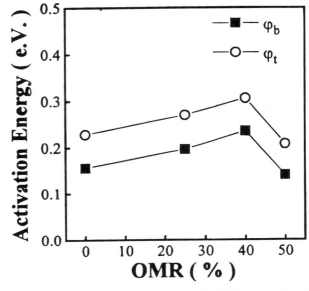

Fig. 18.   The variation of activation energy with OMR for $Ba_{0.5}Sr_{0.5}TiO_3$ films [273].

shown in Figure 18. The φ_B and φ_t increased with increasing OMR, but had a maximum value at 40% OMR. A further increase in OMR decreased the potential barrier height and the depth of trap level. The results show that the barrier height is lowered by high concentration of oxygen vacancies at lower OMR films, which leads to a depth of trap centers level φ_t near the surface (shallow trap level) (see Fig. 18). Therefore, the electrons would be trapped easily, and the PF mechanism dominates the current transportation for higher OMR film (40% OMR), the interface barrier height and the depth of the trap center level are larger (deep trap level), so the SE mechanism dominates the electrical conduction. For 50% OMR film, the smaller interface barrier height and depth of the trap center level and the larger leakage current were attributed to the polarization-enhanced electron transport [293, 298, 300]. Additional mechanisms (e.g., relaxation loss) may also contribute

to the increased conduction beyond 40% OMR. Therefore, the positive leakage current had the minimum value at 40% OMR, because the maximum values of activation energy, potential-barrier height, and trapped level (deep trap level) were seen at 40% OMR. For the same reason, the negative leakage current density also had the minimum value at 40% OMR.

Table X lists the slopes of different linear regions of the plots shown in Figures 14 and 15, the dynamic dielectric constant (ε_d) calculated on the basis of Eqs. (2)–(5), the optical dielectric constant (ε_{op}), which is the square of the refractive index, and the static dielectric constant (ε_s) obtained from capacitance–voltage (C–V) measurement at 100 kHz. Table X shows that the dynamic dielectric constants of 0% OMR films are 154 at SE region I and 113 at PF region I, which are in relatively close to the static dielectric constant 138. Therefore, the authors suggest

Table X. Calculated Slopes of the Various OMR BST Films at Room Temperature Using Various Transport Mechanisms [273]

OMR (%)	SE (region I)		SE (region II)		PF (region I)		Static dielectric constant	Optical dielectric constant
	β_{SE}	Dynamic dielectric constant	β_{SE}	Dynamic dielectric constant	β_{PF}	Dynamic dielectric constant		
0	1.18×10^{-3}	154	4.55×10^{-3}	10	2.76×10^{-3}	113	138	5.37
25	2.08×10^{-3}	50	3.72×10^{-3}	15	2.43×10^{-3}	145	151	5.43
40	2.53×10^{-3}	34	–	–	–	–	192	5.72
50	2.92×10^{-3}	25	–	–	–	–	375	5.48

Positive bias was applied on the top electrode.

that the electronic carrier velocity is slow (polaron-like conduction) due to high concentration oxygen vacancies on 0% OMR films, and thus the static dielectric constant would be used to describe the carrier transport through the films. The dynamic dielectric constants of 40% and 50% OMR films as shown in Table X are reduced, and the PF emission effect of the J–E curve is eliminated because the concentration of oxygen vacancy decreases at high OMR. The electron carrier velocity is higher, so the optical dielectric constant would be used to describe the carrier transport through the films. The dominant conduction mechanism for high-OMR films is the thermionic emission of electrons from the cathode electrode over the potential barrier into the thin film. Hence, the I–V characteristics are influenced by the SE mechanism.

Based on the foregoing observations, Tsai and Tseng [293] indicated that in BST films prepared at low OMR (0%–25%), the SE mechanism dominated below the transition electric field of 490 kV/cm and the PF transport mechanism dominated beyond 490 kV/cm, whereas in BST films prepared at high OMR (40%–60%), the SE mechanism dominated both below and above the transition electric field. The difference in dominant mechanism (bulk and electrode-limited conduction) between the films was ascribed to the concentration variation of the oxygen vacancies in the films.

Dietz et al. [243] studied leakage properties for BST thin films grown using LS-CVD, and the following conclusions were drawn based on their findings. The leakage current through BST films was limited primarily by an interfacial Schottky barrier whose properties showed dependence on the electrode material and deposition conditions. Dietz et al. used a thermionic emission model to interpret the temperature and voltage dependence of BST electrical leakage data. Analysis in terms of a Schottky barrier–limited current flow gives acceptable values for cathode barrier height. Along with the Schottky effect, barrier lowering was observed, which was attributed to a distribution of deep acceptor states in the band gap of the ceramic.

It should be noted that from the foregoing analysis, although the functional form of the leakage current is well described by the equations derived from the simple Schottky barrier model, the parameters (other than barrier height) extracted from the

data analysis are physically unrealistic. From the theory of thermionic emission, a saturation regime for the leakage current determined by the Richardson constant is expected for relatively low fields. The observed currents in this regime are much lower than predicted based on expected values for A^{**} (effective Richardson constant) and increase linearly with the electric field. These are strong indications that in this regime, the influence of the film bulk cannot be disregarded.

Shimada et al. [51] studied the temperature-dependent current–voltage characteristics of fully processed $Ba_{0.7}Sr_{0.3}TiO_3$ thin film capacitors integrated in a charge-coupled device delay-line processor as bypass capacitors. The leakage current measured after completion of the integration process was one to two orders of magnitude higher than that measured after capacitor patterning. The leakage current at low voltages (<1 V, 50 kV/cm) indicated ohmic conduction within a measured temperature range of 300–423 K. At high voltages (>10 V, 500 kV/cm), the SE mechanism plays a dominant role in leakage current, whereas the PF emission begins to contribute to the leakage current as temperature rises.

Properties of interfaces between electrodes and BST (e.g., potential barrier) is dependent not only on the electrode material, but also on the processing, such as deposition conditions and postannealing. Park et al. [196] reported that postannealing at temperatures above 850°C resulted in an increased leakage current density due to the roughening of BST–Pt interfaces. Also, Cha et al. [311] reported that BST films on Pt electrodes deposited at different temperatures showed different leakage current characteristics due to different crystallinity of BST and roughness. Lee et al. [311] reported the electrical properties of BST film capacitor, focusing on the variation of the barrier height at interfaces with top and bottom electrodes and leakage conduction mechanisms according to different deposition power of top Pt electrode. The capacitors having a top Pt electrode with deposition power of 0.2 kW showed SE behavior at both top and bottom electrode interfaces with potential barrier heights of 1.24–1.48 eV and 1.88–2.08 eV, respectively. The barrier height increased with postannealing temperature, and the capacitors with a top Pt electrode with a larger deposition power of 0.5 kW showed SE behavior only at the bottom electrode interface with a barrier of 1.61–1.89 eV. Under negative

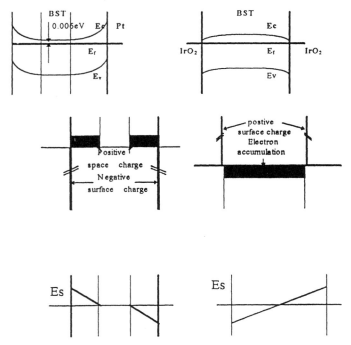

Fig. 19.   Band bending, space charge distribution, and static field distribution for the Pt/BST/Pt and IrO$_2$/BST/IrO$_2$ capacitors [235].

bias, a peculiar $J-V$ behavior indicating a positive temperature coefficient of resistivity (PTCR) effect was observed at the bottom electrode interface. This was attributed to a reduction of the interface potential barrier height between the top Pt and the BST, which was caused by roughening of the top Pt during postannealing. Hwang et al. [53, 255] also found that the Pt/BST/IrO$_2$ capacitor showed a PTCR effect when the Pt electrode was positively biased, whereas the Schottky emission behavior was obtained when the Pt electrode was negatively biased even though the BST was paraelectric. (The PTCR effect is known to be due to abrupt decrease in the dielectric constant of the ferroelectrics, such as BaTiO$_3$ at their Curie temperature [313].) A model based on an upward band bending at the Pt–BST interface and a downward band bending at the BST–IrO$_2$ interface was suggested to explain the asymmetrical electrical conduction behavior and the PTCR effect. The band bending, space charge distribution, and static field distribution for the Pt/BST/Pt and IrO$_2$/BST/IrO$_2$ cases are schematically shown in Figure 19. The electrical conduction mechanisms of the rf-sputter–deposited BST thin films depend on the type of electrode contacts. The Pt–BST contact showed a SE behavior with a potential height of about 1.5–1.6 eV. The magnitude of the potential barrier height is determined largely by surface electron trapping of the BST. Therefore, pinning of the Fermi level at the Pt–BST interface to a place as deep as possible in the band gap is essential to obtain a large potential barrier height and low leakage current. The weak chemical interaction of Pt and BST is the key factor in maintaining a high potential barrier height. The IrO$_2$–BST contact is ohmic in nature. The IrO$_2$/BST/IrO$_2$ capacitor showed ohmic conduction behavior in

the low-voltage region and PF behavior in the high-voltage region, with an electron trapping energy of about 1.23 eV. The much stronger chemical bonds between IrO$_2$ and BST eliminate the surface electron traps, and thus the bulk properties of the two materials are major determinants of contact properties. The small work function for IrO$_2$ appears to produce ohmic contact properties.

The Pt/BST/IrO$_2$ capacitor showed a very distinctive electrical conduction behavior. When the conduction electrons were transferred from Pt to IrO$_2$ through BST, the conduction behavior was consistent with SE at the Pt–BST interface. However, when conduction occurred via the transfer of electrons from IrO$_2$ to Pt through BST, the conduction was neither ohmic nor consistent with Schottky behavior, but rather showed a PTCR effect. Hwang et al. [53, 255] suggested that this may be due to the space charge region within the BST at the Pt–BST contact causing this effect.

It has been reported that leakage currents increased asymmetrically for negative and positive bias voltage with increasing annealing temperature [258]. A model of leakage characteristics based on the SE mechanism has been proposed. Using this model, Maruno et al. [258] predicted that there exist high-density oxygen vacancies, about 10^{20} cm^{-3}, at the Pt–BST interfaces. These oxygen-deficient layers induce large built-in electrical fields within the BST. These electrical fields result in enhanced leakage currents through a reduction of the Schottky barrier height due to image force increase and changes the $J-V$ characteristics with increasing annealing temperature.

During fabrication of top electrode by the sputtering technique, some accelerated sputtering gases struck on the surface of the BST films. This results in out-diffusion of oxygen from the BST films to the platinum top electrode. Oxygen vacancy generated by such a process ($O_O = V_O^{\cdot\cdot} + 2e^- + 1/2O_2$) may act as an electron trap site as so causes a high leakage current of Pt/BST/Pt capacitors [198]. Joo et al. [198] have shown that when Pt/BST/Pt capacitors fabricated by a sputtering process are postannealed under N$_2$ or H$_2$ atmosphere, higher leakage currents flow at the negative bias region than at the positive bias region when bias voltage is applied to the top electrode. Moreover, the enhanced leakage currents are effectively reduced by annealing under O$_2$ atmosphere or adding oxygen into sputtering gas of the platinum top electrode.

The presence of trap states (acceptors and donors) plays a prominent role in the leakage current characteristics of BST capacitors. Using a deep-level transient spectroscopy (DLTS) technique, Wang and Tseng [186] investigated the deep trap levels of rf-sputtered (Ba$_{0.4}$Sr$_{0.6}$)TiO$_3$ films deposited at various temperatures. A single trap was observed at 0.45 eV in 450°C deposited films, whereas two traps, at 0.2 and 0.40 eV, appeared in 550°C deposited films. Figure 20 depicts conventional DLTS signals at various emission rates for the BST films sputtered at 450 and 550°C. The DLTS signals of the 550°C deposited films reveal two downward peaks (Fig. 20b), indicating the existence of two states. Meanwhile, only one downward peak appears in Figure 20a for the films deposited at 450°C (Fig. 20a). Figure 21 illustrates the activation en-

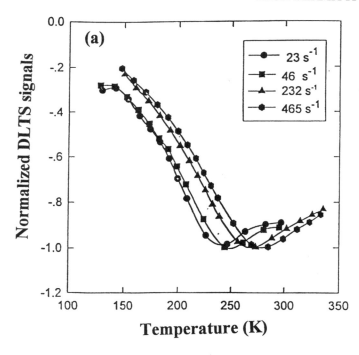

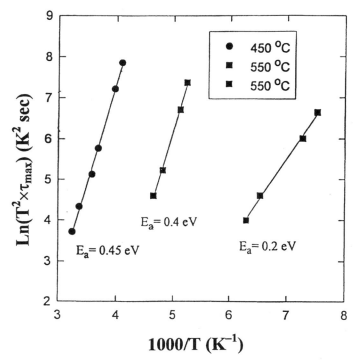

Fig. 21. Arrhenius plot of DLTS spectra for $Ba_{0.4}Sr_{0.6}TiO_3$ films sputtered at 450°C and 550°C [166].

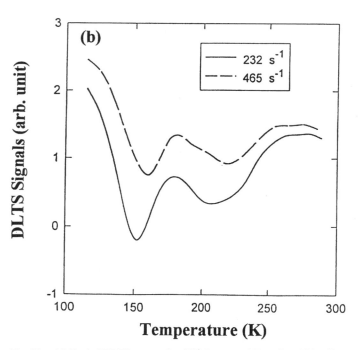

Fig. 20. (a) Typical DLTS spectra for 450°C-sputtered $Ba_{0.4}Sr_{0.6}TiO_3$ films with four indicated emission rates. (b) Typical DLTS spectra for 550°C-sputtered $Ba_{0.4}Sr_{0.6}TiO_3$ films with two indicated emission rates [166].

ergies (E_a) obtained from the slope of the Arrhenius plot of the capacitance change versus temperature data. The activation energies of those traps were 0.2 and 0.40 eV for 550°C deposited films and 0.45 eV for 450°C deposited ones. Trap levels for the $Ba_{0.76}Sr_{0.24}TiO_3$-based PTCR ceramic doped with various oxide additives were investigated using isothermal

capacitance transient spectroscopy and determined to be 0.36–0.46 eV [314].

Further, the $I–V$ characteristics of the films at a temperature range of 298-423 K reveals the presence of two conduction regions, indicating ohmic behavior at low voltage (<1 V) and SE or PF at high voltage (>6 V). The barrier height and trapped level, estimated as 0.46 and 0.51 eV from SE and PF mechanisms, respectively, were in agreement to those obtained from the DLTS technique and hence they concluded that oxygen vacancy seems to be a more reasonable candidate for the source of the trap than cation vacancy. Moreover, the elementary nature of this trap is a prominent factor affecting the conducting properties of BST thin films.

5. DIELECTRIC RELAXATION AND DEFECT ANALYSIS OF BST THIN FILMS

Capacitor performance in DRAM cell applications is affected by charge storage capacity, dissipation of stored charge, dielectric response, and lifetime from quality and reliability standpoints [315–319]. The capacitor in every DRAM cell must store sufficient charges for preserving memorized information even through the cell area and the power supply voltage have been reduced in downsizing trends. Many efforts have been made to achieve sufficiently large capacitance in small areas. Furthermore, reduction of leakage current is also important, because it decreases the electrical charge once stored.

In BST thin film capacitors, capacitance has been observed to have dielectric dispersion, resulting in dielectric relaxation

[315, 320]. The amount of charge stored on a capacitor during the write operation therefore depends on the dielectric dispersion. It has been shown that the dielectric relaxation phenomenon in thin film capacitors greatly affects the capacitors' electrical properties [18, 25, 41, 320, 321]. Dielectric relaxation arises from the dispersive nature of dielectric and increases with increasing value of the power law dependence of capacitance on frequency, which is termed the dispersion parameter [32, 41, 320]. The most crucial influence of dielectric relaxation on DRAM operation is on the pause refresh property [29]. Horikawa et al. [32] estimated that charge loss by the dielectric relaxation for the pause time of 1 second would amount to about 8% of the initially stored charge, which is about two orders of magnitude larger than that produced by dc leakage current.

At least four possible defects—the interface defect, grain boundary defect, shallow trap levels, and oxygen vacancies—may exist in the metal–insulator–metal (MIM) capacitor films, leading to dielectric relaxation as a function of frequency [29, 320, 322, 323]. In the case of grain boundary defect, the grain boundary in dielectric ceramics is considered to represent a resistor, R, and the grain is a thin insulating layer, C. There are many such RC series–equivalent circuits in parallel throughout the ceramic. Equivalent circuit analysis [261, 320, 322–326] indicates that the grain boundary plays a prominent role in the relaxation of ceramics. The grain boundary defect exists within the nonstoichiometric grain boundary, and the dominant defect is the oxygen vacancy. An interface defect exists within the forbidden gap due to the interruption of the periodic lattice structure. Under a dc electrical field, the migration of oxygen vacancies followed by reduction reaction leads to space charge accumulation at the grain boundary and the dielectric–electrode interface reduces the barrier height at the grain boundary and interfaces, and increases the leakage current and forms the dielectric relaxation [320, 323]. The current induced by grain boundary defects is attributed to Poole–Frenkel conduction, and the current induced by interface defect is termed as Schottky emission [38, 46, 186].

The shallow traps can be determined under small-signal ac stress applied at various temperatures. Under small-signal ac stress applied at the Schottky junction, the depletion layer width varies about its equilibrium position due to trapping and detrapping of electrons from the oxygen vacancies or shallow traps, denoted by E_t, where E_t is the trap energy. The shallow trap is located below, near the conduction band, and hence its emission rate is affected by temperature.

Since K. S. Cole and R. H. Cole showed this feature of relaxation behavior with Cole–Cole diagrams [327] in accordance with the Debye equations, the measurement of small-signal ac frequency response of materials over a wide frequency range (i.e., impedance spectroscopy) has become a valuable tool for analyzing and characterizing electroceramic materials, including dielectric and ferroelectric types. A semicircular fit of the data in any of the complex impedance or admittance planes suggests appropriate models, equations, and equivalent circuit representation of the observed dispersion. This approach can reveal the presence of relaxation processes and the relative con-

tribution of defect states to the total ac response under a given set of experimental conditions. Recently several works have been carried out using this technique in thin film capacitors to explain the nature of dielectric relaxation [281, 320, 325, 326].

In addition to direct examination of the grain and/or grain boundary, complex plane analysis is commonly adopted to separate and identify the intergranular/intragranular impedance and also to determine the contribution of defects to the dielectric relaxation. The electrical parameters used to characterize the ac response of a thin film dielectric with parallel plane electrodes are impedance (Z) and admittance (Y). They have the following transformation relationships:

$$Y = \frac{I(\omega)}{V(\omega)} = G_p(\omega) + jB_p(\omega) \tag{6}$$

$$Z = \frac{V(\omega)}{I(\omega)} = R_s(\omega) + jX_s(\omega) \tag{7}$$

$$Y = j\omega \times C(\omega) = j\omega \times \left(C' - jC''\right)$$
$$= j\omega C_0 \times \left(\varepsilon' - j\varepsilon''\right) \tag{8}$$

The following relations can be obtained by comparing Eqs. (6)–(8):

$$G_p(\omega) = \omega C'' = \omega C_0 \varepsilon'' \tag{9}$$

$$B_p(\omega) = \omega C' = \omega C_0 \varepsilon' \tag{10}$$

where ω is the angular frequency, $I(\omega)$ and $V(\omega)$ the electrical current and applied voltage as a function of ω, $G_p(\omega)$ and $B_p(\omega)$ are the parallel relative real admittance and imaginary admittance as a function of ω, $R_s(\omega)$ and $X_s(\omega)$ the series relative real impedance and imaginary impedance as a function of ω, C_0 is the geometric capacitance in free space, ε' and ε'' are the relative real and imaginary dielectric constants, C' and C'' are the real and imaginary capacitance.

Typically, results from impedance spectroscopy measurements are analyzed via complex impedance and admittance plots, in which the imaginary part is plotted against the real part. The resulting curve is parameterized by the applied frequency, with low frequencies at a high real axis intercept and high frequencies at a low one. Depending on the distinctive orders of magnitude for the relaxation time (defined as the time-constant RC in the equivalent circuit to be an indication of the transport process), a series array of parallel RC elements may give rise to independent or overlapping semicircular arcs in the two complex planes. Complex impedance measurements are of great interest, because they allow separation of the contributions of the grains, grain boundaries, and defect states to the total impedance. The shape of the complex-plane plot is usually analyzed using equivalent circuit analysis, using a collection of resistors and capacitors arranged in various combinations of series and parallel circuits to duplicate the experimental spectrum.

Tsai and Tseng [320] investigated the dielectric relaxation and defect analysis of BST thin films through the measurement of dielectric dispersion as a function of frequency (100 Hz ≤ f ≤ 10 MHz) and temperature (27°C ≤ T ≤ 150°C). The frequency dependence of real and imaginary capacitance for BST thin films in the measurement temperature range of

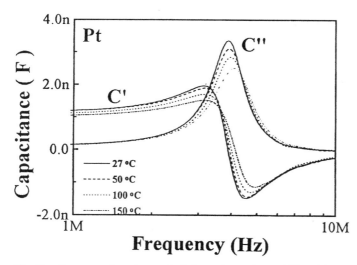

Fig. 22. Frequency dependence of relative real capacitance (C') and imaginary capacitance (C'') of $Ba_{0.47}Sr_{0.53}TiO_3$ films at various temperatures. Reprinted with permission from [300], © 1967, American Physical Society.

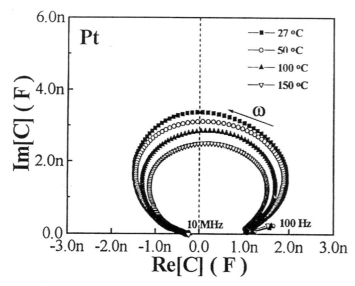

Fig. 23. Complex capacitance plot of $Ba_{0.47}Sr_{0.53}TiO_3$ films at various temperatures. Reprinted with permission from [300], © 1967, American Physical Society.

27–150°C is shown in Figure 22. The capacitance at low frequency (<1 MHz) passes through a minimum. At high frequency (>1 MHz), the capacitance increases, goes through a maximum, and decreases abruptly, passing through zero and attaining, negative values, that is, the BST capacitor film showed a resonance in capacitance at high frequency of 1–10 MHz. Figure 23 shows a complex capacitance plot of BST thin films measured at various temperatures. At high frequencies (1–10 MHz), the resonance phenomenon emerges as a circle in the complex capacitance plane (C^*). The complex capacitance plot in the high-frequency region continues to exhibit an abrupt discontinuity in dispersion with increasing frequency, which indicates the onset of resonance phenomenon. The admittance ($G_p/\omega = C''$) at a given capacitance is constant over

a narrow frequency range on the complex capacitance curve in the C^* plane. Therefore, the locus of the C'' is a circle of diameter $1/R_r$, tangent to the imaginary axis and with its center displaced from the real axis in the positive direction. The lower the value of R_r, the higher the area of the circle [328, 329].

At high frequencies, the films experience completely shorted grain boundary electrical barriers and become electrically conductive, resulting in negative capacitance and associated resonance in capacitance. Similar phenomena were observed in polycrystalline Ta_2O_5 films at frequency >1 MHz [328–331] and were attributed to the type and density of defect states formed at the depletion regions of the grain boundary of the Ta_2O_5 films [328–331].

The resonance frequency for thickness mode is determined by electromechanical coupling factor and should be in the range of 1–10 MHz for dielectric films with a thickness of 0.5 μm. Therefore, the resonance observed in BST films cannot be attributed to piezoelectric resonance [188]. Figures 22 and 23 demonstrate that another resonance can arise from the resonance of an electrical equivalent circuit consisting of resistance, capacitance, and inductance (R-C-L) connected in series, which can be used to duplicate the BST capacitor film [332], where the capacitance of the equivalent circuit is determined mainly by the capacitance of the BST film and the inductance of the equivalent circuit is determined by the conductivity of the electrode and the BST film. But the resonance is not observed in BST films with a different capacitance range [188, 320], which supports the argument that the origin of the resonance is the equivalent circuit. That is, the resonance observed by Tsai and Tseng [320] for BST films in the frequency range 1–10 MHz is an electrical resonance resulting from the electrodes.

The complex capacitance spectral studies reported by Tsai and Tseng [320] clearly demonstrate that the BST thin films exhibit resonance at frequencies above 1 MHz. Therefore, the proposition of R-C-L series equivalent circuit for BST film is reasonable only for measurement frequency ≥ 1 MHz. But the situation is different for measurement frequency in the range of 100 Hz–1 MHz, in which case the impedance (Z) of the equivalent R-C-L circuit can be written as

$$Z = R_{eq} + j\omega L_{eq} + \frac{1}{j\omega C_{eq}} \quad (11)$$

The inductance value (ωL_{eq}) is one to two orders of magnitude smaller than the capacitance value ($1/(j\omega C_{eq})$), for the frequency range 100 Hz $\leq f \leq$ 1 MHz and can be neglected. Hence, the equivalent R-C-L circuit can be simplified to comprise only capacitors (C_{eq}) and resistors (R_{eq}). Figures 24 and 25 show the complex impedance and admittance plane plots of BST deposited on Pt at various temperatures, respectively. On the basis of Z and Y analysis, Tsai and Tseng [320] proposed a practical equivalent circuit for BST capacitors as shown in Figure 26. In the equivalent circuit, the R_{el} represents the electrode resistance, C_f is the frequency dependent capacitance due to the grain, and R_f is the frequency dependent resistance due to the grain boundary and interface. The foregoing equivalent circuit

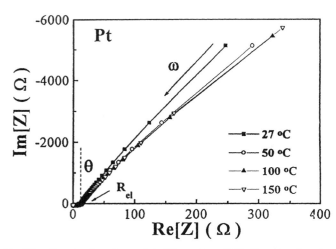

Fig. 24. Complex impedance plot of Ba$_{0.47}$Sr$_{0.53}$TiO$_3$ films at various temperatures. Reprinted with permission from [300], © 1967, American Physical Society.

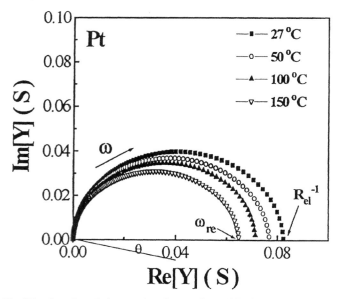

Fig. 25. Complex admittance plot of Ba$_{0.47}$Sr$_{0.53}$TiO$_3$ films at various temperatures. Reprinted with permission from [300], © 1967, American Physical Society.

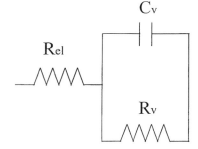

Fig. 26. Schematic equivalent circuit model for Ba$_{0.47}$Sr$_{0.53}$TiO$_3$ film capacitors for frequency range 100 Hz–1 MHz. Reprinted with permission from [300], © 1967, American Physical Society.

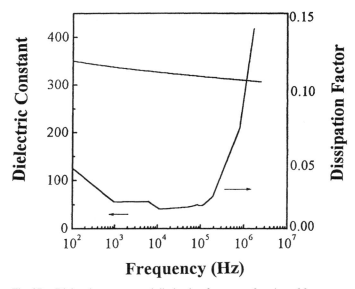

Fig. 27. Dielectric constant and dissipation factor as a function of frequency. Reprinted with permission from [168], © 1995, IEEE.

for BST capacitor film agrees well with the one described by Jonscher [332].

Further, through the measurement of admittance spectral studies in the temperature range of 27–150°C, Tsai and Tseng [320] observed a shallow trap level located 0.005–0.01 eV below the conduction band. Since the measured trap energy values are much smaller than the thermal energy (KT), 25.9 meV at 27°C, they concluded that the effect of shallow trap level can be neglected at the normal temperature range of DRAM operation: 0–70°C ambient and 0–100°C on the chip (at the measurement frequency range <1 MHz) and the equivalent circuit has contributions from only grain, grain boundary, and interface (BST–metal) defects.

Figure 27 shows the dielectric constant and dissipation factor (tan δ) as functions of frequency, as reported by Yoon and

Safari [188]. The capacitance value of the different data points was within 2%, indicating good homogeneity in film thickness (200 nm). The dielectric constant and loss factor for a 200-nm film at a frequency of 100 kHz were 320 and 0.02, respectively. These results were comparable with data reported by Roy and Krupanidhi [50]. However, the dielectric loss shows a large increase above 100 kHz. These phenomena are similar to those of lead zirconate epitaxial films produced by the laser ablation technique [333]. Yoon and Safari [188] also observed resonance in capacitance at a frequency of 10–15 MHz for capacitance of 3–5 nF and explained that this resonance phenomena is the reason for the observed increase in the dielectric loss above 100 kHz. Moreover, the resonance phenomenon was attributed to electrical resonance occurring at 10 MHz and related to the electrode area of the films (0.6 mm diameter). Further, the authors showed the relation between the applied electric field and the charge storage density for BST films (see Fig. 25). The charge storage density, Q_c, can be calculated using the following equation because the BST film is a paraelectric material:

$$Q_c = \varepsilon_0 \varepsilon_r E \qquad (12)$$

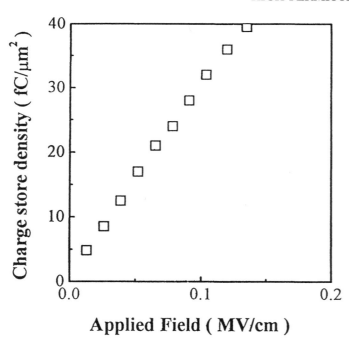

Fig. 28. Charge storage density as a function of applied field. Reprinted with permission from [168], © 1995, IEEE.

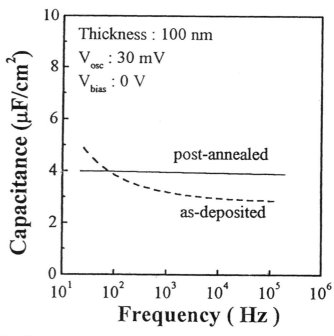

Fig. 29. Frequency dependence of the capacitance of as-deposited and postannealed Pt/Ba$_{0.5}$Sr$_{0.5}$TiO$_3$/Pt capacitors for frequency range 20 Hz–400 kHz. Reprinted with permission from [314], © 1998, American Institute of Physics.

where ε_0 and ε_r are the dielectric constants of vacuum and the BST film, respectively, and E is the applied electric field. The charge storage density was calculated from the C–V measurements on MIM capacitors. The charge storage density at an applied field of 0.15 MV cm^{-1} (leakage current density of 0.8 μA cm^{-2} for a 200-nm-thick film) was about 40 fC μm^{-2}, as shown in Figure 28. It has been reported that for 256-Mb DRAM applications, a lower limit of allowable leakage current density of 3 μA cm^{-2} and a charge storage density of the order of 40 fC μm^{-2} [6, 10] are required for a film thickness of 200 nm.

Dielectric relaxation of (Ba$_{0.5}$Sr$_{0.5}$)TiO$_3$ films deposited by the rf-sputtering method has been investigated by Horikawa et al. [32]. Their observations indicate that the dielectric constant decreased with frequency following the relationship of $d\varepsilon/d(\log_{10} f) \sim -0.01\varepsilon$ and that the dielectric loss was almost constant at less than 1% in the measurement frequency range. The transient current response for a stepwise change of applied voltage was found to be inversely proportional to time in the time range of 1×10^{-4} to 5×10^{2} seconds. Based on the quantitative correspondence between the ε dispersion and the transient response, Horikowa et al. confirmed that the transient current is the absorption current due to dielectric relaxation. The absorption current for the thin films of BST is attributed not to a set of a several distinct dielectric relaxations, but rather to a series of dielectric relaxations distributed continuously throughout the wide frequency range. In DRAM operations, this type of dielectric relaxation is one of the main causes of storage charge dissipation than the film's leakage current.

Basceri et al. [242] investigated the temperature- and field-dependent permittivities of fiber-textured Ba$_{0.7}$Sr$_{0.3}$TiO$_3$ thin films grown by LS-CVD were investigated as a function of film thickness. These films displayed a nonlinear dielectric response, and the behavior was related to the classical nonlinear, nonhysteretic dielectric as described in terms of a power series expansion of the free energy in the polarization, as in the Landau–Ginzburg–Daonshire approach. Curie–Weiss like behavior was observed above the bulk Curie point ($\sim$300 K). Small-signal capacitance measurements of films with different thicknesses (24–160 nm) indicated that only the first term in the power series expansion varied significantly with film thickness or temperature. Fukuda et al. [334] described the mechanism of the dielectric relaxation of BST capacitors postannealed in oxygen ambient. From comparison of the electrical characteristics of the as-deposited and postannealed BST film capacitors, these authors concluded that electrons from oxygen vacancies in the interfacial depletion layer are the origin of the dielectric relaxation phenomenon. Figure 29 shows the frequency dependence of the capacitance of the as-deposited and postannealed BST films for a frequency range of 20–400 KHz measured with an oscillation voltage of 30 mV. The capacitance of the as-deposited film decreases drastically with an increase in frequency. In contrast, that of the postannealed capacitor is almost independent of the frequency for the measured frequency range. Supposing that the observed frequency dependence of the capacitance were attributed to the dielectric aftereffect, the frequency dependence of the capacitance, $C(f)$, could be expressed by $C_\infty + C_0 f^{n-1}$, where C_∞ denotes the capacitance that can respond to high frequency, C_0 denotes the capacitance due to the dielectric relaxation, and f is the frequency. By fitting this equation to the experimental results,

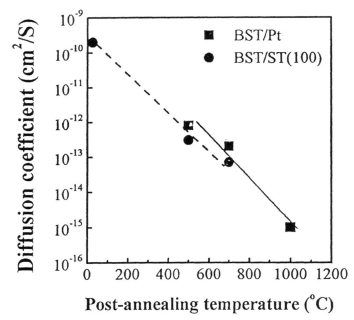

Fig. 30. Postannealing temperature dependence of ^{18}O diffusion coefficient in Ba$_{0.5}$Sr$_{0.5}$TiO$_3$ films [288].

Fukuda et al. [334] found that $C(f)$ of the as-deposited film was $3.0 + 3.6f^{-0.24}$ (μF/cm^2) and that of the postannealed film was $3.1 + 1.0f^{-0.023}$ (μF/cm^2). The values of C_∞ were almost equal for both films. On the other hand, the C_0 of the as-deposited film is 3.6 times larger than that of the postannealed films. Therefore, the authors remarked that the frequency dependence of capacitance for the as-deposited film is attributed to the dielectric relaxation and the effect of this phenomenon can be dramatically suppressed by the postdeposition annealing in the oxygen ambient.

Fukuda et al. [308, 335] also studied the temperature dependence of the dielectric relaxation current of the BST capacitor and found that its thermal activation energy is 0.3 eV. This relatively small activation energy suggests that the electron is most likely responsible for the observed dielectric relaxation. Furthermore, considering that an oxygen vacancy is an electron donor in perovskite oxide [336, 337], the effect of postannealing is assumed to compensate for such vacancies by the introduction of oxygen atoms. To confirm this, the authors measured the diffusion coefficient of the oxygen atoms in the BST films which were postannealed at various temperatures. Since the oxygen vacancy density is considered to be proportional to the oxygen diffusion coefficient [310], the difference in the diffusion coefficients represents the relative difference in oxygen vacancy density. Figure 30 shows the relation between the diffusion coefficient and the postannealing temperature obtained from the BST films deposited on Pt (100 nm)/SiO$_2$ (100 nm)/Si (100) and SrTiO$_3$ (100) substrates. As shown, the higher the postannealing temperature, the smaller the diffusion coefficient. The oxygen vacancy density of the BST film postannealed at 500°C is more than two orders of magnitude lower than that of the as-deposited film.

As occurs with the other perovskite-oxides, the Pt–BST contact forms a Schottky junction at the interfaces [196]. If a step-function voltage is applied in the current-time measurement or an oscillation voltage is applied in the capacitance measurement, the applied voltage drops primarily at a one-sided reversely biased Schottky junction, and as a result, its depletion layer width is modulated with electrons charging or discharging at oxygen vacancies in the depletion layer. These electrons in the voltage-modulated Schottky depletion layer are said to be responsible for the origin of dielectric relaxation [308].

Tsai and Tseng [338] reported on the ionic and electronic conductivity characteristics and the diffusion of oxygen ion in the (Ba$_{0.47}$Sr$_{0.53}$)TiO$_3$ thin films rf-sputtered at 450°C on a Pt bottom electrode at various OMRs. The grain boundary defect and the interface defect—together the total defect density—of BST/metal is considered to be a donor when it becomes neutral or positive by donating two electrons. When an ac voltage is applied, the defect levels move up or down with respect to the valence and conduction bands while the Fermi level remains fixed. A change of charge in the defect occurs when it crosses the Fermi level. Therefore, the defect density calculations can be done from the measurement of real capacitance (C') and the imaginary part of the capacitance (C'') as a function of frequency. Once C'' is known, the defect density can be obtained from the relation $D_{df} = C''/(qA)$, where A is the metal plate area and q is the elemental charge [320, 331, 339, 340]. Defect density decreases with increasing OMR. The defect traps of perovskite titanates often lead to dielectric relaxation as a function of frequency, in which the dielectric constant decreases and the loss tangent increases with increasing frequency.

High-temperature deposition of BST films under a nonoxidizing atmosphere, such as Ar, generally produces oxygen vacancies in the film according to

$$O_O \leftrightarrow V_O^{\cdot\cdot} + 2e^- + 1/2O_2 \qquad (13)$$

where O_O, $V_O^{\cdot\cdot}$, and e^- represent the oxygen ion on its normal site, oxygen vacancy and electron, respectively. As shown by Eq. (13), the BST material tends to show an n-type conductivity, although the conductivity is usually small due to the electrons associated with the formation of oxygen vacancies [38, 293]. The leakage current and defect density decreased with increasing OMR, resulting from the reduced concentration of the oxygen vacancies and the mobile charge (electrons) existing in the high-OMR–sputtered films [38, 293]. Thus, the electrical conductivity of the films is due mainly to the motion of the oxygen vacancies and electrons [by Eq. (13)], and, consequently, the leakage current density decreases with increasing OMR. The electrical conductivity of various OMR BST films at 27°C indicated in Figure 31 shows that conductivity decreased with increasing OMR, which may also result from the reduced concentration of oxygen vacancies and mobile charge (electrons) existing in the high-OMR films [38, 293]. The oxygen vacancies are the major defect in these BST films [10, 39–41], but the conductivity of the BST is due to the electrons associated with the formation of oxygen vacancies by using Eq. (13).

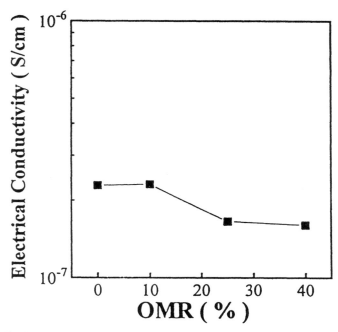

Fig. 31. Electrical conductivity of various OMR $Ba_{0.47}Sr_{0.53}TiO_3$ thin films at 27°C. Reprinted with permission from [318], © 1998, Taylor & Francis, Ltd.

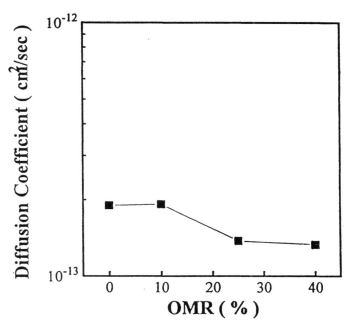

Fig. 32. Diffusion coefficient of oxygen ions in various OMR $Ba_{0.47}Sr_{0.53}TiO_3$ thin films at 27°C. Reprinted with permission from [318], © 1998, Taylor & Francis, Ltd.

Hence, the majority of electrical conductivity is contributed by electrons.

The temperature dependence of the diffusion coefficient and the electrical conductivity of a perovskite structure, such as $BaTiO_3$, when thermodynamic equilibrium of the crystalline material is achieved are given by the following equations [310]:

$$D = D_0 \exp\left[-\frac{(E' + E''/2)}{kT}\right] \quad (14)$$

$$\sigma = \frac{\sigma_0}{T} \exp\left[-\frac{(E' + E''/2)}{kT}\right] \quad (15)$$

Here D and σ are the diffusion coefficients and the electrical conductivity, respectively, and D_0 and σ_0 are constants. The activation energy for the jump of an ion into a vacancy is represented by E', whereas E'' is the work necessary to create a vacancy by moving the ions from the interior to surface of the crystal. The ratio σ/D has the simple form

$$\frac{\sigma}{D} = \frac{N(qZ)^2}{kT} \quad (16)$$

which is known as the Einstein relation. Here N is the number of ions per unit volume of crystal (i.e., charge carrier density) and qZ is the particle charge. The diffusion coefficient of oxygen ions in BST films can be determined using Eq. (16). If the lattice constant of BST is 3.95 Å [341, 342] for cubic structure, then $N = 4.86 \times 10^{22}$ cm^{-3}. The diffusion coefficient of oxygen ions in BST films at 27°C, shown in Figure 32, indicates that diffusion coefficient decreases with increasing OMR. It has been reported [308, 337] that the value of the diffusion coefficient of oxygen ions for BST on Pt at 27°C was 10^{-12}–10^{-13} cm^2/s, which is reasonably close to the values shown in

Figure 32. The oxygen vacancy density is considered to correspond to the oxygen ion diffusion coefficient [308, 310], and the difference in oxygen ion diffusion coefficients due to the variation of vacancy concentration represents the relative difference in oxygen vacancy density. Therefore, the oxygen ion diffusion coefficient increases with increasing defect density (oxygen vacancies).

The activation energies for oxygen ion diffusion in the BST sputtered under various OMRs can be obtained from the slopes of the linear regions of the Arrhenius plots, $\log(D)$ vs. $1000/T$ in the temperature range of 27–150°C. The activation energy for oxygen ion diffusion in 40% OMR BST thin film has the maximum value; hence, the motion of oxygen ions is small. The reduction of oxygen vacancy in the higher OMR [38, 273] deposited films leads to the suppression of oxygen ion motion. Therefore, the ionic conductivity of 40% OMR BST is smaller than that of the others.

The total electrical conductivity, σ_{total}, of a ceramic capacitor can be described as a sum of the electronic conductance, σ_{ele}, and the ionic conductivity, σ_{ion} [343]. The fraction of the total conductivity carried by each charged species is termed the transference number, t_i, where

$$t_i = \frac{\sigma_i}{\sigma_{total}} \quad (17)$$

The ionic conductivity, σ_{ion}, of BST film can be calculated from the number of defect density (oxygen vacancy only), N, and diffusion coefficient data (see Fig. 32) according to Eq. (16). Figure 33 shows the σ_{ion} of various OMR BST films at 27°C. The σ_{ion} is much smaller than the total conductivity (see Fig. 31), so the majority of electrical current is carried by the liberated electrons associated with the formation of oxygen

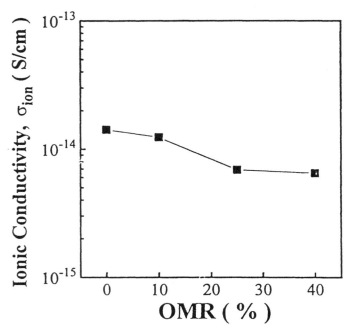

Fig. 33.   Ionic conductivity of various OMR Ba$_{0.47}$Sr$_{0.53}$TiO$_3$ films at 27°C. Reprinted with permission from [318], © 1998, Taylor & Francis, Ltd.

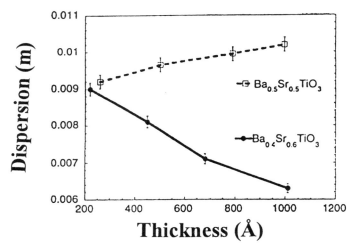

Fig. 34.   Thickness dependence of dispersion (represented by the exponent m). Ba$_{0.5}$Sr$_{0.5}$TiO$_3$ films exhibit increasing dispersion with thickness, whereas Ba$_{0.4}$Sr$_{0.6}$TiO$_3$ films show decreasing dispersion with increasing thickness. Reprinted with permission from [298], © 1967, American Institute of Physics.

vacancies, as shown in Eq. (13), and hence the BST is an electronic conductor ($t_{ele} \sim 1$). The oxygen vacancies are the major defect in BST thin film, but the mobility of oxygen vacancies (ionic conductivity) is much smaller than the mobility of electrons (electronic conductivity). Hence, the majority of electrical conductivity is contributed by electrons.

Capacitance vs. frequency measurements were performed on BST films with varying thicknesses and are described in [319, 320]. Their study shows that the interfacial capacitance is independent of frequency and increases with temperature. Also, the bulk dielectric constant is observed to have a power law dependence on frequency and to decrease with increasing temperature. Based on the bulk and interfacial study, the authors showed that the observed increase in the dispersion with BST film thickness is a volumetric effect and that the measured increase in dispersion with temperature is attributed mainly to decreasing bulk dielectric constant with increasing temperature.

Dielectric dispersion in BST film capacitors is also influenced by the properties of the electrode–dielectric interface in addition to bulk of the film. Balu et al. [180, 318, 344] studied the dielectric dispersion in BST thin films of compositions Ba$_{0.5}$Sr$_{0.5}$TiO$_3$ and Ba$_{0.4}$Sr$_{0.6}$TiO$_3$ deposited on an Ir bottom electrode using rf sputtering. Figure 34 shows the dispersion of BST capacitors as a function of thickness for the two compositions mentioned earlier. In the case of Ba$_{0.5}$Sr$_{0.5}$TiO$_3$ films, an increase in dispersion with increasing thickness was observed, while the Ba$_{0.4}$Sr$_{0.6}$TiO$_3$ films exhibited decreasing dispersion with thickness. Their study showed that dispersion in the Ba$_{0.5}$Sr$_{0.5}$TiO$_3$ film was bulk dominated, whereas dispersion in the Ba$_{0.4}$Sr$_{0.6}$TiO$_3$ was interface controlled, and the BST capacitors of each composition were modeled as a series

of combination of an interfacial and a bulk capacitance. Balu et al. also envisaged various multilayered BST capacitors with combinations of the aforementioned compositions and different individual layer thicknesses and observed that the experimental data agree with the model predictions for films with layer thicknesses greater than 150 Å.

6. RELIABILITY

Yield and reliability of thin dielectric films have been major reliability concerns throughout the history of MOS integrated circuit production. As devices are scaled down, dielectric films become thinner, but despite scaling of power supply voltage, the electric field applied to the dielectric during operation is increased. Because of the high field present in the dielectric, mandated by the aggressive dielectric scaling and the presence of oxide defects, catastrophic failure of the dielectric has always been the predominant oxide reliability concern, which limits scaling. Manufacturing economics and stringent customer requirements demand higher yield and lower failure rates. Dielectric yield and reliability have always been critical parameters for economical integrated circuit production.

Since practical considerations limit the assessment of oxide integrity on actual circuits, the question of oxide integrity is usually abstracted to the breakdown study of capacitors whose area is comparable to or larger than the area of active transistors in the circuit. We need to arrive at an accelerated testing paradigm that allows engineers to extrapolate the lifetime of good product based on a minimal number of short-time, high-field accelerated tests. This section focuses on techniques that allow engineers to determine whether or not their dielectric film technology is sufficiently robust to guarantee the lifetime specification of their product at its intended operating voltage, including today's operating voltages at or below 3.3 V.

Different types of test structures are needed to completely assess dielectric reliability. The most suitable test chip for a particular application of the dielectric is defined by the available chip area, test time, and cost constraints associated with the application [345–347]. The details of generic types of test structures used for gate/tunnel dielectrics and interlevel/double-polysilicon dielectrics were described by Martin et al. [347]. The following measurement methods are commonly used in dielectric yield and reliability assessment [347]:

(i) Constant voltage stress (CVS), used to measure the time to breakdown (T_{BD}) at different stress fields.

(ii) Constant current stress (CCS), used to measure the charge to breakdown (Q_{BD}).

(iii) Ramped voltage stress (RVS), used to measure breakdown fields (E_{BD}) and I–V characteristics.

(iv) Exponentially ramped current stress (ERCS), used to obtain a fast measure of Q_{BD}, breakdown strength (J_{BD} and E_{BD}) and I–V characteristics.

(v) Combined ramped/constant stress measurements: these measurements are used to record extrinsic breakdowns and intrinsic times to breakdown with short measurement times and high resolution.

The CVS method is frequently used measure the T_{BD} and the extrapolation of oxide lifetimes at normal operating conditions from highly accelerated reliability stresses. In the literature, extensive results have been reported using the CVS with fields of 1–12 MV/cm and temperatures of 25–400°C. When low/moderate stress fields are applied to high-quality oxides at room temperature, measurement times range from days to months. In this case, long-term reliability is investigated with CVS and extrinsic as well as intrinsic breakdown recorded. Such long measurements can be done on packaged samples or at the wafer level with a lot of devices stressed in parallel [348]. For wafer-level tests, it is usual to accelerate the field/or the temperature to keep measurement times low, $T_{BD} < 10,000$ seconds. As a consequence, at highly accelerated stress fields, dielectric structures with defects very often will break down at time zero, and it will not be possible to record the extrinsic distribution. Therefore, it is mainly the intrinsic breakdowns that are monitored with CVS measurements at highly accelerated fields [349]. The principle of CVS, which is also referred to as TDDB (time-dependent dielectric breakdown) measurement, is straightforward. A constant voltage is applied to the dielectric under test. The current through the dielectric is measured until breakdown occurs. Plotting the recorded current vs. stress time allows the current-time (I–t) characteristic to be studied. Also, if repeated the experiment at different voltage stress and recording the time to breakdown at each voltage stress enables creation of a plot relating the voltage stress vs. time to breakdown. This plot will show a linear curve that, on extrapolation, can predict the TDDB lifetime of the device. Figure 35 shows the TDDB lifetime for various DRAM dielectrics. TDDB, also referred to as resistance degradation of the dielectrics, shows a slow increase of leakage current under dc field stress. Degradation may take place at a much lower dielectric field than that of

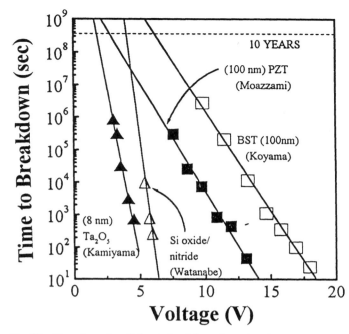

Fig. 35. Time-dependent dielectric breakdown for various DRAM dielectrics [371].

dielectric breakdown or thermal breakdown. Waser et al. [29] proposed an empirical power law between the electric field, E, and the characteristic time constant, t_{ch} (or lifetime), that is, $t_{ch} \propto E^{1.1}$, to determine the resistance degradation of dielectric materials. Typically, when a dc field is applied to a dielectric material, the current varies with time, as shown in Figure 36 [350]. One can observe three regions in the curve: (i) the initial current decay, called the charging current; (ii) leakage current, where the current stabilizes; and (iii) the electrical degradation region. The phenomena of electrical degradation can be ascribed to deterioration of grain boundaries, oxygen vacancy piling up at the electrode, reduction of potential barrier height due to space charge accumulation at grain boundaries and demixing reactions of oxygen vacancies [350].

The most important result of a CVS measurement is T_{BD}. The T_{BD} of a CVS is the most commonly measured reliability indicator, and thus extrapolated or measured data from ramped or combined ramped/constant measurements are usually compared with the T_{BD} of a CVS to validate the extrapolation method or the results of a new measurements [349, 351]. It can be said that the T_{BD} of a CVS is somehow a "reference value" for the oxide quality, since it is assumed that a CVS at constant operating voltage is the worst case of stress in real oxide life. T_{BD} is dependent on several stress parameters, including oxide thickness, process variations, and measurement temperature. The charge to breakdown, Q_{BD}, can be calculated from the recorded current time characteristics. Q_{BD} is the current injected through the oxide integrated over time,

$$Q_{BD} = \int^{T_{BD}} I(t)\, dt \qquad (18)$$

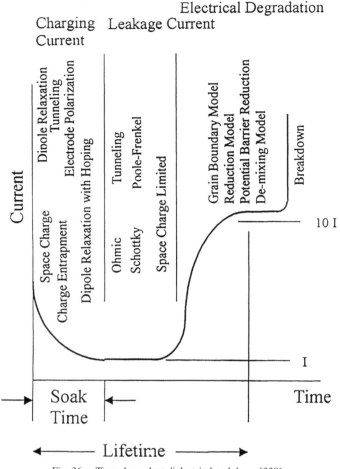

Fig. 36.   Time-dependent dielectric breakdown [330].

Since the CVS current decays with time, the frequency of the current readings influences the quality of the integration of the charge to breakdown.

The CCS was first described by Harari [352]. In a CCS measurement, the current is held constant and the voltage measured. For breakdown detection, continuous voltage monitoring is recommended to achieve good Q_{BD} resolution. In a CCS, T_{BD} is measured and then multiplied by the current density to obtain the charge to breakdown, Q_{BD}, in units of C/cm²,

$$Q_{BD} = T_{BD} J_{CCS} \qquad (19)$$

The RVS is really the workhorse of dielectric measurements. It is a fast measurement method that can provide intrinsic and extrinsic breakdown strengths of the dielectric with good resolution. The principle of the RVS is that a continuously increasing voltage is applied to the dielectric until breakdown is detected. In practice, the form of the voltage ramp used can vary widely. In earlier days, a continuous linear ramp was used [353], but today, some form of staircase ramp [354] that which approximates to a linear ramp is much more common. One of the reasons for this is that dielectric measurements are commonly performed with parametric test equipment, and the staircase ramp is easy to implement and control with the avail-

able parametric test hardware and software. Another type of voltage ramp which is sometimes used is the stepped ramp, which generally has long dwell times and several measurements per step [355]. The time step used can be on the order of minutes to hours.

The ERCS is similar to the RVS. In both cases, the bias is stepped up until breakdown. In the ERCS, the current is kept constant throughout each step of the ramp, and the applied voltage is measured at each step. ERCS is used mainly for the characterization of thin oxides in nonvolatile memory technologies. It is a fast method for the evaluation of charge to breakdown Q_{BD} [356]. In high-volume integrated circuit production, it is desirable to have fast measurement methods that provide maximum information on extrinsic and intrinsic breakdowns. Therefore, many semiconductor manufacturers perform their own custom-designed reliability measurements, such as combined ramped/constant stresses [357]. Constant stresses and ramped stresses both have advantages and disadvantages. It is a target to combine the advantages of both stress methods while trying to avoid their disadvantages [357, 358]. This done in some degree by performing a ramped stress followed by a constant stress. Either current density or field biases can be applied in these combined reliability measurements [359, 360]. In the following paragraphs we present the reliability properties of BST thins films studied by various researchers.

Electrical properties such as a low leakage current, fast dielectric response, low dielectric loss, and long lifetime are essential to ensure the reliability of BST film capacitors. Dielectric relaxation greatly affects the electrical properties of the BST capacitor, such as field-stress leakage current [324], stored-charge loss [32], and pause refresh properties [29]. The details are presented in Section 5. TDDB and stress-induced leakage current (SILC) have been investigated for the reliability of BST thin films [212]. The breakdown was strongly affected by leakage current properties, and did not depend on the dielectric constant. Notably, the 10-year breakdown field for BST was six times larger than that for SiO₂. The T_{BD} and SILC for BST films were measured by Yamamichi et al. [212]. Both the T_{BD} vs. E and T_{BD} vs. $1/E$ plots show universal straight lines, independent of the film thickness, as shown in Figure 37. Both plots project lifetimes longer than 10 years at +1 V for 50-nm BST films with an SiO₂ equivalent thickness (t_{eq}) of 0.70 nm. The breakdown lifetime was correlated with the leakage current, but there was no monotonic relationship between dielectric constant and lifetime. SILC was observed at +1 V in the time domain after a stress charge injection. The thinner films displayed greater endurance for stres, because of the smaller number of traps generated in the thinner BST. Although SILC leads to increased charge loss after long-term operation, a 10-year lifetime for Gb-DRAMs is also predicted by the charge loss vs. stress charge plot shown in Figure 38. It is well known from the work of Horikowa et al. [361] that T_{BD} and the slope of the T_{BD} vs. $1/E$ curves are dramatically worse for an ac stress than for a dc stress. The currents for both polarities of the dc stress decrease with time, most likely due to generation and occupation of traps near the cathode. The trapping time constants

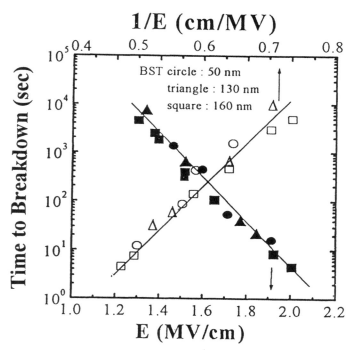

Fig. 37. T_{BD} vs. E and T_{BD} vs. $1/E$ plots for BST of 50, 130, and 160 nm, with a (Ba + Sr)/Ti ratio of 1.05 [192].

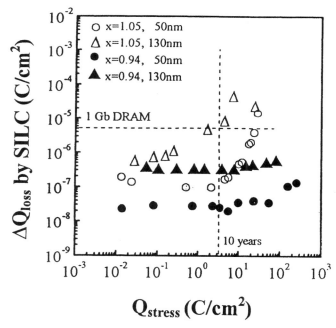

Fig. 38. ΔQ_{loss} as a function of Q stress. ΔQ_{loss} increases with Q_{loss}, especially for thicker films, but 1 year of operation is successfully guaranteed for 1-Gb DRAMs [192].

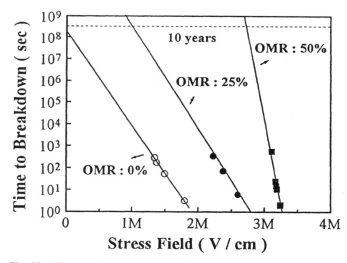

Fig. 39. Time to breakdown of $Ba_{0.5}Sr_{0.5}TiO_3$ films deposited at 450°C at various OMRs, as a function of stress field. Reprinted with permission from [38], © 1997, American Institute of Physics.

have a wide range, from 100 ms to very long times. Trapping impedes current and in consequence improves lifetime. But in the case of ac stress, a part of the charges trapped during one pulse gets detrapped during the following pulse with the opposite polarity. So ac stress effectively prevents occupation of short time-constant traps [362].

Lifetime extrapolation using CVS–TDDB studies (Fig. 39) predicts the 10-year lifetime at a 1-V operating voltage [38]. It indicates that a 50% OMR sample has a longer lifetime than do 25% and 0% OMR samples. TDDB is characteristic of the intrinsic materials, the procedures, and the quality of the processing and electrode materials [6, 360]. Several models are available in literature describing the phenomena of TDDB, such as grain boundary model and the reduction model [363–368]. All other models except the grain boundary model are based on the mobile oxygen vacancies. The grain boundary barrier height reduction model [364–366] assumes that the positively charged oxygen vacancies with a relatively high mobility electromigrate toward the cathode under a dc electric field. The oxygen vacancies then pile up at the front of the cathode. At the anode, in contrast, an electrode reaction leads to the generation of oxygen gas and electrons, leaving oxygen vacancies behind. The net result is that the space charge accumulation at the grain boundaries reduces the grain boundary height and increases the leakage current. The simplest way to improve the TDDB is to reduce the concentration of oxygen vacancies in the materials. In fact, a significant improvement is seen in the 50% OMR sample as compared to the 0% OMR sample (see Fig. 39). Tsai and Tseng [49] reported that BST films deposited on Pt, Ir,

Ir(600°C), and IrO₂/Ir samples have longer TDDB lifetimes at 1-V operating voltage than BST on Ru and RuO₂/Ru (Fig. 40).

Extrapolated lifetimes of $(Ba_{0.7}Sr_{0.3})TiO_3$ ($t_{oxeq} = 6$ Å) films were reported to be in a safe regime for ±1-V operation but not for ±2-V operation [362]. Therefore, the authors predict that further scaling of BST to $t_{oxeq} = 3$ Å and lower will not yield the required reliability [342]. Yamabe et al. [369] investigated time-dependent leakage currents under high voltage stress (±5 V) to BST film at high temperatures (~225°C). Their results indicate that under the application of negative stress, the leakage current increases once and decreases after reaching

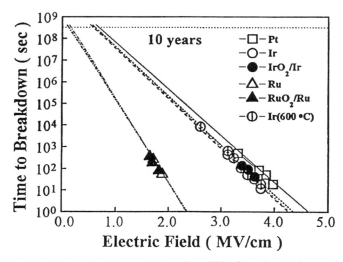

Fig. 40. Time to breakdown of $Ba_{0.47}Sr_{0.53}TiO_3$ films deposited on various bottom electrodes as a function of stress field. Reprinted with permission from [49], © 1999, American Ceramic Society.

the maximum point, whereas the application of positive stress leads to a steep increase in leakage current after a given time. Three possible mechanisms were proposed to be responsible for the above degradation processes. The first mechanism involves a combination of electron and hole trapping in the BST film [370]. The second involves a combination of cation and anion distribution in the BST film. The third consists of a combination of electron trapping and cation distribution or of hole trapping and anion distribution in the BST film.

Numata et al. [35] suggested that a possible cause of the polarity dependence of the $I–V$ characteristics of BST films is the intrinsic inhomogeneous distribution of the oxygen vacancies in the films. Accumulation of electrically drifted oxygen vacancies is thought to be the cause of the resistance degradation of BST. A singular characteristic at the initial stage of the sputtering deposition or reaction between the deposited films and the bottom electrodes might produce a greater concentration of oxygen vacancies near the bottom interface, leading to polarity dependence of the resistance degradation.

Shimada et al. [51] ascribed the time-dependent increase in leakage current of BST films to the change in the conduction mechanism from the interface-controlled Schottky type to the bulk-related space-charge-limited type due to the accumulation of oxygen vacancies near the cathode as a result of interface barrier lowering and the migration of distributed oxygen vacancies across the film. For a 100-nm-thick $(Ba_{0.5}Sr_{0.5})TiO_3$ film, the time required for 50% device failure, T_{50}, was determined to be larger than 10^2 years under a stress voltage of 5 V [60]. Similar TDDB studies on BST films by Ohno et al. [37] and Fujii et al. [117] support the observations of Koyama et al. [60]. Park et al. [26] also reported the high reliability of 20-nm-thick BST thin films.

The preceding discussion on the electrical reliability of BST thin films clearly demonstrates that the lifetime of BST has a quantitative relationship with voltage or current stress, film thickness, and temperature of measurement and only hints at

the physical mechanism responsible for breakdown. During the course of electrical stress, dielectric damage manifests itself as generated interface and bulk traps, stress-induced leakage current, and, finally, catastrophic breakdown. Suitable deposition techniques and the associated postannealing process need to be developed to minimize the charge trapping and thereby reduce the interface instability and enhance the electrical reliability of BST film capacitors for Gb-era DRAMs.

7. KEY TECHNOLOGIES FOR GIGABIT DRAMS

7.1. Cell Structure and Cell Technology

As is well known, the unit cell scheme composed of one transistor and one capacitor is the most important parameter for characterizing DRAMs for die size as well as performance. The unit cell can be laid out in one of two different ways, depending on the bit line arrangement: either folded bit line cell architecture or open bit line cell architecture. The smallest unit cell sizes are $8F^2$ in the folded bit line cell architecture and $6F^2$ in the open bit line architecture, where F denotes the minimum feature size that is normally determined by lithography. Folded bit line cell architecture has been the standard memory cell layout because of its superior noise immunity despite increased area [4, 5, 371]. The folded bit line architecture has evolved into two different cell structures using trench and stacked cells (STC) from the planar cell structure. The STC-type cell has recently evolved into the single or multilayer cylinder type and the single or multilayer fin type. A rugged surface has also been introduced for enlarging the surface areas. These will be important for STCs in the Gb era, as area-enlargement methods approach their limits. Another way to enlarge the surface area of capacitors is to use trench capacitors. In this category, substrate-plate–type cells and stacked-trench–type cells have been introduced. However, it is very difficult to use high-ε ferroelectric films inside the trench capacitors, because these materials contain heavy ions, which act as generation/recombination sites in the Si substrate. In the trench-type cells, therefore, silicon on insulator (SOI)-plus-trench capacitor structure may be the better way to make Gb-era DRAMs, because it gives the maximum capacitor layout in a limited cell region. The cell sizes and die sizes corresponding to densities of Gb DRAMs based on $8F^2$ folded line cell architecture are estimated as a function of minimum feature size F. As feature size decreases, cell size rapidly decreases from 0.27 μm^2 for a 1-Gb DRAM to 0.08 μm^2 for a 16-Gb DRAM. Manufacturing Gb DRAMs economically necessitates a smaller chip size by realizing cell size smaller than $8F^2$ in the folded bit line cell architecture.

Memory cell capacitance is the critical parameter for Gb-density DRAMs in determining sensing signal voltage, sensing speed, data retention times, and endurance against soft error events. It is generally accepted that the minimum cell capacitance is more than 25 fF/cell regardless of density. However, lower supply voltage and increased junction leakage current due to high doping density drive memory cell capacitance toward higher values, more than 25 fF/cell in the Gb-density DRAMs. Capacitance of DRAM cells can be described by parallel plate capacitor which is determined by area of memory cell capacitor, dielectric constant of capacitor dielectric and the thickness of capacitor dielectric. To date, strategies for increasing memory cell capacitance have focused on increasing memory cell capacitor area and decreasing dielectric thickness.

Thin film BST dielectric is the best capacitor material for the Gb-era DRAMs due to its extremely high dielectric constant, around 200–600. High-dielectric BST capacitors are basically formed in an MIM structure with metal electrodes to have strong resistance to native oxide; the details were described in Section 3. Therefore, a MIM cell capacitor using the BST dielectric seems to be the ultimate solution for the Gb era.

7.2. Lithography

The key issues in lithography technology for the Gb era are resolution and alignment accuracy in printing feature size patterns smaller than the wavelength of the exposure tool. The suppression of pattern size variations across a large chip is another concern. These issues become much more critical as the feature size shrinks. A pattern size smaller than the wavelength of the lithography tool requires techniques such as off axis illumination, phase shifting mask, and optical proximity control. At the same time, the decreased depth of focus (DOF) in the resolution enhancement technique should be overcome by providing a flat surface before each lithography step. The flat surface can be achieved by chemical mechanical polishing (CMP) planarization. The lithography technology in the Gb era is expected to be deployed as follows. The 248-nm lithography is believed to be used mainly in the production of the 0.25-μm generation since it was introduced for the 0.30-μm generation. It will be a major lithography in the 0.18-μm generation and is expected to be pushed to the 0.15-μm generation by using proper high-resolution techniques together with higher values of numerical apertures of optics and improved photo-resistant performance. Thus, 248-nm KrF lithography will be the major lithography for 0.25 μm to 0.15 μm. The 193-nm ArF lithography will be first introduced in the 0.13-μm generation and then continuously pushed to the 0.1-μm generation as previous optical lithographies 248 nm (KrF), 365 nm (i-line), and 468 nm (g-line) are pushed to their ultimate limit. Below a 0.1-μm feature size, postoptical lithography (e.g., as X-ray, electron, and ion beam projection) will definitely be necessary. These technologies have much more DOF than optical lithography. Another concern in printing Gb DRAMs is how to reduce the variation of the critical dimension (CD). Keeping the variation of CD within an acceptable range in a large chip area requires extremely tight control of the mask CD and a well-tuned lithography process in terms of antireflecting layer, scan speed of exposure, soft bake and postexposure bake process, and development process [5, 372, 373].

The requirements of etching technology for Gb-scale devices are high etching selectivity, small deep contact hole etching, self-aligned contact (SAC) etching, and new materials etching. Since the vertical etching of layers is strongly dependent on etching selectivity, enhancement of etching selectivity will not only improve the etching process latitude, but also greatly help the vertical scaling. The lateral dimensions of DRAMs have been scaled down by a factor of 0.84 per year. The scaling of vertical dimensions in DRAM has not been practiced as the scaling of lateral dimensions has been done. For instance, the gate stack using polycide is only scaled down from 250 nm (poly, 100 nm; silicide, 150 nm) in 0.35-μm, 64-Mb DRAM to 200 nm (poly, 100 nm; silicide, 100 nm) in 0.18-μm, 1-Gb DRAM, resulting in a vertical scaling factor of only 0.8 with a horizontal scaling factor 0.5. The discrepancy between the vertical and horizontal scaling factors is caused mainly by the low etching selectivity between layers. Insufficient vertical scaling implies that the deposition of layers without generating voids or seams becomes more difficult and the aspect ratio of contact holes becomes larger. Thus, high etching selectivity is indispensable in Gb-era DRAMs [5].

7.3. Device Technology

It is generally accepted that the shallow trench isolation (STI) introduced for the 0.30-μm to 0.25-μm generation will be the main isolation scheme for Gb era due to its superior isolation characteristics [5, 374, 375]. The main advantages of STI are no loss of active width (free from oxide encroachment onto the active region) and suppression of low-isolation punchthrough voltage. Considerable loss of the active region in local oxidation structure (LOCOS) comes from the encroachment of field oxide onto the active region. The loss of active region can be nearly eliminated in STI because high-temperature field oxidation is no longer necessary in STI. Punchthrough limitation encountered in LOCOS below the 0.25-μm minimum feature size can be eliminated in STI by increasing the isolation distance, which is done by simply increasing the depth of trench. Recent progress in STI can meet both the physical (i.e., flat surface, no dishing, no oxide dipping in field oxide) and electrical (i.e., hump-free subthreshold conduction) requirements of Gb-scale isolation. In addition to these advantages, STI has been noted to produce lower junction leakage current and lower junction capacitance compared to LOCOS, resulting in better DRAM cell operation [376, 377].

The two key processes of STI are oxide filling after trench etching and CMP planarization. The filling oxide leaves various topologies depending on the pattern density, determined by the ratio of active area to field area and on pattern structures, such as wide active or wide field. CMP alone seems to be difficult to achieve a flat surface without dishing or field oxide recess. A "dummy" active pattern or an additional masking step is typically used to relieve the burden of CMP [374, 378, 379]. Since the filling properties of deposited oxide are strongly dependent on methods of deposition, different dummy patterns and/or different CMP planarizations, depending on the filling properties of deposited oxide, should be used.

The suppression of crystal defects that might be generated in the process of STI fabrication (e.g., trench etching, densification of filled oxide) is very important. The generated defects during the STI process seem to be further aggravated with subsequent thermal cycles, etching processes like spacer etching, and high-dose ion implantation (e.g., source/drain implantation). Recent investigations into defects related to STI indicate that the proper processes, such as round profiles of trench, densification at elevated temperature around 1200°C [380], and optimized ion implantation, can greatly suppress the generation of defects. By virtue of a well-optimized STI process that is less prone to generate defects, STI will be the main stream of the isolation scheme in the Gb era.

Two different classes of transistors are used in DRAM chips. One class is the n-MOS memory cell transistors, which comprises the memory cell array; the other is periphery CMOS, which comprises the periphery CMOS circuit. The requirements and approaches for the two classes are quite different. The constraints on memory cell transistors are nonscaleable high-threshold voltage around 1 V regardless of feature size, density, and supply voltage to suppress the subthreshold leakage currents and higher-gate oxide thickness to sustain the high voltage. The threshold voltage variations due to active CD and gate CD variations in 0.18-μm, 1-Gb DRAMs are reportedly as high as 150 mV/15 nm and 200 mV/15 nm, respectively [4]. The most important thing in memory cell transistors is that the junction leakage current due to high substrate doping density be greater than 10^{18} cm^{-3}, which is the most critical factor for realizing the Gb-density DRAMs. A substrate doping density higher than 10^{18} cm^{-3} is required to keep the threshold voltage around 1 V and to suppress the short-channel effect. Electric field–assisted tunneling current is the major leakage current in such a high doping regime [379]. To suppress the substrate doping density, the self-aligned channel implantation scheme [381] seems to be a good solution, in which high doping is achieved only in the channel region and substrate doping beneath the source and drain region is prohibited. Reduced doping density underneath the source and drain regions is beneficial in reducing the junction capacitance.

Periphery CMOS transistors can be designed with the guidelines provided by scaling theories [382, 383]. So far, the periphery CMOS transistors used in DRAMs have been developed that focus on low leakage current and high breakdown voltage rather than on high performance (e.g., high driving capability, low parasitic resistances), which is more important in logic application. Also, the cost-effectiveness of DRAMs has driven fabrication of transistors toward simple processes. As a result, a single polysilicon gate or a single polycide gate has been the standard scheme for DRAM transistors with higher gate oxide thicknesses compared to high-performance logic transistors at the same channel length for both n-MOS and p-MOS. This approach is expected to be extended in the Gb era as long as the single gate is used. However, recent movement toward merged technology with logic, as in logic-embedded DRAM or DRAM-embedded logic, demands transistors with high performance (i.e., high driving capability). Various threshold voltages depending on the nature of the circuit will be used to satisfy both fast operation under low supply voltage and low standby current.

7.4. Metallization

It is generally known that the metallization scheme determines the chip yield as well as chip performance in DRAMs. The metallization scheme in DRAMs has been changed from single-metal scheme, which had been used up to 4-Mb DRAM generation, to a double-metal scheme, which was introduced for 16-Mb DRAMs and has been continuously used for up to 64- and 256-Mb DRAM generations. Further extension of the double-metal scheme in the Gb era is no longer possible, because of fine patterns and the global topology. The triple-metallization process introduced for 1-Gb DRAM generation [4, 5] can solve difficult processes due to insufficient DOF and provide a much more relaxed design rule as well as improved performance. The triple metallization will be used for up to 16-Gb DRAMs. Beyond this generation, multiple-level metallization will be necessary for high performance. In multiple-level metallization [384], Cu and low-dielectric technologies become important to reduce power consumption, speed delay, cross-talk, and insulation resistance (IR) drop. Cu technology involves many issues, including fast diffusion in oxide and Si [385] and difficulty in reactive ion etching (RIE) [386]. Low-dielectric technology gives only a small reduction in dielectric constants like SiOF [387]. The typical ohmic and barrier layers used today and expected to be used in the future are the Ti/TiN layers deposited by sputtering [388]. Here Ti serves as the ohmic layer by forming TiSi$_2$, and TiN serves as the barrier layer. The lack of conformality in sputtering deposition has been improved by using the collimation method [388], long-through sputtering [389], and ion metal plasma deposition [390]. The fundamental approaches to deposit ohmic and barrier layer formation involve CVD deposition, in which the conformality is extremely good. Stable and reliable ohmic and barrier layer technologies are necessary for future Gb-era DRAMs.

7.5. Integration Issues

There have been encouraging developments in the integration of BST-based ferroelectrics into high-density DRAMs. One of the primary problems is the need to deposit the BST films under oxidizing conditions and to minimize postdeposition thermal treatments under low-oxygen partial pressures, because the perovskites are susceptible to reduction. This also limits the choice of electrodes for the capacitor stack to either Pt, Ru, or conducting oxides, since forming an insulating oxide at the BST–electrode interface lowers the stack capacitance. The electrode material selection then results in a host of consequences for both processing and integration. NEC uses a stack consisting of polysilicon plug/thin Ti (forms Ti silicide)/TiN/RuO$_2$/BST/TiN/Al [181]. The RuO$_2$ is deposited under conditions that minimize oxidation of the TiN. During BST deposition, this RuO_2 also acts as an oxygen diffusion barrier. The approach appears viable if care is taken to ensure that deposition temperatures are kept as low as possible, which NEC achieves by using the oxygen ECR plasma-assisted CVD system.

Mitsubishi uses a simpler scheme consisting of polysilicon/Ru/BST/Ru [103]. It is believed that the Ru at the BST must partially oxidize, but the oxide has satisfactory conductivity. If the extent of oxidation is kept small, then the large volume change associated with the oxidation apparently is not deleterious. The oxidation is minimized by using low deposition temperatures and a nitrogen atmosphere during the rapid thermal anneal used for the crystallization. However, the ruthenium silicide formation would occur at the processing temperature.

The U.S. DRAM consortium, Samsung [392], and earlier work by NEC [393] used a barrier system based on Pt, for example, a polysilicon/Ti/TiN/Pt/BST/top electrode. Because oxygen diffusion rates along Pt grain boundaries are high, the Pt does not act as an oxygen diffusion barrier. Thus process conditions must be selected to minimize the TiN oxidation, because oxidation results in a rapid increase in resistance.

It appears from recent publications that etching methods have been found for Ru and RuO_2 electrodes [103, 181]. This is expected, because of known volatile Ru oxides and halides. Encouragingly, Mitsubishi has also reported an attractive X-ray lithography procedure for Ru and demonstrated its use for the processing of very fine scale features [95]. However, there has been little evidence of a true reactive ion etch for Pt at room temperature. This presents a serious problem, for in the case of fine scale features with high aspect ratios, the Pt is redeposited on the side of the mask, forming unwanted features. One known solution to the Pt etching problem is to undertake the reactive ion etching at elevated temperatures, which then requires compatible hard masks.

Recently, Kim et al. [394] reported that the critical issues of MIM BST capacitor for 0.15 μm can be solved by many novel processes, including recessed barrier with SiN spacers, Pt-encapsulated Ru storage nodes (modified etch mask structures and controlled etch mask sizes), MOCVD BST deposition, and an Al_2O_3 barrier layer to suppress hydrogen damage. A novel self-aligned electroplating process for fabricating Pt electrodes of integrated high-dielectric capacitors for 1-Gb DRAMs and beyond was developed by Horii et al. [395]. Pt pillars of 210 nm diameter and 650 nm height were fabricated after the SiO_2 wet strip. The leakage current density of the sputtered BST capacitor using electroplated Pt was less than 2×10^{-7} A/cm^2 at ± 1.5 V. The t_{eq} and dissipation factor of 40-nm-thick BST film were 0.70 nm and 0.0080 at 0 V, respectively. More recently, Kiyotoshi et al. [396] successfully developed a new in-situ multistep (IMS) process technology to achieve both conformal step coverage and a high dielectric constant of CVD-BST. IMS is a sequential repetition of low-temperature CVD of BST and its crystallization in a batch-type hot wall reactor that enables uniform deposition over 200-mm wafers. The authors demonstrated that using the IMS CVD process combined with $SrRuO_3$ electrodes, a conformal growth of local epitaxial grown BST film with dielectric constant exceeding 300, and leakage current density of around 10^{-7} A/cm^2 at 1 V were easily attained.

8. OPTICAL PROPERTIES

Ferroelectric thin films of high dielectric constant, such as BST, have been widely investigated for their feasibility in thin film integrated capacitors and storage capacitors in Gb dynamic random access memory (DRAM) applications, and have received much attention for fabricating novel functional devices [397–402]. Nevertheless, there is an active interest in the study of optical properties of BST thin films because of their high electro-optical coefficient, which is highly promising for optical applications [403]. These films are expected to be excellent in realizing various optical applications because of their wide energy gap (>3 eV), large static dielectric constant, high refractive index, and low absorption coefficient. In addition, films or multilayer systems of BST can be used in nonlinear optical devices, such as planar waveguides or optical switches, with minimal optical propagation losses. Most optical losses are directly related to the thin film growth and film morphology that develops [404–406]. It has been demonstrated that for $x \leq 0.12$, BST materials are suitable for applications in optical signal processing, dynamic holography, phase conjugation, and two-beam coupling [407–409]. Bulk $BaTiO_3$ in the tetragonal phase has large optical nonlinearities and has potential applications in thin film electro-optic modulators and waveguide frequency doublers [410]. The new-generation optical and optoelectronic devices require properties of thin films, which cannot be achieved in existing materials. Solid solutions or mixing of one or more compounds allows for the possibility of changing many properties of the thin films, including refractive index, extinction coefficient, microstructure, and surface roughness. BST films can display a paraelectric phase by controlling the Ba/Sr ratio and have a high transparency, hence they can be used as an insulating layer of an electroluminescent device [411, 412]. Current research into the optical properties of BST should attract more interests because of its possible application in flat panel displays [413, 414] and integrated optics.

Very few studies on the optical properties of BST thin films compared to the electrical and dielectric properties have been reported. Investigations have focused on the dependence of the optical properties and film morphology on the deposition parameters of the thin films grown by different methods, such as sputtering, laser ablation, and the sol-gel method on various substrates. The published data on refractive index and dispersion parameters as well as fundamental band gap energies suggest that the optical properties, crystallinity, and morphology are interdependent. Different authors have used different thin film preparation methods, different processing conditions, different optical constant computational methods, and different values for the optical constants. These variations indicate the dependence of optical constants on the thin film preparation history.

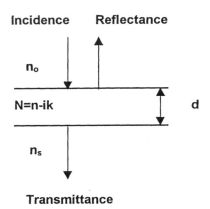

Incidence　　Reflectance

n_o

$N = n - ik$　　　d

n_s

Transmittance

Fig. 41. Schematic diagram representing reflection and transmittance by a thin film.

8.1. Refractive Index and Dispersion Parameters

Optical properties of a medium can be described by the complex index of refraction, $N = n - ik$. The real and imaginary part of the refractive index are termed the refractive index (n) and the extinction coefficient (k). The latter is connected to the absorption coefficient, α, by the equation

$$\alpha = \frac{2\omega k}{c} \tag{20}$$

where ω is the photon frequency and c is speed of light. Here n, k, and α are referred as the optical constants of the medium. The values of these optical constants are dependent on the photon energy, $E = \hbar\omega$; that is, $N = N(E) = n(E) - ik(E)$ and $\alpha = \alpha(E)$. These functions are called *optical dispersion relations*. Figure 41 shows a schematic diagram of a thin film with a complex refractive index, $N = n - ik$, and thickness, d, bound by two media with refractive index n_0 and n_s, one being air and later is substrate. For this type of uniform multilayer structure, an interference pattern is expected in the transmission, reflectance, or absorption spectrum because of the presence of multiple air–film and film–substrate interfaces similar to those shown in Figure 42 [415]. If the thickness of the film is not uniform, or if it is slightly tapered, then all interference effects will be destroyed and the spectrum will be a smooth curve. Thus, the transmission, reflectance, or absorption spectrum also serves as a technique for determining the uniformity of the film. Different methods for determining the refractive index and extinction coefficient from the observed transmittance spectrum or reflectance spectrum have been published [416–420].

Wang et al. [415] reported on a detailed study on the optical properties of BST thin films prepared on transparent substrates by rf sputtering at different substrate temperatures to investigate the effect of substrate temperature on optical properties of BST thin films. Figure 42 shows the spectral transmittance characteristics of bare Corning glass substrate and the $Ba_{0.7}Sr_{0.3}TiO_3$ thin films deposited on it at different substrate temperatures. The interference fringes are a result of the air–film and film–substrate interfaces. Wang et al. [415] calculated the refractive index from this spectrum following Swanepoel's

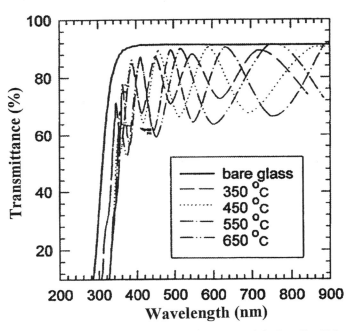

Fig. 42. Transmittance spectra as functions of wavelength for $Ba_{0.7}Sr_{0.3}TiO_3$ films deposited at various substrate temperatures on Corning glass [415].

envelope method [416] by drawing two envelopes through the maxima and minima of the transmittance spectra and using the expressions

$$n = \left[N + \left(N^2 - n_s^2\right)^{1/2}\right]^{1/2} \tag{21}$$

and

$$N = 2n_s\left(\frac{1}{T_{\min}} - \frac{1}{T_{\max}}\right) + \frac{(n_s^2 + 1)}{2} \tag{22}$$

where $T_{\max}$ and $T_{\min}$ are the corresponding transmittance maximum and minimum at a wavelength λ and n_s is the refractive index of the substrate. The refractive index dispersion spectrum of $Ba_{0.7}Sr_{0.3}TiO_3$ thin film deposited on Corning glass ($n_s = 1.51$) at various substrate temperatures is plotted in Figure 43 and has been fitted to the Cauchy formula [415]

$$n = A + \frac{B}{\lambda} + \frac{C}{\lambda^2} \tag{23}$$

(where A, B, and C are the fitting constants), as shown by solid lines. The dispersion curve rises sharply toward shorter wavelengths, displaying the typical shape of a dispersion curve near an electronic interband transition. The strong increase in the refractive index is due to the proximity of the fundamental band gap. Wang et al. [415] observed that refractive index increased from 2.17 to 2.59 (at $\lambda = 410$ nm) as the substrate temperature increased from 350 to 650°C. The increase in the refractive index of the films deposited at higher substrate temperature is due to the increased crystallinity and also to the increased packing density of the thin film. A porous film has a lower refractive index, because the air trapped in the film's pores, which has a lower refractive index, can effectively lower the

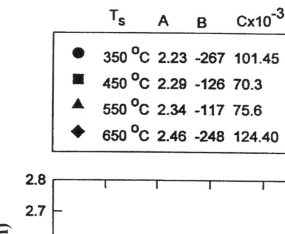

T_s	A	B	$C \times 10^{-3}$
● 350 °C	2.23	-267	101.45
■ 450 °C	2.29	-126	70.3
▲ 550 °C	2.34	-117	75.6
◆ 650 °C	2.46	-248	124.40

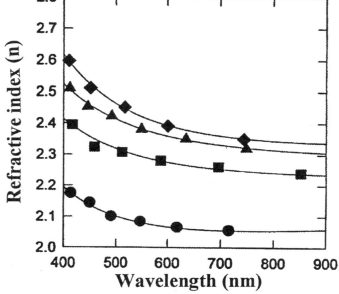

Fig. 43. Variations of refractive index as functions of wavelength for $Ba_{0.7}Sr_{0.3}TiO_3$ films deposited at various substrate temperatures on Corning glass [415].

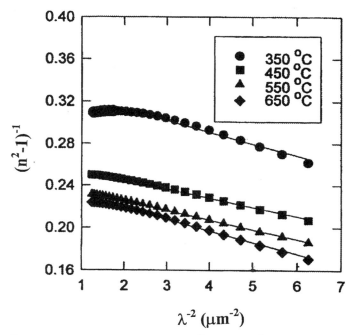

Fig. 44. Plots of refractive index factor $(n^2 - 1)$ vs. λ^{-2} for BST films deposited at various substrate temperatures [415].

refractive index. Postdeposition annealing of the films can considerably increase the refractive index by increasing the film's crystallinity and density by eliminating the pores. The refractive index increased from 2.08 ($\lambda = 550$ nm) for as-deposited film to 2.165 ($\lambda = 550$ nm) after postdeposition annealing at 650°C, which can be explained on the basis of improved crystallinity and increased film density. The dispersion parameters from the experimental refractive index dispersion curve have been obtained by fitting to Wemple's single electronic oscillator model [421, 422], wherein the refractive index as a function of wavelength is expressed as

$$n^2 - 1 = \frac{E_0 E_d}{E_0^2 - (hc/\lambda)^2} \quad (24)$$

where c is the light speed, h is Planck's constant, E_0 is the single-oscillator energy, and E_d is the dispersion energy. E_0 is empirically related to the energy gap of an insulator by $E_0 \sim 1.5 E_{gap}$, whereas E_d is correlated with the coordination of cations with the nearest-neighbor anions and effective

electronic charge by

$$E_d = \beta N_c Z_e N_e \quad (25)$$

where $N_c = 6$ is the coordination of the cations to their nearest-neighbor anions, $Z_e = 2$ is the number of the anions, $N_e = 8$ is the number of valence electrons, β is an empirical factor given as 0.26 ± 0.04 eV [421–423] for most of the oxides. The E_0 and E_d values can be obtained from the intercept and slope of the $(n^2 - 1)^{-1}$ vs. λ^{-2} plot (Fig. 44) [415]. At long wavelength regions, the plot shows a deviation from the linear behavior due to the negative contribution of lattice vibrations to the refractive index [421]. The dispersion parameters obtained from the single-oscillator model fitting for the $Ba_{0.7}Sr_{0.33}TiO_3$ thin films are listed in Table XI. The E_0 values decreased with increasing substrate temperature. The E_d value obtained for BST films is close to the E_d of pure bulk $BaTiO_3$ (23.3 eV) and $SrTiO_3$ (23.32 eV) [421]. A small amount of amorphous phase would result in a higher E_0 value for the lower-temperature deposited films. Apart from substrate temperature, the change in the substrate also changes the film's refractive index. The refractive index of the $Ba_{0.7}Sr_{0.3}TiO_3$ films deposited on Al_2O_3 ($n_s = 1.80$), fused quartz ($n_s = 1.51$), and Corning glass ($n_s = 1.51$) substrates are different. The refractive indexes at $\lambda = 600$ nm of the film are 2.28 on Al_2O_3, 2.35 on fused quartz and 2.4 for films deposited onto the Corning glass, indicating the effect of substrate on the refractive index of the film. This is due mainly to the substrate's influence on the film's microstructure and also due to the film stress (or strain) as a result of a lattice parameter and thermal expansion coefficient mismatch between the film and substrate. The β-value obtained by Wang et al. [415] (see Table XI) is comparable to that of bulk oxides,

Table XI.   The Dispersion Parameters Obtained from Wemple's Single-Oscillator Model Fitting for BST
Films Deposited at Various Substrate Temperatures [415]

Substrate temperature, T_s (°C)	Single-oscillation energy, E_0 (eV)	Dispersion energy, E_d (eV)	Empirical ratio, β (eV)
350	6.89	20.61	0.214
450	6.69	26.14	0.272
550	6.42	26.33	0.274
650	5.87	24.48	0.255

Table XII.   Refractive Index ($\lambda = 600$ nm) of $Ba_{0.7}Sr_{0.3}TiO_3$ on SiO_2/Si
After Annealing at Different Temperatures [423]

Annealing temperature (°C)	Refractive index ($\lambda = 600$ nm)
500	2.025
550	2.04
650	2.15
750	2.24

$\beta = 0.26 \pm 0.04$ eV [421]. Postdeposition processing of the film also has a substantial effect on β-value [423]. The β-value of $Ba_{0.7}Sr_{0.3}TiO_3$ thin films deposited on a SiO_2/Si substrate increased with increasing annealing temperature. A typical β-value [423] for the film annealed at 750°C is 0.23, which is close to that of bulk $BaTiO_3$ (0.26 eV) and of $SrTiO_3$ (0.25 eV) and correlates well with many bulk oxides [421, 422].

The deposition parameters and postdeposition annealing have a marked influence on the refractive index of the thin films. Kuo and Tseng [423] studied the optical properties of rapidly thermal annealed $Ba_{0.7}Sr_{0.3}TiO_3$ thin films deposited on SiO_2/Si substrates prepared by rf-magnetron sputtering. Kuo and Tseng obtained a refractive index from a reflectance spectrum using an n&k analyzer (n&k Technology, Inc.), which fits the observed experimental data using the Forouhi–Bloomer (FB) physical model [417–419]. This physical model describes the optical constants of crystalline semiconductor dielectrics based on the single-electron model assuming the electronic transition between the valence band and the conduction band, which includes a finite lifetime (τ) of the excited electronic state. The dispersion relations are then derived as follows [417, 418]. The extinction coefficient is

$$k(E) = \frac{A(E - E_g)^2}{E^2 - BE + C} \tag{26}$$

where $A = \text{const} \times |\langle \sigma^* | x | \sigma \rangle|^2 \gamma$, $B = 2(E_{\sigma*} - E_\sigma)$, and $C = (E_{\sigma*} - E_\sigma)^2 + \hbar^2 \gamma^2 / 4$. The refractive index dispersion relation is

$$n(E) = n(\infty) + \sum_{i=1}^{q} \frac{B_{0_i} E + C_{0_i}}{E^2 - B_i E + C_i} \tag{27}$$

where q is an integer,

$$B_{0_i} = \frac{A_i}{Q_i} \left[-\frac{B_i^2}{2} + E_g B_i - E_g^2 + C_i \right]$$

$$C_{0_i} = \frac{A_i}{Q_i} \left[(E_g^2 + C_i) \frac{B_i}{2} - 2 E_g C_i \right]$$

$$Q_i = \frac{1}{2} \left(4C_i - B_i^2 \right)^{1/2}$$

In this physical model, $n(\infty) \neq 1$, in contrast to the classical dispersion theory. The reflectance spectrum is simulated using the equation

$$R(E) = \frac{[n(E) - 1]^2 + k^2(E)}{[n(E) + 1]^2 + k^2(E)} \tag{28}$$

These equations indicate that the variation with energy of the real part, n (index of refraction), and imaginary part, k (extinction coefficient), of the complex index of refraction is obtained in terms of five physical parameters [417, 418]:

1. Index of refraction at high energies.
2. Optical band gap energy E_g.
3. The energy difference between the bonding and antibonding states ($E_{\sigma*} - E_\sigma$).
4. The lifetime τ of the electrons involved in the transition.
5. A, a quantity related to the position matrix element and τ.

In this model, the $n(E)$ and $k(E)$ are related by Kramers–Kronig [417] analysis.

Kuo et al. [423] also observed that the refractive index increases as the temperature of annealing is increased, in contrast to as-deposited film (Table XII). The refractive index of the samples annealed at higher temperature is 0.25% (at 500°C) to 10.7% (at 750°C) higher than that of the as-deposited samples. Kuo et al. [423] ascribed the increase in refractive index to increases in the film's crystallinity and packing density.

The dispersion parameters have been deduced using Wemple's single-oscillator [421] model. Their results showed that the E_0 is almost independent of annealing temperature ($E_0 \sim 6$ eV). On the other hand, the annealing temperature significantly affects E_d in a manner qualitatively similar to the refractive index. As the annealing temperature increased, the E_d value also increased, for example, $E_d = 16.6$ eV for the

500°C annealed sample and $E_d = 21.8$ eV for the 750°C annealed sample.

The compositional dependence of the refractive index has been investigated by many authors [424–426]. Panda et al. [424] deposited BST thin films of various compositions by rf-magnetron sputtering on silicon and Corning glass substrates. Through transmittance and reflectance measurements, they showed that the refractive index is a function of thin film composition. They derived the refractive index from the transmittance spectra following the method of Manifacier et al. [420] using

$$n = \left[N + \left(N^2 - n_0^2 n_s^2 \right)^{1/2} \right]^{1/2} \tag{29}$$

where

$$N = \frac{(n_0^2 + n_s^2)}{2} + 2 n_0 n_s \left\{ \frac{T_{\max} - T_{\min}}{T_{\max} T_{\min}} \right\} \tag{30}$$

where n_s is the refractive index of the substrate, n_0 is that of the surrounding medium (air $n_0 = 1$), and $T_{\max}$ and $T_{\min}$ are the maximum and minimum transmittances in the oscillation. The refractive index of $Ba_{0.8}Sr_{0.2}TiO_3$ thin films deposited on Corning glass derived using the foregoing method decreased with increasing wavelength similar to other investigations [415, 423] and were well fitted to the Cauchy equation. The dispersion parameters of refractive index in the low-absorption region has been fitted to the single-term Sellmeier relation [422]

$$n^2(\lambda) - 1 = \frac{S_0 \lambda_0^2}{1 - (\lambda_0/\lambda)^2} \tag{31}$$

where S_0 is the oscillator strength, λ_0 is the average oscillator position, and E_0 is the oscillator energy given by $E_0 = hc/\lambda$. The parameters S_0 and λ_0 in Eq. (31) can be obtained experimentally by plotting $1/(n^2 - 1)$ vs. $1/\lambda^2$. The slope of the resultant straight line gives $1/S_0$, and the infinite wavelength intercept gives $1/(S_0 \lambda_0)$. Panda et al. [424] obtained a single oscillator energy of $E_0 = 6.76$ eV, which is slightly higher than that reported by Kuo et al. [423], and $S_0 = 10^{14}/\text{m}^2$. Panda et al. also investigated the dependence of the refractive index of the BST thin films on film composition. Their results (Fig. 45) show that the refractive index decreases with the increase in Ba percentage in the BST. The refractive index drop is drastic above a Ba content of 80%. These authors also measured the refractive index of BST thin film deposited on silicon by ellipsometry. The refractive index of these samples also decreased with increasing Ba content. Also, the notable difference is the refractive index is lower than that of BST on Corning glass. The decrease in refractive index with increase in Ba-content is due to the difference in the morphology and crystallinity of the film between a Ba-poor and a Ba-rich film. Similarly, the difference in refractive index between the films deposited on two different substrates can also be explained with the similar reasoning of differences in film orientation and morphology with deposition on different substrates.

Suzuki et al. [425] used spectroscopic ellipsometry for the optical characterization of $Ba_{0.7}Sr_{0.3}TiO_3$ thin films prepared by the sol-gel method after postdeposition annealing at various

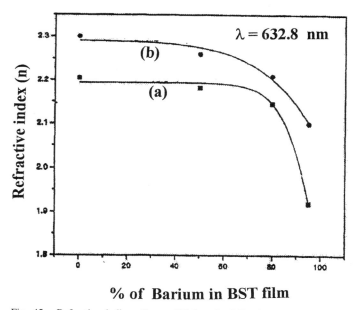

Fig. 45. Refractive indices ($\lambda = 632.8$ nm) of $Ba_x Sr_{1-x}TiO_3$ films as functions of film composition (a) from transmittance measurements (b) from ellipsometric measurements. Reprinted with permission from [424], © 1998, Elsevier Science.

temperatures and atmospheres (O_2 or N_2). Spectroscopic ellipsometry is a sensitive optical technique that has been widely recognized as a reliable tool for characterizing structural and optical properties of thin film materials [425–430]. Ellipsometric measurement involves measuring changes in the amplitude and phase of a linearly polarized monochromatic incident light on an oblique reflection from the sample surface. Experimentally, the measured ellipsometry parameters (φ and Δ), which are related to the optical and structural properties of the samples, are defined by [425, 429, 430]

$$\rho = \frac{R_p}{R_s} = \tan \Psi \cdot \exp(i\Delta) \tag{32}$$

where R_p and R_s are the complex reflection coefficients of the light, polarized parallel (p) and perpendicular (s) to the plane of incidence, respectively. For a layered sample, the measured spectrum from this method can be analyzed using an appropriate fitting model, constructed based on the sample structure. In this model, unknown parameters (e.g., thickness and optical constants of a layer) may be designated as fitting variables. These parameters are then determined by minimizing the difference between the calculated spectrum generated from the fitting model and the experimental spectrum using a nonlinear least squares fitting algorithm [425, 429]. The refractive index dispersion curve obtained by Suzuki et al. [425] for $B_{0.7}Sr_{0.3}TiO_3$ thin film annealed at different temperatures and atmospheres (O_2 and N_2) indicates that the refractive index increased with increasing annealing temperature. The film annealed in O_2 showed a more drastic increase, indicating better crystallinity of the film, which was also confirmed by X-ray diffraction [425]. Also, the film annealed in N_2 atmosphere had a lower refractive index than that annealed in O_2 atmosphere. The refractive

Table XIII.   Dispersion Parameters of $Ba_{0.5}Sr_{0.5}TiO_3$ Thin Films Deposited
on Sapphire Substrate

Material	S_0	E_0 (eV)	E_0/S_0
$BaTiO_3$ [422]	0.91, 0.84	5.88, 5.57	6.4, 6.6
			7.4 [433]
$SrTiO_3$ [422]	0.89	5.74	6.5
			7.4 [433]
$Ba_{0.5}Sr_{0.5}TiO_3$ [432]	1.04	7.16	6.9

Reprinted with permission from [432], © 1997, Elsevier Science.

indexes of amorphous $Ba_{0.67}Sr_{0.33}TiO_3$ thin films obtained by Zhu et al. [431] using ellipsometric spectroscopy varied from 1.95–1.75 in the range of 400–1000 nm, which is much lower than the crystalline BST thin films.

Techeliebou et al. [432] reported the optical properties of $Ba_{0.5}Sr_{0.5}TiO_3$ thin films deposited on sapphire substrates using pulsed laser deposition techniques. The refractive index was derived from the transmittance spectrum following Swanepoel's [416] method. The refractive index that they obtained was around 2.096–2.046 in the wavelength range of 300–800 nm. The dispersion data of the refractive index have been interpreted in terms of an individual dipole oscillator model [421, 422]. Table XIII shows the dispersion parameters of the complex refractive index of single crystals of $BaTiO_3$, $SrTiO_3$ [422, 433], and $Ba_{0.5}Sr_{0.5}TiO_3$ polycrystalline film [432]. The BST thin films showed higher values for both S_0 and E_0 than those of single-crystal $BaTiO_3$ and $SrTiO_3$ [422]. Based on these results, Techeliebou et al. [432] concluded that the assumption that the lowest-energy oscillator is the largest contributor to the refractive index may not be adequate for the analysis of the optical behavior of BST thin films. On the other hand, the ratio of E_0 to S_0 has been predicted to be constant for a wide range of materials in different structural phases and compositions [422]. The E_0/S_0 ratio depends on the characteristics of interband transitions and enters directly into the evaluation of the electro-optical and nonlinear optical constants of the crystal [432]. The refractive index dispersion coefficient, E_0/S_0, is found to be 6.9, which is also higher than the value reported in [422] for single crystals of $BaTiO_3$ and $SrTiO_3$. Mansingh et al. [433] reported a value of 7.4 for microcrystalline $BaTiO_3$ and $SrTiO_3$. The thin film value obtained is between that of single crystalline and microcrystalline bulk $BaTiO_3$ and $SrTiO_3$, which is suggestive of the difference in microstructure.

Thielsch et al. [434] studied the optical properties of laser-deposited $Ba_xSr_{1-x}TiO_3$ thin films grown on a MgO (001) substrate. The refractive indexes of the films were calculated from the reflectance spectrum according to the equation [435]

$$n_f(\lambda) = n_s(\lambda) \frac{1 + \sqrt{R_{max}(\lambda)}}{1 - \sqrt{R_{max}(\lambda)}} \qquad (33)$$

where $n_s(\lambda)$ is the refractive index of the bare substrate (determined from the reflection spectra of the mirror-like polished

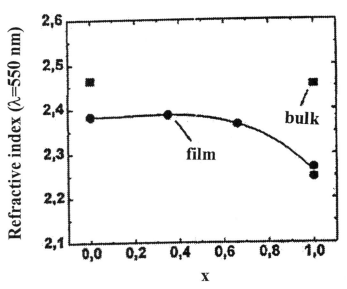

Fig. 46.   Variation of refractive indices n ($\lambda = 550$ nm) of $Ba_xSr_{1-x}TiO_3$ films with various compositions. (■) Bulk values; (●) film values. Reprinted with permission from [434], © 1997, Elsevier Science.

MgO (001) substrate surface [436, 437]) and R_{max} represents the maximum reflection spectrum of the films. The refractive indexes reported by Thielsch et al. [434] for $Ba_{0.65}Sr_{0.35}TiO_3$ deposited on MgO (001) at different substrate temperatures show a mixed trend. The refractive index ($\lambda = 550$ nm) of 760°C ($n \sim 2.25$) deposited film was lower than that of 730°C ($n \sim 2.28$) deposited film; it was at its maximum at 800°C ($n \sim 2.36$) and decreased thereafter (at 840°C, $n \sim 2.24$). The authors explained the observation of maximum refractive index for the 800°C grown film as due to the growth of films in a closely packed columnar grain mode with smooth surfaces of low roughness. For lower and higher deposition temperatures, the difference has been attributed to imperfect growth, resulting in films of reduced packing density at these temperatures [434]. Thielsch et al. [434] also investigated the influence of film composition on the refractive index (Fig. 46) by depositing a series of films of various compositions at a fixed substrate temperature of 800°C. The refractive index showed a decrease as the Ba content in the film composition increased. Similar results were later reported by Panda et al. [424]. The refractive index of the film was lower than the $SrTiO_3$ and $BaTiO_3$ bulk refractive index, as shown in Figure 46.

Thielsch et al. [434] studied the dependence of dispersion parameters E_0 and E_d on the deposition temperature. In the case of optimal growth temperature (800°C), the value of oscillator energy, E_0, was found to be 5.36 eV, which agrees well with reported bulk values for $BaTiO_3$ (5.63 eV) and $SrTiO_3$ (5.58 eV) [421]. These authors [434] also found that E_0 depended only slightly on the deposition temperature for lower growth temperature, but that dispersion energy, E_d, was significantly affected by the deposition temperature in a manner

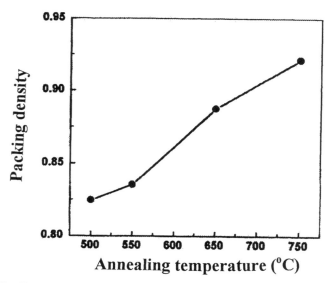

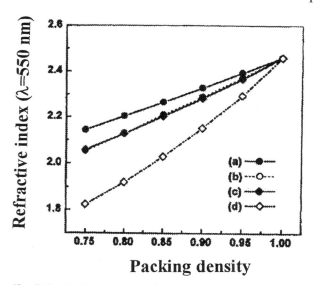

Fig. 47. Packing density of $Ba_{0.7}Sr_{0.3}TiO_3$ films annealed at various temperatures. Reprinted with permission from [423], © 1999, Elsevier Science.

Fig. 48. Refractive index n ($\lambda = 550$ nm) vs. packing density for two microstructures calculated according to the Bragg–Pippard model: (a) close-packed columns with filled voids; (b) close-packed voids with empty voids; (c) columnar growth with filled with voids; (d) columnar growth with empty voids. Reprinted with permission from [423], © 1999, Elsevier Science.

qualitatively similar to that of refractive index. Similar observations were reported by Kuo et al. [423].

8.2. Packing Density of Film

The packing density of thin film can be determined from the experimentally derived refractive index [423, 438]. The packing density is defined as the ratio of average film density (ρ_f) to bulk density (ρ_b),

$$p = \frac{\rho_f}{\rho_b} \qquad (34)$$

The correlation between packing density and refractive index can be expressed as [439, 440]

$$p = \frac{\rho_f}{\rho_b} = \left(\frac{n_f^2 - 1}{n_f^2 + 2}\right) \times \left(\frac{n_b^2 + 2}{n_b^2 - 1}\right) \qquad (35)$$

where n_f denotes the refractive index of the film and n_b denotes the refractive index of the bulk. Kuo et al. [423] calculated the packing density by fixing the bulk refractive index of BST as 2.465 at $\lambda = 550$ nm. This is plotted as a function of annealing temperature in Figure 47, which indicates an increase in packing density with increasing annealing temperature. This implies that the observed increase in refractive index with increasing annealing temperature is due to an increase in the packing density of the film. This is due mainly to the reduction in air-filled pores. Air has the lowest refractive index and effectively reduces the average refractive index of the film.

To understand the growth morphology and relation between the refractive index of the film, n_f, and that of the bulk, n_b, which contain a certain volume part of voids, the effective medium mode of Bragg and Pippard [441], which takes into account the growth morphology of the film [442, 443], was used by Kuo et al. [423] and Thielsch et al. [434]. For a closely packed columnar grain morphology, which contains a certain

part of the voids, the relation between the refractive index of the film n_f and that of the bulk n_b is given by

$$n_f^2 = \frac{n_b^4 p + (2 - p)n_b^2 n_p^2}{(2 - p)n_b^2 + pn_p^2} \qquad (36)$$

and that of the columnar structure of reduced density and thus more widely separated grains is given by

$$n_f^2 = \frac{(1 - p)n_p^4 + (1 + p)n_p^2 n_b^2}{(1 + p)n_p^2 + (1 - p)n_b^2} \qquad (37)$$

where n_p is the refractive index of the voids which is 1 for empty voids or 1.33 in the case of water-filled voids [441–443].

Kuo et al. [423] calculated the refractive index and packing density using the Bragg–Pippard model [441] for two different microstructures and is presented in Figure 48. The figure reveals that a good fitting with the experimental results is obtained by the closely packed columns with empty voids and columnar growth with water-filled voids similar to that reported by Thielsch et al. [434].

8.3. Optical Band Gap

The optical band gap energy (E_{gap}) of the film was deduced from the spectral dependence of the absorption coefficient $\alpha(\lambda)$ by applying the Tauc relation [444]

$$\alpha(\lambda)E \approx A(E - E_{gap})^{1/r} \qquad (38)$$

where A is a constant and $r = 1/2, 2, 3/2,$ or 3 for allowed direct, allowed indirect, forbidden direct, and forbidden indirect electronic transition, respectively [445, 446]. The absorption

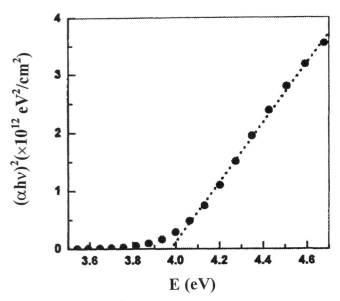

Fig. 49. Plot of $(\alpha E)^2$ vs. photon energy E for 750°C-annealed $Ba_{0.7}Sr_{0.3}TiO_3$ film deposited on SiO_2/Si. Reprinted with permission from [423], © 1999, Elsevier Science.

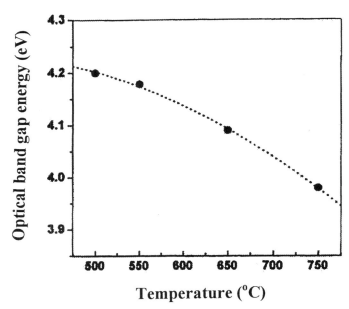

Fig. 50. Optical band gap energies of $Ba_{0.7}Sr_{0.3}TiO_3$ films deposited on SiO_2/Si and annealed at various temperatures. Reprinted with permission from [423], © 1999, Elsevier Science.

coefficient is determined from the extinction coefficient obtained from the experiments using the equation

$$k(\lambda) = \frac{\alpha(\lambda)\lambda}{4\pi} \qquad (39)$$

and the optical band gap can be determined by plotting $(\alpha E)^{1/r}$ vs. E and extrapolating the linear portion of the curve to $(\alpha E)^{1/r} = 0$. Figure 49 shows the $(\alpha E)^2$ vs. E plot for the $Ba_{0.7}Sr_{0.3}TiO_3$ (considering $r = 1/2$) reported by Kuo et al. [423]. In the high-energy region of the curve $(\alpha E)^2$ vs. E plot is linear and in the low-energy region the spectrum deviated from the straight line. Notably, straight line behavior of the $(\alpha E)^2$ vs. E plot in the high-energy range was taken to be the prime evidence for a direct allowed transition between the highest occupied state of the valence band and the lowest unoccupied state of the conduction band. These authors [423] also showed that the optical band gap energy varied with the annealing temperature of the film (Fig. 50). The optical band gap decreased from 4.2 to 3.98 with an increase in annealing temperature up to 750°C. Some studies [447] suggested that the decrease in the optical band gap with the increase in annealing temperature might be attributed to the lowering of the interatomic spacing, which reduced the polarization and electron hole interactions. Wang et al. [415] used the wavelength absorption data for BST thin films prepared on Al_2O_3 and quartz substrates to obtain the fundamental absorption of 3.5 eV, which is close to the energy gap of $BaTiO_3$ films [448, 449] and $SrTiO_3$ bulk materials [450]. Recent investigations involving the dependence of optical band gap energies of polycrystalline-sputtered $BaTiO_3$ films on the grain size suggest a size effect on the optical energy gap [451]. Those investigators demonstrated that optical band gap energy decreased with increasing grain size. The energy gap of $BaTiO_3$ with a grain size of 8 nm is 3.68 eV and

decreases to 3.53 eV for a grain size of 35 nm. Qualitative explanation has been given by Lu et al. [451] following Harrison's [452] description of energy bands. The increase in grain size was accompanied by a decrease in lattice parameters, and so the rise of the highest valence band resulted in a reduction of the energy gap. Moreover, deviation also might be caused by a tensile-stress–induced distortion of the energy band and the effect of the crystal boundary on the band [451]. Kuo et al. [423] showed that the optical band gap energy decreased from about 4.20 eV for a mean grain size of 8.8 nm to 3.98 eV for a grain size of about 10.8 nm, which is higher than the E_{gap} of the polycrystalline-sputtered $BaTiO_3$ film [451] with a 3.68 eV for a mean grain size of 8 nm to 3.53 eV for a grain size of about 35 nm. Thielsch et al. [434] and Panda et al. [424] also observed that the optical band gap energy decreases with increasing barium content in the film (Fig. 51). The values reported by Thielsch et al. [434] are 0.1 to 0.2 eV higher than the corresponding bulk values of $BaTiO_3$ and $SrTiO_3$ ($E_g = 3.2$ and 3.6 eV, respectively [421]). Panda et al. [424] reported slightly lower values in the case of $SrTiO_3$ films and higher values in the case of $BaTiO_3$. The energy gap variation showed a simple composition-related shift toward lower energies as the Ba concentration increased in the composition. Cheng [453] reported an E_g value of 2.8 eV for the laser-deposited $(Sr_{0.2}Ba_{0.8})TiO_3$ thin films, and Techeliebou et al. [432] reported an E_g value of 3.96 eV for the composition $Ba_{0.5}Sr_{0.5}TiO_3$. Different values have been reported for BST films with different compositions and deposited by different techniques and on different substrates.

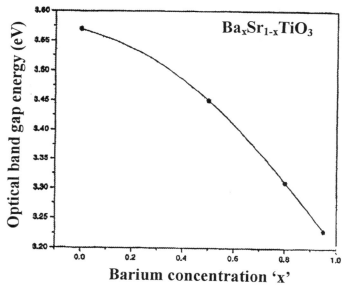

Fig. 51. Optical band gap energies of $Ba_x Sr_{1-x}TiO_3$ films as a function of Ba concentration. Reprinted with permission from [424], © 1999, Elsevier Science.

8.4. Homogeneity of the Film

The optical homogeneity of the films can be analyzed by using the near-normal reflectance of the films and the bare substrate surface. The inhomogeneous refractive index (Δn) along the film thickness was determined using the empirical relation [454]

$$\Delta n = n_o - n_i = n_{av} \frac{R_f - R_s}{4.4 R_s} \qquad (40)$$

where n_o is the refractive index at the air–film interface, n_i is the refractive index at the film–substrate interface, and n_{av} is the average index of the film. R_s and R_f can be computed form the measured reflectance of the bare and deposited substrates, R_{SO} and R_{fo}, using the equations [455]

$$R_{SO} = \frac{2 R_s}{1 + R_s} \qquad (41)$$

and

$$R_{fo} = R_f + \frac{R_s(1 - R_f)^2}{1 - R_s R_f} \qquad (42)$$

where Δn can be positive or negative according to the gradation in the film [454]. Negative Δn values indicate that the refractive index decreases from the substrate to the top surface of the film and vice versa. The degree of inhomogeneity of the films, $\Delta n / n_{av}$, calculated by Wang et al. [415] from the points of reflectance minima around 600–700 nm (Fig. 52), where the R_{SO} (R_s) is almost independent of λ, is tabulated in Table XIV. It is found to vary from -7.3% to -4.3% as the annealing temperature of the film increased from 350°C to 650°C.

The results obtained by various authors indicate that the optical properties of the BST thin film depend on deposition parameters, postdeposition processing methods, annealing temperature, film composition, substrate type, microstructure,

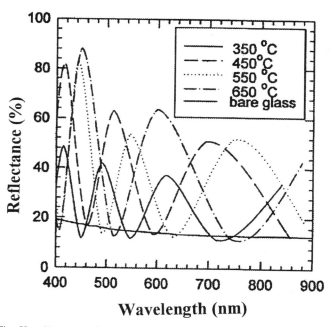

Fig. 52. Observed reflectance spectra as a function of wavelength for bare substrate (R_{SO}) and BST films (R_{fo}) deposited at various substrate temperatures [415].

Table XIV. Optical Inhomogeneity Analysis for Films Deposited at Various Temperatures [415]

Substrate temperature, T_s (°C)	Wavelength (nm)	$\Delta n / n_{av}$
350	715	−7.3
450	593	−3.3
550	630	−3.7
650	739	−4.3

grain size, and the film packing density. Different authors have reported different values because of the variations in these parameters.

9. OTHER POSSIBLE APPLICATIONS

9.1. Hydrogen Gas Sensors

With the increasing global awareness and urgency for environment protection and pollution control, the gas sensor research activities especially the semiconductor based gas sensors have been stimulated extensively in recent years [431, 456]. The applications of the gas sensors range from air-to-fuel ratio control in combustion processes (e.g., automotive engines, industrial furnaces) to leakage detection of inflammable and toxic gas in domestic and industrial environments. Semiconductor gas sensors not only have the advantages small size, simple operation, and high sensitivity, but they also show good potential for integration with electric control into a single Si chip. Such integrated gas sensors would be suitable for large-scale production and could greatly reduce the production cost [431, 456].

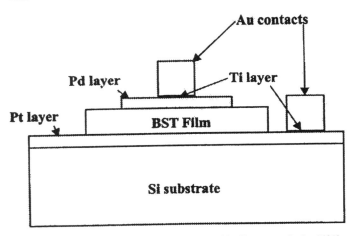

Fig. 53. Schematic diagram of the MFM BST thin film sensor device [431].

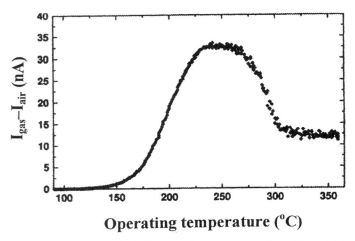

Fig. 54. The difference in dc ($I_{gas} - I_{air}$) vs. temperature for a 125-nm-thick film annealed at 475°C in air for 1 hour. Reprinted with permission from [456], © 1998, American Institute of Physics.

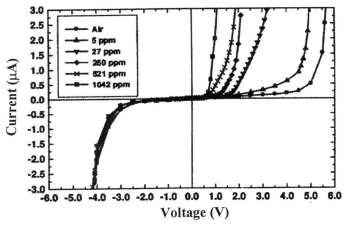

Fig. 55. $I-V$ characteristics as a function of H_2 gas concentration in air for the sample annealed at 475°C in air for 1 hour. Reprinted with permission from [456], © 1998, American Institute of Physics.

Zhu et al. [431, 456] were the first to fabricate hydrogen gas sensors using amorphous BST thin films with a polarization potential as large as 4.5 V at 1000 ppm H_2 in air, which is about seven times larger than the best value reported in the literature [457] for the hydrogen gas response of palladium gate field effect transistor. These authors used the MFM-type thin film sensor schematically illustrated in Figure 53, in which the ferroelectric material is amorphous $Ba_{0.67}Sr_{0.33}TiO_3$. They prepared the ferroelectric film by the sol-gel method and studied the $I-V$ behavior of this type of sensor in air and in different concentrations of H_2 gas mixed in air. They observed that the $I-V$ curve for the film annealed at 475°C showed negligible leakage current of -5 to 5 V, which is typical electric insulating behavior of amorphous ferroelectric materials and is excellent for sensor applications. As the annealing temperature increased, the $I-V$ curves showed increasingly leaky behavior. The increased leakage current after annealing at higher temperature is attributed to the polycrystalline grain growth and the open grain boundaries in the thin film [431, 456]. The optimum operating temperature of the sensor was determined by

sweeping the operating temperature from 100–360°C in both air and 1000-ppm H_2 in air. The difference in dc $I_{gas} - I_{air}$, which is a measure of the hydrogen gas sensitivity of the sensor, is plotted in Figure 54 [431, 456]. The change of current in the hydrogen gas environment increased steadily up to around 250°C, then fell and reached a stable value above 300°C. This phenomenon has been explained by Zhu and co-workers [431, 456] based the fact that the increase in temperature brings an increased dissociation of H_2 molecules into H^+ ions at the interface between the palladium layer and the amorphous BST film. With further temperature increases after the peak, more electrons in this film were thermally activated and neutralized as part of the ionized H^+ ions. Also, the H^+ ions at the interface dislodged with increasing H^+ ion mobility, resulting in the decrease of dc. Figure 54 infers that the high gas sensitivity can be observed in the temperature range of about 230–270°C [431, 456].

Zhu and co-workers [431, 456] studied the hydrogen gas effect on the induced polarization of amorphous ferroelectric $(Ba,Sr)TiO_3$ thin films systematically by the dc $I-V$ measurements after exposing the device to different concentrations of H_2 in air (Fig. 55). In the forward bias, the turn-on voltage shifted consistently toward the lower voltage with increased H_2 concentration in air. The turn-on voltage shifted from 4.5 V in air to 0.5 V in 1000-ppm H_2 in air. Figure 55 also indicates that in the reverse bias, all of the curves converge together indicating that there is no hydrogen gas effect under the reverse bias. Under the forward bias, all curves exhibit the Schottky diode current characteristic, and with increasing hydrogen gas concentration, the turn-on voltage is shifted to the left, indicating that the interfacial polarization potential increases with H_2 concentration in air, then reduces the required turn-on forward bias applied voltage. The voltage shift is about 4.0 V at about 1000-ppm H_2 gas in air and about 0.4 V at 5-ppm H_2 gas in air. The foregoing data reported by Zhu and co-workers [431, 456] gave strong supporting evidence as to the advantage of using amorphous ferroelectric thin films that have relatively high dielectric

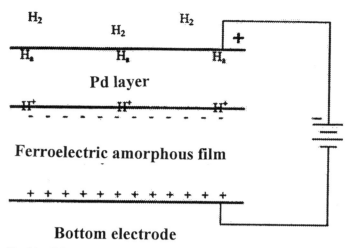

Fig. 56. Schematic model to explain enhanced hydrogen-induced interfacial polarization potential between the Pd layer and the amorphous ferroelectric thin film. Reprinted with permission from [456], © 1998, American Institute of Physics.

constant over the conventional low dielectric constant materials in this type of device.

The annealing temperature of the thin film has a marked effect on H_2 gas sensitivity. Zhu and co-workers [431, 456] reported that the voltage shift is higher for the film annealed at 475°C, which is amorphous, than that for the film annealed at 525°C. This is because amorphous BST thin film (a 475°C annealed sample) can effectively stop the H^+ ion at its interface with the Pd layer and thus enhance the induced dipole polarization at the interface. With increasing annealing temperature, BST films develop the polycrystalline phase with more open grain boundaries, and thus accumulation of H^+ at the interface progressively decreases, because very small-sized H^+ ions can easily diffuse through the polycrystalline thin film with big grains and open grain boundaries. Hence, the voltage shift is correspondingly reduced [431, 456].

The Ti concentration has a marked effect on the gas sensitivity characteristics of the film. The voltage shift is maximum for stoichiometric Ti ($= 1$) concentration and decreases if the concentration is increased or decreased. This is because the stoichiometric composition has the lowest leakage current and the Ti-deficient sample (0.98) has the highest leakage current value. This indicates that the lower the leakage current, the higher the gas sensitivity. With high electric conduction, the probability of recombination of H^+ ions with electrons increases, and the buildup of potential at the interface decreases. The sample with a lower leakage current has less background current, thus improving its H_2 gas sensitivity [431, 456].

The enhanced interface polarization potential has been explained by Zhu and co-workers [431, 456] based on the hydrogen interface blocking model, as shown in Figure 56. The top Pd layer has a positive bias, whereas the bottom electrode is negatively biased. When the H_2 molecules reach the Pd surface, they become hydrogen atoms because of the catalytic effect of Pd on hydrogen. The hydrogen atoms are then ionized under the Pd metal layer and the amorphous ferroelectric BST

film. Hydrogen ions (proton) are blocked at this interface of the crack-free amorphous film that does not provide any connected channels for atomic diffusion, even in the case of a small-sized proton. This accumulation of positive hydrogen at the interface results in the buildup of the charges, which in turn causes a dipole polarization potential across the space charge layer at the interface. Since the dielectric constant of the amorphous ferroelectric BST film is higher than that of SiO_2, the polarization potential is enhanced. In the reverse bias, hydrogen ions are driven out of the Pd layer, and consequently no dipole polarization potential can be built up at this interface between the Pd layer and the amorphous ferroelectric BST film, which is indicated by the merger of all the curves in the reverse bias region of the $I–V$ plot [431, 456].

The foregoing study by Zhu et al. [431, 456] shows yet another possible use of BST thin film in the field of sensors, which may trigger a surge of studies in the near future.

9.2. Pyroelectric Sensors

Infrared uncooled focal plane arrays (UFPAs) have garnered much attention because of the easy realization of thermal imaging and has shown a wide market of great potential [458–460]. The UFPAs based on resistive bolometers have already been developed and are becoming commercially available [458, 461, 462]. The disadvantage of the resistive bolometer is a voltage drop across the detector resistor, making current consumption nonnegligible. This causes a temperature increase and requires a thermal stabilizer [458]. In the manufacture of devices for infrared detection and thermal imaging, uncooled pyroelectric materials such as ferroelectrics are an alternative to the cooled semiconductor detectors commonly used today. Ferroelectric thin films are considered to have great potential, because they would result in a more economical design with better performance than ceramic or single-crystal detectors. The eventual challenge would be to deposit the detector as a thin film integrated into silicon-based circuitry [463]. Ferroelectric BST has been identified as a very promising material because of its composition-dependent Curie temperature [464, 465]. As a result, maximum infrared response can be obtained at room temperature. A high pyroelectric coefficient ($>600\ \mu C/m^2\ K$) coupled with a low dielectric constant (<1200) have been reported in these materials [466]. Ferroelectric materials can be used as both pyroelectric detectors and bolometers. In its application as a pyroelectric detector, the change in polarization charge of the material with respect to the rate of change in temperature is measured. The high value of the dielectric constant is important in a small amount of highly integrated pyroelectric (PE) array sensors to suppress switching noise by a good capacitance match to the preamplifier [458]. In the dielectric bolometer mode, the dielectric constant change against temperature is detected through the capacitance change in the thin film. For ferroelectric bolometers, the ferroelectric material should show a large change in dielectric constant with temperature. Increased dielectric loss and the risk of progressive depoling are the disadvantages of ferroelectrics as pyroelectric detectors.

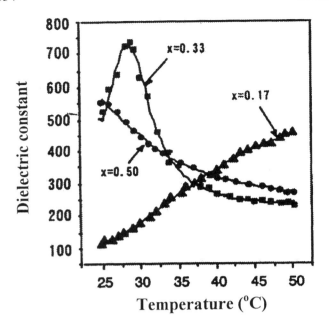

Fig. 57. Dielectric constant vs. temperature of $Ba_{1-x}Sr_xTiO_3$ thin films of $x = 0.17$, 0.33, and 0.5 deposited at $T_s = 500°C$ and 0.1 torr. Reprinted with permission from [458], © 1999, Elsevier Science.

On the other hand, its advantage is its high pyroelectric coefficient at any desired temperature achieved by simply adjusting the composition of the ferroelectrics [466].

Noda et al. [458] deposited crack-free highly oriented BST thin films by pulsed laser deposition on the Pt/Ti/SiO$_2$/Si (100) substrates in a different ambient. Temperature dependence of the dielectric constant of the BST films showed a sharp change with a peak around room temperature (Fig. 57). A peak in dielectric constant exists around 25°C in the film with Ba/Sr = 1 and shifts to above 50°C in the films with Ba/Sr = 6. For Ba/Sr = 2, the peak appears near 27°C and is very sharp. The differential dielectric constant against temperatures is the largest for Ba/Sr = 2, about 100/K, which corresponds to 10% K, where the absolute capacitance is about 640 pF at the pixel capacitor size of about 280 μm square. This value is equivalent to a pyroelectric coefficient of 1.8×10^{-7} C/cm^2 K in thick film (0.5 μm), a capacitance area of 8×10^{-4} cm^2, and a bias field of 20 KV/cm, which is almost same as that of the LiTaO$_3$ single crystal.

Sengupta et al. [466] showed that by adjusting the composition of BST-MgO ceramics, the maximum reversible pyroelectric coefficient can be centered around room temperature and the corresponding Curie temperature can also designed to be at room temperature. Unlike undoped BST, the doped BST-MgO materials have low dielectric loss (<0.01) at operating temperature. This resulted in a higher material figure of merit (FOM), defined as [466]

$$FOM = \frac{p}{d\varepsilon_r \tan\delta} \tag{43}$$

where p is the peak pyroelectric coefficient, d is the thickness of the material, ε_r is the relative dielectric constant, and $\tan\delta$

is the dielectric loss. The pyroelectric coefficient is calculated from the following equation [466]:

$$P = -IA\frac{dt}{dT} \tag{44}$$

where I is the current in amperes, A is the metal contact area, and dT/dt is the rate of change of temperature with time. The pyroelectric coefficient; the corresponding temperature of peak pyroelectric coefficient, T (°C); the Curie temperature, T_c (°C); the pyroelectric coefficient at room temperature, p_{RT}; the dielectric properties, ε_r and $\tan\delta$, at 200 Hz; and the FOM of the tape cast material are listed in Table XV [466]. The table indicates that the pyroelectric coefficient is maximum for a composition of $Ba_{0.64}Sr_{0.36}TiO_3$. The peak in $p(T)$ is at a higher temperature for $Ba_{0.64}Sr_{0.36}TiO_3$ than the $Ba_{0.94}Sr_{0.06}TiO_3$/0.25 wt% MgO ceramic and magnitude is also larger. Since the dielectric constant and the loss tangent are also much higher for the $Ba_{0.64}Sr_{0.36}TiO_3$ sample, the overall FOM of the sample is less than that of the $Ba_{0.94}Sr_{0.06}TiO_3$/0.25 wt% MgO samples. Table XV also gives the pyroelectric and dielectric properties of the material in the thin film form. The figure of merit and pyroelectric coefficient are maximum for $Ba_{0.64}Sr_{0.36}TiO_3$. Also, compared to tape cast samples, thin films exhibited better FOM because of the reduced thickness according to Eq. (45). The grain size of the films accompanied with the orientation of the grains will contribute toward pyroelectric behavior of the thin films [466].

Cheng et al. [467] deposited tetragonal $Ba_{0.8}Sr_{0.2}TiO_3$ film with columnar grains 100–200 nm in diameter on Pt/SiO$_2$/Si substrates using 0.05 M precursor solution by sol-gel processing. They obtained a pyroelectric coefficient of 4.586×10^{-4} C/m^2 K at 33°C, which is larger than that of a LiTaO$_3$ single crystal ($p = 2.3 \times 10^{-4}$ C/m^2 K), which is one of the material currently used for infrared detectors [459]. The maximum value of FOM is around 1.47×10^{-5} C/m^3 K at 33°C. Compared to LiTaO$_3$ single crystal, the BST thin films showed a lower FOM because of high dielectric constant.

The foregoing studies on the pyroelectric sensors based on BST thin films of appropriate composition can be developed to replace the presently used single-crystal LiTaO$_3$-based pyroelectric detectors. However, the low FOM of BST thin films due to the large dielectric constant must be overcome by suitable compositional and processing adjustments.

9.3. A Dielectric Layer in a Thin Film Electroluminescent (TFEL) Device

The TFEL device is a multilayer capacitor structure of metal-electrode/phosphor/insulator/indium tin oxide (ITO) transparent electrode/glass type [468]. The function of the insulator layer is to suppress the avalanche breakdown of the phosphor layer, which are connected serially. This puts an additional requirement on the insulator layer along with high dielectric constant that it be transparent in the phosphor emission region. The voltage drop across the insulating layer causes a decrease in the voltage applied in phosphor layers. The amount of voltage

Table XV. Maximum Pyroelectric Coefficients (p_{peak}), Corresponding Temperatures of Peak Pyroelectric Coefficient T (°C), Curie Temperature T_c (°C), the Pyroelectric Coefficient at Room Temperature (24°C) (p_{RT}), the Dielectric Properties (ε_r, tan δ) at 200 Hz, and the Figure of Merit of the Tape Cast Materials and Thin Films (Thickness = 3 μm)

Material	p_{peak} (μC/m^2 K)		T (°C)		d (μm)	T_c (°C)		p_{RT} (μC/m^2 K)		ε_r		tan δ		FOM (μC/m^3 K)	
	Bulk	Film	Bulk	Film	Bulk	Bulk	Film	Bulk	Film	Bulk	Film	Bulk	Film	Bulk	Film
Ba$_{0.64}$Sr$_{0.36}$TiO$_3$	950	1120	30	32	120	46	36	759	1940	4478	594	0.012	0.024	0.118	45.36
Ba$_{0.94}$Sr$_{0.06}$Zr$_{0.18}$Ti$_{0.82}$O$_3$	165	2000	50	35	150	34	39	96.5	1260	1818	1119	0.008	0.141	0.044	2.66
Ba$_{0.94}$Sr$_{0.06}$TiO$_3$ + 0.25 wt% MgO	580	258	24	40	140	44	34	593	187	2370	505	0.004	0.008	0.447	15.4
Ba$_{0.94}$Sr$_{0.06}$TiO$_3$ + 0.25 wt% MgO	680	310	10	5	110	42	42	550	129	1370	297	0.007	0.015	0.521	9.65

Reprinted with permission from [466], © 1998, Taylor & Francis, Ltd.

155

Table XVI.   Electrical Properties of BaTiO$_3$ (BT) Films Used in TFEL

	Amorphous BT	Polycrystalline BT	Layered BT
Dielectric constant (1 kHz, sine)	17	140	98
Breakdown field (MV C m^{-1})	2.7	0.8	>2.2
Figure-of-merit (μC cm^{-2})	4.1	9.9	>19

Reprinted with permission from [474], © 1998, Elsevier Science.

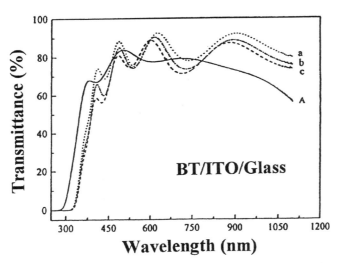

Fig. 58.   Transmission spectra of ITO-coated glass (A) and BaTiO$_3$ films deposited on ITO coated glass and annealed at (a) 650°C for 10 minutes, (b) 650°C for 60 minutes, and (c) 700°C for 60 minutes. Reprinted with permission from [468], © 2000, Elsevier Science.

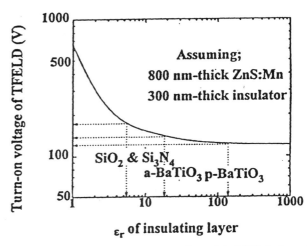

Fig. 59.   Relationship between the turn-on voltage of the TFEL device and the dielectric constant of insulating layer used in the TFEL device. Reprinted with permission from [474], © 1998, Elsevier Science.

drop increases with the decrease of capacitance of insulating layer. To effectively supply a high electric field to the phosphor layer, the voltage drop due to the insulating layers should decrease as much as possible. The dielectric layer with high dielectric constant is very effective to realize TFEL devices with a low driving voltage [469, 470]. Many workers [469, 470] have studied the TFEL device to improve the operating characteristics, but the critical problem of high driving voltage remains unsolved. In general, SiON or Si$_3$N$_4$ has been used as an insulating layer in TFEL panels. The low dielectric constant (5–6) of these insulating layers have resulted in a high driving voltage of 150–200 V.

It has been reported that BaTiO$_3$ has suitable properties for the TFEL device, which needs high transmittance in the visible region, chemical resistivity, and a high dielectric constant [471–473]. Zhang et al. [468] deposited BaTiO$_3$ on ITO-coated glass and showed that in the visible region, the multilayer structure has high transmittance of around 80% (Fig. 58). The relationship between the dielectric constant of the insulating layer in the TFEL device and the turn-on voltage of the TFEL device is shown in Figure 59 [474]. The figure indicates that a higher dielectric constant above 100 is not effective in decreasing the driving voltage of TFEL devices. The BaTiO$_3$ films were investigated in both amorphous and crystalline forms [475, 476].

The amorphous films were highly insulating, whereas the crystalline films were more leaky because of the grain boundaries, which provide one of the mechanisms for dielectric breakdown under the low electric field of 0.5–0.8 MV cm^{-1} [476]. For the practical use of BaTiO$_3$ in the TFEL device, it is desirable that advantages of both amorphous and crystalline film be combined. Some studies [477, 478] have suggested that a multilayered BaTiO$_3$ thin film structure with a high dielectric constant of 98 and a high breakdown strength of 2.2 MV cm^{-1} was practical in TFEL. Song et al. [474] prepared ZnS:Mn TFEL devices using layered BaTiO$_3$ thin film structures by different stacking methods and investigated the effects of the crystallinity of BaTiO$_3$ layers on the operating characteristics of TFEL devices. The electrical and dielectric properties of the BaTiO$_3$ thin films used by Song et al. [474] are listed in Table XVI. The TFEL device using the layered BaTiO$_3$ (amorphous and crystalline layers) thin film structure showed stable efficiency characteristics with operating time as well as low turn-on voltage of ~50 V and high saturated brightness of ~3000 cd m^{-2} [474].

The results of the aforementioned studies indicate that indeed, high-permittivity thin films with very high transmittance in the visible region can act as a insulator layer in thin-film and thick-film electroluminescent devices, there by reducing the driving voltage of the devices and protecting them from breakdown.

Table XVII. Dielectric Constants and Tunabilities of Thin Films and Bulk Materials Measured at 0.5 MHz

Material	Applied field (V/μm)		ε' ($V = 0$)		Tunability %	
	Film	Bulk	Film	Bulk	Film	Bulk
BST	3.0	0.73	1200	3300	23	20
BST/1.5 vol% MgO	3.0	2.32	926	1276	24	16
BST/1.5 vol% ZrO$_2$/MgO	3.0	1.2	1216	2515	20	12
BST/1.5 vol% MgO	3.0	1.7	745	3065	20	19

9.4. Microwave Phase Shifters and Voltage Tunable Devices

Ferroelectrics are a class of nonlinear dielectric materials that exhibit an electric field–dependent dielectric constant [479–481]. This property is being used in a new class of frequency tunable microwave devices that includes phase shifters, tunable filters, steerable antennas, varacters, frequency triplers, and so forth [482–486]. The dielectric constant of BST can be varied by an applied dc electrical field, which makes it suitable for microelectronic device applications and also composition-dependent Curie temperature. The critical properties that need to be optimized for the tunable microwave devices are the magnitude of the change in dielectric constant as a function of applied electric field and dielectric loss at microwave frequencies [487, 488]. An excellent review of novel ferroelectric materials for phased array antennas has been given by Sengupta and Sengupta [489]. A phased array refers to an antenna configuration comprising numerous elements that emit phased signals to form a radio beam. The radio signal can be electronically steered by actively manipulating the relative phasing of the individual antennas [489]. These antennas are currently constructed using ferrite phase shifting elements; however, the circuit and power requirements necessary to operate these present day antennas make them extremely costly, large, and heavy. Therefore, such antennas have been limited to military applications. To make these devices practical for many other commercial and military uses, better antenna materials must be developed [489]. The material characteristics of the dielectric material of choice for phased array applications should be moderate to low dielectric constant at microwave frequencies ($30 < \varepsilon_r < 1500$), low loss tangents ($0.005 < \tan\delta < 0.01$) (high dielectric quality factor $Q = 1/\tan\delta$), and high dielectric voltage tunability (change in dielectric constant with applied voltage >10%). The tunability of a material is measure of by how much the dielectric constant changes with applied voltage, defined as

$$\text{tunability} = \frac{\varepsilon_{max} - \varepsilon_{min}}{\varepsilon_{max}} \quad (45)$$

where ε_{max} and ε_{min} are the maximum and minimum dielectric constants. The degree of phase-shifting ability is directly related to tunability; therefore, higher tunabilities are desired [489]. Higher tunabilities can be achieved by having a low insertion loss material. The loss tangent serves to dissipate or absorb the incident microwave energy and thus is desired to be 0.01 or lower for the specific design of the antennas. BST thin films are of ideal candidates for this type of application, because their dielectric constants and loss tangents can be optimized by changing the composition.

Application of pure BST thin films in microwave devices is limited by these films' high dielectric constant and high dielectric loss. Various methods have been investigated [489] to decrease the dielectric constant and loss by forming composites of BST and oxides such as MgO, ZrO$_2$, and alumina [489, 490]. Table XVII summarizes the dielectric constants and tunability of various BST thin films ($\sim$300 nm on RuO$_2$/sapphire) measured at 0.5 MHz and their bulk forms (1 mm). The dielectric constants of thin films are lower than those of bulk forms; also, tunability is improved slightly [489, 490]. The loss tangent at microwave frequencies is reported to be minimum for the 40–50 vol% MgO + BST composite, but the dielectric constant is effectively lowered for high MgO (>50 vol%)-containing composites [489]. The foregoing studies indicate that tailoring the electronic properties of BST thin films in the low-frequency region is possible through incorporation of metal oxides. The decreased dielectric constant along with high tunability play important roles in the impedance matching of these films into electronic circuits [489].

Im et al. [491] deposited BST thin films of various composition to determine the effect of composition on dielectric properties and tunability. They showed that BST films with high tunabilities, low losses, and high dielectric breakdown fields can be grown using magnetron sputter deposition with judiciously chosen process parameters. Table XVIII summarizes the properties of BST thin film as functions of the composition. The data reveal that the capacitance density, dielectric constant, and dielectric loss at zero bias also exhibited a clear dependence on the total process pressure or the (Ba + Sr)/Ti ratio. The film with (Ba + Sr)/Ti = 0.98 exhibited the highest dielectric constant. A significant reduction in permittivity and dielectric loss at zero bias was observed for films with a (Ba + Sr)/Ti ratio <0.85. The near-stoichiometric BST film with (Ba + Sr)/Ti = 0.98 displayed the largest tunability, approximately 74%, whereas the sample with a (Ba + Sr)/Ti = 0.73 had the lowest dielectric loss, 0.0047 at zero bias [491]. The results show that the defect density contributes to the tunability of the dielectric constant. The larger nonstoichiometry, the higher the defect density.

Table XVIII.   Properties of BST Films Deposited under Various Total Process Pressures at 650°C Substrate
Temperature

P (mT)	(Ba + Sr)/Ti	C/A (fF/μm^2)	ε	Tunability	tan δ	K	Roughness (Å)
22	0.73	17	153	0.476	0.0047	101	20.8
30	0.85	18.5	167	0.488	0.0061	80	20.4
40	0.90	53	420	0.72	0.0131	55.0	28.9
58	0.98	67	530	0.74	0.0187	25.1	31.6

C/A: capacitance density, ε: permittivity, tan δ: loss tangent.

Reprinted with permission from [491], © 2000, American Institute of Physics.

The correlation between tunability and loss is given by the FOM [491, 492],

$$K = \frac{\text{tunability}}{\tan \delta} \qquad (46)$$

As shown in Table XVIII, the highest K value was reported for the sample with a (Ba + Sr)/Ti = 0.73. This shows that the decrease in zero bias permittivity for the highly nonstoichiometric samples is compensated for by their lower loss and high breakdown fields, so they may offer superior performance for some applications [492].

The annealing temperature and the film strain (or stress) has great influence on the dielectric properties of the thin films [493, 494]. The primary source of strain in heteroepitaxial thin films is due to the lattice parameter and thermal expansion coefficient mismatch between the thin film and the substrate. For the film deposited on (100) MgO Chang et al. [493] observed that after postdeposition annealing (1000–1200°C), the dielectric constant decreased and Q (1/ tan δ) increased. The film deposited onto (100) LaAlO$_3$ after postdeposition annealing ($\leq$1000°C) exhibited a significant increase in the dielectric constant and a decrease in the quality factor (Q). The difference in the BST thin film behavior on two different substrates is attributed to a change in film stress, which affects the extent of ionic polarization [493]. Stress is a very significant factor affecting dielectric properties. It has been reported that hydrostatic compression or two-dimensional compression parallel to the electrode of a ferroelectric parallel plate capacitor leads to a decrease in the dielectric constant and the Curie temperature (T_c) [495–497]. It was also reported that application of two-dimensional compression normal to the electrode induced increases in dielectric constant and T_c. The effects of stress on the dielectric constant and the T_c are due to the fact that ionic position and vibrations in a ferroelectric are modified by stress, and these changes are coupled to the polarization mechanism in the ferroelectric [498, 499]. Table XIX [494] shows average dielectric properties for the BST ($x = 0.5$) film at 1–20 GHz as a function of substrate type. The data indicate that the average percent tuning and dielectric constant obtained for the film deposited on (100) MgO is less than that of film deposited on (100) LaAlO$_3$. However, the quality factor Q is higher for film deposited on (100) MgO than that deposited on (100) LaAlO$_3$. In general, authors observed either a large di-

Table XIX.   Average Values of the Dielectric Constant and Dielectric Q for Annealed BST Films (1–20 GHz) as a Function of Substrate Type

	Average	
Substrate	MgO	LaAlO$_3$
ε	100	1500
% tuning	30	50
Q	45	25

ε: film dielectric constant; % dielectric tuning at $E = 67$ kV/cm; and dielectric Q.

Reprinted with permission from [494], © 2000, American Institute of Physics.

electric tuning (50%) or a high dielectric Q (45) but not both at the same time [494], which is attributed to the difference in film strain on two different substrates.

Kim et al. [500] reported a strong correlation between the structure and the microwave properties of epitaxial Ba$_{0.5}$Sr$_{0.5}$TiO$_3$ thin films deposited onto (001) MgO by pulsed laser deposition. Oxygen deposition pressure has a very strong effect on the structure of the epitaxial film deposited; in particular, the lattice parameters varied oxygen deposition pressure varied. The change in the lattice parameters of the film deposited with oxygen deposition pressure of 350–1000 mtorr was relatively small; however, those deposited at lower oxygen pressure (3–50 mtorr) show tetragonal distortion of the unit cell in the form of large changes in both in-plane (a) and surface normal (c) directions. The calculated tetragonal distortion of the film, $D = a/c$, strongly depends on the oxygen deposition pressure. At 3 mtorr, D is 0.996 less than 1 ($c > a$). The deposited film is nearly cubic around 50 mtorr ($D = 1.0004$). D increases monotonically with increasing oxygen pressure to a maximum at 800 mtorr ($D_{800} = 1.003$). At microwave frequencies, BST films with low tetragonal distortion had higher dielectric constants ($\sim$500) and lower dielectric loss (tan $\delta \sim 0.02$) compared to films with higher distortion. The FOM also was maximum for the film with the minimum distortion (at 50 mtorr), with a maximum value $\sim$700. The correlation of microwave properties and film structure is attributed to the stresses and polarization in the films. The BST film grown at an oxygen deposition pressure of 50 mtorr showed a large dielectric constant change concurrent with a low dielectric loss, which corresponds to the film in low stress ($D = 1.0004$). The films with very minimal deviation from the cubic symmetry had the highest quality factor (Q). For tunable microwave applications, BST films with low stress are desirable to achieve both low dielectric loss and high tunability [500]. Chen et al. [501] deposited BST ($x = 0.5$) thin films on (001) LaAlO$_3$ substrates by pulsed laser ablation. They obtained a loss tangent of 0.007 and relatively large dielectric constant of 1430. The tunability of the dielectric constant was reported to be 33% at room temperature for the bias voltage changing from 0 to 35 V at 1 MHz.

Jia et al. [502] proposed the idea of simultaneous tuning by magnetic and electrical techniques through the integration of BST thin films with polycrystalline yttrium iron garnet. With dual tuning, a constant microwave impedance can be maintained while the frequency or propagation velocity is tuned. In the case of a tunable filter, for example, dual tuning enables not only fine tuning of the filter profile to achieve symmetric and optimum filter characteristics electrically, but also broadband magnetic tuning of the filter passband to demonstrate adaptive filter response over a wide frequency range [502]. A typical phased array antenna may have several thousand radiating elements with a tunable phase shifter for every antenna element. Thus dual-tuning phase shifters make global tuning easier. For this type of device, the thin films must be deposited on polycrystalline ferrite substrates. Ferrite materials, such as yttrium iron garnet (YIG), have been used for magnetically tunable microwave devices whose electrical response can be tuned by applying a dc magnetic field. The deposition of device-quality Ba$_{1-x}$Sr$_x$TiO$_3$ on a polycrystalline ferrite YIG substrate can be a formidable challenge, because of the inherent crystallographic incompatibility of the two materials and the lack of an epitaxial template for the growth of well-oriented BST thin films [502]. Jia et al. [502] have deposited biaxially oriented Ba$_{0.6}$Sr$_{0.4}$TiO$_3$ films on polycrystalline YIG substrates by using biaxially oriented MgO as a buffer layer deposited with an ion beam–assisted deposition process. The MgO buffer layer acted not only as a template layer, but also as a diffusion barrier to prevent chemical reaction between the BST and YIG. The BST films were grown on this substrate using pulsed laser deposition [502]. Figure 60 shows the dc bias-voltage dependence of the capacitance and dielectric loss of the multilayer structure at 100 kHz. The capacitance tunability of the coplanar capacitor with a gap of 5 μm is greater than 25% at a dc bias voltage of 40 V. The dielectric loss decreases from <0.02 at zero bias to 0.005 at a dc bias voltage of 40 V [502].

The use of ferroelectric materials as nonlinear dielectrics at microwave frequencies and the integration of tunable dielectrics with conductors with low microwave surfaces (R_S) are currently being investigated for advanced high-frequency applications [503–507]. The thin film multilayer heterostructures of ferroelectric materials with thin film high-temperature superconductors (HTS) provides a means of producing low-loss, voltage-tunable monolithic microwave circuits with reduced length scales [505–507] that operate in the cryogenic temperatures. The HTS/ferroelectric coplanar waveguide structure for voltage-tunable phase shifters and delay lines is being studied extensively [503–507]. The advantage of this type of coplanar

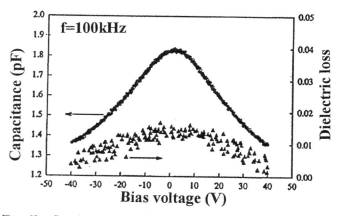

Fig. 60. Capacitance and dielectric loss of the biaxially oriented Ba$_{0.6}$Sr$_{0.4}$TiO$_3$ deposited on polycrystalline YIG substrate as a function of dc bias voltage measured at room temperature. Reprinted with permission from [502], © 1999, American Institute of Physics.

HTS-based waveguide is the possibility of very low insertion loss but at cryogenic temperatures.

10. SUMMARY

This article has briefly reviewed how such factors as processing, film composition, microstructure, film thickness, electrode materials, and interfacial properties affect BST thin film properties. The electrical properties of BST films prepared using various deposition techniques have also been compared. The properties of BST films have been found to be strongly dependent on the fabrication method, nature of the substrates and electrode materials, and postdeposition annealing treatment. Copel et al. [220] demonstrated that acceptor impurities can dramatically increase the depletion width in BST films and contribute to reduced leakage current. However, the physical origins of the doping effect on the films must be further elucidated, and the possible decrease in dielectric constant caused by the doping also solved. In addition, a new postannealing technique is necessary to bring forth improvements in crystallinity and defect density reduction and thus improve the electrical properties of the films without significantly roughening the surface morphology of the upper electrode and BST films.

After the requirements for capacitance and leakage current are satisfied, a remaining relevant issue is reliability in the practical use of the BST films for memory cell capacitors. As commonly observed, leakage current gradually increases with degradation of the insulation when temperature and ac and dc voltage stresses are applied to BST capacitors for a certain time interval. The degradation process limits the lifetime and reliability of BST capacitors. Further understanding of the defect formations and distribution, relaxation, conduction, and breakdown of BST capacitors is required to enhance reliability. On the other hand, recent results indicate high integration of BST films into the DRAM capacitor structure. However, issues related to improving the manufacturing processes and compatibility—such as new etching processes and equipment for BST films and electrodes, a lithography scale of ~0.1 μm, low-temperature processing, and thermal stability of electrodes—also must be addressed.

Researchers are now focusing more on applications, such as sensors, microwave phase shifters, voltage tunable devices, coplanar waveguides, and so forth. There is a large application potential of BST materials in fabricating microwave devices. However, it has been reported that the high inherent material loss and high dielectric constant hinder the application of pure BST in the microwave frequencies. Hence, different methods, such as aliovalent substitution and composite films, may provide a better solution. The challenge in the future is to reduce insertion loss at high frequency ranges (>15 GHz). Sensor applications are another area in which the uses of BST film are yet to be fully. The dependence of the BST film's dielectric properties on temperature is effectively exploited for the dielectric bolometer and pyroelectric sensors, which can be used in uncooled infrared detectors for the thermal imaging device. For the pyroelectric sensor, low-loss materials are highly desirable. Optical properties of these films have been studied considering their possible use in nonlinear optical devices, such as coplanar waveguides. The research work in this field is in the preliminary stage, and more studies are needed to improve understanding the structure property relations to fabricate a feasible device.

The present and future research on BST thin films, after years of research work, remains focused on the finding thin film fabrication technique most compatible with ULSI technology that can yield a highly crystalline or epitaxial BST thin film at a low temperature. Also of interest is continuous improvement of thin film quality and properties through various postannealing processes, such as rapid thermal annealing and plasma annealing, or by modifying composition through isovalent/aliovalent doping or substitution or forming composite thin films. The advancement of thin film technology and the principles of miniaturization/integration will keep BST thin films fertile field of research and development, full of application potential and rich in science.

REFERENCES

1. J. F. Scott, C. A. Paz de Auraujo, B. M. Melnick, L. D. McMillan, and R. Zuleeg, *J. Appl. Phys.* 70, 382 (1991).
2. R. Waser and M. Klee, *Integ. Ferroelectrics* 2, 23 (1992).
3. P. C. Joshi and S. B. Krupanidhi, *Appl. Phys. Lett.* 61, 1525 (1992).
4. K. N. Kim, J. Y. Lee, K. H. Lee, B. H. Noh, S. W. Nam, Y. S. Park, Y. H. Kim, H. S. Kim, J. S. Kim, J. K. Park, K. P. Lee, K. Y. Lee, J. T. Moon, J. S. Choi, J. W. Park, and J. G. Lee, *VLSI Tech. Dig. Tech.* 9 (June 1997).
5. K. Kim, C. G. Hwang, and J. G. Lee, *IEEE Trans. Electron. Dev.* 45, 598 (1998).
6. L. H. Parker and A. H. Tasch, *IEEE Circ. Dev. Mag.* 6, 17 (1990).
7. C. A. Paz de Auraujo, L. D. McMillan, B. M. Melnick, J. D. Cuchiaro, and J. F. Scott, *Ferroelectrics* 104, 241 (1990).
8. R. Moazzami, C. Hu, and W. H. Shepherd, *IEEE Electron. Dev. Lett.* 11, 454 (1990).
9. K. Kashibara, H. Ito, K. Tsukamoto, and Y. Akasaka, "Extended Abstracts of the 1991 International Conference on Solid-State Devices and Materials," Yokohama, 1991, p. 192.
10. S. Yamamichi, T. Sakuma, K. Takemura, and Y. Miyasaka, *Japan. J. Appl. Phys.* 30, 2193 (1991).
11. Y. Miyasaka and S. Matsubara, "Proceedings of the 7th International Symposium on the Applications of Ferroelectrics," 1991, p. 121.
12. H. Kawano, K. Mori, and Y. Nakayama, *J. Appl. Phys.* 73, 5141 (1993).
13. T. Horikawa, N. Mikami, T. Makita, J. Tanimura, M. Kataoka, K. Sato, and M. Nunoshita, *Japan. J. Appl. Phys.* 32, 4126 (1993).
14. K. Abe and S. Komatsu, *Japan. J. Appl. Phys.* 31, 2985 (1992).
15. J. F. Scott, B. M. Melnick, C. A. Paz de Auraujo, L. D. McMillan, and R. Zuleeg, *Integ. Ferroelectrics* 1, 323 (1992).
16. R. Moazzami, C. Hu, and W. H. Shepherd, *IEEE Trans. Electron. Dev.* 39, 2044 (1992).
17. J. F. Scott, B. M. Melnick, L. D. McMillan, and C. A. Paz de Auraujo, *Integ. Ferroelectrics* 3, 129 (1993).
18. X. Chen, A. I. Kingon, L. Mantese, O. Auciello, and K. Y. Hsieh, *Integ. Ferroelectrics* 3, 259 (1993).
19. P. C. Joshi and S. B. Krupanidhi, *J. Appl. Phys.* 73, 7627 (1993).
20. K. Abe and S. Komatsu, *Japan. J. Appl. Phys.* 32, 4186 (1993).
21. W. Antpohler, G. W. Dietz, M. Klee, and R. Waser, "Proceedings of the Electroceramics IV," 1994, p. 169.
22. D. J. Wouters, G. Wilems, G. Groeseneken, H. E. Maes, and K. Brooks, *Integ. Ferroelectrics* 7, 73 (1995).
23. J. F. Scott, M. Azuma, C. A. Paz de Auraujo, L. D. McMillan, M. C. Scott, and T. Roberts, *Integ. Ferroelectrics* 4, 61 (1994).
24. H. Hu and S. B. Krupanidhi, *J. Mater. Res.* 9, 1484 (1994).
25. Y. Fukuda, K. Aoki, K. Numata, and A. Nishimura, *Japan. J. Appl. Phys.* 33, 5255 (1994).
26. T. Tamura, K. Takai, H. Noshiro, M. Kimura, S. Otani, and M. Yamada, *Japan. J. Appl. Phys.* 33, L1697 (1994).
27. J. J. Lee, C. L. Thio, and S. B. Desu, *J. Appl. Phys.* 78, 5073 (1995).
28. G. W. Dietz, W. Antpohler, M. Klee, and R. Waser, *J. Appl. Phys.* 78, 6113 (1995).
29. R. Waser, in "Science and Technology of Electroceramic Thin Films," NATO ASI Series, Vol. 284, p. 223. (O. Auciello and R. Waser, Eds.). Kluwer Academic Publishers, Dordrecht, 1995.
30. H. Kawano, K. Mori, and Y. Nakayama, *J. Appl. Phys.* 73, 5141 (1993).
31. J. F. Roeder, S. M. Bilodeau, R. Carl, P. C. Van Brukirk, C. Basceri, S. K. Streiffer, A. I. Kingon, and J. Fair, "Proceedings of the 9th International Meeting on Ferroelectricity," 1997, p. 218.
32. T. Horikawa, T. Makita, T. Kuroiwa, and N. Mikami, *Japan. J. Appl. Phys.* 34, 5478 (1995).
33. M. Yamamuka, T. Kawahara, T. Makita, A. Yuuki, and K. Ono, *Japan. J. Appl. Phys.* 35, 729 (1996).
34. K. Abe, N. Yanase, S. Komatsu, K. Sano, N. Fukushima, and T. Kawakubo, *IEICE Trans. Electron.* E81-C 505 (1998).
35. K. Numata, Y. Fukuda, K. Aoki, and A. Nishimura, *Japan. J. Appl. Phys.* 34, 5245 (1995).
36. T. Eimori, Y. Ohno, J. Matsufusa, S. Kishimura, A. Yoshida, H. Simitani, T. Maruyama, Y. Hayashide, K. Moriizumi, T. Katayama, M. Asakura, T. Horikawa, T. Shibano, H. Itoh, K. Namba, T. Nishimura, S. Satoh, and H. Miyoshi, *IEDM Tech. Dig.* 631 (1993).
37. Y. Ohno, T. Horikawa, H. Shinkawata, K. Kashihara, T. Kuroiwa, T. Okudaira, Y. Hashizume, K. Fukumoto, T. Eimori, T. Shibano, K. Arimoto, H. Itoh, T. Nishimura, and H. Miyoshi, *Symp. VLSI Tech. Dig.* 149 (1994).
38. M. S. Tsai, S. C. Sun, and T. Y. Tseng, *J. Appl. Phys.* 82, 3482 (1997).
39. B. Gnade, S. R. Summerfelt, and D. Crenshaw, in "Science and Technology of Electroceramic Thin Films," NATO ASI Series, Vol. 284, p. 373. (O. Auciello and R. Waser, Eds.). Kluwer Academic Publishers, Dordrecht, 1995.
40. T. Y. Tseng, *IEDMS* C2-5, 89 (1996).
41. A. I. Kingon, S. K. Streiffer, C. Basceri, and S. R. Summerfelt, *MRS Bull.* 21, 46 (1996).
42. P. C. Fazan, *Integ. Ferroelectrics* 4, 247 (1994).
43. K. W. Kwon, I.-S. Park, D. H. Han, E. S. Kim, S. T. Ahn, and M. Y. Lee, *IEDM Tech. Dig.* 835 (1994).
44. K. W. Kwon, S. O. Park, C. S. Kang, Y. N. Kim, S. T. Ahn, and M. Y. Lee, *IEDM Tech. Dig.* 53 (1993).

45. S. Ezhilvalavan and T. Y. Tseng, *J. Mater. Sci.: Mat. Electron.* 10, 9 (1999).

46. S. Ezhilvalavan and T. Y. Tseng, *J. Appl. Phys.* 83, 4797 (1998).

47. S. Ezhilvalavan and T. Y. Tseng, *J. Am. Ceram. Soc.* 82, 600 (1999).

48. T. Y. Tseng, *Ferroelectrics* 232, 881 (1999).

49. M. S. Tsai and T. Y. Tseng, *J. Am. Ceram. Soc.* 83, 351 (1999).

50. D. Roy and S. B. Krupanidhi, *Appl. Phys. Lett.* 62, 1056 (1993).

51. Y. Shimada, A. Inoue, T. Nasu, Y. Nagano, A. Matsuda, Y. Vemoto, E. Fujii, M. Azuma, Y. Oishi, S. Hayashi, and T. Otsuki, *Japan. J. Appl. Phys.* 35, 140 (1996).

52. T. Kuroiwa, Y. Tsunemine, T. Horikawa, T. Makita, J. Tanimura, N. Mikami, and K. Sato, *Japan. J. Appl. Phys.* 33, 5187 (1994).

53. C. S. Hwang, S. O. Park, H.-J. Cho, C. S. Kang, H.-K. Kang, S. I. Lee, and M. Y. Lee, *Appl. Phys. Lett.* 67, 2819 (1995).

54. W. J. Yoo, J. H. Hahn, H. W. Kim, C. O. Jung, Y. B. Koh, and M. Y. Lee, *Japan. J. Appl. Phys.* 35, 2501 (1996).

55. S. Zaima, T. Furuta, and Y. Yasuda, *J. Electrochem. Soc.* 137, 1297 (1990).

56. H. Shinriki, M. Nakata, Y. Nishioka, and K. Mukai, *IEEE Electron. Dev. Lett.* 10, 514 (1989).

57. N. Rausch and E. P. Burte, *J. Electrochem. Soc.* 140, 145 (1993).

58. J. Shappir, A. Anis, and I. Pinsky, *IEEE Trans. Electron. Dev.* 33, 442 (1986).

59. L. Manchanda and M. Gurvitch, *IEEE Electron. Dev. Lett.* 9, 180 (1988).

60. K. Koyama, T. Sakuma, S. Yamamichi, H. Watanabe, H. Aok, S. Ohya, Y. Miyasaka, and T. Kikkawa, *IEDM Tech. Dig.* 823 (1991).

61. Y. F. Kuo and T. Y. Tseng, *Electrochem. Solid State Lett.* 2, 236 (1999).

62. K. Torii, T. Kaga, and E. Takeda, *Japan. J. Appl. Phys.* 31, 2989 (1992).

63. S. K. Dey, *IEEE Trans. Electron. Dev.* 39, 1607 (1992).

64. R. E. Jones, *Appl. Phys. Lett.* 60, 1022 (1992).

65. R. Moazzami, *IEDM Tech. Dig.* 973 (1992).

66. R. H. Dennard, U.S. Patent 3,387,286 (1968).

67. C. N. Ahlquist, J. R. Breivogel, J. T. Koo, J. L. McCollum, W. G. Oldham, and A. L. Renninger, *IEEE J. Solid State Circ.* 11, 570 (1976).

68. T. C. May and M. H. Wood, *IEEE Trans. Electron. Dev.* 26, 2 (1979).

69. S. Asai, *IEDM Tech. Dig.* 6 (1984).

70. C. Hu, *IEDM Tech. Dig.* 368 (1985).

71. W. M. Smith, *IBM Tech. Disclos. Bull.* 3585 (1973).

72. M. Koyanagi, H. Sunami, N. Hashimoto, and M. Ashikawa, *IEDM Tech. Dig.* 348 (1978).

73. M. Koyanagi, Y. Sakai, M. Ishihara, M. Tazunoki, and N. Hashimoto, *IEEE J. Solid State Circ.* 15, 661 (1980).

74. Y. Takemae, T. Ema, M. Nakano, F. Baba, T. Yabu, K. Miyasaka, and K. Shirai, *ISSCC Dig. Paper* 250 (1985).

75. H. Sunami, T. Kure, N. Hashimoto, K. Itoh, T. Toyabe, and S. Asai, *IEDM Tech. Dig.* 806 (1982).

76. M. Sakamoto, T. Katoh, H. Abiko, T. Shimizu, H. Mikoshiba, K. Hamono, and K. Kobayashi, *IEDM Tech. Dig.* 711 (1985).

77. W. F. Richardson, D. M. Bordelon, G. P. Pollack, A. H. Shah, S. D. S. Malhi, H. Shichijo, S. K. Banerjee, M. Elahy, R. H. Womack, C. P. Wang, J. Gallia, H. E. Davis, and P. K. Chatterjee, *IEDM Tech. Dig.* 714 (1985).

78. N. Lu, P. Cottrell, W. Craig, S. Dash, D. Critchlow, R. Mohler, B. Machesney, T. Ning, W. Noble, R. Parent, R. Scheuerlein, E. Sprogis, and L. Terman, *IEDM Tech. Dig.* 771 (1985).

79. M. Yanagisawa, K. Nakamura, and M. Kikuchi, *IEDM Tech. Dig.* 13 (1986).

80. M. Taguchi, S. Ando, N. Higaki, G. Goto, T. Ema, K. Hashimoto, T. Yabu, and T. Nakano, *IEDM Tech. Dig.* 135 (1986).

81. T. Furuyama, S. Saito, and S. Fujii, *Japan. J. Appl. Phys.* 21, 61 (1982).

82. K. Tsukamoto, M. Shimizu, M. Inuishi, Y. Matsuda, H. Oda, H. Morita, M. Nakjima, K. Kobayashi, Y. Mashiko, and Y. Akasaka, *IEDM Tech. Dig.* 328 (1987).

83. K. P. Muller, B. Flietner, C. L. Hwang, R. L. Kleinhenz, T. Nakao, R. Ranader, Y. Tsunashima, and T. Mii, *IEDM Tech. Dig.* 507 (1996).

84. H. Shinriki, T. Kisu, S. I. Kimura, Y. Nishioka, Y. Kawamoto, and K. Mukai, *IEEE Trans. Electron. Dev.* 37, 1939 (1990).

85. K. Itoh, *IEEE J. Solid State Circ.* 25, 778 (1990).

86. G. Bronner, *IEDM Tech. Dig.* 75 (1996).

87. T. Ema, S. Kawanago, T. Nishi, S. Yoshida, H. Nishibe, T. Yabu, Y. Kodama, T. Nakano, and M. Taguchi, *IEDM Tech. Dig.* 592 (1988).

88. S. Kimura, Y. Kawamoto, T. Kure, N. Hasegawa, J. Etoh, M. Aoki, E. Takeda, H. Sunami, and K. Itoh, *IEDM Tech. Dig.* 569 (1988).

89. W. Wakamiya, Y. Tanaka, H. Kimura, H. Miyatake, and S. Satoh, *VLSI Tech. Symp.* 69 (1989).

90. M. Sakao, N. Kasai, T. Ishijima, E. Ikawa, H. Watanabe, K. Terada, and T. Kikkawa, *IEDM Tech. Dig.* 655 (1990).

91. M. Yoshimaru, J. Miyano, N. Inoue, A. Sakamoto, S. You, H. Tamura, and M. Ino, *IEDM Tech. Dig.* 659 (1990).

92. D. Temmler, *VLSI Tech. Symp.* 13 (1991).

93. H. Watanabe, T. Tarsumi, S. Ohnishi, T. Hamada, I. Honma, and T. Kikkawa, *IEDM Tech. Dig.* 259 (1992).

94. H. K. Kang, K. H. Kim, Y. G. Shin, I. S. Park, K. M. Ko, C. G. Kim, K. Y. Oh, S. E. Kim, C. G. Hong, K. W. Kwon, J. Y. Yoo, Y. G. Kim, C. G. Lee, Y. S. Paick, D. I. Suh, C. J. Park, S. I. Lee, S. T. Ahn, C. G. Hwang, and M. Y. Lee, *IEDM Tech. Dig.* 635 (1994).

95. Y. Nishioka, K. Shiozawa, T. Oishi, K. Kanamoto, Y. Tokuda, H. Sumitani, S. Aya, H. Yabe, K. Itoga, T. Hifumi, K. Marumoto, T. Kuriowa, T. Kawahara, K. Nishikawa, T. Oomori, T. Fujino, S. Yamamoto, S. Uzawa, M. Kimata, M. Nunoshita, and H. Abe, *IEDM Tech. Dig.* 903 (1995).

96. E. S. Yang, "Microelectronic Devices." McGraw-Hill, New York, 1990.

97. M. Aoki, J. Itoh, S. Kimura, and Y. Kawamoto, *ISSCC Dig. Paper* 238 (1989).

98. S. Kimura, Y. Kawamoto, N. Hosegawa, A. Hiraiwa, M. Horiguchi, A. Aoki, T. Kisu, and H. Sunami, *Extended Abs. Int. Conf. Solid State Dev. Mater.* 19 (1987).

99. T. Licata, E. G. Colgun, J. M. E. Harper, and S. E. Luce, *IBM Res. Devel.* 38, 419 (1995).

100. J. Yugami, T. Mine, S. Iijima, and A. Hiraiwa, *Extended Abs. Int. Conf. Solid State Dev. Mater.* 173 (1989).

101. T. Kaga, T. Kure, H. Shinriki, Y. Kawamoto, F. Murai, T. Nishida, Y. Nakagome, D. Hisamoto, T. Kisu, E. Takeda, and K. Itoh, *IEEE Trans. Electron. Dev.* 38, 255 (1991).

102. S. Yamamichi, P.-Y. Lesaicherre, H. Yamaguchi, K. Takemura, S. Sone, H. Yabuta, K. Sato, T. Tamura, K. Nakajima, S. Ohnishi, K. Tokashiki, Y. Hayashi, Y. Kato, Y. Miyasaka, M. Yoshida, and H. Ono, *IEEE Trans. Electron. Dev.* 44, 1076 (1997).

103. A. Yuuki, M. Yamamuka, T. Makita, T. Hotikawa, T. Shibano, N. Hirano, H. Maeda, N. Mikami, K. Ono, H. Ogata, and H. Abe, *IEDM Tech. Dig.* 115 (1995).

104. T. Kanehara, I. Koiwa, Y. Okada, K. Ashikaga, H. Katoh, and K. Kaifu, *IEDM Tech. Dig.* 601 (1997).

105. M. Takeo, M. Azuma, H. Hirano, K. Asari, N. Moriwaki, T. Otsuki, and K. Tatsuuma, *IEDM Tech. Dig.* 621 (1997).

106. K. H. Joo and S. K. Joo, *IEDM Tech. Dig.* 127 (1995).

107. P. C. Van Buskrik, S. M. Bilodeau, J. F. Roeder, and P. S. Kirlin, *Japan. J. Appl. Phys.* 35, 2520 (1996).

108. P. Y. Lesaicherre, S. Yamamichi, H. Yamaguchi, K. Takemura, H. Watanabe, K. Tokashiki, K. Satoh, T. Sakuma, M. Yoshida, S. Ohnishi, K. Nakajima, K. Shibahara, Y. Miyasaka, and H. Ono, *IEDM Tech. Dig.* 831 (1994).

109. M. P. Lepselter and S. M. Sze, *IEEE Circ. Dev. Mag.* 1, 53 (1985).

110. Y. Tarui, *Nikkei Microdev.* 21, 165 (1987).

111. Y. Tarui and T. Tarui, *IEEE Circ. Dev.* 7, 44 (1991).

112. Y. Tarui, *IEDM Tech. Dig.* 7 (1994).

113. B. Nagaraj, T. Sawhney, S. Perusse, S. Aggarwal, R. Ramesh, V. S. Kaushik, S. Zafar, R. E. Jones, J. H. Lee, V. Balu, and J. Lee, *Appl. Phys. Lett.* 74, 3194 (1999).

114. W. J. Lee, I. K. Park, G. E. Jiang, and H. G. Kim, *Japan. J. Appl. Phys.* 34, 196 (1995).

115. L. L. Hench and J. K. West, "Principles of Electronic Ceramics." Wiley, New York, 1990.

116. K. Abe and S. Komatsu, *J. Appl. Phys.* 77, 6461 (1995).

117. E. Fujii, Y. Uemoto, S. Hayashi, T. Nasu, Y. Shimada, A. Matsuda, M. Kibe, M. Azuma, T. Otsuki, G. Kano, M. Scott, L. C. McMillan, and C. A. Paz de Araujo, *IEDM Tech. Dig.* 267 (1992).

118. K. Arita, E. Fujii, Y. Shimada, Y. Uemoto, T. Nasu, A. Inoue, A. Matsuda, T. Otsuki, and N. Suzuoka, *Japan. J. Appl. Phys.* 33, 5397 (1994).

119. K. Arita, E. Fujii, Y. Uemoto, M. Azuma, S. Hayashi, T. Nasu, A. Inoue, A. Matsuda, Y. Nagano, S. Katsu, T. Otsuki, G. Kano, L. D. McMillan, and C. A. Paz de Araujo, *IEICE Trans. Electron.* 392 (1994).

120. H. O. Pierson, "Handbook of Chemical Vapor Deposition: Principles, Technology and Applications." Noyes Publications, Park Ridge, NJ, 1992.

121. A. C. Adams, in "VLSI Technology," Chap. 3, p. 93. (S. M. Sze, Ed.). McGraw-Hill, New York, 1983.

122. W. Kern, *Semiconductor Int.* 8, 122 (1985).

123. S. Wolf and R. N. Tauber, "Silicon Processing for the VLSI Era," Vol. 1. Lattice Press, Sunset Beach, CA, 1986.

124. W. Kern and G. L. Schnable, *IEEE Trans. Electron. Dev.* 26, 647 (1979).

125. P. Singer, *Semiconductor Int.* 5, 72 (1984).

126. R. S. Rosler, *Solid State Technol.* 20, 63 (1977).

127. A. C. Adams, *Solid State Technol.* 4, 135 (1983).

128. B. Gorowitz, T. B. Gorczyca, and R. J. Saia, *Solid State Technol.* 4, 197 (1985).

129. J. Y. Chen and R. Henderson, *J. Electrochem. Soc.* 131, 2147 (1984).

130. R. Solanki, C. Moore, and G. Collins, *Solid State Technol.* 6, 220 (1985).

131. A. Feingold and A. Katz, *Mater. Sci. Eng.* R13, 56 (1994).

132. S. K. Dey and P. V. Alluri, *MRS Bull.* 21, 44 (1996).

133. S. K. Dey, in "Ferroelectric Thin Films: Synthesis and Basic Properties," Vol. 10 (C. Paz de Araujo, J. F. Scot, and G. W. Taylor, Eds.). Gordon and Beach, New York, 1996.

134. P. C. Van Buskirk, J. F. Roeder, and S. Bilodeau, *Integ. Ferroelectrics* 10, 9 (1995).

135. J. B. Wachtman and R. A. Haber, "Ceramic Thin Films and Coatings," p. 208. Noyes Publications, Park Ridge, NJ, 1993.

136. R. Ramesh, H. Gilchrist, T. Sands, V. G. Keramidas, R. Haakenaasen, and D. K. Fork, *Appl. Phys. Lett.* 63, 3592 (1993).

137. R. Dat, J. K. Lee, O. Auciello, and A. I. Kingon, *Appl. Phys. Lett.* 67, 572 (1995).

138. N. Taga, H. Odaka, Y. Shigesato, I. Yasui, and T. E. Haynes, *J. Appl. Phys.* 80, 978 (1970).

139. C. J. Brinker and G. W. Scherer, "Sol-Gel Science: The Physics and Chemistry of Sol-Gel Processing." Academic Press, New York, 1990.

140. G. Yi and M. Sayer, *Ceram. Bull.* 70, 1173 (1991).

141. C. D. Lakeman and D. A. Payne, *Mater. Chem. Phys.* 38, 305 (1994).

142. D. M. Tahan, A. Safari, and L. C. Klein, *J. Am. Ceram. Soc.* 79, 1593 (1996).

143. F. Wang, A. Uusimäki, S. Leppävuori, S. F. Karmanenko, A. I. Dedyk, V. I. Sakharov, and I. T. Serenkov, *J. Mater. Res.* 13, 1243 (1998).

144. S. I. Jang, B. C. Choi, and H. M. Jang, *J. Mater. Res.* 12, 1327 (1997).

145. S. I. Jang and H. M. Jang, *Thin Solid Films* 330, 89 (1998).

146. T. Hayashi, H. Shinozaki, and K. Sasaki, *J. Eur. Ceram. Soc.* 19, 1011 (1999).

147. Z. A. Burhanuddin, M. S. Tomar, and E. Dayalan, *Thin Solid Films* 253, 53 (1994).

148. A. Nazeri, M. Kahn, and T. Kidd, *J. Mater. Sci. Lett.* 14, 1085 (1995).

149. F. M. Pontes, E. B. Arauja, E. Leite, J. A. Eiras, E. Longo, J. A. Varela, and M. A. Pereira-da-Silva, *J. Mater. Res.* 15, 1176 (2000).

150. F. M. Pontes, E. Longo, J. H. Rangel, M. I. Bernardi, E. R. Leite, and J. A. Varela, *Mater. Lett.* 43, 249 (2000).

151. W. G. Luo, A. L. Ding, X. He, B. Qi, and P. Qiu, *Ferroelectrics* 231, 199 (1999).

152. D. Bao, Z. Wang, W. Ren, L. Zhang, and X. Yao, *Ceram. Int.* 25, 261 (1999).

153. E. Dien, J. B. Briot, M. Lejeune, and A. Smith, *J. Eur. Ceram. Soc.* 19, 1349 (1999).

154. R. W. Schwartz, *Chem. Mater.* 9, 2325 (1997).

155. K. A. Vorotilov, M. I. Yanovskaya, L. I. Solovjeva, A. S. Valeev, V. I. Petrovsky, V. A. Vasiljev, and I. E. Obvinzeva, *Microelectron. Eng.* 29, 41 (1995).

156. M. C. Gust, N. D. Evans, L. A. Momodo, and M. L. Mecartney, *J. Am. Ceram. Soc.* 80, 2828 (1997).

157. H. Kozuka and M. Kajimura, *J. Am. Ceram. Soc.* 83, 1056 (2000).

158. V. Y. Shur, E. B. Blankova, A. L. Subbotin, E. A. Borisova, D. V. Pelegov, S. Hoffmann, D. Bolten, G. Gerhardt, and R Waser, *J. Euro. Ceram. Soc.* 19, 1391 (1999).

159. S. Hoffmann and R. Waser, *J. Euro. Ceram. Soc.* 19, 1339 (1999).

160. M. Grossman, R. Slowak, S. Hoffmann, H. John, and R. Waser, *J. Eur. Ceram. Soc.* 19, 1413 (1999).

161. R. Singh, *J. Appl. Phys.* 63, R59 (1988).

162. C. J. Peng and S. B. Krupanidhi, *Thin Solid Films* 223, 327 (1993).

163. T. Horikawa, N. Mikami, H. Ito, Y. Ohno, T. Makita, and K. Sato, *IEICE Trans. Electron.* E77-C, 385 (1994).

164. S. Saha and S. B. Krupanidhi, *Mater. Sci. Eng.* B57, 135 (1999).

165. M. de Keijser and G. Dormans, *MRS Bull.* 21, 37 (1996).

166. T. Kawahara, M. Yamamuka, T. Makita, J. Naka, A. Yuuki, N. Mikami, and K. Ono, *Japan. J. Appl. Phys.* 33, 5129 (1994).

167. M. Yoshida, H. Yamaguchi, T. Sakuma, Y. Miyasaka, P.-Y. Lesaicherre, and A. Ishitani, *J. Electrochem. Soc.* 142, 244 (1995).

168. S. Yamamichi, P.-Y. Lesaicherre, H. Yamaguchi, K. Takemura, S. Sone, H. Yabuta, K. Sato, T. Tamura, K. Nakajima, S. Ohnishi, K. Tokashiki, Y. Hayashi, Y. Kato, Y. Miyasaka, M. Yoshida, and H. Ono, *IEDM Tech. Dig.* 199 (1995).

169. T. Horikawa, N. Mikami, H. Ito, Y. Ohno, T. Makita, and K. Sato, *IEICE Trans. Electron.* E77-C, 477 (1994).

170. T. Horikowa, N. Mikami, H. Ito, Y. Ohno, T. Makita, and K. Sato, *IEICE Trans. Electron.* E77-C, 385 (1994).

171. C. M. Chu and P. Lin, *Appl. Phys. Lett.* 70, 249 (1997).

172. M. Izuha, K. Abe, M. Koike, S. Takeno, and N. Fukushima, *Appl. Phys. Lett.* 70, 1405 (1997).

173. H. J. Cho, H. Horii, C. S. Hwang, J. W. Kim, C. S. Kang, B. T. Lee, S. I. Lee, Y. B. Koh, and M. Y. Lee, *Japan. J. Appl. Phys.* 36, L874 (1997).

174. C. M. Chu and P. Lin, *Appl. Phys. Lett.* 72, 1241 (1998).

175. M. Izuha, K. Abe, and N. Fukushima, *Japan. J. Appl. Phys.* 36, 5866 (1997).

176. S. I. Ohfuji, M. Itsumi, and H. Akiya, *Integ. Ferroelectrics* 16, 209 (1997).

177. P. Bhattacharya, K. H. Park, and Y. Nishioka, *Japan. J. Appl. Phys.* 33, 5231 (1994).

178. P. Bhattacharya, T. Komeda, K. H. Park, and Y. Nishioka, *Japan. J. Appl. Phys.* 32, 4103 (1993).

179. S. Hayashi, M. Huffman, M. Azuma, Y. Shimada, T. Otsuki, G. Kano, L. D. McMillan, and C. A. Paz de Araujo, *VLSI Tech. Dig.* 153 (1994).

180. T. S. Chen, D. Hadad, V. Balu, B. Jiang, S. H. Kuah, P. McIntyre, S. Summerfelt, J. M. Anthony, and J. C. Lee, *IEDM Tech. Dig.* 679 (1996).

181. S. Yamamichi, P.-Y. Lesaicherre, H. Yamaguchi, K. Takemura, S. Sone, H. Yabuta, K. Sato, T. Tamura, K. Nakajima, S. Ohnishi, K. Tokashiki, Y. Hayashi, Y. Kato, Y. Miyasaka, M. Yoshida, and H. Ono, *IEDM Tech. Dig.* 119 (1995).

182. S. Sone, H. Yabuta, Y. Kato, T. Iizuka, S. Yamamichi, H. Yamaguchi, P.-Y. Lesaicherre, S. Nishimoto, and M. Yoshida, *Japan. J. Appl. Phys.* 35, 5089 (1996).

183. M. Yamamuka, T. Kawahara, A. Yuuki, and K. Ono, *Japan. J. Appl. Phys.* 35, 2530 (1996).

184. M. Yamamuka, T. Kawahara, T. Makita, A. Yuuki, and K. Ono, *Japan. J. Appl. Phys.* 35, 729 (1996).

185. Y. C. Choi, J. Lee, and B.-S. Lee, *Japan. J. Apl. Phys.* 36, 6824 (1997).

186. Y. P. Wang and T. Y. Tseng, *J. Appl. Phys.* 81, 6762 (1997).

187. R. Khamankar, B. Jiang, R. Tsu, W. Y. Hsu, J. Nulman, S. Summerfelt, M. Anthony, and J. Lee, *VLSI Tech. Dig.* 127 (1995).

188. S. G. Yoon and A. Safari, *Thin Solid Films* 254, 211 (1995).

189. A. Outzourhit, J. U. Trefny, T. Kito, B. Yarar, A. Naziripour, and A. M. Hermann, *Thin Solid Films* 259, 218 (1995).

190. K. Abe and S. Komatsu, *J. Appl. Phys.* 77, 6461 (1995).

191. T. S. Kim, M. H. Oh, and C. K. Kim, *Japan. J. Appl. Phys.* 32, 2837 (1993).

192. N. Ichinose and T. Ogiwara, *Japan. J. Appl. Phys.* 34, 5198 (1995).

193. S. C. Sun, M. S. Tsai, P. Lin, J. A. Lay, D. C. H. Yu, and M. S. Liang, "Abstracts of the ECS Meeting," May 5–10, 1996, Los Angeles, 96-1, 184 (1996).

194. K. Abe and S. Komatsu, *Japan. J. Appl. Phys.* 33, 5297 (1994).

195. R. Primig, R. Bruchhaus, and G. Schindler, *VLSI Tech. Dig.* 208 (1996).

196. S. O. Park, C. S. Hwang, H. J. Cho, C. S. Kang, H. K. Kang, S. J. Lee, and M. Y. Lee, *Japan. J. Appl. Phys.* 35, 1548 (1996).

197. Y. Fukuda, K. Numata, K. Aoki, and A. Nishimura, *Japan. J. Appl. Phys.* 35, 5178 (1996).

198. J. H. Joo, J. M. Scon, Y. C. Jeon, K.-Y. Oh, J. S. Roh, and J. J. Kim, *Appl. Phys. Lett.* 70, 3053 (1997).

199. J. Lee, Y. C. Choi, and B. S. Lee, *Japan. J. Appl. Phys.* 36, 3644 (1997).

200. T. Horikawa, J. Tanimura, T. Kawahara, M. Yamamuka, M. Tarutani, and K. Ono, *IEICE Trans. Electron.* E81-C4, 497 (1998).

201. M. Kiyotoshi, K. Eguchi, K. Imai, and T. Arikado, *Japan. J. Appl. Phys.* 37, 4487 (1998).

202. S. Yamamichi, H. Yabuta, T. Sakuma, and Y. Miyasaka, *Appl. Phys. Lett.* 64, 1644 (1994).

203. C. Basceri, Ph.D. Dissertation, North Carolina State University, 1997.

204. R. K. Sharma, N. H. Chan, and D. M. Smyth, *J. Am. Ceram. Soc.* 64, 448 (1981).

205. S. Stemmer, S. K. Streiffer, N. D. Browning, and A. I. Kingon, *Appl. Phys. Lett.* 74, 2432 (1999).

206. M. Sedlar and M. Sayer, *Integ. Ferroelectrics* 10, 113 (1995).

207. T. G. In, S. Baik, and S. Kim, *J. Mater. Res.* 13, 990 (1998).

208. D. Hennings, M. Klee, and R. Waser, *Adv. Mater.* 3, 334 (1991).

209. M. Vollman and R. Waser, *J. Am. Ceram. Soc.* 77, 235 (1994).

210. Y. M. Chiang and T. Takagi, *J. Am. Ceram. Soc.* 73, 3278 (1990).

211. Y. F. Kuo, Ph.D. Dissertation, Department of Electronics Engineering, Nationa Chiao-Tung University, Taiwan, ROC, 1999.

212. S. Yamamichi, A. Yamamichi, D. Park, and C. Hu, *IEDM Tech. Dig.* 261 (1996).

213. I. C. Ho and S. L. Fu, *J. Am. Ceram. Soc.* 75, 728 (1992).

214. B. Huybrechts, K. Ishizaki, and M. Takata, *J. Am. Ceram. Soc.* 75, 722 (1992).

215. K. N. Tu, J. W. Mayer, and L. C. Feldman, "Electronic Thin Film Science for Electrical Engineers and Materials Scientists," p. 236. Macmillan, New York, 1992.

216. M. Copel, J. D. Baniecki, P. R. Duncombe, D. Kotecki, R. Laibowitz, D. A. Neumayer, and T. M. Shaw, *Appl. Phys. Lett.* 73, 1832 (1998).

217. S. B. Krupanidhi and C. J. Peng, *Thin Solid Films* 305, 144 (1997).

218. S. Matsubara, T. Sakuma, S. Yamamichi, H. Yamaguchi, and Y. Miyasaka, *Mater. Res. Soc. Symp. Proc.* 200, 243 (1992).

219. S. Paek, J. Won, K. Lee, J. Choi, and C. Park, *Japan. J. Appl. Phys.* 35, 5757 (1996).

220. R. F. Pinizzotto, E. G. Jacobs, H. Yang, S. R. Summerfelt, and B. E. Gnade, *Mater. Res. Soc. Symp. Proc.* 243, 463 (1992).

221. L. A. Knauss, J. M. Pond, J. S. Horwitz, D. B. Chrisey, C. H. Mueller, and R. Treece, *Appl. Phys. Lett.* 69, 25 (1996).

222. D. Y. Noh, H. H. Lee, T. S. Kang, and J. H. Je, *Appl. Phys. Lett.* 72, 2823 (1998).

223. M. H. Yeh, K. S. Liu, and I. N. Lin, *Japan. J. Appl. Phys.* 34, 2247 (1995).

224. W. C. Tsai and T. Y. Tseng, *J. Am. Ceram. Soc.* 81, 768 (1998).

225. T. Kawahara, M. Yamamuka, A. Yuuki, and K. Ono, *Japan. J. Appl. Phys.* 34, 5077 (1995).

226. K. Kashihara, T. Okudaira, H. Itoh, T. Higaki, and K. Abe, "Extended Abstracts of the 53rd Autumn Meeting of the Japanese Society of Applied Physics," 16-p-zv-13, 1992 (in Japanese).

227. K. Kashihara, T. Okudaira, H. Itoh, T. Higaki, and K. Abe, *VLSI Tech. Dig.* 49 (1993).

228. M. Shimizu, M. Sugiyama, H. Fujisawa, and T. Shiosaki, *Japan. J. Appl. Phys.* 33 (1994).

229. G. A. Smolenskii and K. I. Rozgachev, *Zh. Tekh. Fiz.* 24, 1751 (1954).

230. G. Arlt, D. Hennings, and G. Dewith, *J. Appl. Phys.* 58, 1619 (1985).

231. K. Takemura, A. Nakai, T. Samuka, and Y. Miyasaka, "Extended Abstracts of the 55th Autumn Meeting of the Japanese Society of Applied Physics," 341, 1994 (in Japanese).

232. S. S. Lee and H. G. Kim, *Integ. Ferroelectrics* 11, 137 (1995).

233. V. M. Mukhortov, Yu. I. Golovko, V. A. Aleshin, E. Sviridov, Vl. M. Mukhortov, V. P. Dudkevich, and E. G. Fesenko, *Phys. Status Solidi* A78, 253 (1983).

234. K. Abe and S. Komatsu, *Japan. J. Appl. Phys.* 31, 2985 (1994).

235. K. Abe and S. Komatsu, *Japan. J. Appl. Phys.* 32, L115 (1993).

236. K. Abe and S. Komatsu, *Japan. J. Appl. Phys.* 32, 4186 (1993).

237. S. G. Yoon, J. C. Lee, and A. Safari, *J. Appl. Phys.* 76, 2999 (1994).

238. H. Terauchi, Y. Watanabe, H. Kasatani, K. Kamigaki, Y. Yano, T. Terashima, and Y. Bando, *J. Phys. Soc. Jpn.* 61, 2194 (1992).

239. K. Iijima, Y. Yano, T. Terashima, and Y. Bando, *Oyo-Butsuri (Japan)* 62, 1250 (1993).

240. W. J. Lee, H. G. Kim, and S. G. Yoon, *J. Appl. Phys.* 80, 5891 (1996).

241. M. S. Tsai, S. C. Sun, and T. Y. Tseng, *IEEE Trans. Electron. Dev.* 46, 1829 (1999).

242. C. Basceri, S. K. Streiffer, A. I. Kingon, and R. Waser, *J. Appl. Phys.* 82, 2497 (1997).

243. G. W. Dietz, M. Schumacher, R. Waser, S. K. Streiffer, C. Basceri, and A. I. Kingon, *J. Appl. Phys.* 82, 2359 (1997).

244. P. K. Larsen, G. J. M. Dormans, D. J. Taylor, and P. J. Van Veldhoven, *J. Appl. Phys.* 76, 2405 (1994).

245. C. Y. Ting and M. Wittmer, *J. Appl. Phys.* 54, 937 (1983).

246. L. Krusin-Elbaum, M. Wittmer, and D. S. Yee, *Appl. Phys. Lett.* 50, 1879 (1987).

247. M. Wittmer, *J. Vac. Sci. Technol.* A2, 273 (1984).

248. L. Krusin-Elbaum, M. Wittmer, C. Y. Ting, and J. J. Cuomo, *Thin Solid Films* 104, 81 (1983).

249. R. G. Vadimsky, R. P. Frankenthal, and D. E. Thompson, *J. Electrochem. Soc.* 126, 2017 (1979).

250. P. H. Holloway and C. C. Nelson, *Thin Solid Films* 35, L13 (1976).

251. M. L. Green, M. E. Gross, L. E. Papa, K. J. Schnoes, and D. Brasen, *J. Electrochem. Soc.* 132, 2677 (1985).

252. K. P. Lee, Y. S. Park, D. H. Ko, C. S. Hwang, C. J. Kang, K. Y. Lee, J. S. Kim, J. K. Park, B. H. Roh, J. Y. Lee, B. C. Kim, J. H. Lee, K. N. Kim, J. W. Park, and J. G. Lee, *IEDM Tech. Dig.* 907 (1995).

253. T. Nakamura, Y. Nakao, A. Kamisawa, and H. Takasu, *Japan. J. Appl. Phys.* 33, 5207 (1994).

254. R. Ramesh, T. Sands, and V. G. Keramidas, *J. Electron. Mater.* 23, 19 (1994).

255. C. S. Hwang, B. T. Lee, C. S. Kang, J. W. Kim, K. H. Lee, H. J. Cho, H. Horii, W. D. Kim, S. I. Lee, and Y. B. Roh, *J. Appl. Phys.* 83, 3703 (1998).

256. M. S. Tsai, Ph.D. Dissertation, Department of Electronics Engineering, National Chiao-Tung University, Taiwan, ROC, 1999.

257. M. S. Tsai and T. Y. Tseng, *IEEE Trans. Compon. Packag. Tech.* 23, 128 (2000).

258. S. Maruno, T. Kuroiwa, N. Mikami, K. Sato, S. Ohmura, M. Kaida, T. Yasue, and T. Koshikawa, *Appl. Phys. Lett.* 73, 954 (1998).

259. P.-Y. Lesaicherre, S. Yamamichi, H. Yamaguchi, K. Takemura, H. Watanabe, K. Tokashiki, K. Satoh, T. Sakuma, M. Yoshida, S. Ohnishi, K. Nakajima, K. Shibahara, Y. Miyasaka, and H. Ono, *IEDM Tech. Dig.* 831 (1994).

260. L. Kursin-Elbaum and M. Wittmer, *J. Electrochem. Soc.* 135, 2610 (1988).

261. W. Pan and S. B. Desu, *J. Vac. Sci. Technol.* B12, 3208 (1994).

262. S. H. Paek, K. S. Lee, J. Y. Seong, D. K. Choi, B. S. Kim, and C. S. Park, *J. Vac. Sci. Technol.* A16, 2448 (1998).

263. X. Chen, A. I. Kingon, H. N. Al-Shareef, K. R. Bellur, K. Gifford, and O. Auciello, *Integ. Ferroelectrics* 7, 291 (1995).

264. H. N. Al-Shareef, O. Auciello, and A. I. Kingon, in "Science and Technology of Electroceramic Thin Films" p. 133. (O. Auciello and R. Waser, Eds.). Kluwer Academic Publishers, Dordrecht, 1995.

265. T. Nakamura, Y. Nakao, A. Kamisawa, and H. Takasu, *Japan. J. Appl. Phys.* 34, 5184 (1995).

266. L. S. Roblee, M. J. Mangaudis, E. D. Lasinsky, A. G. Kimball, and S. B. Brummer, *Mater. Res. Soc. Symp. Proc.* 55, 303 (1986).

267. J. H. Ahn, W. Y. Choi, W. J. Lee, and H. G. Kim, *Japan. J. Appl. Phys.* 37, 284 (1998).

268. J. A. Rard, *Chem. Rev.* 85, 1 (1985).

269. R. S. Roth, T. Negas, and L. P. Cook, "Phase Diagrams for Ceramists," Vol. IV, p. 9. (G. Smith, Ed.). American Ceramics Society, Columbus, OH, 1981.

270. P.-Y. Lesaicherre, S. Yamamichi, K. Takemura, H. Yamaguchi, K. Tokashiki, Y. Miyasaka, M. Yoshida, and H. Ono, *Integ. Ferroelectrics* 11, 81 (1995).

271. T. Aoyama, A. Murakoshi, M. Koike, S. Takeno, and K. Imai, *Japan. J. Appl. Phys.* 37, L242 (1998).

272. C. B. Eom, R. B. Van Dover, J. M. Philips, D. J. Werder, J. H. Marshall, C. H. Chen, R. J. Cava, R. M. Fleming, and K. K. Fork, *Appl. Phys. Lett.* 63, 2570 (1993).

273. L. A. Wills and J. Amano, *Mat. Res. Soc. Symp. Proc.* 361, 47 (1995).

274. J. X. Jia, X. D. Wi, S. R. Foltyn, and P. Tiwari, *Appl. Phys. Lett.* 66, 2197 (1995).

275. S. Y. Hou, J. Kwo, R. K. Watts, J.-Y. Cheng, and D. K. Fork, *Appl. Phys. Lett.* 67, 1387 (1995).

276. S. Yamamichi, K. Takemura, T. Sakuma, H. Watanabe, H. Ono, K. Tokashiki, E. Ikawa, and Y. Miyasaka, "Proceedings of the 9th International Symposium on Applications of Ferroelectrics," 1994.

277. K. Tokashiki, K. Sato, K. Takemura, S. Yamamichi, P.-Y. Lesaicherre, H. Miyamoto, E. Ikawa, and Y. Miyasaka, "Proceedings of the 15th Symposium Dry Processes," 1994.

278. K. Takemura, T. Sakuma, and Y. Miyasaka, *Appl. Phys. Lett.* 64, 2967 (1994).

279. A. Grill, W. Kane, J. Viggiano, M. Brady, and R. Laibowitz, *J. Mater. Res.* 7, 3260 (1992).

280. K. Yoshikawa, T. Kimura, H. Noshiro, S. Otani, M. Yamada, and Y. Furumura, *Japan. J. Appl. Phys.* 33, L867 (1994).

281. K. Takemura, S. Yamamichi, P.-Y. Lesaicherre, K. Tokashiki, H. Miyamoto, H. Ono, Y. Miyasaka, and M. Yoshida, *Japan. J. Appl. Phys.* 34, 5224 (1995).

282. H. Yamaguchi, T. Iizuka, K. Koga, K. Takemura, S. Sone, H. Yabuta, S. Yamamichi, P.-Y. Lesaicherre, M. Suzuki, Y. Kojima, K. Nakajima, N. Kasai, T. Sakuma, Y. Kato, Y. Miyasaka, M. Yoshida, and S. Nishimoto, *IEDM Tech. Dig.* 675 (1996).

283. B. T. Lee, K. H. Lee, C. S. Hwang, W. D. Kim, H. Horii, H.-W. Kim, H. J. Cho, C. S. Kang, J. H. Chung, S. I. Lee, and M. Y. Lee, *IEDM Tech. Dig.* 50 (1997).

284. R. B. Khamankar, M. A. Kressley, M. R. Visokay, T. Moise, G. Xing, S. Nemoto, Y. Okuno, S. J. Fang, A. M. Wilson, J. F. Gaynor, T. Q. Hurd, D. L. Crenshaw, S. Summerfelt, and L. Colombo, *IEDM Tech. Dig.* 45 (1997).

285. K. R. Milkove and C. X. Wang, *J. Vac. Sci. Tech.* A15, 596 (1997).

286. B. A. Baumert, L. H. Chang, A. T. Matsuda, T. L. Tsai, C. J. Tracy, R. B. Gregory, P. L. Fejes, N. G. Cave, W. Chen, D. J. Taylor, T. Otsuki, E. Fujii, S. Hayashi, and K. Suu, *J. Appl. Phys.* 82, 2558 (1997).

287. H. J. Wintle, *IEEE Trans. Elect. Insulat.* 25, 27 (1990).

288. K. C. Kao and W. Hwang, "Electrical Transport in Solids." Pergamon Press, London, 1981.

289. T. Mihara and H. Watanabe, *Japan. J. Appl. Phys.* 34, 5664 (1995).

290. T. Mihara and H. Watanabe, *Japan. J. Appl. Phys.* 34, 5674 (1995).

291. P. Li and T. M. Lu, *Phys. Rev.* B43, 261 (1991).

292. C. S. Hwang, S. O. Park, C. S. Kang, H. J. Cho, H. K. Kang, S. T. Ahn, and M. Y. Lee, *Japan. J. Appl. Phys.* 34, 5178 (1995).

293. M. S. Tsai and T. Y. Tseng, *J. Electrochem. Soc.* 145, 2853 (1998).

294. J. Frenkel, *Tech. Phys. USSR* 5, 685 (1938).

295. J. Frenkel, *Phys. Rev.* 54, 647 (1938).

296. S. M. Sze, "Physics of Semiconductor Devices," 2nd ed. Wiley, New York, 1981.

297. A. M. Cowly and S. M. Sze, *J. Appl. Phys.* 36, 3212 (1965).

298. S. M. Sze, *J. Appl. Phys.* 38, 2951 (1967).

299. N. F. Mott, *Philos. Mag.* 24, 911 (1971).

300. J. G. Simmons, *Phys. Rev.* 155, 657 (1967).

301. C. A. Mead, *Phys. Rev.* 128, 2088 (1962).

302. S. Banerjee, B. Shen, I. Chen, J. Bohlman, G. Brown, and R. Doering, *J. Appl. Phys.* 65, 1140 (1989).

303. J. C. Schug, A. C. Lilly, Jr., and D. A. Lowitz, *Phys. Rev.* B1, 4811 (1970).

304. J. Antula, *J. Appl. Phys.* 43, 4663 (1972).

305. A. K. Jonscher and R. M. Hill, in "Physics of Thin Films," Vol. 8, p. 222. (G. Hass, Ed.). Academic Press, New York, 1975.

306. Y. Fukuda, K. Aoki, K. Numata, S. Aoyama, A. Nishimura, S. Summerfelt, and R. Tsu, *Integ. Ferroelectrics* 11, 121 (1995).

307. Y. Fukuda, K. Numata, K. Aoki, and A. Nishimura, "Proceedings of the 10th IEEE International Symposium on Applied Ferroelectrics," 1995.

308. Y. Fukuda, H. Haneda, I. Sakaguchi, K. Numata, K. Aoki, and A. Nishimura, *Japan. J. Appl. Phys.* 36, L1514 (1997).

309. Y. Fukuda, K. Numata, K. Aoki, A. Nishimura, G. Fujihashi, S. Okamura, S. Ando, and T. Tsukamoto, *Japan. J. Appl. Phys.* 37, L453 (1998).

310. D. Mapother, H. Crooks, and R. Mauer, *J. Chem. Phys.* 18, 1231 (1950).

311. S. Y. Cha, H. C. Lee, W. J. Lee, and H. G. Kim, *Japan. J. Appl. Phys.* 34, 5220 (1995).

312. K. H. Lee, C. S. Hwang, B. T. Lee, W. D. Kim, H. Horii, C. S. Kang, H. J. Cho, S. I. Lee, and M. Y. Lee, *Japan. J. Appl. Phys.* 36, 5860 (1997).

313. B. M. Kulwicki, in "Advances in Ceramics," Vol. 1, p. 138. (L. M. Levinson and D. C. Hill, Eds.). American Ceramics Society, Columbus, OH, 1981.

314. J. C. Simpson and J. F. Cordaro, *J. Appl. Phys.* 63, 1781 (1988).

315. T. Kaga, M. Ohkura, F. Murai, N. Yokoyama, and E. Takeda, *J. Vac. Sci. Technol.* B 13, 2329 (1995).

316. S. Kamiyama, T. Saeki, H. Mori, and Y. Numasawa, *IEDM Tech. Dig.* 827 (1991).

317. G. Lo, D. L. Kwong, P. C. Fazan, V. K. Mathews, and N. P. Sandler, *IEEE Electron. Dev. Lett.* 14, 216 (1993).

318. V. Balu, T. S. Chen, S. Katakam, J. H. Lee, B. White, S. Zafar, B. Jiang, P. Zurcher, R. E. Jones, and J. Lee, *Integ. Ferroelectrics* 21, 155 (1998).

319. S. Zafar, R. E. Jones, P. Chu, B. White, B. Jiang, D. Taylor, P. Zurcher, and S. Gillespie, *Appl. Phys. Lett.* 72, 2820 (1998).

320. M. S. Tsai and T. Y. Tseng, *Mater. Chem. Phys.* 57, 47 (1998).

321. M. S. Tsai, S. C. Sun, and T. Y. Tseng, *IEDM Tech. Dig.* 100 (1997).

322. M. A. Alim, M. A. Seitz, and R. W. Hirthe, *J. Appl. Phys.* 63, 2337 (1988).

323. C. H. Lai and T. Y. Tseng, *IEEE Trans. Comp. Packag. Manufact. Tech.* 17, 309 (1994).

324. K. Watanabe, J. Tressler, M. Sadamoto, C. Lsobe, and M. Tanaka, *J. Electrochem. Soc.* 143, 3008 (1996).

325. R. Waser, T. Baiatu, and K. H. Hardtl, *J. Am. Ceram. Soc.* 73, 1645 (1990).

326. M. Tomozane, Y. Nannichi, I. Onodera, T. Fukase, and F. Hasegawa, *Japan. J. Appl. Phys.* 27, 260 (1988).

327. K. S. Cole and R. H. Cole, *J. Chem. Phys.* 9, 341 (1941).

328. S. Ezhilvalavan and T. R. N. Kutty, *Appl. Phys. Lett.* 69, 3540 (1996).

329. S. Ezhilvalavan and T. R. N. Kutty, *IETE J. Res. (India)* 43, 233 (1997).

330. G. E. Pike, *Mater. Res. Soc. Proc.* 5, 369 (1982).

331. S. Ezhilvalavan and T. Y. Tseng, *J. Phys. D: Appl. Phys.* 33, 1137 (2000).

332. A. K. Jonscher, "Dielectric Relaxation in Solids." Chelsa Dielectrics Press, London, 1983.

333. J. Lee and A. Safari, Ph.D. Thesis, Rutgers University, 1993.

334. Y. Fukuda, K. Numata, K. Aoki, and A. Nishimura, *Japan. J. Appl. Phys.* 35, 5178 (1996).

335. Y. Fukuda, K. Numata, K. Aoki, and A. Nishimura, "Extended Abstracts of the 56th Autumn Meeting of the Japanese Society of Applied Physics," 26aZG9, 1997.

336. D. M. Smyth, M. P. Harmer, and P. Peng, *J. Am. Ceram. Soc.* 72, 2276 (1989).

337. R. Waser, *J. Am. Ceram. Soc.* 72, 2234 (1989).

338. M. S. Tsai and T. Y. Tseng, *J. Phys. D: Appl. Phys.* 32, 2141 (1999).

339. S. Ezhilvalavan and T. Y. Tseng, *Appl. Phys. Lett.* 74, 2477 (1999).

340. S. M. Sze, "Physics of Semiconductor Devices," 2nd ed. Wiley, New York, 1981.

341. L. A. Knauss, J. M. Pond, J. S. Horwitz, D. B. Chrisey, C. H. Mueller, and R. Treece, *Appl. Phys. Lett.* 69, 25 (1996).

342. H. Kobayashi and T. Kobayashi, *Japan. J. Appl. Phys.* 33, L533 (1994).

343. W. D. Kingery, H. K. Bowen, and D. R. Uhlmann, "Introduction to Ceramics," 2nd ed. Wiley, New York, 1991.

344. B. Jiang, V. Balu, T. S. Chen, S. H. Kuah, J. C. Lee, P. Y. Chu, R. E. Jones, P. Zurcher, D. J. Taylor, and S. J. Gillespie, *Symp. VLSI Tech. Dig.* 26 (1996).

345. J. S. Suehle, P. Chaparala, C. Messick, W. Miller, and K. C. Boyko, *IEEE Int. Integ. Reliabil. Work.* 59 (1993).

346. D. J. Dumin, N. B. Heilemann, and N. Husain, *IEEE Int. Conf. Microelect. Test Struct.* 61 (1991).

347. A. Martin, P. O'Sullivan, and A. Mathewson, *Microelectron. Reliabil.* 38, 37 (1998).

348. J. Prendergast, J. S. Suehle, P. Chaparala, E. Murphy, and M. Stephenson, *IEEE Int. Phys. Symp.* 124 (1995).

349. A. Martin, P. O'Sullivan, and A. Mathewson, *Microelectron. J.* 27, 633 (1996).

350. S. B. Desu and I. K. Yoo, *J. Electrochem. Soc.* 140, L133 (1993).

351. E. Rosenbaum, J. C. King, and C. Hu, *IEEE Trans. Electron. Dev.* 43, 70 (1996).

352. E. Harari, *J. Appl. Phys.* 49, 2478 (1978).

353. C. Fritzsche, *Z. angewandte Physik* 24, 48 (1967).

354. T. Brozek and A. Jakubowski, *Microelectron. Reliabil.* 33, 1637 (1993).

355. H. S. Momose, S. I. Nakamura, Y. Katsumata, and H. Iwai, "Proceedings of the 27th European Solid-State Device Research Conference," 1997.

356. P. Cappelletti, P. Ghezzi, F. Pio, and C. Riva, *IEEE Int. Conf. Microelectron. Test Struct.* 4, 81 (1991).

357. A. Martin, P. O'Sullivan, and A. Mathewson, *Qual. Reliabil. Eng. Int.* 12, 281 (1996).

358. M. Kamolz, *Int. Wafer Level Reliabil. Work.* 121 (1992).

359. A. Martin, J. S. Suehle, P. Chaparala, P. O'Sullivan, and A. Mathewson, *IEEE Int. Reliabil. Phys. Symp.* 67 (1996).

360. P. Hiergeist, A. Spitzer, and S. Rohl, *IEEE Trans. Electron. Dev.* 36, 913 (1989).

361. T. Horikowa, N. Mikami, H. Ito, Y. Ohno, T. Makita, and K. Sato, *IEEE Int. Phys. Symp.* 82 (1997).

362. H. Reisinger, H. Wendt, G. Beitel, and E. Fritsch, *VLSI Tech. Dig.* 58 (1998).

363. K. F. Schuegraf and C. Hu, *Semicond. Sci. Technol.* 9, 989 (1994).

364. H. Y. Lee and K.-L. Lee, *IEEE Trans. Compon. Hybrids Manuf. Technol.* 4, 443 (1984).

365. E. Loh, *J. Appl. Phys.* 53, 6229 (1982).

366. H. Neumann and G. Arlt, *Ferroelectrics* 69, 179 (1986).

367. L. Lehorec and G. A. Shirn, *J. Appl. Phys.* 33, 2036 (1962).

368. C. Sudhamn, J. Kim, J. Lee, V. Chikarmane, W. Shepherd, and E. R. Myers, *J. Vac. Sci. Technol.* B11, 1302 (1993).

369. K. Yamabe, M. Inomoto, and K. Imai, *Japan. J. Appl. Phys.* 37, L1162 (1998).

370. A. Kudo and T. Sakata, *J. Phys. Chem.* 99, 15,963 (1995).

371. K. Itoh, R. Hori, H. Masuda, Y. Kamigaki, H. Kawamoto, and H. Katto, *ISSCC Dig. Tech. Papers* 228 (1980).

372. J. H. Bruning, *Proc. SPIE* 3051, 14 (1997).

373. C. J. Wilson, *Proc. SPIE* 3051, 28 (1997).

374. B. H. Roh, Y. H. Cho, Y. G. Shin, C. K. Hong, S. D. Kwon, K. Y. Lee, H. K. Kang, K. N. Kim, and J. W. Park, *Japan. J. Appl. Phys.* 35, 4618 (1996).

375. A. Bryant, W. Hansen, and T. Mii, *IEDM Tech. Dig.* 671 (1994).

376. B. H. Roh, C. S. Yoon, D. U. Choi, M. J. Kim, D. W. Ha, J. Y. Lee, K. N. Kim, and J. W. Park, *SSDM'96 Tech. Dig.* 830 (1996).

377. S. M. Hu, *VLSI Sci. Tech. Electrochem. Soc.* 722 (1987).

378. B. Davari, C. W. Koburger, R. Schulz, J. D. Warnock, T. Furukawa, M. Jost, Y. Taur, W. G. Schwittek, J. K. DeBrosse, M. L. Kerbaugh, and J. L. Mauer, *IEDM Tech. Dig.* 61 (1989).

379. T. Hamamoto, S. Sugiura, and S. Sawada, *IEDM Tech. Dig.* 915 (1995).

380. K. Ishimaru, F. Matsuoka, M. Takahashi, M. Nishigohri, Y. Okayama, Y. Unno, M. Yabuki, K. Umezawa, N. Tsuchiya, O. Fujii, and M. Kinugawa, *VLSI Tech. Dig. Tech. Papers* 123 (June 1997).

381. H. Ha, J. H. Sim, and K. Kim, *SSDM Tech. Dig.* 514 (1997).

382. R. H. Dennard, F. H. Gaenssln, H. N. Yu, V. L. Rideout, E. Bassous, and A. R. LeBlamc, *IEEE J. Solid State Circ.* SC-9, 256 (1974).

383. P. K. Chatterjee, W. R. Hunter, T. C. Holloway, and Y. T. Lin, *IEEE Electron. Dev. Lett.* EDL-1, 220 (1980).

384. M. T. Bohr, *IEDM Tech. Dig.* 241 (1995).

385. G. Raghavan, C. Chiang, P. B. Anders, S. M. Tzeng, R. Villasol, G. Bai, M. Bohr, and D. B. Fraser, *Thin Solid Films* 262, 168 (1995).

386. J. Li, T. E. Seidel, and J. W. Mayer, *MRS Bull.* XIX, 15 (1994).

387. N. Hayasaka, H. Miyajima, Y. Nakasaki, and R. Katsumata, *SSDM Tech. Dig.* 157 (1995).

388. S. M. Rossnagel, D. Mikalsen, H. Kinoshita, and J. J. Cuomo, *J. Vac. Sci. Technol.* A9, 261 (1991).

389. N. Motegi, Y. Kashimoto, K. Nagatani, S. Takahashi, T. Kondo, Y. Mizusawa, and I. Nakayama, *J. Vac. Sci. Technol.* B13, 1906 (1995).

390. G. A. Dixit, W. Y. Hsu, A. J. Konecni, S. Krishnan, J. D. Luttmer, R. H. Havemann, J. Forster, G. D. Yao, M. Narasimhan, Z. Xu, S. Ramaswami, F. S. Chen, and J. Nulman, *IEDM Tech. Dig.* 357 (1996).

391. R. E. Jones, P. Zurcher, P. Chu, D. I. Taylor, Y. T. Lii, B. Jiang, P. D. Maniar, and S. J. Gillespie, *Microelectron. Eng.* 29, 3 (1995).

392. C. S. Hwang, B. T. Lee, H. J. Cho, K. H. Lee, C. S. Kang, and M. Y. Lee, "International Conference on Solid-State Devices and Materials," Osaka, August 12–24, 1995.

393. P.-Y. Lesaicherre, H. Yamaguchi, Y. Miyasaka, H. Watanabe, H. Ono, and M. Yoshida, *Integ. Ferroelectrics* 8, 201 (1995).

394. K. N. Kim, D. H. Kwak, Y. S. Hwang, G. T. Jeong, T. Y. Chung, B. J. Park, Y. S. Chun, J. H. Oh, C. Y. Yoo, and B. S. Joo, *VLSI Tech. Symp.* T4A-1 (1999).

395. H. Horii, B. T. Lee, H. J. Lim, S. H. Joo, C. S. Kang, C. Y. Yoo, H. B. Park, W. D. Kim, S. I. Lee, and M. Y. Lee, *VLSI Tech. Symp.* T8A-4 (1999).

396. M. Kiyotoshi, S. Yamazaki, E. Eguchi, K. Hieda, Y. Fukuzumi, M. Izuha, S. Niwa, K. Nakamura, A. Kojima, H. Tomita, T. Kubota, M. Satoh, Y. Kohyama, Y. Tsunashima, T. Arikado, and K. Okumura, *VLSI Tech. Symp.* T8A-3 (1999).

397. T. Matsuki, Y. Hayashi, and T. Kunio, *IEEE Tech. Dig.—Int. Electron. Devices Meeting* 691 (1996).

398. C. S. Hwang, *Mater. Sci. Eng.* B56, 178 (1998).

399. K. Okamoto, Y. Nasu, and Y. Hamakawa, *IEEE Trans. Electron. Dev.* ED-28, 698 (1991).

400. A. Yuuki, M. Yamamuka, T. Makita, T. Horikawa, T. Shibano, N. Hirano, H. Maeda, N. Mikami, K. Ono, H. Ogata, and H. Abe, *IEEE Tech. Dig.—Int. Electron. Devices Meeting* 115 (1995).

401. T. Horikawa, M. Tasayoshi, T. Kawahara, M. Yamamuka, N. Hirano, T. Sato, S. Matsuno, T. Shibano, F. Uchikawa, K. Ono, and T. Oomori, *Mater. Res. Soc. Symp. Proc.* 541 (1999).

402. A. Kingon, S. K. Streiffer, C. Basceri, and S. R. Summerfelt, *MRS Bull.* 46 (1996).

403. B. Biharaj, J. Kumar, G. T. Stauf, P. C. Stauf, P. C. Van Burkirk, and C. S. Hwang, *J. Appl. Phys.* 76, 1169 (1994).

404. K. H. Guenther, *Appl. Opt.* 23, 3612 (1984).

405. C. G. Granqvist and O. Hunderi, *Phys. Rev. B* 16, 3513 (1977).

406. W. K. Chen, C. M. Cheng, J. Y. Huang, W. F. Hsieh, and T. Y. Tseng, *J. Phys. Chem. Solids* 61, 969 (2000).

407. J. Zhuang, G. S. Li, X. C. Gao, X. B. Guo, Y. H. Huang, Z. Z. Shi, Y. Y. Weng, and J. Lu, *Optics Comm.* 82, 69 (1991).

408. N. A. Vainos and M. C. Gower, *J. Opt. Soc. Am.* B8, 2355 (1991).

409. M. B. Klein and G. C. Valley, *J. Appl. Phys.* 57, 4901 (1985).

410. B. Bihari, J. Kumar, G. T. Stauf, P. C. Van Buskirk, and C. S. Hwang, *J. Appl. Phys.* 76, 1169 (1994).

411. T. S. Kim, M. H. Oh, and C. H. Kim, *Japan. J. Appl. Phys.* 32, 2837 (1993).

412. P. M. Alt, *Proc. SID* 25, 123 (1984).

413. X. Ouyang, A. H. Kittai, and T. Xiao, *J. Appl. Phys.* 79, 3299 (1996).

414. T. S. Kim, C. H. Kim, and M. H. Oh, *J. Appl. Phys.* 75, 7998 (1994).

415. Y. P. Wang and T. Y. Tseng, *J. Mater. Sci.* 34, 3573 (1999).
416. R. Swanepoel, *J. Phys. E: Sci. Instrum.* 16, 1214 (1983).
417. A. R. Forouhi and I. Bloomer, *Phys. Rev. B* 38, 1865 (1988).
418. A. R. Forouhi and I. Bloomer, *Phys. Rev. B* 34, 7018 (1986).
419. J. Szczyrbowski, K. Schmalzbauer, and H. Hoffmann, *Thin Solid Films* 130, 57 (1985).
420. J. C. Manifacier, J. Gasiot, and J. P. Fillard, *J. Phys. E: Sci. Instrum.* 9, 1002 (1976).
421. S. H. Wemple and M. DiDomenico, Jr., *Phys. Rev. B* 3, 1338 (1971).
422. M. DiDomenico, Jr., and S. H. Wemple, *J. Appl. Phys.* 40, 720 (1969).
423. Y. F. Kuo and T. Y. Tseng, *Mater. Chem. Phys.* 61, 244 (1999).
424. B. Panda, A. Dhar, G. D. Nigam, D. Bhattacharya, and S. K. Ray, *Thin Solid Films* 332, 46 (1998).
425. I. Suzuki, M. Ejima, K. Watanabe, Y. M. Xiong, and T. Saitoh, *Thin Solid Films* 313-314, 214 (1998).
426. C. H. Peng and S. B. Desu, *J. Am. Ceram. Soc.* 77, 1486 (1994).
427. S. Troiller-McKinstry, H. Hu, S. B. Krupanidhi, P. Chindaudom, K. Vedam, and R. E. Newnham, *Thin Solid Films* 230, 15 (1993).
428. S. Troiller-McKinstry, J. Chen, K. Vedam, and R. E. Newnham, *J. Am. Ceram. Soc.* 78, 1907 (1995).
429. P. G. Snyder and Y. M. Xiong, *Surface Interface Anal.* 18, 107 (1992).
430. R. M. A. Azzam and N. M. Bashara, "Ellipsometry and Polarised Light." North-Holland, Amsterdam, 1977.
431. W. Zhu, O. K. Tan, J. Deng, and J. T. Oh, *J. Mater. Res.* 15, 1291 (2000).
432. F. Techeliebou, H. S. Ryu, C. K. Hong, W. S. Park, and S. Baik, *Thin Solid Films* 305, 30 (1997).
433. A. Mansingh and C. V. R. Vasanta Kumar, *J. Mat. Sci. Lett.* 7, 1104 (1988).
434. R. Thielsch, K. Kaemmer, G. Holzapfel, and L. Schultz, *Thin Solid Films* 301, 203 (1997).
435. F. R. Flory, "Thin Films for Optical Systems." Marcel Dekker, New York, 1995.
436. J. M. Bennet and M. J. Borty, *Appl. Optics* 5, 41 (1966).
437. I. Ochidal, K. Navratil, and E. Schmidt, *Appl. Phys.* A29, 157 (1982).
438. H. Demiryont, L. R. Thomson, and G. J. Collins, *Appl. Optics* 25, 1311 (1986).
439. G. Bauer, *Ann. Phys.* 19, 434 (1934).
440. R. Jacobson, *Phys. Thin Films* 8, 51 (1975).
441. W. L. Bragg and A. B. Pippard, *Acta Crystallogr.* 6, 865 (1953).
442. H. A. McLeod, *J. Vac. Sci. Technol.* A4, 418 (1986).
443. R. Theilsch, "Proceedings of the 2nd Conference on Future Applications of Thin Films in Microoptics, Bioelectronics and Optics," 1989.
444. J. C. Tauc, "Optical Properties of Solids." North-Holland, Amsterdam, 1972.
445. N. F. Mott and E. A. Davis, "Electronic Processes in Non-Crystalline Materials," 2nd ed. Clarendon Press, Oxford, UK, 1979.
446. J. C. Tauc and A. Menth, *J. Non-Crystal. Solids* 8–10, 569 (1972).
447. C. V. R. Vasanta Kumar and A. Mansingh, "7th IEEE International Symposium on Applications of Ferroelectrics." IEEE Press, New York, 1990.
448. J. S. Zhu, X. M. Lu, W. Jiang, W. Tain, M. Zhu, M. S. Zhang, X. B. Chen, X. Liu, and Y. N. Wang, *J. Appl. Phys.* 81, 1392 (1997).
449. P. Pasierb, S. Komornicki, and M. Radecka, *Thin Solid Films* 324, 134 (1998).
450. M. N. Kamalasanan, N. D. Kumar, and S. Chandra, *J. Appl. Phys.* 74, 679 (1993).
451. X. M. Lu, J. S. Zhu, W. Y. Zhang, G. Q. Ma, and Y. N. Wang, *Thin Solid Films* 274, 165 (1996).
452. W. A. Harrison, "Electronic Structure and the Properties of Solids." W. H. Freeman, San Franscisco, 1980.
453. H. F. Cheng, *J. Appl. Phys.* 79, 7965 (1996).
454. D. P. Arndt, R. M. Azzam, J. M. Bennett, J. P. Borgono, C. K. Carnuglia, W. E. Case, J. A. Dobrovolski, U. J. Gibson, T. T. Hart, F. C. Ho, V. A. Hodgin, W. R. Klapp, H. A. Macleod, E. Pelletier, M. K. Purvis, D. M. Quinn, D. H. Stome, R. Swenson, P. A. Temple, and T. F. Thoun, *Appl. Optics* 23, 3571 (1984).
455. R. Jacobsson, *Arkiv. Fys.* 24, 17 (1963).
456. W. Zhu, O. K. Tan, and X. Yao, *J. Appl. Phys.* 84, 5134 (1998).
457. Y. Morita, K. Nakamura, and C. Kim, *Sensors Actuators* B33, 96 (1996).
458. M. Noda, K. Hashimoto, R. Kubo, H. Tanaka, T. Mukaigawa, H. Xu, and M. Okuyama, *Sensors Actuators A* 77, 39 (1999).
459. C. Hanson, H. Beratan, R. Owen, M. Corbin, and McKenny, *SPIE Infrared Detectors: State of the Art* 1735, 17 (1992).
460. R. Watton, *Integ. Ferroelectrics* 4, 175 (1994).
461. M. Ueno, O. Kaneda, T. Ishikawa, K. Yamada, A. Yamada, M. Kimata, and M. Nanoshita, *SPIE Infrared Technol.* XXI, 2552, 636 (1995).
462. A. Tanaka, S. Matsumoto, N. Tsukamoto, S. Itoh, K. Chiba, T. Endoh, A. Nakazato, K. Okuyama, Y. Kamazawa, M. Hijikawa, H. Gotoh, T. Tanaka, and N. Teranishi, *IEEE Trans. Electron. Dev.* 43, 1844 (1996).
463. C. Björmander, K. Sreenivas, A. M. Grishin, and K. V. Rao, *Appl. Phys. Lett.* 67, 58 (1995).
464. R. W. Whatmore, P. C. Osbond, and N. M. Shorrocks, *Ferroelectrics* 76, 351 (1987).
465. C. Hanson and H. Beratan, *IEEE Trans.* 657 (1995).
466. L. C. Sengupta, S. Sengupta, D. A. Rees, J. Synowcyznski, S. Stowell, E. H. Ngo, and L. H. Chiu, *Integ. Ferroelectrics* 22, 393 (1998).
467. J. G. Cheng, X. J. Meng, B. Li, J. Tang, S. L. Guo, J. H. Chu, M. Wang, H. Wang, and Z. Wang, *Appl. Phys. Lett.* 75, 2132 (1999).
468. H. X. Zhang, C. H. Kam, Y. Zhou, X. Q. Han, Y. L. Lam, Y. C. Chan, and K. Pita, *Mater. Chem. Phys.* 63, 174 (2000).
469. K. Okamoto, Y. Nasu, and Y. Hamakawa, *IEEE Trans. Electron. Dev.* 28, 698 (1981).
470. R. Fukao, H. Fujikawa, and Y. Hainakawa, *Japan. J. Appl. Phys.* 28, 2446 (1989).
471. P. J. Haroop and D. S. Cambell, *Thin Solid Films* 2, 273 (1968).
472. A. B. Kauftnan, *IEEE Trans. Electron. Dev.* 16, 562 (1969).
473. N. F. Borelli and M. M. Layton, *IEEE Trans. Electron. Dev.* 16, 562 (1969).
474. M. H. Song, Y. H. Lee, T. S. Hahn, M. H. Oh, and K. H. Yoon, *Solid State Electron.* 42, 1711 (1998).
475. P. Li, T. M. Lu, and H. Bakhm, *Appl. Phys. Lett.* 58, 2639 (1991).
476. I. H. Pratt and S. Firestone, *J. Vac. Sci. Technol.* 8, 256 (1971).
477. M. H. Song, Y. H. Lee, T. S. Hahn, M. H. Oh, and K. H. Yoon, *J. Kor. Ceram. Soc.* 32, 761 (1995).
478. M. H. Song, Y. H. Lee, T. S. Hahn, M. H. Oh, and K. H. Yoon, *J. Appl. Phys.* 79, 3744 (1996).
479. L. Davis, Jr., and L. G. Rubin, *J. Appl. Phys.* 24, 1194 (1953).
480. K. M. Johnson, *J. Appl. Phys.* 33, 2826 (1962).
481. K. Bethe and F. Welz, *Mater. Res. Bull.* 6, 209 (1971).
482. F. W. Van Keuls, R. R. Romanofsky, N. D. Varalijay, F. A. Miranda, C. L. Candey, S. Aggarwal, T. Venkatesan, and R. Ramesh, *Microwave Optics Technol. Lett.* 20, 53 (1999).
483. V. N. Keis, A. B. Kozyrev, M. L. Khazov, J. Sok, and J. S. Lee, *Electron. Lett.* 34, 1107 (1998).
484. F. A. Miranda, R. Romanofsky, F. W. Van Keuls, C. H. Mueller, R. E. Treece, and T. E. Rivkin, *Integ. Ferroelectrics* 17, 231 (1998).
485. J. M. Ponds, S. W. Kirchoeffer, W. Chang, J. S. Horwitz, and D. B. Chrisey, *Integ. Ferroelectrics* 22, 317 (1998).
486. F. A. Miranda, F. W. Van Keuls, R. R. Romanofsky, and G. Subramanyam, *Integ. Ferroelectrics* 22, 269 (1998).
487. W. J. Kim, W. Chang, S. B. Qadri, J. M. Pond, S. W. Kirchoefer, J. S. Horwitz, and D. B. Chrisey, *Appl. Phys. A* 70, 313 (2000).
488. C. M. Carlson, T. V. Rivkin, P. A. Parilla, J. D. Perkins, D. S. Ginley, A. B. Kozyrev, V. N. Oshadchy, and A. S. Pavlov, *Appl. Phys. Lett.* 76, 1920 (2000).
489. L. C. Sengupta and S. Sengupta, *IEEE Trans. Ultrason., Ferroelect. Freq. Cont.* 44, 792 (1997).
490. S. Sengupta, L. C. Sengupta, D. P. Vijay, and S. B. Desu, *Integ. Ferroelectrics* 13, 1 (1997).
491. J. Im, O. Auciello, P. K. Baumann, S. K. Streiffer, D. Y. Kaufmann, and A. R. Krauss, *Appl. Phys. Lett.* 76, 625 (2000).
492. A. B. Kozyrev, V. N. Keis, G. Koepft, R. Yandrofski, O. I. Soldatenkov, K. A. Dudin, and D. P. Dovgan, *Microelectron. Eng.* 29, 257 (1995).
493. W. Chang, J. S. Horwitz, A. C. Carter, J. M. Pond, S. W. Kirchoefer, C. M. Gillmore, and D. B. Chrisey, *Appl. Phys. Lett.* 74, 1033 (1999).

494. W. Chang, C. M. Gilmore, W. J. Kim, J. M. Pond, S. W. Kirchoefer, S. B. Qadri, D. B. Chrisey, and J. S. Horwitz, *J. Appl. Phys.* 87, 3044 (2000).
495. W. J. Merz, *Phys. Rev.* 77, 52 (1950).
496. J. C. Slater, *Phys. Rev.* 78, 748 (1950).
497. R. F. Brown, *Can. J. Phys.* 39, 741 (1961).
498. P. W. Forsbergh, Jr., *Phys. Rev.* 93, 686 (1954).
499. W. R. Buessem, L. E. Cross, and A. K. Goswami, *J. Am. Ceram. Soc.* 49, 36 (1966).
500. W. J. Kim, W. Chang, S. B. Qadri, J. M. Pond, S. W. Kirchoefer, D. B. Chrisey, and J. S. Horwitz, *Appl. Phys. Lett.* 76, 1185 (2000).
501. C. L. Chen, H. H. Feng, Z. Zhang, A. Braedeikis, Z. J. Huang, W. K. Chu, C. W. Chu, F. A. Miranda, F. W. Van Keuls, R. R. Romanofsky, and Y. Liou, *Appl. Phys. Lett.* 75, 412 (1999).
502. Q. X. Jia, J. R. Groves, P. Arendt, Y. Fan, A. T. Findikoglu, S. R. Foltyn, H. Jiang, and F. A. Miranda, *Appl. Phys. Lett.* 74, 1564 (1999).
503. J. Lindner, F. Weiss, J. P. Senateur, V. Galindo, W. Haessler, M. Weihnacht, J. Santiso, and A. Figueras, *J. Euro. Ceram. Soc.* 19, 1435 (1999).
504. O. G. Vendik, L. T. Ter-Martirosyan, A. I. Dedyk, S. F. Karmanenko, and R. A. Chakalov, *Ferroelectrics* 144, 33 (1993).
505. D. C. DeGroot, J. A. Beall, R. B. Marks, and D. A. Rudman, *IEEE Trans. Appl. Superconductiv.* 5, 2272 (1995).
506. O. G. Vendik, I. G. Mironenko, and L. T. Ter-Martirosyan, *Microwaves &RF* 33, 67 (1994).
507. S. S. Gevorgian, D. I. Kaparkov, and O. G. Vendik, *IEEE Proc. Microwave Antenn. Propag.* 141, 501 (1994).

Chapter 3

ULTRATHIN GATE DIELECTRIC FILMS FOR Si-BASED MICROELECTRONIC DEVICES

C. Krug, I. J. R. Baumvol

Instituto de Física, Universidade Federal do Rio Grande do Sul, Porto Alegre, RS 91509-900, Brazil

Contents

1. INTRODUCTION

A microelectronic device may be defined as one that relies on electrons (specifically, on their quantum behavior) for functioning. "Micro," although related to physical dimension, stands for a character of building block; an electronic device is usually an assembly of microelectronic components. Generally speak-ing, a device must be controllable to be useful. The possibility of controlling a significant physical quantity, electron "flow," is offered by the band structure of a crystal of semiconducting material suitably doped with an electron donor or acceptor. From a myriad of semiconductors, silicon has been chosen to form the vast majority of microelectronic devices, totally dominating the commercial market. However, if we are currently witnessing

Handbook of Thin Film Materials, edited by H.S. Nalwa
Volume 3: Ferroelectric and Dielectric Thin Films
Copyright © 2002 by Academic Press
All rights of reproduction in any form reserved.

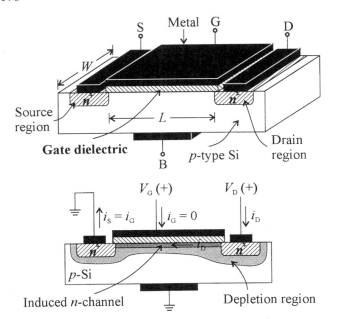

Fig. 1. Three-dimensional sketch and cross section of an *n*-channel MOSFET on *p*-type Si, showing the channel width *W* and length *L*. The active region of the device under bias is also shown.

a silicon age, it is not solely because silicon is a semiconductor. (Gallium arsenide (GaAs), for example, shows superior electron transport properties and special optical properties [1]). It is perhaps mainly due to the fact that silicon has a stable electrical passivating oxide, which can be placed on top of it with a structurally sharp interface of incomparable electronic quality; i.e., there is an extremely low density of interface electronic quantum states. Electrical passivation of the semiconductor crystal surface is the key factor to control, as surface properties can be the dominant influence on the electronic current in the device in opposition to the contribution of charge transport in the bulk. This is undesirable because devices with surface-dominated characteristics usually exhibit poorer performance and stability than devices with bulk-dominated characteristics. Even ideal surfaces can possess a density of electronic states that would make any device uncontrollable [2]. Finally, silicon is very abundant and can conveniently be converted into high-purity single crystals suitable for device manufacturing.

A key component of present-day microelectronics built on the concept of the electrical passivation of silicon by its oxide is the metal-oxide-semiconductor field-effect transistor (MOSFET) [3, 4], shown schematically in Figure 1. It can be structurally divided into gate, oxide, source, drain, and bulk semiconductor. The gate is a conducting electrode used to control the device. The passivating oxide is also called the gate dielectric. The source and drain are islands doped opposite from the substrate. The semiconductor in contact with the gate oxide and between the source and drain forms the active region of the MOSFET. In such a device, an electric field produced by a voltage applied between the gate electrode and the silicon substrate (the gate voltage V_G) can control charge flow (electrical current) from the source to the drain (the drain cur-

rent I_D, produced by an applied drain voltage V_D)—hence the "field effect" denomination. Under favorable gate bias, a channel appears, connecting the source and drain, and charge flows between them (Fig. 1). The metal/oxide (SiO_2)/semiconductor (Si) or MOS structure is, without a doubt, the core structure of modern-day electronics. (Actually, heavily doped polycrystalline silicon (polysilicon, poly-Si) has been favored over metal (aluminum) electrodes since about 1970, eliminating the technological problem of aligning the gate to the insulating film.) This cannot be ascribed to a single characteristic of the MOSFET, but is due to a favorable combination of properties [3]. The two-terminal MOS capacitor (MOS-C) is the structural heart of all MOS devices. There lies the gate dielectric. The MOS-C consists of a parallel plate capacitor with a metallic plate and the semiconductor substrate as electrodes. The two electrodes are separated by a thin insulating layer of SiO_2. The MOS designation is reserved for the metal-SiO_2-Si system; a more general designation, metal-insulator-semiconductor (MIS), is used to identify similar device structures composed of an insulator other than SiO_2 or a semiconductor other than Si.

One key feature of the MOSFET is its planar geometry, which allowed the development of integrated circuits (ICs), i.e., the arrangement of many individual devices on the same semiconductor chip. The vast majority of discrete devices and ICs are based on MOSFETs. In complementary metal-oxide-semiconductor (CMOS) technology, both *n*-channel (*p*-bulk, electron current carrying) MOSFETs and *p*-channel (*n*-bulk, hole current carrying) MOSFETs are fabricated on the same chip. Today, CMOS technology is the dominant semiconductor technology for microprocessors, memories, and application-specific integrated circuits (ASICs). Sales of semiconductor ICs were estimated to be $150 billion for the year 2000. The main advantage of CMOS is its low power dissipation. A CMOS circuit has almost no static power dissipation—power is only dissipated when the circuit actually switches. This makes it possible to integrate many more CMOS gates on an IC, resulting in much better performance.

Significant expressions of MOSFET performance are drain current, charge carrier mobility, and switching speed. In part, these depend on properties more or less under the designer's control [3]. Both drain current and switching speed are proportional to the gate capacitance and to the device aspect ratio, given by the channel length divided by the channel width. The gate capacitance, in turn, is inversely proportional to the gate oxide thickness. The drive for increasing device performance has led to continuous MOSFET scaling in the direction of decreasing gate oxide thickness and increasing aspect ratio. Such scaling, involving such parameters as gate length, gate width, gate oxide thickness, source/drain junction depth, substrate dopant concentration, and supply voltage (Table I), must follow some rules [5] to cope with the "short-channel" effect. Briefly stated, the short-channel effect refers to the tendency toward a lower gate voltage at which drain current is observed (threshold voltage) as the source-to-drain distance (channel length) is reduced, because of the two-dimensional electrostatic effect in-

Table I. MOS Transistor Scaling at a Glance

Design parameter	Scaling factor
Upon	
Transistor dimensions	Decreased by k
Voltage	Decreased by k
Doping	Increased by k
Result	
Circuit area	Reduced by $1/k^2$
Speed	Increased by k
Currents	Reduced by $1/k$
Power per circuit	Reduced by $1/k^2$
Power per unit area	Held constant

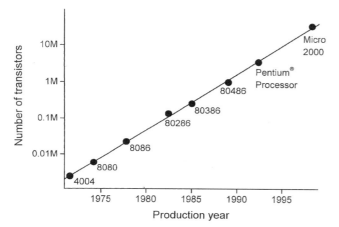

Fig. 2. Moore's law illustrated by the time evolution of the number of transistors in Intel chips.

volving the source, the drain, the gate, and the channel regions. As a rule of thumb, the gate oxide thickness should be about 1/50 to 1/25 of the channel length to keep the short-channel effect under control [6]. So if one takes gate capacitance as the scaling parameter to increase device performance, channel length must be scaled concomitantly. As a result, the whole device shrinks, and more devices can be placed on the same chip area, increasing integration and decreasing the cost per transistor. Gate and drain voltages must be scaled at the same time, and one finds that power consumption is drastically reduced. With regard to the MOSFET, one is tempted to say that "the smaller, the better."

Because of scaling, semiconductor device manufacturing has gone through medium- (MSI), large- (LSI), very large- (VLSI), and now ultralarge-scale integration (ULSI). Computing power has been doubling every 18 months since the 1970s [7], following what became known as Moore's law [8] (Fig. 2). Goals and strategies for developments in semiconductor technology appear in the biannual editions of the Semiconductor Industry Association's roadmap [9].

Reduction of the gate dielectric film thickness is one of the areas of device fabrication that now limits CMOS scaling. In active CMOS devices, the thickness of the gate insulator has been scaled down from 100 nm in the early 1970s to less than 3 nm for current devices with an effective channel length of less than 0.2 μm. Already at the present stage one cannot make a strict distinction between "surface," "bulk," and "interface" of the gate dielectric films, from a structural or compositional point of view. The quality and reliability of these films determine the performance of the ICs that contain them. The drive of ULSI establishes new and very strict material requirements. Outstanding scientific and technological challenges have been created by the downscaling of CMOS devices. Most of them have been surpassed at the cost of massive research, but some remain at the very forefront, limiting further evolution of silicon science and technology [10–12]. Those concerning gate oxides, and more generally gate dielectrics, are close to fundamental limits, where the need for atomic-scale understanding and control becomes ever more critical. A limit to silicon oxide as a gate dielectric has already been foreseen, and alternative materials are now under intense investigation [13].

This chapter is intended to present and discuss the broad view of ultrathin gate dielectric films on silicon. The main purpose is to review the current status of ultrathin gate dielectric science and provide the means for the reader to keep pace with the new scientific literature in this very lively research area. Discussion in depth is restricted for the sake of brevity, with references offered to the interested reader. The material is intended for either continuous or sectional reading. Emphasis is placed on the correlation of dielectric quality, physicochemical issues, and processing parameters. Basic requirements of ultrathin dielectric films to be used in microelectronic devices are listed in Section 2. Film preparation methods are outlined in Section 3. Both electrical and physicochemical methods of characterization of dielectric films are presented in Section 4. The significance and effects of the hydrogen presence in gate dielectrics are presented in Section 5. Section 6 describes silicon oxide films thermally grown on single-crystalline silicon in dry oxygen. They have allowed and promoted semiconductor device development for 40 years. Silicon oxide is the gate dielectric *par excellence*, and all discussions take it as a reference. Understanding and simulating modern processing routes to the formation of gate dielectric films require an understanding of the atomic transport processes responsible for their growth. Atomic transport is the natural way to approach the growth of ultrathin films and so is explored in this text. Because of ever-increasing constraints brought about by integration issues, silicon oxide has been at least partially replaced. Dielectrics generally referred to as silicon oxynitrides are addressed in Section 7, considering a variety of preparation methods. (Silicon nitride finds a number of applications in the semiconductor industry, but not in the form of ultrathin gate dielectric films, and so will not be discussed here.) Recently proposed dielectric materials that are alternatives to these are covered in Section 8. Section 9 summarizes the chapter and gives concluding remarks.

2. REQUIREMENTS OF ULTRATHIN GATE DIELECTRIC FILMS

Ultrathin dielectric films must fulfill a number of electronic, physical, and chemical requirements to be useful in microelectronic devices [14]. A major concern for the gate dielectric is minimization of leakage current, as this ensures the field effect. A large bandgap is desirable for the prevention of leakage and for ensuring a lower sensitivity to radiation. A high dielectric constant (permittivity) is becoming more and more important as integration increases. Uniform and high dielectric strength (high breakdown voltage) is also necessary. To obtain high-performance MOSFETs with fast turn-on characteristics and high carrier mobility, electrically active defects (and the associated charge) encountered in the gate insulator and at its interfaces must be minimized. The dominant criterion is still a low density of electronic states at the insulator–semiconductor interface, which makes SiO_2 the first choice unless one of the other already mentioned aspects cannot be fulfilled.

To make sure that the fabrication yield of ICs is kept acceptable despite the ever-increasing chip size, gate dielectric films must be homogeneous, uniform, and both physically and chemically stable. (Although individual devices keep shrinking, chips are growing because of increasing device integration.) For the sake of device stability, these dielectric films should prevent all foreign species from diffusing into active parts. Foreign species can be doping atoms, ions diffusing from a metal gate electrode, or any type of contaminant introduced during the fabrication process. One very illustrative and disturbing example is diffusion of boron from heavily doped gate electrodes made of polycrystalline silicon. Such characteristics are usually perfected through cleanliness and control of the film growth rate. The domain of ultrathin dielectric films is particularly tricky because as a general rule the first stages of formation are not fully understood.

3. ULTRATHIN GATE DIELECTRIC FILM PROCESSING

In current IC fabrication technology, passivation of the semiconductor surface with a suitable dielectric film is one of the first processing steps. It is therefore considered as a "front end of line" process. With regard to manufacturing, dielectric films can be divided into two groups: "grown" and "deposited." The term "grown" designates films that are formed by means of a chemical reaction (oxidation, nitridation, etc.) involving the semiconductor substrate and whose growth depends on atomic transport, and "deposited" means that the dielectric has been formed without any chemical reaction with the semiconductor [14]. A film unintentionally grown by simple exposure of a clean surface to the ambient is usually called "native." Examples are SiO_2 and Al_2O_3 films formed on clean Si and Al upon exposition to air.

This section addresses different approaches to grown and deposited dielectric film fabrication. The presentation is preceded by considerations on substrate cleaning, which is a key factor regardless of the particular method chosen for the preparation of the passivating film.

3.1. Silicon Wafer Cleaning

Cleaning is the most frequently repeated step in IC production. One of the most critical cleaning steps is that which precedes ultrathin dielectric film growth or deposition. The RCA clean, developed in 1965, still forms the basis of most front-end wet cleans [15–19]. A typical RCA-type cleaning sequence starts with a sulfuric acid–hydrogen peroxide mixture (SPM) step (H_2SO_4/H_2O_2) followed by a dip in diluted HF. The SC1 step ($NH_4OH/H_2O_2/H_2O$) then removes particles, and the SC2 step ($HCl/H_2O_2/H_2O$) removes traces of metals. Environmental concerns and the search for greater cost effectiveness have recently led to much research effort directed toward understanding cleaning chemistries [20].

A variety of organic contaminants can exist in IC processing (human skin oils, pump oil, silicon vacuum grease, cleaning solvents). A potential problem arising from this is the presence of an organic film at the silicon surface, preventing the action of cleaning solutions. Therefore, removal of organic contamination is often the first step in cleaning. SPM (H_2SO_4/H_2O_2) or sulfuric acid–ozone mixture (SOM) (H_2SO_4/O_3) have successfully been applied to this end. Traditional cleaning sequences use mixtures based on H_2SO_4 to grow a sufficiently thick chemical oxide to obtain high particle removal efficiencies in the second step of the cleaning sequence. The effect of organic contamination on SiO_2 film quality was tested by wafer exposure either between cleaning and oxidation or oxidation and poly-Si gate electrode deposition [20]. Charge-to-breakdown (see Section 4.1.2) measurements showed no difference in the intrinsic oxide breakdown, but the extrinsic tail was influenced by the different treatments. The presence of organic contamination is likely responsible for the extrinsic defects in the thin oxide layer.

In the SC2 mixture the H_2O_2 can be left out completely, inasmuch as it has been shown that diluted HCl is as effective in the removal of metals as the standard SC2 solution [20]. Moreover, at low HCl concentrations, particles are kept away from the silicon surface. This is due to the fact that the isoelectric point for silicon and silicon oxide is between pH 2 and 2.5. For most particles in liquid solutions at pH values greater than 2–2.5, an electrostatic repulsion barrier is formed between the particles in the solution and the surface.

An additional step can be added to the cleaning sequence to make the silicon surface hydrophilic. This allows easier drying without the generation of drying spots or watermarks. In the Marangoni dryer [21], the drying is performed by a strong natural force (the Marangoni effect) in cold deionized water, and the wafer is rendered completely dry without the evaporation of water.

Trace amounts of noble-metal ions, such as from Ag, Au, and, especially, Cu, are present in HF solutions and can deposit on the Si surface. An optimized HF/HCl mixture provides protection against metal outplating from the solution. Ca shows a pronounced tendency to deposit on the wafer surface, with a strong effect on the gate oxide integrity [22]. Low pH and high temperature were found to result in lower Ca surface concentration [20]. A transport-limited surface deposition was also observed, so low pH, high temperature, and fast cleaning minimized Ca contamination.

The chemical state of the silicon surface after a cleaning process can be either an oxidized form or oxide-free, bare silicon, terminated by Si—H groups [23]. It has been shown that the presence of a thin chemical oxide (0.6 to 0.8 nm) on the Si surface before deliberate oxidation does not significantly influence final oxide thickness or thickness variation on a wafer. This chemical oxide forms a passivation layer with a dangling-bond defect density in the range of 10^{12} cm^{-2} at the SiO$_2$–Si interface, two orders of magnitude higher than that of device-grade thermal oxides. Wet chemical oxides protect the Si surface from some recontamination or at least render the surface less sensitive. On the other hand, such thin oxides left on the surface are detrimental, for example, to epitaxial silicon deposition and might become increasingly critical for subsequent thermal oxidation as the gate oxide thickness decreases. For such purposes, cleaning sequences that end with an HF step might be preferred, as the surface becomes hydrophobic and the silicon dangling bonds are passivated with hydrogen atoms. However, no significant differences were found between wafers that received HF and SC2 final cleaning, with respect to electrical performance and defect density, provided metal and particle contamination was sufficiently low. So far experimental proof for the superiority of HF-last cleaning has not been reported [24].

Surface roughness of Si wafers received much attention over the last few years as a possible cause of gate oxide defects. It was recently reported that as far as yield loss is concerned, silicon surface roughness for very thin oxides (<6 nm) appears to be negligible, at least at moderate levels of surface roughening [25–27]. In a reevaluation of yield loss due to local roughening and the relationship among Si surface roughness, metal contamination, and oxide defects, the authors were led to believe that yield loss was due not to the observed silicon surface roughening, but rather to locally high levels of iron contamination [20].

For surface science studies, the so-called rapid thermal cleaning of silicon allows reduction of native oxides and some etching [14]. Samples are typically heated to 1000–1150°C for 2–10 s in forming gas (10 vol.% H$_2$ in N$_2$), optionally containing HCl at a small concentration. SiO is volatile above 750°C and is formed upon heating of SiO$_2$ at very low O$_2$ partial pressures. For etching, it is believed that hydrogen reduces silicon to gaseous SiH$_x$ and eliminates metallic impurities in the form of hydrides.

The effect of different silicon wafer cleaning procedures before oxidation on the kinetics of thermal growth was investigated by, among others, Gould and Irene [28], who used cleaning solutions based on the RCA recipe [16], and Ganem et al. [29], who used HF diluted in ethanol. Hahn et al. [30]

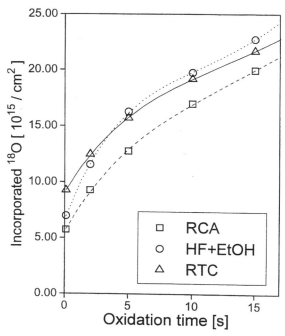

Fig. 3. Oxygen content vs. oxidation time in ^{18}O$_2$ for Si wafers submitted to different cleaning procedures. The top of the scale for O content corresponds to about 5.5 nm of SiO$_2$. The lines connecting experimental data are just guides to the eyes. Reprinted with permission from [31], © 1996, Elsevier Science.

studied the effect of Si wafer cleaning, polishing, and storage time; oxidation parameters; and postoxidation annealings on the roughness of the SiO$_2$–Si interface. Stedile et al. [31] compared Si cleaning methods: standard RCA cleaning, HF etching followed by a rinse in ethanol, and rapid thermal cleaning. Different oxidation kinetics (Fig. 3) and a pronounced change in the thickness of the SiO$_2$–Si interface were observed. The thickest interface was obtained after HF/ethanol cleaning (comprising the equivalent of about 11 monolayers of Si), and the thinnest (eight monolayers of Si), after RTC [31]. (For a description of the SiO$_2$–Si interface, see Section 6.)

3.2. Thermal Growth of Gate Dielectrics

Thermal growth is by far the most widely used method for preparing gate SiO$_2$ because the electrical properties of the films thus produced are vastly superior to those of silicon oxide films deposited by chemical vapor methods (see Section 3.4). For the thermal growth of gate dielectrics, one basically needs an apparatus capable of keeping a sufficient temperature in a controllable (clean) atmosphere. Key factors are uniformity and repeatability. In the domain of ULSI, acceptable parameter fluctuation has become very stringent and process monitoring specially important. Two concepts have proved useful for the thermal growth of dielectric films on silicon: the conventional furnace, which relies on heating due to the Joule effect, and the rapid furnace, which makes use of electromagnetic radiation strongly absorbed by crystalline silicon.

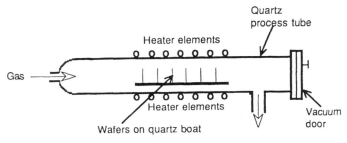

Fig. 4. Schematic cross section of the reaction chamber of a conventional furnace. Reprinted with permission from [14], © 1999, Elsevier Science.

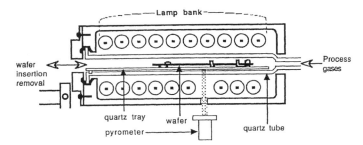

Fig. 5. Schematic cross-section of the reaction chamber of an RTP module. Reprinted with permission from [14], © 1999, Elsevier Science.

3.2.1. Conventional Thermal Processing

Dry oxidation of silicon in so-called conventional furnaces (Fig. 4) is usually performed under a flow of pure O_2 (1 bar) at 900–1150°C. Time and temperature of oxidation can be chosen to grow oxides with reasonable control of the thickness. In the wet oxidation process, up to a few parts per million of water vapor are introduced in the oxidizing ambient. "Wet" films grow faster than "dry" films at the same temperature [32–41], rendering thickness control difficult in the ultrathin oxide film regime.

Besides flow furnaces, static variants have been used for low-pressure processing. Low-pressure oxidation offers an attractive way of growing ultrathin oxides in a controllable manner [42]. Oxide thicknesses between 3 and 14 nm have been achieved in this way at 900–1000°C under 0.2–2.0 mbar of O_2. Diluted O_2 (e.g., in N_2) can also be used. With this method the growth rate is slower, which makes it attractive if controlling the oxide thickness is very critical. The oxides are very uniform, homogeneous, and similar to thicker oxides prepared at atmospheric pressure [43]. Static furnaces are specially useful for growing oxides under an isotopically pure atmosphere ($^{16}O_2$ or $^{18}O_2$), because contamination and waste of the high-purity gases are minimal and recovery after oxide growth for subsequent use is also possible. As shown in Section 4.2, isotopic tracing experiments were essential for an understanding of mechanistic aspects of silicon oxide film growth.

Small HCl or HF concentrations in the oxidation ambient are typically used to passivate contaminant ionic sodium, to improve the dielectric breakdown voltage, and to reduce the amount of impurities and defects in the grown silicon oxide films. To reduce the density of electrically active defects, oxidation is often followed by thermal annealing at the oxidation temperature in an inert ambient (N_2 or Ar). This is often followed by an annealing in forming gas (10 vol.% H_2 in N_2 or Ar) at 450°C (see Section 5).

3.2.2. Rapid Thermal Processing

Rapid thermal processing (RTP) refers to the use of a broad range of energy sources to rapidly heat materials used in semiconductor and IC processing for short periods (1–100 s) [14]. RTP was originally introduced to anneal defects originating from ion implantation, but proved useful for a variety of ad-

ditional applications [44]. (Processes such as source, drain, and poly-Si gate electrode doping are now carried out by ion implantation from a particle accelerator. This introduces a number of defects (atom dislocations and such) in the system, which are at least partially annealed under thermal treatment.) Indeed, as device dimensions decrease, low thermal budget processes are necessary to limit dopant diffusion. Two solutions exist: low-temperature processes enhanced by photons, plasmas, etc. (see Section 3.3) and RTP processes in which high temperatures are applied for short periods. As conventional furnaces, RTP processes have been used to prepare silicon oxide and oxynitride films. Dry cleaning (see Section 3.1) and annealing of wafers can usually also be performed.

Figure 5 shows a schematic cross section of a single-wafer, rapid thermal processing module. Heat is provided by a bank of tungsten-halogen lamps, and the temperature is monitored with an optical pyrometer and a feedback system for controlling the power supplied to the lamp bank. As in a typical conventional furnace, the reaction chamber is a quartz tube. The apparatus allows one to perform several processes sequentially in the same chamber. The small size of the chamber makes it possible to quickly change the reacting gases, and the substrate temperature switches the chemical reactions on and off. As determined from I–V and C–V measurements (Section 4.1), rapid thermal oxide films display better electrical characteristics than oxide films grown in conventional furnaces: higher dielectric strength, lower defect density, and lower density of interface states are observed [14].

3.3. Hyperthermal Processing of Gate Dielectrics

"Hyperthermal processing" refers to processing with the chemical species involved at an energy range above that achieved by heating reactive gases and/or substrate. The energy range of such a hyperthermal regime has been quoted as being between 1 and 1000 eV (Fig. 6) [45]. Compared with thermal processing (Fig. 7), the hyperthermal approach shows a number of advantages for the microelectronics industry. One key point is the use of much lower substrate temperatures for the preparation of dielectric films, reducing dopant diffusion during processing. Hyperthermal processes also make it possible to overcome the thermodynamic barrier to the formation of metastable phases that might be of interest, as in the case of silicon oxynitrides

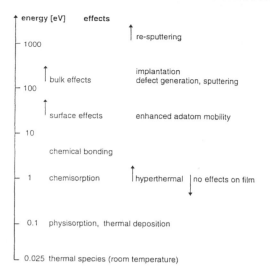

Fig. 6. Effects of low-energy species in the modification of materials. Reprinted with permission from [45], © 1997, Elsevier Science.

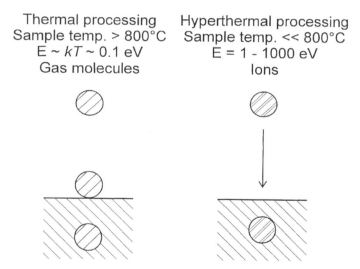

Thermal processing
Sample temp. > 800°C
$E \sim kT \sim 0.1$ eV
Gas molecules

Hyperthermal processing
Sample temp. << 800°C
$E = 1 - 1000$ eV
Ions

Fig. 7. Schematic comparison of thermal and hyperthermal processing.

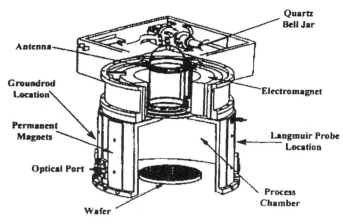

Fig. 8. Cross section of a high-density plasma experimental process module. Reprinted with permission from [46], © 1997, American Institute of Physics.

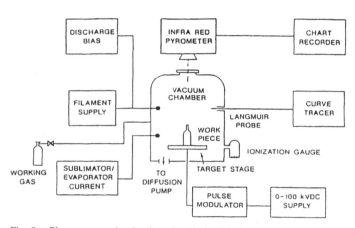

Fig. 9. Plasma source ion implantation device block diagram showing vacuum chamber, power supplies, and diagnostics. Reprinted with permission from [47], © 1987, American Institute of Physics.

(Section 7). Whereas thermal growth makes use of reacting molecules that have to go through adsorption (and dissociation in cases), diffusion, and reaction, hyperthermal processing usually makes use of energetic ions, eliminating adsorption and facilitating the diffusion and the reaction. Actually, substrate temperature plays a role in hyperthermal processing, and this is mainly related to diffusion after most of the ion energy has been released. Ion energy and fluence (i.e., number of ions impinging on the sample) are of primary importance in the process. Together, these parameters allow the modification of materials to various degrees, which makes hyperthermal processing very attractive. One disadvantage of hyperthermal methods is that quite often the interface between insulator and semiconductor is not sharp. Moreover, implantation damage (atom dislocations and such) occurs.

The most straightforward way to produce hyperthermal species is to accelerate charged particles to the desired energy in an electromagnetic field. In the following, three different hyperthermal methods that have been used to produce silicon oxynitrides (Section 7) are outlined. Figure 8 [46] shows an apparatus used for the nitridation of SiO$_2$ that uses a plasma. The plasma source is fed with a precursor (as NH$_3$ or N$_2$) of ions of interest, and the sample is placed within the limits of the plasma sheath. If the substrate is grounded, implantation occurs at floating potential (about 20 V, depending on plasma density). Additional bias can be supplied to the sample to increase implantation energy. Figure 9 [47] shows an apparatus for plasma immersion ion implantation (PIII). In this case the sample is placed away from the plasma sheath and is pulse biased to a negative voltage. As a result, ions are massively accelerated toward the sample and implanted. The interval between pulses allows recomposition of the plasma. Figure 10 shows how a conventional ion implanter can be used for hyperthermal processing. One applies the ion beam extraction voltage but a fixed potential to the sample, reducing the effective ion energy at landing. This approach is different from those directly using a plasma in that the nature and energy of the hyperthermal species

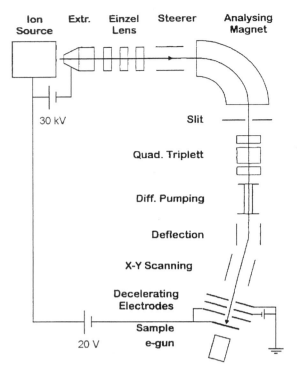

Fig. 10. Schematic of an ion implanter modified to perform hyperthermal modification of materials.

are exactly defined. Plasma methods always produce a mixture of ions arriving at the sample with a distribution of energies.

3.4. Physical or Chemical Deposition of Gate Dielectrics

A myriad of insulating materials can be deposited on silicon from physical or chemical vapor. In physical vapor deposition (PVD), a film is deposited without the involvement of a chemical reaction. This can be done in many different ways. For example, one can simply use a difference in temperature between the substrate and a piece of the material to be deposited if it is sufficiently volatile. An ion beam can also be used to sputter atoms from a suitably chosen target; some of these atoms are then collected on the substrate. In this case, the target can be either an elemental or composite material, and multiple targets can be used. PVD is very well established for the deposition of relatively thick films. It is now being applied [48] in the thickness range of interest for current and future gate dielectrics.

In chemical vapor deposition (CVD), a reaction occurs between chosen precursors and leaves the insulating films as a product. The process consists of exposing a clean substrate surface to a reactive mixture of gases. The substrate surface may play a role as a catalyst, but it is not consumed in the process. Films prepared by CVD are usually not as pure and defect-free as (hyper)thermal ones, so one should not expect to see gate SiO_2 produced by CVD. (Silicon oxide deposited from chemical vapor actually plays an important role in IC fabrication, but as an interlayer dielectric and not at the gate.) Many potentially good insulators, however, cannot be (hyper)thermally grown on

silicon, and CVD quite often offers a reasonable alternative. One major impurity from CVD is hydrogen, a major source of instability in silicon-based microelectronic devices (Section 5).

There are many variants of CVD. The classical process consists of heating the silicon substrate to a few hundred degrees Celsius (the actual temperature varyies according to the material to be deposited and the precursors) and exposing it to a mixture of reacting gases. CVD can be performed at atmospheric pressure (APCVD) or at low pressure (LPCVD). LPCVD is replacing the conventional APCVD because of lower costs and superior film quality. Variants include plasma-enhanced CVD (PECVD) and rapid thermal CVD (RTCVD), the main purpose of which is reduction of the thermal budget during sample processing. Another modification is jet vapor deposition (JVD), which makes use of a supersonic precursor flow [49]. The need to control film thickness and uniformity led to the development of atomic-layer CVD (ALCVD) [50]. This is a technique that uses sequential pulses of precursors and the self-saturating nature of certain surface reactions, allowing the chemical deposition of materials with a monolayer control. Novel gate dielectric materials (Section 8) can often be deposited by metalorganic chemical vapor deposition (MOCVD), in which the precursor is an organometallic compound.

4. CHARACTERIZATION OF ULTRATHIN GATE DIELECTRIC FILMS

Characterization of dielectric films is a challenging, key aspect of present-day microelectronics. Its relevance to the industry stems from the need for process modeling and model testing. More than ever it seems that adequate modeling rather than blind experimentation is the way to obtain the best performance from the materials available. This section describes a few aspects of the electrical characterization of ultrathin dielectric films and briefly presents physicochemical characterization techniques. No attempt is made to fully describe any of the methods; the interested reader should consult the references provided in the text. Electrical characterization is important for both dielectric testing and study; physicochemical description is more important at a fundamental level. This section lays the basis for the discussion of gate dielectric performance and nature in Sections 5, 6, and 7.

4.1. Electrical Characterization

Electrical properties of the MIS system are the bottom line for judging an insulator material intended for use as a gate dielectric. An understanding of some electrical characterization techniques, including interpretation of raw experimental data, is thus highly desirable when dealing in this field. This section addresses the electrical characterization of the MIS system, with emphasis on the insulator layer.

Even when the focus is on gate dielectrics for application in MISFETs, it is the MIS capacitor (MIS-C) that is usually applied for electrical characterization. It has the advantages of simplicity of fabrication and of analysis. When the MOS-C is used to measure properties of the MOS system, the following insulator characteristics can be obtained [2]:

- thickness
- electric field at dielectric breakdown
- charge configurations such as fixed charge and the charge at the interface between SiO_2 and another insulator deposited on top of it
- work function difference relative to the top electrode (gate) material
- charge carrier tunneling (leakage current)
- dielectric constant
- properties of electron and hole traps

Moreover, trap level density for charge carriers at the interface between insulator and semiconductor can be profiled as a function of energy in the semiconductor bandgap. Each of these results depends to a variable degree on previous knowledge about the system under investigation. The SiO_2-Si system has been subjected to the most comprehensive analysis, but the MIS capacitor is also the primary structure for the characterization of insulators other than SiO_2 (such as SiO_xN_y and Al_2O_3) and semiconductors other than Si (such as GaAs and SiC). Analysis to the extent suggested above requires a deep understanding of the physics and adequate modeling of the MOS system. A qualitative discussion is offered below as an introduction to the subject. A primary reference for the interested reader is the text by Nicollian and Brews [2].

In the MISFET, conductivity changes in the channel can be measured to determine charge carrier mobility. The mobility is not directly measured but is inferred from the I_D–V_D characteristics of the MISFET. The capability for measurement of channel mobility is one of the major advantages of MISFETs over MIS-Cs used for electrical characterization.

Fabrication of a MIS-C involves the passivation of a semiconductor surface with a suitable insulator film, which can be either thermally grown (in the case of SiO_2 on Si) or deposited by a variety of physical and chemical methods (see Section 3). Aluminum is a common gate metal because it is easy to evaporate and it adheres strongly to the oxide. The gate is usually evaporated onto the oxide surface in the shape of a circular dot. Heavily doped poly-Si can alternatively be used as an electrode. A suitable contact must also be provided for the back of the silicon wafer. This is an ohmic contact [51], whose two basic properties are as follows: (i) it permits the passage of current into or out of the crystal without placing any resistance of its own in the path; and (ii) passing current through such a contact does not perturb carrier densities within the sample; i.e., equilibrium carrier densities are guaranteed. Such contact can be made if the substrate has a degenerate layer or, if it does not, if the substrate surface is abraded with emery cloth before a metal is evaporated to complete the contact [2].

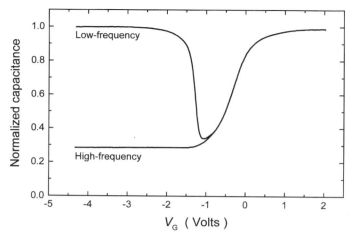

Fig. 11. High- and low-frequency capacitance–voltage characteristics of an MOS-C built on n-type Si.

4.1.1. Capacitance versus Gate Voltage MIS-C Characteristics

Once an insulator prevents d.c. current flow through the MIS-C, one can measure the capacitance of the system [1]. The capacitance is a function of applied gate voltage, and from an experimental capacitance–voltage (C–V) curve one can extract valuable information concerning both the insulator and the semiconductor. C–V characteristics are routinely monitored during device fabrication.

A C–V meter superimposes a small a.c. signal on a preselected d.c. voltage and detects the resulting a.c. current flowing through the sample. The a.c. signal is typically 15 mV rms or less, and a common signal frequency is 1 MHz. The d.c. voltage is slowly ramped so that a continuous C–V characteristic is obtained.

MIS C–V characteristics are found to be useful in the so-called low-frequency and high-frequency limits. These refer to the frequency of the a.c. signal used in capacitance measurements. (An equivalent to low-frequency data can also be collected from quasi-static measurements, which involve only a V_G ramp and no a.c. signal.) High- and low-frequency C–V data for an MOS-C built on n-doped Si are displayed in Figure 11. To explain the observed form of the C–V characteristics, one must consider how the charge in the test structure responds to the applied a.c. signal as the d.c. bias is systematically changed from accumulation, through depletion, to inversion (from right to left in Fig. 11). Under accumulation, the majority carrier concentration is greater near the insulator–semiconductor interface than in the bulk of the semiconductor (majority carriers are those introduced in silicon by the dopants); under depletion, electron and hole concentrations at the interface are both less than in the bulk semiconductor; and under inversion, minority carrier concentration at the interface exceeds the bulk majority carrier concentration.

In accumulation the d.c. state is characterized by the pileup of majority carriers at the insulator–semiconductor interface. If the semiconductor is Si and experiments are run at room temperature, it is reasonable to assume that the device can follow the applied a.c. signal quasi-statically at frequencies of 1 MHz and less, with the small a.c. signal adding or subtracting small amounts of charge close to the edges of the insulator. For either low or high probing frequencies, the behavior is very much that of an ordinary parallel-plate capacitor. Thus the low- and high-frequency $C-V$ characteristics are the same in Figure 11 in the gate bias region corresponding to accumulation.

Under depletion biasing the d.c. state of an n-type MOS structure is characterized by a $-Q$ charge on the gate and a $+Q$ depletion layer charge in the semiconductor. The depletion layer charge is directly related to the withdrawal of majority carriers from the SiO$_2$–Si interface. As in accumulation, only majority carriers are involved in the operation of the device, and the charge state inside the system can be changed very rapidly. When the a.c. signal places an increased negative charge on the gate, the depletion layer inside the semiconductor widens almost instantaneously. For all probing frequencies this situation is analogous to two parallel-plate capacitors in series, and so low- and high-frequency $C-V$ characteristics are the same in Figure 11 in the gate bias region corresponding to depletion.

Once inversion is achieved an appreciable number of minority carriers pile up near the insulator–semiconductor interface in response to the applied d.c. bias. Now the a.c. charge response of the capacitor depends on probing frequency, as shown in Figure 11. At low frequencies minority carriers can be generated or annihilated in response to the applied a.c. signal. As in accumulation, one has a situation where charge is being added or subtracted close to the edges of a single-layer insulator (so the analysis follows that of one parallel-plate capacitor). At high frequencies the generation–recombination process will not be able to supply or eliminate minority carriers in response to the applied a.c. signal. The number of minority carriers in the inversion layer therefore remains fixed at its d.c. value, and the width of the depletion layer simply fluctuates about an average corresponding to the d.c. bias. Similarly to depletion biasing, this is equivalent to two parallel-plate capacitors in series. Intermediate measurement frequencies lead to a mixed behavior.

The way $C-V$ curves are used to probe dielectric films is as follows [1–3, 51]. Accumulation capacitance is used to determine film thickness or dielectric constant, one of them being known or assumed. Because of the current search for dielectric materials alternative to SiO$_2$, one often speaks of an equivalent oxide thickness (EOT) when comparing devices with different materials as gate dielectrics. The EOT is the thickness of an SiO$_2$ film presenting the same accumulation capacitance measured for the alternative sample: $\text{EOT} = t_{\text{alt}}\epsilon_{\text{ox}}/\epsilon_{\text{alt}}$, where t_{alt} is the physical thickness of the alternative film, ϵ_{ox} is the permittivity (or dielectric constant) of SiO$_2$, and ϵ_{alt} is the permittivity of the alternative material. Displacement of the curve in the gate voltage axis is taken as a direct measure of threshold voltage shift. The threshold voltage is the gate voltage at which a MISFET is turned on, i.e., the onset of inversion at the silicon

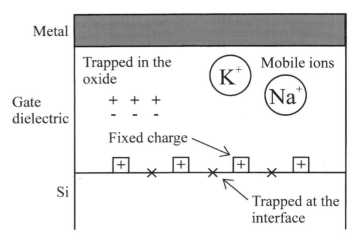

Fig. 12.   Nature and location of charges in the MOS system.

surface. (This would equivalently be described as the voltage at which the conducting channel is formed at the silicon surface.) Another important figure of merit is the flatband voltage, the voltage at which the Fermi level at the semiconductor surface is the same as in its bulk.

The exact $C-V$ characteristics are strongly dependent on the doping level of the semiconductor and on the gate electrode material. Of course, actual threshold voltage measurements require MOSFETs; they are usually extracted from drain conductance versus gate voltage measurements [1]. Threshold voltage shifts are due to parasitic charge in the MIS-C. This charge can be due to mobile ion charge, insulator trapped charge, interface trapped charge, and fixed insulator charge, as shown in Figure 12. The mobile ionic charge is mostly made up of sodium ions mobile at room temperature and above. The insulator trapped charge depends on the nature of the trapped carrier distributed in the bulk of the insulator, and so can be either positive or negative. The interface trapped charge is made up of carriers trapped on interface defects, which, depending on bias, can exchange carriers with the semiconductor. The fixed insulator charge is the sum of all of the other charges that are impervious to changes in bias conditions. Because the thickness of the gate insulator is much reduced in modern devices, the effects of mobile ionic charge and insulator trapped charge are also reduced, but interface trapped charge and insulator fixed charge play a major role.

One partially resolves charge identity by comparing $C-V$ characteristics taken before and after the sample has been submitted to stressing (biased to a gate voltage for a given time and at a certain temperature). Changes are due to mobile charge that has drifted. Charge in the insulator can give rise to significant effects, including large voltage shifts and instabilities. The shape of the low-frequency $C-V$ curve is related to the distribution of charge traps. Comparison of experimental and calculated or experimental low- and high-frequency characteristics yields [2] the trap density profile at the dielectric–semiconductor interface along the semiconductor bandgap. A common manifestation of a significant interfacial trap concentration within an MIS-C is the distorted or spread-out nature of the C–V characteristics. Judged in terms of their wide-ranging and degrading effect on the operational behavior of MIS devices, insulator–semiconductor interfacial traps must be considered the most important nonideality in MIS structures. A device-grade SiO_2 film on Si yields an interface trap level density of about 10^{10} cm^{-2} eV^{-1} at midgap. (Whenever an interface trap level density appears without reference to a position in energy in the semiconductor bandgap, midgap is assumed.)

4.1.2. Gate Current versus Gate Voltage MIS-C Characteristics and Time to Dielectric Breakdown

A rather simple experiment that is of increasing importance for the characterization of ultrathin dielectrics is gate current density versus gate voltage measurement in an MIS-C (I–V or J–V). In this case, one measures the actual ability of the insulator to block d.c. current flow through the gate structure. Leakage in ultrathin dielectrics (Fig. 13) [52] is due to direct quantum tunneling of charge carriers through the insulator film. Tunneling is the process whereby charge carriers overcome a potential barrier in the presence of a high electric field. Direct tunneling (DT) differs from Fowler–Nordheim tunneling (FNT) in that the tunneling barrier changes from triangular to trapezoidal [53]. Both processes are quantum mechanical in nature, and both contribute to the I–V characteristics in Figure 13. The transition between the main conduction modes occurs at the maximum slope in the curves. Quite often in the case of ultrathin dielectric films, high leakage currents prevent adequate characterization by other methods, such as C–V.

Electrons traversing the gate dielectric as leakage can damage the film, creating electron traps and interface states [54–59]. Once the concentration of induced defects reaches a critical value (N_{BD}), the insulating properties of the oxide (catastrophically) fail; i.e., oxide breaks down. This is seen as a sharp increase in leakage current. Dielectric reliability is typically expressed as charge or time to breakdown (Q_{BD} or t_{BD}). For most applications, it is generally accepted that MOS devices are required to work continuously for about 10 years. As testing devices for such long times is impracticable, the breakdown is accelerated by stressing devices to gate biases and temperatures above those of intended use. Under such conditions one measures the so-called stress-induced leakage current (SILC) [10, 58, 60]. SILC represents the relative change in leakage current density $\Delta J / J_0$ due to defect creation of MOS devices under electrical stress (applied bias) conditions (Fig. 14). A major point is then to be able to extrapolate experimental Q_{BD} and t_{BD} obtained under stress to values corresponding to operating conditions. The slope of the linear portion of the SILC curve is a measure of the defect generation rate P_{gen}, defined as the number of defects produced per electron of a given energy traversing the oxide. The generation rate strongly depends on applied bias (electron energy) and temperature [61–64]. Knowing P_{gen} (and its functional dependence on applied bias

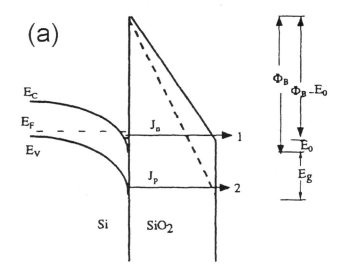

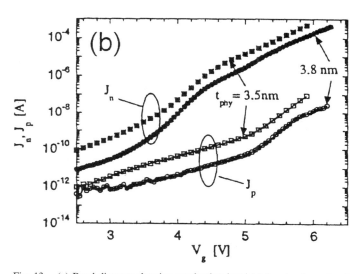

Fig. 13. (a) Band diagram showing conduction band (J_n) and valence band (J_p) electron tunneling paths and (b) conduction band and valence band tunneling currents vs. applied gate voltage. Reprinted with permission from [52], © 1999, Materials Research Society.

and temperature) and N_{BD} (as a function of oxide thickness), one can calculate (for individual devices) charge to breakdown ($Q_{BD} = N_{BD}/P_{gen}$) and time to breakdown ($t_{BD} = Q_{BD}/J$) and extrapolate the parameters to lower (operating) voltages, 1–1.5 V.

Dielectric film breakdown is a statistical phenomenon. It has been established that the statistics of breakdown is described by the Weibull distribution [10, 60]. Figure 15 shows an example of a typical Weibull distribution of Q_{BD} and t_{BD} for a series of capacitors. The data shown is typical of sub-3-nm oxides and oxynitrides. The Weibull distribution is characterized by two parameters: the characteristic Q_{BD} (t_{BD}) value (at 50% or 63% of the distribution) and the slope β. The latter plays a major role in extrapolating the measured lifetime of individual devices (of

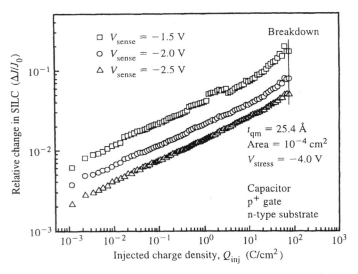

Fig. 14. Example of the generation of defects measured by the relative change in the SILC under three different sense conditions for stress of bias −4 V (t_{qm} is the oxide thickness extracted from C–V data and corrected for quantum mechanical effects in the measurement). Reprinted with permission from [10], © 1999, IBM Technical Journals.

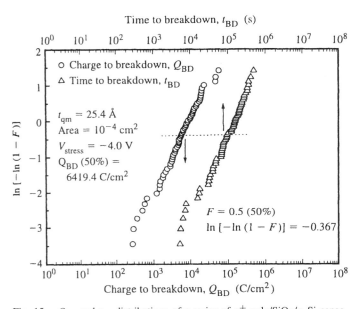

Fig. 15. Q_{bd} and t_{bd} distributions of a series of n^+-poly/SiO$_2$/p-Si capacitors with a cumulative probability of failure F. Reprinted with permission from [10], © 1999, IBM Technical Journals.

area A_{device}) to the lifetime of the chip (of area A_{chip}), which is

$$(Q_{BD})_{chip} = (Q_{BD})_{device} \left(\frac{A_{device}}{A_{chip}} \right)^{1/\beta} \qquad (1)$$

The larger the β, the longer the lifetime of the chip for a given failure rate of the individual device.

4.2. Physicochemical Characterization

Physicochemical characterization of device structures is becoming more and more important for the microelectronics industry as further advances in silicon technology increasingly depend on an atomic-scale description of the correlation among processing, structure, and performance. Such characteristics reveal the physics and chemistry behind the results of electrical measurements.

Continuous thinning of gate dielectrics, as well as the introduction of many insulators besides SiO$_2$ as potentially relevant for silicon technology, makes characterization a very challenging task. Surface science techniques have long been invoked and, in combination with isotopic substitution experiments, have provided much of the currently available mechanistic description concerning film growth and behavior upon thermal annealing [65]. In this section selected techniques found to be useful for the physicochemical characterization of ultrathin dielectrics are presented, in an attempt to provide an understanding of data presented in Sections 5, 6, and 7.

4.2.1. Microscopy

Two microscopies have been used to characterize the topography of ultrathin gate dielectric films. If a cleaved sample is used, both can yield cross-sectional (in-depth) micrographs. One is atomic force microscopy (AFM), one of the so-called scanning probe microscopies (SPMs). A surface is imaged at high (almost atomic) resolution (Fig. 16) [66] by rastering an atomically sharp tip back and forth in close contact with the surface. The major impact of AFM is the possibility of observing atomic-scale structures in real space. AFM can be done at room temperature in the presence of air, requires little sample preparation, and is nondestructive. The AFM has a tip mounted on a cantilever of low elastic constant. The tip scans a surface under investigation. It may touch the surface or not (contact or noncontact mode of operation). The force of interaction between the sample and the tip provokes deflection of the cantilever. AFM uses laser light to measure deflections. The light is focused at the edge of the cantilever; after reflection, it is collected by a detector that measures the variations in beam position and intensity produced by deflections of the cantilever. To make very small scans, the sample is mounted on a piezoelectric ceramic. Increments of less than 1 Å can be achieved. Data collected by the detector are processed with the aid of a computer, rendering bi- or tridimensional images.

The other microscopy currently applied to ultrathin dielectric characterization is transmission electron microscopy (TEM). Atomic resolution can be achieved with its high-resolution variant (HRTEM) (Fig. 17). The microscope consists of an electron source and focusing elements in an evacuated (10^{-5} to 10^{-8} Torr) metal cylinder. Electrons are extracted, accelerated to energies on the order of 100 keV, and made to impinge on a sample. As the electron beam traverses the sample, some electrons are scattered and the remainder are focused

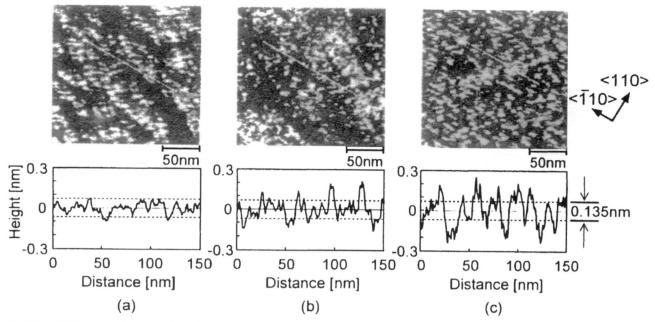

Fig. 16. AFM images and cross sections of oxide films with thicknesses of (a) 0.83 nm, (b) 1.53 nm, and (c) 1.92 nm. Cross-sectional profiles were obtained along the lines in the AFM images. Reprinted with permission from [66], © 1999, Materials Research Society.

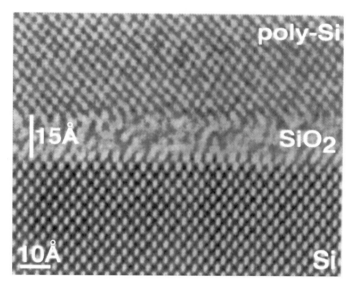

Fig. 17. High-resolution TEM cross section of the poly-Si/SiO$_2$/Si structure with a ~1.5-nm gate oxide (micrograph by M. Gribelyuk and P. Varekamp, IBM). Reprinted with permission from [60], © 2000, Kluwer Academic Publishers.

onto a phosphorescent screen or photographic film to form an image. The limited penetrating power of electrons requires samples in the 50–100 nm thickness range. Contrast in the TEM depends on the atomic number of the atoms in the sample; the higher the atomic number, the more electrons are scattered and the greater the contrast. Unfocused electrons are blocked out by the objective aperture, resulting in an enhancement of the image contrast.

4.2.2. Electron Spectroscopies

Two of the most useful analytical techniques for the physicochemical characterization of ultrathin gate dielectric films have been X-ray photoelectron spectroscopy [XPS, also known as electron spectroscopy for chemical analysis (ESCA)] and Auger electron spectroscopy (AES). They provide qualitative and quantitative elemental analysis. Their unique feature is that the analysis is not restricted to the identity of the elements in the samples, but chemical bonding information is also provided.

In XPS, the sample is illuminated with X-rays (usually from the Mg K_α or Al K_α line), and photoelectrons emitted from the surface region are energy-analyzed and detected. Practically only photoelectrons leaving the sample without interaction (energy loss) can be detected, so the technique is highly surface specific. The inelastic mean free path (up to a few nanometers) depends on the composition of the sample and on the energy of the photoelectrons. Photoelectrons leaving the sample pass through an energy analyzer to reach a channeltron detector. This gives a spectrum (number of photoelectrons vs. photoelectron energy) with a series of peaks (Fig. 18). The kinetic energy of these emitted electrons is characteristic of each element. Each peak in the spectrum is labeled according to the element and orbital from which emission occurred, as in "Si $2p$." The peak areas can be used (with appropriate sensitivity factors) to determine the composition of the sample surface. This is usually expressed as an atomic percentile (at.%) by comparison to a standard. The chemical state of an atom alters the binding energy of a photoelectron and so its kinetic energy upon detection. Bonding information is derived from these chemical shifts and from the shapes of the peaks. XPS is not sensitive to hydrogen or helium, but can detect all other elements with a detection

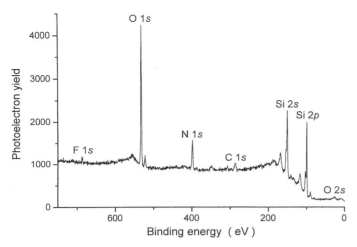

Fig. 18. Survey XPS spectrum from an ultrathin SiO_2 film on Si nitrided with a low-energy ion beam. Selected peaks are identified.

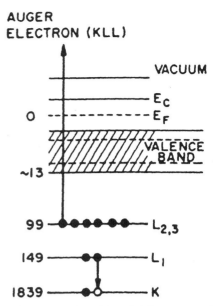

Fig. 19. Schematic representation of the KLL Auger process in a solid. This core hole leads to a contraction of the outer shells. The K hole is filled by an L electron in the transition process, and the excess energy is transferred to another L electron, which is ejected from the atom. The final state is a doubly ionized atom. Reprinted with permission from [68], © 1986, North-Holland.

limit of about 1 at.%. XPS must be carried out in ultra-high-vacuum conditions ($<10^{-8}$ mbar) to minimize sample surface contamination. A monochromator can be used after the X-ray source to improve resolution in the determination of chemical shifts. The availability of synchrotron radiation sources [67] made XPS possible over extremely short sampling depths (less than 1 nm) and with an energy resolution of 0.02 eV. Elemental depth profiling with chemical information is also possible upon repeated analysis of a sample at different photoelectron take-off angles (this actually makes use of the well-behaved variation of sampling depth with experimental geometry), constituting angle-resolved XPS (ARXPS). (The take-off angle is the angle between the sample surface normal and the axis of the electron energy analyzer.)

AES employs a beam of low-energy (3–20 keV) electrons as a surface probe, analyzing and detecting secondary electrons in very much the same way as done in XPS. The primary electrons cause core electrons from atoms contained in the sample to be ejected, resulting in a free electron and an atom with a core hole. The atom then relaxes via electrons with a lower binding energy dropping into the core hole. The energy thus released can be converted into an X-ray, or an electron can be emitted. This electron is called an Auger electron, labeled according to the electronic shells involved in the emission, as in "KLL" (Fig. 19) [68]. Auger electrons characteristic of each element present are emitted from the sample, identifying the element and carrying chemical bonding information (Fig. 20) [69]. (In fact, Auger electrons are always present in XPS spectra. The use of electrons instead of X-rays as the primary energy source in AES, however, increases the Auger electron yield. One fundamental difference between photoelectrons and Auger electrons is that the energy of the latter is fixed (within the range of chemical shifts), whereas that of the former depends on the energy of incident X-rays.) As in XPS, given the limited inelastic mean free path of electrons traversing matter, only those Auger electrons that emerge from the topmost atomic layers contribute to the spectrum. Also in analogy with XPS, AES must be carried

out in ultra-high-vacuum conditions. AES detects all elements other than hydrogen and helium, usually to a sensitivity better than 1 atom percent of a monolayer, and results are usually expressed as atomic percentiles. (Although the number of atoms corresponding to a monolayer depends on the element under consideration, 10^{15} cm^{-2} atoms is a general approximation.) It is most sensitive to elements with a low atomic number. Although angle-resolved analysis is also possible, depth profiling with AES is usually done in combination with sample sputtering. Sputtering involves directing a beam of ions (usually Ar$^+$ ions) at 500 eV to 5 keV at the sample. This process erodes away the sample to reveal structure beneath the surface. If sputtering is used in combination with a rotating sample holder, the depth resolution can be below 1 nm. (In the remainder of this chapter, depth resolution is used as the minimum distance separating two monolayers of a given species such that both can be identified.)

4.2.3. Ellipsometry

Ellipsometry [70–73] is an optical technique that uses polarized light to probe the dielectric properties of a sample. It is based on the measurement of the variation of the polarization state of the light after reflection from a plane surface. Depending on what is already known about the sample, the technique can probe a range of properties, including layer thickness, morphology, and chemical composition. A distinction should be made between single-wavelength and spectroscopic ellipsometries. The first can measure only two parameters; the latter can analyze complex structures such as multilayers and provide information about interface roughness. Characteristics of ellipsometry

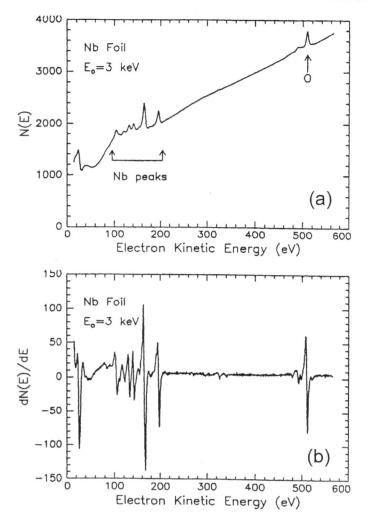

Fig. 20. (a) Auger electron spectrum for Nb, showing the electron yield as a function of electron kinetic energy and (b) its differentiated form. Historically, the spectra were differentiated to enhance the Auger signal, which was relatively weak. Reprinted with permission from [69].

are a nondestructive character, high sensitivity, and large measurement range (from fractions of monolayers to micrometers). In semiconductor research and fabrication, it is mainly used to determine properties of layer stacks of thin films and the interfaces between the layers.

After reflection from a sample surface, a linearly polarized light beam is generally elliptically polarized. The reflected light undergoes phase changes that are different for electric field components polarized parallel (p) and perpendicular (s) to the plane of incidence. Ellipsometry measures this state of polarization. Different measurement techniques of the polarization after reflection exist. All of them use the same optical components: a source, a polarizer, an analyzer, and a detector. To these basic elements other components like modulators or compensators can be added.

If the sample is an ideal bulk, the real and imaginary parts of the complex refractive index may be calculated from the measured ellipsometric parameters (amplitude ratio and phase shift of the p and s components) with a knowledge of the incidence angle. The optical index and thickness of a transparent layer on a known substrate can also be deduced in the same way. This kind of analysis is characteristic of a single-wavelength ellipsometric measurement.

The spectroscopic ellipsometry technique can be used for the analysis of more complex samples. The idea is to measure the two ellipsometric parameters in a large range of wavelengths and to assume that the optical indices of the materials are known. With an optical model it is then possible to extract the different physical parameters of the sample. Ellipsometry is not a direct deductive method, except in one simple case: the case of a bulk material. It is generally necessary to build a priori multilayer models to extract physical information after numerical fitting of experimental data.

4.2.4. Infrared and Raman Spectroscopies

Infrared radiation can be used to qualitatively probe chemical groups in a sample [74]. If a group has a dipole moment, it can absorb infrared light, but only at certain fixed frequencies. Hence, an infrared spectrum of light that traversed the sample or was reflected from its surface will show absorption peaks that are characteristic of the chemical groups in that sample. Silicon is essentially transparent in the spectral region of interest, so that the characteristic high-frequency (>2000 cm^{-1}) stretching vibrations of hydrogen-containing species such as SiH, H_2O, OH, NH_x, and CH_x can be readily probed. Furthermore, the lower-frequency (<1500 cm^{-1}) modes of important atoms such as O, C, N, and F can also be studied, although this frequency region is particularly challenging because of a combination of silicon phonon absorption and lower sensitivity of infrared detectors in this region. Infrared spectroscopy is found to be useful for the analysis of ultrathin dielectric films, especially in the specular reflectance (reflection absorption infrared spectroscopy, RAIRS) and multiple internal reflectance (MIR) modes.

RAIRS provides a nondestructive method of analyzing films on reflective surfaces without requiring any sample preparation. Specular reflectance is a mirror-like reflectance from the surface of the sample. The angle of incidence that is selected for measurement depends upon the thickness of the film that is being analyzed. For ultrathin films in the nanometer range, an 80° angle of incidence would be chosen. Reflectance measurements at this angle of incidence are often referred to as grazing angle measurements.

Semiconducting materials lack the almost perfect reflectivity that metals exhibit in the infrared, so alternative methods to RAIRS have been developed for the analysis of such materials. One of these is MIR, also known as attenuated total reflectance (ATR). In this technique, the beam is internally reflected many (typically 50) times within the sample, which is shaped to behave like a prism; because the silicon interfaces are sampled repeatedly, strong absorption bands can be obtained (Fig. 21) [75]. MIR is generally a powerful technique, but it suffers from two disadvantages. First, the sample requires careful

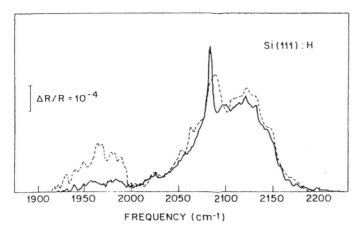

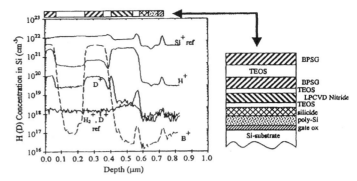

Fig. 21. Example of MIR spectrum: reflectivity change induced by a saturation coverage of hydrogen on the Si(111) surface. The solid line is for a
laser-annealed 1×1 surface with a coverage of 1 monolayer, and the dashed
line is for a thermally annealed 7×7 surface with a coverage of 1.25 monolayer.
Reprinted with permission from [75], © 1983, American Physical Society.

Fig. 22. Elemental depth profiles determined by SIMS in a device structure.
Reprinted with permission from [77], © 1999, American Institute of Physics.

construction and preparation, and second, the long path length
through the sample precludes measurements in any region of
the spectrum where the substrate absorbs, even quite weakly.

About the same information as provided by infrared spectroscopy can be obtained from Raman spectroscopy, which
makes use of a laser beam as the primary excitation source.
In this case, however, it is not a permanent dipole moment
that makes a chemical group detectable, but its polarizability.
Generally, a group that is IR active is not active in Raman and
conversely, so the techniques are complementary.

4.2.5. Secondary Ion Mass Spectrometry

If a surface is bombarded with high-energy ions, then ions
and atoms from the sample will be sputtered from the surface. Direct qualitative and quantitative elemental information
is produced by determining the mass of the ionized fragments
removed, which is the principle of secondary ion mass spectrometry (SIMS) [76].

Primary beam species useful in SIMS include Cs^+, O_2^+, Ar^+,
and Ga^+ at energies between 1 and 30 keV. The bombarding
primary ion beam produces monatomic and polyatomic particles of sample material and resputtered primary ions, along with
electrons and photons. The secondary particles carry negative,
positive, and neutral charges, and they have kinetic energies that
range from zero to several hundred electron-volts. Only a small
proportion of the elements emitted from the surface are ionized. Ion yields vary over many orders of magnitude for the
various elements. As the probability of ionization in SIMS is
highly variable, it is difficult to quantify SIMS data. This problem is largely overcome in sputtered neutral mass spectrometry
(SNMS). The neutral atoms (more than 99% of the sputtered
species) are detected by post-ionizing any atoms that are ejected
from the surface. This post-ionization can be accomplished by
using lasers or electron bombardment of the atoms entering the
mass analyzer, usually a quadrupole mass spectrometer.

Given continuous erosion of the sample by the ion beam,
monitoring the secondary ion count rate of selected elements
as a function of time leads to depth profiles (Fig. 22) [77]. To
convert the time axis into depth, the SIMS analyst uses a profilometer to measure the sputter crater depth. Depth resolution
can be close to 1 nm near the sample surface, depending on
flat bottom craters, but also on atom mixing produced by the
primary ion beam. Primary ions are implanted and mixed with
sample atoms to depths of 1 to 10 nm. When the sputtering rate
is extremely slow, the entire analysis can be performed while
consuming less than a tenth of an atomic monolayer. This slow
sputtering mode is called static SIMS, in contrast to dynamic
SIMS used for depth profiles. Only dynamic SIMS yields quantitative information. Results are usually expressed as volumetric
densities (cm^{-3} atoms).

SIMS has the highest sensitivity of all surface analysis
techniques. Detection limits for most trace elements are between 10^{12} and 10^{16} cm^{-3} atoms. Oxygen, present as residual
gas in vacuum systems, is an example of an element with
background-limited sensitivity. Analyte atoms sputtered from
mass spectrometer parts back onto the sample by secondary
ions constitute another source of background. Mass interfer-
ences also cause background limited sensitivity. Because sputtered ion mass is determined in SIMS, all elements can be
detected, including hydrogen and, similarly, all of the isotopes
of all of the elements.

4.2.6. Ion and Atom Scattering and Recoil Spectrometries

These techniques provide qualitative and quantitative, depth-
resolved elemental analysis of thin films. They are low-energy
ion scattering [LEIS, also known as ion scattering spectrometry (ISS)], medium-energy ion scattering (MEIS), high-energy
ion scattering [HEIS, better known as Rutherford backscattering spectrometry (RBS)], and elastic recoil detection (ERD).
They involve impinging collimated, nearly monoenergetic ion
beams on a sample and detecting backscattered or recoil particles. The energy of primary ions in LEIS is normally 1–10 keV;
in MEIS, 20–200 keV; and in HEIS, 200–2000 keV. ERD makes
use of primary ion beams of 2000 keV and above. The lateral

resolution of the techniques corresponds to the beamspot, usually between 0.5 and 4 mm in diameter.

In ion scattering spectrometries, information is extracted from the energy spectrum of ions scattered by atomic nuclei in the target. In ERD, use is made of the fact that high-energy ions impinging on a sample knock out atoms from the near-surface region; it is the energy spectrum of such recoils that is used to characterize the sample. These analytical techniques can be understood based on four physical quantities: kinematic factor, cross section for ion scattering or elastic recoil, stopping cross section, and straggling constant.

Knowledge of the kinematic factor allows qualitative elemental analysis. For ion scattering, it is defined as the ratio between the ion energies after and before collision with a nucleus in the sample. From the conservation of energy and momentum, such kinematic factors can be calculated for ion scattering as well as for elastic recoil. They depend on the mass of primary ions and target atoms and on the geometry of the experiment (scattering or recoil angle). The kinematic factor varies more strongly between light elements, thus allowing isotope-resolved analysis in this case. Once the energy of the primary beam and kinematic factors are known, one determines the composition of a sample from the distribution in energy of the signals observed in an experimental spectrum.

Ion beam techniques intrinsically provide depth profiles because the ions lose energy as they traverse matter. This energy loss is quantitatively expressed by the stopping cross section. If the scattering or recoil event occurs at the sample surface or at a given depth the detected particles will have different energies (provided the target element is the same). Depending on sample thickness and composition, signals from two or more elements may appear stacked in a backscattering spectrum. Stopping cross sections vary greatly, depending on ion identity, ion energy, and sample composition.

The quantitative character of ion scattering and elastic recoil techniques is due to well-known scattering and recoil cross sections, which indicate the probability of scattering or recoil occurrence. The cross sections depend on the atomic number and mass of primary ions and target nuclei, as well as on the geometry of the experiment. The number of scattered ions or recoils detected with a given energy depends on, in addition to the cross section, the number of primary ions reaching the sample, the solid angle of detection, and the content of the target element in the sample. This content is expressed as an areal density (so one speaks of 2×10^{15} cm^{-2} oxygen atoms, for example), the product of the volumetric density of the target element in the sample and the thickness corresponding to the interval between two (or more) points in the experimental energy spectrum of the scattered ions or recoils.

As ion scattering conveniently offers accurate absolute quantitative elemental information, it is quite often used to study film growth. In this case, the areal density of a given element (say O in SiO$_2$ films) is usually converted to film thickness through the volumetric density of the bulk material. For example, 1×10^{15} cm^{-2} O corresponds to 0.226 nm of thermal SiO$_2$ and 1×10^{15} cm^{-2} N corresponds to 0.188 nm of thermal

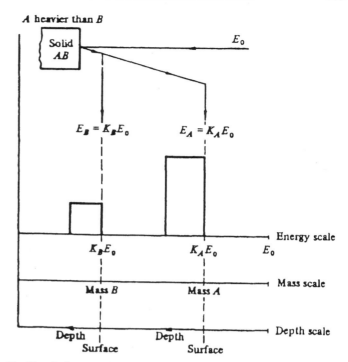

Fig. 23. Scale conversion in ion beam analysis; K_i is the kinematic factor for element i. Reprinted with permission from [78], © 1989, Cambridge University Press.

Si$_3$N$_4$. As the volumetric density of an ultrathin film usually varies with thickness and so cannot be exactly equal to the bulk density of its constituting material, one may find growth kinetics and the like directly expressed in the form of areal densities. Such is the case in Figure 3.

A schematic spectrum from ion beam analysis is presented in Figure 23 [78], showing the detected particle energy scale also as target mass and target depth distribution scales. Mass and depth resolution in ion beam analysis depend on the masses of projectile and target, projectile energy, experimental geometry, and detector resolution. They degrade with increasing depth at which the collision occurs because of statistical fluctuations in the energy loss of ions traversing matter, which constitutes "straggling" and is quantified by the straggling constant.

In the case of ultrathin amorphous dielectric films on single-crystalline silicon, one wants to concentrate on the amorphous region. In such cases, medium- and high-energy ion beam experiments are usually performed in a so-called channeling geometry. The ion beam is aligned along one of the major crystallographic directions in the sample, such that the first atom of each atomic row casts a shadow cone, reducing the contribution of the crystalline substrate to the collected spectrum. If the detection of the backscattered particles is performed at grazing angles (close to 90° relative to the direction of beam incidence), the outgoing path length of the emergent particles can be much larger than for the path corresponding to the usual detection angle of nearly 180°. Such stretching spreads the total number of detected scattering events in a given thickness over a greater energy interval in the collected spectrum. The combined effects of

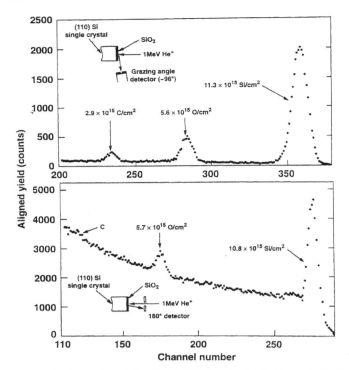

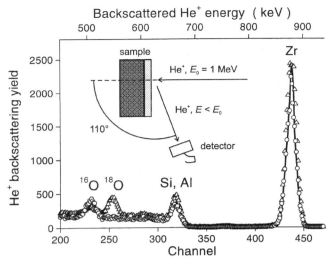

Fig. 24. Backscattering yields of 1-MeV He ions for ⟨110⟩ alignment of a Si crystal (having a {110} planar surface) that was covered by 1.5 nm of oxide. Results from both a grazing angle detector and a 180° detector are shown [79]. Reprinted with permission from [79], © 1980, Elsevier Science.

Fig. 25. RBS spectrum from a film of $ZrAl_xO_y$ on Si recorded with the use of ion channeling and grazing angle detection. The solid line corresponds to the as-deposited sample; the circles, to the sample after thermal annealing in vacuum; and the triangles, to the sample after thermal annealing in $^{18}O_2$.

channeling and grazing angle detection are improved detection limit for elements lighter than Si and increased sensitivity to the sample surface. Jackman et al. [79] investigated in detail the application of this technique to the determination of the amounts of O and Si in silicon oxide films, as shown in Figure 24 [79].

RBS [78, 80] is usually performed with H^+ or He^{2+} ions. It can be distinguished from other ion beam analysis techniques by its excellent ability to extract quantitative data. Typical accuracy is between 3% and 5%. This is due to the precise knowledge of the Rutherford scattering cross sections. Because of a lower detection limit (10^{-2} to 10^{-4} monolayers) and increased sensitivity, RBS is ideally suited for determining the concentration of elements heavier than the major constituents of the substrate and so has found application in the analysis of alternative dielectrics to SiO_2. Its detection limit for light masses is 10^{-1} to 10^{-2} monolayers. Because primary ions have at least the mass of hydrogen, backscattering from this element is not possible, and so it cannot be detected at all by RBS. Depth resolution in RBS can be less than 2 nm with the use of grazing ion incidence or detection. For samples with thicknesses below depth resolution, no profiling can be done, but RBS can still be used to determine the total amount of a given element. One RBS spectrum collected for a $ZrAl_xO_y$ film on Si is shown in Figure 25.

MEIS [81–83] is usually performed with H^+ primary ions of sufficiently low energy for an electrostatic energy analyzer to be used before the ion detector, greatly improving energy resolution in comparison with RBS. Not only the energy

but also the angular distribution of the scattered ions is usually recorded, allowing crystallographic studies [81, 82, 84]. Isotope-sensitive concentration profiles with depth resolutions of up to 0.3 nm [85] and overall sensitivities of about 10^{13} cm^{-2} atoms can be achieved under favorable experimental conditions. This makes MEIS one of the most suitable techniques for the physical characterization of ultrathin dielectric films. Figure 26 shows the nitrogen and oxygen sections of a MEIS spectrum for an ultrathin silicon oxynitride film grown on Si(001) [85]. For each element there is a corresponding peak, with the lighter element appearing at lower backscattering energies. The areas under the peaks are proportional to the total amounts of the elements in the probed region of the sample, and the shape of each peak contains information about the depth distribution of the corresponding element.

ISS [86] is only sensitive to the first atomic layer of the surface. The technique is ideal for certain types of analysis where unique depth resolution, virtually nondestructive analysis, and general ease of interpretation are important requirements. As in MEIS, an energy analyzer is used before the ion detector, yielding high energy resolution (Fig. 27). A drawback of LEIS is that it suffers from complications due to multiple scattering. Scattering cross section increases slowly with target mass; a factor of 10 covers most atomic species. The cross section for the detection of oxygen or carbon is about a factor of 10 lower than for heavy atoms. The detection limit in ISS is about 10^{-4} to 10^{-3} atoms per monolayer, and is better for heavy atoms on a light substrate.

In ERD [87] one determines the yield and energy not of scattered primary ions (as in RBS, MEIS, and ISS), but of particles ejected from the surface region of samples because of collision with high-energy, relatively heavy primary ions (C^{n+}, O^{n+}, Si^{n+}, or others). A consequence of the conservation laws is that the energy of a target atom after collision with a pro-

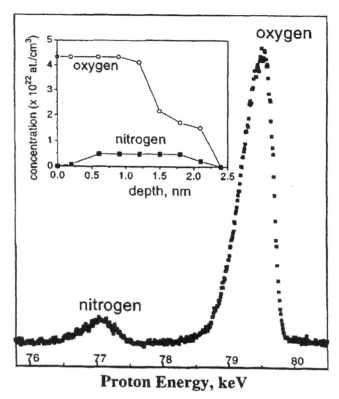

Fig. 26. Oxygen and nitrogen sections of a MEIS spectrum from an ultrathin silicon oxynitride film grown on Si(001); simulated profiles are shown in the inset. Reprinted with permission from [85], © 1996, American Institute of Physics.

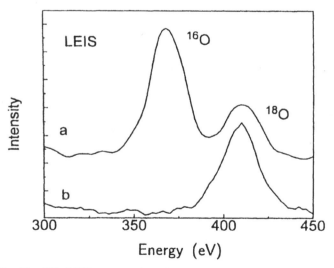

Fig. 27. ISS (LEIS) spectra taken after oxidation of Si (b) at 1080 K for 100 min under ~10^{-3} Torr of $^{18}O_2$ and (a) sequential annealing at 1120 K for 120 min under ~10^{-1} Torr of $^{16}O_2$. The probing ion beam was He$^+$ at 1.0 keV. Reprinted with permission from [82], © 1995, American Physical Society.

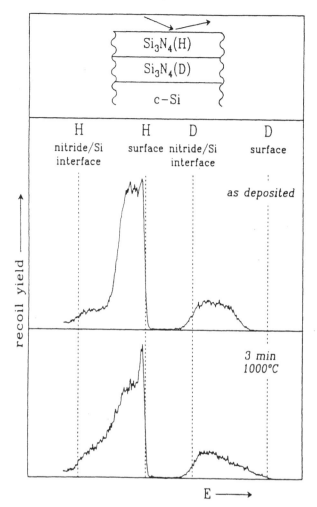

Fig. 28. ERD spectrum (recoil yield vs. recoil energy) of a LPCVD Si_3N_4 layer structure. The top layer (60 nm) was deposited with the use of NH_3, and the bottom layer (95 nm), with the use of ND_3 (D stands for ^2H). Top spectrum: as-deposited. Bottom: after annealing for 3 min at 1000°C. From the raw spectra, interdiffusion of H and D across the original H/D interface is clearly observed. Reprinted with permission from [88], © 1992, Elsevier Science.

analysis—if the film thickness is less than the depth resolution of the technique, the content of a given element in the sample can still be determined. Figure 28 [88] shows a sample ERD spectrum.

4.2.7. Nuclear Reaction Techniques

Charged particles (mainly protons and deuterons) accelerated in the 0.1–2 MeV range can induce nuclear reactions on light nuclides, with potential use in the determination of overall near-surface nuclide amounts. This is the basis of nuclear reaction analysis (NRA). An ion beam is directed to a sample to induce a nuclear reaction, and reaction products are collected with a suitable detector (semiconductor surface barrier detectors for particles, scintillator detectors for gamma rays). The number of reaction products collected for a given ion fluence is converted to nuclide concentration with the use of standard

jectile contains the same kind of information as that of the scattered particle. ERD is unique in allowing hydrogen detection and depth profiling. Depth resolution is about 10 nm, so the same situation as observed for RBS appears in ultrathin film samples or tabulated cross sections. Some particularly useful reactions are $^{16}O(d,p)^{17}O$, $^{18}O(p,\alpha)^{15}N$, $^{14}N(d,\alpha_0)^{13}C$, and $^{15}N(p,\alpha\gamma)^{12}C$. (These are shorthand notations for reactions like $^{16}O + d \longrightarrow ^{17}O + p$; p indicates proton and d deuteron.) As the projectiles lose energy traversing matter and reaction probability is a function of ion energy, analysis is preferentially done in the plateau regions of the cross section curves, respectively around 810, 1000, 1450, and 760 keV [89]. The fact that NRA is nuclide-specific has been of great advantage for isotopic tracing experiments in ultrathin dielectric films [65]. Depending on the reaction used, the sensitivity achieved is better than 10^{14} cm^{-2}, with about 3% accuracy. NRA can yield nuclide depth profiles if used in combination with chemical etchback. In this technique, a sample is analyzed for the amount of a given nuclide, then a surface layer is chemically etched away and the sample is analyzed again, and so on. A plot of the amount of a nuclide in a film as a function of the total number of atoms in the film (film thickness) yields a depth profile through differentiation. The etch rate of SiO_2 films in HF solutions is well known. Etchback combined with NRA can yield depth profiles in the nanometer depth resolution range. Chemical etch of oxynitride films is less homogeneous, resulting in degraded depth resolution. Some of the materials proposed as alternative to SiO_2 can be etched by HF, such as Al_2O_3, but the process is much less understood.

When the cross section curve for a nuclear reaction as a function of projectile energy presents a narrow resonance, the measured yield curve of emitted reaction products (charged particles or gamma rays) as a function of ion beam energy in the vicinity of the resonance energy may give valuable information on the concentration profile of the probed nuclide. This constitutes narrow nuclear resonance profiling (NNRP). Interpretation of experimental data with the use of the stochastic theory of energy loss for ions in matter can lead to depth resolutions of up to 1 nm, with sensitivities of about 10^{13} cm^{-2} atoms [89, 90]. Useful reactions are $^{18}O(p,\alpha)^{15}N$ at 151 keV, $^{15}N(p,\alpha\gamma)^{12}C$ at 429 keV, $^{29}Si(p,\gamma)^{30}P$ at 417 keV, $^{27}Al(p,\gamma)^{28}Si$ at 405 keV, and $^{1}H(^{15}N,\alpha\gamma)^{12}C$ at 6.40 MeV.

The principle of narrow nuclear resonance profiling is illustrated in Figure 29. If a sample is bombarded with projectiles at the resonance energy, the reaction product yield is proportional to the probed nuclide concentration on the surface of the sample. If the beam energy is raised, the yield from the surface tends to vanish as the resonance energy is reached only after the beam has lost part of its energy by inelastic interactions with the sample. Then the reaction yield is proportional to the nuclide concentration at a given depth, roughly the excess energy of the incident beam divided by the average ion energy loss per unit length. The higher the energy of the beam above the resonance energy, the deeper the probed region in the sample. At the heart of NNRP, high depth resolution is a consequence of the narrow resonance acting as an extremely high-resolution energy filter.

A key aspect in NNRP is that the reaction product yield versus beam energy curve (called the excitation curve) is not the probed nuclide concentration profile itself, but is related to it

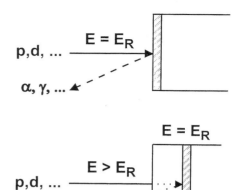

Fig. 29. Principle of narrow nuclear resonance depth profiling. p and d represent protons or deuterons in an incoming ion beam, and α and γ represent detectable nuclear reaction products. E is the ion beam energy, and E_R is the energy of the resonance.

by the integral transform [90],

$$N(\overline{E}_0) = \int_0^\infty C(x) q_0(x; \overline{E}_0)\, dx \tag{2}$$

in which $N(\overline{E}_0)$ is the number of detected reaction products as a function of beam energy (the experimental excitation curve), $C(x)$ is the nuclide concentration profile, and $q_0(x; \overline{E}_0)\, dx$ is the energy loss spectrum at depth x (the probability that an incoming particle produces a detected event in the vicinity dx of x for unit concentration). Through the last factor, the width of which sets the depth resolution of the method, the excitation curve depends on the resonance shape and width, on the ion beam energy spread (an instrumental function), and on the incoming ion slowing down process. This is illustrated for an arbitrary profile in Figure 30 [89]. In the case of light projectiles, straggling (i.e., additional energy spread due to the slowing down process) dominates the interpretation of the excitation curve. Real excitation curves and the corresponding depth profiles are presented in Figure 31 [91].

The stochastic theory of energy loss as implemented in the Spaces program [92] has been used to calculate and accurately interpret experimental data. Nuclide concentration profiles are assigned on an iterative basis, by successive calculation of a theoretical excitation curve for a guessed profile followed by comparison with experimental data. The process is repeated with different guessed profiles until satisfactory agreement is achieved.

Resonance widths are often between 50 and 150 eV for proton energies from 300 to 1200 keV, with comparable instrumental beam energy spreads. This leads to near-surface depth resolutions of a few nanometers, which may be improved with the use of grazing incidence geometry [89]. Under such modified conditions, increased trajectory of incoming and outgoing (when applicable) particles results in improved depth resolution, which may go down to 1 nm. The ultimate resolution is

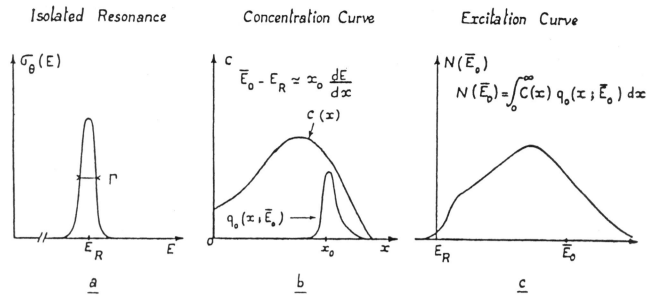

Fig. 30. Relation among (a) resonance peak, (b) nuclide concentration profile and energy loss spectrum, and (c) excitation curve. Reprinted with permission from [90], © 1982, Elsevier Science.

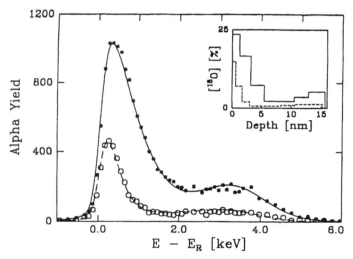

Fig. 31. Experimental (points) and simulated (lines) excitation curves for the nuclear reaction $^{18}O(p,\alpha)^{15}N$ around the resonance energy of 151 keV for an $Si^{16}O_2$ film annealed in $^{15}N^{18}O$ for different times. The inset shows the ^{18}O profiles used in the simulations. Reprinted with permission from [91], © 1998, American Institute of Physics.

limited by energy straggling and angular multiple scattering processes.

5. HYDROGEN AND ULTRATHIN GATE DIELECTRIC FILMS

Hydrogenous species play a key role in the performance of MIS devices. This section addresses aspects relative to hydrogenous species in gate dielectric materials and at the insulator–semiconductor interface. Specifically, the identity and location

of such species in the SiO_2-Si system is addressed, and the so-called giant isotope effect [93] is presented and discussed. Issues concerning hydrogen have been much less explored in other dielectric materials.

Hydrogen is ubiquitous in thin and ultrathin silicon oxide films on silicon. It is present at significant concentrations even in thermal oxides grown in dry, nominally pure oxygen. It is also present in other dielectrics, especially when deposited by CVD and its variants from hydrogenated precursors. The relatively high background level of hydrogen-containing molecules like H_2, H_2O, Si_2H_2, NH_3, HF, and others in semiconductor processing ambients allied to high mobility and significant reactivity also accounts for a significant amount of hydrogen in devices. Whereas other impurities (e.g., alkali or other metals) have largely been eliminated in modern semiconductor fabrication, hydrogen remains.

Hydrogen has both beneficial and detrimental effects in semiconductor devices. As for the former, electrical characterization of the MOS structure and electron paramagnetic resonance (EPR) [94] have shown that hydrogen acts as a passivant for SiO_2–Si interfaces. It ties to silicon dangling bonds, passivating the charge traps existing therein [95]:

$$Si- + H \rightarrow Si-H \tag{3}$$

(Such paramagnetic defects as dangling bonds are usually referred to as P_b centers in the literature.) Figure 32 [96] shows that interface charge trap level in dry SiO_2 on Si treated with a limitless supply of passivant depends basically on the annealing temperature. Higher temperature causes a greater reduction in Q_{it}, and a sample previously annealed at a higher temperature depassivates at a lower temperature, tending toward the Q_{it} corresponding to the lower temperature. It should be mentioned that Q_{it} from electrical measurements comprises more types

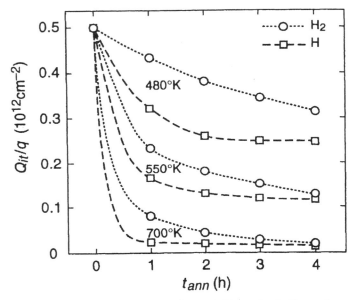

Fig. 32.   Passivation of interface traps by atomic H or molecular H_2 as a function of annealing time with annealing temperature as a parameter. Reprinted with permission from [96], © 1988, The Electrochemical Society, Inc.

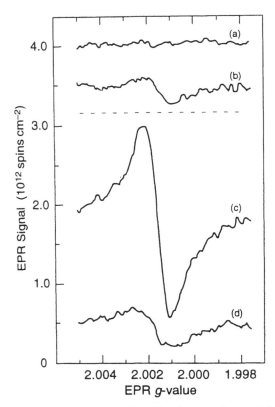

Fig. 33.   Passivation and depassivation of interface traps by atomic hydrogen at 300 K. (a) P_b in as-oxidized wafer; (b) after exposure to H; (c) as-oxidized wafer, dessicated in vacuum; (d) dessicated wafer exposed to H. The EPR g-value is inversely proportional to the magnetic field. Reprinted with permission from [97], © 1993, American Institute of Physics.

of defects than the well-defined P_b center, ideally detected by EPR.

Concerning deleterious effects of hydrogen in devices, its involvement in hot-electron-induced degradation and dielectric breakdown of SiO_2 films was suggested by a number of experiments, notably the observation of substrate dopant passivation and hydrogen redistribution during hot electron stress and enhanced degradation rate of hydrogen-soaked films. This involves a depassivation reaction,

$$Si-H \longrightarrow Si- + H \qquad (4)$$

It was found that the steady-state balance between passivation and depassivation determines the final P_b density [97, 98]. Figure 33 [97] presents illustrative EPR results. The reaction balance is strongly dependent on temperature, as already shown in Figure 32.

The passivation of silicon dangling bonds by hydrogen gave origin to a deliberate operation used in various processing steps in the semiconductor industry. It consists of an annealing in forming gas (10 vol.% H_2 in N_2 or Ar) aimed at passivating electrical activity at the SiO_2–Si interface. It is only after this step that a device-grade interface with a density of electronic states in the low 10^{10} cm^{-2} eV^{-1} is achieved, because P_b centers amount to about 10^{12} cm^{-2}. Further processing and MOS device operation can break the Si−H bonds, releasing hydrogen and depassivating the interface, with consequent degradation of the electrical characteristics of a device [94, 99].

Hydrogen release during device processing is related to the stability of Si−H bonds upon thermal annealing in different ambients and is one of the reasons for reducing the thermal budget during device processing. Hydrogen release during device operation is promoted by bombardment of the interface with electrons at a few electron-volts deviated from the channel (the

so-called hot electrons in light of the correspondence between energy and temperature given by the Boltzmann constant) [97, 100, 101]. It was recently found that replacing hydrogen [(^{1}H)$_2$] with deuterium [(^{2}H)$_2$ or D_2 for short] during the final wafer sintering process in forming gas at 450°C greatly reduces hot electron degradation effects in MOS transistors [93, 102–105]. Given increases in hot electron device lifetime by factors of 10–50 [102], this was called a "giant isotope effect." The isotope effect strongly correlates with the amount of D (more precisely, with the ratio of D to H) at the SiO_2–Si interface [102]. There is no clear evidence that D substitution has any significant impact on device reliability with respect to oxide breakdown (which is different from the reliability with respect to hot electron degradation).

The improved hot electron immunity is believed to be related to the H (D) depassivation from the interface caused by multiple excitations of the Si−H(D) vibrational modes under electrical stress. The depassivation/desorption yield is governed by two competing processes: (i) the excitation rate, which in turn depends on current density and electron transferred energy, and (ii) the rate at which the energy can dissipate by coupling to Si lattice phonons. Assuming that the lifetime is controlled by the Si−H (Si−D) bending modes, and taking into account the vibrational frequencies of the bending modes

(650 cm^{-1} for Si−H and 460 cm^{-1} for Si−D), one notices that the frequency for the Si−D bending mode is close to the frequency of the Si bulk TO phonon state (463 cm^{-1}). Therefore, the Si−D bending mode is expected to be coupled to the Si bulk phonon, resulting in an efficient channel for deexcitation. This enables more efficient energy dissipation of the excited Si−D bond. The result was obtained from first-principles calculations [106], which showed no effect of isotopic substitution on thermal desorption. This indicates that D is as susceptible as H with regard to loss due to the processing thermal budget. Such a result should be due to the simultaneous excitation of Si−H(D) bonds and the crystalline silicon substrate. In addition to pure loss, exchange of H for D in successive processing steps is a major concern.

Deuterated MOS devices can be produced in several ways. Annealings in D_2 after the skeleton MOS has been built are the most common and economical [93, 103, 105]. Another approach is to use deuterated precursors when depositing an interlayer dielectric on top of the MOS structure [104]. Wet oxide growth with the use of D_2O has also been performed [107]. As the giant isotope effect is actually due to the crystalline silicon substrate, it can be expected to hold for insulating films other than SiO_2.

Although it is the most important mode from the point of view of electrical performance, it has been shown that Si−H is not the only bonding mode of H in the MOS system. D_2 was used [108] to mimic the behavior of H_2 interacting with dry thermal SiO_2 on Si(001), and two distinguished configurations were identified: one, hypothesized to involve Si−D bonding, produced D uptake at 300°C and above followed by release at 600°C; the other, identified as O−D, formed readily at 100°C and dissociated at about 800°C. Evidence indicated that incorporation is due to interactions with preexisting defect sites rather than chemical reactions involving breaking of Si−O bonds. D was so seen to incorporate mainly at the film surface and interface regions. Detailed studies indicate that only a small portion of the incorporated H (D) is used to passivate the interface dangling bonds. Most of the H (D) atoms are found to be bonded to Si, O, or H (D) (so remaining in molecular form) close to (but not exactly at) the SiO_2−Si interface. This will later (Section 6) be shown as a transition region characterized by the presence of substoichiometric oxide. The distribution of H in ultrathin SiO_2 films on Si prepared by different methods is well illustrated in Figures 34 [109] and 35 [110] as determined from SIMS and etchback NRA, respectively. Figure 36 [111], independently drawn from NNRP data, summarizes the results. It is worth noting that from such physical characterization data, ultrathin (<6 nm thick) SiO_2 films do not qualify as "bulk oxide."

Hydrogen-related issues have been studied to a far lesser extent in systems other than SiO_2-Si. Information on silicon oxynitride films deposited on silicon by CVD is available [112, 113]. Although the data presented were originally obtained for films above the ultrathin limit, marked features are expected to hold.

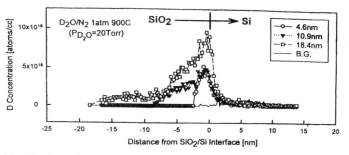

Fig. 34. Deuterium profiles determined by SIMS in thermally grown wet (D_2O) silicon oxide films of various thicknesses. Reprinted with permission from [109].

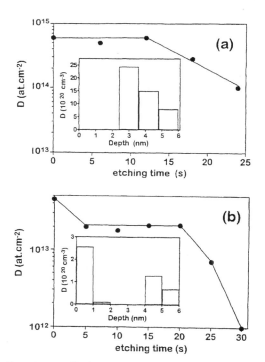

Fig. 35. Deuterium profile determined by etch-back NRA in thermally grown dry silicon oxide (a) after annealing in deuterated forming gas (a mixture of D_2 and N_2) for 45 min at 450°C and (b) after additional annealing in a vacuum for 30 min at 450°C. Reprinted with permission from [110], © 1998, American Institute of Physics.

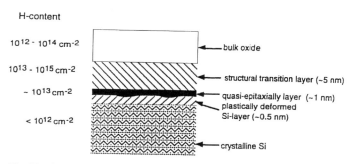

Fig. 36. Hydrogen concentration in the different characteristic regions of the SiO_2−Si structure. Reprinted with permission from [111], © 1996, The Electrochemical Society, Inc.

Hydrogen incorporated into deposited or grown silicon oxynitride films plays a role in their physical, chemical, and electrical stability. In both wet oxidation and nitridation of the Si—O—N system, hydrogen stabilizes intermediate reaction products, allowing multistep reactions to proceed. Interruption of the process results in incorporation of hydrogenated intermediates in the whole film thickness. The presence of hydrogen in these materials is usually associated with deviations from the "ideal" amorphous network structure, and a relation between defects and hydrogen is often emphasized. Most importantly, the presence of hydrogen in these materials is a major cause of their instability, because hydrogen can migrate into and out of the films at relatively low temperatures, which may result in detrimental macroscopic and microscopic defects.

During CVD of silicon oxynitride, hydrogen is incorporated into both N—H and Si—H configurations to concentrations of up to 10 at.%. The amount of incorporated hydrogen depends on both deposition temperature and the oxygen-to-nitrogen ratio in the films. Generally speaking, the higher the deposition temperature, the lower the hydrogen amount. Hydrogen desorption and overall reactivity of the films increase for O/N > 0.5. Samples containing both Si—H and N—H configurations in significant amounts lost hydrogen upon thermal annealings in vacuum or inert gas ambients above the deposition temperature. This was ascribed to a cross-linking effect,

$$\text{Si—H} + \text{N—H} \rightarrow \text{Si—N} + \text{H}_2 \qquad (5)$$

and

$$\text{Si—H} + \text{Si—H} \rightarrow \text{Si—Si} + \text{H}_2 \qquad (6)$$

with blistering and/or cracking of the films. The process is reversible; i.e., hydrogen uptake occurs if a sample is exposed to a hydrogen-rich ambient at elevated temperature.

The general picture emerging from considerations of hydrogen-related issues in ultrathin dielectrics is as follows. Most of the studies performed so far focus on the SiO$_2$-Si system. In this case hydrogen is desirable for passivating electrically active defects at the interface. The release of such hydrogen is of major concern with regard to device stability. Although most of the hydrogen in the SiO$_2$-Si system is not exactly at the interface, there it plays by far its most important role. Significant improvement of hot electron device lifetime has been foreseen upon substitution of deuterium for hydrogen at the SiO$_2$—Si interface (Fig. 37) [114]. Such an effect, however, is not expected to overcome the fundamental roadblocks for oxide scaling (Section 6). Hydrogen in CVD silicon oxynitride is reminiscent of the chemistry of film deposition. Further reactions involving hydrogen can lead to severe mechanical damage.

Analysis of both the SiO$_2$-Si and SiO$_x$N$_y$-Si systems showed significant but distinct roles for hydrogen. Silicon dangling bonds at the interface are most probably passivated when other insulating films are used, as annealings in forming gas are routinely seen to improve their electrical quality. However, no general rule can be offered for deeper considerations.

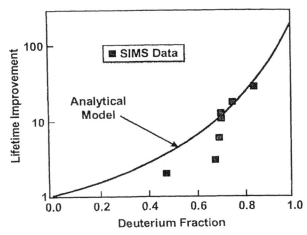

Fig. 37. Device lifetime improvement with regard to hot electron degradation as a function of deuterium fraction at the SiO$_2$–Si interface from experimental SIMS data and according to a simple analytical model. Reprinted with permission from [114], © 1999, IEEE.

6. SILICON OXIDE GATE DIELECTRIC FILMS

The whole of planar electronics processing and the modern IC industry have been made possible by the unique properties of silicon dioxide: the only oxide of a single semiconductor that is stable in water and at elevated temperatures, is an excellent electrical insulator, is a mask to common diffusing species, and is capable of forming a nearly perfect electrical interface with its substrate. This section presents characteristics of ultrathin SiO$_2$ films as gate dielectrics from electrical and physicochemical viewpoints. The ultrathin regime is generally assumed to be where the oxide thickness is less than around 4 nm, but is actually defined by a significant change in behavior as compared with thicker films and so varies from about 7 to 3 nm, depending on the characteristic under consideration.

6.1. Electrical Characteristics

Thinning the gate SiO$_2$ plays an important role in the scaling scheme of MOSFETs [115]. It can suppress short-channel effects by controlling the channel potential through the thinner gate oxide. It can also provide higher drain current because of the larger number of carriers in the channel due to the larger capacitance of the gate SiO$_2$ film. High drain current drive under low power consumption depends on drain voltage reduction. Without thinning of the gate dielectrics, improvement of the drain current drive cannot be expected. Now, further thinning of the gate SiO$_2$ is close to its limit and becomes the most critical issue for the development of next-generation CMOS. In this section, electrical limitations to further thinning of SiO$_2$ are presented and discussed, namely direct-tunneling leakage current, dielectric breakdown, poly-Si gate electrode depletion and inversion layer quantization, and boron penetration.

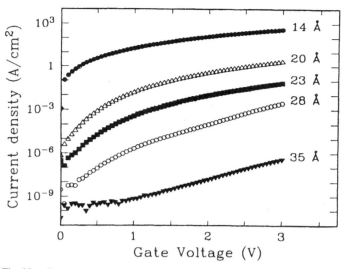

Fig. 38. Current density versus gate voltage for a series of MOS-C with gate dielectric films ranging from 1.4 to 3.5 nm. The SiO_2 films were grown at either 700°C or 750°C in dry O_2. Reprinted with permission from [12], © 1996, The Electrochemical Society, Inc.

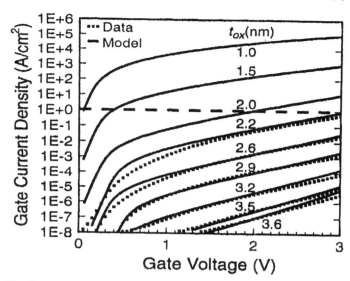

Fig. 39. Gate leakage current as a function of gate voltage for a series of oxide thicknesses. One can see that somewhere between 1.5 and 2.0 nm the gate leakage current density is higher than 1 A cm^{-2} at a gate voltage of 1 V, which may be too high for many applications. Reprinted with permission from [6], © 1999, IEEE.

6.1.1. Direct-Tunneling Leakage Current

It is well known that as the gate oxide thickness is scaled down to 3 nm and below, the current density at working bias conditions becomes substantial and increases exponentially with decreasing thickness. In Figure 38 the current density is shown as a function of applied gate bias for a series of devices with oxide thicknesses ranging from 3.5 to 1.4 nm. At a gate bias of 1.5 V the current density increases by 11 orders of magnitude for a thickness decrease of a factor of 2.5 [12]. The conduction mechanism in this regime is direct quantum mechanical tunneling [116, 117]. It has been shown [118] that the direct-tunneling gate current does not affect the MOSFET operation down to a gate oxide thickness of 1.5 nm for MOSFETs with gates less than 1 μm long. This is due to the fact that the tunneling gate current is, in fact, negligibly small compared with the large drain current provided by short-channel MOSFETs. Thanks to the large capacitance value of the direct-tunneling oxide, extremely high performance (i.e., high drain current and fast switching) has been reported [119].

Even though the gate leakage current for individual transistors may not be a problem for short-channel MOSFETs, the total gate current for an entire chip will become a big problem for low-power applications [118, 120]. If one assumes that the total active gate area per chip is on the order of 0.1 cm^2 for future generation technologies, the maximum tolerable tunneling current will be about 1–10 A cm^{-2} [121]. This figure is to be taken at a gate bias of 1 V. The minimum SiO_2 film thickness needed to meet this criterion as the gate dielectric will be around 2 nm, as shown in Figure 39. To maintain device performance, it is projected that the oxide field during normal operation will stay around 5 MV cm^{-1} and that the minimum supply (drain) voltage will be around 1 V.

Partial use of direct-tunneling gate oxide MOSFETs in a chip is one solution. *In situ* cleaning of HF gas followed by UV Cl before the RTO is reported to be very effective in keeping the uniformity of the SiO_2 film and thus in suppressing the gate leakage current. In that case, 1.4 nm is the limit of thinning [120]. It is also reported that an epitaxially grown Si layer reduces tunneling leakage, presumably because of the improvement of the quality of the Si surface layer [122].

The direct-tunneling gate leakage component at high gate bias and low drain voltage is significant in the MOSFET characteristics when the gate oxide film thickness becomes 1.3 nm, even with an ultra-small gate length of 40 nm [119]. Because of drain-to-gate, source-to-gate, and channel-to-gate tunneling paths, the drain current reduces with thinning of the gate oxide beyond 1.3 nm [123]. Thus, 1.3 nm would be the limit of the gate SiO_2 thinning in terms of d.c. characteristics of MOSFETs. Even if direct-tunneling gate leakage could be suppressed by introducing high-k dielectrics (Section 8), the increase in performance is expected to reach its limit at a gate insulator film thickness somewhere around the equivalent to 1–0.5 nm of SiO_2. This is in terms of capacitance, because of the channel (which is an inversion layer at the silicon surface) and gate electrode capacitances connected in series with that of the gate dielectric [115].

6.1.2. Breakdown Characteristics

Reliability as expressed by the time to dielectric breakdown is one of the biggest concerns together with the total gate leakage in a chip [124]. It has been found that the minimum reliable gate oxide film thickness is about 2.6 nm under a supply (drain) voltage of 1.0 V [63]. This can possibly be improved by improving the SiO_2 thickness uniformity, bringing the limit to

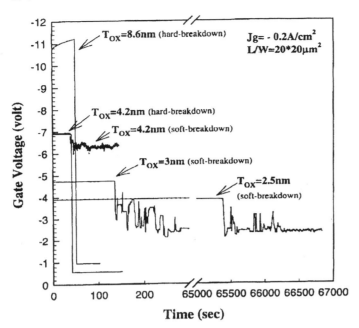

Fig. 40. Typical V_G–t curves for constant-current stressing measurements. The oxide thickness ranges from 8.6 to 2.5 nm. Soft breakdown occurs when it is scaled to 4.2 nm and dominates exclusively when it is below 3 nm. Reprinted with permission from [53], © 1993, Elsevier Science.

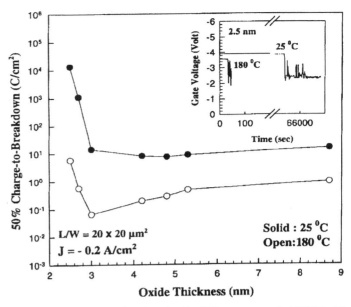

Fig. 41. 50% Q_{BD} as a function of oxide thickness measured at 25°C (solid symbols) and 180°C (open symbols) under gate injection polarity at a constant current of -0.2 A cm^{-2}. The inset shows typical V_G–t curves. Reprinted with permission from [53], © 1999, Elsevier Science.

1.6 nm. On the other hand, a model has been presented that predicts almost infinite time to dielectric breakdown under low supply voltage, which is the case in ULSI [125]. Thus, reliability is still controversial for ultrathin SiO$_2$. Different issues are involved, such as the nature of dielectric breakdown and the effects of stressing temperature and polarity, oxide film thickness, hot-carrier degradation, and device geometry. The discussion includes whether widely accepted methods used for thin oxide characterization are meaningful in the ultrathin film regime.

In ultrathin oxides tested at low voltages, a phenomenon called soft breakdown occurs [126]. Unlike conventional or hard breakdown, which results in large changes in the resistivity of the oxide layer, soft breakdown results in very small changes in resistivity and large changes in the level of current or voltage noise. Soft breakdown can be reliably detected with the use of current or voltage noise. The occurrence of soft breakdown is characteristic of SiO$_2$ films thinner than 6 nm. Its occurrence complicates oxide reliability evaluation. Figure 40 shows typical V_G–t curves generated during charge-to-breakdown measurements using constant current stress, which is a well-known and widely accepted method for evaluating oxide reliability for various gate oxide thicknesses. Whereas the post-breakdown voltage after hard breakdown is around 1 V or less, the voltage after soft breakdown can be more than 1 V in magnitude and represents a characteristic "noisy" behavior. Such behavior could be ascribed to on/off switching events of one or more local conduction spots [127]. Trends in operating voltage, oxide thickness, and device area indicate that soft breakdown will occur in future device operation. Devices that undergo soft breakdown, for the most part, will continue to function. At some

point in time after the initial soft breakdown, the leakage may become too high for the device to operate well.

The oxide breakdown characteristics show strong thickness dependence, as evidenced by Figure 40. For the thickest oxide, only hard breakdown is observed, whereas both soft and hard breakdown events are induced in oxide films with intermediate thickness. For ultrathin oxide, soft breakdown dominates breakdown events. For sub-2-nm oxide films, the voltage drop after soft breakdown is very small because of the extremely large direct-tunneling current, and oxide breakdown becomes very difficult to detect from V_G–t curves.

Stressing temperature has a significant effect on ultrathin oxide Q_{BD} or, more specifically, on the difference between thin and ultrathin oxide films with regard to Q_{BD} [53, 128]. This is shown in Figure 41. At room temperature, it is found that the Q_{BD} under constant current density stressing of an oxide film of thickness 2.6 nm is about three orders of magnitude higher than that of thicker oxides. Such behavior indicates that ultrathin oxides show higher tolerance to direct-tunneling currents because negligible energy is deposited in the oxide layer. However, when the temperature is raised, Q_{BD} of a 2.6-nm-thick oxide is only about one order of magnitude higher than that of thicker films. This implies a significant temperature acceleration effect for ultrathin oxide breakdown under direct-tunneling stress. It is not clear [10] whether the temperature acceleration of Q_{BD} is dominated by N_{BD} or P_{gen} (Section 4.1.2). This may be another reason for preferring low thermal budget processing of devices with ultrathin gate oxides.

Figure 41 incidentally shows that Q_{BD} in the ultrathin film regime increases with decreasing oxide thickness. In this respect, Figure 42 [124] shows Q_{BD} versus gate bias data with

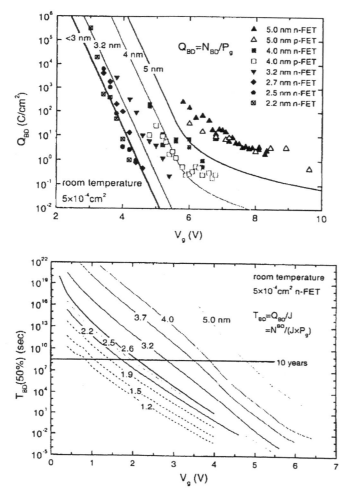

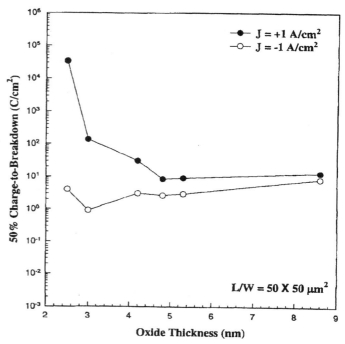

Fig. 43. 50% Q_{BD} measured at 25°C as a function of oxide thickness, both under substrate injection (+1 A cm^{-2}, solid symbols) and gate injection (−1 A cm^{-2}, open symbols). Reprinted with permission from [53], © 1999, Elsevier Science.

Fig. 42. Top: Q_{BD} data and model calculation as a function of gate voltage for devices with ultrathin gate oxides. Open symbols are p-FETs; filled symbols are n-FETs. Bottom: t_{BD} model calculation. Oxide thicknesses from classical C–V extrapolation. Reprinted with permission from [124], © 2000, The Electrochemical Society, Inc.

oxide film thickness as a parameter from measurements using constant-voltage stressing; Q_{BD} is clearly seen to decrease with decreasing oxide thickness for stressing at any fixed gate voltage. Concerning the apparent contradiction arising from a qualitative comparison of Figures 41 and 42, two aspects should be mentioned. One is that once different stressing conditions are used in the measurements, no direct comparison should be attempted at all. Under a theory of dielectric breakdown that is equally accurate in describing constant-current and constant-voltage stressing, one should expect that extrapolation of the data from both graphs to actual operating conditions would result in about the same Q_{BD}. The other is that irrespective of the fact that N_{BD} actually decreases with decreasing film thickness, serious reliability concerns appear after the measured Q_{BD} of individual test devices is extrapolated to the Q_{BD} of a whole chip, as the Weibull slope β (Section 4.1.2) definitely decreases with decreasing oxide thickness.

The polarity dependence of Q_{BD}, which is the Q_{BD} difference between gate injection ($V_G < 0$) and substrate injection

($V_G > 0$) stressing, is shown in Figure 43. It has been well documented that the polarity dependence increases with decreasing oxide thickness for oxides thicker than 4 nm. This is ascribed to the different properties of the polysilicon–SiO$_2$ and SiO$_2$–Si interfaces. As shown in Figure 43, as oxide is further scaled down, the polarity dependence becomes even more dramatic. This is mainly due to the rapid rise in Q_{BD} under substrate injection polarity as oxide is thinned down.

Hot carrier-induced oxide degradation has also received much attention for scaled oxides. Figure 44 shows hot carrier degradation results for 2.5 and 4.2 nm thick oxide films. The hot carrier stress was performed with constant drain voltages of 3 V for the 2.5 nm oxide and 4 V for the 4.2 nm oxide, with appropriate gate bias to ensure maximum substrate current injection. Results confirm that the degradation in threshold voltage is much smaller for the thinner oxide. Thus the ultrathin gate oxide does exhibits higher hot carrier resistance, despite the fact that it is biased under a higher electric field stress due to a not proportionally scaled drain voltage.

As described above, the evaluation of ultrathin oxide reliability becomes much more complicated as the oxide is scaled down. The choice of stress condition (constant current, constant voltage, or constant field) is very important for comparison of the different oxide thicknesses. In addition, several recent studies have pointed out that oxide breakdown is a strong function of device geometry. As such, the conventional use and its interpretation of Q_{BD} for comparing different MOS processes may lead to erroneous conclusions. By the same token, traditional use of large-area samples (i.e., capacitors) for evaluating oxide

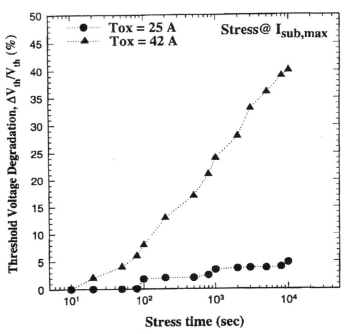

Fig. 44.   Threshold voltage shift after hot carrier stressing at maximum substrate current injection for devices with 2.5- and 4.2-nm-thick gate oxides. Reprinted with permission from [53], © 1999, Elsevier Science.

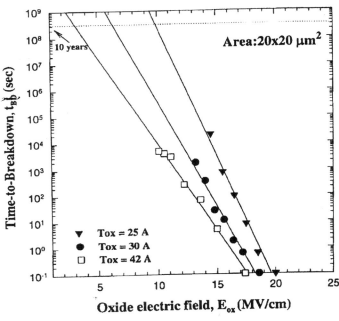

Fig. 45.   Time-dependent dielectric breakdown (TDDB) versus oxide electric field for various oxide thicknesses. (In the absence of charges in the oxide, the electric field is constant and equal to the gate voltage divided by the film thickness.) Reprinted with permission from [53], © 1996, Elsevier Science.

reliability may also lead to an erroneous conclusion for device-level applications.

According to the percolation model for oxide breakdown [62, 63, 129, 130], the critical concentration of defects for breakdown decreases rapidly with decreasing oxide thickness. It was experimentally measured that whereas approximately 10^{12} cm^{-2} defects are produced before a 6-nm SiO$_2$ film breaks down, for a 2–3 nm film the value is about five orders of magnitude lower, i.e., oxide fails much faster [63]. A systematic study of the parameters N_{BD}, P_{gen}, and β (Section 4.1.2) led to the projection that for near-future generation devices gate oxides thinner than approximately 2 nm will not pass the reliability lifetime requirement of at least 10 years of chip operation before gate oxide breakdown [63]. Even for very uniform films, the value of β decreases with oxide thickness, reaching the limiting value of unity at approximately 2 nm. Regardless of uncertainties in lifetime extrapolation, it is very clear that the combined effect of the increasing leakage current and decreasing N_{BD} and β on oxide breakdown may limit the use of ultrathin SiO$_2$ films in the 1.5–2 nm range. This may happen before the oxide scaling reaches the leakage current limit.

The basic relationship for the calculation of charge to breakdown ($Q_{BD} = N_{BD}/P_{gen}$) has been shown to be a valid description of the breakdown of silicon oxide in the ultrathin regime [63, 130, 131]. However, Q_{BD} tests have been criticized as a tool for evaluating ultrathin oxide reliability [132]. The constant voltage stressing (CVS) method, which represents a more realistic situation for practical applications, has been presented as possibly more suitable for evaluating ultrathin oxide reliability [133]. Figure 45 illustrates the time-dependent di-

electric breakdown (TDDB) characteristics for various oxide thicknesses; t_{BD} tests under different oxide voltages at high stressing fields are often employed for predicting the oxide lifetime under a normal operating field (e.g., 5 MV cm^{-1}). As shown in Figure 45, TDDB improves with decreasing oxide thickness. This can be ascribed to less trapped charges as well as reduced interface state generation after electrical stress as the oxide is scaled to the direct-tunneling regime. By extrapolating the data from Figure 45, an electric field of over 9 MV cm^{-1} at room temperature is projected for a 10-year lifetime. It is important to recall that oxide lifetime is strongly dependent on gate area. Whereas the total gate area on a chip is on the order of 0.1 cm^2 for future generation technologies, the projected lifetime from Figure 45 may be overestimated.

Perhaps the clearest expression of the reliability concern for ultrathin gate oxides is Figure 46, which shows the maximum operating voltage for an oxide, so that no more than one device out of 10,000 will fail in a chip (a standard 0.01% percentile) within the 10-year lifetime as a function of oxide film thickness [134]. The results are scaled to a chip area of 0.1 cm^2. The solid line with open symbols corresponds to specifications in the Semiconductor Industry Association's International Technology Roadmap for Semiconductors (SIA ITRS) [9], which take into account requirements of power dissipation and circuit speed for successive technology generations. Projections intersecting the SIA roadmap [63, 135] imply that ultrathin oxides cannot be made sufficiently reliable. As shown, however, there are also optimistic projections, based on higher than anticipated Weibull slopes β and break-

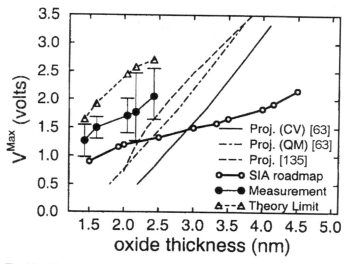

Fig. 46. Maximum device operating voltage as a function of gate oxide thickness from different sources. Reprinted with permission from [134], © 2000, The Electrochemical Society, Inc.

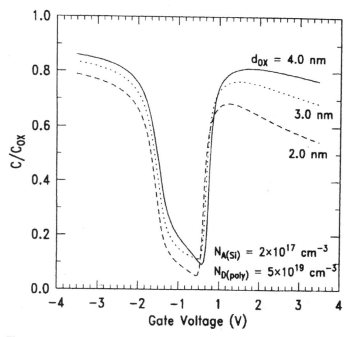

Fig. 47. Normalized capacitance as a function of gate bias calculated for an n-FET with different gate dielectric thicknesses, taking poly-Si depletion and inversion layer quantization into account. Reprinted with permission from [12], © 1996, The Electrochemical Society, Inc.

down acceleration factors due to stressing at voltages above operation conditions [134]. Thus it is clear that there is no single analysis leading to the limiting gate oxide thickness for application in microelectronic devices, as judged from oxide reliability.

6.1.3. Polysilicon Depletion Effects

As CMOS device dimensions are scaled down at a different rate compared with the gate bias, the electric field at the silicon surface continues to increase. One of the consequences is that while the silicon substrate is in inversion (when the MOSFET is turned on) the poly-Si gate may begin to be depleted of carriers [12, 53]. Such a poly-Si depletion effect is caused by insufficient active dopant concentration near the polysilicon–SiO_2 interface. Moreover, high surface fields reveal quantization of the silicon inversion layer.

Poly-Si depletion and quantization of the silicon inversion layer are additional complications for the continued scaling of gate SiO_2. Their combined effect is shown in Figure 47, where normalized capacitance is shown as a function of gate bias for three different dielectric thicknesses: 2, 3, and 4 nm. For these simulations (based on a 1-D self-consistent solution of Schrödinger and Poisson equations) the poly-Si and substrate doping were 5×10^{19} and 2×10^{17} cm^{-3}, respectively. On the inversion side of the $C-V$ curve, the maximum capacitance drops from a little over 80% of the dielectric capacitance for the 4.0 nm film to less than 70% for the 2.0 nm film. This trend indicates that the drain current gained by using a thinner oxide is partly compensated for by poly-Si depletion and inversion layer quantization. A solution to this problem would be to increase the gate dopant implant dose and/or the thermal budget for the gate dopant activation. This, however, would put at risk the steepness of dopant profiles required by deep submicron technologies (i.e., CMOS devices with a channel length less than 0.2 μm). Alternatives proposed at the research level to reduce or eliminate poly-Si depletion effects include poly-$Si_{1-x}Ge_x$, TiN, or WN_x as gate electrode materials.

6.1.4. Boron Penetration

For deep-submicron CMOS technologies, p^+-polysilicon (poly-Si heavily doped with boron) gates are indispensable for p-channel devices, as they offer superior short-channel behavior. However, boron diffusion from the gate into the thin gate oxide and the underlying channel region causes p-channel devices to display flatband voltage shift, threshold voltage instability, drain current reduction, and gate oxide degradation, to mention a few [53]. Boron penetration becomes more severe with increasing temperature, thus constituting one of the major drives for a lower thermal budget in CMOS processing. The situation is more critical for devices with ultrathin gate oxides, because the thermal budget must be sufficient to achieve adequate gate dopant activation to avoid performance loss caused by gate (poly-Si) depletion. Fluorine and hydrogen enhance boron diffusion and so their levels must be kept as low as possible in the process flow. Alternatives to reducing boron penetration include the use of gate materials such as α-Si, stacked poly-Si layers, or poly-$Si_{1-x}Ge_x$. Resistance to boron penetration has also been achieved by nitridation of the gate SiO_2 film (Section 7).

6.2. Physicochemical Characteristics and the Silicon Oxidation Process

This section presents physicochemical characteristics of ultrathin silicon oxide films on silicon and discusses the silicon oxidation process. Discussion is limited to thermal growth in dry O_2, as oxide films prepared by such processes present superior quality. Emphasis is placed on understanding the interaction between device characteristics and the electrical and chemical properties of the oxide layer. These are all intimately related to the oxide growth process, although the details of this relationship are not always clear.

The discussion on silicon oxidation is divided into oxidation kinetics, oxide composition and structure, and oxidation mechanisms. Oxidation kinetics deals with the quantitative prediction of the thickness of oxide grown at a given temperature and pressure for a given length of time, so it is of primary practical concern. Experimental kinetics data and both compositional and structural information obtained from ultrathin SiO_2 films on Si have contributed much to the development of models for the mechanistic description of silicon oxidation. Insight into silicon oxidation mechanisms leads to control of the characteristics of thermally grown oxide, which is the ultimate need for technological applications.

6.2.1. Silicon Oxidation Kinetics

In 1965, Deal and Grove published a seminal paper [33] in which they presented results on silicon (dry and wet) oxidation and proposed a model to describe the growth kinetics of oxide films. The model of Deal and Grove considers diffusion of an oxidant species through a silicon oxide film from the surface toward the SiO_2–Si interface. The process is described using constant diffusivity and assuming (i) a steady-state regime, where the gradient of the oxidant species is constant over the oxide film, and (ii) reaction between the oxidant species and Si at a sharp oxide–Si interface. The result is a well-known linear–parabolic growth law [33]. Interestingly, Deal and Grove noted that the initial oxidation regime was anomalously fast, and their model could not be successfully applied to oxides thinner than about 30 nm (Fig. 48). It is precisely this "anomalous regime" that is now used and studied.

Description of the growth kinetics in the lower thickness range was addressed by many authors, within the framework of Deal and Grove, by adding new terms to the linear–parabolic expression. Although they fit experimental data and so provided useful analytical expressions capable of reproducing the whole growth thickness interval, the extra terms added to the law of Deal and Grove did not have a well-defined physical meaning, even though their dependence on some processing parameters, like temperature, for instance, could be explored. The solutions were obtained by either assuming a nearly steady state and/or a sharp interface [136], similar to the solution of Deal and Grove, or assuming variable diffusivities or reaction rates [136–143].

The thermal growth of silicon oxide films on Si has recently been modeled as a dynamical system, assuming only that it is

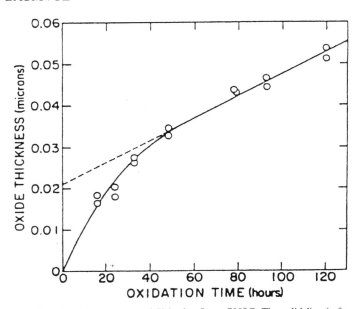

Fig. 48.   Oxidation kinetics of Si in dry O_2 at 700°C. The solid line is for guidance through the experimental data. The dashed line represents the linear–parabolic growth law of Deal and Grove. Reprinted with permission from [33], © 1965, American Institute of Physics.

basically a reaction–diffusion phenomenon with constant reaction rate and diffusivity [144]. Because the steady-state regime was not imposed, an initial oxide thickness was not required. Moreover, the structure of the interface between the oxide film and silicon was obtained as a result rather than given as an assumption of the model. The formulation is based on a system of coupled partial differential equations used to model the time and space dependence of the volume density of Si and of the oxidizing species (assumed to be O_2). The volume density of SiO_2 does not appear explicitly in the equations, but is in direct connection with that of Si. Figure 49 shows the good agreement between growth kinetics calculated from this model and experimental data in the sub-20 nm film thickness regime. Kinetic curves taken at different temperatures, under different O_2 partial pressures and for different periods of time, collapsed to a single curve because of the symmetry in the form of the reaction–diffusion model equations together with initial and boundary conditions. Finally, the model yielded a varying gradient for the oxidizing species in the oxide film and a graded interface between SiO_2 and Si, as shown in Figure 50, contrasting with the assumptions made by Deal and Grove. Experimental evidence of a graded interface between SiO_2 and Si is shown below. It was taken together with the agreement between experimental and calculated kinetics shown in Figure 49 to validate the proposed model. Practical application of the dimensionless single oxidation kinetics curve obtained from the model (in the sense of predicting the thickness of an oxide film grown on Si) requires knowledge of the diffusivity of the oxidant species in the oxide film and of the reaction rate between the oxidant species and Si under given experimental conditions (temperature, O_2 pressure, etc.). Both parameters were assumed to be constant during the oxidation process in the original for-

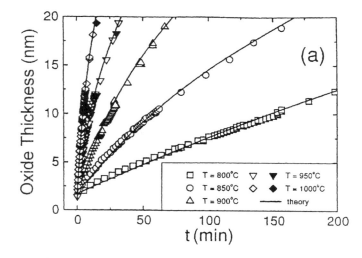

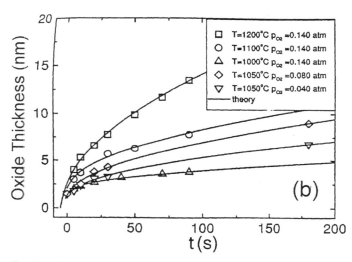

Fig. 49. Calculated (line) and experimental (symbols) silicon oxidation kinetics for (a) conventional furnace processing under 1 atm of O_2 and (b) rapid thermal processing. Reprinted with permission from [144], © 2000, American Physical Society.

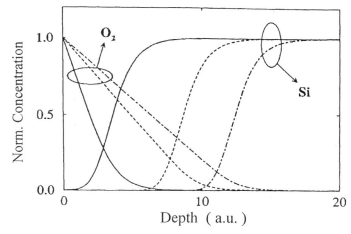

Fig. 50. Calculated normalized Si and O_2 (taken as the oxidant species) profiles in the solid phase at a given temperature and O_2 pressure, for different oxidation times. The solid, dashed, and dot-dashed lines represent, respectively, the profiles at increasing oxidation times. According to the model of Deal and Grove, O_2 profiles would be straight lines and Si profiles would be step-like. Reprinted with permission from [144], © 2000, American Physical Society.

transmission electron microscopy (XTEM), X-ray photoelectron spectroscopy (XPS), ellipsometry) differ from each other (depending on the parameters used; e.g., photoelectron mean free path in XPS and refractive index in ellipsometry) and are slightly different from the "electrical" thickness (as measured by C–V and other electrical tools). The electrical thickness is generally larger because of poly-Si depletion and quantum effects in the silicon substrate. Measured values usually differ by 0.2–0.5 nm, but the difference can be as high as 1 nm.

6.2.2. Composition and Structure

Essentially all thermally grown (and deposited) oxides in semiconductor processing are amorphous. In special circumstances, notably when the silicon surface is flat on an atomic scale, it appears possible to grow a crystalline layer, rather than an amorphous layer, though the precise crystalline form is in doubt. Oxide structure, especially in the near interfacial region, appears to be one of the critical parameters that determines device performance. The electrical reliability of the SiO_2-Si(001) system, as compared with SiO_2-Si(111), is caused by the difference in oxide quality or oxide structure (even when the interface microroughness is of the same level). Si(001) substrates are preferred in manufacturing lines for yielding reduced parasitic charge and density of electronic states at the SiO_2–Si interface, a result that has been confirmed for ultrathin oxide films [145]. Hattori [146] describes in detail the chemical changes occurring during interface formation between SiO_2 and Si(001) or Si(111). It has been shown experimentally that thin thermal oxides are strained in a zone that extends 1.5 to 3 nm away from the dielectric–semiconductor interface and differs from bulk SiO_2 [147]. It appears that oxidation carried out at high temperatures leads to an increase in thermal stresses in the underlying semiconductor and to a degradation of its properties.

mulation of the model; values extracted from curve fitting (as in Fig. 49) are presented in the original work [144]. Finally, it should be noted that the linear–parabolic law of Deal and Grove was obtained as an asymptotic solution of the reaction–diffusion equations.

Independently of the physical model or fitting expression used to represent the kinetics of thermal oxidation of Si, one should be aware that in the thickness range 0–6 nm the actual result is strongly dependent on processing conditions, like silicon wafer cleaning, substrate orientation, rapid or conventional thermal processing, water vapor content, and others. This is a major concern for those seeking predictive power.

Also of relevance to the modeling of silicon oxidation kinetics is the fact that accurate determination of oxide thickness is difficult in the ultrathin regime. As gate oxide thickness shrinks below 5 nm, "physical" measurements of oxide thickness (as monitored by techniques such as cross-sectional

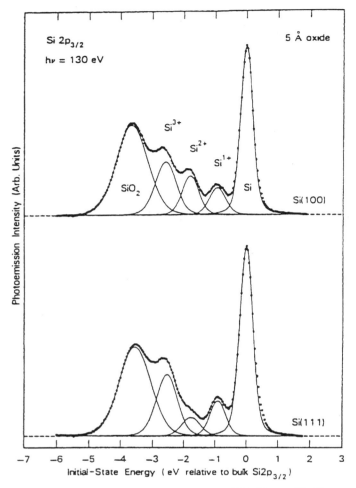

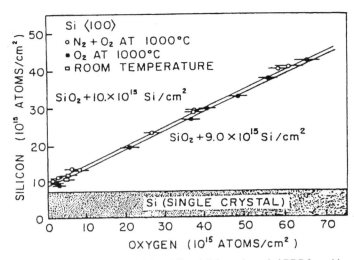

Fig. 52. Measured areal densities of Si and O from channeled RBS for oxides grown on Si(001) under different conditions. The straight line fits to the data assume stoichiometric SiO2 and indicate slightly different values of Si excess for different growth parameters. Reprinted with permission from [79], © 1980, Elsevier Science.

Fig. 51. Si $2p$ photoelectron spectra for 0.5-nm-thick silicon oxide films on Si(001) and Si(111). Reprinted with permission from [148], © 1988, American Physical Society.

The amount of microdefects propagating from the Si substrate into silicon dioxide during oxidation decreases for thin oxides. Thus, in MOS-C testing devices, thin oxides display a better reliability than thick oxides, and their breakdown field is higher.

Compositional characterization of ultrathin silicon oxide films indicates that stoichiometric SiO_2 is predominant in the "bulk" and surface of films with thicknesses above 2 to 5 nm. For thinner films or near the interface between silicon oxide films of any thickness and silicon, nonstoichiometric (oxygen-deficient) oxides exist, helping to accommodate the transition between materials. Figure 51 [148] shows Si $2p$ photoelectron spectra for 0.5 nm-thick silicon oxide films thermally grown in dry O_2 on reconstructed Si(001) and Si(111) surfaces, indicating intermediate oxidation states for Si. The absolute intensities of the peaks corresponding to intermediate oxidation states decrease with increasing film thickness. The existence of intermediate oxidation states at the oxide surface was ruled out by these studies. Observation of intermediate oxidation states in reconstructed surfaces [obtained by long-lasting ther-

mal annealing of the clean (native oxide free) crystalline silicon substrates in ultrahigh vacuum] sets a lower limit for the thickness of the suboxide region. In a nonreconstructed Si surface, not only do the relative intensities of the different oxidation states change; the distance from the oxide–silicon interface at which they subsist changes as well. In addition to depending on the quality of the Si substrate, the amount of intermediate oxidation states depends on oxidation conditions.

Structural transition layers are formed on both sides of the SiO_2–Si interface to relax stress originating from lattice mismatch and different thermal expansion coefficients. They include deviations of the Si—O—Si bond angle from that of bulk SiO_2, Si interstitials and precipitates, oxide microcrystals, and others. It has been found that SiO_2–Si interface microroughness, which reflects initial silicon surface flatness, is one of the important parameters determining the electrical properties of the SiO_2-Si system, such as maximum breakdown field, charge to breakdown, and carrier mobility. A general denomination of "intermediate layer" has been used for the adaptation region. Its thickness was determined by various techniques, many of them model-dependent and relying on simulations. Figure 52 illustrates the results obtained with high-energy ion scattering (or Rutherford backscattering, RBS) under channeling of the ion beam into the crystalline silicon substrate. Linear fits to the data indicate (i) SiO_2 stoichiometry (from the slope) and (ii) excess of Si at the SiO_2–Si interface (from the line intercepts at null oxygen content). Such an excess corresponds to a silicon layer not aligned to the single-crystalline substrate. Its thickness was determined to be between 2 and 4 nm for oxides produced in conventional furnaces [79] and 1 to 2 nm for rapid thermal oxides, depending on the Si wafer cleaning procedure [31]. The thickness of the transition region was also determined by MEIS. Figure 53 indicates a graded interface between SiO_2 and Si. The thickness estimated for the compositional transi-

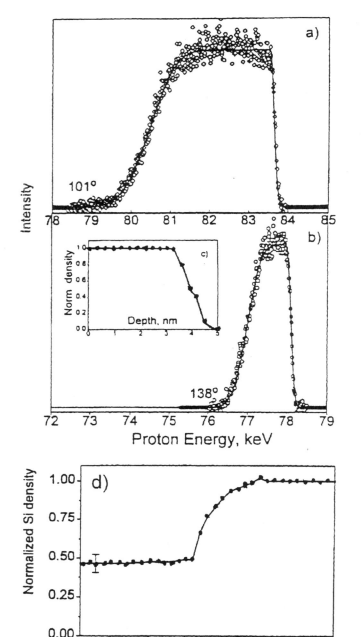

Fig. 53. (a) Oxygen section of an MEIS spectrum for a 4.5-nm-thick oxide on Si(001) taken with a scattering angle of 101°. (b) As in (a), with a scattering angle of 138°. (c) Oxygen profile producing the simulations shown as solid lines in (a) and (b). (d) Silicon profile obtained from the corresponding section of the spectra [82, 149]. Parts (a)–(c) reprinted with permission from [82], © 1995, American Physical Society. Part (d) reprinted with permission from [149], © 1996, Elsevier Science.

tion region from MEIS experiments was 1.2 ± 0.4 nm [149]. Experiments using grazing-incidence X-ray diffraction from a synchrotron source yielded transition regions with thicknesses of up to 7 nm [150]. It is well timed to compare such findings with the picture incidentally provided by Figure 36. The SiO_2–

Si interface has also received much attention from a theoretical approach [151–157].

In summary, it is generally agreed that there is a transition region (of altered structure and stoichiometry) between crystalline silicon and SiO_2. However, the thickness of the region has been reported to vary from 0.5 to 3 nm (or even 7 nm). Although such a large scatter can be attributed to differences in oxidation procedure and oxide thickness, more important factors are that the width of the transition region depends strongly on both the probing technique and the definition of the transition region. Despite extensive work, neither the atomic-scale structure nor the composition (or gradient) in the transition region is well understood—currently, there is no universally accepted model.

6.2.3. Growth Mechanisms

Oxide films grown in pure oxygen and in steam have different chemical, mechanical, and electrical properties. Silicon oxide grown in steam is actually SiO_2 doped with water to some variable degree. For oxidation in "dry" oxygen (i.e., containing up to a few ppm of H_2O), the stoichiometric chemical reaction producing a silicon oxide film is

$$Si(solid) + O_2 \rightarrow SiO_2(solid) \tag{7}$$

The equation describes the overall reaction between oxygen and silicon; there may be elementary reactions in which intermediate species are produced during the oxidation process.

Because the silicon surface is highly reactive, a layer of oxide rapidly forms on exposure to an oxidant gas. The rate of oxide growth is limited by the availability of the reactants, namely, oxidant molecules and Si—Si bonds. However, as the oxide film becomes thicker, its rate of formation becomes diffusion-limited because the silicon and the oxidizing ambient are separated by the oxide layer. Silicon oxidation thus raises a whole series of questions [158, 159], such as where oxidation takes place, what the nature of diffusing species is, how oxidation affects interface structure, and how these phenomena depend on oxidation parameters, such as temperature, pressure, and postoxidation treatment. An overview of the present understanding of silicon oxidation mechanisms is offered in the following.

It was not known a priori whether the oxidation proceeded at the SiO_2–Si interface or at the gas–SiO_2 interface. It has been demonstrated by experiment that the oxidizing species diffuses through the oxide layer and reacts with the silicon when it arrives at the SiO_2–Si interface. Evidence for the inward migration of the oxidizing species and reaction at the SiO_2–Si interface is given by isotope tracing methods. The basic experiment consists of (i) growing an oxide layer in O_2 containing the natural abundance of isotopes (about 99.8 at.% ^{16}O and 0.2 at.% ^{18}O) and (ii) continuing the growth in O_2 enriched in the ^{18}O isotope. This sequence is preferred because ^{18}O can conveniently be profiled with the use of NNRP (Section 4). To illustrate the mechanistic evolution of silicon oxidation with respect to the oxide thickness, Figure 54 shows observed ^{18}O

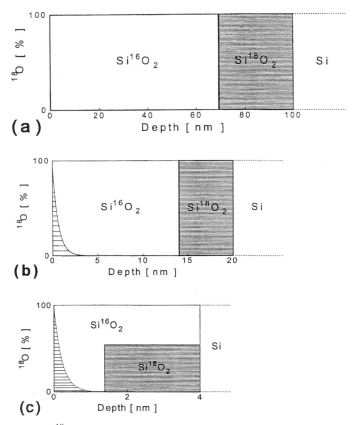

Fig. 54. ^{18}O profiles as determined by NNRP for silicon oxide films of different thicknesses sequentially grown in natural (quoted ^{16}O) and ^{18}O-enriched O_2. The profiles are representative of original (Si$^{16}O_2$) oxide films (a) thicker than 30 nm; (b) between 10 and 30 nm; and (c) thinner than 10 nm. Reprinted with permission from [31], © 1996, Elsevier Science.

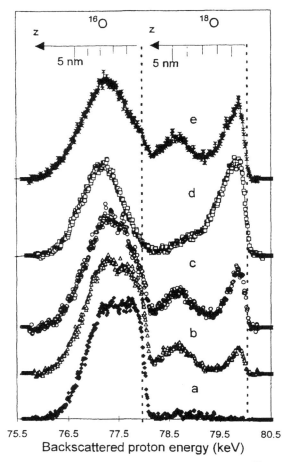

Fig. 55. Oxygen section of MEIS spectra for (a) an initial Si$^{16}O_2$ film sequentially exposed to $^{18}O_2$ at (b) 800°C, 1 Torr, 21.5 h; (c) 800°C, 1 Torr, 40 h; (d) 900°C, 0.1 Torr, 5 h; (e) 900°C, 1 Torr, 5 h. Reprinted with permission from [160], © 1995, American Institute of Physics.

profiles determined by NNRP in SiO$_2$ films originally grown in natural O$_2$ for ranges thicker than 30 nm, between 10 and 30 nm, and thinner than 10 nm [31]. In the thickest film, most of the ^{18}O (>95%) is found near the SiO$_2$–Si interface, as proposed in the model of Deal and Grove. In the film of intermediate thickness, ^{18}O is also found near the gas–SiO$_2$ interface. In this region it is distributed according to a complementary error function, a profile typical of diffusing species. In the thinnest film, enrichment of SiO$_2$ in the ^{18}O isotope at the interface with Si is no longer 100% and ^{18}O becomes incorporated in the "bulk" oxide. Figures 55 [160] and 56 show extensive isotope intermixing after $^{16}O_2$, $^{18}O_2$ sequential oxidations.

Figure 54 provides the basis for a "physical" definition of thick, thin, and ultrathin oxide film. Although the mechanisms responsible for the distinct features in Figure 54c as compared with its counterparts are certainly active during the growth of thicker films, their contribution to oxide growth is increasingly negligible. Continuing with reference to where silicon oxidation takes place, these mechanisms are now explored.

The first systematic experimental investigation of isotopic substitution in ultrathin films was made by Rochet et al. [161]. Growth was promoted in ultra-dry O$_2$ (less than 1 ppm of H$_2$O),

in the temperature range from 810°C to 1090°C, to total film thicknesses from 2 to 7 nm. Oxidations performed in the $^{16}O_2$, $^{18}O_2$ gas sequence showed most (between 98% and 80%) of the ^{18}O incorporated in the near-surface region of the samples. This is illustrated in Figure 57, which shows results from etch-back NRA (Section 4). The steep dashed lines in the plots indicate the enhanced incorporation of ^{18}O in the near-surface of the oxide film. Such ^{18}O is found to be incorporated in two different modes, namely either in exchange for ^{16}O originally in the samples or contributing to oxide film growth. This brings two new ideas not contemplated within the framework of the model of Deal and Grove: reaction of oxidizing species not at the SiO$_2$–Si interface, but near the gas–SiO$_2$ interface and resulting in (i) O exchange and (ii) film growth. Isotopic exchange between the oxygen atoms from the gas phase and those already existent in the oxide film was shown to be present at all temperatures above 800°C. At all oxidation temperatures, the contribution to oxide film growth of ^{18}O incorporated near the surface decreases with the increase in the total film thickness. Experiments also showed that the less dry the O$_2$, the less favored is the fixation of ^{18}O in the near-surface of the sam-

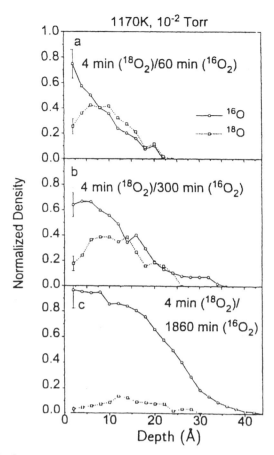

Fig. 56.   Isotope depth distributions from MEIS measurements on silicon oxide films sequentially grown in $^{18}O_2$ and $^{16}O_2$. Reprinted with permission from [82], © 1995, American Physical Society.

ples. This could be due to a fast-diffusing species combining H and ^{18}O, like $H_2^{18}O$. Experiments involving oxide films in the thickness range between 20 and 300 nm are in agreement with such results [162, 163].

Figure 58a shows the ^{18}O profile in a silicon oxide film thermally grown on Si in a $^{16}O_2$, $^{18}O_2$ gas sequence, as measured by NNRP (Section 4). The near-surface ^{18}O distribution follows a complementary error function (erfc). Figure 58b shows the ^{18}O profile after a third oxidation step in $^{16}O_2$. As expected, a region enriched in ^{16}O is found in the near-surface. Also shown is a significant loss of ^{18}O previously existing close to the gas–oxide interface, accompanied by modification of the original erfc-like profile. Figure 59 shows the experimentally determined erfc-like and final ^{18}O profiles, together with a calculated (labeled "predicted") final profile. The good agreement with experimental data was taken as a confirmation of the hypotheses used in the calculation, namely (i) the defects responsible for near-surface O incorporation can be thought of as diffusing in a semi-infinite medium; (ii) the gas–oxide interface constitutes a permeable interface; and (iii) the transport of the oxidizing species in the near-surface region follows a diffusion-exchange mechanism, related to step-by-step motion of network oxygen

atoms [164]. The latter has also been inferred from previous two-step sequential oxidation experiments [165].

There is more than one model capable of incorporating the experimental evidence found so far concerning silicon oxidation [158]. One picture is the reactive layer model [166]. The idea is that the oxide nearest to the silicon is not fully oxidized, so that species (perhaps O_2) that diffuse interstitially through the outer oxide react within this layer, before reaching the silicon itself. Diffusion continues by some other mechanism. The reactive layer is therefore opaque to interstitial O_2. The reaction does not have to be at the outside (stoichiometric oxide side) of the reactive layer, but this side will receive the largest flux of interstitial oxidizing species. Figure 60 shows a pictorial sketch of the basic idea announced in the reactive layer model [167]. Based on the reactive layer model, Stoneham et al. [166] were able to explain the results given in Figure 57 and equivalent data, concluding that the thickness of the reactive layer is between 1.5 and 2.0 nm. From the point of view of experimental evidence and confirmation, the reactive layer model is solidly based on results concerning growth kinetics, structure, composition, and isotope-exchange data.

One possibility according to the reactive layer model that has not been addressed in this chapter so far is the mobility of species involving Si. The long-range mobility of Si was rendered very improbable by the initial results on oxygen isotopic substitution. Indeed, direct evidence of Si immobility was given by Si isotopic substitution experiments. Results of a ^{31}Si isotopic substitution experiment with a rather poor depth resolution (approximately 50 nm) [168] were confirmed by a ^{29}Si isotopic substitution experiment of much better resolution (approximately 0.7 nm near the sample surface) [169]. The profiles of ^{29}Si before and after thermal oxidation in $^{18}O_2$ and of ^{18}O revealed that no Si is lost and that it is immobile during oxide growth, in the sense that it does not diffuse across the growing oxide to react with O_2 at the gas–oxide interface. These results do not exclude short-range Si transport from the substrate into the near oxide–silicon interface, as hypothesized in the reactive layer model. It is important to note, however, that Si mobility is not at the basis of the model: "whilst it is convenient to talk of out-diffusion of silicon, it suffices if interstitial oxygen stops at the outside of the reactive layer. Even diffusion through the layer would give an equivalent effect provided it is a vacancy mechanism or one involving exchange" [167].

The overall atomic transport picture in the growth of ultrathin oxide films on silicon comes from the relative thicknesses of the different regions relevant to the process: (i) the thickness of the reactive oxide layer, approximately 2 nm from the oxide–silicon interface [166]; (ii) the thickness of the oxygen-excess region, approximately 3–4 nm from the gas–oxide interface [170, 171]; and (iii) the oxide film thickness. Experimental findings and models discussed above revealed the atomic transport mechanisms acting in the ultrathin regime. Oxygen either (i) diffuses through the oxide network without interaction (interstitially) to react within the reactive layer or (ii) follows a diffusion-exchange mechanism, related to step-by-step motion of network atoms, and is fixed at the oxygen-excess region. On

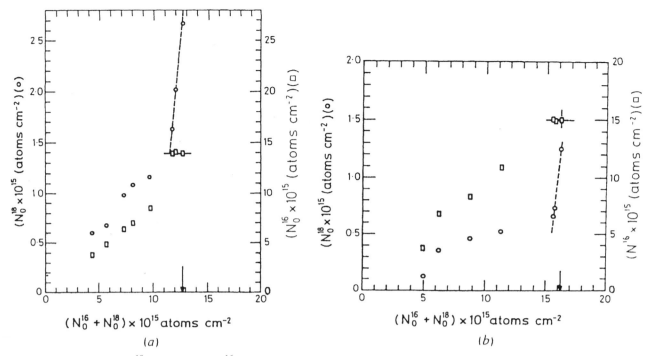

Fig. 57. Areal densities of ^{18}O (circles) and of ^{16}O (squares) as a function of the total amount of O from etch-back NRA in silicon oxide films grown at 930°C in ultradry gas. (a) 30 min in ^{16}O$_2$ followed by 30 min in ^{18}O$_2$; (b) 330 min in ^{16}O$_2$ followed by 30 min in ^{18}O$_2$. The arrows indicate the gas–oxide interface. The steep dashed lines indicate enhanced ^{18}O incorporation at the sample surface. Reprinted with permission from [161], © 1986, Taylor and Francis.

the basis of atomic transport, the ultrathin oxide film regime is conveniently defined as that at which film thickness becomes comparable to the sum of the reactive layer and the oxygen-excess zone, or equivalently, that at which these two regions overlap. Oxygen diffusing step-by-step in the oxygen-excess zone can then reach the reactive layer to react therein. As a direct consequence, the contribution to film growth of oxygen incorporated near the surface increases.

The identity of O-containing species diffusing during silicon oxide growth is now addressed. Atomic transport during thermal oxidation of Si in dry O$_2$ or water vapor was first described by Deal and Grove [33] as steady-state interstitial diffusion of molecular oxygen (O$_2$) or water (H$_2$O) across the growing oxide and subsequent reaction with Si at a sharp SiO$_2$–Si interface. This has been supported by both experiment and quantum chemical calculations [159]. However, recent experimental results have shown that atomic oxygen is also a possible candidate for the transported species, and it is now claimed that while traditional arguments for molecular oxygen being the transported species are valid for atomic oxygen as well, more recent experimental results support atomic oxygen as the transported species [172]. The new results concern (i) the dissociation rate of oxygen molecules at the SiO$_2$ surface compared with the oxidation rate [173]; (ii) coupling between the dissociation rate and oxidation kinetics [174]; and (iii) oxygen exchange at the oxide–silicon interface [175]. As for the step-by-step motion of oxygen atoms in the near surface, peroxyl bridges (O—O bonds, equivalent to an excess-oxygen intersti-

tial) have been identified as probably being responsible [165]. This is in good agreement with experimental EPR data concerning defects associated with Si depletion or O excess centers (the so-called EX centers) [170, 171]. The depth distribution of EX centers is remarkably similar to that of oxygen found in the near-surface region of an oxide film after the second step of a sequential ^{16}O$_2$, ^{18}O$_2$ oxidation. Oxygen transport through peroxy bridge defects in silicon oxide has been explored with the use of molecular dynamics and Monte Carlo combined to first-principles calculations [154, 176].

6.3. Remarks and Limitations Concerning Ultrathin Silicon Oxide Films as Gate Dielectrics

Silicon dioxide has remained the gate dielectric of choice because it has close to ideal properties: its dielectric strength is large and the SiO$_2$–Si interface contains very few defects. The scaling method with some modifications has succeeded in downsizing MOSFETs for 30 years, to the 0.18-μm technology node with gate lengths of 0.12 μm. By thinning of the gate oxide to less than 2 nm, various advantages in the MOSFET performance were confirmed. Further downscaling, however, is being simultaneously threatened by different parameters. Among them, gate SiO$_2$ thinning is thought to be the most severe. First, the presence of large quantum mechanical tunneling current is a serious scaling limitation in terms of standby power consumption. Second, breakdown characteristics for ultrathin oxides become even more critical because of

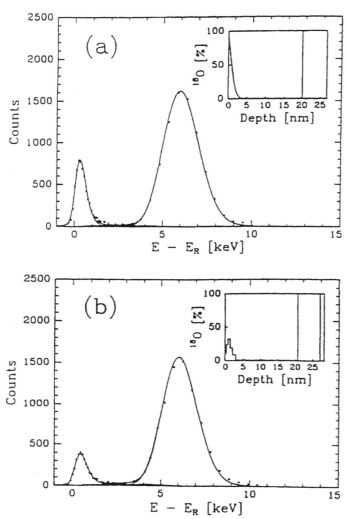

Fig. 58. Excitation curves of the $^{18}O(p,\alpha)^{15}N$ nuclear reaction around the resonance at 151 keV for a silicon oxide film grown in (a) $^{16}O_2$, $^{18}O_2$ sequence and (b) $^{16}O_2$, $^{18}O_2$, $^{16}O_2$ sequence. Corresponding profiles are shown in the insets. Reprinted with permission from [164], © 1997, American Institute of Physics.

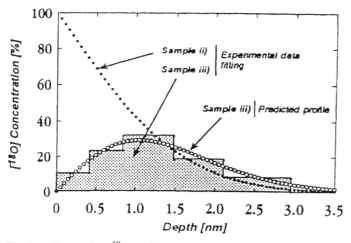

Fig. 59. Near-surface ^{18}O profiles from Figure 58 and the predicted (calculated) profile to be compared with Figure 58b. Reprinted with permission from [164], © 1997, American Institute of Physics.

the dramatic increase in electric field across the oxide during normal device operation. Third, poly-Si gate depletion effects are known to get worse with oxide scaling, as operating gate voltage normally does not scale proportionally to oxide thickness. Furthermore, as gate oxide thickness decreases, process integration issues emerge as new challenges. Boron penetration from p^+-polysilicon gates into the thin gate oxide and the channel region in p-MOSFETs is one of the major concerns for CMOS technologies.

It is now well established that ULSI reliability and electrical properties are strongly dependent on the quality of the SiO_2–Si interface region. Channel mobility, leakage current, time-dependent breakdown, and hot electron-induced effects have all been correlated with the oxide structure and defects at the SiO_2–Si interface. Although the electrical defects are controlled by fabrication conditions, oxidation ambient, etc., relatively little is known about the atomic configuration of these defects, especially for ultrathin oxide films. The mechanism of breakdown of ultrathin dielectrics is also not fully understood. An atomic-scale description of silicon oxidation in the ultrathin oxide film regime is also still to be developed.

7. SILICON OXYNITRIDE GATE DIELECTRIC FILMS

At this time, silicon oxynitrides (SiO_xN_y or, more accurately, nitrogen-doped SiO_2) are the leading candidates for replacing pure SiO_2 in ultrathin gate dielectric films [177]. Oxynitride films are of great interest because they retain favorable features of both silicon oxide and silicon nitride while minimizing their drawbacks [14]—one takes advantage of the passivating and masking properties of Si_3N_4 while retaining the excellent electrical properties of the SiO_2–Si interface. The three main reasons for the attractiveness of silicon oxynitrides as a replacement for pure SiO_2 are (i) very good diffusion barrier properties (particularly against boron penetration from p^+-polysilicon gates); (ii) a slightly higher dielectric constant that is reflected in some reduction of leakage current; and (iii) enhanced reliability. Even small amounts of N (1×10^{14} cm^{-2} or more) in the SiO_2 network significantly improve its diffusion barrier properties. The dielectric constant of oxynitrides linearly increases with N content from $\epsilon_{SiO_2} = 3.9$ to $\epsilon_{Si_3N_4} = 7.8$ [178].

At first, nitrogen was introduced in thicker gate oxide films [179–181], increasing their reliability. It was shown [56] that N incorporation results in a reduced defect generation rate (Section 4.1.2). Many ultrathin (<3 nm) SiO_2 films are now nitrided to reduce hot electron effects, limit boron penetration, and, under certain conditions, prolong device lifetime. Nitrogen incorporation into SiO_2, however, can also lead to deleterious effects, depending on the amount and profile. In the following, an overview of ultrathin silicon oxynitride gate dielectric films is presented. Different approaches to silicon oxynitride formation are briefly reviewed, and electrical and physicochemical characteristics of the resulting films are discussed. These characteristics can be strongly dependent on the method of film

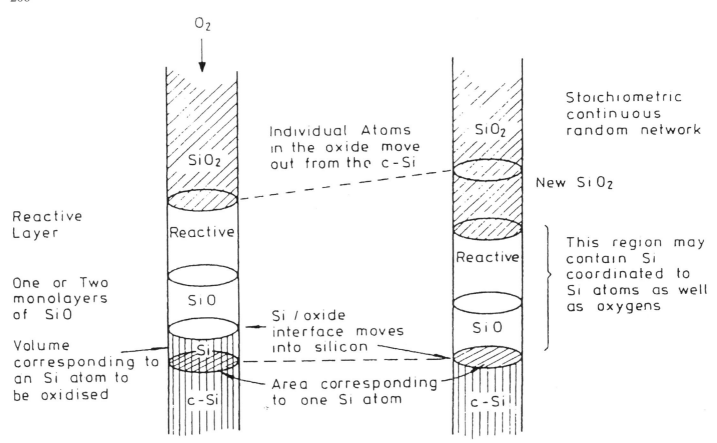

Fig. 60.   Pictorial sketch of the reactive layer model. A column of silicon atoms is shown both before and after oxidation by one oxygen molecule. Reprinted with permission from [167], © 1986, Taylor and Francis.

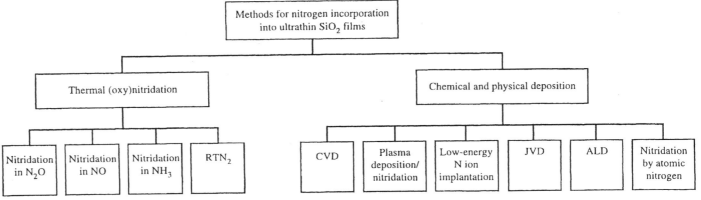

Fig. 61.   Different methods for nitrogen incorporation into ultrathin SiO_2 films. Reprinted with permission from [177], © 1999, IBM Technical Journals.

preparation and on the parameters adopted in a given method. General trends are presented as much as possible.

7.1. Preparation Methods

In contrast to the case of SiO_2, which for gate dielectric application is exclusively thermally grown, many different methods have been developed to produce SiO_xN_y films (Fig. 61). Nitrogen may be incorporated into SiO_2 by either conventional or rapid thermal oxidation/annealing [182–195] or chemical and physical deposition [196–201] methods. In both classes the approach can be either nitridation of an ultrathin oxide film or direct growth/deposition of the oxynitride on Si.

Thermal methods are particularly attractive because of their similarity to standard silicon oxidation methods, the availability of both single-wafer and batch tools (and low processing cost), and, most importantly, hydrogen-free and damage-free processing [202]. Historically, thermal nitridation of SiO_2 in

NH_3 was studied first, following an approach used to grow Si_3N_4 on Si. Silicon oxide layers can be nitrided by direct exposure to gaseous NH_3 at high temperature (800–1200°C), at high or low pressures of ammonia (1 bar, 10^{-1} to 10^{-6} mbar). Thermal nitridation of silicon oxide in NH_3 can be of great interest because it allows simultaneous incorporation of N and D near the oxynitride–Si interface [203]. Nitrous oxide (N_2O) and later nitric oxide (NO) have also been used in the nitridation of SiO_2, with increased benefit to device reliability as compared with processing in NH_3 due to the absence of H. It was also found that oxynitrides could be directly grown on Si with the use of N_2O and NO.

Chemical and physical deposition methods have been used to modify SiO_2 mainly with N_2^+ and N^+ as nitriding species, with the use of plasmas or ion implantation in the hyperthermal energy range (Section 3.3). Deposition from chemical vapor with the use of a variety of precursors has shown great flexibility at producing oxynitrides in the whole range of possible stoichiometries. Ion beam or plasma nitridation of bare Si followed by oxidation has also been used. Finally, these methods can be applied either alone or in sequence. For different reasons that will be made clear, annealing in O_2 ("reoxidation") is particularly frequent after either thermal or physical/chemical nitridation. It becomes evident that there are numerous ways of producing oxynitride films. Next the electrical and physicochemical characteristics of representative samples are used to show the effectiveness of these methods at producing ultrathin films suitable for use as gate dielectrics.

7.2. Electrical Characteristics

Nitridation of silica in gaseous ammonia [14] introduces a substantial amount of hydrogen in the film, which in turn enhances the density of electron traps, results in a fixed-charge buildup, and causes the leakage current to increase. The reoxidation of nitrided SiO_2 films has proved effective in reducing the density of incorporated H atoms. It has been investigated as a technique for minimizing bulk electron trapping and reducing interface states without affecting other properties. Nitridation of SiO_2 in N_2O or NO, however, is generally preferred. The obtained films display a greater resistance to high-field stresses (specially to carrier injection) and possess a greater resistance to dielectric breakdown. Electrical properties of NO-grown oxynitrides appear to be better than those of N_2O-grown films.

High-frequency C–V curves reveal that a nitrided oxide film behaves like the original oxide, except for its capacitance in the accumulation mode and for a flatband voltage shift. Enhancement in the film capacitance is explained by an increase in the effective dielectric constant of the nitrided oxide or film growth during nitridation. Increasing the dielectric constant can be advantageous in reducing leakage current as compared with pure SiO_2. There is a trade-off involved, however: as the dielectric constant of SiO_xN_y increases from that of SiO_2 (3.9) to that of Si_3N_4 (7.8), the bandgap decreases from $\sim$8 eV to $\sim$5 eV. The incorporation of large amounts of N leads to the reduction of boron penetration, but also causes threshold-voltage shifts

and degradation of charge carrier mobility, which are dependent upon both the nitrogen concentration and profile. This is due at least partially to positive charge that results from the nitrogen incorporation into the SiO_2 matrix.

The dielectric strength of nitrided oxide films varies from 2 to about 30 MV cm^{-1}, depending on nitridation conditions. On average, it is greater than that for SiO_2. Compared with silica films, nitridation modifies, besides fixed oxide charge and flatband voltage, the density of interface states in a complex manner, which depends on the nitrogen concentration at the SiO_2–Si interface. At the level of 4 at.% at the SiO_2–Si interface, N led to an increase in the density of interface states observed with n^+-polysilicon gate and p-type Si substrate [204]. For p^+-polysilicon gates nitridation of silica is effective in preventing boron diffusion, thus preventing strong electrical degradation of the dielectric–Si interface.

A few examples are taken from the recent literature to illustrate these properties of ultrathin silicon oxynitride films intended for use as gate dielectrics. Electrical characteristics are shown first for oxynitride films grown on Si with the use of high-pressure thermal processing in NO [205], then for oxide films nitrided by plasma immersion ion implantation [206], and finally for a silicon nitride/oxide stack produced by CVD [207]. The impact of nitridation by a variety of methods on device characteristics and reliability has been studied by Hook et al. [208].

Figure 62 shows gate leakage current densities of control SiO_2 samples and SiO_xN_y films grown on Si with the use of high-pressure thermal processing in NO (HPNO). Both nitrided MOS-C and MOSFET present an order of magnitude improvement relative to a thermal oxide grown by rapid processing. As the samples presented the same equivalent oxide thickness (EOT, Section 4.1.1), the authors attributed such improvement to an increased dielectric constant of the SiO_xN_y film as compared with SiO_2.

Figure 63 shows I–V characteristics of oxynitride films produced by plasma immersion ion implantation (PIII) of nitrogen into 2.0 nm-thick SiO_2 films followed by rapid thermal annealing. All nitrided samples present reduced leakage as compared with the control oxide film. Although an increased permittivity of the nitrided dielectrics could be invoked (as before) to explain this result, the authors found that the reductions in tunnel currents were due only to the effective increase in film thickness. Thickness measurements were performed by spectroscopic ellipsometry, and spectra obtained after implantation required the inclusion of an amorphous Si layer at the SiO_xN_y–Si interface into the model to obtain good fits. Rapid thermal annealing was seen to be essential in repairing such damage to the implanted substrate.

High-frequency (1 MHz) C–V characteristics of the oxynitride films produced by PIII after annealing are shown in Figure 64. The increased thickness quoted above appears as reduced capacitance under accumulation. A shift of the C–V curves toward higher gate voltages is noticeable for the low-energy nitrogen plasma implants compared with the unimplanted control oxide. As the implant energy is increased, the

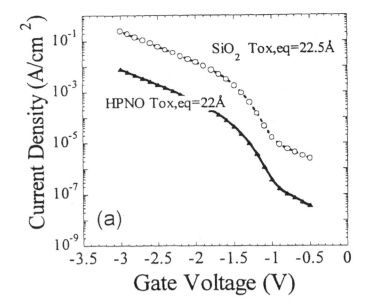

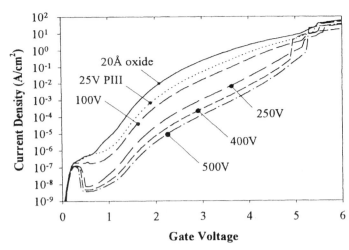

Fig. 63. *I–V* characteristics of gate oxynitrides produced by plasma immersion ion implantation at the bias voltages shown. The characteristics of a pure SiO_2 film are included for comparison. Reprinted with permission from [206], © 1999, Materials Research Society.

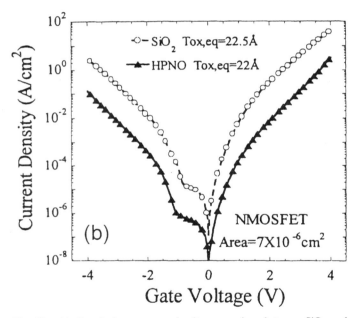

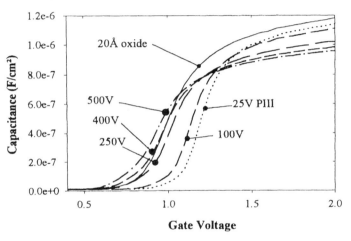

Fig. 64. *C–V* characteristics of the dielectric films corresponding to Figure 63. Reprinted with permission from [206], © 1999, Materials Research Society.

Fig. 62. (a) Gate leakage current density comparison between SiO_2 and HPNO with the use of n^+-poly NMOS capacitors. (b) As in (a), with the use of n^+-poly NMOSFETs. A more than 10-fold reduction is leakage current is observed with SiO_xN_y as compared with SiO_2 in the bias range between 1.5 and 2.5 V. Reprinted with permission from [205], © 1999, Materials Research Society.

C–V curve shifts back toward lower voltages, most likely because of incorporation of nitrogen into the silicon substrate. This nitrogen, contributing with donor states, would make the substrate more *n*-type.

Figure 65 shows *C–V* and *I–V* characteristics of gate dielectric films produced by remote plasma-enhanced CVD of Si_3N_4 on ultrathin SiO_2. *C–V* measurements were used to monitor the suppression of boron diffusion from p^+-polysilicon gate electrodes. In Figure 65a, both quasi-static and high-

frequency *C–V* curves for the devices with pure oxide dielectrics are shifted to positive voltages compared with device simulations (not shown), indicating significant penetration of boron into the channel region. Based on the simulations, the relative positions of the *C–V* curves for the devices with the stacked nitride/oxide dielectrics demonstrate that the top nitride layer is effective in suppressing boron diffusion out of p^+-polysilicon gates. In addition, as the density of electronic states at the dielectric–Si interface determines the deviation between high- and low-frequency (or, equivalently, quasi-static) *C–V* curves at the onset of inversion (see labeled area in Fig. 65a), this deviation can be used to compare the electrical quality of the interfaces between the dielectric films and Si. Although for high substrate doping in ultrathin oxide devices the conventional *C–V* technique is no longer effective for extracting accurate values of interface states, it is clear that the

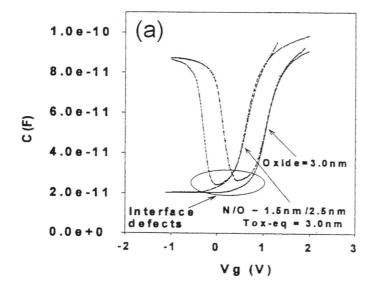

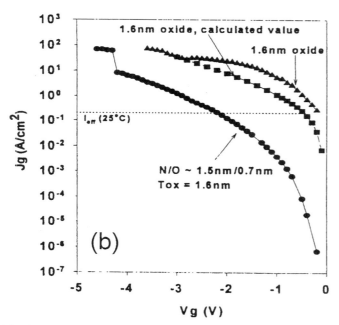

Fig. 65. $C-V$ and $I-V$ characteristics of silicon nitride/silicon oxide gate dielectric stacks compared with pure SiO_2 of the same EOT. Reprinted with permission from [207], © 1999, Materials Research Society.

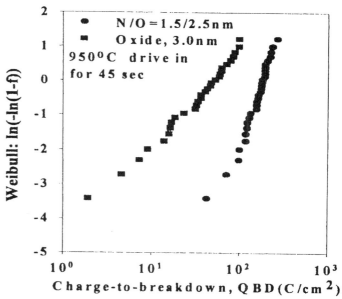

Fig. 66. Weibull plots of Q_{BD} for capacitors with stacked Si_3N_4/SiO_2 and pure SiO_2 dielectrics with the same EOT under constant current stressing (substrate injection at 500 mA cm^{-2}). Reprinted with permission from [207], © 1999, Materials Research Society.

deviations are smaller for the stacked dielectrics, indicating a reduced density of states at the interface. This was attributed to incorporation of N at the SiO_2–Si interface during the annealing that followed nitride deposition. Figure 65b compares tunneling current through 100 μm × 100 μm capacitors for 1.6-nm SiO_2 and Si_3N_4/SiO_2 stacks with 1.6-nm equivalent oxide thickness. Leakage through the stack is about two orders of magnitude lower at -1 V. Compared with calculated data, the oxide shows higher leakage due to incorporation of diffusive boron.

Figure 66 shows Weibull plots for MIS-Cs subjected to a uniform injection stress of 500 mA cm^{-2}. An improvement of Q_{BD}

of about an order of magnitude is observed in devices with stacked dielectrics. The difference in reliability between devices with pure oxide and stacked dielectric films was attributed to defects near the dielectric–Si interface associated with the presence of boron.

Taking into account both desirable and deleterious effects of nitrogen incorporation into SiO_2 gate dielectric films, an "ideal" profile has been devised for MIS devices [202]. This shows nitrogen peaks at the SiO_2–Si interface, to improve hot carrier resistance, and at the polysilicon–SiO_2 interface, to prevent the penetration of boron from the heavily doped gate electrode. More nitrogen is required at the polysilicon interface, because the boron flux can be very large, depending upon the thermal budget. At the Si interface, only enough nitrogen to improve the hot carrier resistance is needed, as more might also lower the channel mobility. Very little nitrogen in the interior of the dielectric is desired, because a large total amount of incorporated nitrogen might raise the fixed charge. The actual ideal amounts of nitrogen required at each interface are not known, but typical concentrations range from 1 to 5 at.% at each interface [10, 202].

7.3. Physicochemical Characteristics and Mechanistic Aspects

This section begins with considerations of the thermal nitridation of SiO_2 films in NH_3. It then moves to the direct growth of oxynitride films or nitridation of SiO_2 in NO_2 and NO. Finally, hyperthermal and deposition methods are discussed. Particularly for the latter, no attempt is made at a full description, but general trends are presented.

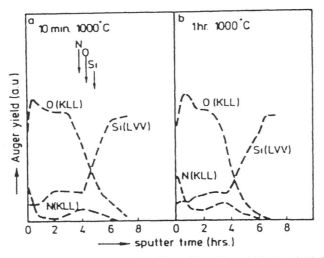

Fig. 67. Auger depth profiling of a 20-nm-thick silicon oxide film nitride in NH$_3$ at 1000°C for 10 and 60 min. Reprinted with permission from [209], © 1982, American Institute of Physics.

7.3.1. Thermal Nitridation of SiO$_2$ in NH$_3$

One of the techniques used to produce oxynitride films is nitridation of SiO$_2$ in gaseous NH$_3$. The obtained results depend on the nitridation time and temperature and on the initial oxide thickness. The basic mechanism behind the process is the replacement of O with N atoms coming from the nitriding gas [14]. Annealing of SiO$_2$ films in NH$_3$ leads to accumulation of nitrogen in the near-surface and near-interface regions. For long annealing times and high temperatures, nitrogen is also incorporated in the bulk of the oxide films. Along with nitrogen, significant amounts of hydrogen are incorporated. Figure 67 [209] shows the N, O, and Si profiles obtained in typical NH$_3$-nitrided oxides. These profiles were interpreted as resulting from the diffusion and reaction of ammonia-like species (NH$_x$, $x = 1, 2, 3$) through the SiO$_2$ film. The reaction occurs preferably in the near-surface region because the water created can escape from the material into the gas ambient or near the interface, from where it oxidizes the silicon substrate. XPS performed with synchrotron radiation confirmed the formation of one monolayer of Si$_3$N$_4$ at the SiO$_2$–Si interface at the initial stages of nitridation [210]. SIMS profiling of ^{16}O, ^{18}O, and ^{14}N in bilayered Si^{16}O$_2$/Si^{18}O$_2$ thin films nitrided in NH$_3$ indicated that thermal nitridation of the oxide is accomplished by transport of nitrogenous species from the surface to the interface of the films, concomitantly with an autodiffusion of the network oxygen atoms and replacement of the oxygen atoms in the SiO$_2$ network by nitrogenous species (Fig. 68) [211]. The fact that N at the interface is reduced for thicker oxides indicates that the nitridation process in NH$_3$ is limited by the diffusion of NH$_x$ species.

A nitridation mechanism in two steps has been proposed [212, 213]. It includes diffusion of NH$_x$ to interstitial sites that are uniformly occupied, followed by migration to substitutional sites. Nitridation could proceed via the following

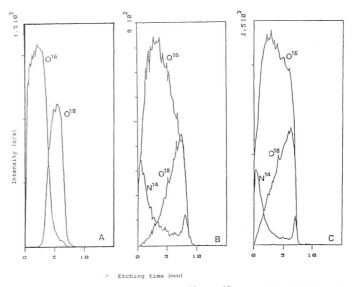

Fig. 68. SIMS profiles of a bilayered Si^{16}O$_2$/Si^{18}O$_2$ thin film (A) before nitridation, (B) after 10 min nitridation in NH$_3$ at 1100°C, and (C) after 30 min of nitridation. Reprinted with permission from [211], © 1991, Elsevier Science.

reaction:

$$2SiO_2(s) + 2NH_3(g) \rightarrow Si_2N_2O(s) + 3H_2O \qquad (8)$$

Thin nitrided oxide films display lower oxidation rates than thermal oxides. Resistance to oxidation increases with nitridation time and temperature and, thus, with nitrogen content. The oxidation rate of samples whose surface has been partially etched in HF is found to be practically identical to that of nonetched samples, showing that the nitrogen-rich interfacial layer is the most likely contributor to oxidation resistance [214].

Initial nitridation produces electrically active defects (strained bonds, broken Si—O bonds, possibly accompanied by displaced Si and O and Si dangling bonds) that are all sources of positive charge, i.e., which all produce negative threshold voltage shift. These defects are due to in-diffusion of hydrogenous species. The shift has been correlated to Si—H and Si—OH groups by SIMS and Raman spectroscopy. It is then reduced as Si—N groups form, reducing silicon dangling bonds or redistributing existing defects. After long nitridation, the negative voltage shift is greatly reduced, as the amount of incorporated nitrogen keeps increasing.

Oxides nitrided by rapid thermal processing show properties that are similar to those of oxides nitrided in a conventional furnace. However, rapid nitridation makes it possible to minimize the duration of the nitridation step and thus permits one to reduce both strain and impurity redistribution at the SiO$_2$–Si interface.

Besides NH$_3$ and disregarding nitrogen oxides (discussed in the next section), gaseous N$_2$ does react, to a slight extent, with silicon at the SiO$_2$–Si interface, but high temperatures (>1000°C) and long nitridation times (several days) must be used to obtain a sizable effect. The binding energy of N$_2$

molecules is fairly high, whereas the standard free energy of the reaction between silicon and nitrogen is low. The fact that nitridation of SiO$_2$ is possible with NH$_3$ but nearly impossible with N$_2$ alone points to a role for hydrogen in breaking the Si—O bonds.

Oxide films nitrided in NH$_3$ are often reoxidized. This reoxidation yields insulating films whose properties are half-way between those of silicon oxide and those of nitrided oxide films. During reoxidation, the top nitrided layer is quickly reoxidized, a thin oxide layer grows beneath the interfacial oxynitride layer, and the hydrogen content in the film is much reduced.

7.3.2. Thermal Nitridation of SiO$_2$ and Direct Oxynitride Growth in N$_2$O and NO

One of the drawbacks of nitridation in NH$_3$ is the inevitable formation of hydrogenated species in the film, which leads to high densities of electron traps. Oxynitride films can be grown directly by submitting a Si substrate or an SiO$_2$ film to a gaseous N$_2$O or NO ambient. This process permits better control of the nitridation step and minimizes the amount of incorporated hydrogenous species.

The bulk phase diagram of the Si—N—O system is shown in Figure 69 [177, 202]. Four phases appear: Si, SiO$_2$ (cristobolite and tridymite), Si$_3$N$_4$, and Si$_2$N$_2$O. The three compound phases have similar structural units: SiO$_4$ tetrahedra for SiO$_2$, SiN$_4$ tetrahedra for Si$_3$N$_4$, and slightly distorted SiN$_3$O tetrahedra for Si$_2$N$_2$O, implying that the phases can be converted from SiO$_2$ to Si$_2$N$_2$O and finally to Si$_3$N$_4$ by replacing oxygen with nitrogen. However, the nitride and the oxide phases

never coexist. They are always separated by Si$_2$N$_2$O, which is the only thermodynamically stable and crystalline form of silicon oxynitride. According to chemical equilibrium, N should not incorporate at all into an SiO$_2$ film grown in almost any partial pressure of oxygen, depending upon temperature (Fig. 69).

At least two reasons for the presence of nitrogen in SiO$_2$ films processed at the usual temperatures and pressures can be suggested. First, nitrogen atoms may simply be kinetically trapped at the reaction zone near the interface (i.e., the nitrogen is present in a nonequilibrium state, where the rate of the transition to equilibrium is slow and some nitrogen is trapped). Alternatively, the nitrogen at the interface may indeed be thermodynamically stable, because of the presence of free energy terms that are not yet understood. For example, if the nitrogen plays a role in lowering the interfacial strain, there may be a strain free energy contribution when nitrogen is incorporated at the interface. After oxynitridation with N$_2$O or NO, one typically observes the incorporated nitrogen to be segregated very close to the SiO$_2$–Si interface, consistent with a special, stabilizing role at the interface. The N peak can be shifted to other areas of the film, but the processing required to accomplish this must take into account the metastability of the incorporated nitrogen in the SiO$_2$. Even when nitrogen is implanted into silicon, it tends to migrate to the SiO$_2$–Si interface and be incorporated into the SiO$_2$.

Figure 70 is a schematic illustration of the incorporation of N into SiO$_2$ films by thermal processing in NO and N$_2$O. It is shown in Figure 70a that the exposure of silicon to NO produces an oxynitride film with an approximately uniform distribution

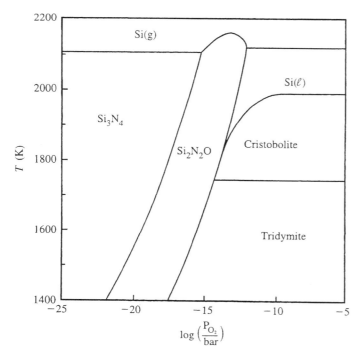

Fig. 69. Thermodynamic phase diagram of the Si–N–O system. Based on [249].

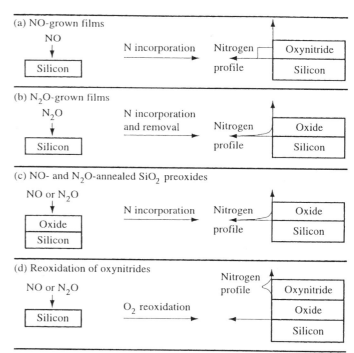

Fig. 70. Schematic diagram illustrating nitrogen depth distributions for different thermal nitridation sequences. Reprinted with permission from [10], © 1999, IBM Technical Journals.

of nitrogen. At a given temperature, the oxidation of Si with NO is essentially a self-limiting growth, because the increasingly N-rich oxide layer acts as a diffusion barrier. As in Figure 70b, when N_2O is used there is both incorporation and removal of nitrogen, which results in a profile peaked close to the SiO_2–Si interface. The peaked profile results from the reduction of N_2O in the gas phase to primarily NO and O radicals. The NO efficiently incorporates the nitrogen into the film as it grows, while the oxygen radicals remove N from the films, especially at its outer surface. When either N_2O or NO is used to anneal a previously grown SiO_2 film as in Figure 70c, the N concentration is peaked near the SiO_2–Si interface, with NO producing substantially higher N concentrations for a given thermal cycle. Only with the reoxidation in O_2 of an oxynitride film can the peak of the nitrogen profile be displaced from the interface, as in Figure 70d.

Oxynitridation mechanisms for thermal methods based in N_2O and NO, just like those based in NH_3, are fairly well understood, as presented in recent reviews [65, 177, 200, 215, 216]. Both N_2O and NO promote film growth. The former leads to incorporation of 0.1–1 at.% of N; the latter, to substantially higher amounts [184, 190, 191, 217–230]. The reason for this is in the gas phase chemistry of N_2O. As N_2O decomposes when heated to the temperatures at which (oxy)nitridation takes place ($>800°C$), the actual nitriding species is NO, but in an environment containing N_2, O_2, and atomic oxygen, O. The first is essentially inert; O_2 can promote some film growth, but at reduced rates if compared with the continued oxidation of SiO_2/Si structures because of the inhibiting effect of nitrogen incorporated into the film; atomic oxygen, in turn, is responsible for the partial removal of nitrogen from the film, as stated above. It has been found [231] that in a heated conventional furnace N_2O rapidly decomposes into about 60 mol% N_2, 30 mol% O_2, and 10 mol% NO. Rapid thermal oxynitridation with N_2O leads to slightly different results, as the gas decomposes only upon reaching the heated sample surface [223].

The growth kinetics of thermal silicon oxynitride films on Si in NO are self-limited to 2.5 nm at any temperature below 1100°C. At the initial stages, NO adsorbs dissociatively on Si(001) as well as in Si(111), forming one monolayer of Si_3N_4 at the dielectric–Si interface, followed by several monolayers of subnitrides (nitrogen-defective silicon nitrides) and, most probably, suboxides as well. Isotopic substitution studies clearly indicate that NO diffuses toward the oxynitride–Si interface. The presence of one Si_3N_4 monolayer or even a fraction of a monolayer at the interface largely prevents the migrating NO molecules from reacting with Si atoms of the substrate. So direct thermal growth of silicon oxynitride films on Si in NO proceeds within a very limited atomic transport scenario, because of the diffusion barrier properties of a layer with an appreciable concentration of N at and near the oxynitride–Si interface. Most of the activity consists of replacement of N originally incorporated by O, as well as completion of the suboxide/subnitride network. The degree of replacement depends on processing time, NO pressure, and film thickness as a whole.

Silicon oxynitride films can also be produced by thermal nitridation of silicon oxide films. When the nitriding gas is NO, nitrogen is introduced only in the near-interface region. Contrary to what is observed in the direct thermal growth of oxynitrides in NO, the nitrogen concentration at and near the interface increases with increasing annealing time. No appreciable film growth (thickness increase) occurs. XPS analysis indicates that nitrogen is predominantly bonded to Si, as in stoichiometric Si_3N_4. Only a minor portion of the N atoms presents a Si=N—O bond structure. Nitridation of SiO_2 films in N_2O leads to similar results, except that significant film growth can occur. This should be due to the presence of O_2 as a product of N_2O decomposition, as discussed above.

Isotopic substitution results [222] indicate that the nitridation of silicon oxide films in NO takes place by two atomic transport mechanisms occurring in parallel, as in the case of dry oxidation of silicon oxide films in O_2: (i) NO diffuses through the silica network and reacts at the SiO_xN_y interface to fix both N and O (this mechanism involves, in fact, a minor fraction of the NO molecules entering the oxynitride network) and (ii) step-by-step motion of network oxygen atoms induced by the presence of network defects leads to incorporation of O in the near-surface. As (ii) involves only O, one can presume that N is released in the form of a nonreacting molecule, like N_2.

Figure 71 [232] shows SIMS depth profiles for silicon oxynitride samples grown in a rapid furnace at 1050°C for 1 min with the use of (a) N_2O or (b) a mixture of NO and O_2. Profiles are not reliable within the top 0.5 nm because of the presence of an adventitious carbon surface layer. The film grown in N_2O shows a characteristic nitrogen pile-up at the interface with the Si substrate, whereas the other sample has the nitrogen uniformly distributed throughout the oxynitride layer. The latter is typical of oxynitridation with NO. As the samples were prepared by rapid thermal treatment, the result is explained by the decomposition of N_2O being restricted to the sample surface, in small amounts, so that a continuous supply of O radicals was available during film growth.

The nitrogen and oxygen profiles in a sample nitrided in NO and subsequently annealed in O_2 at high pressure and low temperature are shown in Figure 72. This sample corresponds to the I–V characteristics shown in Figure 62. Processing parameters are detailed in the figure. One observes relatively high nitrogen incorporation in the film (>5 at.%) at low processing temperatures. Moreover, the N profile is shifted to the sample surface as compared with the usual feature for an SiO_xN_y film grown in NO. These are all desirable characteristics for a nitrided gate dielectric.

Figure 73 presents in schematic form a sequence devised to produce the ideal nitrogen profile in a gate oxynitride film. It consists of an oxynitridation step of Si in NO followed by oxidation in O_2 and repeated oxynitridation in NO. The first step results in an oxynitride film with a given concentration of N at the interface with Si. The second step promotes growth of an underlying oxide film, displacing the N distribution in the direction of the sample surface. The third step introduces N at the new dielectric–silicon interface. One expects to be able

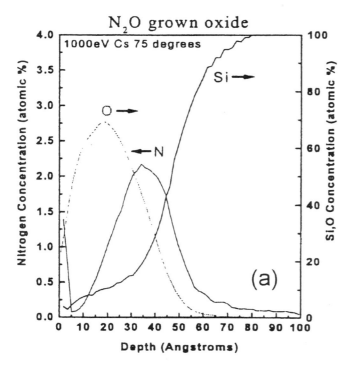

N$_2$O grown oxide

1000eV Cs 75 degrees

(a)

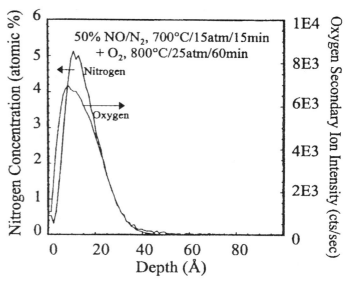

50% NO/N$_2$, 700°C/15atm/15min
+ O$_2$, 800°C/25atm/60min

Nitrogen

Oxygen

Fig. 72. SIMS profile of high-pressure NO oxynitride with subsequent high-pressure O$_2$ annealing. Reprinted with permission from [205], © 1999, Materials Research Society.

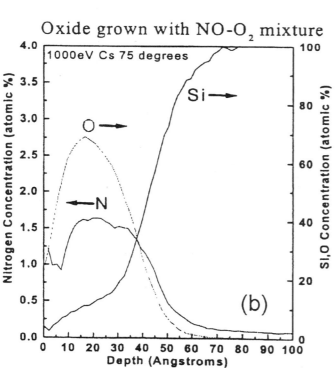

Oxide grown with NO-O$_2$ mixture

1000eV Cs 75 degrees

(b)

Fig. 71. SIMS depth profiles for silicon oxynitride samples grown in a rapid furnace at 1050°C for 1 min with (a) N$_2$O or (b) a mixture of NO and O$_2$. Reprinted with permission from [232], © 1999, Materials Research Society.

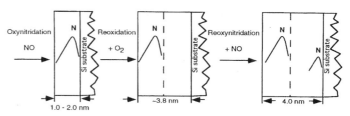

Fig. 73. Depiction of the process flow for creating the ideal nitrogen profile with NO gas. Reprinted with permission from [202], © 1998, Kluwer Academic Publishers.

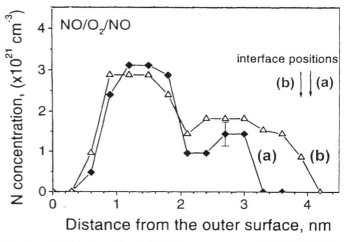

Fig. 74. MEIS depth profiles for Si(001) samples processed in (a) NO (800°C, 5 min) followed by O$_2$ (900°C, 52 min) with final annealing in NO (900°C, 5 min), and (b) NO (800°C, 5 min) followed by O$_2$ (900°C, 48 min) with final annealing in NO (900°C, 40 min). Reprinted with permission from [217], © 1998, American Institute of Physics.

to tailor N concentration at the sample surface and interface through the annealing temperature in NO in the first and third steps. The successful result of such an approach is shown in Figure 74.

7.4. Hyperthermal and Deposition Methods

Thermal nitridation of SiO_2 in NO or N_2O generally results in a relatively low concentration of nitrogen in the films, on the order of 10^{15} cm^{-2} N atoms [189, 202, 230, 233]. Because nitrogen content increases with temperature, thermal oxynitridation is typically performed at high temperatures (i.e., >800°C). Other nitridation methods, such as with the use of energetic nitrogen particles (plasma, nitrogen atoms, or ions) [46, 206, 234–247], can be used. These nitridation methods can be performed at lower temperatures, ~300–400°C. However, low-temperature deposition methods may result in nonequilibrium films, and subsequent thermal processing steps are often required to improve film quality and minimize defects and induced damage [199, 248]. Because the thermodynamics [249, 250] of the SiO_xN_y system and the kinetics [184, 188, 194, 202, 203, 223, 228, 229, 233] of nitrogen incorporation are rather complex, these different methods produce oxynitride films with different total nitrogen concentrations and depth distributions. From a scientific viewpoint, the addition of N to the Si—O system opens a number of questions concerning microstructure, defects, and reaction mechanisms.

At this point it is revealing to consider the nitridation of silicon with hyperthermal ion beams, which has been studied (and modeled) in great detail by Hellman et al. [235]. In this case it was observed that ion energy, ion fluence, and substrate temperature during nitridation all play significant roles in the process. The kinetic energy provided to the ions allows their insertion into the substrate at no cost in thermal energy, which is then used to activate the chemical reaction. Once the reaction is complete (i.e., a stoichiometric phase has been formed) at a depth corresponding to the ion range in the sample, temperature becomes fundamental for the diffusion of nitriding species until a region comprising substoichiometric nitride is reached. Although such results may constitute an interesting parallel to the nitridation of SiO_2 films with hyperthermal species, thermodynamic considerations mentioned above should be taken into account, as should the presence of the SiO_2–Si interface.

Figure 75 shows the nitrogen profiles observed in ultrathin SiO_2 films on Si submitted to plasma immersion ion implantation. I–V and C–V characteristics of these films were presented in Figures 63 and 64. The profile observed in an SiO_xN_y film produced by thermal oxynitridation of Si with N_2O is presented for comparison. The samples prepared by PIII clearly show higher N content. Among the plasma implants, the one performed at 25 V bias places more nitrogen in the bulk of the oxide and less at the interface with Si. Nearly 10 times more nitrogen is seen in the sample nitrided at the highest bias; however, the tail of the nitrogen distribution extends through the interface and into the Si substrate, which is not a desirable characteristic but may be tolerable to some extent.

SIMS profiles of oxynitride films produced by treating oxide samples with different nitrogen plasma sources (remote plasma or helicon plasma source) are shown in Figure 76 [232]. Resulting nitrogen contents are similar among the films. In the

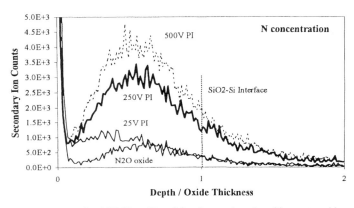

Fig. 75. Normalized SIMS profiles of the nitrogen introduced into gate oxides upon plasma immersion ion implantation at the bias voltages shown. The profile corresponding to a thermal oxynitride grown in N_2O is included for comparison. The quoted oxide thickness is 2 nm. Reprinted with permission from [206], © 1999, Materials Research Society.

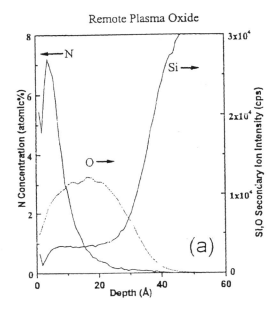

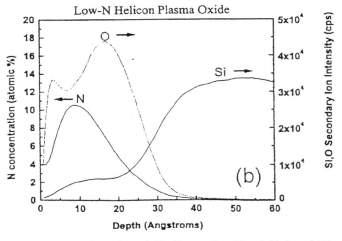

Fig. 76. SIMS profiles of oxynitride films produced by nitridation of SiO_2 (a) with a remote nitrogen plasma and (b) with a helicon plasma source. Reprinted with permission from [232], © 1999, Materials Research Society.

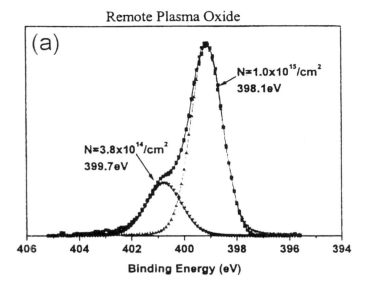

Remote Plasma Oxide

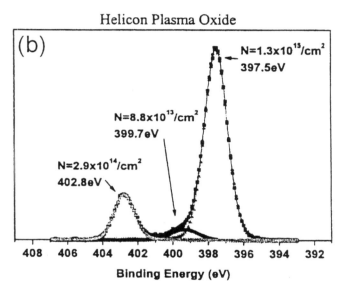

Helicon Plasma Oxide

Fig. 77. High-resolution N 1s photoelectron spectra taken for ultrathin SiO₂ films nitrided (a) with a remote plasma and (b) with a helicon plasma source. Reprinted with permission from [232], © 1999, Materials Research Society.

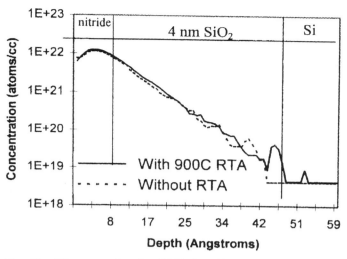

Fig. 78. Nitrogen profiles by SIMS for silicon nitride/silicon oxide (0.8 nm/4.0 nm) gate dielectrics before and after 30 s, 900°C postdeposition rapid thermal annealing. Reprinted with permission from [207], © 1999, Materials Research Society.

remote plasma sample the nitrogen is concentrated in a sharp surface peak confined to about 1 nm at the sample surface. The helicon source produces nitrogen buried in the oxide substrate. Areal densities extracted from the profiles range from 1 to 3×10^{15} cm^{-2} N, enough to form up to 0.3 nm of Si₃N₄.

XPS characterization of the nitrogen in plasma samples described above is shown in Figure 77. Both spectra are dominated by N(−Si)₃ peaks. Together with high nitrogen content, this indicates the presence of stoichiometric Si₃N₄ in the helicon plasma sample. In addition to this major peak, the remote plasma sample also presents a peak at 399.7 eV, characteristic of O−N(−Si)₂. The helicon plasma film shows a lower O−N(−Si)₂ peak as well as a third peak at 402.8 eV. The relationship between the N 1s peak position and the number of

N−O bonds has been used to suggest that the 402.8 eV peak is due to (O−)₂N−Si. Thus, the plasma-nitrided films show more complex bonding, including bonds to oxygen, than that usually exhibited by thermally nitrided samples.

To produce oxynitride films with higher nitrogen content than that available from hyperthermal nitridation, deposition methods such as chemical vapor deposition (CVD) [200] with different precursors and its low-pressure (LP) and/or rapid thermal (RT) variants [198] can be used. Jet vapor deposition (JVD) [251] and atomic layer deposition (ALD) [252] can also be considered. In this case, a mechanistic description of film deposition depends on the precursors chosen. The simplest choice, SiH₄ and NH₃, usually leads to hydrogen incorporation into the oxynitride films. Gases such as N₂O and NO have also conveniently replaced NH₃ in this case.

Figure 78 shows SIMS profiles of N for a silicon nitride/silicon oxide stack before and after a postdeposition rapid thermal annealing. The annealing step is performed to promote chemical and structural relaxation, as well as to reduce the density of electronic states at the dielectric–silicon interface. The annealing drives the N atoms to the SiO₂–Si interface, as shown by a distinctive feature in Figure 78 ("tailing" of the nitrogen signal into the oxide alone was reported as an artifact of SIMS). Interfacial nitrogen concentrations achieved in this manner are reported to be much less than one monolayer. Replacing Si−O bonds at the interface, Si−N is believed to relieve interface strain because of the smaller effective size of N atoms and has proved to yield a physically smoother interface.

7.5. Remarks and Limitations Concerning Ultrathin Silicon Oxynitride Films as Gate Dielectrics

Silicon oxynitrides present significant advantages over silicon oxide for application as gate dielectrics: increased resistance to boron penetration, reduced leakage current (due to increased dielectric constant), and increased reliability. Such properties combined with a low density of electronic states at the interface with Si make silicon oxynitrides (and nitride/oxide stacks) the materials chosen to replace silicon oxide in current state-of-the-art and near-future technologies. A large number of processing schemes have been devised to produce SiO_xN_y, and a significant knowledge of reaction mechanisms exists.

Although there is possibly room for improvements in the current status of silicon oxynitrides as gate dielectrics, limitations of these materials are already evident. As the offered increase in the dielectric constant is only marginal if compared with SiO_2, dielectric films cannot be made much thicker than the current ones. If one further considers that nitrogen must be kept away (except for very small beneficial concentrations) from the interface with silicon—otherwise threshold voltage shifts, degraded channel mobility, and increased density of interface states take over—it becomes clear that silicon oxynitrides do not offer a significant margin for continuing device scaling. Such might be the case for another material, however.

8. ALTERNATIVE (HIGH-k) GATE DIELECTRIC FILMS

The search for an alternative to SiO_2 as the main material for gate dielectrics constitutes a new and very lively research area [253, 254] because the exponential increase in tunnel current with decreasing film thickness is a fundamental limit on the scaling of gate dielectrics [10]. Scaling of CMOS suggests that the use of SiO_2-based dielectrics appears to be limited to a range somewhere between 2.0 and 1.5 nm. For 2 nm-thick SiO_2 films, for instance, leakage current densities can rise to a few amperes per square centimeter [64, 255]. To reduce transistor area while keeping leakage current at acceptable levels and maintaining the same gate capacitance, an alternative gate dielectric film made with a material of higher dielectric constant is required. If such a material can be used then the thickness of the gate dielectric can be increased proportionally to the increase in dielectric permittivity with respect to that of SiO_2 ($\epsilon_{ox} = 3.9$).

Many materials with higher dielectric constants (commonly referred to as high-k materials) have been suggested as replacements for SiO_2, like Ta_2O_5, TiO_2, Al_2O_3, and many double (e.g., zirconium silicate) and triple (e.g., barium strontium titanate, BST) oxides [256–261]. (As pointed out by Buchanan [10], "the term 'high-k' refers to a material with dielectric constant significantly higher than that of SiO_2. The Greek letter ϵ is more commonly used in the semiconductor industry to represent the relative permittivity of a material. The letter k is found more often in the fields of chemistry and physics." Even though the materials are referred to as high-k, in this chapter the symbol ϵ is used for the relative permittivities.) The ITRS roadmap [9] foresees gate dielectrics scaling and materials as indicated in Figure 79. However, a

higher dielectric constant than that of SiO_2 (high-k) is not sufficient for an alternative dielectric. A large bandgap, comparable to that of SiO_2 (9 eV), and a large energy barrier from the conduction band to the gate electrode are also mandatory requirements, otherwise leakage current will be unacceptably high despite the higher dielectric constant. This is a rather restrictive requirement, because for most high-k oxides mentioned in Figure 79, as the dielectric constant increases the band energy gap decreases, following approximately an $E_g \approx 1/\epsilon^2$ law [262], as illustrated for simple dielectric materials in Figure 80. Al_2O_3 is one of the very few exceptions to this rule, as its bandgap is comparable to that of SiO_2, and therefore this material and double oxides involving aluminum (like $ZrAl_xO_y$) are serious candidates for alternative gate dielectrics, even though their dielectric constants are not exceptionally high.

Furthermore, it might be desirable as well as necessary to preserve the outstanding benefits of the low density of electronic states at the SiO_2–Si interface as compared with other dielectric materials, which has been instrumental so far in the success of this technology. Therefore, an ultrathin interlayer of silicon oxide, nitride, or oxynitride may still be required between the silicon substrate and the high-k dielectric to minimize interface states and/or to be a diffusion barrier between the layers [60, 263]. The interlayer should be very thin—just enough to have a good-quality interface and diffusion barrier, if needed—to minimize the effect of the serial association of the two capacitances, namely that of the interlayer and that of the high-k dielectric. This is illustrated in Figure 81 [256] for the case of Ta_2O_5 films deposited on thermally grown SiO_2 interlayers on Si(001). One can clearly see that for an electrical equivalent oxide thickness of the stack to be below, for example, 1.5 nm, the thickness of the interlayer must stay below 1 nm.

For any candidate material to be an alternative gate dielectric it must be capable of retaining its properties when submitted to thermal annealing [253], which is inherent to further processing steps after gate dielectric deposition, such as dopant activation, contact silicide formation, and others. The primary concern is chemical stability of the gate oxide on silicon: under thermal annealing the gate dielectric must not change its stoichiometry, undergo deleterious reactions with silicon, or phase segregate, and the amorphous candidates must not crystallize. If cations from the gate dielectric react or diffuse into the silicon channel, the electrical properties will suffer. On the other hand, previous investigations have widely demonstrated [257–261] that a postdeposition annealing at moderate temperatures, in dry O_2 or N_2O, of several of the above-proposed gate dielectric materials can bring the leakage current and the density of interface states down to well acceptable limits, without a significant lowering of the dielectric constant due to the formation of a SiO_2 layer between the alternative dielectric and the Si substrate.

Thus, the technologically imperative search for an alternative material to replace SiO_2 as a gate dielectric is a complicated task, requiring information on reaction with silicon, oxygen

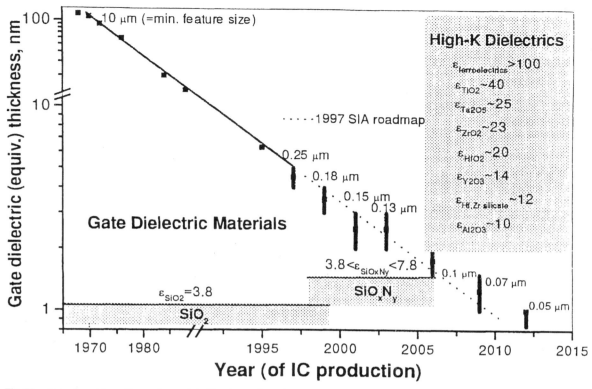

Fig. 79.   Gate dielectric scaling and materials. The figure shows how (equivalent) oxide thickness shrinks during ULSI device evolution. It also lists dielectric materials and their dielectric constants. Reprinted with permission from [60], © 2000, Kluwer Academic Publishers.

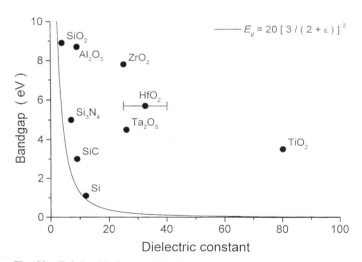

Fig. 80.   Relationship between bandgap and permittivity for several commonly studied gate dielectric materials and semiconductors. Reprinted with permission from [262], © 1997, IEEE.

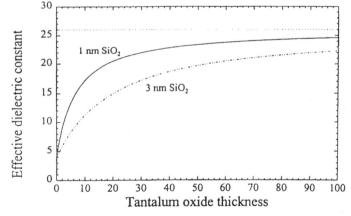

Fig. 81.   Effective dielectric constant of the double-layered Ta_2O_5/SiO_2 structure vs. Ta_2O_5 thickness calculated assuming two capacitors in series with the dielectric constant of silicon oxide and tantalum pentoxide for 1 nm (solid line) and 3 nm (dashed line) of interfacial SiO_2. Reprinted with permission from [256], © 1998, Elsevier Science.

diffusion, stoichiometry, film crystallization, and phase segregation. One clear illustration of the difficulties is given in Figure 82, showing that the use of BST—a triple oxide that can have very high permittivity—as a gate dielectric for CMOS applications may be prevented because of eventual minor changes in stoichiometry during thermal annealing in vacuum or in

oxygen, leading to a very high decrease in the permittivity.

Nevertheless, improvements in device electrical characteristics are being reported for capacitors and transistors made with dielectrics alternative to SiO_2. A few illustrations are given in the next section.

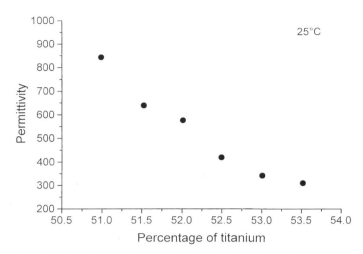

Fig. 82. Plots of permittivity of $(Ba_{0.7}Sr_{0.3})TiO_3$ (BST) produced by met-alorganic chemical vapor deposition (MOCVD) as a function of the $(Ba+Sr)/Ti$ stoichiometry. Reprinted with permission from [253], © 2000, Macmillan.

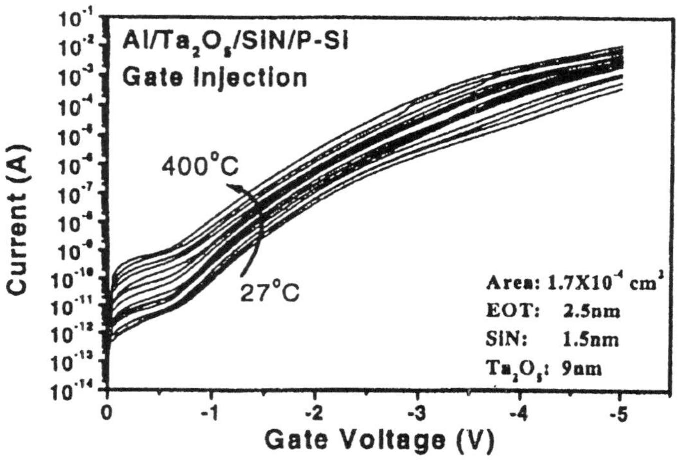

Fig. 83. I–V characteristics of an $Al/Ta_2O_5/Si_3N_4/Si$ capacitor where the Si_3N_4 is an ultrathin buffer layer in each sample. Reprinted with permission from [263], © 2000, The Electrochemical Society, Inc.

8.1. Electrical Characteristics

Different aspects determine the electrical characteristics of high-k dielectrics on Si; the most important ones concern the deposition process (method, substrate temperature, and chemical precursors), intermediate layer nature and thickness (intentionally grown or not), and postdeposition annealing. Other factors also may be decisive, such as the nature of the gate electrode (poly-Si, Al, Au, etc.).

Figure 83 shows the I–V characteristics of capacitors made with 9 and 12 nm Ta_2O_5 films deposited by rapid thermal chemical vapor deposition (RTCVD) on Si(001) substrates in which 1.5 nm intermediate silicon nitride layers were thermally grown by rapid thermal nitridation (RTN) in NH_3 [263]. There is a noteworthy reduction in leakage current as compared with capacitors made with SiO_2 of the same equivalent thickness (EOT) [255]. Figure 84 shows I–V and C–V characteristics for MIS devices made with Ta_2O_5 films and approximately 0.5 nm silicon oxide or nitride intermediate layers deposited by plasma-enhanced chemical vapor deposition (PECVD) methods [264]. The C–V data for these Ta_2O_5 devices with plasma-processed interfaces (oxide or nitride intermediate layers) gave no evidence of fixed charge at the dielectric–Si interface or at the internal interface of the stacked structure, whereas devices prepared on HF-last Si displayed negative flatband shifts corresponding to a fixed positive charge of mid-10^{11} cm^{-2}. Interface nitridation produced 10-fold reductions in tunneling current, as it did in combination with SiO_2 dielectrics.

The C–V characteristics of $HfSi_xO_y$ and $ZrSi_xO_y$ films approximately 5 nm thick, deposited by reactive sputtering or reactive e-beam evaporation, are shown in Figure 85 [265]. C–V curves for an $Au/Hf_6Si_{29}O_{65}/n^+$-Si structure are shown in Figure 85a, where the silicate film was deposited directly on Si at 500°C (no intentional intermediate layer) and post-deposition annealed at 450°C in forming gas. The largest value of measured capacitance density in accumulation corresponds

to $\epsilon \approx 11$ or EOT = 1.78 nm. The C–V characteristics for an $Au/Zr_4Si_{31}O_{65}/n^+$-Si structure are shown in Figure 85b, corresponding to EOT = 2.08 nm. The results shown in Figure 85 also demonstrate that the Au electrode, with a work function $\Phi_B = 5.3$ eV, creates a flatband condition at zero bias. A comparison of the C–V curves with ideal ones indicates that the interface density of electronic states is $D_{it} \approx$ (1–5) × 10^{11} cm^{-2} eV^{-1}, a figure that can be substantially reduced to close to the SiO_2 values by optimization of the post-deposition annealing. The corresponding I–V characteristics are shown in Figure 86, where one notices that both alternative dielectrics, $HfSi_xO_y$ and $ZrSi_xO_y$, display leakage currents several orders of magnitude smaller than SiO_2 films with the same EOT. Devices that were voltage-ramped to hard breakdown showed breakdown fields of $E_{BD} > 10$ MV cm^{-1}.

C–V characteristics of Al-gate capacitors and 6.5 nm-thick Al_2O_3 films deposited on n-type Si without an intentional intermediate layer are shown in Figure 87. It can be seen that the D_{it} is rather low. The quasi-static C–V measurements indicate a small d.c. leakage for the larger voltages. By changing the ramp rate, it is verified that this leakage does not distort the quasi-static C–V in the relevant interval -1 to 0 V. The high-frequency C–V was ramped from -2 to $+2$ V and back. Very little hysteresis was observed in this voltage range. Scanning to larger voltages indicated the occurrence of some electron trapping, evidenced by a flatband shift toward more positive voltages.

Finally, I–V characteristics are shown in Figure 88 for 12.5 nm Gd_2O_3 films directly deposited on Si(001) by reactive e-beam evaporation [48]. The as-deposited film is leaky, but a 10-min annealing in O_2 at 500°C or 700°C resulted in a dramatic improvement. Estimates of the thickness of the intermediate SiO_2 layers formed after annealing in O_2 obtained from C–V characteristics are shown in Figure 89 [48] along with the average dielectric constants of the films.

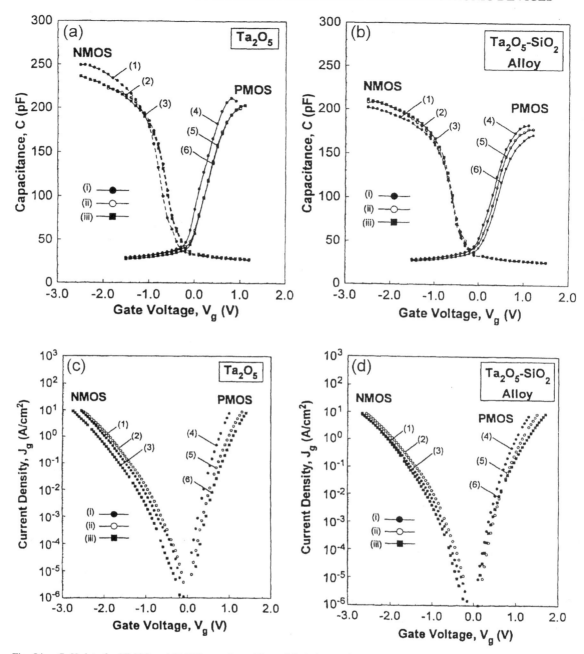

Fig. 84. $C-V$ data for NMOS and PMOS capacitors. The solid circles are for direct deposition on HF-last Si, the open circles for remote plasma-assisted oxidized (RPAO), and the solid squares for RPAO/remote plasma-assisted nitrided interfaces. (a) Ta_2O_5 dielectrics with the following EOT: (1) 1.09 nm, (2) 1.20 nm, (3) 1.17 nm, (4) 1.19 nm, (5) 1.30 nm, and (6) 1.29 nm. (b) Ta silicate alloys: (1) 1.40 nm, (2) 1.35 nm, (3) 1.33 nm, (4) 1.45 nm, (5) 1.51 nm, and (6) 1.58 nm. The areas of the devices are all the same, 10^{-4} cm^{-2}. (c) and (d) show $I-V$ data, respectively, for the MOS-C devices in (a) and (b). Reprinted with permission from [264], © 2000, The Electrochemical Society, Inc.

Electrical characteristics of several other alternative dielectric films are becoming available in the literature, some of them constituting excellent candidates to replace SiO_2. Many different illustrations of the effects of the above-mentioned aspects like deposition process, intermediate layer, postdeposition annealing, and gate electrode on the electrical characteristics are also being explored. In this section some illustrative ex-

amples were given, mainly indicating that physicochemical stability against thermal annealing must be fully understood before any of these alternative dielectrics can be incorporated into Si-based device technology. One word of caution: the strict requirements of breakdown lifetime will also apply to novel high-k gate dielectrics. Because the breakdown properties are very material-dependent, systematic studies of reliability of the

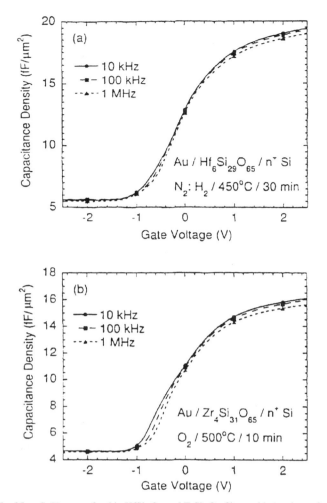

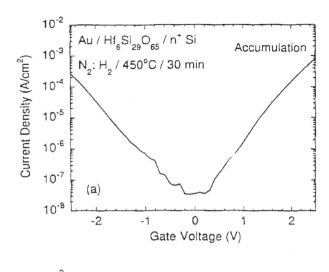

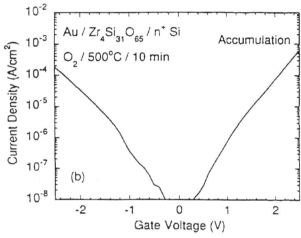

Fig. 85. $C-V$ curves for thin HfSi$_x$O$_y$ and ZrSi$_x$O$_y$ films, with Au electrodes ($A = 1.76 \times 10^{-4}$ cm^2). (a) 50-Å Hf$_6$Si$_{29}$O$_{65}$ film on n^+-Si, deposited at 500°C, and subsequently annealed in forming gas at 450°C for 30 min. The C_{max}/A value in accumulation yields EOT = 17.8 Å. (b) 50 Å Zr$_4$Si$_{31}$O$_{65}$ film on n^+-Si, deposited at 25°C, and subsequently annealed in O$_2$ at 600°C for 10 min. The capacitance density in accumulation yields EOT = 20.8 Å. These films have some dispersion near zero bias, which indicates the presence of interface traps. Reprinted with permission from [265], © 2000, American Institute of Physics.

Fig. 86. $I-V$ curves for films shown in Figure 85. (a) Au/50 Å Hf$_6$Si$_{29}$O$_{65}$/n^+-Si. (b) Au/50 Å Zr$_4$Si$_{31}$O$_{65}$/n^+-Si. These films show extremely low leakage currents, which are below 2×10^{-6} A cm^{-2} at 1.0-V gate bias in accumulation. The $I-V$ curves are well behaved and appear nearly symmetrical about zero bias. This suggests nearly equal barrier heights in the two polarities. Reprinted with permission from [265], © 2000, American Institute of Physics.

novel high-k materials will be needed to evaluate their applicability. Despite the reduced leakage, some of the new materials may not pass the reliability (lifetime) criterion.

8.2. Physicochemical Characteristics

The structure and composition of alternative, high-k dielectric films of potential interest as replacements for SiO$_2$ that have been investigated so far may have rather different characteristics: (i) they can be amorphous or epitaxial with the c-Si substrate. Amorphous films will most probably be initially used, whereas epitaxial films will constitute a further improvement in latter stages; (ii) stoichiometry and chemical bonds can be different, depending on deposition methods and parameters; (iii) the sharpness of the dielectric–Si interface is a

critical aspect, similar to what happens in the case of the SiO$_2$–Si interface; (iv) the as-deposited structures will be submitted to thermal annealing in vacuum, forming gas, or O$_2$, and the resulting structures and compositions will be a direct consequence of the physicochemical stability of the different thin film materials on Si; (v) atomic transport and chemical reaction may be inhibited by ultrathin, intermediate oxide, nitride, or oxynitride layers. They can be intentionally grown or deposited on the c-Si substrate before high-k film deposition or be present unintentionally, as a result of the formation of silicon oxide on the c-Si surface after cleaning or during high-k film deposition. A few examples are given here with the only aim of illustrating the variety of cases and the need for detailed investigation of the above-mentioned characteristics.

Figure 90 shows high-resolution transmission electron microscopy (HRTEM) micrographs of Ta$_2$O$_5$ films deposited by

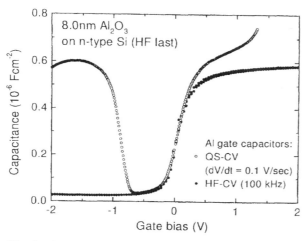

Fig. 87. Quasi-static (open symbols) and high-frequency (solid symbols) $C-V$ characteristics of an Al/Al$_2$O$_3$/n-Si structure. Reprinted with permission from [266], © 2000, American Institute of Physics.

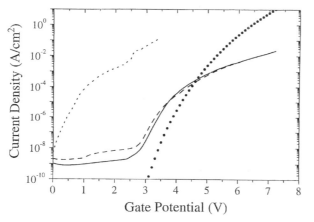

Fig. 88. $I-V$ data for a 12.5-nm-thick Gd$_2$O$_3$ film annealed for 10 min in oxygen: solid line, 700°C; dashed line, 500°C; dotted line, as-deposited; circles, calculated Fowler–Nordheim current for ideal SiO$_2$ layer 2.05 nm thick. Reprinted with permission from [48], © 2001, The Electrochemical Society, Inc.

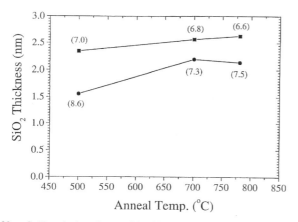

Fig. 89. $C-V$ analysis estimate of the thickness of the SiO$_2$ interface layer as a function of annealing temperature for 8-nm-thick Gd$_2$O$_3$ films with (circles) and without (squares) a previous 10-min vacuum anneal at 700°C. The anneals were done in oxygen for 10 min. The average dielectric constants are shown in parentheses at the thickness data points. Reprinted with permission from [48], © 2001, The Electrochemical Society, Inc.

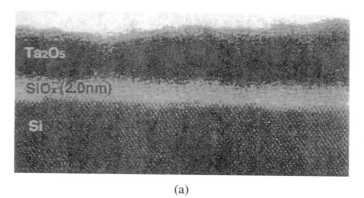

(a)

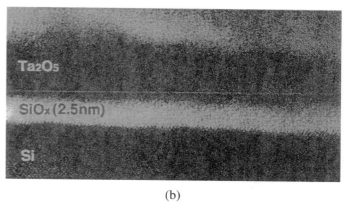

(b)

Fig. 90. High-resolution TEM micrographs of tantalum oxide and interfacial region after (a) a low-temperature plasma anneal and (b) a rapid thermal anneal in oxygen at 800°C. Crystalline regions and growth of the interfacial region can be observed after the thermal anneal. Reprinted with permission from [258], © 1998, American Institute of Physics.

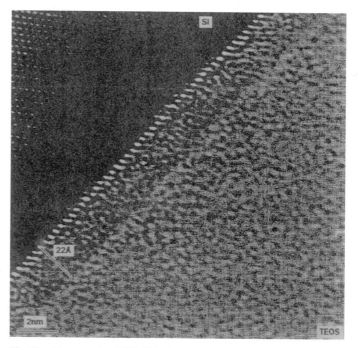

Fig. 91. Cross-sectional HRTEM of an Al$_2$O$_3$ film deposited on HF-treated Si. Reprinted with permission from [266], © 2000, American Institute of Physics.

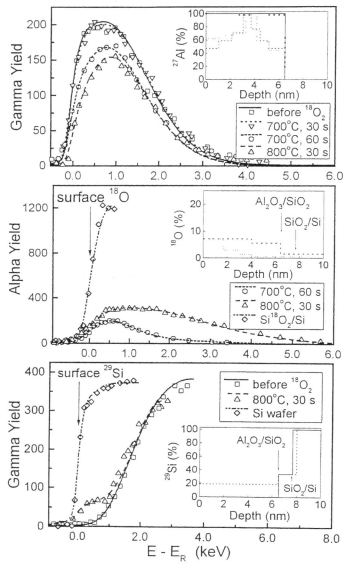

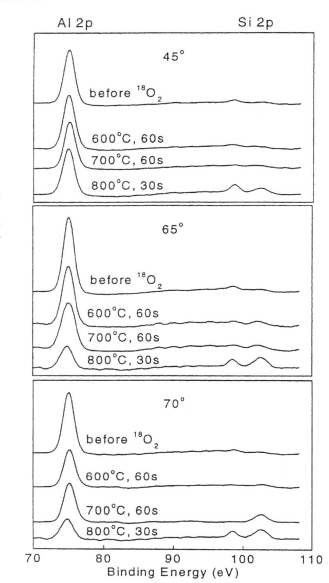

Fig. 92. Excitation curves of the $^{27}Al(p,\gamma)^{28}Si$ (top), $^{18}O(p,\alpha)^{15}N$ (center), and $^{29}Si(p,\gamma)^{30}P$ (bottom) nuclear reactions around resonance energies, with the corresponding profiles in the insets for ultrathin $Al_2O_3/SiO_2/Si$ structures submitted to rapid thermal annealing in $^{18}O_2$. One hundred percent of ^{27}Al, ^{18}O, and ^{29}Si correspond, respectively, to their concentrations in Al_2O_3, $Al_2{}^{18}O_3$, and Si. Excitation curves from a standard $Si^{18}O_2$ film and from a virgin Si wafer are shown as a reference to the points on the energy scales corresponding to the nuclides at the sample surface. The arrows in the insets indicate the positions of the interfaces before rapid thermal annealing. Reprinted with permission from [267], © 2000, American Physical Society.

Fig. 93. XPS spectra showing the Al 2p and Si 2p photoelectron peak regions for different take-off angles of $Al_2O_3/SiO_2/Si$ structures submitted or not to rapid thermal annealing in $^{18}O_2$. Reprinted with permission from [267], © 2000, American Physical Society.

CVD at temperatures below 400°C [258]. Si substrates were cleaned in HF solution before deposition to remove any native oxide, and even so the as-deposited films showed a 2 nm interfacial region with stoichiometry close to SiO_2 and 7 nm of amorphous Ta_2O_5 on top. Thermal annealings at and above 800°C induced thickening of the SiO_2 layer and partial crystallization of the Ta_2O_5 layer.

Al_2O_3 films deposited by atomic layer chemical vapor deposition (ALCVD) on HF-cleaned Si substrates display atomi-

cally sharp interfaces and no detectable intermediate SiO_2 layer, as shown in Figure 91 [266]. When a 0.5 nm SiO_2 layer is intentionally grown on the surface of Si before Al_2O_3 deposition, the Al and Si profiles are as shown in Figure 92 [267]. Thermal annealing in isotopically enriched oxygen ($^{18}O_2$) leads to incorporation of ^{18}O and transport of Al and Si as shown in Figure 92. Angle-resolved X-ray photoelectron spectroscopy (ARXPS) of Si 2p electrons indicates the presence of SiO_2 in the near-interface region in the as-deposited samples, and the formation of Si-Al-O compounds in the near-surface region after thermal annealing in O_2, as shown in Figure 93.

Figure 94 illustrates the dependence of the morphology of TiO_2 films deposited by CVD on the c-Si substrate tempera-

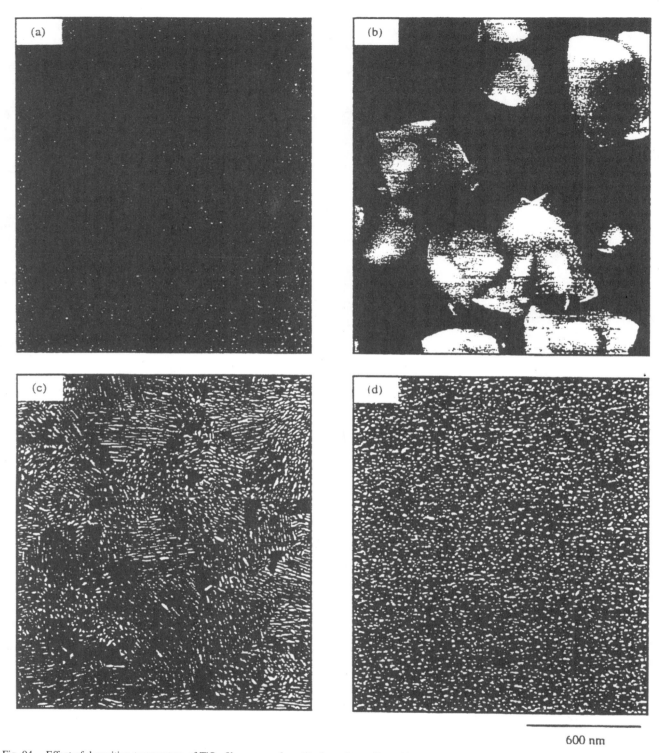

Fig. 94. Effect of deposition temperature of TiO$_2$ films grown from titanium nitrate. Deposition temperatures were (a) 170°C, (b) 210°C, (c) 340°C, and (d) 490°C. Reprinted with permission from [268], © 1999, IBM Technical Journals.

ture during deposition [268]. Although they are much thicker than the approximately 5 nm films that will be necessary for FET insulators, these thick films allow relatively easy electron microscopy of the grain structure. At the lowest substrate tem-

perature (~170°C), completely amorphous films were obtained (Fig. 94a). At 200°C, anatase crystals in an amorphous background were observed (Fig. 94b). Increasing the temperature further led to rough anatase films (Fig. 94c). At 500°C the

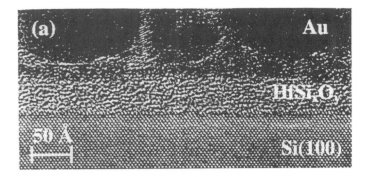

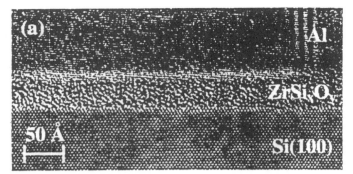

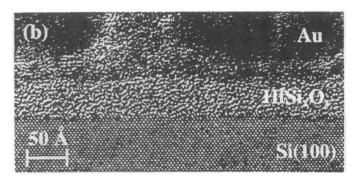

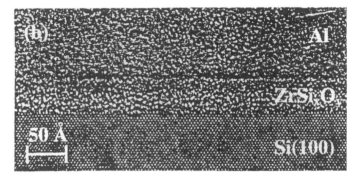

Fig. 95.   TEM cross section of a 50-Å $Hf_6Si_{29}O_{65}$ film with an Au cap (a) as deposited at 500°C and (b) after annealing at 800°C for 30 min in N_2. The silicate film remains amorphous and stable, with no sign of reaction after the anneal. Reprinted with permission from [265], © 2000, American Institute of Physics.

Fig. 96.   TEM cross section of a 50-Å $Zr_4Si_{31}O_{65}$ film with an Al cap (a) as deposited at 500°C and (b) after annealing at 800°C for 30 min in N_2. The silicate film remains amorphous and stable, but the slight contrast difference at the Al interface indicates a possible reaction. Reprinted with permission from [265], © 2000, American Institute of Physics.

films became considerably smoother (Fig. 94d), with a columnar structure and a grain width of about 10 to 20 nm. Films grown at a substrate temperature above 600°C were primarily rutile.

Finally, Figures 95 and 96 show HRTEM images of Hf and Zr silicate films, respectively, deposited on HF-cleaned Si by reactive sputtering or e-beam evaporation [265]. There are no visible intermediate layers, and the films are amorphous with sharp interfaces with the Si substrates. Hf $4f$ and Zr $3d$ XPS analyses [265] of these films are shown in Figure 97, indicating the presence of Hf—O or Zr—O bonds and no Hf—Si or Zr—Si bonds.

8.3. Remarks and Perspectives Concerning Alternative (High-k) Materials as Gate Dielectrics

A great deal remains to be learned before identification of the most attractive alternative gate dielectrics [253]. Interesting compositions have been identified, but none has yet been able to overcome all of the associated difficulties. Most certainly, this corresponds to insufficient understanding of the materials in question, the interfaces they form, and the technologies required to produce the sensitive processing equipment needed for accomplishing these goals. It seems, however, that the issues mentioned above are being addressed systematically, and the results of these fundamental investigations

represent real progress, as documented in a recent review article [269].

9. FINAL REMARKS

Very few areas of human activity have possibly experienced such an accelerated evolution as that of microelectronics. Because of silicon and its formidable oxide—not to mention the MOSFET device—originally large, massive, and expensive machines (which are nonetheless marvels of creation) have become portable and cheap to the point of giving to almost anyone on Earth a computing power never dreamed of by early generations. This astonishing evolution, which has been both pulling and pushing itself for more, is now seriously threatened.

The remarkable performance of SiO_2 in passivating Si is now limited by such diverse and central issues as leakage current, reliability with respect to dielectric breakdown, and masking capability. Although atomic-scale smoothness and deuterium incorporation into device structures may be of some help, the need for an alternative dielectric has finally become evident. Silicon oxynitrides emerged as potentially useful materials for application in gate dielectric films not very long ago. After significant efforts, much has been learned about them, to the point where they are now state of the art for high-performance applications. Fundamental improvements relative

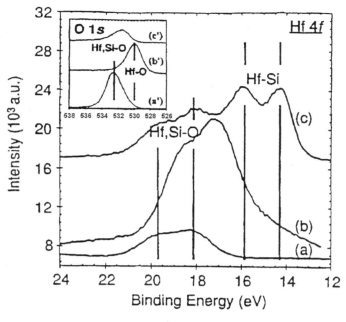

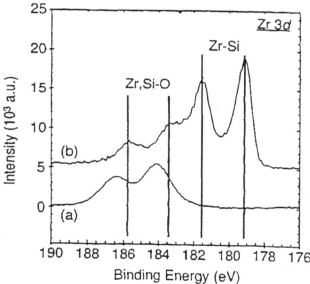

Fig. 97. (Top) Hf 4f XPS features for (a) thin Hf$_6$Si$_{29}$O$_{65}$ film, (b) thick HfO$_2$ film, and (c) thick O-rich HfSi$_2$ film. No evidence of Hf—Si bond formation is observed for the Hf-silicate film. (Inset) Oxygen 1s XPS features for (a') thin Hf$_6$Si$_{29}$O$_{65}$ film, (b') thick HfO$_2$ film, and (c') thick O-rich HfSi$_2$ film. Silicate formation (vs. metal oxide) is confirmed by the O 1s feature. (Bottom) XPS data with Zr 3d features of (a) a Zr$_5$Si$_{31}$O$_{64}$ film and (b) an O-rich ZrSi$_2$ film. No evidence of Zr—Si bond formation is observed for the Zr-silicate film. Reprinted with permission from [265], © 2000, American Institute of Physics.

to SiO$_2$, however, are marginal, and a limit to the applicability of oxynitrides has already been foreseen. One can expect them to be in use until 2005 or so. High-k materials for application as gate dielectrics constitute at present a very active research area. From the significant wealth of knowledge already available, it could be said that ZrSiO$_4$, given its thermodynamic stability, is a serious candidate as a substitute for silicon oxynitrides. It

could be followed by Ta$_2$O$_5$ and TiO$_2$, and finally by a single crystal dielectric. Among the speculations, it seems to be that an ultrathin silicon oxynitride layer will continue to provide the interface to silicon, now composing a stack with a high-k material.

Finally, it should be said that it is also possible that instead of a continued evolution the next step may be a revolution, like that of solid-state devices about 40 years ago. Revolutionary or not, gate dielectrics will surely take part in its preparation.

ACKNOWLEDGMENTS

The authors express their gratitude to their working team at UFRGS for conclusive support. Special thanks are due to Prof. F. C. Stedile, Prof. T. D. M. Salgado, Prof. J. Morais, Prof. R. M. C. da Cunha, Prof. H. I. Boudinov, Prof. J. P. de Souza, E. B. O. da Rosa, C. Radtke, R. P. Pezzi, R. Brandão, and L. Miotti.

REFERENCES

1. R. F. Pierret, "Semiconductor Device Fundamentals." Addison-Wesley, Reading, MA, 1996.
2. E. H. Nicollian and J. R. Brews, "MOS (Metal Oxide Semiconductor) Physics and Technology." Wiley, New York, 1982.
3. R. M. Warner, Jr., and B. L. Grung, "MOSFET Theory and Design." Oxford Univ. Press, New York, 1999.
4. J. W. Mayer and S. S. Lau, "Electronic Materials Science: For Integrated Circuits in Si and GaAs." Macmillan, New York, 1988.
5. R. H. Dennard, F. H. Gaensslen, H. N. Yu, V. L. Rideout, E. Barsous, and A. R. LeBlanc, *IEEE J. Solid-State Circuits* SC-9, 256 (1974).
6. Y. Taur, *IEEE Spectrum* 36, 25 (1999).
7. P. S. Peercy, *Nature (London)* 406, 1023 (2000).
8. G. E. Moore, *Electronics* 38, 114 (1965).
9. Semiconductor Industry Association, "International Technology Roadmap for Semiconductors," 1999 edn. International Sematech, Austin, TX, 1999.
10. D. A. Buchanan, *IBM J. Res. Dev.* 43, 245 (1999).
11. L. Feldman, E. P. Gusev, and E. Garfunkel, in "Fundamental Aspects of Ultrathin Dielectrics on Si-based Devices," NATO Science Series, Vol. 47 (E. Garfunkel, E. P. Gusev, and A. Vul, Eds.), pp. 1–24. Kluwer, Dordrecht, the Netherlands, 1998.
12. D. A. Buchanan and S.-H. Lo, in "The Physics and Chemistry of SiO$_2$ and the Si–SiO$_2$ Interface" (H. Z. Massoud, E. H. Poindexter, and C. R. Helms, Eds.), Vol. 3, pp. 3–14. Electrochemical Society, Pennington, NJ, 1996.
13. D. A. Buchanan and S.-H. Lo, *Mater. Res. Soc. Symp. Proc.* 567 (1999).
14. B. Balland and A. Glachant, in "Instabilities in Silicon Devices" (G. Barbottin and A. Vapaille, Eds.), Vol. 3, Chap. 1. Elsevier, Amsterdam, 1999.
15. W. Kern, Ed., "Handbook of Semiconductor Wafer Cleaning Technology." Noyes, Park Ridge, NJ, 1993.
16. W. Kern and D. A. Puotinen, *RCA Rev.* 31, 187 (1970).
17. M. Meuris, P. M. Mertens, A. Opdebeeck, H. F. Schmidt, M. Depas, M. M. Heyns, and A. Philipossian, *Solid State Technol.* 38, 109 (1995).
18. T. Hattori, T. Osaka, A. Okamoto, K. Saga, and H. Kuniyasu, *J. Electrochem. Soc.* 145, 3278 (1998).
19. T. Ohmi, *J. Electrochem. Soc.* 143, 2957 (1996).
20. M. M. Heyns, T. Bearda, I. Cornelissen, S. De Gendt, R. Degraeve, G. Groeseneken, C. Kenens, D. M. Knotter, L. M. Loewenstein, P. W. Mertens, S. Mertens, M. Meuris, T. Nigam, M. Schaekers, I. Teerlinck, W. Vandervorst, R. Vos, and K. Wolke, *IBM J. Res. Dev.* 43, 339 (1999).

21. A. F. M. Leenars, J. A. M. Huethorst, and J. J. Van Oekel, *Langmuir* 6, 1701 (1990).

22. S. Verhaverbeke, M. Meuris, P. W. Mertens, M. M. Heyns, A. Philipossian, D. Gräf, and A. Schnegg, in "International Electron Devices Meeting Technical Digest," p. 71. New York, 1991.

23. H. F. Okorn-Schmidt, *IBM J. Res. Dev.* 43, 351 (1999).

24. M. Depas, T. Nigam, K. Kenis, M. M. Heyns, H. Sprey, and R. Wilhelm, in "Proceedings of the 3rd International Symposium on Ultra Clean Processing of Silicon Surfaces" (M. Heyns, M. Meuris, and P. Mertens, Eds.), p. 291. Acco, Leuven, 1996.

25. M. Hirose, M. Hiroshima, T. Y. Asaka, and S. Miyazaki, *J. Vac. Sci. Technol., A* 12, 1864 (1994).

26. C. J. Sofield and A. M. Stoneham, *Semicond. Sci. Technol.* 10, 215 (1995).

27. R. I. Hedge, M. A. Chonko, and P. J. Tobin, *J. Vac. Sci. Technol., B* 14, 3299 (1996).

28. G. Gould and E. A. Irene, *J. Electrochem. Soc.* 134, 1031 (1987).

29. J.-J. Ganem, S. Rigo, I. Trimaille, and G.-N. Lu, *Nucl. Instrum. Methods Phys. Res., Sect. B* 64, 784 (1992).

30. P. O. Hahn, M. Grundner, A. Schnegg, and H. Jacob, *Appl. Surf. Sci.* 39, 436 (1989).

31. F. C. Stedile, I. J. R. Baumvol, I. F. Oppenheim, I. Trimaille, J.-J. Ganem, and S. Rigo, *Nucl. Instrum. Methods Phys. Res., Sect. B* 118, 493 (1996).

32. F. P. Fehlner, *J. Electrochem. Soc.* 119, 1723 (1972).

33. B. E. Deal and A. S. Grove, *J. Appl. Phys.* 36, 3770 (1965).

34. B. E. Deal, *J. Electrochem. Soc.* 127, 979 (1980).

35. A. C. Adams, T. E. Smith, and C. C. Chang, *J. Electrochem. Soc.* 127, 1787 (1980).

36. P. J. Caplan, E. H. Poindexter, B. E. Deal, and R. R. Razouk, *J. Appl. Phys.* 50, 5847 (1979).

37. E. H. Poindexter, P. J. Caplan, and R. R. Razouk, *J. Appl. Phys.* 52, 879 (1981).

38. H. Z. Massoud, J. D. Plummer, and E. A. Irene, *J. Electrochem. Soc.* 132, 1745 (1985).

39. H. Z. Massoud, J. D. Plummer, and E. A. Irene, *J. Electrochem. Soc.* 132, 2685 (1985).

40. H. Z. Massoud, J. D. Plummer, and E. A. Irene, *J. Electrochem. Soc.* 132, 2693 (1985).

41. L. N. Lie, R. R. Razouk, and B. E. Deal, *J. Electrochem. Soc.* 129, 2828 (1982).

42. E. H. Nicollian and A. Reisman, *J. Electron. Mater.* 17, 4 (1987).

43. E. A. Irene, *J. Appl. Phys.* 54, 5416 (1983).

44. H. Fukuda, T. Arakawa, and S. Ohno, *IEEE Trans. Electron Devices* 39, 127 (1992).

45. W. Ensinger, *Nucl. Instrum. Methods Phys. Res., Sect. B* 127/128, 796 (1997).

46. R. Kraft, T. P. Schneider, W. W. Dostalik, and S. Hattangady, *J. Vac. Sci. Technol., B* 15, 967 (1997).

47. J. R. Conrad, J. L. Radtke, R. A. Dodd, F. J. Worzala, and N. C. Tran, *J. Appl. Phys.* 62, 4951 (1987).

48. D. Landheer, J. A. Gupta, G. I. Sproule, J. P. McCaffrey, M. J. Graham, K.-C. Yang, Z.-H. Lu, and W. N. Lennard, *J. Electrochem. Soc.* 148, G29 (2001).

49. T. P. Ma, *IEEE Trans. Electron Devices* 45, 680 (1998).

50. T. Suntola, *Appl. Surf. Sci.* 100/101, 391 (1996).

51. R. M. Warner, Jr., and B. L. Grung, "Semiconductor-Device Electronics." Holt, Rinehart and Winston, Philadelphia, 1991.

52. S. Okhonin, A. Ils, D. Bouvet, P. Fazan, G. Guegan, S. Deleonibus, and F. Martin, *Mater. Res. Soc. Symp. Proc.* 567, 253–258 (1999).

53. C.-Y. Chang, C.-C. Chen, H.-C. Lin, M.-S. Liang, C.-H. Chien, and T.-Y. Huang, *Microelectron. Reliability* 39, 553 (1999).

54. D. J. DiMaria, *J. Appl. Phys.* 65, 2342 (1989).

55. D. A. Buchanan and D. J. DiMaria, *J. Appl. Phys.* 67, 7439 (1990).

56. E. Cartier, D. A. Buchanan, and G. J. Dunn, *Appl. Phys. Lett.* 64, 901 (1994).

57. D. A. Buchanan, D. J. DiMaria, C.-A. Chang, and Y. Taur, *Appl. Phys. Lett.* 65, 1820 (1994).

58. D. A. Buchanan, J. H. Stathis, E. Cartier, and D. J. DiMaria, *Microelectron. Eng.* 36, 329 (1997).

59. D. J. DiMaria and J. H. Stathis, *Appl. Phys. Lett.* 70, 2708 (1997).

60. E. P. Gusev, in "Defects in SiO$_2$ and Related Dielectrics: Science and Technology," NATO Science Series II: Mathematics, Physics and Chemistry, Vol. 2 (G. Pacchioni, Ed.), pp. 1–23. Kluwer, Dordrecht, the Netherlands, 2000.

61. R. Degraeve, P. Roussel, H. E. Maes, and G. Groeseneken, *Microelectron. Eng.* 36, 1651 (1996).

62. R. Degraeve, J. L. Ogier, R. Bellens, P. Rousel, G. Groeseneken, and H. E. Maes, in "Proceedings of the International Reliability Physics Symposium." IEEE, New York, 1996.

63. J. H. Stathis and D. J. DiMaria, in "International Electron Devices Meeting Technical Digest," p. 167. IEEE, New York, 1998.

64. D. A. Buchanan and S. H. Lo, *Microelectron. Eng.* 36, 13 (1997).

65. I. J. R. Baumvol, *Surf. Sci. Rep.* 36, 1 (1999).

66. T. Hattori, M. Fujimura, and H. Nohira, *Mater. Res. Soc. Symp. Proc.* 567, 163 (1999).

67. H. Winick, Ed., "Synchrotron Radiation Sources," Series on Synchrotron Radiation Techniques and Applications, Vol. 1. World Scientific, Singapore, 1994.

68. L. C. Feldman and J. W. Mayer, "Fundamentals of Thin Film and Surface Spectroscopy." North-Holland, New York, 1986.

69. R. H. Schaus and R. J. Smith, "Auger Electron Spectroscopy." Montana State Univ., Bozeman, MT, 1992.

70. Available at http://www.sopra-sa.com/ellipso.htm.

71. Available at http://www.beaglehole.com/elli_intro/elli_intro.html.

72. H. G. Tompkins and W. A. McGahan, "Spectroscopic Ellipsometry and Reflectometry: A User's Guide." Wiley, New York, 1999.

73. H. G. Tompkins, "A User's Guide to Ellipsometry." Academic Press, New York, 1993.

74. P. Hollins, *Vacuum* 45, 705 (1994).

75. Y. J. Chabal, G. S. Higashi, and S. B. Christman, *Phys. Rev. B: Solid State* 28, 4472 (1983).

76. A. Benninghoven, F. G. Rüdenauer, and H. W. Werner, "Secondary Ion Mass Spectrometry: Basic Concepts, Instrumental Aspects, Applications, and Trends." Wiley, New York, 1987.

77. P. J. Chen and R. M. Wallace, *J. Appl. Phys.* 86, 2237 (1999).

78. W. A. Grant, in "Methods of Surface Analysis" (J. M. Walls, Ed.), Chap. 9. Cambridge Univ. Press, Cambridge, U.K., 1989.

79. T. E. Jackman, J. R. MacDonald, L. C. Feldman, P. J. Silverman, and I. Stensgaard, *Surf. Sci.* 100, 35 (1980).

80. W. K. Chu, J. W. Mayer, and M.-A. Nicolet, "Backscattering Spectrometry." Academic Press, New York, 1978.

81. M. Copel, *IBM J. Res. Dev.* 44, 571 (2000).

82. E. P. Gusev, H. C. Lu, T. Gustafsson, and E. Garfunkel, *Phys. Rev. B: Solid State* 52, 1759 (1995).

83. J. F. van der Veen, *Surf. Sci. Rep.* 5, 199 (1985).

84. P. Bailey and T. Noakes, "User Manual for the Daresbury MEIS Facility," available at www.dl.ac.uk/ASD/MEIS, 1998.

85. H. C. Lu, E. P. Gusev, T. Gustafsson, E. Garfunkel, M. L. Green, D. Brasen, and L. C. Feldman, *Appl. Phys. Lett.* 69, 2713 (1996).

86. M. P. Murrell, *Vacuum* 45, 125 (1994).

87. W. M. Arnold Bik and F. H. P. M. Habraken, *Rep. Prog. Phys.* 56, 859 (1993).

88. F. H. P. M. Habraken, *Nucl. Instrum. Methods Phys. Res., Sect. B* 68, 181 (1992).

89. G. Amsel, J. P. Nadai, E. D'Artemare, D. David, E. Girard, and J. Moulin, *Nucl. Instrum. Methods* 92, 481 (1971).

90. B. Maurel, G. Amsel, and J. P. Nadai, *Nucl. Instrum. Methods* 197, 1 (1982).

91. I. J. R. Baumvol, J.-J. Ganem, L. G. Gosset, and S. Rigo, *Appl. Phys. Lett.* 72, 2999 (1998).

92. I. Vickridge and G. Amsel, *Nucl. Instrum. Methods Phys. Res., Sect. B* 45, 6 (1990).

93. J. W. Lyding, K. Hess, and I. C. Kizilyalli, *Appl. Phys. Lett.* 68, 2526 (1996).

94. J. H. Stathis, in "Fundamental Aspects of Ultrathin Dielectrics on Si-Based Devices," NATO Science Series, Vol. 47 (E. Garfunkel, E. P. Gusev, and A. Vul, Eds.), pp. 325–333. Kluwer, Dordrecht, the Netherlands, 1998.

95. C. R. Helms and E. H. Poindexter, *Rep. Prog. Phys.* 57, 791 (1994).

96. L. DoThanh and P. Balk, *J. Electrochem. Soc.* 135, 1797 (1988).

97. E. Cartier, J. H. Stathis, and D. A. Buchanan, *Appl. Phys. Lett.* 63, 1510 (1993).

98. J. H. Stathis and E. Cartier, *Phys. Rev. Lett.* 72, 2745 (1994).

99. C. R. Helms and B. E. Deal, Eds., "The Physics and Chemistry of SiO_2 and the $Si–SiO_2$ Interface," Vol. 2, Chap. VIII. Plenum, New York, 1993.

100. E. Cartier and D. J. DiMaria, *Microelectron. Eng.* 22, 207 (1993).

101. E. Cartier, D. A. Buchanan, J. H. Stathis, and D. J. DiMaria, *J. Non-Cryst. Solids* 187, 244 (1995).

102. K. Hess, I. C. Kizilyalli, and J. W. Lyding, *IEEE Trans. Electron Devices* 45, 406 (1998).

103. R. A. B. Devine, J.-L. Autran, W. L. Warren, K. L. Vanheusdan, and J.-C. Rostaing, *Appl. Phys. Lett.* 70, 2999 (1997).

104. T. G. Ference, J. S. Burnham, W. F. Clark, T. B. Hook, S. W. Mittl, K. M. Watson, and L.-K. K. Han, *IEEE Trans. Electron Devices* 46, 48 (1999).

105. H. C. Mogul, L. Cong, R. M. Wallace, P. J. Chen, T. A. Rost, and K. Harvey, *Appl. Phys. Lett.* 72, 1721 (1998).

106. C. J. Van de Walle, *J. Vac. Sci. Technol., A* 16, 1767 (1998).

107. H. Kim and H. S. Hwang, *Appl. Phys. Lett.* 74, 709 (1999).

108. S. M. Myers, *J. Appl. Phys.* 61, 5428 (1987).

109. K. Muraoka, S. Takagi, and A. Toriumi, "Extended Abstracts of the 1996 International Conference on Solid State Devices and Materials," Yokohama, 1996, p. 500.

110. I. J. R. Baumvol, E. P. Gusev, F. C. Stedile, F. L. Freire, Jr., M. L. Green, and D. Brasen, *Appl. Phys. Lett.* 72, 450 (1998).

111. J. Krauser, A. Weidinger, and Bräuning, in "The Physics and Chemistry of SiO_2 and the $Si–SiO_2$ Interface" (H. Z. Massoud, E. H. Poindexter, and C. R. Helms, Eds.), Vol. 3, pp. 184–195. Electrochemical Society, Pennington, NJ, 1996.

112. F. H. P. M. Habraken, E. H. C. Ullersma, W. M. Arnoldbik, and A. E. T. Kuiper, in "Fundamental Aspects of Ultrathin Dielectrics on Si-based Devices," NATO Science Series, Vol. 47 (E. Garfunkel, E. P. Gusev, and A. Vul, Eds.), pp. 411–424. Kluwer, Dordrecht, the Netherlands, 1998.

113. F. H. P. M. Habraken and A. E. T. Kuiper, *Mater. Sci. Eng.* R12, 123 (1994).

114. W. F. Clark, P. E. Cottrell, T. G. Ference, S.-H. Lo, J. G. Massey, S. W. Mittl, and J. H. Rankin, in "International Electron Devices Meeting Technical Digest." IEEE, New York, 1999.

115. H. Iwai, H. S. Momose, and S.-I. Ohmi, in "The Physics and Chemistry of SiO_2 and the $Si–SiO_2$ Interface" (H. Z. Massoud, I. J. R. Baumvol, M. Hirose, and E. H. Poindexter, Eds.), Vol. 4, pp. 3–17. Electrochemical Society, Pennington, NJ, 2000.

116. K. F. Schuegraf, C. Park, and D. Hu, in "International Electron Devices Meeting Technical Digest," p. 609. IEEE, New York, 1994.

117. M. Hirose, *Mater. Sci. Eng.* B41, 35 (1996).

118. H. S. Momose, M. Ono, T. Yoshitomi, T. Ohguro, S. Nakamura, M. Saito, and H. Iwai, in "International Electron Devices Meeting Technical Digest," pp. 593–596. IEEE, New York, 1994.

119. G. Timp, J. Bude, K. K. Bourdelle, J. Garno, A. Ghetti, H. Gossmann, M. Green, G. Forsyth, Y. Kim, R. Kleiman, F. Klemens, A. Komblit, C. Lochstampfor, W. Mansfield, S. Moccio, T. Sorsch, D. M. Tennant, W. Timp, and R. Tung, in "International Electron Devices Meeting Technical Digest," pp. 55–58. IEEE, New York, 1999.

120. S.-H. Lo, D. A. Buchanan, Y. Taur, L.-K. Han, and E. Wu, in "Symposium on VLSI Technology Technical Digest," pp. 149–150. IEEE, New York, 1998.

121. Y. Taur and E. J. Nowak, in "International Electron Devices Meeting Technical Digest," p. 789. IEEE, New York, 1998.

122. H. S. Momose, T. Ohguro, E. Morifuji, H. Sugaya, S. Nakamura, T. Yoshitomi, H. Kimijima, T. Morimoto, F. Matsuoka, Y. Katsumata,

H. Ishiuchi, and H. Iwai, in "International Electron Devices Meeting Technical Digest," pp. 819–822. IEEE, New York, 1999.

123. H. Z. Massoud, J. P. Shiely, and A. Shanware, *Mater. Res. Soc. Symp. Proc.* 567, 227–239 (1999).

124. J. H. Stathis and D. J. DiMaria, in "The Physics and Chemistry of SiO_2 and the $Si–SiO_2$ Interface" (H. Z. Massoud, I. J. R. Baumvol, M. Hirose, and E. H. Poindexter, Eds.), Vol. 4, pp. 33–44. Electrochemical Society, Pennington, 2000.

125. K. Okada and K. Yoneda, in "International Electron Devices Meeting Technical Digest," pp. 445–448. IEEE, New York, 1999.

126. B. E. Weir, P. J. Silverman, G. B. Alers, D. Monroe, M. A. Alam, T. W. Sorsch, M. L. Green, G. L. Timp, Y. Ma, M. Frei, C. T. Liu, J. D. Bude, and K. S. Krisch, *Mater. Res. Soc. Symp. Proc.* 567, 301–306 (1999).

127. E. Miranda, J. Suñé, R. Rodriguez, M. Nafria, and X. Aymerich, in "Proceedings of the 36th International Reliability Physics Symposium," p. 42. IEEE, New York, 1998.

128. C. C. Chen, C. Y. Chang, C. H. Chien, T. Y. Huang, H. C. Lin, and M. S. Liang, *Appl. Phys. Lett.* 74, 3708 (1999).

129. R. Degraeve, P. Roussel, H. E. Maes, and G. Groeseneken, *Microelectron. Reliability* 36, 1651 (1996).

130. J. H. Stathis, *Microelectron. Eng.* 36, 325 (1997).

131. D. J. DiMaria, *Microelectron. Eng.* 36, 317 (1997).

132. T. Nigam, R. Degraeve, G. Groeseneken, M. M. Heyns, and H. E. Maes, in "Proceedings of the 36th International Reliability Physics Symposium," p. 62. IEEE, New York, 1998.

133. E. Wu, E. Nowak, J. Aitken, W. Abadeer, L. K. Lan, and S. Lo, in "International Electron Devices Meeting Technical Digest," p. 187. IEEE, New York, 1998.

134. M. A. Alam, B. Weir, P. Silverman, J. Bude, A. Ghetti, Y. Ma, M. M. Brown, D. Hwang, and A. Hamad, in "The Physics and Chemistry of SiO_2 and the $Si–SiO_2$ Interface" (H. Z. Massoud, I. J. R. Baumvol, M. Hirose, and E. H. Poindexter, Eds.), Vol. 4, pp. 365–376. Electrochemical Society, Pennington, NJ, 2000.

135. R. Degraeve, N. Pangon, B. Kaczer, T. Nigam, G. Groeseneken, and A. Naem, in "Symposium on VLSI Technology Technical Digest," p. 59. IEEE, New York, 1999.

136. K.-Y. Peng, L.-C. Wang, and J. C. Slattery, *J. Vac. Sci. Technol., B* 14, 3316 (1996).

137. V. R. Mhetar and L. A. Archer, *J. Vac. Sci. Technol., B* 16, 2121 (1998).

138. Y.-L. Chiou, C. H. Sow, and K. Ports, *IEEE Electron Device Lett.* 10, 1 (1989).

139. L. Verdi, A. Miotello, and R. Kelly, *Thin Solid Films* 241, 383 (1994).

140. T. K. Whidden, P. Thanikasalam, M. J. Rack, and D. K. Ferry, *J. Vac. Sci. Technol., B* 13, 1618 (1995).

141. P. Thanikasalam, T. K. Whidden, and D. K. Ferry, *J. Vac. Sci. Technol., B* 14, 2840 (1996).

142. S. Dimitrijev and H. B. Harrison, *J. Appl. Phys.* 80, 2467 (1996).

143. G. F. Cerofolini, G. La Bruna, and L. Meda, *Mater. Sci. Eng., B* 36, 104 (1996).

144. R. M. C. de Almeida, S. Gonçalves, I. J. R. Baumvol, and F. C. Stedile, *Phys. Rev. B: Solid State* 61, 12992 (2000).

145. T. Ohmi, K. Matsumoto, N. Nakamura, K. Makihara, J. Takano, and K. Yamamoto, *J. Appl. Phys.* 77, 1159 (1995).

146. T. Hattori, in "The Physics and Chemistry of SiO_2 and the $Si–SiO_2$ Interface" (H. Z. Massoud, I. J. R. Baumvol, M. Hirose, and E. H. Poindexter, Eds.), Vol. 4, pp. 392–405. Electrochemical Society, Pennington, NJ, 2000.

147. A. Fargeix and G. Ghibaudo, *J. Appl. Phys.* 54, 2878 (1983).

148. F. J. Himpsel, F. R. McFeely, A. Taleb-Ibrahimi, J. A. Yarmoff, and G. Hollinger, *Phys. Rev. B: Solid State* 38, 6084 (1988).

149. E. P. Gusev, H. C. Lu, T. Gustafsson, and E. Garfunkel, *Appl. Surf. Sci.* 104/105, 329 (1996).

150. E. Hasegawa, A. Ishitani, K. Akimoto, M. Tsukiji, and N. Ohta, *J. Electrochem. Soc.* 142, 273 (1995).

151. A. Pasquarello and M. S. Hybertsen, in "The Physics and Chemistry of SiO_2 and the $Si–SiO_2$ Interface" (H. Z. Massoud, I. J. R. Baumvol,

M. Hirose, and E. H. Poindexter, Eds.), Vol. 4, pp. 271–282. Electrochemical Society, Pennington, NJ, 2000.

152. A. Stirling, A. Pasquarello, J.-C. Charlier, and R. Car, in "The Physics and Chemistry of SiO$_2$ and the Si–SiO$_2$ Interface" (H. Z. Massoud, I. J. R. Baumvol, M. Hirose, and E. H. Poindexter, Eds.), Vol. 4, pp. 283–294. Electrochemical Society, Pennington, NJ, 2000.

153. A. Markovits and C. Minot, in "Fundamental Aspects of Ultrathin Dielectrics on Si-based Devices," NATO Science Series, Vol. 47 (E. Garfunkel, E. P. Gusev, and A. Vul', Eds.), pp. 131–146. Kluwer, Dordrecht, the Netherlands, 1998.

154. K.-O. Ng and D. Vanderbilt, *Phys. Rev. B: Solid State* 59, 10132 (1999).

155. T. Yamasaki, C. Kaneta, T. Uchiyama, T. Uda, and K. Terakura, in "The Physics and Chemistry of SiO$_2$ and the Si–SiO$_2$ Interface" (H. Z. Massoud, I. J. R. Baumvol, M. Hirose, and E. H. Poindexter, Eds.), Vol. 4, pp. 295–306. Electrochemical Society, Pennington, NJ, 2000.

156. T. Watanabe and I. Ohdomari, in "The Physics and Chemistry of SiO$_2$ and the Si–SiO$_2$ Interface" (H. Z. Massoud, I. J. R. Baumvol, M. Hirose, and E. H. Poindexter, Eds.), Vol. 4, pp. 319–330. Electrochemical Society, Pennington, NJ, 2000.

157. K. Raghavachari, A. Pasquarello, J. Eng, Jr., and M. S. Hybertsen, *Appl. Phys. Lett.* 76, 3873 (2000).

158. A. M. Stoneham and C. J. Sofield, in "Fundamental Aspects of Ultrathin Dielectrics on Si-based Devices," NATO Science Series, Vol. 47 (E. Garfunkel, E. P. Gusev, and A. Vul', Eds.), pp. 79–88. Kluwer, Dordrecht, the Netherlands, 1998.

159. A. M. Stoneham, in "The Physics and Chemistry of SiO$_2$ and the Si–SiO$_2$ Interface" (C. R. Helms and B. E. Deal, Eds.), Vol. 2, pp. 3–6. Plenum, Pennington, NJ, 1993.

160. H. C. Lu, T. Gustafsson, E. P. Gusev, and E. Garfunkel, *Appl. Phys. Lett.* 67, 1742 (1995).

161. F. Rochet, S. Rigo, M. Froment, C. D'Anterroches, C. Maillot, H. Roulte, and G. Dufour, *Adv. Phys.* 35, 237 (1986).

162. C.-J. Han and C. C. R. Helms, *J. Electrochem. Soc.* 135, 1824 (1988).

163. I. Trimaille and S. Rigo, *Appl. Surf. Sci.* 39, 65 (1989).

164. J.-J. Ganem, I. Trimaille, P. André, S. Rigo, F. C. Stedile, and I. J. R. Baumvol, *J. Appl. Phys.* 81, 8109 (1997).

165. I. Trimaille, F. C. Stedile, J.-J. Ganem, I. J. R. Baumvol, and S. Rigo, in "The Physics and Chemistry of SiO$_2$ and the Si–SiO$_2$ Interface" (H. Z. Massoud, E. H. Poindexter, and C. R. Helms, Eds.), Vol. 3, p. 59. Electrochemical Society, Pennington, NJ, 1996.

166. A. M. Stoneham, C. R. M. Grovenor, and A. Cerezzo, *Philos. Mag. B* 55, 201 (1987).

167. N. F. Mott, S. Rigo, F. Rochet, and A. M. Stoneham, *Philos. Mag.* 60, 189 (1989).

168. R. Pretorius, W. Strydom, J. W. Mayer, and C. Comrie, *Phys. Rev. B: Solid State* 22, 1885 (1980).

169. I. J. R. Baumvol, C. Krug, F. C. Stedile, F. Gorris, and W. Schulte, *Phys. Rev. B: Solid State* 60, 1492 (1999).

170. A. Stesman and F. Scheerlinck, *Phys. Rev. B: Solid State* 50, 5204 (1994).

171. A. Stesman and F. Scheerlinck, *J. Appl. Phys.* 75, 1047 (1994).

172. T. Åkermark, *J. Electrochem. Soc.* 147, 1882 (2000).

173. T. Åkermark and G. Hultquist, *J. Electrochem. Soc.* 144, 1456 (1997).

174. T. Åkermark, *Oxid. Met.* 50, 167 (1998).

175. T. Åkermark, L. G. Gosset, J.-J. Ganem, I. Timaille, and S. Rigo, *J. Electrochem. Soc.* 146, 3389 (1999).

176. D. R. Hamann, *Phys. Rev. Lett.* 81, 3447 (1998).

177. E. P. Gusev, H. C. Lu, E. L. Garfunkel, T. Gustafsson, and M. L. Green, *IBM J. Res. Dev.* 43, 265 (1999).

178. D. M. Brown, P. V. Gray, F. K. Heumann, H. R. Philipp, and E. A. Taft, *J. Electrochem. Soc.* 115, 311 (1968).

179. G. J. Dunn, *Appl. Phys. Lett.* 53, 1650 (1988).

180. M. Dutoit, P. Letourneau, N. N. J. Mi, J. Manthey, and J. S. d. Zaldivar, *J. Electrochem. Soc.* 41, 549 (1993).

181. S. S. Krisch and C. G. Sodini, *J. Appl. Phys.* 76, 2284 (1994).

182. H. Hwang, W. Ting, B. Maiti, D. L. Kwong, and J. Lee, *Appl. Phys. Lett.* 57, 1010 (1990).

183. H. Fukuda, T. Arakawa, and S. Ohno, *Japan. J. Appl. Phys.* 29, L2333 (1990).

184. E. C. Carr and R. A. Buhrman, *Appl. Phys. Lett.* 63, 54 (1993).

185. H. T. Tang, W. N. Lennard, M. Zinke-Allmang, I. V. Mitchell, L. C. Feldman, M. L. Green, and D. Brasen, *Appl. Phys. Lett.* 64, 64 (1994).

186. M. Bhat, L. K. Han, D. Wristers, J. Yan, D. L. Kwong, and J. Fulford, *Appl. Phys. Lett.* 66, 1225 (1995).

187. Z. Q. Yao, H. B. Harrison, S. Dimitrijev, and Y. T. Yeow, *IEEE Electron Device Lett.* 16, 345 (1995).

188. M. L. Green, D. Brasen, L. C. Feldman, W. Lennard, and H. T. Tang, *Appl. Phys. Lett.* 67, 1600 (1995).

189. M. L. Green, D. Brasen, K. W. Evans-Lutterodt, L. C. Feldman, K. Krisch, W. Lennard, H. T. Tang, L. Manchanda, and M. T. Tang, *Appl. Phys. Lett.* 64, 848 (1994).

190. R. I. Hedge, P. J. Tobin, K. G. Reid, B. Maiti, and S. A. Ajuria, *Appl. Phys. Lett.* 66, 2882 (1995).

191. Z. H. Lu, S. P. Tay, R. Cao, and P. Pianetta, *Appl. Phys. Lett.* 67, 2836 (1995).

192. D. G. J. Sutherland, H. Akatsu, M. Copel, F. J. Himpsel, T. Callcott, J. A. Carlisle, D. Ederer, J. J. Jia, I. Jimenez, R. Perera, D. K. Shuh, L. J. Terminello, and W. M. Tong, *J. Appl. Phys.* 78, 6761 (1995).

193. Z. Q. Yao, *J. Appl. Phys.* 78, 2906 (1995).

194. J. J. Ganem, S. Rigo, I. Trimaille, I. J. R. Baumvol, and F. C. Stedile, *Appl. Phys. Lett.* 68, 2366 (1996).

195. M. Copel, R. M. Tromp, H. J. Timme, K. Penner, and T. Nakao, *J. Vac. Sci. Technol. A* 14, 462 (1996).

196. D. Landheer, Y. Tao, D. X. Xu, G. I. Sproule, and D. A. Buchanan, *J. Appl. Phys.* 78, 1818 (1995).

197. S. V. Hattangady, H. Niimi, and G. Lucovsky, *Appl. Phys. Lett.* 66, 3495 (1995).

198. W. L. Hill, E. M. Vogel, V. Misra, P. K. McLarty, and J. J. Wortman, *IEEE Trans. Electron Devices* 43, 15 (1996).

199. G. Lucovsky, in "Fundamental Aspects of Ultrathin Dielectrics on Si-based Devices," NATO Science Series, Vol. 47 (E. Garfunkel, E. P. Gusev, and A. Vul', Eds.), pp. 147–164. Kluwer, Dordrecht, the Netherlands, 1998.

200. F. H. P. M. Habraken and A. E. T. Kuiper, *Mater. Sci. Eng. Rep.* R12, 123 (1994).

201. S. V. Hattangady, H. Niimi, and G. Lucovsky, *J. Vac. Sci. Technol., A* 14, 3017 (1996).

202. M. L. Green, D. Brasen, L. Feldmand, E. Garfunkel, E. P. Gusev, T. Gustafsson, W. N. Lennard, H. C. Lu, and T. Sorsch, in "Fundamental Aspects of Ultrathin Dielectrics on Si-based Devices," NATO Science Series, Vol. 47 (E. Garfunkel, E. P. Gusev, and A. Vul', Eds.), pp. 181–190. Kluwer, Dordrecht, the Netherlands, 1998.

203. I. J. R. Baumvol, F. C. Stedile, J. J. Ganem, I. Trimaille, and S. Rigo, *Appl. Phys. Lett.* 70, 2007 (1997).

204. D. J. DiMaria and J. R. Abernathey, *J. Appl. Phys.* 60, 1727 (1986).

205. S. C. Song, C. H. Lee, H. F. Luan, D. L. Kwong, M. Gardner, J. Fulford, M. Allen, J. Bloom, and R. Evans, *Mater. Res. Soc. Symp. Proc.* 567, 65–70 (1999).

206. Y. Ono, Y. Ma, and S.-T. Hsu, *Mater. Res. Soc. Symp. Proc.* 567, 39–44 (1999).

207. Y. Wu and G. Lucovsky, *Mater. Res. Soc. Symp. Proc.* 567, 101–106 (1999).

208. T. B. Hook, J. S. Burnham, and R. J. Bolam, *IBM J. Res. Dev.* 43, 393 (1999).

209. A. E. T. Kuiper, M. F. C. Willemsen, A. M. S. Theunissen, W. M. van de Wijgert, F. H. P. M. Habraken, R. H. G. Tijhaar, W. F. van der Weg, and J. T. Chen, *J. Appl. Phys.* 59, 2765 (1982).

210. K. Yamamoto and M. Nakazawa, *Japan. J. Appl. Phys.* 33, 285 (1994).

211. A. Serrari, J. L. Chartier, R. Le Bihan, S. Rigo, and J. C. Dupuy, *Appl. Surf. Sci.* 51, 133 (1991).

212. T. Ito, T. Nozaki, and H. Ishikawa, *J. Electrochem. Soc.* 127, 2053 (1980).

213. Y. Hayafuji and K. Kajiwara, *J. Electrochem. Soc.* 129, 2102 (1982).

214. S. S. Wong, S. H. Kwan, H. R. Grinolds, and W. G. Oldham, in "Proceedings of the Symposium on Silicon Nitride Thin Insulating Films"

(V. J. Kapoor and H. J. Stein, Eds.), Vol. 83-8, pp. 346–354. Electrochemical Society, Pennington, NJ, 1983.

215. G. Lucovsky, *IBM J. Res. Dev.* 43, 301 (1999).

216. K. A. Ellis and R. A. Buhrmann, *IBM J. Res. Dev.* 43, 287 (1999).

217. E. P. Gusev, H. C. Lu, E. Garfunkel, T. Gustafsson, M. L. Green, D. Brasen, and W. N. Lennard, *J. Appl. Phys.* 84, 2980 (1998).

218. K. A. Ellis and R. A. Buhrman, *Appl. Phys. Lett.* 68, 1696 (1996).

219. P. J. Tobin, Y. Okada, S. A. Ajuria, V. Lakjhotia, W. a. Feil, and Hedge, *J. Appl. Phys.* 75, 1811 (1994).

220. R. I. Hedge, B. Maiti, and P. J. Tobin, *J. Electrochem. Soc.* 144, 1081 (1997).

221. Y. Okada, P. J. Tobin, and V. Lakhotia, *J. Electrochem. Soc.* 140, L87 (1993).

222. Y. Okada, P. J. Tobin, and S. A. Ajuria, *IEEE Trans. Electron Devices* 41, 1608 (1994).

223. E. C. Carr, K. A. Ellis, and R. A. Buhrman, *Appl. Phys. Lett.* 66, 1492 (1995).

224. K. A. Ellis and R. A. Buhrman, *Appl. Phys. Lett.* 70, 545 (1997).

225. K. A. Ellis and R. A. Buhrman, *J. Electrochem. Soc.* 145, 2068 (1997).

226. E. Garfunkel, E. P. Gusev, H. C. Lu, T. Gustafsson, and M. L. Green, in "Fundamental Aspects of Ultrathin Dielectrics on Si-based Devices," NATO Science Series, Vol. 47 (E. Garfunkel, E. P. Gusev, and A. Vul', Eds.), pp. 39–48. Kluwer, Dordrecht, the Netherlands, 1998.

227. Z. H. Lu, R. J. Hussey, M. J. Graham, R. Cao, and S. P. Tay, *J. Vac. Sci. Technol., B* 14, 2882 (1996).

228. H. C. Lu, E. P. Gusev, T. Gustafsson, and E. Garfunkel, *J. Appl. Phys.* 81, 6992 (1997).

229. H. C. Lu, E. P. Gusev, T. Gustafsson, M. L. Green, D. Brasen, and E. Garfunkel, *Microelectron. Eng.* 36, 29 (1997).

230. E. P. Gusev, H. C. Lu, T. Gustafsson, E. Garfunkel, M. L. Green, and D. Brasen, *J. Appl. Phys.* 82, 896 (1997).

231. M. J. Hartig and P. J. Tobin, *J. Electrochem. Soc.* 143, 1753 (1996).

232. S. W. Novak, J. R. Shallenberger, D. A. Cole, and J. W. Marino, *Mater. Res. Soc. Symp. Proc.* 567, 579–586 (1999).

233. M. L. Green, in "Advances in Rapid Thermal and Integrated Processing" (F. Roozeboom, Ed.). Kluwer, Dordrecht, the Netherlands, 1995.

234. W. DeCoster, B. Brijs, H. Bender, J. Alay, and W. Vandervorst, *Vacuum* 45, 389 (1994).

235. O. C. Hellman, N. Herbots, and O. Vancauwenberghe, *Nucl. Instrum. Methods Phys. Res., Sect. B* 67, 301 (1992).

236. R. Hezel and N. Lieske, *J. Electrochem. Soc.* 129, 379 (1982).

237. H. Kobayashi, T. Mizokuro, Y. Nakato, K. Yoneda, and Y. Todokoro, *Appl. Phys. Lett.* 71, 1978 (1997).

238. J. S. Pan, A. T. S. Wee, C. H. A. Huan, H. S. Tan, and K. L. Tan, *Appl. Surf. Sci.* 115, 166 (1997).

239. K. H. Park, B. C. Kim, and H. Kang, *Surf. Sci.* 283, 73 (1993).

240. Z.-M. Ren, Z.-F. Ying, X.-X. Xiong, M.-Q. He, F.-M. Li, Y.-C. Du, and L.-Y. Cheng, *Appl. Phys. A* 58, 395 (1994).

241. A. G. Schrott and S. C. Fain, *Surf. Sci.* 111, 39 (1981).

242. A. G. Schrott, Q. X. Su, and S. C. Fain, *Surf. Sci.* 123, 223 (1982).

243. J. A. Taylor, G. M. Lancaster, A. Ignatiev, and J. W. Rabalais, *J. Chem. Phys.* 68, 1776 (1978).

244. Y. Saito and U. Mori, *Mater. Res. Soc. Symp. Proc.* 567, 33–37 (1999).

245. S. K. Kurinec, M. A. Jackson, K. C. Capasso, K. Zhuang, and G. Braunstein, *Mater. Res. Soc. Symp. Proc.* 567, 265–270 (1999).

246. I. J. R. Baumvol, C. Krug, F. C. Stedile, M. L. Green, D. C. Jacobson, D. Eaglesham, J. D. Bernstein, J. Shao, A. S. Denholm, and P. L. Kellerman, *Appl. Phys. Lett.* 74, 807 (1999).

247. I. J. R. Baumvol, T. D. M. Salgado, C. Radtke, C. Krug, and F. C. Stedile, *J. Appl. Phys.* 83, 5579 (1998).

248. G. Lucovsky, A. Banerjee, B. Hinds, B. Claflin, K. Koh, and H. Yang, *Microelectron. Eng.* 36, 207 (1997).

249. M. Hillert, S. Jonsson, and B. Sundman, *Z. Metallkd.* 83, 648 (1992).

250. H. Du, R. E. Tressler, and K. E. Spear, *J. Electrochem. Soc.* 136, 3210 (1989).

251. T. P. Ma, *Appl. Surf. Sci.* 117/118, 259 (1997).

252. H. Goto, K. Shibahara, and S. Yokoyama, *Appl. Phys. Lett.* 68, 3257 (1996).

253. A. I. Kingon, J.-P. Maria, and S. K. Streiffer, *Nature (London)* 406, 1032 (2000).

254. M. C. Gilmer, T.-Y. Luo, H. R. Huff, M. D. Jackson, S. Kim, G. Bersuker, P. Zeitzoff, L. Vishnubhotla, G. A. Brown, R. Amos, D. Brady, V. H. C. Watt, G. Gale, J. Guan, B. Nguyen, G. Williamson, P. Lysaght, D. Torres, F. Geyling, C. F. H. Gordran, J. A. Fair, M. T. Schulberg, and T. Tamagawa, *Mater. Res. Soc. Symp. Proc.* 567, 323–341 (1999).

255. G. Timp, K. K. Bourdelle, J. E. Bower, F. H. Baumann, T. Boone, R. Cirelli, K. Evans-Lutterodt, J. Garno, A. Ghetti, H. Gossmann, M. L. Green, D. Jacobson, S. Moccio, D. A. Muller, L. E. Ocola, M. L. O'Malley, J. Rosamilia, J. Sapjeta, P. J. Silverman, T. Sorsch, D. M. Tennant, W. Timp, and B. E. Weir, in "International Electron Devices Meeting Technical Digest," p. 615. IEEE, New York, 1998.

256. C. Chaneliere, J. L. Autran, R. A. B. Devine, and B. Balland, *Mater. Sci. Eng.* R22, 269 (1998).

257. I. Kim, S.-D. Ahn, B.-W. Cho, S.-T. Ahn, J. Y. Lee, J. S. Chun, and W.-J. Lee, *Japan. J. Appl. Phys.* 33, 6691 (1994).

258. G. B. Alers, D. J. Werder, Y. Chabal, H. C. Lu, E. P. Gusev, E. Garfunkel, T. Gustafsson, and R. S. Urdahl, *Appl. Phys. Lett.* 73, 1517 (1998).

259. J.-L. Autran, R. Devine, C. Chaneliere, and B. Balland, *IEEE Electron Device Lett.* 18, 447 (1997).

260. K. A. Son, A. Y. Mao, Y. M. Sun, B. Y. Kin, F. Liu, and R. N. Vrities, *Appl. Phys. Lett.* 72, 1187 (1998).

261. G. B. Alers, R. M. Fleming, Y. H. Wong, B. Dennis, A. Pinczuk, G. Redinbo, R. Urdahl, E. Ong, and Z. Hasan, *Appl. Phys. Lett.* 72, 1308 (1998).

262. S. A. Campbell, D. C. Gilmer, X. Wang, M. Hsieh, H.-S. Kim, W. L. Gladfelter, and X. Yan, *IEEE Trans. Electron Devices* 44, 104 (1997).

263. T. P. Ma, in "The Physics and Chemistry of SiO$_2$ and the Si–SiO$_2$ Interface" (H. Z. Massoud, I. J. R. Baumvol, M. Hirose, and E. H. Poindexter, Eds.), Vol. 4, pp. 19–32. Electrochemical Society, Pennington, NJ, 2000.

264. H. Niimi, R. S. Johnson, G. Lucovsky, and H. Z. Massoud, in "The Physics and Chemistry of SiO$_2$ and the Si–SiO$_2$ Interface" (H. Z. Massoud, I. J. R. Baumvol, M. Hirose, and E. H. Poindexter, Eds.), Vol. 4, pp. 487–494. Electrochemical Society, Pennington, NJ, 2000.

265. G. D. Wilk, R. M. Wallace, and J. M. Anthony, *J. Appl. Phys.* 87, 484 (2000).

266. E. P. Gusev, M. Copel, E. Cartier, I. J. R. Baumvol, C. Krug, and M. A. Gribelyuk, *Appl. Phys. Lett.* 76, 176 (2000).

267. C. Krug, E. B. O. da Rosa, R. M. C. de Almeida, J. Morais, I. J. R. Baumvol, T. D. M. Salgado, and F. C. Stedile, *Phys. Rev. Lett.* 85, 4120 (2000).

268. S. A. Campbell, S.-H. Kim, D. C. Gilmer, B. He, T. Ma, and W. L. Gladfelter, *IBM J. Res. Dev.* 43, 383 (1999).

269. G. D. Wilk, R. M. Wallace, and J. M. Anthony, *J. Appl. Phys.* 89, 5243 (2001).

Chapter 4

PIEZOELECTRIC THIN FILMS: PROCESSING AND PROPERTIES

Floriana Craciun, Patrizio Verardi

Istituto di Acustica "O.M. Corbino," Consiglio Nazionale delle Ricerche, 00133 Rome, Italy

Maria Dinescu

National Institute for Lasers, Plasma and Radiaton Physics, Institute of Atomic Physics, Bucharest, Romania

Contents

1. INTRODUCTION

The direct piezoelectric effect, discovered by Pierre and Jacques Curie in 1880, consists of the generation of a macroscopic polarization in certain dielectric materials when subjected to stress. Vice versa, the application of an external electric field produces in these materials a strain through the converse effect. For small deformations, both effects are linear, unlike the electrostrictive effect where the obtained strain is proportional to the square of the amplitude of the applied electric field.

Not all materials can be piezoelectric. In crystals, symmetry properties are crucial in determining if and how a material is piezoelectric [1]. For example, a crystal which possesses inversion symmetry (centrosymmetric) cannot be piezoelectric because in this case its polarization, which must change sign when the axes are inverted, should be necessarily zero. For a given symmetry class the piezoelectricity is determined by the balance between two contributions, ionic and electronic, whose microscopic origin will be described later.

Piezoelectric materials can be classified according to different criteria. Usually they are divided into 70 polar piezoelectric materials (which possess a net dipole moment) and nonpolar piezoelectric materials (whose dipole moments summed on the different directions give a null total moment). Polar materials

Handbook of Thin Film Materials, edited by H.S. Nalwa
Volume 3: Ferroelectric and Dielectric Thin Films

ISBN 0-12-512911-4/$35.00

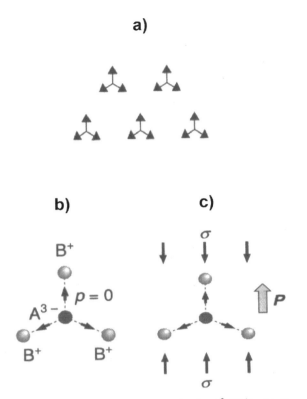

Fig. 1. Sketch of a crystal constituted by molecules $A^{3-}(B^+)_3$ (a). The conformation of each molecule is shown in the absence (b) and in the presence (c) of an applied external stress σ. In the first case (b) the total dipole electric moment is null; in the second case (c) the stress produces a finite macroscopic polarization.

are also called pyroelectric because they show production of electric charge as a consequence of uniform heating. Among the 20 noncentrosymmetric piezoelectric classes, 10 have a spontaneous polarization. Furthermore polar materials can be classified as ferroelectrics [2] (if their dipole moment can be switched by a finite electric field) and nonferroelectrics. From the applications point of view the most important piezoelectrics belong to the ferroelectric class.

Even though piezoelectric phenomena had been well known since the end of the nineteenth century, a correct microscopic approach was attempted much later [3, 4], when it was demonstrated that piezoelectricity was a bulk property and did not depend on surface effects [3], as had been previously believed. To roughly understand the bulk mechanisms which are responsible for the piezoelectric effect, let us first consider the case of a purely ionic system, where the electrons are rigidly bonded to the ions. It is easy to understand how the application of a uniaxial stress can produce a net dipolar electric moment, even if, in the absence of stress, the moment is zero due to symmetry. Figure 1 shows such an example where every unit cell contains a planar group of ions formed by an A^{3-} ion surrounded by three B^+ ions (a). In the absence of an applied stress, there is a symmetry axis for rotations of $\pm 2\pi/3$ and the dipole moments associated with every bond (indicated by arrows) give

a null total moment (b). When an external stress σ is applied along one of the bonds, the result of the sum of the dipole moments is no longer zero, but it gives a net polarization $\mathbf{P}$ in the indicated direction (c). In this model with rigid ions the polarization is therefore simply obtained as a sum of the ion charges multiplied by their displacement. Even if this description is very simplified, it explains qualitatively the behavior of certain ionic dielectrics. It has been demonstrated that, together with the purely ionic contribution, the polarization contains also an electronic contribution, due to the fact that the electronic shell is not rigid but can be polarized. In semiconductors the electronic contribution is nearly equal in absolute value to the ionic contribution and tends to cancel it, so that the piezoelectric constants of semiconductors are much smaller than those of ionic dielectrics [3].

In ferroelectrics the piezoelectric effect is simply produced by a variation of the intrinsic net dipole moment as a consequence of an applied stress. The large piezoelectric response in these materials is mainly due to the large relative displacement of cationic and anionic sublattices induced by the macroscopic strain [4].

Another type of approach to piezoelectric phenomena in crystalline materials is the macroscopic description [1]. This is based on the series development of thermodynamic potential in the strain tensor S and electric field E (or other pair of thermodynamic variables), up to the quadratic term. From this expression the stress σ and dielectric induction D (or other pair of variables) can be obtained as the derivatives of the thermodynamic potential. The choice of independent variables leads to four types of such equations:

$$D_l = e_{l\mu}S_\mu + \varepsilon_{lm}^S E_m \qquad \sigma_\lambda = c_{\lambda\mu}^E S_\mu - e_{m\lambda}E_m \qquad (1)$$

$$D_l = d_{l\mu}\sigma_\mu + \varepsilon_{lm}^T E_m \qquad S_\lambda = s_{\lambda\mu}^E \sigma_\mu + d_{m\lambda}E_m \qquad (2)$$

$$E_l = -h_{l\mu}S_\mu + \beta_{lm}^S D_m \qquad \sigma_\lambda = c_{\lambda\mu}^D S_\mu - h_{m\lambda}D_m \qquad (3)$$

$$E_l = -g_{l\mu}S_\mu + \beta_{lm}^T D_m \qquad S_\lambda = s_{\lambda\mu}^D \sigma_\mu + g_{m\lambda}D_m \qquad (4)$$

Summation over indices appearing twice in any product is understood. The symbols occurring in these equations are defined in Table I. These equations define the *piezoelectric constant tensor* d (or, briefly, the piezoelectric tensor) as the strain produced in a medium under constant stress by an electric field, or as the dielectric induction produced by a strain in conditions of constant electric field. The term piezoelectric *strain* coefficients for the $d_{l\mu}$, $g_{l\mu}$, and *stress* coefficients for the $e_{l\mu}$, $h_{l\mu}$, indicate whether strain or stress is the dependent variable in the converse effect. Summation over the subscripts μ and λ extends from 1 to 6, and over the subscripts m and l from 1 to 3. In the absence of any symmetry, each pair of equations (1)–(4) represents a set of 9 linear equations and the coefficients form a 9×9 symmetric matrix containing 21 elastic, 6 dielectric, and 18 piezoelectric coefficients. The symmetry properties of the crystal limit the number of independent components of these tensors and the classes where the piezoelectric effect is possible: for instance, in the centrosymmetric classes all the components of

Table I. Definition of the Main Tensors Involved in the Piezoelectric Effect

Name	Symbol	Definition	IS units
Stress component	σ	Force/area	N/m^2
Strain component	S_μ	Displacement/distance	Numeric
Electric field component	E_l	Electric potential difference/distance	V/m
Electric displacement component	D_l	Electric flux/area	C/m^2
Elastic stiffness constant	$c_{\lambda\mu}$	$\partial T_\lambda / \partial S_\mu$	N/m^2
Elastic compliance constant	$s_{\lambda\mu}$	$\partial S_\lambda / \partial T_\mu$	m^2/N
Dielectric permittivity	ε_{lm}	$\partial D_l / \partial E_m$	F/m
Dielectric constant	K_{lm}	$\varepsilon_{lm}/\varepsilon_0$ (ε_0 is vacuum permittivity)	Numeric
Dielectric impermeability	β_{lm}	$\partial E_l / \partial D_m$	m/F
Piezoelectric strain coefficients	$d_{l\mu}$	$(\partial D_l / \partial T_\mu)_E = (\partial S_\mu / \partial E_l)_T$	C/N = m/V
Piezoelectric strain coefficients	$g_{l\mu}$	$-(\partial E_l / \partial T_\mu)_D = (\partial S_\mu / \partial D_l)_T$	$(V \cdot m)/N = m^2/C$
Piezoelectric stress coefficients	$e_{l\mu}$	$(\partial D_l / \partial S_\mu)_E = (\partial T_\mu / \partial E_l)_S$	$C/m^2 = N/(V \cdot m)$
Piezoelectric stress coefficients	$h_{l\mu}$	$-(\partial E_l / \partial S_\mu)_D = -(\partial T_\mu / \partial D_l)_S$	V/m = N/C

the piezoelectric tensor must be zero, because the inversion operation $a_{ij} = -\delta_{ij}$ applied to the transformation $d'_{ijk} = a_{il}a_{jm}a_{km}d_{lmn}$ (where the uncontracted notation has been used) yields $d'_{ijk} = -d_{ijk}$, whereas $d'_{ijk} = d_{ijk}$, which is possible only if $d_{ijk} = 0$.

The constants occurring in Eqs. (1)–(4) are interrelated by the following equations [5]:

$$e_{m\lambda} = d_{m\tau}c_{\tau\lambda}^E \qquad d_{m\lambda} = e_{m\tau}s_{\tau\lambda}^E \qquad (5)$$

$$h_{m\lambda} = g_{m\tau}c_{\tau\lambda}^D \qquad g_{m\lambda} = h_{m\tau}s_{\tau\lambda}^D \qquad (6)$$

$$e_{m\lambda} = \varepsilon_{tm}^S h_{t\lambda} \qquad d_{m\lambda} = \varepsilon_{tm}^T g_{t\lambda} \qquad (7)$$

$$h_{m\lambda} = \beta_{tm}^S e_{t\lambda} \qquad g_{m\lambda} = \beta_{tm}^T d_{t\lambda} \qquad (8)$$

$$c_{\lambda\mu}^D - c_{\lambda\mu}^E = h_{t\lambda}e_{t\mu} \qquad s_{\lambda\mu}^D - s_{\lambda\mu}^E = -g_{t\lambda}d_{t\mu} \qquad (9)$$

$$\beta_{lm}^S - \beta_{lm}^T = h_{l\tau}g_{m\tau} = h_{m\tau}g_{l\tau}$$
$$\varepsilon_{lm}^S - \varepsilon_{lm}^T = -e_{l\tau}d_{m\tau} = -e_{m\tau}d_{l\tau} \qquad (10)$$

For the sake of convenience other dependent coefficients called *electromechanical coupling factors* are used to characterize the conversion of energy between electrical and elastic forms. The square of the coupling factor k measures the fraction of stored energy in the conversion process and is defined by the following relationships [5]:

$$k_{m\lambda}^2 = \frac{d_{m\lambda}^2}{\varepsilon_{mm}^T s_{\lambda\lambda}^E} = \frac{h_{m\lambda}^2}{\beta_{mm}^S c_{\lambda\lambda}^E}$$
$$= \frac{\varepsilon_{mm}^T - \varepsilon_{mm}^S}{\varepsilon_{mm}^T} = \frac{s_{\lambda\lambda}^E - s_{\lambda\lambda}^D}{s_{\lambda\lambda}^E} \qquad (11)$$

$$\frac{k_{m\lambda}^2}{1 - k_{m\lambda}^2} = \frac{e_{m\lambda}^2}{\varepsilon_{mm}^S c_{\lambda\lambda}^E} = \frac{g_{m\lambda}^2}{\beta_{mm}^T s_{\lambda\lambda}^D} \qquad (12)$$

It must be mentioned that coupling factors are not the components of a tensor.

Piezoelectric materials are of great importance in applications such as transducers for the electrical–mechanical signal conversion used for the generation and reception of acoustic waves, transducers for microdisplacements, and electrooptic devices [2, 6]. They have important applications in echography and in nondestructive testing as well as in scanning tunneling microscopes (STM) and devices based on surface acoustic waves (SAW).

Thin piezoelectric films are important for applications such as microsensors, micromotors, and integrated electrooptic devices [7–9]. Their development has been stimulated by the need for piezoelectric transducers of very high frequency [7], which cannot normally employ bulk materials, because frequencies of about 1 GHz would normally require a layer of a few microns which cannot be obtained other than by thin film deposition techniques.

The first deposited piezoelectric films, such as CdS, ZnS, and ZnO, were simple binary compositions, obtained by evaporation techniques. For a historical review on this subject we send the reader to [10]. The introduction of other techniques such as sputtering, pulsed laser deposition (PLD), molecular beam epitaxy (MBE), chemical vapor deposition (CVD), and the sol–gel method allowed an increase in complexity of the deposited films to materials such as $LiNbO_3$, $LiTaO_3$, $PbTiO_3$, and $Pb(Zr,Ti)O_3$ and allowed high-quality AlN, GaN, and ZnO thin films to be obtained.

In this chapter we present an overview of the various piezoelectric thin films together with the growth techniques which allowed them to be obtained and the characterization methods applied to measure their properties. Section 2 discusses finite-size effects on these properties. Section 3 gives a review of the main deposition techniques together with a discussion of some significant growth aspects. Section 4 is devoted to the main characterization techniques used to obtain the physical properties. Section 5 is concerned with the physical properties of the piezoelectric thin films and the most important parameters on which they depend.

2. PIEZOELECTRICITY IN THIN FILMS: SIZE EFFECTS

It has been shown that material properties change when the particle size is decreased to the nanometer range [11]. These changes can be attributed to quantum size effects, surface (interface) effects, variations in the cell parameters and lattice symmetry, etc. [12]. The equilibrium lattice parameters and crystal stability are controlled by short- and long-range forces. Whereas short-range repulsive forces are given by near-neighbor interactions and hence the influence of crystal size on them is negligible, long-range forces and cooperative phenomena are nevertheless affected by size effects. In some oxides an anisotropic expansion of the unit cell was observed as the size was decreased; in these materials the lattice becomes more symmetric. The size-induced lattice distortions lead to important changes in many physical properties and tend to destroy the cooperative long-range order. Therefore large deviations in critical temperature, order parameter, and related properties with respect to the corresponding bulk, when the size is decreased to a few tens of nanometers, have been generally registered in all nanosized systems and in nanocrystalline ferroelectrics, in particular. The understanding of finite size effects in ferroelectric piezoelectrics is essential because they limit the miniaturization of devices based on these materials. In a study of ferroelectric phase transitions in nanocrystalline $PbTiO_3$ Chattopadhyay et al. [13] reported a decrease in critical temperature as well as a broadening of the dielectric peak with decreasing particle size (the similarity with relaxor ferroelectric behavior is evident and can be attributed to the same basic phenomenon, that is, size dependence of ferroelectric state). The critical size below which the ferroelectric state becomes unstable was determined to be 10.7 nm by Raman measurements [14]. Finite size effects have been also numerically evaluated in $BaTiO_3$, $PbTiO_3$, and $Pb(Zr_{0.5}Ti_{0.5})O_3$ [15]. Surface or interface effects should strongly affect polarization stability in nanosized ferroelectrics because the translational invariance of the system is broken on the surface and the local symmetry and behavior of soft modes is changed. The nonpolarized surface layers tend to disturb the normal correlation of dipoles and influence the stability of polar phase. Moreover the correlation forces of dipoles in ferroelectrics are anisotropic; the parallel alignment of dipoles along the polar axis is due to a strong long-range interaction and leads to the formation of long dipole chains with a transverse optic phonon character. In directions normal to the polar axis the interaction force which is responsible for their alignment is weaker and has a short-range character. Therefore, an anisotropic behavior with size reduction is expected. In [15] mean-field theory was used to examine ferroelectric phase instability in perovskite ferroelectrics with consideration of anisotropy. A critical thickness of 4 nm for $PbTiO_3$ cell and 8 nm for $Pb(Zr_{0.5}Ti_{0.5})O_3$ cell was predicted. Calculations based on other models (Landau phenomenological theory) found size-driven phase transitions in ferroelectric nanoparticles [16]. Many experimental results have confirmed

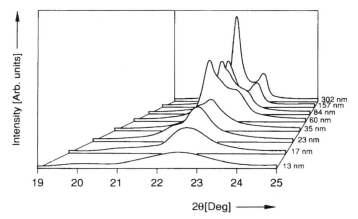

Fig. 2. X-ray diffraction patterns of the (100) and (001) reflections of $PbTiO_3$ films with different thicknesses. Reprinted with permission from [21], © 1991, American Institute of Physics.

the existence of size effects in ferroelectric size-reduced systems [12, 13, 17–19].

In the last years many studies on size effects in piezoelectric films have been also performed [20–31]. The meaning is different from that of bulk materials, because they include not only crystallite size but also film thickness. The separation of the two effects has been obtained in epitaxial films, but as we will see further, a true finite size effect due to only thickness reduction in thin films is questionable and evidence exists against or sustaining it, both theoretically and experimentally. In the following the influence of finite size effects on the properties of thin films will be discussed; the most explored are variation of lattice parameters and tetragonality ratio, changes in ferroelectric properties, changes in dielectric constant, and changes in the phase transition and critical temperature.

2.1. Variation of Lattice Parameters and Asymmetry

Effects of thickness reduction on lattice constants (a, c) and tetragonality ratio c/a have been found in $PbTiO_3$ thin films grown epitaxially on (001) MgO single crystalline substrates [21]. Lead titanate $PbTiO_3$ (PT) is a ferroelectric material which has a relatively large tetragonal distortion $c/a \sim 1.065$. A series of PT films with thickness ranging from ~ 2.5 to ~ 300 nm was grown by metalorganic chemical vapor deposition (MOCVD). X-ray diffraction (XRD) studies on different films evidenced only $(00l)$ and $(h00)$ reflections, corresponding to the preferential orientations. In Figure 2 XRD patterns show the variation of the spectrum with the film thickness. A remarkable change in the XRD pattern is observed when the film thickness is decreased from about 300 nm to 20 nm. At the high limit both the (001) and (100) reflections are present. When the film thickness is decreased, a gradual shift of the (001) peak to a higher angle and of the (100) to a lower angle is observed. At a film thickness of about 35 nm the (001) and (100) reflections become a single peak. The value of the lattice constant of about 4 Å for the thinnest film corresponds to the value for the pseudocubic $PbTiO_3$ phase. Average crystallite sizes in the

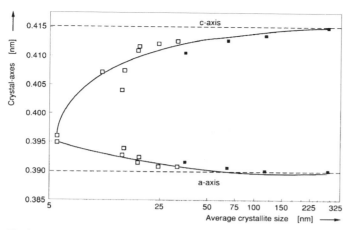

Fig. 3. Measured a- and c-axis lattice constants vs. average crystallite size for PbTiO$_3$ thin films on MgO substrates. Reprinted with permission from [21], © 1991, American Institute of Physics.

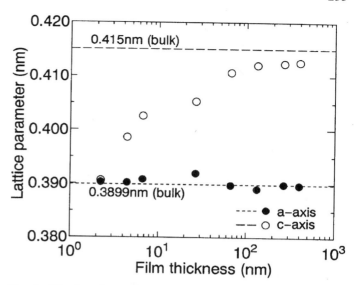

Fig. 4. The dependence of a- and c-axis lattice parameters on the film thickness for PbTiO$_3$ thin films on Pt/MgO(100) substrates. Reprinted with permission from [22], © 1996, Institute of Pure and Applied Physics.

films were calculated with Scherrer's equation using the width of the (001) reflection. A gradual growth of the average crystallite size with film thickness has been observed. For example, films of 15 nm thickness have a crystallite size of about 15 nm, whereas films with 200 nm thickness have a crystallite size of about 25 nm. The changes of the a- and c-axis lattice constants [obtained from (h00) and (00l) reflections] as a function of the average crystallite size are presented in Figure 3. A gradual variation in lattice parameters is observed down to a critical average crystallite size of about 25 nm, followed by a fast change below 25 nm. On the same graph, data corresponding to the microcrystalline PT powder are also shown. A good correlation is observed between powder and thin film results. The decrease of the c lattice parameter and the increase of a with decreasing crystallite size reduce the tetragonality ratio and consequently the lattice asymmetry.

Thickness size effects in ultrathin films also have been investigated by Fujisawa et al. [22] on epitaxial PbTiO$_3$ thin films grown on Pt(100)/MgO(100) by radiofrequency (rf) sputtering. Both the crystalline structure and lattice parameters were found to be dependent on the film thickness, which was varied between 2.2 and 398 nm. The PT films showed c-axis orientation, when their thickness was less than 26.5 nm. However, as the thickness increased, the formation of some a-domains was observed. From the measurement of θ rocking curves, it was found that a-domains tilted from the MgO(100) planes by 2.6°. Figure 4 presents the film thickness dependence of the lattice parameters of a and c axes, calculated from the PT(100) peaks in the in-plane XRD profiles at $\omega = 0°$ and from the PT(001) and (002) peaks in vertical XRD profiles. The variations of lattice parameters shows trends similar to those reported in [21]. The lattice parameter of the a axis remained constant at about 0.390 nm. On the other hand, the lattice parameter of the c axis was strongly dependent on the film thickness, showing very low value in ultrathin films. Consequently the tetragonality ratio drops to 1, which corresponds to a cubic perovskite and therefore to the absence of ferroelectricity and piezoelectric-

ity. This effect has been attributed to the presence of a tensile misfit strain near the electrode which is about 0.61% at room temperature and which produces an in-plane tensile stress. Consequently a compressive stress is produced perpendicular to the plane, which shortens the c-axis parameter. For thicker films the strain is relaxed by the formation of a-domains and the lattice parameters approach the values of the bulk crystal.

Similar results have been reported in [23] for epitaxial lead zirconate titanate (PZT) thin films on Mg(100) deposited by MOCVD, with thicknesses between 20 and 200 nm. The tetragonality ratio hardly decreased with thickness reduction from 100 nm to 20 nm. This effect has been attributed to the in-plane tensile strain introduced by the positive misfit between substrate and film lattices, which is 4.36% at room temperature.

A rather different thickness dependence of lattice parameters and tetragonality ratio has been reported in [24] for Pb(Zr,Ti)O$_3$ (PZT) thin films epitaxially deposited on SrRuO$_3$(SRO)(100)/SrTiO$_3$(STO)(100) by MOCVD. It has been found that the a-axis parameter decreased and the c-axis parameter increased; therefore the tetragonal ratio was enhanced with thickness reduction from 400 nm to 40 nm. The different behavior of PZT/SRO with respect to PZT/MgO or PT/Pt consists of the sign of the misfit in lattice constants between PZT and SRO (which is negative because $a_{SRO} < a_{PZT}$, so a compressive strain exists) and that between PZT and MgO or PT/Pt (positive, so a tensile strain exists). Hence, the PZT lattice in PZT/SRO is compressed along the interface plane. The compression stress produces the decrease of the a constant in ultrathin films and it is accompanied by a tensile strain on the direction perpendicular to the plane, which enhances the c constant. As thickness increases, the epitaxial strain is relaxed by different mechanisms (domain formation etc.); therefore lattice constants reach their bulk single crystal values. Similar results have been reported by Zhao et al. [25] for BaTiO$_3$ thin films grown epitaxially on SrTiO₃(001) by laser molecular beam epitaxy, for different thicknesses varying in the range 10–400 nm. It must be mentioned that, in this case, the misfit between substrate and film also gives an in-plane compressive strain and a tensile strain on the direction of the c axis. This produces the increase of the c/a ratio in ultrathin films (where the misfit stress was not relaxed by introducing dislocations or by other mechanisms). Other similar results have been reported by Nagarajan et al. [26] concerning the tetragonality ratio gradually increasing in epitaxial PZT 20/80 films grown by PLD on (001)LaAlO₃ with LSCO electrodes, with thicknesses ranging between 60 and 400 nm: the misfit between film and electrode that introduces high compressive in-plane strain is considered responsible for an increasing c/a ratio. The stress is gradually released in thicker films by a-domain formations, as shown by transmission electron microscopy (TEM) investigations.

It must be mentioned that true thickness size effects are very difficult to observe because they require an ultrathin film which is completely free from the influence of the substrate. In a TEM study of ultrathin free-standing BaTiO₃ thin films the observation of ferroelectric domain weakening has been suggested as due to a transition from ferroelectric to paraelectric phase of the BaTiO₃ [27].

2.2. Change of Ferroelectric Properties

Piezoelectric thin films with ferroelectric properties are characterized by a typical hysteresis loop polarization P and electric field E [2]. Piezoelectric and ferroelectric phenomena are closely related because the piezoelectric response in these materials is given by the polarization variation under an applied mechanical stress. A size dependence of the hysteresis loop has been observed in many thin films with reduced dimensions. Measurements on epitaxial PZT/MgO(100) thin films with a thickness of 20 nm show very small hysteresis with nearly zero remanent polarization [23]. On the other hand clear hysteresis loops have been reported for 10-nm-thick tetragonal epitaxial PZT 25/75 films deposited by reactive evaporation on SRO(001)/BTO/ZrO₂/Si heterostructure [28] and for 4-nm-thick epitaxial PZT 20/80 grown on Nb-doped (001)STO by off-axis rf magnetron sputtering [29]. This behavior is closely correlated to that of the tetragonality ratio because the onset of ferroelectricity in displacive ferroelectrics is accompanied by the cell distortion [2].

Effects of film thickness reduction on the remanent polarization and coercive field of epitaxial PZT films grown in different conditions are shown in Figure 5. The data have been taken from different references. The set of data indicated by × for the remanent polarization P_r and by open circles for the coercive field E_c has been taken from [30] and represents values measured on epitaxial tetragonal thin films with thicknesses between 100 and 2000 nm grown on (100)Pt/(100)MgO substrates by MOCVD. The other set of data, indicated by small crosses for P_r and solid squares for E_c, comes from [24] and has been obtained for epitaxial PZT films grown on SRO/STO(100) by MOCVD, with thicknesses ranging from 40 to 438 nm. The

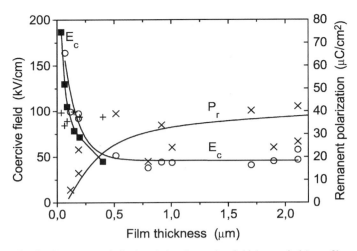

Fig. 5. Remanent polarization (×) and coercive field (open circle) vs. film thickness for PZT films. The data have been taken from [30]. On the same graph, corresponding values [indicated by plus signs (P_r) and solid squares (E_c)] measured on similar films in [24] have been reported.

data for P_r shows a larger dispersion, but a general trend of decreasing with thickness can be observed, as expected for strained films. Coercive field data from the two sets show similar behavior: a huge increase when film thickness is reduced below 100 nm.

An increase of polarization with decreasing thickness has been found in epitaxial PZT 20/80 films grown by PLD on (001)LaAlO₃ with LSCO electrodes for thickness ranging from 60 to 400 nm [26, 31]. This agrees with the tetragonality ratio enhancement due to the misfit between film and electrode which introduces high compressive in-plane strain.

2.3. Influence on Dielectric Constant and Its Temperature Dependence

The influence of thickness reduction on dielectric properties has been reported in [23] for films with thicknesses between 20 and 200 nm; the samples have been epitaxial Pb(Zr,Ti)O₃ thin films grown by MOCVD on single-crystal MgO(100). The dielectric constant slightly decreased when thickness varied between 100 and 10 nm. This decrease occurred simultaneously with that of the tetragonality ratio, previously discussed; therefore the decreases were correlated.

Effects of film thickness reduction on the dielectric constant of epitaxial PZT films grown in different conditions are shown in Figure 6. The data have been taken from different references. The set of data plotted with open circles has been taken from [30] and represents values measured on epitaxial tetragonal thin films with thicknesses between 100 and 2000 nm grown on (100)Pt/(100)MgO substrates by MOCVD. The other set of data, plotted with solid squares, is borrowed from [24] and was obtained for epitaxial PZT films grown on SRO/STO(100) by MOCVD, with thicknesses ranging from 40 to 438 nm. A sudden decrease of the dielectric constant while the film thickness was varying in the range 100–40 nm was observed. The continuous and dashed lines are calculated from a model which

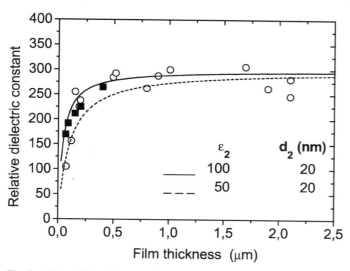

Fig. 6. Effect of film thickness reduction on the dielectric constant of epitaxial PZT films grown in different conditions. The data have been taken from [30] (open circles) and [24] (solid squares).

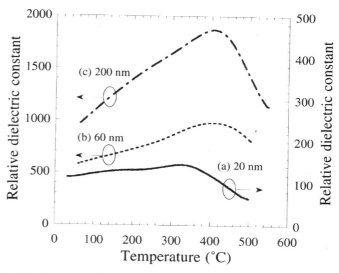

Fig. 7. Temperature dependence of the dielectric constant of PZT films with different thicknesses. Reprinted with permission from [23], © 1999, Institute of Pure and Applied Physics.

assumes that an ultrathin layer (about 20 nm) with low dielectric constant (50 or 100) exists connected in series with the PZT layer [30]. Since chemical and physical analyses did not evidence the presence of other phases, nor the electrical field used in measurements was so high to produce a space charge layer which could decrease the dielectric constant, the origin of the low-dielectric-constant layer has been attributed to the compressive stress existing in the ultrathin epitaxial films or to a true size effect. The first hypothesis is supported also by other results obtained in the same experiment, as previously discussed.

Finite-size effects on the dielectric constant variation with temperature have been found in nanocrystalline $PbTiO_3$ particles for crystallite size ranging between 26 and 80 nm [13]. In normal perovskite ferroelectrics the dielectric constant shows a narrow maximum at the critical temperature where the crystal transforms from paraelectric (cubic) to ferroelectric (tetragonal or orthorhombic). The critical temperature T_c does not depend on the frequency. On the other hand, relaxor ferroelectrics (disordered ferroelectrics, where the ferroelectric order is preserved only in very small nanosized regions of 10–20 nm diameter [32]) show broad peak in dielectric constant vs temperature, which shifts with frequency toward higher temperatures. This is exactly what was found in nanocrystalline $PbTiO_3$ powder [13]: a decrease in the particle size makes the ferroelectric transition increasingly diffuse, with a broad dielectric peak. This broadening has been attributed to the critical temperature dependence on the particle size, which in a powder has not a unique value but ranges around a mean value. A similar T_c dependence on particle size has been evidenced by Raman measurements in [14] for $PbTiO_3$ particles ranging between 14 and 40 nm: when the particle size decreases to 14 nm, the critical temperature decreases to 440°C, which is about 50°C lower than its bulk value. These measurements (correlated with XRD measurements of the tetragonality ratio) allowed prediction of a critical size of about 10.7 nm where the ferroelectric state might

be unstable, due to the weakening of dipole–dipole interaction in a confined medium.

Size effects on the dielectric constant, originating from both reduced film thickness and crystallite dimension, have been reported in polycrystalline $BaTiO_3$ ferroelectric thin films prepared by rf sputtering [20]. There it has been evidenced by scanning electron microscopy (SEM) that the size of the crystallites increased from 15 nm to 300 nm along with an increase of the film thickness from 4 nm to 2000 nm. By measuring the dielectric constant ε vs temperature for every film a clear correlation has been found between the critical temperature and the width of the phase transition and the size of coherent scattering regions d. In particular no phase transition peak could be detected in the $\varepsilon(T)$ curves below a certain $d \cong 30$ nm. This result has been interpreted in terms of the lattice-dynamical theory of the size effect in ferroelectrics: the coherent scattering regions are slightly disoriented blocks inside the crystallites. Due to their disorientation, the lattice vibrations of neighboring blocks are incoherent, and consequently it can be supposed that the dielectric properties of the sample are determined only by the lattice vibrations of a separate block. Therefore, the cyclic Born–von Karman condition for the lattice vibrations is replaced in the finite crystal by the boundary condition on the surface, leading to the discreteness of the values of the phonon wavevectors and the splitting-off from the crystal vibration spectrum of the vibrations with $k < \pi/d$, where d is the smallest size of the crystal [20]. The splitting of vibrations with $k < \pi/d$ influences only phonon branches with strong dispersion near the center of the Brillouin zone, such as the soft ferroelectric mode. Hence, in ferroelectrics the decrease of the crystal size must influence the lattice vibration spectrum and, consequently, the dielectric properties.

Investigation of the structural phase transition temperature of PZT films showed a clear dependence of its character on thickness [23]. Figure 7 shows the temperature dependence of the dielectric constant of the epitaxial PZT thin films deposited on MgO for thicknesses 20, 60, and 200 nm. A very broad dielectric constant peak due to structural phase transition is observed at 328°C, confirming the existence of ferroelectricity at the thickness of 20 nm. As the film thickness increased to 60 nm, T_c increased to about 410°C. This shift was attributed to the changes in tetragonality in ultrathin films due to the inner expanding strain in the in-plane direction. The dielectric constant peak became sharper as the film thickness increased.

Size influence on phase transition properties also has been studied for other systems such as $BaTiO_3/SrTiO_3$ multilayered thin films with periodicities varying in the range 1.6–32 nm and total thickness 400 nm prepared by pulsed laser deposition [33]. Broadening of the dielectric constant peak, strong frequency dispersion of the dielectric constant, and shift of critical temperature toward lower values have been observed when the periodicity was decreased. This clear relaxational behavior has been attributed to size effects that hinder long-range ferroelectric ordering in these films.

In conclusion, evidence has been presented for and against the existence of true finite thickness effects in ultrathin films. The problem is still under investigation, waiting for theoretical models free of strong approximations and for clear experimental evidence free of possible spurious effects.

In this section the discussion was limited to finite size effects in piezoelectric thin films with ferroelectric order, because this aspect may be most perturbed by the intrinsic effect of size reduction in the mesoscopic range. In thin films (of any type) there are other extrinsic effects (strains at interface, chemically different interface layers, clamping effect of the substrate, etc.) which become influential when thickness decreases and which, therefore, affect the effective properties of the structure. Some of these aspects have been only incidentally touched in this section whereas they will be treated in Section 5 dedicated to the properties of piezoelectric thin films.

3. GROWTH TECHNIQUES

Deposition techniques used for piezoelectric thin films are the same as those developed for dielectrics and include sputtering, pulsed laser deposition, the sol–gel technique, and chemical vapor deposition. The use of a particular technique depends on many factors and in general there is not a single choice for the growth of a given material; similar results can be obtained with different deposition methods and different sets of deposition parameters. Each main technique has been further improved by making particular modifications to the basic system. Generally, the effect of a particular modification is limited to the experimental setup where it has been employed and cannot be easily extended to another system. The aim of this section is to give an overview and to ease the understanding of the most important deposition parameters (temperature, pressure, growth rate, etc.) for the growth of films with desired composition and structure (crystalline structure degree, microstructure, and orientation) by using different growth techniques. Other factors influencing

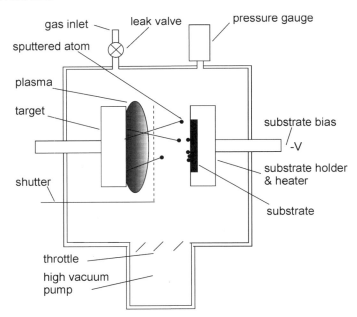

Fig. 8.   General setup of sputtering system.

the growth, such as substrate, electrode, buffer, and template layers, will be considered. The discussion will include the most important piezoelectric thin films grown by each technique.

Our main interest is only in laboratory systems for research. The analysis of industrial systems for mass production is outside the scope of this chapter. A detailed description of thin film deposition systems can be found in [34].

3.1. Sputtering

This technique is widely used for the deposition of piezoelectric thin films and has reached a high degree of maturity during over three decades of investigations. The reason arises from the fact that sputtering satisfies the major requirements for thin film fabrication: strong adhesion to the substrate; smooth surface; uniformity of the properties and thickness over large areas; low deposition temperature, etc. Even though the basic mechanism is quite simple, not all the other processes involved are well understood.

Basically a sputtering system consists of a stainless steel chamber connected with a high vacuum pumping system able to reduce the residual pressure below 10^{-6} mbar (Fig. 8). Inside the chamber, a circular electrode on which is bonded the material to be sputtered (target) is placed in front of another electrode supporting on its surface the substrate to be coated. A proper voltage applied between these electrodes ionizes a gas introduced into the chamber and held at a suitable pressure. The ions of the gas bombard the target, producing the ejection of target particles toward the substrate. If the gas is inert (i.e., Ar) the species deposited are the same as those of the target, while a proper mixture of inert and reacting gasses determines the compound that will result as a combination of the reactive gas and the target species. Other basic elements of the system are as follows: a throttle valve for controlling the pumping speed;

pressure gauges; leak valves controlled by a gas flow meter system; substrate heater; shutter; power supply [direct current (dc) or rf] for plasma; matching network for rf operation; dc supply for substrate biasing. Reactive sputtering is usually preferred for piezoelectric materials such as oxides and nitrides because a metal target offers the advantages that it can be machined and bonded easily to a cooled plate. A metal target has good thermal conductivity and thus can manage high power density without damage.

Various configurations and modes of operation are possible for a sputtering system, depending on the specific application. Here we consider the diode, triode, and magnetron sputtering systems, with either dc or rf power.

Diode sputtering is the oldest and simplest system used. It allows a uniform erosion of the target but has the disadvantages of low deposition rate and overheating of the substrate, due to the bombardment of the growing film by high-energy electrons. The target usually consists of a disk 5–15 cm in diameter bonded on a water-cooled plate. A grounded shield is placed around the target to avoid sputtering from its edges and from the backing plate.

Triode sputtering consists of a modified diode where a hot filament is added to supply electrons for supporting the discharge. In this way the discharge can be sustained at much lower gas pressure and thus a significant increase of the deposition rate can be achieved. The use of a triode configuration has the disadvantage that reactive gas can rapidly destroy the filament.

In magnetron sputtering the increase of the deposition rate is obtained by introducing a static magnetic field crossing the plasma region near the target. With a proper choice of magnetic field shape, strength, and orientation, the secondary electrons are confined in the region near the target. As a consequence the probability of ionizing Ar atoms is increased. The plasma discharge can be more intense at lower pressure and the deposition rate increases significantly. Among the various possible configurations, the most used is the planar magnetron with circular target. Its major disadvantages are the nonuniform erosion of the target and poor film uniformity at low target–substrate distances. An analysis of the design of sputtering systems and performance comparison between various configurations are reported in [35].

Microstructure of deposited films depends strictly on growth parameters. Microstructure control is necessary for the optimization of relevant film properties for the applications. Adatom mobility is the main factor influencing the microstructure; it depends mainly on substrate temperature and particle bombardment. Following the modified model proposed by Thornton [36, 37] for sputtered films, microstructure is classified in four zones as shown in Figure 9. The microstructure is represented as a function of the gas pressure and the ratio T/T_m, which represents the substrate temperature normalized with respect to the melting point temperature of the deposited compound. Analyzing the best results in the literature for reactively sputtered ZnO and AlN, the highest piezoelectric coupling factors have been obtained in the conditions

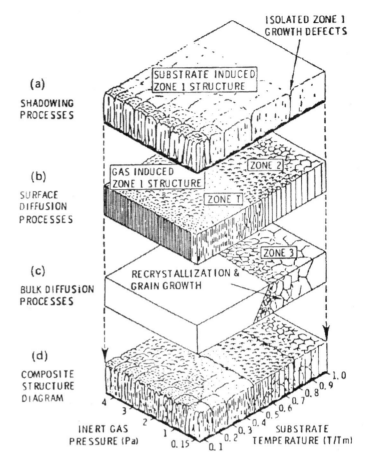

Fig. 9. Microstructure zone model as a function of sputtering pressure and temperature [36]. Reprinted with permission from the *Annual Review of Materials Science*, Volume 7, © 1977, Annual Reviews, www.AnnualReviews.org.

falling in zones T and 2, where the film has a dense columnar structure with smooth surface. For a detailed discussion of growth zones see [34, 36–38]. Vincett [39] established that the optimal substrate temperature should be one-third of the deposited compound boiling point and associated with this an optimum deposition rate value. Figure 10 is a plott of this rate–temperature relationship for ZnO growth data in the literature, as reported by Hickernell [40]. As can be seen, there is a good agreement between the data reported for magnetron sputtering and for Vincett theory. On the other hand if the boiling point is considered equivalent to the melting point and the pressure is not taken into account, Vincett and Thornton predictions are in agreement.

Another important factor influencing the orientation and the morphology of the growing film is the ion bombardment. Depending on their energy, the effect can be so strong that the orientation of the c axis can be parallel or perpendicular to the substrate. This effect is used for obtaining films with adequate orientation for the generation of shear waves that require crystallites tilted either 40° or 90° with respect to substrate normal. The effect of ion bombardment during AlN film growth has been analyzed in [41]. The orientation of c axis plotted as a

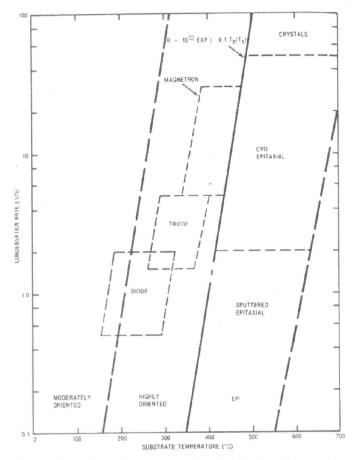

Fig. 10. Dependence of ZnO condensation rate with substrate temperature. Reprinted with permission from [40], © 1985, IEEE.

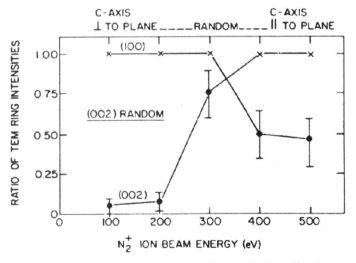

Fig. 11. Microstructural transition from c axis perpendicular to film plane to c axis parallel to film plane, with N_2^+ ion beam energy. Reprinted with permission from [41], © 1983, American Institute of Physics.

shear wave excitation. One of the most used, but hardly reproducible, consists of tilting the substrate plane with respect to the target. An analysis of the film texture in such conditions is given in [42]. The most reproducible method seems that described in [43] in which an external bias was applied.

An extensive study of the microstructure evolution of sputtered ZnO with film thickness is reported in [44]. From the comparison of crystallographic and microstructural characteristics the authors divided the growth behavior of ZnO into three different regions with thickness increasing from 0.15 to 7.3 μm. At the initial stage they found a smooth columnar structure with c axis perpendicular to the substrate; in the intermediate stage the orientation tended to randomize and large and elongated crystallites appeared; in the final growth stage the resulting c axis was parallel to the substrate and crystallites of various size and shape were mixed (Fig. 12). The structure modification was attributed to the effect of bombardment with highly energetic species on the top of the growing film surface.

An original approach for improving the c-axis orientation of magnetron sputtered AlN is reported in [45]. Using a roof-shaped target, the ion bombardment of the substrate has been reduced. Comparisons between AlN films deposited with conventional planar magnetron and this technique indicates the latter to be effective for improving c-axis orientation.

In many laboratories there is a trend to use sputtering configured in a pulsed dc reactive magnetron. The power supply generates a series of dc pulses of a few tens of kilohertz modulated with a lower frequency. With respect to conventional dc sputtering, it allows a higher deposition rate for dielectric films to be obtained; moreover the cost of the generator is considerably lower than that of a radio frequency generator. An extensive comparative analysis of the properties of AlN deposited by this sputtering configuration and by PLD is presented in [46]. The authors report better results with sputtering in terms of oxygen contamination and residual stresses in the film. Other authors report [47] on the growth of AlN with excellent properties using the same sputtering technique in similar conditions. The film has been employed for the realization of a thin film bulk acoustic wave resonator (TFBAR) operating at 3.6 GHz with a coupling factor $k = 0.23$.

The desired texture of AlN thin films for SAW applications is (100) with respect to (001) because of its superior coupling coefficient. Control of preferential orientation on reactively sputtered AlN has been analyzed in [48]. The authors used reactive dc magnetron sputtering with no substrate heating. They obtained preferential (100) orientation at long substrate–target distance (L) and high sputtering pressure. The relative growth rate of (100)/(001) has been varied by changing the deposition unit from atoms to Al–N dimer. That is, when the mean free path of Al and N atoms is greater than L, they deposit directly on the substrate and (001) film grows; when it is smaller than L the collision of Al and N atoms forms Al–N dimers and growth with (100) orientation is enhanced.

A low-temperature deposition of highly c-axis-oriented AlN by dc reactive magnetron sputtering is reported in [49]. The deposition atmosphere was pure nitrogen and the rate was

function of the nitrogen ion energy is reported in Figure 11. As can be seen the c axis is perpendicular to the film plane at low ion energies, whereas above 400 eV it is parallel to the plane. Various methods have been used to obtain a tilted c axis for

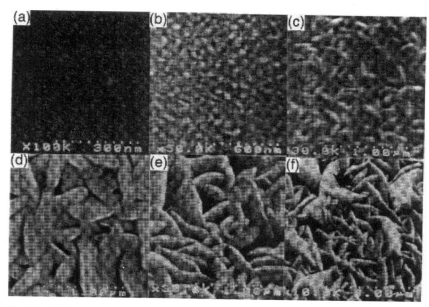

Fig. 12. SEM micrographs of ZnO films with thickness: (a) 30 nm, (b) 0.15 μm, (c) 0.55 μm, (d) 3.3 μm, (e) 4.4 μm, (f) 7.3 μm. Reprinted with permission from [44], © 1998, Materials Research Society.

60 Å/min to minimize the bombardment of the substrate. Various characterizations of the properties confirm the good quality of the film for applications in bulk acoustic wave (BAW) devices at microwave frequencies.

The correlation between acoustic performance related to the square of the electromechanical coupling coefficient (k^2) of sputtered ZnO film and deposition parameters (substrate temperature and sputtering pressure) has been analyzed in [50]. It has be found that, at a pressure of 2 mTorr, k^2 is independent of temperature whereas, at a pressure of 10 mTorr, k^2 decreases when temperature increases and vanishes at 360°C. As a first approach, this can be considered a general guideline.

It should be pointed out that the substrate temperature during deposition is determined not only by the heater but also by the energy of the incoming plasma particles that bombard the substrate. This fact has been analyzed in [51], where it was observed that the effective piezoelectric thickness of the film is less than its physical thickness. This fact has been attributed to the modification of growth conditions for the first layers, which generally grow at lower temperature and thus are randomly oriented.

A different approach to improving nucleation of the ZnO grains consists of periodically chopping the sputtering process [52]. The technique consists of rotating the substrate holder on an axis shifted from the center of the target. The modified sputtering was an rf magnetron and the target a sintered ZnO plate. A highly c-axis-oriented ZnO film on SiO$_2$/Si(100) has been obtained with a proper choice of the ratio of deposition to pause time.

Ferroelectric materials form another group of interesting piezoelectric materials in the form of thin film. The most attractive are lead titanate (PT) and lead zirconate titanate (PZT) in pure form or doped with La$_2$O$_3$ (PLT and PLZT). One of

the first examples of SAW propagation on a PZT thin film is described in [53]. The deposition was made by dc reactive magnetron sputtering of a multielement (Pb, Zr, Ti) metal target with sector areas optimized to obtain the desired film composition. A PLT thin film resonator operating in the gigahertz range is reported in [54]. The film was deposited by rf magnetron sputtering in Ar + O$_2$ atmosphere at 640°C starting from a PLT target.

3.2. Pulsed Laser Deposition

Pulsed laser deposition is an emerging technique for thin film deposition. It consists of material removal by bombarding the surface of a target with short energetic pulses of a focalized laser beam of proper wavelength (laser ablation). This process takes place in a vacuum chamber where a gas is held at constant pressure (Fig. 13). Due to the high power density of the beam, plasma having a plume shape perpendicular to the target surface is generated at the incident point. The plasma contains ions of the target and of the gas atmosphere. The substrate to be coated is placed few centimeters apart from the target facing the top of the plasma plume. Condensation of the particles ejected from the target produces the growth of a thin film on the substrate surface. The deposited film composition can be the same as the target, if the gas atmosphere is inert, or a compound depending on the reactive gas introduced into the deposition chamber. The film can be amorphous or crystalline with preferential orientation determined by various parameters such as substrate temperature, gas pressure, and laser fluence. During the deposition process the target must be continuously rotated and translated to avoid crater formation and to obtain uniform erosion.

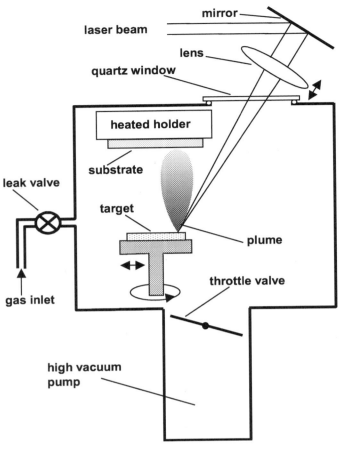

Fig. 13.   Schematic setup of a PLD system.

The setup can be enriched with different equipment for "*in situ*" analysis: Auger electron spectroscopy (AES), spectroellipsometry, etc.

The possibility of transferring complicated stoichiometries directly from the target to a collector situated at a certain distance and position with respect to the target pushed this method into a top position starting with the discovery of high-temperature superconductors, in the second part of the 1980s. Different classes of materials, such as metals, semiconductors, and dielectrics, were successfully deposited by pulsed laser deposition in vacuum or in different inert or chemically active atmospheres. The technique itself is simple and versatile and has some particular advantages [55, 56]; (i) it can synthesize metastable materials that cannot be produced by "standard" techniques; (ii) it can fabricate films from species that are generated only during pulsed laser ablation; (iii) it has the possibility of transferring complicated stoichiometries from bulk into a thin film (congruent ablation), as in the case of high-T_c superconductors, piezoelectric and ferroelectric compounds, biocompatible materials (hydroxyapatite), etc.; (iv) the laser energy source is external to the system and therefore it is a "clean" reactor; (v) the laser beam is incident on a small zone on the target surface, which results in high efficiency, control, and flexibility of the process; (vi) no charge effects appear; (vii) it is a no-"memory" reactor with the possibility of depositing heterostructures by changing the target and/or the reactive gas; (vii) in some cases it is a single-step process, combining synthesis and deposition (GaN from Ga in N_2 [57], ZnO from Zn in O_2 [58–60] Si_3N_4 from Si in NH_3 [63]); (viii) it decreases the deposition temperature and, sometimes, no subsequent thermal treatments are required. The main drawback of the technique comes from the formation of particulates (droplets) on the substrate and film surface. To eliminate this, different solutions have been proposed [55].

3.2.1. Zinc Oxide

ZnO thin films were first deposited by PLD in the early 1980s. Shankur and Cheung [62] reported obtaining crystalline, (002) oriented films by CO_2 ($\lambda = 10.6 \, \mu m$) laser ablation of a ZnO target in oxygen reactive atmosphere in 1983. The laser sources used were related to the general evolution of lasers. The main part of reported results were obtained using excimer lasers with different wavelengths such as ArF ($\lambda = 193$ nm) [63], KrF ($\lambda = 248$ nm) [64–68], and Nd–YAG lasers ($\lambda = 1.06 \, \mu m$) [58–60]. The incident laser fluence varied in the range 1–4 J/cm^2 in the case of excimer lasers, being much higher, 25–30 J/cm^2, for IR (Nd–YAG) lasers [58–60]. Laser frequency was set at 5 or 10 Hz. Mainly high-purity (99.999%) ZnO pellets prepared by different methods were used as targets. The most frequently used are hot sintered [62], pressed [64, 65], and hot pressed [66, 69] ones. Very few results are reported concerning the deposition by laser ablation of a pure (99.99%) Zn target in oxygen atmosphere [58–60]. The target–collector distance was of the order of few centimeters (3–5 cm) in all experiments. For all experiments a reactive gas was introduced in the deposition

The substrate temperature is monitored and controlled via a thermocouple in conjunction with an electronic power regulator. The heater block should have a shape ensuring a good uniform distribution of the temperature in the range 100–800°C.

A turbomolecular pump drained by a rotary pump generally constitutes the pumping system. Because it is required to operate at much higher deposition pressure with respect to other techniques the turbomolecular pump should be able to work up to 10^{-1} mbar. This problem can be partially overcome when the reacting gas is oxygen by using a throttle valve. In this case the operating pressure of the pump can be lower than the chamber pressure. This expedient cannot be used for nitrides deposition, where the residual oxygen gas pressure must remain very low to avoid contamination and thus the pump must operate at maximum flow rate.

The focalization of the laser beam is obtained through a proper optical system with characteristics depending on the laser wavelength. The converging lens or objective is usually placed outside the chamber and the beam reaches the target after passing through a window. Because the spot cross-area on the target decides the laser fluence (J/cm^2), particular care should be taken in checking the laser power and the area of the spot.

Table II.   Typical Experimental Conditions Used for ZnO Film Growth by PLD

Laser, wavelength	Temperature (°C)	Oxygen pressure (Torr)	Fluence (J/cm^2)	Substrate	Reference
CO_2, 10.6 μm	50–450	5×10^{-4}–10^{-2}		Si, sapphire, quartz, GaAs	[62]
KrF, 248 nm	20–500	5–2×10^{-2}	0.5–5	Si, Corning glass, GaAs	[64, 65]
	600–700	10^{-2}		Quartz, sapphire	[71]
ArF, 197 nm	500–600	O_2 with 8% O_3, 2–6×10^{-4}	1	Glass	[63]
KrF, 248 nm	300–800	10^{-5}–10^{-1}	2–3	Sapphire	[66, 67, 70]

chamber. Most frequently this gas was oxygen whose pressure varied in a large range: 5×10^{-6}–10^{-1} mbar. In some cases, the oxygen pressure was varied during the deposition process, to obtain better control of the early stages of film growth: thus the oxygen pressure was initially set at 10^{-4} mbar for the first 100–500 Å and then increased to 10^{-1} mbar [70]. To avoid the oxidation of GaN buffer layer surface an initial (20 Å thin) layer was deposited at 10^{-6} mbar; then the pressure was increased until 10^{-5}–10^{-4} mbar [67]. O_3 was sometimes added to O_2, due to its strong ability to oxidize [63].

As expected for all the materials deposited by PLD, the substrate temperature was found to strongly influence the orientation and quality of the grown films. On the other hand ZnO has a high tendency to develop a columnar structure, with the c axis perpendicular to the film surface, quasi-independent of the substrate temperature. As previously underlined, PLD allows one to obtain high-quality films at lower substrate temperature because of the high energy of particles arriving on the substrate. This energy depends on the incident laser fluence: high fluence increases this energy, and thus a smaller substrate temperature is necessary. This effect was confirmed in [58–60], where c-axis-oriented films with good piezoelectric properties were obtained at only $T = 250°C$, working at high incident laser fluence (25 J/cm^2). However, the substrate temperatures used are in a very wide range: from 50 to 450°C in the case of CO_2 laser [62], 200–500°C [64, 65], or up to 800°C [68]. The deposited film thickness varies, according to the purpose, between tens of nanometers and several microns. Different experimental conditions leads to deposition rates between 0.2 and 1.8 Å per pulse. Table II summarizes the experimental conditions used for ZnO film deposition.

Special attention has been paid to the substrate and bottom electrode–substrate combination, due to its influence on the initial growth stages resulting from the mismatch between the film and substrate lattices and the quality of the interlayer. Different substrates have been used: Si(100) [58, 62, 64]; Si(111) [62]; GaAs(100) [62, 65]; GaAs(111) [62]; Corning glass [58, 62–64]; sapphire [62, 66–68, 70, 71]; quartz [71]; and gold- and titanium-plated substrates. Usually the irradiation geometry used was with the substrate parallel to the target surface [58, 59, 62, 63].

As just mentioned, the growth process depends not only on the deposition parameters but also on the type of bottom electrode and/or buffer layer deposited on the substrate. Sankur and Cheung [62] describe the phenomena which take place on the target and on the substrate during irradiation with a CO_2 laser beam of a ZnO target at different oxygen pressures [62]. They measured and calculated the target temperature inside the focal spot as being in the range 950–1025°C, when ZnO decomposes into Zn and O_2. A congruent evaporation occurs, as inferred from the analysis of the target which did not evidence Zn enrichment and also from the presence in the residual gas of only oxygen and Zn. ZnO thin film grows as a result of a reaction which takes place on the substrate between Zn and oxygen. It was experimentally observed that the oxidation of Zn adatoms occurs with a high rate, resulting in stoichiometric layers for a deposition rate up to 20 Å/s. On the other hand the deposition rate was found to depend on the reevaporation rate of Zn atoms from the grown interface, which in addition depends on the substrate temperature. The sticking coefficient of Zn to a ZnO surface decreases with temperature, reducing the deposition rate. So, a compromise among the substrate temperature, incident laser fluence, and oxygen pressure should be established to have a high enough deposition rate for films with desired stoichiometry. A more detailed description of the different stages of the growth process is presented in [63] for a ZnO film grown on an amorphous substrate. Using a reflection high-energy electron diffraction (RHEED) technique a gradual transformation from an amorphous phase to a c-axis-oriented layer has been observed. Figure 14 describes the succession of the film growth stages. First there is evidence for an island structure (a few nanometers thick) which needs large formation energy but no crystallization occurs. Further, after "coalescence" of the islands and the formation of a continuum layer, the growth of the ZnO layer changes from heteromode (high-energy process) to homomode (low-energy process). Thus enough energy remains for crystallization, and a polycrystalline ZnO layer grows (5–10 nm thick) on the amorphous one. Subsequently, a third layer, c-axis oriented, starts growing on the polycrystalline film.

Different buffer layers were tested for their influence on the ZnO layer properties. For films deposited on GaAs wafers, SiO_2 was found to alleviate stress problems and promote a good texture, as was demonstrated by X-ray diffraction analysis. For films grown on sapphire the best buffer layer was found to be GaN, which reduces the lattice mismatch from 16.7% to 1.9% [67].

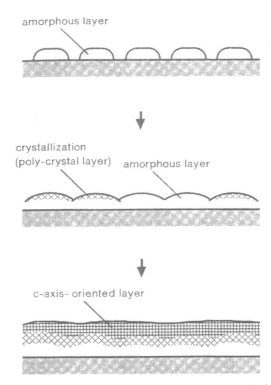

Fig. 14.   The evolution of the growth mode of ZnO films deposited by PLD:
(a) initial (heteromode) stage of growth with thickness between 0 and 5 nm;
(b) homomode stage of growth, with thickness of 5–10 nm; (c) final stage consisting of growth of *c*-axis-oriented layer. Reprinted with permission from [63],
© 1996, American Institute of Physics.

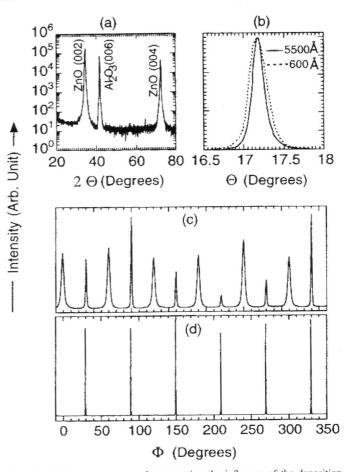

Fig. 15.   XRD measurements demonstrating the influence of the deposition
conditions on the crystallinity and orientation of ZnO films deposited by PLD:
(a) θ–2θ spectrum of a film exhibiting only {00l} family of planes; (b) rocking
curves of two films with different thickness deposited at 750°C; (c) Φ scan
spectra of a film deposited at 500°C using a laser frequency of 15 Hz; (d) Φ
scan spectra of a film deposited at 750°C using a laser frequency of 10 Hz.
Reprinted with permission from [66], © 1996, American Institute of Physics.

The composition and properties of the deposited layers
have been investigated by different techniques such as X-ray
diffraction, reflection high-energy electron diffraction, scanning electron microscopy, X-ray photoelectron spectroscopy
(XPS), Rutherford backscattering (RBS), optical absorption
spectroscopy, spectroellipsometry, ion channeling techniques,
and atomic force microscopy (AFM).

The main results concern the orientation of the *c* axis with
respect to the film surface. For films deposited by PLD the usual
orientation is the *c* axis normal to the film plane. No results have
been reported for the *c* axis inclined to the normal or lying in the
film plane. A typical XRD spectrum of a film deposited by laser
ablation is presented in Figure 15 [66]. It reveals only the (00l)
family of planes of ZnO film and proves the *c*-axis orientation
of the films. Indeed, the only ZnO peaks are (002) at $2\theta = 34.5°$
and (004) ZnO at $2\theta = 73°$. As can be seen, the substrate
temperature influences the in-plane orientation: it changes from
two in-plane orientations with a rotation of 30° between them
for films deposited on sapphire at 500°C substrate temperature to a perfect alignment with a single in-plane orientation
for films deposited at 750°C. This behavior can be correlated
with the increase of adatom energy as proposed by the model
in [62]. The adatom energy enhancement reduces the formation
of the intermediate-interface noncrystalline or polycrystalline
layers. The optimum deposition pressure range identified for
obtaining the previous results was 10^{-5}–10^{-4} mbar. Similar

results concerning the strong perpendicular *c*-axis orientation
have been reported also in [58, 59, 62–65]. The full width at half
maximum (FWHM) of the (002) plane diffraction peak varies
between 1.9–4° in [63], 2–3° in [64, 65], and 0.17° in [66]. The
degree of orientation and the lattice constant strongly depend
on the growth conditions, as was also evidenced by Hayamitzu
et al. [63]. For high substrate temperature the FWHM of the
rocking curve decreases for large film thickness, but a great degree of fluctuation in the *c*-axis orientation was evidenced for
thin layers. The lattice constant is close to that of bulk material
and is independent of the film thickness.

The columnar aspect of the deposited films was investigated
by cross-section SEM [58, 64, 72] and by TEM [60]. A dense,
columnar structure of a ZnO film deposited at low deposition
temperature (275°C) by laser ablation of a Zn target in O_2 reactive atmosphere is presented in Figure 16. The columns are
regular, with dimensions on the order of 50–100 nm thickness
and up to 0.6–0.7 microns long as can be seen in Figure 17a.

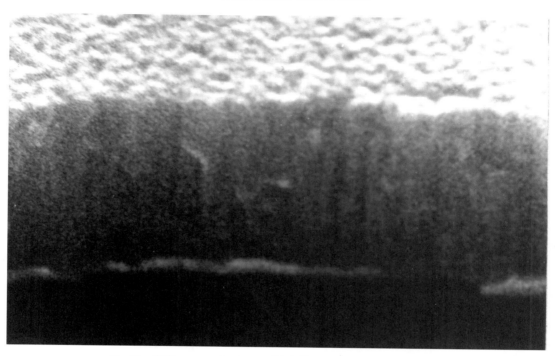

Fig. 16. SEM cross-section image of ZnO film grown on Au/Cr/Si structure.

The diffraction pattern (Fig. 17b) taken on a large selected area reveals a good crystallinity of ZnO film, with a hexagonal structure ($a_0 = 3.249$ Å, $c_0 = 5.205$ Å) (ASTM 5-664). Figure 17c presents the diffraction pattern of the columnar group from Figure 17a. The 002 diffraction spot is elongated due to the slope of the diffraction object, so that we can conclude that the columns are grown along the c-hexagonal axis of ZnO.

Similar results have been obtained by Seabury et al. [73]; they reactively deposited ZnO using a shorter wavelength (248 nm) with respect to [72] and use the film in a microwave BAW resonator. A detailed report on growth and characterization of PLD deposited ZnO is described in [74]. Here the films were obtained starting from a pure ZnO sintered target.

The quality of film surface is very important for further application in building transducers because of the losses at the interface between the ZnO film and the top electrode.

3.2.2. Aluminum Nitride

Growth of aluminum nitride (AlN) by PLD [75–84] is more difficult than growth of other materials such as, for example, oxide compounds because it requires a background atmosphere almost free of oxygen, but at the same time it can be performed at lower substrate temperatures compared with other techniques. As in the case of other piezoelectric materials, decreasing the substrate temperature during deposition is a mandatory requirement for applications compatible with silicon technology. It has been evidenced that many techniques traditionally used for AlN thin film deposition have some disadvantages: Most of them can cause thermal or plasma damage to the substrate or substrate–film

interface [75]. Because laser ablation using high-energy UV lasers is a nonequilibrium process, metastable configurations may develop at low substrate temperature where surface rearrangement is restricted by the low surface mobility of the ablated species [76]. PLD itself leads to low-temperature processing, because the average energy of particles in the laser-evaporated plume (10 eV) is considerably higher than the thermal evaporation energy (1 eV). The additional energy during laser ablation is used in recrystallyzation of the thin film [77].

Growth rate increases nonlinearly with the laser fluence because laser ablation is a multiphoton process [75]. Therefore a film of 1 μm thickness can be grown in 10 min at 4 J/cm^2 laser fluence. This is an extremely rapid growth compared to that typical for rf sputtering (4–5 nm/min) and remote plasma deposition (4 nm/min) [75]. The main experimental conditions used for pulsed laser deposition of AlN films are summarized in Table III. It was demonstrated that N$_2$ pressure, substrate temperature, and laser fluence play an important role in the crystalline structure quality and surface morphology of the film. It has been found also that a dense target is mandatory for reducing the density of particulates on the film surface.

An important parameter for AlN growth by PLD is the deposition pressure, which can be varied in a wide range between high vacuum and 10^{-1} mbar. Films deposited in vacuum, at temperatures of 670°C on sapphire substrates [76], are oriented mainly with the c axis perpendicular to the substrate surface, the (0002) plane of the film being parallel to the ($1\underline{1}02$) plane of the substrate. When they are grown in nitrogen atmosphere at 5×10^{-2} mbar, for the same substrate temperature, the films are always single phase, but with another orienta-

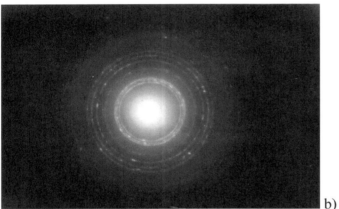

Fig. 17.   TEM and SAED images of ZnO films deposited by laser ablation of a Zn target in oxygen reactive atmosphere: (a) column groups; (b) SAED image of a fragment of the film; (c) SAED image of the columnar group from (a): the (002) diffraction spot is elongated due to the slope of diffraction object, so it can be concluded that the columns are grown along the c-hexagonal axis of ZnO. Reprinted with permission from [60], © 1999, Elsevier Science.

tion: the $(10\underline{1}0)$ plane of the film is parallel with the $(1\underline{1}02)$ plane of the substrate. At higher nitrogen pressure (0.4 mbar) a mixture of grain orientations is evidenced. Similar results are reported in [78] for films deposited in vacuum. An optimum pressure of 0.1 mbar nitrogen for growing of cubic AlN films is reported in [79]. The epitaxial orientation relationships are (100)AlN∥(100)Si and [010]AlN∥(100)Si for Si(100) substrates and (111)AlN∥(111)Si and [011]AlN∥[011]Si for

Si(111) substrates, respectively. Microtwins have been observed in (111)AlN planes. Single crystalline films have been obtained by laser ablation of the AlN target in vacuum [80]. Increasing the nitrogen pressure from 3×10^{-7} to 5×10^{-4} mbar produces a threefold drop of the intensity of the (0002) diffraction peak and a corresponding broadening of the rocking curve from $1.1°$ to $1.9°$ [77]. The influence of the nitrogen pressure on the quality of the film is explained by the increasing number of collisions within the plasma plume with increasing gas pressure, which results in a higher number of particulates generated during the evolution of the plasma plume. These clusters condense on the substrate, damaging the epitaxial quality of the film.

Another important parameter is the substrate temperature, which has been varied between room temperature and 850°C. High-purity, oxygen-free films have been obtained by laser ablation at room temperature [75] but crystalline films were produced only at temperatures greater than 375°C [81, 82] or 550–600°C [76–80]. Compensating effects of temperature and nitrogen pressure were observed in [76]. Thus films deposited in vacuum at 670°C were similar to those deposited at only 500°C for a nitrogen pressure of 0.04 mbar. The best alignment of the (0001)AlN plane with those of the surface is reported in [77, 79] for substrate temperature as high as 750°C (Fig. 18).

The epitaxial growth of AlN layers on different substrates with large lattice mismatch can be explained by the domain matching epitaxy [77, 78], where the domain dimension becomes the unit distance across which the lattice matching is maintained. Thus m lattice parameters of the substrate match with n in the epilayer and the residual domain mismatch becomes $f_d = 2(nb - ma)/(nb + ma)$, where m and n are integers. For this reason even if the direct mismatch between the lattices of Si and AlN is high, a domain misfit of only 1.2% allows the epitaxial growth of AlN on silicon. The same explanation is also valid for films grown on sapphire, where in-plane alignments are $AlN[1\underline{2}10]∥Al_2O_3[0\underline{1}10]$ and $AlN[10\underline{1}0]∥Al_2O_3[\underline{2}110]$ [78]. This indicates also a 30° rotation of the AlN film orientation with respect to the substrate, which leads to a residual domain misfit f_d of 0.7% between the Al atoms in the AlN and Al_2O_3, where eight interatomic distances in AlN matche with nine in Al_2O_3.

No influence of the laser frequency was observed in the range 5–30 Hz [76, 77], whereas the laser fluence was found to be important for the film crystallinity and the surface roughness as well [77, 78].

High-quality films have been obtained for laser fluences in the range 2–4 J/cm². For higher values (>10 J/cm²) Al-rich layers have been obtained, with high density of grain boundaries [78]. The Al is coming from AlN decomposition due to the high incident laser fluence.

A high surface roughness and the presence of many particulates is evidenced for films deposited at higher fluence: this is also accompanied by increasing deposition efficiency at values of 1.12 Å/pulse. The high laser fluences reported in [81, 82] are partially compensating the effect of low substrate temperature. At the same time, taking into account that the target

Table III.   Main Experimental Conditions Used for Pulsed Laser Deposition of AlN Thin Films

No.	Target	Collector	$D_{T\text{-}C}$ (cm)	T (°C)	p_i/p_{N_2}	ν_L	$\lambda/E/S$	(00l)	Reference
1	Many 0.6-mm-thick samples	Sapphire	4	500–670	2×10^{-6} 5×10^{-2}	50	248 2 J/cm^2	W (0002)	[76]
2	AlN powder tablets	KBr, suprasil, quartz, glass, GaAs	3 perpend. geom.	20	10^{-3}	10	248, 193 0.2–4 J/cm^2	W	[75]
3	Hot pressed AlN	Si(111)	5	550–650– 750	10^{-7}	15	248 2–10 J/cm^2	W (0002)	[77]
4	AlN compacting powder	Si(111), Si(100) Best Si(111)	3–4	600–680 680	$(3–5) \times 10^{-6}$ 0.1–0.3 0.1	4	248 2 J/cm^2	Cubic Possible sphalerite as NaCl	[79]
5	Hot pressed AlN	Sapphire		750–800	5×10^{-7}	15	248 2 J/cm^2 2–3	W (000l)	[78]
6	Pressing powder	(0001) sapphire	3	493–487	10^{-9} 1×10^{-7} 5×10^{-3} N$_2$	1	3.4–3.5	W (000l)	[84]
7	AlN	(000l) sapphire 6H–SiC Si(111)	5	100–850	10^{-7}–10^{-4}	10–30	248 2–15 J/cm^2 2	W (000l)	[80]
8	Al	(0001) Sapphire, Si(100), (111)	4–7	350–380	2×10^{-6} 5×10^{-2}	10	1060 25 J/cm^2	W (110), (0001)	[82]

is pure Al and the reactive gas is nitrogen, the balance energy for plasma formation together with the breaking of the nitrogen molecule requires a higher amount of energy. The difference also comes from the mechanisms responsible for plasma formation and expansion for a laser working in the ultraviolet or infrared range.

Experimental conditions similar to those reported in [77, 78, 80] are used in [83] for preparing AlN films which were subsequently tested for surface acoustic wave devices. An interdigital transducer was built on an (110)AlN film deposited on a sapphire substrate.

3.2.3. Gallium Nitride

Gallium nitride is another III–V compound which was discovered to exhibit good piezoelectric properties [85] several decades ago. GaN thin film properties have been reported in [85, 86]. Its application as a material for light-emitting diodes in the UV–visible range has opened a new interest in GaN thin film preparation.

The investigation of GaN film deposition by laser ablation is concentrated on the comparison between ammonia and nitrogen as reactive gases and the advantages generated by the ablation from the liquid phase. Due to the naturally liquid phase

of gallium at temperatures above 29.8°C the laser ablation can overcome the drawbacks generated by the deterioration of the target surface and particulate splashing on the film. The deposition of a ZnO buffer layer [87, 88] was found to favor the c-axis orientation of the GaN layer due to the small mismatch between the two lattice parameters.

NH$_3$ is used as reactive gas in [87] and nitrogen in [88, 89]. Different laser sources have been used: ArF (λ = 197 nm) in [87]; KrF (λ = 248 nm) in [89]; and Nd–YAG (λ = 1.06 μm) in [88]. The fluence varies between 5 J/cm^2 in [89], 10 J/cm^2 in [87], and 20–25 J/cm^2 in [88]. Some of the experimental conditions are summarized in Table IV.

The use of amorphous substrate (fused silica) required the deposition of a buffer layer for deposition of the GaN film. ZnO was used as buffer layer because it is a material isomorphic with wurtzite GaN [87] and the lattice mismatch in the basal plan is 2.2%. X-ray diffraction spectra evidenced that films grown directly on fused silica substrate are polycrystalline, whereas the presence of the ZnO buffer layer allows the growth of c-axis single crystalline phase GaN films. The substrate temperature was also critical for film crystallization: below 600°C only amorphous or polycrystalline films have been obtained, as reported in [87].

Table IV.   Experimental Conditions for the PLD Growth of GaN

	Laser, wavelength	Fluence (J/cm^2)	Frequency (Hz)	Temperature (°C)	Gas pressure (mbar)	Substrate	Reference
1	ArF, 193 nm	10	10	600	NH$_3$, 1	Fused silica	[87]
2	KrF, 248 nm	5–9	12	300–800	N$_2$, 4×10^{-4}	Passivated Si(001)	[89]
3	Nd–YAG, 1060 nm	20–25	10	350	N$_2$, 5×10^{-2}	Si(100) (001)-sapphire	[88]

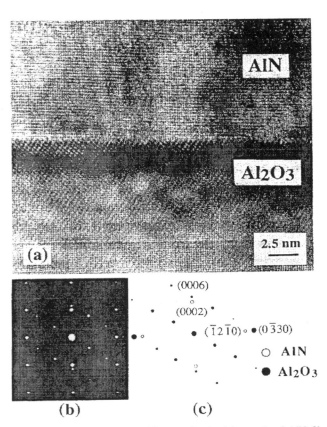

Fig. 18. Electron microscopy evidences of epitaxial growth of AlN films on (0001) sapphire substrate by the PLD technique: (a) high-resolution TEM image; (b) selected area electron diffraction pattern; (c) simulated diffraction pattern of AlN/sapphire (0001). Reprinted with permission from [78], © 1995, American Institute of Physics.

Using high incident laser fluence (20–25 J/cm^2) it was possible to obtain c-axis-oriented GaN films at lower deposition temperatures (350°C) and working in nitrogen reactive atmosphere [88]. The presence of ZnO buffer layer on the sapphire substrate improved the film crystallinity.

A special deposition system was used in [89]. To improve the reactivity of the nitrogen gas, this was introduced in the deposition chamber through a nozzle as pulsed N$_2$ flux directed onto the Si substrate. The synchronization of the nitrogen "pulse" with the laser pulse allows the Ga plasma plume to cross the gas expansion jet. In this way excited nitrogen species are produced and then atomic nitrogen results from the breaking of

N$_2$ molecules. The excited species propagates (adiabatically) toward the film surface due to the pulsed character of the N$_2$ flux.

3.2.4. Lead Titanate Zirconate

The piezoelectric multicomponent materials such as lead titanate zirconate Pb(Zr$_x$Ti$_{1-x}$)O$_3$ (PZT), BaTiO$_3$, PbTiO$_3$, and SrTiO$_3$ are quite difficult to grow in the form of thin films because small variations in the film stoichiometry results in a strong degradation of piezoelectric and ferroelectric properties. Ferroelectric compositions frequently contain a volatile component (e.g., Pb, Li, Bi, or K) and small deviations from the correct stoichiometry can lead to films with undesirable, nonferroelectric phases. Control over the process is required to avoid loss of volatile components and formation of metastable phases. For many applications, the growth of oriented (or even epitaxial) films would be of high interest. Other important film qualities are dense and smooth surface morphology, uniform film thickness, and good ferroelectric and piezoelectric properties [55].

An "*in situ*" deposition technique for PZT and other piezoelectric materials with complex stoichiometries is desirable because postprocessing of amorphously deposited materials requires high substrate temperatures and routinely leads to the production of large grain boundaries, impurity phases, and randomly oriented materials [90].

PLD shows some advantages with respect to other techniques used for preparing such films. The most important is that epitaxial thin films can be deposited at low substrate temperature and with a high deposition rate, over a large range of target phases and compositions and with a relatively low number of experimental parameters to be optimized [55].

Each of the three steps of PLD process—the interaction of the laser beam with the target and plasma formation, the plasma expansion and its physical and chemical interaction with the surrounding reactive gas, and the growth of the film on the heated substrate—plays an important role in the further composition and crystallinity of the film. There are several categories of parameters which can be controlled during PLD process: composition, density, and structure of the target; laser pulse characteristics (wavelength, fluence, frequency, duration); the content and the pressure of the gas; and the substrate–electrode combination [55, 90–107].

As we have mentioned before, film stoichiometry is mandatory for piezoelectric and ferroelectric applications. One of the

Table V.   Main Experimental Conditions Used for the Deposition of PZT by PLD

	Laser, wavelength	Fluence (J/cm^2)	Frequency (Hz)	Temperature (°C)	Oxygen (gas) pressure (mbar)	Substrate	Reference
1	KrF, 248 nm	2	5	20–650	0.2–0.3	SrTiO$_3$	[90]
						MgO	
						Pt/Ti/Si	
						(GaAs)	
2	ArF, 197 nm	2–10	5	20–590	0.133	MgO(100)	[91]
3	XeCl, 308 nm	3.3	3	560	0.2	Si(100)	[94]
4	ArF	10	1	450–530	+0.8% O$_3$	SrTiO$_3$(100)	[96]
5	Nd–YAG, 532 nm	10	2–12	560		MgO	[97]
6	XeCl, 308 nm	25	0.3–3	20	10^{-1}–10^{-5}	Si(100)	[101]
7	XeCl, 308		4	20–550	Vacuum 0.2 mbar		[98]
8	Nd–YAG, 1060 nm	10	1	20	10^{-4}	Glass	[106]

main problems is the control of the lead content and its distribution in the film. In [90] it was demonstrated that the films deposited at substrate temperature higher than 400°C and oxygen pressure lower than 0.1 mbar are lead deficient. Films deposited in vacuum at room temperature preserve the target stoichiometry whereas a decrease of the lead content was observed when the temperature increased by 20% from the initial value. For a constant substrate temperature of 550°C, the lead content in the film increased with increasing oxygen pressure in the deposition chamber until it reached the value of 0.42 (from 0.5) at a pressure of 0.2–0.3 mbar. This behavior can be explained by complex contributions of the processes related to the vapor thermalization by collision with the ambient gas and to the PbO vapor pressure. The oxygen presence is compulsory for the chemistry of the process: experiments done in argon in similar conditions result in films with very low Pb content. Another explanation for the deficient lead content of films is given in [96], for films prepared from targets subjected to prolonged ablation at small laser fluence. The evaporation of the volatile lead component occurs as a result of the dominant thermal ablation mechanism. This is not the case at higher fluences, when a prevailing plasma regime is responsible for the evolution and this effect was not observed.

Generally, low Pb content in films prepared by PLD is not due only to the low reactivity of the Pb atoms with oxygen [55, 90]; oxygen also exhibits a low chemical reactivity because of high dissociation energy. The solution for avoiding Pb loss during the deposition process is to increase the oxygen pressure [100] and/or to use a more reactive oxidant ambient, such as ozone [96] or reactive species resulting from a discharge. On the other hand, increasing the deposition pressure too much produces a dramatic decrease in the deposition rate; an optimum range of values was found to be (0.1–0.4) mbar.

Another effect that has been studied is the spatial distribution of lead content in the film [98]. Films deposited at room temperature under vacuum conditions exhibit a dip in the lead distribution, at the target surface normal position. This dip is explained by the revaporization effects due to the high incident laser fluence which generates high energetic species. When the pressure is increasing, the kinetic energy of particles decreases due to the intensification of the collisions inside the laser-induced plasma.

Although the Pb/Ti ratio was found to be dependent on either the deposition pressure or the laser fluence, the Zr/Ti ratio was found to depend only on the laser fluence but not on the deposition pressure. A solution for incorporating Pb in the film is proposed in [19], where an anion assisted PLD technique is used to prepare stoichiometric PZT films. The electrons produced attach easily to the ambient oxygen molecules, yielding molecular anions, which are guided by the electric field to the film surface. The oxygen anions react faster than molecular oxygen with the Pb atoms and increase the probability of Pb incorporation into the film.

The substrate temperature, together with its orientation and/or the buffer or electrode layer strongly influences the crystalline structure and orientation of the PLD deposited films. The substrate choice is important because epitaxial growth results easily when the mismatch between the film and substrate is small and their symmetry is similar. Different substrate–electrode combinations have been used (Table V). The single-crystal substrates were found to promote epitaxial growth of perovskite PZT films even at low substrate temperature. Among them, the most often used are MgO and SrTiO$_3$ [90, 96]: (100) epitaxial PZT layers have been obtained despite the quite large mismatch between the lattice constant of film and substrate. On MgO, the pyrochlore phase formation was frequently noticed even under optimum conditions used to deposit perovskite PZT on SrTiO$_3$. Thus, for SrTiO$_3$ substrate, at 550°C in 0.2 mbar oxygen a single phase oriented PZT is obtained; the pyrochlore phase is still present for films grown on MgO substrate [90]. The greater difficulty associated with deposition of perovskite on MgO may be related to the larger mismatch between the lattice parameters of PZT and MgO (1–8%, depending upon composition) relative to SrTiO$_3$ (<1%). The difference in film growth can also be the result of a nucleation effect: the perovskite substrate (cubic) promotes the formation of the perovskite film, whereas the MgO (NaCl) structure may not. Obviously one should carefully choose the substrate and its orientation for getting a particular ferroelectric phase and orientation of the deposited film [55]. However, the film has a slightly distorted structure due to the stress induced at the interface.

By introducing an electrode between the substrate and the deposited layer and using as substrates Si and GaAs, the phase composition and the crystallinity dramatically change. Films deposited in the same experimental conditions as those on MgO and SrTiO$_3$ were found to contain mainly pyrochlore for deposition temperatures $T < 550°C$ and $p = 0.3$ mbar O$_2$. The optimum deposition temperature was found to be 600°C, when the perovskite phase was estimated to be 70% for GaAs and 90% for Si, respectively. The further increase of the substrate temperature induces the diffusion of Ti (used as intermediate layer) into the Pt (electrode) layer and films contain less perovskite than pyrochlore, which can be interpreted also as a decrease of the Pb content due to PbO evaporation [90].

Pt-coated or Pt/SiO$_2$-coated MgO and Si substrates have been also used [99, 102]. The presence of Pt on the MgO substrate produces a shift of (002)–(200) peaks toward higher values [99]. Differences have been observed also between films grown on coated MgO and coated Si substrates: However, in both cases a polycrystalline structure was evidenced. Another structure which promotes epitaxial growth and (00l) preferential orientation is LSCO/Pt/TiN/Si [100]. TiN is a barrier layer which growths heteroepitaxially on Si (domain matching epitaxy) and avoids the interdiffusion between Pt layer and Si substrate. LSCO is a perovskite type material which allows a sharp interface with PZT film, free of oxygen vacancies, to be obtained.

The deposition of a SiO$_2$ buffer layer, with thickness in the range 40–100 Å was also found to promote an epitaxial growth of (100) oriented thin films on Si substrates [102].

The use of Au as bottom electrode was found to promote the (111)PZT preferential growth of the films, even at low substrate temperature [108–110]. Thus, highly oriented PZT films have been deposited on Au/Si(100) and Au/Pt/NiCr/glass structures at 400°C. This could be a consequence of the high matching between the PZT and Au lattice constant with (111) orientation.

The deposition temperature was found to strongly influence the phase present in the film as well as the crystalline structure. Thus, for $T < 250°C$, only amorphous films are deposited (despite the single crystalline structure of the substrate) and subsequent thermal annealing at 550°C induces mainly pyrochlore crystalline phases, whereas by increasing the temperature to 600°C a randomly oriented pseudocubic perovskite phase is formed [90]. Films deposited at temperatures between 350 and 700°C exhibit (100) and (111) oriented perovskite phase accompanied by a pyrochlore phase. The lead-deficient films were found to contain mainly pyrochlore phases. It has been observed also that films deposited at room temperature were predominantly amorphous [106]. The subsequent thermal treatment at 450°C improves the crystallinity, films becoming polycrystalline with both tetragonal and rhombohedral phases.

By increasing the annealing temperature to 650°C only the tetragonal phase is observed, together with other phases of new (unidentified) compounds. The disappearance of the rhombohedral phase is attributed to the Zr depletion by reaction with the glass substrate.

Postdeposition annealing processes are often used to improve the film crystallinity [94, 97]. A high-rate heating process (100°C/min) to 700°C was performed in an oxygen atmosphere in [94], whereas a longer treatment (30 min) at only 570°C in an oxygen atmosphere was used in [97].

Special mention should be devoted to a complex PLD method, reported in [93]. A multitarget system containing the three base oxides ZrO$_2$, TiO$_2$, and PbO is used. Alternative ZrO$_2$, PbO, TiO$_2$, and PbO structures, with different thicknesses, have been deposited at 200°C substrate temperature. The annealing process (at 600°C) transforms the multilayer structure dominated by (001)PbO orientation into PZT(100).

3.2.5. Lithium Niobate

Lithium niobate (LiNbO$_3$) is a widespread material in modern technology due to its high electromechanical coupling and electrooptic coefficients, high pyroelectric coefficients, etc. In the form of thin films it has been used in surface acoustic wave devices, optical waveguides, electrooptic devices, etc.

The application of pulsed laser deposition technique for LiNbO$_3$ film growth is quite recent [111–117]: the deposited layers were found to have properties appropriate for SAW device applications [112, 115].

Some experimental conditions are summarized in Table VI. The three parameters which influence the deposition of single-phase, stoichiometric LiNbO$_3$ layers are the Li/Nb ratio in the target, substrate temperature during the film deposition, and type and gas pressure inside the deposition chamber.

As for other materials, the nature and pressure of the gas influence the process of the film growth. Thus, it was found [111] that the addition of argon is mandatory for the growth of stoichiometric thin films. The optimum argon/oxygen ratio was found to be 400 mTorr Ar/100 mTorr O$_2$; for a smaller Ar concentration traces of LiNb$_3$O$_8$ are present in the film whereas by increasing the Ar beyond 450 mTorr a mixture of different phases appears. A critical oxygen pressure of 10^{-3} Torr was identified in [112]: a small increase to 5×10^{-2} Torr produces a decrease of the Li content in the film. This is explained by the collisions between the Li atoms and the oxygen species which result in the decrease of the number of Li atoms arriving on the substrate surface. Another important feature is the color of the layer. Despite the presence in the X-ray spectra of peaks attributed to single-phase LiNbO$_3$ crystal, the brown color of films prepared at pressure lower than 10^{-3} mTorr is due to the mixed valence states of Nb ions or to the presence of oxygen vacancies. An annealing process at 700°C in an oxygen atmosphere transforms the film color from brown to transparent. Similar effects are obtained by working in a more oxidizing atmosphere containing 8% O$_3$ [112, 115]. A different behavior is observed in [116], where the oxygen pressure was found to

Table VI.   Main Experimental Conditions for the Deposition of LiNbO₃ by PLD

	Laser, wavelength	Fluence (J/cm²)	Frequency (Hz)	Temperature (°C)	Oxygen (gas) pressure (mbar)	Substrate	Reference
1	ArF, 193 nm	2.3	5	600	Vacuum O₂ + Ar O₂ 0.5	Si(100)	[111]
2	ArF, 193 nm		15	500–800	O₂ + 8% O₃ 10^{-3}–5×10^{-2}	Sapphire (001)	[112]
3	KrF, 248 nm	3–5		650–800	0.1–1 O₂	α-Sapphire	[113]
4	XeCl, 308 nm	0.8–1.3		525–825	0.001–0.1 O₂	MgO/GaAs	[114]
5	ArF, 193 nm		10–45	500–800	0.001–0.05 O₂ + 8% O₃	Sapphire (001), (110)	[115]
6	Nd–YAG, third harmonic, 355 nm	1.4–10	10	350–550	0.002–0.4 O₂	Sapphire (0001)	[116]
7	ArF, 193 nm	0.4–2.5	4	20	10^{-5}–10^{-1}	Si(100)	[117]

be less important for layer properties. However, a postdeposition annealing process in oxygen atmosphere was found to be important for single-phase growth. Strongly textured (012) and (024)LiNbO₃ films have been deposited at high oxygen pressure (1 Torr) on sapphire substrates: pressure decreases induce the appearance of new crystallographic phases [113].

Films deposited at a substrate temperature in the range 20–850°C were found to have a preferential orientation or single phase composition for appropriate combination of two other parameters: Li/Nb ratio and gas pressure. Thus, the optimum combination identified in [113] for single-phase, textured surface film perpendicular to the substrate was 730°C and 1 Torr O₂. The temperature decrease at 700°C together with the oxygen pressures in the range 100–600 mTorr results in polycrystalline films. A pressure of 10^{-2} Torr O₂ was found to promote LiNbO₃ single-phase growth for a substrate temperature of 500°C: films deposited at higher temperature (800°C) contain Li-deficient phases, such as LiNb₃O₈, due to the vaporization of the adsorbed Li on the surface as a result of the high temperature [113]. The same effect was observed in [115]. A much stronger temperature dependence is mentioned in [116]: a critical temperature of 450°C was identified for single-phase (0006)–(00$\underline{12}$) oriented LiNbO₃ film growth. X-ray diffraction spectra of films deposited at 500 and 550°C exhibit peaks coming both from LiNbO₃ and (Li-deficient) LiNb₃O₈ phases, whereas films deposited at 350°C give no other spurious peaks [116].

The effect of Li/Nb ratio in the target on the composition of the film was investigated together with the influence of the other two parameters: gas pressure and substrate temperature. Li-enriched targets with Li/Nb ratio between 1 and 3 are used to compensate the Li losses as an effect of the ablation process. Single-phase, stoichiometric LiNbO₃ films have been deposited for a Li/Nb ratio of 2, at 500°C and 10^{-2} Torr O₂ pressure [112, 115], at 825°C [114] and at 20°C and 6×10^{-2} mbar O₂ pressure [117].

Epitaxial films have been grown on sapphire substrates by choosing an optimum set of experimental parameters [112, 115]. Thus, for (001) sapphire substrate the (001) planes of the film were found to be parallel with the substrate surface. From pole figure investigations it is evident that films are completely epitaxial, untwined, with [110] axis of LiNbO₃ parallel to [110] axis of sapphire [115].

3.3. Sol–Gel Method

Sol–gel is one of the most used methods for the deposition of different materials in thin film form, due to its advantages over other techniques, such as control of composition over a large area, low initial processing temperature, high purity, and low cost. It consists of the mixing of liquid material components in adequate proportion and the subsequent transformation of the sol into a gel by hydrolysis reaction. The gel is further transformed by heating treatment in a powder or film material. To obtain a thin film the synthesized solution is first spun onto a substrate and then is subjected to the heating treatment [118]. The as-deposited films are usually amorphous and need postdeposition annealing treatments for crystallization. The number of ceramic thin films prepared by this technique is limited only by the lack of suitable starting precursors.

3.3.1. Lead Zirconate Titanate

An exhaustive review concerning the description of this technique and its application to the deposition of lead zirconate titanate (PZT) thin films is presented in [119]. The most important types of the technique are: (i) sol–gel starting from 2-methoxy ethanol as reactant and solvent; (ii) a hybrid process using a chelatic compound (acetic acid); and (iii) metalorganic decomposition using carboxylate compounds. The influence of the chemical interaction between the starting type of precursors on the species which are formed and on their evolution until the final compound formation is described for the three solution–deposition routes. Thus, when the process starts from carboxylate compounds, the low reactivity of the precursors results in a small interaction between species. On the contrary, when alkoxide compounds are used, a large reactivity is observed, the resulting species containing more than one type of cation.

A commonly used solution for deposition is that starting from 2-methoxy ethanol precursors. The process advantage is a high reproducibility; however, it has also disadvantages resulting from the complicated, high-level chemistry involved together with toxicity of the precursors. The hybrid process has a smaller degree of reproducibility due to (i) complex chemistry of solution preparation and (ii) further reactivity of the precursors after the synthesis, which results in the modification of the precursors over time. The use of metalorganic compounds allows an accurate control of the composition because of the water insensitivity of the precursor species. The limitation of the process comes from cracking during film deposition as a result of the large organic ligands of the starting reagents and the small reactivity of initial precursors.

The synthesized solution is then used for film deposition by applying a spin-coated process. A few drops of solution are deposited on an electrode plated wafer then flooded during the static dispensing before spinning at 1000–8000 rpm. A shrinkage process takes place at room temperature, up to 50–70% in the thickness direction: thus, for obtaining films with thickness appropriate for applications, hundreds of nanometers up to microns, a multistep deposition–crystallization process is considered. The further steps in film preparation are pyrolysis and crystallization. Two ways are proposed: a single-step process, rapidly heating at the crystallization temperature; and a two-step process, pyrolysis at 400–450°C followed by annealing at temperatures in the range 500–750°C, with rates between 5°C/min and 7500°C/min.

The phase evolution is described in detail, depending on the reaction route, annealing treatment, and electrode composition and structure. As a general trend, it is assumed that the transformation steps are as follows: amorphous phase, pyrochlore phase, and, finally, perovskite phase. It was demonstrated by intermediary analysis that the activation energy for phase transformation for the two processes are similar. Other routes, depending on the material and orientation of electrode and substrate, are summarized in [120]. Different explanations concerning the direct perovskite phase growth were proposed: (i) the lattice match between Pt(111) and PZT(111); (ii) the perovskite nucleation on the surface defects (crystallites, stress hillocks) of the Pt electrode; (iii) the promoting of PZT nucleation by Ti diffusion through the Pt layer at the Pt–Ti interface and formation of a Ti-enriched phase which favors the perovskite growth; (iv) the formation of an intermediate layer, with the composition Pt_xPb, etc. Independent of the bottom layer, it is believed that the crystallization of the perovskite phase is a nucleation-controlled process [119, 121].

The parameters which influence PZT thin film structure deposited by the sol–gel method were found to be solution composition and chemistry, Zr/Ti ratio, PbO content and substrate–electrode combination.

The high-temperature annealing treatment is the main drawback of this technique; it is quite difficult to reduce the temperature below 600°C due to the formation of the pyrochlore phase. This high temperature is a drawback for the PZT integration into silicon technology (the selection of suitable barrier layers, etc.). Some of the main results concerning the deposition of high-quality PZT layers by the sol–gel method are summarized here.

The chemistry and homogeneity of PZT precursor solutions were found to strongly influence the crystallization temperature of PZT layers [122]. The optimization of the content of the acetic acid (chelatic acid) results in the homogeneity of the precursor solution by generating a dense polymeric network with a high mixing rate. The precursors used were Zr-n-propoxide, Ti-isopropoxide, and lead acetate as starting material and n-propanol as solvent. The acetic acid was added before adding water in different molar ratios with respect to the Ti-isopropoxide: 2.5, 13, 25, 75, and 250. The best results have been obtained for a ratio of 25. The lead depletion during the deposition process was compensated by increasing by 10% the amount of lead in the solution.

Another combination of precursors is reported in [123]: lead acetate trihydrate [$Pb(CH_3COO)_2 \cdot 3H_2O$], with 10% extra amount added during preparation; zirconium oxyacetate [$ZrO(CH_3COO)_2$]; and tetrabutyl titanate [$Ti(C_4H_9O_4$]. The lead acetate trihydrate and zirconium oxyacetate were dried by distillation in 2-methoxyethanol and cooled at room temperature before addition of tetrabutyl titanate–methoxyethanol mixture. The solution was then stirred 2 h at room temperature and further diluted in 2-methoxymethanol.

A combined precursor solution was used in [124]. Thus, lead acetate and titanium and/or zirconium n-propoxide were dissolved in methoxyethanol ($CH_3OC_2H_4OH$); the solution was refluxed for 12 h at 110°C and then condensed at 0.4 M by distilling off the by-product and solvent. Subsequently the stock solution was partially hydrolyzed (2 : 1 molar ratio) under 0.1 NHO_3. The deposition was performed only after 24 h aging of the hydrolyzed solution.

Starting from the aforementioned precursors different deposition conditions have been used for obtaining stoichiometric, uniform, crack- and pore-free layers. Thus, in [122] the precursor solution was deposited by spin-coating and the as-deposited layer was dried on a plate heated at 150°C in air, for evaporating all volatile products such as water, alcohol, and organics. The sequence spin-coating and drying was repeated several times; then the film was pyrolized, annealed in an oxygen atmosphere at temperature in the range 500–650°C in a preheated furnace, and then cooled in air at room temperature. High rpm value (2800 with respect to 1300 in [125]) are used for the spin-coating process in [124], together with a high annealing temperature (700°C for 90 s). A more complex recipe is used in [123], where a two-step spin-coating process, once at 600 rpm for 9 s and subsequently at 3000 rpm for 30 s, is used. The crystallization was promoted using a rapid thermal annealing process in an oxygen atmosphere, at 600–700°C, for 2 min. An improvement of the thermal treatment conditions is reported in [126]; after drying at 200–300°C, the films were spin-coated at 3000 rpm and then annealed at 460°C in a multistep process conducted in conjunction with X-ray diffraction studies. Generally, the annealing temperature is high, 650–750°C, as reported in [127, 128], thus avoiding the appearance of pyrochlore phase. The PZT films were included in heterostructures for smart thin film devices, as reported in [129].

The effect of the postdeposition treatments on the electrical properties of the PZT layer have been investigated. Thus, the electrical characteristics of the films were found to be improved by exposure to the O_2 plasma [130]. The authors performed a study on the changes produced in the remanent polarization and the leakage current densities of the samples after O_2-plasma treatment for different time durations. The plasma treatment was done by exposing the PZT films to 0.7 mTorr O_2 plasma in a radiofrequency plasma reactor, at room temperature. It was found that the remanent polarization increases slowly with O_2 exposure time up to 90 min, being saturated thereafter.

The influence of the forming gas (4% H_2/balance N_2 or 100% N_2) on the film properties was also investigated [131]: despite the constancy of the texture and crystallization of the film, as observed in X-ray diffraction spectra, the electrical properties dramatically changed, up to the disappearance of the remanent polarization. Because the oxygen loss was found to be small enough to induce any change in the perovskite crystalline structure, another proposed explanation for the degradation of ferroelectric properties is the hydrogen incorporation in the film in the form of (OH) hydroxyl groups, as evidenced by Raman spectra. The hydrogen is supposed to occupy, in the perovskite cell, the tetrahedral sites or the sites between apical oxygen ions and Ti [131].

Much attention was paid to the influence of the electrode–substrate combination on the mechanisms of PZT thin film growth and, consequently, on their properties. The interdiffusion at the bottom electrode–film interface was found to determine the further film orientation and the possible degradation of electrical properties of the film. The most used combination is Pt/Ti/SiO_2/Si [120–132]. Combinations such as YBCO/$SrTiO_3$(001) [131], Pt/Ti/SiO_2/Si [127], ITO/fused silica (glass) [123], $LaAlO_3$ with LSCO electrode [131], and ZrO_2 passivated silicon [132] were also used. An interdiffusion zone at the Ti–Pt interface and an enlargement of the PZT–Pt interface have been observed [121].

The formation of a Pt_xPb transient intermetallic phase was evidenced by many authors, with x having values in the range 3–6 [120, 121, 126]. This phase was supposed to reduce the mismatch between the PZT(111) and Pt(111) and to facilitate the growth of perovskite PZT.

The early stages of the structural development of films also are described in detail in [120]. The authors demonstrate that the intermetallic layer is Pt_3Pb. The amount of Pt_3Pb metastable phase was found initially to increase with annealing time and to decay after reaching a maximum. The kinetics of the growth and decay processes was simulated by using the Avrami

equation (as had been done also for studying the kinetics of perovskite PZT growth in [126]). The Avrami coefficient n and growth rate constant k were determined by comparing the experimental results and the simulated curves, from which activation energies of 40 and 145 kJ/mol were obtained for the growth and decay of the intermetallic phase, respectively. The perovskite PZT was found (by using TEM analyses) to nucleate epitaxially on top of the Pt_3Pb phase. Obviously the Pt_3Pb phase plays a major role in determining the crystallite orientation at the nucleation stage of the perovskite PZT. This was demonstrated to depend strongly on the annealing temperature and time. Thus, for films dried at 200°C and then annealed at 440°C for 7 × 60 s the pyrochlore phase is evidenced together with a strong presence of Pt_3Pb phase and a small amount of PZT perovskite. After 13 × 60 s annealing, the (111)PZT orientation together with a small pyrochlore phase appears, and the Pt_3Pb is completely absent, due to the transformation in Pt. Then, after annealing at 440°C for 20 × 60 s a perovskite columnar structure is evidenced, increasing in amount after 30 × 60 s annealing time. It was also evidenced that films can crystallize even at 400°C if the annealing time is long enough (315 × 60 s), but two main orientations appear: (100) and (111). The same effect was observed by increasing the temperature to 420°C, when a similar crystalline structure was obtained after only 150 × 60 s. With increasing the temperature to 460 and 480°C, the (100) orientation disappears and the film becomes entirely (111) oriented. It was then concluded that the higher annealing temperature inhibits the (100) phase growth. The treatment time was found to have a similar effect: a longer time leads to reduction of the (100) phase development. An important question is related to the condition of the Pt_3Pb layer formation. The proposed mechanism is based on the reduction of Pb^{2+} through reaction with C from the organic species, in the as-deposited film, before the dried-annealing treatment initiation. Then, a competition between the two processes: formation of the Pt_3Pb and the decomposition into Pb^{2+} and Pt takes place, governed by the C amount and oxidizing conditions.

A detailed discussion about different phases which appear at the electrode–PZT layer is presented in [126], for a film with 70/30 composition. The as-deposited films were subjected to a series of annealing stages as follows. First, the sample was put onto a hot plate in air and kept at constant temperature for a certain time (e.g., 10 s). After cooling on a copper block, the film was investigated by X-ray diffraction. The same piece of the thin film was annealed further for a longer time (e.g., 20 s) and the structure was investigated again by XRD. This process was repeated many times. The film was investigated by cross-section transmission electron microscopy to determine how much PZT had been transformed into perovskite at the end of the earlier stages.

The authors used the Avrami equation $x(t) = 1 - \exp(-kt^n)$, where $x(t)$ is the transformed fraction at the time t, to model the transformation from pyrochlore to perovskite. The transformed fraction should be proportional to the integrated area under the PZT(111) diffraction peak, which was normalized to the same peak area for the fully transformed PZT thin films. n is the Avrami coefficient whose value normally varies from 1 to 3, and, providing that there is no change in nucleation mechanism, n is independent of temperature. On the other hand, k is dependent on the nucleation and growth rates and is very sensitive to temperature.

The XRD investigations performed during the annealing stages revealed that the thin films were fully transformed into perovskite after annealing for 5400 s at 460°C. The authors investigated the formation and evolution of the metastable Pt_3Pb phase as a function of annealing time at temperatures of 400, 440, 480, and 500°C. A strong Pt_3Pb peak was observed for films heated for no more than 5 s at 400°C or higher temperatures. The Pt_3Pb phase developed after putting the films onto 350°C hot plate for 12 s and onto a 330°C hot plate for 30 s. However, this intermetallic phase was not observed when the film was annealed at 300°C, even after 600 s, so that the authors concluded that the lowest temperature for the formation of this phase in their material system was 330°C. The formation of the Pt_3Pb intermetallic phase was found to be greatly influenced by the film drying temperature. The perovskite phase formation was identified after annealing at 440°C for 15×60 s for a three-layer PZT thin film. Concerning PZT growth kinetics, it is believed that PZT starts nucleating and growing on top of the intermetallic phase Pt_3Pb, rather than directly on Pt (as demonstrated by cross-section transmission electron microscopy). The benefit for PZT to nucleate on Pt_3Pb rather than on Pt is that there is a better lattice match between PZT ($a_0 = 4.035$ Å) and Pt_3Pb ($a_0 = 4.050$ Å) than between PZT and Pt ($a_0 = 3.9231$ Å). This fact can help the nucleation process. This assumption was supported by the evidence that well-(111)-oriented perovskite PZT was produced at temperatures as low as 440°C.

A particular behavior of the Ti at the interface is also evidenced in [121]: thus, a Ti-rich layer is formed, followed by a slight depletion in film and an increase of the Zr content. This Ti-rich layer influences the columnar perovskite growth of the film. The effect was evidenced for different compositions: 30/70, 53/47. However, the intermetallic layer consisting of Pt_xPb compound is always present, with a variation of stoichiometry depending on the firing and/or annealing temperature.

3.3.2. Zinc Oxide

Different precursor solutions and routes have been used for ZnO thin film deposition. Thus, in [134] aqueous precursors consisting of 6 g of $Zn(NO_3)_2 \cdot 6H_2O$ were mixed with 2.25 g of glycine (H_2NCH_2COOH) and dissolved in 20 g deionized water. The obtained solution was stirred at 80°C for 1 h, cooled, and nucleopore filtered and then deposited by spin-casting (3000 rpm) on silica or silicon for 50 s at room temperature. The annealing temperature was set at 400°C. Another route is proposed in [135], where zinc acetate [$Zn(CH_3CO_2) \cdot 2H_2O$] is dissolved in anhydrous ethanol or methanol. The maximum solubility was 6 wt% in ethanol and 12 wt% in methanol, without lactic acid addition. The deposition was performed using the

drain-coating method: the substrate was fixed and the solution was drained from the vessel. This requires a careful control of the concentration, viscosity, and surface tension of the solution, as well as vapor pressure, temperature, and humidity above the coating bath. The as-prepared film is introduced into an electrically heated furnace and annealed at temperatures in the range 300–500°C, in air or in vacuum, using a heating rate no higher than 10°C/min. Because the obtained average thickness in a cycle was very small, the films were the results of several deposition–annealing cycles.

The films were polycrystalline in both previous cases; the peaks identified in the spectra are the same as for the ZnO powder. The annealing process induces the conversion from an amorphous to a crystalline phase, but increasing the temperature does not result in improvement of c-axis orientation. The growth mechanism proposed in [134] is based on the coordination of the zinc cations in an organic glass forming matrix along with nitrate counterions. Due to the solvent removal during the spin-casting process, a uniform distribution of Zn^{2+} and nitrate anion inside the amorphous glycine matrix will be reached. The further heating induces the oxidation of the organic radicals by nitrite and the formation of a pure zinc oxide phase. Two features have been observed: the c-axis orientation coexists with a random polycrystalline phase and the linewidths are broader, which indicates the presence of lattice defects. The films deposited on silica were found to be stable at temperatures below 800°C. At 1000°C the wurtzite phase is no longer present in the XRD spectra; instead, the peaks assigned to compounds as Zn_2SiO_4 appear.

3.4. Chemical Vapor Deposition

Chemical vapor deposition is one of the most known and used techniques for thin film growth. The synthesis method is based on the reaction of the constituents in the vapor phase and the formation of the solid film on the substrate surface [136]. It has some important advantages with respect to other techniques: large-scale processing, step coverage, and control of composition. Among the disadvantages can be mentioned the high-temperature processing and the corrosive chemical compounds.

It consists of several steps, which involve physical and chemical processes as well. These steps are summarized in [136] and consist of chemical reactions in gas phase and on the substrate surface, transport phenomena (mass transfer, fluid flow, and heat transfer), and the film growth. Each step is complex and requires an interdisciplinary treatment and interpretation.

The processes in a CVD reactor are generally heterogeneous, the (film) deposition taking place on the substrate; a homogeneous reaction results in formation of powders. The most frequent chemical reactions which take place are pyrolysis, reduction, oxidation, hydrolysis, nitride formation, carbide formation, chemical transport reaction, etc. Very often combined reactions take place, as is the case of III–V compound deposition, where transport reaction of the metal is followed by hydride (V) decomposition and the deposition reaction.

Even if CVD is far from thermodynamic equilibrium, the thermodynamic consideration are mandatory for process control and prediction starting from some parameters such as temperature and pressure in the reactor and the initial reactant proportion. The thermodynamic equilibrium assumed for the chemical reactions is established by using some models: the most known are the optimization method and the nonlinear equation method [136]. The first method is based on the calculation of the number of moles from each gaseous and solid species from the minimum free energy condition of the system. This requires knowing the free energy values of formation of all components of the system. The second method allows the calculation of partial pressure of all gaseous species from the mixture, for an established set of experimental conditions: total pressure, temperature, and input reactant concentration. The CVD kinetics involves several processes that control the film growth: these are related to the surface and the reactant gas. The most important are as follows: impinging and adsorption of reactant on the surface; desorption of species from the surface and their diffusion away; and, very important, the surface chemical reactions and modification. The deposition rate depends also on the substrate temperature and orientation.

The CVD reactor is usually constructed as a function of the film characteristics and requirements related to the uniformity in thickness and composition and the efficient growth rate. Because the transport phenomena essentially influence the film properties, the careful control of the gas flow and temperature distribution inside the reactor is compulsory. The most important parameters are gas (vapor) characteristics, flow velocity, mixture pressure, and system geometry.

The film nucleation and growth mechanisms are classical: (i) three-dimensional island growth (Volmer–Weber), two-dimensional full-monolayer growth (Frank–van der Merwe), and (iii) two-dimensional growth of full monolayer followed by nucleation and growth of three-dimensional islands (Stranski–Krastinov) [137]. These models are based on statistical mechanics and chemical bonding characteristics of solid surfaces. The surface sites with strong bonds act as nucleation sites: when a nucleus reaches a dimension appropriate for growth rather than evaporation, it can contact the neighbors and coalesce to form a film. The structure of the film strongly depends on the deposition parameters. Epitaxial films are quite difficult to obtain because the parameter window is very narrow. Most often the films are polycrystalline or amorphous: the polycrystalline films are mainly deposited at low substrate temperature and high gas phase concentration. The species mobility on the surface is small and the resulted nuclei create differently oriented grains. Further decreasing the temperature can result in an amorphous structure.

Different categories of piezoelectric materials have been deposited in the form of thin films by CVD [138–143]: n-type GaN films were deposited on 800°C heated sapphire substrate starting from trimethyl gallium [$(CH_3)_3Ga$] and ammonia [138]. AlN films were grown on sapphire substrate at a much higher temperature, 1200°C, from trimethyl aluminum [$(CH_3)_3Al$] and ammonia [138]. LiNbO$_3$ was also deposited by CVD starting from different precursors. Thus, in [141] niobium pentaethoxide [$Nb(OC_2H_5)_5$] and dipivaloylmethane [$Li(C_{11}H_{19}O_2)$] [Li(DPM)] were used. Li(DPM) is a hygroscopic powder with melting point 280°C whereas $Nb(OC_2H_5)_5$ is liquid at room temperature. Special precautions have been taken: thus, each compound was kept at room temperature in a stainless steel vessel and the vapors were transported in Ar gas and then mixed with O$_2$ and Ar through a tube heated at 210°C to avoid condensation. The film deposition was performed on a LaTiO$_3$ single crystal placed inside an Inconel susceptor heated between 600 and 700°C by radiofrequency. The total pressure inside the reactor was 13 mbar (1300 Pa) and the Li/Nb ratio was controlled by varying the Nb source temperature between 135 and 145°C; the chemical processes involved were pyrolysis and/or oxidation. High-quality epitaxial films have been obtained for substrate temperature 700°C and (001)LaTiO$_3$ substrate; these films were found also to exhibit the best piezoelectric properties [141].

Lead niobium titanate Pb(NbTi)O$_3$ piezoelectric films were prepared by metalorganic chemical vapor deposition [142]. The precursors used were [$Pb(H_{11}C_{19}O_2)_2$], pentaethoxy niobium [$Nb(O{-}C_2H_5)_5$] and tetraisopropoxy titanium [$Ti(O{-}i{-}C_3H_7)_4$]. Different substrates such as Pt/Ti/SiO$_2$/Si(100), PbTiO$_3$/Pt/Ti/SiO$_2$/Si(100), and (100)SrRuO$_3$ plated (100)LaAlO$_3$, heated at temperatures in the range 500–620°C, have been used. The oxidant gas was oxygen, the carrier gas Ar, and the pressure inside the reactor was 665 Pa (6.65 mbar). The film composition was found to depend on the input gas flow rates and on the ratio between the amount of different precursors, but was independent of temperature. The increasing of niobium content in the film (controlled by the flow rate of Nb-containing gas with an excess of Pb) resulted in the appearance of a pyrochlore phase together with the perovskite one. From the variation of the c/a ratio of lattice constants a solubility limit of Nb in Pb(Nb,Ti)O$_3$ phase of 5% was found.

Single-phase ZnO films have been deposited by metalorganic chemical vapor deposition in a rotating disc reactor [143]. The carrier gas was a mixture of N$_2$ and Ar. The precursor, diethyl zinc [DE Zn; $(C_2H_5)_2Zn$], and the oxygen were injected into the reactor separately, to avoid the mixing of reactant in the gas phase. The substrates were C-plane and R-plane sapphire heated at temperatures between 250 and 600°C and the pressure inside the reactor was 24–60 mbar. Epitaxial films, compatible with applications in surface acoustic wave (SAW) devices, have been grown on R-plane sapphire: they exhibit a dense and smooth columnar structure, with a semicoherent interface. The postdeposition annealing treatment at temperatures as high as 1000°C improved the crystallization but generated a spinel interlayer (Al$_2$ZnO$_4$) of 15 nm thickness between the ZnO film and the sapphire substrate. An upper temperature limit of 750°C for the annealing process was found.

Primary beam	Secondary signal	Technique	Acronym	Application
Photon	X-ray	X-ray diffraction	XRD	Crystalline structure
Photon	Electron	X-ray photoelectron spectroscopy	XPS (ESCA)	Surface composition
Laser	Ion	Laser microprobe		Composition of the deposited film
Ion	Ion	Secondary ion mass spectroscopy	SIMS	Film composition versus depth
Ion	Ion	Rutherford backscattering	RBS	Composition versus depth
Electron	Electron	Scanning electron microscopy	SEM	Surface morphology
Electron	Electron	Auger electron spectroscopy	AES	Surface film composition
Electron	Electron	Transmission electron microscopy	TEM	High-resolution (crystalline) structure
Electron	Electron	Electron microprobe	EMP	Film composition

4. CHARACTERIZATION METHODS

The analytical techniques used for solid characterization are based on the interaction between electromagnetic radiation, electrons or ions, and the solid, followed by examination of the emitted secondary particles or radiation. Three classes of structures are investigated by these techniques: external surfaces, thin films (or layered structures), and interfaces. Some combinations of incident and secondary emitted particles and radiation used for film characterization are presented in Table VII [144, 145].

Each of these techniques gives important information about the film composition, texture, crystallinity, and uniformity and about the surface roughness and contamination at the interface between the substrate (electrode or buffer layer) and the film.

The technique and the corresponding instruments used are based on fundamental physical phenomena. Thus, it is known that crystalline materials are organized in ordered periodic arrays which can scatter the incident wave via diffraction processes. The atomic planes can diffract the beam of wavelength λ comparable with their interatomic distances according to the Bragg condition $2d_{hkl} \sin \theta = n\lambda$, where d_{hkl} is the interplanar distance, h, k, and l are the Miller indices of the diffracting plane, n is the order of diffraction, and θ is the angle that the primary beam makes with the diffracting plane. This is valid for X-ray beam and also for electron beam: according to De Broglie's theory, the wavelength associated with a charged-particle beam accelerated by a potential difference V is given by $\lambda = h/(2mVe)^{1/2}$, where h is Planck's constant and m is the particle mass. X-ray diffraction and transmission electron microscopy give, to a certain extent, similar information concerning the crystallite orientation in the film. Thus X-ray diffraction analysis yields information about preferential orientation, grain size, and compound identification. Electron diffraction spectra contain a periodic array of spots, which are few for a single-crystal material and become diffraction rings for a polycrystalline film. Each spot in the ring is produced by a different crystallite which is properly oriented for the particular Bragg condition [144].

Other characterization techniques are aimed at investigating the piezoelectric, dielectric, and elastic properties of thin films. These methods are based on piezoelectric interaction of electrical and elastic fields in the film, on dielectric absorption, elastic wave propagation, acoustooptic interaction, etc.

4.1. X-Ray Analysis

As will be discussed in Section 5, the texture and the crystalline phases present in the film directly influence the piezoelectric and electromechanical coupling coefficient, dielectric constant, remanent polarization, etc. X-ray diffraction, with its facilities such as rocking curve measurements, grazing incidence, and polar figure, is the most widely used technique for piezoelectric thin film characterization because it allows obtaining direct, sometimes rough information about the degree of crystallinity, grain orientation, and piezoelectric or nonpiezoelectric phase. The simplest analysis consists of peak identification, which allows identification of the crystalline phases of the film and of the presence of amorphous or secondary phases, as well as structure randomness or orientation (texture). A particular case is that of films grown on good matching (or "domain matching") single-crystalline substrates. X-ray analysis helped to find the most relevant deposition parameters for the epitaxial growth. Thus, in [90, 97] $(00l)$ epitaxial PZT film growth on $SrTiO_3$ and MgO single crystals by PLD has been reported. It is important to underline that the films have been deposited at the same temperature and oxygen pressure ($T = 550$–$560°C$ and $p(O_2) = 0.25$ mbar) but with different laser wavelength and fluence. There is an indication that the growth mechanisms of PZT perovskite film is essentially governed by the substrate temperature and the oxygen pressure. A similar epitaxial growth is identified in [100] for P(N)ZT on LSCO/Pt/TiN/Si multilayer structure.

An exclusive c-axis orientation of ZnO films grown by PLD on Si(100) single-crystal substrate was evidenced by X-ray diffraction completed by rocking curve measurements [62, 64]. The single-crystal sapphire substrate was also found to promote the single-phase growth of $LiNbO_3$ [115] as well as AlN [78]

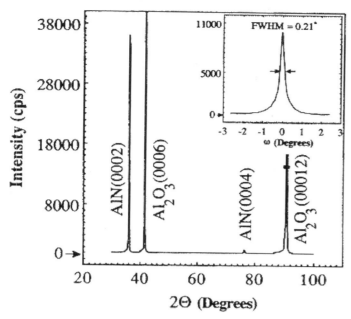

Fig. 19. XRD pattern of epitaxial AlN film deposited by laser ablation on sapphire substrate, in vacuum, $T_s = 800°C$ and $E_s = 2$ J/cm^2. The rocking curve FWHM of the (002) peak (upper right corner) is 0.21. Reprinted with permission from [78], © 1995, American Institute of Physics.

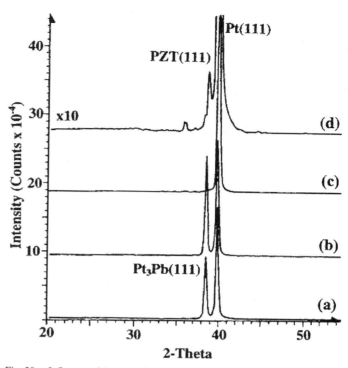

Fig. 20. Influence of the annealing time on the phase formation for PZT films deposited by sol–gel, dried at 200°C, and annealed at 400°C for: (a) 10 s, (b) 150 s, (c) 780 s, (d) 5400 s. The presence of Pt$_3$Pb intermetallic phase promoting the PZT(111) growth is evidenced. Reprinted with permission from [126], © 1999, American Institute of Physics.

by PLD. Again the gas pressure and the substrate temperature were found to essentially influence the growth process. Thus, for LiNbO$_3$ layers three parameters act simultaneously and complementarily for promoting the correct stoichiometry and orientation. If a high substrate temperature (800°C) favors the development of the crystalline structure, it simultaneously produces Li depletion. On the other hand the oxidant atmosphere (enriched by adding 8% O$_3$) pressure, combined with an appropriate Li/Nb ratio in the target, allows the completely epitaxial growth of the film in three directions, with the [110] axis of LiNbO$_3$ parallel to the [001] sapphire axis. This is evidenced by pole figure of the {012} planes of the LiNbO$_3$ and the (300) planes for the substrate. AlN films deposited at medium laser fluence, in vacuum, were found to have an improved crystallinity with increasing substrate temperature. FWHM of the rocking curve reflects the distribution of the crystals with a definite orientation with respect to the film surface normal. The corresponding rocking curve, which is broad for temperatures below 700°C, had 1.2° width at 750°C and only 0.21° at 850°C (Fig. 19). Good alignment, but not single-phase growth, is demonstrated to be obtained at the much lower temperature of only 550°C, but for a higher laser fluence, 12 J/cm^2.

X-ray diffraction studies followed the step-by-step growing process in the case of the sol–gel technique. It allowed one to check the best way to deposit crystalline, (111) preferentially oriented PZT films [122]. PZT layers deposited at low temperature were found to be amorphous: subsequent direct annealing treatment at temperatures between 500 and 650°C promoted an increase of the (111) peak intensity, as was visible from X-ray spectra. It was also observed that the (111) perovskite

peak appeared at 500°C, and, at 650°C, a relative peak intensity $I(111)/[(I(111) + I(100)]$ of 0.96 was obtained. Films pyrolyzed at 400°C and then annealed at 550°C exhibited lower (111) orientation degree and less performant electric properties.

Other important information given by X-ray spectra is identification of compounds generated at the film–substrate interface. Thus, in [122] an intermetallic compound Pt$_3$Ti was identified to form at the interface between the PZT layer and the Pt/Ti/SiO$_2$/Si structure due to the Ti underlayer diffusing through the Pt layer. The lattice matching of this compound with (111)PZT perovskite generates nucleation sites for PZT(111) growth. X-ray spectra of films deposited on substrates with different underlayers without Ti, such as (Pt–Rh/Pt–RhO$_x$/Pt–Rh or Pt/glass) evidenced the presence of pyrochlore phase at an annealing temperature of 550°C. A similar evolution of an intermetallic compound grown at the interface between the PZT film and the bottom electrode, for PZT films grown by sol–gel, was investigated by X-ray diffraction studies [126]. The temperature threshold for this peak growth was found to be 330°C and it nucleates from the Pt–PZT interface toward the middle of the Pt layer. Its X-ray diffraction peak intensity can be correlated with those of PZT(111) versus annealing temperature. The growth kinetics of PZT(111) perovskite phase consist of the nucleation on top of this intermetallic phase and the appearance of the (111)PZT peak coincides with the extinction of the (111)Pt$_3$Pb peak (Fig. 20). Systematic X-ray diffraction analysis performed

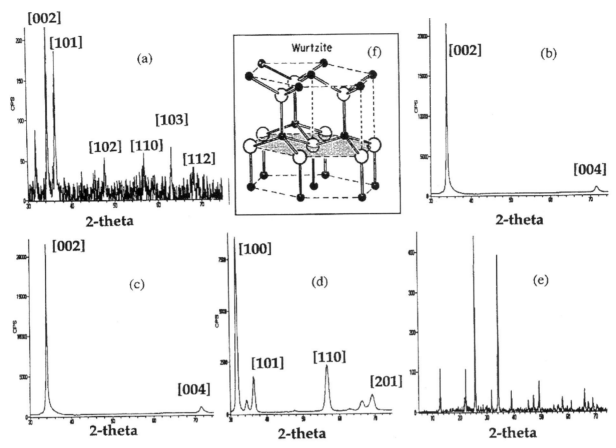

Fig. 21. Influence of the preparation method on the crystalline structure of ZnO thin films: (a) solution deposited; (b) pulsed laser deposited; (c) rf sputtering deposited; (d) 8 mm thick film deposited by rf sputtering; (e) solution deposited, annealed at 1000°C for 1 h in air; (f) crystalline structure of ZnO hexagonal wurtzite phase. Reprinted with permission from [134], © 1995, Elsevier Science.

on pulsed laser deposited PZT films grown using SiO_2 as buffer layer has been reported in [102]. The optimum range of SiO_2 thickness was identified, and it was correlated with the formation of small crystallites at the interface between Si and SiO_2 layer. These crystallites, present in high proportion for a layer in the range 40–100 Å, promote the (111) phase nucleation of PZT compound. A thicker layer has an amorphous structure which generates a randomly oriented PZT and a thinner layer cannot avoid the Si–PZT interdiffusion.

The secondary phases in the ZnO films obtained by different techniques can also be identified by X-ray diffraction (Fig. 21) [134]. As can be seen, the sol–gel deposited layer is polycrystalline, with a slight c-axis orientation, whereas films deposited by sputtering and pulsed laser deposition exhibit a highly oriented wurtzite phase, with the c axis perpendicular to the surface.

4.2. Electron and Atomic Force Microscopy

Transmission electron microscopy is a more precise and sophisticated technique for the analysis of the crystalline phases present in the deposited film. It gives images of microstructural features at magnifications between 1000 and 450,000, with a

resolution smaller than 1 nm, identifying crystal structure and orientation for dimensions greater than 30 nm and having a lattice imaging of crystals with interplanar distance greater than 0.12 nm. The electron beam path through a transmission electron microscope is similar to that of light in the ground glass lenses of an optical microscope working in a transmitted light mode, except that the electrons follow a spiral path through the lenses. The images that result only from transmission electrons are known as bright field images whereas those formed using specific diffracted (*hkl*) beams are called dark field images. Selected area diffraction (SAED) patterns can be used to determine the crystal structure of a given phase, identify the structure and the orientation of a given crystal, and determine the orientation relationship between different coexisting phases [145].

This method was used for the identification of the phases present in the deposited layer. Thus, in [60] the wurtzite phase was identified. The spots from SAED images corresponding to (002) diffraction planes are elongated: they give information about the fact that ZnO film deposited by laser ablation of a Zn target in an oxygen atmosphere developed a columnar structure with the c axis perpendicular to the film surface.

Cross-section TEM is a more appropriate technique for determining the interface behavior, in-plane orientation of the

Fig. 22. SEM image of a cross section of a PZT film with (100) preferred orientation. Reprinted with permission from [128], © 2000, Elsevier Science.

layer with respect to the substrate, and crystalline phases present in the film. The SAED image of a cross-sectional specimen prepared from a AlN layer deposited by PLD at 750°C, in vacuum ($p = 3 \times 10^{-7}$ mbar) [77] shows that the film is single crystal. The in-plane orientation which was identified to be [2110]AlN plane is aligned with Si[011]: for the other two directions AlN[0110]‖Si[422] and AlN[0001]‖Si[111]. Using the same technique, the evolution of the Pt–PZT interface for films deposited by the sol–gel method as a results of different intermediate thermal treatments has been observed [120]. Thus, for films dried at 200°C and subsequently annealed at 440°C for 7×60 s, a pyrochlore phase was evidenced near the PZT layer as can be observed from the presence of white dots just near the Pt film. Corroborating the images with the X-ray spectra, it was deduced that the intermetallic layer of Pt_3Pb formed also at this interface and it was covered by a small amount of crystalline PZT. It was observed that the perovskite grains were growing in the form of a uniform plane front, not in the form of islands, as evidenced for PZT layers directly deposited on MgO. This can be explained by an epitaxial nucleation and growth due to the very small lattice mismatch between Pt_3Pb and PZT(111). Increasing the annealing time to 13×60 s, the cross-section TEM images indicate that the Pt_3Pb layer disappeared, whereas the presence of Pt grains having the same orientation as the PZT grains was identified. Thus, PZT grew heterogeneously directly on the Pt grains resulting from the Pt_3Pb decomposition during the annealing process. For 20×60 s annealing time TEM images evidenced that PZT growth developed in a "plane frontier" and the perovskite crystals had a columnar structure. The absence of any chemical reaction during the interdiffusion process was evidenced by the bright field TEM image of a LSCO/PNZT/LSCO/Pt/TiN/Si structure deposited by PLD [100]. This indicated that the Pt layer acted as a good barrier for oxygen diffusion and LSCO is an appropriate buffer-electrode layer for perovskite growth, its interfaces with the

piezoelectric film being quite sharp and free of reaction products.

The configuration of atoms in the crystalline lattice can be established by high-resolution TEM (HR-TEM) analysis. In the case of multilayer structures, the interface characteristics are the main features to be examined. Such an image of a ZnO layer deposited by pulsed laser deposition on a (0001) sapphire substrate using a GaN buffer layer shows that the interface is sharp and the lattice planes of ZnO are perfectly aligned with those of the GaN layer [67].

The application of piezoelectric materials to device construction implies that the surface roughness is low enough to allow the deposition of top electrodes for insertion in complex structures. The techniques frequently used for surface investigations are scanning electron microscopy and atomic force microscopy. In the SEM technique the surface image is the result of the backscattered electrons detected when an electron beam irradiates a sample surface. The general aspect of the surface can be evidenced from the SEM images. For example, the presence of droplets on the surfaces of the films grown by PLD is frequently proved by means of this technique. To avoid droplet formation, UV laser sources are used instead of IR laser sources. For an automated PZT laser ablation deposition process [93] the SEM pictures revealed a smoother surface when already prepared PZT targets were laser ablated than in the case of metallic oxides used in the form of high-density targets. Cross-section SEM images evidenced the columnar structure of ZnO thin films deposited by laser ablation [58] as well as for PZT thin films prepared by both PLD and sol–gel [99, 128] (Fig. 22).

AFM is another technique used for surface roughness studies, with a higher sensitivity than to SEM. Comparative AFM studies evidenced that the surface roughness was lower for ZnO films deposited by PLD relative to those deposited by sol–gel or sputtering [134]; they also proved that successive layer addition in the sol–gel process does not increase the surface roughness.

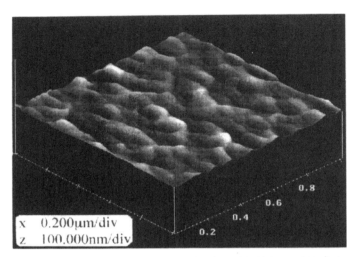

Fig. 23. AFM image of a Pb(Zr$_{0.7}$Ti$_{0.3}$)O$_3$ film with thickness 5000 Å deposited by PLD at 500°C substrate temperature. Reprinted with permission from [96], © 1996, Elsevier Science.

The AFM images of a PZT film surface prepared by PLD [102] revealed higher grain dimensions for high deposition temperature. The annealing process of films deposited at high substrate temperature induced the enhancement of the grain size from 950 Å to 1600 Å for films annealed for 15 min at 800°C. AFM also allowed the observation that very smooth surfaces are obtained for Pb(Zr$_{0.7}$Ti$_{0.3}$)O$_3$ films deposited by PLD [96] (Fig. 23). AFM images allowed the identification of small pyrochlore islands on the surface of the films deposited by sol–gel and annealed at 550°C [122].

4.3. Compositional Analysis

Ion-based techniques have been used for the quantitative compositional analysis of thin films, characterization of the uniformity in depth profile of elements inside the film, and characterization of the interface layer. The most widely used are Rutherford backscattering and secondary ion mass spectroscopy.

Rutherford backscattering spectrometry gives information about composition, but not about chemical bonding between elements. It is based on the two-body elastic collision process which occurs during the interaction between an ion beam and the (heavy) atoms of a target. When the incident ion energy is known, the measurement of the scattered ion energy is directly related to the mass of the target atom where scattering takes place [145]. Thus the backscattered energy signal can be translated in a mass spectrum. The energy loss measurements give the depth where the atom scattering the incident ions is placed. This technique was used for the characterization of the piezoelectric films deposited by different methods. From a RBS spectrum of a PZT film deposited by the sol–gel method [146] it was found that the Zr/Ti ratio in a film treated at only 350°C for 15 min is the same as in the precursor solution, that is, 52/48. The estimated thickness was confirmed by other (spectroellipsometric) measurements. The magnitude of

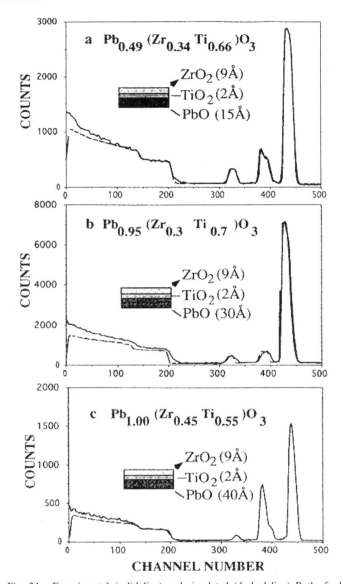

Fig. 24. Experimental (solid line) and simulated (dashed line) Rutherford backscattering spectra of PZT thin films deposited by the automatic PLD technique. The influence of the PbO layer on the film stoichiometry is evidenced. Reprinted with permission from [93], © 1993, American Institute of Physics.

the Si energy step in the spectrum of a film annealed at 400°C corresponds to a dense amorphous layer having a step profile, without any interaction at the PZT–Si interface [146]. Important information concerning the influence of the oxide layer thickness and annealing temperature on the stoichiometry of the PZT films deposited by an automatic PLD technique is given in [93]. Thus, values of 9 Å for ZrO$_2$, 2 Å for TiO$_2$, and 40 Å for PbO have been identified to be the most appropriate for growing PZT layers with the desired (near morphotropic phase boundary) stoichiometry (Fig. 24).

The channeling measurements indicate that the epitaxial growth by PLD of ZnO films on (0001) sapphire substrate is strongly influenced by the oxygen pressure during deposi-

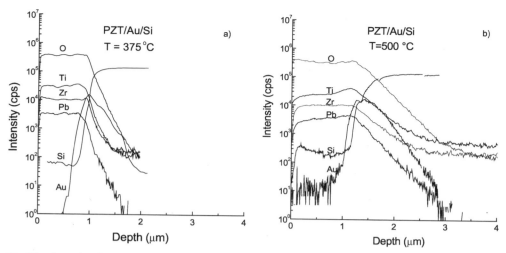

Fig. 25. Secondary ion mass spectroscopy of PZT films deposited by PLD on Au/Si substrate at different temperatures: the increase of the interdiffusion layer is observed as a result of the substrate temperature.

tion [70]. For better control, a two-step process is suggested: during the first step the nucleation layer develops in low oxygen pressure (10^{-4} mbar); subsequently, the pressure increases to 10^{-1} mbar so that the second layer will be grown epitaxially, using the first layer as a template layer.

During SIMS analyses a high-energy ion beam interacts with a target in a vacuum or ultrahigh vacuum chamber. The momentum transferred to the target surface produces the sputtering of atoms and molecules, which are ejected with positive or negative charges (secondary ions). These ions are analyzed by a double focusing mass spectrometer or by an energy filtered quadrupole spectrometer [144, 145]. The method enables us to obtain information about the surface, near surface, and interface as well as about the uniformity in the depth profile of an element in a film. The technique was successfully used to evidence the uniform Al, N, and Ga depth profile in AlN [81, 82] and GaN [57] thin films deposited by laser ablation of pure metallic targets in a nitrogen reactive atmosphere. It also allowed us to establish the low oxygen contamination level for a nitrogen pressure of the order of 10^{-2} mbar, after a pumping down until a base pressure of 10^{-6} mbar. SIMS was used to characterize the interface between the deposited layer and the bottom electrode. The SIMS spectra demonstrate that the interface between the PZT layer and the Au bottom electrode strongly depends on the deposition temperature, as can be observed in Figure 25. Thus, for 500°C a strong interdiffusion between the Au layer, the silicon substrate, and the PZT film was observed, whereas for films deposited at 375°C the profile was sharper and the interdiffusion less significant [109].

A more rapid technique for investigation of film composition is energy dispersive X-ray analysis (EDX). This technique is based on the spectral analysis of the X-rays emitted by samples subjected to bombardment with an electron beam. Using this technique the PZT film composition dependence on oxygen pressure for films deposited at room temperature by PLD was studied in [101]. The Pb/Ti, Zr/Ti, and Zr/Pb ratios were found to be independent of the laser fluence, whereas the lead content was found to increase with the deposition pressure. The multistep sol–gel PZT deposited films exhibit a cyclic composition variation of the metallic elements corresponding to the interfaces between the successive layers grown, pyrolyzed, and annealed [121].

Electron spectroscopy studies are used for obtaining information concerning film composition and depth compositional profiling. The technique uses electrons or X-rays as probing source and information about the surface chemistry is obtained by analyzing the electrons emitted from the surface [144, 145]. Auger electron spectroscopy and X-ray photoelectron spectroscopy (XPS) consist of precise determination of the number of secondary electrons as a function of kinetic energy [144, 145]. The AES two-step process involves the emission of an inner shell electron (K, L, M) followed by the emission of an Auger electron due to the excited state of the atom as a result of the first electron emission. XPS takes advantage of the photoelectron emission as a result of the irradiation with an X-ray beam. In this process, both photoelectrons and Auger electrons are generated. The two techniques were successfully used for chemical composition analyses of different piezoelectric films. Generally, the investigated depth in both techniques is of the order of several angstroms to several tens of angstroms. For in-depth information ion sputtering is performed for material removal. In this way a step-by-step analysis is possible. XPS and AES studies performed in conjunction with TEM evidenced the variation in the film composition for (70/30) PZT layers obtained by the sol–gel method on Pt/Ti/SiO$_2$/Si structures [121]. To avoid film cracking, successive layers have been added to grow a thicker film. After each spin-coating process the structure was heated until 510°C for water and organic solvent evaporation. XPS depth profile revealed some important features. The Pb content, higher at the surface, decreases rapidly and then keeps constant throughout the film depth. Ti and Zr concentration are complementary and exhibit a cyclic fluctuation (Fig. 26). For a single PZT layer the same concentration complementarity was observed, with a higher value of Zr at

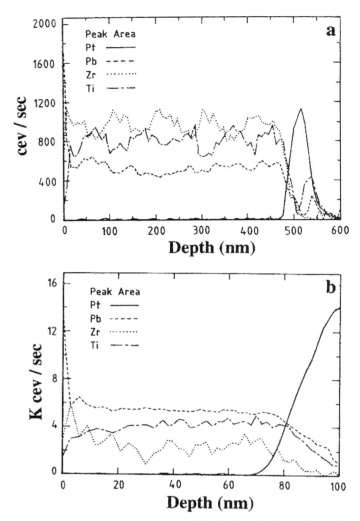

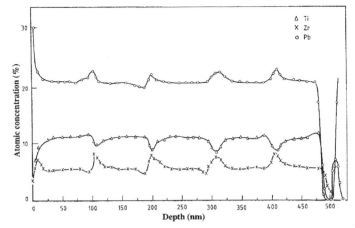

Fig. 27. Compositional depth profile measured by Auger electron spectroscopy of a PZT film deposited by five subsequent sol–gel processes. The compositional variation at the interface is evidenced. Reprinted with permission from [121], © 2000, American Institute of Physics.

Fig. 26. Compositional fluctuation of (a) PZT film deposited by multistep sol–gel process and (b) single layer of PZT deposited by sol–gel, as resulting from XPS depth profile. Reprinted with permission from [121], © 2000, American Institute of Physics.

the surface. Ti concentration slowly increases from the surface toward the Pt layer, whereas Zr reaches a minimum value at the half of the film thickness. For multilayer films, this minimum of the Zr concentration is reached at the edge of each layer. XPS studies have been performed to check the stoichiometric uniformity of films deposited at 370°C by pulsed laser deposition using a Nd–YAG laser [147]. Going from the center of the substrate toward the edges the Zr/Ti ratio was found to be constant, whereas a decrease in lead content was observed. Simultaneously, lead bonded in a (pyrochlore) compound different than stoichiometric PZT was identified. It can be assumed that it is a plasma plume evolution effect, where the spatial distribution of different species is determined by the laser fluence–gas pressure binomial.

A systematic investigation of the composition of a PZT target used for laser ablation is reported in [105]. The evolution of the Pb/(Zr + Ti) and Zr/(Zr + Ti) ratios was studied by XPS depth profile in conjunction with an etching process: the val-

ues were compared with the results obtained by a calibration equation. The investigation of the target surface showed a lead excess and a lack of Zr. From bonding energy of Pb $4f$ in PZT and PbTiO$_3$ it was concluded that the PbTiO$_3$ was formed on the surface due to the Pb absorbed by the dangling bonds of a Ti-rich surface layer.

Auger depth profile describes more accurately the elemental concentration evolution because of the small area the signal comes from (600 μm^2) [121]. The cyclical evolution evidenced by XPS is confirmed (Fig. 27) together with the complementary distribution of Zr and Ti.

4.4. Piezoelectric, Dielectric, and Elastic Characterization

4.4.1. Piezoelectric Measurements

Characterization of piezoelectric properties of thin films can be based on either the direct or the converse piezoelectric effect [148]. In direct measurements the voltage produced by an applied stress is measured; in indirect measurements the displacement produced by an applied electric field is measured through different methods. There are two main classes of direct film measurements: quasi-static measurements and dynamic measurements. The first is typically used for nonresonant applications (sensors and actuators); the second is typical for those applications where the piezoelectric thin film is used for the generation or detection of high-frequency bulk acoustic waves or surface acoustic waves.

Various methods have been described for the measurement of the piezoelectric coefficients. Usually, for the measurement of the longitudinal piezoelectric coefficient d_{33}, a calibrated stress is applied to the film and the generated charge is collected. This requires the application of an electrode on top of the film surface and, moreover, it is not easy to implement as a reliable quantitative method. Nevertheless, a preliminary quantitative check of the quality of the piezoelectric film can be useful just after its deposition and before the realization of

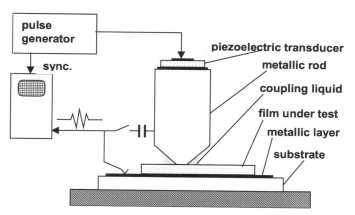

Fig. 28. Schematic of the device for direct d_{33} measurement.

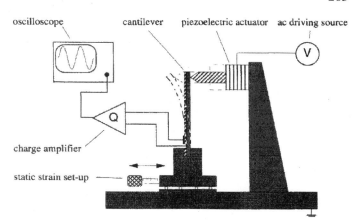

Fig. 29. Schematic view of the setup for the measurement of the transverse piezoelectric coefficient. Reprinted with permission from [151], © 1999, Elsevier Science.

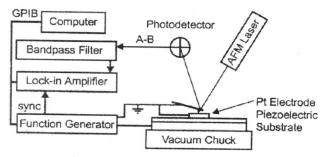

Fig. 30. Schematic of the setup for piezoelectric measurements through an AFM. Reprinted with permission from [152], © 1998, American Institute of Physics.

the complete device. For this purpose many techniques have been described. An interesting one, used by the authors, was described in [149]. A sketch of the arrangement is shown in Figure 28. The low-frequency transducer (2 MHz), bonded on one end of a metallic rod, when driven with a short pulse (0.1–1 ns) produces a longitudinal wave propagating in the rod. The other end of the rod, shaped as a truncated cone with a small flat tip, is put in contact with the film using a droplet of water. This ensures a good mechanical energy transfer from the rod to the film. The consequent voltage generated in the stressed region is picked up from the bottom metal layer and the rod that acts as ground electrode. This voltage is fed to an oscilloscope and measured. To evaluate the d_{33} constant, the measured voltage is compared with that obtained from a known thin piezoelectric plate measured in the same conditions. An important and original feature of this method is the possibility to trace a map of the d_{33} over the deposited area without depositing any top electrode. The sensitivity of the system can be estimated to be less than 1 pC/N, that is sufficient for the common piezoelectric thin films.

The evaluation of the transverse e_{31} piezoelectric constant requires different techniques. An example of a simple method is described in [150]. Here the film is deposited on a cantilever beam and two electrodes are deposited on the free film surface. The free edge of the cantilever is first statically displaced, causing a bending of the cantilever, and then it is suddenly released. The consequent damped oscillation of the cantilever at its natural frequency can be revealed through the voltage appearing between the electrodes and displayed on an oscilloscope. It can be demonstrated that the first maximum of the damped oscillation is proportional to the e_{31} constant. More precisely, because the deformation is not a pure elongation, the exact value should be calculated considering the bulk tensor properties; this implies the knowledge of other material constants. A similar technique consists of measuring the electrode charge not at the resonant frequency of the cantilever but at a frequency imposed by an external exciter moving the free edge of the beam. An example is reported in Figure 29 [151].

The piezoelectric constant d_{33} can be measured also through the converse effect. The order of magnitude of the displacement

that should be measured on piezoelectric thin films is in the range of units or a fraction of an angstrom. Sensitive methods able to detect such small deformations are interferometers and more recently AFM.

AFM has been used for this purpose in various configurations in order to minimize the effect of the applied electric field with the microscope tip. An example is given in [152], where results on various piezoelectric materials are presented; the analysis of error sources gives the guidelines for developing this promising technique. The experimental setup is shown in Figure 30.

Among the various categories of optical interferometers, an interesting and reliable one is the double-beam Mach–Zender type in which the difference of the displacements of the two faces of the sample is measured [153, 154]. A comparative study of this kind of interferometer with respect to a single-beam one is presented in [155]. Referring to the scheme of this arrangement (Fig. 31) the main difference and peculiarity consist of eliminating the contribution of substrate bending motion. The same technique has been employed [156] for the measurement of piezoelectric displacement on PZT film. In this case the piezoelectric coefficients were estimated even from permittivity and polarization data, obtaining good agreement with the interferometric data.

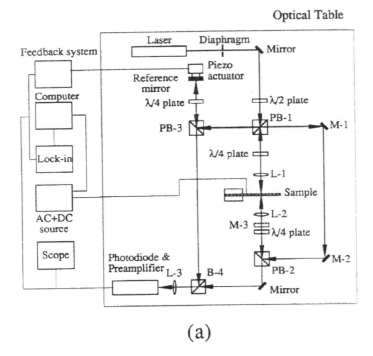

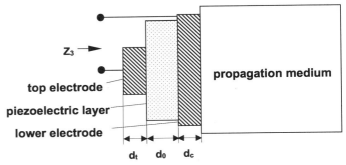

Fig. 32.   BAW thin film transducer structure.

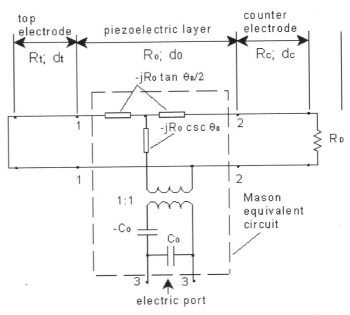

Fig. 33.   Electrical equivalent circuit for the transducer of Figure 32.

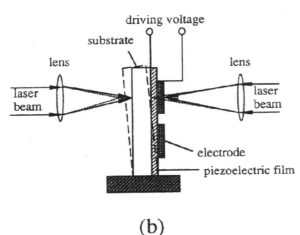

Fig. 31.   Schematic of the double-beam laser interferometer (a) and principle of elimination of bending motion of the substrate (b). Reprinted with permission from [155], © 1996, American Institute of Physics.

As a typical example of dynamic measurement we consider a simple BAW device employing one transducer as represented in Figure 32. We can distinguish two metallic electrodes, the active layer, and the acoustic propagation medium. In practical applications this is the case for an acoustooptical cell, a probe of an acoustic microscope, or a single-port resonator. For the following considerations we assume that the wavelength is much smaller than the lateral dimension of the transducer and only a pure mode with propagation in the thickness direction can be excited [157]. This implies that the symmetry of the active layer is appropriate to generate longitudinal or shear mode. Moreover the effect of the external contact on the top electrode is considered negligible. The physical dimensions of the top electrode

over the plane determine the initial cross section shape of the acoustic beam radiated into the propagation medium.

This device is well represented by the equivalent electric circuit presented in Figure 33. The dashed line confines Mason's equivalent circuit [158] of the transducer. Here the transducer is represented as a three-port circuit where port 3 is electrical whereas 1 and 2 are acoustic ports. Port 2 is short-circuited because no radiation of acoustic power takes place in air; R_0, R_D, R_t, and R_c are respectively the electric unit equivalents of the piezolayer, the propagation medium, and the metallic electrodes. C_0 is the static capacitance of the piezolayer. The electrical impedance Z_3 seen at port 3 has the expression

$$Z_3 = \frac{1}{j\omega C_0} + Z_a$$

where $Z_a = R_a + jX_a$ is the radiation impedance. The electrical power flowing in the real part R_a, named radiation resistance, represents the mechanical power flowing in the propagation medium. When the excitation frequency equals the ratio between the velocity in the piezolayer over 2 times its thickness, that is, $f_0 = \omega_0/2\pi = v_0/2t_0$, and the electrode

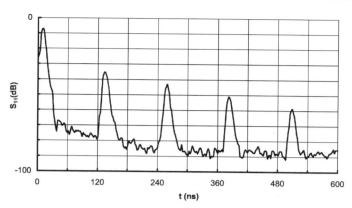

Fig. 34. Time-domain response of a ZnO BAW transducer (1.9 GHz) on 0.5 mm Si substrate.

thickness is negligible with respect to the acoustic wavelength, the parts of Z_a assume the simple expression

$$R_a = \left(4k^2 \bigg/ \pi \frac{R_D}{R_0}\right)(1/\omega_0 C_0), \qquad X_a = 0$$

where k is the electromechanical coupling coefficient. As can be seen, the measurement of this quantity implies a measurement of R_a that is not direct because of the presence of the reactance due to C_0. To overcome the presence of the static capacitance a series inductance can be introduced at port 3 so that, at f_0, $Z_3 = R_a$. In these conditions an impedance meter can easily perform the measurement of the radiation resistance. For the evaluation of k the knowledge of the value of C_0 is required. This can be estimated separately at a frequency much lower than the resonance frequency.

Another important parameter characterizing the efficiency of the transducer is the conversion loss T, which has the expression [159] $T = P_{ML}/P_L$, where $P_{ML} = V_s^2/4R_s$ is the electrical power that the generator can deliver to a matched load and P_L is the power launched in the propagation medium.

If $k < 0.3$ it can be shown that

$$T(\omega) = \frac{\pi r_D}{16k^2} \frac{1 + (\omega C_0 R_s)^2}{\omega C_0 R_s} M(\omega)$$

where $M(\omega)$ is called the band shape factor and has unity value for $\omega = \omega_0$.

The measurement of the conversion loss can be performed in an indirect way using a classical pulse-echo technique [160–162] or better with a network analyzer operating in time domain [72]. Figure 34 shows a typical time-domain response of a BAW transducer operating at 1.9 GHz. The one-way conversion loss can be obtained from the round trip loss by subtracting acoustic propagation losses and dividing by 2 [163]. Propagation losses include attenuation of the propagation medium, diffraction of the acoustic beam, and reflection loss due to absorption from the face of the medium opposite the transducer. The round trip loss corresponds to the difference between the peak at $t = 0$ and the first echo whereas the propagation loss is obtained from the difference of the amplitude between two consecutive echoes. The technique can be applied even

to piezoelectric thin films used for SAW devices on nonpiezoelectric substrates. The measured k^2 can be compared with the calculated value. The last can be plotted as a function of the ratio between the film thickness and the acoustic wavelength and for a given structure of the interdigital transducer. In fact the transducer can be deposited and patterned on the free surface of the film or buried at the film–substrate interface. As an example, in [164] reported calculated plots of k^2 vs h/λ of AlN films for various transducer configurations on different nonpiezoelectric substrates and compared with experimental values obtained from sputtered AlN.

4.4.2. Dielectric Characterization

The characterization of ferroelectric thin films usually requires evaluation of the dielectric constant and a plot of the ferroelectric hysteresis loop. The measurement of the dielectric constant is performed with an impedance meter. Some measurements are performed as a function of temperature and frequency to evidence the type of phase transitions and obtain the critical temperatures. The film should be prepared with a well-defined geometry of the electrodes and its thickness precisely measured by a profilometer. The sample under test is placed in a temperature controlled chamber.

The ferroelectric hysteresis loop can be obtained with the well-known technique introduced by Sawyer-Tower in 1935. By means of this circuit the polarization as a function of the applied voltage can be displayed on an oscilloscope. On this principle many circuits have been proposed to achieve a response insensitive to stray capacitances and resistive losses in the sample under test. For this purpose commercial instruments are now available. An interesting approach is reported in [165], where a constant current generator is used to drive the sample capacitance. In the classical Sawyer-Tower circuit the equation describing the time evolution of the current flowing in the capacitance as a result of the applied voltage $v_s(t)$ is

$$I(t) = \frac{A\varepsilon}{L} \frac{dv_s(t)}{dt} + A \frac{dP(t)}{dt}$$

where A is the capacitance area and L its thickness. In the proposed method $I(t)$ is kept constant and thus the measure of $dP(t)/dt$ is obtained by the derivative of the measured $v(t)$. The advantage is that only the measurement of a voltage is required and no current or charge amplifiers are needed. A typical hysteresis loop obtained on a $Pb(Zr_{0.53}Ti_{0.47})O_3$ thin film [179] is shown in Figure 35.

4.4.3. Elastic Characterization

Complete characterization of piezoelectric thin films requires the knowledge of the set of elastic dielectric constants. These can be determined from SAW velocity dispersion curves that can be obtained in several ways. One method consists of measurement of the phase velocity of acoustic modes through an acoustooptic probing based on measurement of the diffraction angle of the light scattered by the acoustic waves. Figure 36

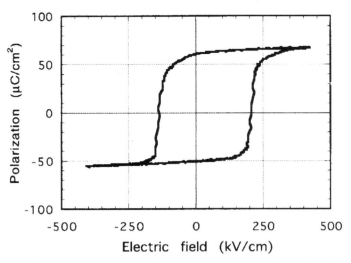

Fig. 35. Polarization–electric field hysteresis loop of a Pb($Zr_{0.53}Ti_{0.47}$)O_3 thin film epitaxially grown on Pt/MgO by rf magnetron sputtering. The measurement has been performed at 1 kHz. Reprinted with permission from [179], © 1997, American Institute of Physics.

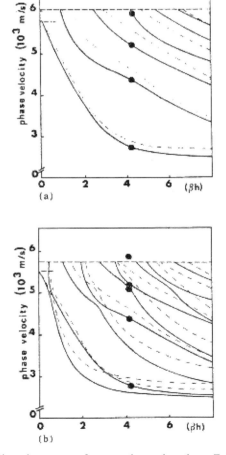

Fig. 36. Dispersion curves for acoustic modes along ZnO layer on (a) zy-Al_2O_3 and (b) zx-Al_2O_3. Dotted lines are referred to ZnO single-crystal and solid lines to effective constants. Reprinted with permission from [166], © 1987, American Institute of Physics.

presents dispersion curves for acoustic modes propagating along a ZnO layer grown on Al_2O_3 for two propagation directions. With this method the set of elastic and piezoelectric constants of the films have been obtained [166]. The experimental determination of the dispersion curves of the acoustic modes can be performed by Brillouin scattering of light by thermally excited acoustic phonons [167]. Elastic constants of sputtered AlN have been determined with this method and are reported in [168].

5. PROPERTIES OF PIEZOELECTRIC THIN FILMS

Piezoelectric materials are divided into two types: polar (which possess a net dipole moment) and nonpolar (whose dipole moments summed on the different directions give a null total moment). Among the 20 noncentrosymmetric piezoelectric classes, 10 have a spontaneous polarization. Polar materials can be further classified as ferroelectric (if the dipole moment can be switched by a finite electric field) and nonferroelectric. The primary feature distinguishing ferroelectrics from other polar materials is that the spontaneous polarization can be reversed at least partially with an applied electric field. When the polarization is measured as a function of an applied electric field a hysteresis loop is obtained [2].

The discussion of properties of piezoelectric thin films, therefore, will be devoted to the most important materials belonging to the two major divisions: ferroelectric and nonferroelectric piezoelectric thin films. Among ferroelectric thin films, attention will be directed to those types which have strong piezoelectric properties and are consequently most used in applications such as microelectromechanical systems, acoustic wave transducers, sensors, and resonators. In addition we have considered the new relaxor–ferroelectric systems, even if below the morphotropic phase boundary they are only electrostrictive, because they promise to revolutionize the field of transducer applications due to their huge piezoelectric constants. Among nonferroelectric piezoelectric thin films, with the exception of some historical mentions, only those materials that have been the object of systematic research in the last decades of the twentieth century will be considered.

As a general rule, piezoelectric materials of great interest for applications will be discussed. Some of the piezoelectric materials, substrates, and electrodes commonly used in electroacoustic devices are listed in Table VIII.

Physical properties of piezoelectric thin films depend on many parameters, including composition, crystalline structure, orientation (texture), film thickness and microstructure, internal stress, presence of secondary phases, dopants, interface layers, misfit strain, and mechanical constraints. In addition, the piezoelectric response in ferroelectric thin films is affected by displacement of domain walls. These problems will be discussed in the second part of the section dedicated to each material, whereas in the first part an overview of the main results reported in the last decades of the twentieth century will be given.

Table VIII. Main Characteristics of Piezoelectric Materials, Substrates, and Electrodes Commonly Used in Electroacoustic Devices

Material	Direction	Density (kg/m^3)	Velocity (m/s)	Impedance [10^6 kg/(m^2·s)]	Attenuation (dB/μs @ 1 GHz)
Piezoelectric					
AlN	*c* axis	3,270	10,400	34.0	
CdS	*c* axis	4,820	4,465	21.52	0.0939
LiNbO$_3$	*c* axis	4,640	7,320	33.96	0.2199
LiNbO$_3$	*x* axis	4,640	6,590	30.58	
LiNbO$_3$	*y* 35° rot.	4,640	7,400	34.34	
LiTaO$_3$	*a* axis	7,454	5,630	41.97	
LiTaO$_3$	*c* axis	7,454	6,337	47.24	
ZnO	*c* axis	5,680	6,330	35.95	8.3176
Substrates					
Si		2,332	8,429	19.66	8.3000
GaAs	[100]	5,307	4,921	26.19	
MgO	[100]	3,580	9,230	33.04	2.8521
CaF$_2$	[100]	3,180	7,550	24.01	
SiO$_2$	*x* axis	2,651	5,730	15.19	2.4983
SiO$_2$	*z* axis	2,651	6,350	16.83	1.7208
SiO$_2$	Fused	2,194	5,970	13.10	8.9968
Al$_2$O$_3$	*a* axis	3,970	11,130	44.19	0.1887
Al$_2$O$_3$	*c* axis	3,970	11,170	44.34	0.8950
MgAl$_2$O$_4$	[100]	3,780	8,800	33.26	0.2024
MgAl$_2$O$_4$	[111]	3,780	10,600	40.07	
MgAl$_2$O$_4$	[110]	3,780	10,200	38.56	
YAl$_5$O$_{12}$	[100]	4,550	8,560	38.95	0.1712
Electrodes					
Al	[100]	2,695	6,350	17.2	13.0175
Au	[100]	19,300	3,210	61.95	
Au	[111]	19,300	3,400	65.62	56.7800
Ag	[100]	10,500	3,440	36.12	86.000
Cr		6,000	5,000	30.00	
Cu		8,940	4,900	43.81	20.5800
W		19,200	5,200	99.84	1.0400
Zn		7,100	4,210	29.89	9.2620

Data taken from [7].

5.1. Ferroelectric Thin Films

5.1.1. Lead Titanate Zirconate

Among ferroelectrics, a very important group is formed by perovskites (called after the mineral perovskite CaTiO$_3$). The perfect perovskite structure is an extremely simple one with the general formula ABO$_3$, where A is a monovalent or divalent metal and B is a tetravalent or pentavalent one. The elementary cell is cubic with the A atoms at the cube corners, the B atoms at the cube center and the oxygens at the face centers. Among ferroelectric perovskites, the most studied is lead titanate zirconate Pb(Zr$_x$Ti$_{1-x}$)O$_3$ (PZT) due to its strong piezoelectric properties. The ferroelectric and piezoelectric properties of PZT are sensitive to both crystallographic structure and compositional ratio of Zr/Ti. They are maximum near the morphotropic phase boundary (MPB) at Zr/Ti = 53/47, which separates the rhom-

bohedral (Zr rich) and tetragonal (Ti rich) symmetries in the PZT phase diagram [169]. Above the Curie temperature the material has a paraelectric perovskite cubic phase.

Pb(Zr,Ti)O$_3$ thin films have been studied mostly for non-volatile ferroelectric memory applications due to their large remanent polarization and low coercive field. MPB compositions show high piezoelectric constants and electromechanical coupling factors; therefore they have found applications in the field of microelectromechanical systems and other piezoelectric microdevices.

5.1.1.1. Overview of the Main Results

Significant efforts have been directed toward the development of novel processing techniques for the deposition of PZT thin films [170–206]. The first papers devoted to PZT film deposition have concentrated mostly on their ferroelectric behavior whereas piezoelectric properties have been less investigated. Systematic research on the structure and physical properties of PZT thin films for bulk and surface acoustic wave transducers and their variation with compositional Zr/Ti ratio has been performed by Sreenivas et al. [170, 171]. PZT films have been deposited by dc magnetron sputtering from multielement targets on glass substrates covered with a conductive layer. Single-phase PZT was formed by a diffusion controlled process during annealing in air/oxygen. It has been found that near-MPB compositions (Zr/Ti = 54/46) yield films with high dielectric constants of 800 but the rhombohedral phase compositions PZT 58/42 resulted in better ferroelectric properties. Well-saturated ferroelectric hysteresis loops were observed with a remanent polarization of 30 μC/cm^2 and a coercive field of 25 kV/cm. Because the films were polycrystalline and unoriented, a poling treatment was necessary to achieve a polar axis. The value of the poling field as determined from measurement of coercive field in the hysteresis curve was about 35 kV/cm and the poling duration was about 2 hours at 170°C. SAW characterization has been performed by using deposited interdigital transducers (IDT) at around 35 MHz. The SAW velocity was found to be around 2100 m/s. The electromechanical coupling coefficient k^2 determined from impedance measurements was in the range 0.57–0.79%.

The preparation of PZT films by a dc sputtering technique with a postdeposition rapid thermal annealing treatment at 650°C for 10 s has been reported in [172]. The films exhibited good structural, dielectric, and ferroelectric properties compared to conventional furnace-annealed films. The measured dielectric constant and loss tangent at 1 kHz were 900 and 0.04 and the remanent polarization and coercive field values were 10 μC/cm^2 and 23 kV/cm, respectively.

Low-temperature metalorganic chemical vapor deposition (MOCVD) of PZT thin films has been reported in [173]. PZT thin films with perovskite structure were prepared on sapphire disks, Pt/Ti/SiO$_2$/Si and RuO$_x$/SiO$_2$/Si substrates at temperatures as low as 550°C. As-deposited films were dense and showed uniform and fine grain size. Films annealed at 600°C showed a spontaneous polarization of 23.3 μC/cm^2.

The piezoelectric properties of sol–gel-derived PZT 53/47 thin layers have been investigated by Li et al. [174]. Typical film thicknesses were in the range 0.3–0.5 μm, and the films exhibited a significant degree of preferred alignment parallel to (111) perovskite. The piezoelectric properties of films were investigated by interferometry as a function of frequency and dc electric bias. It has been found that the value of d_{33} increased with increasing bias and reached a maximum value of nearly 180 pC/N at a bias level of 8×10^6 V/m. Piezoelectric loss was approximately 12% at low frequency and increased to nearly 20% at 10^4 Hz. The influence of polarization reversal on the d_{33} coefficient at 10^3 Hz showed strong hysteresis effects. In addition the hysteresis loop appeared to be significantly asymmetric, with the center displaced from the origin by about 1.5×10^6 V/m.

The obtaining of ferroelectric La–Sr–Co–O/PZT/La–Sr–Co–O heterostructures on silicon via template growth was reported by Ramesh et al. [175]. This heterostructure was grown on [001]Si with an yttria stabilized zirconia (YSZ) buffer layer. A layered perovskite "template" layer of c-axis-oriented Bi$_4$Ti$_3$O$_{12}$ grown between the YSZ buffer layer and the bottom La–Sr–Co–O electrode assured the required (001) orientation of the subsequent layers. The use of these electrode layers allowed the growth temperatures to be decreased by 60–150°C and to obtain structures with large remanent polarization and small coercive field.

The ferroelectric and piezoelectric properties of PZT thin films deposited on Pt-coated Si(100) wafers by electron cyclotron resonance (ECR) sputtering have been discussed in [176]. Measurements of the polarization hysteresis curve have shown that the remanent polarization was about 16.5 μC/cm^2 (for a 410-nm thick film), which is smaller than the value corresponding to bulk PZT (>30 μC/cm^2). From the piezoelectric resonance of a device based on a vibrating Si beam the piezoelectric constant e_{31} was evaluated and was found to be 2.1 C/m^2 without poling treatment, whereas $d_{31} = -33$ pC/N. These values are smaller than for bulk PZT ($e_{31} = 5.2$ C/m^2, $d_{31} = -150$ pC/N), but are similar to other results obtained on thin films.

The effect of poling on the piezoelectric properties of PZT thin films was studied by Watanabe et al. [177]. The samples were deposited by planar rf magnetron sputtering on Pt-coated Si wafers. A remanent polarization $P_r = 30$ μC/cm^2 and a coercive field of 100 kV/cm were found. The piezoelectric properties were obtained from measurement of the resonance amplitude obtained on a cantilever structure. After poling at about 5 kV/cm the d_{31} obtained value was about -100 pC/N, which is in the same order of magnitude as bulk values.

Maeder et al. [178] demonstrated the possibility of obtaining PZT films on reactive metal substrates, opening new possibilities for PZT integration on micromechanical devices. They used a special RuO$_2$/Cr buffer as an efficient electrically conductive barrier on Zr membranes. All films in the structure were deposited by magnetron sputtering. The piezoelectric activity of the films, even if not quantified, was demonstrated by measuring the piezoelectric deflection vs frequency spectrum by using a laser interferometer.

Tanaka et al. [96] studied the controlling factors on the synthesis of Pb(Zr$_x$Ti$_{1-x}$)O$_3$ thin films [deposited on (100)SrTiO$_3$ (STO) substrates using ArF excimer laser ablation] as a function of substrate temperature and oxygen pressure. The authors discussed also the film formation diagram for PZT films for Zr compositions of $x = 0$, 0.3, 0.5, 0.7, and 1, deposited under various oxygen pressures and substrate temperature values. It has been found that crystalline Pb(Zr$_{0.5}$Ti$_{0.5}$)O$_3$ (PZT 50/50) and Pb(Zr$_{0.7}$Ti$_{0.3}$)O$_3$ (PZT 70/30) films are formed in the region where T_s is about 490°C and P_{O_2} is about 50 mTorr; it has been pointed out that in all the cases the formation of the perovskite phase is located near the conditions for PbO crystallization. The dielectric constant of each PZT film was measured at room temperature and at the frequency of 1 kHz. It has been found that the PZT 70/30 film has the largest dielectric constant whereas in bulk materials the maximum is at 55/45. The orientation of PZT film could be controlled by using a buffer layer. Another finding was related to the possibility of controling the a-axis orientation of PZT 30/70 film by using a PZT 70/30 film as a buffer layer on the STO substrate.

The piezoelectric properties of c-axis-oriented PZT thin films at MPB also have been investigated by Kanno et al. [179]. The films were prepared by rf magnetron sputtering on (100)Pt-coated MgO single-crystal substrates. Electrical measurements showed that the relative dielectric constant of the films was 150–200, with a dissipation factor less than 2%. P–E hysteresis loops indicated strong ferroelectricity with a large remanent polarization, more than 50 $\mu C/cm^2$, and a coercive field E_c about 150–200 kV/cm. The shape of the hysteresis loop was found to be asymmetric and slightly shifted toward positive values of P and E. Piezoelectric properties were evaluated by microfabricating cantilever structures covered with PZT films with a thickness of 3.6 μm. It has been found that $d_{31} = -100$ pC/N, which is almost consistent with that of bulk PZT ceramic. It should be noted that the PZT films were not subjected to a poling treatment and furthermore the operating electric fields were less than the coercive field of the PZT films. This result indicated that the as-deposited PZT films were intrinsically poled and that their c-axis orientation allowed good piezoelectric properties to be obtained without poling.

The possibility of using PZT thin films in the ultrahigh-frequency (UHF) range has been demonstrated in [180]. PZT thin films have been grown by the sol–gel method on a Pt/Ti/SiO$_2$/Si substrate. The crystallization of the films, whose final thickness after repeated deposition operations was 800 nm, was achieved at 650°C. BAW transducers were obtained by depositing Al electrodes on the film and the resonator was supported on a silicon wafer fabricated by anisotropic etching of (100) silicon. Transducer performance was evaluated by measuring ultrasonic pulse echo trains coming from the PZT thin film and reflected at the bottom surface of the Si substrate while a polarizing electric field was applied with a dc power supply. Ultrasonic pulse echo trains were detected at 1.7 GHz by thickness vibration. Mechanical factor Q_m was estimated to be 237 and electromechanical coupling factor k_t was estimated to be about 31%. The bandwidth was about 8% at 1.7 GHz.

PLD deposition of PZT 53/47 thin films on SrRuO$_3$ and La$_{0.5}$Sr$_{0.5}$CoO$_3$ grown on (100)LaAlO$_3$ single-crystal substrates was reported by Cho and Park [181]. PZT films with a thickness of 3600 Å were deposited at a substrate temperature of 600°C with 200 mTorr of oxygen pressure. All the films including the oxide electrode layers were obtained by PLD. After the deposition the thin PZT films were cooled down to room temperature with ~600 Torr of oxygen pressure. The in-plane orientation for all films shows heteroepitaxial growth regardless of the substrate temperature. Good ferroelectric properties have been obtained. The remanent polarization was found to be about 35 $\mu C/cm^2$ when a driving voltage of 10 V was applied; coercive field E_c was about 100 kV/cm.

The deposition of PZT 50/50 thin films over a lead titanate seeding layer was reported in [182]. The films have been deposited by cosputtering Pb(Zr$_{0.5}$Ti$_{0.5}$)O$_3$, PbO, and TiO$_2$ targets at a substrate temperature of 200°C and crystallized by rapid thermal annealing at 630°C for 30 s. A relationship between the distance of the (100) lattice plane and the Zr/Ti ratio in the PZT thin films was found. From these results the composition of PZT thin films could be estimated. SEM cross-sectional images taken on PZT films 300 nm thick revealed a columnar grain structure with grain sizes ranging between 100 and 150 nm. The remanent polarization and coercive field values were about 42 $\mu C/cm^2$ and 46 kV/cm, respectively.

Results obtained from the preparation and characterization of sol–gel derived PZT thin films have been reported in [183]. Crack-free PZT thin films with thickness of 3 μm for microactuators have been deposited on Pt/Ti/SiO$_2$/Si substrates. As has been shown in the previous paragraphs, the sol–gel process provides high purity, large deposition area, and easy composition control but it generally has limits in the thickness of a crack-free single-layer film which is about 0.1 μm. For applications in microactuators a much higher thickness is required. Multiple depositions permit the build-up of thicker films, but usually 1 μm is the limit where multilayers crack. The authors reported the obtaining of PZT films with thickness up to 3 μm. The film consisted of the perovskite PZT phase without pyrochlore phase and showed strong (100) preferred orientation. Different pyrolysis process temperatures allowed the crystal orientations to be controlled, with the preferred orientations in the direction of the (111) plane and (100) plane. The dielectric constant and loss value of the films measured at 1 kHz were approximately 1250 and 0.04, respectively; the remanent polarization and the coercive field were 45.5 $\mu C/cm^2$ and 58.5 kV/cm, respectively.

The same technique has been employed by Cho et al. [184] to deposit Pb(Zr$_{0.52}$Ti$_{0.48}$)O$_3$ thin films on Pt/TiO$_2$/SiO$_2$/Si substrates. The films exhibit some preferential orientation in the [110] and [111] directions. The piezoelectric properties have been obtained using a laser interferometer which detected the piezoresponse to a small alternating-current (ac) voltage (0.1 V at 1 kHz) superimposed on a dc field which was varied in steps. The longitudinal piezoelectric coefficient d_{33} was obtained as a function of the bias electric field. A high-temperature poling was required to enhance the piezoelectric response of the thin film. A significant increase of d_{33} was obtained after a poling treatment with a dc voltage applied at 100°C for 10 min. In all films, a typical butterfly curve of the d_{33} value as a function of dc applied voltage was observed. The maximum value of d_{33} after poling was about 130 pm/V. This value is less than that of bulk PZT ceramics, but is consistent with those previously reported for PZT films. The lower response of the film is believed to be due to the substrate constraints as well as to the restriction on the reorientation of ferroelectric domains. Acoustic emission tests were carried out on PZT films in a variety of amplifier and package conditions. Their response has been compared with a commercial sensor. The films showed a faster rise time and a shorter ring downtime than the commercial sensor.

The deposition of PZT thin films at remarkably low substrate temperature (375–400°C) by pulsed laser deposition has been reported in [109, 185–187]. The films were deposited on different types of electrodes (Pt/Si, Au/Pt/glass, TiN/Si, etc.) and generally presented (111) orientation. They showed an intrinsic polarization with a d_{33} constant of about 40 pC/N and a dielectric constant of about 500. The remanent polarization was about 15 μC/cm^2 and the coercive field showed a very high value, 100 kV/cm.

Lian and Sottos [188] studied the effect of thickness on the piezoelectric and dielectric properties of lead zirconate titanate thin films. They have deposited PZT thin films with a Zr/Ti ratio of 52/48 on platinized silicon substrates by a sol–gel method. The crystallized films showed (111) or (100) orientation. The piezoelectric constants d_{33} and the field-induced strain of the films with different thickness and preferred orientation were measured by a laser Doppler heterodyne interferometer as a function of film thickness. Both the piezoelectric properties and the dielectric constant were found to increase with film thickness. Films with (100) preferred orientation were found to have higher piezoelectric constants and lower dissipation factors than films with (111) preferred orientation. The dependence of properties on thickness was correlated with the residual stress state in the films.

The piezoelectric properties of rhombohedral PZT thin films with different orientations [(100), (111), and random] have been studied by Taylor and Damjanovic [189]. Sol–gel derived Pb(Zr$_{0.6}$Ti$_{0.4}$)O$_3$ films with (111) and "random" orientation were grown on a TiO$_2$/Pt/TiO$_2$/SiO$_2$/Si substrate. The same type of substrate with an addition of a 10 nm thick PbTiO$_3$ seeding layer was used to deposit the (100)-oriented film. The thickness of the films was about 0.9 μm. All piezoelectric measurements were made under a dc electric bias field of 6 MV/m. The purpose of the dc bias was to polarize and keep films in the same polarization state during measurements. The largest d_{33} was found in (100)-oriented films and the smallest along the polarization direction in (111)-oriented films. The value of the piezoelectric coefficient of the randomly oriented film ($d_{33} \sim 77$ pm/V) was between those for (100) and (111) orientations. The field dependence of d_{33} was also investigated as a function of crystallographic orientation of the films. It was found that (100)-oriented films with the highest piezoelectric coefficient exhibited the weakest nonlinearity. Observed variation in the piezoelectric nonlinearity with film orientation have been explained by taking into account domain wall contributions, which are dependent on film orientation. The experimental dependence of d_{33} on orientation confirmed the theoretical predictions by Du et al. [190, 191] which will be discussed in the next subsection. In Table IX the main physical properties of PZT films, together with the corresponding bulk values, have been listed.

5.1.1.2. Influence of Structure and Composition on Physical Properties

Whereas the growth process and the relationships between deposition parameters and the obtaining of a definite structure were treated in Section 3, here we mainly investigate the physical properties of the deposited films and their relationship with structure, orientation composition, internal stress, poling treatment, etc. It must be pointed out that the field of piezoelectric thin films is continuously expanding and not all the aspects of the correlation between structure and physical properties have been clarified or investigated. Obviously the following discussion will include only those aspects which have been object of systematic research in this field.

5.1.1.2.1. Piezoelectric Properties

(a) Influence of Orientation.   The influence of orientation of the structure on the piezoelectric properties has been theoretically predicted in two fundamental papers for tailoring and developing high-quality PZT films [190, 191]. A phenomenological model was used to calculate the orientation dependence of piezoelectric properties of single-crystal PZT with various compositions around MPB. The piezoelectric constants in any orientations can be obtained by applying the tensor operations to the corresponding values calculated with this model. Figure 37 shows the piezoelectric constants d_{33} in the [001] and [111] orientations for different compositions. For tetragonal PZT the piezoelectric constant d_{33} for [001] orientation is larger than the piezoelectric constant d_{33} for [111] orientation. However, the difference between $d_{33[001]}$ and $d_{33[111]}$ is much larger for rhombohedral PZT. Near the morphotropic phase boundary on the rhombohedral site the $d_{33[001]}$ rapidly increases whereas the $d_{33[111]}$ increase is less strong; therefore a large enhancement of their ratio is obtained. A comparison with the experimental values of d_{33} for PZT ceramics plotted in the same graph shows that the d_{33} of the ceramic in the tetragonal phase is lower than that of the single crystal due to the averaging on the randomly oriented grains. For the rhombohedral phase experimental ceramic d_{33} values are larger than the theoretical $d_{33[111]}$ of the single crystal because the rhombohedral phase crystal orientation away from the spontaneous polarization direction [111] will enhance the effective d_{33}. In Figures 38 and 39 the orientation dependence of d_{33} and of the electromechanical coupling factor k_{33} for the tetragonal composition PZT 40/60 and rhombohedral PZT 60/40 have been represented. In these figures the distance between the surface of the graph and the origin represent the values of d_{33} and k_{33} in that orientation. For tetragonal PZT d_{33} has the maximum value in the spontaneous polarization direction [001] and monotonously decreases as the crystal cutting angle from the direction [001] increases. For rhombohedral PZT (Fig. 39), d_{33} has the maximum value in a direction 56.7° away from the polarization direction [111]. It must be noted that the directions with maximum d_{33} are very close to the perovskite [001] directions. Consequently, it has been suggested that PZT epitaxial thin films used for actuators and sensors should adopt rhombohedral compositions near

Table IX.   Main Physical Properties of PZT Thin Films as Reported in Different References

Composition	Deposition technique, substrate	Orientation	Relative dielectric const. ε_{33} (1 kHz)	Dielectric loss factor	P_r (μC/cm^2)	E_c (kV/cm)	d_{33} (pC/N)	d_{31} (pC/N)	Reference
PZT 80/20	Sol–gel Pt/Ti/SiO$_2$/Si(100)	Random	481		19	28	64		[192]
PZT 70/30	PLD SrTiO$_3$(001)	(001)	600						[181]
PZT 68/32	Dc magnetron sputtering Conducting glass	Random	424		14.1	33.1			[170]
PZT 65/35	Metalorganic decomposition Ti/Pt/Ti/SiO$_2$/SiO(110)	(111) + (100)	1080					−160	[193]
PZT 65/35	Metalorganic decomposition Ti/Pt/Ti/SiO$_2$/SiO(110)	(100)	1040					−100	[193]
PZT 65/35	Metalorganic decomposition Pt/Ti/SiO$_2$/SiO(110)	(111)	1170					−160	[193]
PZT 60/40	Sol–gel Pt/Ti/SiO$_2$/Si(100)	Random	608		16.5	28	90		[192]
PZT 60/40	Dc magnetron sputtering Conducting glass	Random	507		29.2	23.1			[170]
PZT 60/40	Sol–gel Pt/TiO$_2$/SiO$_2$/Si(100)	(111)	1200	0.04			40		[47]
PZT 58/42	Dc magnetron sputtering Conducting glass	Random	562		29.6	23.6			[170]
PZT 56/44	Sol–gel Pt/Ti/SiO$_2$/Si(100)	Random	1065		28	32	115		[192]
PZT 55/45	Sol–gel Pt(111)/Ti/SiO$_2$/Si	(100)	1250	0.04	45.5	58.5			[183]
PZT 53/47	Sol–gel (111)Pt/Ti/SiO$_2$/Si(100)	(111)	750	0.03	20	50	180		[174]
PZT 53/47	PLD SrRuO$_3$/(100)LaAlO$_3$	(001)	500		35				[181]
PZT 53/47	PLD PT/Pt/Si PT/Pt/diamond	(001)	500–650 (100 kHz)				50–350		[202]
PZT 53/47	PLD Pt/Ti/SiO$_2$/Si	(111)	410 (100 kHz)	0.021	12	38			[122]
PZT 53/47	Sol–gel Pt/Si						400		[203]
PZT 53/47	MOCVD Pt/Si						200		[203]
PZT 53/47	Sol–gel Pt/TiO$_2$/SiO$_2$/Si(100)	(111)	1300	0.05			70		[47]
PZT 52/48	Sol–gel Pt/Ti/SiO$_2$/Si(100)	Random	1310		36	32	194		[192]
PZT 52/48	Sol–gel (111)Pt/Si	(111)	840.6	0.003			86.4	−33.9	[128]
PZT 52/48	Sol–gel (111)Pt/Si	(100)	734.7	0.003			97.5	−33.9	[128]
PZT 52/48	Sol–gel Pt/TiO$_2$/SiO$_2$/Si(100)	(110) + (111)			30	47	130		[184]
PZT 52/48	Planar rf magnetron sputtering	Random			31	110		−102	[177]

Table IX.   Continued

Composition	Deposition technique, substrate	Orientation	Relative dielectric const. ε_{33} (1 kHz)	Dielectric loss factor	P_r (μC/cm^2)	E_c (kV/cm)	d_{33} (pC/N)	d_{31} (pC/N)	Reference
PZT 52/48	Pt/Ta/Si$_3$N$_4$/Si(100) Rf planar magnetron sputtering	Random	494	0.072	100	0.6			[204]
PZT 52/48	Pt/Si Dc magnetron sputtering	Random	787		13.6	34.6			[170]
PZT 52/48	Conducting glass Sol–gel		900						[180]
PZT 50/50	Pt/Ti/SiO$_2$/Si(100) Rf magnetron sputtering	(001)	200	0.02	50	200		−100	[179]
PZT 49/51	MOCVD	(111)	1000		20	25	50		[156]
PZT46/54	Ir/SiO$_2$/Si Sol–gel	Random	1004		22.5	39	126		[192]
PZT 45/55	Pt/Ti/SiO$_2$/Si(100) Dc magnetron sputtering	Random	385		4	54.2			[170]
PZT 45/55	Conducting glass Sol–gel	(111)	1100	0.03			50		[47]
PZT 45/55	Pt/TiO$_2$/SiO$_2$/Si(100) Reactive sputtering	(111)	900	0.03			55		[47]
PZT 45/55	Pt/TiO$_2$/SiO$_2$/Si(100) Reactive sputtering	(100)	650	0.03			80		[47]
PZT 40/60	Pt/TiO$_2$/SiO$_2$/Si(100) Sol–gel	Random	715		26	48	101		[192]
PZT 40/60	Pt/Ti/SiO$_2$/Si(100) Rf magnetron sputtering	(111)			22	50			[205]
PZT 20/80	Pt(111)/TiN/Ti/SiO$_2$/Si Sol–gel	Random	319		17.5	77	81		[192]
PZT 20/80	Pt/Ti/SiO$_2$/Si(100) PLD	(001) epitaxial	230	0.04	26.5	100			[206]
Bulk PZT 52/48	YBCO/(001)LaAlO$_3$		730 (ε_{33}) 1180 (ε_{11})		34		223 494 (d_{15})	−93.5	[169]

Note: For comparison bulk values have been also included.

MPB with perovskite [001] orientation [190, 191]. In this case an enhancement of the effective d_{33} more than four times the already reported experimental values for the PZT films may be obtained.

An experimental investigation of the dependence of piezo-electric properties on the crystallographic orientation has been reported also by Taylor and Damjanovic [189] for rhombohe-dral PZT 60/40 thin films with (111), (100), and random orien-tation, prepared by sol–gel. The largest piezoelectric response ($d_{33} \sim 100$ pm/V) was observed in films with (100) orientation. The piezoelectric coefficient in the (111)-oriented films (along spontaneous polarization direction [111]) was much smaller (63 pm/V). The value for the randomly oriented film is interme-diate (77 pm/V). These results are in good qualitative agreement with the theoretical predictions of [190, 191] previously pre-

sented. When compared with the results reported in Figure 37, a discrepancy between the calculated single-crystal value and measured film can be observed, which is much higher for (100) orientation (50%) than for (111) orientation (10%). The differ-ence between single film and crystal values are explained by the clamping effect of the substrate. This difference is much smaller for (111)-oriented thin films because in this case there is a strong contribution from domain walls which compensate for the reduction in the piezoelectric effect due to film clamp-ing [189].

The dependence of piezoelectric constant on film orientation has been observed also on PZT 52/48 thin films deposited by the sol–gel method on Pt(111)/Si substrates [188]. The inves-tigation has been performed on two series of films, one with (100) and the other with (111) preferred orientation. Cross-sectional SEM photographs indicated that both groups of films had well-aligned columnar grains. The measured piezoelectric d_{33} constant was about 86 pm/V for (111)-oriented films with respect to 98 pm/V for (100)-oriented films. This is in qualita-

tive agreement with the tendency shown in Figure 37, even if a great discrepancy between the experimental and predicted values is again observed, which is due to the different stress and constraint conditions of the film with respect to single crystal. Yet other experimental results which confirm the predicted dependence on orientation in Figure 37 come from [184], where it has been reported that tetragonal near MPB PZT 52/48 films with predominant (100) orientation have better piezoelectric response ($d_{33} \sim 85$ pm/V) relative to films with predominant (111) orientation ($d_{33} \sim 60$ pm/V) at zero bias voltage.

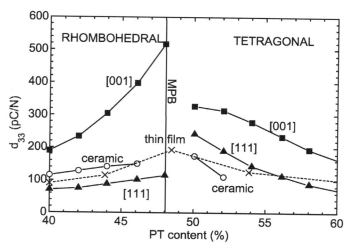

Fig. 37. Crystal orientation dependence of piezoelectric d_{33} constant for PZT around MPB. Curves indexed [001] and [111] represent calculated d_{33} values for different rhombohedral and tetragonal compositions with these orientations, taken from [191]. Experimental data for PZT ceramic (open circles) and random polycrystalline thin films ($\times$) with different compositions, taken from [192], are also shown.

(b) Influence of Composition. A systematic investigation of the relationship between composition and the longitudinal piezoelectric constant d_{33} in PZT thin films has been reported in [192]. A number of $PbZr_xTi_{1-x}O_3$ compositions with x ranging from 0 to 0.8 were prepared by the sol–gel technique on Pt/Ti/SiO$_2$/Si(100) substrates. The film thickness was 1 μm. Because the samples were randomly oriented polycrystalline films they were subjected to a poling treatment under 150 kV/cm dc field for 2 min at room temperature. Figure 37, discussed in the previous paragraph, shows also the effective d_{33} constant of these films as a function of composition (the term "effective" refers to the maximum value of the constant obtained under an applied dc bias field during the measurement). On the same graph the dependence of corresponding bulk values on composition has been also plotted. The maximum value of 194 pC/N is obtained at MPB, as for the bulk

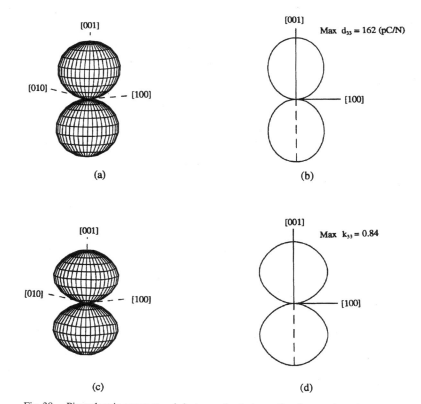

Fig. 38. Piezoelectric constant and electromechanical coupling factor orientation dependence for tetragonal PZT 40/60. Reprinted with permission from [190], © 1997, Institute of Pure and Applied Physics.

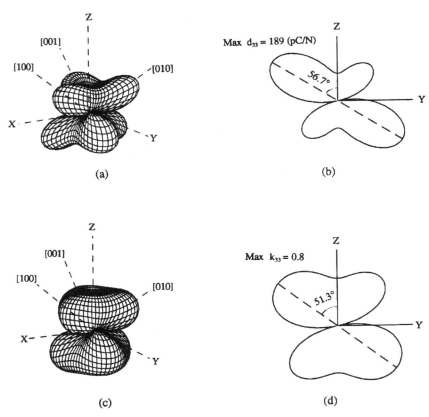

Fig. 39. Orientation dependence of piezoelectric constant and electromechanical coupling factor for rhombohedral PZT 60/40. Reprinted with permission from [190], © 1997, Institute of Pure and Applied Physics.

ceramics. The observed d_{33} values for PZT films and that of bulk ceramics show similar trends. As in bulk ceramics, the maximum of piezoelectric properties at MPB is due to the presence of both tetragonal and rhombohedral phases and therefore to the enhanced possibilities of dipole orientation.

The influence of composition on the piezoelectric d_{31} constant has been observed in PZT films deposited by a metalorganic decomposition process, for the compositions 45/55 (tetragonal), 55/45 (nearly MPB), and 65/35 (rhombohedral) [193]. The films have mixed [111]/[100] orientations. The thickness of all three types of films was 600 nm. The largest d_{31} constant (160 pm/V) was obtained for films with near-MPB composition, as expected.

(c) Influence of Poling Field. In polycrystalline unoriented films the polarization under high dc electric field is crucial because the piezoelectric effect cannot be obtained in the absence of a polar axis. The optimization of the piezoelectric constants of the PZT thin films by poling treatment has been reported in [184] for PZT 52/48 thin films obtained by a multiple deposition sol–gel method, on Pt(111)/TiO₂/SiO₂/Si(100). The authors studied the behavior at poling in a field of about 20 kV/cm determined from the hysteresis loop. The effective piezoelectric coefficients d_{33} have been plotted as a function of dc voltage in Figure 40. The d_{33} value of the unpoled thicker film (virgin state) was 54 pm/V and the maximum obtained

value was 105 pm/V at 4 V dc bias voltage. After poling the d_{33} value increased to 84 pm/V (virgin state) and the maximum obtained value when a dc field was applied for measurement was 130 pm/V at 3 V. So poling of the films leads to a substantial increase of their piezoelectric properties due to the dipole orientation during the poling treatment. These values are less than those of bulk ceramics (223 pm/V) of similar composition. The authors attributed this difference to the substrate constraints and to the restriction on the reorientation of ferroelectric domains.

A much higher electric field of about 35 kV/cm (whose size was determined from the coercive field in the hysteresis curve) applied at 170°C for 2 h allowed an electromechanical coupling coefficient k^2 of about 0.57–0.79% to be obtained for surface acoustic wave generation at about 35 MHz, for PZT thin films with rhombohedral structure obtained by dc magnetron sputtering [194].

A polarizing electric field of about 5 kV/cm applied on PZT thin films deposited by planar rf magnetron sputtering on Pt-coated Si wafers allowed a d_{31} value of −100 pC/N to be obtained which is in the same order of magnitude as bulk values [177]. It has been shown by Cattan et al. [195] that not only the size of the applied electric field but also the poling procedure can modify the piezoelectric properties and the piezoelectric constant hysteresis loop. A poling field whose intensity was increased in steps allowed a piezoelectric constant e_{31} of about −4 C/m² to be obtained.

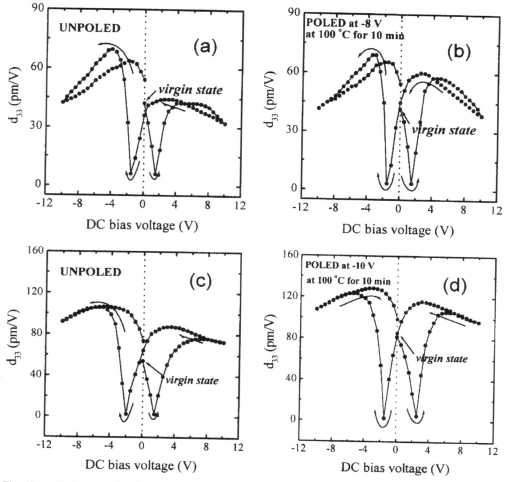

Fig. 40. Effective piezoelectric constant d_{33} as a function of dc voltage for PZT films of different thicknesses: (a, b) 0.32 μm; (c, d) 0.48 μm. Reprinted with permission from [184], © 1999, Institute of Pure and Applied Physics.

The influence of a dc electric field on the piezoelectric properties was evidenced also on MPB PZT53/47 obtained by the sol–gel method [174]. Typical film thicknesses were in the range 0.3–0.5 μm, and the films exhibited a significant degree of preferred alignment parallel to (111) perovskite. The piezoelectric properties of films were investigated by interferometry as a function of frequency f and dc electric bias E. The real part of the piezoelectric response d_{33} and the corresponding piezoelectric loss factor are shown in Figure 41 as a function of f and E. In the initial state the sample had a nonzero d_{33} value, possibly indicative of inherent poling due to the preferential (111) orientation. The value of d_{33} then increased with increasing bias and reached a maximum value of nearly 180 pC/N at a bias level of 8×10^6 V/m. At higher bias levels the value of d_{33} started to decrease. The d_{33} coefficient was also frequency dependent, with a 10% relaxation between 100 Hz and 10 kHz. Figure 41b illustrates the behavior of the piezoelectric loss factor $\tan \delta$ as a function of frequency and field. Tan δ was approximately 12% at low frequency and increased to nearly 20% at 10 kHz. The hysteretic behavior for d_{33} with E had a typical butterfly-like pattern,

consistent with polarization reversal by 90° domain-wall movement (Fig. 42).

(d) Influence of Film Thickness and Substrate Constraint. The dependence of piezoelectric response (piezoelectric displacement and field-induced strain) on film thickness was observed on PZT 52/48 thin films deposited by the sol–gel method on Pt(111)/Si substrates [188]. As can be observed in Figure 43, as the film thickness decreased from 2 to 0.5 μm the maximum displacement decreased significantly for both orientations. The piezoelectric d_{33} constants show a similar behavior for both series: it decreases from 86.4 pm/V for 2 μm to 57.6 pm/V for 0.5 μm thickness in the case of (111)-oriented films, and from 97.5 pm/V (at 1.6 μm) to 63.5 pm/V (at 0.5 μm) for (100)-oriented films. Possible explanations for this behavior are changes in the residual stress with thickness and the presence of an interfacial layer at the electrodes whose influence on the effective properties is higher for thinner films.

A similar dependence was reported in [184], whose results are displayed in Figure 40. Graphs (a) and (b), which refer to

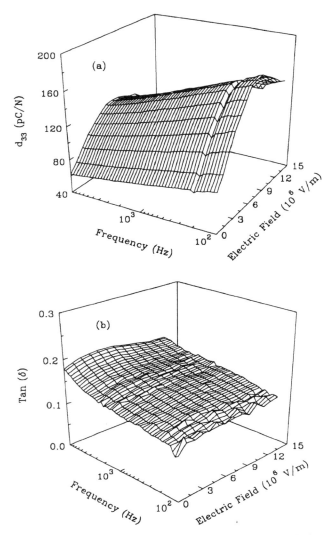

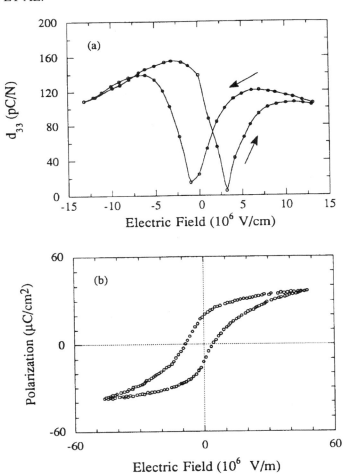

Fig. 41. The complex piezoelectric response [(a) the real part; (b) the imaginary part] for PZT 53/47 thin films as a function of bias field. Reprinted with permission from [174], © 1994, American Institute of Physics.

Fig. 42. Hysteresis of the piezoelectric coefficient (a) and polarization (b) in PZT 53/47 thin films. Reprinted with permission from [174], © 1994, American Institute of Physics.

a film of 0.32 μm thickness, show much lower d_{33} values than (c) and (d), which represent a film of 0.48 μm thickness.

The piezoelectric constant d_{33} depends also on substrate constraint through the relationship $d_{33f} = d_{33} - 2s_{13}/(s_{11} + s_{12})d_{31}$ [31]. It has been observed [31] that separating the epitaxial film in stripes along the [001] direction with width smaller than the film thickness transforms the biaxial constraint to a uniaxial one. The piezoelectric constant is now given by the relationship $d_{33f} = d_{33} - s_{13}/s_{33}d_{31}$. The cutting of the ferroelectric film along two orthogonal directions such that the lateral dimensions are smaller than the film thickness further reduces the constraint. The effect was observed in PZT films 28/68 doped with Nb (PNZT) with a thickness of 120 nm deposited by sol–gel processing onto Si substrates, with LSCO bottom electrode. Test capacitors of 0.25×0.25 μm in plane dimensions were separated by focused Ga ion beam milling and their piezoelectric response was measured by scanning force microscopy. The results show that the piezoelectric constant of

the 0.25×0.25 islands is two times higher than that of the continuous film [31].

(e) Influence of Dopants. It is generally expected that addition of dopants in PZT have effects similar to those in bulk materials. This has been verified by Haccart et al. [196], who investigated PZT 54/46 thin films doped with various Nb concentrations between 0 and 7 at%. The films were deposited on Pt(111)/Ti/SiO$_2$/Si substrates by sputtering. PNZT films showed (111) preferential orientation. The maximum d_{33} constant (115 pm/V) was obtained for 1 at% Nb whereas for the base composition the obtained value was two times lower. It decreased also at higher concentrations; for example, it was found to be only 90 pm/V at 3 at% Nb.

5.1.1.2.2. Ferroelectric Properties

(a) Influence of Composition. The ferroelectric properties of PZT films showed a strong dependence on composition, as can be observed from Figure 44 [170]. The films were prepared by dc magnetron sputtering. High values of the remanent polariza-

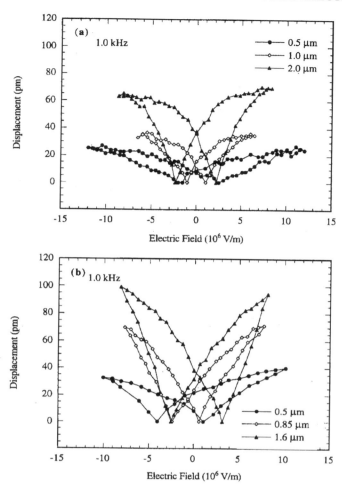

Fig. 43. Piezoelectric response of PZT 52/48 thin films with different thicknesses and different orientations: (a) (111) preferred orientation; (b) (100) preferred orientation. Reprinted with permission from [188], © 2000, American Institute of Physics.

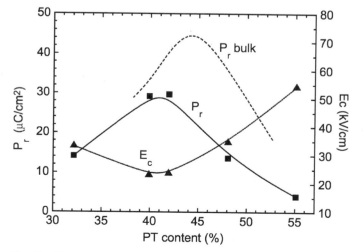

Fig. 44. Dependence of remanent polarization and coercive field on composition for PZT films with different PT content. The experimental data were taken from [170].

tion P_r and low values of coercive field E_c have been found only for the rhombohedral phase compositions, whereas for the tetragonal compositions P_r was low and E_c was high. These differences from bulk values for tetragonal compositions have been attributed by the authors to the effects of small grain size obtained in these films.

A systematic investigation of the relationship of composition and hysteresis loop, remanent polarization, and coercive field has been reported by Chen et al. [192]. A number of $PbZr_xTi_{1-x}O_3$ compositions with x ranging from 0 to 0.8 were prepared by the sol–gel technique on Pt/Ti/SiO$_2$/Si(100) substrates. The film thickness was 1 μm. A maximum remanent polarization of 36 μC/cm^2 was found for films at MPB, whereas the coercive field increased systematically with the content of lead titanate in the films.

The influence of composition on the ferroelectric properties has been reported also in [197]. PZT films deposited on various substrates and with Zr content ranging between $x = 0.2$ and $x = 0.9$, with submicron thicknesses (600–900 nm), have been obtained by rf magnetron sputtering. For films deposited on Pt/MgO which were (001) oriented, E_c decreased and P_r increased with increasing Zr content. An abrupt change in P_r observed at about $x = 0.55$ corresponds to MPB and indicates that the PZT films in the rhombohedral zone have soft ferroelectricity.

(b) Domain Orientation. For achieving the best control of ferroelectric properties, besides crystallite orientation, the role of 90° domain configurations is extremely important [198], because there is limited electrical switching of these domains. The important role of misfit strain and thermal stress in domain orientation was pointed out by Tuttle and Schwartz in [198]. It has been observed that, after cooling from the deposition temperature through the critical temperature of paraelectric–ferroelectric phase transformation, two identical sol–gel PZT films 53/47 but deposited on Si and on Pt/MgO show different behavior. Due to the different misfit strain the film deposited on Pt/MgO is highly c-domain oriented, whereas the film deposited on Si is highly a-domain oriented. The c-domain-oriented film shows much higher remanent polarization (55 μC/cm^2) compared to the a-domain-oriented film, whose remanent polarization is two times lower.

(c) Thickness Dependence. Thickness dependence of ferroelectric properties has been observed in PZT 55/45 films obtained by the sol–gel technique [183]. The films have been grown on Pt(111)/Ti/SiO$_2$/Si substrates and with strong (100) texture. Films with 1.2 and 3.3 μm thickness show rather different ferroelectric properties even though their microstructure was similar (columnar cross section, with grain size of about 30–50 nm which does not change with thickness). The PZT films with higher thickness had better ferroelectric properties: a remanent polarization of 45.5 μC/cm^2 with respect to only 12.5 μC/cm^2 measured on the thinner film. Also the coercive field increases with thickness, from 29 kV/cm for the 1.2 μm thick film to about 58.5 kV/cm for the 3.3 μm thick film,

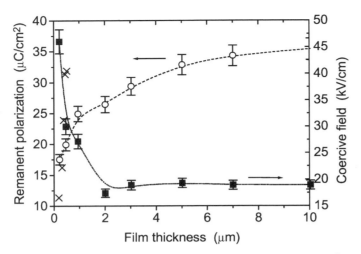

Fig. 45. Dependence of remanent polarization and coercive field on film thickness. The experimental data (P_r, open squares; E_c, solid squares) have been taken from [199] and from [200] (P_r, ×).

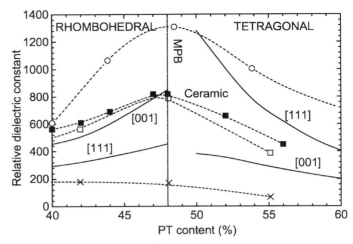

Fig. 46. Dielectric constant for PZT with different orientations and PT content: (continuous lines) theoretical predictions from [191]; (× and open squares) experimental data obtained in [170] for films of 1 and 3 μm thickness, respectively, as compared with bulk values (solid squares); (open circles) experimental dielectric constant relative values taken from [192].

whereas usually the coercive field is higher at smaller thickness. This effect was attributed by the authors to the increased pyrolysis duration for thicker films, which increases the probability of chemical reaction of the film and substrate.

Kurchania and Milne [199] studied the thickness dependence of the ferroelectric properties of sol–gel prepared PZT 53/47 films in the thickness range 0.25–10 μm. The films were deposited on (111)Pt/Ti/SiO$_2$/Si substrates and were completely free of pyrochlore phase. Values of remanent polarization and coercive fields are plotted against film thickness in Figure 45. It can be observed that the coercive field increases sharply at a thickness of about 1 μm. The remanent polarization decreases more gradually with thickness between 2 and 1 μm. The effects of PZT film thickness on dielectric and ferroelectric properties has been interpreted as being due to the proportionally greater contribution from interfacial effects as sample thickness decreases.

A systematic investigation of the thickness dependence of the ferroelectric properties of PZT thin films was reported by Udayakumar et al. [200]. PZT films up to a thickness of 6000 Å have been prepared by using the sol–gel method on substrate structures consisting of Ti/Pt/[100]Si. It has been found that the polarization has a sudden decrease with decreasing thickness. The measured values for polarization are represented by × symbols in Figure 45.

5.1.1.2.3. Dielectric Properties

(a) Influence of Orientation The influence of orientation of the structure on the dielectric properties was theoretically predicted in [190, 191] on the basis of a thermodynamic phenomenological model. The results are shown in Figure 46. For both tetragonal and rhombohedral phases the dielectric constant monotonously increases as the crystal tilting angle from the spontaneous polarization direction increases. Thus for the rhombohedral compositions the maximum dielectric constant is

on the direction (001), whereas for the tetragonal compositions this is much higher on the (111) direction than on the polar axis direction.

The influence of orientation on the dielectric constant has been experimentally observed in [193] in PZT 55/45 films (rhombohedral near MPB) deposited by a metalorganic decomposition process on different electrodes. Different precursor solutions and electrodes allowed films to be obtained with [111], [100], and mixed [111]/[100] orientations. The thickness of all three types of films was 600 nm. The largest dielectric constant (1170) was obtained for films with [111] orientation, whereas lower values (1040) were obtained for films with [100] orientation, which is in contrast with the predictions of [190, 191]. For PZT 60/40 rhombohedral films grown by sol–gel [189] the dielectric constant values obtained were about 740 for [111]-oriented films and 800 for [100]-oriented films. Even if $\varepsilon_{[100]} > \varepsilon_{[111]}$ as predicted by [190, 191] both values are much higher than those predicted for single crystals. This difference has been attributed to the contribution of domain walls to permittivity.

(b) Influence of Composition. Theoretical predictions for the relative dielectric constant as a function of composition for single-crystal PZT can be seen in Figure 46. A maximum of the dielectric constant is observed at MPB for all orientations.

The dependence has been experimentally verified in PZT films deposited by magnetron sputtering [170]. A strong dependence on composition has been obtained, as can be observed in Figure 46. The dielectric constant of films of adequate thickness (for which almost bulk properties have been reached), represented by open square symbols in Figure 46, behave like those of similar compositions in bulk (represented by solid squares); that is, they reach a maximum at MPB.

The influence of composition on the dielectric constant has been analyzed also by Sumi et al. [193] in PZT films deposited

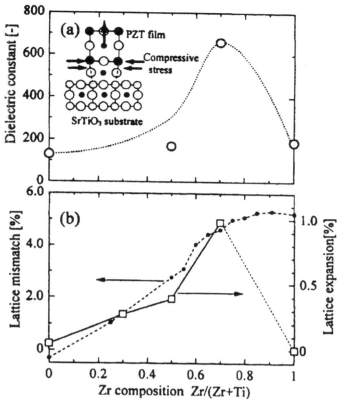

Fig. 47. (a) Dielectric constant of PZT films vs Zr content. The inset shows schematically the deformed crystal structure of PZT due to strain introduced by the SrTiO$_3$ substrate. (b) Relationship between the calculated lattice mismatch (solid circles) and c-axis lattice expansion (open squares). Reprinted with permission from [96], © 1996, Elsevier Science.

by a metalorganic decomposition process, for three compositions 45/55, 55/45, and 65/35. The films have mixed [111]/[100] orientations. The largest dielectric constant (1080) was obtained for films with near-MPB composition, as expected.

A systematic experimental investigation of the relationship between composition and dielectric constant has been reported by Chen et al. [192] for a number of PbZr$_x$Ti$_{1-x}$O$_3$ compositions with x ranging from 0 to 0.8 prepared by the sol–gel technique on Pt/Ti/SiO$_2$/Si(100) substrates. Figure 46 shows the dielectric constant for films of 1 μm thickness (represented by open circle) as a function of composition. The maximum value (1310) was obtained at MPB, as for the bulk ceramics, but the maximum value of bulk ceramics is much lower (850). It is not evident why polycrystalline thin films randomly oriented should have a higher dielectric constant than corresponding ceramics.

Figure 47 shows the influence of Zr variation in Pb(Zr$_x$Ti$_{1-x}$)O$_3$ film composition on the dielectric constant measured at room temperature and at a frequency of 1 kHz [96]. The films were deposited on SrTiO$_3$(100) substrates using excimer laser ablation. Their thickness was about 200 nm. All the films at every composition have been c-axis oriented. It can be observed that the film with $x = 0.7$ has the largest dielectric constant, whereas in bulk materials the maximum of the

dielectric constant is at MPB. The shift of the maximum of the dielectric constant to higher Zr composition in thin films was explained by the large expansion of the film lattice in the c direction due to the increased film–substrate mismatch for PZT with increased Zr concentration (see inset in the figure). The large lattice mismatch reached for high Zr concentration (which, for $x = 1$ is 5.2%) induces large compressive stress, which shrinks the lattice of the film along the substrate plane (a–b axis plane) and extends along the growth direction (c-axis direction). As a result the ionic displacements which contribute to the dielectric constant along the c axis are favored and therefore the relative dielectric constant is increased for high Zr concentration. The decrease at higher concentration is due to the fact that too large a mismatch introduces misfit dislocations to relieve large stress at the interface so that these lattices are not deformed.

The influence of composition on the dielectric properties has been considered also by Ijima et al. [197]. PZT films deposited on various substrates and with Zr content ranging between $x = 0.2$ and 0.9 and with submicron thicknesses (600–900 nm) have been obtained by rf magnetron sputtering. A maximum of the dielectric constant of about 900 was obtained for films deposited on Pt/sapphire and about 500 for those deposited on MgO, all at $x = 0.55$ (near MPB). This is a confirmation that the variation of the piezoelectric properties in PZT solid solution thin films behaves like it does in bulk.

(c) Thickness Dependence. Thickness dependence of dielectric properties has been found in different PZT films grown by different techniques. Films of PZT 55/45 grown on Pt(111)/Ti/SiO$_2$/Si substrates with 1.2 and 3.3 μm thickness have shown rather different dielectric constants and loss factors for the two cases: 1150 and 0.03 for the first and 1250 and 0.04 for the second. The bulk value constants are 1200 and 0.02 [173]. Strong thickness dependence of dielectric constant has been found also for films with different compositions obtained by dc magnetron sputtering [170] as can be observed in Figure 46. Bulk PZT values (about 800) could be achieved only beyond a film thickness of 3 μm, whereas films with thickness 1 μm yield a dielectric constant of only 194 for MPB compositions.

Kurchania and Milne [199] studied the thickness dependence of the dielectric properties of sol–gel prepared PZT 53/47 films in the thickness range 0.25–10 μm. Values of the dielectric constant and dissipation factor are plotted against film thickness in Figure 48. It can be observed that the dielectric constant decreases rapidly with film thickness decreasing below 1 μm whereas the decrease of the dissipation factor was slower. The effects of PZT film thickness on dielectric properties have been interpreted as being due to the enhanced contribution from interfacial effects as sample thickness decreases. On the same graph data obtained on PZT films prepared by sol–gel on Ti/Pt/Si with thicknesses up to 600 nm [200] have been represented also. The relative dielectric constant increases with thickness, saturating at a value of 1300 for film thicknesses of 0.32 μm and above. A concomitant drop in dissipation factor is noted with increasing film thickness. These variations have

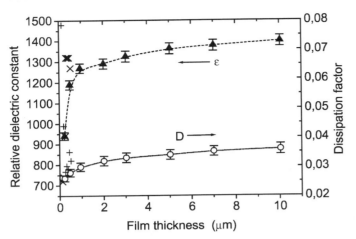

Fig. 48. The dependence of dielectric constant and dissipation factor on thickness. The experimental data have been taken from [199] (solid triangles for ε and open squares for D) and from [200] ($\times$ for ε and $+$ for D).

been attributed to the presence of growth stresses and thermal stresses, which increased with decreasing thickness.

(d) Influence of Dopants. It is expected generally that addition of dopants in PZT has effects on the dielectric constant similar to those in bulk materials. It has been shown by Haccart et al. [196] that Nb doping in PZT films leads to strong dielectric constant variation with respect to the base composition. Nb has been added in concentrations between 0 and 7 at% to PZT 54/46 composition and different films have been sputtered on Pt(111)/Ti/SiO$_2$/Si substrates. The dielectric constant increased with Nb concentration from 0 to 2 at%, reaching the maximum value of 1100 for a doping of 2 at%. At a higher doping level a decrease in the dielectric constant is observed. The variation of the dielectric loss was also measured. For undoped films $\tan \delta$ is of the order of 1.8%. This value does not change in the range 1–7%.

In bulk ceramics, addition of La transforms the ferroelectric PZT into a relaxor over a certain value which depends on the ratio Zr/Ti. Ferroelectric and dielectric properties of La-doped PZT (PLZT) epitaxial films grown on sapphire by rf magnetron sputtering have been reported by Adachi et al. [201].

5.1.2. Lead Titanate

Lead titanate PbTiO$_3$ (PT) belongs to the perovskite cubic m3m class at high temperatures and transforms to a low-temperature tetragonal (4mm) ferroelectric phase at the critical temperature $T_c = 493°C$. Piezoelectric properties of PT have been extensively investigated in view of high-frequency sensing applications. Lead titanate is considered as one of the best candidates for this purpose because both piezoelectric and dielectric properties can be tailored by doping with suitable elements such as Ca, Sm, and La. In particular, it has been found that longitudinal piezoelectric coefficient d_{33} significantly increases with Ca content, whereas the dielectric constant remains low up to a Ca concentration of about 20%. As a result the piezoelectric

voltage coefficient defined as $g_{33} = d_{33}/\varepsilon_0\varepsilon_{33}$ was found to increase in Ca-modified PT (PCT). PbTiO$_3$ thin films have been intensively investigated in the last decades [207–212]. They have been applied in different piezoelectric devices at high frequency due to the high electromechanical coupling factor and high quality factor Q.

5.1.2.1. Overview of the Main Results

The piezoelectric properties of c-axis-oriented PbTiO$_3$ thin films have been investigated by Kushida and Takeuchi [207]. Films with thicknesses of 1–2 μm have been deposited by rf magnetron sputtering on (100)SrTiO$_3$ single-crystal plate, which has the same crystal structure as PT and similar a-lattice constant. A patterned (111)Pt electrode was fabricated on this substrate in an embedded form, with the intention of letting the epitaxially grown PT film on the ST single crystal overgrow laterally in order to cover the Pt electrode. The PT film was found to be highly c-axis oriented. The small portion of film covering the Pt electrode was polycrystalline with small grains of 0.2 μm, whereas all the rest was epitaxial; when the film grew to 1.5 μm thick the polycrystal portion was completely covered with the epitaxial film. The dielectric constant of the film was about 110. From the impedance characteristics, values of the electromechanical coupling factor k_t and mechanical quality factor Q have been obtained in the frequency range 200–700 MHz. The k_t estimated from modes whose frequencies are less than 350 MHz was as large as 0.8, but it was found to be smaller for higher frequencies probably due to the higher dielectric loss, which increased with frequency. More recently Yamada et al. [208] demonstrated the successful fabrication of piezoelectric PT films on GaAs substrates for use in BAW resonators in the gigahertz band. PT films were deposited by rf magnetron sputtering on Pt/Ti/SiO$_2$/GaAs substrates. The film was highly oriented along the $\langle 111 \rangle$ crystal direction. The resonance behavior and piezoelectric properties were investigated on the constructed BAW resonator, whose main resonance occurred at 1.9 GHz. By fitting the electrical parameters in the equivalent circuit of the resonator to this resonance behavior, the electromechanical coupling constant k^2 was estimated to be 9.4%.

The microstructural evolution and the related electromechanical properties of Ca-added PT thin films have been investigated as a function of Ca content and porosity of the films [209]. The porosity was tailored by changing the heating rate during the final crystallization annealing. It was found that piezoelectric properties of porous films heated at slow rates are superior to those of dense films crystallized using higher heating rates. This was explained as a result of the constraining effect of the substrate, which apparently reduces the piezoelectric and strain responses in dense films. When the porosity of the film increases, the clamping effect of the substrate is reduced because the grains can deform more independently in the lateral direction. Electromechanical properties are also improved with Ca addition due to the decrease of tetragonality and ease of 90° domain rotation. Charge piezoelectric coefficients in porous films are close to those of PCT ceramics of the same composition, whereas the voltage piezoelectric coefficients were significantly greater for PCT films than for ceramics [$g_{33} = d_{33}/(\varepsilon_0\varepsilon) \cong 130 \times 10^{-3}$ mV/N for PCT15 film]. These results, combined with the low dielectric constant, make PCT films an attractive material for microelectromechanical applications.

Yamada et al. [54] studied the piezoelectric properties of La-doped PT films for rf resonators: 5-wt% La$_2$O$_3$-doped PbTiO$_3$ piezoelectric films were deposited by rf magnetron sputtering onto the bottom electrodes on CVD-deposited SiN films. The SiN films were used as support films in the membrane structure of the BAW resonators, on which successive Ti/PtTi/Pt layers had been deposited. The intermediate PtTi layer was added to prevent Ti diffusion from the adhesive layer. The PT films on the resonators were polarized under 150 MV/m for 15 min at 200°C. All PT films were strongly oriented in the ⟨111⟩ direction and no other diffraction peaks were observed. The extensional mode resonance of the BAW resonator occurred at 2 GHz. The electromechanical coupling factor k_t^2 and Q values were calculated from the values of the series and parallel resonances and the values of equivalent circuit of the resonator. k_t^2 was estimated to be 11% at 1.4 GHz and 9.4% at 2 GHz and Q was about 70. Radiofrequency devices using La-doped PT films could obtain a broader bandwidth in comparison with the devices using ZnO films, because k_t^2 of the La-doped PT films is superior to the k_t^2 value of 7.8% of ZnO films. For the realization of rf devices with PT films it is necessary to improve the Q value.

Single-domain, single-crystal PbTiO$_3$ (PT) thin films with (001) orientation have been obtained by sputtering on a miscut (001)SrTiO$_3$ (ST) substrate [210]. It has been observed that the films showed a strained deformation due to lattice mismatch, and the tightly bonded interface consequently modifies ferroelectric properties by inducing a decrease of polarization, an increase of coercive field, and a diffused temperature anomaly.

Table X presents the main physical properties of PT thin films, together with the corresponding bulk values.

5.1.2.2. Influence of Structure on Physical Properties

5.1.2.2.1. Piezoelectric Properties

(a) Dependence on Remanent Polarization. The piezoelectric properties of PbTiO$_3$ thin films, like those of other ferroelectric thin films, are dependent on the remanent polarization. A calculation based on the method of effective piezoelectric medium [213] predicted this dependence. Figure 49 shows the variations of the piezoelectric charge coefficient d_{33} with the remanent polarization in PbTiO$_3$ bulk ceramics; the fourth curve shows the effective piezoelectric response d_{33} of a polycrystalline PbTiO$_3$ thin film. The difference between the corresponding piezoelectric constants of the polycrystalline bulk ceramic and polycrystalline thin film is due to the substrate

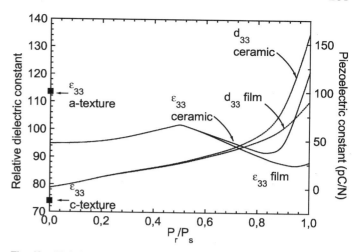

Fig. 49. Variations of the piezoelectric constant d_{33} (dashed lines) and the dielectric constant (continuous lines) for PT polycrystalline films and bulk ceramics, as predicted in [213]. The two solid squares indicate the values of the dielectric constant for oriented films with *a*- and *c*-texture.

clamping. It can be observed that d_{33} increases with the remanent polarization and that the substrate clamping has the effect of decreasing the piezoelectric constant of the film with respect to that of the substrate.

(b) Influence of Dopants. The effect of Ca dopants on the piezoelectric properties of PbTiO$_3$ thin films was considered by Kholkin et al. [209]. Films with thicknesses of 500–700 nm have been prepared by sol–gel with different Ca concentrations ranging from 0 to 15%. Films deposited on Pt/TiO$_2$/SiO$_2$/Si substrates showed (111) to be the preferred orientation. The tetragonal distortion was found to decrease with increasing Ca concentration, thus allowing a more efficient poling to be achieved. Therefore the piezoelectric constant d_{33} (measured after poling with 400 kV/cm) and the electric-field-induced strain (maximum value under 600 kV/cm) increase with Ca concentration. The electric-field-induced strain is mainly determined by 90° domain rotations; by increasing Ca concentrations the tetragonality decreases, thus favoring 90° domain rotation and consequently the increase of the piezoelectric properties.

La doping was found to increase the electromechanical coupling factor, as reported in [54]. The electromechanical coupling factor was increased in 5-wt% La$_2$O$_3$-doped PbTiO$_3$ almost three times with respect to the base composition [54].

(c) Influence of Porosity. The effect of porosity on the piezoelectric properties of Ca-doped PT thin films was also considered in [209]. Even if it is difficult to clearly separate the effects of Ca doping from the effects of increasing porosity, some interesting features have been evidenced. The porosity degree was varied by changing the ramp of the final heat treatment, when the crystallization temperature of 650°C was reached at different heating rates: 1, 6, and 60°C/s. This was found to have a strong effect on the microstructure of the films: whereas the

Table X. Main Physical Properties of PT Thin Films as Reported in Different References (Last Column)

Composition	Deposition technique, substrate	Orientation	Relative dielectric const. (1 kHz)	Dielectric loss factor	P_r (μC/cm^2)	E_c (kV/cm)	d_{33} (pC/N)	d_{31} (pC/N)	Q_m	k^2	Reference
PbTiO$_3$	Rf magnetron sputtering Pt/Ti/SiO$_2$/GaAs	(111)	160	0.008					65	9.4% (1.9 GHz)	[208]
PbTiO$_3$	Rf magnetron sputtering Pt/SrTiO$_3$(100)	(001)								$K_t \sim 0.8$ (below 350 MHz)	[207]
PbTiO$_3$	Sol–gel Pt/fused quartz	Random	285	0.02							[211]
PbTiO$_3$	Sol–gel Pt/Ti/SiO$_2$/Si	Random	80	0.028	97.1						[222]
PbTiO$_3$	Rf magnetron sputtering Pt/Ti/SiO$_2$/Si	(111)	160 (100 kHz)						70 (1.4 GHz)	4.2% (1.4 GHz)	[54]
PbTiO$_3$	Reactive sputtering (100)Pt/(100)MgO	(001)	192	0.036							[221]
PbTiO$_3$	Sol–gel Pt/TiO$_2$/SiO$_2$/Si	Random					40				
5 wt% La$_2$O$_3$-doped PbTiO$_3$	Rf magnetron sputtering Pt/PtTi/Ti/Si$_3$N$_4$/Si	(111)	320 (100 kHz)		11				70 (1.4 GHz) 65 (2 GHz)	11% (1.4 GHz) 9.4% (2 GHz)	[54]
Pb$_{1.03-x}$La$_x$Ti$_{1-x/4}$O$_3$, $x = 0.1$	Sol–gel Pt/Ti/SiO$_2$/Si	Random	220	0.03	10	55					[222]
Pb$_{1.03-x}$La$_x$Ti$_{1-x/4}$O$_3$, $x = 0.2$	Sol–gel Pt/Ti/SiO$_2$/Si	Random	450	0.05	2	20					[222]
Pb$_{1.03-x}$La$_x$Ti$_{1-x/4}$O$_3$, $x = 0.21$	On-axis sputtering SrRuO$_3$/SrTiO$_3$	(001) single domain, single crystal epitaxial	310	0.02	15	400–500					[210]
1 mol% Nb–PbTiO$_3$	Reactive sputtering (100)Pt/(100)MgO	(001)	170	0.02							[221]
2 mol% Nb–PbTiO$_3$	Reactive sputtering (100)Pt/(100)MgO	(001)	200	0.04							[221]
3 mol% Nb–PbTiO$_3$	Reactive sputtering (100)Pt/(100)MgO	(001)	190	0.036	20	200					[221]
Pb$_{1-x}$Ca$_x$TiO$_3$ $x = 5\%$	Sol–gel Pt/TiO$_2$/SiO$_2$/Si	Random					50				[209]
Pb$_{1-x}$Ca$_x$TiO$_3$ $x = 10\%$	Sol–gel Pt/TiO$_2$/SiO$_2$/Si	Random					57				[209]
Pb$_{1-x}$Ca$_x$TiO$_3$ $x = 15\%$	Sol–gel Pt/TiO$_2$/SiO$_2$/Si	Random					59				[209]
Bulk single crystal PbTiO$_3$			37 (ε_{33}) 110 (ε_{11})				160	−26			[212]

Note: For comparison bulk values have been also included.

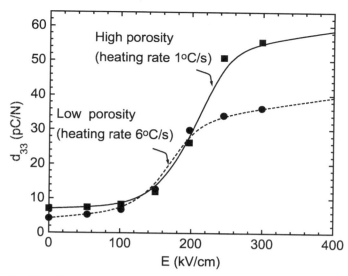

Fig. 50. The variation of piezoelectric constant with poling field for films heated with different ramps. The experimental data have been taken from [209].

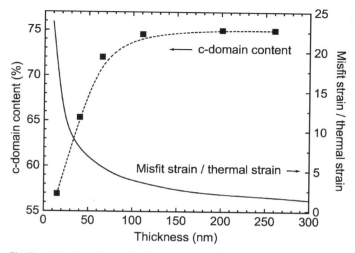

Fig. 51. The dependence of c-domain content and ratio of misfit strain to thermal strain on film thickness. The data are taken from [214].

film heated at 60°C/s showed a dense microstructure with large columnar grains, the film heated at 1°C/s had a high porosity (25%) and small grain size (30–70 nm). Figure 50 compares the variation of d_{33} coefficient with poling field for the films processed at 1 and 6°C/s. The piezoelectric constant d_{33} is higher for more porous films, which is contrary to what is expected from the effective behavior of a porous medium. When the porosity of the film increases, the clamping effect of the substrate is reduced because the grains can deform more independently in the lateral direction. This may compensate the decrease due to mixing with nonpiezoelectric phase.

5.1.2.2.2. Ferroelectric Properties

(a) Domain Orientation. The ferroelectric behavior is closely related to the domain switching behavior and domain structure. Lee and Baik [214] studied the relaxation elastic misfit strains introduced by heteroepitaxy and its effect on the domain formation of epitaxial $PbTiO_3$ thin films grown on MgO(001) substrates. These effects have been studied as a function of film thickness considering thermodynamic equilibrium relief of coherency strain at the initial growth stage. Thickness-dependent evolution of twin domain structures was characterized by two-dimensional reciprocal space mapping using synchrotron X-ray. $PbTiO_3$ thin films have been grown epitaxially at temperatures between 550 and 750°C. At these temperatures the films are in a cubic paraelectric phase. When the films are very thin, the coherency with the substrate lattice is maintained due to the elastic straining of the film lattice. When the thickness of the film increases, the elastic coherency strain energy increases too. If it will exceed the energy required for misfit dislocation generation, dislocations will be formed, thus reducing the strain energy. Once the misfit dislocations are generated, they will create a screening effect, thus reducing the substrate effect by the amount of dislocation density.

This effect can be described by introducing an effective substrate lattice parameter b_{eff} and an effective misfit strain residual in the film, which will depend on the film thickness. The effective misfit strain was calculated for $PbTiO_3$ and it has been found that fully coherent film is under tension below the critical thickness (about 5 Å); then it rapidly decreases when the film thickness grows to 1000 Å and, in the larger thickness region, approaches a constant value but never disappears completely. This effect of thickness-dependent relaxation of elastic misfit strain on the ferroelectric properties (domain formation) was tested experimentally with epitaxial $PbTiO_3$ thin films grown on MgO(001) single-crystalline substrates [214]. $PbTiO_3$ thin films were prepared by PLD on annealed MgO(001) single-crystalline substrates with varying film thickness ranging from 150 to 2500 Å. The thickness of the film was measured by Rutherford backscattering spectrometry and the evolution of twin domain structures with the film thickness were obtained by two-dimensional reciprocal space mapping for $PbTiO_3$ (001) and (100) reflections in the synchrotron X-ray diffraction beam line. The quantitative c-domain abundance was estimated by the ratio of the integrated intensities for (001) and (100) reflections defined as $\alpha = \sum I(001)/\sum[I(001) + I(100)]$. Figure 51 shows α as a function of film thickness. It can be observed that α is critically dependent on film thickness for films of few hundred angstroms thickness; then it rapidly increases with the film thickness, saturating at a value of about 0.75. On the same graph the calculated ratio between the misfit strain normalized to the thermal strain, accumulated during cooling from the growth temperature down to Curie temperature, was represented too. It is clearly seen that the domain formation in ultrathin films (below 1000 Å) is conditioned by the misfit strain.

The problem of strain relaxation by domain formation in epitaxial $PbTiO_3$ films grown on the (001) surface of single crystals of $KTaO_3$, $SrTiO_3$, and MgO has been considered by Kwak et al. [215, 216]. Considerable stress has been revealed by X-ray diffraction and micro-Raman scattering in

$PbTiO_3$ films with thickness 0.6–1.2 μm deposited by sol–gel on Si(100) [217].

The dependence of ferroelectric properties of $PbTiO_3$ thin films on the domain nucleation and orientation has been considered by Wang et al. and Ren et al. [218–220]. They have pointed out that the ferroelectric properties are related to the ferroelectric domain structure and to the domain-wall mobility which in ferroelectric polycrystalline thin films depend on the grain size. Although the strength of the coercive field is determined by the ease of domain nucleation and domain-wall motion, permittivity is related to the density of domain walls and their mobility at low fields. An accurate study of ferroelectric domain structure in free-standing $PbTiO_3$ films formed by grains 60–1000 nm in size has been performed by TEM. A $PbTiO_3$ film about 100 nm thick with tetragonal perovskite structure was deposited on NaCl substrate by the sol–gel method. A free-standing film was obtained by dissolving away NaCl substrate in water. The film was polycrystalline randomly oriented with a grain size of about 60 nm. The unsupported film free from the influence of substrate was directly observed under TEM without further thinning. By removing the condenser lens aperture of the TEM and converging the electron beam, a local growth of the grains could be obtained due to the heating by the high-intensity electron beam. The grain size growth was controlled by changing heating time and beam convergence, so that grains with sizes ranging from 60 to 1000 nm could be obtained in different places in the same film. The grain growth was predominantly in the plane of the film, thus the thickness of the film did not increase. It has been evidenced that the ferroelectric domain structure changes from multidomain to single domain at a grain size of ~150 nm. This finding helps to understand the dependence of ferroelectric properties of thin polycrystalline films on thickness, because different thicknesses often mean different grain sizes. In very thin films small single-domain grains are energetically more stable than those that have been split into domains. As a result the nucleation of new domains under an external field becomes much more difficult and therefore the coercive field increases. Moreover, due to the lack of domain walls the polarization change due to domain-wall motion at low fields is very small and consequently the dielectric constant is very small. Therefore the intrinsic reasons for the sharp increase in coercive field and decrease of dielectric constant in very thin polycrystalline ferroelectric films are related to the stabilization of single-domain grains at small size.

(b) Influence of Misfit Strain in Single-Domain, Single-Crystal Films. The influence of misfit strain on $PbTiO_3$ thin film ferroelectric properties was evidenced in [210]. The films with (001) orientation were synthesized on a miscut (001)SrTiO$_3$ (ST) substrate by sputtering. Single-domain, single-crystal PT thin films with thickness about 125 nm have been characterized. These films showed a strained deformation due to the slight lattice mismatch. The interface between sputtered PT films and the substrates was coherent and the films were tightly bonded to the substrates, without any interfacial structure. Therefore the misfit strain in these films was not relaxed as in other epitaxial films

by dislocations and domain formation. The measurement of the P–E hysteresis curve showed that the coercive field E_c for these single-domain, single-crystal films was 400–500 kV/cm, which is an order of magnitude higher than other PT films which do not have this structure. The saturation polarization P_s was one-half that of the bulk values. It is considered that the low value of P_s is due to the strained deformation of the film which limits the displacements of the atoms. The high value of E_c can be due to the difficulties in domain orientation in this clamped structure.

5.1.2.2.3. Dielectric Constant and Loss Factor

(a) Influence of Poling. Pertsev et al. calculated the dependence of the dielectric properties of $PbTiO_3$ thin films on poling by using a model based on the method of effective piezoelectric medium [213]. Figure 49 shows the variations of the dielectric constant ε_{33} with the remanent polarization in $PbTiO_3$ bulk ceramics and in polycrystalline $PbTiO_3$ thin film. The difference between the corresponding dielectric constants of the polycrystalline bulk ceramic and polycrystalline thin film is due to the substrate clamping. The dielectric constant shows a local maximum at $P_r/P_s \sim 0.5$ both for the polycrystalline bulk ceramic and for the polycrystalline untextured film. Black squares on the figure show the dielectric constants of unpolarized *c*- and *a*-textured $PbTiO_3$ film. It can be observed that the crystal texture has a strong impact on the dielectric response of ferroelectric thin films. This effect may be attributed to the high anisotropy of dielectric properties of $PbTiO_3$ single crystals.

(b) Influence of Misfit Strain. The influence of misfit strain on $PbTiO_3$ thin film properties was put in evidence by Wasa et al. [210]. The effect was studied on sputtered single-domain, single-crystal PT thin films with thickness about 125 nm. The films with (001) orientation were synthesized on a miscut (001)SrTiO$_3$ (ST) substrate with a strained deformation due to the slight lattice mismatch, as previously discussed. The effect of this strain on the dielectric properties was a diffused temperature anomaly: the broadening of the peak of dielectric constant at the critical temperature. This behavior was observed also in the loss factor. It is considered that the effect is due to the fact that the displacements of the atoms in this film are clamped.

(c) Influence of Doping. Different additives in PT have been generally used to enhance the resistivity but more recently they have been used also to improve other properties. Ibrahim et al. [221] studied the effect of Nb additive on the dielectric properties of PT thin films epitaxially grown on (100)-cut MgO single crystals and on Pt/MgO by reactive sputtering. The samples were doped at a Nb concentration of approximately 1, 2, 3, or 4%. All samples were highly (001) oriented, but low-intensity (100) diffraction peaks corresponding to 90° domain formation have been also observed. The epitaxial growth of PT and Pt bottom electrode on the (100)MgO substrates was verified by the in-plane energy dispersive total reflection X-ray

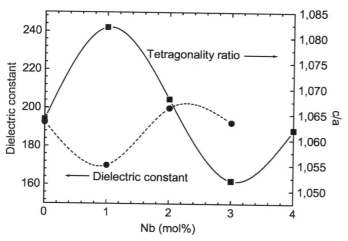

Fig. 52. The variation of the dielectric constant and tetragonality ratio on Nb concentration in PT thin films. The data are taken from [221].

diffraction (ED-TXRD) methods, whereas the lattice parameters have been obtained from the (001) diffraction (c-lattice parameter) and from the in-plane diffractions by ED-TXRD (a-lattice parameter). Being closely related, the variation of the tetragonality ratio c/a and of the dielectric constant are represented in Figure 52. A maximum of the tetragonality ratio is obtained at Nb concentration of about 1 mol%. At the same value the permittivity shows a minimum, due the fact that increasing the already strong anisotropy of PT makes the orientation of 90° domains more difficult.

The effects of La doping in thin PT films have been considered in [222]. The dielectric constant and the dissipation factor increased gradually with La addition. Moreover the Curie temperature was found to decrease monotonically with increasing La doping up to 25 mol%.

5.1.3. Lithium Niobate

5.1.3.1. Overview of the Main Results

Lithium niobate $LiNbO_3$ (LN) has been the object of intensive study both in bulk and in thin film form due to its high piezoelectric and electrooptical constants and its use in many signal-processing devices such as low-loss wide-bandpass filters and high-efficiency convolvers. LN is uniaxial at all temperatures, with a structural phase transition from paraelectric to ferroelectric at 1200°C. Although LN does not have the perovskite structure it has an ABO_3 lattice with oxygen octahedra. The basic room-temperature structure consists of a sequence of distorted oxygen octahedra joined by their faces along a trigonal polar axis [2].

LN thin film deposition was first reported by Foster [223], who succeeded in forming piezoelectric polycrystalline LN films on sapphire substrates and evaluated their piezoelectric properties. LN films have been also successfully deposited by rf sputtering on silicon and sapphire substrates by Rost et al. [224]. The films were oriented polycrystalline.

Furushima et al. [225] used a rf planar magnetron sputtering system to prepare LN thin films on α-Al_2O_3(001), (012), and (110) and $LiTaO_3$(110). Films with different orientations were epitaxially grown on these substrates. Although the waveguide optical properties were good and optical propagation loss was low, no piezoelectric response could be obtained from interdigital transducers deposited on the LN thin films, probably due to the surface roughness (which could cause defects in the transducer pattern) or to the multidomain structure.

The preparation and characterization of LN thin film by CVD was reported in [226]. The films were grown epitaxially on $LiTaO_3$ (LT). It has been found that the crystallinity of LN greatly depends on substrate temperature and surface orientation of the LT substrate. The properties of surface acoustic wave propagation have been improved by using LT as substrate. The electromechanical coupling factor k^2 was found to be 0.52% and the sound velocity 3200 m/s.

The preparation of textured LN thin films by off-axis laser deposition was reported by Kaemmer et al. [227]. The films were deposited on (012) and (001) sapphire and $LaAlO_3$ single-crystalline substrates. The presence of a secondary phase $LiNb_3O_8$ was also observed in some conditions, but it could be eliminated by optimizing the deposition parameters. The dielectric polarization and domain structure of the films were obtained by electric scanning force microscopy.

The epitaxial growth of $LiNbO_3$ thin films on sapphire substrates by using an excimer laser ablation technique and the evaluation of the surface acoustic wave properties of these films have been reported by Shibata et al. [112, 115, 228]. Their results demonstrated that thin films of LN on sapphire can be used in high-frequency applications such as high-frequency SAW filters due to their high velocities of propagation (5300–6600 m/s) and good temperature coefficients (−34 to −80 ppm/°C). By controlling the ratio Li/Nb in the target, the substrate temperature, the composition, and the gas pressure they obtained epitaxial films without twin structure with good crystallinity and surface flatness. The rocking curve of the (012) reflection of the LN single-phase film had half width 0.17°, which was very close to that of sapphire substrate (0.15°). It has been found that the as-grown LN films on sapphire substrates are sufficiently piezoelectric to act as SAW filters. Moreover SAW properties were observed over the entire film surface, suggesting that the films consist of a single domain without poling.

The growth of LN thin films by PLD has been reported also in [229], where oriented LN thin films were deposited on fused silica substrates. The as-prepared films showed good surface quality. During the deposition a low electric field was applied *in situ* on the film through a copper grid used as the top electrode and positioned between the target and the substrate, whereas the metallic substrate was the electrically grounded bottom electrode. A drastic influence of the substrate temperature on the stoichiometry of the thin films and their orientation was found. It was observed that the application of the electric field during the deposition helped to achieve a complete (001) orientation.

The structural evolution of epitaxial LN thin films deposited on sapphire (001) during growth by pulsed laser deposition was studied by Veignant et al. [230]. Their structure was investigated at different stages of growth by TEM. It was found that it changed from the early islands stage to the continuous films stage. The large lattice mismatch (8.6% at the temperature of deposition and 7.7% at room temperature) between Al_2O_3 ($a = 4.79$ Å) and LN ($a = 5.22$ Å) induces considerable stress in the LN films and misfit dislocations to accommodate it. Nanocracks preferentially oriented along the boundaries are introduced in films 100 nm thick, but they are not sufficient to relax the elastic deformation, and the lattice parameters remain different from those in bulk crystal. The elastic deformation is fully relaxed in the 220 nm thick films where the cracks have a mean width of 60 nm. Twinnings are also produced during cooling stage. These defects influence the optical properties of the lithium niobate films. The optical waveguiding was not obtained in thicker films which displayed a multitude of cracks of about 60 nm.

The deposition of (012)-oriented LN thin films on glass substrates has been reported by Park et al. [231]. The addition of K to the target improved the mobility of the sputtered materials through the reevaporation of K on the surface of the substrate during crystallization of LN. The enhanced mobility enabled the growth of a LN film with a minimum surface energy for the (012) plane which promoted the (012) preferred orientation on glass substrates.

LN thin film deposition on sapphire substrates by MOCVD was discussed in [232]. The LN films deposited on (001) sapphire substrates showed strong (001) orientation involving twin structures. Highly oriented (100) films were obtained on the (012) sapphire substrates. The FWHM of the X-ray rocking curves of (001) and (100) LN films were 0.16° and 0.21°, respectively. Very interesting morphological features have been revealed on the (100)-oriented films, where long and narrow grains aligned in the (012) plane of the sapphire with the c axis of LN parallel to the short axis of grains have been observed.

The first successful deposition of high-quality textured LN thin films on Si substrates was reported by Ghica et al. [233]. This result is of great importance for the use of LN thin films in devices integrated on Si. By adjusting the deposition parameters, the presence of undesired Li-deficient phases was avoided. The measurement of the refractive indices demonstrate good agreement with the corresponding single-crystal bulk values in all the visible spectral range. Growth of oriented LN thin films on Si was reported also in [234], where chemical beam epitaxy was employed.

Lee et al. [235] studied thin film $LiNbO_3$ on diamond-coated Si substrates for SAW applications. The goal of their project was to produce gigahertz-range electronic filters with low insertion loss for front-end radiofrequency SAW devices. Theory predicts that LN thin films on diamond/Si substrates have high electromechanical coupling coefficients and SAW propagation velocities because of the interaction with the high acoustic velocity diamond layer. Radiofrequency magnetron sputtering technology was used to grow LN films on these substrates, thus allowing the SAW filters to be integrated with electronic circuits. Interdigital transducers deposited on the LN/diamond/silicon films allowed the SAW properties to be studied. The thickness of the diamond film was about 20 μm and the surface was polished. For this thickness surface acoustic waves are confined in the LN and diamond layers and no SAW energy can penetrate into the Si substrate, thus allowing high quality devices to be obtained.

5.1.4. Lead Magnesium Niobate–Lead Titanate

The new single-crystal relaxor ferroelectric materials such as $Pb(Mg_{1/3}Nb_{2/3})O_3$–$PbTiO_3$ (PMN-PT) exhibit a piezoelectric effect that is 10 times larger than in conventional ceramics and can therefore revolutionize the field of applications in ultrasonic devices, medical imaging, miroelectromechanical systems, etc. [236]. Relaxor ferroelectric materials possess several unique properties which make them attractive for dielectric, piezoelectric, and electrostrictive applications: namely a high dielectric constant and a broadened dielectric maximum at the transition temperature as compared to normal ferroelectric materials [237–239]. PMN is the most known relaxor ferroelectric material which exhibits an abnormally high dielectric constant, high electrostrictive coefficient, and diffuse phase transition near −15°C. To increase the Curie temperature solid solutions of PMN-PT have been synthesized. The end members of the family PMN-PT are on one side PT, which is a highly anisotropic ferroelectric with large tetragonal distortion, typical ferroelectric–paraelectric transition at $T_c = 490$°C and long-range spontaneous ferroelectric order below T_c, and on the other side PMN, which is a relaxor ferroelectric characterized by a maximum of dielectric constant of extremely high value ($\varepsilon_{max} \cong 15,000$ at $T_{max} = -15$°C) associated with a significant frequency dispersion. However, in PMN no phase transition takes place at T_{max} and the dielectric dispersion does not follow the classical relaxation model. Ferroelectric relaxors are characterized by the existence of a ferroelectric order limited to regions of nanometer dimension (nanodomains), isolated by disordered zones which prevent the long-range ferroelectric order from developing. Application of an electric field can, however, induce a ferroelectric phase with formation of polar macrodomains (field-enforced phase transition). The solid solutions between PMN and PT are expected to combine properties of both relaxor PMN and ferroelectric PT. Indeed single crystals of PMN-PT have been reported to exhibit extremely large piezoelectric strain (>1%) and a very high electromechanical coupling factor ($k_{33} > 90$%). The ferroelectric solid solution of PMN-PT covers a wide range of applications which depend on the amount of PT. A large induced electrostrictive strain can be obtained from compositions which contain small amounts of PT, whereas the piezoelectric effects become significant for compositions near the morphotropic phase boundary of the PMN-PT system, which occurs at a PT content of about 35%. As the PT content is increased above 40% this material behaves like a normal ferroelectric. The high electromechanical properties and large electrostrictive strain make PMN-PT a promising candidate for actuator and transducer applications. A significant advance in the field of materials with high piezoelectric efficiency is marked by the growth of single-crystal solid solutions of PMN-PT. Single-crystal relaxor ferroelectrics exhibit an ultrahigh piezoelectric coefficient d_{33} of about 2500 pC/N, very high electromechanical coupling factors of 90%, strain level of 1.7%, and low hysteresis. These solid solutions have a rhombohedral structure with polarization along the $\langle 111 \rangle$ direction below MPB and tetragonal above MPB. Optimized electromechanical properties are observed for compositions just on the rhombohedral side of the MPB, when these single crystals were measured along $\langle 100 \rangle$-type directions, even though the polar axis lies along $\langle 111 \rangle$. As Fu and Cohen [240] observed, the most likely explanations for the giant piezoelectric response are the following: (i) the reorientation of polar domains by the applied field (in this hypothesis the fact that these materials are solid solutions with a relaxor end member is crucial but it is hard to understand why they must be single crystals); (ii) polarization rotation where the large response does not depend in an essential way on mesoscopic structure or ordering. As Fu and Cohen demonstrated, this polarization rotation can drive a giant piezoelectric response. The polarization rotation from the $\langle 111 \rangle$ direction to the $\langle 001 \rangle$ direction in a rhombohedral single crystal with a huge strain (as would be a PT with a "rhombohedral" ground phase) would drive a colossal mechanical strain at normal applied electric field. In single crystals the local polarizations will be rotated coherently from $\langle 111 \rangle$ to $\langle 001 \rangle$ directions whereas the effect will cancel in a randomly oriented ceramic.

5.1.4.1. Overview of the Main Results

PMN-PT thin films with different compositions have been obtained by using different deposition techniques [241–260]. Udayakumar et al. [244] reported the deposition of PMN-PT 65/35 thin films by the sol–gel technique on Ti–Pt/SiO$_2$/Si(100) substrate. The films were perovskite polycrystalline randomly oriented and free of pyrochlore. They showed a high dielectric constant of 2900 and low dissipation factor of 0.02. Film thickness was about 0.4 μm. The measured temperature dependence of the dielectric constant showed a broad maximum with a Curie point of about 148°C.

Jiang et al. [245] reported the deposition of PMN-PT with different PT content by rf magnetron sputtering on Si(100) and (111)Pt/Ti/SiO$_2$/Si(100). The films were polycrystalline randomly oriented. The dielectric properties of the as-deposited PMN-PT films were measured at room temperature. All films had the same thickness of ~0.5 μm. The dielectric constant ε increased from 520 for PMN-PT 10/90 to 1300 for PMN-PT 50/50 and then decreased to 1018 for PMN-PT 70/30. The variation in ε with respect to the change of composition from 50/50 to 70/30 was not consistent with the tendency reported for bulk PMN-PT ceramics. This inconsistency must be related to the presence of a TiO$_2$ interface layer which had a low dielectric constant of ~100. Because $\varepsilon_{\text{PMN-PT}} \gg \varepsilon_{\text{TiO}_2}$ an increase in thickness of the TiO$_2$ layer would cause a significant decrease of the apparent dielectric constant of the film. The ferroelectric

characteristics of the same series of films were also investigated. The remanent polarization P_r and coercive field E_c were found to decrease with increasing PMN content of the films.

PMN-PT 70/30 epitaxial thin film deposition was reported in [246]. The 70/30 ratio was chosen because it lies about 5 mole% to the rhombohedral side of the morphotropic phase boundary where the composition dependence of the electrical properties is the least dramatic. PMN-PT (70/30) epitaxial thin films with SrRuO$_3$ bottom electrodes were prepared on (001)LaAlO$_3$ substrates by pulsed laser deposition. PMN-PT targets were prepared by the columbite method and then an additive mass of PbO was mixed with the powder and sintered. Because the films were rich in Pb, a mixture of epitaxial perovskite and PbO$_x$ is present. In addition to the epitaxial perovskite, a small peak corresponding to the (110) orientation is present at 31.5°. When large quantities of excess PbO$_x$ are present in the films small volumes of nonepitaxial perovskite (typically <1 vol%) were unavoidable. Azimuthal scans and rocking curves showed that all the samples were oriented in-plane with peak widths between 0.6° and 0.8° in Φ and between 0.4° and 0.7° in ω. Phase-pure samples were found to have insufficient resistivity due to lead deficiency; therefore it was preferred to grow samples with measurable amounts of PbO$_x$. The deposited films showed electrical properties similar to those of single crystals.

Lu et al. [247] reported the fabrication of PMN-PT thin film for optical waveguides. Highly (110)-oriented and (100) nearly epitaxial PMN-PT 70/30 thin films were deposited on (1012) sapphire and (100)LaAlO$_3$ substrates, respectively, by using the sol–gel method. The films grown on both sapphire and LaAlO$_3$ substrates were free of pyrochlore. Electrooptic properties of the film were evaluated by measuring the phase retardation using 632.8-nm He–Ne laser light transmitted normal to the film. They exhibited a large quadratic electrooptic effect with a value of 0.75×10^{-16} (m/V)^2, which is comparable to the best values reported for PLZT (9/65/35) [1×10^{-16} (m/V)^2].

More recently Stemmer et al. [248] discussed the microstructure of epitaxial PMN-PT thin films grown by MOCVD. Dielectric and electromechanical properties (electric-field-induced strain) in PMN films were studied by Catalan et al. [249]. The electromechanical strain response was found to depend on the deposition conditions for each sample. The maximum value for the tensile electrostrictive strain was 0.2%. Compressive strains were also found and they reached maximum values of 0.35%.

Structural and electrooptic properties of PMN-PT thin films have been investigated by Lu et al. [250]. An influence of PT content on lattice constants and electrooptic properties was found. Low propagation loss was obtained for films near MPB. Propagation loss of about 3 dB/cm was found, which is much smaller than those of ferroelectric materials. The increased quadratic electrooptic coefficient at MPB was attributed to the increasingly strong coupling between the tetragonal and rhombohedral phases through the long-range order spontaneous polarization. This composition range corresponds to structural changes from polar nanodomains ($x < 0.3$) to tweedlike domains ($x > 0.34$).

The preparation of epitaxial PMN thin films by MOCVD was reported in [251]. The films were deposited on (001)SrTiO$_3$ (ST) and on SrRuO$_3$ (SR)/ST substrates. A cube-on-cube orientation relationship between the thin film and the ST substrate was found, with a (001) rocking curve width of 0.1° and an in-plane rocking curve width of 0.8°. The root mean square surface roughness of a 200 nm thick film on ST was 2–3 nm as measured by scanning probe microscopy. The zero bias dielectric constant and loss measured at room temperature and 10 kHz for a 200 nm thick film on SR/ST were approximately 1100 and 2%, respectively. A remanent polarization of approximately 16 μC/cm^2 was measured for this sample, despite the fact that, as a relaxor ferroelectric, PMN measured at room temperature is typically expected to demonstrate only a very slim loop response. The observed ferroelectric-like behavior may be a result of the imposed tetragonality as observed by XRD. A similar effect was reported in [252] in PMN films deposited by PLD on Au/Pt/NiCr/glass. A substantial piezoelectric effect in the absence of an applied electric field was measured at room temperature. The samples were polycrystalline and partially oriented. The effect has been attributed to the very small grain column width (50–60 nm) that helped to fix the orientation of nanodomains which could occur during growth due to an existing electric field on the substrate.

Table XI shows the main physical characteristics of PMN-PT thin films, as reported in the references listed in the last column.

5.1.4.2. Influence of Composition and Structure on Physical Properties

5.1.4.2.1. Dielectric and Ferroelectric Properties

(a) Influence of Composition. The properties of PMN-PT thin films have been studied as a function of the PMN content variation between 10 and 70% in [245]. Thin films with perovskite structure were deposited by rf magnetron sputtering on Si(100) and on Pt/Ti/SiO$_2$/Si(100). The dielectric properties of the as-deposited thin films were measured at room temperature. The dielectric constant for different compositions varied between 500 for PMN-PT 10/90 and a maximum of 1300 which was reached for 50/50. The dependence of ferroelectric hysteresis loop of a 0.5 μm thick PMN-PT thin film on the variation of composition, which changes from 10 to 70% PMN content, is shown in Figure 53. With the increasing of PMN content the ferroelectric properties gradually decreased. The remanent polarization P_r decreased from 23.5 to 8 μC/cm^2 and E_c from 82.5 to 60 kV/cm. The unsaturated loop for PMN 70/30 could be explained by the presence of a relatively thick layer of TiO$_2$ which was formed during deposition. Therefore the voltage drop on the PMN-PT layer might not be large enough to polarize the layer so as to have a well-developed P–E hysteresis loop. The understanding of the ferroelectric behavior of the PMN-PT thin films with composition comes from

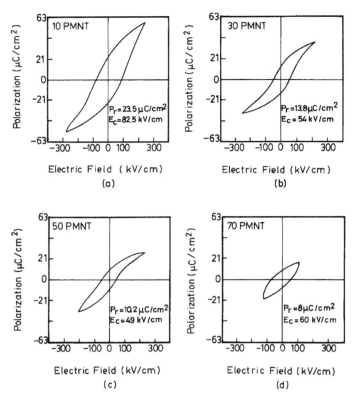

Fig. 53. The dependence of ferroelectric hysteresis loop of PMN-PT thin films on the variation of composition: (a) PMN-PT 10/90; (b) PMN-PT 30/70; (c) PMN-PT 50/50; (d) PMN-PT 70/30. Reprinted with permission from [245], © 1995, Institute of Pure and Applied Physics.

the complex transformations which occur in these materials when the relative content of the ferroelectric and relaxor phase is changed. Ye and Dong [253] discuss the domain structure variations and the phase transitions with PT content in single crystals of PMN-PT. An interesting aspect in these crystals is the mesoscopic domain structures and phase transition. The enhanced properties of the piezocrystals are nevertheless related to the morphotropic phase boundary effects and the formation of macrodomain states resulting from the partial substitution of Ti^{4+} for the complex ions (Mg$_{1/3}$Nb$_{2/3}$)$^{4+}$ on the B site. The MPB is located at 65/35, which separates the rhombohedral phase from the tetragonal phase. However, X-ray or neutron diffraction usually indicates an average cubic symmetry and could not reveal the morphotropic features of the solid solution systems. Therefore in [253] the investigation of the mesoscopic domain structures and phase transitions was aimed at understanding the microscopic mechanisms of the relaxor ferroelectric properties and the enhanced piezoelectric performance of the relaxor-based single crystals. Crystal optical studies of the mesoscopic domain structures and phase transitions have revealed that the substitution of Ti^{4+} ions for the B site complex (Mg$_{1/3}$Nb$_{2/3}$)$^{4+}$ ions results in development of long-range ferroelectric order and a symmetry breaking from relaxor PMN associated with the formation of birefringent macrodomains. Although PMN-PT 50/50 crystals exhibit domain structures of the tetragonal 4mm phase with strong bire-

Table XI.   Main Physical Properties Characteristic of PMN-PT Thin Films

Composition	Deposition technique, substrate	Orientation	Relative dielectric const. ε_{33} (1 kHz)	Dielectric loss factor	P_r (μC/cm^2) (at 25°C)	E_c (kV/cm) (at 25°C)	d_{33} (pC/N)	d_{31} (pC/N)	Reference
PMN	Sol–gel TiO$_2$/Pt/TiO$_2$/SiO$_2$/Si	(111)	2,200	>0.01	0 10 (−160°C)	0 40 kV/cm (−160°C)	65 (25°C, 70 kV/cm)		[257]
PMN	MOCVD (001)SrRuO$_3$/ (001)SrTiO$_3$	(001) Epitaxial	1,100	0.02	16				[251]
PMN	PLD LSCO/MgO(100)	(001)	3,800	0.08					[255]
PMN	PLD PZT/Pt/NiCr/ Si(111)	Partial (211)	500 (effective, PMN + low ε interf. layer)				60–80 (25°C, no bias)		[252]
PMN-PT 90/10	PLD LSCO/MgO(100)	(001)	2,800	0.09					[255]
PMN-PT 70/30	Rf magnetron sputtering (111)Pt/Ti/SiO$_2$/Si	Random	71,000	0.02–0.05	8	60			[245]
PMN-PT 70/30	PLD SrRuO$_3$/ (001)LaAlO$_3$	(001) Epitaxial	73,500	<0.08	17	20			[246]
PMN-PT 70/30	PLD LSCO/MgO(100)	(001)	1,100	0.06					[255]
PMN-PT 68/32	PLD LSCO/MgO(100)		550–750	0.04–0.06	4.5	25	24		[258]
PMN-PT 68/32	PLD LSCO/MgO(100)	(001)	1,300	<0.05	20	28–32			[254]
PMN-PT 65/35	Sol–gel (111)Pt/SiO$_2$/ Si(100)	Random	2,900 (10 kHz)	0.02	11	11			[244]
PMN-PT 65/35	PLD LSCO/MgO(100)	(001)	2,200	0.05					[255]
PMN-PT 60/30	PLD LSCO/MgO(100)	(001)	1,200	0.07					[255]
PMN-PT 50/50	Rf magnetron sputtering (111)Pt/Ti/SiO$_2$/Si	Random	1,300	0.02–0.05	10.2	49			[245]
PMN-PT 30/70	Rf magnetron sputtering (111)Pt/Ti/SiO$_2$/Si	Random	1,000	0.02–0.05	13.8	54			[245]
PMN-PT 10/90	Rf magnetron sputtering (111)Pt/Ti/SiO$_2$/Si	Random	500	0.02–0.05	23.5	82.5			[245]
Bulk single crystal PMN-PT (APC Int. Ltd.)			4,200				2200	−520	[259]
Bulk PMN-PT ceramic (PMN-15)			20,000	<0.008	26 (saturation polarization, 25°C, 35 kV/cm)		700 (25°C, 6.5 kV/cm)	−230 (25°C, 6.5 kV/cm)	[260]

fringence reflecting the symmetry of PT and transforming into the cubic phase at $T_C = 220°C$, single crystals of PMN-PT 65/35 exhibit the morphotropic phase boundary characteristics. The complex morphotropic domain structures are composed of the rhombohedral R3m and the tetragonal P4mm phases intimately mixed, with multiple orientation states. The morphotropic domain structures seem to prove the assumption that the enhanced piezoelectric properties of the PMN-PT crystals are due to the multidomain states with the presence of both the rhombohedral and tetragonal polarizations [253].

(b) Influence of Structural Defects. By examining the ferroelectric properties and the phase transition temperature in PMN-PT 70/30 films Maria et al. [246] found a dependence on the crystallite size or scattering domain dimension in XRD. The (001)-oriented epitaxial films were deposited by PLD on SRO/LaAlO$_3$ substrates. Generally a very reduced ferroelectric behavior was found, with a slim hysteresis loop and small remanent polarization. These samples behaved like relaxors in the vicinity of T_{max}. By introducing the sample into liquid nitrogen and measuring the hysteresis curve at this temperature a well-developed hysteresis loop was found with a remanent polarization of 16 $\mu C/cm^2$. Evidence of a reduced transition temperature was also found in the temperature dependence of the dielectric constant. A temperature T_m of approximately 50°C at 1 kHz was found instead of a transition temperature of 150°C. It has been also observed that films prepared at slightly higher substrate temperatures showed dielectric constant, remanent polarization, and transition temperature similar with single-crystal values. Comparison of the X-ray patterns showed that the PMN-PT films deposited at higher temperatures (660°C) have a larger coherent X-ray scattering dimension. A dimension of about 50 nm was obtained from the analysis. Instead, in films deposited at lower temperature (620°C) the scattering size was found to be 10 nm. This analysis gives a semiquantitative picture of the subgrain structure. Combining these data with the permittivity temperature dependence one finds that these smaller scattering regions are accompanied by temperature-shifted transition, broadened and more dispersive dielectric maxima, and deviation from Curie–Weiss behavior. Interestingly, the 10 nm dimension is consistent with the 10 nm diameter ordered regions observed in ceramic relaxors of several compositions. The broadened X-ray reflections in the lower temperature deposited samples are produced by two factors: larger variations in interplanar spacing in the crystal and a mosaic subgrain structure. The more pronounced mosaic structure of the lower temperature deposited sample would result in greater concentrations of dislocations which occur at the interface of the slightly misoriented subgrains. The higher defect concentrations are believed to interrupt the long-range order of the lattice and to produce electrical behavior more consistent with that of relaxor ferroelectrics. It seems that if deposition temperatures could be further increased or if a better lattice matched substrate–electrode–film system were used, electric properties similar to those of single crystals could be realized.

Similar influence has been discussed also by Tyunina et al. [254]. Highly oriented perovskite films of PMN-PT with compositions near MPB 70/30 were formed by PLD on LSCO/MgO(100). The dielectric properties of the films were studied over the frequency range 100 Hz to 1 MHz and in the temperature range 20–350°C. The room-temperature polarization of the films and the dielectric permittivity of 250 nm thick films were close to those of bulk ceramics. The films exhibited relaxor-type behavior with temperature of the dielectric peaks corresponding to those in bulk. Instead it has been found that the width of the transition peak was larger than in bulk ceramics. The films demonstrated ferroelectric behavior at room temperature. The polarization was about 20 $\mu C/cm^2$ and the coercive field 28–32 kV/cm, respectively. A broad peak in dielectric permitivity was observed in PMN-PT 90/10. The smearing of the Curie temperature has also been found in normal ferroelectric PZT films. A broad peak was found in these films. Its width was two times larger than in corresponding PMN-PT single crystas. Such smearing has been found also in thin strained epitaxial ferroelectric films, but in these films the strain was relaxed by the formation of domains. The most acceptable reason for the additional broadening of the transition in the PMN-PT films was the small grain size. The broadening of the transition in nanocrystalline ferroelectrics was seen as a size effect (see Section 2). In highly oriented epitaxial films the grain boundaries could be treated as extended defects. The small grain size is then accompanied by a high concentration of defects.

(c) Influence of Pyrochlore Phase. As in other lead-based ferroelectric materials the presence of pyrochlore phase influences the properties of PMN-PT thin films as has been found by Tantigate et al. [255]. Thin films of PMN-PT with compositions 90/10, 70/30, 65/35, and 60/40 were deposited on MgO(100) substrates by PLD. Heterostructures of PMN-PT/LSCO were prepared on MgO(100) to evaluate the dielectric properties. The films were about 500 nm thick. It was found that the appearance of pyrochlore phase in different compositions of PMN-PT greatly decreased the value of the dielectric constant. The highest value was found in PMN films with 99% perovskite and it was 3800. This maximum value is still below the bulk single crystal value, probably due to the effect of a low dielectric constant layer near the electrode.

(d) Influence of Misfit Strain. It has been shown in previous sections that the existence of an unrelaxed misfit strain in ultrathin films changes their properties dramatically. Antiferroelectric materials such as PbZrO$_3$ when grown in very thin films could become ferroelectric as reported by Ayyub et al. [256]. Bai et al. [251] studied pure PMN thin films deposited by MOCVD. The films were deposited on (001)STO and on SrRuO$_3$/STO substrates. A tetragonal structure was found for the PMN film with out-of-plane and in-plane lattice parameters of 0.406 and 0.404 nm, respectively. This tetragonal distortion is most likely the result of temperature-dependent misfit strain (i.e., lattice mismatch and difference in thermal

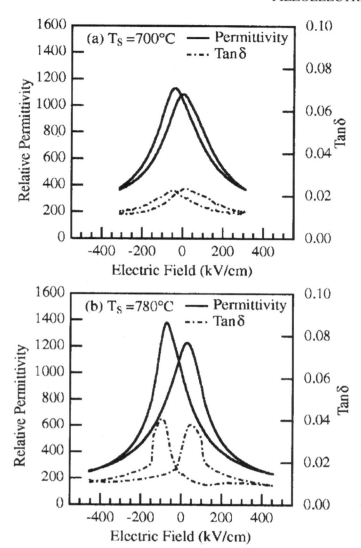

Fig. 54. Dielectric constant vs. electric field curves measured at 10 kHz for two PMN films of different thicknesses: (a) 300 nm; (b) 200 nm. Reprinted with permission from [251], © 2000, American Institute of Physics.

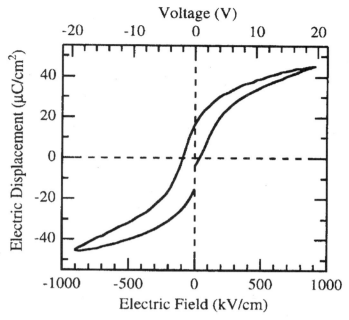

Fig. 55. Electric displacement vs. electric field hysteresis loop for a 200-nm PMN film. Results from Argonne National Lab., Chicago Univ. & US-DOE under Contract No. W-31-109-ENG-38. Reprinted with permission from [251], © 2000, American Institute of Physics.

expansion coefficient) induced by the substrate and which has been only partially relieved. Extremely weak, broad superlattice reflections were found at the (1/2, 1/2, 1/2) reciprocal lattice positions of the PMN, indicating a very small degree of B-site ordering. Zero-bias relative permittivities of approximately 1000 were found, still substantially lower than that obtained in bulk materials or by other deposition techniques. Interestingly, substantial hysteresis is observed even at room temperature in the forward and reverse sweeps (Fig. 54). This is also reflected in electric displacement vs. electric field measurements (Fig. 55). A remanent polarization of approximately 16 $\mu C/cm^2$ was measured for this sample, despite the fact that, as a relaxor ferroelectric, PMN measured at room temperature is typically expected to demonstrate only a very slim loop response. Although a time-dependent transition from the relaxor state to an electric-field-induced ferroelectric state has been demonstrated in bulk materials, the measurements described

here are outside the bounds of reported phase diagrams. Therefore the observed ferroelectric-like behavior may be a result of the imposed tetragonality as observed by XRD.

5.2. Nonferroelectric Piezoelectrics

5.2.1. Zinc Oxide

Zinc oxide is one of the most used nonferroelectric piezoelectrics. Due to its high piezoelectricity and electromechanical coupling factor, ZnO has been used in many practical applications such as ultrasonic transducers, SAW devices and sensors, resonators, and filters [261–263]. The crystal structure of zinc oxide is hexagonal (wurtzite type). Each Zn atom is tetrahedrally coordinated with four O atoms and the zinc d electrons hybridize with the oxygen p electrons. Although stoichiometric ZnO has a very high resistivity, it usually contains excess zinc atoms, which influence electrical conductivity, piezoelectricity, and defect structure. Wurtzite structure is polar and a mechanical compression in the c-direction will produce either a positive or negative voltage in this direction, depending on the polarity of the c axis. Along the polar axis Zn atom layers define the positive face (0001 direction) and the O atom layers the negative face (000$\underline{1}$ direction). These two crystal faces behave differently regarding their surface energies, growth rates, and etch and abrasive characteristics. The oxygen face has a lower surface energy and slower growth rate and it abrades and etches more rapidly than the Zn face. In films ZnO shows a strong tendency to grow with its c axis perpendicular to the substrate surface even if the substrate is amorphous. However, the degree of orientation is influenced by growth conditions such as temperature, background gas composition and pressure, particle energy, substrate type, and surface condition. Because structural differences occur during growth, it is important to understand how and to what extent they will influence the physical properties of the films such as piezoelectric, dielectric, and elastic constants, electrical conductivity, and acoustical loss. These problems will be discussed in the second part of this section, whereas the first part gives an overview of the main results reported in the last decades of the twentieth century.

5.2.1.1. Overview of the Main Results

As for other materials, the introduction of sputtering techniques marked the turning point for obtaining high-quality ZnO thin films. The first successful use of this technique for piezoelectric ZnO films was reported in 1965 by Wanuga et al. [264]. Since then many results have been reported [265–275] on the deposition of ZnO films by sputtering and other techniques. In one of the first systematic studies of zinc oxide properties, Dybward [265] reported the influence of the substrate type (crystalline or amorphous) on the polar axis orientation (000$\underline{1}$ or 0001) and suggested that heat conductivity of the substrates was responsible for this phenomenon.

ZnO films with good crystal orientation and surface flatness grown with high deposition rate by rf planar magnetron sputtering have been reported in [266]. A Li-doped ZnO ceramic target was used and the temperature of the substrate was about 300–350°C. The obtained films had the c axis perpendicular to the substrate and a value of the standard deviation angle σ of the c-axis orientation distribution smaller than 0.5°. Transparent ZnO films with a thickness up to 48 μm have been reproducibly prepared with high crystal orientation, surface flatness, and optical transmittance, within a circular area of 40 mm in diameter. The surface roughness of a 30 μm thick film was smaller than 0.2 μm. The measured electromechanical coupling factor was of more than 95% of the theoretical value.

The successful deposition of c-axis-oriented thin films of piezoelectric ZnO on unheated substrates by rf magnetron sputtering was reported by Krupanidhi and Sayer [267]. The target was pure zinc which was sputtered in a reactive ambient of 100% oxygen over a pressure range from 5 to 70 mTorr. It was found that below 10 mTorr the film structure, orientation, and physical properties were dependent on the lateral position of the substrates with respect to the target. This was attributed to the role of high-energy neutral atoms in resputtering the film in the region facing the target erosion zone. At higher pressures when the atom mean free path decreases the film structure and properties are uniform. It has been found also that at 700 W sputtering power the optimum pressure lies between 20 and 30 mTorr, leading to highly oriented ZnO with a dielectric constant of 11 (bulk value) and Rayleigh wave velocity of 2600 m/s.

Two review papers by Hickernell [40, 268] summarized the main achievements in the field of sputtered ZnO films coming from the author's own experience and from other reported results. A comparison between the performances of the main sputtering systems—dc diode, rf diode, dc triode, rf magnetron sputtering—for ZnO film deposition was presented. Systematic investigation of the structural defects and surface roughness effects on acoustical loss was also reported for the first time [268].

The first complete set of constants of ZnO films was reported by Carlotti et al. [166]. The properties of ZnO films deposited by rf magnetron sputtering on Al_2O_3 were analyzed by an acoustic technique. The phase velocities of the spectrum of acoustic modes propagating along the structure were measured and the results used for the determination of the complete set of elastic, piezoelectric, and dielectric constants of the film. The effective elastic constants of the films were found to be lower than those of the bulk material by amounts which depend on the elastic constants considered (from -1.2% for c_{33} to -24.8% for c_{11}). The values obtained were used for determining the dispersion curves of acoustic modes propagating along ZnO layers deposited on other substrates (fused quartz and silicon) and showed good agreement with experimental results, demonstrating that they can be taken as a reference set.

Deposition of c-axis-tilted ZnO films by reactive dc magnetron sputtering and their use in a transducer for the simultaneous generation of both longitudinal and shear waves has been reported in [269]. At a c axis tilted at approximately 16° the electromechanical coupling factors k^2 for longitudinal and shear waves are almost equal (0.25) and the generated waves are close to pure modes. It has been shown that by varying the operating frequency the ratio of the excitation strength between the longitudinal and shear waves can be adjusted.

Deposition of highly oriented, dense, and fine-grain polycrystalline ZnO films with an excellent surface flatness has been reported in [270]. The deposition was performed by reactive magnetron sputtering on Al-coated substrates [Si(111), oxidized Si(111), oxidized Si(100), and 7059 Corning glass]. The films showed (002) preferred orientation, with only minor presence of (100) orientation in some cases. The (002) peak width was 0.05 degrees, corresponding to a crystallite size along the c axis of 0.16 μm estimated using the Scherrer formula. The role of the substrate in determining the degree of orientation of the c axis as well as secondary nucleation and poor film texture was evidenced. It was found that the quality and the orientation of the c axis of the films depends strongly on the initial conditions of deposition of the first few monolayers. The piezoelectric properties of the films were investigated by measuring the acoustic delay–line insertion loss. From the insertion loss measurement a value of 0.21 was obtained for the electromechanical coupling constant, somewhat lower than the bulk value of 0.28 for single-crystal ZnO.

ZnO films with very high resistivity have been obtained by Moeller et al. [271], who noticed that periodic interruption of column grain growth by nucleation-like layers can greatly increase their resistivity, due to the fact that the nucleation layer has a high electrical resistivity, whereas column grain kernels

Table XII.   Elastic, Dielectric, and Piezoelectric Constants of ZnO Thin Films

Composition	Deposition technique, substrate	Orientation	Density (10^3 kg/m^3)	Relative dielectric constant ε_{33} (1 kHz)	c_{11} (10^{10} N/m^2)	c_{13} (10^{10} N/m^2)	c_{33} (10^{10} N/m^2)	c_{55} (10^{10} N/m^2)	c_{66} (10^{10} N/m^2)	e_{31} (C/m^2)	e_{33} (C/m^2)	e_{15} (C/m^2)	d_{33} (pC/N)	Ref.
ZnO	Rf magnetron sputtering z-cut Al$_2$O$_3$	c-axis oriented	5.72		15.7	8.3	20.8	3.8	3.4	−0.51	1.22	−0.45		[166]
ZnO	Dc magnetron sputtering Al/7059 glass	c-axis oriented		11										[282]
ZnO	PLD Au/Cr/Si(100)	c-axis oriented											8	[58, 72]
Bulk ZnO			5.665	$\varepsilon_{33} = 11$ $\varepsilon_{11} = 9.3$	20.9	10.46	21.06	4.23	4.42	−0.57	1.321	−0.48	12.3	[275]

usually have a lower resistivity, due to the existence of free Zn atoms. A ramp-shaped sputtering cycle was used for growing films as thick as 20 μm, with a resistivity of about 10^{11} Ω cm.

An interesting result was obtained by Onodera et al. [272], who investigated the temperature dependence of dielectric constants, specific heat, and D–E hysteresis loops of Li-doped ZnO. A ferroelectric phase transition was found for the first time in a II–VI semiconductor. Moreover the electrical resistivity due to carriers was improved by the introduction of Li ions, which is suitable for measurements of dielectric activity in a Li-doped ZnO system. It has been suggested that the small Li ions (0.6 Å) which substitute Zn atoms (0.73 Å) in the lattice occupy off-center positions therefore giving rise to local permanent electric dipoles. At low temperatures, the off-center ions become ordered and the crystal undergoes a ferroelectric phase transition. The measurements of the D–E hysteresis loop revealed a spontaneous polarization of about 0.044 μC/cm^2 at room temperature, which is two orders of magnitude smaller than those of typical ferroelectrics but is nevertheless an evidence of the ferroelectric activity in Li-doped ZnO.

Verardi and Dinescu [72] and Dinescu and Verardi [58] in a series of papers reported the successful deposition of ZnO films by laser ablation of a Zn target in an oxygen reactive atmosphere. The films were crystalline with the c axis perpendicular to the substrate surface and exhibited high optical transmission and a high piezoelectric coefficient.

A detailed study of deposition parameters which influence crystallite orientation in-plane or perpendicular to the substrate and correlations with the photoconductive properties of ZnO films was made by Zhang and Brodie [273]. The epitaxial growth of ZnO on LiNbO$_3$ substrate and the SAW properties on ZnO grown on GaAs substrates have been reported [274]. The physical properties of ZnO films as reported in different references are listed in Table XII.

5.2.1.2. Influence of Structure on Physical Properties

The physical properties of zinc oxide films are highly dependent on their structure. Crystalline ZnO films obtained by different deposition techniques can have different c-axis orientation and spreading, polar direction and local polar inversion, internal stress, structural defects, etc. In the following, some results coming from reported systematic studies will be discussed. The role of crystallinity degree in achieving good properties is well known; therefore it is only mentioned here.

5.2.1.2.1. Piezoelectric Properties

(a) c-Axis Tilting. Usually crystalline textured ZnO films with c axis oriented perpendicular to the substrate are obtained. If for some reason (the relative position of target and substrate, background gas pressure, substrate condition, etc.) the obtained film has a tilted c axis, the global piezoelectric properties can be very different, as can be observed in Figure 56, where the electromechanical coupling factor k^2 is represented as a function of direction in the zx plane of ZnO for conversion of energy to a longitudinal wave and to one shear wave. If the coupling to the longitudinal wave is considered, it can be observed that by tilting the c axis the electromechanical coupling factor decreases from its maximum value (0.28). Moreover the excited mode is no more a pure mode, because excitation of the shear wave is now possible in the same configuration. This can be also a positive aspect if the possibility of efficiently exciting different types of waves in the same structure is desired; the deposition of a ZnO film with the c axis tilted by 16° for this purpose was reported by Jen et al. [269]; 30°-tilted ZnO films have been deposited and used for SAW generation by Carlotti et al. [276].

(b) c-Axis Angular Spreading. A systematic experimental study of the relationship between the angular spreading of crys-

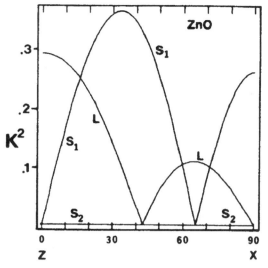

Fig. 56. The electromechanical coupling constant along the *zx* plane of ZnO. Reprinted with permission from [269], © 1988, Acoustical Society of America.

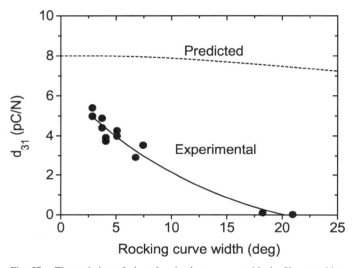

Fig. 57. The variation of piezoelectric d_{31} constant with the X-ray rocking curve width. The predicted and measured values have been taken from [277].

tallite *c* axis, as reflected in the full width at half maximum (FWHM) of the rocking curve, and the piezoelectric d_{31} constant has been reported by Gardeniers et al. [277]. The films were deposited by rf planar magnetron sputtering on *p*-doped Si(001) substrates covered with SiO_2, Si_3N_4, or W. All the films were *c*-axis oriented. The d_{31} constant was measured on cantilever structures and its variation with X-ray rocking curve FWHM is shown in Figure 57. To evaluate to what extent a misalignment of the *c* axis could influence the piezoelectric constant, a model was proposed which allows the evaluation of the effective piezoelectric constants as a function of the single-crystal properties and the crystallite spreading distribution width [as reflected in the FWHM of the rocking curve of the (0002) peak]. To calculate the piezoelectric constant for a ZnO film with a certain spreading in the *c*-axis orientation, the properties of a grain with the *c* axis misaligned

with respect to the *z* axis have been obtained by rotating the matrices with an angle Φ about an axis perpendicular to the *z* axis, and with the in-plane rotation angle α and integrating with respect to the both Φ and α. A gaussian distribution of the misalignment angle Φ whose width is related to the FWHM of the rocking curve was considered. Analytical expressions have been found for the dependence of the effective compliances s_{11} and s_{12}, piezoelectric constants d_{31} and d_{32}, and dielectric constant ε_{33} on FWHM of the X-ray rocking curve of (0002) peak. The result thus obtained for the piezoelectric constant is presented in Figure 57. It can be observed that the predicted decrease is much softer than the experimental one. It has been supposed that this great difference can be produced by the different polarizations of the grains in the film, that is, a large fraction (15–30%) of the grains has the polar *c* axis directed opposite to that of the majority of grains.

(c) Polarity Inversion. Such defects as inversion of polarity regions called inversion domain boundaries (IDB) as well as other structural defects such as stacking faults and their relation with the (0002) peak broadening have been examined by TEM in [278]. In the hexagonal wurtzite structure of ZnO with lattice parameters $a = 0.3253$, $c = 0.5213$, and $u = 0.382$ the oxygen stacking is similar to an hcp structure and the cations occupy one-half of the tetrahedral sites. The stacking sequence along [0001] of the wurtzite structure may be described as AaBbAaB..., where the capital letters indicate oxygen atoms and the small letters indicate Zn atoms. Locally the hexagonal arrangement may transform to a sphalerite (cubic) structure giving rise to one or two layers having the wrong stacking sequence such as AaBbAa/CcBbCc.... This defect is a stacking fault and is found in materials having wurtzite structure. For this defect there is no change of polarity. In the case of inversion domain boundaries there is also a change of polarity. The regions on both sides of the boundary plane are related by an inversion. IDBs have been observed in ZnO ceramics that contain dopants or some impurities; therefore it has been supposed that the impurities stabilize the polarity reversal. IDBs contribute strongly to the reduction of piezoelectric coefficients because grains or regions of inverse polarity are symmetric with respect to inversion and therefore have a null piezoelectric activity.

(d) Lattice Distortion. Lattice distortions have been observed in many *c*-axis-oriented ZnO films deposited on different substrates. They have been measured from the shift of the (0002) peak position toward lower or higher values. This indicates that the ZnO lattice is deformed along to the (0001) direction due to a stress. Both compressive and tensile stresses have been observed. They can be produced due to lattice misfit with the substrate or during growth, as reported in [267], where bombardment with high-energy atoms during sputtering at low pressure produced a compressive stress which manifests as a shift in the diffraction (0002) peak.

A model proposed by Qian et al. [279] for calculating elastic and piezoelectric constants of ZnO films and based on the equivalent atomic charge approach showed that the influence of lattice distortions on these constants is much greater than the influence of grain misalignment.

(e) Grain Interconnections. Shear piezoelectric constant d_{15} which couples an electrical field applied on the c axis of the textured film to a shear deformation is obviously highly dependent on the boundary interconnections between grains. If the structure is poorly connected on the direction perpendicular to the c axis the effective d_{15} constant is very much decreased. In [280] the relationship between the coupling factor of a piezoelectric film layer and the relative contributions of d_{15} and d_{33} constants for a textured film was discussed in terms of the lateral strength of the columnar structure. It has been pointed out that a soft connection of columns even if perfectly oriented can decrease the efficiency of the layer by 4–5 times.

5.2.1.2.2. Elastic Constants

(a) Influence of Internal Strain. A set of experiments reported by Carlotti et al. [167] and aimed at obtaining the set of elastic constants of ZnO films deposited on different substrates with thicknesses ranging from 20 nm to 20 μm revealed that the same set of constants did not fit the results obtained from the entire range of thicknesses, but only those coming from films with thickness above 200 nm. It has been hypothesized that the difference could come from strains in ultrathin films and a set of experiments was performed to clarify this. Accurate measurements of the dispersion curves of Rayleigh acoustic modes in ZnO films of different thicknesses in the range 20–320 nm have been performed by Brillouin scattering. The measurement is based on the photon–phonon inelastic interaction and allows the simultaneous determination of the frequency and wavevector of the absorbed phonon. The frequency is determined from the frequency shift in the Brillouin spectra and the wavevector from the angle of scattering. The measurements were performed on ZnO films deposited by reactive rf diode magnetron sputtering on amorphous SiO_2 and on (001)-cut p-doped silicon. Rayleigh mode propagation in oriented polycrystalline films with the c axis perpendicular to the substrate was investigated. The physical constants of the film were evaluated from measurements of the phase velocity of Rayleigh modes over a large range of qh, where q is the wavevector and h the film thickness, by means of the fit of experimental data to the theoretical dispersion curves. Although the dispersion curves of films deposited on amorphous substrates could be fitted with the same set of constants as used for thicker films, those obtained from films deposited on crystalline silicon showed a different behavior. The behavior of the first Rayleigh mode has been interpreted in terms of a change in the effective elastic constants of the film in a region close to the ZnO–Si interface caused by the lattice misfit between the two media. The hypothesis of a very thin film with modified constants just near the interface helped to improve the fitting and new effective constants were obtained whose values

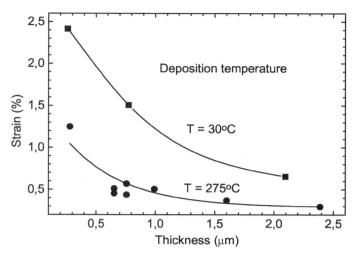

Fig. 58. Variation of strain in ZnO films as a function of film thickness (the data have been taken from [281]).

were dependent on thickness. For example, for a film of 75 nm thickness the relative decrease of its constants with respect to thicker films ranged between -3% for c_{13} and -19% for c_{55}.

Thin films frequently develop large internal stresses during growth which are dependent on the deposition conditions, nature of deposited material, etc. It has been shown by Hickernell and Hickernell [281] that the problem is present for the deposition of those materials which have high melting points; therefore the deposition temperature could not be sufficiently high to allow the arrangement of particles in the lowest energy states on the surface. Therefore during growth large internal stresses can develop. An intrinsic tensile strain was present in ZnO films obtained by sputtering [281], which produced a variation in their elastic constants. The SAW velocity was reduced in such films. This effect is highly undesired because it changes the parameters of high-frequency devices in which the ZnO film is incorporated. Moreover these parameters can vary in time due to uncontrolled stress relaxation. Higher strain values have been found in thinner ZnO films. The dependence of internal strain on film thickness is shown in Figure 58 for two deposition temperatures. A progressive decrease in strain with increasing thickness is observed. The internal strain is leveling at a value of 0.25%, which corresponds to a stress level of about 10^9 N/m^2, whereas the highest value reached for a film of about 0.3 μm is approximately 2.4%. Therefore thin films are more likely to be subjected to variation of elastic constants, as was also found in [167].

5.2.1.2.3. Electrical Conductivity

(a) Role of Additives. Based on its wide bandgap (3.3 eV) ZnO should have a very high resistivity. Practically, ZnO film resistivity usually is many orders of magnitude lower (10^6–10^8 Ω cm). Generally as-deposited films show n-type conduction due to uncompensated Zn atoms. In such films efficient piezoelectric transduction can only be obtained at high frequencies of the electrical field applied across the films. One of the

first systematic investigations of ZnO film conductivity was reported by Hada et al. [282] on polycrystalline films prepared by sputtering on amorphous substrates. Some samples were prepared by a special process of cosputtering with Al or Cu. It was found that the addition of Al increased the conductivity by over three orders of magnitude. On the contrary, the copper addition decreased the conductivity in approximately the same ratio. The relatively low electrical conductivity observed in these films may result not only from a small carrier concentration but also from a low mobility caused by the scattering of carriers on the crystallite boundaries. Taking the electrical conductivity of $8 \times 10^{-5} \ \Omega^{-1} \text{cm}^{-1}$ and Hall mobility of $0.4 \text{ cm}^2/(\text{V s})$ for the films, the concentration of zinc interstitials was estimated at about 1015 cm^{-3} under the assumption that all are fully ionized. The increase in conductivity observed in the films containing aluminum suggests that the aluminum is introduced as a donor, and from the temperature dependence of the carrier concentration a donor level at about 0.08 eV below the conduction band was obtained. In contrast to the effect of aluminum the reduced conductivity observed in the films containing copper suggests that copper forms deep traps and behaves as an acceptor as one might expect. Similar results for Cu-doped ZnO films have been reported also by Hayamizu et al. [63]. They investigated the conductivity of crystalline ZnO films deposited on amorphous glass substrates by PLD with an ArF excimer laser. The resistivities obtained were very low (10^2–$10^4 \ \Omega \text{ cm}$). The reason for these lower values was attributed to the existence of a primary layer at the interface which has a very low resistivity. The resistivity in the films was improved by about two orders of magnitude by postannealing in oxygen gas at 1 atm due to the decrease of the oxygen defects. Some samples have been Cu-doped and in this case the resistivity was above $10^7 \ \Omega \text{ cm}$. It is considered that Cu was incorporated at the Zn site as an acceptor impurity and carriers were decreased by recombination.

Electrical properties of ZnO films have been reported in [283] for single-crystal films epitaxially grown on the (0010) and (0112) planes of sapphire by rf diode sputtering. Both pure ZnO films and Li-doped films were examined. Undoped ZnO films had very low resistivities which prevented them from being used in SAW devices. The resistivity was increased by doping with Li because Li atoms act as acceptors that compensate the excess Zn atoms. In addition an annealing in air at 600°C for 30 min improved the resistivity, but at the same time increased the surface roughness.

(b) Oxygen Vacancies. Krupanidhi and Sayer [267] investigated the electrical resistivity of ZnO layers deposited at room temperature by rf magnetron sputtering as a function of oxygen pressure and position of the substrate with respect to the target. The substrates were conducting glass or Pt. The pure oxygen pressure was varied between 5 and 70 mTorr. Below 10 mTorr film structure, orientation, and physical properties depended on the position of the substrates with respect to the target. This was attributed primarily to resputtering of the films in the region below the target erosion area by high-energy neutral oxygen. At higher pressures where the mean free path

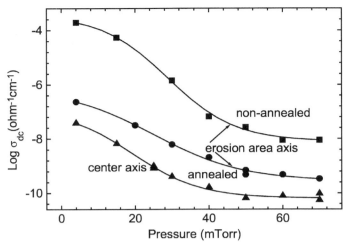

Fig. 59. Direct-current conductivity of ZnO films grown in different experimental conditions that produce variation of oxygen vacancies. The measured values have been taken from [267].

inhibits such effects the film structure and properties were uniform. The dark dc conductivity was measured on films with thicknesses between 1 and 3 μm. The dc conductivity of film sputtered in position A (in front of the target center) and B (in front of the erosion zone) under different pressures is reported in Figure 59 (lower and upper curves, respectively). The conductivity shows a decrease with increasing pressure up to about 30 mTorr and then tends to saturate with further increase in sputtering pressure. This behavior was followed by both samples, but the differences in conductivity were much higher at low oxygen pressure. On the whole, films deposited in B position (upper curve) are more conducting. The middle curve shows data for B films subjected to postdeposition annealing at 500°C for 4 h. The conductivity decreased very much for films deposited at low pressure and to some extent for those deposited at high pressures (50 mTorr). This behavior indicates that (a) the conductivity in sputtered ZnO films is caused by the nonstoichiometric oxygen deficiencies in general and (b) the oxygen vacancies are more pronounced at B having undergone bombardment by energetic neutrals. The conductivity of B films even after postdeposition annealing is higher than those of as-grown films prepared at A. This observation reveals that the intrinsic stoichiometry in the as-deposited films can provide much higher electrical resistivity than that achieved by subjecting nonstoichiometric films to postdeposition annealing. It is often proposed that high-resistivity ZnO films can be achieved by increasing the annealing temperature. However, such a procedure often leads to cracks and macroscopic imperfections in the films due to mismatch of the coefficients of thermal expansion of the film and substrate. The temperature dependence of dc conductivity was measured under constant dc bias. From the Arrhenius plot of dc conductivity versus temperature the activation energy was estimated to be 0.65 eV. This value is close to that reported for donor levels in single-crystal ZnO films. The increase of the dark conductivity with temperature may be attributed to the release of thermally activated trapped donors at the grain boundaries.

(c) Conduction along Columnar Grain Kernels. A conduction model has been proposed by Blom et al. [284] for polycrystalline ZnO films, based on models for polycrystalline semiconductors. The film consists of columnar grains with an internal semiconducting kernel and a completely depleted boundary region. Generally the grain boundaries contain surface states, due to defects or absorbed ions, which can trap free carriers from the bulk material. This leads to band bending near the grain boundary and depletion of the boundary material. The conduction across the boundary is described in terms of a Schottky barrier with thermoionic emission as the dominant conduction mechanism. The model has been verified on ZnO films grown on chemically vapor deposited SiO_2 and on aluminum.

The increase in resistivity by interrupting the conduction through columnar grain kernels was obtained by Moeller et al. [271], who used a ramp-shaped sputtering cycle. It was observed that the films presented a significant electrical resistivity only in the nucleation zone, whereas columnar oriented grains have a much lower resistivity especially in their internal zone (kernel) due to the existence of free Zn atoms. Therefore by interrupting periodically the process and reinitiating growth the columns are interrupted by high-resistivity nucleation layers. In this way the resistivity of ZnO films was increased to 10^{11} Ω cm.

5.2.1.2.4. Dielectric Constant. Only few papers have reported investigations of dielectric constant of ZnO films as a function of relevant parameters. A study regarding the influence of high-energy neutrals on ZnO films during sputtering was reported in [267]. The dielectric constant of films was measured in the frequency range 0.1–100 kHz. Figure 60 shows the dielectric constant measured at 100 kHz for films at A and B positions (see the previous section for the experiment description) as a function of pressure during the sputtering process. At A an increase in dielectric constant is observed from 9.6 to 11 as the pressure is increased from 5 to 30 mTorr. The value of 11 is close to that reported in single-crystal ZnO. Films deposited at B showed poor dielectric constant at low pressure which increased to about 9.5 as the pressure increased to 50 mTorr. However, the dielectric behavior of these films increased markedly when they were subjected to postdeposition annealing. Figure 60 shows that the dielectric constant of ZnO films is clearly influenced by the defects introduced into the structure by the high-energy-atom bombardment. This behavior is similar to that normally observed in other defect oxide dielectric films.

5.2.1.2.5. Acoustical Loss. It has been observed that acoustical loss in thin zinc oxide films is much higher than that in single crystals, because single-crystal bulk zinc oxide is intrinsically a low propagation loss material. The problem of loss mechanisms in sputtered ZnO films has been considered by

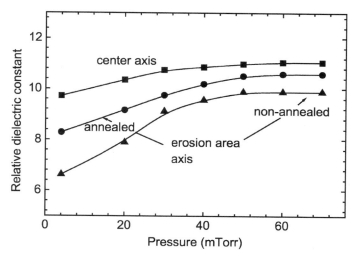

Fig. 60. Dielectric constant of ZnO films grown in different conditions and with different resulting oxygen vacancies. The measured values have been taken from [267].

Hickernell [285], who reported results obtained on SAW propagation loss. For bulk wave propagation in ZnO, losses for longitudinal wave propagation along the c axis are 14 dB/cm and for shear mode propagation normal to the c axis 7 dB/cm. An estimation of the SAW attenuation gives approximately 9.4 dB/cm at 1 GHz. This loss level is similar to that of GaAs and approximately three times higher than in lithium niobate and tantalate. In reality such low surface losses have not been observed with ZnO crystals or films. They were approximately 20 dB/cm at 0.4 GHz, which is 12 times higher than that predicted. The quality of the crystals may be responsible for this high loss level. Observed loss levels for sputtered ZnO have been much higher than predicted. Because the shear mode contribution dominates the SAW propagation loss structural properties of the films which influence this term could be the major source of loss. Data appearing in the literature on the SAW propagation loss for ZnO films on amorphous substrates have shown the following trends: it is 8 to 30 times higher than the predicted intrinsic loss level and the loss level follows a frequency-squared dependence; it is process, substrate, and microstructure dependent. For films deposited by dc triode sputtering, at 500 MHz the loss was 8 dB/cm whereas the predicted loss at the same frequency was 2–3 dB/cm. Regarding substrates, crystal quartz gives much lower losses with respect to oxidized silicon and fused quartz. At 630 MHz for a ZnO film deposited by dc triode the losses were 7.5 dB, with respect to the predicted value of 3.5 MHz [285].

(a) Influence of Bulk Structural Defects. The presence of structural defects in a film greatly increases the acoustic wave propagation loss. In a series of experiments Hickernell [285] reported the dependence of the SAW attenuation (measured at 220 MHz) on film thickness for several ZnO films deposited on fused quartz. All the films were of good quality with smooth surfaces and grown under similar deposition conditions. It can be observed that the loss increases with the film thickness. The

evidence seems to support the possibility of bulk structural defects influencing the observed loss levels. A theoretical model for the scattering of Rayleigh waves by point defects on or near the propagation surface shows that the scattering is sensitive to the defect configuration and also to its position relative to the surface. For defects right at the surface, scattering into other Rayleigh waves is the strongest process. For deeper defects most of the scattered energy goes into bulk waves. The scattering into transverse bulk waves is stronger than the scattering into longitudinal waves. The scattering by point defects situated at the surface is predicted to vary as f^5. If the point defects are randomly but uniformly dispersed through the volume the scattering varies as f^4. Horizontal line defects at the surface (i.e., scratches) scatter as f^4. Vertical line defects through the volume scatter as f^3 and vertical defect sheets as f^2 [285]. Therefore the type of defects which show a frequency characteristic like that observed with ZnO films would be volume-type defects such as collections of vertical lines or vertical sheets of defects. Natural questions arise as to the existence of such defects, their density, and the experimental evidence about their role in losses. Hickernell evidenced by surface etching the defects in the structure of a series of sputtered films [285]. The rf diode films have a very irregular surface with little evidence of distinct and separate defect sites. The dc triode films have a small number of identifiable conic-shaped etch pits, whereas magnetron sputtered films have a surface similar to that of triode films but with a higher density of defects. It has been found that the attenuation loss varies with a cubic law with the density of defects of the films. Then the source of the frequency-squared loss factor is considered to be bulk defects running vertically in the film, representing pores or structurally weak boundaries between the polycrystalline grains. The influence of the initial layer of growth can be also a source of loss. The lowest loss films have been grown on crystalline substrates which should represent a better base for growing material with lower defect density. Also the cleanliness of the substrate is very important.

(b) Surface Roughness. Surface roughness is a factor which has been given consideration as a source of loss. In general as ZnO film thickness increases, surface roughness also increases. For films of 5–6 μm thickness deposited by sputtering Hickernell [285] reported a correlation between the attenuation and the peak-to-peak roughness value but at a rate that is less than linear. It is clear the that the losses due to roughness are not the only ones responsible for the total losses in ZnO films. Theoretically the attenuation by roughness was predicted to increase as the square of the peak-to-peak value of the roughness. The data observed show a much weaker relationship. On the other side high losses have been observed in films with very smooth surfaces.

5.2.2. Aluminum Nitride

Aluminum nitride AlN belongs to the hexagonal class (wurtzite structure). It has a good piezoelectric constant, 5.4 pC/N, together with high electrical resistivity, 10^{14} Ω cm. Thin films of AlN have been developed especially for high-frequency transducers, because this material has a very high velocity of propagation of bulk acoustic waves (10,400 m/s) which allows film thicknesses of 2.5–0.5 μm to cover the range from 2 to 10 GHz in the fundamental resonance mode. Compared with ZnO, AlN has a lower coupling constant but it has the advantage of a much higher electrical resistance. It also shows a high breakdown voltage and low dielectric loss. Moreover low thermal expansion coefficients may allow small thermal drifts of the characteristics to be achieved. AlN thin films also can be employed in microelectromechanical systems, because the small piezoelectric coefficient e_{31}, which is 9 times smaller than that corresponding to PZT, is balanced by the dielectric constant, which is 100 times smaller than that of PZT. Therefore AlN thin films can be expected to be competitive in sensor, actuator, and ultrasound applications where low loss, low thermal drift, and high signal-to-noise ratios are requested. Based on high electromechanical coupling coefficient and superior optical qualities, AlN layers have been considered to be among the most promising materials for optical waveguides, acoustoelectric applications, acoustooptic interaction media, and surface acoustic wave and bulk acoustic wave transducers. AlN has been characterized as a single-crystal epitaxial film and as an oriented polycrystalline film [168, 286]. The published data originate from SAW devices with AlN films grown on sapphire or silicon. It has been found that with some small differences the same set of data can represent high-quality films (single-crystal epitaxial films or highly oriented films) obtained by various techniques.

5.2.2.1. Overview of the Main Results

One of the first papers reporting the deposition of AlN thin films [287] presented an accurate investigation of their electrical properties (resistivity, dielectric constant, and loss factor) as a function of temperature and at different frequencies. The films were prepared by reactive sputtering and were found to show dielectric properties superior to those of bulk polycrystalline material. A series of successive papers in the following years reported the growth of single-crystal epitaxial films on various substrates (usually sapphire and silicon) and by different methods [288–303]. It was then realized that for many applications highly oriented AlN films offer performance similar to epitaxial films.

Onishi et al. reported the obtaining of colorless, transparent, c-oriented AlN films grown at low temperature by a modified sputter gun [293]. A very strong magnetic field made it possible to sputter at low pressures. This high-vacuum sputtering improves c-orientation, surface smoothness, film color, transparency, deposition rate, and thermal expansion coefficient matching. The high magnetic field of the sputter gun makes possible the growth of c-axis-oriented, colorless, transparent mirror-surface AlN films up to 14 μm thick. In addition, the obtained AlN films were completely free from electron bombardment effects.

Testing of AlN films on SAW devices with AlN on silicon was reported in [294]. Reactive rf planar magnetron sputtering was used at substrate temperatures below 300°C to deposit highly oriented piezoelectric AlN films on silicon for surface acoustic wave device applications. The substrates were (100)-oriented, *n*-type silicon with and without a thermally grown oxide. The resulting AlN film was strongly *c*-axis oriented with thickness of about 1.6 μm. The films were piezoelectric and different devices have been constructed on it to test the properties. Growth of AlN epitaxial films for zero temperature coefficient SAW delay lines has been presented by Tsubouchi and Mikoshiba [295]. The authors studied single-crystal AlN films on sapphire and Si obtained by epitaxial growth using MOCVD. They have shown that AlN/Al$_2$O$_3$ or AlN/Si combinations have remarkably attractive SAW characteristics of high frequency, low dispersion, zero temperature coefficient, and low propagation loss. From SAW measurements material constants of AlN films were obtained by computer fitting with experimental data. Single-crystal AlN films have a negative temperature coefficient of delay (TCD) of about -30 ppm/°C and so it is expected to obtain a zero TCD by combination with appropriate positive TCD substrates. The propagation loss of SAW on AlN films on sapphire substrates in the frequency range 490–1200 MHz was also measured. The propagation loss of SAW on the as-grown AlN surface obtained at 990°C was about 15 dB/cm at $f = 1.2$ GHz and $kh = 1.26$. Instead measured propagation loss on AlN/basal-plane sapphire was found to be much lower than that of AlN/R-plane sapphire. Propagation loss on these AlN films seems to be due to defects in the films and to surface roughness, but the detailed mechanisms of loss and its frequency dependence are not fully understood.

The effect of oxygen impurities in AlN films deposited by sputtering have been discussed by Liaw et al. [296]. The sputtered films were typically 1 μm thick. A 200 nm thick layer of Si$_3$N$_4$ was used between the AlN and Si substrate as an electrical insulator. Oxygen concentrations in AlN were estimated by Auger spectroscopy. The sputtered AlN films exhibited columnar growth in a (002) direction perpendicular to the main substrate surface. A correlation among the preferred orientation of the grains, oxygen concentration in the films, and the surface roughness was evidenced.

SAW characteristics of sputtered AlN on Si and GaAs have been reported in [297]. Different interface layers and AlN film thickness were examined. Contrary to expectations, it was found that the thicker AlN films gave transducer and propagation losses poorer than thin AlN films. This is mainly due to the fact that well-oriented grain alignment was degraded as film thickness increased because in thicker films there may be adverse effects due to increases in stress and decreases in grain preferred orientation.

The deposition of highly oriented polycrystalline AlN films and their successful use in SAW devices was reported also in [164]. The films were deposited on various substrates by rf reactive diode magnetron sputtering. From the measurements on a SAW delay line, phase velocities were obtained. Acoustic propagation loss was also evaluated.

Polycrystalline AlN films deposited on Si covered with Si$_3$N$_4$ have been characterized by Brillouin scattering by Carlotti et al. [168]. The complete set of elastic constants was determined. Only small differences were found with the corresponding elastic constants of epitaxial films previously reported in [286].

The possibility of controlling the preferential orientation of AlN films would allow one to obtain films with (100) orientation which are more efficient for SAW generation; in [48] Ishihara et al. demonstrated that (100)-oriented films can be obtained by reactive dc magnetron sputtering. The preferential orientation of the AlN films was controlled by the distance between target and substrate, the gas pressure, and the sputtering power. It was found that the (100)-oriented AlN film is obtained by increasing the distance L, by increasing the pressure, and by decreasing the sputtering power. The authors have given a theoretical explanation of the preferential orientation. Nucleations grow randomly on a substrate but only those survive whose fastest growth directions are closest to the normal of the substrate (geometric selection). The equilibrium form of the wurtzite is made of (100) and (001) planes. The difference in growth rate of these planes decides the orientation. The relative growth rate of (100)/(001) is varied by changing the deposition unit from atoms (Al, N) to dimers such as Al–N. When the mean free path of the atoms is longer than the distance L, Al and N atoms deposit directly on the substrate and the (001)-oriented films grow. When it is shorter than the distance L, the collisions of Al and N atoms occur in the space between the target and the substrate and the Al–N dimers are formed and deposit on the substrate. In this case the (100) orientation is enhanced.

Dubois and Muralt [47] studied the properties of AlN thin films at very high frequency. They prepared polycrystalline AlN thin films with reactive pulsed dc magnetron sputtering, deposited on (111)Pt/Ta/Si. Pt was chosen because the (111) plane of its cubic structure has a hexagonal symmetry like the (002) plane of AlN, and because it is stable in N$_2$ at the deposition temperature. Extremely well *c*-axis-oriented films with a columnar microstructure were obtained. The $e_{31,f}$ piezoelectric coefficient was measured with the cantilever method. The $d_{33,f}$ piezoelectric coefficient, defined for a thin film clamped on a substrate as $d_{33,f} = e_{33}/c_{33}$, was determined at the same frequency with a double-beam laser interferometer. The obtained value was 87% of the single-crystal coefficient (3.9 pC/N). This reduction could be due to the existence of a few columnar grains with an antiparallel polarization. The coefficient has been found to be independent of the applied electric field. Electromechanical coupling coefficient was evaluated from measurements performed on thin film bulk acoustic resonators (TFBAR). A coupling coefficient k of 0.23 was calculated from the main resonance. The acoustic wave propagation velocity v_s in AlN was calculated from the reflection condition of a standing wave in a $\lambda/2$ layer. The velocity extracted from the wavenumber was $v_s = 11,428$ m/s. A value of 11,400 m/s was obtained after the removal of the SiO$_2$ layer of the membrane.

Ruffner et al. [298] studied the effect of substrate composition and deposition parameters on the piezoelectric response of

Table XIII.   Measured Physical Properties of AlN Thin Films

Composition	Deposition technique, substrate	Orientation	Relative dielectric constant ε_{33} (1 kHz)	Dielectr. loss	c_{11} (10^{10} N/m^2)	c_{13} (10^{10} N/m^2)	c_{33} (10^{10} N/m^2)	c_{55} (10^{10} N/m^2)	c_{66} (10^{10} N/m^2)	e_{31} (C/m^2)	e_{33} (C/m^2)	e_{15} (C/m^2)	d_{31} (pC/N)	d_{33} (pC/N)	Ref.
AlN	Dc magnetron sputtering Si$_3$N$_4$/Si(100)	c-axis oriented			36	12.3	41	11.6	11.9						[168]
AlN	Dc magnetron sputtering (0001)Al$_2$O$_3$	Epitaxial	$\varepsilon_{33} = 10.7$ $\varepsilon_{11} = 9.1$		34.5	12	39.5	11.8	11	−0.58	1.55	−0.48	−2.65	5.53	[286, 295]
AlN	Rf magnetron sputtering Pt/Si	c-axis oriented	10.4	0.002			42			−1.02				3.4	[47]
AlN	Rf reactive sputtering Ru/Ti/SiO$_2$/Si	c-axis oriented											−3.5 to +4.2		[298]
Bulk single crystal AlN			$\varepsilon_{33} = 9.3$		41	9.89	38.85	12.5	13.1				−2	5	[275, 303]

reactively sputtered AlN thin films. They observed piezoelectric response values ranging from −3.5 to +4.2 pC/N for 1 μm thick AlN films deposited onto Ti/Ru electrode stacks. This substantial variation in the piezoelectric response occurred despite the fact that all of the AlN thin films exhibited the correct crystallographic orientation for piezoelectric activity [(0002) crystallographic planes parallel to the substrate]. This has been attributed to the different conditions of oxidation of the electrode layer which favored the partial inversion of polarity, thus changing the response from negative to positive values.

Tomabechi et al. [299] studied the development of high-quality AlN epitaxial film for a 2.4-GHz front-end SAW matched filter. AlN films were grown epitaxially by the MOCVD method. AlN deposition on off-angle substrates and the AlN deposition at high temperature were investigated for eliminating the problems of cracks on the surface of epitaxial AlN films. AlN films deposited on an Al$_2$O$_3$ surface which was −4° off-angle from the c axis were free of cracks as shown by SEM observations. The propagation loss was evaluated from the characteristics of time-domain impulse response of 2.4 GHz matched filters. The propagation loss of crackless AlN films was drastically improved by one order of magnitude compared with that of cracked AlN films.

High-quality optoelectronic-grade epitaxial AlN films grown on α-Al$_2$O$_3$, Si, and 6H-SiC by pulsed laser deposition were reported by Vispute et al. [80]. The laser fluence and the substrate temperature have been found to be the main parameters that control the process. The X-ray rocking curve width of epitaxial films on sapphire and SiC was about 0.06–0.07°. The measured bandgap was found to be 6.1 eV, the electrical resistivity 5–6 × 10^{13} Ω cm, and the breakdown field 5 × 10^6 V/cm.

The effect of an external magnetic field applied during growth on c-axis orientation and residual stress in AlN films has been discussed in [300]. The AlN films were deposited by dc planar magnetron sputtering on borosilicate glass, which has a similar thermal expansion coefficient. Large crystal grains, a high degree of orientation, and large tensile residual stresses were obtained at large values of the applied external magnetic field.

Detailed studies of crystallinity and microstructures of AlN/Si(111) and AlN/Si(100) have been reported by Liaw et al. and by Chubachi et al. [301, 302]. Table XIII reports the measured physical properties of AlN films, as obtained by different authors for films grown in different conditions.

5.2.2.2. Influence of Structure on Physical Properties

5.2.2.2.1. Piezoelectric Properties

(a) Degree of Preferred Orientation.   A systematic study of the influence of preferential orientation degree on the properties of AlN films deposited by reactive sputtering on Si$_3$N$_4$/Si(100) was presented in [296]. The sputtered films were typically 1 μm thick. The preferred (002) orientation of the AlN grains was evaluated by the XRD diffraction intensities of the line at $2\theta = 36.1°$. The parameters of the X-ray generator were kept constant for each measurement so that the diffraction intensity depended only on the preferred orientation and comparison could be made between different samples. The sputtered AlN films exhibited columnar growth in a (002) direction perpendicular to the substrate surface. If a SAW transducer were deposited on the film its piezoelectric properties could be estimated from the transducer loss: the lower the loss, the higher the piezoelectric constants and electromechanical coupling factor. Therefore SAW transducers were fabricated on the AlN by forming Al interdigital electrodes on the top surface. The insertion loss of the IDT transducers constructed on the different films has been evaluated at 150 MHz and is presented in Fig-

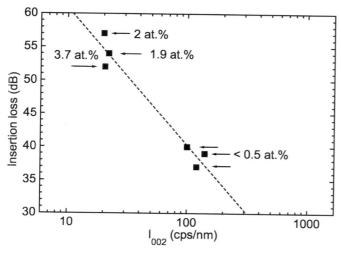

Fig. 61. The variation of insertion loss measured on AlN films with different (002) diffraction intensities, corresponding to different oxygen content. The experimental data have been taken from [296].

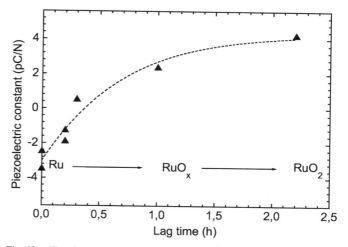

Fig. 62. The piezoelectric response of AlN films as a function of the lag time between deposition of Ru electrode and AlN films. The variation is attributed to the different states of Ru oxidation which causes partial reversal of dipoles in the AlN films. The data are taken from [298].

ure 61 vs. XRD intensity of (002) planes. It can be observed that the smallest insertion loss (highest electromechanical coupling $k^2 = 0.5$–0.8%) is obtained in films with the highest degree of orientation, as expected. The decrease in the degree of orientation can be due to other orientations of AlN grains or to the appearance of spurious phases. As the only orientation of the grains was with the c axis perpendicular to the substrate, the oxygen content of the films was considered. Oxygen concentrations in AlN were estimated by Auger spectroscopy. The detection limit of the technique was approximately 0.5 at%. The oxygen content of every film is shown in Figure 61. Because Al has a strong chemical affinity to oxygen for the formation of Al_2O_3 it is supposed that films with higher oxygen content will have also a higher concentration of Al_2O_3, and therefore a lower content of the piezoelectric phase with preferred orientation.

(b) Polarity Inversion. If different regions in AlN film have different orientations of the polar axis (with Al atom or N atom toward the upper surface) the overall piezoelectric effect will be decreased in proportion to these inversion domains. Ruffner et al. [298] put in evidence the crucial role of polarity inversion (and ultimately of the substrate surface condition that produced it) in determining the size and the sign of the piezoelectric constant in reactively sputtered AlN thin films. Different thin film samples with thickness about 1 μm have been prepared on Ru/Ti/SiO$_2$/Si substrates. The length of time between deposition of Ru electrode layer and AlN layer was different for each film, varying between 0 and 2.2 h, to determine how the contamination of the underlying electrode film affected the properties of the overlying AlN film. Piezoelectric response was measured by sectioning the samples into cantilever beams of known area and measuring deflection as a function of applied voltage at 100 Hz. The piezoelectric response for each of the Ti/Ru/AlN samples is shown in Figure 62 as a function

of lag time between deposition of the Ru and AlN thin film layers. The variation in the piezoelectric response cannot be attributed to a change in crystallographic orientation because all of the AlN thin film samples exhibited very strong preferential (0002) orientation when analyzed using θ–2θ XRD. The change in piezoelectric response must result from a reversal of the dipoles within that crystallographic configuration. As can be seen from Figure 62, for samples in which there was a relatively long lag time (>0.5 h) the piezoelectric response has positive values. Exposition of Ru surface to residual gases in the vacuum chamber produced its oxidation. Compositional depth profiling using SIMS performed on some samples with small and large lag time showed a high oxygen-to-aluminum concentration ratio at the Ru–AlN interface of samples with large lag time, suggesting the presence of a RuO$_2$ layer. Although the RuO$_2$ layer was probably limited to the first several monolayers of the surface, it appears to have been sufficiently thick to affect crystalline and dipole orientation of the subsequently deposited AlN. For samples in which there was a minimal lag time (<20 min) between deposition of the Ru and AlN thin film layers, the piezoelectric response has negative values. This result suggests that the composition of the Ru surface affects the dipole orientation of the subsequently deposited AlN thin film, which determines the sign of the piezoelectric response. The dependence of piezoelectric response on the composition of the Ru surface can be explained in terms of sputtering kinetics and chemical bonding between the thin film materials. During the reactive sputter deposition of the AlN thin film, the plasma contains Al^{3+} and N^{3-} ions. The sticking coefficient of these ions to the Ru thin film is dependent upon their relative charges. The electrode layer which is exposed to residual oxygen gas will develop a RuO$_2$ layer with oxygen atoms at the surface. When this surface is bombarded with Al^{3+} and N^{3-} ions during the sputtering process, the Al^{3+} ions will preferentially bond to the O^{2-} ions at the surface. Therefore, the first monolayer of atoms

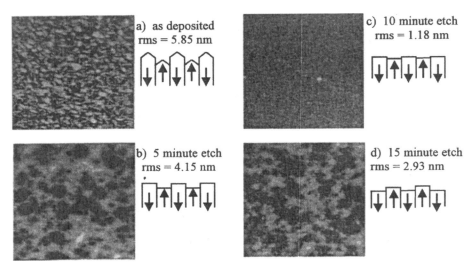

Fig. 63. AFM images of an AlN film with equal mixture of the two dipole orientations for different etching times. (Reprinted from [298], © 1999, with permission from Elsevier Science.)

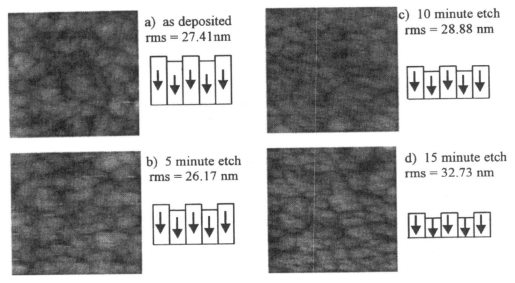

Fig. 64. AFM images of an AlN film with one dipole configuration for different etching times. Reprinted with permission from [298], © 1999, Elsevier Science.

consist almost entirely of Al. Nitrogen is highly reactive with Al and therefore the second monolayer of atoms consist mostly of nitrogen atoms. In this way the thin film is deposited such that dipoles of the resultant AlN molecules are all oriented in the same direction (away from the RuO_2 surface). It seems that this mechanism is a very effective orientation technique as the piezoelectric response (4.2 pC/N) for an AlN film deposited in this manner approaches the bulk value (5.4 pC/N). On the other hand, a pristine Ru surface will preferentially bond with the N^{3-} ions in the plasma, resulting in a first monolayer consisting mostly of nitrogen. The second monolayer will consist mostly of Al^{3+} and so on. The dipole orientation of the resulting AlN molecules will be reversed compared to those for the oxidized Ru surface. The negative piezoelectric response values for samples with a pristine Ru/AlN interface support this

explanation. In addition progressive contamination at the interface would explain the gradual shift from negative to positive values of the piezoelectric response as the lag time between Ru and AlN deposition was increased, as shown in Figure 63. The near-zero piezoelectric response in films with partially oxidized electrodes suggests a nearly equal mixture of the two dipole orientations. To verify this hypothesis etch rate experiments were conducted on these samples to determine the dipole orientation as a function of deposition conditions. Due to the different bonding conditions for the opposite dipole orientations, one dipole configuration will etch more rapidly than the other. A series of images showing this aspect as seen by AFM of a sample with 0.2 h lag time and −1.29 pm/V piezoelectric response at different etching times are presented in Figure 64. The evolution of rms can be explained by supposing that opposite polarities are present in almost equal parts: at the beginning the as-deposited film is relatively rough (rms 5.85 nm) because the partially oxidized surface favored the formation of opposite dipoles, which have different growth rates (a); after 10 min etching the surface become more smooth (rms 1.18 nm) (b and c) but then returns to being rough (rms 2.93 nm) due to the different etch rates of the opposite dipoles (d). By contrast, a sample with unoxidized electrode (Fig. 64) favored the formation of a single polarity, whose etching does not modify the roughness significantly because the etching rate is the same everywhere.

5.2.2.2.2. Elastic Constants. Role of Oxygen Impurities.

Oxygen impurities can hardly modify the elastic constants of AlN. In AlN films deposited on Si_3N_4/Si by reactive sputtering [296] samples with different oxygen content have been examined The SAW phase velocity hardly decreased in samples with high oxygen content. The lower phase velocity was attributed to a lower stiffness factor resulting from a less rigid Al–N bonding in the presence of oxygen or Al–O bonds. Jagannadham et al. [304] found very high compressive stress in AlN films with oxygen impurities deposited by pulsed laser deposition. To avoid the formation of particulates when ablating a metallic target, an AlN target was used. Therefore, even if the deposition chamber was previously evacuated down to 2×10^{-7} mbar, release of oxygen impurities from the sintered polycrystalline AlN target was possible, as confirmed by SIMS analysis of the obtained films. The films were deposited on Si(111), which is favorable for domain matching epitaxy AlN (0002)[11$\underline{2}$0]//Si(111)[1$\underline{1}$0]. The domain mismatch between four atomic spacings along Si[110] and five of AlN [11$\underline{2}$0] is evaluated to give a compressive strain of -1.28% in the film. Normally these stresses are partially or totally relaxed for films deposited at 600°C, but it has been observed that the films have an increased compressive stress with respect to that attributed to lattice misfit. From the Fourier transform IR absorption peak shift the compressive residual stress was estimated to be about 30 GPA. This was attributed to oxygen interstitials. These compressive stresses could only be relieved by diffusion of the oxygen ions from the lattice, which does not take place because of the high affinity of oxygen for aluminum.

5.2.2.2.3. Breakdown Strength.

The electrical characteristics of AlN films deposited by dc planar magnetron sputtering on different substrates have been analyzed in [49]. The films where highly c-axis oriented with average crystallite size [as determined by the (0002) peak width] of 27 nm. The films were free from any impurities. A dielectric constant 8.7 (the same as for a single crystal) was measured.

Unlike other materials (perovskite ceramics) the average breakdown strength of AlN was found to increase with decreasing thickness. A similar phenomenon was observed in silicon dioxide and other dielectric materials. When measuring the current–field characteristics a tunneling characteristic is obtained at the injection threshold field. For carrier injection

from the electrode at fields higher than this threshold, the electrical conduction is filamentary because the electrode surface is not microscopically identical in asperity and the surface condition varies from domain to domain. Thus there may be one or more microregions in which the potential barrier has a profile more favorable for carrier injection than in other regions. Furthermore, the material itself is never microscopically homogeneous. It is believed that the current filament which leads to final destructive breakdown is the filament carrying the majority of the total current and that destructive breakdown is initiated by thermal instability followed by impact ionization leading to a sharp increase in current at the breakdown field. The increasing of breakdown strength for thinner films can be attributed to the easier heat transfer in thin samples than in thick ones.

5.2.2.2.4. Acoustical Loss.

Propagation loss of surface acoustic waves on AlN films on different substrates have been evaluated by Tsubouchi and Mikoshiba [286]. The films were epitaxially grown on sapphire substrates. It was found that the loss increased with film thickness. This was attributed to bulk defects and surface roughness. At 1 GHz the loss was about 8 dB/cm at $kh = 1$.

The propagation loss in highly oriented polycrystalline AlN films has been reported by Liaw et al. [296] in AlN films deposited by reactive sputtering on Si_3N_4/Si. The sputtered films were typically 1 μm thick. SAW transducers were fabricated on the AlN by forming Al interdigital electrodes on the top surface. The propagation loss of the acoustic wave was derived from the slope of the plot between the transducer insertion loss and transducer separation distance. A loss which is comparable with that obtained in epitaxial AlN films [286] was measured up to 1 GHz.

5.2.3. Gallium Nitride

Generally, the interest in the piezoelectric properties of thin film materials has been given by applications such as microactuators, microsensors, and ultrasonic motors, but in the case of piezoelectric semiconductors such as gallium nitride there is an increased interest because their piezoelectric properties affect their semiconducting properties. Indeed the piezoelectric field in strained layers has been used to modify the operating characteristics of devices fabricated from these materials. Development of technology and understanding of the growth mechanism in heteroepitaxial growth of nitrides on highly mismatched substrates allowed the growth of high-quality GaN, AlGaN, and GaInN and their quantum well structures. GaN and InGaN are the most promising materials for use in blue–UV integrated solid-state light-emitting devices due to the ability to tailor the direct bandgap between 2 and 3.4 eV and their high chemical bond strength [305].

GaN belongs to the hexagonal class (wurtzite structure). Many physical properties measured on films of widely varying quality have been reported. However there is only a limited amount of measured data for piezoelectric coefficients. Willmott and Antoni [306] discussed the growth of GaN(0001) thin films on Si(001). Wurtzite GaN films were grown by reactive crossed-beam PLD at 248 nm (KrF) using a liquid Ga target and a synchronous N_2 pulse on atomically flat reconstructed Si(001) substrates. The films were (0001) single phase for substrate temperatures between 200 and 700°C and also grew in a twinned epitaxial manner with the crystallites oriented parallel to the (110) and ($\underline{1}$10) in-plane directions. The films were subsequently investigated *ex situ* by X-ray diffraction, reflection high-energy electron diffraction, and photoluminescence. Although single-phase GaN grows down to 200°C, the (0002) reflex intensity per unit film thickness drops sharply with temperature, signifying an increase in the amorphous contribution to the film.

An estimate of the crystalline size in the growth direction is made from the width of the GaN(0002) line and using Scherrer's formula. For the investigated temperature range the authors found $0.14° < FWHM_{GaN(0002)} < 0.3°$, which corresponds to a crystallite size in the growth direction of 200–95 nm, respectively. It has been observed that GaN(0001) grows in a locked twinned mode, with crystallites oriented parallel to the equivalent Si(110) azimuths. There will therefore be two crystallite orientations with their hexagonal basal nets at 30° to one another. Their interface is incommensurate, where large defect densities and possible amorphous growth will occur. The optical properties will therefore be affected by the average in-plane crystallite size. Although large crystallites will reduce grain boundary defect densities there will also be an increase in strain and defect densities within the crystal due to the additional incommensurability of the substrate–film interface.

Muensit and Guy [307] reported the first determination of the d_{33} piezoelectric coefficient of wurtzite GaN by using a laser interferometer. The GaN films used were grown by chemical vapor deposition on n-(100)Si. XRD confirmed that the films consisted of the wurtzite phase, with the c axis perpendicular to the film surface. The film thickness was around 1 μm. Aluminum was deposited on the top surface of the film to act as an electrical contact, the substrate providing contact to the other face. An electric field, applied via these contacts, produced piezoelectrically induced changes in the film thickness. These changes were measured using an optical interferometer. The arrangement thus measures the converse piezoelectric effect, the change in strain resulting from a change in field. Several configurations of optical interferometer have been used to measure piezoelectric displacements. The piezoelectric response could be measured for driving fields as low as 0.2 kV/mm and the response was found to be reasonably linear up to the highest fields used, which were about 60 kV/mm. The variation of d_{33} with frequency in the range 1–100 kHz was also measured. The value of the piezoelectric constant was found to be substantially constant over the full range except at frequencies near 100 kHz where bending mode vibrations are present. Sample mounting was found to play an important part in the observed response. For rigidly mounted samples the measured d_{33} was 2.0 ± 0.1 pm/V. Because the samples are polycrystalline in nature, it is expected that this value would be lower than the intrinsic value for a single GaN crystal. Several factors act to reduce the piezoelectric coefficient in a film, such as interfacial stress, defects, and variation in c-axis alignment.

6. CONCLUSIONS

In this article we have reviewed the available experimental and theoretical studies of piezoelectric thin films. The most important techniques of deposition have been presented, with emphasis on the particular growth conditions for obtaining pieozoelectric thin films of high technological interest. The most used characterization techniques have been also presented and specific devices for piezoelectric and dielectric measurements have been described.

Our approach in reviewing the main results reported in this field was to discuss, on one hand, the influence of growth conditions on the microstructure and crystalline properties of the obtained thin films; on the other hand, the relationship between thin film structure and physical properties, such as piezoelectric, dielectric and elastic constants, electrical conductivity, breakdown strength etc.

Theoretical and experimental results regarding the peculiar aspects of piezoelectric phenomena in thin films as reported in literature have been also presented.

Acknowledgments

We would like to thank our colleagues for useful discussions during the preparation of this chapter, and the authors of numerous reviewed papers who helped us with useful suggestions and kindly agreed to use their results and original figures.

Italian Ministry for Foreign Affairs, Progetto Finalizzato MADESS and NATO SfP Program 97 1934 are acknowledged for support in our activity.

REFERENCES

1. J. F. Nye, "Physical Properties of Crystals." Oxford University Press, Oxford, 1979.
2. M. E. Lines and A. M. Glass, "Principles and Applications of Ferroelectrics and Related Materials." Clarendon, Oxford, 1977.
3. R. M. Martin, *Phys. Rev. B* 5, 1607 (1972).
4. G. Saghi-Szabo, R. E. Cohen and H. Krakauer, *Phys. Rev. Lett.* 80, 4321 (1998).
5. W. R. Cook, Jr., and H. Jaffe, in Landolt-Börnstein, "Numerical Data and Functional Relationships in Science and Technology," Vol. 11, p. 287. Springer-Verlag, Berlin, 1979.
6. K. Uchino, "Piezoelectric Actuators and Ultrasonic Motors." Kluwer Academic, Boston, 1997.
7. S. K. Krishnaswamy, B. R. McAvoy, and M. H. Francombe, in "Physics of Thin Films" (M. H. Francombe and J. L. Vossen, Eds.), Vol. 17, p. 145. Academic Press, Boston, 1993.
8. M. H. Francombe, in "Physics of Thin Films" (M. H. Francombe and J. L. Vossen, Eds.), Vol. 17, p. 225. Academic Press, Boston, 1993.
9. M. Sayer, "Proc. 1991 Ultrasonics Symposium," p. 595.
10. N. F. Foster, in "Handbook of Thin Film Technology" (L. Z. Maissel and R. Glang, Eds.). McGraw-Hill, New York, 1970.

11. A. N. Goldstein, "Handbook of Nanophase Materials." Dekker, New York, 1997.

12. P. Ayyub, V. R. Palkar, S. Chattopadhyay, and M. Multani, *Phys. Rev. B* 51, 6135 (1995) and references therein.

13. S. Chattopadhyay, P. Ayyub, V. R. Palkar, A. V. Gurjar, R. M. Wankar, and M. S. Multani, *Phys. Rev. B* 52, 13177 (1995).

14. K. Ishikawa, T. Nomura, N. Okada, and K. Takada, *Japan. J. Appl. Phys.* 35, 5196 (1996).

15. S. Li, J. A. Eastman, J. M. Vetrone, C. M. Foster, R. E. Newnham, and L. E. Cross, *Japan. J. Appl. Phys.* 36, 5169 (1997).

16. W. L. Zhong, Y. G. Wang, P. L. Zhang, and B. D. Qu, *Phys. Rev. B* 50, 698 (1994).

17. K. Ishikawa, K. Yoshikawa, and N. Okada, *Phys. Rev. B* 37, 5852 (1988).

18. B. Jiang, J. L. Peng, L. A. Bursill, and W. L. Zhong, *J. Appl. Phys.* 87, 3462 (2000).

19. X. G. Tang, Q. F. Zhou and J. X. Zhang, *J. Appl. Phys.* 86, 5194 (1999).

20. V. P. Dudkevich, V. A. Bukreev, V. M. Mukhortov, Yu. V. Golovko, Yu. G. Sindeev, V. M. Mukhortov, and E. G. Fesenko, *Phys. Status Solidi A* 65, 463 (1981).

21. M. de Keijser, G. J. M. Dormans, P. J. van Veldhoven, and D. M. de Leeuw, *Appl. Phys. Lett.* 59, 3556 (1991).

22. H. Fujisawa, M. Shimizu, T. Horiuchi, T. Shiosaki, and K. Matsushige, *Japan. J. Appl. Phys.* 35, 4913 (1996).

23. H. Okino, T. Nishikawa, M. Shimizu, T. Horiuchi, and K. Matsushige, *Japan. J. Appl. Phys.* 38, 5388 (1999).

24. H. Fujisawa, S. Nakashima, K. Kaibara, M. Shimizu, and H. Niu, *Japan. J. Appl. Phys.* 38, 5392 (1999).

25. T. Zhao, F. Chen, H. Lu, G. Yang, and Z. Chen, *J. Appl. Phys.* 87, 7442 (2000).

26. V. Nagarajan, I. G. Jenkins, S. P. Alpay, H. Li, S. Aggarwal, L. Salamanca-Riba, A. L. Roytburd, and R. Ramesh, *J. Appl. Phys.* 86, 595 (1999).

27. F. Tsai and J. M. Cowley, *Appl. Phys. Lett.* 65, 1906 (1994).

28. T. Maruyama, M. Saitoh, I. Sakai, T. Hidaka, Y. Yano, and T. Noguchi, *Appl. Phys. Lett.* 73, 3524 (1998).

29. T. Tybell, C. H. Ahn, and J. M. Triscone, *Appl. Phys. Lett.* 75, 856 (1999).

30. Y. Sakashita, H. Segawa, K. Tominaga, and M. Okada, *J. Appl. Phys.* 73, 7857 (1993).

31. A. L. Roytburd, S. P. Alpay, V. Nagarajan, C. S. Ganpule, S. Aggarwal, E. D. Williams, and R. Ramesh, *Phys. Rev. Lett.* 85, 190 (2000).

32. K. Uchino and S. Nomura, *Ferroelectrics Lett.* 44, 55 (1982).

33. B. D. Qu, M. Evstigneev, D. J. Johnson, and R. H. Prince, *Appl. Phys. Lett.* 72, 1394 (1998).

34. J. L. Vossen and W. Kern, Eds., "Thin Film Process II." Academic Press, San Diego, 1991.

35. S. Maniv, *Vacuum* 23, 215 (1983).

36. J. A. Thornton, *Ann. Rev. Mater. Sci.* 7, 256 (1977).

37. J. A. Thornton, *J. Vac. Sci. Technol.* 11, 666 (1974).

38. W. D. Westwood, in "Physics of Thin Films" (M. H. Francombe and J. L. Vossen, Eds.), Vol. 14, pp. 1–79. Academic Press, San Diego, 1989.

39. P. S. Vincett, *J. Vac. Sci. Technol.* 21, 972 (1982).

40. F. S. Hickernell, *IEEE Trans. Sonics Ultrason.* SU-32/5, 621 (1985).

41. J. M. E. Harper, H. T. G. Hentzell, and J. Cuomo, *Appl. Phys. Lett.* 43, 547 (1983).

42. I. Červeň, T. Lacko, I. Novotný, V. Tvarožek, and M. Harvanka, *J. Cryst. Growth* 131, 546 (1993).

43. S. V. Krishnaswamy, B. R. McAvoy, and R. A. Moore, "Proc. IEEE Ultrasonics Symp.," Dallas, 1984, p. 411.

44. Y. E. Lee, Y. J. Kim, and H. J. Kim, *J. Mater. Res.* 13, 1260 (1998).

45. K. Tominaga, Y. Sato, I. Mori, K. Kusala, and T. Hanabusa, *Japan. J. Appl. Phys.* 37, 5224 (1998).

46. K. Jagannadham, A. K. Sharma, Q. Wei, R. Kalyanraman, and J. Narayan, *J. Vac. Sci. Technol. A* 16, 2804 (1998).

47. M. A. Dubois and P. Muralt, *Appl. Phys. Lett.* 74, 3032 (1999).

48. M. Ishihara, S. J. Li, H. Yumoto, K. Akashi, and Y. Ide, *Thin Solid Films* 316, 152 (1998).

49. D. Liufu and K. C. Kao, *J. Vac. Sci. Technol. A* 16, 2360 (1998).

50. S. V. Krishnaswamy and B. R. McAvoy, *J. Wave–Material Interaction* 4, 43 (1989).

51. B. Wacogne, M. P. Roe, T. J. Pattinson, and C. N. Pannell, *Appl. Phys. Lett.* 67, 1674 (1995).

52. B. M. Han, S. Chang, and S. Y. Kim, *Thin Solid Films* 338, 265 (1999).

53. K. Sreenivas, M. Sayer, D. J. Baar, and M. Nishioka, *Appl. Phys. Lett.* 52, 709 (1988).

54. A. Yamada, C. Maeda, F. Uchikawa, K. Misu, and T. Honma, *Japan. J. Appl. Phys.* 38, 5520 (1999).

55. D. B. Chrisey and G. K. Hubler, Eds., "Pulsed Laser Deposition of Thin Films." Wiley, New York, 1994.

56. D. Bäuerle, "Laser Processing and Chemistry." Springer-Verlag, Berlin/Heidelberg/New York, 1996.

57. M. Dinescu, P. Verardi, C. Boulmer-Leborgne, C. Gerardi, L. Mirenghi, and V. Sandu, *Appl. Surf. Sci.* 127/129, 559 (1998).

58. M. Dinescu and P. Verardi, *Appl. Surf. Sci.* 106, 149 (1996).

59. P. Verardi, M. Dinescu, and A. Andrei, *Appl. Surf. Sci.* 96/98, 827 (1996).

60. P. Verardi, N. Nastase, C. Gherasim, C. Ghica, M. Dinescu, R. Dinu, and C. Flueraru, *J. Cryst. Growth* 197, 523 (1999).

61. S. A. Studenikin, N. Golego, and M. Cocivera, *J. Appl. Phys.* 83, 2104 (1998).

62. H. Shankur and J. T. Cheung, *J. Vac. Sci. Technol., A* 1, 1806 (1983).

63. S. Hayamizu, H. Tabata, H. Tanaka, and T. Kawai, *J. Appl. Phys.* 80, 787 (1996).

64. V. Craciun, J. Elders, J. G. E. Gardeniers, and I. W. Boyd, *Appl. Phys. Lett.* 65, 2963 (1994).

65. V. Craciun, J. Elders, J. G. E. Gardeniers, J. Geretovsky, and I. W. Boyd, *Thin Solid Films* 259, 1 (1995).

66. R. D. Vispute, V. Talyansky, Z. Trajanovic, S. Choopun, M. Downes, R. P. Sharma, T. Venkatesan, M. C. Woods, R. T. Lareau, K. A. Jones, and A. A. Iliadis, *Appl. Phys. Lett.* 70, 2735 (1997).

67. R. D. Vispute, V. Talyansky, S. Choopun, R. P. Sharma, T. Venkatesan, M. He, X. Tang, J. B. Halpern, M. G. Spencer, Y. X. Li, L. G. Salamalanca-Riba, A. A. Iliadis, and K. A. Jones, *Appl. Phys. Lett.* 73, 348 (1998).

68. P. L. Washington, H. C. Ong, J. Y. Dai, and R. P. H. Chang, *Appl. Phys. Lett.* 72, 3261 (1998).

69. H. Cao, J. Y. Wu, H. C. Ong, J. Y. Dai, and R. P. Chang, *Appl. Phys. Lett.* 73, 572 (1998).

70. S. Choopun, R. D. Vispute, W. Noch, A. Balsamo, R. P. Sharma, T. Venkatesan, A. Iliadis, and D. C. Look, *Appl. Phys. Lett.* 75, 3497 (1999).

71. V. Srikant and D. R. Clarke, *J. Appl. Phys.* 81, 6357 (1997).

72. P. Verardi and M. Dinescu, "Proc. IEEE Ultrason. Symp.," 1995, pp. 1015–1018.

73. C. W. Seabury, J. T. Cheung, P. H. Kobrin, and R. Addison, "Proc. IEEE Ultrason. Symp.," 1995, pp. 909–911.

74. S. L. King, J. G. E. Gardeniers, and I. W. Boyd, *Appl. Surf. Sci.* 96/98, 811 (1996).

75. K. Seki, X. Xu, H. Okabe, J. M. Frye, and J. B. Halpern, *Appl. Phys. Lett.* 60, 2234 (1992).

76. M. G. Norton, P. G. Kotula, and B. Carter, *J. Appl. Phys.* 70, 2871 (1991).

77. R. D. Vispute, J. Narayan, H. Wu, and K. Jagannadham, *J. Appl. Phys.* 77, 4724 (1995).

78. R. D. Vispute, H. Wu, and J. Narayan, *Appl. Phys. Lett.* 67, 1549 (1995).

79. W.-T. Lin, L.-C. Meng, G.-J. Chen, and H.-S. Liu, *Appl. Phys. Lett.* 66, 2066 (1995).

80. R. D. Vispute, J. Narayan, and J. D. Buday, *Thin Solid Films* 299, 94 (1997).

81. P. Verardi, M. Dinescu, C. Stanciu, C. Gerardi, L. Mirenghi, and V. Sandu, *Mater. Sci. Eng. B* 50, 223 (1997).

82. P. Verardi, M. Dinescu, C. Gerardi, L. Mirenghi, and V. Sandu, *Appl. Surf. Sci.* 109/110, 371 (1997).

83. J. Meinschien, G. Behme, F. Falk, and H. Stafast, *Appl. Phys. A* 69, S683 (1999).

84. D. Feiler, R. S. Williams, A. A. Talin, H. Yoon, and M. S. Goorsky, *J. Cryst. Growth* 171, 12 (1997).

85. G. D. O'Clock, Jr., and M. T. Duffy, *Appl. Phys. Lett.* 23, 55 (1973).
86. I. L. Guy, S. Muensit, and E. M. Goldys, *Appl. Phys. Lett.* 75, 4133 (1999).
87. R. F. Xiao, H. B. Liao, N. Cue, X. W. Sun, and H. S. Kwok, *J. Appl. Phys.* 80, 4226 (1996).
88. M. Dinescu, P. Verardi, C. Boulmer-Leborgne, C. Gerardi, L. Mirenghi, and V. Sandu, *Appl. Surf. Sci.* 127/129, 559 (1998).
89. P. R. Willmott and F. Antoni, *Appl. Phys. Lett.* 73, 1394 (1998).
90. J. S. Horwitz, K. S. Grabovski, D. B. Chrisey, and R. E. Leuchtner, *Appl. Phys. Lett.* 59, 1565 (1991).
91. I. Kidoh, T. Ogawa, H. Yashima, A. Morimoto, and T. Shimizu, *Japan. J. Appl. Phys.* 30, 2167 (1991).
92. R. Ramesh, W. K. Chan, B. Willkens, H. Gichrist, T. Sands, J. M. Tarascon, D. K. Fork, J. Lee, and A. Safari, *Appl. Phys. Lett.* 61, 1537 (1992).
93. O. Auciello, L. Manteze, J. Duarte, X. Chen, S. H. Rou, A. I. Kingon, A. F. Schreiner, and A. R. Kraus, *J. Appl. Phys.* 73, 5197 (1993).
94. A. Iembo, F. Fuso, M. Allegrini, E. Arimodo, V. Beradi, N. Spinelli, F. Lecabue, B. E. Watts, G. Franco, and G. Chiorboli, *Appl. Phys. Lett.* 63, 1194 (1993).
95. D. J. Lichtenwalner, O. Auciello, R. Dat, and A. I. Kingon, *J. Appl. Phys.* 74, 7497 (1993).
96. H. Tanaka, H. Tabata, T. Kawai, Y. Yamazaki, S. Oki, and S. Gohda, *Thin Solid Films* 289, 29 (1996).
97. G. C. Tyrrell, T. H. York, L. G. Coccia, and I. W. Boyd, *Appl. Surf. Sci.* 96/98, 769 (1996).
98. C. S. Ma, S. K. Hau, K. H. Wong, P. W. Chan, and C. L. Choy, *Appl. Phys. Lett.* 69, 2030 (1996).
99. J. Lappalainen, J. Frantti, and V. Lantto, *J. Appl. Phys.* 82, 3469 (1997).
100. B. Yang, S. Aggarwal, A. M. Dhote, T. K. Song, R. Ramesh, and J. S. Lee, *Appl. Phys. Lett.* 71, 356 (1997).
101. M. Tyunina, J. Levoska, and S. Lepavuori, *J. Appl. Phys.* 83, 5489 (1998).
102. Y. Lin, B. R. Zhao, H. B. Peng, B. Xu, H. Chen, F. Wu, H. J. Tao, Z. X. Zhao, and J. S. Chen, *Appl. Phys. Lett.* 73, 2781 (1998).
103. M. Tyunina, J. Wittborn, C. Bjormander, and K. V. Rao, *J. Vac. Sci. Technol. A* 16, 2381 (1998).
104. S. G. Mayr, M. Moske, K. Samwer, M. E. Taylor, and H. A. Atwater, *Appl. Phys. Lett.* 75, 4091 (1999).
105. O. Sugiyama, S. Saito, K. Kato, S. Osumi, and S. Kanero, *Japan. J. Appl. Phys.* 38, 5461 (1999).
106. B. E. Watts, F. Leccabue, G. Bocelli, G. Calestani, F. Calderon, O. de Melo, P. P. Gonzales, L. Vidal, and D. Carillo, *Mater. Lett.* 11, 183 (1991).
107. Y. Shybata, K. Kaya, and K. Akashi, *Appl. Phys. Lett.* 61, 1000 (1992).
108. P. Verardi, M. Dinescu, F. Craciun, and C. Gerardi, *Thin Solid Films* 318, 265 (1998).
109. P. Verardi, M. Dinescu, F. Craciun, and A. Perrone, *Appl. Surf. Sci.* 127/129, 457 (1998).
110. P. Verardi, M. Dinescu, F. Craciun, R. Dinu, and M. F. Ciobanu, *Appl. Phys. A* 69, S667 (1999).
111. S. B. Ogale, R. Nawathey-Dikshit, S. J. Dikshit, and S. M. Kanektar, *J. Appl. Phys.* 71, 5718 (1992).
112. Y. Shibata, K. Kaya, K. Akashi, M. Kanai, T. Kawai, and S. Kawai, *Appl. Phys. Lett.* 61, 1000 (1992).
113. A. M. Marsh, S. D. Harkness, F. Qian, and R. K. Singh, *Appl. Phys. Lett.* 62, 952 (1993).
114. D. K. Fork and G. B. Anderson, *Appl. Phys. Lett.* 63, 1029 (1993).
115. Y. Shibata, K. Kaya, K. Akashi, M. Kanai, T. Kawai, and S. Kawai, *J. Appl. Phys.* 77, 1498 (1995).
116. S. H. Lee, T. K. Song, T. W. Noh, and J. H. Lee, *Appl. Phys. Lett.* 67, 43 (1995).
117. C. N. Afonso, J. Gonzalo, F. Vega, E. Dieguez, J. C. C. Wong, C. Ortega, J. Siejka, and G. Amsel, *Appl. Phys. Lett.* 66, 1452 (1995).
118. S. H. Risbud, in "The Engineering Handbook" (R. C. Dorf, Ed.), p. 1738. CRC Press, Boca Raton, FL, IEEE Press, New York, 1997.
119. B. A. Tuttle and R. W. Schwartz, *MRS Bull.* 1996 (June), 49 (1996).
120. Z. Huang, Q. Zhang, and R. W. Whatmore, *J. Appl. Phys.* 86, 1662 (1999).
121. S. A. Impey, Z. Huang, A. Patel, R. Beanland, N. M. Shorrocks, R. Watson, and R. W. Whatmore, *J. Appl. Phys.* 83, 2202 (2000).
122. Y. S. Song, Y. Zhu, and S. B. Desu, *Appl. Phys. Lett.* 72, 2686 (1998).
123. J. Zeng, C. Lin, K. Li, and J. Li, *Appl. Phys. A* 69, 93 (1999).
124. J. H. Jang and K. H. Yoon, *Appl. Phys. Lett.* 75, 130 (1999).
125. N. Ykarashi, *Appl. Phys. Lett.* 73, 1955 (1998).
126. Z. Huang, Q. Zhang, and R. W. Whatmore, *J. Appl. Phys.* 85, 7355 (1999).
127. J. C. Shin, C. S. Hwang, H. J. Kim, and S. K. Park, *Appl. Phys. Lett.* 75, 3411 (1999).
128. L. Lian and N. R. Sottos, *J. Appl. Phys.* 87, 3941 (2000).
129. P. G. Mercado and A. P. Jardine, *J. Vac. Sci. Technol., A* 13, 1017 (1995).
130. H. K. Jang, S. K. Lee, C. E. Lee, S. J. Noh, and W. I. Lee, *Appl. Phys. Lett.* 76, 882 (2000).
131. S. Agarwal, S. R. Perusse, C. W. Tripton, R. Ramesh, H. D. Drew, T. Venkatesan, D. B. Romero, V. B. Podobedov, and A. Weber, *Appl. Phys. Lett.* 73, 1973 (1998).
132. B. Xu, R. G. Polcawich, S. Trolier-McKinsky, Y. Ye, L. E. Cross, J. J. Bernstein, and R. Miller, *Appl. Phys. Lett.* 75, 4180 (1999).
133. Z. Xie, E. Z. Luo, J. B. Xu, I. H. Wilson, H. B. Peng, L. H. Zhao, and B. R. Zhao, *Appl. Phys. Lett.* 76, 1923 (2000).
134. G. J. Exarhos and S. K. Sharma, *Thin Solid Films* 270, 27 (1995).
135. W. Tang and D. C. Cameron, *Thin Solid Films* 238, 83 (1994).
136. J. L. Vossen and W. Kern, Eds., "Thin Films Processes," p. 257. Academic Press, San Diego, 1978.
137. J. S. Horwitz and J. S. Sprague, in "Pulsed Laser Deposition of Thin Films" (D. B. Chrisey and G. H. Hubler, Eds.), p. 229. Wiley, New York, 1994.
138. G. D. O'Clock, Jr., and M. T. Duffy, *Appl. Phys. Lett.* 23, 55 (1973).
139. W. P. Li, R. Zhang, Y. G. Zhou, J. Yin, H. M. Bu, Z. Y. Luo, B. Shen, Y. Shi, R. L. Jiang, S. L. Gu, Z. G. Liu, Y. D. Zheng, and Z. C. Huang, *Appl. Phys. Lett.* 75, 2416 (1999).
140. L. Guy, S. Muensit, and E. M. Goldys, *Appl. Phys. Lett.* 75, 4133 (1999).
141. Y. Sakashita and H. Segawa, *J. Appl. Phys.* 77, 5995 (1995).
142. T. Matsuzaki and H. Funakubo, *J. Appl. Phys.* 86, 4559 (1999).
143. C. R. Gorla, N. W. Emanetoglu, S. Liang, W. E. Mayo, Y. Lu, M. Wraback, and H. Shen, *J. Appl. Phys.* 85, 2595 (1999).
144. S. M. Sze, Ed., "VLSI Technology." McGraw–Hill, New York, 1988.
145. R. C. Dorf, Ed., "The Engineering Handbook." CRC Press, IEEE Press, 1996.
146. S. K. Dey and R. Zuleeg, *Ferroelectrics* 108, 37 (1990).
147. P. Verardi, F. Craciun, L. Mirenghi, M. Dinescu, and V. Sandu, *Appl. Surf. Sci.* 138/139, 552 (1999).
148. F. S. Hickernell, "Proc. IEEE Ultrason. Symp.," 1996, pp. 235–242.
149. V. A. Vyun, V. N. Umashev, and I. B. Yakovkin, *Sov. Phys. PTE* 6, 192 (1986).
150. E. Cattan, T. Haccart, G. Vélu, D. Rémiens, C. Bergaud, and L. Nicu, *Sens. Actuators A* 74, 60 (1999).
151. M.-A. Dubois and P. Muralt, *Sens. Actuators* 77, 106 (1999).
152. J. A. Christman, R. R. Woolcott, Jr., A. I. Kingon, and R. J. Nemanich, *Appl. Phys Lett.* 73, 3851 (1998).
153. Q. M. Zhang, S. J. Jung, and L. E. Cross, *J. Appl. Phys.* 65, 2807 (1989).
154. W. Y. Pan and L. E. Cross, *Rev. Sci. Instrum.* 60, 2701 (1989).
155. A. L. Kholkin, Ch. Wütchrich, D. V. Taylor, and N. Setter, *Rev. Sci. Instrum.* 67, 1935 (1996).
156. H. Maiwa, J. A. Christman, S.-H. Kim, D.-J. Kim, J.-P. Maria, B. Chen, S. Streiffer, and A. I. Kingon, *Japan. J. Appl. Phys.* 38, 5402 (1999).
157. N. F. Foster, G. A. Coquin, G. A. Rozgonyi, and F. A. Vanatta, *IEEE Trans. Sonics Ultrason.* SU-15, 28 (1968).
158. W. P. Mason, "Physical Acoustics," Vol. 1A, pp. 233–242. Academic, New York, 1964.
159. T. M. Reeder, *Proc. IEEE (Lett.)* 55, 1099 (1967).
160. H. Meitzler and E. K. Sittig, *J. Appl. Phys.* 40, 4341 (1969).
161. D. L. Denburg, *IEEE Trans. Sonics Ultrason.* 18, 31 (1971).
162. A. H. Fahmy and E. L. Adler, *IEEE Trans. Sonics Ultrason.* 19, 346 (1972).

163. T. M. Reeder and D.K. Winslow, *IEEE Trans. Microwave Theory Tech.* MTT-17, 927 (1969).

164. C. Caliendo, G. Saggio, P. Verardi, and E. Verona, "Proc. IEEE Ultrason. Symp.," 1993, pp. 249–252.

165. J. A. Giacometti, C. Wisniewski, W. A. Moura, and P. A. Ribeiro, *Rev. Sci. Instrum.* 70, 2699 (1999).

166. G. Carlotti, G. Socino, A. Petri, and E. Verona, *Appl. Phys. Lett.* 51, 1889 (1987).

167. G. Carlotti, D. Fioretto, L. Palmieri, G. Socino, L. Verdini, and E. Verona, *IEEE Trans. Ultrason., Ferroelect., Freq. Contr.* 38, 56 (1991).

168. G. Carlotti, F. S. Hickernell, H. M. Liaw, L. Palmieri, G. Socino, and E. Verona, "Proc. IEEE Ultrason. Symp.," 1995, pp. 353–356.

169. B. Jaffe, W. R. Cook, Jr., and H. Jaffe, "Piezoelectric Ceramics." Academic Press, London, 1971.

170. K. Sreenivas, M. Sayer, C. K. Jen, and K. Yamanaka, "Proc. 1988 Ultrasonics Symposium," p. 291.

171. K. Sreenivas, M. Sayer, D. J. Baar, and M. Nishioka, *Appl. Phys. Lett.* 52, 709 (1988).

172. C. V. R. Vasant Kumar, M. Sayer, R. Pascual, D. T. Amm, Z. Wu, and D. M. Swanston, *Appl. Phys. Lett.* 58, 1161 (1991).

173. C. H. Peng and S. B. Desu, *Appl. Phys. Lett.* 61, 16 (1992).

174. J. F. Li, D. D. Viehland, T. Tani, C. D. E. Lakeman, and D. A. Payne, *J. Appl. Phys.* 75, 442 (1994).

175. R. Ramesh, H. Gilchrist, T. Sands, V. G. Keramidas, R. Haakenaasen, and D. K. Fork, *Appl. Phys. Lett.* 63, 3592 (1993).

176. M. Toyama, R. Kubo, E. Takata, K. Tanaka, and K. Ohwada, *Sens. Actuators A* 45, 125 (1994).

177. S. Watanabe, T. Fujiu, and T. Fujii, *Appl. Phys. Lett.* 66, 1481 (1995).

178. T. Maeder, P. Muralt, L. Sagalowicz, I. Reaney, M. Kohli, A. Kholkin, and N. Setter, *Appl. Phys. Lett.* 68, 776 (1996).

179. I. Kanno, S. Fujii, T. Kamada, and R. Takayama, *Appl. Phys. Lett.* 70, 1378 (1997).

180. N. Hanajima, S. Tsutsumi, T. Yonezawa, K. Hashimoto, R. Nanjo, and M. Yamaguchi, *Japan. J. Appl. Phys.* 36, 6069 (1997).

181. J. H. Cho and K. C. Park, *Appl. Phys. Lett.* 57, 549 (1999).

182. T. Sakoda, K. Aoki, and Y. Fukuda, *Japan. J. Appl. Phys.* 38, 3600 (1999).

183. Z. Wang, R. Maeda, and K. Kikuchi, "Proc. Symp. Design, Test and Microfabrication of MEMS and MOEMS," Paris, France, 1999, *SPIE* 3680, 948.

184. C. R. Cho, L. F. Francis, and M. S. Jang, *Japan. J. Appl. Phys.* 38, L751 (1999).

185. F. Craciun, P. Verardi, M. Dinescu, F. Dinelli, and O. Kolosov, *Thin Solid Films* 336, 281 (1998).

186. F. Craciun, P. Verardi, M. Dinescu, and G. Guidarelli, *Thin Solid Films* 343/344, 90 (1999).

187. F. Craciun, P. Verardi, M. Dinescu, L. Mirenghi, and F. Dinelli, in "Piezoelectric Materials: Advances in Science, Technology and Applications" (C. Galassi, M. Dinescu, K. Uchino, and M. Sayer, Eds.), pp. 273–284. Kluwer Academic, Dordrecht, 2000.

188. L. Lian and N. R. Sottos, *J. Appl. Phys.* 87, 3941 2000).

189. D. V. Taylor and D. Damjanovic, *Appl. Phys. Lett.* 76, 1615 (2000).

190. X. H. Du, U. Belegundu, and K. Uchino, *Japan. J. Appl. Phys.* 36, 5580 (1997).

191. X. H. Du, J. Zheng, U. Belegundu, and K. Uchino, *Appl. Phys. Lett.* 72, 2421 (1998).

192. H. D. Chen, K. R. Udayakumar, C. J. Gaskey, and L. E. Cross, *Appl. Phys Lett.* 67, 3441 (1995).

193. K. Sumi, H. Qiu, M. Shimada, S. Sakai, S. Yazaki, M. Muray, S. Moriya, Y. Miyata, and T. Nishiwaki, *Japan. J. Appl. Phys.* 38, 4843 (1999).

194. K. Sreenivas and M. Sayer, *J. Appl. Phys.* 64, 1484 (1988).

195. E. Cattan, T. Haccart, and D. Remiens, *J. Appl. Phys.* 86, 7017 (1999).

196. T. Haccart, E. Cattan, D. Remiens, S. Hiboux, and P. Muralt, *Appl. Phys. Lett.* 76, 3292 (2000).

197. K. Ijima, I. Ueda, and K. Kugimiya, *Japan. J. Appl. Phys.* 30(9B), 2149 (1991).

198. B. A. Tuttle and R. W. Schwartz, *MRS Bull.* 21, 49 (1996).

199. R. Kurchania and S. J. Milne, *J. Mater. Res.* 14, 1852 (1999).

200. K. R. Udayakumar, P. J. Schuele, J. Chen, S. B. Krupanidhi, and L. E. Cross, *J. Appl. Phys.* 77, 3981 (1995).

201. H. Adachi, T. Mitsuyu, O. Yazamaki, and K. Wasa, *J. Appl. Phys.* 60, 736 (1986).

202. H. Du, D. W. Johnson, Jr., W. Zhu, J. E. Graebner, G. W. Kammlott, S. Jin, J. Rogers, R. Willett, and R. M. Fleming, *J. Appl. Phys.* 86, 2220 (1999).

203. K. Lefki and G. J. M. Dormans, *J. Appl. Phys.* 76, 1764 (1994).

204. C. C. Chang and C. S. Tang, *J. Appl. Phys.* 87, 3931 (2000).

205. B. Ea-Kim, P. Aubert, F. Ayguavives, R. Bisaro, F. Varniere, J. Olivier, M. Puech, and B. Agius, *J. Vac. Sci. Technol., A* 16, 2876 (1998).

206. R. Ramesh, A. Inam, W. K. Chan, F. Tillerot, B. Wilkens, C. C. Chang, T. Sands, J. M. Tarascon, and V. G. Keramidas, *Appl. Phys. Lett.* 59, 3542 (1991).

207. K. Kushida and H. Takeuchi, *Appl. Phys. Lett.* 50, 1800 (1987).

208. A. Yamada, C. Maeda, T. Uemura, F. Uchikawa, K. Misu, S. Wadaka, and T. Ishikawa, *Japan. J. Appl. Phys.* 36, 6073 (1997).

209. A. Kholkin, A. Seifert, and N. Setter, *Appl. Phys. Lett.* 72, 3374 (1998).

210. K. Wasa, Y. Ai, Y. Ichikawa, D. G. Schlom, S. Troiler-McKinstry, Q. Gang, and C. B. Eom, "Proc. 1999 IEEE Ultrasonics Symposium," p. 999.

211. A. Patel, N. M. Shorrocks, and R. W. Whatmore, *IEEE Trans. Ultrason., Ferroelect., Freq. Contr.* 38, 672 (1991).

212. E. G. Fesenko, V. G. Gavrilyacenko, and A. F. Semenchev, "Domain Structure of Multiaxial Ferroelectric Crystals." Rostov University Press, Rostov on Don, 1990.

213. N. A. Pertsev, A. G. Zembilgotov, and R. Waser, *J. Appl. Phys.* 84, 1524 (1998).

214. K. S. Lee and S. Baik, *J. Appl. Phys.* 87, 8035 (2000).

215. B. S. Kwak, A. Erbil, B. J. Wilkens, J. D. Budai, M. F. Chisholm, and L. A. Boatner, *Phys. Rev. Lett.* 68, 3733 (1992).

216. B. S. Kwak, A. Erbil, J. D. Budai, M. F. Chisholm, L. A. Boatner, and B. J. Wilkens, *Phys. Rev. B* 49, 14865 (1994).

217. E. Ching-Prado, A. Reynes-Figueroa, R. S. Katiyar, S. B. Majumder, and D. C. Agrawal, *J. Appl. Phys.* 78, 1920 (1995).

218. Y. N. Wang, S. B. Ren, J. S. Liu, Z. Zhang, F. Yan, C. J. Lu, J. Zhu, and H. M. Shen, *Ferroelectrics* 231, 1 (1999).

219. S. B. Ren, C. J. Lu, J. S. Liu, H. M. Shen, and Y. N. Yang, *Phys. Rev. B* 54, R14337 (1996).

220. S. B. Ren, C. J. Lu, H. M. Shen, and Y. N. Yang, *Phys. Rev. B* 55, 3485 (1997).

221. R. C. Ibrahim, T. Shiosaki, T. Horiuchi, and K. Matsushige, "Proc. 1999 IEEE Ultrasonics Symposium," p. 1005.

222. R. W. Schwartz, B. A. Tuttle, D. H. Doughty, C. E. Land, D. G. Goodnow, C. L. Hernandez, T. J. Zender, and S. L. Martinez, *IEEE Trans. Ultrason., Ferroelect., Freq. Contr.* 38, 677 (1991).

223. N. F. Foster, *J. Appl. Phys.* 40, 420 (1969).

224. T. A. Rost, H. Lin, T. A. Rabson, R. C. Bauman, and D. L. Callahan, *J. Appl. Phys.* 72, 4336 (1992).

225. Y. Furushima, T. Nishida, M. Shimizu, and T. Shiosaki, "Proc. 1993 Ultrasonics Symposium," p. 263.

226. Y. Sakashita and H. Segawa, *J. Appl. Phys.* 77, 5995 (1995).

227. K. Kaemmer, K. Franke, B. Holzapfel, D. Stephan, and M. Weihnacht, "Proc. 1996 IEEE Ultrasonics Symposium," p. 243.

228. Y. Shibata, N. Kuze, K. Kaya, M. Matsui, M. Kanai, T. Kawai, and S. Kawai, "Proc. 1996 IEEE Ultrasonics Symposium," p. 247.

229. Z. C. Wu, W. S. Hu, J. M. Liu, M. Wang, and Z. G. Liu, *Mater. Lett.* 34, 332 (1998).

230. F. Veignant, M. Gandais, P. Aubert, and G. Garry, *J. Cryst. Growth* 196, 141 (1999).

231. S. K. Park, M. S. Baek, S. C. Bae, K. W. Kim, S. Y. Kwun, Y. J. Kim, and J. H. Kim, *Japan. J. Appl. Phys.* 38(Part 1), 4167 (1999).

232. K. Shiratsuyu, A. Sakurai, K. Tanaka, and Y. Sakabe, *Japan. J. Appl. Phys.* 38(Part 1), 5437 (1999).

233. D. Ghica, C. Ghica, M. Gartner, V. Nelea, C. Martin, A. Cavaleru, and I. N. Mihailescu, *Appl. Surf. Sci.* 138/139, 617 (1999).

234. V. A. Joshkin, P. Moran, D. Saulys, T. F. Kuech, L. McCaughan, and S. R. Oktyabrsky, *Appl. Phys. Lett.* 76, 2125 (2000).

235. J. T. Lee, N. Little, T. Rabson, and M. Robert, "Proc. 1999 IEEE Ultrasonics Symposium," p. 269.

236. S. E. Park and T. R. Shrout, *J. Appl. Phys.* 82, 1804 (1997).

237. L. E. Cross, *Ferroelectrics* 76, 241 (1987).

238. L. E. Cross, *Ferroelectrics* 151, 305 (1994).

239. K. Uchino, *Ferroelectrics* 151, 321 (1994).

240. H. Fu and R. E. Cohen, *Nature* 403, 281 (2000).

241. V. E. Wood, J. R. Busch, S. D. Ramamurthi, and S. L. Swartz, *J. Appl. Phys.* 71, 4557 (1992).

242. L. F. Francis and D. A. Payne, *J. Am. Ceram. Soc.* 74, 1360 (1991).

243. K. Okuwada, S. Nakamura, M. Iami, and K. Kakuno, *Japan. J. Appl. Phys.* 29, 1553 (1990).

244. K. R. Udayakumar, J. Chen, P. J. Schnele, L. E. Cross, V. Kumar, and S. B. Krupanidhi, *Appl. Phys. Lett.* 60, 1187 (1992).

245. M. C. Jiang, T. J. Hong, and T. B. Wu, *Japan. J. Appl. Phys.* 33, 6301 (1994).

246. J.-P. Maria, W. Hackenberger, and S. Trolier-McKinstry, *J. Appl. Phys.* 84, 5147 (1998).

247. Y. Lu, G. H. Jin, M. Cronin-Golomb, S. W. Liu, H. Jiang, F. L. Wang, J. Zhao, S. Q. Wang, and A. J. Drehman, *Appl. Phys. Lett.* 72, 2927 (1998).

248. S. Stemmer, G. R. Bai, N. D. Browning, and S. K. Streiffer, *J. Appl. Phys.* 87, 3526 (2000).

249. G. Catalan, M. H. Corbett, R. M. Bowman, and J. M. Gregg, *Appl. Phys. Lett.* 74, 3035 (1999).

250. Y. L. Lu, B. Gaynor, C. Hsu, G. Jin, M. Cronin-Golomb, F. Wang, J. Zhao, S. Q. Wang, P. Yip, and A. J. Drehman, *Appl. Phys. Lett.* 74, 3038 (1999).

251. G. R. Bai, S. K. Streiffer, P. K. Baumann, O. Auciello, K. Ghosh, S. Stemmer, A. Munkholm, C. Thompson, R. A. Rao, and C. B. Eom, *Appl. Phys. Lett.* 76, 3106 (2000).

252. P. Verardi, M. Dinescu, F. Craciun, R. Dinu, and I. Vrejoiu, "Proc. E-MRS 2000, Symposium D," D-VII.5.

253. Z. G. Ye and M. Dong, *J. Appl. Phys.* 87, 2312 (2000).

254. M. Tyunina, J. Levoska, A. Sternberg, and S. Leppävuori, *J. Appl. Phys.* 86, 5179 (1999).

255. C. Tantigate, J. Lee, and A. Safari, *Appl. Phys. Lett.* 66, 1611 (1995).

256. P. Ayyub, S. Chattopadhyay, R. Pinto, and M. S. Multani, *Phys. Rev. B* 57, 5559 (1998).

257. Z. Kinghelman, D. Damjanovic, A. Seifert, L. Sagalowicz, and N. Setter, *Appl. Phys. Lett.* 73, 2281 (1998).

258. A. Sternberg, M. Tyunina, M. Kundzinsh, V. Zauls, M. Ozolinsh, K. Kundzinsh, I. Shorubalko, M. Kosec, L. Pardo, M. L. Calzada, M. Alguero, R. Kullmer, D. Bäuerle, J. Levoska, S. Leppävuori, and T. Martoniemi, in "Ferroelectric Thin Films, Proc. COST 514 European Concerted Action Workshop" (F. Leccabue, B. E. Watts, and G. Bocelli, Eds.), Parma, Italy, 1997, p. 95.

259. http://www.americanpiezo.com/html/pmn-pt.htm

260. http://www.trsceramics.com/electrostrictors.html

261. H. Ieki and M. Kadota, "Proc. 1999 IEEE Ultrasonics Symposium," p. 281.

262. P. Osbond, C. M. Beck, C. J. Brierley, M. R. Cox, S. P. Marsh, and N. M. Shorrocks, "Proc. 1999 IEEE Ultrasonics Symposium," p. 911.

263. K. M. Lakin, "Proc. 1999 IEEE Ultrasonics Symposium," p. 895.

264. S. Wanuga, J. Midford, and J. P. Dietz, "Proc. IEEE Ultrasonics Symposium," Boston, 1965.

265. G. L. Dybward, *J. Appl. Phys.* 42, 5192 (1971).

266. T. Yamamoto, T. Shiosaki, and A. Kawabata, *J. Appl. Phys.* 51, 3113 (1980).

267. S. B. Krupanidhi and M. Sayer, *J. Appl. Phys.* 56, 3308 (1984).

268. F. S. Hickernell, "Proc. 1980 Ultrasonics Symposium," p. 785

269. C. K. Jen, K. Sreenivas, and M. Sayer, *J. Acoust. Soc. Am.* 84, 26 (1988).

270. G. Perluzzo, C. K. Jen, and E. L. Adler, "1989 Ultrasonics Symposium Proceedings," p. 373.

271. F. Moeller, T. Vandhal, D. C. Malocha, N. Schwesinger, and W. Buff, "Proc. 1994 Ultrasonics Symp.," p. 403.

272. A. Onodera, N. Tamaki, Y. Kawamura, T. Sawada, and H. Yamashita, *Japan. J. Appl. Phys.* 35, 5160 (1996).

273. D. H. Zhang and D. E. Brodie, *Thin Solid Films* 251, 151 (1994).

274. Y. Kim, W. D. Hunt, F. S. Hickernell, and R. J. Higgins, "Proc. 1992 IEEE Ultrasonics Symposium," p. 413.

275. Landolt-Börnstein, "Numerical Data and Functional Relationships in Science and Technology," Group III, Vol. 11. Springer-Verlag, Berlin, 1979.

276. G. Carlotti, D. Fioretto, G. Socino, L. Palmieri, A. Petri, and E. Verona, "Proc. 1990 Ultrasonics Symposium," p. 449.

277. J. G. E. Gardeniers, Z. M. Rittersma, and G. J. Burger, *J. Appl. Phys.* 83, 7844 (1998).

278. L. Sagalowicz and G. R. Fox, *J. Mater. Res.* 14, 1876 (1999).

279. Z. Qian, X. Zhang, M. Zhao, X. Wu, and Y. Lin, *IEEE Trans. Sonics Ultrason.* SU-32, 630 (1985).

280. R. S. Wagers and G. S. Kino, *IEEE Trans. Sonics Ultrason.* SU-21, 209 (1974).

281. F. S. Hickernell and T. S. Hickernell, "Proc. 1983 IEEE Ultrasonics Symposium," p. 311.

282. T. Hada, K. Wasa, and S. Hayakawa, *Thin Solid Films* 7, 135 (1971).

283. T. Mitsuyu, S. Ono, and K. Wasa, *J. Appl. Phys.* 51, 2464 (1980).

284. F. R. Blom, F. C. M. van de Pol, G. Bauhuius, and Th. J. A. Popma, *Thin Solid Films* 204, 365 (1991).

285. F. S. Hickernell, "Proc. 1982 IEEE Ultrasonics Symposium," p. 325.

286. K. Tsubouchi and N. Mikoshiba, *IEEE Trans. Sonics Ultrason.* SU-32, 634 (1985).

287. A. J. Noreika, M. H. Francombe, and S. A. Zeitman, *J. Vac. Sci. Technol.* 6, 194 (1969).

288. J. K. Liu, K. M. Lakin, and K. L. Wang, *J. Appl. Phys.* 46, 3703 (1975).

289. T. Shiosaki, T. Wamamoto, T. Oda, and A. Kawabata, *Appl. Phys. Lett.* 36, 643 (1980).

290. M. Morita, S. Isogai, N. Shimizu, K. Tsubouchi, and N. Mikoshiba, *Japan. J. Appl. Phys.* 20, L173 (1981).

291. M. Morita, N. Uesugi, S. Isogai, K. Tsubouchi, and N. Mikoshiba, *Japan. J. Appl. Phys.* 20, 17 (1981).

292. K. Sato, "Proc. 1985 IEEE Ultrasonic Symposium," p. 192.

293. S. Onishi, M. Eschwei, S. Bielaczy, and W. C. Wang, *Appl. Phys. Lett.* 39, 643 (1981).

294. L. G. Pearce, R. L. Gunshor, and R. F. Pierret, *Appl. Phys. Lett.* 39, 878 (1981).

295. K. Tsubouchi and N. Mikoshiba, "1983 Ultrasonics Symposium," p. 299.

296. H. M. Liaw, W. Cronin, and F. S. Hickernell, "Proc. 1993 Ultrasonics Symposium," p. 267.

297. H. M. Liaw and F. S. Hickernell, "1994 Ultrasonics Symposium Proc.," p. 375.

298. J. A. Ruffner, P. G. Clem, B. A. Tuttle, D. Dimos, and D. M. Gonzales, *Thin Solid Films* 354, 256 (1999).

299. S. Tomabechi, K. Wada, K. Saigusa, H. Matsuhashi, H. Nakase, K. Masu, and K. Tsubouchi, "Proc. 1999 IEEE Ultrasonics Symposium," p. 263.

300. K. Kusaka, T. Ao, T. Hanabusa, and K. Tominaga, *Thin Solid Films* 332, 247 (1998).

301. H. M. Liaw, R. Doyle, P. L. Fejes, S. Zollner, A. Konkar, K. J. Linthicum, T. Gehrke, and R. F. Davis, *Solid State Electronics* 44, 747 (2000).

302. Y. Chubachi, K. Sato, and K. Kojima, *Thin Solid Films* 122, 259 (1984).

303. L. E. McNeil, M. Grimsditch, and R. H. French, *J. Am. Ceram. Soc.* 76, 1132 (1993).

304. K. Jagannadham, A. K. Sharma, Q. Wei, R. Kalyanraman, and J. Narayan, *J. Vac. Sci. Technol. A* 16, 2804 (1998).

305. I. Akasaki and H. Amano, *Japan. J. Appl. Phys.* 36, 5393 (1997).

306. P. R. Willmott and F. Antoni, *Appl. Phys. Lett.* 73, 1394 (1998).

307. S. Muensit and I. L. Guy, *Appl. Phys. Lett.* 72, 1896 (1998).

Chapter 5

FABRICATION AND CHARACTERIZATION OF FERROELECTRIC OXIDE THIN FILMS

Jong-Gul Yoon

Department of Physics, University of Suwon, Hwaseung, Kyung-gi-do 445-743, Korea

Tae Kwon Song

Department of Ceramic Science and Engineering, Changwon National University, Changwon, Kyungnam 641-773, Korea

Contents

1. INTRODUCTION

In last three decades have brought a tremendous increase in the investigation of ferroelectric materials in different forms, including single crystals, ceramics, composites, liquids, polymers, and thin films. Ferroelectrics is characterized by a spontaneous polarization that has two or more orientation states in the absence of an electrical field and can be switched to another of these states by an electrical field [1]. Any two of the orientation states are identical in crystal structure and differ only in their electric polarization vector at zero electrical field. The polar character of the orientation states represents an absolutely stable configuration at zero field. In perovskite ferroelectric materials, such as $BaTiO_3$ and $PbTiO_3$, ferroelectric spontaneous polarization comes from small displacements of ions in crystal structures. This displacement is susceptible to external electric fields, and the electrical, mechanical, and optical properties of ferroelectric materials are coupled with each other. These couplings of physical properties result in piezoelectric and electro-optic properties that make ferroelectric materials useful in various applications. The physical properties of ferroelectrics depend on chemical composition and crystal structures.

Ferroelectric materials possess a multitude of potentially useful properties and a variety of new applications of these materials have been developed for use in memories, sensors, and displays. The technical importance of multicomponent oxide ferroelectric thin films is reflected in the wide range of applications in microelectronic hybrid and discrete devices with active and passive components of semiconductor signal processors, which are currently under investigation. These devices are de-

Handbook of Thin Film Materials, edited by H.S. Nalwa
Volume 3: Ferroelectric and Dielectric Thin Films

ISBN 0-12-512911-4/$35.00

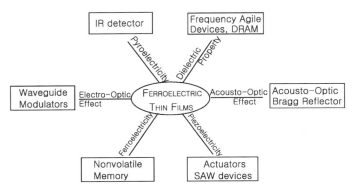

Fig. 1.   Applications of ferroelectric thin films.

signed to combine the unique and/or optimized properties of ferroelectrics with the advantages of semiconductor integrated circuits. Promising devices include nonvolatile memory [2], dynamic random access memory (DRAM) [3, 4], pyroelectric infrared sensors [5], surface acoustic wave filters [6], microacutators [7], electro-optic modulators/switches [8], and second harmonic generators [9]. Some examples of thin film devices using the properties of ferroelectric materials are shown in Figure 1.

The potential utilization of ferroelectric materials has driven intensive research efforts into establishing suitable synthesis/ processing–microstructure property relationships of multicomponent oxide thin films. It is worth noting that the use of oxide films integrated with semiconductors has been limited to the simple oxides. The majority of ferroelectric thin film growth involved materials of perovskite structure composed of oxygen octahedra with a transition metal element at its center. Recently, ferroelectric thin films of much more complex structures than simple perovskite structure, such as $SrBi_2Ta_2O_9$, have been fabricated [10]. Thin film techniques and knowledge accrued in thin film synthesis of high-temperature superconductors are readily applied to ferroelectric film growth. However, there remain many difficulties in achieving the reliable production of device-quality films directly on large semiconductor substrates.

The main goal of thin film technology is to obtain bulk properties of ferroelectric materials in thin film form. To optimize the useful properties of ferroelectric materials in thin film, many factors affecting the physical properties of ferroelectric thin films should be considered. The various techniques adopted in fabricating ferroelectric thin films have produce various structural, electrical, optical, and mechanical characteristics depending on the deposition conditions as well as on the substrate materials. The optimization of deposition conditions requires a good understanding of the basic phenomena related to these techniques and their influence on film composition, microstructure, and properties. Physical properties of thin films are closely related to the structure of deposited films and interactions with the substrates although the exact relationship is not yet clear. These concerns involve the mechanisms of thin film growth, interface formation, thermodynamic reaction between the film and substrate and so forth.

Advances in thin film technology have made it possible to obtain high-quality ferroelectric thin films on foreign substrates. Historically, materials of perovskite structure began to be synthesized in thin films after World War II. The first report on successful vapor growth of epitaxial films appeared in 1963 by Muller et al. [11] using a grain-by-grain evaporation technique. By the early 1970s, the newly developed rf-sputtering technique was adopted successfully for synthesis of $BaTiO_3$ [12], $SrTiO_3$ [13], and $LiNbO_3$ [14]. Subsequently, the rf-sputtering method of growth was commonly used for ferroelectric film growth. Materials of somewhat more complex structure than simple perovskite-type material (e.g., $Bi_4Ti_3O_{12}$) were synthesized [15], and such problems as nonferroelectric phase formation and composition control were encountered [5, 15, 16]. During the early to mid 1980s, various synthesis techniques, including magnetron sputtering, sol-gel, and metalorganic decomposition (MOD), were developed. In recent years other techniques, including metal-organic chemical vapor deposition (MOCVD), pulsed-laser deposition (PLD), and activated reactive evaporation [or molecular beam epitaxy (MBE)] have been introduced. Much useful information on the growth of oxide thin films has been transferred from the thin film research on high-temperature superconductors of perovskite-like structures to the synthesis of ferroelectric film growth.

The emphasis in this review chapter is on currently applied techniques for the synthesis of ferroelectric oxide thin films and the characterization of the films to inform synthesis/processing– microstructure property relationships. Although there are many deposition techniques, typical physical and chemical deposition techniques are discussed. The chapter also addresses the analysis of deposition processes and basic characterization of the deposited films for ferroelectric oxides within limited compositions and structures. In Section 2, basic physical properties of ferrolectrics are briefly reviewed. Brief descriptions of the typical deposition techniques, including physical and chemical deposition, are given in Section 3. Characterization and determination of processing–microstructure property relationships are also discussed in Sections 3 and 4.

2. OVERVIEW OF BASIC PHYSICAL PROPERTIES OF FERROELECTRIC OXIDES

Ferroelectric properties were found in Rochelle salt crystals in 1920 [17]. A ferroelectric crystal is defined as a crystal belonging to the pyroelectric family and in which the direction of spontaneous polarization can be reversed by an electric field [1]. The usual procedure for examining whether a crystal is ferroelectric uses the Sawyer–Tower circuit [18], in which a parallel condenser made of a crystal is inserted. One of the simplest such circuits is shown in Figure 2a. If the crystal is made of ferroelectric crystal with its plate perpendicular to ferroelectric axis, then a polarization-electric field (P–E) hysteresis loop appears, as shown in Figure 2b. This method is also used in ferroelectric characterization of thin films.

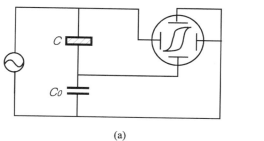
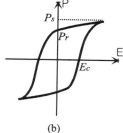
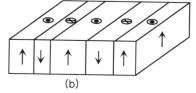

(a)   (b)

Fig. 2.   (a) Sawyer–Tower setup and (b) P–E hysteresis loop. P_s is the saturated polarization, P_r the remnant polarization, and E_c is the coercive field required to reverse the polarization direction.

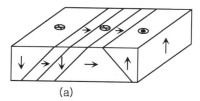

(a)   (b)

Fig. 3.   Schematic representation (a) 90° and (b) 180° domains. The arrows indicate the direction of spontaneous polarization. The circle with a cross represents the negative end of the dipole, and the circle with a dot represents the positive end of the dipole.

Usually the ferroelectric crystal has a twin structure consisting of regions that have spontaneous polarization with negative and positive polarity, respectively. Such a twin structure is the domain structure. The origin of the ferroelectric domain is not yet clearly understood. From a phenomenological stand point, a possible explanation is that the system with polydomain is in a state of minimum free energy. Microscopically, the domains may be caused by a change in the electrostatic forces acting on the crystal faces due to the spontaneous polarization. Also, the ferroelectric domain structure may be caused by defects and internal stresses in a crystal that exhibits piezoelectricity. The structure of the domain depends on the structure of the ferroelectric crystal. In ferroelectric crystals, the variety of domain patterns and the number of types of domain walls depend on the number of all conceivable orientations of the dipole moment when the spontaneous polarization occurs from the prototype high-symmetry phase. For example, BaTiO$_3$ is a cubic crystal that transforms to a tetragonal crystal below the Curie temperature ($\sim$120°C) with its polar tetragonal c-axis. Since any of the three mutually perpendicular fourfold axes of the cubic phase can be the c-axis, the spontaneous polarization must have six possible directions. Thus the angle between the polarization vectors of the domains can be either 90° or 180° (Fig. 3). The domains with a mutually perpendicular direction of the spontaneous polarization are called the 90° domains and have 90° domain walls separating them. Similarly, there can be 180° domains and 180° domain walls. The allowed 90° domain walls coincide with the tetragonal {101} planes [19]. A strong electric field may reverse the spontaneous polarization of the domain. This dynamic process of domain reversal is called "domain switching."

2.1. Structure of Ferroelectric Oxides

Useful properties of ferroelectrics (i.e., ferroelectric, antiferroelectric, pyroelectric, and piezoelectric properties) are related directly to the arrangement of the ions in the unit cell of the crystal and to the modification of the lattice symmetry caused by distortions of structural units. According to the nature of their chemical bond, crystalline ferroelectrics may be classified into compounds with hydrogen bonds and those without hydrogen [20]. Most ferroelectric oxides in thin film studies are compounds without hydrogen bonds. Among them, we limit the structural description to those compounds with oxygen octahedra and hexagonal manganites.

2.1.1. Ferroelectrics with the Perovskite Structure

A very important group of ferreoelectrics is the perovskites, named from the mineral perovskite calcium titanate (CaTiO$_3$). Most of the useful ferroelectrics, including barium titanate (BaTiO$_3$), lead titanate (PbTiO$_3$), lead zirconate titanate (PbZr$_{1-x}$Ti$_x$O$_3$), lead lanthanium zirconate titanate (PLZT), and potassium niobate (KNbO$_3$), have a perovskite-type structure.

The perfect perovskite structure is extremely simple with the general chemical formula ABO$_3$, where O is oxygen, A represents a cation (a monovalent or divalent metal) with a larger ionic radius, and B is a cation (a tetravalent or pentavalent metal) with a smaller ionic radius. It is cubic, with the A ions at the cube corners, B ions at the body centers, and the oxygen ions at the face centers. A perovskite-type structure is essentially a three-dimensional network of BO$_6$ octahedra, as shown in Figure 4; it also may be regarded as a set of BO$_6$ octahedra arranged in a simple cubic pattern and linked together by shared oxygen atoms, with the A atoms occupying the octahedral interstitial positions.

In a certain temperature range, these ions are at their equilibrium positions, and the center of positive charge does not coincide with the center of negative charge. $BaTiO_3$, the first ferroelectric perovskite discovered, transforms successively to three ferroelectric phases: first to tetragonal below 120°C, then to orthorhombic at about 5°C, and finally to a trigonal phase below −90°C. Lattice constants at tetragonal phase are $a = b = 3.992$ Å and $c = 4.036$ Å. The polar axis in the three ferroelectric phases is [001], [011], and [111] respectively. In the tetragonal phase, Ti^{4+} and O^{2-} ions move relative to Ba at the origin from their cubic positions [i.e., Ti at (1/2, 1/2, 1/2) and three oxygens at (1/2, 1/2, 0), (1/2, 0, 1/2), and (0, 1/2, 1/2)] to Ti at (1/2, 1/2, 1/2 + dz_{Ti}), O_I at (1/2, 1/2, dz_{O_I}), and O_{II} at (1/2, 0, 1/2 + $dz_{O_{II}}$) and (0, 1/2, 1/2 + $dz_{O_{II}}$), where $dz_{Ti} = 0.05$ Å, $dz_{O_I} = -0.09$ Å, and $dz_{O_{II}} = -0.06$ Å [21], thereby creating a dipole below 120°C, as shown in Figure 5a. The oxygen octahedra are slightly distorted in each of the ferroelectric

phases. The structural instability of ABO_3-type perovskites has been explained by the soft-mode concept [22, 23]. When a negative or positive electric field is applied on opposite faces of a crystal, the small Ti^{4+} ions are displaced up or down while the O^{2-} ions move down or up, resulting in the polarization switching.

$KNbO_3$ is in many ways qualitatively similar to $BaTiO_3$. It undergoes the same series of ferroelectric transitions in the same sequence as the temperature is lowered. The cubic–tetragonal transition occurs at 435°C; the tetragonal–orthorhombic at 225°C; and the orthorhombic–rhombohedral at −10°C. Lattice constants at orthorhombic phase are $a = 3.973$ Å, $b = 5.695$ Å, and $c = 4.036$ Å. The static structures determined in each of the ferroelectric phases show that the oxygen octahedron distortion is considerably less than in $BaTiO_3$ and that an essentially rigid oxygen cage is displaced relative to the reference K ion at each of the phase transitions [24].

$PbTiO_3$ undergoes a first-order transition at 493°C from cubic perovskite to a tetragonal ferroelectric phase isomorphic with tetragonal $BaTiO_3$. However, no additional transitions are reported. The displacement of Ti and O ions in the ferroelectric phase of $PbTiO_3$ is much greater than that of $BaTiO_3$. Room-temperature crystal structure of tetragonal $PbTiO_3$ was determined with the displacement of ions, $dz_{Ti} = 0.17$ Å and $dz_{O_I} = dz_{O_{II}} = 0.47$ Å [25]. Lattice constants in the tetragonal phase are $a = b = 3.904$ Å and $c = 4.151$ Å. The oxygen octahedra suffer no distortion in going to the ferroelectric phase, and, unlike in $BaTiO_3$ and $KNbO_3$, the oxygens and B-cations are shifted in the same direction in $PbTiO_3$ relative to the A-cations (see Figs. 5a and b). The large ionic shifts in $PbTiO_3$ lead to a particularly large room-temperature spontaneous polarization [26].

When Ti^{4+} ions in $PbTiO_3$ are partially replaced by Zr^{4+} with a molar ratio x, a solid solution of $PbTi_{1-x}Zr_xO_3$ [lead zirconate titanate (PZT)] binary system is formed. This system sometimes contains a number of other components. Substitution of Zr^{4+} for Ti^{4+} in $PbTiO_3$ reduces the tetragonal distortion and ultimately causes the appearance of another ferroelectric phase of rhombohedral symmetry [27, 28]. Figure 6 is

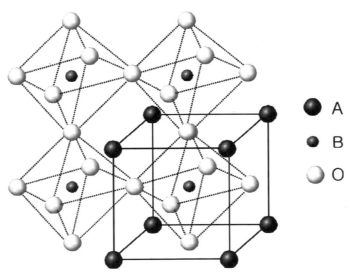

Fig. 4. A cubic ABO_3 perovskite-type unit cell and the three-dimensional network of BO_6 octahedra.

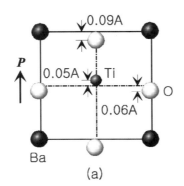

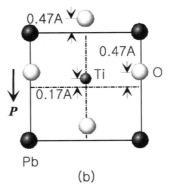

Fig. 5. A comparison of the unit cells of (a) $BaTiO_3$ and (b) $PbTiO_3$ with the origin at the A-cations in the tetragonal ferroelectric phase. The direction of spontaneous polarization is indicated by arrows.

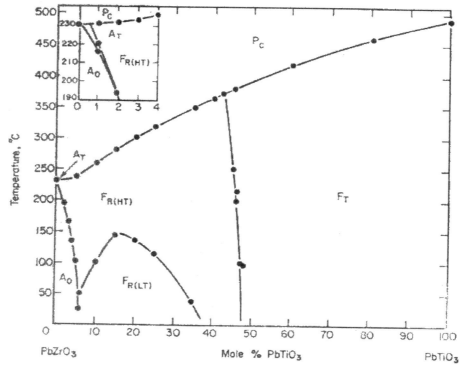

Fig. 6. The subsolidus phase diagram for $PbTiO_3$-$PbZrO_3$. Reprinted with permission from B. Jaffe, W. R. Cook, Jr., and H. Jaffe, "Piezoelectric Ceramics." Academic Press, London, 1971. Copyright 1971, Academic Press.

the temperature–composition $(T–x)$ phase diagram of the PZT system, where the T_C-line is the boundary between the cubic paraelectric phase and the ferroelectric phase [29]. The boundary between the tetragonal and rhombohedral phases is nearly independent of temperature and is called the *morphotropic boundary*. At room temperature, the morphotropic boundary is at the point Zr/Ti = 53/47. The rhombohedral ferroelectric phase actually divides into two phases; the low-temperature phase has a multiple rhombohedral cell [30]. In the region where Zr/Ti lies between 100/0 and 94/6, the solid solution is in an antiferroelectric orthorhombic phase. Physical properties of PZT near the morphotropic boundary are of special interest. The piezoelectric coefficient, electromechanical coupling coefficients, and dielectric constant pass through a maximum at the morphotropic boundary, whereas the mechanical quality and the coersive field pass through a minimum [29, 31, 32]. The remnant polarization also reaches a maximum value of about 47 $\mu C/cm^2$ for a rhombohedral composition near the phase boundary [32]. Since the position of the phase boundary is almost independent of temperature, the temperature dependence of the physical properties is not highly sensitive to small compositional variations.

2.1.2. Ferroelectrics with a Layer-Structured Perovskite

In recent years, a bismuth layer-structured ferroelectric, $SrBi_2Ta_2O_9$ (SBT), has become the most important material in the field of ferroelectric memory because of its low fatigue endurance and low switching voltage [10, 33]. SBT is a

member of the family of Aurivillius-structure–layered bismuth compounds [34–36] with the formula $Bi_2A_{m-1}B_mO_{3m+3} = [Bi_2O_2]^{2+}[A_{m-1}B_mO_{3m+1}]^{2-}$, with A = Na^+, K^+, Ba^{2+}, Ca^{2+}, Pb^{2+}, Sr^{2+}, Bi^{3+}, ... and B = Fe^{3+}, Ti^{4+}, Nb^{5+}, Ta^{5+}, W^{6+}, ... The pseudo-perovskite (perovskite-like) $[A_{m-1}B_mO_{3m+1}]^{2-}$ layers in the compounds alternate with bismuth-oxide layers, $[Bi_2O_2]^{2+}$. These layered compounds have mBO_6 octahedra between two neighboring Bi_2O_2 layers, which gives $(m-1)$ pseudo-perovskite ABO_3 units in a pseudo-perovskite layer. SBT is the structure of $m = 2$, whereas bismuth titanate, $Bi_4Ti_3O_{12}$ (BTO), is that of $m = 3$ (Figs. 7a and b).

SBT undergoes a ferroelectric phase transition at about $T_C = 335°C$ with a slight orthorhombic distortion from a high-temperature tetragonal structure [37]. It has been customary to define the a-axis as a polar axis direction of ferroelectricity. Lattice constants at the orthorhombic phase are $a = 5.512$ Å $\cong b$ and $c = 25.00$ Å. The main contribution to the polarization arises from the off-center position of the Ta^{5+} at the B-site ion relative to its octahedron of surrounding oxygens [38]. The spontaneous polarization is also derived by several displacing mechanisms involving rigid displacement of the Sr^{2+} ions at the A-site of a pseudo-perovskite unit along the ferroelectric a-axis, as well as rotations of an octahedron around the polar a-axis and the c-axis.

BTO transforms from the tetragonal to the ferroelectric monoclinic phase at about $T_C = 675°C$ [39]. Ferroelectricity in BTO involves very large polarization along the a-axis but small

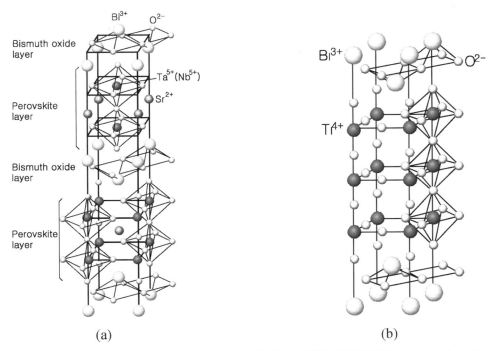

Fig. 7.   The structures of bismuth layer-structured ferroelectrics (a) SBT and (b) BTO. The structure shown for BTO is half of the BTO unit cell.

polarization along the c-axis, because the polarization vector of BTO is tilted in the monoclinic a–c-plane with the major component along the a-axis. Although the symmetry of BTO at room temperature is in fact monoclinic [39], the crystal structure can be regarded as nearly orthorhombic because the monoclinic plane coincides with the orthorhombic plane [35, 36, 40]. Lattice constants of the pseudo-orthorhombic phase are $a = 5.411$ Å, $b = 5.448$ Å, and $c = 32.83$ Å.

2.1.3. Ferroelectrics with a Distorted Perovskite Structure: Lithium Niobate and Lithium Tantalate

Lithium niobate ($LiNbO_3$) and lithium tantalate ($LiTaO_3$) are ferroelectric below 1210°C [41, 42] and 620°C [43], respectively. They have similar crystal structures. In view of their excellent piezoelectric, pyroelectric, and optical properties, they are well-known ferroelectric crystals and have been extensively studied for various device applications. Their structure is shown in Figure 8a [42] and may be regarded as a strongly distorted perovskite structure. The structure of both crystals is trigonal (rhombohedral) in the ferroelectric phase at room temperature. Sometimes, it is more convenient to choose a hexagonal cell when describing their structure. At room temperature, lattice parameters for hexagonal unit cell are $a_H = 5.148$ Å and $c_H = 13.863$ Å in $LiNbO_3$ [41], and $a_H = 5.154$ Å and $c_H = 13.784$ Å in $LiTaO_3$ [42]. Figure 8a shows the sequence of distorted octahedra along the polar c-axis with Nb (or Ta) at the origin. The Nb-occupied octahedron shares a face with the second, vacant octahedron and also with the preceding Li-occupied octahedron. The cation arrangement follows the sequence Nb (or Ta), vacancy, Li, Nb (or Ta), vacancy, Li, ... In the paraelectric phase, Nb^{5+} ions lie between the

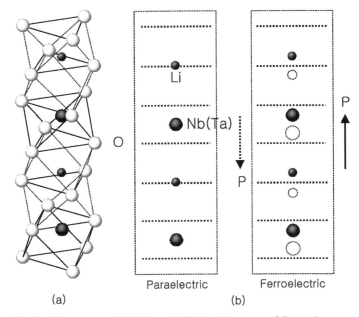

Fig. 8.   The structure of $LiNbO_3$ and $LiTaO_3$: (a) sequence of distorted oxygen octahedra along the polar c-axis; (b) schematics for the locations of cations and the oxygen layers.

octahedra while Li^+ ions are located in the plane defined by oxygen atoms at the surface of the octahedron. Below Curie temperature, observable cation displacement occurs relative to the oxygen layer along the c-axis. On transition to the ferroelectric phase, Li^+ ions leave the common plane of the oxygen octahedra while Nb^{5+} ions leave the center of the octahedra. The direction of displacement of Nb^{5+} ions coincides with that of Li^+ ions (Fig. 8b).

2.1.4. Ferroelectrics with a Tetragonal Tungsten-Bronze Structure

There are many tungsten-bronze-type ferroelectric crystals. Some of the niobate ferroelectric crystals of tungsten-bronze structure have many useful properties—in particular, large electro-optic effects or large optical nonlinearities [44, 45]—and thus have great potential as materials for laser modulation and laser frequency multiplication.

A tetragonal tungsten-bronze structure is constructed from straight oxygen octahedra chains that lie parallel to the four-fold axis [46]. As in perovskite, the octahedra of the adjacent chains are connected through their corners. The oxygen octahedral framework of tungsten-bronze–type niobates is shown in Figure 9. The structure has tetragonal, pentagonal, and triangular tunnels and corresponding vacancies. The general chemical formula of the tungsten-bronze compounds can be written as $(A_1)_4(A_2)_2(C)_4(B_1)_2(B_2)_8O_{30}$, where A_1, A_2, and C are the ions in the pentagonal, tetragonal, and triangular tunnels, respectively. B_1 represents the ions in the less strongly distorted octahedra; B_2, those in the more strongly distorted octahedra (see Fig. 9). The sites A_1, A_2, C, B_1, and B_2 can be either partially or fully occupied by different cations. In the case of niobates or tantalates, the B_1 and B_2 sites are occupied by Nb^{5+} or Ta^{5+}, while the A_1, A_2, and C sites are occupied by alkaline earth ions, alkali ions, or both. Usually the triangular tunnels are unoccupied, and the formula is $(A_1)_4(A_2)_2(B_1)_2(B_2)_8O_{30}$ or $A_6B_{10}O_{30}$. Some AB_2O_6 compounds $(A_5B_{10}O_{30})$ have other tunnels in addition to the triangular tunnels that have vacancies. $(Sr_{1-x}Ba_x)Nb_2O_6$ (SBN) is one example. This material has a wide range of solid solubility along the $SrNb_2O_6$–$BaNb_2O_6$ binary combination [47]. Many other ferroelectrics of tetragonal tungsten-bronze structure are described in the literature [1, 48]. Generally, the origin of ferroelectricity lies in the displacement of the cations at A and B sites from the centrosymmetric oxygen planes in the paraelectric phase with temperature decrease below T_C. At this stage, spontaneous polarization occurs in the direction of the c-axis.

2.1.5. Ferroelectrics with a Hexagonal Manganite Structure

Rare-earth and yttrium manganites with the general formular $RMnO_3$ have a hexagonal lattice for R with smaller ionic radius (R = Ho, Er, Tm, Yb, Lu, Y), whereas the compounds with R of larger radius (R = La, Ce, Pr, Nd, Sm, Eu, Gd, Tb, or Dy) are orthorhombic [49–51]. The hexagonal $RMnO_3$ compounds belong to the class of ferroelectromagnet materials [52] characterized by the coexistence of magnetic and ferroelectric ordering, while the orthorhombic compounds are not ferroelectric [53]. The hexagonal lattice is constructed from layers of corner-sharing MnO_5 bipyramids with a triangular base of nonequivalent oxygen atoms [50], as shown in Figure 10 for the ferroelectric phase (below 640°C) of $YMnO_3$. The hexagonal lattice parameters at room temperature are $a = 6.130$ Å and $c = 11.505$ Å. The MnO_5 bipyramids form a two-dimensional array that is perpendicular to the c-axis. Layers of the bipyramids are separated from each other by layers of yttrium ions or ions of rare-earth elements. The sixfold c-axis of these ferroelectrics is the polar axis. In ferroelectric phase, the axial Mn—O bonds of bipyramids are slightly tilted with respect to the c-axis by small rotation around axes that are parallel to one of the base-triangular sides and pass through the Mn site [54].

Oxygen octahedra

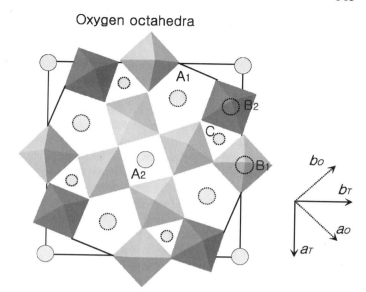

Fig. 9. Schematic diagram showing octahedral framework of the tungsten-bronze structure on the (001) plane. The axes a_T and a_O are for tetragonal and orthorhombic unit cells, respectively.

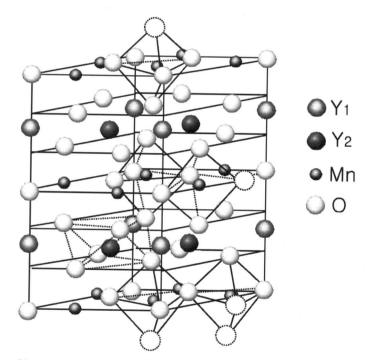

Fig. 10. Unit cells of $YMnO_3$ in its ferroelectric phase. Empty circles are atoms outside the unit cell.

2.2. Physical Properties of Ferroelectric Oxides

The physical properties of ferroelectric materials are related to dielectric, elastic, piezoelectric, pyroelectric, ferroelectric, acoustic, optical, electro-optic, nonlinear optic, and other parameters. These physical parameters are affected by structural symmetry of a crystal [55]. Depending on their symmetry with respect to a point (crystal classes), crystals are classified into 32 crystal classes, 11 of which have a center of symmetry. Of the remaining 21 noncentric crystal classes, all except one class exhibit electrical polarity when subject to stress (and also its converse, the production of strain by application of an electric field). The effect is termed the *piezoelectric* effect and is linear, with reversal of the stimulus resulting in a reversal of the response. Among the 20 piezoelectric crystal classes, 10 crystal classes have only one unique polar axis. Such crystals are called *polar* crystals, since they display *spontaneous polarization* or electric moment per unit volume. The spontaneous polarization is in general temperature dependent, and a variation of the charge density can be detected by observing the flow of charge to and from the surfaces on change of temperature. This is the *pyroelectricity* in crystals that exhibit spontaneous polarization. Most pyroelectric crystals exhibit spontaneous polarization, P_s, in a certain temperature range, and the direction of P_s can be reversed by an external electric field. Such crystals are called *ferroelectric* crystals.

Ferroelectric materials have piezoelectric, pyroelectric, and ferroelectric properties. Ferroelectric materials are also very often dielectrics. Devices using the dielectric, piezoelectric, and pyroelectric properties of ferroelectrics have received much attention. However, in recent years, the interest in the ferroelectric nature of the material—namely, the large reversible spontaneous polarization—has increased. Ferroelectrics are prime candidates for nonvolatile digital memory devices, since their bistable polarization offers the potential for binary memory and they can remain stable after removal of electrical power. Although attempts to make memory devices were made in the early 1950s [56, 57], practical problems (e.g., reliable switching characteristics, fatigue) diminished the enthusiasm for these after a few years. In the past decade, the situation has changed, due largely to the recent development of thin film technologies.

Ferroelectrics have also been exploited in the field of optics. The crystalline anisotropy of ferroelectrics generally gives rise to a large optical birefringence. In addition, ferroelectric materials (including crystals, ceramics, and thin films) have electro-optic, nonlinear optic, photorefractive, and photoelastic effects. Based on these useful properties of ferroelectric materials, a number of useful optical devices have been developed, including spatial light modulators, integrated optical waveguides, light-beam deflectors, optical frequency multipliers, and holographic optical strorage media. Ferroelectric materials also have played an important role in the development of laser techniques and optical communications.

The polarization of ferroelectrics can be described by the relation

$$\mathbf{D} = \mathbf{D}(\text{linear}) + \mathbf{D}(\text{nonlinear})$$

Generally, both parts of the electric displacement vector $\mathbf{D}$ depend on the applied field, the stress, and the crystal temperature. The linear part is related to ferroelectrics' dielectric, piezoelectric, and pyroelectric properties. The nonlinear part includes the electro-optic, photo-elastic, and thermo-optic effects. The photoelastic effect is of significant practical importance, since it allows for the interaction of acoustic and optic waves and enables the acousto-optic modulation of light. The thermal term is also important, because it gives rise to the thermal self-focusing of light beams. However, we restrict attention to the electro-optic effect.

2.2.1. Dielectric, Piezoelectric, and Pyroelectric Properties of Ferroelectrics

The large dielectric and piezoelectric constants of ferroelectrics immediately made these materials attractive candidates for a variety of applications. For most such applications, the dielectric constant and dielectric loss are important parameters. The dielectric constant is defined by the relation between electric field, E, and electric displacement, D. The relative dielectric constant, ε_r, is a function of measuring frequency, ω, and is written as a complex number,

$$\varepsilon_r(\omega) \equiv \frac{\varepsilon(\omega)}{\varepsilon_0} = \varepsilon'(\omega) - j\varepsilon''(\omega)$$

where $\varepsilon'(\omega)$ and $\varepsilon''(\omega)$ are the real part and the imaginary part of the dielectric constant, respectively, and ε_0 is the dielectric permittivity of vacuum. The latter represents the dielectric loss. Instead of $\varepsilon''(\omega)$, $\tan\delta$ is most frequently used by engineers to express this loss; its definition is

$$\tan\delta = \frac{\varepsilon''}{\varepsilon'}$$

Generally, dielectric permittivity is represented by a tensor, $\varepsilon_{ij} = \partial D_i / \partial E_j$, because of the crystalline anisotropy and is dependent on temperature. For instance, by using the *a*-cut (or the *c*-cut) plate of an anisotropic tetragonal crystal, the dielectric constant $\varepsilon_{11}/\varepsilon_0$ (or $\varepsilon_{33}/\varepsilon_0$) is obtained. If the dielectric constant is measured at a far lower frequency than the sample's piezoelectric resonance frequency, then the "free" dielectric permittivity (the case of constant stress), ε_{ij}^T, is obtained. On the other hand, if the dielectric constant is measured at a far higher frequency than the sample's piezoelectric resonance frequency, then the "clamped" dielectric permittivity (the case of constant strain), ε_{ij}^S, is obtained.

For piezoelectric applications, several properties of ferroelectric materials are important. These include the dielectric constant, the piezoelectric constant, the elastic constant (coefficient), and the electromechanical coupling factor. The piezoelectric and elastic constants are defined by the piezoelectric strain equations and Hook's law, respectively. The electromechanical coupling factor, k_p, is defined as the conversion ratio of electric energy to mechanical energy in a specific vibration mode of a piezoelectric vibrator; k_p can be expressed in terms of the other three constants.

The piezoelectric constant, d_{ij}, is defined by the tensor relations between dielectric displacement **D**, stress **T**, strain **S**, and electric field **E** as

$$d_{ij} = \frac{\partial D_i}{\partial T_j} = \frac{\partial S_i}{\partial E_j}$$

A high d constant is desirable for materials intended to develop motion or vibration, such as sonar or ultrasonic transducers. Another frequently used piezoelectric constant is g, which gives the field produced by a stress. The g constant is related to the d constant by the permittivity

$$g = \frac{d}{\varepsilon'} = \frac{d}{\varepsilon_r \varepsilon_0}$$

A high g constant is desirable in materials intended to generate voltages in response to a mechanical stress, as in a phonograph pickup. Additional piezoelectric constants that are used only occasionally are e, which relates stress to field, and h, which relates strain to field. Rigorous development of the relationships can be found elsewhere [58, 59]. The elastic coefficient s_{ij} is defined as $s_{ij} = \partial S_i / \partial T_j$.

Ferroelectric crystals can develop an electric charge when heated uniformly, owing to a change in magnitude of the spontaneous polarization with temperature. Under the conditions of constant field and constant stress, the pyroelectric coefficient, p_i, in ferroelectrics is equal to the negative temperature derivative of the spontaneous polarization, P_s:

$$p_i^{S,E} = -\left(\frac{dP_s}{dT}\right)_{S,E}$$

where the minus sign makes p_i positive when dT is positive. The $p_i^{S,E}$ is called the *first* pyroelectric coefficient. If thermal expansion of the sample occurs when the sample is being heated, then an additional change of polarization may be caused by the piezoelectric effect,

$$p_i^{T,E} = p_i^{S,E} + d_{ij}\alpha_{ij}^E / s_{ij}^E$$

where α_{ij}^E is the thermal expansion coefficient and the second term on the right-hand side of the equation is the *second* pyroelectric coefficient. Properties of some important ferroelectric crystals and ceramics can be found in the literature [1, 48].

2.2.2. Optical and Electro-Optical Properties of Ferroelectrics

Ferroelectric crystals are optically anisotropic materials that can be classified as optically uniaxial ($n_1 = n_2 = n_o$ and $n_3 = n_e$, see below) crystals and optically biaxial ($n_1 \neq n_2 \neq n_3$) crystals. Optically uniaxial crystals belong to tetragonal, trigonal, or hexagonal crystal classes, whereas optically biaxial crystals belong to orthorhombic, monoclinic, or triclinic crystal classes. In the case of ferroelectric ceramics, optical properties are isotropic due to the random grains. In the visible range, the principal refractive indices $n_i = (\varepsilon_i / \varepsilon_0)^{1/2}$ and the optical anisotropy of a crystal is characterized by an index-ellipsoid (or

indicatrix) defined as

$$\frac{x^2}{n_1^2} + \frac{y^2}{n_2^2} + \frac{z^2}{n_3^2} = \sum \varepsilon_0 \kappa_{ij} x_i x_j = 1$$
$$(x_1 = x, \; x_2 = y \text{ and } x_3 = z)$$

where $\kappa = \varepsilon^{-1} = \varepsilon_0/\mathbf{n}^2$ is the dielectric impermeability tensor and a coordinate system (x, y, z) is chosen to coincide with the three principal axes of a crystal. Usually high values of optical relative dielectric constants ($\varepsilon/\varepsilon_0$) in ferroelectrics give high refractive index values.

When a steady electric field **E** is applied, the index ellipsoid is modified. A linear electro-optic (Pockels) coefficient r_{ijk} and a quadratic electro-optic (Kerr) coefficient s_{ijk} are conventionally defined by the relation

$$\Delta\kappa_{ij} \equiv \kappa_{ij}(E) - \kappa_{ij}(0) = r_{ijk}E_k + s_{ijk}E_j E_k$$

where the frequency of the applied electric field lies below the vibrational frequencies of the crystal. Higher orders than quadratic are neglected in the expansion. The electro-optic effect is the change in the refractive index resulting from the application of a dc or low-frequency electric field. The number of subscript indices in r_{ijk} can be reduced; i.e., the first and second indices are replaced by a single index running from 1 to 6, while the third index is fixed. A plane wave propagating through the crystal polarized linearly along one of the principal axis of the index ellipsoid emerges from the crystal as a linearly polarized wave with a field-dependent phase due to the electro-optic effect. The linear electro-optic coefficient and the transverse effective electro-optic coefficient of some important ferroelectric crystals and ceramics are well summarized in the literature [1, 48]. The transverse electro-optic effect is defined as $r_c = r_{33} - (n_1^3/n_3^3)r_{13}$ when the field is applied in the direction of the uniaxial optical axis and the propagation of the light beam is perpendicular to the axis.

3. DEPOSITION OF FERROELECTRIC OXIDE THIN FILMS

In the last decade, unprecedented advances have been made in the processing of perovskite oxide thin film materials, including high-temperature superconductors and ferroelectrics, due to growing demand for compatibility with integrated circuit technology. Desirable features of a deposition technique for producing multicomponent oxide thin films and heterostructures include (i) the capability of producing stoichiometric films, (ii) a high deposition rate, (iii) the ability to produce conformal deposition, and (iv) scalability to cover large-area substrates with uniform composition and thickness. The common thin film processing techniques for these multicomponent systems are classified into physical vapor deposition, chemical vapor deposition, and chemical solution deposition. Although these methods have not yet been optimized, remarkable understanding of many basic processes for the different deposition techniques has been obtained.

3.1. Physical Vapor Deposition

There have been many review papers and book chapters on physical vapor deposition (PVD) for ferroelectric thin film growth. They include general description of PVD methods [60–62], ferroelectric thin film deposition and integration [2, 63–75], and device applications of ferroelectric thin films [72, 75]. Currently, a large part of research is directed at establishing suitable synthesis methods for ferroelectic thin films, and at investigating process–microstructure property relationships. PVD techniques, such as plasma/ion-beam sputter deposition and pulsed laser deposition (PLD), are used extensively in synthesizing multicomponent ferroelectric thin films. In this review on PVD techniques, we discuss progress in ferroelectric thin film research, with particular emphasis on ferroelectric/dynamic random access memory(FRAM/DRAM) applications. Because the conditions for the integration of ferroelectric materials in Si-based memory technology are so strict, PVD methods developed for memory applications can be easily applied to other applications, such as electro-optical, infrared, microelectromechanical systems (MEMS), and microwave applications.

3.1.1. Current Issues in Ferroelectric Thin Film Research for Ferrodynamic Random Access Memory

To obtain high-density FRAMs, an one-transistor/one-capacitor (1T-1C) cell with a stacked capacitor structure is needed [76]. In terms of technologic aspects, there are some major reliability concerns, including uniformity control, thermal stability, fatigue, and aging of the FRAMs. Several issues must be considered for integration of ferroelectric capacitors into a commercially viable memory technology, including (i) an appropriate electrode technology, (ii) a suitable ferroelectric layer, (iii) scalable film-deposition processes with good conformal coverage of nonplanar surfaces, (iv) appropriate lithography and reactive ion-etching processes to produce submicron patterns for high-density memories, and (v) suitable back-end and encapsulation processes [72, 75].

Figure 11 shows a cross-section of an integrated ferroelectric memory cell with a capacitor stacked over a contact plug. Deposition processes expose the bottom electrode to a high-temperature oxygen-containing atmosphere. Any oxidation of the electrode material to form an insulation layer can seriously degrade the capacitor performance. To avoid oxidation, the electrode must be an oxidation-resistant metal (e.g., Pt), a highly conductive metal oxide, or a metal that forms such a conductive oxide. Pt is the most commonly used electrode. The adhesion layer (e.g., Ti) is commonly deposited on the substrate before the electrode layer. Interaction between Pt and Ti must be controlled, or the Ti can diffuse into the Pt/ferroelectric interface and negatively influence the properties of the ferroelectric layer. Pt and Ti will interdiffuse readily at crystallization temperatures of ferroelectrics in inert atmospheres. In an oxygen atmosphere, there is a competing diffusion of oxygen through the Pt. Also, TiO_2 is formed within the electrodes, although it

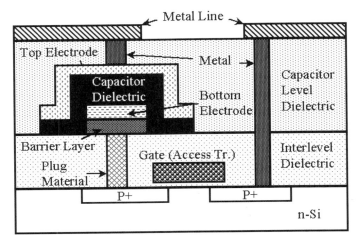

Fig. 11. Schematic of a high-density DRAM showing electrical contact between bottom electrode and a plug, via a barrier layer. Reprinted with permission from H. N. Al-Shareef and A. I. Kingon, in "Ferroelectric Thin Films: Synthesis and Basic Properties," Chap. 7 (C. P. de Araujo, J. F. Scott, and G. W. Taylor, Eds.). Gordon and Breach, Amsterdam, 1996. Copyright 1996, Gordon and Breach Publishers.

appears to remain relatively localized. The final stable profile depends on the detailed thermal cycle, as well as on the relative Ti to Pt thickness.

Deposition of capacitor-level dielectrics over the completed capacitors is required for isolation before the metal interconnect deposition. But this process encounters one of the most fundamental problems in the integration of ferroelectrics, that ferroelectrics are easily damaged by exposure to hydrogen.

Above all of these integration issues, long-term reliability issues, such as fatigue and retention of FRAM devices, which are related to defect chemistry in ferroelectric thin films, are most important for commercial applications [77]. Fatigue is defined as the loss of polarization in ferroelectrics resulting from repeated polarization switching. Retention is the ability to retain a certain level of polarization in ferroelectrics after a given duration from the write operation. These long-term reliability issues are partly overcome by using conducting oxide electrodes and Bi-layered structure ferroelectrics.

3.1.2. PVD Techniques

Any thin film deposition process involves three main steps: (i) production of the appropriate atomic, molecular, or ionic species; (ii) transport of these species to the substrate through a medium; and (iii) condensation on the substrate, either directly or via chemical and/or electrochemical reaction, to form a solid deposit [62]. Formation of a thin film takes place via nucleation and growth processes.

The growth process can be summarized as consisting of a statistical process of nucleation, surface-diffusion controlled growth of the three-dimensional nuclei, formation of a network structure, and this structure's subsequent filling to give a continuous film. Depending on the thermodynamic parameters of the deposit and the substrate surface, the initial nucleation and growth stages may be classified as (i) island type (Volmer–Weber type), (ii) layer-by-layer type (Frank–van der Merwe type), or (iii) mixed type (Stranski–Krastanov type).

There are many deposition techniques for the material formations. Vapor deposition methods can be classified into two main reaction mechanisms: PVD and chemical vapor deposition (CVD). Some factors that distinguish PVD from CVD are [78]: (i) reliance on solid or molten sources, (ii) physical mechanisms (evaporation or collisional impact) by which source atoms enter the gas phase, (iii) reduced pressure environment through which the gaseous species are transported, and (iv) general absence of chemical reactions in the gas phase and at the substrate surface (reactive PVD processes are exceptions).

Conceptually, the simplest PVD method is thermal evaporation, by which material from the bulk is transported onto a thin film. Thermal deposition techniques are defined as those in which the kinetic energies of the impinging particles are the order of 0.1 eV. This technique was the first method used for film growth, and it is still used today in a refined manner in the form of molecular beam epitaxy (MBE). However, established thermal growth techniques may be unable to accommodate two or more complex chemical systems on the same substrate; for example, the growth conditions needed for one system can result in the thermal decomposition of the other. The ability to deposit films under nonthermal conditions may alleviate such difficulties. Among the PVD techniques, sputtering, ion-beam, and PLD techniques are nonthermal methods. The nonthermal methods have proven advantageous for congruent transfer (i.e., where the elemental ratios in the original material are preserved in the deposited film) of multicomponent materials which, if thermally heated, would decompose before sufficient vapor pressure was reached [79].

One advantage of PVD over other deposition techniques is its ability to grow oriented or epitaxial ferroelectric thin films. Since ferroelectric materials are anisotropic in nature, orientation control of depositing thin films is important to optimize the physical properties of ferroelectrics for various applications. Also, unusual or artificial heterostructures of ferroelectric materials can be deposited by PVD methods. In nonthermal PVD, the dynamics of thin film formation cannot be described by equilibrium thermodynamics. Thus it is possible to grow unusual heterostructures of ferroelectric materials that are not obtainable in single-crystal form by nonthermal PVD methods. Multilayer or superlattice structures of ferroelectric materials are very interesting for applications as well as for scientific research.

Figure 12 shows a schematic of the PVD process. High energy (①) is applied on the target (②) surface. The target material is evaporated or sublimed (③), makes plasma (④), and then arrives on the substrate (⑤) surface. In the deposition vacuum chamber, gases are introduced. Argon is used as a sputtering gas in sputtering deposition method. Argon gas is ionized by the high energy rf electromagnetic source and incident on the target surface for the sputtering. For the deposition of oxides, oxygen gas is also introduced. Since these gases col-

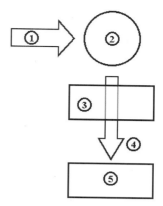

Fig. 12. Schematics of PVD.

lide with the species evaporated or ablated from the target, the gas partial pressures are important parameters of the deposition.

Energy sources (①), such as high-frequency electromagnetic energy in sputtering or high-energy pulse laser beam in PLD, are applied on the surface of the target (②). Various physical deposition methods are classified by these energy sources. The applied energy scales vary from a few eV to MeV ranges, depending on the method used.

3.1.2.1. Comparison of PVD Techniques

Each PVD technique has its own strengths and weaknesses. The production of device-compatible high-quality films and heterostructures with specific properties (i.e., large remanent polarization value, fatigue-free, long retention time, etc.) is required for each technique. A process compatible with integrated device processing is also required.

Evaporation technology is one of the oldest PVD techniques used for depositing thin films. A vapor is first generated by evaporating a source material in a vacuum chamber, then transported from the source to the substrate and condensed to a solid film on the substrate surface. MBE is a sophisticated, finely controlled method for growing epitaxial films in an ultrahigh vacuum at 10^{-8} to 10^{-10} torr. The most widely studied materials are epitaxial layers of III–V semiconductor compounds and SiGe compounds. However, problems associated with MBE that make it difficult to use as manufacturing process include relatively low deposition rates, a questionable capability for covering large-area substrates with uniform films, difficulty of integration with other processing steps, and the relatively high cost of equipment. This thermal technique has not been extensively used to investigate the synthesis of ferroelectric films and heterostructures. New MBE methods (e.g., laser MBE) have been proposed to grow ferroelectric thin films.

A promising technique is sputtering. Although differing in their details, all sputtering techniques use a process whereby material is dislodged and ejected from the surface of targets by momentum exchange associated with surface bombardment by energetic particles. The target is therefore eroded, or "sputtered," and a beam of target atoms with energy of 10–100 eV is produced. The generated vapor of the material is then deposited on the substrate. The bombarding energetic particles are generally ions of a heavy inert gas generated by plasma discharge. Deposition of ferroelectric films is achieved by sputtering targets exposed to a dc or rf plasma discharge in a high-vacuum chamber backfilled with an inert (commonly Ar) or oxidant (generally O_2) gas, or a mixture of both, to a pressure of 0.5–200 mtorr. Most studies on plasma sputter deposition of ferroelectric films have been performed in the 5–50 mtorr pressure range. The substrates are positioned in front of the target so that they intercept the flux of sputtered atoms. The plasma–target–substrate interaction has been known to affect the deposition process, and thus various target–substrate geometries have been used. Unlike evaporation, sputtering processes are very well controlled and are generally applicable to all materials, including metals, insulators, semiconductors, and alloys. Various sputtering methods have been developed. However, sputtering suffers from problems of impurities, mainly because (i) a pressure range of about 0.5–200 mtorr of working gas is need to ignite the electrical discharge, and (ii) the sputtering plasma also erodes surfaces other than the intended target.

The PLD method is relatively a new deposition method developed for the deposition of high-temperature superconductor materials. This technique has become one of popular thin film deposition techniques, primarily in oxide thin film deposition applications, through the demonstration of the growth of very high quality thin films and heterostructures. PLD is conceptually and experimentally a simple technique. Thin films of the desired material and composition are deposited by ablation of the materials from solid target surface by a short laser pulse. Typically, an ultraviolet (193- or 248-nm) excimer laser or frequency-tripled YAG laser is used with a pulse width of a few nanoseconds. The instantaneous interaction of the pulsed laser beam with the target produces a plume of material that is very forward directed and is transported toward the heated substrate, which is placed directly in the line of this plume. The impact of the laser beam on the target surface results in various complicated processes, including ablation, melting, and evaporation of materials and production of a plasma due to excitation and ionization of the species ejected from the target by the laser photons. All of these processes are triggered by the transformation of electromagnetic energy into electronic excitation, followed by a transformation into thermal, chemical, and mechanical energy. The spatial and temporal evolutions of the ablated plume are quite complicated and are not discussed in this review.

Under appropriate deposition conditions, the composition of the film is the same as that of the target within a small central area of the film, but is nonstoichiometric outside this central area. The ability to produce stoichiometric films using multicomponent oxide targets is perhaps the main intrinsic advantage of PLD in comparison with other techniques, such as sputtering. The compositional congruency in the central part of the ablated plume is translated to the film depending on the substrate temperature, the substrate–film interface chemistry, and the oxygen partial pressure during deposition. Since useful ferroelectric materials, such as $BaTiO_3$, are mostly oxides and they have very complicated compositions, high oxygen partial pressure is needed for the deposition. On the other hand, some PVD methods, such as MBE, which needs ultra-high vacuum, cannot be used without modification.

3.1.2.2. Sputtering Methods

One of the most elementary means of forming a plasma is in the form of a dc diode discharge [61]. To allow the use of insulating materials as either sputtering targets or as samples, and to increase the level of ionization in the plasma, rf power is often applied to the electrodes in a diode discharge. While most commercially available systems operate at a frequency of 13.56 MHz because of government communications regulations, plasmas can be powered at any practical frequency up into the GHz range.

In the practical realm of power supplies, one must deal with the various impedances in the plasma, transmission lines, and power supply. In many cases, the plasma is shown as an electrical circuit, with an effective resistance, capacitance, and inductance. Most power supplies are designed with an output impedance of 50 ohms. The practical problem, then, is how to couple the highest fraction of this power into the plasma. The most common approach is to use a "matchbox," which consists of two tunable capacitors and a fixed inductor. The matchbox impedance is adjusted such that half of the total power is applied to the plasma and the remaining half is lost in the matchbox. Most matchboxes are configured with self-tuning circuits, which adjust the level of the two capacitors to find the best impedance match.

The class of sputter sources called magnetrons has a magnetic field of about 50–500 gauss parallel to the target surface, which in combination with the electric field causes the secondary electrons to drift in a closed circuit, or "magnetron tunnel," in front of the target surface. This electron confinement significantly increases the efficiency; as a result, a magnetron can operate at low pressures and low voltage. The current density at the cathode of a magnetron is peaked where the magnetic field lines are tangent to the surface of the cathode. Therefore, the erosion of the target is nonuniform [80].

The plasma sputtering deposition technique has been developed to produce multicomponent oxide thin films, including high-temperature superconductors, ferroelectrics, and electrooptic materials. In this technique, the interaction of plasmas with targets and substrates is important. The plasma–material interaction and transport of species through the plasma determine the composition, microstructure, and properties of films. The deposition of ferroelectric materials by sputtering methods, entails several problems, however. Stoichiometry control is one such problem. Since ferroelectric materials are complex in composition and structure, the control of composition and structure is important in obtaining ferroelectric properties (which are very sensitive to both composition and structure) for device applications. Several problems are related to controlling film stoichiometry. The well-known preferential sputtering phenomena (i.e., different sputtering rates of compositional elements) occurring during ion bombardment of multicomponent materials, particularly oxides, can alter the surface composition of the target and result in films with different stoichiometry than that of the target. Some components (e.g., Pb and Bi) are highly volatile. For sputter-deposited PZT films, enrichment of Pb in the PZT target is necessary to compensate for the preferential sputtering. The development of an ion bombardment–induced surface topography can also affect the sputtered flux and, consequently, the film thickness and composition uniformity. Moreover, the impact of plasma ions on multicomponent oxide targets results in the emission of a large amount of O^- ions, which are accelerated by the plasma sheath in front of the target, acquiring enough kinetic energy to produce, upon impact on the films, resputtering and compositional changes (the "negative ion effect") [65]. In addition to the negative ions effect, films are also bombarded by secondary electrons emitted from the oxide targets, and these electron can also produce compositional alterations through the electron bombardment–induced oxide breakdown phenomenon.

A second problem is related to the stoichiometry problem. Since the sputtering rates of composition elements of the target vary, the composition of the target is changed before and after deposition. As a result, controlling film composition for a long enough period is incredibly difficult. Another problem related to the target is the fact that the ferroelectric target is easily broken during sputtering because of thermal damage. During the sputtering deposition, energetic particles collide against the target, heating the target. In the sputtering gun, a cooling system helps solve this problem. But although this cooling is effective for metal or semiconductor target materials, insulating target materials, including ferroelectric materials, are generally also thermally insulating, so that the heat generated at the ferroelectric material surface cannot be easily cooled down by a water cooling system. The difference in the temperature of the target surfaces induces mechanical stress and can break the target into pieces, which can change the optimum deposition conditions.

Despite these problems, sputtering deposition methods for the ferroelectric thin films have been studied and developed for decades. Sputtering method is a well-developed PVD method and viable in CMOS technologies for the deposition of polysilicon, silicon nitrides, and metals. Also, sputtering is a scalable deposition method. Silicon wafers as large as 12 inches are used in recent Si technology. Ceramic targets of well-known ferroelectric materials are now commercially available. In comparison with PLD, targets as large as about 4 inches in diameter are used in sputtering deposition.

Magnetron sputtering has be used to produce PZT films with controlled stoichiometry and properties, using disk-shaped targets consisting of metallic Pb, Zr, and Ti sectors of different sizes to control the amount of each material deposited on the substrate at a particular temperature [81]. Investigations into the basic plasma–target–substrate interaction and their effects on the deposition process have revealed that the growth using a particular magnetron sputtering deposition method was governed by three main processes: (i) formation of a reproducible oxide layer on the target surface, (ii) the stability of oxide species formed during transport through the plasma toward the substrate, and (iii) the nucleation and growth of the film on the substrate surface.

From a practical stand point, deposition pressure, electric source power, and the ratio between argon and oxygen partial pressures are the most important deposition conditions affecting thin film properties such as microstructure, composition, deposition rate, and electrical properties. Although information about the basic sputtering process for simple materials such as metals is readily available, the sputtering process for ferroelectric materials is not yet well established. Optimal deposition conditions differ among studies.

A reactive sputtering method, the ion-beam sputter deposition (IBSD) technique, has been used to synthesize multicomponent oxide ferroelectric thin films [67, 69, 70]. The IBSD technique is a suitable alternative to plasma sputtering deposition because many of the aforementioned undesirable effects (e.g., negative ion effect, impurity introduction) are not present or are much smaller in this technique. The ISBD technique involves single ion beam and multiple ion beam target systems. In a single-beam system, a rotating target holder sequentially positions elemental material targets in front of the sputter beam to control film stoichiometry. The surface topographic phenomenon, which can influence film composition and thickness uniformity, can be minimized by rotating the target continuously during deposition. The multiple ion beam target system is a quite flexible IBSD technique, although the need for simultaneous control of various sources makes the hardware assembly complicated. Since different single-element targets are sputtered individually by independently operated ion sources, the muliltiple IBSD has the capability of varying and controlling the composition of depositing films. An advantage of using elemental target materials over the use of multicomponent oxide targets is that the preferential sputtering phenomenon is eliminated in sputtering of single elements, resulting in better control of film composition. Compared to plasma sputtering deposition techniques, the IBSD offers several advantages, including (i) the ability control film composition, (ii) reduction of impurities, (iii) better control of deposition rate, and (iv) no negative ion effect.

3.1.2.3. Pulsed Laser Deposition

PLD has been used extensively by many groups to produce multicomponent oxide thin films. The versatility of this technique is demonstrated by the fact that it has been used to deposit close to 128 different materials in thin film form [82]. The impact of a laser beam on the target produces localized melting and resolidification, which for some target materials produce topographical and compositional changes on the impacted area of the target, which affects the reproducibility of the film characteristic after several hundred pulses. This problem can be solved by continuous rotation of the target and/or laser beam scanning on the surface to permanently expose a fresh area to the laser beam.

PLD has many merits in ferroelectric thin film deposition. Since a highly energetic laser beam (usually ultraviolet; KrF [248 nm], XeCl [308 nm]) is incident on the target surface, all the target material sublimes suddenly so that target composition is easily transferred to the film. In other words, in PLD the film and the target material have almost the same composition. This is an important advantage in ferroelectric thin film deposition, because ferroelectric material is complex in composition and structure and its properties are strongly dependent on them. Another merit of PLD is that there is no reactive component in the chamber. In MBE, where filaments are easily contaminated, an ultra-high vacuum must be maintained during deposition. Because there are no such reactive components in the PLD chamber (the laser beam comes through a window), deposition with relatively high oxygen partial pressure is possible in PLD. This is crucial for ferroelectric oxide film deposition. The growth rate of ferroelectric thin films in PLD is relatively high. Much research on ferroelectric thin films has involved PLD, because PLD has many advantages over other PVD techniques. Several problems are associated with PLD, however, including scalability. Since the uniform laser flume region is so narrow, large and uniform deposition is very difficult with PLD. Several researchers have tried to solve this problem by using a scanning laser beam. Other problems related to PLD, such as particulate problems, also remain to be solved.

As in sputtering, in PLD achieving optimal conditions for deposition is an important issue. Partial oxygen pressure and laser fluence are critical factors in PLD. The distance between the target and the substrate, laser wavelength, substrate temperature and postannealing conditions also affect ferroelectric thin film composition and structure. For epitaxial film growth, selection of substrates and laser repetition rates are also important. Compared with oxygen partial pressure, laser power requires some precautions. Most laser systems provide laser power indication, but the laser power on the target surface is different than the laser power of laser system. Some optical components (e.g., mirrors, focusing lenses, ultraviolet windows) affect laser power. It is important to periodically check the laser power at the target surface. In an excimer laser system, excimer gas is consumed and laser power decreases; thus the laser power should be checked at every deposition. Since target material is also deposited on the window surface where the laser beam enters, the window surfaces must be cleared regularly. Another problem on the laser fluence is related with target surface. Since the laser beam is focused on a small area of the target, if high-power laser beam is incident on a certain point of the target, then the point is damaged during the deposition, because the target material is sublimed as it is evaporated. This may cause changes in composition and deposition conditions. To avoid this problem, the target holder is usually spun or scanned during deposition but it is not enough so that a groove is made after the deposition. Before each deposition, the target surface is ground and cleaned. After these processes, optical alignment should be rearranged carefully since laser fluence is an important factor.

The affect of laser fluence on the electrical properties of PLD-deposited SBT films was reported by Bu et al. [83]. Re-manent polarizations of the films depended on laser fluence, the optimal range of which was 1.0–1.5 J/cm^2 for the system. Bi content was closely related to remanent polarization and laser fluence. The physical processes in PLD are highly complex and interrelated, and depend on the laser pulse parameters and properties of the target materials [79]. The rationale for using PLD in preference to other deposition techniques lies primarily in its "pulse nature," the possibility of carrying out surface chemistry far from thermal equilibrium, and, under favorable conditions, the ability to reproduce thin films with the same elemental ratios of even highly chemically complex bulk ablation targets.

Film growth and chemistry may be enhanced or modified by performing PLD in an ambient background gas. Gases are often used to thermalize the plasma species through multiple collisions or to compensate for the loss of an elemental component of the target through incongruent ablation. Compound targets are often used in PLD, particularly in the production of dielectric and ceramic films and other multielemental oxides and nitrides. Congruent ablation is not always guaranteed, however, and preparation of compound targets carries its own set of problems. Films deposited at various fluences and at various partial pressure of oxygen on postannealing showed variations in stoichiometry that seem to affect orientation in PZT thin films [84]. Film composition is clearly dependent on the fluence in low-energy densities. Beyond a fluence of 2 J/cm^2, the composition of the films showed no dependence on the ablation fluence. In the absence of substrate temperature effects, low energy and low ablation pressures seem to produce stoichiometric films.

One interesting development of PLD is laser MBE [85]. Using laser MBE, a series of BaTiO$_3$ thin films with thicknesses of 10–400 nm was epitaxially grown under various oxygen pressures from 2×10^{-4} to 12 Pa on (001) SrTiO$_3$ substrates. A new method for the growth of solid-solution thin films has been developed, in which a predetermined ratio of the components can be deposited through the controlled use of a single rotating target. This variable compositional control is achieved by altering the radial position of either the target, which incorporates wedge-shaped segments of the solid-solution components, or the laser beam. KTa$_{1-x}$Nb$_x$O$_3$ thin films have been grown using this method [86].

3.1.2.4. Growth Mechanisms, Optimal Conditions, and in situ Characterization

To get the optimal conditions for the ferroelectric film growth, *in situ* characterization is more important and useful than *ex situ* characterization. With *in situ* characterization, ferroelectric film growth mechanisms have been studied using various methods. *In situ* optical emission spectroscopy measurements showed a correlation between the evolution of characteristic atomic emission line intensities and the thin film composition during the rf-magnetron sputtering deposition of PZT from a metallic target in a reactive argon/oxygen gas mixture [87]. The reactive magnetron sputtering process of a multiatomic target was shown to consist of two or three steps of transitions (oxide-metal mode), corresponding to the different binding energies for Pb—O, Ti—O, and Zr—O as rf power density increased. When the flow rate of reactive gas and the rf power density were high enough, the reactions at the substrate surface were shown to be very significant. A high reactive sputtering deposition rate could be obtained in the quasi-metallic mode when the oxygen deposition rate was maximum, and this gave a good compositional PZT thin film.

In situ, real-time studies of layered SBT film growth processes have been performed using a time-of-flight ion scattering and the recoil spectroscopy technique [88]. These studies revealed two important features related to the synthesis of SBT films via ion-beam sputter deposition: (i) atomic oxygen originating from a multicomponent SBT target during the sputtering process is incorporated in the growing film more efficiently than molecular oxygen, and (ii) the SBT surface appears to be determined in an incomplete $(Bi_2O_2)^{2+}$ layer with a top surface of oxygen atoms. A more well-developed mass spectroscopy of recoiled ions technique, which is suitable for monolayer-specific surface analysis of thin films during growth, has been used to study the effect of different bottom electrode layers on metallic species and oxygen incorporation in the early stages of SBT film growth via ion-beam sputter deposition. It was found that the incorporation of Bi in sputter-deposited SBT films depended critically on bottom electrode surface composition and on growth temperature [88–91].

To clarify the fundamental mechanisms of the PLD process, Mach–Zehnder interferometry and laser scattering spectroscopy were applied to $YBa_2Cu_3O_7$ (YBCO) ablation plasma plume in atmospheric pressure. The yield of particulates in the ablation plume, which was assumed to correspond to the surface roughness of deposited film, increased with increasing laser energy density and the laser wavelength [92].

Adsorption-controlled conditions have been identified and used to grow epitaxial BTO thin films by reactive MBE [93]. Growth of stoichiometric, phase-pure, *c*-axis-oriented epitaxial films was achieved by supplying a large overabundance of bismuth and ozone continuously to the surface of the depositing film. Titanium was supplied to the film in the form of shuttered bursts, with each burst containing a three-monolayer dose of titanium to grow one formula unit of $Bi_4Ti_3O_{12}$. The surface morphology of (001) $Bi_4Ti_3O_{12}$ grown on (001) $SrTiO_3$ by reactive MBE has been examined using atomic force microscopy (AFM). A Stranski–Krastonov growth mode was observed [94].

The growth rate and composition of films deposited by laser ablation of $Pb(Zr_{0.65}Ti_{0.35})O_3$, both in vacuum and in ambient oxygen and argon with laser fluences in the range $0.3–3.0$ J/cm^2, have been studied [90]. The film growth rate was shown to increase with an increase in laser fluence and to demonstrate two modes of behavior on the addition of gas. The growth rate decreased at low laser fluence and increased nonmonotonously at high laser fluence on adding gas; however, high pressure of gas decreased the deposition rate regardless of laser fluence. Deposition in vacuum resulted in Pb-deficient films, with the Pb/Ti ratio decreasing with increasing laser fluence, while deposition in a gas resulted in highly increased Pb content, with the Pb/Ti ratio increasing faster un-

der strong laser irradiation. Changes in the film growth rate and composition were similar on the addition of either oxygen or argon. The deposition of single-layer electroceramic ($BaTiO_3$, $PbZr_{0.52}Ti_{0.48}O_3$) thin films by pulsed excimer laser radiation (248 nm) on (111) Ti/Pt/Si substrates has been investigated as a function of laser parameters (e.g., fluence, repetition rate, beam geometry) and processing variables (e.g., pressure and composition of the processing gas, the target–substrate arrangement) under conditions of various temporal and spatial properties of the involved vapor and/or plasma states represented in the type, number, ionization degree, momentum, and energy of the ensemble of species generated [95–98].

3.1.3. *Progress in PVD Techniques*

3.1.3.1. *Oriented/Epitaxial Growth and Multilayer/Superlattice System*

Since many physical properties of ferroelectrics are anisotropic in nature, oriented growth of ferroelectric thin film is necessary to optimize thin film device performance. In the case of the $Pb(Zr,Ti)O_3$ system, *c*-axis-oriented film has shown higher remanent polarization. For the Bi-layer system, such as $SrBi_2Ta_2O_9$, its ferroelectric axis is within the *ab* plane, and many researchers have tried to grow the SBT films epitaxially in *ab* planes. Attempts have been made to deposit non-*c*-axis-oriented SBT thin films, since polycrystalline SBT thin films had been deposited by sol-gel deposition and showed good long-term reliability with Pt electrodes. Compared with *c*-axis-oriented films [99–101], non-*c*-axis-oriented SBT films showed better ferroelectric properties [102–108].

Superlattice structures have received much attention, not only in applications, but also from a theoretical standpoint. Superlattice structures of dielectric and/or ferroelectic materials are of interest and may open a new research area. As techniques of dielectric or ferroelectric thin film deposition are developed, deposition of ferroelectric superlattice has been attempted. The fabrication of artificially designed superlattice structures can provide a strained crystal structure and give rise to unusual properties.

In 1989, deposition of *c*-axis-oriented $BaTiO_3$-$SrTiO_3$ double-layer thin film with rf-magnetron sputtering was reported by Fujimoto et al. [114]. The strain in the film was studied. $SrTiO_3/BaTiO_3$ superlattices were deposited with PLD by Tabata et al. [109]. Tabata and Kawai [110] also deposited $(Sr_{0.3}Ba_{0.7})TiO_3/(Sr_{0.48}Ca_{0.52})TiO_3$ dielectric superlattice. It was reported that the crystal structure was controlled with atomic-order accuracy and that a large lattice stress of $0.5–1$ GPa could be introduced periodically at the interfaces because of the lattice mismatch of the constituent layers.

$KTaO_3/KNbO_3$ and $PbTiO_3/SrTiO_3$ strained-layer superlattices of variable periodicity were grown by PLD [111–114] and MBE [115]. The effects of strain from substrate in epitaxial ferroelectric thin film have been studied experimentally and theoretically. A giant permittivity associated with the motion of domain walls was predicted and reported in epitaxial heterostructures with alternating layers of ferroelectric and non-ferroelectric oxides [116, 117].

A phenomenological thermodynamic theory of ferroelectric thin films epitaxially grown on cubic substrates has been developed using a new form of the thermodynamic potential, that corresponds to the actual mechanical boundary conditions of the problem. From this theory, changes in the sequence of the phases and the appearance of phases forbidden in the bulk crystal were predicted [118, 119]. By the application of Devonshire thermodynamic formalism to the epitaxial ferroelectric film, a shift of the Curie point or an apparent absence of ferroelectricity was also predicted [120–122].

3.1.3.2. Variations of Basic PVD Methods

An interesting result of the modification of the PVD method has been reported by Jiang et al. [115]. They grew $PbTiO_3/SrTiO_3$ superlattices by reactive MBE and showed atomically sharp interface structure. Molecular beams of the constituent elements were supplied to the film surface from thermal sources and purified ozone was used as the oxidant, supplied to the film via a directed inlet nozzle.

In the plasma sputtering deposition method, it is more difficult to change the composition of thin films than in reactive sputtering using metal targets, since a new composition target should be prepared for the new composition film. One solution to this problem is the use of two targets. Tsai et al. [123] have deposited ferroelectric $Sr_{0.8}Bi_{2.5}Ta_{1.2}Nb_{0.9}O_{9+x}$ thin films using two-target off-axis rf-magnetron sputtering. They used two different targets of $Sr_{0.8}Bi_{2.5}Ta_{1.2}Nb_{0.8}O_9$ and Bi_2O_3 simultaneously. The Bi_2O_3 target was used to compensate volatile Bi atoms. Another example of off-axis co-sputtering was performed by Cukauskas et al. [124] for the deposition of $Ba_{1-x}Sr_xTiO_3$ thin films with targets of $SrTiO_3$ and $BaTiO_3$. Reaction of the sputtered species in the plasma resulted in the growth of $Ba_{1-x}Sr_xTiO_3$ thin films. Film composition could be controlled by adjusting the relative power supplied to the $SrTiO_3$ and $BaTiO_3$ targets. In these two co-sputtering systems, off-axis geometry was used, because arranging two targets in the vacuum chamber proved difficult. As a result, the growth rate was very low; the growth rate was 43 nm/h in the case of $Ba_{1-x}Sr_xTiO_3$ film. On-axis double-target rf-magnetron sputtering was performed to deposit $Ba_{1-x}Sr_xTiO_3$ film by rotating the substrate above the targets [97], but the growth rate (about 75 nm/h) was not high enough.

Off-axis sputtering for the ferroelectric $Pb(Zr_{0.2}Ti_{0.8})O_3$ thin film was investigated for growing atomically smooth films with thicknesses ranging from a few unit cells to 800 Å. A stable ferroelectric polarization was observed in films down to a thickness of 40 Å [125]. This result is interesting in view of the size or thickness effects in ferroelectric thin films. Similar modification of the deposition method was performed for the PLD. Droplet-free and epitaxial $SrTiO_3$ thin films were grown by eclipse PLD in $(O_2 + Ar)$ ambient gas [126]. Another reported modification of PLD is on the $Ba_{0.5}Sr_{0.5}TiO_3$ thin films grown

by ultraviolet-assisted PLD. Enhanced electrical performance of high-dielectric–constant thin film was reported [127].

The kinetics of the processes in facing target sputtering of multicomponent oxide films were studied by Nathan et al. [128]. Discharge diagnostics performed using optical emission spectroscopy revealed strong dependence of plasma parameters on process conditions of YBCO thin films. Deposition of multicomponent oxide thin films involved several problems arising primarily from the inhomogeneous distribution of plasma density, angular ejection of sputtered atoms, differential sputtering yield of individual component, and differential transport of sputtered atoms.

3.1.3.3. Electrodes and Barrier Layers

An electrode can influence the behavior of a ferroelectric material in many ways, from changing the chemical nature of the material by diffusion into and chemical reaction with the ferroelectric to introducting strain into the material, causing spontaneous polarization of the ferroelectric. The principal influence is the chemical interaction caused by diffusion between the electrode and ferroelectric at the interface. This includes the diffusion of cations from the electrode into the ferroelectric, causing doping of the material, and the diffusion of oxygen through the electrode or gettering of oxygen from the ferroelectric, causing oxidation of the electrode at the interface [129, 130].

It has been suggested that the formation of oxygen defects in the material and the accumulation of oxygen defects at electrodes (via diffusion) may be a mechanism for fatigue. The use of oxide electrodes provides a possible way to overcome this problem. The advantage of a conducting oxide electrode is that oxygen diffusion and reaction are reduced because of a less abrupt oxygen gradient at the interface. The conductivity of the oxide electrode is less sensitive than that of metallic electrodes to further oxidation during annealing of ferroelectric thin films.

There are a number of criteria for choosing electrodes. Electrode materials must be not only of low resistance, but also chemically compatible with the semiconductor. Other requirements, such as morphologically stability during the process and electrically conducting diffusion barrier for oxygen or other species, are related to integration issues in Si technology. The electrode materials that have been studied for use with PZT ferroelectric thin films are grouped into three categories: metallic (Pt), simple oxide (RuO_2, IrO_2), and complex oxide (($LaSr)CoO_3$, YBCO) electrodes.

A 1T-1C memory cell structure requires that the bottom electrode of the capacitor be in direct electrical contact with the source/drain of the memory cell transistor. A high-density FRAM process requires the integration of conducting barrier layers to connect the drain of the pass-gate transistor to the bottom electrode of the ferroelectric stack. Conventionally, this is accomplished through a highly doped polysilicon plug. There has been a report on the growth of polycrystalline thin film capacitors of $Pb(Nb_{0.04}Zr_{0.28}Ti_{0.68})O_3$ with top and bottom

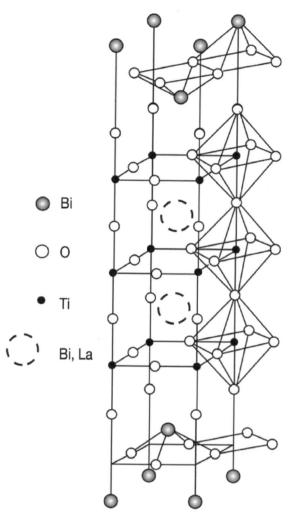

Fig. 13. Lattice structure of the La-substituted bismuth titanate. The compound studied is $Bi_{3.25}La_{0.75}Ti_3O_{12}$ (BLT), in which La ions can be substituted by other rare-earth ions such as Pr, Nd, or Sm. Reprinted with permission from B. H. Park et al., *Nature* 401, 682 (1999). Copyright 1999, Nature Publishing Group.

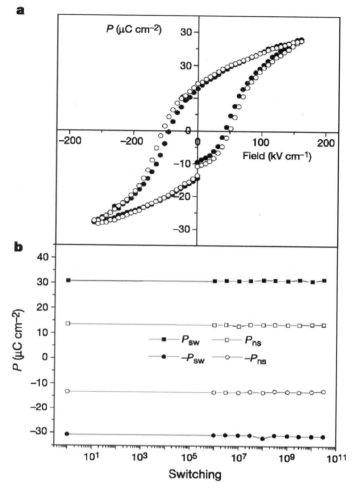

Fig. 14. Results of the fatigue test at 1 MHz. (a) $P–E$ hysteresis loops for the Au/BLT/Pt/Ti/SiO$_2$/Si films before (filled circles) and after (open circles) being subnected to 3×10^{10} read/write cycles. (b) Variation of P_{sw}, P_{ns}, $-P_{sw}$, and $-P_{ns}$ versus number of cycles. Here $-P_{sw}$ and $-P_{ns}$ denote a switching polarization and linear nonswitching polarization, respectively, when a negative read voltage is applied to the capacitor. Reprinted with permission from B. H. Park et al., *Nature* 401, 682 (1999). Copyright 1999, Nature Publishing Group.

electrodes of $La_{0.5}Sr_{0.5}CoO_3$ on polysilicon using TiN/Pt as conducting barriers [131].

Conducting oxide thin films, such as $La_{0.5}Sr_{0.5}CoO_3$, have the same perovskite structure as ferroelectric materials, and the oxygen stoichiometry effects on the electrical properties of $La_{0.5}Sr_{0.5}CoO_3$ have been studied [132]. Different electrode effects of (La,Sr)CoO$_3$ and LaCoO$_3$ were investigated studied by Lee et al. [133]. These oxide materials work not only as electrodes, but also as seed layers for ferroelectric materials. It was found that $La_{0.25}Sr_{0.75}CoO_3$ bottom electrodes with a cubic structure strongly promoted the formation of (001) $Pb(Ta_{0.05}Zr_{0.48}Ti_{0.47})O_3$ films and improved the fatigue and retention properties of the capacitors [134]. Similar results were shown for SrRuO$_3$ electrodes [135].

A novel Pt-Rh-O$_x$/Pt-Rh/Pt-Rh-O$_x$ electrode-barrier structure, which acts as an electrode as well as a diffusion barrier for integration of the ferroelectric capacitors directly onto sil-

icon deposited using an *in situ* reactive rf-sputtering process, was proposed by Bhatt et al. [136]. Oxide electrode materials (e.g., RuO$_2$, IrO$_2$, SnO$_2$ [137], SrRuO$_3$ [138], BaRuO$_3$ [139], (Ca,Sr)RuO$_3$, and (Ba,Sr)RuO$_3$ [140] have been explored as replacements for Pt, which might not be an adequate electrode material due to its difficulty in creating patterns with submicron scales. Oxide electrodes of SrRuO$_3$ and $La_{0.5}Sr_{0.5}CoO_3$ were compared by Cho et al. [141]. Other kinds of oxide electrodes, including LaNiO$_3$ [142] and $Sr_{0.7}NbO_3$ [143], have also been studied [144–152].

3.1.3.4. New Materials and New Structures

A recent breakthrough with PLD is the synthesis of lanthanum-substituted bismuth titanate reported by Park et al. [153, 154]. They reported that lanthanum-substituted bismuth titanate thin films could be deposited at temperatures of ~650°C and that their values of P_r were larger than those of SBT films. $Pb(Zr,Ti)O_3$ and $SrBi_2Ta_2O_9$ systems have been studied for the application of FRAM with a 1T-1C structure. With this structure, the stored information is destructively read out. For the $Pb(Zr,Ti)O_3$ system, some donor dopants (e.g., La^{3+} and Nb^{5+}) were introduced to improve electrical and long-term reliability properties. The doped system—$(Pb,La)(Zr,Ti)O_3$ and $Pb(Nb,Zr,Ti)O_3$—showed better results. The effects of dopants in conjuction with structural and microstructural properties have been studied extensively. For the $SrBi_2Ta_2O_9$ system, isomorphs (e.g., $CaBi_2Ta_2O_9$, $SrBi_2Nb_2O_9$, and $SrBi_2TaNbO_9$) have been studied. One of the interesting results on the doping effects is the lanthanum-substituted bismuth titanate. Substituting lanthanum in a Bi-layered system, bismuth titanate ($Bi_4Ti_3O_{12}$), produced excellent electrical properties for FRAM applications, i.e., fatigue-free with metal electrodes, high remanent polarization, and relatively low deposition temperature.

A new memory structure of field effect transistors (FETs) has been proposed in which the stored information could be nondestructively read out and memory structure is simple with only one transistor (1T). One of the candidate materials is $YMnO_3$ [155–158]. In such metal-ferroelectric-semiconductor (MFS) FET devices, ferroelectric thin films directly face the semiconductor silicon. Since the interface between semiconductor and oxide materials has not yet been well studied, it is very difficult to control the interface. Because rare-earth elements have powerful deoxidizing abilities, $RMnO_3$ (were R denotes rare-earth elements), thin films are expected to be the deoxidizing agent for native oxide on Si. Moreover, $RMnO_3$ with low dielectric permittivity has the greatest advantage for a device directly on Si, because the applied voltage for ferroelectric thin films does not decrease much, because of the existence of the layer with low dielectric permittivity.

Highly c-axis-oriented ferroelectric $YMnO_3$ thin films were deposited by Lee et al. [157] using rf sputtering on Y_2O_3/Si and Si substrates. Ferroelectric $SrBi_2Ta_2O_9$ thin films were deposited on top of $YMnO_3$ for the metal/ferroelectric/insulator/semiconductor (MFIS) structure using PLD [159]. Relaxor ferroelectrics, including $Pb(Mg_{1/3}Nb_{2/3})O_3$-$PbTiO_3$ and $Pb(Yb_{1/2}Nb_{1/2})O_3$-$PbTiO_3$, deposited by PLD have been studied for applications in FRAM and piezoelectric devices [154, 160–163].

3.2. Chemical Vapor Deposition

In chemical vapor deposition (CVD) processes, a solid material is deposited from the vapor via chemical reactions occurring either on or near the substrate. A characteristic of CVD, that is of technological importance is that a very good step coverage can be achieved. As integrated circuit design dimensions decrease, processes capable of conformal coverage are required for the substrates with steps, contact holes, and trenches with large aspect ratios. Other advantageous characteristics of CVD include use of relatively high reactant (e.g., oxygen or hydrogen) partial

pressure during film growth, relatively low deposition temperatures, high deposition rates, and amenability to large-scale and large-area depositions.

For the deposition of films, volatile compounds of the elements to be incorporated into the solid, the so-called precursors, are transported via the gas phase to the region where deposition take place. If metal-organic compounds are used as precursors in a CVD process, then the technique is referred to as metal-organic CVD (MOCVD). Metal-organic compounds usually, but not always, have higher volatility than inorganic compounds, and also have relatively low decomposition temperatures. This enables higher deposition rates at low temperatures. Also, physical and chemical properties can be controlled by introducing small changes in the ligands. Thus deposition of thin films of high purity is possible with the proper choice of precursors and deposition conditions.

3.2.1. MOCVD

MOCVD recently has been applied successfully in the deposition of oxide thin films such as high-temperature superconductors and ferroelectrics. MOCVD has the advantage over most other techniques in that a high oxygen partial pressure can be applied during deposition. The technique allows direct formation of thin films in the proper crystalline phase and even epitaxial films for lattice-matched substrates. MOCVD uses metal-organic compounds, which are characterized by the presence of direct bonds between the central metal atom and carbon atoms of the surrounding organic ligands.

3.2.1.1. Metal-Organic Precursors

In a CVD process, using precursors of sufficient vapor pressure and mass transport as well as thermal stability is critical. The precursors for MOCVD of multicomponent oxides (titanates, niobates, tantalates) and high-temperature superconductors are predominantly metal alkoxides ($M(OR)_n$), β-diketonates [or chelates; $M(R_1COCHCOR_2)_n$], and to a lesser extent, alkyl (MR_n) or cyclopentadienyl derivatives [$M(C_5H_5)_n$]. Here M is the metal with valency n, and R is an alkyl group or other organic ligands. The formulations of these perovskite-type materials are based on large elements, alkaline-earth metals, lead or bismuth, and early transition metals or lanthanides. Major drawbacks to date include the limited studies on volatile alkaline-earth (the precursor for Ba-, Sr-, and Pb-based perovskites) and lanthanide compounds for MOCVD, the high cost of starting chemicals, and environmental safety concerns. Additional issues of concern include the low volatility of available precursors, the high toxicity of volatile precursors, and carbon contamination in films. These precursors can be grouped into two general categories: those that bond oxygen to the metal center (β-diketonates and alkoxides) and those that bond carbon to the metal center (alkyls and cyclopentadienyl derivatives).

Metal alkoxides are the predominant precursors in the MOCVD of metal oxides. For the metal alkoxides, compounds with very low or very high vapor pressure are known. Ti and Zi alkoxides have shown sufficient volatility and ease of decomposition, but Pb, Ba, Sr, and La alkoxides behave differently. Indeed, a low charge on a metal ion combined with a large ionic radius (e.g., Ba^{2+}) causes considerable problems when attempting to achieve volatility. Due to their large atomic radius and tendency for higher coordination numbers (e.g., 4, 5, and 6 in $Pb(OR)_2$), strong intermolecular associations occur with Pb alkoxides even with bulky ligands such as the t-butoxide, $^tOC(CH_3)_3$. For example, $Pb(^tOC(CH_3)_3)_2$, which has been found to be trimeric [164, 165], has the advantage of having less carbon than Pb-β-diketonate, but has much lower volatility and much greater air sensitivity.

The volatility of alkoxides depends on their degree of oligomerization. If two or more alkoxides cluster together, then vapor pressure will decease due to the increased molecular weight of the cluster. This tendency of clustering is correlated to the incomplete coordination of the central metal atom by the ligands and to the positive charge on the metal, which readily attracts negatively charged ligands of neighboring alkoxides.

Clustering (agglomeration) of the alkoxides can be suppressed by introducing bulky ligands, causing steric hindrance around the metal atom. In this way, the positive charge on the central metal atom is shielded from the negative charge on oxygen atoms in the ligands. This causes the phenomenon in which alkoxides with larger molecular weights (which in general cause vapor pressure to decrease) can have higher vapor pressures than the smaller alkoxides. For example, for tetravalent metal alkoxides, volatility increases from the methoxides to the ethoxides and propoxides. If the ligand becomes too large, then volatility decreases again, due to the increased molecular weight of the alkoxide. The demands for large ligands and low molecular weights are conflicting and not easily achieved. The steric hindrance effect can also be obtained by using branched alkoxides to increase volatility. For alkoxides of Zi, volatility increases, going from n-butoxide ($OCH_2CH_2CH_2CH_3$) via i-butoxide [$OCH_2CH(CH_3)_2$] to t-butoxide [$OC(CH_3)_3$], which is a monomeric liquid with a vapor pressure of ~ 0.1 mbar at 50°C.

The main difficulty in handling the metal alkoxides is their ease of hydrolysis by even very small amounts of water or alcohols and stringent precautions are required to prevent this deterioration of the precursor. On the other hand, this reactivity towards water may be used for a low temperature decomposition of these precursors.

Double-alkoxide compounds have been applied for the deposition of the corresponding double oxides. These precursors ensure a stoichiometric supply of metals to the deposition region [166]. Oxides can be formed from the alkoxide compounds by simple decomposition; no additional oxidizing species is necessary.

The β-diketonates [$M(R_1COCHCOR_2)_n$] and their adducts are quite useful and frequently used precursors for the deposition of oxide thin films, especially in metals such as Ba that do not form any other type of volatile compounds. The schematic structure of some β-diketonate precursors is presented in Figure 15a. It is well known that steric hindrance on the α-carbon

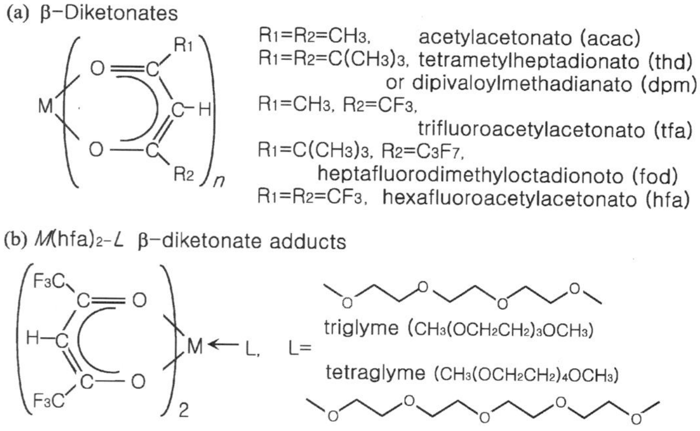

(a) β-Diketonates

$R_1=R_2=CH_3$,   acetylacetonato (acac)
$R_1=R_2=C(CH_3)_3$, tetrametylheptadionato (thd) or dipivaloylmethadianato (dpm)
$R_1=CH_3$, $R_2=CF_3$,   trifluoroacetylacetonato (tfa)
$R_1=C(CH_3)_3$, $R_2=C_3F_7$, heptafluorodimethyloctadionoto (fod)
$R_1=R_2=CF_3$, hexafluoroacetylacetonato (hfa)

(b) $M(hfa)_2$-L β-diketonate adducts

triglyme ($CH_3(OCH_2CH_2)_3OCH_3$)

tetraglyme ($CH_3(OCH_2CH_2)_4OCH_3$)

Fig. 15. Schematic structural diagram of (a) typical β-diketonate precursors and (b) $M(hfa)_2 \cdot L$ adducts.

atom favors volatility. 2,2,6,6-Tetramethyl-3,5-heptanedionate $M(O_2C_{11}H_{19})_n$ [or $M(thd)_n$] bases on β-diketones, where $R_1 = R_2 = t$-butyl, thus is usually the best choice for nonfluorinated derivatives [167]. These are also often called dipivaloylmethanates, $M(dpm)_n$. Examples, include the barium derivative barium bis(2,2,6,6-tetramethyl-3,5-heptanedionate), $Ba(O_2C_{11}H_{19})_2$, and $Ba(thd)_2$ [168]. However, because $Ba(thd)_2$ and $Sr(thd)_2$ are gradually decomposed by moisture and carbon dioxide, these materials should be handled in inactive gas in a glove box. A major problem associated with these solid compounds is that their volatility is often low, and fairly high temperatures are needed to cause sublimation. Substitution of fluorine for hydrogen—for example, Pb or Ba bisheptafluorodimethyloctadione [Pb or $Ba(O_2C_{10}H_{10}F_7)_2$; Pb or $Ba(fod)_2$]—increases volatility [169] but can lead to the deposition of metal fluorides. The β-diketonate precursors are stable up to about 200°C, and they are nonhygroscopic and nonpyrophoric. The toxicity of these compounds is generally low. Deposition of oxide thin films from β-diketonates is by thermal decomposition, and theoretically, no oxidant is needed.

Recently, there has been progress in metal β-diketonate chemistry. Similar to the alkoxides, their volatility depends strongly on the ability of the ligands to prevent the metal atoms from increasing their coordination number by forming intermolecular bonds that shield off the positive charge on the central metal atom. This spatial shield was studied by Gardiner et al. [170]. Tetraglyme [or 2,5,8,11,14-pentaoxopentadecane ($C_{10}H_{22}O_5$)] adducts of Ba β-diketonate were synthesized to obtain a Ba source with thermal stability and high vapor pressure for MOCVD. Complexation has been widely used as a means of obtaining a more stable "second generation" of precursors. Complexes of the type $M(hfa)_2 \cdot L$ [hfa = 1,1,1,5,5,5-hexafluoro-2,4-pentanedionato or hexafluoroacethylacetonate; M = Ca, Sr, Ba; L = tetraglyme, triglyme ($C_8H_{18}O_4$)] (Fig. 15b) have been reported [171] and found to exhibit high, stable vapor pressures [170, 172]. The superior vapor pressure characteristics and thermal stability of these second-generation $M(hfa)_2 \cdot L$ complexes have been exploited in the MOCVD growth of thin films of high-temperature superconductors [172–174]. Also, M(hfa)$_2$·L complexes [L = CH$_3$O(C$_2$H$_4$O)$_5$C$_2$H$_5$ or CH$_3$O(C$_4$H$_4$O)$_6$-n-C$_4$H$_9$] have been used with titanium tetraethoxide as MOCVD precursors to epitaxial BaTiO$_3$ films [175]. However, these can promote contamination of the deposits by barium fluoride. Heat treatment of the films in an H$_2$O-O$_2$ atmosphere is then necessary for defluorination. Various approaches used to tailor volatility with barium derivatives have been investigated [176].

Metal alkyls MR$_n$ with high vapor pressures are available for those metals of interest for ferroelectric thin films. The compounds are extremely volatile and have been used for the MOCVD of metals, semiconductors, nitrides, and oxides. For example, tetraethyllead [Pb(C$_2$H$_5$)$_4$] has been used for lead-based perovskite thin films. The metal alkyls are extremely sensitive to oxygen and water, and are very toxic. Cyclopentadienyls, M(C$_5$H$_5$)$_n$, are those compounds of metal π complexes. The π complexing ligands donate electrons to the central metal atoms. This electron donation promotes volatility of the complexes by satisfying the metals, electronically and coordinately, and preventing intermolecular associations with adjacent ligands. In contrast to the oxygen-containing precursors, the formation of oxides from the alkyls or cyclopentadienyls occurs by oxidation rather than by thermal decomposition. Therefore, an additional oxidant (e.g., oxygen) is necessary to obtain the oxide, as in, for example, the following reaction:

$$2Pb(C_2H_5)_4 + 3O_2 \rightarrow 2PbO + 8C_2H_4 + 4H_2O$$

The metal–carbon bond implies a risk of carbon incorporation into the solid.

To modulate volatility, decomposition temperature, and/or reactivity of precursor, it may be desirable to use a combination of ligands. For example, alkyllead alkoxide with the formula Pb(C$_2$H$_5$)$_3$(OR), where R = C(CH$_3$)$_3$ or CH$_2$C(CH$_3$)$_3$, has been used as the Pb source in MOCVD of PZT [177]. To obtain reproducible oxide ferroelectric thin films at low temperature, precursors must be handled under moisture-free conditions. Also, for multicomponent oxide films, it is desirable to use component precursors with similar volatility and adequate mass transport characteristics at the lowest possible temperature.

3.2.1.2. The MOCVD Process

MOCVD involves a number of physical and chemical subprocesses: evaporation and transport of precursor, chemical reactions in the gas phase, diffusion of reactants to the surface, adsorption and desorption, surface diffusion, chemical reaction at the surface, incorporation into the growing films, and diffusion of reaction products from the surface to the gas phase. A complete description of the process is very complicated, since it depends strongly on the precursors involved, the reactor geometry, and process conditions. Various textbooks treat this subject in detail [178, 179]. A general approach to get an impression of the process is to study the deposition rate as a function of such parameters as the precursor partial pressure,

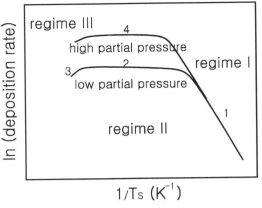

Fig. 16. Schematic representation of an Arrhenius plot showing three different regimes. Segment (1) is reaction-rate controlled regime, and segments (2) and (4) are diffusion-limited at two different partial pressures of reactant gas precursor.

substrate temperature, and reactor pressure. An Arrhenius plot of the deposition rate is the usual way to describe the deposition behavior as a function of temperature. A qualitative description of the plot, as shown in Figure 16, illustrates several aspects involved in the deposition process.

At low temperatures (regime I), where the plot yields a steep straight line, thin film growth is controlled by chemical reaction, e.g., the decomposition of a precursor at the surface. The deposition rate depends exponentially on the deposition temperature and is proportional to the precursor partial pressure. This strong temperature dependence requires very accurate control of substrate temperature over the whole substrate area to get a uniform layer thickness and, in the case of a multicomponent layer, a uniform composition. Under these conditions, films exhibit relatively low growth rates and have strong texture and relatively small grain size, due to the low growth temperature. Since the deposition rate depends only on the substrate temperature and not on the diffusion of species, a conformal step coverage may be obtained.

An increase in temperature produces a transition area in the Arrhenius plot, after which the slope is lower and the deposition rate is no longer limited by these reactions, but rather by the diffusion (mass-transport) of species through the gas phase to the substrate. This regime is often called a diffusion-limited (or mass-transfer-controlled) deposition regime (regime II). In this regime, the reaction rate is fast enough so that it does not control the overall rate. Rather, the arrival rate of the reactants (and hence the surface concentration) controls the rate. Also, decreasing the total pressure of a reactor increases the mean free path and increases the diffusion coefficient of the reactants. In this regime, the deposition rate is weakly dependent on temperature, because the diffusion coefficients of the chemical species increase only slightly with temperature. It is advantageous to deposit layers in this regime, since the temperature uniformity over the substrate is no longer critical. Films grown in this regime exhibit excellent property control. The gas flow rate can be better controlled than substrate temperature, and hence a high and uniform growth rate can be achieved simultaneously. However, film conformality is poor with this growth regime, because mass transfer through various features, such as trenches, can be different for different gases and the deposition rate also varies. Microstructurally, the higher temperature allows for larger grain sizes. However, very fast growth rates combined with high temperature can result in high film stresses. The transition from regime I to regime II depends not only on temperature, but also on reaction pressure, gas-flow rate, precursors, reactor design, and so on. This allows an extra degree of freedom in designing the process.

At very high temperatures, the deposition rate often decreases. Such a decrease can be attributed to the onset of film reevaporation (desorption) or parasitic side reactions. These reactions may cause the formation of clusters of molecules or even particles in the gas phase that do not contribute to the deposition, resulting in decreased growth efficiency.

From these considerations, it is clear that film deposition in either regime I or II is advantageous, depending on the desired properties of the layer. In addition, the deposition temperature may be restricted to some maximum temperature to suppress substrate degradation or to a minimum temperature to obtain a sufficiently crystallized film.

In MOCVD of oxide materials, a number of the available precursors of essential elements have a rather low vapor pressure at room temperature. This implies that the temperature at which these precursors must be evaporated must be raised (typically to 50–150°C). Higher partial pressures of the precursors cannot be obtained simply by raising the temperature of the bubbler only, since in this case the precursor material would condense in the cooler parts of the system in between the bubbler and the reactor cell. These factors make the hardware more complicated, although a standard MOCVD system is simple. A simple solution is to wrap the gas system with heater tape. The main disadvantage to this approach is that obtaining uniform heating appears to be very difficult. It is harder to heat components like valves. In a custom-built system, coaxial tubes are used for uniform heating of transport lines; hot oil flows through the outer tube, heating the inner tube. Uniform heating of valves is also accomplished by mounting the components on metal blocks that are heated by the same oil that flows through the lines. Various MOCVD setups are found in the literature [179].

The control of the composition is a matter of concern in the case of lead-based (ferroelectric) thin films. This is due to the relatively high vapor pressure of both lead and lead oxide, resulting in relatively fast evaporation of these compounds from the growing surface. In the case of MOCVD, the lead loss can be compensated for by adjusting the lead precursor partial pressure. The high vapor pressures of lead and lead oxide turn out to be of benefit, because the stoichiometry of the film is not very sensitive to variations in deposition temperature and precursor partial pressures (within a certain range) [180, 181]. A process window has been observed with respect to both the gas-phase composition $(Pb/Ti)_g$ and the temperature, in which stoichiometric $PbTiO_3$ forms. At 670°C, the

composition of $PbTiO_3$ film turns out to be stoichiometric and independent of the gas-phase composition within an interval of $0.2 < (Pb/Ti)_g < 0.6$ [181]. Hence film composition is independent of small temperature variations over the wafer and of fluctuations in precursor partial pressures. This is important for a reproducible and uniform deposition of such films over large wafers. The size of the process window depends on, for example, reactor geometry, substrate temperature, and growth rate [181].

This process window may be explained by a mechanism involving a fast surface reaction of PbO with TiO_2 (and ZrO_2) to form $PbZr_xTi_{1-x}O_3$ in competition with the evaporation of the relatively volatile PbO. This mechanism is expected to be independent of the type of substrate. The growth rate, ratio of the precursor partial pressures, and substrate temperature are the important parameters. This mechanism regulating the composition may not be specific to MOCVD of lead-based films, and comparable observations are expected for other in situ deposition techniques and other materials. An example is $Bi_4Ti_3O_{12}$, in which the presence of volatile Bi_2O_3 may induce a comparable mechanism.

3.2.1.3. Synthesis of Ferroelectric Thin Films by MOCVD

The discussion that follows is restricted to the growth of the perovskite ferroelectric oxides [$SrTiO_3$, $BaTiO_3$, $(Pb,La)(Zr,Ti)O_3$ and $Pb(Sc,Ta)O_3$] and bismuth-oxide layered perovskites [$SrBi_2Ta_2O_9$ and $Bi_4Ti_3O_{12}$].

Perovskite Ferroelectrics. $BaTiO_3$ has been grown by MOCVD using the β-diketonate complex of $Ba(thd)_2$ and $Ti(^iOC_3H_7)_4$ [182, 183]. Highly resistive dielectric layers were obtained on silicon substrates at 600°C [182] and on Pt/MgO [183] stacks at 800°C with deposition rates of about 1 μm/h. However, the $Ba(thd)_2$ evaporates at high temperatures about 250°C, and the evaporation temperature is close to the region where $Ba(thd)_2$ starts to decompose. Moreover, this compound degrades in the bubbler with time, resulting in decreased vapor pressure. The evaporation temperature of the barium precursor could be reduced down to 105–125°C by using a $Ba(hfa)_2$-tetraglyme precursor [184]. Unfortunately, the layers then were found to be contaminated with fluorine. Introduction of water vapor is used to remove the fluorine as HF, which results from the fluorinated barium precursor [185]. Argon was used as the carrier gas, and O_2 bubbled through deionized water was used as the reactant gas. $Ba(thd)_2 \cdot (tpa)_2$ (tpa = tetraethylenepentamine; $C_8H_{23}N_5$) also has been used for the synthesis of $BaTiO_3$ thin film [186]. At 600°C, carbon incorporation in the form of $BaCO_3$ was observed in thermally grown $BaTiO_3$. Using plasma enhancement, films can be grown at 600°C and 680°C without any incorporation of carbon [187]. The precursor $Ti(^iOC_3H_7)_4$, which has a higher vapor pressure, was evaporated at temperatures of 25–60°C. Instead of $Ti(^iOC_3H_7)_4$, $TiO(thd)_2$ has been used to improve step-coverage characteristics [188]. Recently, step coverage of the $(Ba,Sr)TiO_3$ films using a new Ti source,

$Ti(^iOC_4H_9)_2(thd)_2$, was found to be 80% and 70% at aspect ratios of 3.3 and 5, respectively [189].

The precursors used for the deposition of $SrTiO_3$ were $Ti(^iOC_3H_7)_4$ and $Sr(thd)_2$, evaporated at 200–230°C and 50°C, respectively [190, 191]. Single-phase layers with a preferred orientation were deposited at 800°C on sapphire. When using H_2O as an oxidant, the layers were found to be contaminated with carbon. This could be effectively suppressed by a supplying additional oxygen. As discussed previously, the source $Sr(thd)_2$ has some weak points, including high sublimation temperature, decomposition of the compound around the sublimation temperature, and sensitive reaction with H_2O and CO_2. Novel Sr sources for MOCVD have been synthesized with adducts. Among about 50 materials tested, $Sr(thd)_2 \cdot (tta)_2$ (tta = trietylenetetramine; $C_6H_{18}N_4$) and $Sr(thd)_2 \cdot (tpa)_2$ were selected as MOCVD sources [192]. They have a low melting point (43–75°C) and are in the liquid phase at a vaporization temperature of 120–130°C. A single phase of $SrTiO_3$ film with (110) orientation could be deposited on a $Pt/Ta/SiO_2/Si$ substrate at 550°C [192].

In terms of high dielectric constant materials, $(Ba,Sr)TiO_3$ (BST) thin films have been synthesized by MOCVD [188]. The same Ba (Sr) sources for $BaTiO_3$ ($SrTiO_3$) films are used for the solid-solution films. Most investigations have used $Ba(thd)_2$, $Sr(thd)_2$, and $Ti(^iOC_3H_7)_4$. Recently, the use of $Ba(thd)_2 \cdot (phen)_2$ and $Sr(thd)_2 \cdot (phen)_2$ (phen = phenanthroline) sources has been reported [193]. Oxygen gas was used as an oxidant. BST films grown on Pt/MgO at 650°C showed a low leakage current density of 10^{-7} A/cm^2 at 3 V and a room-temperature dielectric constant of 400–590, and the breakdown voltage was above 500 kV/cm [193]. The use of $Ba(thd)_2 \cdot (tpa)_2$ and $Sr(thd)_2 \cdot (tta)_2$ also has been tried, but remarkable differences from the use of $Ba(thd)_2$ were not observed [194].

Attempts to deposit thin films of complex perovskite ferroelectric oxide, $Pb(Sc,Ta)O_3$ (PST) directly by MOCVD using the precursors $Pb(^iOC_4H_9)_2$, $Ta(OC_2H_5)_5$, and $Sc(fod)_3$ proved unsuccessful [195]. In these attempts first the cubic phase of $ScTaO_4$ was obtained by annealing Sc_2O_3 and Ta_2O_5 thin films, and then PbO was diffused into the film from a surface layer to produce the perovskite phase of $Pb(Sc,Ta)O_3$. Later, Ainger et al. [196] succeeded in growing PST films directly by MOCVD using $Pb(fod)_2$, $Sc(fod)_3$, and $Ta(OC_2H_5)_5$ as precursors. The addition of water vapor to the reactant gas stream was very important, because it significantly enhanced the breakdown of the (fod)-type precursors to the oxide. The oxygen content of the gas needed to be at least 30% to form perovskite [196]. Recently highly (100) textured PST thin films were grown on $LaNiO_3/Pt/Ti$ electrode-coated Si substrate by MOCVD at 685°C using the precursors $Pb(thd)_2$, $Sc(thd)_3$, and $Ta(OC_2H_5)_5$ [197].

Lead Titanate Family. Since the first report on the deposition of $PbTiO_3$ by CVD using chlorides as precursor materials [198], a number of papers on the growth of $PbTiO_3$, $PbZr_xTi_{1-x}O_3$, $(Pb,La)TiO_3$, and $(Pb,La)(Zr,Ti)O_3$ have appeared. After a report by Okada et al. [199], $Pb(C_2H_5)_4$ became the most popular precursor for lead. $PbTiO_3$ was deposited by MOCVD with a system operating at atmospheric pressure using $Pb(C_2H_5)_4$ and $Ti(^iOC_3H_7)_4$ as precursors. Since the $Pb(C_2H_5)_4$ does not contain oxygen, oxygen was needed to form the lead oxide. The deposition rate of PbO was saturated if the oxygen partial pressure exceeded a certain critical value, whereas that of TiO_2 was independent of the oxygen partial pressure. Oxidation of $Pb(C_2H_5)_4$ was proposed to explain the effect of oxygen partial pressure on the PbO deposition rate. The deposition rates of both PbO and TiO_2 first increased with substrate temperature, indicative of a reaction-controlled process, and then showed a decrease or a saturation. Deposition of $PbTiO_3$ was achieved by matching the deposition rates of PbO and TiO_2:

$$PbO + TiO_2 \rightarrow PbTiO_3$$

Single-phase perovskite-type $PbTiO_3$ was obtained on (001) Si, sapphire, and (001) MgO. The deposition temperature was 500–600°C.

Low-pressure MOCVD for $PbTiO_3$ deposition was reported by Kwak et al. [200] using $Pb(C_2H_5)_4$ and $Ti(^iOC_3H_7)_4$. Nitrogen dioxide (NO_2) was introduced in the photo-enhanced MOCVD as an oxidizing gas source to obtain high-quality film at low growth temperature [201]. The N—O bond in NO_2 has a lower dissociation energy than the O—O bond in O_2, and efficient photolysis of NO_2 occurs with absorption of UV light. Perovskite $PbTiO_3$ films were grown on sapphire at substrate temperatures above 530°C with $Pb(C_2H_5)_4$ and $Ti(^iOC_3H_7)_4$ sources [201]. Hirai et al. [202] reported on the deposition of $PbTiO_3$ by the alternate introduction of $Pb(C_2H_5)_4$ and $Ti(^iOC_3H_7)_4$ into the deposition chamber. In combination with the use of ozone as an oxidant, a low deposition temperature of 510°C was possible. The use of $Pb(fod)_2$ and $Ti(C_6H_{10}O_3)_2(^iOC_3H_7)_2$ as precursors for $PbTiO_3$ films has been reported [196].

$PbZr_xTi_{1-x}O_3$ (PZT) shows promising characteristics for future applications, and a number of studies on the MOCVD of this material have been reported. The first MOCVD of $PbZr_xTi_{1-x}O_3$ was reported by Okada et al. [203] using a low-pressure deposition system. These authors used $Pb(C_2H_5)_4$ and $Ti(^iOC_3H_7)_4$ for Pb and Ti sources, respectively. $Zr(^iOC_3H_7)_4$ and $Zr(thd)_4$ were used and evaluated as Zr sources. With both precursors, deposition of single-phase perovskite-type $PbZr_xTi_{1-x}O_3$ was obtained. Using $Zr(thd)_4$, a higher preferential orientation on (001) MgO was obtained. In comparison with the deposition of $PbTiO_3$, a smoother surface topography was observed. The high zirconium content resulted in a denser film. Highly oriented [($h00$) and/or ($00l$)] PZT films were grown on Pt-coated 10-cm-diameter Si substrates at 700°C using $Pb(C_2H_5)_4$, $Ti(^iOC_3H_7)_4$, and $Zr(^iOC_4H_9)_4$ as precursors; the MOCVD-grown PZT films showed good electrical and ferroelectric properties [204]. Because $Pb(C_2H_5)_4$ is poisonous, $Pb(thd)_2$ was studied as a nonpoisonous alternative lead precursor, evaporated at 100–150°C, in combination with $Zr(thd)_4$ and $Ti(^iOC_3H_7)_2(thd)_2$ [205, 206]. Oxygen gas was used to produce an oxidative atmosphere. Films grown at reduced pressure showed better crystallinity and thickness uniformity [206]. A low processing temperature was achieved using either $(C_2H_5)PbOCH_2C(CH_3)_3$ (triethyl n-pentoxy lead) as a Pb precursor or NO_2 as an oxidizing gas [207, 208]. Tetragonal and rhombohedral PZT films were grown at substrate temperatures above 470°C and 505°C, respectively [207]. In addition, use of a liquid Pb precursor, $(C_2H_5)PbOCH_2C(CH_3)_3$, resulted in improved reproducibility of the composition and growth rate. Variations in film thickness of less than ±1.5% were obtained on a 15-cm Si wafer [207].

As in the case of $PbTiO_3$ [181], there exists a process window where the film stoichiometry is insensitive to partial precursor pressures. Although some differences are found between the reports, the composition of film, $[Pb/(Zr + Ti)]_f$, is not very sensitive to variations in both the deposition temperature and the gas supply ratio, $[Pb/(Zr + Ti)]_g$, within a certain range [207, 209, 210].

Thin films of (Pb,La)TiO_3 (PLT) are very interesting for the pyroelectric infrared detector because of their large pyroelectric coefficient and small dielectric constant. The preparation of PLT thin films using MOCVD was first reported by Tominaga et al. [211]. The synthesis of PLT thin films was described using the precursors $Pb(C_2H_5)_4$, $Ti(^iOC_3H_7)_4$ and $La(thd)_3$. The study of the deposition of the individual components found that the formation of La_2O_3 involved oxidation of the $La(thd)_3$, comparable with PbO formation from $Pb(C_2H_5)_4$. The deposition rate of La_2O_3 leveled off at about 20% oxygen in the gas phase. When the deposition rates of the individual components were matched, single-phase perovskite-type PLT thin films could be obtained. Using the same precursors, highly textured PLT films were grown on Si [212]. The tetragonality, c/a, of the PLT films was found to decrease as the La concentration increased. Substrate dependence in the growth of epitaxial PLT films was investigated in a subsequent study [213]. An initial TiO_2 layer on MgO promoted growth of three-dimensional epitaxial PLT films. Recently, PLT films were deposited by MOCVD using a solid delivery technique with the source precursors of $Pb(thd)_2$, $La(thd)_3$, and $Ti(^iOC_3H_7)_2(thd)_2$ [214].

(Pb,La)(Zr,Ti)O_3 (PLZT) films deposited by MOCVD were first reported by Okada et al. [215] using the precursors $Pb(C_2H_5)_4$, $La(thd)_3$, $Zr(thd)_4$, and $Ti(^iOC_3H_7)_4$. Single-phase preovskite-type PLZT was deposited on (001) MgO, Pt/(001) MgO and Pt/SiO_2/Si again by adjusting the deposition rates of the individual oxides. The film properties were examined as a function of the La content for $Zr/(Zr + Ti) < 0.5$. The values for the dielectric permittivity were found to depend on the $Zr/(Zr + Ti)$ and to increase with increasing La content up to 5 at%. Growth of PLZT was also done using $(C_2H_5)PbOCH_2C(CH_3)_3$, $La(thd)_2$, $Zr(^tOC_4H_9)_4$, and $Ti(^iOC_3H_7)_4$ as precursors and O_2 as an oxidizing gas [207]. Improved fatigue properties compared with the PZT(54/46) thin films were observed. High-quality PLZT films were deposited on Pt-coated Si wafers at 650°C by direct liquid injection MOCVD using $Pb(thd)_2$, $La(thd)_3$, $Zr(thd)_4$, and $Ti(OC_2H_5)_4$ as precursors [216]. The precursors were dissolved in a so-

lution of tetrahydrofuran, isopropanol, and tetraglyme in a molar ratio of 8 : 2 : 1 to form a source solution, which was injected into the heated (250°C) vaporization chamber. The mixed precursor vapors and the oxidant (O_2) were carried into the reactor separately. Singh et al. [217] deposited PLZT films using a dual-spectral source rapid thermal processing–asssisted MOCVD system. Halogen quartz lamps were used to provide optical and thermal energy, and deuterium lamp was used as the source of optical energy. These authors used [thd] sources for all metals (Pb, La, Zr, Ti) in the deposition process and a mixture of N_2O and O_2 gases for oxidation.

Bismuth-Oxide Layered Perovskites. Bismuth-oxide layered perovskite ferroelectrics have attracted much attention recently for the potential applications to nonvolatile memory device. $Bi_4Ti_3O_{12}$ and $SrBi_2Ta_2O_9$ are the most widely studied materials. Single-phase BTO layers were deposited onto silicon substrates at 600–750°C by low-pressure MOCVD at a deposition rate of 2.4–4.8 μm/h [191]. The precursors used were triphenyl-bismuth, $Bi(C_6H_5)_3$ (at evaporation temperature 130–170°C), and $Ti(^iOC_3H_7)_4$ (at 40°C). Wang et al. [218] deposited dense BTO films at 550°C using $C_{16}H_{36}O_4Ti$ as the Ti source and N_2 and O_2 gases as a carrier and an oxidizing gas, respectively. It was found that the Bi/Ti ratio decreased with increasing deposition temperature, probably due to the evaporation of relatively volatile bismuth oxide (BiO_x) from the growing film [191, 218]. There have been reports on the growth of high-quality and fatigue-free BTO thin films at a low substrate temperature of 400°C [219–221]. It was found that the BiO_x layer effectively lowered the crystallization temperature and controled the composition of BTO thin films [220]. They used $Bi(^oC_7H_7)_3$ and $Ti(^iOC_3H_7)_3$ as source materials. After low-temperature annealing at 400°C in N_2 atmosphere, films with good ferroelectric properties were obtained.

SBT is known as a new ferroelectric material with low intrinsic defect density suitable for nonvolatile memory [10]. The first successful deposition of SBT thin films by MOCVD was reported by Li et al. [221]. With precursors of $Sr(thd)_2$, $Bi(C_6H_5)_3$, and $Ta(OC_2H_5)_5$, crystalline films could be obtained on Pt/Si and sapphire substrates above 650°C using a liquid-delivery system. Hendrix et al. [222] showed that $Bi(thd)_2$ is a superior source of Bi than $Bi(C_6H_5)_3$, offering a wide decomposition window with compatible Sr and Ta precursors, $Sr(thd)_2$-tetraglyme and $Ta(^iOC_3H_7)_4(thd)$. Films with 90% conformality have been grown on 0.6-μm structures with an aspect ratio of 1. Recently, new liquid source materials, $Bi(CH_3)_2$ and $Sr[Ta(OC_2H_5)_6]_2$, which have high vapor pressure and thermal stability, were introduced for the use in conventional MOCVD [223]. $Bi(CH_3)_2$ and $Sr[Ta(OC_2H_5)_6]_2$ were vaporized at 20°C and 140°C, respectively. An almost single-phase of SBT was deposited at 670°C.

Miscellaneous. The first MOCVD growth of $LiNbO_3$, using a lithium β-diketonate and niobium alkoxide precursors, was reported in 1975 [224]. Feigelson [225] reported growth of high-quality epitaxial $LiNbO_3$ films for waveguiding applications on sapphire and LiTaO$_3$ substrates by solid-source MOCVD. A mixture of 65% Li(thd) and 35% Nb(thd)$_5$ powder sources was used for single-phase LiNbO$_3$ films. Epitaxial LiNbO$_3$ films were prepared on LiTaO$_3$, and sapphire substrates were prepared by MOCVD using Li(thd) and Nb(OC$_2$H$_5$)$_5$ precursors [226, 227]. Recently, a high-vacuum chemical beam epitaxy technique was introduced to grow LiNbO$_3$ with alkoxide precursors [Li(^{i}OC$_4$H$_9$) and Nb(OC$_2$H$_5$)$_5$] [228]. By depositing alternating layers of single metal oxides, highly textured LiNbO$_3$ can be obtained directly on Si substrate.

Strontium-barium niobate [(Ba,Sr)Nb$_2$O$_6$] thin films have been grown by MOCVD with the precursors Ba(thd)$_2$, Sr(thd)$_2$, and Nb(OC$_2$H$_5$)$_5$ on various substrates in a hot wall reactor at 550°C [229]. YMnO$_3$ films were grown at 450°C by MOCVD with Y(thd)$_3$ and (CH$_3$C$_5$H$_4$)Mn(CO)$_3$ as precursors for Y and Mn, respectively [159]. The YMnO$_3$ films were used as buffer layers for SBT film growth on Si.

3.2.2. Plasma-Enhanced CVD

Most authors report deposition temperatures in the range of 600–700°C for typical ferroelectrics to have acceptable electrical properties. It is necessary to lower the deposition temperatures as much as possible so as to not damage underlying structures for future microelectronic devices. One way to decrease the deposition temperature is to use additional external energy sources, such as plasmas or photons. The use of such sources is expected to have an influence on crystal quality and thus on the ferroelectric properties of the oxide thin film. As far as the MOCVD of PbZr$_x$Ti$_{1-x}$O$_3$ is concerned, a few papers have described the use of such external power sources. Photo-enhanced deposition of PbTiO$_3$ was done using the precursors Pb(C$_2$H$_5$)$_4$, Ti(^{i}OC$_3$H$_7$)$_4$, and O$_2$ in combination with a Xe-Hg lamp of 500 W [201]. Using this photo-enhancement, improved crystallinity was achieved and the deposition temperature of PbTiO$_3$ was lowered to 530°C. When the O$_2$ was replaced by the more reactive NO$_2$, improvements in the dielectric properties of the photo-irradiated PbTiO$_3$ films deposited at temperatures as low as 550°C were observed [201].

The advantages of low-temperature plasma-enhanced CVD (PECVD) are (i) previously deposited layers, interfaces, or doping profiles are not perturbed; (ii) solid-state diffusion and defect formation is minimized; (iii) compatibility with low-melting point substrates/films is maintained; and (iv) complex and metastable compounds may be synthesized. PECVD enables a direct, one-to-one correspondence between gas and solid composition, in contrast to the results from single-thermal CVD experiments. Although the chemistry and physics of a plasma glow discharge are complex, the plasma performs two key functions: (i) reactive chemical species generated by electron-impact overcome kinetic (activation) barriers for a particular reaction, and (ii) activated gas atoms and molecules, energetic neutrals and ions, metastable species, electrons, and ions bombard and stimulate surface processes. These effects are equivalent to raising the temperature of the reaction without

substantially raising the temperature of the substrate. The combination of physical processes and chemical reactions results in material properties and etching profiles and rates unachievable by either process acting alone.

PECVD of PbZr$_x$Ti$_{1-x}$O$_3$ using 13.56-MHz rf plasma and the precursors Pb(C$_2$H$_5$)$_4$, Zr(^{i}OC$_4$H$_9$)$_4$, Ti(^{i}OC$_3$H$_7$)$_4$, and O$_2$ has been reported [230]. Films were deposited at substrate temperatures as low as 425–500°C. At a pressure of 0.3 torr and low rf power, the pyrochlore-type PbZr$_x$Ti$_{1-x}$O$_3$ was formed. Increasing the rf power and lowering the reactor pressure to 0.16 torr yielded up to 92% of perovskite-type PbZr$_x$Ti$_{1-x}$O$_3$ in the thin films deposited on a platinized substrate. Typical deposition rates of 0.5–0.6 μm/h were obtained, which are in the same range as for thermal MOCVD. Composition of PECVD PbTiO$_3$ thin films was intensively influenced by the input flow rate ratio of precursors but was independent of the deposition temperatures [231]. Recently, PECVD-grown SrTiO$_3$ and BST thin films were reported [232, 233]. For the SrTiO$_3$ thin films, Sr(thd)$_2$ and Ti(^{i}OC$_3$H$_7$)$_4$ were used as precursors. Deposition at higher rf power ($>$180 W) resulted in a poor crystalline structure. The leakage current density decreased as the deposition temperature increased in the temperature range of 400–550°C [232]. The BST precursors were Ba(thd)$_2$-trietherdiamine, Sr(thd)$_2$-trietherdiamine, and Ti(^{i}OC$_3$H$_7$)$_2$(thd)$_2$ [233]. It was shown that the reactive oxygen radicals produced by microwave plasma were involved more in breaking O−C bonds than in substituting the precursor ligands for the film formation. All BST films grown at temperatures of 560–675°C were highly textured on Pt-coated Si [233].

Although SBT films have high potential for nonvolatile memory device application, the desired electrical properties can be obtained only by annealing at temperatures above 750°C. Recently, SBT thin films were deposited at a low temperature (about 550°C) by plasma-enhanced MOCVD (PE-MOCVD) [234]. For the SBT growth, Sr(hfa)$_2$-tetraglyme, Bi(C$_6$H$_5$)$_3$, and Ta(OC$_2$H$_5$)$_5$ were used as precursors for Sr, Bi, and Ta, respectively. A BTO (Bi$_4$Ti$_3$O$_{12}$ or Bi$_2$Ti$_4$O$_{11}$) phase formed at the interface between the SBT films and Pt/Ti/SiO$_2$/Si during the SBT deposition at 550°C [235]. The BTO phase decreased the leakage current density of the SBT films.

3.2.3. Aerosol-Assisted CVD

A major aspect in the development of MOCVD processes is the controlled introduction of relatively involatile precursors to the reactor. For precursors of low volatility, viable alternatives to the classical transport methods have been proposed [236]. Dissolution of solid precursors in an appropriate solvent can be an alternative to the limitation of solid precursors. In a liquid-source MOCVD technique, liquid precursors are transported with or without a solvent and are evaporated or injected by a nebulizer just before entering the deposition reactor. These options may prove useful for processes using liquid precursors with very low vapor pressures or solid precursors [188, 201, 237, 238], and the technique using a liquid-delivery system is a common MOCVD process. For these emerging delivery technologies, liquid precursors or solutions of precursors in organic solvents are used.

In aerosol-assisted CVD (AACVD), which includes spray pyrolysis, a solution of metal complexes is transported directly via aerosol formation onto the substrate. Complicated reactions between precursors and deposited molecules of the solution occur before the deposit is formed. These reactions include evaporation of solvent melting of precursors, vaporization, pyrolysis, and nucleation of solid film. Criteria for selecting solvents for the nebulization process in AACVD are high solubility of the precursors, low vapor pressure, and low viscosity. Various alcohols or acetylacetone have often been used [239]. 2-Methoxyethyl ether, $CH_3(OC_2H_4)_2OCH_3$ (diglyme; boiling point $160°C$), and other glymes are also of interest. The solvent is an undesirable ballast in a CVD process; thus it should contain few carbon atoms, because the partial pressure of the carbon dioxide generated by the decomposition of the solvent and/or precursors is generally the limiting factor in film growth rates.

Solubility becomes a critical issue for AACVD, whereas volatility is the most important issue for classical MOCVD. Compounds with poor volatility (e.g., acetates) have been used as precursors in AACVD experiments. Another important issue in AACVD is the long-term stability of the solution. Evolution of the solubility properties and precipitation may occur during the deposition process, especially during storage, as a result of impurities (hydrolysis-generated by trace amounts of water) or by reactions between different types of precursors used for multicomponent systems. Ligand exchange reactions or formation of mixed metal species can occur if metal alkoxides are associated with β-diketonates or carboxylates in the feed solution; this can modify the homogeneity of the system. Readily synthesized, low-cost chemicals that can be manipulated on open benches are also an important objective. The requirements of precursors for classical MOCVD and for AACVD have been discussed in detail by Hubert-Pfalzgraf and Guillon [176].

There have been some reports on the deposition of ferroelectric thin films by the AACVD technique. $BaTiO_3$ thin films were deposited at $550–700°C$ using barium-diethylhexanoate $[Ba(O_2C_2HC_2H_5C_4H_9)_2]$ and $Ti(^iOC_3H_7)_4$ as the precursors and acetylacetone as the solvent [240]. Growth of $PbTiO_3$, PLT, and PZT thin films by AACVD has also been reported [241]. Pb and La acetates and Ti and Zr isopropoxides were used as precursors and taken in proper proportions in methanol solvent. The films were deposited on $SrTiO_3$ substrates at about $350°C$ and annealed at the same temperature for 12 hours in air. In this study, the $PbTiO_3$, PLT, and PZT thin films showed a weak $P–E$ hysteresis. Epitaxial $LiNbO_3$ and $LiTaO_3$ thin films were grown on sapphire substrates by AACVD [242, 243]. A bimetallic complex of Li(thd) and $Nb(OC_2H_5)_5$ [or $Nb(OC_2H_5)_5$] had been formed to prepare a single-source precursor by reacting in refluxing toluene under Ar, as Li(thd) is insoluble in toluene in the absence of the Nb alkoxide. $LiNbO_3$ films were deposited at substrate temperatures of $490–590°C$ [242]. $LiTaO_3$ thin films grown at $452°C$ were

amorphous and were crystallized by annealing at $700°C$ [243]. Other groups also reported on AACVD $LiTaO_3$ thin films [244–246]. A $LiTa(^tOC_4H_9)_6$ single-source precursor was dissolved into toluene and injected in a pulsed mode [244]. The films were grown at the temperatures of $550–800°C$ on a sapphire substrate. Textured $LiTaO_3$ films with preferred c-axis orientation were grown on SiO_2-coated Si (111) substrates at the temperature range of $600–650°C$ [245]. The precursor solution was prepared by mixing $Ta(OC_2H_5)_5$ and Li(acac) in methanol (used as a solvent) at a ratio of $Li/Ta = 0.7$ [245]. Recently, (116) oriented $LiTaO_3$ thin films were deposited by AACVD on MgO-buffered Si substrates [246]. $Li(OC_2H_5)$ and $Ta(OC_2H_5)_5$ were reacted in 2-methoxyethanol by refluxing and distilled to prepare precursor solution. Addition of small amounts of acetylacetone to the precursor increased the deposition rate under the same deposition conditions. The $LiTaO_3$ thin films were deposited at $430°C$ and annealed at $650°C$ for 1 hour. Recently, inductively coupled plasma was used to facilitate the AACVD process [247]. $PbTiO_3$ thin films were deposited using $Pb(thd)_2$ and $Ti(^iOC_3H_7)_4$ dissolved in an organic solvent (heptane).

3.3. Chemical Solution Deposition

During the past decade, numerous solution-based chemical processing techniques for the synthesis of electroceramic thin films have been developed. Unique names are used to describe techniques, depending on the manner in which the solution is prepared (e.g., sol-gel method, MOD, hybrid sol-gel method) and the way in which the lay-down is accomplished (e.g., spin-on, dip-in, mist coating, electrochemical formation). Chemical solution deposition (CSD) uses colloidal solutions, oligomeric and polymeric solutions, and complex and simple solutions. One problem with the use of CSD methods is the lack of fundamental understanding of the effects of solution chemistry variations and structural evolution behavior on film dielectric and ferroelectric properties. Reported results for films prepared by CSD show significant differences in dielectric and ferroelectric properties, crystallization temperatures, and thin film microstructures. Since the final film properties can conceivably be affected by variations in the preparation process, each step of the overall processing route, from solution preparation to heat treatments must be considered.

3.3.1. Sol-Gel Method

Sol-gel processing is an important and useful processing technique for growing metal oxide–based ferroelectric thin films. Since the largest group of ferroelectric materials includes compounds of metal oxides, the sol-gel process has been particularly useful and there is a strong effort in using this technique to fabricate oxide-based ferroelectric films. Advantages of the sol-gel processing technique include good homogeneity, ease of accurate composition control (which is especially important for multicomponent systems), large-area thin films, high possibility of epitaxy, ease of process integration with standard semiconductor manufacturing technology, and comparatively lower cost than other techniques. The first solution deposition process introduced to the fabrication of oxide-based ferroelectrics was the work on fabricating $BaTiO_3$ conducted by Fukushima et al. [248]. Subsequent reports of the use of the sol-gel process for the PZT family of ferroelectric thin films followed [249–251]. The sol-gel process involves (i) preparation of a homogeneous solution containing the precursor for the ferroelectric, usually metal alkoxides; (ii) deposition of the solution on the substrate, producing a wet-gel thin film; (iii) drying or pyrolysis of the film to form an amorphous thin film; and (iv) crystallization of the amorphous thin film at elevated temperatures. The first step (i.e., preparation of solutions) includes tailoring of solution characteristics, such as viscosity. In the deposition step, hydrolysis and polycondensation occur simultaneously. Also, pyrolysis and crystallization may be accomplished by a single-step heat treatment. Despite the apparent ease and versatility of the scheme, the properties of oxides are extremely sensitive to the different conditions of their preparation. Changes in any of the processing steps could potentially affect not only the nature of the species obtained in the subsequent step, but also the final film microstructure and electrical properties.

3.3.1.1. Preparation of Precursor Solutions

The sol-gel method begins with the preparation of liquid precursors. A *sol* is a suspension or dispersion of discrete colloidal particles or polymeric species in a solution, and a *gel* is a colloidal or polymeric solid that has an internal network structure with mixed or highly dispersed solid and fluid components. The sol is prepared by dispersing the starting materials in an organic solvent. Preparation of the initial solution becomes an especially important stage for solutions used in film applications. The starting materials can be metal-organics (e.g., alkoxides), acetates, or inorganics (e.g., metal hydroxide, nitrites). Probably the best starting materials for sol-gel precursors are the materials known as metal alkoxides. The alkoxides of most metals can be synthesized [252] and are convenient starting materials with respect to availability and cost. All metals form alkoxides, and these can be represented by the general formula $M(OR)_n$, where M is the metal or metalloid with valency n and R is an alkyl group. Table I shows various types of alkyl groups. These can be considered to be derivatives of alcohol, ROH, in which H has been replaced by M.

All metal alkoxides, with two notable exceptions (the alkoxides of silicon and phosphorous), are rapidly hydrolyzed to the corresponding hydroxide or oxide. The overall reaction (hydrolysis) can be represented as follows:

$$M(OR)_n + nH_2O \rightarrow M(OH)_n + nROH$$
$$2M(OH)_n \rightarrow M_2O_n + nH_2O$$

or, for partial hydrolysis,

$$M(OR)_n + mH_2O \rightarrow M(OH)_m(OR)_n + mROH$$

The byproduct, ROH, is an aliphatic alcohol that is readily removed by volatilization. For film deposition, the precursor so-

	Table I. Alkyl and Alkoxy Groups		
Alkyl		Alkoxy	
Methyl	$\bullet CH_3$	Methoxy	$\bullet OCH_3$
Ethyl	$\bullet CH_2CH_3$	Ethoxy	$\bullet OCH_2CH_3$
n-Propyl	$\bullet CH_2CH_2CH_3$	n-proxy	$\bullet O(CH_2)_2CH_3$
Iso-propyl	$H_3C(\bullet C)HCH_3$	Iso-proxy	$H_3C(\bullet O)CHCH_3$
n-Butyl	$\bullet CH_2(CH_2)_2CH_3$	n-butoxy	$\bullet O(CH_2)_3CH_3$
Sec-butyl	$H_3C(\bullet C)HCH_2CH_3$	Sec-butoxy	$H_3C(\bullet O)CHCH_2CH_3$
Iso-butyl	$\bullet CH_2CH(CH_3)_2$	Iso-butoxy	$\bullet OCH_2CH(CH_3)_2$
Tert-butyl	$\bullet C(CH_3)_3$	Tert-butoxy	$\bullet OC(CH_3)_3$

Note: $\bullet$ indicates the bonding site.

lutions may be partially hydrolyzed beforehand. Modification of the precursor solution is discussed later in the chapter. There is also a limited class of compounds known as double alkoxides, which contain two different metals in the same compound and have the general formula

$$M'_x M''_y (OR)_m$$

where M' and M'' are metals and x, y, and m are integers. Double-metal alkoxides have advantage of retaining exact molecular stoichiometry between metals, which can be used to deposit stoichiometric multicomponent ferroelectric oxide films. The modern state of chemistry of these compounds has been reviewed [253]. The simplest example of a double-metal alkoxide for ferroelectrics is the precursor for $LiNbO_3$. Preparation of a $LiNbO_3$ precursor solution involves making a solution of $Li(OC_2H_5)$ in a suitable organic solvent (ethanol or 2-methoxyethanol), then reacting the solution with $Nb(OC_2H_5)_5$ to form a $LiNb(OC_2H_5)_6$ precursor solution [254–257]. Recently, triple-alkoxide percursors, $Sr[BiTa(OC_2H_4OCH_3)_9]_2$ and $Sr[BiTa(OC_2H_5)_9]_2$ were also synthesized for the deposition of SBT thin films [258].

Recent progress in sol-gel technology is associated with substitution of metal alkoxides by methoxyethoxides, $M(OC_2H_4OR)_n$ (R = CH_3, C_2H_5). Methoxyethanol proved to be the most convenient solvent for the preparation of the initial solution, which is more and more widely used instead of aliphatic alcohol. One of its advantages lies in its ability to dissolve other organic and inorganic derivatives of metals along with alkoxides, which greatly extends the potential uses of the sol-gel method. Dissolution of metals in methoxyethanol may be used for the synthesis of solutions containing two or three elements. Besides, methoxyethoxides are somewhat more stable to hydrolysis and oxidation. The use of methoxyethanol instead of aliphatic alcohol as the solvent produced films with much improved density and morphology for $LiNbO_3$ [257, 259] and PZT [260]. Methoxyethanol-based precursors were found to be more stable with respect to hydrolysis, a result attributed to the steric and electrostatic differences between the alkoxy ligands [257].

For some metals, using alkoxides is inconvenient because of preparation problems or unavailability, and alternative starting materials must be found. Metal salts (acetates and nitrates) provide a viable alternative provided that they can be readily converted to the oxide form by thermal or oxidative decomposition and (preferably) are soluble in organic solvents. Nitrates are really the only suitable inorganic salts, because others, such as chlorides or sulfates, are more thermally stable, and removing the anion may be difficult. A major drawback to the use of a metal nitrate precursor is the tendency for nitrates to crystallize during dehydration, which tends to disturb the homogeneity of the system [261]. Salts of organic acids (particularly acetates) are potential candidates. A disadvantage of using acetates is that they do not thermally degrade as cleanly as nitrates and can be a source of carbonaceous residues. Further, in view of the weakly acid character of the parent acid, solutions of many acetates are basic but this can be partly overcome by buffering the solution with acetic acid.

Use of a combination of metal alkoxide and metal acetates is also possible. Sol-gel preparations involving salts are usually more complex than those with only alkoxides because of the more facile hydrolyzability of the alkoxides. It was found that certain acetates react with some alkoxides to form metallometaloxane derivatives with the liberation of alkyl acetates [262]. The first step in the reaction is

$$M(OR)_n + M'(OAc)_x$$
$$\rightarrow (RO)_{n-1}M-O-M'(OAc)_{x-1} + ROAc$$

where Ac represents the group $-COCH_3$. The reaction continues with further reaction of acetate and alkoxide groups, leading to higher-molecular-weight products. Heating the reaction mixture containing more alkoxide than acetate groups in the absence of solvent distills out about 60–80% of the ester. Often this partially hydrolyzed alkoxide derivative is soluble in organic solvents, or else its solubility can be enhanced to some extent by refluxing the same with alcohol. Thus, it is often a convenient precursor for the sol-gel process. It is preferable that the molar ratio of alkoxide to acetate be greater than 1, this leads to soluble and hydrolyzable products that can be subsequently used in sol-gel preparation similarly to a double alkoxide. Examples of this type of combination include BaTiO₃ [263], PbTiO₃ [264, 265], and PZT precursors [266]. Precursor complexes in the PZT system with varying Zr/Ti ratios were synthesized under dry nitrogen atmosphere using standard Schlenk technique by the following reaction of lead acetate with Zr and Ti methoxyethoxide [266]:

$$Pb(OAc)_2 + xZr(OR)_4 + (1-x)Ti(OR)_4$$
$$\rightarrow Pb(Zr_xTi_{1-x})O_2(OR)_2 + 2ROAc$$

where $R = CH_3OCH_2CH_2$. The Ti and Zr alkoxides were refluxed separately, thereby deliberately avoiding the formation of a double alkoxide. The reactions of the individual alkoxides with lead precursor promotes the replacement of the acetate ligands with alkoxide ligands. This process minimizes the number of acetate groups, which do not undergo hydrolysis, and thereby gives a cleaner pyrolysis.

Preparation of the organometallic precursor solution is usually carried out under dry inert gas (Ar or N_2) using an air-sensitive handling apparatus, since the metal alkoxides hydrolyze vigorously on contact with water. All metal alkoxides decompose even in the presence of traces of water in solvents or in the atmosphere. The standard Schlenk technique was used to synthesize organometalic precursor complexes; the details of the experimental setups for atmospheric and vacuum distillation have been reported earlier [267].

For the PZT precursor, the vapor temperature is monitored by thermometer during distillation to indicate completion of alcoholysis reactions for formation of the Ti and Zr methoxyethoxides. Methoxyethoxides of Ti and Zr are created by the alcohol exchange reactions of methoxyethanol with Ti-isopropoxide and Zr-butotoxide. A constant vapor temperature (122–124°C) indicates distillation of only methoxyethanol and demonstrates that further exchange of alkoxide groups with methoxyethanol is not possible. In the preparation of the PZT alkoxide precursor, lead acetate trihydrate is first dissolved in 2-methoxyethanol. To form lead methoxyethoxide, dehydration of lead acetate trihydrate and subsequent alcoholysis reaction of anhydrous lead acetate with methoxyethanol or ROH is necessary. However, an incomplete alcoholysis reaction generally occurs [268]. Stoichiometric amounts of Zr and Ti methoxyethoxides are mixed with the lead precursor, and the resulting solution is refluxed, followed by redilution [266]. Excess Pb precursor is often added at this stage, as necessary.

A distinct advantage of sol-gel processing over conventional film-forming techniques is the ability to precisely control the microstructure of the deposited film, i.e., pore volume, pore size, and surface area of the deposited film. Because the pore structure reflects to some extent the size and topology of the solution-grown polymers, microstructure tailoring is achieved by controlled polymer growth in solution before film deposition. Before deposition, water and catalysts are added to the sol solution to initiate a series of hydrolysis and polycondensation reactions that produce an oxide gel network. The network generation in the precursor solution increases the solution viscosity. The hydrolysis step must be controlled so that precipitation is avoided while polymerization proceeds. The partial hydrolysis products condense to form metal–oxygen–metal linkages that form the basis for the final crystalline structure. The conditions of partial hydrolysis (i.e., alkoxide concentration, relative amount of water, additive) have important effects on the deposition characteristics, final macrostructure quality, and microstructure development in the layer on heat treatment. The possible structures of the solution precursors range from weakly branched polymeric species to discrete porous clusters to fully dense colloidal uniform particles depending on the hydrolysis conditions [269]. Solution conditions also affect the relative rates of condensation and evaporation during deposition. Both polymer topology and the relative rates of condensation and evaporation influence the evolution of structure during film formation.

For control of the hydrolysis conditions, different additives (e.g., CH_3COOH and NH_4OH as acid and base catalysts, respectively) were added to the solution. Acid and base catalysts can influence both the hydrolysis and condensation reaction rates and thus the structure of the condensed product. The effects of the catalysts for silicon alkoxides have been well investigated [269]. In silicate solutions prepared at low pH (<2.5) and with understoichiometric addition of water, polymer growth is biased toward linear or randomly branched chains during polycondensation reaction. In addition, the condensation rate is very low under these conditions. The linear chains are interpenetrable [270] and produce denser film structures. If the pH is increased slightly above 2.5, then the condensation rate is greatly increased, resulting in the formation of weakly condensed, spherically expanding clusters whose density decreases radially from the center of mass. Porosity is created at an early stage of the deposition process, and the resulting pore volume is quite large. Further increases in pH and/or over stoichiometric additions of water increase the rate of depolymerization in silicate systems. This leads to the formation of highly condensed species through bond breakage and reformation. A high-pH solution often yields particulate film structures [269].

It is not clear whether these catalyst effects are applicable to other organometallic precursors. NH_4OH, CH_3COOH, and HNO_3 were added in PZT solution [260, 271], and the resulting microstructures were not so sensitive to the catalysts. HNO_3 and NH_4OH were used in $PbTiO_3$ solution [272], and it was found that acidic gels seemed more capable of polymeric rearrangement during drying, yielding denser amorphous structures with microcrystalline regions in $PbTiO_3$ films. It was suggested that the acid-catalyzed precursor consisted of entangled polymer chains that allowed a dense gel to form [272]. Additives of NH_4OH, HNO_3, and benzoic acid were investigated in $Pb(Mg,Nb,Ti)O_3$ (PMNT) [273]; benzoic acid promoted the formation of the most uniform, dense microstructures in PMNT films. Also, dense $(Sr,Ba)Nb_2O_6$ (SBN) thin films and dense $KNbO_3$ films can be made using the alkoxide solution modified by 2-ethylhexonaic acid [274]. Other types of chemical additives, such as glycol and β-diketones, can also be used. Their addition results in partial substitution of OR groups by acetic or acetylacetonate groups, condensation with formation of oxo bridges, and elimination of ester. The use of ethylene glycol and glycerol can reduce the pore size of the gel films. Gel films prepared from solutions containing glycerol are transparent after drying, indicating that the pore size of the gel films is very small or that pores may be totally absent. Such films prepared can be crystallized into the perovskite structure at temperatures as low as 450°C [275].

The mechanism for the use of ethylene glycol and glycerol in precursor solutions for PZT films is still under study. These substances do not increase the viscosity of the solution, but may simply act as solvents in the presence of water. It was also observed that solutions containing ethylene glycol and glycerol gel at higher temperatures [275]. Since the ethylene glycol and glycerol have more than one functional group, they also act as chelating agents to form chelated derivatives with metal complexes:

$$M(OR)_n + HO-CH_2-CH_2-OH$$

$$\leftrightarrow (RO)_{n-2}M \begin{array}{c} O \\ \diagdown \\ O \end{array} \begin{array}{c} CH_2 \\ | \\ CH_2 \end{array} + 2ROH$$

$$M(OR)_n + HO-CH_2-\underset{\underset{OH}{|}}{CH}-CH_2-OH$$

$$\leftrightarrow (RO)_{n-2}M \begin{array}{c} O-CH_2 \\ \diagup \\ \diagdown \\ O-CH_2 \end{array} CHOH + 2ROH$$

Since the chelated derivatives are more resistant to hydrolysis than their counterpart metal complexes, formation of the chelated derivatives may be the reason why such solutions form gels at high temperatures. Also, the number of the alkoxide groups taking part in the subsequent hydrolysis is thus considerably limited. Thus, it is possible to change qualitatively the properties of the oxide materials obtained by variation of the hydrolysis conditions. Organically modified sol-gel precursors have been prepared in many different ways [276–278].

Another important factor is the effect of the correlation between the solution structure and the final crystal structure of sol-gel-derived systems. It was realized that the structure of the alkoxides in the solution has a significant bearing on the early stage of crystallization. Homogeneity at the molecular level allows better compositional control and lower crystallization temperature. It was also found that a close resemblance of the local atomic arrangement of the alkoxide complexes in the solution can play a critical role in the crystallization processes. It enables epitaxial growth of crystalline thin films on selected substrates. Such similarity of structure allows crystallization and epitaxy without long-distance diffusion, which in turn plays a crucial role in the sol-gel process. From such considerations, the crystallization temperature of $LiNbO_3$ has been significantly lowered [279, 280].

3.3.1.2. Thin Film Deposition

Sol-gel deposition is usually done on unheated substrates, whereas chemical vapor deposition is done at elevated temperatures due to the need for a hot substrate surface for condensing films. The fluid sol or solution is ideal for preparing thin films by such common processes as dipping, spinning, and spraying. The dipping and spinning processes have been examined with respect to such parameters as withdrawal rate, spin speed, viscosity, surface tension, and evaporation rate. In the deposition process, specific attention is given to the consequences of the overlap of the deposition and drying stages [269].

The dip coating process can be divided into five stages: immersion, startup, deposition, drainage, and evaporation [281]. With volatile solvents, evaporation normally accompanies the startup, deposition, and drainage steps. How the deposited film thickness depends on measured rheological parameters is not well accounted for by the available approximate theories. In real life situations, the slower the substrate speed, the thinner the film and the greater the overlap of the deposition and drying stages. Since condensation continues during sol-gel film formation, the relative condensation and evaporation rates will dictate the extent of further cross-linking that accompanies the deposition and drainage stages. The overlap of the deposition and drying stages can significantly affect the drying profile. However, it is unclear what effect these phenomena have on the structure of the depositing films.

Major uses of spin coating have been in microelectronic manufacturing, photoresister coating, and magnetic memory technology. The technique has been popularly applied to the deposition of ferroelectric oxide thin films by CSD methods. Spin coating is divided into four stages [282]: deposition, spin-up, spin-off, and evaporation (although evaporation accompanies the other stages, as discussed for dip coating). An excess of liquid is delivered to the substrate at rest or while the substrate is spinning slowly in the first stage. In the second stage, the liquid flows radially outward, driven by the centrifugal force generated by the rotating substrate. An advantage of spin coating is that a film of liquid tends to become uniform in thickness during spin-off. This tendency arises due to the balance between centrifugal force and viscous force. The thickness of a spin coating is the outcome of the delicate transition by which evaporation takes over from spin-off. According to a model of spin coating by Meyerhofer [283] that separates the spin-off and evaporation stages, the final thickness is proportional to $(\eta m/\omega^2)^{1/3}$, where η is the viscosity of solution, m is the evaporation rate of solvent, and ω is the angular velocity.

Solvent is evaporated during the coating processes, and wet gel film is formed on the substrates. Polymer growth during the deposition probably occurs through a process similar to cluster-cluster aggregation [284]. Gelation is referred to the moment when the condensing network is sufficiently stiff to withstand flow due to gravity or centrifugal force and yet is still filled with solvent. Further evaporation may collapse the film or generate porosity within the film. The overlap of the deposition and evaporation stages establishes a competition between evaporation and the continuing condensation reaction. The evaporation compacts the structure of film, whereas the condensation reactions stiffen the structure, resulting in increased resistance to compaction. Because aggregation, gelation, and drying occur in the short time span of deposition, films experience much less aging than bulk gels, resulting in more compact dried structures. The term "aging" describes the process of change in structure and properties after gelation. Aging may involve further condensation, dissolution, and precipitation of monomers or oligomers, or phase transformations within the solid or liquid phases. These wet gel films are dried on a hot plate or fired at elevated temperatures to obtain amorphous films. During drying and firing, gel films are constrained by the substrate, and shrinkage of film volume by solvent evaporation often leads to cracking of the film, especially in thick films. To avoid film cracking, previously mentioned additives can be used. In wet gel films, the initial polymeric network is highly stretched and expanded by the liquid. If the liquid phase in the gel is not constrained as it is removed, then the network will spontaneously shrink to release this tensile stress.

3.3.1.3. Heat Treatments

Since the dry gel films contain organic residues such as carbon species, pyrolysis is necessary to remove the organic components. Thermal analysis (i.e., thermogravimetric analysis and differential thermal analysis) is used to study the decomposition of organometallic molecules into metal oxides. Most of the organic components are decomposed and removed below about 400–450°C [275, 277]. However, for PZT films derived from acetates, it has been shown that substantial amounts of organic residues were found at higher temperatures [277]. Thus, sufficient duration of pyrolysis in oxygen atmosphere may be necessary. In the pyrolysis stage, the films are still amorphous. To crystallize the films, annealing at higher temperatures should be followed. Sometimes the pyrolysis step can be omitted.

Unlike the various vapor deposition techniques, film deposition for sol-gel process is not done atom-by-atom, and therefore relatively long-distance diffusion is inevitable if the local arrangement does not resemble that of the crystalline state. As a result, the mechanisms of crystallization and epitaxy are expected to be differ from conventional vapor deposition methods. Solid-state crystallization for sol-gel-derived films thus requires relatively high-temperature postannealing process. The microstructural evolution occurs after each processing step of heat treatments. Changes in thickness, refractive index, and band gap of films annealed at different temperatures are correlated with chemical modifications of the precursor solution, as well as with structural and microstructural changes occurring during the formation of final thin films [285]. Determining the best temperature schedule for the control of microstructure and properties of sol-gel-derived films is necessary. The temperature schedule is especially important for thick films, to prevent cracking due to volume shrinkage [275]. In this process, a metastable intermediate phase can be formed, which is in most cases undesirable.

A proper combination of deposition and heat-treatment processes was found to be effective for obtaining high-quality films. Recently, BST films with good ferroelectricity have been obtained using a highly diluted precursor solution (0.05 M solution) [286]. After each layer was spin coated to 8-nm thickness, the film was given a pyrolysis heat treatment at 350°C for 5 minutes to remove residual organics, and then annealed at 750°C for 10 minutes in air. Ferroelectricity and dielectric properties of the films were strongly dependent on the concentration of the precursor solution, as shown in Figure 17.

For those processes involving postdeposition annealing, crystallization in a conventional furnace was conducted at temperatures above 650°C for 30 minutes to several hours. This conventional furnace annealing process led to undesirable results, such as Pb losses in PZT thin films and film–substrate interface reactions. As an alternative method of crystallization, the rapid thermal annealing (RTA) technique has been applied

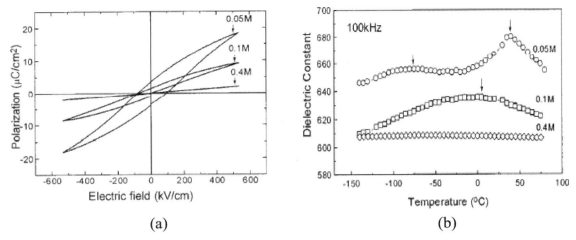

Fig. 17.   (a) $P–E$ hysteresis of sol-gel-derived $Ba_{0.8}Sr_{0.2}TiO_3$ thin films from 0.4, 0.1, and 0.05 M precursor solutions, respectively. (b) Temperature dependence of the dielectric constant for sol-gel-derived $Ba_{0.8}Sr_{0.2}TiO_3$ thin films from 0.4, 0.1, and 0.05 M precursor solutions, respectively. Reprinted with permission from J.-G. Cheng et al., *Appl. Phys. Lett.* 75, 2132 (1999). Copyright 1999, American Institute of Physics.

to ferroelectric thin film process [287–289]. RTA has several advantages over conventional furnace annealing. It can reduce the thermal budget by minimizing the processing time to only a few seconds and minimize the interfacial reaction and loss of such volatile elements of films as Pb and Bi even at annealing temperatures above 700°C. More importantly, for the PZT system, the absence of secondary phases (e.g., a pyrochlore phase) is a key factor in quality control of films. The metastable pyrochlore phase has a nonferroelectric oxygen-deficient cubic fluorite structure [290] and is often observed in films processed at low temperatures (400–600°C) or in Pb-deficient films [287, 291, 292]. The pyrochlore phase is therefore undesirable for the applications under consideration. The same aspects of pyrochlore phase formation have been observed in $Bi_4Ti_3O_{12}$ films [293]. Transformation from the pyrochlore phase to the perovskite phase requires activation energy, and thus it is necessary to minimize crystallization of undesirable metastable phases. RTA is an effective method for this purpose. Theoretical investigation has been reported on the effects of the RTA process on thin film crystallization in the perovskite phase without formation of a metastable pyrochlore phase [294]. The transformation from the pyrochlore to the perovskite phase appeared to be influenced by ramping rate and holding time [295]. Interestingly, a lower ramping rate was effective to the transformation.

Recently, Hu et al. [296] have reported that the preferentially (100)-oriented SBT films prepared by a layer-by-layer RTA method show a very large remanent polarization (28.6 $\mu C/cm^2$) at an annealing temperature of 750°C [296]. Low-pressure (50 torr) RTA under oxygen ambient was shown to be effective in lowering the crytallization temperature of sol-gel-derived PZT films [297]. This effect was ascribed to the effectiveness in removing residual gases (e.g., acetone, CO_2, H_2O) at low pressure. A new face-to-face annealing method has been proposed to control composition of the volatile Bi element during the crystallization process, in which a sol-gel SBT film de-

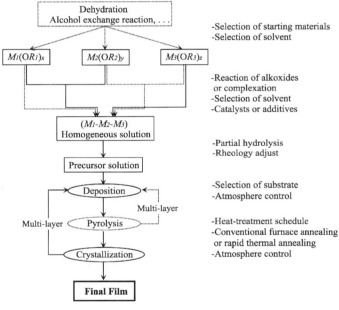

Fig. 18.   Schematic diagram for the preparation of precursor solution, deposition, and heat-treatments. Key focusing points are described. The dotted and solid lines are for optional procedures.

posited on a substrate is placed directly (with the film side down) on the other SBT film during crystallization [298]. The samples were crystallized using a RTA furnace at 750°C for 30 minutes in an oxygen flow and successively annealed at 400°C in a 6.2 wt% ozone flow by the face-to-face annealing method. A remanent polarization value as large as 23 $\mu C/cm^2$ was obtained, and the saturated leakage current density was below 1×10^{-9} A/cm^2 at an electrode field of 65 kV/cm [298]. The overall processes for the deposition of ferroelectric oxides by the sol-gel method are shown schematically in Figure 18.

3.3.2. MOD and the Hybrid Sol-Gel Method

MOD and hybrid sol-gel methods are similar to the sol-gel method except in the preparation of precursor solutions. In contrast to sol-gel approach, MOD eliminates the hydrolysis step through careful selection of the precursors and involves direct pyrolysis of the dried films. The method is more straightforward and simpler than the sol-gel method. The MOD method typically uses carboxylate precursors, $M(OOCR)_n$, with large organic moieties, such as 2-ethylhexanoate ($R = C_7H_{15}$) and neodecanoate ($R = C_9H_{19}$) ligands. The precursors are typically water insensitive. To avoid disturbing the chemistry, no water is added, and a hydrophobic solvent can even be used. In MOD, gellation is obtained by the evaporation of solvent within a more compliant film, whereas in the sol-gel process, gellation is obtained by hydrolysis.

The MOD films can accommodate to the substrate and release stress during drying, resulting in less cracking than occurs in sol-gel films. However, the use of precursors with smaller organic groups is desirable, because cracking problems are frequently encountered in films with high organic content. To optimize both the chemical properties and the rheological and drying properties of the solution, combining the sol-gel and MOD methods is necessary. This hybrid solution deposition method generally uses low-molecular-weight carboxylate and alkoxide precursors, namely acetates and propoxides or butoxides. In the hybrid sol-gel method, less attention is given to control of the hydrolysis conditions because of the use of chemical modifiers, which diminish the hydrolysis rates of the alkoxides. More often, the hybrid sol-gel method is simply referred to MOD.

BST thin films have been fabricated by MOD using Ba-acetate, St-acetate, and Ti-ammonium-lactate as starting materials for precursors [299]. The RTA process increased the grain size of the films. SBT films also have been deposited by MOD [300]. Precursor solutions were prepared by dissolving $Sr(^iOC_3H_7)_2$ and $Ta(OC_2H_5)_5$ in 2-ethylhexanoic acid and mixing Bi-2-ethylhexanoate toluene solution. Xylene was used as solvent. Solution of 20% Sr-deficient and 10% Bi-excess composition resulted in better electrical properties than found the stoichiometric solution. Octylates of bismuth and titanium with xylene solvent were used for the synthesis of $Bi_3Ti_4O_{12}$ thin films [301], a two-step MOD process controlled the growth orientation.

3.3.3. Synthesis of Ferroelectric Thin Films by CSD

Hundreds of reports have been published on the fabrication of ferroelectric oxide films using the CSD. In this review, details of every aspect are not discussed. Table II summarizes recently reported source materials used in the preparation of precursor solutions for limited ferroelectric materials. Many factors should be considered in controlling the quality of ferroelectric thin films. In the CSD techniques, precursor chemistry is an important factor. Although there is no absolute guide to the preparation of the precursor solution, it is preferable to choose precursor solutions that can feasibly control hydrolysis and condensation as well as rheological properties. Also, precursor solutions containing as little organic content as possible are desirable, to reduce residual carbon contamination, which often deteriorates the electrical properties of film. It has been observed that the chemistry of precursor preparation has significant effects on microstructure, texture development, and physical properties [314, 316, 318, 326]. Proper choice and modification of precursors enables low-temperature crystallization of films [258, 312, 322, 323, 329, 330]. Deposition process and heat treatment also affect film crystallization temperature and physical properties [286, 299, 301, 308, 311, 313, 324]. It has been shown that the density of sol-gel-derived films can change on repetition of the deposition process [331]. Void formation at the film–substrate interface during solid-state crystallization may be responsible for this density change.

Composition control in ferroelectric thin films is considered an important issue for materials containing volatile elements such as Pb and Bi. This issue is common to other deposition techniques, including PVD and CVD methods. Generally, excessive amounts of such volatile elements are supplied not only to obtain stoichiometric films, but also to prevent such non-ferroelectric metastable phases as the pyrochlore phase. The effects of excessive Bi in SBT films has been studied [332]. In SBT films composed of fluorite and Bi-layered structure grains, the Bi-layered structure grains had a higher Bi content than the fluorite grains. Closely stoichiometric $Sr_{0.9}Bi_{2.1}Ta_2O_9$ films showed no fatigue even after undergoing 3×10^{12} switching cycles. Also, the films deposited using precursors of Sr-deficient and Bi-excess composition yielded better electrical properties than films deposited using precursors of stoichiometric composition [300]. The 20% Sr-deficient and 10% Bi-excess compositions showed maximum remanent polarization value (as high as 7.9 $\mu C/cm^2$) and good fatigue endurance. The results were attributed to Bi substitution of Sr sites, and it was suggested that the Sr deficiency prevented Bi vaporization [300, 333].

3.4. Epitaxy

Royer [334], recognizing the importance of lattice matching, introduced the term "epitaxy" (meaning "arrangement on") to denote the phenomena of the oriented growth of one substance on the crystal of a foreign substrate. The crystallographic order of the film is significantly influenced by that of the substrate as a result of some degree of matching between the two along the interface. By far the most extensive and sophisticated applications of epitaxial growth are in the semiconductor field. Superlattices and heterostructures, quantum wells, compositionally graded interfaces, and two-dimensional wells for electron gases are grown with almost atomic perfection.

One of the most important issues in the study of heterostructures is the nature of the film–substrate interaction. These

Table II.   Source Materials for the Preparation of Precursor Solutions

Material	Starting reagents	Solvents (additives)	Characteristics	References
$BaTiO_3$	Ba- and Ti-methoxyethoxide	2-Methoxyethanol	Nonferroelectric nanocrystalline $BaTiO_3$	[302]
	$Ba(OH)_2$, $Ti(^iOC_3H_7)_4$	Methanol (ethylene glycol)	Highly (h00)-oriented films	[303]
	Ba-acetate, $Ti(^tOC_4H_9)_4$	Ethanol (acetic acid, ethylene glycol)	Hf- and Zr-modified films	[304]
		Methanol (acetic acid)	$BaTiO_3$/$LaNiO_3$, electrical properties	[305]
	Ba-acetate, $Ti(^iOC_3H_7)_4$	2-Methoxyethanol	Fatigue	[306]
$(Ba,Sr)TiO_3$	Ba- and Sr-acetate, Ti-ammonium lactate	Acetic acid (acetylacetone)	MOD, RTA effect on grain growth	[299]
	Ba- and Sr-acetate, $Ti(^tOC_4H_9)_4$	2-Methoxyethanol	Formation of large grains from highly dilute spin-on solutions (0.05 M)	[286]
$KNbO_3$	$K(^tOC_4H_9)$, $Nb(OC_2H_5)_5$		Ferroelectric-like amorphous films	[307]
$PbTiO_3$	Pb-acetate, $Ti(^tOC_4H_9)_4$	Butanol (acetylacetone)	Film thickness effect on texturing	[308]
		Methanol, 2-methoxyethanol (acetic acid)	Effects of solvent on film properties	[309]
P(L)ZT	Pb-acetate, $Ti(^iOC_3H_7)_4$, $Zr(OC_3H_7)_4 \cdot C_3H_7OH$	2-Methoxyethanol	Crystallization of films, structure–property relationships	[267]
	Pb-acetate, $Ti(^iOC_3H_7)_4$, $Zr(^nOC_3H_7)_4$	n-propanol (acetic acid)	Lowed crystallization temperature from excessive Pb	[310]
			Nitrogen annealing	[311]
		Ethanol	Low-temperature (550°C) crystallization	[312]
		Ethanol	O_2-H_2O ambient annealing, low-temperature crystallization	[313]
	Pb-acetate, $Ti(^iOC_3H_7)_4$, $Zr(^nOC_4H_9)_4$-(C_4H_9OH)	Methanol, acetic acid, acetylacetone	Inverted mixing order method, solution chemistry	[314, 315]
		Methoxyethanol, methoxybutanol, acetic acid	Material–property relationship	[316]
	Pb-acetate, $Ti(^iOC_3H_7)_4$, $Zr(^nOC_4H_9)_4$	2-Methoxyethanol (acetylacetone)	Optical properties	[317]
		Methoxyethanol, buthoxyethanol	Precursor chemistry, microstructure, texturing	[318]
	Pb-acetate, $Ti(^iOC_3H_7)_4$, $Zr(^nOC_4H_9)_4$, $La(^iOC_3H_7)_3$	2-Methoxyethanol	Thick films for optical applications	[319]
$LiNbO_3$	$Li(OC_2H_5)$, $Nb(OC_2H_5)_5$	Ethanol		[254–256, 259]
		2-Methoxyethanol		[257, 259]
	$LiNb(OC_3H_7)_6$	Isopropanol		[320]
$LiTaO_3$	$Li(OC_2H_5)$, $Ta(OC_2H_5)_5$	2-Methoxyethanol		[259]
$SrBi_2Ta_2O_9$ (SBT)	Bi 2-ethylhexanoate, $Ta(OC_2H_5)_5$, $Sr(OC_3H_7)_2$	2-Ethylhexanoic acid, xylene	Piezoelectric property	[321]
	Bi 2-ethylhexanoate, $Ta(OC_2H_5)_5$, Sr-acetate	2-Ethylhexanoic acid, acetic acid, 2-methoxyethanol	Complete perovskite phase at 650°C annealing	[322]
	Sr, $Bi(OC_2H_5)_3$, $Ta(OC_2H_5)_5$	Ethanol, 2-methoxyethanol (diethanolamine)	Precursor chemistry, low-temperature crystallization	[258, 323]
	$Sr(OC_4H_9)_2$, Bi-acetate, $Ta(OC_2H_5)_5$	Acetic acid, pyrimidine	Layer-by-layer annealing, (200) preferred orientation	[324]
$Bi_4Ti_3O_{12}$ (BTO)	Bi- and Ti-octylates	Xylene	Two-step MOD, film orientation control	[301]
	$Bi(NO_3)_3$, $Ti(OC_4H_9)_4$	Acetic acid (ethanolamine)	Precursor chemistry	[325]
	Bi-acetate or Bi-nitrate, $Ti(C_5H_7O_2)_2(OC_3H_7)_2$	2-Methoxyethanol	Bi-acetate is preferable	[326]
	Bi- and Ti-naphthenates	Toluene	Epitaxial growth	[327]
$YMnO_3$	$Y(^iOC_3H_7)_3$, Mn(III)-acetylacetonate	2-Methoxyethanol (acetylacetone)	(001) preferred orientation	[328, 329]
	Y-acetate or $Y(^iOC_3H_7)_3$, Mn-acetate	2-Methoxyethanol (diethanolamine)	Effectiveness of $Y(^iOC_3H_7)_3$ for low-temperature crystallization	[330]

studies may improve the general understanding of nucleation and growth mechanisms. Film–substrate interactions can modify the final structure and properties of the film in epitaxial heterostructures. The interactions may also produce interfacial defects, pseudomorphic growth, and domain formation, resulting in film structures and properties that are different from those of bulk phase.

The essential requirements for epitaxial growth are (i) a single-crystal substrate, (ii) nucleation and growth initiated on the substrate surface, and (iii) a single energetically favorable nucleation orientation (usually, but not necessarily, determined by the minimum lattice mismatch between film and substrate). The single-crystal substrate has a dominant influence on the oriented growth of the film. The orientation of the film depends on the orientation and the crystal structure of the substrate. Some symmetry relation exists (although it is not always obvious) between the contacting planes of the two materials. Although a lattice misfit is expected to be the key to the understanding of epitaxy, the relative positioning of the atoms, rather than the best geometrical fit of the two lattice, is the significant variable. Substrate temperature is another important factor in epitaxy. There seems to exist a critical temperature, called the "epitaxial temperature," above which epitaxy is perfect and below which it is imperfect [335]. A high substrate temperature lowers supersaturation and thus allows the dilute gas of adatoms sufficient time to reach the equilibrim positions, provides activation energy for adatoms to occupy the positions of potential minima, and enhances recrystallization due to the coalescence of islands by increasing surface and volume diffusions. These factors are dependent on other deposition conditions, making it difficult to define a precise value of the epitaxial temperature. An adatom should have sufficient time to jump to an equilibrium position of an ordered state by surface diffusion before it interacts with another adatom. For a given system with fixed atomic mobility, this requirement is satisfied if the deposition rate of a monolayer is less than the jump frequency of adatoms.

Epitaxial ferroelectric thin films have been shown to offer unique characteristics compared to randomly oriented films [336, 337], as expected. Forthcoming optical technology will require epitaxial growth of oxides [337]. Especially for optical waveguide devices (e.g., acousto-optic deflectors, electro-optic switches, second harmonic generators), single-crystal-like heteroepitaxial thin films are required because of low optical propagation loss, their near-single-crystal properties, and compatibility with integration to semiconductor lasers. In general, vapor-phase growth processes (physical vapor deposition and chemical vapor deposition) afford greater ease of epitaxial film growth. Oriented or epitaxial PZT thin films, fabricated by reactive sputtering [338], PLD [339, 340], MOCVD [341], and sol-gel methods [342], have been reported from the early 1990s. It should be noted that the term "epitaxial" is also used to describe films that could be more accurately described as having some preferential orientation not conforming to the more strict definition of epitaxy.

3.4.1. Epitaxial Ferroelectric Oxide Films on Single-Crystal Oxides

The success of oxide thin film devices will depend on their ability to deposit films of high crystalline quality and to control their microstructure, composition, and properties. One approach to addressing these requirements is creating lattice- and chemistry-matched heterostructures of ferroelectric with single-crystal oxide substrates to yield single crystalline films. There have been many studies on ferroelectric epitaxial films grown on oxide single crystals such as MgO, $SrTiO_3$, and sapphire.

Epitaxial $BaTiO_3$ thin films have been grown by various techniques, including rf sputtering, reactive evaporation (or MBE), PLD, and MOCVD on (100) MgO [183, 343–348], (100) $SrTiO_3$ [85, 349–352], and (100) $LaAlO_3$ [184, 353]. The epitaxial relationship has been found to be cube on cube. Excellent crystalline quality of $BaTiO_3$ thin films was obtained even up to thickness of a few microns by PLD [348]. Waveguide losses of 2.9 dB/cm were demonstrated. It has been shown that the phase composition and epitaxial quality were sensitive to the reactant partial pressures and growth temperature [184]. In general, it has been reported that PLD results in c-axis-oriented films [343, 346, 348, 350], whereas MOCVD produces a-axis-oriented films [183, 184, 343–345, 353]. It has not yet been well established what factors determine the crytallographic orientation of the films. The crystallographic phase and lattice distortion could be explained by the theories developed by Speck et al. [354], who posited that growth stresses resulting from supersaturated point defect populations may be associated with the growth orientation. The strain relaxation mechanism [354] could explain the observed thickness and oxygen-pressure-dependent structural characteristics of the $BaTiO_3$ thin films [85]. Also, it has been shown that ion size and electrostatics at the first atomic layers of the interface between simple perovskites and MgO influences the heteroepitaxial growth of these oxides [355]. Using the TiO_2-truncated MgO surface, optical-quality $BaTiO_3$ thin film can be grown by MBE to the μm thicknesses required for application in electro-optic devices. Interfacial energy minimization at the first atomic layers is the basis for a commensurate, unit-cell stability. Film deposition rates were also shown to affect the phase of the films; relatively higher deposition rates yielded multiphase films [185].

Chemical matching at the interface was also investigated in the growth behaviors of $SrTiO_3$ on $LaAlO_3$ substrates [356]. For ABO_3-type perovskite, the AO (BO_2) layer was favorably wet on the BO_2 (AO) layer to minimize interfacial energies. The electrically compensated initial TiO_2 layer in the LaO-terminated substrate provided a chemically well-matched interface, resulting in strained layer. This allowed a strain-induced roughening, a transition from layer-by-layer growth to island growth. The noncompensated initial layer in the AlO_2-terminated substrate seemed to have many defects, such as vacancies and interstitials, but relaxed the strain quickly enough to allow layer-by-layer growth after a few monolayers.

Epitaxial ferroelectric $PbTiO_3$ thin films grown on $KTaO_3$ substrate by MOCVD have been shown to have domains with the a-axis near the substrate surface normal (a-domain) and domains with the c-axis near the normal (c-domains) for thicknesses greater than 30 nm [357]. TEM revealed that the films consisted of four different symmetry-equivalent twinned crystals. Each type of twinned crystals contained periodic a- and c-domains in an alternating sequence separated by a 90° domain wall that formed an angle of about 45° with the surface of the substrate. It has been suggested that the interfacial strain is accommodated by the formation of a periodic domain pattern in the overlayer [357]. In (001) $PbTiO_3/SrTiO_3$, which exhibits an excellent lattice match between respective a lattice parameters, the films existed as a single c-domain [358]. The strain-accommodating mechanisms in these heterostructures depended on both the lattice and the thermal expansion–coefficient mismatch between the film and the substrate. Formation of dislocations and/or domains could be observed above a critical thickness about 250 nm [357]. Heteroepitaxial growth of PZT on $SrTiO_3$ using MOCVD has been reported covering most of the phase diagram ($x \leq 0.9$) [359]. The heteroepitaxial films with a tetragonal structure ($x < 0.53$) had a microstructure characterized by the presence of a-axis-oriented regions penetrating the film with misfit dislocations at the interface. For the films with a rhombohedral structure ($x > 0.53$), the films consisted of regions with a slightly different in-plane orientation respective to each other with amorphous zones near the interface. The solid-solution diagram of the $PZT/SrRuO_3/SrTiO_3$ system for thin films was different from the bulk, due to epitaxy-induced strains and interfacial defect formation [360]. The total strains may be relieved either by misfit generation or by domain formation. Also, high values of remnant polarization (32–35 $\mu C/cm^2$) were observed for most ferroelectric compositions. The coercive field decreased with increasing Zr concentration. Differential thermal expansion, cooling rate, and applied electric fields may be used to effectively control domain structure in the PZT system [361]. The thermal expansion difference between the substrate and the film is the relevant parameter controlling the domain formation for ferroelectric films with thicknesses exceeding critical thickness. Enhanced formation of c-domains of $PbTiO_3$ was observed under the influence of a dc bias field during sputtering [361]. A giant dielectric permittivity, as high as 420,000 at low frequencies, has been reported in epitaxial structures with alternate layers of ferroelectric and nonferroelectric oxides, superlattices of $PbTiO_3/Pb_{0.72}La_{0.28}TiO_3$ [117]. The observation was interpreted as a result of the motion of a pinned domain wall lattice at low electric fields and a sliding motion at high electric fields. Epitaxial PLT thin films were also grown on (100) MgO and (0001) Al_2O_3 (C-sapphire) substrates [213]; the preferred orientations of the epitaxial PLT thin films were (100) and (111) for (100) MgO and (0001) Al_2O_3, respectively.

Epitaxial growth of bismuth-layer structured perovskite ferreoelectrics has been reported. Epitaxial BTO thin films have been successfully grown by PLD on (100) $SrTiO_3$ [362, 363], (100) MgO [364, 365], and (0001) Al_2O_3 [366]. Under the usual deposition conditions, they showed highly (001)-oriented structure, indicating that underlying lattice structure significantly affects growth orientation of the BTO films. The (110) $SrTiO_3$ substrate was found to be useful in controlling the growth orientation of BTO film, resulting in (117)-oriented grains [363]. Also, BTO films grown on (110) $SrTiO_3$ and (110) MgO substrates showed large quadratic electro-optic coefficients [363, 365]. BTO film grown on (0001) Al_2O_3 substrate showed highly oriented films with (104) planes normal to the substrate's c-axis [366].

For SBT films, (001)- and (116)- oriented epitaxial $SrBi_2Ta_2O_9$ thin films were deposited by MOCVD on (100) $SrRuO_3 \| (100)$ $SrTiO_3$ substrates at 750°C and on (110) $SrRuO_3 \| (110)$ $SrTiO_3$ substrates at 820°C, respectively [367]. The (116)-oriented SBT films showed ferroelectricity with the remnant polarization of 11.4 $\mu C/cm^2$ and the coercive field of 80 kV/cm, while the (001)-oriented films showed no ferroelectricity. Since it is supposed that the ferroelectric polarization direction is in the ab-plane, a- or b-axis-oriented SBT films are desirable for many applications, and the control of growth orientation is important. Recently, control of growth orientation has been attempted through the orientation of the substrates [102]. Growth of (100)-oriented epitaxial SBT films on (100) $LaAlO_3$ and (100) YSZ substrates has been reported although the crystalline phase may be a nonferroelectric cubic fluorite one [368]. Also, (110)-oriented SBT film was reported on a (100) $LaSrAlO_4$ substrate, which has rectangular surface mesh with a width and length each approximately half of the dimensions of the rectangular surface mesh of with (100) SBT [104]. Thin films of a-/b-axis and c-axis oriented SBT were grown on (110) MgO and (100) MgO substrates, respectively [106]. The orthorhombic SBT phase was confirmed by TEM and infrared reflectance.

It has been argued that epitaxial films are more easily achieved for ionically bonded materials than for convalently bonded materials [369]. For a covalently bonded material, the preferred growth orientation of the film itself must be considered, as must the film–substrate interface. On the other hand, for ionically bonded films, the controlling feature is the crystallographic relationship to the substrate. The (0001) and $(10\bar{1}0)/(11\bar{2}0)$ growth orientations of $LiNbO_3$ and $LiTaO_3$ are useful for electro-optic applications, as well as for electrical and piezoelectric applications. For optical signal modulation, orientations with the polar axis in the plane, i.e., $(10\bar{1}0)$ and $(11\bar{2}0)$ orientations, are desired, whereas second harmonic generation with quasi-phase matching requires that the polar axis be normal to the plane, the (0001) orientation. Growth orientation of ionically bonded $LiNbO_3$ films could be controlled by paying attention to the sharing between octahedra in the structure and to the formation of the octahedra containing Li and Nb ions [369]. $LiNbO_3$ films deposited on R-cur sapphire showed that the $(01\bar{1}2)$ textured film changed to $(10\bar{1}0)$ through $(11\bar{2}0)$ by increasing the Li concentration in the films, despite their large lattice misfit. Based on the degree of freedom in sharing of octahedra containing Li and Nb, sufficient Li concentration and the effect of interfacial restriction were considered to promote the formation of the $(10\bar{1}0)$ epitaxial film. Epitaxial growth of $LiNbO_3$ and $LiTaO_3$ films using various deposition techniques has been reported [9, 225–227, 242, 244, 370–374]. The ability to avoid twinning and high-angle grain boundaries is of prime importance for ensuring optical quality.

Epitaxial growth of the sol-gel-derived films can occur during film crystallization. The films are either amorphous or microcrystalline before high-temperature annealing. Properly controlled, high-temperature annealing can convert an amorphous film to an epitaxial film. This is essentially a solid-phase epitaxy process. In contrast, high-temperature annealing of a microcrystalline film can only achieve either a randomly or a preferentially oriented film, because the growth and coalescence of the microcrystalline result in polycrystalline grains with various orientations. The ideal process requires epitaxial nucleation on the substrate surface, but no random nucleation in the amorphous film. The processing parameters for preparation of epitaxial films by the sol-gel process have been suggested by Rou et al. [375]: (i) An amorphous film with atomic-scale compositional homogeneity on a single-crystal substrate is required, (ii) a high annealing temperature is preferred to minimize the probability of some homogeneous nucleation occurring, and (iii) a rapid ramping rate to the annealing temperature is preferred, to increase the relative stability of heterogeneous nuclei over the homogeneous nuclei.

3.4.2. Epitaxial Ferroelectric Thin Films on Semiconductor Substrates

Although it is feasible to grow epitaxial ferroelectric thin films on single-crystal oxide substrates, growing the epitaxial oxide layer on a semiconductor is generally necessary to optimize the electronic and optical properties of the oxides which to form hybrid-type devices. This allows possible integration of various useful physical properties of ferroelectric oxides directly with semiconductor-based electronic and optoelectronic devices. The growth of oxide films on semiconductor substrates has been investigated for more than 20 years, with much of the early work carried out in Japan. In 1978, Ishida et al. [376] reported the growth of preferentially oriented ferroelectric PLZT films on (001) GaAs by rf sputtering and found evidence of significant Pb diffusion into the semiconductor. In the case of $SrTiO_3$, epitaxy on (001) Si was achieved by first deposition of a Sr layer [377] to reduce the native oxide, whereas films grown directly on (001) Si displayed evidence of interfacial reaction [378]. Many other attempts to achieve deposition of ferroelectric oxide directly on semiconductor substrates have produced polycrystalline films. The origin of this polycrystallization has not been well studied. Chemical reactions and the native semiconductor oxide probably can play an important role in the initial stage of epitaxy. Also, structural and chemical differences between oxides and semiconductors can give rise to complex and unexpected orientation relations.

Comparatively few epitaxial oxides have been grown directly on semiconductors. Interest in the growth of oxide-based devices on semiconductors has led to the use of epitaxial

MgO films to provide a chemically stable buffer layer [379, 380], because MgO is chemically stable and its crystal structure is similar to that of many oxides. Growth of highly textured MgO film on semiconductor substrates using various techniques has been reported. Depending on the growth conditions, MgO thin films of (100), (110), and (111) preferred orientations have been grown on semiconductors [346, 379–384]. There have been several reports on the use of a MgO buffer layer for the growth of ferroelectric thin films. $BaTiO_3$ films of (001) orientation were grown epitaxially on MgO/GaAs by PLD [380]; the in-plane epitaxial relationship was $BaTiO_3[100]\|MgO[100]\|GaAs[100]$, despite of a large lattice mismatch (25.5%) between MgO and GaAs. The crystallographic orientation of the $BaTiO_3$ film could be affected by the thermal expansion mismatch between the film and the underlying substrate, with a-axis-oriented films formed on MgO/GaAs and c-axis-oriented films formed on (001) MgO single crystal [346]. Also, the $BaTiO_3$ films were weakly ferroelectric, probably due to the heterogeneous strain fields produced by the dislocation distribution in the $BaTiO_3$ films. Partially $(h00)$- or $(00l)$-textured $BaTiO_3$ thin films have been grown on (100) MgO/Si(100) using rf magnetron sputtering [385].

(100)-oriented $PbTiO_3$ ferroelectric thin films were grown on (100) MgO/(100) GaAs [386]. La-modified $PbTiO_3$ [$Pb_{0.9}La_{0.1}TiO_3$ (PLT)] thin films with the preferred c- and a-axis orientation were grown by rf-magnetron sputtering on (100) MgO/(111) Si [387]. Growth of highly (100)-oriented PZT films on oxidized (100) Si substrates using a thin (7–70 nm) MgO buffer layer also has been reported [158]. This two-layer-buffer structure (i.e., MgO/SiO_2) was expected to provide a template for the growth of oriented PZT films and a diffusion barrier between ferroelectric and a substrate during device fabrication, protecting the SiO_2–Si interface.

Epitaxial thin film growth of $LiNbO_3$ family on semiconductor substrates has been observed for their potential application to electro-optic devices. Epitaxial (0001) $LiNbO_3$ film was grown on (111) GaAs by PLD with and without intermediate epitaxial (111) MgO buffer layers [388]. Epitaxial $LiTaO_3$ film with the preferred [0001] direction was grown on (111) MgO/(111) GaAs [389]. Although MgO and $LiNbO_3$ (or $LiTaO_3$) have different crystal structures, they have the same oxygen ion framework in their (111) and (0001) planes, thus favoring an epitaxial relationship. Presence of high twin density was revealed in the epitaxial film of $LiTaO_3$ by X-ray diffraction ϕ-scan [375]. Highly textured ferroelectric $LiNbO_3$ thin films with the preferred (0001) orientation have been grown on (111) MgO-buffered (100)/(111) Si by the sol-gel method [390]. Diffusion of MgO into $LiNbO_3$ seemed to affect the crystalline quality of the films [391]. On the other hand, vapor-deposited $LiTaO_3$ film on (100) MgO-buffered Si showed a highly (116)-oriented structure [246]. An epitaxial relationship of higher-order commensuration may play an important role in the structure of preferred orientation.

The use of other oxide layers besides MgO as buffer layers or electrodes has been reported. CeO_2, Y_2O_3, LSCO,

SrRuO$_3$, LaNiO$_3$, and RuO$_2$ have been used as templates for the growth of oriented ferroelectric oxide films. SrRuO$_3$, LaNiO$_3$, and RuO$_2$ have been used as oxide-conducting electrodes and buffer layers. Occasionally, nitride compounds (e.g., TiN and Si$_3$N$_4$) have been used. PbTiO$_3$ films grown by PLD showed (100) + (001) mixed texture on CeO$_2$/(100) Si and polycrystalline on Y$_2$O$_3$/(100) Si [392]. PZT films were epitaxially grown on (111) Si substrates using Y$_2$O$_3$ layers [393]. Strongly (101)-oriented PZT films grew on the Y$_2$O$_3$/(111) Si structures at a substrate temperature of 700°C. Completely (111)-textured Pb(Ta,Zr,Ti)O$_3$ (PTZT) films were grown on Pt/TiO$_2$/SiO$_2$/(001) Si substrates by PLD using SrRuO$_3$ as a buffer layer [135]. It has been argued that the small lattice mismatch in (111) plane between Pt and SrRuO$_3$ and the compatible perovskite structures for SrRuO$_3$ and PLZT are responsible for the complete (111) orientation of PTZT films. This heterostructure with low leakage current density exhibited excellent resistance to bipolar fatigue and quite good retention characteristics.

Amorphous SiN$_x$ layers were found to be effective for the growth of c-axis-oriented LiNbO$_3$ film [394]. The local arrangement of SiN$_4$ tetrahedra in the sputtered SiN$_x$ may be related to the nucleation of the (0001) LiNbO$_3$ films. Large-area epitaxial ferroelectric films of Ba$_2$Bi$_4$Ti$_5$O$_{18}$ were grown by PLD on LaNiO$_3$-buffered (100) Si [395]. The ferroelectric layer showed excellent c-axis orientation, although the films showed very small remnant polarization, probably due to their c-axis-oriented grains.

4. CHARACTERIZATION OF FERROELECTRIC THIN FILMS

4.1. Structural Characterization

The range of analytical tools useful to the structural properties of thin films is expansive. This section provides a brief introduction to some major methods used for structural characterizations. These techniques encompass microscopies and spectroscopies for ascertaining specific information of thin film structure, surfaces, interfaces, and compositions.

4.1.1. X-Ray Analysis

Structural characterization is important for understanding the relationship between the structure and properties of materials. One of the most powerful structural analysis tools is X-ray diffraction (XRD), through which such characteristics as phase chemistry can be directly and easily identified in crystalline materials. Quantitative information can be easily obtained from a diffraction pattern and expressed in terms of a variety of parameters, e.g., volume fractions of phases, lattice parameters, atomic site occupancies, residual stress, particle sizes, microstrains, and dislocation densities. Structural properties of thin films are investigated by XRD techniques, including θ–2θ scan, ω scan, and X-ray pole figure measurements. These X-ray techniques are complementary with one another.

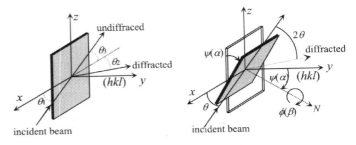

θ–2θ scan : $\theta_1 = \theta_2 = \theta_3 = \theta$, $\theta_2 + \theta_3 = 2\theta$
Rocking curve (ω scan): $\theta_1 = \omega$ scan, $\theta_2 + \theta_3 = 2\theta$ fixed
Pole figure : ψ and ϕ scan (θ–2θ fixed)

Fig. 19.   Plane of arrangements for structural analysis of thin films using X-ray techniques.

In most experiments, to meet the Bragg reflection condition, the angle of incidence is θ and the reflected X-rays are detected at an angle 2θ with respect to the incoming beam, a θ–2θ measurement. The θ–2θ scan probes crystalline grains whose atomic planes are parallel to the substrate surface. In addition, XRD measurements provide information on the crystal structure and texture of (poly)crystalline thin films. Information on the quality of the texture (i.e., out-of-plane alignment of the crystallites) can be obtained by measuring a rocking curve. In this geometry, the angles of the sample and detector are first tuned to a Bragg reflection angle. Next, the sample is rotated over an angle ω while the detector remains at a fixed position, and the intensity is measured as a function of the angle ω. The geometry of the measurements is shown in Figure 19. The rocking curve measurement provides information on how well a given axis of the grains are aligned. The width of a ω scan is a direct measure of the range of orientation present in the films.

The X-ray pole figure gives a lateral registry between the crystal axes of the thin film and in-plane vectors of the substrate, as well as in-plane alignment of the crystallites. The pole figure measures diffraction intensity at a particular 2θ (i.e., a particular plane) as a function of orientation. Here θ–2θ geometry is used, so that the diffracting planes are always at the same orientation relative to the beam. The X-ray source and the detector are fixed, and the sample is tilted and rotated. The plane spacing of interest is chosen by setting up the initial diffraction condition for the appropriate Bragg angle, as shown in Figure 19. To make a pole figure, the sample is tilted through an angle ψ (occasionally the notation α is used) about the x-axis and rotated through ϕ (or β) about the sample normal (see Fig. 19). After the sample is tilted through an angle ψ, the sample normal is ψ away from the diffracting plane normal (hkl). A pole figure is plotted on a stereographic projection, centered around the sample normal. For each ψ and ϕ, the intensity recorded by the detector is marked on the pole figure. Equal-intensity contours or shaded regions may then be drawn. If the measurement were performed at a fixed ψ while ϕ scanning 2π, then it corresponds to the ϕ scan. High-diffraction intensity at some angle ψ means that the planes of interest are oriented ψ from the sample normal. For epitaxial films, the diffraction intensity distribution reflects the symmetry of a single crystal. Distribution of the diffracting planes is not radially symmetric, and this sample is described as having in-plane texture. However, for random polycrystalline films or for oriented films without in-plane texture (fiber texture), the diffraction intensity distribution in pole figure is radially symmetric. In this case, the ϕ-scan data do not show sharp intensity peaks, but rather are dispersive.

The crystallization of PZT films sputtered at low temperature (~200°C) has been studied as a function of annealing temperature and time using XRD [291]. As-deposited amorphous PZT films were first crystallized into a pyrochlore phase for annealing temperatures as low as 450°C for 10 hours. Single-phase perovskite appeared for thermal processing conditions of 600°C for 30 minutes. For annealing conditions between 450°C/10 hours and 600°C/30 minutes, both pyrochlore and perovskite phases were observed. The initial formation of perovskite phase occurred by annealing at 550°C for 30 minutes. X-ray studies on the pyrochlore to perovskite phase transformation has also been reported in sol-gel-derived PZT [396].

Fine characterization by XRD of structural modifications, such as cell parameters and domain microstructure, as a function of film preparation method and the nature of its substrate has been reported [397]. $PbZr_{0.2}Ti_{0.8}O_3$ cell tetragonality c/a, which was always smaller than that measured for the PZT powder or bulk ceramic, was shown to be related to PZT grain size, meaning that tetragonality is influenced more by internal stress than by external stress induced by lattice mismatch. Epitaxial stress did not induce any change in PZT cell parameters and tetragonality, showing that relaxation process occurred by domain formation. On the other hand, ferroelectric domain microstructure modification of the PZT grains was directly related to the substrate, because of thermal treatment, which induces external stress between the PZT film and the substrate. A change of the domain microstructure could be deduced by measuring the XRD intensity ratio of two doublet peaks of PZT film. For a free unpolarized ferroelectric PZT film, the peak intensity ratio of $(001)/\{(010)(100)\}$ doublet is $1/3$ and that of $\{(101)(011)\}/(110)$ doublet is $2/3$ in the diffractogram [398], because it has a random distribution of all domains and an equal volume of 180° and 90° domains in each grain. An electric field and an external stress can modify the relative volumes of 180° and 90° domains, and the effect appears in the diffractogram as a modification of the peak intensity ratio of a doublet, since XRD allows differentiation of 180° and 90° domains. The sign of the difference between the PZT film and substrate thermal dilatation coefficient $\alpha_{PZT} - \alpha_{sub}$ and the microstructure type were correlated. For a negative difference, the domain microstructure is of a-domain type, for a zero difference, it is normal, and for a positive difference, it is of c-domain type. An improvement of the electrical properties was observed for dense PZT film with (100)-oriented grains whose domain microstructure was of c-domain type.

It is often the case that the microstructure of films is still polycrystalline even though XRD is of a highly oriented structure. To determine the layer perfection, the rocking curve full width at half maximum (FWHM) of Bragg reflections and the pole figure should be measured. The FWHM is a direct mea-

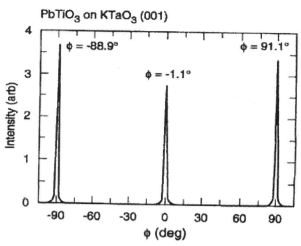

Fig. 20. ϕ scan of X-ray diffraction pattern from the (220) $PbTiO_3$ reflection of the a-domains showing the excellent in-plane epitaxial relations. $\phi = 0$ is aligned with the in-plane [100] $KTaO_3$ direction. Reprinted with permission from B. S. Kwak et al., *Phys. Rev. B* 49, 14865 (1994). Copyright 1994, American Physical Society.

sure of the range of orientation present in the irradiated area of the films. The rocking curve width for ideal single crystal is about 0.003°; however, most crystals exhibit widths 10 to 100 times greater. Epitaxial $BaTiO_3$ films grown on (100) $LaAlO_3$ showed FWHM in the range of 0.6–1.0° [399]. The crystalline quality of PLD-grown $SrTiO_3$ on (001) MgO substrates was found to be significantly improved in the oxygen partial pressure range of 0.5–1 mtorr compared to the films deposited at higher pressures of 10–100 mtorr [400]. The XRD rocking curves for the films grown at oxygen pressure of 1 mtorr and 100 mtorr yielded FWHM of 0.7° and 1.4°, respectively. Thermal annealing of the $SrTiO_3$ films in oxygen further improved the quality, and the 1-mtorr films gave FWHM of 0.13°. Similar results have been reported for $SrTiO_3$ films grown in TiN-buffered Si substrate [184]. The crystallinity of the epitaxial films was improved with increasing substrate temperature and decreasing ambient oxygen pressure, whereas the film surface morphology was degraded with increases in either of the two parameters. For $Bi_4Ti_3O_{12}$ thin film on (100) $SrTiO_3$, the FWHM was about 0.3° for the (006) film peak, indicating that grains were well aligned with their c-axes normal to the substrate [363].

The in-plane epitaxial alignment of films is investigated using either ϕ scans or pole figure analysis. The corresponding ϕ scan or pole figure from the substrate also can be obtained. By doing so, the in-plane orientation relationship between the films and the substrates can be investigated. Epitaxial heterostructures of $PbTiO_3$ films on various single-crystal substrates have been investigated by ϕ scan [358], as shown in Figures 20 and 21. A ϕ scan of the (220) $PbTiO_3$ reflection scattering from a-domains showed that the in-plane orientation was such that the principal a and b axes of the films and substrate were not exactly aligned, even though the film grown on $KTaO_3$ was indeed three-dimensionally epitaxial (see Fig. 20). The interfacial steps resulting from the miscuts (2°) in the substrate surface was

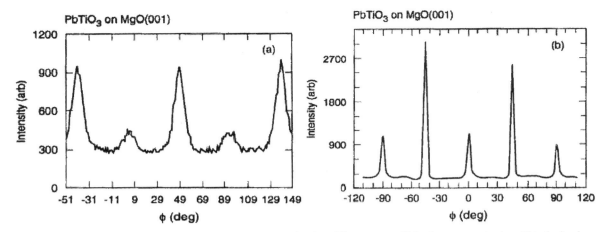

Fig. 21. (a) ϕ scan from a-domains in PbTiO$_3$ on (001) MgO showing different types of island arrangements. $\phi = 49$ is the in-plane [100] MgO direction. (b) ϕ scan from c-domains in PbTiO$_3$ on (001) MgO showing different types of c-domains. $\phi = 0$ is the in-plane [100] MgO. Reprinted with permission from B. S. Kwak et al., *Phys. Rev. B* 49, 14865 (1994). Copyright 1994, American Physical Society.

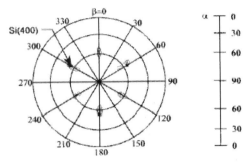

Fig. 22. Result of LiNbO$_3$\{012\} pole figure analysis of c-axis-oriented LiNbO$_3$ thin film grown on MgO/(111) Si. Filled squares are the poles of Si (400) peak. Reprinted with permission from J.-G. Yoon and K. Kim, *Appl. Phys. Lett.* 68, 2523 (1996). Copyright 1996, American Institute of Physics.

considered as a nucleation site for twin boundaries. Also, for a- and c-domains of PbTiO$_3$/(100) MgO epitaxial heterostructure, ϕ scans from the (202) reflection showed two types of both a- and c-domains, one having the in-plane alignment between [100] axes of the film and the substrate and the other with the [100] direction of the film forming an angle of 45° with the [100] direction of the substrate (see Fig. 21).

Figure 22 presents results of pole figure experiments showing the order of in-plane epitaxial alignment of a sol-gel-derived LiNbO$_3$ thin film grown on MgO-buffered Si [390]. For the c-axis-oriented LiNbO$_3$ films, the threefold symmetry should give only three \{012\} poles separated by an angle of 120° at an appropriate ψ (or α). The six poles in the figure indicate the presence of twins of grains (90° or 180° twins). If the film were a single crystal, then the angular distribution of poles should be very narrow.

Epitaxial films deposited on lattice-mismatched substrates are often subject to large coherency strains. The diffraction peaks of highly strained films are broadened considerably. Broadening of the diffraction peaks is the sum of contributions from grain (domain) size and strain. The contribution of local strain to the FWHM of the θ–2θ Bragg peaks is given by

$$\Gamma_s^2 = \frac{4\ln 2}{C^2}\tan^2\theta = K_s\tan^2\theta$$

and the contribution of grain size is given by

$$\Gamma_L^2 = \frac{4\ln 2}{\pi L^2}\frac{\lambda^2}{\sin^2 2\theta} = K_L\frac{\lambda^2}{\sin^2 2\theta}$$

where K_s and K_L are the strain and grain coefficients, respectively, λ is the wavelength of X-ray radiation, C is a constant, and L is the coherently diffracting grain size. Hence the broadening is given by

$$\Gamma_{\theta-2\theta}^2 = \Gamma_s^2 + \Gamma_L^2 = K_s\tan^2\theta + K_L\frac{\lambda^2}{4\sin^2\theta\cos^2\theta}$$

and

$$\Gamma_{\theta-2\theta}^2\sin^2\theta\cos^2\theta = K_s\sin^4\theta + K_L$$

From the plot of $\sin^2\theta\cos^2\theta\,\Gamma_{\theta-2\theta}^2$ versus $\sin^4\theta$, the relative contributions of strain and grain size to the FWHM are determined from the slope and the intercept of the plots, respectively. In this way, microstrain in epitaxial films of BaTiO$_3$ on MgO has been assessed by XRD [401]. The brodening was due predominantly to strain. The magnitude of the microstrain decreased sharply with increasing film thickness.

Information on the layer thickness of a thin film or, in case of a multilayer, the multilayer period can also be obtained from XRD measurements. Recently, ferroelectric superlattices have begun to be exploited [113]. Figure 23 shows the high-angle XRD result for a KTaO$_3$/KNbO$_3$ multilayer sample grown on KTaO$_3$ substrate by PLD. Satellite peaks appear, characteristic of multilayer thin films, around the substrate (002) peak. The satellite peaks can be indexed and the superlattice periodicity Λ accurately determined. The peak positions are given by Bragg condition $\sin\theta_n = n\lambda_X/2\Lambda$, where n is an integer and λ_X is the X-ray wavelength, and so Λ is given by the separation of the adjacent peaks, $\Lambda = \lambda_X/2\sin\theta_{n+1} - \sin\theta_n$.

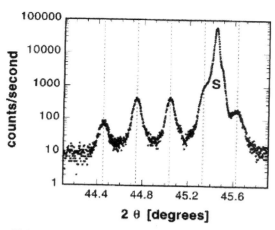

Fig. 23. High-angle XRD pattern of $KTaO_3/KNbO_3$ superlattice. X-ray θ–2θ scan through the (002) Bragg peak for a $KTaO_3/KNbO_3$ superlattice showing satellite peaks, from which the periodicity of the superlattice can be determined to be 33.8 nm. S indicates the substrate (002) peak. Reprinted with permission from H.-M. Christen et al., *Appl. Phys. Lett.* 72, 2535 (1998). Copyright 1998, American Institute of Physics.

When θ is typically below 50 mrad, the technique is sometimes called *glancing-incidence X-ray analysis* (GTXA). The specular X-ray reflectivity represents the interference pattern of the reflected X-rays from the surface and the interface, which is described by [402]

$$S_{\text{spec}}(\overline{q}) = \frac{A}{q_z^2}\left[\rho_1^2 e^{-q_z^2\sigma_1^2} + (\rho_2 - \rho_1)^2 e^{-q_z^2\sigma_2^2}\right.$$
$$\left. + 2\rho_1(\rho_2 - \rho_1)\cos(q_z d)e^{-q_z^2(\sigma_1^2 + \sigma_2^2)/2}\right]\delta(\overline{q}_{\parallel})$$

where q_z ($q_{\parallel}$) is X-ray momentum transfer along the surface normal (in-plane) direction, ρ_1 (ρ_2) is the electron density of films, σ_1 (σ_2) is the root-mean-squared surface (interface) roughness, and d is the average film thickness. Film thickness can be determined by the period of the intensity oscillations Δq in the reflectivity curve by $2\pi/\Delta q$. X-ray reflectivity measurement is a powerful method for studing the morphological properties of buried interfaces. Roughness at the interfaces between various layers reduces specular reflection and gives rise to diffuse scattering in all directions. Thus, when the height of the peaks is analyzed, information can be obtained about the interface quality.

The crystallization of amorphous (Ba,Sr)TiO_3 thin films was studied in a synchrotron X-ray scattering experiment [403]. The reflectivity decreased dramatically as the films were annealed to the temperature at which a nanocrystalline phase was formed, indicating that the interface became rough. The reflectivity intensity oscillation was also disappeared at the same temperature. A buried SiO_2 layer, formed during the annealing of sol-gel-derived MgO film on Si, could be detected by the same method [404].

4.1.2. Rutherford Backscattering Spectrometry

The Rutherford backscattering spectrometry (RBS) technique is increasingly used in thin film and thin layer analysis. It uses

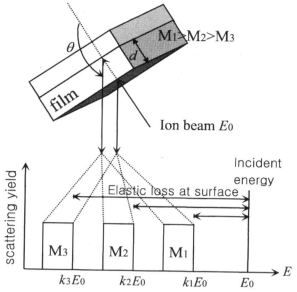

Fig. 24. Schematic backscattering spectrum for a thin homogeneous film of a compound with elements of different atomic masses ($M_1 > M_2 > M_3$).

1- to 3-MeV ions, usually ^{4}He, to analyze the surface and the outer 0.5- to 3.0-μm depth of materials under investigation. The penetration depth increases with increasing energy of the ion beam. The results produce information on composition, as well as qualitative and quantitative information on the distribution of atoms within the vicinity of the surface. The principles of RBS are shown in Figure 24, which provides insight into the composition of a sample. Details of RBS are given in the literature [405].

For the x-axis of the backscattering spectrum, the so-called kinematic factor k states where exactly the signal of an element of any given mass has its high-energy edge (i.e., leading edge). The factor k is defined as the ratio of the projectile energy after the elastic collision to that before the collision, and depends on the scattering angle and the masses of the ion and the target. The locations of the high-energy edges are indicated by the length of the arrows labeled $k_i E_0$ below the energy axis of the spectrum of Figure 24. When a particle of energy E_0 collides elastically with a stationary particle of M_i in the sample, energy is transferred from the moving particle to the stationary particle. In elastic scattering, less energy is transferred for heavier target atoms. Thus heavy masses go to high energies of the leading edge, and light masses go to low energies.

Similarly, the scattering cross-section gives the scaling factor for the yield axis of different elements, because the probability of a collision is given by the differential cross-section. The differential cross-section is proportional to the square of the atomic number. Thus the relative concentration ratio of elements transforms into relative yields by a ratio given essentially by the cross-sectional ratio of the elements. High atomic numbers give high yields, and low atomic numbers give low yields. Information on the depth profile of film composition can also be obtained. In real situations, some corrections must be applied. In analysis of the RBS spectrum, a program called RUMP, based on the basic theory of RBS and some corrections, is often used for simulation [406].

Measurement of film thickness is an obvious use of backscattering spectrometry. Since the area under each signal is proportional to the total number of atoms in the film, the film thickness from the area of a signal can also be obtained. The energy difference, ΔE, between particles scattered from the surface of the film and those scattered from the film–substrate interface is related to film thickness. It should be remembered that the exact measuring quantity is not the physical tnickness, but rather the quantity corresponding to the number of atoms per square centimeter, Nd, where N and d are the volume density of atoms and the film thickness, respectively. RBS does not identify voids in films, so the calculated film thickness is often different from that measured by other techniques, such as optical methods and electron microscopy.

The epitaxial nature of films is easily confirmed by use of channeling effects. The channeling effect arises because rows or planes of atoms can steer energetic ions by means of a correlated series of gentle small-angle collisions. In terms of backscattering spectrometry, channeling effects produces strikingly large changes in the yield of backscattered particles as the orientation of the single-crystalline films is changed with respect to the incident beam. Indeed, there can be a 100-fold decrease in the number of backscattered particles when the film is rotated so that the beam is incident along axial or planar directions rather than along the direction viewing a random collection of atoms.

Stoichiometry of $LiNbO_3$ films grown by rf sputtering has been analyzed by RBS [407]. Since Li is a light element, the Li signal cannot be observed in the RBS spectra. By comparing the areas under the curves for O and Nb and dividing by the square of the atomic number, the ratio of O to Nb in the sample was determined to be 3.0 to the resolution of the measurement. This indicates that the film may be stoichiometric $LiNbO_3$, since Li-excessive or Li-deficient films have different Nb/O ratios. Sufficient oxygen partial pressure was necessary for stoichiometric $LiNbO_3$ films.

Figure 25 shows channeling yield χ_{min} measurements for PLD-grown $SrTiO_3/MgO$ film. The measurements were made from the RBS spectra of 1.8-MeV $^4He^+$ ions. The minimum channeling yields, χ_{min}, as measured by the Sr signal are 1.7% for the annealed sample, and <5% for the as-deposited film grown under the same conditions [399]. The minimum channeling yields for $PbTiO_3$ thin films grown on $KTaO_3$ and $SrTiO_3$ were compared [358], improved channeling in $PbTiO_3$ films grown on $SrTiO_3$ was attributed to the single-domain nature of the overlayer. De Keijser et al. [359] obtained good agreements between the random and simulated spectra of stoichiometric $PZT/SrTiO_3$. They deduced the heteroepitaxial quality of the films from the RBS channeling experiments. These authors also studied the epitaxial quality of the films as a function of the Zr concentration. Despite the rather large mismatch of rhombohedral PZT with the (001) $SrTiO_3$, values as low as 4% for the minimum channeling yield were obtained [359].

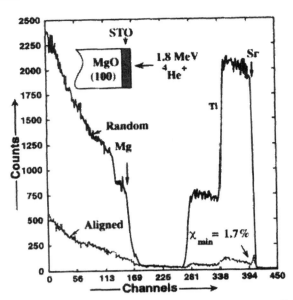

Fig. 25. RBS spectrum for the $SrTiO_3/MgO$ film grown at 1 mtorr P_{O_2} and subsequently annealed. The Sr channeling yield κ_{min} is 1.7%. Reprinted with permission from R. Kalyanaraman et al., *Appl. Phys. Lett.* 71, 1709 (1997). Copyright 1997, American Institute of Physics.

4.1.3. Transmission Electron Microscopy, Scanning Electron Microscopy, and Reflection High-Energy Electron Diffraction

Transmission electron microscopy (TEM) is the primary investigative tool for the investigation of the structural properties of crystals, thin films, and interfaces [408]. While XRD yields valuable information averaged over large areas, high-resolution transmission microscopy (HRTEM) provides information on an atomic scale. Details of atomic arrangements and defects can be examined with TEM. With lattice imaging and localized diffraction, phases and crystallographic information can be identified from small localized areas. Information is typically obtained from very small regions of a 3-mm-diameter sample using a submicron-size electron probe and is mostly qualitative in nature.

In TEM, a focused high-energy electron beam is incident on a thin sample. The TEM signal is extracted from directed and diffracted electron-beam components that penetrate the sample. TEM offers two mechanisms for observing the sample: diffraction and image. In the diffraction mode, the electron diffraction patterns correspond to an XRD analogue, and they provide information on crystallinity and crystal orientation. The image mode produces a representation of the entire sample depth. The image contrast results from several mechanisms, including diffraction contrast (the scattering of the electrons by structural inhomogeneities), mass contrast (caused by the spatial separations and orientations of constituent atoms), thickness contrast (caused by nonuniformities in film thickness), and phase contrast (the result of coherent elastic scattering). Each of these modes provides specific information about a selected sample area.

The crystallization temperature of amorphous $BaTiO_3$ has been investigated by TEM [409]. For the sample annealed at 400°C for 1 minute by rapid thermal process, the electron diffraction pattern showed a very diffuse band, which is typical for a sample with an amorphous structure. However, in the sample annealed at 500°C, the diffraction pattern showed a superposition of a diffuse band and scattered diffraction spots. The electron micrograph for the sample showed a uniform amorphous background containing ellipsoidal microcrystallites. Figure 26a is a cross-sectional TEM image of a-/b-axis-oriented $SrBi_2Ta_2O_9$ (SBT) film grown on (110) MgO substrate by rf sputtering [106] showing a clear interface between the film and the substrate. Figures 26b, c, and d show selected area electron diffraction patterns of the SBT film, the (110) MgO substrate, and the interface covering both the film and the substrate, respectively. They show that the grown film has an orthorhombic structure with orientation relationship of SBT [001]∥MgO [001] and SBT [010]∥MgO [1$\bar{1}$0] (SBT [100]∥MgO [1$\bar{1}$0]). The a-/b-axis-oriented SBT films are expected to show better electrical properties compared to those of random or c-axis-oriented SBT films because SBT has structural and electrical anisotropy [367].

Ferroelectric superlattice grown by the MBE technique showed beautiful TEM images of the superlattice. Because of the different structural factors of the constituent ferroelectric layers, the multilayer structure is clearly visible in cross-sectional TEM observation. Figure 27 is a HRTEM image of the $[(PbTiO_3)_{10}/(SrTiO_3)_{10}]_{15}$ superlattice prepared by reactive MBE [115]. The individual $PbTiO_3$ and $SrTiO_3$ layers show highly perfect single-crystal structures and an extremaly sharp interface between the layers. In the superlattice region of the film, the $PbTiO_3$ layers of a tetragonal structure are

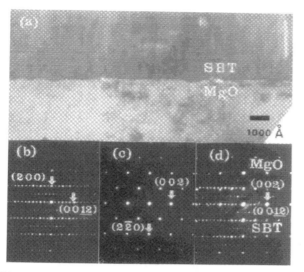

Fig. 26. (a) Cross-sectional bright-field TEM image of the a-/b-axis-oriented SBT film on (110) MgO, and selected area diffraction patterns obtained from the (b) film layer, (c) MgO substrate, and (d) interface covering the film layer and the MgO substrate, respectively. Reprinted with permission from S. E. Moon et al., *Appl. Phys. Lett.* 75, 2827 (1999). Copyright 1999, American Institute of Physics.

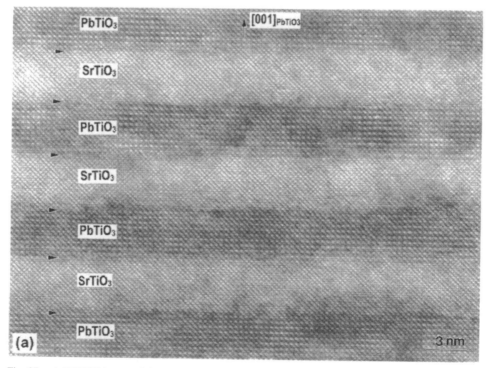

Fig. 27. A HRTEM image of the $[(PbTiO_3)_{10}/(SrTiO_3)_{10}]_{15}$ superlattice structure taken with the electron beam parallel to the [100] axis of $SrTiO_3$. Reprinted with permission from J. C. Jiang et al., *Appl. Phys. Lett.* 74, 2851 (1999). Copyright 1999, American Institute of Physics.

c-axis-oriented and free of twin boundaries, contrary to the general trends of domain formation in PbTiO$_3$-based heteroepitaxial thin films [357, 358]. It has been argued that small value of lattice mismatch can be accommodated through elastic distortion, rather than being relaxed through the formation of dislocations [115]. Such clear interfaces without interfacial reaction layer were also observed for BaTiO$_3$/SrTiO$_3$ superlattice system grown by a laser MBE [410]. The multilayers were perfectly *c*-axis oriented and the epitaxial crystalline structure showed the orientation relations of (001) BaTiO$_3$∥(001) SrTiO$_3$ and (100) BaTiO$_3$∥(100) SrTiO$_3$. These results show the possible growth of high-quality artificial crystalline oxide superlattices with full control over the composition and structure at the atomic level, making it possible to study predictions of enhanced dielectric performance in nanoscale dielectric and ferroelectric superlattices [116, 411].

Scanning electron microscopy (SEM) is the traditional instrument used to observe topography or topology on the microscale [412]. In addition to its capabilities for detailed surface, microstructural, and microchemical analysis, SEM can be used to determine crystallographic orientation information using electron-backscattered patterns (EBSP) from single crystals and grains. Essentially these are electron diffraction patterns wherein the various crystallographic planes develop different intensity levels, forming an image with crystallographic contrast, called a pattern of Kikuchi bands.

The availability of ultra-high-vacuum instruments has made it possible to study the structure of surfaces under clean conditions by reflection high-energy electron diffraction (RHEED). RHEED can be used to determine the lateral arrangement of atoms in the topmost layers of a surface, including the structure of adsorbed layers. RHEED is far superior to low-energy electron diffraction (LEED) for investigations of rough surfaces, because the fast electrons can penetrate the asperities and produce a transmission diffraction pattern. When the growing surface is smooth, the RHEED pattern is streaked and displays channeling effects that arise from good surface order. *In situ* monitoring of two-dimensional epitaxial growth of ferroelectric thin films by monitoring the RHEED specular spot intensity was shown to be possible [356, 413–415]. Two-dimensional growth can be classified as either layer-by-layer growth or step-flow growth. The former involves nucleation and subsequent growth and coalescence of islands on atomically flat terraces due to high supersaturation, which can be caused by either a high deposition rate or a relatively low deposition temperature. This can be monitored *in situ* by measuring RHEED specular spot intensity oscillations [356, 413, 414]. The RHEED oscillation period corresponds to the height of one unit cell, which satisfies the chemical composition and electrical neutrality of the oxide [413]. Also, strain-induced roughening of film surface has been deduced from the observation of RHEED oscillation amplitude damping, indicating a transition from layer-by-layer growth to island growth [356]. On the other hand, step-flow growth occurs if the growth temperature is high enough to allow adsorbed precursors to migrate to step edges on a viscinal surface before nucleation can occur on the terraces. A direct

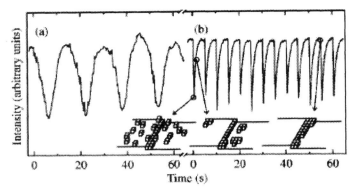

Fig. 28. RHEED specular spot intensity during homoepitaxy of SrTiO$_3$ at (a) 800°C and 2 Hz (laser pulse rate) and (b) 1200°C and 0.5 Hz. Reprinted with permission from M. Lippmaa et al., *Appl. Phys. Lett.* 74, 3543 (1999). Copyright 1999, American Institute of Physics.

evidence of step-flow growth during PLD of SrTiO$_3$ was observed by *in situ* RHEED, and it was shown that dramatically improved dielectric properties could be obtained by step-flow growth [415]. At temperatures below 900°C, the SrTiO$_3$ film growth proceeded in layer-by-layer mode, as indicated by the oscillatory nature of the specular RHEED spot intensity time dependence, as shown in Figure 28a. At 1200°C, film growth was found to proceed in the step-flow mode. The corresponding RHEED intensity behavior is shown in Figure 28b. After every laser pulse, the RHEED intensity dropped sharply, due to a momentary increase of surface roughness. The RHEED intensity recovered completely after every pulse up to the end of a deposition run, showing that surface smoothness was maintained. The intensity recovery time was found to be strongly temperature dependent [415]. The dielectric constant of the SrTiO$_3$ film grown by the step-flow mode was much larger than that grown by the layer-by-layer growth mode and could be tuned by 80% by applying a bias voltage of ±1 V at 4.2 K. The improved dielectric properties were attributed to the high crystallinity of the films grown in the step-flow mode at very high temperature.

4.1.4. Scanning Probe Microscopy

Ultra-high-resolution surface topography and surface imaging of materials can be accomplished with scanning probe microscopy (SPM). These instruments can be especially useful for nanoscale microstructures. SPM is a relatively new technique based on a very simple principle of measuring the height change of a nanoprobe (i.e., an atomically sharp probe), mounted on a cantilever, as a function of interaction with the sample, typically topographic changes. There are two basic scanning probe instruments, scanning tunneling microscopy (STM) and atomic force microscopy (AFM). STM requires a conductive sample because it monitors the tunneling current to maintain a constant probe tip-to-sample distance using a piezoelectric transducer to monitor probe height changes in the subangstrom range. AFM is based on the fact that there is an interatomic force between scanning probe and the sample surface. In operation, the interatomic force is maintained constant; thus a constant probe

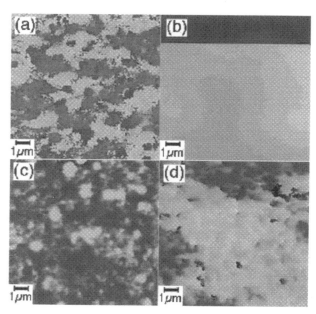

Fig. 29. Orientations of the polarization for a (a) "as-prepared" and (b) poled virgin ferroelectric thin film capacitor. The upper dark part of (b) was obtained by the application of the opposite field. The bright areas in (c) and (d) represent the frozen polarization after 10^5 and 10^7 cycles of P–E hysteresis measurement, respectively. Reprinted with permission from E. L. Colla et al., *Appl. Phys. Lett.* 72, 2763 (1998). Copyright 1998, American Institute of Physics.

Fig. 30. Polarization orientation in a fatigued sample after application of the two opposite poling fields. The frozen regions of opposite orientation are randomly distributed. Reprinted with permission from E. L. Colla et al., *Appl. Phys. Lett.* 72, 2763 (1998). Copyright 1998, American Institute of Physics.

tip to sample spacing, to provide a feedback mechanism to the piezoelectric scanner that moves the probe to maintain a constant tip-to-sample distance. A laser beam system is used to record the height variations of the scanner. The complexity of the technique results from the various interactions with the probe tip and the sample that lead to the creation of image contrast. The forces involved can include van der Waals, electrostatic, frictional, and magnetic, and instruments working on each of these force mechanisms, including magnetic force microscopes (MFMs) and electrostatic force microscopes (EFMs), are available.

In studies on ferroelectric thin films, the AFM technique has been modified to detect ferroelectric polarization states or domains in ferroelectric thin films. Electric field imaging [416], friction microscopy [417], and piezoelectric detection techniques [418–420] have been developed. The degrees of fatigue in PZT thin films have been studied by analyzing the piezoelectric vibration phase images [419]. Figure 29 shows orientations of the polarization for PZT thin films. The dark and bright regions represent polarization of opposite orientation in the "as-prepared" state (Fig. 29a). After electrical treatment (i.e., after the first polarization loop in P–E hysteresis), the polarization can be polarized quite homogenously in both electric field directions (Fig. 29b). The phase images after 10^5 and 10^7 cycles of P–E hysteresis loops are shown in Figures 29c and d, respectively; these were obtained after the application of $+17$ V on the bottom electrode. The dark regions correspond to the still-switchable parts of the film, whereas the bright regions correspond to the frozen parts along the preferred direction. Af-

ter 10^7 cycles, the degree of fatigue calculated from the frozen domains was 70%. An homogeneous phase signal could be obtained by poling in the opposite direction (-17 V on the bottom electrode). Thus it was concluded that the fatigue is due to "region-by-region" or "grain-by-grain" freezing of switching polarization, and that the frozen polarization can have a preferential orientation [419]. However, no preferential orientation existed in some samples, probably due to inhomogeneity of the film or the surface quality and higher fatiguing field (Fig. 30).

Retention loss of remnant polarization has been investigated by the piezoelectric response generated in PZT thin film [420]. It was demonstrated that polarization reversal occurs under no external field (i.e., loss of remnant polarization) via a dispersive continuous-time random work process, identified by a stretched exponential decay of the remnant polarization. The piezoresponse images given in Figures 31d and e show time-evolution snapshots of the change in the domain structure of the central grain (grain 1) that occurred after the removal of the dc field. The central grain was fully switched on application of a pulse (Fig. 31c). Figure 31d is an image recorded 4 hours after pulse switching; the first stages of backward switching, which started at the grain boundary, are discernible. This is direct evidence of the role played by grain boundaries in initiating spontaneous polarization reversal through the electrostatic interaction across the grain boundary between grain 2 of opposite polarization. Once the reversal begins, it proceeds through the sidewise expansion of the reversed portion of the grain, and reversed domain stabilization requires a grain boundary.

4.2. Electrical Characterization

4.2.1. Ferroelectric Properties: Polarization Switching, Fatigue, and Retention

Investigation of fast polarization reversal in thin films can be done by integral experimental methods. The most popular such method is the measurement of transient current. A fundamental aspect of the method is the correlation between the rate of switching identified by the maximum switching current, i_{max}, and the applied field E. In general, this has been shown to have an exponential dependence as $i_{max} = i_0 \exp(-\alpha/E)$. The re-

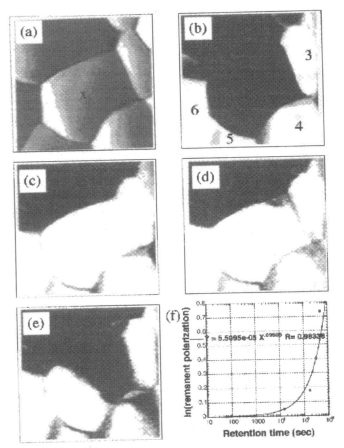

Fig. 31. AFM images showing loss of polarization in PZT thin films. AFM retention experiments conducted on PZT(20/80) grown on LSCO/TiN/Si using 6 V, 200 ms write pulse: (a) topography (200×200 nm^2), (b) piezoresponse of the as-grown surface with 2.3 V$_{ac}$, (c) piezoresponse after dc poling grain 1, (d) piezoresponse after 4 hours with no external field, (e) piezoresponse after 140 hours, and (f) stretched exponential fit of the SFM studies plotted with the logarithm of the remanent switched grain area vs time. Reprinted with permission from A. Gruverman et al., *Appl. Phys. Lett.* 71, 3492 (1997). Copyright 1997, American Institute of Physics.

lation governs the switching dynamics through the magnitude of α, the activation field of nucleation, which is a function of temperature [421]. The polarization switching in ferroelectric materials includes the nucleation of new domains, their growth, and the combination of these domains. Switching kinetics are important in ferroelectric thin films, because the speed of switching limits the speed of read and write performance in memory devices.

For the FRAM application, electric pulses are used for read and write operations instead of sinusoidal continuous electric fields, for which a ferroelectric hysteresis loop is obtained. As an important quantity of ferroelectricity, the value at zero electric field (remanent polarization) in hysteresis loop is not so meaningful in the FRAM device; instead, pulse polarization is used. Even though the hysteresis loop parameters, such as remanent polarization (P_r) and coercive field (E_c), are related with pulse polarization, it is necessary to understand their relationship in detail. Standardized commercial testers (e.g., Radiant's

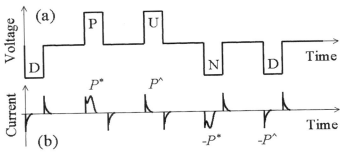

Fig. 32. (a) Double bipolar electric pulse train used to simulate the read and write operation of FRAM devices. Pulses P, U, N, and D denote positive, up, negative, and down, respectively. (b) Typical switching current responses of a ferroelectric capacitor with a double bipolar pulse train.

RT66A) are generally used for the measurement of ferroelectric properties of thin film capacitors for FRAM devices. Since FRAM devices are operated at clock rate above 1 GHz, the ferroelectric capacitor should be switched within less than 1 ns. High-speed switching properties can be measured from switching current responses. To simulate read-write operations in a FRAM device, a double-bipolar pulse train is used, as shown in Figure 32. The pulses P, U, N, and D denote positive, up, negative, and down, respectively. The stored information "1" written with the first D pulse is read by the P pulse, so that the switched pulse polarization (P^*) could be measured with the P pulse. This P pulse also writes information with "0." The U pulse reads the stored information of "0." The nonswitched pulse polarization ($P^\wedge$) is measured with the U pulse. The same is true for the negative electric fields with the N and D pulses. A pulse about 1 μs wide has generally been used up to now. Pulse polarization studies with pulses shorter than 1 ns become important, because FRAM devices are operated at very high clock rates. It was reported that P^* increased but $P^\wedge$ decreased with increasing pulse width. As a result, $\Delta P = P^* - P^\wedge$ increases with increasing pulse width, so ΔP measured with 1 ns pulse would be expected to be much smaller than with a 1-μs pulse. It is generally accepted that $2P_r$ is almost same as $P^* - P^\wedge$ in the relationship between the hysteresis loop and pulse polarizations. This relation is not true, however, because there is retention loss between measuring pulses, so that $P^* - P^\wedge$ is less than $2P_r$ by the amount of retention loss in 1 second.

It has been reported that the width dependence of pulse polarizations is strongly related to the structural and microstructural properties of ferroelectric material. The activation field, α, of ferroelectric material, which is measured from the switching current responses with different electric fields, represents switching properties. The strong pulse width dependence was shown with the higher α value, and this α value was reported to depend linearly on tetragonality (c/a) of perovskite PZT system in tetragonal phase.

As a standard measurement, the delay time between measuring pulses are set with at 1 second, which is related to retention loss properties of the ferroelectric capacitor cell. As a long-term reliability property, retention loss is the ability to retain information for a long time. For the retention loss measurement, pulse polarizations of $P*$ and $P^\wedge$ are measured as the delay time between pulses are changed. It was reported that retention losses occurred in two time regimes shorter and longer than 1 second. In the shorter regime, the depolarization effect plays an important role, and about 20–30% of the polarization is lost. This retention loss property also depends on writing pulse width. The shorter pulse width results in greater retention loss. This pulse width dependence is also related to the structural and microstructural properties of the ferroelectric material [421–425].

For nonvolatile memory applications, the fatigue-free property is very important. Some ferroelectric thin films suffer a serious loss of switchable polarization after bipolar switching, i.e., fatigue [64]. Although there is considerable debate regarding the mechanisms that cause fatigue, fatigue failure in ferroelectric materials are in general thought to be induced by defect charges, such as oxygen vacancies, present inside the materials or near electrodes [426–428]. In perovskite-type ferroelectrics, oxygen vacancies are relatively mobile and may segregate to either electrode of domain walls to impede domain switching. Also, charge carriers can influence fatigue. Numerous attempts have been made to improve fatigue resistance, including efforts to reduce movement and concentration of oxygen vacancies using oxide ectrodes (see Section 3.1.1.3), doping with a small amount of impurities that help getter oxygen [429]. For SBT thin films, it is widely accepted that the Bi_2O_3 layers play an important role in preventing fatigue failure [10]; the Bi_2O_3 layers have net electrical charges, and their positioning in the lattice is self-regulated to compensate for space charges near electrodes. It has been reported that the oxygen ions at the metal–oxygen octahedra are much more stable than those at the Bi_2O_3 layer [430].

4.2.2. Dielectric Properties

Ferroelectric (i.e., strictly paraelectric) high-dielectric-constant materials, such as $(Ba,Sr)TiO_3$, have attracted much attention as dielectric materials in capacitor structures for applications such as dynamic random access memory (DRAM), where the availability of avoiding the deeply trenched or exfoliated structures is required to achieve sufficient capacitance with SiO_2 dielectrics. Recently, the materials promise to bring about a new generation of such high-frequency devices as tunable capacitor [431], microwave phase shifter [432], and electro-optic modulators [433]. However, when thin film dielectrics are used as capacitor structures, the high dielectric constant values in bulk crystals tend to be significantly reduced, so that the desired performance of the device incorporating them may not be reached. The reduction may be due to a low quality of the film or to other effects (e.g., a size effect). The high dielectric loss at high frequencies is a major barrier to the widespread use of thin film dielectrics in a number of high-frequency applications. Thus, as device capabilities increase, measuring these properties over an increasingly broad band of frequencies is becoming more important.

There are many techniques for quantitatively measuring the dielectric permittivity of thin film samples. Usually, thin film capacitors allow measurement of the in-plane component [434] as well as the normal component of the permittivity tensor. Dielectric resonators [435] and Corbino measurements [436] have also been used at high frequencies in the rf and microwave range.

The low-frequency dielectric response of dielectric thin films is usually measured using a low-frequency inductance-capacitance-resistance (LCR) meter. In view of structure–property relationships, epitaxial or high-quality crystalline thin films show better dielectric properties than amorphous or polycrystalline films. Surface roughness can also affect the dielectric properties of thin films [400, 437]. Furthermore, interfaces between dielectric thin films and electrodes alter the observed dielectric properties. Jia et al. [438] showed that epitaxial (50/50) BST thin film with a $SrRuO_3$ bottom electrode had superior electrical and dielectric properties [438] compared to epitaxial BST with $YBa_2Cu_3O_{7-x}$ as a bottom electrode [439]. The crystalline thin films with a configuration of $Ag/BST/SrRuO_3/LaAlO_3$ had a dielectric constant around 500, a dielectric loss of less than 0.01 at 10 kHz, and a leakage current density of less than 50×10^{-8} A/cm^2 at a field intensity of 2×10^6 V/cm [438]. Also, thin film capacitors consisting tatally of perovskite oxides [$SrRuO_3$/BST (20 nm)/$SrRuO_3$] have been fabricated on Si and $SrTiO_3$ by rf-magnetron sputtering [440]. The relative dielectric constants for polycrystalline and single-crystal epitaxial capacitors were found to be 274 and 681, respectively. Compared with the electrical properties of samples with Pt top electrodes, large differences in the values of dielectric constant were observed. The difference were explained in terms of differences in the interface between the dielectric and the top electrode, and the possible existence of a low-dielectric layer at the interface between the Pt top electrode and BST due to ion vacancies, interface levels, and/or local lattice distortion [440]. Recently, structure–property relationships were reviewed for high-permittivity $BaTiO_3$, $SrTiO_3$, and BST ceramic thin films [441, 442].

The size effect refers to a dependence of the dielectric constant of a thin dielectric film on the film thickness. Nanosized particles and films of ferroelectric perovskites exhibit variations in polarization behavior with changes in applied voltage or electric field, measurement temperature, or particle size or film thickness [441, 442]. Frequently, a significant decrease in permittivity is reported if film thickness is reduced below a critical value. For $SrTiO_3$, the critical thickness value below which the permittivity drop occurs is about 30–70 nm [443, 444]. A good understanding of size effect on the dielectric and ferroelectric properties in such films of high-permittivity material is essential for successful application. Many factors can cause the size effect, the most obvious possibilities have been discussed by Waser [441]. These include a low-dielectric interfacial layer, a microstructure/grain size effect, porosity or second phases, a stress/strain effect, stoichiometry variation, mean free path length of phonons, high-field space charge depletion layers, and measurement artifact. Recently, the temperature- and field-dependent permittivities of fiber-textured BST thin films were investigated as a function of film thickness (24–160 nm) [445]. The thickness dependence of dielectric permittivity was discussed in terms of a nonlinear dielectric interfacial layer. The nature of size effect has also been discussed based on the spatial correlation of the ferroelectric polarization and boundary condition of the ferroelectric polarization on electrodes [446]. It was shown that the $SrRuO_3/BST/SrRuO_3$ structure could be characterized by free boundary conditions for the ferroelectric polarization, providing the highest value of the effective dielectric constant of a dielectric layer in the sandwich structure. Apart from the interfacial layer effect, an intrinsic "dead-layer effect" on the surface of dielectric film that reduced the effective dielectric constant of the film was suggested [447].

It has been suggested that the formation of 90° domain walls in ultra-thin films grown epitaxially on appropriate substrates can increase its dielectric permittivity markedly because of the field-induced translational vibrations of the walls [116]. This expectation was realized in $PbTiO_3/PLT$ superlattices that have both a- and c-domains [117]. At low frequencies (below 1 kHz), permittivities as high as 420,000 were found. The dielectric behavior of the system, including frequency and bias voltage dependencies, was explained by a model based on the rigid-body motion of a domain wall lattice. It is noteworthy that the superlattice with short period, which may experience smaller strains, showed different dielectric behavior. The superlattice with periodicity of 10 nm, which was found to be paraelectric, had a much lower dielectric constant (about 750) and no significant frequency dependence in the frequency range of 300 Hz–1 MHz [117].

The formation of dielectric and ferroelectric superlattices provided a powerful method for creating new high-dielectric materials [110]. The superlattice structure and the periodic stress at the interface between the constituent $SrTiO_3$ and $BaTiO_3$ layers were found to play an important role in the enhancement of the dielectric constant. Enhanced tetragonality in ferroelectric layers, which was induced by in-plane lattice-mismatch strain, may produce the large dielectric constant. Figure 33 shows the dielectric constant of the $SrTiO_3/BaTiO_3$ superlattices and the solid-solution as a function of total thickness. The superlattice with a longer periodicity shows better dielectric behavior. The $(Sr_{0.3}Ba_{0.7})TiO_3/(Sr_{0.48}Ca_{0.52})TiO_3$ superlattice, optimized by adjusting the composition and the periodicity of the strained superlattice, showed a dramatically large dielectric constant (900) even at a film thickness of 50 nm. The optimum pressure for inducing tetragonality in the BST layers and enhanced dielectric properties occurred at a lattice mismatch of 2.5–3% [110].

A large dielectric constant and a high nonlinear dielectric response make ferroelectric materials attractive for microwave and radio frequency applications, e.g., field-tunable capacitor elements. The combination of $YBa_2Cu_3O_{7-x}$ and $SrTiO_3$ (or SBT) offers the potential of unpredicted performance in tunable planar circuits, e.g., phase shifters for steerable beam antennas at 20–30 GHz [448]. However, the dielectric loss in ferroelectric thin films is still too large to allow these films to be useful

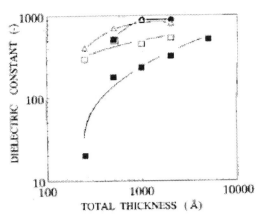

Fig. 33. The dielectric constant of the STO/BTO superlattices as a function of the total thickness. The dielectric constant of the STO/BTO superlattices and solid solution SBTO against the total thickness between 250 and 4000 Å. □, 1/1 unit stacking; ●, 2/2 unit stacking; △, 4/4 unit stacking; ■, SBTO solid solution. Reprinted with permission from H. Tabata and T. Kawai, *Appl. Phys. Lett.* 70, 321 (1997). Copyright 1997, American Institute of Physics.

in real devices. Therefore, it is important to understand the microwave properties of ferroelectric thin film in conjunction with the films' processing and microstructure.

High-frequency capacitance and loss information for $SrTiO_3$ thin film have been extracted from resonances in $YBa_2Cu_3O_{7-x}$ microstrip resonators fabricated on $LaAlO_3$ substrates with $SrTiO_3$-filled gaps at the superconducting strips [449]. The obtained dielectric constants at 11 GHz and 4 K were 920 at zero bias and 330 at 40 V (although the uncertainty could be as great as a factor of 2). The complex dielectric function of $SrTiO_3$ was also measured as a function of temperature and electric field bias using a microwave ring resonator and a flip-chip technique [450]. The films with large grains in the plane, which were grown by PLD with an oxygen pressure of 600 mtorr, had the highest dielectric constant. The in-plane component of the dielectric constant of thin films in the configuration of coplanar capacitors or interdigitated capacitors has been extracted using a conformal mapping technique [451, 452]. In this technique, the dielectric constant of thin films is calculated from the capacitance, electrode geometry, film thickness, and dielectric constant of the substrate. Petrov et al. [453] have shown that the dielectric constant of $SrTiO_3$ thin film is independent of frequency to at least 40 GHz, while the losses exhibit no or weak frequency dependence. Multilayers with eight intermediate oxygen relaxation during $SrTiO_3$ film deposition had the lowest losses ($\tan \delta < 0.005$) and the highest dielectric constant (1800 at 20 GHz) with 25% tunability of the effective dielectric constant at 20 K and an applied field of 19 kV/cm [453]. Oxygen relaxation involves the deposition (at 760°C) and cooling of a 50-nm $SrTiO_3$ sublayer to 300°C at 20°C/minute in 0.9 atm of oxygen, followed by heating and deposition of a 50-nm $SrTiO_3$ sublayer, and so on. Recently, it has been shown that the degree of crytallinity, film strain, and surface roughness can affect the microwave properties of BST films through a postdeposition annealing effect [454, 455]. The tetragonal distortion (i.e., ratio of in-plane and surface normal lattice parameters,

$D = a/c$) also has been shown to affect the microwave properties of SBT(50/50) thin films grown on MgO by PLD [456]. The distortion D varied by oxygen pressure ranging from 0.996 at 3 mtorr to 1.003 at 800 mtorr. At microwave frequencies (1–20 GHz), BST films with low distortion had a higher dielectric constant ($\sim$500) and lower dielectric loss (tan $\delta \sim 0.02$) than films with higher distortion.

Recently, techniques of submicron resolution have been applied to the studies on microwave properties of ferroelectric thin films [457–459]. The techniques give quantitative results consistent with conventional thin film measurements performed with contact interdigital electrodes at microwave frequencies [460]. Mapping of microwave dielectric properties was performed using a scanning evanescent microwave microscope at 1 GHz for an epitaxial thin film ternary composition spread of $(Ba_{1-x}Sr_xCa_y)TiO_3$ ($0 < x < 1$ and $0 < y < 1$) fabricated on an equilateral-triangle-shaped $LaAlO_3$ substrate [458]. The composition region $Ba_{0.12-0.25}Sr_{0.35-0.47}Ca_{0.32-0.5}TiO_3$ was found to have desirable properties for electronic applications such as DRAM. Steinhauer et al. [459] measured tunability of BST thin films using a near-field scanning microwave microscope with a spatial resolution of 1 μm. Permittivity images and local hysteresis loops were also demonstrated at 7.2 GHz. The technique is able to measure changes in relative permittivity $\Delta\varepsilon_r$ as small as 2 at $\varepsilon_r = 500$, and changes in dielectric tunability as small as 0.03/V. These authors also observed the role of annealing in the recovery of dielectric tunability in a damaged region of the thin film. A new optical probe of GHz polarization dynamics in ferroelectric thin films has been reported [461]. The technique uses the electro-optic response of thin films. A microwave voltage, derived from the pulse train of a mode-locked Ti:sapphire laser (frequency $f_1 = 76$ MHz), was applied to BST film via interdigitated electrodes. The laser pulses were focused to a diffraction-limited spot on the sample. The change in polarization state of the reflected light could be related to the ferroelectric polarization at one particular phase of the microwave field. An understanding of the physical origin of the local phase shifts observed may help reduce dielectric loss in ferroelectric thin films [461].

4.2.3. Piezoelectric and Pyroelectric Properties

Knowledge of piezoelectric response in ferroelectric thin films can be useful for understanding materials property issues and developing applications and integration of ferroelectric microelectromechanical systems (MEMSs). Currently, several techniques, including interferometry [462–467], AFM [467–470], direct methods [471, 472], and measurement of the displacement or resonance of cantilever [473–476], are used to evaluate piezoresponse.

Interferometry has the advantage of being able to measure the piezoelectric effect with high resolution [462–467]. A modified Michelson–Morley interferometer has been used to make ac field-induced shape change measurements. For a film on a substrate, a bending contribution to the sample face displacement occurred. To compensate for this bending, a double-beam

interferometer was used, in which the measurement beam was reflected off the front and back of the bimorph, thus allowing determination of the thickness change only. An asymmetry in butterfly-like strain-field hysteresis curves was observed and was interpreted in terms of residual interfacial stresses between the substrate and the films [464–466]. The interfacial stress could be relieved electrically, and strength of the stress field was inferred to depend on the microstructure of films [465]. Also, the piezoelectric constant and electromechanical coupling coefficient of PZT were found to be frequency dependent in the rf range [464], whereas the piezoelectric response of the bulk ceramic was found to be independent of frequency. These phenomena were ascribed to the relaxation of the electrically induced strain and piezoelectric response, resulting in decreased d_{33} value with increased frequency [465]. Recently, the piezoelectric properties of rhombohedral PZT films were found to depend on their texture [477]. The highest piezoelectric coefficients was observed in (100)-oriented films, which had smaller polarization than films with (111) preferred orientation, in agreement with the predictions in the calculation for single crystals [478]. AFM has been used to determine the piezoelectric properties of PZT films in the morphotropic phase boundary [469].

An AFM tip was used as a top electrode to apply a voltage to polarize the film and to apply an oscillating field to obtain piezoelectric coefficients and piezoelectric loops from the inverse piezoelectric effect induced on the film. This method can provide insight into the structure and the dielectric, ferroelectric, and piezoelectric properties of distinct nanoregions of films [469]. It was found that non-180° domains did not contribute to the measured piezoelectric coefficient for the stress levels applied, and that the mobility of ferroelectric domain walls was low for PZT films [469]. In this technique, accurate determination of the electric field is necessary to obtain the absolute magnitude of the effective longitudinal piezoelectric coefficient (d_{33}). Also, it should be pointed out that the piezoelectrically exited region may be constrained by the surrounding material. Maiwa et al. [462] reported on the piezoelectric response of PZT thin films determined using the interferometric method, and compared the obtained d_{33} values to those measured by AFM. They showed that the interferometry and AFM techniques are very complementary when used together. Extrinsic contributions to the piezoelectric displacement can be greatly suppressed in the measured film by considering "intrinsic" permittivity, film orientation, and substrate clamping [467].

The material property of main interest in MEMS is the piezoelectric coefficient, d_{31}. Application of an electric field in the direction perpendicular to the film surface causes the piezoelectric film to shrink or expand in the transverse direction parallel to the film surface. For this measurement, microfabricated cantilevers were used with piezoelectric thin films. Poling the polycryalline PZT film proved to be as effective as poling bulk PZT ceramics [473, 476]. The piezoelectric coefficient d_{31} of c-axis-oriented film was directly measured from the transverse expansion or deflection of a cantilever beam [474, 475]. The measurement revealed that the PZT films were naturally polarized and had a piezoelectric coefficient, d_{31}, as large as about 100×10^{-12} m/V without poling.

Pyroelectric materials in point detectors or arrays have many applications in temperature-sensing systems for fire and intruder detection, air-conditioning control, and thermal imaging [479]. Pyroelectric detectors formed from ferroelectric materials are of great interest for application in noncryogenic, low-cost, high-performance thermal imaging systems. An effective pyroelectric coefficient, $p = dP_s/dT$, can be deduced from the temperature dependence of remanent polarization. Alternatively, the pyroelectric coefficients of a film is determined by measuring the spontaneous thermally induced current in a diagnostic resistance under the conditions of uniform heating, constant stress, and low electric field. The pyroelectric coefficient can be obtained from the temperature gradient of pyroelectric current ($p = I_p/\{A(dT/dt)\}$, where I_p is the pyroelectric current and A is the electrode area). The dynamic Chyoweth method [480] is also used with modifications. In this method, samples are periodically heated and cooled with a given frequency, and the resulting pyroelectric current is measured.

Ferroelectric BST has been identified as a very promising material due to its composition-dependent transition temperature (30–400 K) [481], which allows the maximum infrared response to be obtained at room temperature. However, the pyroelectric coefficients of thin films have been found to be much smaller than those of ceramics and single crystals. Sol-gel-derived (80/20) BST thin films showed a pyroelectric coefficient value of about 0.046 μC/cm^2 K at 33°C [482]. Compositionally graded BST thin films have attracted interest for their charge pumping effect [483, 484]. The compositionally graded BST thin films have a continuous composition change from $BaTiO_3$ to $Ba_{0.7}Sr_{0.3}TiO_3$. On interrogating capacitor-like structures with a strong periodic electric field (active mode operation), free charge was found to preferentially accumulate on one electrode of the device, a result of the inherent asymmetries provided by the internal self-biases created by the compositional gradient [483–485]. The "pumping action" of the graded ferroelectric devices leads to offsets, or shifts, along the displacement axis in otherwise conventional displacement versus electric field ferroelectric hysteresis loops. Effective (pseudo) pyroelectric coefficients as large as 0.06 μC/cm^2 K were obtained from the active operation mode (charge pumping) [484]. In contrast, a value of only -0.003 μC/cm^2 K was measured for the conventional pyroelectric coefficient [484]. Recently, a high pyroelectric coefficient of 0.24 μC/cm^2 K was reported for (66/34) BST thin films grown by rf sputtering [486]. The sputtering target of SrO-excess (2–4%) films yielded almost stoichiometric films of larger grains and showed a higher tendency to grow along the (100) orientation compared to the films grown without excess SrO.

Pb-based perovskite ferroelectrics also have attracted interest for application to infrared detectors because of excellent ferroelectricity. Pyroelectric properties of sol-gel-derived $PbTiO_3$ thin films have been reported and compared with previous works [487]. La- or Ca-modified $PbTiO_3$ thin films have

also been studied [387, 488–490]. Such A-site dopants have been used to obtain higher pyroelectric coefficients by lowering the Curie temperature. The RTA process at high temperature (800°C) was shown to be effective in the preparation of preferentially c-axis-oriented Ca-modified $PbTiO_3$ and for the enhancement of pyroelectric properties [490]. For good voltage response [479], it is necessary to maximize p and lower the relative permittivity, ε_r, to increase the FOM, $F_v = p/\varepsilon_r$. For current detection [479], the electric loss (tan δ) becomes important, and the FOM is $F_1 = p/c_v\sqrt{\varepsilon_r}\tan\delta$ (where c_v denotes heat capacity). Incorporation of void space was done in an attempt to lower ε_r [489]. The heating rate of sol-gel-derived films was found to have a profound effect on the microstructure development of the films [489]. Relatively porous films with low ε_r could be obtained at slow heating rates. The obtained values of the FOM for voltage detection were 3–4 times higher, and those for current detection were 1.5–2 times higher, than the results for PZT compositions [489].

4.2.4. Capacitance–Voltage and Current–Voltage Characteristics

Analysis of small-signal capacitance–voltage (C–V) characteristics offers the switching dielectric properties of ferroelectric thin films. The ferroelectric permittivity is very large in a small region about the coercive field value (E_c), due to the switching of the domains. This occurs because during switching, a very small change in the applied electric field results in a very large change in the ferroelectric polarization. Thus, the ideal C–V characteristics of ferroelectric thin film capacitors include a hysteresis with maximum permittivity peaks at $\pm E_c$. Typical C–V characteristic were reported for sol-gel-derived ferroelectric PLT thin films with varying a concentrations [491]. Film capacitance increased with increasing La content, and exhibited a strong nonlinearity on an applied voltage. The C–V curves showed the hysteresis that resulted from the switching of ferroelectric domains, and were symmetric about the zero bias voltage. The hysteretic behaviors decreased as the La content increased. A domain model consistent with the measured C–V characteristic of PZT capacitors was proposed [492]. The interaction between domain properties and electrical doping has been explored. The C–V behavior and the switching of the polarization can be modulated by the space charge produced by ionization of trap states in the surface region of the ferroelectrics due to immense electric fields, induced by large polarization, at the surface of the ferroelectric crystal [493]. The metallic electrodes in a ferroelectric capcitor form Schottky contacts. The space charge accumulation at the interface is due to the band bending caused by the Schottky potential and to the applied bias [494]. The depolarizing field in the ferroelectrics is produced by the difference in space charge concentrations at the two junctions. The ferroelectric capacitor is a three-layer dielectric sandwich with high-permittivity layer between two low-permittivity layers. A mathematical model of the small-signal characteristics of the ferroelectric capacitor has shown that the C–V behavior is largely determined by the space charge concentrations at the ferroelectric–contact interface [495]. The $C-V$ characteristics and consequently the device performances are strongly dependent on the interfacial properties of the electrode and ferroelectric thin films, as well as on the intrinsic electrical properties of the film.

The $C-V$ characteristics for paraelectric (50/50) BST thin films showed a shift of capacitance maximum toward the negative bias region by annealing under high vacuum conditions [496]. This fact was attributed to a built-in electric field, probably due to oxygen vacancies formed at the interface between the film and the top Pt electrode during annealing. This electric field was thought to result in the enhanced leakage currents through reduction of the Schottky barrier height due to image force increases and altered $C-V$ characteristics with increasing annealing temperature [496]. Recently, thickness dependence of electrical properties of SBT thin films was reported [497]. The apparent coercive field obtained from the hysteretic $C-V$ curve was different from that obtained from the $P-E$ hysteresis loop. The result may be due to mobile ionic defects in the films because the sweep time of the bias voltage was much slower than that of the $P-E$ measurement. The dielectric constant measured at zero bias was strongly dependent on the film thickness. This result was explained by an interfacial layer adjacent to the electrode [497]. The $C-V$ characteristics of ferrelectric thin films are known to be affected by fatigue [498]. The zero bias capacitance decreased by about 20% after fatigue. The repeated domain switching during fatigue may cause a significant increase in the impeding forces against the domain wall motion. It has been argued that the effect may lead to a reduction in domain wall mobility and/or in the total amount of the active domain walls participating in the switching process [498]. The coercive field was also found to be decreased, which was attributed to the decrease in the space charge layer capacitance or the increase in its thickness [498]. The effects of constant voltage stress on ferroelectric PZT thin-film capacitors have been studied [499]. The stress effects were revealed by the reduction of capacitance and a voltage shift of the $C-V$ curve and the effect were found to depend on the stress electric field and the injected charge fluence. The reduction of capacitance and the voltage shift were correlated with electron trapping inside the film.

A ferroelectric nonvolatile data storage element is commonly a capacitor consisting of a thin ferroelectric film sandwiched between two conductive electrodes [63]. The direction of polarization vector is set by a voltage pulse applied to the capacitor. The stored bit is read by applying another voltage pulse and determining whether or not the polarization switched direction. During the read process, the data are destroyed (destructive readout), and thus the bit must be reprogrammed after each read. On the other hand, the metal/ferroelectric/semiconductor field-effect transistor (MFSFET) offers a significant advantage over the ferroelectric capacitor approach: the data-reading process does not destroy the stored data, and the ferroelectric gate offers simple circuits [500]. The ferroelectric field effect is the modulation of the surface potential of semiconductor by the spontaneous polarization of a ferroelectric that is in intimate contact with the semiconductor material. The ferroelectric thin films do not need to have large spontaneous polarization, because the spontaneous polarization directly controls the channel conductivity of the FET. The information is read by sensing the FET channel current, which depends on the polarization state of the ferroelectric gate. MFS gate structures have serious problems, including poor interface, high leakage current, retention, and fatigue [328, 501, 502]. In a typical MFSFET configuration, the ferroelectric materials make direct contact with silicon, often resulting in interdiffusion or reaction and thus a high density of interface states (donors or acceptors) that can trap charges. These trapped charges can be ionized to form space charges by a strong polarization-induced field, which can degrade device performance. An alternative solution has been adopted through inserting an insulating buffer layer between ferroelectrics and silicon, producing a metal/ferroelectric/insulator/semiconductor (MFIS) structure. Fabrications of MFSFETs or MFISFETs with limited success have been reported using various material systems, including $BaTiO_3$ [503], $LiNbO_3$ [504, 505], $Bi_4Ti_3O_{12}$ [500, 506, 507], P(L)ZT [508–513], SBT [514–518], and $YMnO_3$ [156, 328, 519–521]. The insulating layers used were SiO_2, CeO_2, Y_2O_3, $SrTiO_3$, and other oxides compatible with semiconductors and ferroelectric oxides. Results have shown the better to avoid an insulating layer of low dielectric constant, because a relatively high electric field would be applied to the insulating layer when a bias voltage is applied across the gate, and this may cause a charge injection to the insulator layer. Alternatively, ferroelectric materials of low dielectric constant are preferred to reduce the electric field applied to the insulating layer.

The $C-V$ characteristics of ferroelectric thin films in an MFS or MFIS structure provides useful information on the ferroelectricity of films, defects in films, interface properties, and other parameters. Most of the arguments applied to metal/oxide/semiconductor (MOS) devices [522] can be used. In MFS or MFIS structures, the effects of ferroelectric polarization should be considered for the analysis of the $C-V$ characteristics.

Ferroelectricity of thin films in MF(I)S structures is often confirmed via a ferroelectric switching-type $C-V$ hysteresis loop. If the gate is ferroelectric, then the direction of the hysteresis is clockwise (counterclockwise) in the $C-V$ curve for p-type (n-type) semiconductor substrates. Figure 34 shows the $C-V$ curves for $YMnO_3$ thin films grown by CSD on n-type Si substrate. The hysteresis in $C-V$ curves is consistent with that arising from ferroelectric switching. However, cautious analysis is necessary to confirm whether the origin of the hysteresis is the ferroelectricity of the film. There have been reports that the width of $C-V$ hysteresis (memory window) is dependent on the sweep rate of the gate bias voltage and the sweep width of the bias voltage. Ionic conduction in nonferroelectric gate materials can apparently cause the same $C-V$ characteristics as those caused by ferroelectric gate materials [523, 524]. Mobile ions in the gate layer are responsible for the sweep rate dependence of the memory window, as well as for the observed stretchout in $C-V$ curves and shifts in threshold voltage [524]. The effects of

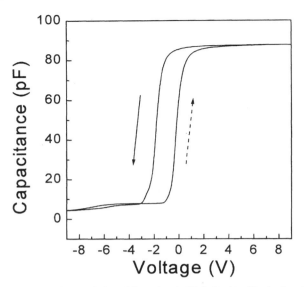

Fig. 34. *C–V* characteristics of ferroelectric YMnO$_3$ thin film in the structure of Au/YMnO$_3$/*n*-Si(100). The *C–V* curves were obtained at a measuring frequency of 1 MHz.

mobile charges from ferroelectric gates have been investigated by observing the temperature dependence of *C–V* curves [328]. Charge injection in general causes a *C–V* hysteresis of a different direction then that caused by polarization switching. If charge injection from semiconductor substrates occurs, then the memory window induced by polarization switching can be reduced or diminished. Also, the memory window may be caused by interfacial polarization and rearrangement of the space charge [521]. In an MFIS structure, interfacial polarization may be generated because of the difference in the charges in ferroelectric layer and insulator layer. The interfacial charge is difficult to distinguish from the remanent polarization. The discharge time of the interfacial polarization is on the order of minutes. This slow polarization can also be observed by the rearrangement of space charges. Recently, problems of conventional *C–V* measurement for evaluating MFIS capacitor were discussed [521]. In the case of the MFIS capacitor, because the ferroelectric materials should have lower remanent polarization than the MFM capacitor, conventional *C–V* characteristics are influenced more by interfacial polarization and rearrangement of the space charge. It was shown that pulsed *C–V* measurement was effective in evaluating the MFIS capacitor. In this method, bidirectional step voltage pulses were applied to the gate. After a short time pulse voltage was applied to polarize the ferroelectric layer (±9 V/0.1 second), the capacitance was measured after a holding time (0.2 second) with a voltage corresponding to gate bias voltage. Bidirectional step voltage was needed to eliminate the accumulation of charge due to the slow polarization by repeated pulses with the same sign.

The leakage current of ferroelectric thin films is a primary concern in estimating the capacitor charge retention. Various mechanisms responsible for the leakage current in electroceramic thin film and bulk ceramic have been proposed [525–528]. Essentially, the trap states play a prominent role in the

leakage current and capacitance dispersion of capacitors. Due to the existence of a grain boundary, bulk structural/electronic defects, or a heterojunction, the existence of trap states is suspected for many ferroelectric thin films. The current transportation in ferroelectric thin films at high fields is normally ascribed to either the Poole–Frenkel or the Schottky emission conduction mechanism [528, 529]. The two conduction mechanisms are quite similar and cannot be easily distinguished [530]. Compared with SiO$_2$ (with a band gap of about 9 eV), most of ferroelectrics have much smaller band gaps (3–4 eV), so that Schottky emission into band states could be a serious source of leakage conduction. Recently, it has been reported that the Schottky barrier is controlled mainly by Fermi-level pinning at the interface rather than by the work function of electrode materials [531]. However, it seems that the leakage conduction can originate from many different intrinsic and extrinsic factors. There has been a report that the dielectric and electrical leakage characteristics of BST thin films are dependent on film thickness as well as on the fabrication processes, especially when a Pt electrode is used [532]. Atmosphere control during annealing also has been shown to affect leakage conduction characteristics [533–535]. Schottky emission, thermionic tunneling, and Fowler–Nordheim tunneling mechanisms have been investigated for the electrical behavior of sol-gel-derived PZT films with transition metal and noble metal electrodes [536]. Space-charge-limited currents has also been proposed as an important factor in the description of leakage current and breakdown [2, 533, 537]. Space-charge-limited currents can occur from any microscopic transport mechanism (e.g., Schottky, Fowler–Nordheim, Poole–Frenkel) or a combination of several such mechanisms, whenever current densities reach a certain value such that current is not limited by surface or interface details.

4.3. Optical Characterization

Infrared (IR) and Raman spectroscopies have been widely used to study ferroelectric materials, especially on ferroelectric phase transition in conjunction with soft modes. In ferroelectric thin films, however, these optical spectroscopic methods have not been well used compared to single crystal. Recently, these methods were used to identify the ferroelectric phase of materials and to study strain effects in thin films [368, 538, 539]. However, compared to single crystal, these optical measurements encounter some difficulties because of substrate effects. The effects of the film–substrate interface and an interpretation method of IR spectra of SrTiO$_3$ thin films on MgO were introduced by Song et al. [540]. The optical properties of PbTiO$_3$ have been studied in the form of thin films and compared with the results from single crystal [541]. A shift in the transition temperature was observed from the soft mode behavior, these are related to effective hydrostatic pressure due to the clamping of grains by their neighbors in the distorted tetragonal structure. Similar results were reported in BaTiO$_3$ thin films deposited by rf-magnetron sputtering [542]. These spectroscopic methods are very useful tools for studying phase transition behaviors of ferroelectric materials in thin films.

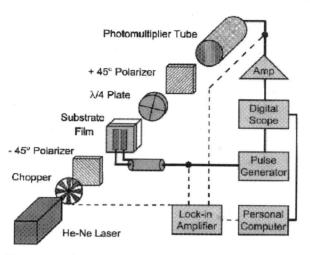

Fig. 35. A schematic representation of the electro-optic measurement system. Solid lines indicate the electronic circuit used to measure the response up to 150 MHz. Dashed lines indicate the electronic connections used to measure r_{eff} values at 100 kHz. Reprinted with permission from B. H. Hoerman et al., *Appl. Phys. Lett.* 75, 2707 (1999). Copyright 1999, American Institute of Physics.

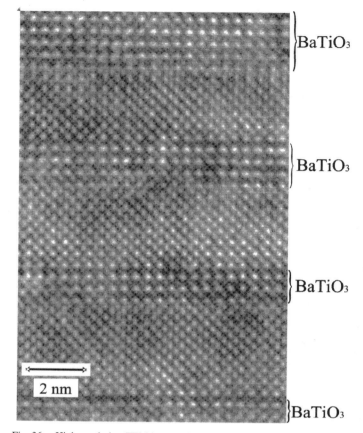

Fig. 36. High-resolution TEM image of graded $BaTiO_3/SrTiO_3$ superlattice. Courtesy of D. G. Schlom.

Along with spectroscopic characterization, optical waveguide and electro-optic properties of ferroelectric thin films have been investigated. Film optical loss is often measured by the prism coupling technique [9, 225, 348]. Optical losses in ferroelectric oxide thin films were discussed extensively by Fork et al. [543]. A thin-film waveguide electro-optic modulator was demonstrated with epitaxial $BaTiO_3$ thin films [433]. The electro-optic effect of ferroelectric thin films is usually observed in a transverse mode with a geometry of normal incidence. There have been reports on electro-optic effect with large linear and quadratic coefficients [336, 365, 544]. Figure 35 is a schematic representation of the electro-optic measurement system adopted by Hoerman et al. [545]. The dynamic response (from 10 μs to 10 ns) of the electro-optic coefficient of epitaxial $KNbO_3$ thin films was measured, dispersion of the electro-optic coefficient was observed to follow a power law, $\sim At^m$, which may be a consequence of a relaxation process [545].

5. SUMMARY AND CONCLUDING REMARKS

In this review chapter, we have tried to provide information on prevalent thin film deposition techniques for ferroelectric oxides and thin film characterizations. Key issues in fabricating multicomponent ferroelectric oxide thin films include controlling the stoichiometry of films and enhancing structural properties to preserve useful physical properties of the materials. Exact control of film stoichiometry can reduce electronic defects, including oxygen vacancies, in the films and optimize physical properties of the films. This control depends on various deposition techniques and process conditions. Reliable deposition techniques and well-established process conditions should be provided for mass production of the films with other requirements discussed in Section 3. Along with stoichiometry control,

interface problems should be resolved to enhance structural properties of films. From the standpoint of film growth mechanisms nucleation and subsequent growth of films may govern the microstructural evolution of films. Minimization of structural defects can enhance the physical properties of films. Also, control of growth orientation may be possible and thus enable optimization of the physical properties of films for various applications. These issues require basic understanding of thin film science relevant to process–structure–property relationships.

With advances in thin film technology, a challenging issue is the growth of commensurate crystalline oxides on semiconductors. Recently, McKee et al. [546] reported growth of commensurate crystalline $SrTiO_3$ on Si. They showed that alkaline-earth and perovskite oxides could be grown in perfect registry on the (001) face of silicon, totally avoiding the amorphous silica phase that ordinarily forms when silicon is exposed to an oxygen-containing environment. This result introduced the possibility of an entirely new device physics based on use of the anisotropic response of crystalline oxides grown commensurately on a semiconductor. Another challenging issue is the fabrication of artificial superlattice structure consisting of oxide materials. A graded $BaTiO_3/SrTiO_3$ superlattice structure (Fig. 36) has been fabricated [547]. Although the number of reported superlattic structures is limited and the physical properties of the systems have not yet been explored in detail, this system is expected to have an impact on material science and to provide new scientific research issues in oxide electronics.

Many research reports have been published on thin film material processing, characterizations, device applications, and device integration issues. All of these issues are being explored very rapidly, and it is difficult to follow-up on all of the recent research results in a single review chapter. The Proceedings of the International Symposium on Integrated Ferroelectrics (ISIF), or International Symposium on Application of Ferroelectrics (ISAF), and related publications may be helpful sources of information on recent progress in ferroelectric thin films.

REFERENCES

1. M. E. Lines and A. M. Glass, "Principles and Applications of Ferroelectrics and Related Materials." Clarendon, Oxford, UK, 1982.
2. J. F. Scott, *Ferroelectrics Rev.* 1, 1 (1998).
3. R. H. Dennard, U.S. Patent 3387286, 1968.
4. P. C. Fazan, *Integ. Ferroelectrics* 4, 247 (1994).
5. K. Iijima, Y. Tomita, R. Takayama, and I. Ueda, *J. Appl. Phys.* 60, 361 (1986); *ibid.* 60, 2914 (1986).
6. Y. Shibata, K. Kaya, A. Akashi, M. Kanai, T. Kawai, and S. Kawai, *Appl. Phys. Lett.* 61, 1000 (1992).
7. K. R. Carroll, J. M. Pond, D. B. Chrisey, J. S. Horwitz, and R. E. Leuchtner, *Appl. Phys. Lett.* 62, 1845 (1993).
8. D. K. Fork and G. B. Anderson, *Appl. Phys. Lett.* 63, 1029 (1993).
9. H. Xie, W.-Y. Hsu, and R. Raj, *J. Appl. Phys.* 77, 3420 (1994).
10. C. A. Paz de Araujo, J. D. Cuchiaro, L. D. McMillian, M. C. Scott, and J. F. Scott, *Nature* 374, 627 (1995).
11. E. K. Muller, B. J. Nicholson, and G. E. Turner, *J. Electrochem. Soc.* 110, 969 (1963).
12. I. H. Pratt and S. Firestone, *J. Vac. Sci. Technol.* 8, 256 (1971).
13. W. B. Pennebaker, *IBM J. Res. Dev.* 686 (1969).
14. N. F. Foster, *J. Appl. Phys.* 40, 420 (1969).
15. W. J. Takei, N. P. Formigoni, and M. H. Francombe, *J. Vac. Sci. Technol.* 7, 442 (1969).
16. R. A. Roy, K. F. Etzold, and J. J. Cuomo, *Mater. Res. Soc. Symp. Proc.* 200, 77 (1990).
17. J. Valasek, *Phys. Rev.* 15, 537 (1920); *ibid.* 17, 475 (1921).
18. C. B. Sawyer and C. H. Tower, *Phys. Rev.* 35, 269 (1930).
19. F. Jona and G. Shirane, "Ferroelectric Crystals." Pergamon Press, Oxford, UK, 1962.
20. I. Bunget and M. Popescu, "Physics of Solid Dielectrics." Elsevier, Amsterdam, 1984.
21. J. Harada, T. Pederson, and Z. Barnea, *Acta Crystallogr.* A 26, 336 (1970).
22. W. Cochran, *Phys. Rev. Lett.* 3, 412 (1959); *Adv. Phys.* 9, 387 (1960).
23. P. W. Anderson, "Fizika Dielektrikov" (G. I. Skanavi, Ed.). Akad. Nauk SSSR, Moscow, 1959.
24. A. W. Hewat, *J. Phys. C* 6, 2559 (1973).
25. G. Shirane, R. Pepinsky, and B. C. Frazer, *Acta Crystallogr.* 9, 131 (1956).
26. J. P. Remeika and A. M. Glass, *Mater. Res. Bull.* 5, 37 (1970).
27. G. Shirane and K. Suzuki, *J. Phys. Soc. Jpn.* 7, 333 (1952).
28. E. Sawaguchi, *J. Phys. Soc. Jpn.* 8, 615 (1953).
29. B. Jaffe, W. R. Cook, Jr., and H. Jaffe, "Piezoelectric Ceramics." Academic Press, London, 1971.
30. C. Michel, J.-M. Moreau, G. D. Achenbach, R. Gerson, and W. J. James, *Solid State Commun.* 7, 865 (1969).
31. B. Jaffe, R. S. Roth, and S. Marzullo, *J. Res. Nat. Bur. Stand.* 55, 239 (1955).
32. D. Berlincourt, C. Smolik, and H. Jaffe, *Proc. IRE* 48, 220 (1960).
33. P. C. Joshi, S. O. Ryu, X. Zhang, and S. B. Desu, *Appl. Phys. Lett.* 70, 1080 (1997).
34. B. Aurivillius, *Ark. Kemi.* 1, 463 (1950).
35. B. Aurivillius, *Ark. Kemi.* 1, 499 (1950).
36. B. Aurivillius, *Ark. Kemi.* 2, 519 (1951).
37. A. D. Rae, J. G. Thompson, and R. L. Withers, *Acta Crystallogr., Sect. B: Struct. Sci.* 48, 418 (1992).
38. R. E. Newnham, R. W. Wolfe, R. S. Horsey, F. A. Diaz-Colon, and M. I. Kay, *Mater. Res. Bull.* 8, 1183 (1973).
39. S. E. Cummins and L. E. Cross, *J. Appl. Phys.* 39, 2268 (1968); A. Fouskova and L. E. Cross, *ibid.* 41, 2834 (1970).
40. A. D. Rae, J. G. Thompson, R. L. Withers, and A. C. Willis, *Acta Crystallogr., Sect. B: Struct. Sci.* 46, 474 (1990).
41. K. Nassau, H. J. Levinstein, and G. M. Loiacono, *J. Phys. Chem. Solids* 27, 983 (1966).
42. S. C. Abrahams, W. C. Hamilton, and J. L. Bernstein, *J. Phys. Chem. Solids* 27, 997 (1966); S. C. Abrahams, W. C. Hamilton, and J. M. Reddy, *ibid.* 27, 1013 (1966); S. C. Abrahams and J. L. Bernstein, *ibid.* 28, 1685 (1967).
43. Z. I. Shapiro, S. A. Fedulov, Y. N. Venevtsev, and L. G. Rigerman, *Sov. Phys. Cryst.* 10, 725 (1966).
44. P. V. Lenzo, E. G. Spencer, and A. A. Ballman, *Appl. Phys. Lett.* 11, 23 (1967).
45. A. Watanabe, Y. Sato, T. Yano, and I. Kitahiro, *J. Phys. Soc. Jpn.* 28, Suppl. 93 (1970).
46. P. B. Jamieson, S. C. Abrahams, and J. L. Bernstein, *J. Chem. Phys.* 48, 5048 (1968).
47. A. A. Ballman and H. Brown, *J. Cryst. Growth* 1, 311 (1967).
48. Y. Xu, "Ferroelectric Materials and Their Applications," Chap. 6. Elsevier, Amsterdam, 1991.
49. M. A. Gilleo, *Acta Crystallogr.* 10, 161 (1957).
50. H. L. Yakel, W. C. Koehler, E. F. Bertant, and E. F. Forrat, *Acta Crystallogr.* 16, 957 (1963).
51. F. Moussa, M. Hennion, J. Rodriguez-Carvajal, H. Moudden, L. Pinsard, and A. Revcolevschi, *Phys. Rev. B* 54, 15149 (1996).
52. G. A. Smolenskii and I. E. Chupis, *Sov. Phys. Usp.* 25, 475 (1982).
53. F. Bertant, F. Forrat, and P. Fang, *C. R. Acad. Sci.* 256, 1958 (1963).
54. M. N. Iliev, H.-G. Lee, V. N. Popov, M. V. Abrashev, A. Hamed, R. L. Meng, and C. W. Chu, *Phys. Rev. B* 56, 2488 (1997).
55. J. F. Nye, "Physical Properties of Crystals." Clarendon Press, Oxford, U.K., 1964.
56. J. R. Anderson, *Elec. Eng.* 71, 916 (1952).
57. W. Kaenzig, *Phys. Rev.* 98, 549 (1955).
58. W. P. Mason, "Piezoelectric Crystals and Their Application to Ultrasonics." Van Nostrand, New York, 1950.
59. H. Jaffe and D. A. Berlincourt, *Proc. IEEE* 53, 1372 (1965).
60. J. L. Vossen and W. Kern, "Thin Film Processes." Academic Press, Boston, 1978.
61. J. L. Vossen and W. Kern, "Thin Film Processes II." Academic Press, Boston, 1991.
62. K. Wasa and S. Hayakawa, "Handbook of Sputter Deposition Technology—Principles, Technology and Applications." Noyes Publications, Park Ridge, NI, 1992.
63. J. F. Scott and C. A. Paz de Araujo, *Science* 246, 1400 (1989).
64. C. A. Paz de Araujo, L. D. McMillan, B. M. Melnick, J. D. Cuchiaro, and J. F. Scott, *Ferroelectrics* 104, 241 (1990).
65. R. A. Roy, K. F. Etzold, and J. J. Cuomo, *Mater. Res. Soc. Symp. Proc.* 200, 77 (1990); *ibid.* 200, 141 (1990).
66. W. A. Geideman, *IEEE Trans. Ultrason., Ferroelectrics, Freq. Cont.* 38, 704 (1991).
67. O. Auciello and A. I. Kingon, in "Proceedings of the 8th International Symposium on Applications of Ferroelectrics," p. 320 (M. Liu, A. Safari, A. Kingon, and G. Haertling, Eds.), IEEE, Piscataway, NJ, 1992.
68. T. Mihara, H. Yoshimori, H. Watanabe, and C. A. P. de Araujo, *Japan. J. Appl. Phys.* 34, 5233 (1995).
69. O. Auciello, A. R. Krauss, and K. D. Gifford, in "Ferroelectric Thin Films: Synthesis and Basic Properties," Chap. 10 (C. P. de Araujo, J. F. Scott, and G. W. Taylor, Eds.). Gordon and Breach, Amsterdam, 1996.

70. S. B. Krupandhi, H. Hu, and G. R. Fox, in "Ferroelectric Thin Films: Synthesis and Basic Properties," Chap. 11 (C. P. de Araujo, J. F. Scott, and G. W. Taylor, Eds.). Gordon and Breach, Amsterdam, 1996.

71. O. Auciello, R. Dat, and R. Ramesh, in "Ferroelectric Thin Films: Synthesis and Basic Properties," Chap. 13 (C. P. de Araujo, J. F. Scott, and G. W. Taylor, Eds.). Gordon and Breach, Amsterdam, 1996.

72. *MRS Bull.* 21, 6/7 (1996).

73. O. Auciello, J. F. Scott, and R. Ramesh, *Phys. Today* 51, 22 (1998).

74. "Science and Technology of Electroceramic Thin Films" (O. Auciello and R. Waser, Eds.). Kluwer, Dordrecht, 1994.

75. *Integ. Ferroelectics* 27, 1 (1999).

76. A. K. Sharma, "Semiconductor Memories; Technology, Testing, and Reliability." IEEE Press, New York, 1997.

77. G. H. Haertling, *Ceram. Trans.* 25, 1 (1992).

78. C. Y. Chang and S. M. Sze, "ULSI Technology." McGraw-Hill, New York, 1996.

79. P. R. Willmott and J. R. Huber, *Rev. Modern Phys.* 72, 315 (2000).

80. K. Iijima, I. Ueda, and K. Kugimiya, *Ceram. Trans.* 25, 33 (1992).

81. A. Crouteau and M. Sayer, in "Proceedings of the 6th International Symposium on Applications of Ferroelectrics," p. 606 (V. E. Wood, Ed.). IEEE, Press New York, 1986.

82. F. Beech and I. W. Boyd, in "Photochemical Processing of Electronic Materials," p. 387 (I. W. Boyd and R. B. Jackman, Eds.). Academic Press, New York, 1991.

83. S. D. Bu, B. H. Park, B. S. Kang, S. H. Kang, T. W. Noh, and W. Jo, *Appl. Phys. Lett.* 75, 1155 (1999).

84. D. Roy, S. B. Krupanidhi, J. P. Dougherty, and L. E. Cross, *Ceram. Trans.* 25, 121 (1992).

85. T. Zhao, F. Chen, H. Lu, G. Yang, and Z. Chen, *J. Appl. Phys.* 87, 7442 (2000).

86. H.-M. Christen, D. P. Norton, L. A. Géa, and L. A. Boatner, *Thin Solid Films* 312, 156 (1998).

87. F. Ayguavives, B. Ea-kim, P. Aubert, B. Agius, and J. Bretagne, *Appl. Phys. Lett.* 73, 1023 (1998).

88. J. Im, A. R. Krauss, A. M. Dhote, D. M. Gruen, O. Auciello, R. Ramesh, and R. P. H. Chang, *Appl. Phys. Lett.* 72, 2529 (1998).

89. K. Wasa, Y. Haneda, T. Satoh, H. Adachi, and K. Setsune, *J. Vac. Sci. Technol. A* 15, 1185 (1997).

90. M. Tyunina, J. Levoska, and S. Leppavuori, *J. Appl. Phys.* 83, 5489 (1998).

91. O. Auciello, A. R. Krauss, J. Im, D. M. Gruen, E. A. Irene, R. P. H. Chang, and G. E. McGuire, *Appl. Phys. Lett.* 69, 2671 (1996).

92. Y. Yamagata, K. Shingai, A. M. Grishin, T. Ikegami, and K. Ebihara, *Thin Solid Films* 316, 56 (1998).

93. C. D. Theis, J. Yeh, D. G. Schlom, M. E. Hawley, G. W. Brown, J. C. Jiang, and X. Q. Pan, *Appl. Phys. Lett.* 72, 2817 (1998).

94. G. W. Brown, M. E. Hawley, C. D. Theis, J. Yeh, and D. G. Schlom, *Thin Solid Films* 357, 13 (1999).

95. E. W. Kreutz, J. Gottmann, M. Mergens, T. Klotzbücher, and B. Vosseler, *Surf. Coat. Technol.* 116–119, 1219 (1999).

96. T. Hase and T. Shiosaki, *Japan. J. Appl. Phys.* 30, 2159 (1991).

97. K. Abe and S. Komatsu, *J. Appl. Phys.* 77, 6461 (1995).

98. M. Brazier, M. McElfresh, and S. Mansour, *Appl. Phys. Lett.* 72, 1121 (1998).

99. S. B. Desu, D. P. Vijay, X. Zhang, and B. He, *Appl. Phys. Lett.* 69, 1719 (1996).

100. S. B. Desu, H. S. Cho, and P. C. Joshi, *Appl. Phys. Lett.* 70, 1393 (1997).

101. J. S. Lee, H. H. Kim, H. J. Kwon, and T. W. Jeong, *Appl. Phys. Lett.* 73, 166 (1998).

102. J. H. Cho, S. H. Bang, J. Y. Son, and Q. X. Jia, *Appl. Phys. Lett.* 72, 665 (1998).

103. J. Lettieri, C. I. Weber, and D. G. Schlom, *Appl. Phys. Lett.* 73, 2057 (1998).

104. J. Lettieri, Y. Jia, M. Urbanik, C. I. Weber, J.-P. Maria, D. G. Schlom, H. Li, R. Ramesh, R. Uecker, and P. Reiche, *Appl. Phys. Lett.* 73, 2923 (1998).

105. J. S. Lee, H. J. Kwon, Y. W. Jeong, H. H. Kim, S. J. Hyun, and T. W. Noh, *Appl. Phys. Lett.* 74, 2690 (1999).

106. S. E. Moon, T. K. Song, S. B. Back, S.-I. Kwun, J.-G. Yoon, and J. S. Lee, *Appl. Phys. Lett.* 75, 2827 (1999): S. E. Moon, S. B. Back, S.-I. Kwun, T. K. Song, and J.-G. Yoon, *J. Korean Phys. Soc.* 35, S1206 (1999).

107. T. Nagahama, T. Manabe, I. Yamaguchi, T. Kumagai, T. Tsuchiya, and S. Mizuta, *Thin Solid Films* 353, 52 (1999).

108. J. Lettieri, M. A. Zurbuchen, Y. Jia, D. G. Schlom, S. K. Streiffer, and M. E. Hawley, *Appl. Phys. Lett.* 76, 2937 (2000).

109. H. Tabata, H. Tanaka, T. Kawai, and M. Okuyama, *Japan. J. Appl. Phys.* 34, 544 (1995).

110. H. Tabata and T. Kawai, *Appl. Phys. Lett.* 70, 321 (1997).

111. B. D. Qu, M. Evstigneev, D. J. Johnson, and R. H. Prince, *Appl. Phys. Lett.* 72, 1394 (1998).

112. E. D. Specht, H.-M. Christen, D. P. Norton, and L. A. Boatner, *Phys. Rev. Lett.* 80, 4317 (1998).

113. H.-M. Christen, E. D. Specht, D. P. Norton, M. F. Chisholm, and L. A. Boatner, *Appl. Phys. Lett.* 72, 2535 (1998).

114. K. Fujimoto, Y. Kobayashi, and K. Kubota, *Thin Solid Films* 169, 249 (1989).

115. J. C. Jiang, X. Q. Pan, W. Tian, C. D. Theis, and D. G. Schlom, *Appl. Phys. Lett.* 74, 2851 (1999).

116. N. A. Pertsev, G. Arlt, and A. G. Zembilgotov, *Phys. Rev. Lett.* 76, 1364 (1996).

117. A. Erbil, Y. Kim, and R. A. Gerhardt, *Phys. Rev. Lett.* 77, 1628 (1996).

118. N. A. Pertsev, A. G. Zembilgotov, and A. K. Tagantsev, *Phys. Rev. Lett.* 80, 1988 (1998).

119. N. A. Pertsev and V. G. Koukhar, *Phys. Rev. Lett.* 84, 3722 (2000).

120. G. A. Rossetti, Jr., L. E. Cross, and K. Kushida, *Appl. Phys. Lett.* 59, 2524 (1991).

121. N. A. Pertsev, A. G. Zembilgotov, S. Hoffmann, R. Waser, and A. K. Tagantsev, *J. Appl. Phys.* 85, 1698 (1999).

122. S. H. Oh and H. M. Jang, *Appl. Phys. Lett.* 72, 1457 (1998).

123. H.-M. Tsai, P. Lin, and T.-Y. Tseng, *Appl. Phys. Lett.* 72, 1787 (1998).

124. E. J. Cukauskas, S. W. Kirchoefer, W. J. DeSisto, and J. M. Pond, *Appl. Phys. Lett.* 74, 4034 (1999).

125. T. Tybell, C. H. Ahn, and J.-M. Triscone, *Appl. Phys. Lett.* 75, 856 (1999).

126. M. Tachiki, M. Noda, K. Yamada, and T. Kobayashi, *J. Appl. Phys.* 83, 5351 (1998).

127. A. Srivastava, V. Craciun, J. M. Howard, and R. K. Singh, *Appl. Phys. Lett.* 75, 3002 (1999).

128. S. S. Nathan, G. M. Rao, and S. Mohan, *Thin Solid Films* 347, 14 (1999).

129. J. M. Bell, P. C. Knight, and G. R. Johnston, in "Ferroelectric Thin Films: Synthesis and Basic Properties," Chap. 4 (C. P. de Araujo, J. F. Scott, and G. W. Taylor, Eds.). Gordon and Breach, Amsterdam, 1996.

130. H. N. Al-Shareef and A. I. Kingon, in "Ferroelectric Thin Films: Synthesis and Basic Properties," Chap. 7 (C. P. de Araujo, J. F. Scott, and G. W. Taylor, Eds.). Gordon and Breach, Amsterdam, 1996.

131. B. Yang, S. Aggarwal, A. M. Dhote, T. K. Song, R. Ramesh, and J. S. Lee, *Appl. Phys. Lett.* 71, 356 (1997).

132. S. Madhukar, S. Aggarwal, A. M. Dhote, R. Ramesh, A. Krishnan, D. Keeble, and E. Poindexter, *J. Appl. Phys.* 81, 3543 (1997).

133. J. Lee, C. H. Choi, B. H. Park, T. W. Noh, and J. K. Lee, *Appl. Phys. Lett.* 72, 3380 (1998).

134. J. Yin, T. Zhu, Z. G. Liu, and T. Yu, *Appl. Phys. Lett.* 75, 3698 (1999).

135. J. Yin, Z. G. Liu, and Z. C. Wu, *Appl. Phys. Lett.* 75, 3396 (1999).

136. H. D. Bhatt, S. B. Desu, D. P. Vijay, Y. S. Hwang, X. Zhang, M. Nagata, and A. Grill, *Appl. Phys. Lett.* 71, 719 (1997).

137. T. Maeder, P. Muralt, L. Sagalowicz, *Thin Solid Films* 345, 300 (1999).

138. I. Stolichnov, A. Tagantsev, N. Setter, S. Okhonin, P. Fazan, J. S. Cross, and M. Tsukada, *J. Appl. Phys.* 87, 1925 (2000).

139. C. M. Chu and P. Lin, *Appl. Phys. Lett.* 72, 1241 (1998).

140. B.-S. Kim, S.-H. Oh, S.-Y. Son, K.-W. Park, Z. R. Dai, and F. S. Ohuchi, *J. Appl. Phys.* 87, 4425 (2000).

141. J. H. Cho and K. C. Park, *Appl. Phys. Lett.* 75, 549 (1999).

142. T. F. Tseng, R. P. Yang, K. S. Lin, and I. N. Lin, *Appl. Phys. Lett.* 70, 46 (1997).

143. R. M. Bowman, D. O'Neill, M. McCurry, and J. M. Gregg, *Appl. Phys. Lett.* 70, 2622 (1997).

144. J. Lee, R. Ramesh, V. G. Keramidas, W. L. Warren, G. E. Pike, and J. T. Evans, Jr., *Appl. Phys. Lett.* 66, 1337 (1995).

145. W. L. Warren, D. Dimos, G. E. Pike, B. A. Tuttle, M. V. Raymond, R. Ramesh, and J. T. Evans, Jr., *Appl. Phys. Lett.* 67, 866 (1995).

146. G. E. Pike, W. L. Warren, D. Dimos, B. A. Tuttle, R. Ramesh, J. Lee, V. G. Keramidas, and J. T. Evans, Jr., *Appl. Phys. Lett.* 66, 484 (1995).

147. H. N. Al-Shareef, B. A. Tuttle, W. L. Warren, D. Dimos, M. V. Raymond, and M. A. Rodriguez, *Appl. Phys. Lett.* 68, 272 (1996).

148. R. Ramesh and V. G. Keramidas, *Ann. Rev. Mater. Sci.* 25, 647 (1995).

149. S. Aggarwal, B. Yang, and R. Ramesh, "Thin Film Ferroelectric Materials and Devices," p. 221. Kluwer, Boston, 1997.

150. F. Wang and S. Leppävuori, *J. Appl. Phys.* 82, 1293 (1997).

151. W. Wu, P. W. Chan, K. H. Wong, J. T. Cheung, X.-G. Li, and Y. H. Zhang, *Physica C* 282–287, 619 (1997).

152. J. P. Wang, Y. C. Ling, Y. K. Tseng, K. S. Liu, and I. N. Lin, *J. Mater. Res.* 13, 1286 (1998).

153. B. H. Park, B. S. Kang, S. D. Bu, T. W. Noh, J. Lee, and W. Jo, *Nature* 401, 682 (1999).

154. A. Kingon, *Nature* 401, 658 (1999).

155. N. Fujimura, T. Ishida, T. Yoshimura, and T. Ito, *Appl. Phys. Lett.* 69, 1011 (1996).

156. T. Yoshimura, N. Fujimura, and T. Ito, *Appl. Phys. Lett.* 73, 414 (1998).

157. H. N. Lee, Y. T. Kim, and Y. K. Park, *Appl. Phys. Lett.* 74, 3887 (1999).

158. N. A. Basit, H. K. Kim, and J. Blachere, *Appl. Phys. Lett.* 73, 3941 (1998).

159. K.-J. Choi, W.-C. Shin, J.-H. Yang, and S.-G. Yoon, *Appl. Phys. Lett.* 75, 722 (1999).

160. J.-P. Maria, W. Hackenberger, and S. Trolier-McKinstry, *J. Appl. Phys.* 84, 5147 (1998).

161. V. Bornand and S. Trolier-McKinstry, *J. Appl. Phys.* 87, 3958 (2000).

162. V. Bornand, S. Trolier-McKinstry, K. Takemura, and C. A. Randall, *J. Appl. Phys.* 87, 3965 (2000).

163. V. Bornand and S. Trolier-McKinstry, *Thin Solid Films* 370, 70 (2000).

164. R. Papiernik, L. G. Hubert-Pfalzgraf, and M.-C. Massiani, *Inorg. Chim. Acta* 165, 1 (1989).

165. S. C. Goel, M. Y. Chang, and W. E. Buhro, *Inorg. Chem.* 29, 4640 (1990).

166. D. C. Bradley, *Chem. Rev.* 89, 1317 (1989).

167. L. G. Hubert-Pfalzgraf, *Appl. Organometal. Chem.* 27, 627 (1992).

168. R. C. Mehrotra, R. Bohra, and D. P. Gaur, "Metal β-Diketonenates and Allied Derivatives." Academic Press, New York, 1978.

169. H. Holzschuh, S. Oehr, H. Suhr, and A. Weber, *Mod. Phys. Lett. B* 2, 1253 (1988).

170. R. Gardiner, D. W. Brown, P. S. Kirlin, and A. L. Rheingold, *Chem. Mater.* 3, 1053 (1991).

171. K. Timmer, K. I. M. A. Spee, A. Mackor, H. A. Meinema, A. L. Spek, and P. v. d. Sluis, *Inorg. Chim. Acta* 190, 109 (1991).

172. G. Malandrino, D. S. Richeson, T. J. Marks, D. C. DeGroot, J. L. Schindler, and C. R. Kannewurf, *Appl. Phys. Lett.* 58, 182 (1991).

173. S. J. Duray, D. B. Buchholz, S. N. Song, D. S. Richeson, J. B. Ketterson, T. J. Marks, and R. P. H. Chang, *Appl. Phys. Lett.* 59, 1503 (1991).

174. J. M. Chang, B. W. Wessels, D. S. Richeson, T. J. Marks, D. C. DeGroot, and C. R. Kannewurf, *J. Appl. Phys.* 69, 2743 (1991).

175. D. A. Neumayer, D. B. Studebaker, B. J. Hinds, C. L. Stern, and T. J. Marks, *Chem. Mater.* 6, 878 (1994).

176. L. G. Hubert-Pfalzgraf and H. Guillon, *Appl. Organometal. Chem.* 12, 221 (1998).

177. G. J. M. Dormans, D. de Keijser, P. J. van Veldhoven, D. M. Frigo, J. E. Holewijn, G. P. M. van Mier, and C. J. Smit, *Chem. Mater.* 5, 448 (1993).

178. G. B. Stringfellow, "Organometallic Chemical Vapor Phase Epitaxy: Theory and Practice." Academic Press, San Diego, 1989.

179. S. Sivaram, "Chemical Vapor Deposition: Thermal and Plasma Deposition of Electronic Materials." Van Nostrand Reinhold, New York, 1995.

180. G. J. M. Dormans, P. J. van Veldhoven, and M. de Keijser, *J. Cryst. Growth* 123, 537 (1992).

181. M. de Keijser, G. J. M. Dormans, P. J. van Veldhoven, and P. K. Larsen, *Integ. Ferroelectrics* 3, 131 (1993).

182. B. S. Kwag, K. Shang, E. P. Boyd, A. Erbil, and B. J. Wilkens, *J. Appl. Phys.* 69, 767 (1991).

183. H. Nakazawa, H. Yamane, and T. Hirai, *Japan. J. Appl. Phys.* 30, 2200 (1991).

184. L. A. Wills, B. W. Wessels, D. S. Richeson, and T. J. Marks, *Appl. Phys. Lett.* 60, 41 (1992).

185. L. A. Wills, B. W. Wessels, D. L. Schulz, and T. J. Marks, *Mater. Res. Soc. Symp. Proc.* 243, 217 (1992).

186. D. Nagano, H. Funakubo, K. Shinozaki, and N. Mizutani, *Appl. Phys. Lett.* 72, 2017 (1998).

187. P. C. van Buskirk, R. Gardiner, P. S. Kirlin, S. Krupanidhi, and S. Nutt, *Mater. Res. Soc. Symp. Proc.* 243, 223 (1992).

188. T. Kawahara, M. Yamamuka, T. Makita, J. Naka, A. Yuuki, N. Mikami, and K. Ono, *Japan. J. Appl. Phys.* 33, 5129 (1994).

189. T. Kawahara, S. Matsuno, M. Yamamuka, M. Tarutani, T. Sato, T. Horikawa, F. Uchikawa, and K. Ono, *Japan. J. Appl. Phys.* 38, 2205 (1999).

190. W. A. Feil, B. W. Wessels, L. M. Tonge, and T. J. Marks, *J. Appl. Phys.* 67, 3858 (1991).

191. L. A. Wills, W. A. Feil, B. W. Wessels, L. M. Tonge, and T. J. Marks, *J. Cryst. Growth* 107, 712 (1991).

192. T. Kimura, H. Yamauchi, H. Machida, H. Kokubun, and M. Yamada, *Japan. J. Appl. Phys.* 33, 5119 (1994).

193. Y. Takeshima, K. Shiratsuyu, H. Tagaki, and Y. Sakabe, *Japan. J. Appl. Phys.* 36, 5830 (1997).

194. H. Funakubo, D. Nagano, A. Saiki, and Y. Inagaki, *Japan. J. Appl. Phys.* 36, 5879 (1997).

195. C. J. Brierley, C. Trundle, L. Considine, R. W. Whatmore, and F. W. Ainger, *Ferroelectrics* 91, 181 (1989).

196. F. W. Ainger, C. J. Brierley, M. D. Hudson, C. Trundle, and R. W. Whatmore, *Mater. Res. Soc. Symp. Proc.* 200, 37 (1990).

197. C. H. Lin, S. W. Lee, H. Chen, and T. B. Wu, *Appl. Phys. Lett.* 75, 2485 (1999).

198. T. Nakagawa, J. Yamaguchi, M. Okuyama, and Y. Hamakawa, *Japan. J. Appl. Phys.* 21, L655 (1982).

199. M. Okada, H. Watanabe, M. Murakami, A. Nishiwaki, and K. Tomita, *J. Ceram. Soc. Jpn.* 96, 676 (1988).

200. B. S. Kwak, E. P. Boyd, and A. Erbil, *Appl. Phys. Lett.* 53, 1702 (1988).

201. T. Katayama, M. Fujimoto, M. Shimizu, and T. Shiosaki, *Japan. J. Appl. Phys.* 30, 2189 (1991).

202. T. Hirai, T. Goto, H. Matsuhashi, S. Tanimoto, and Y. Tarui, *Japan. J. Appl. Phys.* 32, 4078 (1993).

203. M. Okada, K. Tominaga, T. Araki, S. Katayama, and Y. Sakashita, *Japan. J. Appl. Phys.* 29, 718 (1990).

204. G. J. M. Dormans, M. de Keijser, and P. J. van Veldhoven, *Mater. Res. Soc. Symp. Proc.* 243, 203 (1992).

205. H. Yamazaki, T. Tsuyama, I. Kobayashi, and Y. Sugimori, *Japan. J. Appl. Phys.* 31, 2995 (1992).

206. H. Tomonari, T. Ishiu, K. Sakata, and T. Takenaka, *Japan. J. Appl. Phys.* 31, 2998 (1992).

207. M. Shimizu and T. Shiosaki, *Mater. Res. Soc. Symp. Proc.* 361, 295 (1994).

208. M. Shimizu, T. Katayama, M. Sugiyama, and T. Shiosaki, *Japan. J. Appl. Phys.* 32, 4074 (1994).

209. M. de Keijser, P. J. van Veldhoven, and G. J. M. Dormans, *Mater. Res. Soc. Symp. Proc.* 310, 223 (1993).

210. I.-S. Chen and J. F. Roeder, *Mater. Res. Soc. Symp. Proc.* 541, 375 (1999).

211. K. Tominaga, M. Miyajima, Y. Sakashita, H. Segawa, and M. Okada, *Japan. J. Appl. Phys.* 29, L1874 (1990).

212. Z. C. Feng, B. S. Kwak, A. Erbil, and L. A. Boatner, *Appl. Phys. Lett.* 64, 2350 (1994).

213. Y. Kim, A. Erbil, and L. A. Boatner, *Appl. Phys. Lett.* 69, 2187 (1996).

214. J. C. Shin, J.-W. Hong, J. M. Lee, C. S. Hwang, H. J. Kim, and Z.-G. Khim, *Japan. J. Appl. Phys.* 37, 5123 (1998).

215. M. Okada and K. Tominaga, *J. Appl. Phys.* 71, 1955 (1992).

216. W. Tao, S. B. Desu, C. H. Peng, B. Dickerson, T. K. Li, C. L. Thio, J. J. Lee, and W. Hendricks, *Mater. Res. Soc. Symp. Proc.* 361, 319 (1995).

217. R. Singh, S. Alamgir, and R. Sharangpani, *Appl. Phys. Lett.* 67, 3939 (1995).

218. W. Wang, L. W. Fu, S. X. Shang, S. Q. Yu, X. L. Wang, Z. K. Lu, and M. H. Jiang, *Mater. Res. Soc. Symp. Proc.* 243, 213 (1992).

219. T. Kijima, S. Satoh, H. Matsunaga, and M. Koba, *Japan. J. Appl. Phys.* 35, 1246 (1996).

220. T. Kijima, M. Ushikubo, and H. Matsunaga, *Japan. J. Appl. Phys.* 38, 127 (1999).

221. T. Li, Y. Zhu, S. B. Desu, C.-H. Peng, and M. Nagata, *Appl. Phys. Lett.* 68, 616 (1996).

222. B. C. Hendrix, F. Hintermaier, D. A. Desrochers, J. F. Roeder, G. Bhandari, M. Chappuis, T. H. Baum, P. C. Buskirk, C. Dehm, E. Fritsch, N. Nagel, W. Honlein, and C. Mazure, *Mater. Res. Soc. Symp. Proc.* 493, 225 (1998).

223. H. Funakubo, N. Nukaga, K. Ishikawa, and T. Watanabe, *Japan. J. Appl. Phys.* 38, L199 (1999).

224. B. J. Curtis and H. R. Brunner, *Mater. Res. Bull.* 10, 515 (1975).

225. R. S. Feigelson, *J. Cryst. Growth* 166, 1 (1996); S. Y. Lee and R. S. Feigelson, *ibid.* 186, 594 (1998).

226. Y. Sakashita and H. Segawa, *J. Appl. Phys.* 77, 5995 (1995).

227. K. Shiratsuyu, A. Sakurai, K. Tanaka, and Y. Sakabe, *Japan. J. Appl. Phys.* 38, 5437 (1999).

228. V. A. Joshkin, P. Moran, D. Saulys, T. F. Kuech, L. McCaughan, and S. R. Oktyabrsky, *Appl. Phys. Lett.* 76, 2125 (2000).

229. A. Greenwald, M. Horenstein, M. Ruane, W. Clouser, and J. Foresi, *Mater. Res. Soc. Symp. Proc.* 243, 457 (1992).

230. W. T. Petuskey, D. A. Richardson, and S. K. Dey, in "Proceedings of the 3rd International Symposium on Integrated Ferroelectrics," p. 571, 1991.

231. W. G. Lee, S. I. Woo, J. C. Kim, S. H. Choi, and K. H. Oh, *Appl. Phys. Lett.* 63, 2511 (1993).

232. Y. B. Han and D. O. Kim, *J. Vac. Sci. Technol. A* 17, 1982 (1999).

233. Y. Gao, C. L. Perkins, S. He, P. Alluri, T. Tran, S. Thevuthasan, and M. A. Henderson, *J. Appl. Phys.* 87, 7430 (1999).

234. N.-J. Seong, S.-G. Yoon, and S.-S. Lee, *Appl. Phys. Lett.* 71, 81 (1997).

235. N.-J. Seong, C.-H. Yang, W.-C. Shin, and S.-G. Yoon, *Appl. Phys. Lett.* 72, 1374 (1998).

236. S. K. Dey and P. V. Alluri, *MRS Bull.* 21, 44 (1996).

237. P. C. van Buskirk, R. Gardiner, P. S. Kirlin, and S. B. Krupandhi, in "Proceedings of the 8th Symposium on Applications of Ferroelectrics," p. 340 (M. Liu, A. Safari, A. Kingon, and G. Haertling, Eds.). IEEE Press, Piscataway, NJ, 1992.

238. J. F. Roeder, S. M. Bilodeau, P. C. van Buskirk, V. H. Ozguz, J. Ma, and S. H. Lee, in "Proceedings of the 9th Symposium on Applications of Ferroelectrics," p. 687 (R. K. Pandey, M. Liu, and A. Safari, Eds.). IEEE Press, Piscataway, NJ, 1994.

239. M. Langlet, D. Walz, P. Marage, and J. C. Joubert, *Thin Solid Films* 221, 44 (1992).

240. I.-T. Kim, S. J. Chung, and S. J. Park, *Japan. J. Appl. Phys.* 36, 5840 (1997).

241. A. R. Raju and C. N. R. Rao, *Appl. Phys. Lett.* 66, 896 (1995).

242. A. A. Wernberg, H. J. Gysling, A. J. Filo, and T. N. Blanton, *Appl. Phys. Lett.* 62, 946 (1993).

243. A. A. Wernberg, G. Braunstein, G. Paz-Paujalt, H. J. Gysling, and T. N. Blanton, *Appl. Phys. Lett.* 63, 331 (1993).

244. H. Xie and R. Raj, *Appl. Phys. Lett.* 63, 3146 (1993).

245. V. Bornand, P. Papet, and E. Philippot, *Thin Solid Films* 304, 239 (1997).

246. J.-G. Yoon, Y. S. Nam, and J. H. Oh, *J. Korean Phys. Soc.* 35, S1219 (1999).

247. N. Wakiya, S. Nagata, M. Higuchi, K. Shinozaki, and N. Mizutani, *Japan. J. Appl. Phys.* 38, 5326 (1999).

248. J. Fukushima, K. Kodaira, and T. Matsushita, *Am. Ceram. Soc. Bull.* 55, 1064 (1976).

249. E. Wu, K. C. Chen, and J. D. McKenzie, *Mater. Res. Soc. Symp. Proc.* 32, 169 (1984).

250. J. Fukushima, K. Kodaira, and T. Matsushita, *J. Mater. Sci.* 19, 595 (1984).

251. K. D. Budd, S. K. Dey, and D. A. Payne, *Br. Ceram. Soc. Proc.* 36, 107 (1985).

252. D. C. Bradley, R. C. Mehrotra, and D. P. Gaur, "Metal Alkoxide," Chap. 2. Academic Press, London, 1978.

253. K. G. Caulton and L. G. Hubert-Pfalzgraf, *Chem. Rev.* 90, 969 (1990).

254. M. I. Yanovskaya, E. P. Turevskaya, A. P. Leonov, S. A. Ivanov, N. V. Kolganova, S. Y. Stefanovich, N. Y. Turova, and Y. N. Venevtsev, *J. Mater. Sci.* 23, 395 (1988).

255. S. I. Hirano and K. Kato, *Adv. Ceram. Mater.* 3, 503 (1988).

256. S. Balbaa and G. Gowda, *J. Mater. Sci. Lett.* 5, 751 (1986).

257. D. J. Eichorst and D. A. Payne, *Mater. Res. Soc. Symp. Proc.* 121, 773 (1988).

258. K. Kato, *Japan. J. Appl. Phys.* 37, 5178 (1998).

259. K. Nashimoto, *Mater. Res. Soc. Symp. Proc.* 310, 293 (1993).

260. C.-C. Hsueh and M. L. Mecartney, *Mater. Res. Soc. Symp. Proc.* 200, 219 (1990).

261. G. J. McCarthy and R. Roy, *J. Am. Ceram. Soc.* 639 (1971).

262. I. M. Thomas, in "Sol-Gel Technology for Thin Films, Fibers, Performs, Electronics and Specialty Shapes," Part I, p. 2 (L. C. Klein, Ed.). Noyes Publications, Park Ridge, NJ, 1988.

263. G. Tomandl, H. Rosch, and A. Steigelschmitt, *Mater. Res. Soc. Symp. Proc.* 121, 281 (1988).

264. S. Gurkovich and J. Blum, in "Ultrastructure Processing of Ceramics, Glasses and Composites," p. 152 (L. L. Hench and D. R. Ulrich, Eds.). Wiley, New York, 1984.

265. R. W. Schwartz, C. D. E. Lakeman, and D. A. Payne, *Mater. Res. Soc. Symp. Proc.* 180, 335 (1990).

266. S. K. Dey and R. Zuleeg, *Ferroelectrics*, 108, 37 (1990).

267. S. S. Dana, K. F. Etzold, and J. Clabes, *J. Appl. Phys.* 69, 4398 (1991).

268. S. K. Dey, in "Ferroelectric Thin Films: Synthesis and Basic Properties," Chap. 9 (C. P. de Araujo, J. F. Scott, and G. W. Taylor, Eds.). Gordon and Breach, Amsterdam, 1996.

269. C. J. Brinker and G. W. Scherer, "Sol-Gel Science: The Physics and Chemistry of Sol-Gel Processing," Chap. 13. Academic Press, London, 1990.

270. T. A. Witten and M. E. Gates, *Science* 232, 1607 (1986).

271. C.-C. Hsueh and M. L. Mecartney, *Mater. Res. Soc. Symp. Proc.* 243, 451 (1992).

272. K. D. Budd, S. K. Dey, and D. A. Payne, *Mater. Res. Soc. Symp. Proc.* 72, 317 (1986).

273. L. F. Francis and D. A. Payne, *Mater. Res. Soc. Symp. Proc.* 200, 173 (1990).

274. Y. Xu and J. D. Mackenzie, *Integ. Ferroelectrics* 1, 17 (1992); and references therein.

275. G. Yi and M. Sayer, in "Proceedings of the 8th International Symposium on Applications of Ferroelectrics," p. 289 (M. Liu, A. Safari, A. Kingon, and G. Haertling, Eds.). IEEE Press, Piscataway, NJ, 1992.

276. S. P. Faure, P. Barboux, P. Gaucher, and J. Livage, *J. Mat. Chem.* 2, 713 (1992).

277. G. Yi, Z. Wu, and M. Sayer, *J. Appl. Phys.* 64, 2717 (1988).

278. P. Gaucher, J. Hector, and J. C. Kurfiss, in "Science and Technolgy of Electroceramic Thin Films," p. 147 (O. Auciello and R. Waser, Eds.). Kluwer, Dordrecht, 1994.

279. S. Hirano and K. Kato, *Mater. Res. Soc. Symp. Proc.* 155, 181 (1989).

280. R. Xu, Y. H. Xu, and J. D. Mackenzie, in "Proceedings of the 3rd International Symposium on Integrated Ferroelectrics," p. 561, 1991.

281. L. E. Scriven, *Mater. Res. Soc. Symp. Proc.* 121, 717 (1988).

282. D. E. Bornside, C. W. Macosko, and L. E. Scriven, *J. Imaging Techol.* 13, 122 (1987).

283. D. Meyerhofer, *J. Appl. Phys.* 49, 3993 (1978).

284. P. Meakin, *Phys. Rev. Lett.* 51, 1119 (1983).

285. M. N. Kamalasanan, N. D. Kumar, and S. Chandra, *J. Appl. Phys.* 76, 4603 (1994).

286. J.-G. Cheng, X.-J. Meng, B. Li, J. Tang, S.-L. Guo, J. Wang, H. Wang, and Z. Wang, *Appl. Phys. Lett.* 75, 2132 (1999).

287. B. A. Tuttle, R. W. Schwartz, D. H. Doughty, and J. A. Voigt, *Mater. Res. Soc. Symp. Proc.* 200, 159 (1990).

288. R. W. Schwartz, Z. Xu, D. A. Payne, T. A. DeTemple, and M. A. Bradley, *Mater. Res. Soc. Symp. Proc.* 200, 167 (1990).
289. H. Hu, L. Shi, V. Kumar, and S. B. Krupandhi, *Ceram. Trans.* 25, 113 (1991).
290. H. Megaw, "Ferroelectricity in Crystals," p. 142. Methuen, London, 1957.
291. C. Kwok, S. Desu, and L. Kammerdiner, *Mater. Res. Soc. Symp. Proc.* 200, 83 (1990).
292. K. Iijima, Y. Tomita, R. Takayama, and I. Ueda, *J. Appl. Phys.* 60, 361 (1986).
293. W. J. Takei, N. P. Formigoni, and M. H. Francombe, *J. Vac. Sci. Tech.* 7, 442 (1969).
294. E. K. F. Dang and R. J. Gooding, *Phys. Rev. Lett.* 74, 3848 (1995).
295. E. M. Griswold, L. Weaver, I. D. Calder, and M. Sayer, *Mater. Res. Soc. Symp. Proc.* 361, 389 (1995).
296. G. D. Hu, I. H. Wilson, J. B. Xu, W. Y. Cheng, S. P. Wong, and H. K. Wong, *Appl. Phys. Lett.* 74, 1221 (1999).
297. Y. Fujimori, T. Nakamura, and H. Takasu, *Japan. J. Appl. Phys.* 38, 5346 (1999).
298. K. Aizawa, E. Tokumitsu, K. Okamoto, and H. Ishiwara, *Appl. Phys. Lett.* 76, 2069 (2000).
299. Y. Wu, E. G. Jacobs, R. F. Pinizzotto, R. Tsu, H.-Y. Liu, S. R. Summerfelt, and B. E. Gnade, *Mater. Res. Soc. Symp. Proc.* 361, 269 (1995).
300. T. Noguchi, T. Hase, and Y. Miyasaka, *Japan. J. Appl. Phys.* 35, 4900 (1996).
301. S. Okamura, Y. Yagi, K. Mori, G. Fujihashi, S. Ando, and T. Tsukamoto, *Japan. J. Appl. Phys.* 36, 5889 (1997).
302. M. H. Frey and D. A. Payne, *Appl. Phys. Lett.* 63, 2753 (1993).
303. S.-J. Lee, K.-Y. Kang, J.-W. Kim, S. K. Han, and S.-D. Jeong, *Mater. Res. Soc. Symp. Proc.* 541, 495 (1999).
304. J. Thongrueng, T. Tsuchiya, Y. Masuda, S. Fujita, and K. Nagata, *Japan. J. Appl. Phys.* 38, 5309 (1999).
305. A. Li, C. Ge, P. Lü, D. Wu, S. Xiong, and N. Ming, *Appl. Phys. Lett.* 70, 1616 (1997).
306. H. B. Sharma, H. N. K. Sarma, and A. Mansingh, *J. Appl. Phys.* 85, 341 (1999).
307. R. F. Xiao, H. D. Sun, H. S. Siu, Y. Y. Zhu, P. Yu, and G. K. L. Wong, *Appl. Phys. Lett.* 70, 164 (1997).
308. G. Guzman, P. Barboux, and J. Perrière, *J. Appl. Phys.* 77, 635 (1995).
309. A.-D. Li, D. Wu, C.-Z. Ge, P. Lü, W.-H. Ma, M.-S. Zhang, C.-Y. Xu, J. Zuo, and N.-B. Ming, *J. Appl. Phys.* 85, 2145 (1999).
310. T. Hirano, H. Kawai, H. Suzuki, S. Kaneko, and T. Wada, *Japan. J. Appl. Phys.* 38, 5354 (1999).
311. C. J. Kim, T.-Y. Kim, I. Chung, and I. K. Yoo, *Mater. Res. Soc. Symp. Proc.* 541, 399 (1999).
312. Y. J. Song, Y. Zhu, and S. B. Desu, *Appl. Phys. Lett.* 72, 2686 (1998).
313. H. Suzuki, M. B. Othman, K. Murakami, S. Kaneko, and T. Hayashi, *Japan. J. Appl. Phys.* 35, 4896 (1996).
314. R. W. Schwartz, R. A. Assink, D. Dimos, M. S. Sinclair, T. J. Boyle, and C. D. Buchheit, *Mater. Res. Soc. Symp. Proc.* 361, 377 (1995).
315. R. W. Schwartz, J. A. Voigt, B. A. Tuttle, D. A. Payne, T. L. Reichert, and R. S. DaSalla, *Mater. Res.* 12, 444 (1997).
316. J. Kim, R. Khamankar, B. Jiang, P. Mania, R. Moazzami, R. E. Jones, and J. C. Lee, *Mater. Res. Soc. Symp. Proc.* 361, 409 (1995).
317. X. Meng, Z. Huang, H. Ye, J. Cheng, P. Yang, and J. Chu, *Mater. Res. Soc. Symp. Proc.* 541, 723 (1999).
318. L. Fè, B. Malic, G. Norga, M. Kosec, D. J. Wouters, T. A. Bartics, and H. E. Maes, *Mater. Res. Soc. Symp. Proc.* 541, 369 (1999).
319. M. Linnik, O. Wilson, Jr., and A. Christou, *Mater. Res. Soc. Symp. Proc.* 541, 735 (1999).
320. S. Wang, Q. Su, M. A. Robert, and T. A. Rabson, *Mater. Res. Soc. Symp. Proc.* 541, 717 (1999).
321. A. L. Kholkine, K. G. Brooks, and N. Setter, *Appl. Phys. Lett.* 71, 2044 (1997).
322. P. C. Joshi, S. O. Ryu, X. Zhang, and S. B. Desu, *Appl. Phys. Lett.* 70, 1080 (1997).
323. K. Kato, C. Zheng, J. M. Finder, S. K. Dey, and Y. Torii, *J. Am. Ceram. Soc.* 81, 1869 (1998).

324. G. D. Hu, I. H. Wilson, J. B. Xu, W. Y. Cheung, S. P. Wong, and H. K. Wong, *Appl. Phys. Lett.* 74, 1221 (1999).
325. H. Gu, C. Dong, P. Chen, D. Bao, A. Kuang, and X. Li, *J. Crystal Growth* 186, 403 (1998).
326. C. J. Kim, C. W. Chung, and K. S. Lee, *Mater. Res. Soc. Symp. Proc.* 493, 255 (1998).
327. K.-S. Hwang, B.-H Kim, T. Manase, I. Yamaguchi, T. Kumagai, and S. Mizuta, *Japan. J. Appl. Phys.* 38, 219 (1999).
328. W.-C. Yi, J.-S. Choe, C.-R. Moon, S.-I. Kwun, and J.-G. Yoon, *Appl. Phys. Lett.* 73, 903 (1998).
329. W.-C. Yi, C.-S. Seo, S.-I. Kwun, and J.-G. Yoon, *Appl. Phys. Lett.* 77, 1044 (2000).
330. H. Kitahata, K. Tadanaga, T. Minami, N. Fujimura, and T. Ito, *Japan. J. Appl. Phys.* 38, 5448 (1999).
331. J.-G. Yoon, S. Y. Kim, and S.-I. Kwun, *Ferroelectrics* 152, 207 (1994).
332. I. Koiwa, Y. Okada, J. Mita, A. Hashimoto, and Y. Sawada, *Japan. J. Appl. Phys.* 36, 5904 (1997).
333. T. Atsuki, N. Soyama, T. Yonezawa, and K. Ogi, *Japan. J. Appl. Phys.* 34, 5069 (1995).
334. L. Royer, *Bull. Soc. Franc. Mineral.* 51, 7 (1928).
335. L. Brück, *Ann. Physik* 26, 233 (1932).
336. T. M. Graettinger, S. H. Rou, M. S. Ameen, O. Auciello, and A. I. Kingon, *Appl. Phys. Lett.* 58, 1964 (1991).
337. H. Adachi and K. Wasa, *Mater. Res. Soc. Symp. Proc.* 200, 103 (1990).
338. T. Okamura, M. Adachi, T. Shiosaki, and A. Kawabata, *Japan. J. Appl. Phys.* 30, 1034 (1991).
339. R. Ramesh, A. Inam, W. K. Chan, F. Tillerot, B. Wilkens, C. C. Chang, T. Sands, J. M. Tarascon, and V. G. Keramidas, *Appl. Phys. Lett.* 59, 3542 (1991).
340. J. S. Horwitz, K. S. Grabowski, D. B. Chrisey, and R. E. Leuchtner, *Appl. Phys. Lett.* 59, 1565 (1991).
341. Y. Sakashita, T. Ono, and H. Segawa, *J. Appl. Phys.* 69, 8352 (1991).
342. K. Nashimoto and S. Nakamura, *Japan. J. Appl. Phys.* 33, 5147 (1994).
343. M. G. Norton, C. Scarfone, J. Li, C. B. Carter, and J. W. Mayer, *J. Mater. Res.* 6, 2022 (1991).
344. H. A. Lu, L. A. Wills, and B. W. Wessels, *Appl. Phys. Lett.* 64, 2973 (1994).
345. D. L. Kaiser, M. D. Vaudin, L. D. Rotter, Z. L. Wang, J. P. Cline, C. S. Hwang, R. B. Marinenko, and J. G. Gillen, *Appl. Phys. Lett.* 66, 2801 (1995).
346. V. Srikant, E. J. Tarsa, D. R. Clarke, and J. S. Speck, *J. Appl. Phys.* 77, 1517 (1995).
347. S. Kim, S. Hishita, Y. M. Kang, and S. Baik, *J. Appl. Phys.* 78, 5604 (1995).
348. L. Beckers, J. Schubert, W. Zander, J. Ziesmann, A. Eckau, P. Leinenbach, and Ch. Buchal, *J. Appl. Phys.* 83, 3305 (1998).
349. K. Iijima, T. Terashima, Y. Bando, K. Kamigaki, and H. Terauchi, *J. Appl. Phys.* 72, 2840 (1992).
350. J. Gong, M. Kawasaki, K. Fujito, U. Tanakam, N. Ishizawa, M. Yoshomoto, H. Koinuma, M. Kumagai, K. Hirai, and K. Horiguchi, *Japan. J. Appl. Phys.* 32, L687 (1993).
351. H. Shigetani, K. Kobayashi, M. Fujimoto, W. Sugimura, Y. Matsui, and J. Tanaka, *J. Appl. Phys.* 81, 693 (1997).
352. Y. Yoneda, T. Okabe, K. Sakaue, H. Terauchi, H. Kasatani, and K. Deguchi, *J. Appl. Phys.* 83, 2458 (1998).
353. C. S. Chern, J. Zhao, L. Luo, P. Lu, Y. Q. Li, P. Norris, B. Kear, F. Cosandey, C. J. Maggiore, B. Gallios, and B. J. Wilkens, *Appl. Phys. Lett.* 60, 1144 (1992).
354. J. S. Speck, A. Seifert, W. Pompe, and R. Ramesh, *J. Appl. Phys.* 76, 477 (1994).
355. R. A. McKee, F. J. Walker, E. D. Specht, G. E. Jellison, Jr., L. A. Boatner, and J. H. Harding, *Phys. Rev. Lett.* 72, 2741 (1994).
356. D.-W. Lee, D.-H. Kim, B.-S. Kang, T. W. Noh, D. R. Lee, and K.-B. Lee, *Appl. Phys. Lett.* 74, 2176 (1999).
357. B. S. Kwak, A. Erbil, B. J. Wilkens, J. D. Budai, M. F. Chisholm, and L. A. Boatner, *Phys. Rev. Lett.* 68, 3733 (1992).

358. B. S. Kwak, A. Erbil, J. D. Budai, M. F. Chisholm, L. A. Boatner, and B. J. Wilkens, *Phys. Rev. B* 49, 14865 (1994).

359. M. de Keijser, J. F. M. Cillessen, R. B. F. Janssen, A. E. M. De Veirman, and D. M. de Leeuw, *J. Appl. Phys.* 79, 393 (1996).

360. C. M. Foster, G.-R. Bai, R. Csencsits, J. Vetrone, R. Jammy, L. A. Wills, E. Carr, and J. Amano, *J. Appl. Phys.* 81, 2349 (1997).

361. S. Matsubara, N. Shohata, and M. Mikami, *Japan. J. Appl. Phys.* S24, 10 (1985).

362. R. Ramesh, K. Luther, B. Wilkens, D. L. Hart, E. Wang, J. M. Tarascon, A. Inam, X. D. Wu, and T. Venkatesan, *Appl. Phys. Lett.* 57, 1505 (1990).

363. W. Jo, G.-C. Yi, T. W. Noh, D.-K. Ko, Y. S. Cho, and S.-I. Kwun, *Appl. Phys. Lett.* 61, 1516 (1992).

364. H. Buhay, S. Sinharoy, W. H. Kasner, M. H. Farancombe, D. R. Lampe, and E. Stepke, *Appl. Phys. Lett.* 58, 1470 (1991).

365. W. Jo, H.-J. Cho, T. W. Noh, B. I. Kim, D.-Y. Kim, Z. G. Kim, and S.-I. Kwun, *Appl. Phys. Lett.* 63, 2198 (1993).

366. W. Jo and T. W. Noh, *Appl. Phys. Lett.* 65, 2780 (1994).

367. K. Ishikawa and H. Funkubo, *Appl. Phys. Lett.* 75, 1970 (1999).

368. S. J. Hyun, B. H. Park, S. D. Bu, J. H. Jung, and T. W. Noh, *Appl. Phys. Lett.* 73, 2518 (1998).

369. N. Fujimura, M. Kakinoki, H. Tsuboi, and T. Ito, *J. Appl. Phys.* 75, 2169 (1994).

370. H. Matsunaga, H. Ohno, Y. Okamoto, and Y. Nakajima, *J. Cryst. Growth* 99, 630 (1990).

371. Y. Saito and T. Shiosaki, *Japan. J. Appl. Phys.* 30, 2204 (1991).

372. K. Nashimoto and M. J. Cima, *Mater. Lett.* 10, 348 (1991).

373. Y. Shibata, K. Kaya, K. Akashi, M. Kanai, T. Kawai, and S. Kawai, *Japan. J. Appl. Phys.* 32, L745 (1993).

374. T. N. Blanton and D. K. Chatterjee, *Thin Solid Films* 256, 59 (1995).

375. S. H. Rou, T. M. Graettinger, A. F. Chow, C. N. Soble, D. J. Lichtenwalner, O. Auciello, and A. I. Kingon, *Mater. Res. Soc. Symp. Proc.* 243, 81 (1992).

376. M. Ishida, S. Tsuji, K. Kimura, H. Matsunami, and T. Tanaka, *J. Cryst. Growth* 45, 393 (1978).

377. M. Mori and H. Ishiwara, *Japan. J. Appl. Phys.* 30, L1415 (1991).

378. H. Nagata, T. Tsukahara, S. Gonda, M. Yoshimoto, and H. Koinuma, *Japan. J. Appl. Phys.* 30, L1136 (1991).

379. D. K. Fork, F. A. Ponce, J. C. Tramontana, and T. H. Geballe, *Appl. Phys. Lett.* 58, 2294 (1991).

380. K. Nashimoto, D. K. Fork, and T. H. Geballe, *Appl. Phys. Lett.* 60, 1199 (1992).

381. E. J. Tarsa, M. De Graef, D. R. Clarke, A. C. Gossard, and J. S. Speck, *J. App. Phys.* 73, 3276 (1993).

382. R. Huang and A. H. Kitai, *Appl. Phys. Lett.* 61, 1450 (1992).

383. E. Fujii, A. Tomozawa, S. Fujii, H. Torii, R. Takayama, and T. Hirao, *Japan. J. Appl. Phys.* 33, 6331 (1994).

384. J.-G. Yoon and K. Kim, *Appl. Phys. Lett.* 66, 2661 (1995).

385. S. Kim and S. Hishita, *J. Mater. Res.* 12, 1152 (1997).

386. W.-Y. Hsu and R. Raj, *Appl. Phys. Lett.* 60, 3105 (1992).

387. S. Fujii, A. Tomozawa, E. Fujii, H. Torii, R. Takayama, and T. Hirao, *Appl. Phys. Lett.* 65, 1463 (1994).

388. D. K. Fork and G. B. Anderson, *Appl. Phys. Lett.* 63, 1029 (1993); *ibid.*, *Mater. Res. Soc. Symp. Proc.* 285, 355 (1993).

389. L. S. Hung, J. A. Agostinelli, J. M. Mir, and L. R. Zheng, *Appl. Phys. Lett.* 62, 3071 (1993).

390. J.-G. Yoon and K. Kim, *Appl. Phys. Lett.* 68, 2523 (1996).

391. J.-G. Yoon, *J. Korean Phys. Soc.* 29, S648 (1996).

392. Y.-M. Wu and J.-T. Lo, *Japan. J. Appl. Phys.* 37, 4943 (1998).

393. B.-E. Park, S. Shouriki, E. Tokumitsu, and H. Ishiwara, *Japan. J. Appl. Phys.* 37, 5145 (1998).

394. S. Tan, T. E. Schlesinger, and M. Migliuolo, *Appl. Phys. Lett.* 68, 2651 (1996).

395. A. R. James, A. Pignolet, D. Hesse, and U. Gösele, *J. Appl. Phys.* 87, 2825 (2000).

396. C. K. Kwok and S. B. Desu, *Appl. Phys. Lett.* 60, 1430 (1992).

397. N. Floquet, J. Hector, and P. Gaucher, *J. Appl. Phys.* 84, 3815 (1998).

398. C. Valot, J. F. Berar, C. Courteous, M. Maglione, M. Mesnier, and J. C. Niepce, *Ferroelectr. Lett.* 17, 5 (1994).

399. R. Kalyanaraman, R. D. Vispute, S. Oktyabrsky, K. Dovidenko, K. Jagannadham, J. Narayan, J. D. Budai, N. Parikh, and A. Suvkhanov, *Appl. Phys. Lett.* 71, 1709 (1997).

400. M. B. Lee and H. Koinuma, *J. Appl. Phys.* 81, 2358 (1997).

401. S. Chattopadhyay, A. Teren, and B. W. Wessels, *Mater. Res. Soc. Symp. Proc.* 541, 489 (1999).

402. S. K. Sinha, M. K. Sanyal, S. K. Satija, C. F. Majkzak, D. A. Neumann, H. Homma, S. Szpala, A. Gibaud, and H. Morkoc, *Physica B* 198, 72 (1994).

403. D. Y. Noh, H. H. Lee, T. S. Kang, and J. H. Je, *Appl. Phys. Lett.* 72, 2823 (1998).

404. J.-G. Yoon, Y. J. Kwag, and H. K. Kim, *J. Korean Phys. Soc.* 31, 613 (1997).

405. W.-K. Chu, J. W. Mayer, and M.-A. Nicolet, "Backscattering Spectrometry." Academic Press, London, 1978.

406. L. R. Doolittle, *Nucl. Instrum. Meth. Phys. Res. B* 9, 344 (1985).

407. T. A. Rost, H. Lin, T. A. Rabson, R. C. Baumann, and D. L. Callahan, *J. Appl. Phys.* 72, 4336 (1992).

408. P. R. Buseck, M. M. Cowley, and L. Eyring, "High Resolution Transmission Elelectron Microscopy and Associated Techniques." Oxford University Press, New York, 1988.

409. P. Li, J. F. McDonald, and T.-M. Lu, *J. Appl. Phys.* 71, 5596 (1992).

410. N. Wang, H. B. Lu, W. Z. Chen, T. Zhao, F. Chen, H. Y. Peng, S. T. Lee, and G. Z. Yang, *Appl. Phys. Lett.* 75, 3464 (1999).

411. S. Li, J. A. Eastman, J. M. Vetrone, R. E. Newnham, and L. E. Cross, *Philos. Mag. B* 76, 47 (1997).

412. L. Reimer, "Scanning Electron Microscopy." Springer-Verlag, Berlin, 1985.

413. T. Terashima, Y. Bando, K. Iijima, K. Yamamoto, K. Hirata, K. Hayashi, K. Kamigaki, and H. Terauchi, *Phys. Rev. Lett.* 65, 2684 (1990).

414. G. J. H. M. Rijnders, G. Koster, D. H. A. Blank, and H. Rogalla, *Appl. Phys. Lett.* 70, 1883 (1997).

415. M. Lippmaa, N. Nakagawa, M. Kawasaki, S. Ohashi, Y. Inaguma, M. Itoh, and H. Koinuma, *Appl. Phys. Lett.* 74, 3543 (1999).

416. F. Saurenbach and B. D. Terris, *Appl. Phys. Lett.* 56, 1703 (1990).

417. R. Lüthi, H. Haefke, K.-P. Meyer, E. Meyer, L. Howald, and H.-J. Güntherodt, *J. Appl. Phys.* 74, 7461 (1993).

418. P. Güthner and K. Dransfeld, *Appl. Phys. Lett.* 61, 1137 (1992); T. Hidaka, T. Maruyama, M. Saitoh, N. Mikoshiba, M. Shimizu, T. Shiosaki, L. A. Wills, R. Hiskes, S. A. Dicarolis, and J. Amano, *ibid.* 68, 2358 (1996); A. L. Gruverman, O. Auciello, and H. Tokumoto, *ibid.* 69, 3191 (1996).

419. E. L. Colla, S. Hong, D. V. Taylor, A. K. Tagantsev, N. Setter, and K. No, *Appl. Phys. Lett.* 72, 2763 (1998).

420. A. Gruverman, H. Tokumoto, A. S. Prakash, S. Aggarwal, B. Yang, M. Wuttig, R. Ramesh, O. Auciello, and T. Venkatesan, *Appl. Phys. Lett.* 71, 3492 (1997).

421. W. J. Merz, *J. Appl. Phys.* 27, 938 (1956).

422. P. K. Larsen, G. L. M. Kampschoer, M. J. E. Ulenaers, G. A. C. M. Spierings, and R. Cuppens, *Appl. Phys. Lett.* 59, 611 (1991).

423. G. J. M. Dormans and P. K. Larsen, *Appl. Phys. Lett.* 65, 3326 (1994).

424. D. J. Taylor, P. K. Larsen, G. J. M. Dormans, and A. E. M. de Veirman, *Integ. Ferroelectrics* 7, 123 (1995).

425. T. K. Song, S. Aggarwal, A. S. Prakash, B. Yang, and R. Ramesh, *Appl. Phys. Lett.* 71, 2211 (1997).

426. W. L. Warren, B. A. Tuttle, and D. Diamos, *Appl. Phys. Lett.* 67, 1426 (1995).

427. H. M. Duiker, P. D. Beale, J. F. Scott, C. A. Paz de Araujo, B. M. Melnick, J. D. Cuchiaro, and L. D. McMillan, *J. Appl. Phys.* 68, 5783 (1990).

428. I. K. Yoo, S. B. Desu, and J. Xing, *Mater. Res. Soc. Symp. Proc.* 310, 165 (1993).

429. J. F. Scott, C. A. Paz de Araujo, B. M. Melnick, L. D. McMillan, and R. Zuleeg, *J. Appl. Phys.* 70, 382 (1991).

430. B. H. Park, S. J. Hyun, S. D. Bu, T. W. Noh, J. Lee, H.-D. Kim, T. H. Kim, and W. Jo, *Appl. Phys. Lett.* 74, 1907 (1999).

431. S. K. Streiffer, C. Basceri, A. I. Kingon, S. Lipa, S. Bilodeau, R. Carl, and P. C. Van Buskirk, *Mater. Res. Soc. Symp. Proc.* 415, 219 (1996).

432. J. S. Horwitz, J. M. Pond, B. Tadayan, R. C. Y. Auyeung, P. C. Dorsey, D. B. Chrisey, S. B. Qadri, and C. Muller, *Mater. Res. Soc. Symp. Proc.* 361, 515 (1995).

433. D. M. Gill, C. W. Conrad, G. Ford, B. W. Wessels, and S. T. Ho, *Appl. Phys. Lett.* 71, 1783 (1997).

434. B. H. Hoerman, G. M. Ford, L. D. Kaufmann, and B. W. Wessels, *Appl. Phys. Lett.* 73, 2248 (1998).

435. A. T. Findikoglu, C. Doughty, S. M. Anlage, Q. Li, X. X. Xi, and T. Venkatesan, *J. Appl. Phys.* 76, 2937 (1994).

436. J. C. Booth, D. H. Wu, and S. M. Anlage, *Rev. Sci. Instrum.* 65, 2082 (1994).

437. S. M. Bilodeau, R. Carl, P. C. Van Buskirk, J. F. Roeder, C. Basceri, S. E. Lash, C. B. Parker, S. K. Streiffer, and A. I. Kingon, *J. Korean Phys. Soc.* 32, S1591 (1998).

438. Q. X. Jia, X. D. Wu, S. R. Foltyn, and P. Tiwari, *Appl. Phys. Lett.* 66, 2197 (1995).

439. S. G. Yoon, J. C. Lee, and A. Safari, *J. Appl. Phys.* 76, 2999 (1994).

440. M. Izuha, K. Abe, and N. Fukushima, *Japan. J. Appl. Phys.* 36, 5866 (1997).

441. R. Waser, *Integ. Ferroelectrics* 15, 39 (1997).

442. R. Waser and S. Hoffmann, *J. Korean Phys. Soc.* 32, S1340 (1998).

443. S. Yamamichi, T. Sakuma, K. Takemura, and Y. Miyasaki, *Japan. J. Appl. Phys.* 30, 2193 (1991); K. Abe and S. Komatsu, *ibid.* 32, 4186 (1993).

444. P.-Y. Lesaicherre, H. Yamaguchi, Y. Miyasaka, H. Watanabe, H. Ono, and M. Yoshida, *Integ. Ferroelectrics* 8, 201 (1995).

445. C. Basceri, S. K. Streiffer, A. I. Kingon, and R. Waser, *J. Appl. Phys.* 82, 2497 (1997).

446. O. G. Vendik, S. P. Zubko, and L. T. Ter-Martirosayn, *Appl. Phys. Lett.* 73, 37 (1998).

447. C. Zhou and D. M. Newns, *J. Appl. Phys.* 82, 3081 (1997).

448. F. W. Van Keuls, R. P. Romanofsky, D. Y. Bohman, M. D. Winters, F. A. Miranda, C. H. Mueller, R. F. Treece, T. V. Rivkin, and D. Galt, *Appl. Phys. Lett.* 71, 3075 (1997).

449. D. Galt, J. C. Price, J. A. Beall, and R. H. Ono, *Appl. Phys. Lett.* 63, 3078 (1993).

450. M. J. Dalberth, R. E. Stauber, J. C. Price, C. T. Rogers, and D. Galt, *Appl. Phys. Lett.* 72, 507 (1998).

451. H.-D. Wu, F. S. Barnes, D. Galt, J. C. Price, and J. A. Beall, *Proc. SPIE* 2156, 131 (1994).

452. S. S. Gevorgian, T. Martinsson, P. L. J. Linnér, and E. L. Kollberg, *IEEE Trans. Microwave Theory Technol.* 44, 896 (1996).

453. P. Kr. Petrov, E. F. Carlsson, P. Larsson, M. Friesel, and Z. G. Ivanov, *J. Appl. Phys.* 84, 3134 (1998).

454. W. Chang, J. S. Horwitz, A. C. Carter, J. M. Pond, S. W. Kirchoefer, C. M. Gilmore, and D. B. Chrisey, *Appl. Phys. Lett.* 74, 1033 (1999).

455. C. M. Carlson, T. V. Rivkin, P. A. Parilla, J. D. Perkins, D. S. Ginley, A. B. Kozyrev, V. N. Oshadchy, and A. S. Pavlov, *Appl. Phys. Lett.* 76, 1920 (2000).

456. W. J. Kim, W. Chang, S. B. Qadri, J. M. Pond, S. W. Kirchoefer, D. B. Chrisey, and J. S. Horwitz, *Appl. Phys. Lett.* 76, 1185 (2000).

457. Y. Cho, A. Kirihara, and T. Saeki, *Rev. Sci. Instrum.* 67, 2297 (1996).

458. H. Chang, I. Takeuchi, and X.-D. Xiang, *Appl. Phys. Lett.* 74, 1165 (1999).

459. D. E. Steinhauer, C. P. Vlahacos, F. C. Wellstoood, S. M. Anlage, C. Canedy, R. Ramesh, A. Stanishevsky, and J. Melngalis, *Appl. Phys. Lett.* 75, 3180 (1999).

460. C. Gao and X.-D. Xiang, *Rev. Sci. Instrum.* 69, 3846 (1998).

461. C. Hubert and J. Levy, *Rev. Sci. Instrum.* 70, 3684 (1999).

462. J.-F. Li, P. Moses, and D. Vieland, *Rev. Sci. Instrum.* 66, 215 (1995).

463. A. L. Kholkine, C. Wuetchrich, D. V. Taylor, and N. Setter, *Rev. Sci. Instrum.* 67, 1935 (1996).

464. J.-F. Li, D. Vieland, T. Tani, C. D. E. Lakeman, and D. A. Payne, *J. Appl. Phys.* 75, 442 (1994).

465. J.-F. Li, D. Vieland, and D. A. Payne, *J. Korean Phys. Soc.* 32, S1311 (1998).

466. A. Sternberg, M. Tyunina, M. Kundzinsh, V. Zauls, M. Ozolinsh, K. Kundzinsh, I. Shorubalko, M. Mosec, L. Calzada, L. Pardo, M. Alguero, R. Kullmer, D. Bäuerle, J. Levoska, S. Leppävuori, and T. Martoniemi, *J. Korean Phys. Soc.* 32, S1365 (1998).

467. H. Maiwa, J. A. Christman, S.-H. Kim, D.-J. Kim, J.-P. Maria, B. Chen, S. K. Streiffer, and A. I. Kingon, *Japan. J. Appl. Phys.* 38, 5402 (1999).

468. A. Gruverman, O. Auciello, and H. Tokumoto, *J. Vac. Sci. Technol. B* 14, 602 (1996).

469. G. Zavala, J. H. Fendler, and S. Trolier-McKinstry, *J. Appl. Phys.* 81, 7480 (1997); *ibid.*, *J. Korean Phys. Soc.* 32, S1464 (1998).

470. J. A. Christman, R. R. Woolcott, Jr., A. I. Kingon, and R. J. Nemanich, *Appl. Phys. Lett.* 73, 3851 (1998).

471. K. Lefki and J. M. Dormans, *J. Appl. Phys.* 76, 1764 (1994).

472. F. Xu, F. Chu, J. F. Shepard, Jr., and S. Trolier-McKinstry, *Mater. Res. Soc. Symp. Proc.* 493, 427 (1998).

473. S. Watanabe, T. Fujiu, and T. Fujii, *Appl. Phys. Lett.* 66, 1481 (1995).

474. S. Fujii, I. Kanno, T. Kamada, and R. Takayma, *Japan. J. Appl. Phys.* 35, 6065 (1997).

475. I. Kanno, S. Fujii, T. Kamada, and R. Takayma, *Appl. Phys. Lett.* 70, 1378 (1997); *ibid.*, *J. Korean Phys. Soc.* 32, S1481 (1998).

476. J. G. E. Gardeniers, A. G. B. J. Verholen, N. R. Tas, and M. Elwenspoek, *J. Korean Phys. Soc.* 32, S1573 (1998).

477. A. L. Kholkin, K. G. Brooks, D. V. Taylor, N. Setter, and A. Safari, *Mater. Res. Soc. Symp. Proc.* 541, 623 (1999).

478. X. Du, U. Belegundu, and K. Uchino, *Japan. J. Appl. Phys.* 36, 5580 (1997).

479. R. W. Whatmore, *Rep. Prog. Phys.* 49, 1335 (1986).

480. A. G. Chynoweth, *J. Appl. Phys.* 27, 78 (1956).

481. R. W. Whatmore, P. C. Osbond, and N. M. Shorrocks, *Ferroelectrics* 76, 351 (1897).

482. J.-G. Cheng, X.-J. Meng, J. Tang, S.-L. Guo, and J.-H. Chu, *Appl. Phys. Lett.* 75, 3402 (1998).

483. F. Jin, G. W. Auner, R. Naik, N. W. Schubring, J. V. Mantese, A. B. Catalan, and A. L. Micheli, *Appl. Phys. Lett.* 73, 2838 (1998).

484. M. S. Mohammed, G. W. Auner, R. Naik, J. V. Mantese, N. W. Schubring, A. L. Micheli, and A. B. Catalan, *J. Appl. Phys.* 84, 3322 (1998).

485. N. W. Schubring, J. V. Mantese, A. L. Micheli, A. B. Catalan, and R. J. Lopez, *Phys. Rev. Lett.* 68, 1778 (1992).

486. J. S. Lee, J.-S. Park, J.-S. Kim, J.-H. Lee, Y. H. Lee, and S.-R. Hahn, *Japan. J. Appl. Phys.* 38, L574 (1999).

487. K. K. Dab, K. W. Bennett, and P. S. Brody, *J. Vac. Sci. Technol. A* 13, 1128 (1995).

488. Y.-K. Tseng, K.-S. Liu, J.-D. Jiang, and I.-N. Lin, *Appl. Phys. Lett.* 72, 3285 (1998).

489. A. Seifert, P. Muralt, and N. Setter, *Appl. Phys. Lett.* 72, 2409 (1998).

490. T. Imai, M. Maeda, and I. Suzuki, *J. Korean Phys. Soc.* 32, S1485 (1998).

491. S. J. Lee, S. Y. Kang, S. K. Han, M. S. Jang, B. G. Chae, Y. S. Yang, and S. H. Kim, *Appl. Phys. Lett.* 72, 299 (1998).

492. F. K. Chai, J. R. Brews, R. D. Schrimpf, and D. P. Birnie III, *J. Appl. Phys.* 82, 2505 (1997).

493. R. R. Mehta, B. D. Silverman, and J. T. Jacobs, *J. Appl. Phys.* 44, 3379 (1973).

494. V. M. Fridkin, "Ferroelectric Semiconductors," Chap. 3. Consultants Bureau, New York, 1980.

495. C. J. Brennan, in "Proceedings of the 3rd International Symposium on Integrated Ferroelectrics," p. 354, 1991.

496. S. Maruno, T. Kuroiwa, N. Mikami, K. Sato, S. Ohmura, M. Kaida, T. Yasue, and T. Koshikawa, *Appl. Phys. Lett.* 73, 954 (1998).

497. D. Wu, A. Li, H. Ling, T. Yu, Z. Liu, and N. Ming, *J. Appl. Phys.* 87, 1795 (2000).

498. H. Basnatakumar, H. N. K. Sarma, and A. Mansingh, *J. Appl. Phys.* 85, 341 (1999).

499. H.-M. Chen and J. Y.-M. Lee, *Appl. Phys. Lett.* 73, 309 (1998).

500. S. Y. Wu, *IEEE Trans. Electron. Dev.* ED-21, 499 (1974).

501. T. Nakamura, Y. Nakao, A. Kamisawa, and H. Takasu, *Ferroelectrics* 11, 161 (1995).

502. Y. T. Kim and C. W. Lee, *Japan. J. Appl. Phys.* 35, 6153 (1996).

503. J. Zeng, H. Wang, M. Wang, S. Shang, Z. Wang, and C. Lin, *J. Phys. D: Appl. Phys.* 31, 2416 (1998).

504. T. A. Rost, H. Lin, and T. A. Rabson, *Appl. Phys. Lett.* 59, 3654 (1991).

505. C. H. J. Huang, H. Lin, and T. A. Rabson, *Integ. Ferroelectrics* 4, 191 (1994).

506. K. Sugibuchi, Y. Kurogi, and N. Endo, *J. Appl. Phys.* 47, 2877 (1975).

507. T. S. Kalkur, G. J. Kulkarni, Y. C. Lu, M. Rowe, W. Y. Han, and L. Kammerdiner, *Ferroelectrics* 116, 136 (1991).

508. Y. Higuma, Y. Matsui, M. Okuyama, T. Nakagawa, and Y. Hamakawa, *Jpn. J. Appl. Phys. Suppl.* 17, 209 (1978).

509. Y. Nakao, T. Nakamura, A. Kamisawa, and H. Takasu, *Integ. Ferroelectrics* 6, 23 (1995).

510. E. Tokumitsu, R. Nakamura, and H. Ishiwara, *J. Korean Phys. Soc.* 29, S640 (1996).

511. J. Yu, Z. J. Hong, W. Zhou, G. Cao, J. Xie, X. Li, S. Li, and Z. Li, *Appl. Phys. Lett.* 70, 490 (1997).

512. J. Senzaki, K. Kurihara, N. Nomura, O. Mitsunaga, Y. Iwasaki, and T. Ueno, *Japan. J. Appl. Phys.* 37, 5150 (1998).

513. Y. Lin, B. R. Zhao, H. B. Peng, Z. Hao, B. Xu, and Z. X. Zao, *J. Appl. Phys.* 86, 4467 (1999).

514. Y. T. Kim and D. S. Shin, *Appl. Phys. Lett.* 71, 3507 (1997).

515. M. Noda, H. Sugiyama, and M. Okuyama, *Japan. J. Appl. Phys.* 38, 5432 (1999).

516. S.-M. Yoon, E. Tokumitsu, and H. Ishiwara, *IEEE Electron. Dev. Lett.* 20, 229 (1999); *ibid.* 20, 526 (1999).

517. K. Sakamaki, T. Hirai, T. Uesugi, H. Kishi, and Y. Tarui, *Japan. J. Appl. Phys.* 38, L451 (1999).

518. E. Tokumitsu, G. Fujii, and H. Ishiwara, *Japan. J. Appl. Phys.* 39, 2125 (2000).

519. T. Yoshimura, N. Fujimura, N. Hokayama, S. Tsukui, K. Kawabata, and T. Ito, *J. Korean Phys. Soc.* 32, S1632 (1998).

520. H. N. Lee, Y. T. Kim, and S. H. Cho, *Appl. Phys. Lett.* 76, 1066 (2000).

521. T. Yoshimura, N. Fujimura, D. Ito, and T. Ito, *J. Appl. Phys.* 87, 3444 (2000).

522. E. H. Nicollian and J. R. Brews, "MOS (Metal Oxide Semiconductor) Physics and Technology." Wiley, New York, 1982.

523. E. M. Ajimine, F. E. Fagaduan, M. M. Rahman, C. Y. Yang, and H. Inokawa, *Appl. Phys. Lett.* 59, 2889 (1991).

524. J. Qiao. E. M. Ajimine, P. P. Patel, G. L. Giese, C. Y. Yang, and D. K. Fork, *Appl. Phys. Lett.* 61, 3184 (1992).

525. H. M. Chen, J. M. Lan, J. L. Chen, and J. Y. Lee, *Appl. Phys. Lett.* 69, 1713 (1996).

526. T. Zhang, I. K. Yoo, and L. C. Burton, *IEEE Trans. Compon. Hybrids Manuf. Technol.* 12, 613 (1989).

527. R. Moazzami, C. Hu, and W. H. Shepherd, *IEEE Trans. Electron. Dev.* 39, 2044 (1992).

528. T. Mihara and H. Watanabe, *Japan. J. Appl. Phys.* 34, 5664 (1995).

529. Y. Shimada, A. Inoue, T. Nasu, K. Arita, Y. Nagano, A. Matsuda, Y. Uemoto, E. Fujii, M. Azuma, Y. Oishi, S. I. Hayashi, and T. Otsuki, *Japan. J. Appl. Phys.* 35, 140 (1996).

530. S. M. Sze, "Physics of Semiconductor Devices." Wiley, New York, 1981.

531. I. Stolichnov, A. Tagantsev, N. Setter, J. S. Cross, and M. Tsukada, *Appl. Phys. Lett.* 75, 1790 (1999).

532. J. C. Shin, J. Park, C. S. Hwang, and H. J. Kim, *J. Appl. Phys.* 86, 506 (1999).

533. H.-J. Cho, W. Jo, and T. W. Noh, *Appl. Phys. Lett.* 65, 145 (1994).

534. J. H. Joo, J. M. Seon, Y. C. Jeon, K. Y. Oh, J. S. Roh, and J. J. Kim, *Appl. Phys. Lett.* 70, 3053 (1997).

535. Y. Shimamoto, K. Kushida-Abdelghafar, H. Miki, and Y. Fujisaki, *Appl. Phys. Lett.* 70, 3096 (1997).

536. J. J. Lee, P. Alluri, and S. K. Dey, *Appl. Phys. Lett.* 65, 2027 (1994).

537. J. C. Shin, C. S. Hwang, H. J. Kim, and S. O. Park, *Appl. Phys. Lett.* 86, 3411 (1999).

538. I. Fedorov, J. Petzelt, V. Zelezný, G. A. Komandin, A. A. Volkov, K. Brooks, Y. Huang, and N. Setter, *J. Phys.: Condens. Matter* 7, 4313 (1995).

539. J. F. Scott and B. Pouligny, *J. Appl. Phys.* 64, 1547 (1988).

540. T. K. Song, J. S. Ahn, H. S. Choi, T. W. Noh, and S.-I. Kwun, *J. Korean Phys. Soc.* 30, 623 (1997).

541. Yu. I. Yuzyuk, R. Farhi, V. L. Lorman, L. M. Rabkin, L. A. Sapozhnikov, E. V. Sviridov, and I. N. Zakharchenko, *J. Appl. Phys.* 84, 452 (1998).

542. B. Wang, L. D. Zhang, L. Zhang, Y. Yan, and S. L. Zhang, *Thin Solid Films* 354, 262 (1999).

543. D. K. Fork, F. Armani-Leplingard, and J. J. Kingston, *Mater. Res. Soc. Symp. Proc.* 361, 155 (1995).

544. S. Schwyn, H. W. Lehmann, and R. Widmer, *J. Appl. Phys.* 72, 1154 (1992).

545. B. H. Hoerman, B. M. Nichols, M. J. Nystrom, and B. W. Wessels, *Appl. Phys. Lett.* 75, 2707 (1999).

546. R. A. McKee, F. J. Walker, and M. F. Chisholm, *Phys. Rev. Lett.* 81, 3014 (1998).

547. W. Tian, J. H. Haeni, X. Q. Pan, and D. G. Schlom (unpublished).

Chapter 6

FERROELECTRIC THIN FILMS OF MODIFIED LEAD TITANATE

J. Mendiola, M. L. Calzada

Instituto Ciencia de Materiales de Madrid (CSIC), Cantoblanco, 28049 Madrid, Spain

Contents

1. INTRODUCTION

Ferroelectric materials can be phenomenologically defined as dielectric materials with spontaneous polarization, P_s, in a range of temperature. This is produced by a lack of symmetry of the crystal structure with respect to the higher symmetry of the paraelectric phase. Spontaneous polarization decreases with the increase of temperature up to where the phase transition occurs. This polarization changes with temperature and pressure (pyroelectric and piezoelectric effect, respectively). It can also be reversed by an external electric field, causing a change in the charge distribution of the unit cell and turning the crystal into electric twins or ferroelectric domains separated by walls. This reversal character of the polarization permits tracing of P–E hysteresis loops in these materials. Reversal of polarization occurs at a very high rate (switching time).

Microelectronic devices including ferroelectric thin films integrated with silicon substrates have created much interest during the last years due to the possibility of using their ferro, piezo, and pyroelectric properties at a microscale level [1]. Among ferroelectric materials, the family of perovskites has been the most studied since the discovery of ferroelectricity.

The most popular and well-known $BaTiO_3$ and $PbTiO_3$ compounds belong to this family. As bulk ceramic, the commercial material, most widely used in numerous applications, such as piezoelectric devices, is PZT. It is a solid solution of lead, zirconium, and titanium oxides; in fact, it is a modification of lead titanate, extensively studied in the past [2]. PZT compositions have been considered the best candidates for nonvolatile memories, piezoelectric devices, or infrared sensors [3]. Recently, other compounds, such as layered perovskite oxides, have also been considered as competitive materials for the fabrication of ferroelectric nonvolatile memories due to their low fatigue and good retention [4].

$PbTiO_3$ is a perovskite with a large spontaneous polarization that makes it interesting for technological applications. However, $PbTiO_3$ has undesirable mechanical properties due to its large tetragonal strain ($c/a = 1.064$). It also has a high coercive field that makes poling difficult. In this perovskite, substitution of Pb by isovalent cations, such as Ca, or cations with other valences, such as Sm, Nd, or La, produces a decrease in the coercive field and in the tetragonality of the material and reduces the risk of cracking, maintaining, however, considerable values of remanent polarization [5]. These

Handbook of Thin Film Materials, edited by H.S. Nalwa
Volume 3: Ferroelectric and Dielectric Thin Films

ISBN 0-12-512911-4/$35.00

elements have been used as dopants in lead titanate bulk ceramics. In these ceramics, substitution of lead by 20–30% of calcium or by ~8% of samarium induces a high piezoelectric anisotropy that is useful in ultrasonic transducers [6, 7]. Low percentages of lanthanum, 5–10%, also lead to materials with electro-optic and pyroelectric responses, whereas higher amounts of La, ~25%, produce ceramics with a relaxor behavior. This facilitates the use of these materials in applied electronic devices. One expects a similar behavior of these materials deposited as thin films. However, properties of films largely depend on substrate, microstructure, texture, interfaces, and thickness.

$(Pb,Ca)TiO_3$, $(Pb,La)TiO_3$, and $(Pb,Sm)TiO_3$ thin films on platinum/silicon, platinum/MgO, and platinum/SrTiO_3 substrates have been reported to be good candidates for piezoelectric devices and infrared sensors [8–11]. Besides, $(Pb,Ca)TiO_3$ thin films on platinum/silicon substrates have appropriate values of retention and fatigue for their use in non-volatile memories (NVFERAM) [12].

Early in the 1980s and during the last decade, several groups, including ours, were devoted to obtaining thin films of pure lead titanate by chemical and physical deposition methods. Chemical and structure analyses and ferroelectric properties were reported [13–21]. One micrometer crack-free films obtained by sol-gel with a single coating were reported by Phillips et al. [17]. For pyroelectric IR detectors, La-modified lead titanate films are the most studied. We must cite the early work of Okuyama et al. [22] describing a two-dimensional array sensor, the work of Choi et al. [23, 24] on films epitaxially grown onto MgO substrates, and the work of Wang et al. [25] giving the performances of point detectors onto platinized silicon substrates. For more information, see the review published by our group [26]. Attempts have also been made to get advantages of the modification of lead titanate by La and Ca together, with promising results [27, 28]. In the present work we mainly deal with Ca-modified $PbTiO_3$ (PTCa) thin films deposited by sol-gel [a chemical solution deposition (CSD) method].

The excellent ferroelectric performances of PTCa bulk ceramics motivated the research of our group on PTCa thin films onto electroded silicon wafers, as an alternative material to PZT in applications such as microelectromechanical devices (sensors and actuators), nonvolatile ferroelectric random-access memories (NVFERAM), and IR detectors. These devices always consist of an array of different densities of capacitors formed by the ferroelectric film and two electrodes, the common bottom electrode joined to the silicon substrate and the top electrode. Both are usually of platinum. Now, one of the main issues is to make the use of Pt electrodes on silicon substrates compatible with a low degradation of the ferroelectric properties of the film [29]. For PZT films, oxygen vacancies and charge injection at the ferroelectric–electrode interfaces, which are reported as the main causes of degradation, can be controlled by using conductive oxide electrodes (see references quoted by Auciello et al. [29]), as well as by other alternatives that will be described later. Reliability of films with these compositions is based on the retention of polarization and on their switchability, which mainly concerns degradation phenomena (fatigue and aging).

This review will mainly deal with PTCa films fabricated by sol-gel. We think that this is the most promising modification of lead titanate. Ca is introduced stoichiometrically in the Pb lattice sites, without creating an appreciable amount of vacancies that spoil the ferroelectric properties. On the other hand, the sol-gel is an inexpensive method for processing films that permits the control of composition and reduces the temperature of crystalization.

Two deposition techniques have mainly been used to fabricate $Pb_{1-x}Ca_xTiO_3$ (PTCa) thin films. RF-magnetron sputtering has been used by Yamaka et al. [30]. They make a complete work that included the processing of a wide range of compositions with different Pb amounts substituted by Ca ($x = 0$–40) and chemical, microstructural, and structural analyses. They used different substrates, such as Pt/MgO and Pt/StTiO_3, and reported on the improvement of the performances of a pyroelectric detector. The films obtained had a fixed polarization pointing to the film surface. This was a consequence of the relative ion displacements due to the thermal treatments during the film formation. Although no data of switching behavior, neither fatigue or retention, were reported, this work can be considered a pioneer on the matter. Several years later, Maiwa et al. [31] reported the ferroelectric properties of PTCa films on $Pt/Ti/SiO_2/Si$ substrates prepared by multicathode rf-magnetron sputtering. They studied the hysteresis loop dependency on the substrate temperature. Our group has also used pulsed laser ablation deposition (PLD) for preparing these films [32]. Preliminary parameters of the technique were established to obtain the ferroelectric activity of the films. But, the most used deposition technique of PTCa films has been the chemical solution method (CSD), mainly the sol-gel. With this technique, Tsuzuki et al. [33] reported on films deposited on platinized fused silica, and Torii et al. [34] obtained c-axis oriented films on single-crystal MgO substrates with a pyroelectric coefficient of $3.1 \times 10^{-8}\,C\,cm^{-2}\,K^{-1}$. At the same time, our group [35] reported on a new sol-gel method based on the use of diols as solvents that permitted us to obtain crack-free ferroelectric PTCa films up to 700 nm. Further contributions of the authors will be described in detail in this work. In the last decade, some other authors have reported on PTCa films prepared with the same technique [36–38]. Seifert et al. [39, 40] developed by CSD a way for a controlled incorporation of porosity into the film with the result of a matrix–void composite with a 3-0 connectivity. They claimed to control nucleation and growth during the rapid thermal processing (RTP). As a consequence of the large porosity of the film, they measured a low permittivity and a high pyroelectric coefficient, after an adequate electric poling of the film. This led to an improvement of the material figure of merit of interest for IR pyroelectric detectors.

Table I.   Most Common Deposition Techniques of Lead Titanate-Based Perovskite Films

Deposition technique	Lead titanate-based perovskite films prepared by the technique	Devices	Deposition temperature (°C)	Stoichiometry	Epitaxy	Cost	Problems of the technique
Sputtering	$Pb(Zr,Ti)O_3$, $(Pb,Ca)TiO_3$, $(Pb,La)TiO_3$	DRAMs, NVFERAMs, MEMS, infrared sensors, SAW	500–700	5	8	High	Negative ions, target surface, uniformity
Evaporation	$Pb(Zr,Ti)O_3$	MEMS, infrared sensors, SAW	500–700	4	7	High	Deposition velocity
Laser ablation	$Pb(Zr,Ti)O_3$, $(Pb,Ca)TiO_3$, $(Pb,La)TiO_3$, $(Pb,Sm)TiO_3$	DRAMs, NVFERAMs, MEMS, infrared sensors, SAW	500–700	6	9	High	Uniformity
MOCVD	$Pb(Zr,Ti)O_3$	DRAMs, NVFERAMs, MEMS, infrared sensors, SAW	600–700	7	5	High	High substrate temperature
CSD	$Pb(Zr,Ti)O_3$, $(Pb,Ca)TiO_3$, $(Pb,La)TiO_3$, $(Pb,Sm)TiO_3$	MEMS, infrared sensors, SAW	450–750	9	4	Low	Nonconformal deposits, multiple coatings

2. CHEMICAL SOLUTION DEPOSITION OF MODIFIED LEAD TITANATE THIN FILMS

2.1. Processing Considerations

Different deposition techniques are described in the literature for the preparation of ferroelectric thin films [1]. Table I shows the most common deposition techniques used for the preparation of lead titanate perovskite films.

Deposition techniques have been classically divided into physical deposition methods (sputtering, evaporation, and laser ablation in Table I) and chemical deposition methods [MOCVD (metal organic chemical vapor deposition) and CSD in Table I]. Physical methods and MOCVD use expensive vacuum equipments but, however, can provide epitaxy films highly interesting for some applications. CSD methods include sol-gel and metal organic decomposition (MOD). These are much cheaper than the others and permit a high compositional control of the film. But, it is quite difficult to obtain epitaxy films and conformal deposits by CSD methods. Advantages and disadvantages of physical and chemical deposition techniques are collected in Table I.

The modified lead titanate compositions deposited as thin films that have been well studied as the partial substitution of lead by calcium, lanthanum, or samarium [5–12]. Applications of films with these compositions have been described in the Introduction. Lanthanum-modified lead titanate thin films have been prepared by means of magnetron sputtering [9], laser ab-

lation [10], or spin-on solution deposition [11]. Percentages of lanthanum between 5 and 28% have been incorporated into the perovskite films, researching some of the mentioned applications. The preparation of samarium lead titanate thin films was first reported by some of the authors of this work [8] and later by others [41, 42]. In both cases, spin-on solution deposition was used. In the case of calcium, $PbTiO_3$ perovskite films containing 20 to 30% Ca have been prepared by sputtering, laser ablation, and sol-gel [30, 43, 44].

Among the different techniques for deposition of thin films, chemical solution deposition (CSD) facilitates stoichiometric control of complex mixed oxides and is relatively rapid, inexpensive, and compatible with semiconductor technologies. The fabrication of films by CSD involves three steps: (1) synthesis of the precursor solution, (2) deposition of this solution on a substrate by spin or dip-coating, and (3) thermal treatment of the as-deposited amorphous layer. For solution deposition methods, differences stem from the chemical route used for the synthesis of the solution.

The most extensively used method in the preparation of ferroelectric thin films is the 2-methoxyethanol sol-gel route [45]. This method uses transition metal alkoxides, lead acetate, and methoxyethanol as solvent. It offers controllable and reproducible chemistry. However, the toxicity of methoxyethanol and the eagerness toward water of the reagents involved in the process are handicaps of this route.

The replacement of 2-methoxyethanol by other solvents such as diols has been reported [46]. This route takes advantage of the reactivity of glycols with alkoxides, forming derivatives that are more resistant toward hydrolysis than the counterpart metal alkoxide [47]. Also, it has been reported [48] that the use of a solution process that combines diol solvents, titanium alkoxides modified with β-diketonates, and lead acetate, provides Pb-Ti gels capable of redissolving in water [49]. This process makes possible the incorporation into the system of modifiers or dopants as some of their water-soluble compounds, simplifying the preparation of multicomponent solutions.

Applications of film devices prepared with the techniques of Table I strongly depend on the film composition, but also on the microstructural and heterostructural properties of the final device. Microstructure and heterostructure of the device are related with the nucleation and growth on a substrate of the crystalline ferroelectric phase from the amorphous layer [50]. For CSD methods, nucleation and growth are related with the steps (2) and (3) above. Nucleation of the desired phase requires a thermal treatment in an appropriate temperature range and with an appropriate heating rate, thus avoiding the crystallization of nonferroelectric phases such as pyrochlores. In films, nucleation is also affected by the substrate. The film/substrate interface is a site where heterogeneous nucleation is facilitated. Competition between homogeneous and heterogeneous nucleation occurs during the thermal treatment of the film and can be controlled by means of the heating rate and the type of substrate. This competition is especially important when films with a particular orientation are required. Modified lead titanate films with preferred orientation in the c-direction of the perovskite are desired for piezo- and pyroelectric applications, since this orientation provides a spontaneous polarization of the film without the necessity of applying an electric field.

In the following sections, we will focus on the preparation by a diol-based CSD method of modified lead titanate thin films. The physico-chemical properties of the synthesized solutions will be presented. The precursor solution will be deposited by spin-coating onto substrates. The effect of the substrate and of the thermal treatment on the nucleation of crystalline phases, texture, and microstructure of the films, will be shown. These characteristics of the films will be related with their electrical properties. Structural, microstructural, and electrical properties will be presented, mainly for calcium-modified lead titanate thin films with a composition of $Pb_{0.76}Ca_{0.24}TiO_3$ (PTCa). Additional information about samarium- or lanthanum-modified lead titanate thin films can be found in [51] and [26].

2.2. Diol-Based Sol-Gel Method for the Fabrication of Modified Lead Titanate Thin Films

2.2.1. Synthesis of Precursor Solutions

Figure 1 is the general scheme of synthesis of modified lead titanate precursor solutions. Lead acetate trihydrate, $Pb(OOCCH_3)_2 \cdot 3H_2O$, was refluxed in air for 1 hour in 1,3-propanediol, $OH(CH_2)_3OH$. Then, titanium-diisopropoxide-bisacetylacetonate, $(OC_3H_7)_2(CH_3COCHCOCH_3)_2$, was added to the former mixture. Partial distillation of byproducts was carried out after promoting complexation with 8 hours of reflux in

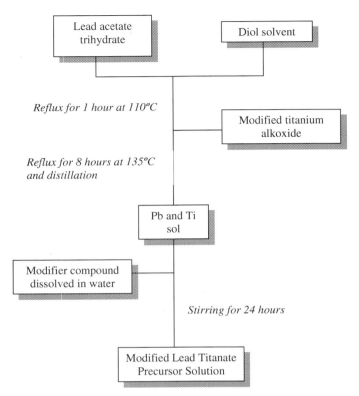

Fig. 1. Scheme of the sol-gel process used for the synthesis of modified lead titanate precursor solutions.

air. Other diol solvents with a carbon number from 2 to 5 have also been tested. Results obtained with these diols can be seen in [48]. Solvent was used in a ratio of five moles of diol per mole of titanium. Molar ratios of lead to titanium were a function of the desired nominal composition. Solutions without excess of Pb and with a 10 mol% of Pb excess were prepared [52].

Modifier cations were added to the Pb-Ti sol as a water-dissolved salt: acetates or nitrates. The nominal composition of the films mainly considered in this review is going to be $Pb_{0.76}Ca_{0.24}TiO_3$. So, calcium was added to the sol with a ratio of 0.24 of calcium to titanium. Films with appropriate properties were obtained when the calcium was incorporated as a water solution of calcium acetate, $Ca(OOCCH_3)_2 \cdot xH_2O$ [8]. Air stable solutions with a density of ~ 1.2 g cm^{-3} and concentration of ~ 0.9 mol L^{-1} were obtained. Dilution of this solution was carried out with water.

2.2.1.1. Synthesis Mechanisms of the Solutions

Major problems of other CSD methods used for the preparation of lead titanate films are related to the large reactivity toward moisture of the used metal alkoxides, the low affinity among the alkoxides and the solvents, and the toxicity of the raw materials. The chemical process described above solves many of these problems by using a titanium alkoxide modified with a β-diketonate and diols as solvents.

The chelating ligand acetylacetonate has been used in sol-gel as a stabilizing agent for metal alkoxide compounds [53].

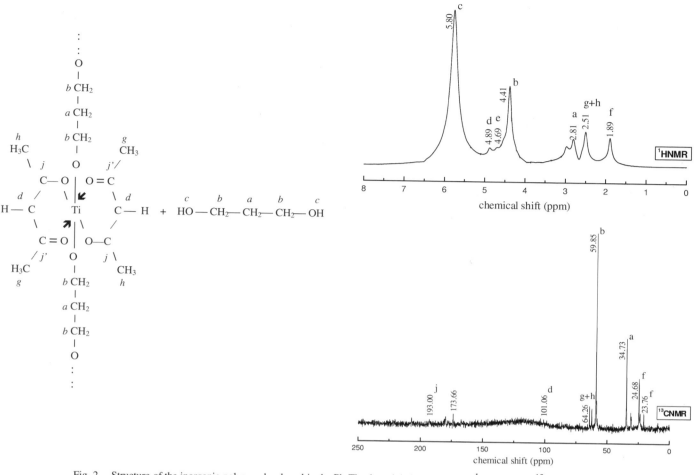

Fig. 2. Structure of the inorganic polymer developed in the Pb-Ti sols and deduced from the ^{1}H NMR and ^{13}C NMR spectra.

Besides, in the 1960s some authors reported the reactions of dihydroxy alcohols and titanium alkoxides [54]. The titanium compounds obtained turned out to be more stable than the corresponding titanium alkoxide [47].

^{1}H and ^{13}C NMR experiments were made on the synthesized solution and on the raw materials to have an approximation to the structure of the inorganic polymer developed in solution. A summary of the results obtained from these measurements is collected in Figure 2. This figure shows the ^{1}H and ^{13}C NMR spectra of the stock solution. The polymer structure depicted in the figure can be deduced from the spectra. In this structure, titanium remains bonded to the acetylacetonate groups in the solution. Both ^{1}H and ^{13}C NMR spectra show the chemical shifts corresponding to the acetylacetonate ligands of the raw $Ti(OC_3H_7)_2(CH_3COCHCOCH_3)_2$ [$\delta_{Hd} \sim 4.89$ ppm, $\delta_{Hg,h} \sim 2.51$ ppm, $\delta_{Cd} \sim 101.06$ ppm, $\delta_{Cg,h} \sim 64.26$ ppm, and $\delta_{Cj} \sim 193.00$]. However, reflux makes possible the rupture of the titanium–isopropyl bonds and the formation of new bonds between titanium and diol groups coming from the solvent [$\delta_{Ha} \sim 2.81$ ppm, $\delta_{Hb} \sim 4.41$ ppm, $\delta_{Ca} \sim 34.73$ ppm, and $\delta_{Cb} \sim 59.85$ ppm]. Formation of these new

bonds is supported by the IR analysis results of Table II. This analysis indicates that the byproducts of the reaction contain isopropanol, water, and acetone. A complete set of the analyses carried out on the raw materials, the synthesized solution, and the byproducts can be found in [55]. These analyses point to the fact that in the solution, the isopropyl ligands of the raw $Ti(OC_3H_7)_2(CH_3COCHCOCH_3)_2$ have been exchanged by solvent molecules and the association of these species has given rise to a polymerized solution, in accordance with the reaction sequence of Figure 3. Because of the steric hindrance, the inorganic polymer developed during reflux of the solutions shows low sensitivity toward moisture, and partially polymerized solutions can be prepared and stored without controlled atmosphere conditions. The Pb-Ti gels formed after reflux and distillation of byproducts can be redissolved in water excess by water adsorption and dissociative chemisorption, in a reaction known as rehydration [49, 56] (see Fig. 3).

The rates of hydrolysis and condensation reactions change with the number of carbons of the diol solvent. The diols tested in this work have been ethylenglycol, 1,3-propanediol, 1,4-butanediol, and 1,5-pentanediol. More than five carbons

Table II.   Summary of the IR Analysis Made on the Synthesized Solutions and on the Byproducts

Byproducts	H_2O		$\nu(O-H) \sim 3424$ cm^{-1}
			$\delta(HOH) \sim 1622$ cm^{-1}
	$(CH_3)_2CHCOOCH_3$		$\nu(C=O) \sim 1714$ cm^{-1}
	C_3H_7OH		$\nu(C-H) \sim 2970-2882$ cm^{-1}
			$\nu(O-H) \sim 3424$ cm^{-1}
			$\nu(C-O) \sim 947$ cm^{-1}
Synthesized solutions	[[HO-R-O-Ti-O-R-OH]··· Pb ···]$_q$ with acac	Acetylacetonate ligands	$\nu(C=O) \sim 1578$ cm^{-1}
			$\nu(C-C) \sim 1418$ cm^{-1}
		Diol groups	$\nu(C-O) \sim 1000-900$ cm^{-1}
			$\nu(Ti-O) \sim 660, 620$ cm^{-1}
	$HO(CH_2)_3CH_2$ solvent		$\nu(C-H) \sim 2958-2876$ cm^{-1}
			$\nu(O-H) \sim 3348$ cm^{-1}
			$\nu(C-O) \sim 986, 940$ cm^{-1}

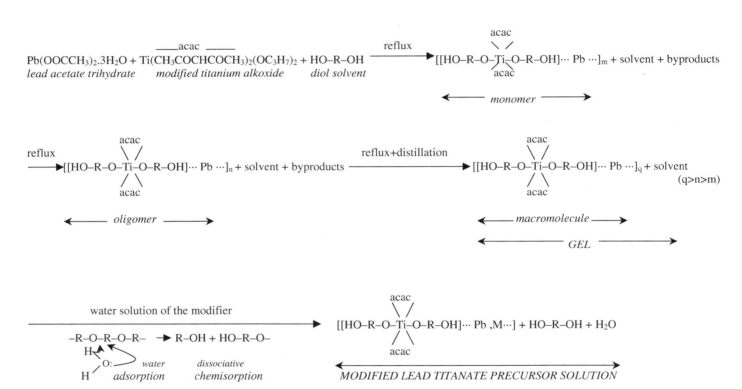

Fig. 3.   Scheme of the mechanisms of synthesis involved in the preparation of the Pb-Ti sols.

leads to solid dihydroxyalcohols. Gels obtained after condensation of the synthesized solutions can be redissolved using molar ratios of H_2O/diol > 1. Table III shows the gelation behavior of gels prepared with the diols above indicated. Note the differences among the solutions prepared with ethylenglycol and with the other diols. There is not enough experimental data to establish the origin of these differences, but they have to be related with the structure of the inorganic polymer developed in solution and with its steric hindrance (see Fig. 2).

Thus, addition of large amounts of water does not lead to gelation and rehydration of the ethylenglycol-derived solutions. So, these solutions were not used as precursors of modified lead titanate compositions, since the incorporation of the modifier in water solution is not feasible. The other glycols are equally useful for the preparation of modified lead titanate precursor solutions. But, the 1,3-propanediol was chosen as solvent, due to its lower number of carbons. Low carbon content leads to simple decomposition processes of the deposited film during

Table III. Gelation and Rehydration of the Solutions Obtained from Different Diols

	H$_2$O/diol molar ratio				
	1	2	3	4	5
Ethyleneglycol	Sol	Viscous sol	Viscous sol	Viscous sol	Viscous sol
1,3-Propanediol	Gel	Cloudy solution	Clear solution	Clear solution	Clear solution
1,4-Butanediol	Gel	Cloudy solution	Clear solution	Clear solution	Clear solution
1,5-Pentanediol	Gel	Cloudy solution	Clear solution	Clear solution	Clear solution

thermal treatment of crystallization.

The capability of rehydration of these Pb-Ti gels makes possible the addition of salts of modifier cations dissolved in water, obtaining, by an easy way, stable solutions of modified lead titanate compositions.

2.2.1.2. Thermal Decomposition of the Solutions

As described in Section 2.1, thermal treatment of films derived from solutions is a compulsory step for the preparation of crystalline films from CSD methods. Pyrolysis of organic compounds, sintering of the amorphous film, rearrangement of the cations, and crystallization of the film occur during this stage [50]. Decomposition of the compounds of the solution, taking place at different temperatures and consuming different amounts of energy, can lead to heterogeneities in the system that, later on, appear in the ceramic films as structural and microstructural defects and that ruin their electrical properties.

Aliquots of the precursor solutions were dried in air at 100°C for 24 hours. Dried samples were subjected to simultaneous thermogravimetric and differential thermal analysis (TGA/DTA). Evolved gas analysis (EGA) coupled to TGA/DTA and with mass spectrometry, was used to determine the gases eliminated during the thermal decomposition of the samples. Data were recorded in air between room temperature and 800°C, with a heating rate of 10°C/min. Other heating rates were also tested for recording data. A delay of all the energetic processes is produced when heating rates higher than 10°C/min were used for these analyse.

As indicated in Section 2.2.1, the salts used for the addition of the modifiers to the Pb-Ti sol have been acetates and nitrates. It will be shown that the nitrate and acetate groups have a strong effect on the thermal decomposition of these systems; complementary information about this can be found in [57]. In this reference, the thermal decomposition of Sm- and La-modified PbTiO$_3$ compositions can also be found.

Figure 4 shows the TGA/DTA and the EGA curves of the Pb$_{0.76}$Ca$_{0.24}$TiO$_3$ precursor solutions. Curves for the incorporation of Ca as acetate or as nitrate are presented.

The first weight loss measured in the two systems at temperatures lower than 200°C is due to volatilization of physically bonded water. Weight losses produced between 200 and 500°C are accompanied by a considerable exothermic process. Mass spectroscopy analysis of the gases evolved in this step shows the leaving of water, H$_2$O, carbon dioxide, CO$_2$, and acetone, CH$_3$COCH$_3$, for the calcium incorporated as acetate (Fig. 4a). For the nitrate system, nitrogen monoxide, NO, is also detected in the EGA curve (Fig. 4b). Combustion of organics occurs in this temperature interval. This combustion produces H$_2$O, CO$_2$, and CH$_3$COCH$_3$. The latter and some of the CO$_2$ are side products of the acetate groups [57–59]. For the nitrate system, literature indicates that, in hydrated calcium nitrates, the NO$_3^-$ group starts to decompose at temperatures lower than that of the vaporization of the crystal water [60]. This is observed in the EGA curve (Fig. 4b). The energetic processes of the nitrate system occurring during pyrolysis are more complex than those of the acetate system. The major exothermic peak is observed in the former system at temperatures over 500°C, whereas in the latter the elimination of the majority of the organics has finished at this temperature. However, residues of CaCO$_3$ remain in both systems up to high temperatures. These carbonate residues are responsible for the small weight loss observed in the TGA curves at temperatures above 750°C. The existence of carbonates at these temperatures is confirmed by the elimination of CO$_2$ in the EGA measurements and by the two main bands at ~1446 and ~874 cm^{-1} detected in the IR spectra of powders treated over 750°C (see [57]). These are characteristic bands of carbonate groups [61].

CaCO$_3$ is always formed during the decomposition of the solutions containing calcium, even if this cation is incorporated through a salt different from the acetate [57]. It is due to its highly negative enthalpy of formation and to the large thermal stability of the alkaline-earth carbonates [62]. These carbonates remain in the system up to temperatures higher than 700°C. This probably indicates that all the crystalline PTCa films would contain carbonate residues, since their crystallization and integration with silicon substrates has to be carried out at temperatures <700°C. We will show later on if the presence of carbonate residues in the PTCa films has effect on their electrical properties.

Crystalline films prepared from the solutions where calcium was incorporated as a nitrate have large craters that render them unfit for applications (Fig. 5). These craters are due to the exothermic reactions produced during the vaporization of crystal water. Hydrated calcium nitrates behave differently from other hydrated inorganic salts when deposited on a solid surface [60]. Liberation of crystal water in this salt is produced through an explosive desorption accompanied by a surface ex-

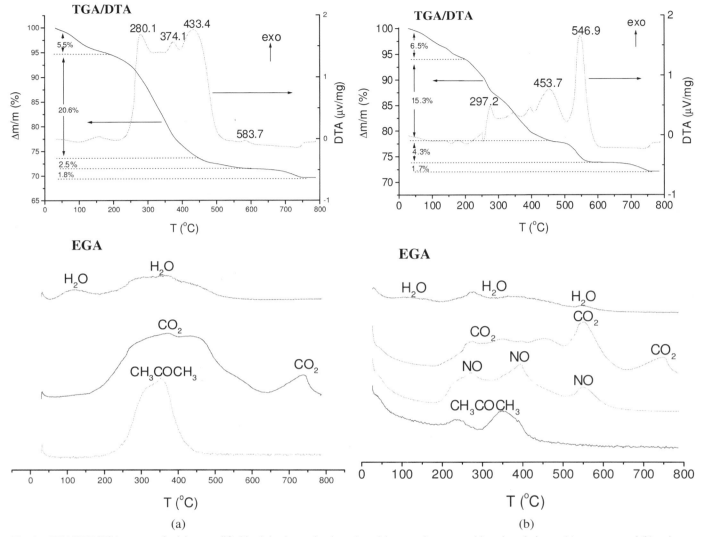

Fig. 4. TGA/DTA/EGA spectra of calcium-modified lead titanium gels where the calcium was incorporated into the solution as (a) an acetate and (b) a nitrate.

plosion process. This produces the damage of the ceramic film and ruins its electrical properties. So, the films studied in this work were prepared from solutions derived from the acetate system.

Lead titanate perovskite-based materials have problems of compositional control due to the loss of Pb by volatilization during thermal treatments of crystallization. This phenomenon is produced in all materials of these compositions (bulks, films, etc.), but it is critical in thin films where the surface/thickness ratio is very large. Pb excess in solutions is usually used in the CSD methods to balance Pb losses during annealing. In our work, we tested different Pb excesses in the precursor solutions. Composition of these solutions was checked by induced coupled plasma analysis (ICP). Final composition of the crystalline films after annealing was determined by energy dispersive spectroscopy (EDS). This study was carried out on films with a thickness >1000 nm. These thick layers were used to minimize the interference of the substrate in the EDS measurements. The analysis was made on areas with diameters of ~0.06 μm

and ~100 μm, to obtain local and average composition, respectively. The lines used for the analysis were the Pb M-lines, Ti K-lines, and Ca K-lines. The La L-lines and Sm L-lines were also used for the analysis of the films modified with La or Sm.

Table IV shows the composition of the crystalline films with expected composition of $Pb_{0.76}Ca_{0.24}TiO_3$. It can be seen that films prepared without Pb excesses are lead deficient. The desired composition is only obtained when a 10 mol% excess of Pb is incorporated into the precursor solution. This Pb excess seems to be the appropriate amount to compensate the Pb losses by volatilization in these films.

2.2.2. Films Derived from the Solutions

Solutions were deposited by spin-coating onto substrates using a velocity of 2000 rpm/min for 45 s. Wet films were dried on a hot plate at 350°C for 60 s. Then, crystallization was carried out by thermal treatment at 650°C. Films with thickness up to ~1 μm could be obtained by a single deposition, drying, and

Table IV.   Atomic Ratios Measured in the Films by EDS

| | mol% excess of Pb in solution | | | |
| | 0% | | 10% | |
	Local composition (over 0.06 μm diameter spot)	Average composition (over 100 μm diameter area)	Local composition (over 0.06 μm diameter spot)	Average composition (over 100 μm diameter area)
Pb/Ti[a]	0.68 ± 0.02	0.69 ± 0.02	0.80 ± 0.02	0.77 ± 0.02
Ca/Ti[a]	0.23 ± 0.02	0.24 ± 0.02	0.26 ± 0.02	0.22 ± 0.02

[a] Desired composition of the perovskite film: $Pb_{0.76}Ca_{0.24}TiO_3$.

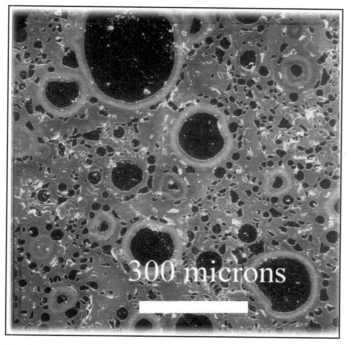

Fig. 5.   SEM image of the surface of a PTCa film derived from a solution where the calcium was incorporated as a nitrate.

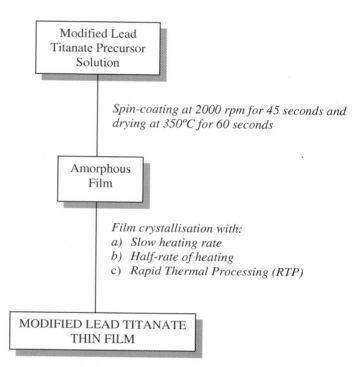

Fig. 6.   Scheme of the preparation of the films.

crystallization, using the CSD method described in this work. Figure 6 shows the scheme of preparation of the films.

Films with a preferred orientation on different substrates were deposited from solutions with a 10 mol% excess of Pb and crystallized by RTP (see scheme of Fig. 6; treatment c). Four depositions on the substrate of a solution with a concentration of ~0.3 mol/L, followed by the drying and crystallization of each layer, were made. These films are going to be described in Section 2.2.2.1.

A single layer of solutions with concentration of ~0.9 mol/L and without Pb excess or with a 10 mol% of Pb excess, were deposited on silicon-based substrates for preparing the films described in Section 2.2.2.2.

2.2.2.1. Selection of Substrates

The type of substrate mainly used has been Pt(~1000 Å)/ TiO$_2$(~500 Å)/SiO$_2$/(100)Si, although other substrates have also been tested: Pt(~1000 Å)/MgO [63], Pt(~1000 Å)/ SrTiO$_3$ [7], Ti(~50 Å)/Pt(~1000 Å)/Ti/SiO$_2$/(100)Si, and Pt(~1000 Å)/Ti(~500 Å)/SiO$_2$/(100)Si [64]. The Pt, Ti, and TiO$_2$ layers were deposited onto the wafers by rf-magnetron sputtering. SiO$_2$ is produced on the top surface of the silicon wafer by its spontaneous oxidation. The Pt layer of all the substrates works as the bottom electrode. It permits the further polarization of the material with an electric field.

The type of substrate used for the preparation of the films was related with the application of the final device, because the substrate has a major effect on the ferroelectric properties of the film. Silicon-based substrates are preferable for microelectronic applications, and other substrates have only been chosen when a desired preferred orientation has to be induced in the ceramic film.

Orientation of films on a substrate is strongly affected by stresses. Most of the stresses in CSD films result from the wetting of the substrate by the solution and the pyrolysis of organic groups during the amorphous to crystalline transformation. A film cannot contract in the plane of the substrate, so all the shrinkage caused during these processes has to be accommodated by decreasing the film thickness [50]. This leads to tensile stresses in the film parallel to the interface with the substrate. These stresses decrease with the film thickness, and there is a critical thickness for the crack of the film. After thermal treatment, stresses are generated on cooling down, because of the thermal expansion mismatch between the film and the substrate. In addition, strains resulting from the paraelectric-to-ferroelectric phase transformation upon cooling have also to be considered in ferroelectric thin films. It has been reported that lead zirconate titanate (PZT) films deposited on substrates with low thermal coefficients of expansion, such as silicon, are preferentially a-domain oriented, whereas those on substrates with high thermal coefficients of expansion are c-domain oriented [65]. In the former case, a tensile transformation stress is produced, whereas in the latter stresses are compressive. c-Oriented films are preferable for piezo- and pyroelectric applications since this orientation provides a considerable spontaneous and, in some cases, nonswitchable polarization very adequate for these devices. The preferential a-domain orientation of lead titanate perovskite films on silicon, whenever easily switchable, can be adequate for memory devices. The percentage of orientation in the a-direction of the films on silicon can be decreased taking into account factors different from stresses. Template layers can be used on silicon substrates to promote a preferred orientation of the perovskite film in its polar direction [66]. In this case, the lattice mismatch between film and substrate has to prevail over the tensile stresses developed during deposition and thermal treatment of the film.

Heterostructure of Silicon Substrates.   Heterostructure of silicon-based substrates is designed to minimize diffusion and to improve properties of the ferroelectric film. In this way, buffer and adhesive layers containing titanium have been used in our laboratory. We have mainly used a TiO_2 layer. It improves adhesion of the Pt bottom electrode to the Si substrate and highly decreases diffusion.

The effect of this layer as a buffer can be observed in the TEM micrographs of Figure 7. Figure 7a, b, and c shows the PTCa film deposited on (100)Si, Pt/(100)Si, and Pt/TiO$_2$/SiO$_2$/(100)Si, respectively. In all cases, formation of a thin layer of SiO$_2$ is observed. Figure 7c shows the cross-section image where a minimum diffusion and reaction between substrate and film is detected. The others develop rough interfaces produced by the interdiffusion of elements of the film and the substrate. This interdiffusion has been detected in our films by means of EDS and XPS [67]. The latter analysis shows the formation of silicates when the films are directly deposited on (100)Si. Formation of lead silicate has been reported for lead titanate films deposited on silicon substrates [68]. Besides, calcium silicates have been also observed in the PTCa films of this work. The reaction of lead titanate-based perovskites with Pt-coated silicon, as observed in the PTCa films on Pt/(100)Si of Figure 7b, has also been previously reported [69].

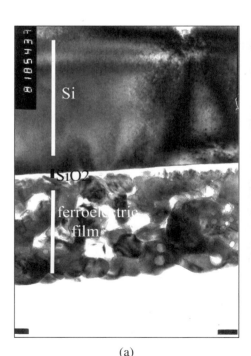

(a)

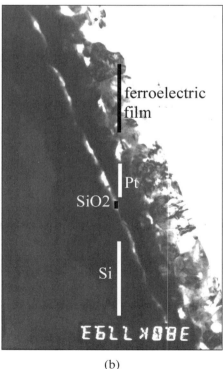

(b)

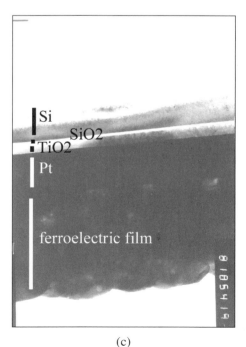

(c)

Fig. 7.   TEM image of cross sections of the PTCa films onto (a) (100)Si, (b) Pt/(100)Si, and (c) Pt/TiO$_2$/(100)Si.

Efficacy of TiO_2 as buffer layer has been also shown by other authors [70]. It avoids migration of silicon to the ferroelectric layer and migration of elements of the film to the silicon substrate. However, interaction between film and Pt bottom electrode is difficult to prevent [69]. It can be minimized by using RTP treatments. This will be discussed later.

PTCa films deposited on the multilayer structure of $Pt/TiO_2/SiO_2/(100)Si$ have adequate ferroelectric properties and are especially interesting for switchable applications [12].

Template Layers and Other Nonsilicon Substrates. Piezoelectric and pyroelectric responses of ferroelectric films supported on a substrate depend on their remanent polarization in the direction perpendicular to the substrate. Preparation of films with a preferred orientation in the polar direction of the perovskite is desired in these materials, since this obviates the application of an electric field to the film for its polarization. Besides, more stable devices are obtained if this spontaneous polarization is not switchable [71].

Preferred orientation of films depends on nucleation mechanisms of the crystals of the film on the substrate. In films, nucleation is always facilitated by the substrate surface that is a site for heterogeneous nucleation [50]. However, during crystallization of films, a competition between growth of crystals heterogeneously nucleated at the film/substrate interface and of crystals homogeneously nucleated through the bulk film occurs. Epitaxial films have been mainly fabricated by physical deposition techniques. But, CSD methods have also been used for the preparation of films with a high degree of orientation, if appropriate preparative parameters and substrates are selected [72]. Using CSD methods, oriented films have even been obtained on noncrystalline substrates [73]. However, the degree of orientation of crystalline films increases with the similarity between the structure of the substrate and that of the film [74–76].

Orientation improves in CSD films by depositing very thin layers. This is due to the reduction of the volume in which homogeneous nucleation can occur. So, multiple deposition and crystallization of thin films have been used for the fabrication of PTCa films with a high preferred orientation in the polar direction. This procedure makes the first perovskite thin layer to replicate the structure of the surface substrate. Following perovskite layers are going to replicate the structure of their underlayer, again. In this way, tensile stresses decrease and the desired orientation can be obtained on silicon substrates if appropriate template layers are placed at the substrate surface.

Here, we present two silicon substrates of all those tested in our laboratory for promoting orientation in the PTCa films. They are $Ti/Pt/Ti/SiO_2/Si(100)$ and that presented in the former section, $Pt/TiO_2/SiO_2/Si(100)$. PTCa films were prepared from ~0.3 mol/L solutions with a 10 mol% excess of Pb and by multiple spin-coating deposition, drying, and RTP crystallization of layers with an average thickness, after crystallization, of ~70 nm. Orientation of the crystalline films was studied by X-ray diffraction, using the Bragg–Brentano geometry. In the patterns of these materials, the 111 peaks of the $Pb_{0.76}Ca_{0.24}TiO_3$ film and of the Pt appear overlapped.

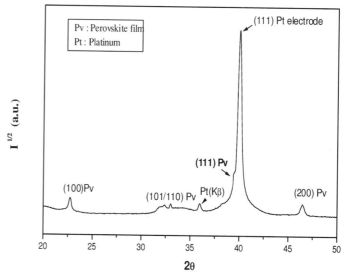

Fig. 8.   XRD pattern of a PTCa film onto $Ti/Pt/Ti/SiO_2/(100)Si$.

To minimize this overlapping, θ and 2θ were misaligned an angle of ~3° for recording the patterns. This experimental setting allowed us to separate the two peaks and to obtain semiquantitative values of preferred orientations in the films [64]. Orientation of the films was estimated from the I_{001}/I_{100}, I_{101}/I_{110}, and I_{111} perovskite doublets and peaks with Miller indexes of 001/100, 101/110, and 111, respectively.

The XRD pattern obtained for the film deposited on $Ti/Pt/Ti/SiO_2/Si(100)$ is shown in Figure 8. A shoulder is detected on the left of the 111 Pt peak. It corresponds to the 111 peak of the $Pb_{0.76}Ca_{0.24}TiO_3$ perovskite. This indicates that the film grows with a preferred ⟨111⟩ orientation, since the intensity of this peak is much higher than the other peaks of the perovskite. Table V shows the results of preferred orientation of the PTCa films onto different substrates. This table also shows the film thickness and the measured spontaneous pyroelectric coefficient. Spontaneous pyroelectricity of the films is a clear indication of their growth with the polar direction of the perovskite perpendicular to the substrate plane. The ⟨111⟩ orientation of these films has a component in the 001 axis of the perovskite that provides the spontaneous polarization of the films. Peak profiles of the films onto $Pt/TiO_2/SiO_2/Si(100)$ and onto $Ti/Pt/Ti/SiO_2/Si(100)$ are shown in Figure 9. Note the preferred ⟨111⟩ orientation of the PTCa film deposited on $Ti/Pt/Ti/SiO_2/Si(100)$.

A preferred ⟨111⟩ orientation in the $Pb_{0.76}Ca_{0.24}TiO_3$ film is expected to be induced by the (111)Pt bottom electrode of the substrate. There is a reasonable match between the cubic Pt and the tetragonal $Pb_{0.76}Ca_{0.24}TiO_3$. However, many other orientations are obtained. Actually, the $Pb_{0.76}Ca_{0.24}TiO_3$ on the $Pt/TiO_2/SiO_2/Si(100)$ substrate develops a preferred ⟨100⟩ orientation (see Fig. 9). Aoaki et al. [77] indicated that Pt(111) is not the ideal substrate to nucleate perovskites.

It has been reported that PZT thin films grow on platinized silicon substrates with a preferred ⟨111⟩ orientation, whenever

Table V. Preferred Orientations of $Pb_{0.76}Ca_{0.24}TiO_3$ Thin Films on Different Substrates

Substrate	Preferred orientation	Strain	$\gamma_s \times 10^9$ (C cm^{-2} K^{-1})	Thickness (nm)
$Pt/TiO_2/SiO_2/Ti$	$\langle 100 \rangle$	tensile	~5	~300
$Ti/Pt/Ti/SiO_2/Si$	$\langle 111 \rangle$	tensile	~7	~300
Pt/MgO	$\langle 001 \rangle$	compressive	~25	~200
$Pt/SrTiO_3$	$\langle 001 \rangle$	compressive	~11	~200

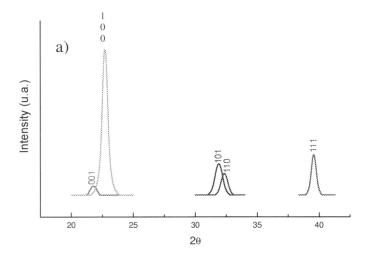

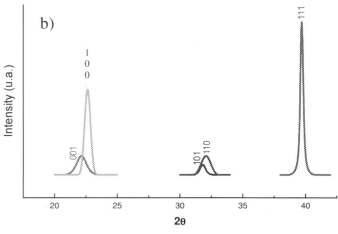

Fig. 9. Peak profiles of the PTCa films onto (a) $Pt/TiO_2/SiO_2/Si(100)$ and (b) $Ti/Pt/Ti/SiO_2/Si(100)$.

the formation of perovskites enriched in TiO_2 at the substrate surface [78]. TiO_2 came from the oxidation of the Ti layer on Pt, prior to the film deposition or during the film pyrolysis. Again, this first layer of TiO_2-rich lead titanate perovskite has a good matching with the $Pb_{0.76}Ca_{0.24}TiO_3$ film and makes easy the $\langle 111 \rangle$ growing. The last hypothesis proposes that the preferred $\langle 111 \rangle$ orientation of these films is caused by the formation of a $(111)Pt_xPb$ textured phase [79]. This would be an intermediate phase formed at the substrate/film interface during the thermal treatment of crystallization of the films.

Other nonsilicon substrates have also been tested, namely Pt/MgO and $Pt/SrTiO_3$. The PTCa films grow on both substrates with a $\langle 001 \rangle$ preferred orientation. Similarity between structures of $Pb_{0.76}Ca_{0.24}TiO_3$ and $SrTiO_3$ is evident. However, nucleation of the perovskite occurs on the $(111)Pt$ surface and not on the $SrTiO_3$ or MgO. According to Tuttle et al. [65], compressive transformation stresses produced during the cooling of the thermal treatment of crystallization have to be responsible for this $\langle 001 \rangle$ orientation. Orientation degree of these films together with their spontaneous pyroelectric coefficients are also shown in Table V. Figure 10a and b shows the GIXRD patterns of the $Pb_{0.76}Ca_{0.24}TiO_3$ on Pt/MgO and $Pt/SrTiO_3$, respectively. The Bragg–Brentano geometry has not been used for recording these data because peaks of the substrates and the $Pb_{0.76}Ca_{0.24}TiO_3$ are overlapped. Other more appropriate techniques have been used for the determination of the texture of the oriented films. Pole figures measured in these films have been reported in [80].

2.2.2.2. Thermal Treatment of Crystallization of the Films

The films, as deposited, are amorphous and have to be thermally treated so as to obtain the desired crystalline phase. Competition between sintering and crystallization occurs during thermal treatment in CSD films. One prevails against the other depending on heating rate of the thermal treatment used for the crystallization of the films [81]. It has been indicated in Section 2.2.1.2 that the increase of the heating rate delays the energetic processes produced during the thermal treatment of the solutions. So, crystallization is also delayed. Raising the heating rate favors a complete sintering of the amorphous layer prior to its crystallization. So, rapid heating rates provide denser films than films obtained with slow heating rates [50].

Here, results on $Pb_{0.76}Ca_{0.24}TiO_3$ films deposited on $Pt/TiO_2/SiO_2/(100)Si$ are going to be presented. Three types of thermal treatments have been used for the crystallization of the

Ti is present on the platinum surface [66]. This orientation seems to be facilitated by the matching between the structures of the perovskite film and that of the platinum surface. However, formation of different compounds at the platinum surface has been predicted for these substrates. Some authors indicate the formation of a Pt_3Ti cubic phase [75]. The Ti–Ti distance in the (111) plane is 5.55 Å in this structure, and in the $Pb_{0.76}Ca_{0.24}TiO_3$ perovskite it is 5.52 Å. Matching between both structures is good enough for the growing of the perovskite film with the $\langle 111 \rangle$ orientation. Other authors defend

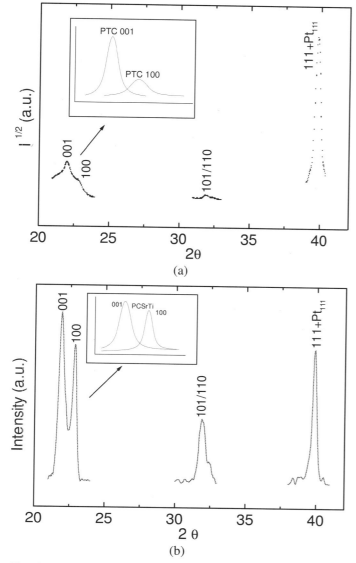

Fig. 10. Peak profiles of the PTCa films onto (a) Pt/MgO and (b) Pt/SrTiO₃.

composition and to the formation of lead-deficient phases. Besides, during annealing, an interdiffusion process may occur between the substrate and the modified lead titanate film. These phenomena spoil the final material and its electrical response. The nature of the treatment has a decisive influence on the crystallinity of lead titanate-based films and on the interlayer diffusion [82]. There seems to be unanimity as regards the advantages of rapid treatments, which minimize the formation of undesired phases and the diffusion [50, 83].

Characterization of the crystalline films was focused on the determination of composition, crystalline phases, structure, and microstructure of the films. Electrical characterization of these films will be considered in Section 3. The films prepared for electrical characterization have an average thickness of ~250–350 nm. Thickness was measured by profilemetry with a Taylor Hobson Form Talysurf-50S2 equipment.

Surface microstructures and cross sections of the films were obtained with a field emission scanning electron microscopy (SEM) Hitachi S800.

A Siemens D500 powder diffractometer with a Cu anode and with the grazing incidence X-ray asymmetric Bragg geometry (GIXRD) was used to monitor crystal structure [84]. An incidence angle of $\alpha = 2°$ was used to record the data. This assures that signal from the film with a minimum interference of the substrate would be obtained. The same diffractometer with the Bragg–Brentano geometry (XRD) was used for the determination of preferred orientations in the films.

Profile composition was determined by means of Rutherford backscattering spectroscopy (RBS) and X-ray photoelectron spectroscopy (XPS). These techniques allowed us to obtain the different layers and interfaces forming the heterostructure of the materials.

For the RBS measurements, data were collected with a 3.1 Van de Graaff accelerator. A 1.6-MeV He⁺ beam was collimated and directed on the sample down to an area of 1 mm². Backscattered ions were detected by two surface barrier detectors of energy resolutions of 12 and 18 keV at 180° and 140° to the beam direction, respectively. Solid angles were 3.8° and 22° for the 180° and 140° detectors. Samples were set in a vacuum chamber perpendicular to the ion beam. This experimental setting sample provides a general information of the heterostructure of the material (film, interfaces, substrate, electrodes, buffer layers, etc.) [85].

XPS combined with Ar⁺ depth profiling was carried out with a PHI 5700 equipment. Survey and multiregion spectra were recorded of Pb 4f, Ca 2p, Ti 2p, O 1s, Pt 4f, Si 2p, and C 1s photoelectron peaks and Pb MNN, Ti LMV, and O KVV Auger peaks. The atomic concentrations were calculated from the photoelectron peak areas, using Shirley background subtraction [86] and sensitivity factors provided by the spectrometer manufacturer PHI [87]. All binding energies refer to Ca 2p₃/₂ at 347.0 eV. Depth profiling was carried out by 4-keV Ar⁺ bombardment and a current density of ~3 μA/cm².

Structure and Microstructure of the Films. Crystalline phases and surface microstructure developed in the films after a ther-

films: (a) a thermal treatment obtained by heating the films at 650°C for 3600 s with a slow heating rate of 10°C/min, (b) a thermal treatment at 650°C for 720 s obtained by direct insertion of the sample into a preheated tubular furnace; this process provides a heating rate of ~8°C/s that can be considered as a half-rate of heating, and (c) a rapid thermal processing (RTP) obtained by heating the film at 650°C for 50 s with a rapid heating rate of ~30°C/s. This was carried out using infrared heat lamps in a commercial RTP system, model JETSTAR 100T of JIPELEC. The films presented in this section have been obtained by a single deposition of a precursor solution with a concentration of 0.9 mol/L. Solutions used for the deposition of the films prepared with the treatments (a) and (b) do not have Pb excess. The films used for the treatment (c) (RTP) were deposited from solutions with a 10 mol% of Pb excess.

Volatilization of lead occurs during annealing of lead-perovskite thin films. This leads to an unbalance in the film

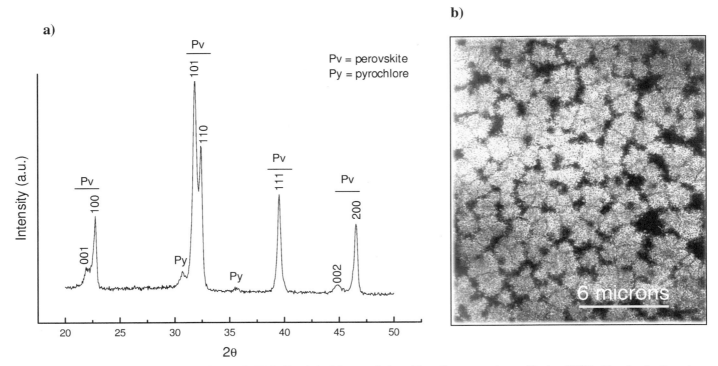

Fig. 11.   (a) GIXRD pattern and (b) SEM image of a PTCa film derived from a solution without Pb excess and crystallized at 650°C with a slow heating rate.

mal treatment with a slow heating rate of 10°C/min are shown in Figure 11. Two microstructural and crystalline phases are observed in the film. GIXRD analysis of this film shows the development of the perovskite phase accompanied by a minor second phase (Fig. 11a). The peaks recorded for this second phase seem to indicate that its structure is close to a pyrochlore structure [83]. A SEM image of this film shows a heterogeneous microstructure formed by ensembles of small grains surrounded by a weakly emitting background (Fig. 11b). Usually, the ensembles of grains correspond to the perovskite phase, whereas the fine-grain background is a lead-deficient phase. EDS analysis made on a film crystallized under these conditions indicated that both microstructural phases had a close composition. These results have been presented in Table IV of Section 2.2.1.1. In this section, we showed how the addition of a 10 mol% excess of Pb to the precursor solution led to films with the desired $Pb_{0.76}Ca_{0.24}TiO_3$ composition. Films with Sm and La modifiers develop microstructures in which two microstructural and compositional phases are also detected. In these cases, the groups of grains appear as rosettes of perovskite surrounded by a matrix of fine grains of pyrochlore where a considerable lead deficiency was measured [8]. The rosette-shaped regions are textured ensembles of nanosized single crystals with a preferred orientation in the a-axis of the perovskite. The pyrochlore does not have any preferred orientation [6].

A second phase is not detected in the GIXRD patterns of the films derived from the same solutions as the former film, but treated at 650°C with a heating rate of ~8°C/s (Fig. 12a). The microstructure of this film is formed by clusters or associations of grains with inter- and intracluster porosity (Fig. 12b).

For films deposited from a solution with a 10 mol% of Pb excess and crystallized with rapid heating rates of ~30°C/s (RTP), second phases are also not detected in the GIXRD pattern of Figure 13a. The SEM image of this film (Fig. 13b) also shows a microstructure formed by clusters. But in this case, the clusters and grains are larger than those of the former film and porosity is lower. Besides, the latter film shows a preferred ⟨100⟩ orientation.

Combination of Pb excess and rapid heating rates produces a decrease of second phases and the obtainment of films with the desired perovskite composition. It is well known that rapid RTP processes minimize the content of second phases by circumventing the temperatures in which they are stable [88]. As we have discussed before, RTP films are denser than others treated with slower heating rates because the sintering of the amorphous film is completed before the crystallization starts [50]. These RTP films grow on Pt/TiO₂/SiO₂/(100)Si with a preferential ⟨100⟩ orientation because rapid heatings facilitate the heterogeneous nucleation of the first crystals over their homogeneous nucleation [50].

Profile Composition of the Films.   Profile composition of a thin film is conditioned by the underlying substrate and the deposition method [89].

The problems of silicon for use as substrate of ferroelectric layers are related to the deposition temperatures of these films. In general, all the deposition techniques use temperatures over 500°C. Diffusion of silicon is strong at these temperatures. This diffusion makes easy the reaction between the ferroelectric

a)

Pv = perovskite

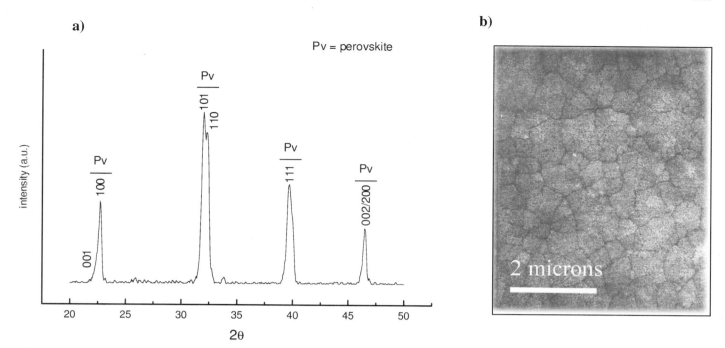

Fig. 12.   (a) GIXRD pattern and (b) SEM image of a PTCa film derived from a solution without Pb excess and crystallized at 650°C with a half-rate of heating.

a)

Pv = perovskite

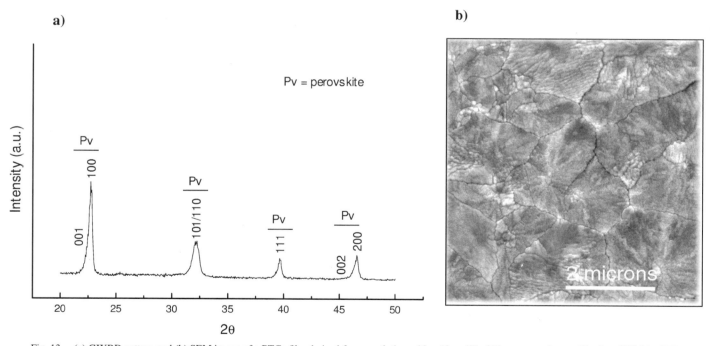

Fig. 13.   (a) GIXRD pattern and (b) SEM image of a PTCa film derived from a solution with a 10 mol% of Pb excess and crystallized at 650°C by RTP.

layer and the substrate. This reaction produces non-ferroelectric interfaces that damage the response of the device.

All the silicon devices containing ferroelectric films also have bottom and top electrodes to form the capacitor. Top electrodes are deposited when the deposition process has finished. Bottom electrodes are part of the underlying substrate and have to be taken into account during the film deposition. The most used bottom electrode is platinum. Adhesion of platinum to silicon is not good. So, adhesive layers are usually placed between platinum and silicon [70]. It is intended that adhesive layers also work as buffers that minimize diffusion, as is the case of TiO_2 (Section 2.2.2.1). However, interaction between the Pt bottom electrode and the CSD film cannot be avoided during the thermal treatment. This interaction is a function of the temperature and the duration of the treatment [90]. So, RTP treatments are preferable for the crystallization of these

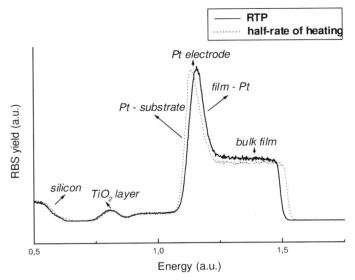

Fig. 14. RBS spectra of PTCa films crystallized at 650°C with different heating rates.

films. They are short-time processes that decrease the diffusion phenomena between film and substrate. This minimizes the formation of undesired film/substrate interfaces. Also, migration of elements of the films through their thickness is brought down with RTP treatments. This should provide films with more homogeneous composition profiles.

For the study of the profile of these films, the Pt electrode/ferroelectric film interface and the bulk ferroelectric layer have been studied by RBS and XPS.

RBS spectra were simulated from the data recorded by the RBS technique and using the RUMP software [91]. The spectra of Figure 14 correspond to the experimental data recorded for the PTCa films deposited from solutions without Pb excess and with a 10 mol% of Pb excess and thermally treated with a half-rate of heating of ∼8°C/s and RTP, respectively (see Section 2.2.2.2). Films prepared with other excesses of Pb and with slower heating rates have been also studied by RBS. Results of these films can be seen in [22, 24].

We have considered some assumptions for making the simulations. First, calcium and titanium cannot be calculated from the spectra because the Pt signal overlaps the Ca and Ti signals. So, calcium and titanium concentrations have been considered unchanging and equal to those of the nominal perovskite. Second, the simulations provide values of thickness of the layers in atom/cm^2. To change these units to nanometers, the bulk density of the layer has to be considered. For these calculations, we have used a density equal to that obtained from the atomic densities of each element of the $Pb_{0.76}Ca_{0.24}TiO_3$ perovskite. This is not the density of the bulk film, since the possible existence of porosity, strains, second phases, and other possible defects in the films have not been considered. We also consider that layers with variable compositions can have been formed during processing [52].

Regions in the spectra from which information has been inferred are marked in Figure 14. This figure shows the to-

tal multilayer structure of the device: film/film–Pt interface/Pt/ Pt–substrate interface/TiO$_2$/SiO$_2$/(100)Si. However, the study of the film/film–Pt interface is going to be emphasized here, because it is the part of the device that has more effect on the ferroelectric properties [92]. As discussed previously, these results also indicate that the film thermally treated with a slower heating rate has a larger thickness than the other due to its larger porosity, ∼440 nm and ∼400 nm, respectively [93]. About composition, the former film has a lower amount of Pb than the latter, Pb/Ca of ∼0.66/0.24 and ∼0.77/0.24, respectively. This is inferred from the lower height of the Pb signal in the film treated with a half-rate of heating. The thickness of the film–Pt interface can be estimated from the slope of the straight line joining the Pt electrode and the bulk film. Higher slopes indicate thicker interfaces. Note that the thickness of this interface is smaller in the RTP film. Thicknesses of the film–Pt interface calculated from the RBS spectra are ∼70 nm and ∼60 nm for the films treated with a half-rate of heating and the RTP, respectively.

Figure 15 shows the XPS spectra obtained for the same films analyzed by RBS. The XPS analysis indicated that both films have a large contamination of carbon at their top surfaces, about 30%. This concentration decreases after 1 min of Ar$^+$ bombardment. Then, the film surfaces appear clean and no signal of carbon is detected inside the bulk films. This means that the carbonate residues observed by TGA/DTA/EGA/IR in the samples of the precursor solutions treated over 750°C (Section 2.2.1.2) are not detected in the bulk films. This could be attributed to an easy elimination of the organic groups in the films due to their large surface/thickness ratio. However, we do not have experimental data to justify this behavior. The Pb/Ca ratios of the films have been calculated just after the elimination of the carbon-contaminated surface of the films. A composition close to the expected $Pb_{0.76}Ca_{0.24}TiO_3$ perovskite is obtained in the film treated by RTP (Pb/Ca ∼ 0.76/0.24). This result is in accordance with that obtained by RBS. The defect of Pb of the other film (Pb/Ca ∼ 0.68/0.32) is due to the lack of Pb excess in the precursor solution and to the slower heating rate used for its crystallization. The longer duration of the thermal treatment of the film treated at ∼8°C/s compared for the RTP film gives opportunity for a higher volatilization of Pb.

Prolonged Ar$^+$ bombardment causes a preferential loss of O and Pb, a reduction of Pb^{2+} to the metal state, and a reduction of Ti^{4+} to lower oxidation states [94]. So, quantitative compositions cannot be calculated in the steady-state situation of the XPS spectra. These data only provide information about the profile distribution of the cations in the bulk film and about the Pt–film interface. The RTP film is homogeneous in depth (Fig. 15b). However, the longer duration of the other thermal treatment provides enough time for the mobility and volatilization of elements of the film (Fig. 15a). So, its profile is less homogeneous than that of the RTP film. Also, a thicker Pt–film interface is formed in this film than in the RTP film. Then, a heating rate of ∼8°C/s seems to be rapid enough to inhibit the formation of undesirable second phases (see Section 2.2.2.2), but not to avoid Pb volatilization and diffusion.

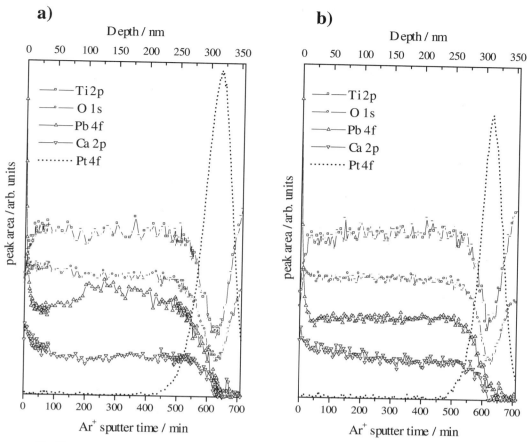

Fig. 15.   XPS spectra of PTCa films crystallized at 650°C with (a) a half-rate of heating and (b) by RTP.

The thickness trend obtained by XPS is in accordance with that deduced from RBS. However, values obtained by both techniques are different. In the case of XPS analyses, deviations came from the preferential sputtering of Pb and O, and that these thicknesses were calculated with respect to the sputter rate of the Ta_2O_5 at 4 keV Ar^+ [95]. Neither are the thicknesses obtained by RBS exact because the real bulk density of the film cannot be considered in the simulations.

2.3. Remarks About Preparation of Ca-Modified Lead Titanate Thin Films

In the study of the preparation of Ca-modified lead titanate thin films from a sol-gel method, the following conclusions can be drawn:

1. Precursor solutions of Ca-modified lead titanate thin films have been synthesized from the diol-based sol-gel route. The modifier (Ca) is incorporated into the chemical system as calcium acetate in water. The solutions are stable in air due to the structure of the inorganic polymer developed in the sol during its synthesis. Steric hindrance to water of this polymer permits decreasing the hydrolysis and condensation rates of the solutions. Thermal decomposition of the gels derived from these solutions takes place below 500°C. Carbonate residues

are formed during pyrolysis that are eliminated above 700°C. However, these residues have not been found in the crystalline films derived from the solutions. A 10 mol% excess of Pb has to be added to the precursor solutions to obtain the desired composition in the crystalline film ($Pb_{0.76}Ca_{0.24}TiO_3$).

2. Texture of the crystalline films is determined by the underlaying substrate. Platinized MgO or $SrTiO_3$ substrates promote a preferred ⟨001⟩ orientation in the PTCa perovskite films. A preferred ⟨111⟩ orientation is developed in films onto platinized silicon substrates with some Ti on the platinum surface. Both orientations, ⟨001⟩ and ⟨111⟩, provide a net spontaneous polarization in the films, highly interesting for the integration of these PTCa films into piezoelectric and pyroelectric devices. Films on platinized silicon substrates without Ti on the platinum surface grow with a random orientation.

3. Heating rate of the thermal treatment used for the crystallization of the films has a strong effect on the composition, structure, and multilayer structure of the film. Diffusion between substrate and film decreases with the increase of the heating rate. Films with a minimum interaction between the ferroelectric layer and the substrate are obtained by rapid thermal processings (RTP). The composition of the film in profile is unbalanced when long-duration thermal treatments are used. Besides, these treatments favor the formation of undesired lead-deficient second phases. Films with a homogeneous distribution

of the cations in profile are also obtained by RTP crystallizations.

3. FERROELECTRIC CHARACTERIZATION TECHNIQUES

3.1. Measurement Techniques

The principal ferroelectric properties of thin films concern the value of the electrical polarization, P_s, and its ability to be switched by an electric field, the time spent to commute, its endurance against the repeated electric switching, and the time of degradation. In order to approach the characterization of ferroelectric thin films, the main experimental techniques currently used will be described here. Attempts to make a standard test methodology for ferroelectric nonvolatile memory applications have been published by authors belonging to an American company consortium [96]. The ferrolectric parameters usually used in the field of ferroelectric thin films were defined, as well as the conditions of measurement. Definitions and terminology consistent with the IEEE Standard Committee on ferroelectric thin films referenced to hysteresis loops were also described, as well as specific guidelines of testing procedures. All of these are concerned with switching current measurements, with possible causes of errors, with the effect of rise time of the pulses used (order of ns), and with the capacitor size and its preconditioning.

Thin films are generally formed by joined crystallite grains randomly oriented that do not show a remanent value of electric polarization, because the sum of the projection along the normal direction to the film plane of the elementary polarization of the grains is zero. Nevertheless, in some occasions, a texture can be promoted through the processing method and a net value of polarization is possible. Therefore, a poling process is usually mandatory to reveal the main ferroelectric properties.

Among ferroelectric measurements, the most representative ones are switching currents and hysteresis loops. For these measurements, a sequence of square pulses is usually applied as shown in Figure 16a. Two reading pulses of the same amplitude are used for the measurement of current curves, after applying to the sample several poling pulses of few volts and 200 μs width, separated by 20-μs intervals. Rise time of the pulses should be <10 ns (lower than the expected switching time, or the time spent to invert the direction of the ferroelectric domains) and the time elapsed between the polarization and the measurement may be 300 ms. Figure 16b shows schematically the measuring circuit. The switched charge, Q_s, is obtained with the first measuring pulse. This is computed by integration of the current throughout the measuring time. The nonswitched charge, Q_{ns}, is obtained with the second pulse because this does not find a switchable charge (it only saves a small contribution due to the recovery of the switched charge during the time elapsed between both reading pulses). Therefore, the linear effect of the capacitor is measured. Usually, two more reading pulses of opposite sign are applied to the film to obtain a similar type of current with opposite sign that gives information about

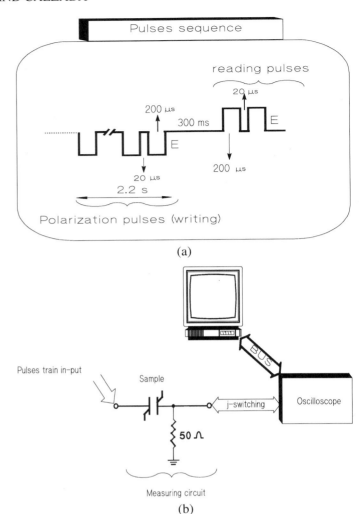

Fig. 16. (a) Pulse sequence used to measure switching currents (a five-pulses method). (b) Schematic equivalent circuit and setup of measuring.

the symmetry of the phenomenon and, consequently, about the nature and mechanism involved in the origin of charges (this method is occasionally called a five-pulse method). The remanent polarization, P_{sw}, is calculated from these charges and the area of the electrode, S, using the equation

$$P_{sw} = S(Q_s - Q_{ns})/2 \qquad (1)$$

The switching times (time of the maximum of the current curve, t_m, and time corresponding to 10% of the maximum value of the current, t_s, which is the true switching time) can be also calculated from the above current curves. Usually t_m and t_s are the most important parameters on memory applications since they define the velocity of writing/reading of data to be stored.

Hysteresis loops may be measured with a modified Sawyer–Tower circuit (Fig. 17). This is the most conventional method used. Nowadays, a Radiant Technology RTA 66a Standardized Ferroelectric Test System is frequently used in the virtual ground mode with a triangular signal of few volts of am-

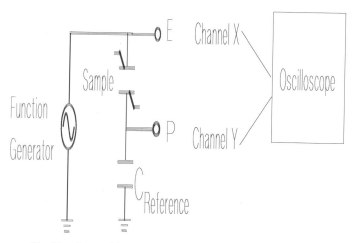

Fig. 17. Sawyer–Tower circuit and setup to trace hysteresis loops.

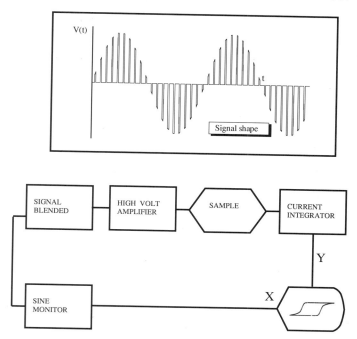

Fig. 18. Block diagram of the electrical circuit setup to measure hysteresis loops by an alternative method. The inset shows the employed sinusoidal pulsed signal of low frequency. Reprinted with permission from [98], copyright 1999, American Institute of Physics.

plitude at different frequencies, although a sinusoidal signal is the most common in the classical setups. Hysteresis loops can be corrected by approximately compensating the contribution of ohmic and capacitive currents with a nonperturbing method [97], except for leakage currents of diverse origins that would still have to be accounted for. Generally, values of remanent polarization, $\pm P_r$, can be deduced from the intersection of the loop with the vertical axis, for $\pm E_{ap} = 0$ ($\pm E_{ap}$ is the applied electric field). The coercive field, E_c, can also be obtained from the two intersections of the loop with the horizontal axis, for $P = 0$.

As stated above, in tracing hysteresis loops, corrections by contributions of non-ferroelectric charge are a common practice to get true results. This is mandatory with lossy thin films. But sometimes, corrections are not possible due to diverse causes. It is known that the electrode–film interfaces produce states of electrical degradation that cause breakdown at low electric fields, due to the increase of the leakage currents. This can mask the true ferroelectric charge and leads to the disappearance of the hysteresis loop. In these cases, compensation, on the base of an ohmic current of the sample, is not good and a wrong estimation of the ferroelectric hysteresis loop would be obtained. Furthermore, since the leakage contributions strongly depend on the time that the electric field is applied, the hysteresis loops have to be traced at high frequencies to avoid the effects of leakages. This also reduces the risk of the film degradation by breakdown. These problems can be reduced by using an alternative method recently proposed by our group [98]. This method combines the advantages of the hysteresis loop method and the use of short pulses (low perturbing). In this alternative method, short pulses of variable amplitude separated by relatively long times are generated, followed by a sinusoidal wave of low frequency, as shown schematically in Figure 18. Electrical and mechanical relaxation of the film are produced during the time elapsed. Thus, a sinusoidal pulsed electric field can be produced by blending a sinusoidal signal of different frequencies with a pulse signal of variable period and width. A small fraction of the true sinusoidal voltage (the root mean square

voltage, V_{rms}) is used with this measurement technique. Its effect on the sample is not disturbing, since it is applied for only a short part of the period of the whole cycle.

Leakage currents are drawbacks for practical uses of ferroelectric films. Therefore, they have to be controlled by measuring the density current curves, j_l, as a function of time, at polarization and depolarization regimes. To know their origin, they are usually measured at several dc voltages of both signs and at several temperatures. For this, a step voltage technique is used with an electrometer Keithley model 6512, and a function generator HP3325B.

For memory applications, it is very convenient to know the degradation of the ferroelectric polarization when the film is subjected to switching cycles (fatigue) and its evolution with time (retention). These measurements can be carried out using the mentioned Radiant RTA 66A Ferroelectric Test System. Alternate pulses of few volts of amplitude and variable width and frequency are used for the fatigue measurements. The five-pulses method, above described, is used to obtain polarization values. This method is also used to measure the decrease of polarization with time.

Pyroelectric coefficient, $\gamma = dP/dT$, is the most characteristic property to qualify ferroelectric films for IR detectors, together with the monitoring of the poling degree. There are several methods well established for the measurement of γ. The dynamic method permits one to calculate γ from the pyroelectric currents obtained after applying to the film a triangular thermal wave of few degrees of amplitude and at low frequency. This procedure gets an effective gradient of temperature of

3°C/min. Theoretically, the response of the intensity current to the triangular thermal wave should be a square-shape signal and γ should be easily calculated.

Strains are inherent to the nature of thin films. They come from the necessary substrate used to support the film and from the phenomena involved in the film processing, as it will be described later. The measurement by profilemetry of the curvature radii of the film surface from the deflection across the sample is one of the methods used to obtain strains of films. Successive measurements of the curvature of the film sample after different steps of the processing make possible the calculation of stresses using the modified Stoney equation [99]:

$$\sigma_{ij} = \frac{E_s}{6(1 - \nu_s)} \frac{t_s^2}{t_f} \left[\frac{1}{R_j} - \frac{1}{R_i} \right] \quad (i, j = 0, 1, 2) \quad (2)$$

where σ_{ij} is the film stress, E_s is the Young's module of the substrate, ν_s is the Poison ratio for the substrate, t_s and t_f are the thickness of the substrate and the film, respectively, and R is the curvature radius.

The methods described here for the characterization of ferroelectric thin films, together with the analytical methods treated in the first part of this review, have allowed us to carry out the studies on the PTCa films that will be explained in the next section. First, we will describe the effects of conditioning electrical or thermal treatments on the ferroelectric properties of the films. Afterwards, we will describe the diverse handling to promote improvements of the final performances of the films, emphasizing those that lead to PTCa films with adequate properties for applications. Mechanisms involved in these processes will be suggested to aid in understanding the origin of the behavior and final properties of the films.

3.2. Conditioning Effects

In general, final ferroelectric properties of thin films depend on the parameters of the processing method used for their fabrication. Strains and charged defects developed in the films during their preparation influence their ferroelectric response. These effects may be reduced or even eliminated by an adequate electrical or thermal treatment. These types of treatments and their effects will be described for sol-gel derived-PTCa thin films onto Pt/TiO$_2$/SiO$_2$/(100)Si substrates.

3.2.1. Electrical Treatment

It is observed that the remanent polarization calculated from hysteresis loops, P_r, normally depends on the type of wave signal used to trace the loop [99–101], and on the history of the electrical signal applied to the films for their stabilization [102], deaging, or conditioning [103]. Values of P_r are usually overestimated from the loops [104]. They differ from those obtained from switching currents measurements [105]. Differences may be important, depending on the preparation conditions [106]. Klee et al. [107] and Yoo and Desu [108] have associated the ferroelectric degradation of thin films of Pb compounds to physical defects (lead and oxygen vacancies) and to their interaction

with the substrate. Based on these findings, Ramos et al. [109] developed an experimental strategy to act on the defective state of PTCa samples that also have lead and oxygen vacancies produced during their processing. This strategy consists of applying to the sample, prior to any ferroelectric measurement, a square wave signal of 1 Hz and ±320 kV/cm of amplitude for 60 seconds. Ferroelectric properties before and after the electric treatment were compared. Differences were quite apparent: before any electrical treatment, the film does not show any hysteresis response, since the loop is buried under a considerable amount of leakages. But, after subjecting the film to the electrical signal described, the same film shows a well-defined hysteresis loop and the remanent polarization and the coercive field can be calculated. Switching currents are also enhanced after this electrical treatment. In the same work [109], an important dependence of the hysteresis loops with the frequency is reported for the samples electrically treated. Hysteresis loops are lower at higher frequencies, but the contribution of leakages also decreases. This is explained on the basis of the role of Pb and O vacancies of moderate mobility, measured in these compounds by the authors [52]. As stated by Carl and Härtl [110] and Larsen et al. [111], a small grain size together with the presence of charged defects may hinder the inversion of domains in films in their virgin state. This is due to a certain degree of stabilization of the domain configuration that would explain the poor ferroelectric character of the films before the application of the electrical signal. The charge involved in the measurements could almost be exclusively due to leakage currents. However, the high electric fields applied during the electrical treatment (as large as ±500 kV/cm) would cause the movement of charged defects, in a similar way to that reported by Larsen et al. [103]. They called this phenomenon a deaging effect. In fact, leakage currents are reduced and domains are unclamped from defects, walls being more free to be switched. Hysteresis loops are better defined and coercive fields are more exactly measured.

As it has been pointed out before, values of P_{sw} (polarization obtained by integration of switching current curves) are usually lower than those obtained from the hysteresis loops, P_r. The authors propose an explanation based on the description of PZT films reported by Larsen et al. [111]. These authors state that a distribution of switching times would be expected for polycrystalline films with a grain size lower than the film thickness (ceramic morphology) [112]. So, pulses would have to be applied for a longer time than the expected switching time, in order to switch all the domains. This is the case of the samples studied by Ramos et al. [109] (pulses of width of 200 μm are used and the expected switching times are $<10^3$ ns). Therefore, the values of P_{sw} should be as high as possible, without any other charge contribution than the ferroelectric one. On the contrary, values of P_r calculated from hysteresis loops have other contributions, such as leakage currents, due to the much longer integration times, which also justify the decrease of P when the frequency is high. Thus, an electrical treatment prior to any ferroelectric measurement on PTCa thin films (with Pb and O vacancies generated during processing) is strongly recommended to minimize the effects of non-ferroelectric charges.

3.2.2. Thermal Treatment

As it has already been described in the first part of this work, addition of an excess of PbO to the precursor solution is usually needed for compensating Pb volatility during the heating process of the solution-derived PTCa perovskite films. The films are a part of a multilayer structure formed by materials with different mechanical and thermal properties. Therefore, the final performances of the films will be conditioned by the behavior of the whole stack during the film processing (wetting, drying, and crystallization mechanisms), as reported by Spiering et al. [113]. This means that the films just processed have a certain amount of charge defects and a strain distribution that are susceptible to change by providing energy in an adequate way, such as heating. This will cause variations in their properties [114, 115]. This will be called a thermal treatment and results in a conditioning of the sample.

Regarding PTCa films, our group has extensively studied this topic in samples crystallized by RTP [116]. Three types of samples were electrically characterized: (a) the crystalline films just prepared, named V, (b) the crystalline films subjected to a post-annealing or thermal treatment that consisted of a heating at 650°C for 60 minutes and cooling down to room temperature, with a heating and cooling rate of 10°C/min, called VT, and (c) the post-annealed films with top electrodes that are heated again at 500°C for 5 minutes and quenched down to room temperature, hereinafter referred for as VTQ. The effects of these thermal treatments on the traced hysteresis loops are depicted in Figure 19. A reduction of the leakage contributions and a lower coercive field are observed for the samples VT and VTQ compared to the sample V, resulting in a more realistic value of the remanent polarization, P_r, and in a small shift of the hysteresis loop along the electric field axis. This means that there is a reduced presence of charged defects in the samples responsible for internal electric fields. Furthermore, the switching process is increased as a consequence of the treatment, as shown in Figure 20, although the quenching does not show the same effect (see the broadening of the VTQ curve). Here, larger values of remanent polarization, P_{sw}, and a slower switching are calculated. It was reported in [117] that these thermal treatments of the crystalline films do not affect their composition or their crystal structure. So, the changes measured in the ferroelectric properties of these films should be associated to some other factors, such as intrinsic charge carriers (vacancies) originated during the fabrication of the films.

The reduction of the coercive field of the film after the thermal treatment may justify the more effective switching. This can be explained as follows: the charged defects created during the RTP crystallization, at first trapped at the grain boundaries and domain walls, are liberated and redistributed during the thermal post-treatment of the film, leading the domains to be more free to move, as stated by Waser et al. [118]. The same reason could justify the changes in the switching current curves

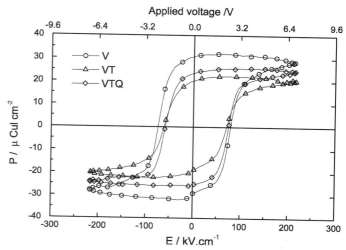

Fig. 19. P–E hysteresis loops of a PTCa film obtained by RTP process (V) and later subjected to a conditioning thermal pretreatment (VT) and further quenching with top electrodes (VTQ). Measurement conditions: 7 V and 100 Hz. Reprinted with permission from [116], copyright 1998, Elsevier Science.

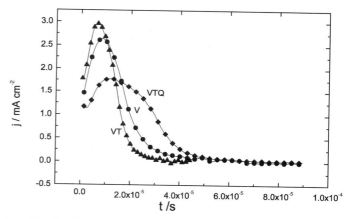

Fig. 20. Switching currents measured on PTCa films after the three types of conditioning treatments. Reprinted with permission from [116], copyright 1998, Elsevier Science.

of Figure 20. Once again, the measured switching time is more than an order of magnitude higher than that expected. This has already been justified as an effect of the film microstructure. According to Larsen et al. [111], a switching time distribution should be expected in this type of materials which would lead to a broadening of the current curves (Fig. 20), since switching is limited by grain boundaries,

For the RTP films, moderately higher values of permittivity were measured compared with those obtained in samples prepared with slower heating rates [119]. The authors associate this effect to a reduction of the non-ferroelectric interlayer placed between the platinum bottom electrode and the (Pb,Ca)TiO$_3$ film. Thicknesses of this interface were obtained from RBS and XPS analyses (see Section 1). But, another consequence of the RTP heating is the higher T_{max} values measured in the films. This is because of the larger stresses induced in the RTP films than in the others, as was previously demonstrated by the authors [120]. These strains are reduced by the effect of the post-annealing and quenching of the films, resulting in a slight reduction of the T_{max}.

Another important topic to identify the origin of the non-ferroelectric charges involved in the experiments is the study of the leakage currents. A certain degree of asymmetry is observed in the direct measurement of leakages of the three types of samples. This can also be deduced from the hysteresis loops. This asymmetry is related with the different nature of the top and bottom electrode-film interfaces. Each electrode was deposited following different deposition procedures. According to Hon-Ming and Lee [121], these interfaces act as traps for the electric carriers and will modify the film properties. The post-annealing treatment (sample VTP) moves the carriers from the bulk to the interfaces, liberating the ferroelectric domains. They become easier to switch with the applied electric field. But, the new charge distribution could also lead to the pinning of some of the ferroelectric domains close to the interfaces, leading to lower polarization values. Furthermore, the lower degradation of the sample VTQ, as a result of the quenching of the top electrodes, could be explained by a decrease of the trapped charges at the interfaces that go back to the bulk, reducing, again, the domain mobility by a pinning effect. On the other hand, the same authors have deduced from the linear dependence of the logarithm of the leakage currents against $E^{1/2}$ that a Schottky emission should be the responsible mechanism, whereas the low asymmetry of the leakage currents could be due to an additional Frenkel effect [122].

3.3. Improvements of Properties

3.3.1. Causes and Effects of Strains

Among the effects affecting properties of films, the stresses developed during the preparation process must be emphasized. This topic has been analyzed by Foster et al. [123] for epitaxially grown $PbTiO_3$ films. They define three types of strains: (a) those resulting from the different lattice mismatch between the film and the substrate, (b) those that come from the phase transformation upon cooling from the paraelectric to the ferroelectric phase, and (c) those developed during the cooling down of the sample and that came from the thermal expansion mismatch between the film and the substrate. For the same compounds, Li and Desul [124] observed a continuous decreasing of compressive stresses of the films during cooling, from the crystallization temperatures, above 500°C, to room temperature. An example of how preparation process produces stresses on films is given in the paper of Spierings et al. [113], where Pt/PZT/Pt thin film stacks prepared by sol-gel were studied. They carried out measurements of changes in the radius of curvature of the wafer after distinct steps of the preparation process. They calculated the stresses from these curvature radii, reporting that an annealing of the film with top electrodes has influence on the final stress of the PZT films, because it produces changes in the poling direction of the ferroelectric layer. More significant are the works of Scherer [50] and Lewis et al.

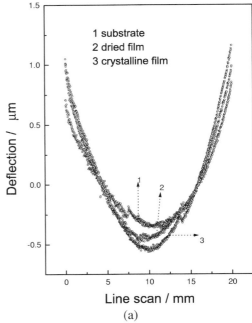

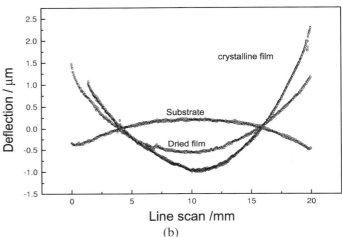

Fig. 21.   Deflection profiles obtained by profilemetry on the surface of the substrate, the dried film, and the crystalline film. Scans corresponding to PTCa films obtained from a solution with a concentration of (a) 0.9 M (PTC1) and (b) 0.5 M (PTC2). Reprinted with permission from [120], copyright 1987, American Institute of Physics.

[126] about drying of sol-gel films on rigid substrates. Large shifts of the phase transition temperature of PTCa thin films prepared by rf-sputtering were first reported by Yamaka et al. [30]. These shifts were explained by the stresses developed during deposition [113].

In order to know the stresses and their effects, the authors [120] approached the study of strains developed in the PTCa thin films, by measuring the curvature of the surfaces of the samples after several steps of the film preparation. Figure 21 shows the deflection scans obtained for two films derived from solutions with different concentration: (a) 0.9 mol/L and (b) 0.5 mol/L. Note that the adhesion of the film to the underlying substrate after the drying step is quite remarkable, as

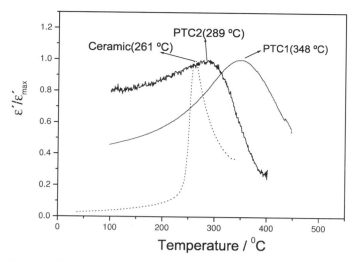

Fig. 22. Variation of permittivity with temperature at 1 kHz for PTCa films derived from solutions with concentrations of (a) 0.9 M (PTC1) and (b) 0.5 M (PTC2). Measurements of a bulk ceramic of the same nominal composition are also shown for comparison. Observe the important shift of the maximum corresponding to the more stressed film. (a) reprinted with permission from [120], copyright 1987, American Institute of Physics.

deduced from the curvature data. Three scans were recorded for each sample, corresponding to the bare substrate (R_0), to the dried film (R_1), and to the crystalline film (R_2). Stresses were calculated using Eq. (1). For the silicon substrate $E_s = 130 \times 10^9$ Pa, $\nu_s = 0.3$, and $t_s = 0.5 \times 10^{-3}$ m. As can be observed, the films obtained are always under a tensile stress (positive stress value). Two different contributions to the stress state were measured. The first one is that produced during the deposition and the further drying of the film, and the second one is that produced during the thermal treatment of crystallization of the film.

To obtain information about the preferred orientation of the films, the GIXRD technique with an incidence angle of $\alpha = 2°$ was used. For this incidence, a penetration of the beam into the sample of approximately 290 nm was calculated, considering that the diffracted X-ray intensity is e^{-1} times the intensity of the incident X-ray. The degree of preferred orientation, α_c, is obtained from the expression

$$\alpha_c = \frac{I_{001} R_t}{I_{001} R_t + I_{100}} \qquad (3)$$

According to Kim et al. [127], a value of α_c of 0.5 corresponds to a random orientation.

Permittivities versus temperature for the films and bulk ceramics with a nominal composition of $(Pb_{0.76}Ca_{0.24})TiO_3$ were also measured (Fig. 22). The transition temperature, T_c, measured as the maximum of the permittivity with temperature, is always higher for the films than for the bulk material in a similar way to that described for PTZ.

Here, the origin of the measured strains is discussed. The a and c lattice parameters obtained for the films show a deviation from the theoretical parameters of the perovskite that cannot be justified only by considering the Pb losses measured by

EDS, RBS, and XPS (about 10 mol%) [52]. The mean strain of $0.8 \pm 0.4\%$, obtained from these parameters, fits well with that calculated from the Hooks law: $\varepsilon = \sigma / E$, where $E = 80$ GPa [127] and considering a value of $\sigma = 800 \pm 10\%$ MPa (the experimental average stress measured in the films). This result seems to indicate that the films are subjected to stresses induced by the substrate.

From experimental results, two different processes seem to be involved in the development of stresses in the sol-gel PTCa thin films deposited on $Pt/TiO_2 TSiO_2/(100)Si$: (1) the wetting of the substrate by the precursor solution joined to the drying of the wet layer, and (2) the thermal processing used for the crystallization of the amorphous film.

The mechanism of strain generation during the wetting and drying of the film can be described as follows: the spin-coating deposition of the solution onto the substrate gives rise to a good adhesion between the solution layer and the underlaying substrate. The subsequent drying produces a large shrinkage in the film due to the volatilization of solvent and to the condensation reactions in the gel layer. Scherer [50] indicates that the drying produces a compressive stress on the substrate, since the only free dimension in the gel film is that perpendicular to the substrate. This creates a tensile stress in the film plane that is proportional to the viscosity of the solution and that explains the curvatures measured in the films after drying, $1/R_1 - 1/R_0$. Results also confirm the increase of the tensile stresses, σ_{01}, with the viscosity of the solutions. Viscosity is directly proportional to the concentration of the solution. So, the conclusion is that this type of stress generated on these films is inherent to the sol-gel deposition method. It is not present in other alternative deposition methods such as sputtering or laser ablation.

The stresses generated during the crystallization of the films can be described as follows: the three types of stresses described by Foster et al. [123] can be developed in the films treated by RTP up to a temperature of 650°C, well above the expected Curie temperature of the $Pb_{0.76}Ca_{0.24}TiO_3$ perovskite film (~260°C). Furthermore, a polynucleation of the perovskite on the substrate has to be expected, since the lattice matching between the film and the (111)Pt bottom electrode is not possible. This leads to films with a random orientation, as deduced from the GIXRD data ($\alpha_c \sim 0.5$). This means that the contribution of this process to the stresses should be negligible. On the other hand, the thermal processing also causes a volume shrinkage (~−36%) of the amorphous film due to the loss of volatiles, the perovskite crystallization, and the sintering of the ceramic layer. Besides, during cooling down a lattice expansion is produced, in particular during the paraelectric–ferroelectric phase transition. Finally, the third effect coming from the differences of the linear thermal expansion coefficients of the silicon substrate and the Ca-modified lead titanate have to be also considered. The authors express the difficulty of quantifying the contribution to the tensile stresses, σ_{12}, of the two last factors due to the large number of assumptions that would have to be considered.

Raman spectra were also measured for a PTCa thin film and a bulk ceramic with the same composition [120]. The spectra showed broad peaks and a large background, as expected for polycrystalline samples. They also showed, for the film, shifts of the frequencies corresponding to the soft mode E(1TO). This supports the presence of a stressed state in the films considering the differences between the measured Raman frequencies of the bulk ceramic and the film, $\nu_{tf} - \nu_{ce}$, and the linear pressure coefficients, $d\nu/dP$, reported by Sanjurjo et al. [128] for pure $PbTiO_3$ under hydrostatic pressure. An increase in the Raman frequencies of the film would indicate a negative pressure or a lattice expansion. Although the pressure sustained by the films is not hydrostatic, qualitatively but not quantitatively, tensile stresses can be considered equivalent to that. The Raman frequencies measured show an average lattice expansion in the film with respect to that of the bulk ceramic. Negative pressures are estimated for these films that agree with the tensile stresses measured by profilemetry. This means that Raman spectroscopy appears to be an adequate alternative technique to study stresses in ferroelectric thin films.

3.3.2. Selection of Parameters for the Optimization of the Ferroelectric Properties

We have described above the influence of the conditioning treatments on the performances of the thin films. Improvements can also be obtained acting during the sol-gel processing of the films, by controlling their composition, microstructure, texture, and strain state. Thus, concentration, viscosity, and stoichiometry of the precursor solutions, and thermal treatment conditions of crystallization of the initially amorphous layer are the major parameters that have to be controlled to obtain films with appropriate ferroelectric properties. As will be described, thickness and stresses of the films, which are related each other, are also dependent on the preparation conditions. Our group [119], following a feedback methodology, have prepared thin films by a sol-gel method [35] testing different PbO excesses in the precursor solution and different heating rates of crystallization. By this way, the best performances of the films were obtained. A more detailed description of this has been given in the first part of this review. In the following, the ferroelectric properties of the improved PTCa films as well as the discussion of these properties made by the authors will be described.

Figure 23 shows the $P-E$ hysteresis loop traced in a PTCa film at 1000 Hz, by means of a modified Sawyer–Tower circuit. This loop has been corrected by compensating the contribution of the ohmic and the capacitive currents. The $j-E$ density currents of the films are depicted in the same figure. A moderated contribution of leakages is observed on both curves that could be due to remaining charge defects in the films. Once again, calculations lead to an overestimated P_r value as compared with the value obtained from the switching current curves, using the alternative method of pulses and Eq. (1).

Results of fatigue and retention of polarization are shown in Figure 24. This figure also shows switching times. The fatigue curve shows a decrease of the initial polarization of about 50% of its initial value, after 5×10^{11} cycles (write and read). This is

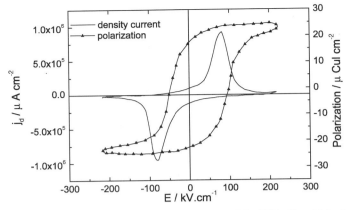

Fig. 23. $P-E$ hysteresis loop and $j-E$ curve traced for PTCa film obtained by optimizing the affecting parameters. Reprinted with permission from [119], copyright 1998, Elsevier Science.

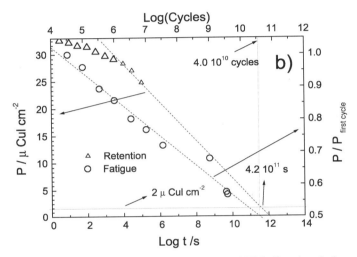

Fig. 24. Fatigue and retention behavior of optimised PTCa films deposited on platinized silicon substrates. Values extrapolated are included. Reprinted with permission from [119], copyright 1998, Elsevier Science.

a promising result for this type of films on Pt electrodes. The endurance of polarization to aging or retention is also quite good, since δP remains over 2 $\mu C/cm^2$ after $\sim 10^{11}$ s (extrapolated value). This limit is considered valid for memory operation. Both results are much better than those obtained for films with the same composition but prepared under other conditions. The pyroelectric coefficient was also measured in these films, obtaining a value as high as $\gamma \sim 3 \times 10^{-8}$ $C\,cm^{-2}\,K^{-1}$. This value is even higher than that reported for bulk ceramics [5]. This is an important finding that shows the potential of these films for IR detectors. The good switching properties measured also show the applicability of these films in memory applications.

3.3.3. Effect of the Substrate

The fabrication of uncooled pyroelectric infrared detectors for thermal imaging and intruder alarms, among other devices, is, at present, highly interesting. Ferroelectric thin films deposited on silicon could compete with semiconductor devices that are more expensive and that have serious difficulties of integration (see Whatmore [129] and Coutures et al. [130]). For these applications, it is recommended to fabricate ferroelectric thin films highly oriented along the polar axis or with any other orientation which provides a large component of the spontaneous polarization perpendicular to the film surface. This fact prevents the usual poling and favors the retention of the electric polarization. This has been recently observed in PTCa textured films [131]. As previously described, other substrates that promote a favorable film texture can also be used instead of silicon, such as MgO single crystals [22].

Nowadays, technologies of pyroelectric devices for infrared detection and thermal imaging face three main issues, in order to improve the performance of the device: (1) type of material, (2) feasibility of integration with silicon readout circuits, and (3) assembly process. Concerning material, a figure of merit is defined as

$$F_M = \gamma / C^E (\varepsilon' \cdot \varepsilon_0 \cdot \tan \delta) \qquad (4)$$

where C^E is the specific heat, ε' is the relative permittivity, ε_0 is the permittivity of free space, and γ and $\tan \delta$ are the pyroelectric coefficient and the tangent of loss angle, respectively. The value of F_M must be as high as possible. This value is used to compare different pyroelectric materials. A control of composition, structure, microstructure, and texture of the film is needed to optimize, conveniently, this response.

Pure lead titanate was considered one of the best candidates for pyroelectric sensors, since it has a large spontaneous polarization, P_s, and a relatively low dielectric constant, ε. But, its large coercive field, E_c, makes it unsuitable for its use at modern microelectronic technologies that require materials that can work with low voltages (3 V or even less). Two ways to get appropriate ferroelectric films can be proposed: to lower the coercive field and thus, reduce the poling field, or to grow highly oriented films along the polar axis. The way to decrease the E_c of pure PbTiO₃ is the partial substitution of Pb²⁺ by another cation, such as Ca²⁺, which causes a considerable reduction of E_c, but maintains high values of P_s and γ. Furthermore, oriented films in the polar direction can be obtained if the PTCa film is deposited onto certain substrates, such as MgO.

For growing lead-based perovskite films, different deposition techniques such as rf-sputtering, CVD, laser ablation, MOCVD, and sol-gel have been reported (see the review paper of Okuyama and Hamakawa [132]). Among these techniques, which are undergoing continuing improvement, rf-sputtering [133] and sol-gel [8, 134] are, at present, the most extensively used, at least such that conformal deposits would not be required. Physical deposition techniques seem to be more appropriate for obtaining oriented or epitaxial thin films [1]. Films of PbZr$_x$Ti$_{1-x}$O₃, PbTiO₃, (Pb,La)TiO₃, and (Pb,Ca)TiO₃ on (111)Pt/(100)Si substrates have also been obtained by sol-gel with ⟨111⟩ and ⟨100⟩ preferred orientations [135–137]. However, the most appropriate orientation of these films for pyroelectric devices is the ⟨001⟩ preferred orientation.

As mentioned before, it is generally assumed that the main condition for obtaining oriented films along the polar direction is a good lattice match between substrate and film. On this basis, epitaxial lead-based perovskite films have been prepared on (100)MgO and (100)SrTiO₃ by rf-sputtering and laser ablation [133, 138, 139]. PTCa films have also been deposited by rf-sputtering on (100)Pt/(100)MgO, with a strong ⟨001⟩ preferred orientation [30]. Nevertheless, well-oriented films of LiNbO₃ onto sapphire have been reported by Joshi et al. [73], despite the lattice mismatch between the substrate and the film being larger than 7%. The possibilities of the formation of favorable textures have been described in detail in the first part of this work. But, mechanisms controlling orientation in this type of films are still waiting for an explanation.

The authors [140] have prepared sol-gel PTCa films on (111)Pt/(100)MgO and have crystallized them by RTP. Films with a low porosity and a grain size of ~100 nm were obtained. These films have a c-axis orientation degree of $\alpha = 0.82$, calculated with the expression of Kim and Baif [126]. Curvature radii of the surfaces of the (111)Pt/(100)MgO substrate and of the deposited films were measured following the experimental strategy described for the study of strains of the PTCa onto Si substrates [141]. The stress generated in this film throughout the whole preparation process was calculated using Eq. (2). For the MgO substrate, values of $E_s = 135 \times 10^9$ Pa, $v_s = 0.34$, and $t_s = 10^{-3}$ m were used. A compressive stress of -1300 MPa was calculated.

For the films on MgO, the sought objectives, according to the authors, were the preparation of PTCa films with a high degree of orientation along the c-axis, for pyroelectric applications. But, a complete characterization was also carried out and the result concerning fatigue is noteworthy. Fatigue is not good enough, since remanent polarization decreases about 50% after applying 10^7 cycles at 43 kHz. Despite this, the polarization value is almost recovered after an electric refreshment. This could be an indication of a degradation of the film so soft that its response can be easily repaired with the refreshment. This seems to be a practical finding for applications of these films. But, the most important results of these films are their pyroelectric response. In fact, these oriented films show spontaneous pyroelectric activity, since values between 1.5×10^{-8} and 2.5×10^{-8} C cm^{-2} K^{-1} were measured prior to poling. Figure 25 shows how the spontaneous pyroelectric coefficient remains without change versus time. These values can be increased by an electric poling, alternatively along each of both senses, but, the pyroelectric coefficient decreases to a steady value of $\sim 1.5 \times 10^{-8}$ C cm^{-2} K^{-1}, after 3×10^3 s. The steady value is close to the net spontaneous pyroelectric coefficient. This time evolution can also be observed in Figure 25. The figure of merit, F_M, for this film is 3.0×10^{-6} (N m^{-2})$^{-0.5}$. The large net pyroelectric coefficient of this film is correlated with its high c-axis orientation perpendicular to the surface of the film. This coefficient is of the same order as that reported by other authors for PbTiO₃, La-modified PbTiO₃, and Ca-modified PbTiO₃ on Pt-coated MgO(100) substrates [24, 30, 142], but deposited by rf-sputtering and with a Pt highly

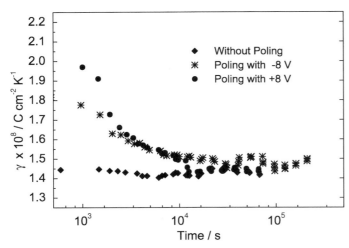

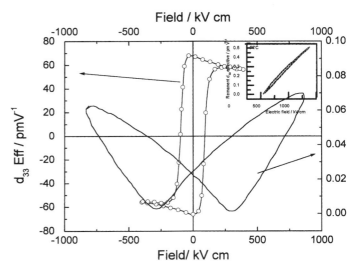

Fig. 25. Time evolution of the pyroelectric coefficient, γ, for the bare sample and after the electrical poling with both signal sense. The steady value coincides after 10^5 seconds. Reprinted with permission from [63], copyright 1999, Elsevier Science.

Fig. 26. Piezoelectric hysteresis loop, $d_{33}-E$ and strain-electric field hysteresis loop (butterfly shape) are shown for a PTCa film. Good linearity strain dc-field is shown in the inset. Reprinted with permission from [145], copyright 1996, American Institute of Physics.

oriented in the $\langle 100 \rangle$ direction, instead of the (111)Pt used in this work. This fact reinforces the suspicion that to promote a heterogeneous nucleation in the film, the lattice matching between the film and the under-film layer (Pt in this case) is not an exclusive condition. Some other reasons have to be considered. In this way, the role of stresses should be important, assuming the strong compliance anisotropy of PTCa, since Quo et al. [142] reported that $S_{33}^E \gg S_{11}^E$. Tuttle et al. [65] also reported the importance of stresses for the orientation of PZT films. Furthermore, the negligible aging is attributed to the absence of externally oriented 90° domains which are usually responsible for this phenomenon. This is clearly observed on the decay of the samples polarized, where 90° domains are surely switched.

In [140], the authors attempted to explain the c-axis orientation of the PTCa films on Pt/MgO based on the controlling of stresses generated during film processing. According to Kweon et al. [143], besides the strains described by Foster et al. [123], a fourth type of stress, an intrinsic stress, should also be considered in CSD films. It is the effect of wetting and it has already been described for Si substrates. To minimize this stress, two thin layers of PTCa solution were deposited on the Pt/MgO substrate, allowing each layer to reduce its stress during crystallization and, probably facilitating the promotion of a preferred orientation following a mechanism close to that described by Miller and Lange [143]. These authors state that the strongest contribution to the total stress of these PTCa films should come from the thermal strain. A compressive stress of 700 MPa can be calculated using published data on thermal expansion of PTCa and MgO and the equation given by Foster et al. [123]:

$$\sigma_{th} = E_s\big[\alpha(\text{parael.}) - \alpha_s\big][T_{process} - T_{transition}] \\ + E_s\big[\alpha(\text{ferroel.}) - \alpha_s\big][T_{transition} - T_{room\ temp.}] \quad (5)$$

This stress would induce a $\langle 001 \rangle$ preferred orientation in the film according to Tuttle et al. [65].

3.4. Piezoelectric Measurements

Micromechanical applications require knowledge of the piezoelectric properties of the material that cannot be measured on thin films by the usual resonance method employed for single crystals or bulk ceramics. Piezoelectric characterization of PTCa thin films has been carried out by Kholkin et al. [145] using a sensitive double-beam laser interferometer. This method was developed by Kholkin et al. [146] and it measures the electrical induced strains. From them, an effective longitudinal piezoelectric coefficient $(d_{33})_{eff}$ can be deduced.

Figure 26 shows the piezoelectric hysteresis loop (d_{33} versus dc applied electric field, E) of a PTCa film on Pt/TiO$_2$/SiO$_2$/(100)Si substrate. It exhibits a change of sign of d_{33}, as a consequence of the polarization switching. Values larger than ± 70 pm V^{-1} are obtained. They are similar to those of bulk ceramics with the same composition [147, 148]. The stability of the piezoelectric response after removing the dc electric field indicates that these films are very promising materials for their use in piezoelectric sensors and actuators. A typical butterfly curve (strain versus dc field) is shown in the same figure. This curve has been traced under a bipolar driving condition. It shows the switching of ferroelectric domains, again. The inset of the figure also shows the strain versus electric field under a unipolar driving condition. It exhibits a good linearity very useful for micropositioning devices, since a minimum hysteresis is an important requirement. The electric field-induced strain, close to 0.8% for 1 MV/cm of the applied electric field, is larger than that reported for PZT thin films [149].

4. CONCLUSIONS AND TRENDS

1. Sol-gel Ca-modified lead titanate thin films were prepared onto platinized substrates. These films have a good

performance for switching applications (NVFERAM). They have a moderate fatigue and a good value of retention. Their piezoelectric response is also appropriate for piezoelectric applications.

2. The large preferred orientation of the PTCa films onto Pt-coated MgO substrates makes them useful for the fabrication of IR detectors without the necessity of a previous poling with an electric field. These films maintain their properties for an infinite time.

3. The piezoelectric properties of the films make possible their use as substrates for the propagation on their surfaces of elastic waves (SAW). These waves are produced by an inter-digital electrode (IDE) deposited on the piezoelectric film and collected by a second IDE, where an electrical signal is produced. This response is used in many devices for selection of radiofrequency. For this, the ferroelectric film has to have as large a piezoelectric effect as that of the lead titanate films modified with a low amount of Ca or La (Ca < 30% or La < 15%). Furthermore, when Pb is substituted by a large amount of Ca or La (Ca > 40% or La > 15%), the films have a large permittivity, close to 1000, that makes them useful for dinamic random access memories (DRAM).

Acknowledgments

This work has been supported by the Spanish CICYT Projects MAT95-0110 and MAT98-1068. The authors gratefully acknowledge Dr. C. Alemany, Dr. R. Jiménez, Dr. R. Sirera, Dr. P. Ramos, and Dr. M. J.Martín for their help in the experimental work and in the discussion of results.

REFERENCES

1. O. Auciello and R. Ramesh, *MRS Bull.* 21(6), 21 (1996).
2. B. Jaffe, R. S. Roth, and S. Marzullo, *J. Res. Nat. Bur. Standar.* 55[5], 2626 (1955).
3. S. L. Swartz and V. E. Wood, *Condensed Matter News* 1(5), 1108 (1985).
4. C. A. Paz de Araujo, J. D. Cuchiaro, L. D. McMillan, M. C. Scott, and J. F. Scott, *Nature* 374, 627 (1995).
5. J. Mendiola, B. Jiménez, C. Alemany, L. Pardo, and L. del Olmo, *Ferroelectrics* 94, 183 (1989).
6. M. L. Calzada, R. Sirera, J. Ricote, and L. Pardo, *J. Mater. Chem.* 8, 111 (1998).
7. R. Jiménez, M. L. Calzada, and J. Mendiola, Proceedings of the Eleventh IEEE International Symposium on Applications of Ferroelectrics (E. Colla, D. Damjanovic, and N. Setter, Eds.), p. 155, 1998.
8. R. Sirera and M. L. Calzada, *Mater. Res. Bull.* 30(1), 11 (1995).
9. K. Ijima, R. Takayama, Y. Tomita, and I. Ueda, *J. Appl. Phys.* 60(8), 2914 (1986).
10. Y. M. Kang, J. K. Ku, and S. Baik, *J. Appl. Phys.* 78(4), 2601 (1995).
11. Y. Shimizu, K. R. Udayakumar, and L. E. Cross, *J. Am. Ceram. Soc.* 74(12), 3023 (1991).
12. R. Jiménez, M. L. Calzada, C. Alemany, P. Ramos, and J. Mendiola, *Bol. Soc. Esp. Cerám. Vidrio* 38, 477 (1999).
13. A. Ariizumi, K. Kawamura, I. Kikuchi, and I. Kato, *Japan. J. Appl. Phys.* 24, Suppl. 24-3, 7 (1985).
14. S. G. Yoon and H. G. Kim, *IEEE Trans. Ultrasonics Ferroelectrics Freq. Control* 37(5) 333 (1990).
15. K. Kushida and H. Takeuchi, *Ferroelectrics* 108, 3 (1990).
16. C. C. Li and S. B. Desu, *Mat. Res. Soc. Symp. Proc.* 243, 387 (1992).
17. N. J. Phillips, M. L. Calzada, and S. J. Milne, *J. Non-Cryst. Solids* 147–148, 285 (1992).
18. K. Y. Kim, H. I. Hwang, J. Y. Lee, and W. K. Choo, *Mat. Res. Soc. Symp. Proc.* 243, 197 (1992).
19. M. Okuyama, J. Asano, T. Imai, D. Lee, and Y. Hamakawa, *Japan. J. Appl. Phys.* 32, Part I, No 9b, 4107 (1993).
20. M. L. Calzada, F. Carmona, R. Sirera, and B. Jiménez, *Ferroelectrics* 152, 19 (1994).
21. B. Pachaly, R. Bruchhaus, D. Pitzer, H. Huber, W. Wersing, and F. Koch, *Integrated Ferroelectrics* 5, 333 (1994)
22. M. Okuyama and Y. Hamakawa, *Integrated Ferroelectrics* 118, 261 (1991).
23. J. R. Choi, D. H. Lee, and S. M. Choo, *Integrated Ferroelectrics* 5, 119 (1994).
24. J. R. Choi, D. H. Lee, H. J. Nam, S. M. Choo, J. H. Lee, and K. Y. Kim, *Integrated Ferroelectrics* 6, 241 (1995).
25. C. W. Wang, Y. C. Chen, M. S. Lee, C. C. Chiu, and Y. T. Huang, *Japan. J. Appl. Phys.* 38, 5A, 2831 (1999).
26. L. Pardo, J. Ricote, M. Algueró, and M. L. Calzada, in "Handbook of Low and High Dielectric Constant Materials and their Applications. Materials and Processing" (H. S. Nalwa, Ed.). Academic Press, New York, 1999.
27. L. Zhen, A. M. Grishin, and K. V. Rao, *J. Phys. IV France* 8, Pr9-143 (1998).
28. L. Zheng, P. Yang, W. Xu, C. Lin, W. Wu, and M. Okuyama, *Integrated Ferroelectrics* 20, 73 (1998).
29. O. Auciello, J. F. Scot, and R. Ramesh, *Phys. Today* 50[7], 22 (1998).
30. E. Yamaka, H. Watanabe, H. Kimura, H. Kanaya, and H. Ohkuma, *J. Vac. Sci. Technol. A* 6(5), 2921 (1988).
31. H. Maiwa and N. Ichinose, *Japan. J. Appl. Phys.* 36, Part I, No. 9B, 5825 (1997).
32. J. Mendiola, M. J. Martin, P. Ramos, and C. Zaldo, *Microelect. Eng.* 29, 209 (1995).
33. A. Tsuzuki, H. Murakami, K Kani, K. Watari, and Y. Torii, *J. Mat. Sci. Lett.* 10, 125 (1991).
34. Y. Torii, A. Tsuzuki, and H. Murakami, *J. Mat. Sci. Lett.* 13, 1364 (1994).
35. R. Sirera, M. L. Calzada, F. Carmona, and B. Jiménez, *J. Mat. Sci. Lett.* 13, 1804 (1994).
36. J. Shyu and K. Mo, *J. Mat. Sci. Lett.* 15, 620 (1996).
37. G. Teowee, K. C. McCarthy, T. P. Alexander, T. J. Bukowski, and D. R. Uhlmann, "Proceedings of the Tenth IEEE International Symposium on Applications of Ferroelectrics—ISAF'96," Piscataway, NJ (B. M. Kyiwicki, A. Amnand and A. Safari, Eds.), p. 487, 1996.
38. S. Chewasatn and S. J. Milne, *J. Mat. Sci.* 32, 575 (1997).
39. A. Seifert, P. Muralt, and N. Setter, *Appl. Phys. Lett.* 72(19), 2409 (1998).
40. A. Seifert, L. Sagalowicz, P. Muralt, and N. Setter, *J. Mat. Res.* 14[5], 2012 (1999).
41. C. L. Fan and W. Huebner, "Proceedings of the Ninth International Symposium on Applications of Ferroelectrics" (R. K. Pandeny, M. Li, and A. Safari, Eds.), p. 512. Univ. Park, PA, 1994.
42. D. S. Paik, A. V. Prasadarao, and S. Somarneni, *Mater. Lett.* 32, 97 (1997).
43. A. Tsuzuki, H. Murakami, K. Kani, K. Watari, and Y. Torii, *J. Mater. Sci. Lett.* 10, 125 (1991).
44. M. J. Martín, Ph.D. Thesis. Univ. Autónoma de Madrid, Spain, 1996.
45. K. D. Budd, S. K. Dey, and D. A. Payne, *Brit. Ceram. Proc.* 36, 107 (1985).
46. N. J. Phillips and S. J. Milne, *J. Mater. Chem.* 1(5), 893 (1995).
47. D. C. Bradley, R. C. Mehrotra, and D. P. Gaur, in "Metal Alkoxides," p. 183. Academic Press, London, 1978.
48. M. L. Calzada and R. Sirera, *J. Mater. Electr.* 7, 39 (1996).
49. C. J. Brinker and G. W. Scherer, in "Sol-Gel Science. The Physics and Chemistry of Sol-Gel Processing," p. 649. Academic Press, London, 1990.
50. G. W. Scherer, *J. Sol-Gel Sci. Techn.* 8, 353 (1997).
51. R. Sirera, Ph.D. Thesis. Univ. Autónoma de Madrid, Spain, 1997.

52. M. L. Calzada, M. J. Martín, P. Ramos, J. Mendiola, R. Sirera, M. F. da Silva, and J. C. Soares, *J. Phys. Chem. Solids* 58(7), 1033 (1997).

53. D. M. Puri, K. C. Pando, and R. C. Mehrotra, *J. Less-Common Met.* 4, 393 (1962).

54. D. M. Puri and R. C. Mehrotra, *Indian J. Chem.* 5, 448 (1967).

55. M. L. Calzada, R. Sirera, F. Carmona, and B. Jiménez, *J. Am. Ceram. Soc.* 78(7), 1802 (1995).

56. C. J. Brinker, R. J. Kirkpatrick, D. R. Tallant, B. C. Bunker, and B. Montez, *J. Non-Cryst. Solids* 99, 418 (1999).

57. M. L. Calzada, B. Malic, R. Sirera, and M. Kosec, *J. Sol Gel Sci. Techn.* (in press).

58. P. R. Coffman, C. K. Barlingay, A. Gupta, and S. K. Dey, *J. Sol-Gel Sci. Techn.* 6, 83 (1996).

59. B. Malic, M. Kosec, K. Smolej, and S. Stavber, *J. Eur. Ceram. Soc.* 19, 1334 (1999).

60. H. Asada, M. Udaka, and H. Kawano, *Thin Solid Films* 252, 49 (1994).

61. K. Nakamoto, in "Infrared and Raman Spectra of Inorganic and Coordination Compounds," p. 252. Wiley, New York, 1986.

62. J. E. Macintyre, Ed., "Dictionary of Inorganic Compounds," Chapman & Hall, London, 1992.

63. R. Jiménez, M. L. Calzada, and J. Mendiola, *Thin Solid Films* 348, 253 (1999).

64. A. González, R. Poyato, R. Jiménez, J. Mendiola, L. Pardo, and M. L. Calzada, *Surf. Interf. Anal.* 29, 325 (2000).

65. B. A. Tuttle, T. J. Garino, J. A. Voigt, T. J. Headley, D. Dimos, and M. O. Eatough, in "Science and Technology of Electroceramic Thin Films" (O. Auciello and R. Waser, Eds.), p. 117. Kluwer Academic, Netherlands, 1995.

66. T. Tani, Z. Xu, and D. A. Payne, *Mater. Res. Soc. Symp. Proc.* 310, 269 (1993).

67. D. Leinen, E. Rodríguez-Castellón, R. Sirera, and M. L. Calzada, *Surf. Interf. Anal.* 29, 612 (2000).

68. G. Guzmán, P. Barboux, J. Livage, and J. Perriere, *J. Sol-Gel Sci. Tech.* 2, 69 (1994).

69. C. D. E. Lakeman, D. J. Guistolise, T. Tany, and D. A. Payne, *Br. Ceram. Proc.* 52, 69 (1994).

70. K. D. Klissurska, T. Maeder, K. G. Brooks, and N. Setter, *Microelect. Eng.* 29, 297 (1995).

71. A. L. Kholkin and N. Setter, *Appl. Phys. Lett.* 71(19), 2854 (1997).

72. R. W. Schwartz, J. A. Voigt, B. A. Tuttle, D. A. Payne, T. L. Reichert, and R. S. DaSalla, *J. Mater. Res.* 12(2), 444 (1997).

73. V. Joshi, D. Roy, and M. L. Mecartney, in "Ferroelectric Thin Films III" (E. R. Myers, B. A. Tuttle, S. B. Desu, and P. K. Larsen, Eds.), p. 287. Pittsburgh, PA, 1993.

74. K. Kushida, K. R. Udayakumar, S. B. Krupanidhi, and L. E. Cross, *J. Am. Ceram. Soc.* 76(5), 1345 (1993).

75. T. Tani and D. A. Payne, *Br. Ceram. Proc.* 52, 87 (1994).

76. O. Auciello, H. N. Al-Shareef, K. D. Gifford, D. J. Lichtenwalner, R. Dat, K. R. Bellur, A. I. Kingon, and R. Ramesh, *Mater. Res. Soc. Symp. Proc.* 341, 341 (1994).

77. K. Aoaki, Y. Fukuda, K. Numata, and A. Nishimura, *Japan. J. Appl. Phys. Part 1* 34, 192 (1995).

78. P. Muralt, T. Maeder, L. Sagalowicz, S. Hiboux, S. Scalese, D. Naumovic, R. G. Agostino, N. Xauthopoulos, H. J. Mathieu, L. Patthey, and E. L. Bullock, *J. Appl. Phys.* 83(7), 3835 (1998).

79. S. Y. Chen and I. W. Chen, *J. Am. Ceram. Soc.* 77(9), 2332 (1994).

80. J. Ricote, D. Chateigner, L. Pardo, M. Algueró, J. Mendiola, and M. L. Calzada, *Ferroelectrics* 241, 167 (2000).

81. J. L. Keddie, P. V. Braun, and E. P. Giannelis, *J. Am. Ceram. Soc.* 77(6), 1592 (1994).

82. S. A. Mansour, D. A. Binford, and R. W. Vest, *Integrated Ferroelectrics* 1, 43 (1992).

83. H. Hu, L. Shi, V. Kumar, and S. B. Krupanidhi, *Ceram. Trans. Ferroelectric Films* 25, 113 (1992).

84. J. Mendiola, M. L. Calzada, R. Sirera, and P. Ramos, "Proceedings of the 4th International Conference on Electronic Ceramics & Applica-

tions" (R. Waser, S. Hoffmann, D. Bonnenberg, and Ch. Hoffmann, Eds.), Vol. 1, p. 327. Aachen, Germany, 1994.

85. M. J. Martín, M. L. Calzada, J. Mendiola, M. F. da Silva, and J. C. Soares, *J. Sol-Gel Sci. Techn.* 13, 843 (1998).

86. D. A. Shirley, *Phys. Rev. B* 5, 4709 (1998).

87. Physical Electronics, 6509 Flying Cloud Drive, Eden Prairie, MN 55344.

88. C. D. E. Lakeman and D. A. Payne, *J. Am. Ceram. Soc.* 75(11) 3091 (1992).

89. O. Auciello and R. Ramesh, *MRS Bull.* 21(7), 29 (1996).

90. D. Leinen, R. Sirera, E. Rodríguez-Castellón, and M. L. Calzada, *Thin Solid Films* 354, 66 (1999).

91. L. R. Doolittle, *Nucl. Inst. Methods* B9, 334 (1984).

92. P. K. Larsen, R. Cuppens, and G. A. C. M. Spierings, *Ferroelectrics* 128, 265 (1992).

93. J. Chen, K. R. Udayakumar, K. G. Brooks, and L. E. Cross, *J. Appl. Phys.* 71(9), 4465 (1992).

94. D. Leinen, A. Fernández, J. P. Espinós, and A. González-Elipe, *Appl. Phys.* A63, 237 (1996).

95. J. F. Moulder, W. F. Stickle, P. E. Sobol, and K. D. Bomben, in "Handbook of X-ray Photoelectron Spectroscopy" (J. Chastras, Ed.). PHI, Eden Prairie, MN, 1979.

96. S. Bernacki, L. Jack, Y. Kisler, S. Kollins, S. D. Berstein, R. Hallock, B. Armstrong, J. Shaw, J. Evans, B. Tuttle, B. Hammetter, S. Roger, J. Henderson, J. Benedetto, R. Moore, C. R. Pugh, and A. Fennelly, *Integrated Ferroelectrics* 3, 97 (1993).

97. L. Pardo, J. Mendiola, and C. Alemany, *J. Appl. Phys.* 64, 5092 (1988).

98. C. Alemany, R. Jiménez, J. Revilla, J. Mendiola, and M. L. Calzada, *J. Phys. D: Appl. Phys.* 32, L79 (1999).

99. G. G. Stoney, *Proc. R. Soc. London, Ser. A* 82, 172 (1909).

100. P. Ramos, J. Mendiola, F. Carmona, M. L. Calzada, and C. Alemany, *Phys. Stat. Sol. (A)* 156, 119 (1996).

101. W. Bennett, P. S. Brody, B. J. Rod, L. P. Cook, and P. K. Schenick, *Integrated Ferroelectrics* 4, 191 (1993).

102. L. Shi, S. B. Krupanidhi, and G. H. Haertling, *Integrated Ferroelectrics* 1, 111 (1992).

103. P. K. Larsen, G. J. M. Dormans, D. J. Taylor, and P. J. Van Velhoven, *J. Appl. Phys.* 76, 2405 (1994).

104. R. D. Pugh, M. J. Sabochick, and T. E. Luke, *J. Appl. Phys.* 72, 1049 (1992).

105. A. Sheikholeslamani and P. G. Gulak, *Microelectronic Eng.* 29, 141 (1995).

106. R. D. Nasby, J. R. Swank, M. S. Rodgers, and S. L. Miller, *Integrated Ferroelectrics* 2, 91 (1992).

107. M. Klee, A. De Veirman, D. J. Taylor, and P. K. Larsen, *Integrated Ferroelectrics* 4, 197 (1994).

108. I. K. Yoo and S. B. Desu, *Phys. Stat. Sol. (A)* 133, 565 (1992).

109. P. Ramos, J. Mendiola, F. Carmona, M. L. Calzada, and C. Alemany, *Phys. Stat. Sol. (A)* 56, 119 (1996).

110. K. Carl and K. H. Hädtl, *Ferroelectrics* 17, 473 (1978).

111. P. K. Larsen, G. L. M. Kampschoer, M. B. Van der Mark, and M. Klee, "International Symposium on Applications of Ferroelectrics, ISAF'92" (M. Liu, A. Safari, A. Kingon, and G. Haerthings, Eds.), p. 217. Greenville, USA, 1992.

112. L. F. Francis and D. A. Payne. *J. Am. Ceram. Soc.* 74(2), 3000 (1991).

113. G. A. C. M. Spierings, G. J. M. Dormans, W. G. J. Moors, M. J. E. Ulenaers, and P. K. Larsen, *J. Appl. Phys.* 78, 1926 (1995).

114. K. Nashimoto, D. K. Fox, and G. B. Anderson, *Appl. Phys. Lett.* 66(7), 822 (1995).

115. R. Ramesh, H. Gilchrist, T. Sands, V. G. Keramidas, R. Haakenasen, and D. K. Fork, *Appl. Phys. Lett.* 63, 3592 (1993).

116. R. Jiménez, M. L. Calzada, and J. Mendiola, *Thin Solid Films* 335[1–2], 292 (1998).

117. J. Mendiola, P. Ramos, and M. L. Calzada. *J. Phys. Chem. Solids* 59 [9], 1571 (1998).

118. R. Waser, in "Science and Technology of Electronics," p. 223. 1995.

119. M. L. Calzada, R. Jiménez, P. Ramos, M. J. Martin, and J. Mendiola, *J. Phys. IV* 8, 9 (1998).

120. J. Mendiola, M. L. Calzada, P. Ramos, M. J. Martin, and F. Agullo-Rueda, *Thin Solid Films* 315, 195 (1998).

121. C. Hon-Ming and J. Ya-Min Lee, *J. Appl. Phys.* 82(7), 3478 (1987).

122. H. N. AlShareef and D. Dimos, *J. Am. Ceram. Soc.* 80(12), 3127 (1997).

123. C. M. Foster, Z. Li, M. Bucckett, D. Miller, P. M. Baldo, L. E. Rehn, G. R. Bai, D. Guo, H. You, and L. Merkle, *J. Appl. Phys.* 78(4), 2607 (1995).

124. C. C. Li and S. B. Desu, *J. Vac. Sci. Technol. A* 14(1), 1 (1996).

125. J. A. Lewis, K. A. Blackman, A. L. Odgen, J. A. Payne, and L. F. Francis, *J. Am. Ceram. Soc.* 79(12), 3225 (1996).

126. S. Kim and S. Baik, *Thin Solid Films* 266, 205 (1995).

127. B. Jiménez, J. Mendiola, C. Alemany, L. Del Olmo, L. Pardo, E. Maurer, J. De Frutos, A. M. Gonzalez, and M. C. Fandiño. *Ferroelectrics* 87, 97 (1988).

128. J. A. Sanjurjo, E. Lopez-Cruz, and G. Burns, *Phys. Rev. B* 28, 7260 (1983).

129. R. W. Whatmore, *Ferroelectrics* 118, 241 (1991).

130. J. L. Coutures, J. Leconte, and G. Boucharlat, *J. Phys. IV France* 8, 9 (1998).

131. R. Jiménez, R. Poyato, L. Pardo, C. Alemany, and J. Mendiola, *Bol. Soc. Esp. Ceram. Vidr.* 38[5], 482 (1999).

132. M. Okuyama and Y. Hamakawa, *Ferroelectrics* 118, 261 (1991).

133. S. Kim and S. Kaik, *Thin Solid Films* 266, 205 (1995).

134. Y. Huang, M. Daglish, I. Reaney, and A. Bell, *Third Euro-Ceramic* 2, 699 (1993).

135. S. Y. Chen and I. W. Chen, *J. Am. Ceram. Soc.* 81(1), 97 (1998).

136. J. Moon, J. A. Kerchner, J. LeBleu, A. A. Morrone, and J. H. Adair, *J. Am. Ceram. Soc.* 80(10), 2613 (1997).

137. M. Alguero, M. L. Calzada, and L. Pardo, "Proceedings of the Tenth IEEE," p. 797. East Brunswick, NJ, 1996.

138. A. Masuda, Y. Yamanaka, M. Tazoe, Y. Yonezawa, A. Morimoto, and T. Shimizu, *Japan. J. Appl. Phys.* 34, 5154 (1995).

139. C. B. Eom, R. B. Dover, J. M. Phillips, R. M. Fleming, R. J. Cava, J. H. Marshall, D. J. Werder, C. H. Chen, and D. K. Fork, *Mat. Res. Soc. Symp.* 310, 145 (1993).

140. R. Jiménez, M. L. Calzada, and J. Mendiola, *Thin Solid Films* 338, 1 (1999).

141. K. Iijima, Y. Tomita, R. Takayama, and I. Veda, *J. Appl. Phys.* 60, 361 (1986).

142. A. B. Qu, D. Kong, W. Zhong, P. Zhang, and Z. Wang, *Ferroelectrics* 145, 39 (1993).

143. S. Y. Kweon, S. H. Yi, and S. K. Choi, *J. Vac. Sci. Technol. A* 15(1), 57 (1997).

144. K. T. Miller and F. F. Lange, *J. Mater. Res.* 6(11), 2387 (1991).

145. A. L. Kholkin, M. L. Calzada, P. Ramos, J. Mendiola, and N. Setter, *Appl. Phys. Lett.* 69, 3602 (1996).

146. A. L. Kolkin, Ch. Wüthrich, D. V. Taylor, and N. Setter, *Rev. Sci. Instrum.* 67, 1935 (1996).

147. Y. Yamashita, K. Yokoyama, H. Honda, and T. Takahashi, *J. Appl. Phys.* 20, 183 (1981).

148. J. de Frutos and B. Jiménez, *Ferroelectrics* 126, 341 (1992).

149. J. F. Li, D. Viehland, L. Tani, C. D. E. Lakeman, and D. A. Payne, *J. Appl. Phys.* 75, 1435 (1995).

Chapter 7

POINT DEFECTS IN THIN INSULATING FILMS OF LITHIUM FLUORIDE FOR OPTICAL MICROSYSTEMS

Rosa Maria Montereali

ENEA C.R. Frascati, Applied Physics Division, 00044 Frascati (RM), Italy

Contents

Handbook of Thin Film Materials, edited by H.S. Nalwa
Volume 3: Ferroelectric and Dielectric Thin Films
Copyright © 2002 by Academic Press

ISBN 0-12-512911-4/$35.00

1. PREFACE

Point defects (impurity ions, color centers, etc.) in insulating crystals represent one of the most investigated fields in solid-state physics connected with optical properties [1]. Alkali halides (AH) containing color centers (CCs) are among the best-known materials in the form of bulk crystals, pure and doped [2]. Their spectroscopic properties have been deeply investigated from the fundamental point of view as prototypes of other kinds of defects in more complex insulating materials. They have found successful applications as suitable active media for the development of optically pumped tunable solid-state lasers operating in the visible and in the near infrared (NIR) [3, 4].

On the other hand, the area of growth and characterization of thin films has seen a considerable expansion in the last decades, due to the increased interest in device miniaturization realized by the successful progress obtained in microelectronics, and, more recently, in the field of integrated optics [5].

On the contrary, research activities concerning CCs in AH films started recently. The first application as media for frequency domain optical storage was proposed [6] by hole-burning spectroscopy of aggregate CCs in polycrystalline doped lithium fluoride (LiF) films.

In recent years, interest has been growing for novel materials exhibiting active optical properties. The opportunities to realize new light-emitting materials by multilayered structures of colored AH films simultaneously active in wide spectral regions extending from the visible to the NIR have also been investigated [7] to obtain predesigned emission properties.

However, the most promising results have been obtained in the generation [8], amplification [9], and waveguiding [10] of visible light in LiF films, where the efficient formation of stable laser active CCs by low-energy electron-beam irradiation induces also a local increase of the refractive index. So through a single-step process, it is possible to create an active waveguiding region [11]. The use of electron-beam lithography (EBL) techniques allows the definition of colored channel waveguides at the surface of LiF crystals and films and opens interesting perspectives for the realization of broadband optical amplifiers and waveguide tunable lasers.

The realization of integrated optical devices is nowadays an object of increasing interest. The peculiar properties of LiF and the use of film growth methods combined with well-developed lithography techniques will favor future opportunities of integration in more complex miniaturized systems, as well as the design and realization of new device configurations. Recently, the fabrication of red-emitting planar microcavities based on low-energy electron-beam irradiated LiF films evaporated on Bragg reflectors has been achieved [12].

A survey of the current state of the art in the investigation of point defects in LiF thin films is presented with particular emphasis on their optical behavior and opportunities of exploitation in novel optical miniaturized systems. The exciting perspectives opened by the most recent results offered by advanced optical microscopy techniques for controlling and monitoring their formation at mesoscopic scale will be highlighted.

Table I. Fundamental Parameters of Low-Refractive-Index Fluoride Crystals (LiF, NaF, MgF_2) as Reported in the Literature

	LiF	NaF	MgF_2
Nearest neighbor distance (Å)	2.013	2.3166	—
Melting point (°C)	848.2[a]	996[a]	1263[a]
Density ρ (g/cm^3)	2.640[a]	2.78[a]	3.148[a]
Molecular weight	25.939[a]	41.988[a]	62.302[a]
Refractive index n at 640 nm and RT	1.3912[b]	1.32460[b]	$n_o = 1.37688$[b] $n_e = 1.38865$[b]
Solubility (g/100 g of water at 25°C)	0.134[a]	3.97[a]	0.013[a]
Thermal expansion (10^{-6}/°C)	37[c]	36[c]	18.8 parallel to c-axis 13.1 perp. to c-axis[c]
Thermal conductivity 10^{-4} cal/(cm s °C) at 41°C	270[d]	220[d]	360[d]
Transmission range (μm)	0.12–7[c]	0.13–12[c]	0.11–8[c]

[a]Ref. [14].

[b]Ref. [15].

[c]Ref. [16].

[d]Ref. [17].

2. LITHIUM FLUORIDE: MATERIAL PROPERTIES

Among AH crystals, LiF stands apart as far as its physical and optical properties are concerned. In spite of its typical ionic structure, the crystal is relatively hard and resistant to moisture. The cation–anion distance is the shortest, and the Li$^+$ ion, as well as the F$^-$ one, has the smallest radius among the alkali and halide ions, 0.06 and 0.136 nm, respectively. Its melting point is lower than the expected values for the series of alkali fluorides.

LiF possesses the largest band gap, greater than 14 eV [13], of any solid, with the possible exception of exotic materials such as solid rare gases. It is optically transparent from $\cong$120 nm to 7 μm, and for this reason it is a widely used window material, in particular in the ultraviolet (UV). In Table I some fundamental parameters of LiF [14], including the optical constants [15], are reported in comparison with those of another face-centered cubic (fcc) fluoride crystal, sodium fluoride, NaF [16], belonging to the same AH family, and of an alkaline earth fluoride crystal of lower refractive index, magnesium fluoride, MgF_2 [17].

3. COLOR CENTERS IN LITHIUM FLUORIDE CRYSTALS

The investigation of lattice defect properties in AH includes both small concentration of interstitial or substitutional dopants

Table II. Values of the Parameters of Optical Absorption and Emission Bands

Center	E_a (eV) λ_a (nm)	HW_a (eV)	E_e (eV) λ_e (nm)	HW_e (eV)
F	5.00[a] 248	0.76[a]		
F_2	2.79[b] 444	0.16[b]	1.83[b] 678	0.36[b]
F_3^+	2.77[b] 448	0.29[b]	2.29[b] 541	0.31[b]
F_2^+	1.92[c] 645	0.433[c]	1.36[c] 910	0.29[c]
F_2^-	1.29[c] 960	0.21[c]	1.11[c] 1120	0.17[c]
F_3 (R_1)	1.92[d] 316	0.52[e]		
F_3 (R_2)	3.31[d] 374	0.66[e]		
F_3^- (R_1^-)	1.88[d] 660			
F_3^- (R_2^-)	1.55[d] 800	0.190[d]	1.38[d] 900	0.331[d]
F_4 (N_1)	2.40[f] 517	0.21[e]		
F_4 (N_2)	2.26[f] 547	0.22[e]		
F_4-like	1.91[g] 650	0.19[g]	1.69[g] 735	0.16[g]

Note: (E = peak position, W = width at half maximum) at RT of F and F-aggregate Centers in LiF as derived from the literature.

[a] Ref. [20].

[b] Ref. [22].

[c] Ref. [3].

[d] Ref. [25].

[e] Ref. [26].

[f] Ref. [28].

[g] Ref. [30].

and intrinsic localized electronic defects. As far as the last ones are concerned, contrary to the other AH crystals, the usual techniques of additive coloration [4] are not effective in their formation in LiF, and it can be colored only by irradiation with ionizing radiation, as X-rays, γ-rays, elementary particles, and ions. This sensitivity has been exploited in radiation dosimetry [18] mainly in doped materials.

Table II reports a list of the main defects found in the literature for colored LiF crystals together with the typical values of the peak position and half-width of their absorption and emission bands at room temperature (RT). One of the crucial problems concerning irradiated LiF crystals is the coexistence of several kinds of aggregate defects with often overlapping absorption bands, which makes it difficult to clearly individuate

and measure the actual contribution due to the individual centers.

The F center, consisting of an anion vacancy occupied by an electron, is the simplest one and plays a crucial role as a primary defect that can aggregate to form complexes [19]. Its absorption is located in the near-UV, at around 248 nm [20]. It is worthwhile to note that, although the investigation started long ago [21], up to now no luminescence originating unambiguously from the F center in LiF has been detected [22].

More complex F-aggregates, such as F_2 and F_3^+ centers (two electrons bound to two and three anion vacancies, respectively), possess almost overlapping optical absorption bands, around 450 nm, generally called M band [23], and a spread in the values of their main parameters can be found in the literature due to their superposition. An accurate determination of the spectroscopic parameters of their absorption and emission bands is reported in [24] also at liquid nitrogen temperature (LNT). Their distinct emissions are located in the red and in the green spectral range [23].

The 315- and 375-nm peaks [25] are due to the R_1 and R_2 transitions of the F_3 centers, consisting of three F centers along the (111) plane. Their half-widths, estimated in [26], are comparable among them, as expected.

The same applies for the 515- and 545-nm absorption bands [27], attributed to the N_1 and N_2 transitions of the F_4 defects (four associated F centers).

The half-width values of absorption bands associated to more complex defects, such as larger F-aggregates, are sometimes absent in the literature. This is because their formation is not negligible only at high doses of irradiation. The exceptions are due to the F_2^+ (ionized F_2), F_3^-, and F_2^- centers (F_2 and F_3 defects capturing an electron, respectively), described in detail in [28] and utilized as laser active media [3, 29] in the NIR region.

Recently [30], a luminescence band centered at 735 nm, observed under excitation at around 650 nm, has been detected in colored LiF crystal and tentatively attributed to an aggregate of four F centers, that is, an F_4-like defect. However, the possible model of an F_2 center stabilized by a monovalent cation, called F_{2A}, has been also evaluated. The presence of impurities in LiF crystals and their influence on the overall spectroscopic features of aggregate defects is a wide and still open field of investigation.

Figure 1 shows the absorption and emission bands, outlined as normalized gaussian curves, of active CCs in nominally pure LiF crystals, whose photoluminescence has been assessed at RT.

4. LASER ACTIVE COLOR CENTERS IN LITHIUM FLUORIDE CRYSTALS

AH crystals containing CCs have been extensively studied for the realization of solid-state vibronic lasers [31] operating in the visible and in the near-infrared spectral ranges, generally

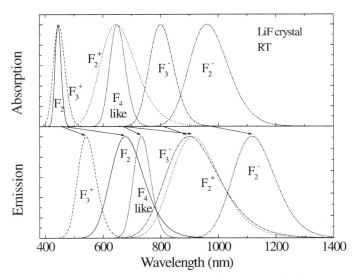

Fig. 1. Absorption and emission bands, outlined as normalized gaussian curves, of the known color centers which possess photoluminescence in LiF at RT.

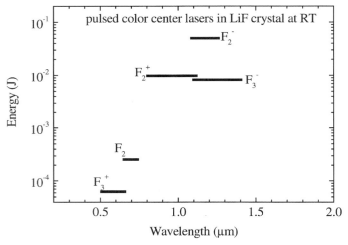

Fig. 2. Average energy and tunability range versus wavelength of the color center lasers operating at RT in LiF crystals in pulsed regime.

Table III. Spectroscopic Properties and Optical Gain Coefficient at RT of Laser Active Color Centers in LiF using $N_{RES} = 10^{17}$ cm^{-3} and $n = 1.39$

CCs	λ_a (nm)	λ_e (nm)	$\Delta\nu$ (10^{13} Hz)	τ_{eff} (ns)	η (%)	g (cm^{-1})
F_3^+	448	541	6.5	11.5	100	7.6
F_2	444	678	7.5	17	100	7.0
F_3^-	800	900	8.0	10	10	1.9
F_2^+	645	910	7.0	15	50	7.6
F_2^-	960	1120	8.0	55	30	1.6

and F_3^+ CCs at RT with a single pumping wavelength at 450 nm, which is located at the middle of their almost overlapping absorption bands.

A sun-color solid-state laser with LiF:F_2^- crystal as active element has been reported [33] to provide simultaneous superbroadband laser radiation in the 1.1–1.25-μm spectral interval in the NIR and oscillation in the visible spectral range, from 0.55 to 0.62 μm, by second-harmonic generation in a nonlinear crystal at the same time.

4.1. Optical Gain Coefficients

The main parameter describing a laser medium is the optical gain G [4], defined as the amplification factor of a light wave of intensity I_{in} which passes through an active medium of gain length l:

$$G = \frac{I_{out}}{I_{in}} = \exp(g - \alpha)l = \exp(\gamma l) \qquad (1)$$

In Eq. (1) it is assumed that the net gain coefficient γ (cm^{-1}) is independent of the position along the medium in the direction of amplification, and it is expressed as the difference between the gain coefficient g, dependent on the luminescence process, and the losses α, due both to the optical quality of the active media as well as to other processes, such as reabsorption. In a homogeneous medium containing CCs, g is due only to the interaction between the defect and the lattice and is described by the same shape of the emission band. The gain coefficient [34] is

$$g(\nu_e) = N_{RES}\sigma_e = N_{RES}\frac{\lambda_e^2 \cdot \eta}{8\pi \cdot n^2 \cdot 1.064 \cdot \tau_{eff} \cdot \Delta\nu} \qquad (2)$$

where N_{RES} is the density of centers in the upper level of the laser transition and σ_e is the emission cross section. For a gaussian band it can be expressed as in Eq. (2), where λ_e and ν_e are the resonance wavelength and frequency, respectively, $\Delta\nu$ is the width at half maximum of the emission band, n is the refractive index of the medium at the excitation wavelength, τ_{eff} is the luminescence decay time, which could be shorter than the luminescence lifetime τ due to the existence of nonradiative processes, and $\eta = \tau_{eff}/\tau$ is the luminescence quantum efficiency.

Table III summarizes the emission properties of laser active CCs in LiF at RT, together with the values of g calculated with Eq. (2), using $N_{RES} = 10^{17}$ cm^{-3} and $n = 1.39$.

at LNT. Their characteristic four-level scheme makes it possible to obtain population inversion even with low pumping power. Moreover, their peculiar properties, such as spectral pure emissions, wide continuous tunable bands, very short pulsed regimes, and reasonably high powers, have been exploited in applied and fundamental research [4]. Among AH crystals, LiF is of particular interest because it can host point defects which are laser active even at RT [3, 29]. Color center lasers (CCLs) based on LiF crystals have been operating at RT in the visible and in the near-infrared for many years. Figure 2 shows the tunability and the energy of LiF crystal-based CCLs as a function of the emission wavelength.

Simultaneous laser emission in the green and in the red spectral range [32] has been obtained in LiF crystals containing F_2

5. LITHIUM FLUORIDE FILMS

Dielectric films have been studied for decades, the main interest being in their optical and electrical applications. Due to their transparency, the first application was in interference multilayer components. The application of LiF films was restricted to antireflecting coatings because of its low index of refraction, and, for this purpose, it has been recently substituted by other fluoride materials.

A first systematic study of AH films was performed in the early 1960s, when films of various materials grown on substrates at RT were analyzed by X-ray diffraction, dielectric loss, and electron diffraction techniques [35].

Recently, renewed attention has been paid to fluoride films for their potential applications in optics [36], due to their low refractive index, in electronics, as insulators on semiconductors [37], and in integrated optics, for the realization of both passive and active waveguiding devices [38]. Moreover, very thin LiF interlayers between the low work function metal cathode and the electron-transport/emitter layer in organic light-emitting diodes (OLED) result in improved device performance [39].

In the last years, the integration of electronic and optical components on the same substrates has been an area of intense research, and a large effort has been directed toward the development of devices compatible with the silicon technologies. Recently, LiF films have been proposed as nuclear sensors [40] for neutrons, grown on silicon, as well as novel photocathode materials [41], grown on germanium, while LiF films treated by low-energy electrons could offer interesting opportunities as miniaturized light sources on Si [42] and on SiO$_2$ on silicon substrates [43].

The present trend toward device miniaturization calls for an exhaustive investigation of the properties of point defects in LiF films.

6. COLOR CENTER FORMATION BY LOW-PENETRATING PARTICLES

Irradiation with low-penetrating particles, such as low-energy electrons or ion beams, involves only thin surface layers, with thickness of the order of several micrometers or less, of the bombarded material.

Ion implantation gives rise along the particle tracks to an energy transfer which is not uniform, being strongly dependent on the velocity of the incident ions [25]. Instead, energy dissipation of low-energy electrons is almost constant up to the end of the particle tracks [44].

Both these approaches are extensively employed for materials modifications and investigations, including the case of LiF [26, 27, 45]. They involve the simultaneous generation of isolated F centers and of more complex defects formed by F-aggregation, such as F$_2$ and F$_3$ or those formed by their ionization. However, low-energy electrons have the peculiarity of interacting mainly with the electrons in the solid; on the contrary, ions can substantially modify the host matrix by atom displacement, depending on their mass and energy, and other phenomena, apart from point defects generation, take place and become predominant [46].

For this reason, and in order to limit the current review to optical properties of point defects in LiF films and their exploitation in optical microsystems, we will treat in detail only this type of low-penetrating ionizing radiation.

7. COLORATION OF LiF BY LOW-ENERGY ELECTRON BEAMS

Low-energy electron-beam irradiation produces different types of CCs located at the surface of LiF, in the form of crystal [47] and film [8].

The electron penetration depth in AH is determined by the beam energy according to the following semiempiric relation [48]:

$$d = \frac{0.064}{\rho} E^{1.68} \tag{3}$$

where d is the maximum electron penetration depth in μm, ρ is the compound density in g/cm^3, and E is the electron-beam energy in keV. In accordance with the density value reported in Table I, in LiF it ranges from 0.1 to 3.7 μm for electron energies between 2 and 20 keV.

A typical RT absorption spectrum of a LiF crystal irradiated at RT by a 12-keV electron beam with a dose of 8×10^{19} eV/cm^2 is reported in Figure 3. The most conspicuous bands are found at 248 nm (attributed to F centers) and at 450 nm (due to the unresolved F$_3^+$ and F$_2$ bands). Other minor contributions due to the absorption bands of more complex aggregate defects can be detected (see Table II).

The features of the spectrum, including the intensity ratios of its various components, reproduce quite well those observed

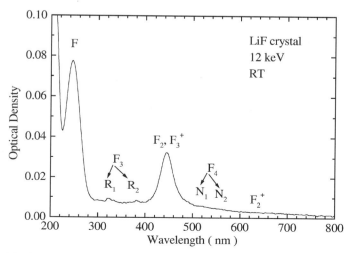

Fig. 3. RT optical absorption spectrum of a LiF crystal irradiated by a 12-keV electron beam at RT.

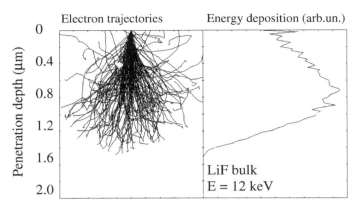

Fig. 4. Monte Carlo simulation for LiF irradiated by a 12-keV electron beam: the figure at left shows a plot of the electron trajectories; the one at right provides a plot of the electron depth dose.

Table IV. Maximum Electron Penetration Depth in LiF Calculated by Semi-empiric Relation (3) and Values Obtained by Monte Carlo Simulations at Several Electron Energies in the Range 2–20 keV

E (keV)	d (μm) $\propto E^{1.68}$	d (μm) Monte Carlo
3	0.15	0.13
12	1.58	1.50
13	1.81	1.72
15	2.30	2.18
16	2.56	2.50
18	3.12	3.03
19	3.42	3.37
20	3.73	3.68

for thin layers of LiF irradiated with 5-keV electrons [49], with 3-keV electrons [50], and with secondary electrons emitted by a microwave discharge [51]. The limited penetration of the used charged particles favors the aggregation of F_2 and F_3^+ CCs, and an F-to-M band intensity ratio of about 3 is obtained. This value is higher by more than one order of magnitude than the ratio observed in LiF crystals in the case of high-penetrating radiation, where the M band is much weaker than the F one, and this peculiarity will be discussed in the next section.

The rate of energy dissipation as a function of penetration for low energy electrons in AH has been investigated by several authors [27]. The electron energy distribution along the penetration depth is not exactly constant, and it ends with a smooth tail into the material. Monte Carlo-based simulation software [52] can be used to determine the electron trajectories and their depth dose, and to verify Eq. (3). In Figure 4 the simulation results for a LiF bulk irradiated by 12-keV electrons have been reported. In Table IV the maximum penetration depth calculated by Eq. (3) and those obtained by Monte Carlo simulations have been listed for LiF at several electron energies in the investigated range, and the agreement in the estimated values is quite satisfactory.

8. KINETICS OF LOW-ENERGY ELECTRON-INDUCED COLOR CENTER FORMATION

Because of its relevance to both the understanding of the fundamental properties of condensed matter and the development of materials and devices suitable for technological application, the production of CCs in AH treated by ionizing radiation has been deeply investigated [1, 2, 53] both experimentally and theoretically. Several models of the mechanisms which lead to the formation of F and F_2 defects in the lattice of AH crystals have been proposed and tested by comparison with the great amount of data collected in the measurements.

It is well known that the F-coloring process consists of three different stages, occurring at different dose intervals. In the initial stage I, observed at low doses, a very efficient increase of the F-center concentration with irradiation takes place. The fast saturation of this growth leads to a substantially constant density of defects, typical of stage II. As the doses increase, a further steep increase in their amount corresponds to stage III, which generally ends when there is a final saturation at higher doses.

Among the many possible channels for the energy transfer from the incoming beam to the ionic material, the halide ionization is the relevant one for CC generation by low-energy electron beams (<50 keV) [54]. The halide ionization is followed by the formation of a V_k center, that is, an X_2^- molecule, where X is a halide atom, which forms a self-trapped exciton (STE) by capturing an electron in an excited state. A primary mechanism of creation of anionic Frenkel pairs via nonradiative exciton decay is generally accepted to explain the F-center generation. Actually only 1/100 of the F-H pairs survives [55], since there is a great probability of mutual annihilation due to the high mobility of the H centers. On the other hand, trapping the mobile H center in dislocations and/or impurities initially present in the crystal, prevents the destruction of the F center and allows the fast growth of the F concentration in the first stage of coloration. After a very detailed study of the effects of X-rays in LiF [56], a model based on competitive recombination of interstitial halogen ions with vacancies and traps, consisting in aggregate of interstitials, was proposed [57] to account for a few features of the F-coloring curve. A unified model describing all stages of the coloration kinetics has been presented [58] and later improved [59]. Competition between recombination with F centers and capture by empty traps and clusters is again postulated, but with two new assumptions: a heterogeneous nucleation of interstitial clusters and the peculiar instability of a short-lived aggregate, formed by two trapped interstitials. The predictions of this most comprehensive theoretical model have found a satisfactory agreement with the experimental results regarding the F-center kinetics in LiF crystals irradiated by 3-keV electrons [50]. The detailed investigation of the coloring process has been carried out on the basis of optical absorption measurements and was restricted to the F and M bands. The dependence of the absorption spectra of LiF crystals

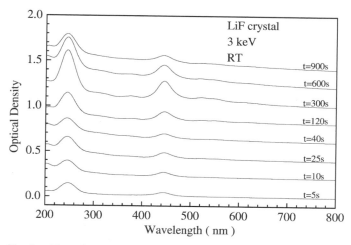

Fig. 5. Absorption spectra of LiF crystals colored at RT by a continuous 3-keV electron beam at different irradiation times with a current density of 1000 μA/cm^2. Each curve is shifted upwards by a constant value with respect to the previous one for the sake of clarity.

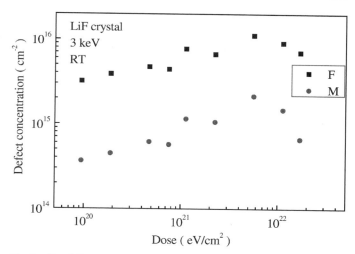

Fig. 6. F and F$_2$ centers concentration as a function of the absorbed dose in LiF crystals colored at RT by a continuous 3-keV electron beam with a current density of 1000 μA/cm^2.

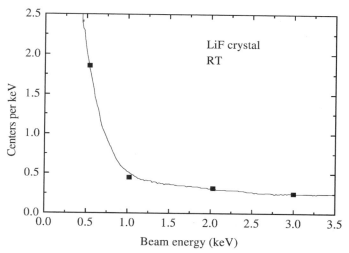

Fig. 7. Number of F centers produced in LiF crystal per energy unit as a function of the electron-beam energy, according with [50, 60].

irradiated at RT by continuous 3-keV electrons on the irradiation times is reported in Figure 5. Only the F and M bands are detected at the beginning, while at higher doses the growth of the two main absorption bands is accompanied by the appearance of other features due to more complex aggregate defects.

From the absorption band associated with a given type of defects, it is possible to calculate the center concentration N by using the Dexter modification of the Smakula formula [2]:

$$Nf = 0.87 \times 10^{17} \left[n/(n^2 + 2)^2 \right] \alpha W \qquad (4)$$

where f is the oscillator strength, n is the refractive index of the host material, W is the absorption band full-width at half maximum, and α is the absorption coefficient at the band peak.

From such spectra, the dose dependence of the concentrations of F and F$_2$ defects has been obtained by the Smakula formula, assuming for the F center an f value of 0.56 [60] and a reasonable oscillator strength of 0.28 for the F$_2$ centers [1] (such a parameter being unknown for the F$_3^+$ defects). The results are presented in a log–log plot in Figure 6 in the case of a current density of 1000 μA/cm^2. The estimated electron penetration is 0.15 μm.

The F-coloring curve exhibits the three stages of behavior. The slope evaluation for the two linear rises of the F-coloring curve during stages I and III allows one to determine the coloration yield m in the formula (5):

$$N = k(dose)^m \qquad (5)$$

where N is the concentration of F centers and k is a proportionality constant. The values of m, rising from 0.3 to 0.6 with increasing current for stage I, are quite similar to those found in LiF for other low-penetrating radiation, such as α-particles, protons, and deuterons [25, 45], which give a mean value of 0.5. Instead, the average value found at different currents, for the slope of the linear zone of the F-coloring curve during stage III, is 0.77, which is in excellent agreement with the calculation

performed in [59] on the basis of a theoretical model, which looks suitable to describe in general the behavior of AH under irradiation.

An evident saturation at high doses is reached in the center production for doses higher than 10^{26} eV/cm^3. The defect concentrations no longer increase and eventually begin to decrease due to the formation of metal colloids by F-center aggregation.

The mean energy needed to produce an F center, which is an estimate of the coloration efficiency, calculated for stage I, has been found to be 3 keV, a value in very good agreement with the data reported in Figure 7 for LiF crystals colored by low-energy electrons [50, 60].

The curve related to the M absorption in Figure 6 shows an analogous three-stage behavior, with an appreciable shift toward higher doses of the separation between different stages. No attempt has generally been made to separate the contributions of the overlapping absorptions due to F$_2$ and F$_3^+$ defects to the M band in low-energy electron-irradiated LiF crystals, with the exception of [24], confirming the well-known preferred formation of F_2 with respect to the F_3^+ ones when the irradiation is performed at RT [61]. The direct aggregation between two V_k centers and their stabilization by the capture of two electrons, together with the simultaneous aggregation of two F centers to produce the F_2 defects, at the high density of electronic excitation reached in a limited spatial region in the case of low-penetrating electrons, have been suggested in [51] as possible explanation for the higher M-to-F observed ratio.

9. REFRACTIVE INDEX MODIFICATION INDUCED BY COLOR CENTERS IN LiF

Besides the efficient defects formation, the electron bombardment of LiF by a beam of energy in the range of a few kiloelectron volts induces a modification of the refractive index of the irradiated layer. The ellipsometric measurements performed on a LiF crystal, heavily colored by a 3-keV electron beam, show an increase of about 4% of the real part of the refractive index in the same spectral range where the F_2 and F_3^+ emissions are located [62], and so the colored layer can act as a guiding region with respect to the uncolored one.

A simple dipole–field interaction model [26] has been recently developed to describe the modification of the complex refractive index induced by CC formation in a host dielectric medium. It has been applied with success to the case of thin colored layers produced by low-energy electron-beam irradiation at the surface of LiF crystals. Starting from a near-normal incidence transmittance measurement, the model can quantitatively evaluate the single contributions that arise from different kinds of electronic defects to the overall refractive index change of the irradiated medium, and predicts an increase in the real part of the refractive index in the colored LiF layer in the visible range. It was assumed that CCs in LiF can be considered as elementary two-level quantum systems, and that the Hamiltonian of the interaction between these systems and the impinging electromagnetic field is of the dipole type, as suggested by both classic and quantum theory. The wavelength-dependent dielectric constant of LiF is modified by the presence of absorbing point defects because they contribute to the dielectric susceptibility of the material in which they are embedded, and depends on three parameters only: the peak energy of the absorption bands, their half-width, and their amplitude. As a consequence, the colored layer can be characterized only by a best fit of a single spectral transmittance measurement. The accuracy of this characterization technique has been successfully checked by comparing the measured complex refractive index of a large surface-colored area of a LiF crystal heavily irradiated by 3-keV electrons with that measured by means of another quite standard optical characterization, that is, a combination of spectrophotometry and ellipsometry [26].

Figures 8 and 9 show the refractive index and extinction coefficient dispersion curves, as calculated from the transmittance

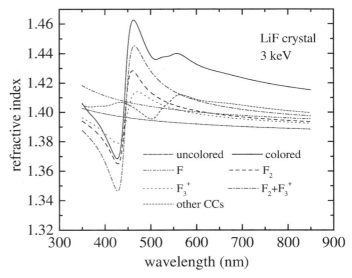

Fig. 8. Refractive index of a 3-keV heavily colored LiF crystal as deduced by the best fit of the spectral transmittance measurements. The partial contributions of several types of color centers are evidenced. The refractive index of uncolored LiF is also reported for comparison.

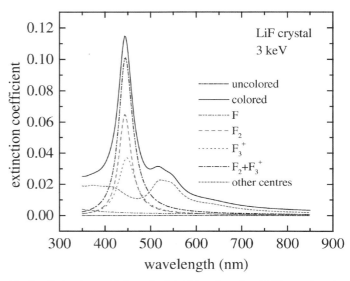

Fig. 9. Extinction coefficient of a 3-keV heavily colored LiF crystal as deduced by the best fit of the spectral transmittance measurements. The partial contributions of several types of color centers are evidenced. The extinction coefficient of uncolored LiF is also reported for comparison.

best-fit parameters, together with the contributions due to different types of defects. At the used dose, the total increase of the real part of the refractive index amounts to 4% with respect to the uncolored value in the same spectral range where the F_2 and F_3^+ visible emissions are located. It is shown in Figure 8 that the main contribution to the refractive index change at longer wavelength is due to the F centers, which absorb light in the UV range. Its amount is comparable with those due to the presence of strong almost-overlapping F_2 and F_3^+ absorption bands, peaked at the wavelengths of 444 and 448 nm, respectively. The measured conspicuous modification is not surprising, if one estimates the concentration of the produced defects, in the range between 10^{20} and 10^{21} centers/cm^3, with respect to 6.1×10^{22} halogen sites/cm^3. As a matter of fact, at the used dose, even the influence of more complex larger aggregate defects is not negligible. Their consistent contribution to the modification of the real part of the refractive index is reported in Figure 8; however, Figure 9 shows that these defects increase considerably the extinction coefficient in the red and green spectral region, where the F_2 and F_3^+ emissions are located, introducing also significant optical losses.

10. WHAT ABOUT "THIN FILMS"?

In order to fully understand the peculiar properties of point defects in LiF films, the differences between thin layers containing a high concentration of CCs induced by low-penetrating ionizing particles into AH single crystals and the same kind of defects host in a polycrystalline matrix directly grown as a thin film by several deposition methods on different types of substrates should be taken into account for a proper understanding of the word "film."

We will generally refer to the second condition, which implies a good knowledge of the used deposition processes and an accurate investigation of the structural, morphological, and optical properties of the as-grown material.

However, the coloration method selected for the film activation by CC formation plays a fundamental role in the determination and control of optical properties of the irradiated films. Apart from the nature and the energy of the impinging particles, the main parameters influencing defects formation and stabilization are the total dose and the dose rate, as well as the sample temperature during irradiation and successive thermo-optical treatments.

11. GROWTH OF LITHIUM FLUORIDE FILMS

LiF films are generally grown by physical vapor deposition (PVD) methods. Thermal evaporation [63] as well as electron-beam assisted PVD [64] have been widely utilized for the growth of alkali fluorides films on different types of substrates, amorphous and crystalline, in a wide range of thickness extending from a few nanometers to several microns.

As a matter of fact, the ionic nature of AH seems not easily compatible with more sophisticated film deposition processes, although the NaF film growth by chemical vapor deposition (CVD) [65] and recently the LiF film production by pulsed laser deposition (PLD) [66] have been achieved.

The structural, morphological, and optical properties of the LiF films are strongly dependent on the main deposition parameters, that is, the substrate temperature T_s during the growth and the total thickness t, as well as on the deposition rate, which plays a fundamental role in the growth process. However, other aspects, such as the evaporation geometry, can be relevant in the determination of the main characteristics of the as-grown LiF films.

11.1. LiF Films Grown on Amorphous Substrates

The growth of LiF films on amorphous transparent substrates, such as glass and fused silica, has been reported by many authors [36, 63, 67]. The films are generally polycrystalline and their structural, morphological, and optical properties are strongly dependent on the deposition parameters, and in particular the influence of T_s on the features of films of different thickness has been investigated.

Spectrophotometric measurements performed to compute the complex refractive index of LiF films thermally evaporated on glass [63] show significant differences in the optical properties of the deposited films, depending on the growth conditions, especially T_s and t. The film complex refractive index was computed from transmittance T and reflectance R measurements at quasi-normal incidence. The thickness t and the complex refractive index $n - ik$ of the film were the unknown quantities [68]. The characterization was carried out over the wavelength range 350–2500 nm, in order to refine the thickness and to obtain information about the film model (homogeneous or inhomogeneous) [69]. Independent mechanical measurements of the LiF film thickness have been performed by stylus profilometer and showed uniform film thickness with sharp edges and smooth surfaces, which were almost the replicas of the substrate ones.

In Figure 10 the mean refractive indices of the LiF films are reported as a function of substrate temperature for different film thickness intervals. Up to 10% differences in the real part of the refractive index value could be obtained by simply varying these deposition parameters. In all the samples, the extinction coefficient, k, was below 10^{-3}, and the real part, n, was always lower than that of the bulk material. Raising the substrate temperature T_s induced an increase of the refractive index n, whereas n was generally lowered by increasing the film thickness, although in this latter case poor film adhesion and fracturing represented the major problem.

By assuming that the void presence reduces the film compactness and consequently n, the relation between a dielectric

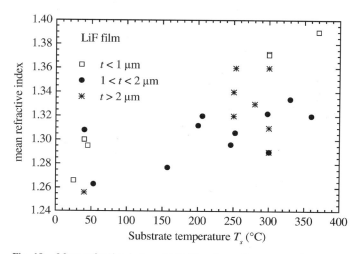

Fig. 10. Mean refractive index of LiF films thermally evaporated on amorphous glass substrates as a function of substrate temperature T_s for different film thickness intervals.

film's refractive index and its density [70] can be estimated in a simplified way by considering the film as an aggregate of material "grains" separated by air interstices. If we define the packing density p as the fraction of total volume of the film occupied by the material, it is possible to compute the value of p, reported in Table V, which reproduces the measured n values with the effective medium approximation formula (EMA) [71]. Assuming for n and ρ of LiF bulk the values reported in Table I,

Table V shows the thickness and the refractive index $n - ik$, deduced from T and R spectra, of two typical couples of LiF films of different thickness, as grown on amorphous substrates at two different T_s, and the values obtained for their mean density in the EMA approximation.

As a matter of fact, the refractive index is strongly related to the density of the deposited film, which is in turn dependent on the film structure. Higher substrate temperatures increase the mobility of the film molecules and thus favor the formation of more tightly packed microcrystals. The recording of direct diffraction pole figures on LiF films grown by thermal evaporation on glass substrate at normal incidence show that the LiF crystallites present a single crystal texture, despite their polycrystalline structure, with an average grain size from 100 to 150 nm at lower temperature and from 200 to 250 nm for higher deposition temperatures, as obtained from the SEM micrographs shown in Figure 11. The LiF films are formed by an ensemble of crystallites, all having the same orientation with respect to the substrate plane, and show different crystallite textures depending on T_s during evaporation [63]. It was observed [72] that for LiF films thermally evaporated on glass

Table V. Refractive Indices at 632.8 nm of Two Couples of LiF Films of Two Different Thicknesses Thermally Evaporated on Glass Substrates at Two Growth Temperatures, Together with the Values of Packing Density p and Mean Density, ρ_{EMA}, Calculated by EMA Approximation

T_s (°C)	t (μm)	n	k	p (%)	ρ_{EMA} (g/cm^3)
30	1.120	1.307 ± 0.002	$(5 \pm 1) \times 10^{-4}$	79	2.072
300	1.055	1.372 ± 0.002	$(5 \pm 1) \times 10^{-4}$	95	2.510
30	1.690	1.263 ± 0.002	$(6 \pm 1) \times 10^{-4}$	67	1.775
300	1.735	1.320 ± 0.002	$(5 \pm 1) \times 10^{-4}$	82	2.159

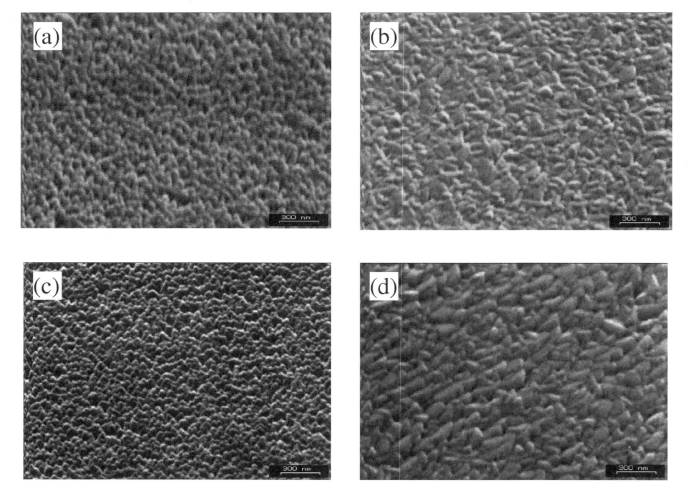

Fig. 11. SEM micrographs of LiF films 1 μm thick thermally evaporated on glass substrates at (a) $T_s = 30°C$, (b) $T_s = 300°C$, and of two LiF films of thickness 1.75 μm grown at (c) $T_s = 30°C$, and (d) $T_s = 300°C$.

substrates, the higher the deposition temperature, the more the $\langle 100 \rangle$ LiF crystallographic direction approached the normal to the substrate plane; that is, a close packed stacking configuration is achieved with a corresponding increase in the film refractive index.

Figure 12 shows the (200)-pole figure recorded on a thin LiF film deposited at low substrate temperature ($T_s = 40°C$, $t = 0.55 \mu m$). The (200)-scattered intensity is well confined in a small region of the pole figure, attesting that the film possesses a nearly single-crystal grain texture. The intensity maximum is located 21° apart from the origin, corresponding to the normal to the substrate plane. It can be deduced that the film is constituted by crystallites all oriented with a $\langle 100 \rangle$ crystallographic direction forming an angle of about 21° with the normal to substrate plane. This angle is reduced down to 10° for high substrate temperature films [72].

The SEM micrographs of Figure 11 show that the high-temperature films retain the low-temperature globular structure, accompanied by grain coarsening. The globular structure suggests that, at all the temperatures, the mobility of the incoming LiF molecules was, in all directions, sufficiently high to avoid the formation of columnar or dendritic structures. The atomic force microscopy (AFM) image of the surface of a LiF film of still higher thickness, $t = 3.3 \mu m$, thermally evaporated on a glass substrate at $T_s = 280°C$, is reported in Figure 13. It is

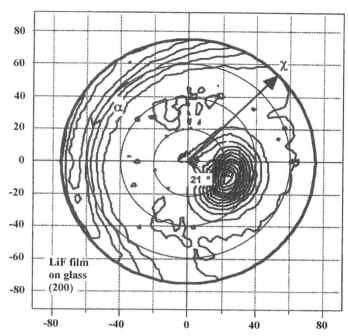

Fig. 12. LiF (200)-pole figure recorded on a LiF film of thickness $t = 0.55 \mu m$ thermally evaporated on a glass substrate at $T_s = 40°C$. Used wavelength: Cu K_α; tilt rotation $\chi \rightarrow 0–75°$; azimuthal rotation $\alpha \rightarrow 0–360°$; $\Delta\alpha, \chi = 5°$; $2\theta = 45°$.

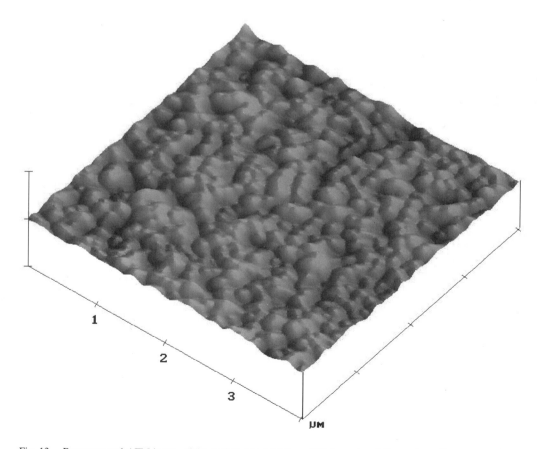

Fig. 13. Reconstructed AFM image of the surface of a LiF film of thickness $t = 3.3 \mu m$ thermally evaporated on a glass substrate at $T_s = 280°C$.

characterized by a low root-mean-square value of the surface roughness, over an area of $4 \times 4\ \mu m^2$, of about 11 nm, which looks like a consequence of the regularity of the grain dimensions.

11.2. LiF Films Grown on Crystalline Substrates

The choice of an oriented substrate plays a dominant role in the determination of the properties of the evaporated LiF film. In Figure 14, the AFM image of the surface of a LiF film thermally evaporated on a polished LiF single-crystal substrate at $T_s = 280°C$ has been reported. The grain dimensions are quite irregular, and the root mean square of the surface roughness over an area of $4 \times 4\ \mu m^2$ is about 47 nm. This picture can be directly compared with that reported in Figure 13, because the samples have been grown in the same evaporation run.

From the absence of an interference fringe pattern in the optical transmission spectrum of the LiF film on the LiF bulk two-layer structure of Figure 14, it could be deduced that the film refractive index was close to the bulk one. Conversely, the LiF film on glass of Figure 13 has a refractive index of 1.34, which, according to EMA approximation, corresponds to a packing density $p \approx 0.87$. Its thickness was 3.3 μm, while the thickness is lower, $t = 3.0\ \mu m$, in the sample grown on LiF substrate. From the thickness ratio, a value of $p \approx 0.91$ is

obtained, assuming the value of p equal to unity for the above-mentioned thinner film grown on bulk on the basis of its bulk-like refractive index. The differences among these packing density values are negligible, taking into account the adopted simplified model. The more packed structure can be ascribed to an epitaxial growth of the film on the LiF (100)-single crystal substrate, as confirmed by the only appearance of an intense (200)-reflection in its XRD Bragg–Brentano pattern, and the perfect superposition of the substrate and film pole figure. As mentioned above, the LiF sample thermally evaporated on glass still presents a single-crystal texture, but with a ⟨100⟩ crystallographic axis tilted 10° with respect to the normal to the substrate plane.

The above examples allow one to better clarify the relevance of the substrate nature to the properties of the deposited films.

For silicon substrates, the textural analysis revealed that also in this case the thermally evaporated LiF films have a "single-crystal" texture, shown in Figure 15, with all the crystallite having the ⟨100⟩ crystallographic direction perfectly parallel to the normal direction to the substrate plane. However, it should be pointed out that the growth of LiF films on silicon cannot be considered epitaxial due to the presence of the thin native oxide layer on the silicon surface.

The ordered growth orientation on silicon substrates is associated to a very uniform and smooth surface with regular grains,

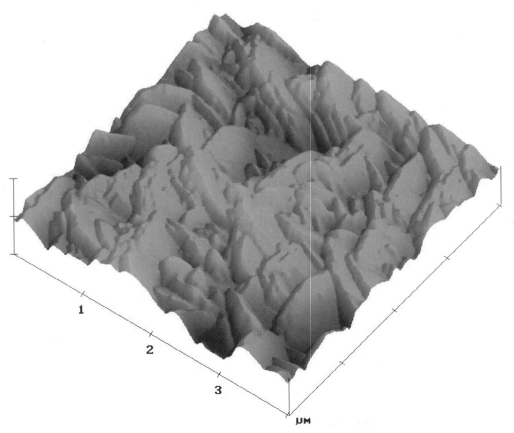

Fig. 14. Reconstructed AFM image of the surface of a LiF film of thickness $t = 3.0\ \mu m$ thermally evaporated on a polished LiF (100) single-crystal substrate at $T_s = 280°C$.

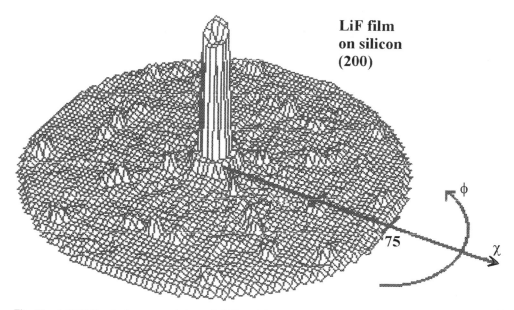

LiF film on silicon (200)

Fig. 15. LiF (200)-pole figure recorded on a LiF film of thickness t = 1.8 μm thermally evaporated on a silicon substrate at T_s = 250°C. Used wavelength: Cu K$_\alpha$; tilt rotation χ → 0–75°; azimuthal rotation α → 0–360°; $\Delta\alpha$, χ = 5°; 2θ = 45°.

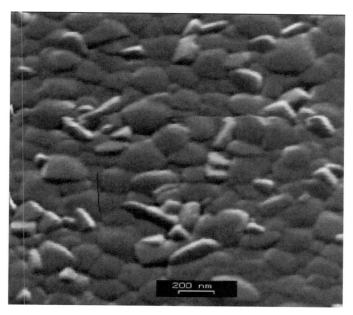

Fig. 16. SEM micrograph of the surface of a LiF film 1.8 μm thick thermally evaporated on a silicon substrate at T_s = 250°C.

whose mean dimensions are enlarged with respect to the films grown in the same evaporation run on a glass substrate. An example is shown for a LiF film of thickness t = 1.8 μm grown on a silicon substrate at T_s = 250°C, whose SEM image is shown in Figure 16. Also in this case, as for glass, raising the substrate temperature causes an enlargement in the dimensions of the grains constituting the films, accompanied by an increase in the surface roughness [73].

11.3. A Special Substrate: NaF Films for Passive Optical Waveguides

Waveguides, by definition, consist of a region of high refractive index surrounded by lower-index material. Due to the low value of the LiF refractive index, 1.3913 at 632.8 nm [15], it is not an easy task to find common substrate materials, which allow the realization of optical waveguides in single-layer structures formed by a LiF film grown on substrates of lower refractive index.

Several two-layer LiF/NaF films, in which the combined films have the highest possible thickness without impairment of their optical characteristics or direct cracking, have been produced by both e-beam-assisted and thermal evaporation techniques. In practice, a LiF film is deposited on a lower-index NaF film, which in turn can be grown on a silica or glass substrate, kept at a constant prefixed temperature T_s. The investigated structures consist of a LiF film, $\cong$1.5 μm thick, on a NaF film, $\cong$2 μm thick, sequentially deposited in the same evaporation run onto a glass substrate kept at fixed temperature.

Guided light propagation of several modes (from 2 to 4, depending on the film refractive indices and thickness) at the wavelength of 632.8 nm has been observed [38] by prism coupling techniques [74] with propagation losses from 6 to 30 dB/cm depending on T_s, which are still high, but quite promising for materials at their infancy in this field of application. The obtained results are summarized in Table VI, for two series of LiF/NaF two-layer systems grown at several T_s and with different thicknesses. The best performances have been obtained at 200°C, despite the results obtained on single-layer LiF deposition, which indicate a still higher packing density for a substrate temperature of about 300°C.

Table VI.　Propagation Losses, Measured at the Wavelength of 632.8 nm, for Two Series of Two-Layer LiF/NaF Structures Grown by e-Beam Evaporation on Glass Substrates at Several T_S and with Different Thickness

Total thickness (μm)	Substrate temperature T_S (°C)	Losses at $\lambda = 632.8$ nm (dB/cm)
3.8	100	6 (4 modes)
3.5	300	30 (4 modes)
3.3 (2.0 NaF + 1.3 LiF)	100	6 (2 modes)
3.1 (1.6 NaF + 1.5 LiF)	200	9 (2 modes)
3.1 (1.6 NaF + 1.5 LiF)	300	30 (2 modes)

Previous studies have evidenced a close relationship between the preferred orientation (texture) of the crystallites and the deposition substrate temperature both for LiF [63, 72] and for NaF [64] materials evaporated onto amorphous substrates. Since the growth conditions have shown a strong influence on the optical properties of thin fluoride films, an accurate study of the structure of the two-layer LiF/NaF system has been performed by X-ray diffraction techniques and has been correlated to their propagation properties. In particular, the preferred orientation of the "isolated" LiF film deposited both onto bare glass substrates and onto NaF buffer layers has been compared. The LiF(111)–NaF(100) pole figure for the two-layer structure grown at $T_s = 200$°C by e-beam evaporation shows an almost "single-crystal" texture of LiF and NaF films, characterized for both materials by a $\langle 100 \rangle$ crystallographic direction close to the normal to the substrate plane. For temperatures higher than 200°C, the X-ray analysis points to a less pronounced order in the NaF films, which causes the absence of a well-defined texture of the LiF guiding film deposited on it. The high losses measured for LiF/NaF structures deposited at 300°C can be, therefore, ascribed to the random film configuration.

The texture analysis has proved that the quality and the waveguiding properties of the LiF/NaF system depend on the structure of the buffer NaF layers, which in turn influences the growth of the guiding LiF films. At deposition temperatures ranging between 100 and 200°C, both the LiF and NaF films grown on glass substrates present a "single-crystal" texture, with a $\langle 100 \rangle$ crystallographic direction almost perpendicular to the substrate plane. In this temperature range, the same structure is conserved by the LiF guiding films deposited onto the NaF buffer layer. It is worth noticing that the two-layer systems grown at $T_s = 300$°C, at which high-quality LiF films on glass substrates have been obtained [63], present a poor LiF texture with a consequent loss of waveguiding capability. This effect could be more likely ascribed to the mismatch of the lattice parameters combined with the difference in the thermal expansion coefficients between LiF and NaF, rather than to the original texture of the thick NaF buffer layer. However, accurate investigations are under way to better clarify the influence of the growth conditions on the film quality [73].

Further improvements in the propagation characteristics of the LiF/NaF planar waveguides are expected by a global optimization of the structure design, taking into account the single-layer thickness and its refractive index, in order to eliminate

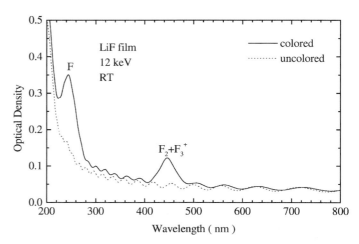

Fig. 17. RT absorption spectra of a LiF film of thickness $t = 1.8$ μm thermally evaporated on a silica substrate at $T_s = 250$°C, inside (solid) and outside (dashed) a colored area irradiated by 12-keV electrons at RT.

the optical losses toward the substrate, which seem to be still present due to the insufficient thickness of the NaF isolating layer, as suggested by the losses measured for samples with a NaF buffer of increasing thickness (see Table VI).

12. OPTICAL ABSORPTION OF COLORED LiF FILMS

Low-energy electron-beam irradiation at RT of LiF films induces the formation of stable primary F centers and several aggregate intrinsic defects, mainly F_2 and F_3^+. Typical absorption spectra, measured at RT in a LiF film of thickness $t = 1.8$ μm thermally evaporated on a silica substrate kept at $T_s = 250$°C during the growth, inside and outside a colored area irradiated at RT by 12-keV electrons, are shown in Figure 17. The absorption bands are superimposed to the interference fringes due to the refractive index differences between the substrate, $n \sim 1.45$, and the film, $n \sim 1.35$, and only the main spectral features can be easily recognized. In particular, it is possible to identify the F and M bands, clearly attributed in the case of low-energy electron-irradiated LiF crystals, as shown in Figure 3.

In the case of a glass substrate, whose transmission is negligible below 300 nm, the F-center absorption cannot be detected. Moreover, the interference fringes contrast is further

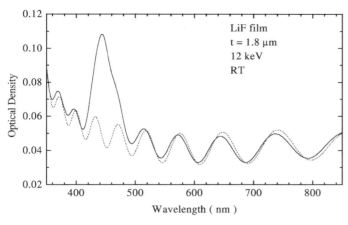

Fig. 18. RT absorption spectra of a LiF film of thickness t = 1.8 μm thermally evaporated on a glass substrate at T_s = 250°C, inside (solid) and outside (dashed) a colored area irradiated by 12-keV electrons at RT.

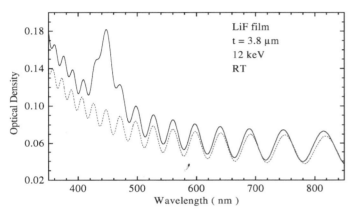

Fig. 19. RT absorption spectra of a LiF film of thickness t = 3.8 μm thermally evaporated on a glass substrate at T_s = 250°C, inside (solid) and outside (dashed) a colored area irradiated by 12-keV electrons at RT.

enhanced with respect to the fused silica substrate, due to the higher difference of glass refractive indices with respect to the LiF one [63]. A typical absorption spectrum, measured at RT on a LiF film grown on a glass substrate with the identical evaporation parameters of Figure 17, colored at RT by 12-keV electrons in the same conditions, is shown in Figure 18, together with the absorption of the same film in the uncolored region. It is possible to identify only the M band, peaked around 450 nm, due to the almost overlapping absorption of the F_2 and F_3^+ centers.

The same measurements are reported in Figure 19 for a LiF film of higher thickness, t = 3.8 μm, grown at the same T_s and irradiated in the same conditions. In this case, the situation is still more complex, because the separation between maxima and minima in the interference pattern of the uncolored film is smaller, due to its higher thickness. In this case, even the absorption bands due to the F_2 and F_3^+ centers can be hidden from the complex spectral features of this simple structure.

The commonly used procedure of a direct subtraction between the spectra of the colored region and the uncolored one, generally adopted for colored LiF crystals, often does not allow one to better isolate the M band in LiF films grown on op-

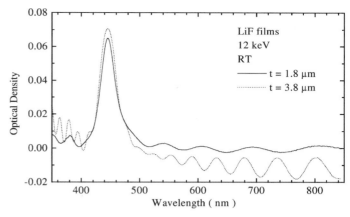

Fig. 20. RT absorption spectra of Figures 18 and 19 roughly corrected from the interference oscillations, of two LiF films, of thickness t = 1.8 μm and 3.8 μm, thermally evaporated on glass substrates at the same temperature T_s = 250°C and colored by 12-keV electrons at RT in the same conditions.

tically transparent substrates such as glass and silica, mainly because the wavelength-dependent refractive index changes induced by the defect formation influence the overall features of the measured interference patterns. Moreover, the presence of less-pronounced bands due to aggregate defects absorbing in the visible can introduce further effects. However, in order to roughly estimate the concentration of the F_2 and F_3^+ centers, a subtraction was performed after a translation of the wavelength axis of the spectrum of the uncolored region in order to match the position of the maxima closest to the absorption band located around 450 nm. The results are shown in Figure 20, where the subtracted spectra are reported for the two samples of Figures 18 and 19. Within the uncertainties due to the presence of fringes, the absorption values due to the overlapping F_2 and F_3^+ centers absorption bands are comparable in the two LiF films of different thickness grown on glass substrate at the same T_s and colored by electron bombardment in the same conditions.

By assuming the defect densities to be constant along the estimated penetration depth of 1.57 μm for 12-keV electrons, a high concentration value of the order of 10^{19} (F_2 + F_3^+)/cm^3 for these aggregate defects is obtained in the case of the colored LiF films of Figure 20.

13. PHOTOLUMINESCENCE OF COLORED LiF FILMS

Stable and intense visible photoluminescence in the green–red region of the spectrum has been measured at RT and LNT in low-energy electron-irradiated LiF films for the first time [8] by exciting the samples with weak argon laser lines between 458 and 514 nm. The emission spectra of a LiF film 1 μm thick thermally evaporated on a fused silica substrate at T_s = 250°C and colored at RT under continuous irradiation by 3-keV electrons are reported in Figure 21a and b both at RT and at LNT. The photoluminescence spectra, excited by the 458- and 476-nm lines of an Ar laser, consist of two bands peaking at around

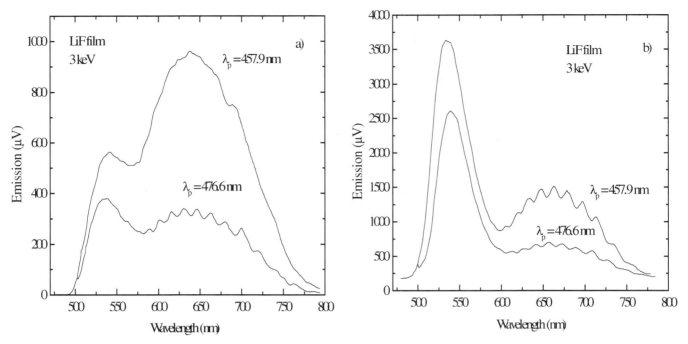

Fig. 21. Photoluminescence spectra of a LiF film of thickness t = 1.0 μm thermally evaporated on a fused silica substrate at T_s = 250°C and colored by 3-keV electrons at RT, excited by the 458- and 476-nm lines of an argon laser at (a) RT and (b) LNT.

540 and 670 nm, which correspond to the known positions of the F_3^+ and F_2 centers (see Table II). The small periodic oscillations superimposed on the broad emission band in the red suggest complex optical interference phenomena of the emitted light in the colored and uncolored thin layers constituting the LiF film. The green emission is more intense at LNT rather than at RT, as expected on the basis of a narrowing of the F_3^+ absorption band with decreasing temperature. On the contrary, the intensities of the red emission appear to be relatively much higher at RT than at LNT. This result is not surprising if we take into account the proper spectral parameters of the F_3^+ and F_2 absorption bands and their specific values at the investigated temperatures, as reported in Table VII, derived from [24] in the case of γ-irradiated LiF crystals. It is evident that the red shift with increasing temperature of the peak position of the F_2 absorption band is more pronounced than that of the F_3^+ one, and then a lower pumping efficiency at 458 nm, that is, 2.71 eV, is obtained. Moreover, the decrease at RT of the F_3^+ emission intensity with respect to that at LNT is also influenced by the existence of a metastable state [75] in its optical cycle, which nonradiatively traps a sizeable fraction of the F_3^+ centers.

Is it really correct to use the optical band parameters measured on entirely colored LiF single crystals to describe the same type of defects hosted in a polycrystalline matrix, like a film? As far as F_3^+ and F_2 photoluminescence is concerned, the experimental results seem to support this kind of approximation. However, in order to obtain the known peak positions for the F_3^+ and F_2 emission bands, it is necessary to correct the measured spectra for the detection system instrumental response. The instrumental calibration is mandatory in order to estimate correctly the peak positions and half-widths for these

Table VII. Values of the Parameters of Optical Absorption and Emission Bands (E = peak position, W = width at half maximum) of F_2 and F_3^+ Centers as Found in Entirely Colored LiF Crystals at RT and LNT

Center	E_a (eV) λ_a (nm)	E_e (eV) λ_e (nm)	W_a (eV)	W_e (eV)	Temperature
F_2	2.82 *440*	1.84 *674*	0.13	0.29	LNT
F_2	2.79 *444*	1.83 *678*	0.16	0.36	RT
F_3^+	2.77 *448*	2.33 *532*	0.23	0.22	LNT
F_3^+	2.77 *448*	2.29 *541*	0.29	0.31	RT

kinds of centers, and its lack is probably the main reason for the spread of the emission peak values reported in the literature [24] for these defects in LiF crystals. For this reason, a direct comparison of the measured photoluminescence spectra taken with different experimental setups is not possible. A typical RT emission spectrum, collected in a collinear geometry between pumping source and detector, is shown in Figure 22 for the film whose absorption spectrum has been shown in Figure 18. The emission spectrum corrected for the used detection system response is also reported. A consistent shift in the peak positions and a remarkable change in the intensities ratio of the two emission bands is obtained. The corrected photoluminescence spectrum has been resolved into the sum of two gaussian bands, as is shown in Figure 23, with peaks and half-widths in

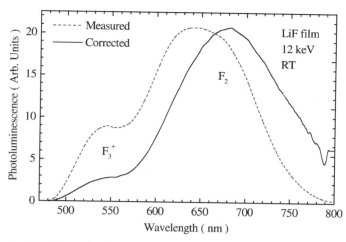

Fig. 22. RT photoluminescence spectrum of the same colored LiF film of Figure 18 excited by the 458-nm line of an argon laser at RT, as measured (dashed) and corrected (solid) for the detection system response.

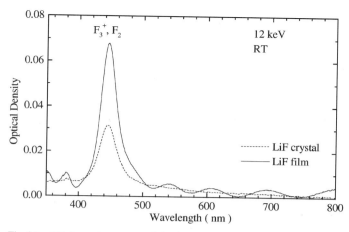

Fig. 24. RT absorption spectra of the LiF crystal (solid) of Figure 3 and of the LiF film (dashed) of Figure 18, corrected from the absorption of the unirradiated part, colored by a 12-keV electron beam in the same irradiation conditions at RT.

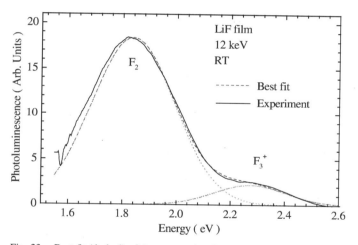

Fig. 23. Best fit (dashed) of the corrected emission spectrum (solid) of Figure 22 as sum of two gaussian bands with spectral parameters consistent with those of F_3^+ (dash-dotted) and F_2 (dotted) emission bands.

good agreement with literature, as in Table II. So, apart from the formation efficiency, which will be dealt with in the next section, the main spectroscopic features of this kind of laser active defects created by low-energy electron beams in polycrystalline films do not show any substantial difference with those measured in crystals in the investigated conditions.

14. INFLUENCE OF LiF FILM STRUCTURE ON COLOR CENTER FORMATION

Although the main spectroscopic features, peak position and half-width, of the emission bands of visible emitting F_3^+ and F_2 defects are similar to the ones observed in bulk crystals, their formation efficiency is influenced by the peculiar structure of the LiF films [76] and depends on their growth condi-

tions, which in turn determine the surface-to-volume ratio, the void presence, and the preferred orientation of crystallites.

In Figure 24, the RT absorption spectrum of the LiF crystal colored by 12-keV electrons of Figure 19, corrected for the absorption of the nonirradiated part of the sample, is compared with that of the film reported in Figure 18, corrected by the interferential oscillations. The irradiation conditions are the same for the two samples. By comparing the absorption spectrum of the colored LiF film with that of the bulk one, it is evident that the intensity of the M absorption band is two to three times greater in films. This means a laser defect density higher than 10^{19} cm^{-3} at the used doses in the LiF film sample. On the contrary, the intensity of the M band in the absorption spectra of Figure 20, referring to two LiF films of different thickness thermally evaporated on glass substrates at the same temperature during the growth, shows comparable values, as expected on the basis of similarity of the film growth conditions [76].

These results have been confirmed also by photoluminescence measurements [77]. In order to compare the formation efficiency of electron-induced CCs in LiF films characterized by different deposition conditions, films of different thickness grown at several T_s have been irradiated by low-energy electron beams, whose energy was selected in order to penetrate the entire film thickness without coloring the substrate. By pumping these samples with the 458-nm line of an argon laser, the F_3^+ and F_2 emissions can be simultaneously observed. In Figure 25a the emission spectrum of a LiF film of t = 1.0 μm grown at $T_s = 300°C$ irradiated by 9-keV electrons is reported, together with the spectra of a sample of the same thickness evaporated at $T_s = 30°C$ and of a LiF crystal irradiated in the same conditions.

The same behavior is also apparent in the emission spectra reported in Figure 25b, which refers to a single crystal and films of the same thickness, t = 1.7 μm, and different T_s irradiated by 12-keV electrons. These photoluminescence spectra clearly indicate that the F_2-center concentration is two

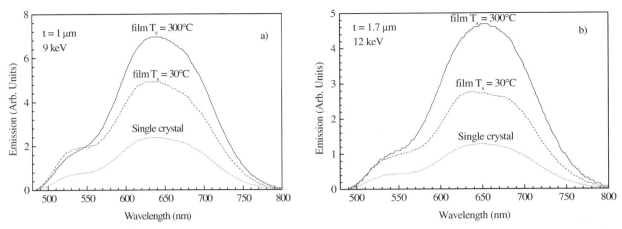

Fig. 25. RT photoluminescence spectra of a LiF single crystal and of two sets of films, thermally evaporated on glass at two different T_s, for two thicknesses and different electron energies: (a) t $=$ 1 μm irradiated at RT by 9-keV electrons. and (b) t $=$ 1.7 μm irradiated at RT by 12-keV electrons with the same dose. The excitation is performed by the 458-nm line of an argon laser.

to five times greater in films than in bulk and that the films evaporated at higher temperatures reach a higher defect density with respect to the films grown at lower temperatures. For the F_2 centers in low-energy electron-irradiated LiF crystals, no quenching is expected on the basis of the preliminary results reported in [77], even at the high concentrations reached in these samples, and the emission intensity can be directly related to the density of defects. However, quenching phenomena have been observed in LiF single crystals entirely colored by 3-MeV electrons [78], as well as in miniaturized samples obtained by medium-energy (0.25–0.6 MeV) electron irradiation of LiF single crystals [79].

For the F_3^+ center, a direct comparison is complicated by the amazing behavior of this type of center [75], whose time-dependent emission intensity is extremely sensitive to pumping power and temperature, due to the presence of a triplet state in its optical cycle. Indeed, in stationary conditions the F_3^+ luminescence increases sublinearly as a function of the exciting power [47]. Moreover, concentration quenching effects should be better investigated for this aggregate defect. Anyway, even in this case, a higher density of F_3^+ centers can be inferred in films.

The formation efficiency of F-aggregate centers is higher in LiF films than in single crystals and it depends on the film growth conditions. It is generally accepted that the main parameters influencing the CCs formation are the surface-to-volume ratio and the void presence as well as the direction of the particle beam with respect to the crystal axes. All these effects contribute to explain the differences between single crystals and polycrystalline films. However, the results shown in Figure 25, obtained on sets of LiF films of different thickness grown at several substrate temperatures and irradiated with low-energy electron beams at several accelerating voltages, suggest that the surface-to-volume ratios and the film compactness play a relevant role because they establish the density of the grain boundaries, which act as a source of vacancies during the CC formation and stabilization processes.

15. NONLINEAR OPTICAL PROPERTIES OF COLORED LiF FILMS

Besides laser emissions, also the generation of degenerate four-wave mixing [80] and/or optical phase conjugation [81] have been pursued by taking into account the nonlinear optical properties of the F_3^+ centers. As already mentioned, this peculiar center, whose absorption band is centered at $\sim$450 nm, almost coincident with the one of the F_2 defect, shows an intense green broad luminescence at $\sim$535 nm. Contrary to the usual luminescent phenomena, the intensity of the F_3^+ emission does not increase linearly with the intensity of the excitation, but it rather displays some kind of saturation effect versus power [82]. It has been shown with systematic measurements of one-photon and two-photon absorption [83, 84], excited state absorption, and emission as a function of power and temperature [47, 85], that a long-lived triplet state is involved in the optical cycle of the F_3^+ center, and as such it can trap a sizeable fraction of the F_3^+ population during excitation [75]. This interesting phenomenon is reflected in the kinetic behaviors of its absorption and emission, and also the F_2 optical cycle is perturbed due to its close connection to F_3^+ through their almost overlapping absorption bands. Although there are some minor effects which do not easily fit the general picture of the F_3^+ and F_2 theoretical model, the majority of their optical properties, which are so important for basic and applied research, are well understood in bulk crystals of LiF.

The peculiar properties of the F_3^+ center derive from its optical cycle, depicted in Figure 26. While the typical four-level scheme of CCs explains the usual absorption–emission behavior, the new level $^3E_1$ introduces a perturbation element in the cycle, which produces an effective quenching of the emission in particular conditions. These nonlinear properties have been addressed by accurate spectral investigations in low-energy electron-irradiated LiF films thermally evaporated on glass [86]. Up to now, very little has been known about the effects, if any, of the polycrystalline structure of the host LiF

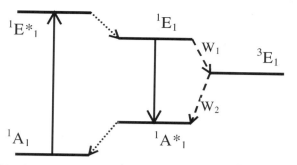

Fig. 26. Optical cycle of the F_3^+ center in LiF showing the radiative (——), nonradiative (- - -), and relaxation (· · ·) transitions.

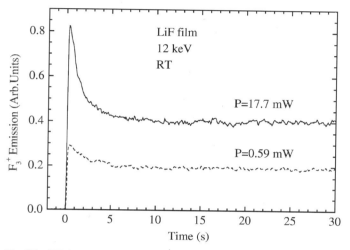

Fig. 27. RT time evolution of the F_3^+ center photoluminescence recorded at 520 nm under c.w. excitation at 458 nm, in a LiF film thermally evaporated on glass and colored by 12-keV electrons at RT, using two different pumping powers.

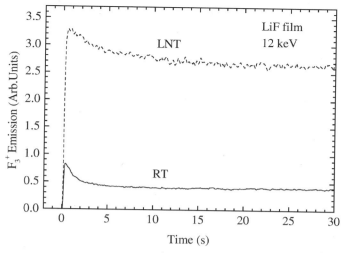

Fig. 28. Time evolution of the F_3^+ center photoluminescence recorded at 520 nm under c.w. excitation at 458 nm, at RT and LNT, in a LiF film thermally evaporated on glass and colored by 12-keV electrons at RT.

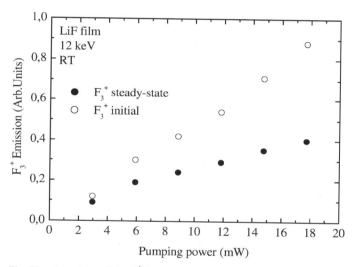

Fig. 29. Intensities of the F_3^+ initial and steady-state emissions at RT measured at 520 nm in a LiF film thermally evaporated on glass and colored by 12-keV electrons at RT, as a function of pumping power under c.w. excitation at 458 nm.

films on the dynamical optical behaviors of point defects embedded in it. From dynamic photoluminescence measurements, it has been possible to measure the two probabilities W_1 and W_2, as defined in Figure 26.

Figures 27 and 28 display the whole dynamic evolution of the F_3^+ green luminescence at 520 nm under a quasi-c.w. optical excitation at 458 nm, starting at t = 0. More precisely, Figure 27 shows its evolution at RT for two different pumping powers, while Figure 28 shows the same behavior at a fixed excitation power for two different temperatures. From the time evolution of the luminescence, it is possible to extract both the initial and final intensity values, and the typical single-exponential decay time. For a more comprehensive description, the initial and final values of F_3^+ emission at RT as a function of pumping power are reported in Figure 29. As far as it is possible to guess at a first analysis, these behaviors are similar to the ones already observed in entirely colored LiF crystals and their accurate analysis, according to [76], allows one to derive the numerical values of the transition probabilities W_1 and W_2, assuming the same values for the F_3^+ luminescence lifetime.

They have been reported in Figure 30 together with those obtained in [76] for a γ-irradiated LiF crystal containing only $\sim 10^{16}$ F_3^+ centers/cm^3, that is, the amount commonly used for optical measurements in bulk. We have found that the values at low temperature agree fairly well with the bulk ones, while those at RT show a rather evident discrepancy for W_1. Indeed, the value of W_1 for the film is almost twice that of the bulk crystal. This slight difference, which increases the trapping properties of the triplet state, should not surprise, because in this case we are dealing with a new type of material, a polycrystalline LiF film with F_3^+ concentrations as high as 10^{18} centers/cm^3. In these extreme conditions, variations of fundamental parameters such as W_1, W_2, lifetime, and emission efficiency are not excluded at all.

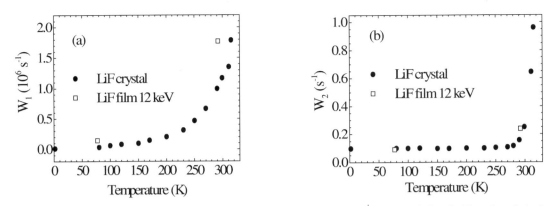

Fig. 30. Temperature dependence of the (a) W_1, and (b) W_2 parameters of the F_3^+ center optical cycle. The values derived for a LiF film (squares), thermally evaporated on glass and colored by 12-keV electrons at RT, are reported together with values relative to an entirely colored LiF crystal (circles).

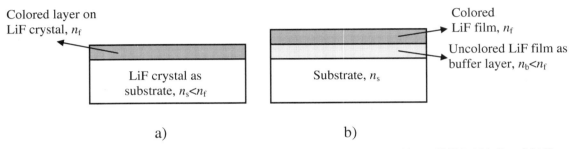

Fig. 31. Outline of the guiding structures investigated for light propagation in a thin colored layer of LiF, in (a) bulk, and (b) films.

16. DESIGN OF ACTIVE WAVEGUIDES IN LiF

The realization of active waveguides in LiF single crystals and films is critically dependent on the modification of the index of refraction induced by low-energy electron-beam bombardment. Apart from the modification of the extinction coefficient of the colored LiF with respect to the uncolored one, the defects formation induces also a change of the real part of the material refractive index. Due to the electron limited penetration depth, from 0.1 to 3.7 μm in the used energies range (see Table IV), through a single-step process it is possible to define an active waveguiding region at the surface of the investigated material, bulk or film. In the case of a LiF crystal, the true waveguide consists of a densely colored layer of LiF located at its surface (single-layer structure, Fig. 31a). As a matter of fact, in a LiF crystal, it could be possible to realize an active waveguide by single-layer structures consisting of colored regions of thickness negligible with respect to the dimension of the crystal. The depth of the colored region is a function of the electron energy, and independently the choice of the refractive index variation is performed via the dose.

In the case of LiF films, a two-layer structure (Fig. 31b) is exploited, in which the guiding upper colored region is optically isolated from the substrate by the uncolored part of the LiF film itself. By a proper choice of the total film thickness, electron energy, and dose, guided propagation of light can be foreseen on any kind of substrate. In other words, the upper layer, the colored region, is the true waveguide, and the lower layer, the uncolored part of the film, acts as an optical buffer. If the buffer layer thickness is high enough, the guiding region is "optically" isolated from the substrate and guided propagation of light is possible even by using high-index substrates.

In every case, to design an active optical waveguide based on LiF, crystal or film, it is important to determine the influence of the coloration conditions, (that is, dose, depth of coloration, irradiation temperature, dose-rate) on the CCs formation and on the induced refractive index modification. Moreover, in the case of polycrystalline LiF films grown on amorphous and crystalline substrates, several properties, in particular refractive index, density, structure, and microstructure, and their influence on the CCs formation induced by low-energy electron-beam bombardment should be taken into account.

17. ELECTRON-BEAM LITHOGRAPHY FOR PATTERN REALIZATION

The use of EBL techniques allows the local modification of the optical properties in predefined patterns at the LiF film and crystal surface by the formation of stable F and F-aggregate centers. Well-developed EBL techniques permit the realization of precise geometrical patterns with various scanning speeds and programmable repetition cycles. The total dose, which is proportional to the density of CCs, can be easily controlled. However, the choice of the specific irradiation conditions, such as electron energy, beam current, scanning speed, apart from

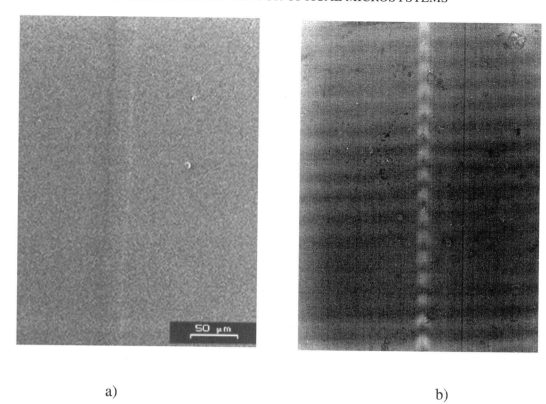

a)   b)

Fig. 32.   (a) SEM image of a 25-μm-wide strip defined by a 12-keV electron beam on a LiF film of thickness t = 3.2 μm thermally evaporated on a glass substrate at T$_s$ = 50°C. (b) Polarizing microscope image of a 100-μm-wide strip defined by a 12-keV electron beam on the same LiF film with a four times higher dose.

their influence on the selective formation of several types of defects, are critical for the mechanical damage of the samples.

In order to realize a channel waveguide, several strips from 2 to 150 μm wide were defined on LiF film surfaces by 12-keV electron-beam irradiation. In Figure 32a, a SEM image of a 25-μm-wide strip defined on a thick LiF film is shown. The photograph was taken on the uncoated sample just after the electron irradiation with the same SEM used for the bombardment. The observed contrast between the irradiated and unirradiated areas is probably due to a local charging of the insulating surface of LiF. The strip width measured from this type of picture is comparable with that obtained in an optical microscope, where the channel is recognizable from the typical yellow coloration due to the M absorption band. In Figure 32b the polarizing microscope image of a 100-μm-wide strip defined on the same thick LiF film by 12-keV electrons shows a fringe pattern modification in the irradiated area, which is attributed to a change in the light optical path, that is, in the refractive index.

Combining the picture reported in Figure 32b with the image obtained in a phase contrast microscope and by a direct comparison with the pictures taken on well-characterized waveguides, the observed changes can be ascribed to an increase in the film refractive index. This result has been recently confirmed in a direct measurement, performed by dark-line spectroscopy technique, on an irradiated strip e-beam written by a 12-keV electron beam on a thick LiF film [87]. As already discussed, el-

lipsometric measurements on a LiF crystal heavily irradiated by 3-keV electrons have shown an increase up to 4% of the real part of the refractive index of the colored region. For the electron-induced channels defined on LiF films, the evaluation of the refractive index difference between the irradiated and nonirradiated area has been accomplished in an indirect way and it could provide only a rough estimate of this quantity. The experiment is performed by using the well-known dark-line spectroscopy setup with the laser beam tightly focused on the film surface in order to discriminate between the area inside and outside the strip. Then, it is possible to clearly observe the difference in the propagation constants inside and outside the irradiated strip through the shift of the coupling dark-line. Such a quantity, though it does not directly yield the refractive index difference, is strictly related, and, in the specific experimental conditions, almost equal to it. The measurements, performed on a channel defined in a LiF film of t = 3.2 μm grown on a glass substrate at T$_s$ = 250°C by 12-keV electrons, gave an effective index difference of 5×10^{-3}, and we estimate this value to be a reliable assessment at least of the order of magnitude of the index change induced by the e-beam irradiation at the used doses.

The RT photoluminescence spectra of 25-μm-wide colored strips defined by 12-keV electrons at the surface of a LiF film and a crystal irradiated at RT in the same conditions, are reported in Figure 33. The samples are excited by the 458-nm

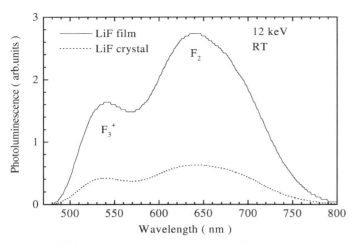

Fig. 33. RT photoluminescence spectra, excited by the 458-nm line of an argon laser, of a LiF film (dashed) and a LiF crystal (solid) irradiated at RT by 12-keV electrons in a strip region 25 μm wide at the dose of 6×10^{-3} C/cm^2.

line of an argon laser and the spectra are collected in a collinear geometry in the same experimental situations. Also in the case of restricted areas, it is possible to recognize the typical F_3^+ and F_2 emission bands, and again we observe signals two to five times higher in the film with respect to the bulk for both kinds of defects. So photoluminescence measurements provide the same information obtained from the optical absorption measurements on laser active CCs formation efficiency in different types of samples, but in addition they allow one to separate the contributions of the two different kinds of defects, and to easily characterize special restricted geometries. Again the formation efficiency of F-aggregate centers is higher in polycrystalline films than in single crystals and depends on the film growth conditions, confirming the results obtained on large irradiated areas.

18. CCs IN ALKALI HALIDE FILMS FOR PASSIVE OPTICAL FUNCTIONS

18.1. Optical Mass Memories

The first application as media for frequency domain optical storage was tested [88] by hole-burning spectroscopy in extremely dense thin films of aggregate CCs confined into a layer a few microns deep produced by ion implantation and low-energy electron bombardment at the surface of alkali fluoride crystals. However, the advantages of the thin film geometry, consistent with the depth of focus of laser beams defined by the diffraction limited optics, were completely exploited for such storage applications in X-ray colored polycrystalline doped thick LiF films [6], which offer the possibility to directly relate the inhomogeneous line-width of the zero-phonon lines (ZPL) to the local environment in the host material and to increase it by appropriate thin-film preparation conditions. In particular, the spectroscopic properties of the ZPL line of the R' center, known as F_3^- according to modern nomenclature, centered at 830 nm in LiF single crystal, have been measured

also in vacuum-evaporated 20-μm-thick doped polycrystalline LiF films. In both types of samples, the CCs were produced by X-ray irradiation at RT. The full-width at half-maximum (FWHM) of the inhomogeneous broadened line, measured by conventional absorption and emission spectroscopy at 2 K, was found to vary from 0.5 nm in the single crystal to 3.0 nm in the polycrystalline sample. This additional broadening was attributed to energy level shifts caused by the extra strain and a substantial number of structural defects such as grain boundaries and stacking faults in the polycrystalline sample. The concentration of the F_3^- defects, estimated by using the Smakula formula by assuming the same oscillator strength for the two samples, was found to be 90 times higher in the film than in the single crystal, according with the role of surface and grain boundaries as well-established vacancy sinks. Also the homogeneous line-width of the burned holes, that is, the natural line-width of the pure electronic transition, increases moving from the single crystal to the polycrystalline sample, and the erasing behavior shows different time dependences and scale, confirming that the nature of the photochemical hole-burning process is influenced by the degree of crystallinity of the LiF sample.

18.2. Selective Optical Filters

The opportunity to introduce higher concentrations of different impurities in AH host materials in the form of film with respect to bulk simply by co-evaporation growth methods, combined with their wide optical transparency window, in which well-defined absorption bands can be tailored by proper selection of the host material and type of defects, has been suggested for the realization of selective optical filters [89]. In the UV spectral region, the intense absorption band centered at 245 nm, due to the F-centers formation, locally induced by using direct focused low-energy electron beams onto LiF crystals and thin films, was proposed as a new type of photomask for photolithography with both the g mercury line and KrF deep-UV lasers [90].

19. CCs IN ALKALI HALIDE FILMS FOR ACTIVE OPTICAL FUNCTIONS

19.1. Broadband-Emitting Novel Materials

The spectroscopic properties of optically active CCs created by low-penetrating ionizing radiation in physical evaporated thin films of alkali halides have been investigated, with some emphasis on their thermo-optical stability [91]. On the other hand, the opportunities to easily realize new light-emitting materials by the growth of multilayered structures of colored AH films simultaneously active in wide spectral regions extending from the visible to the near-infrared have also been studied [92] to obtain predesigned emission properties, controlling the deposition conditions and the irradiation parameters. The growth technique consists of successively evaporated thin films of two original materials, that is, LiF and NaF. The ability to color only a thin dielectric layer of controlled thickness in the structure by using low-penetrating and focalized particle beams allows one to combine the emission properties of the defects host in each of the two fluorides forming the multilayered systems. Such a combination of properties cannot be achieved in mixed crystals. Controlling the deposition parameters and the irradiation conditions in order to influence the CCs spatial distribution makes it possible to obtain green-yellow-red simultaneous emitting materials based on multilayered LiF/NaF structures [92, 93], as well as all-solid novel structures alternately emitting in the visible and in the NIR based on LiF/KCl films [94], and a combination of the above optical properties in a simplified LiF/NaF/KCl film-based system [95].

19.2. Active Optical Channel Waveguides

Well-developed lithography techniques allow one transfer predefined geometric patterns onto the LiF surface. The proper choice of the irradiation conditions, primarily dose and dose-rate once the electron energy, and consequently the depth of the irradiated layer, are fixed allows one to support at least one propagating mode at the emission wavelengths of the point defects laser active in the visible. The waveguiding properties of photoluminescent strips, a few tens of microns wide and more than 1 cm long, realized by electron-beam lithography, have been recently demonstrated for the first time in LiF single crystals [96], LiF films thermally evaporated on LiF bulk [10], and on glass substrates [97].

The realization of a broadband-emitting single-mode channel waveguide on a polished LiF single crystal has been achieved by 12-keV electron-beam irradiation. By using a tent-shaped glass coupling prism and the well-known dark-line spectroscopy technique, the presence of guided modes in the strip has been observed and the corresponding propagation constants have been measured. It appeared that the strip was supporting a single mode at the different test wavelengths, from the blue–green lines of the argon laser to the red line of the He-Ne laser. The dark line, however, was not always sharp, and some measurements were affected by larger error; the effective indices, calculated at various wavelengths from the measured coupling angles, are shown in Figure 34, where a best-fit curve of the chromatic dispersion is also drawn. Optical propagation losses have not yet been measured, but they do appear to be fully acceptable. The usual photoluminescence spectra, consisting of two broad emission bands peaking at 540 and 670 nm, where the F_3^+ and F_2 emissions are located, have been measured by exciting the sample by the 458-nm line of an Ar laser. The intense green–red emitted light is also visible to the naked eye.

Single-mode light propagation together with intense broad photoluminescence in the visible have been also measured in a colored channel induced by electron-beam lithography on a LiF film, 3.0 μm thick, grown by thermal evaporation on a polished LiF single crystal, kept at $T_s = 280°C$ during the deposition. The growth rate, monitored by an oscillator quartz, was around 1.5 nm/s. The electron irradiation parameters, that is, electron energy, beam current, total dose, and dose-rate, have been selected equal to those used successfully in the case of LiF single crystals [96].

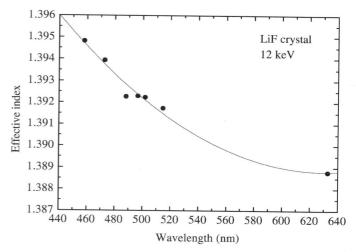

Fig. 34. Dispersion curve of the measured effective index of the single mode supported by a colored channel induced in LiF crystal by a 12-keV electron beam; the continuous line is a second-order polynomial best-fit curve.

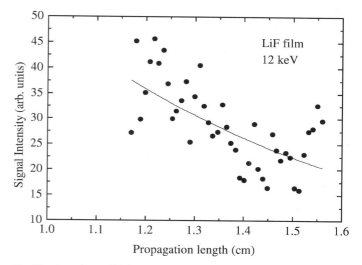

Fig. 35. Loss data at 632.8 nm for a 145-μm-wide channel waveguide induced by a 12-keV electron beam at the surface of a LiF film 3.0 μm thick grown on a LiF single-crystal substrate at $T_s = 280°C$.

It appeared that the strip was supporting a single mode at the 632.8-nm line of an He-Ne laser, and optical losses have been measured by using a scattering detection technique. A plot of some measured data, together with the exponential best fit of the loss measurements, are reported in Figure 35.

As already discussed, though the main spectroscopic features of these defects are similar to the ones in bulk crystals, the formation efficiency of F-aggregate centers is higher in polycrystalline films than in single crystals, even in these restricted geometries, and it is dependent on the peculiar nature of the films, whose structural and morphological properties are strongly influenced by the deposition conditions [76]. The surface-to-volume ratio and the film compactness play the major role because they establish the density of the grain boundaries, which act as a source of vacancies during the CC formation and stabilization processes. So the realization of a light waveguide is not automatically assured by the simple transfer of the irradiation conditions from the LiF crystal to the LiF film-based structure.

Moreover, apart from the optical absorption due to the presence of the defects themselves, further contributions to the losses can be foreseen due to the polycrystalline nature of the film, such as scattering due to grain boundaries, possible lack of homogeneity along the thickness of the film, and the quality of the interface between substrate and film. The measured losses, although not negligible, should be compared with the sizeable optical gain coefficients, greater than 25 dB/cm, measured in colored strips defined by 12-keV electrons at a similar dose in LiF film grown on glass [87].

19.3. Silicon-Compatible Photo-Emitting Structures

The flexibility offered by the thin film geometry has been exploited to easily realize photo-emitting structures, similar to those previously presented, based on colored LiF films grown on silicon. Intense broad green–red photoluminescence has been obtained from colored areas on LiF films thermally evaporated on a Si(100) wafer [42], whose high reflectivity in the visible enhances the brightness of the irradiated regions under illumination with the 458-nm line of an argon laser [98].

The opportunity to exploit the waveguiding properties has been investigated on active channels in LiF films deposited on thick SiO_2 layers grown on Si. The photoluminescence spectrum measured at RT on one of these structures is reported in Figure 36. An interference pattern, which is congruent with the total thickness of the SiO_2 layer plus the LiF one, is superimposed to the known F_3^+ and F_2 broad emission bands.

These examples emphasize the compatibility of colored alkali halide films with the well-assessed silicon technology, and can be seen as a simple way for controlling the distribution of the photo-emitted radiation, by a proper choice of the used substrate. In particular, passive optical elements, such as mirrors or gratings, can be efficiently coupled to the active regions to

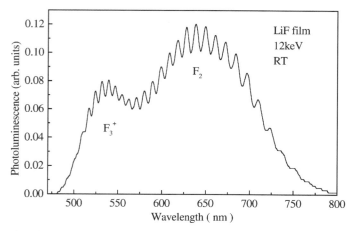

Fig. 36. RT photoluminescence spectrum of a colored strip induced by a 12-keV electron beam in a LiF film thermally evaporated on a thick SiO_2 layer on silicon. The c.w. excitation is at 458 nm.

design new compact emitting devices, as described in the next section.

19.4. Optical Microcavities Based on Electron-Irradiated LiF Films

In recent times, a great deal of attention has been devoted to the investigation of spontaneously emitted radiation modifications inside resonators having at least one size dimension comparable with the emitted wavelength. Starting from the first structures working in the millimetric and near-infrared ranges [99], the interest has naturally turned toward the scale of optical wavelengths [100], where communications and optoelectronics industries work. Inside a microcavity, the optical confinement in one or more dimensions changes the distribution of the electromagnetic field, thus enabling one to control the emission properties of emitting materials placed inside such resonators. For example, alterations of the radiated-power spatial and spectral distributions [101] and enhancement or inhibition of the spontaneous emission rate [102] have been reported in optical microcavities based on insulating layers containing active rare-earth ions, as well as organic light-emitting layers, optically and electrically pumped [103, 104]. Besides their interest as new physical systems, optical microcavities hold technological promise for constructing novel kinds of light-emitting devices such as high-directional LEDs and efficient, high-speed, low-threshold lasers.

Recently, the fabrication of all-solid, multidielectric planar microcavities based on low-energy electron-beam-irradiated thin LiF films evaporated on Bragg reflectors has been achieved. The cavity ends are constituted by Bragg reflectors elaborated by plasma-assisted techniques, consisting of a stack of dielectric layers of a quarter-wavelength optical thickness and alternately high and low refractive index. Over the last high-index layer of the first reflector, the active medium is a half-wavelength LiF film, which has been evaporated and subsequently irradiated by EBL to form color centers inside it, and finally covered by the second mirror. A scheme of the fabricated structures is shown in Figure 37.

Several structures have been designed to perform an electromagnetic constructive interference along the direction orthogonal to the layers for the red photoluminescence generated by optical excitation of laser active F_2 defects. The optical confinement achieved in such a peculiar configuration changes the distribution of the electromagnetic field, thus modifying the emission properties of the active material placed inside the structure, such as the radiated-power spatial and spectral distributions and the spontaneous emission rate.

Spectral and angular-resolved photoluminescence measurements show the modifications of the visible spontaneous emission of F_2 color centers. The resonator induces a narrowing of the emission spectrum and a related increase in the directionality and intensity of the emission along the cavity axis. The resonance effect of the cavity is clearly demonstrated in Figure 38, where the RT photoluminescence spectra, collected in the normal direction, are reported for two colored LiF-based micro-

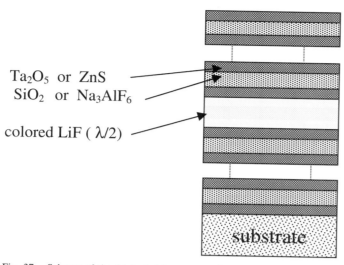

Ta$_2$O$_5$ or ZnS
SiO$_2$ or Na$_3$AlF$_6$

colored LiF (λ/2)

substrate

Fig. 37. Scheme of the fabricated low-energy electron-irradiated LiF film-based optical dielectric microcavities.

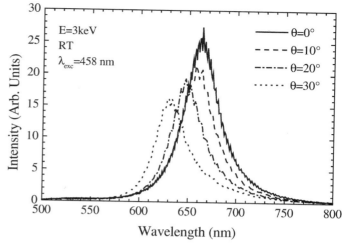

Fig. 39. RT emission spectra of a planar microcavity with a colored LiF film irradiated by a 3-keV electron beam as active spacer, as a function of the detection angle θ with respect to the normal to the sample, under c.w. excitation at 458 nm.

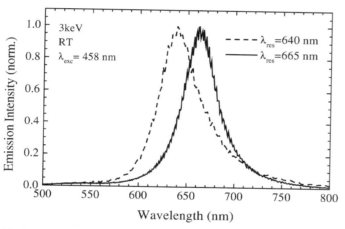

Fig. 38. RT photoluminescence spectra of two planar microcavities, resonating at different wavelengths, with the same colored LiF film irradiated by a 3-keV electron beam as active spacer. The c.w. excitation is at 458 nm and the spectra are collected in the direction perpendicular to the multilayer structures.

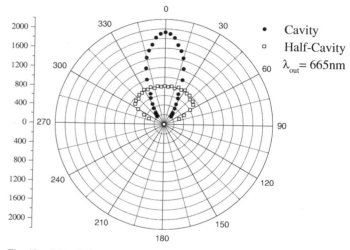

Fig. 40. RT radiation patterns at the resonant wavelength of 665 nm, of a full microcavity and a half cavity, with the same colored LiF film, irradiated by a 3-keV electron beam, as active spacer under c.w. excitation at 458 nm.

cavities resonating at two different wavelengths. The effect of the resonator on the luminescence is twofold. First, a significant change arises in the spectral distribution of the F$_2$ centers emission. The broad emission of the F$_2$ defects has a full-width at half-maximum of about 160 nm (see Table II), while the emission from the microcavities is narrower, in these cases 45 nm, and peaked around the respective resonating wavelengths, overlapping the cavity mode of the Fabry–Perot structures. In addition to these spectral modifications, we can also observe differences in the spatial properties of the light emitted from these kinds of structures. The emission peak shifts to shorter wavelengths when the detection angle is increased, measured with respect to the normal to the multilayer. This well-known behavior of multilayers is shown in Figure 39, for the red-emitting microcavity resonating close to the peak of F$_2$ emission. Figure 40 shows the angular distributions of the radiation emitted at the resonant wavelength of 665 nm for the full microcavity

and a half-cavity. The output light is confined in a narrow cone of 20° around the axis. A significant increase of the emission intensity, of a factor 2.5 at the resonating wavelength, is observed in the normal direction.

All the modifications of the F$_2$ centers/spontaneous emission which have been observed can be ascribed to the resonant nature of the structures where they are placed. By a proper design of the microcavity, the electromagnetic field modes are redistributed and, at the resonant wavelength, they result in a strong confinement of the emitted light in a direction orthogonal to the layer surface. The broad emission band of CCs in LiF allows one to realize optically active structures emitting at different wavelengths based on the same luminescent material simply by modifying the design of the upper reflector of the microcavities. Moreover, a good overlap between the electromagnetic field distribution of the resonant mode of the cavity and the concentration depth profile of the active centers can be achieved by an appropriate choice of the electron irradiation energy.

19.5. Miniaturized Optical Amplifiers and Lasers

Solid-state lasers based on entirely colored LiF crystals, even if compact, do not lend themselves to an easy integration with optical fibers and channel waveguides. A few approaches have been investigated to accomplish this goal. As an example, an evanescent field amplifier was proposed which uses a silica waveguide deposited on top of a LiF:F_2^- crystal [105]. In another device, a waveguide coupler was introduced inside the cavity of a LiF:F_2^- color center laser, in order to efficiently extract the laser light and make easier the coupling to an optical fiber [106].

A sizeable optical gain in a broad wavelength interval in the visible spectral range has been measured at RT on colored strips induced by 12-keV electron beams on LiF films thermally evaporated on glass substrates [107] by the amplified spontaneous emission (ASE) technique [87, 108].

At constant power density of the exciting beam, the photoemission spectra are collected in a perpendicular geometry between pumping source and detector, as a function of the pumped length of the colored strip, which has been varied by moving a screen perpendicular to the exciting beam. The pumping beam was modulated by a chopper and the signal detected by a photomultiplier connected to a lock-in amplifier.

Spontaneous luminescence was amplified (stimulated luminescence) as it passed through the excited volume to the edge of the sample. The laser beam was focused by a cylindrical lens to form a line on the sample surface. This operation allows one to treat the excited medium as a simple one-dimensional optical amplifier and to apply to the detected output intensity signal I the following formula [34]:

$$I(l) = \frac{A}{g}[\exp(gl) - 1] \qquad (6)$$

where A is a constant for a given pump intensity and l is the length of the pumped region. In the micrometric-sized colored strips patterned by EBL at the LiF surface, the one-dimensionality of the excited volume is assured by their peculiar geometry; however, the argon laser beam at 458 nm was focused to reach a higher pumping power density.

In Figure 41, the RT photoluminescence spectrum of a 25-μm-wide strip defined by a 12-keV electron beam on a LiF film of t $= 1.8$ μm thermally evaporated on a glass substrate at $T_s = 250°C$, collected in a standard collinear geometry between pumping source and detector, has been reported together with an ASE spectrum collected for an excited length of 2 mm.

The optical gain has been determined by monitoring the ASE signal as a function of the pumped length. The output signals, measured around the peak of the F_3^+ emission, at 540 nm, and around the peak of the F_2 emission, at 680 nm, are shown in Figure 42. The numerical values of the gain coefficients have

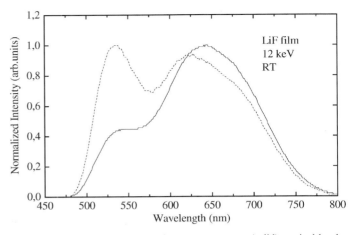

Fig. 41. RT normalized photoluminescence spectrum (solid), excited by the 458-nm line of an argon laser, of a LiF film irradiated at RT by 12-keV electrons in a strip 25 μm wide at the dose of 6×10^{-3} C/cm^2 together with the normalized ASE spectrum (dashed) of the same sample measured for an excitation length of 2 mm.

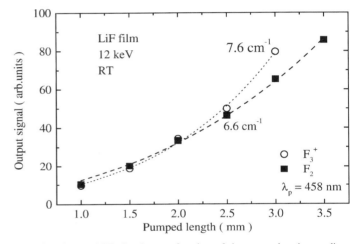

Fig. 42. Output ASE signals as a function of the pumped active medium length in a 25-μm-wide strip induced by 12-keV electron irradiation in a LiF film of t $= 1.8$ μm and grown at $T_s = 250°C$ on a glass substrate, for F_3^+ (circle) and F_2 (square) centers. The numerical values of the gain coefficients, obtained by a best fit (dashed lines) of the experimental measurements with Eq. (6), are also reported.

been obtained performing the best fit of the experimental values by Eq. (6).

The gain coefficients were computed in all the visible spectral range covered by the F_3^+ and F_2 emissions. Sizeable optical gain coefficients have been obtained in the wavelength intervals between 510 and 590 nm and 620 and 710 nm. The obtained results are reported in Figure 43 superimposed to the measured ASE spectrum for a pumped length of 3 mm, and to the same spectrum corrected for the detection system response. As it is expected, the gain effects are observed in the wavelength intervals centered around peak emissions of the laser active defects. In particular, for the F_2 center, this spectral region is located around the center of the emission band in the corrected spectrum. Obviously, outside of the wavelength range where a size-

Table VIII. Numerical Values of the Gain Coefficients Measured by ASE Techniques in LiF Crystals and Films Colored in Different Conditions

LiF sample	Coloration	Centers $\times$ cm^{-3}	F_3^+ (cm^{-1})	F_2 (cm^{-1})
25-μm-wide surface colored strip on a LiF film	12-keV electrons at RT	—	7.6[a]	6.6[a]
Surface colored crystal	3-keV electrons at RT	5×10^{18}	4.4[b]	5.5[b]
Surface colored LiF film	3-keV electrons at RT	5×10^{19}	4.0[b]	7.0[b]
Entirely colored crystal	MeV electrons below RT	$(2.5–4.2) \times 10^{17}$	3–5.5[c]	—
Entirely colored crystal	X-ray at RT	$(3–5) \times 10^{17}$	—	6–15[d]

[a]Ref. [107].

[b]Ref. [9].

[c]Ref. [109].

[d]Ref. [110].

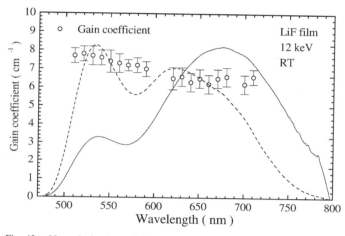

Fig. 43. Numerical values of the optical gain coefficients, obtained by a best fit of ASE measurements, in a colored strip 25 μm wide induced by 12-keV electrons at the dose of 6×10^{-3} C/cm^2 in a LiF film of t = 1.8 μm and grown at T$_s$ = 250°C on a glass substrate. They are superimposed to the measured ASE spectrum (dashed) for a pumped length of 3 mm, and to the same spectrum corrected for the detection system response (solid).

able gain coefficient was unambiguously determined, a linear behavior of the output signal as a function of the pumped length was observed.

The optical gain coefficients of the F_3^+ and F_2 active centers, obtained in 12-keV electron-induced colored strips on a LiF film, are reported in Table VIII together with the values obtained in crystals and films colored at the surface with a lower electron energy (3 keV) on large irradiated regions [9] and the values reported in literature for LiF crystals entirely colored by other types of penetrating ionizing radiation [109, 110]. The listed values of the active defect densities are those provided by the authors and should be considered as an indication, because their exact determination is influenced by the specific choice of many parameters, including oscillator strengths and half-widths. The optical gain coefficients have been generally measured on the peak of the green and red emissions. In the low-energy electron-irradiated LiF crystals and films, their numerical values are comparable with those obtained by the

same ASE technique in entirely colored bulk crystals, containing mainly only one type of center with typical concentration of about 10^{17} cm^{-3}. So, the high concentrations of active centers reached in low-energy electron-irradiated samples affect only slightly the optical gain, while the power threshold appears considerably decreased.

However, the optical behavior is not completely clear and most probably several complex mechanisms of losses play a role in it, such as singlet–triplet transitions and interactions between different types of centers, as well as transformation of active species in other kinds of defects.

As a matter of fact, an efficient laser oscillation can be obtained with a specific active defect only in the case that losses in the optical pumping cycle are kept to a negligible minimum, that is, no interfering deviations of the pumping cycle from a four-energy-level diagram should occur. The most common problems encountered in the achievement of stable RT laser operation are:

(a) excited state absorption (ESA) into higher bound or conduction states, often connected with a destruction of the active centers (optical bleaching);

(b) transition from the upper level of the laser transition to lower-lying metastable states;

(c) pumping power loss, due to other CCs, whose absorption bands overlap with that of the active defects;

(d) reabsorption in the fluorescence range due to the presence of other types of defects in the active media;

(e) quenching of the fluorescence due to competing radiationless transitions at high defects densities.

Information about the parameters affecting the relative concentrations of the laser active centers, not available in detail for LiF films, could allow for a selective generation of the requested kind of defects and therefore for an improvement of light emission performance.

In particular, the stability of the centers under intense optical pumping, as required in laser systems, and in samples with high concentrations of defects are the critical subjects which await full investigation.

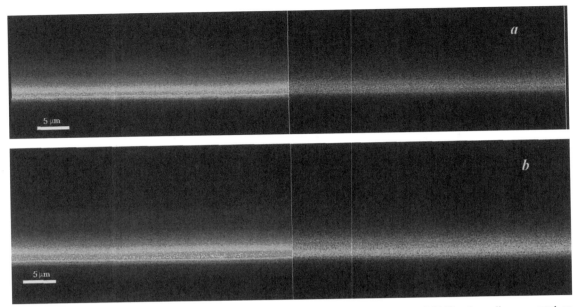

Fig. 44.   CLSM x–z images for the (a) 12-keV and (b) 16-keV electron-induced strip on a LiF film thermally evaporated on glass, optically sectioned in the middle of its width, obtained for excitation at 488 nm for the red (left) and for the green (right) photoluminescence. The electron-irradiated surface is at the bottom of the picture.

20. OPTICAL MICROSCOPY ON LiF-BASED MICROSTRUCTURES

Despite its age and limitations, optical microscopy is currently the most used technique for microscopic investigations. In fact, optical microscopy is fast and low-cost and allows to obtain a great deal of information without the need of particular preparations of the sample. This is very important when the sample must not undergo any modification. In viewing the sample through a conventional microscope, the image is made by the light coming from the whole depth of the sample itself. The formation of the laser active aggregate defects absorbing around 450 nm in LiF is associated with an apparent yellow coloration of the irradiated area, which can be recognized by a standard optical microscope. A confocal light-scanning microscope (CLSM), instead, allows one to view only a selected section of a transparent sample [111].

The high efficiencies of green and red photoemissions from the F_3^+ and F_2 defects, respectively, when simultaneously excited in their almost overlapping absorption bands located around 450 nm, allow one to utilize a CLSM in fluorescence mode [112] to reconstruct the laser active defect distributions along the depth for narrow striplike colored regions induced by EBL on LiF films thermally evaporated on glass substrates. The obtained results are compared with Monte Carlo simulations of electron dose depth profiles.

The low-energy electron-irradiated LiF films, directly grown on a glass cover slip, were firmly attached by nail polish to a microscope slide with the electron-irradiated surface strictly pressed toward it for a proper mounting in a commercial CLSM system equipped with an argon laser and a Nikon microscope. The objective was immersed in oil. The nominal spatial resolu-

tion was 530 nm in depth and 270 nm in plane. An image can be reconstructed by scanning this focal volume through the sample and registering the signal as a function of spatial coordinates. This kind of system allows the specimen to be sectioned optically by scanning a series of planes, eliminating interference from adjacent, overlying, and underlying fluorescence [112]. The sample was analyzed by scanning for two channels simultaneously using the combination fluorescence plus fluorescence with the argon excitation lines at 458 and 488 nm. For dual fluorescence, the signals are recorded simultaneously by two detectors with proper emission filters to select the red and green emissions.

Due to the mounting geometry in the CLSM system, a scanning along the optical axis generates directly x–z images, where the z direction coincides with the depth and the x direction is along the strip length. Typical x–z images are reported in Figure 44a and b for the 12- and 16-keV strips, optically sectioned in the middle of their width. A typical slice image, like the one shown in Figure 45, can be obtained by scanning x–y planes as a function of z. In all the images, the fluorescence signal intensities are converted into colors. Figure 45 shows a practically uniform color center distribution along the width, with the exception of side-boundaries, which look more colored; however, they are regular, well defined, and parallel between them.

According to the color palette, the red luminescence intensity distribution along the z optical axis can be reconstructed. Assuming the red fluorescence intensity proportional to the F_2 center concentration, the F_2 defects spatial profile along the depth has been obtained. The same can be done for the green emission signal, due to the F_3^+ centers [113]. Although the CLSM acquisition software applies a first-order approximation for depth corrections [114], taking into account the ratio be-

Fig. 45. CLSM typical x–y slice images for the 12-keV electron-induced strip of Figure 44 obtained for excitation at 458 nm for the red (left) and for the green (right) photoluminescence. The fluorescence signal intensities are converted into colors according to the palette reported on the right, with increasing values moving from black to white.

tween the refractive indices of the immersion oil and the investigated medium, the major problem encountered in this reconstruction is the clear identification of the sample physical boundaries, due to the high noise level. This could be ascribed to the polycrystalline nature of the investigated sample, which causes high scattering of the photo-emitted light at the grain boundaries and at the film–glass interfaces.

For 12-keV electrons, the energy depth distribution, reported in Figure 46, is characterized by a maximum penetration depth of ~1.5 μm, the depth of maximum deposition rate being at ~0.8 μm. The above values can be directly compared with the normalized spatial profile for the green and red photoluminescence signals, that is, the F_3^+ and F_2 defects depth distribution, reported in the same figure, obtained from the CLSM measurements under optical excitation at 488 nm. The formation of the F_3^+ and F_2 aggregate defects appears restricted to the electron penetration and proportional to the deposited energy in the investigated conditions.

For 16-keV electrons, the energy depth distribution, reported in Figure 47 together with the normalized spatial profile for the green and red photoluminescence signals, is characterized by a maximum penetration depth of ~2.5 μm, the depth of maximum deposition rate being at ~1.3 μm. This confirms an empirical rule which gives a simplified profile of the electron energy deposition rate along with depth. Several computations showed that, for a given material, it is possible to construct a universal simplified curve of energy deposition versus penetration. It was noted [115] that the maximum deposition rate is located at $f \cdot d$, where d is the maximum penetration depth and f is a constant related to the investigated material that, in accordance with the Monte Carlo simulations, in the case of LiF has value ~0.5.

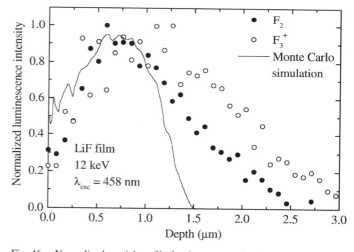

Fig. 46. Normalized spatial profile for the green and red photoluminescence signals, that is, the F_3^+ and F_2 defects depth distribution, obtained for the 12-keV electron-induced strip of Figure 44 from the CLSM measurements excited at 458 nm.

Considering Figure 47, also in this case the F_3^+ and F_2 defects depth distribution appears restricted to the electron penetration and proportional to the deposited energy in the investigated conditions.

Similar results for the F_2 center depth distribution in LiF films irradiated by 15- and 30-keV electron beams have been obtained experimentally by means of optical absorption measurements of multilayered LiF/KBr films [116], and, with the same technique, in multilayered LiF/NaF films irradiated at 25 keV [117]. However, due to the superposition of the F_2 and

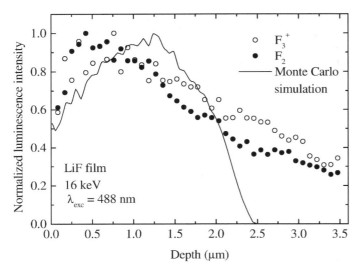

Fig. 47. Normalized spatial profile for the green and red photoluminescence signals, that is, the F_3^+ and F_2 defects depth distribution, obtained for the 16-keV electron-induced strip of Figure 44 from the CLSM measurements excited at 488 nm.

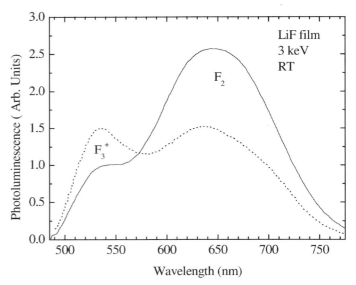

Fig. 48. RT photoluminescence spectra of LiF films thermally evaporated on glass substrates in the same conditions and colored by 3-keV electrons at RT (solid) and $-60°C$ (dashed) with different doses. The c.w. excitation is at 458 nm.

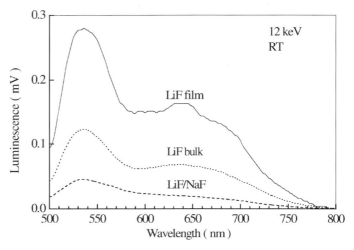

Fig. 49. RT photoluminescence spectra of a LiF crystal, a LiF film grown on glass, and a LiF:NaF planar passive waveguide, all three samples irradiated at RT by 12-keV electrons with the same dose in a strip region 25 μm wide, excited by the 458-nm line of an argon laser.

F_3^+ absorption bands, it was not possible to separate the contribution of these aggregate defects. Moreover, the proposed method appears like very powerful to reconstruct not only the depth profile of the investigated laser active defects emitting in the visible in LiF films irradiated by low-energy electron beams, but also their full spatial distribution in a tridimensional space. It could allow the check and optimization of broadband-emitting single-mode channel waveguide lasers and amplifiers, as well as other active optical miniaturized devices based on this material.

21. PHOTOLUMINESCENCE FOR OPTICAL MICROSYSTEM DEVELOPMENTS

The potentiality of confocal light-scanning microscopy technique in fluorescence mode, described in the previous section, is very attractive. In general, photoluminescence represents a simple, powerful, and versatile tool for the investigation of the spectroscopic properties and device developments of LiF films irradiated by EBL techniques and characterized by restricted geometries and/or nontransparent substrates, which preclude the use of conventional optical absorption measurements. The information obtained by spectral resolved stationary photoluminescence measurements about the enhanced formation efficiency of aggregate CCs in polycrystalline LiF films with respect to single crystal has been already discussed. Apart from the influence of evaporation parameters, also the relevance of the irradiation conditions on the selective formation of different types of active defects in LiF samples should be highlighted. In Figure 48, the photoluminescence spectra of thermally evaporated LiF films, grown in the same conditions and irradiated by 3-keV electrons at RT and $-60°C$ with increasing doses in order to obtain comparable optical absorption at the maximum

of the M band, have been reported. According with similar behavior observed in entirely colored LiF crystals, the F_3^+ concentration is clearly higher when the sample is kept at low temperature during electron bombardment. Similar results have been obtained also at higher electron energies.

In order to investigate the possibility of defining channel waveguiding pathways on LiF-based structure, a 25-μm-wide strip of length greater than 10 mm has been "written" by a 12-keV electron beam on a LiF single crystal, on a LiF film 3.2 μm thick thermally evaporated on a glass substrate, and on a passive LiF:NaF planar waveguide e-beam evaporated on glass. Although the irradiation dose was the same for all the examined samples, the obtained density of induced F-aggregate centers was different in each sample, as it is inferred from the photoluminescence spectra shown in Figure 49. By pumping these samples with the 458-nm line of an argon laser, the F_3^+ and F_2 emissions can be simultaneously observed. The emission intensity is $\cong 2.5$ times higher in the LiF film on glass with respect to bulk crystal, confirming the behavior already found for films of lower thickness. On the contrary, in the passive planar LiF:NaF waveguide, the defect formation efficiency was lowered by more than a factor 5 with respect to the LiF films grown on glass due to the presence of the NaF buffer.

22. CONCLUSIONS

This review describes the interesting characteristics of polycrystalline LiF films and the peculiar properties of point defects induced by low-penetrating ionizing radiation in this material, with the aim of developing miniaturized optical systems.

LiF is unique among all known materials because of its large spectral transparency. Wide band-gap materials are special in many aspects. First of all, they are transparent to visible light and this peculiarity allows control of color. The control over the refractive index on a scale comparable with the wavelength of light, combined with the ease of integration of conventional passive and active optical functions in LiF film-based structures, as in optical microcavities, makes it possible to be optimistic about the perspectives of application in integrated optical circuits. The results already obtained for passive and active waveguides (with propagation losses lower than 6 dB/cm) and for the gain coefficients measured in colored strips on LiF crystals and films (higher than 24 dB/cm) are quite promising.

In photonic applications, control is critical in terms of the performance of the selected instrumentation and methods. The use of versatile, well-assessed, and low-cost fabrication techniques consisting of physical vapor deposition of LiF films on different kinds of substrates combined with low-energy electron-beam direct-writing lithographic techniques allows the realization of optically confined active microstructures, such as optical waveguides and microcavities, and other patterns and configurations can be foreseen.

LiF films are polycrystalline and, as such, many of their properties, such as optical transparency and light scattering, can be expected to be strongly influenced by their grain structure. However, the development and optimization of the proper technological processes for their realization require the careful investigation of several properties, in particular refractive index, density, structure, and microstructure, of polycrystalline LiF films grown by thermal evaporation on amorphous and crystalline substrates. There could be a control of order and disorder on the crystalline but porous LiF films; its straight influence on the main features of the CCs formation induced by low-energy electron beam has been discussed. In particular, the electron energy depth distribution has been investigated on the basis of a Monte Carlo simulation software and of experimental optical measurements on macroscopic and microscopic scale in or-

der to clarify the interaction between the investigated insulating material and the used ionizing radiation.

Changing transparency by charge transfer with low-penetrating ionizing radiation is a simple and efficient way to exploit the potentiality of a wide band-gap material in several areas [118]. In particular, the influence of the polycrystalline structure and morphology of the LiF films on the formation, aggregation, and stabilization of different kinds of aggregate defects still requires a deep investigation of their properties and interactions. The surfaces possess very important features and ionicity can ensure that only certain faces occur; surfaces can be external and internal.

CCs in bulk LiF crystals are optically active point defects with peculiar properties, such as broadband emissions and high efficiencies, and their usefulness in CCLs at RT has been widely proved. In recent years, we have attempted to transfer these properties in low-dimensionality structures such as films, and the results obtained up to now look very promising. Indeed, the sizeable gain coefficients measured on colored regions induced by low-energy electron beams in LiF films grown on glass, as well as other types of substrates, open interesting perspectives for the realization of low-threshold miniaturized lasers. However, while the concentration of point defects is increased in order to obtain more efficient light confinement and more powerful miniaturized optical sources, new phenomena are appearing which require a deeper and more careful spectroscopic investigation. Among them, concentration quenching effects, interactions among different CCs, triplet state, and related transition rates should be well studied under different pumping regimes. Moreover, interactions between different kinds of centers can be reduced by an accurate choice of the irradiation conditions and of the evaporation parameters. In particular, the influence of polycrystalline structure of the films on centers formation and stability should be understood and taken into proper account before optimizing the design of operating photonic devices. Theoretical modeling is important for the progress of the experiments; however, nowadays applications are crucial for the progress of science.

Acknowledgments

I am indebted to Prof. L. C. Scavarda do Carmo for his brilliant suggestions. I would like to thank Dr. G. Baldacchini, Prof. E. Burattini, Prof. P. Picozzi, and Dr. G. C. Righini for valuable discussions. I thank Dr. M. Piccinini for helpful suggestions and a critical reading of the manuscript. I greatly appreciated the precious discussions with Dr. S. Martelli and the useful contribution of Dr. S. Bigotta, F. Bonfigli, M. Cremona, L. Fornarini, A. Mancini, F. Menchini, M. Montecchi, E. Nichelatti, S. Pelli, and F. Somma. The kind contribution of Dr. S. Gagliardi and A. Lai is also acknowledged. Many thanks are due to A. Pace for his skillful and precious help and to A. Grilli and A. Raco for their technical assistance. I warmly acknowledge the patience of my family.

REFERENCES

1. J. H. Schulman and W. D. Compton, in "Color Centers in Solids" (R. Smoluchowski and N. Kurti, Eds.), Pergamon Press, Oxford, 1963.

2. W. B. Fowler, in "Physics of Color Centers" (W. B. Fowler, Ed.), Academic Press, New York and London, 1968.

3. V. V. Ter-Mikirtychev and T. T. Tsuboi, *Prog. Quantum Electr.* 20(3), 219 (1996).

4. W. Gellermann, *J. Phys. Chem. Solids* 52(1), 249 (1991).

5. T. Tamir, in "Integrated Optics" (T. Tamir, Ed.), Springer Verlag, Berlin, 1977.

6. C. Ortiz, C. N. Afonso, P. Pokrowsky, and G. C. Bjorklund, *Appl. Phys. Lett.* 43(12), 1102 (1983).

7. F. Somma, M. Cremona, R. M. Montereali, M. Passacantando, P. Picozzi, and S. Santucci, *Nucl. Instr. Methods B* 116, 212 (1995).

8. R. M. Montereali, G. Baldacchini, and L. C. Scavarda do Carmo, *Thin Solid Films* 201, 106 (1991).

9. G. Baldacchini, M. Cremona, R. M. Montereali, and L. C. Scavarda do Carmo, in "Defects in Insulating Materials" (O. Kanert and J. M. Spaeth, Eds.), p. 176. World Scientific, Singapore, 1993.

10. L. Fornarini, S. Martelli, A. Mancini, R. M. Montereali, S. Pelli, and G. C. Righini, "Proceedings of the 9th European Conference on Integrated Optics," 1999, p. 343.

11. R. M. Montereali, *Mat. Science Forum* 239–241, 711 (1997).

12. F. Menchini, A. Belarouci, B. Jacquier, H. Rigneault, R. M. Montereali, and F. Somma, in "Abstracts INFMeeting '99," 1999, p. 130.

13. M. Piacentini, *Solid State Commun.* 17, 697 (1975).

14. "Handbook of Chemistry and Physics," 78th edition. CRC Press Inc., Cleveland, OH, 1997.

15. E. D. Palik, ed., "Handbook of Optical Constants of Solids." Academic Press, New York, 1997.

16. BDH Crystan Crystal Products, BDH Chemical Ltd, Poole.

17. Harshaw High Power Laser Optics, Technical Paper. The Harshaw Chemical Company, Solon.

18. A. R. Lakshmanan, U. Madhusoodanan, A. Natarajan, and B. S. Panigrahy, *Phys. Stat. Sol. A* 153, 265 (1996).

19. E. Sonder and W. A. Sibley, in "Point Defects in Solids" (J. H. Crawford and L. M. Slifkin, Eds.), Chap. 4. Plenum Press, New York, 1972.

20. E. Hughes, D. Pooley, H. U. Rahman, and W. A. Runciman, Harwell Atomic Research Establishment R-5604, 1967.

21. F. Seitz, *Rev. Mod. Phys.* 26, 7 (1954).

22. G. Baldacchini, M. Cremona, U. M. Grassano, V. Kalinov, and R. M. Montereali, in "Defects in Insulating Materials" (O. Kanert and J. M. Spaeth, Eds.), p. 1103. World Scientific, Singapore, 1993.

23. J. Nahum and D. A. Wiegand, *Phys. Rev.* 154, 817 (1967).

24. G. Baldacchini, E. De Nicola, R. M. Montereali, A. Scacco, and V. Kalinov, *J. Phys. Chem. Solids* 61, 21 (2000).

25. A. Perez, J. Davenas, and C. H. S. Dupuy, *Nucl. Instrum. Methods* 132, 219 (1976).

26. M. Montecchi, E. Nichelatti, A. Mancini, and R. M. Montereali, *J. Appl. Phys.* 86(7), 745 (1999).

27. N. Seifert, H. Ye, N. Tolk, W. Husinsky, and G. Betz, *Nucl. Instrum. Methods B* 84, 77 (1994).

28. J. Nahum, *Phys. Rev.* 158, 814 (1967).

29. T. T. Basiev, S. B. Mirov, and V. V. Osiko, *IEEE J. Quantum Electron.* 24, 1052 (1988).

30. V. V. Ter-Mikirtychev and T. T. Tsuboi, *Canad. J. Phys.* 75, 813 (1997).

31. L. F. Mollenauer, "Tunable Lasers" (L. F. Mollenauer, Ed.). Springer Verlag, Berlin, 1987.

32. L. X. Zheng and L. F. Wan, *Optics Comm.* 55(4), 277 (1985).

33. T. T. Basiev, P. G. Zverev, V. V. Fedorov, and S. B. Mirov, *Appl. Optics* 36(12), 2515 (1997).

34. P. Fabeni, G. P. Pazzi, and L. Salvini, *J. Phys. Chem. Solids* 52, 299 (1991).

35. C. Weaver, *Adv. Phys.* 42, 83 (1962).

36. U. Kaiser, N. Kaiser, P. Weibbrodt, U. Mademann, E. Hacker, and H. Muller, *Thin Solid Films* 217, 7 (1992).

37. S. Sinaroy, *Thin Solid Films* 187, 231 (1990).

38. R. M. Montereali, S. Martelli, G. C. Righini, S. Pelli, C. N. Afonso, A. Perea, D. Barbier, and P. Bruno, "9th CIMTEC-World Forum on New Materials, Symposium X—Innovative Light Emitting Materials" (P. Vincenzini and G. C. Righini, Eds.), p. 321. Techna s.r.l., 1999.

39. R. Schlaf, B. A. Parkinson, P. A. Lee, K. W. Nebesny, G. Jabbour, B. Kippelen, N. Peyghambarian, and N. R. Armstrong, *J. Appl. Phys.* 84, 6729 (1998).

40. F. Cosset, A. Celelier, B. Barelaud, and J. C. Vareille, *Thin Solid Films* 303, 191 (1997).

41. D. A. Lapiano-Smith, E. A. Eklund, F. J. Himpsel, and L. J. Terminello, *Appl. Phys. Lett.* 59(17), 2174 (1991).

42. G. Baldacchini, E. Burattini, L. Fornarini, A. Mancini, S. Martelli, and R. M. Montereali, *Thin Solid Films* 330, 67 (1998).

43. L. Fornarini, A. Mancini, S. Martelli, R. M. Montereali, P. Picozzi, and S. Santucci, *J. Non-Crystalline Solids* 245, 141 (1999).

44. E. J. Kobetich and R. Katz, *Phys. Rev.* 170, 391 (1968).

45. L. H. Abu-Hassan and P. D. Townsend, *J. Phys. C: Solid State Phys.* 19, 99 (1986).

46. C. Trautmann, K. Schwartz, and O. Geiss, *J. Appl. Phys.* 83, 3560 (1998).

47. G. Baldacchini, M. Cremona, R. M. Montereali, L. C. Scavarda do Carmo, R. A. Nunes, S. Paciornik, F. Somma, and V. Kalinov, *Optics Comm.* 94, 139 (1992).

48. C. A. Andersen, "The Electron Microprobe" (T. D. McKinley, K. F. Heinrich, and D. B. Wittry, Eds.), John Wiley, New York, 1964.

49. F. Fischer, *Z. Phys.* 154, 534 (1959).

50. G. Baldacchini, G. d'Auria, R. M. Montereali, and A. Scacco, *J. Phys.: Condens. Matter* 10, 857 (1998).

51. V. V. Ter-Mikirtychev, T. Tsuboi, M. E. Konyzhev, and V. P. Danilov, *Phys. Stat. Sol. (B)* 196, 269 (1996).

52. D. C. Joy, *J. Microsc.* 147, 51 (1987).

53. F. Agullo-Lopez, C. R. A. Catlow, and P. D. Townsend, "Point Defects in Materials," Academic, London, 1988.

54. N. Itoh, *Adv. Phys.* 31, 491 (1982).

55. N. Itoh, *Rad. Eff. Def. Solids* 110, 19 (1989).

56. P. Durand, Y. Farge, and M. Lambert, *J. Phys. Chem. Solids* 30, 1353 (1969).

57. Y. Farge, *J. Phys. Chem. Solids* 30, 1375 (1969).

58. F. Agullo-Lopez and F. Jacque, *J. Phys. Chem. Solids* 34, 1949 (1973).

59. M. Aguilar, F. Agullo-Lopez, and F. Jacque, *Rad. Eff.* 61, 215 (1982).

60. R. Bate and C. V. Heer, *J. Phys. Chem. Solids* 7, 14 (1958).

61. G. Baldacchini, E. De Nicola, G. Giubileo, F. Menchini, G. Messina, R. M. Montereali, and A. Scacco, *Nucl. Instrum. Methods B* 141, 542 (1998).

62. G. Baldacchini, R. M. Montereali, M. Cremona, E. Masetti, M. Montecchi, S. Martelli, G. C. Righini, S. Pelli, and L. C. Scavarda do Carmo, in "Advanced Materials in Optics, Electro-Optics and Communication Technologies" (P. Vincenzini and G. C. Righini, Eds.), p. 425. Techna s.r.l., Faenza, 1995.

63. R. M. Montereali, G. Baldacchini, S. Martelli, and L. C. Scavarda do Carmo, *Thin Solid Films* 196, 75 (1991).

64. R. A. Nunes, A. P. da Silva Sotero, L. C. Scavarda do Carmo, M. Cremona, R. M. Montereali, M. Rossi, and F. Somma, *J. Lumin.* 60&61, 552 (1994).

65. L. J. Lingg, A. D. Berry, A. P. Purdy, and K. J. Ewing, *Thin Solid Films* 209, 9 (1992).

66. A. Perea, J. Gonzalo, C. N. Afonso, S. Martelli, and R. M. Montereali, *Appl. Surf. Sci.* 138-139, 533 (1999).

67. M. Cremona, M. H. P. Mauricio, L. V. Fehlberg, R. A. Nunes, L. C. Scavarda do Carmo, R. R. Avillez, and A. O. Caride, *Thin Solid Films* 333, 157 (1998).

68. H. A. Macleod, "Thin Film Optical Filters," Macmillan, New York, 1986.

69. M. Montecchi, E. Masetti, and G. Emiliani, *Pure Appl. Optics* 4, 15 (1995).

70. H. K. Pulker, *Appl. Optics* 18, 1969 (1979).

71. D. A. G. Bruggeman, *Ann. Phys (Leipzig)* 24, 636 (1935).

72. P. E. Di Nunzio, L. Fornarini, S. Martelli, and R. M. Montereali, *Phys. Stat. Sol. (A)* 164, 747 (1997).

73. G. C. Righini, S. Pelli, E. Bellini, R. M. Montereali, S. Martelli, and M. Cremona, in "Integrated Optics Devices IV" (G. C. Righini and S. Honkanen, Eds.), Vol. 3936, p. 209. Proc. SPIE 2000.

74. M. Olivier, in "New Directions in Guided Wave and Coherent Optics" (D. B. Ostrowsky and E. Spitz, Eds.), Vol. II, p. 639. Martinus Nijhoff, The Hague, 1984.

75. G. Baldacchini, M. Cremona, G. d'Auria, R. M. Montereali, and V. Kalinov, *Phys. Rev. B* 54(24) 17508 (1996).

76. G. Baldacchini, M. Cremona, G. d'Auria, S. Martelli, R. M. Montereali, M. Montecchi, E. Burattini, A. Grilli, and A. Raco, *Nucl. Instrum. Methods B* 116, 447 (1996).

77. R. M. Montereali, *J. Lumin.* 72–74, 4 (1997).

78. G. Baldacchini, F. Menchini, and R. M. Montereali, in "Abstracts International Conference on Defects in Insulating Materials," 2000, p. 114.

79. E. F. Martinovich, V. I. Barishnikov, V. A. Grigorov, and L. I. Shcepina, *Sov. J. Quantum Electron.* 18(1), 26 (1988).

80. H. E. Gu, *Chin. J. Infrared Millimetric Waves* 10, 7 (1990).

81. T. T. Tsuboi and H. E. Gu, *Rad. Eff. Def. Solids* 134, 349 (1995).

82. S. Paciornik, R. A. Nunes, J. P. Von Der Weid, L. C. Scavarda do Carmo, and V. S. Kalinov, *J. Phys. D: Appl. Phys.* 24, 1811 (1991).

83. V. V. Ter-Mikirtychev and T. T. Tsuboi, *Phys. Stat. Sol. (B)* 190, 347 (1995)

84. T. T. Basiev, I. V. Ermakov, and K. K. Pukhov, in "Tunable Solid State Lasers" (W. Strek, E. Lukowiak, and B. Nissen-Sobocinska, Eds.), Vol. 3176, p. 160. SPIE 1997.

85. G. Baldacchini, M. Cremona, G. d'Auria, V. Kalinov, and R. M. Montereali, *Rad. Eff. Def. Solids* 34, 425 (1995).

86. G. Baldacchini, A. Mancini, F. Menchini, and R. M. Montereali, in "9th CIMTEC-World Forum on New Materials, Symposium X—Innovative Light Emitting Materials" (P. Vincenzini and G. C. Righini, Eds.), p. 337. Techna s.r.l., 1999.

87. G. Baldacchini, A. Mancini, R. M. Montereali, S. Pelli, and G. C. Righini, in "Proceedings of the 8th European Conference on Integrated Optics," 1997, p. 373.

88. C. Ortiz, R. M. Macfarlane, R. M. Shelby, W. Lenth, and G. C. Bjorklund, *Appl. Phys.* 25, 87 (1981).

89. L. Oliveira, C. M. G. S. Cruz, M. A. P. Silva, and M. Siu Li, *Thin Solid Films* 250(2) 73 (1994).

90. R. A. Nunes, S. Paciornik, and L. C. Scavarda do Carmo, "Proceedings of Microlithography '92 Conference," Vol. 1674, p. 552. SPIE 1992.

91. A. Ercoli, A. Scacco, F. Somma, M. Cremona, R. M. Montereali, S. Martelli, G. Petrocco, L. Scopa, R. A. Nunes, and L. C. Scavarda do Carmo, *Rad. Eff. Def. Solids* 132, 143 (1994).

92. F. Somma, R. M. Montereali, S. Santucci, L. Lozzi, M. Passacantando, M. Cremona, M. H. P. Mauricio, and R. A. Nunes, *J. Vac. Sci. Technol. A* 15(3), 1750 (1997).

93. F. Somma, E. De Nicola, A. Grilli, R. M. Montereali, and A. Raco, "Proceedings of the Sixth International Conference on Luminescent Materi-

als" (C. R. Ronda and T. Welker, Eds.), Vol. 97-29, p. 153. Electrochemical Society Proceedings, 1998.

94. M. Cremona, A. Grilli, R. M. Montereali, M. Passacantando, A. Raco, and F. Somma, *J. Lumin.* 72-74, 652 (1997).

95. F. Somma, E. De Nicola, A. Mancini, R. M. Montereali, M. Cremona, M. H. P. Mauricio, L. H. Fehlberg, and L. C. Scavarda do Carmo, in "9th CIMTEC-World Forum on New Materials, Symposium X—Innovative Light Emitting Materials" (P. Vincenzini and G. C. Righini, Eds.), p. 329. Techna s.r.l., 1999.

96. R. M. Montereali, A. Mancini, G. C. Righini, and S. Pelli, *Opt. Comm.* 153, 223 (1998).

97. M. Piccinini, Tesi di Laurea, University of Rome "La Sapienza," 1999.

98. J Bell, ed. IOP Magazines, The Haltings, UK. *Opto & Laser Europe* 55, 17 (1998).

99. P. Goy, J. M. Raimond, M. Gross, and S. Haroche, *Phys. Rev. Lett.* 50, 1903 (1983).

100. F. De Martini, G. Innocenti, G. R. Jacobovitz, and P. Mataloni, *Phys. Rev. Lett.* 59, 2995 (1987).

101. H. Rigneault, C. Amra, S. Robert, C. Begon, F. Lamarque, B. Jacquier, P. Moretti, A. M. Jurdyc, and A. Belarouci, *Opt. Mat.* 11, 167 (1999).

102. E. Yablonvitch, T. J. Gmitter, and R. Bhat, *Phys. Rev. Lett.* 61, 2546 (1988).

103. N. Tessler, G. J. Denton, and R. H. Friend, *Nature* 382, 695 (1996).

104. S. Tokito, T. Tsutsui, and Y. Taga, *J. Appl. Phys.* 86(5), 2407 (1999).

105. V. A. Kozlov, A. S. Svakhin, and V. V. Ter-Mikirtychev, *Electron. Lett.* 30, 42 (1994).

106. V. V. Ter-Mikirtychev, E. L. Arestova, and T. Tsuboi, *J. Lightwave Techn.* 14, 2353 (1996).

107. G. Baldacchini, E. Burattini, L. Fornarini, A. Mancini, S. Martelli, M. Montecchi, and R. M. Montereali, in "Proceedings of the 6th International Conference on Luminescent Materials," Vol. 97-29, 1998, p. 78.

108. K. L. Shanklee and R. F. Leheny, *Appl. Phys. Lett.* 18, 475 (1971).

109. L. Zheng, S. Guo, and L. Wan, *Appl. Phys. Lett.* 48, 381 (1986).

110. T. Kulinski, F. Kaczmarek, M. Ludwiczak, and Z. Blaszczak, *Opt. Comm.* 35, 120 (1980).

111. M. Minski, U.S. Patent 3,013,467, 1961.

112. J. B. Pawley, "Handbook of Biological Confocal Microscopy" (J. B. Pawley, Ed.). Plenum Press, New York, 1990.

113. R. M. Montereali, S. Bigotta, M. Piccinini, M. Gianmatteo, P. Picozzi, and S. Santucci, *Nucl. Instrum. Meth. B* 166-167, 764 (2000).

114. CLSM Manual, Vol. 2, "Set-up and Scanning."

115. Y. A. Jammal, D. Pooley, and P. D. Townsend, *J. Phys. C: Solid State Phys.* 6, 247 (1973).

116. L. G. Jacobsohn, R. A. Nunes, and L. C. Scavarda do Carmo, *J. Appl. Phys.* 81(3), 1192 (1997).

117. M. Cremona, L. G. Jacobsohn, H. Manela, M. H. P. Mauricio, L. C. Scavarda do Carmo, and R. A. Nunes, *Material Science Forum* 239-241, 725 (1997).

118. N. Itoh and M. Stoneham, "Materials Modification by Electronic Excitation" (N. Itoh and M. Stoneham, Eds.). Cambridge University Press, 2000.

Chapter 8

POLARIZATION SWITCHING OF FERROELECTRIC CRYSTALS

Lung-Han Peng

Department of Electrical Engineering, National Taiwan University, Taipei, Taiwan, Republic of China

Contents

1. INTRODUCTION

The abundance of novel mechanical, acoustical, electro-optical, and nonlinear optical properties in ferroelectrics has made study of this material system an important course in condensed matter physics [1]. The ferroelectric crystal is defined as a crystal belonging to the pyroelectric family upon which the direction of its spontaneous polarization P_s is subjected to a temperature change. The process leading to the reorientation of all ferroelectric domains in one polarization direction is called as poling, polarization switching, or domain reversal in the literature. The ability of a ferroelectric crystal to switch between its polarization states and to make a small area of reversed domains with fast switching have made ferroelectrics attractive for applications on gigabyte data storage [2–4]. For example, a domain size as small as 50 nm in diameter can be realized on Pb(Zr,Ti)O_3 (PZT) [5] using the nanodomain processing technique to be discussed in Section 3.5. It is noted that in the reverse-poled state of PZT, the corresponding surface charge density is a nearly constant. Combination of these two advantages makes study of the polarization switching process on the PZT material system promising for storage application [6]. Moreover, ferroelectric crystals also exhibit electro-optical and nonlinear activity that make them extremely valuable in the application of modulation, deflection, switching, and frequency generation [7]. These basic tensor properties all hinge upon the status of the spontaneous polarization in the crystal. Study of the ferroelectric polarization switching process therefore not only plays an important role in revealing the fundamental physics but also is crucial for further device applications.

The presence of a homogeneous spontaneous polarization vector P_s in a ferroelectric state will induce in the crystal a depolarization field opposite to P_s [8]. The ferroelectric crystal in this configuration is said to be in a high-energy state and will be split into domains with opposite polarization to minimize the free energy [9, 10]. As a result, the ferroelectric crystal

Handbook of Thin Film Materials, edited by H.S. Nalwa
Volume 3: Ferroelectric and Dielectric Thin Films

ISBN 0-12-512911-4/$35.00

Table I. Characteristics of a Few Typical Ferroelectric crystals

	Curie point T_c (°C)	Coercive field E_c (kV/mm)	Spontaneous polarization P_s (μC/cm²)
TGS	47	0.05	2.8
BaTiO₃	126	0.05	26
KTP	940	2.5	17
LiNbO₃	1145 (Congruent-grown)	21 (Congruent-grown)	70 (Congruent-grown)
	1200 (Stoichiometric)	4.5 (Stoichiometric)	70 (Stoichiometric)
LiTaO₃	600 (Congruent-grown)	21 (Congruent-grown)	50 (Congruent-grown)
	690 (Stoichiometric)	1.7 (Stoichiometric)	50 (Stoichiometric)

would be characterized by a state consisting of P_s having positive and negative polarity, that is, a multidomain structure [11]. The twin polarization component is called the *domain* and the boundary between the domains is called the *wall*. The total free energy of such system then is a sum of the wall energy and the electrostatic energy associated with the depolarization field. Due to the minimization of free energy in the crystal, the domain wall normally retains a width of one or two lattice constants [1]. Moreover, by applying an external electric field (E), the wall can be displaced so that one of the twin domains with P_s parallel to E will be expanded at the expense of the other domain of opposite polarity. If E is larger than the so-called coercive field (E_c), the whole crystal will have the P_s realigned and parallel to the applied field. One therefore obtains a single domain crystal. For conventional ferroelectric device applications, it is desirable to have a single domain crystal, but for modern electro-optics and nonlinear-optics applications, a periodical domain reversal structure is preferred. In both cases, one has to reorient the P_s according to the application requirement. This process of polarization switching, or poling, will constitute the main theme of discussion of this chapter.

A theory of second-order phase transition, proposed by Landau and developed by Ginzberg [12] and Devonshire [13], has been used to describe the status of spontaneous polarization in ferroelectrics. It can also lead to a size effect on the dielectric response of the ferroelectric thin film [14]. In bulk, the characteristics of ferroelectricity disappear due to the thermal motion above a phase transition temperature known as the Curie point (T_c). When the crystal is heated above T_c, it will change into an unpolarized or the so-called paraelectric state. In general, the spontaneous polarization P_s of a ferroelectric state scales with its T_c and so does the coercive field. In Table I we list the relevant parameters of (T_c, E_c, P_s) for several representative ferroelectrric crystals whose polarization switching characteristics will be discussed in this chapter.

We pass by noting several relevant references on the ferroelectric crystals. To explore the basic dielectric, electromechanical, and thermodynamic properties of ferroelectric crystals, one can refer to the books by Känzig [1], Jona and Shirane [15], Fatuzzo and Merz [16], and Burfoot [17]. Summary of the physical and chemical properties of a specific ferroelectric crystal

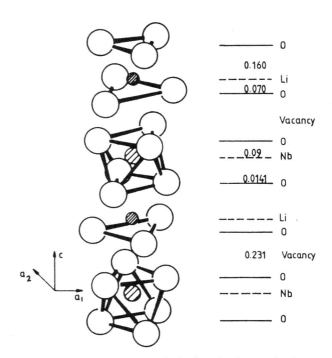

Fig. 1. Crystal structure of LiNbO₃ in the ferroelectric state showing an ordered site sequence containing the oxygen planes and the Li, Nb, and vacancy along the C-axis. The spacing between each atomic plane is shown in the figure in unit of nanometer. Reprinted with permission from [33], copyright 1986, American Institute of Physics.

can be found on lithium niobate (LiNbO₃) [18–21], lithium tantalate (LiTaO₃), [22–25] and potassium titanyl phosphate (KTP) [26–29]. The book series of Landolt–Börnstein also provides basic information on the ferroelectrics [30]. For a recent account of the synthesis of ferroelectric thin film, one can refer to the book by Araujo et al. [31].

Note that many of the tensor properties of the ferroelectric crystals mentioned above originated from the spontaneous polarization P_s in an ordered structure [32]. Let us take LiNbO₃, which exhibits large pyroelectric, piezoelectric, electro-optic, and nonlinear-optic effects, for example. Schematically shown in Figure 1 is the crystal structure of LiNbO₃ in the ferroelectric state that contains the oxygen planes and the Li, Nb, and vacancy sites along the crystal C-axis [33]. Note the polarity of the ferroelectric domain is determined by the direction of the Li ions with respect to the oxygen plane. In a poled LiNbO$_3$ crystal, there exists an ordered structure of a Li–Nb–vacancy– Li–Nb–vacancy site sequence, and in the reverse poled state a Li–vacancy–Nb–Li–vacancy–Nb sequence, respectively. The unique atomic site sequence in a ferroelectric state then results in a spontaneous polarization P_s along the crystal C-axis. A primary feature of the ferroelectric crystal is the ability to reorient the direction of P_s and thus the sign change of its odd-rank tensors. This novel characteristic can lead to numerous applications on acoustics, electro-topics, and nonlinear optics upon which the tensor properties are involved in the action. The article reviewed by Kielich contains significant theoretical and experimental achievements in the optical and electro-optical properties of ferroelectrics up to the 1960s [34]. To explore some of the basic applications of ferroelectrics, the readers can refer to the books by Lines et al. [35], Smolenskii et al. [36], and Taylor et al. [37].

The major interest of this chapter, however, is to focus on the techniques that can lead to the change of the polarity of the spontaneous polarization, i.e., a process called polarization switching or domain reversal, in the ferroelectric crystals. The research activity in this subject has prevailed so much that it has greatly revived technical development in the fields of data storage and optoelectronics. In particular, we will focus on methods that lead to the construction of (i) single domain crystal and (ii) layered structures with P_s anti-parallel or opposite to each other. Here we recognize that the achievement of single domain ferroelectric crystals in the 1980s has inspired research activities in the field of integrated optics, nonlinear optics, and electro-optics. A review on the thin-film guided second- or third-order nonlinear effects in such a material system is provided by Stegeman et al. [38]. What is worth mentioning is the development of the periodical domain poling techniques in the 1990s. It has greatly revived technical development for high-density data storage, and the research activity and applications on nonlinear optics and optoelectronics. We will show that advancergents in poling techniques, crystal quality control, and optical damage resistance have commercialized periodically poled domain reversed ferroelectric devices. Great efficiency improvement in infrared parametric generation [39], short wavelength harmonic conversion [40], ultrafast optical sampling [41], data compression [42], narrow-band terahertz generation [43], integrated electro-optic components [44], ultrasonic resonance and filtering [45], just name a few, are among many other promising applications that benefit from the availability of periodical domain reversal structures.

To be more specific, for a layered ferroelecrtric structure whose domain structure is periodically reversed, there is also a periodic modulation in the sign of all of the odd-rank tensors. This in turn will influence the applications, for example, in which the third-rank tensors such as nonlinear-optic, electro-optic, and piezoelectric effects are involved in the dynamic response. To illustrate this principle, let us consider the effects of periodical domain reversal on the nonlinear second harmonic generation (SHG) and resonant ultrasonic transmission. Here we note that a modulation in the sign of the nonlinear optical and piezoelectric coefficients, respectively, will take place in the corresponding response function.

1.1. Second Harmonic Generation in a Periodical Domain Reversal Structure

The SHG procedure consists of a nonlinear wave mixing process in a noncentral symmetric crystal to generate a wave doubling of the frequency of the fundamental beam. When one considers the conservation of energy and momentum in the SHG process, high conversion efficiency can be obtained only if there is a phase-matching (PM) interaction between the fundamental and the second harmonic waves. In the conventional practice, the phase-matching condition can be realized by using either temperature or angle tuning via the birefringence effects of the uniaxial or biaxial crystals [46]. However, the use of the conventional PM techniques is limited to the beam walk-off problems due to the perpendicular polarized beams involved in a nonlinear wave mixing process using the nonlinear coefficient d_{31}. In 1962, Bloembergen et al. proposed a new concept to enhance the efficiency of nonlinear wave mixing, i.e., by quasi-phase matching (QPM) through the spatial modulation of the nonlinear optical susceptibility [47]. In practice, the QPM device consists of a periodical 180° domain reversal structure with the alignment of the spontaneous polarizations P_s anti-parallel to each other as shown in Figure 2 [48]. In the nonlinear wave conversion procedure of QPM, the periodical domain reversal structure provides a phase-matching mechanism to compensate the refractive index dispersion by

$$\mathbf{k}_2 - 2\mathbf{k}_1 - \mathbf{K} = 0 \tag{1}$$

where $\mathbf{K} = 2m\pi/\Lambda$ is the wave vector of the domain reversal structure with a period of Λ and $\mathbf{k}_1$ and $\mathbf{k}_2$ are the wave vectors of the fundamental and second harmonic waves, respectively. The QPM condition requires the width of the inverted domain to be an odd number of the coherence length l_c to compensate the refractive index dispersion between the fundamental and harmonic wave

$$l_c = \lambda/4(n_2 - n_1) \tag{2}$$

The QPM scheme can be applied to the nonbirefringent as well as the conventional birefringent crystals as will be discussed in Section 4. Advantages of the QPM process on the application of nonlinear optics include the access to the largest nonlinear coefficient d_{33}, avoiding the beam walk-off problems between the mixed waves, and therefore the enhancement of the wave mixing efficiency. Note that a typical ratio of d_{33}/d_{31} on most of the nonlinear ferroelectric crystals is on the order of 5–10. As an example, the enhancement factor of a first order ($m = 1$) QPM device on the SHG efficiency is $I_{QPM}/I_{PM} = [(2/\pi) \cdot d_{33}/d_{31}]^2$ over that of the conventional PM device. However, one has to be aware of that that the period of the reversed domain decreases with the interaction wavelength. This imposes a fair amount of challenge in making fine-pitched

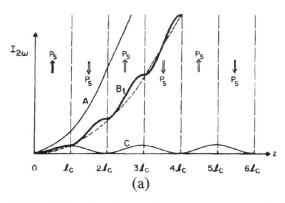
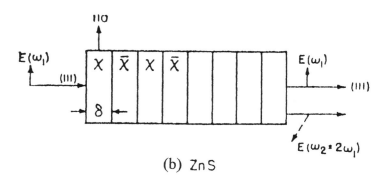

(a)   (b) ZnS

Fig. 2. (a) Variation of the second harmonic power generation in the case of A: conventional birefringent phase matching, B: first-order QPM, and C: non-phase matching interaction. l_c is the coherent length defined in Eq. (2). Reprinted with permission from [48], copyright 1962, American Physical Society. (b) QPM interaction in a semiconductor which may lack optical birefringence. Reprinted with permission from [47], copyright 1968, IEEE.

periodically poled QPM structures on ferroelectric nonlinear crystals as will be discussed in Section 3.

The realization of periodically poled QPM domain reversed structures in the late 1990s has greatly revised the research activities on nonlinear optics and electro-optics. Bulk periodically poled QPM ferroelectric devices suitable for linear and nonlinear optical applications are now commercially available through several vendors. For example, the periodically poled LiNbO$_3$ is now available from Crystal Technology in California (USA), the periodically poled RTA (PPRTA) from Crystal Associates in New Jersey (USA), and the periodically poled KTP (PPKTP) from Raicol Crystals in Israel.

1.2. Ultrasonic Resonance in a Periodical Domain Reversal Structure

It is known that thin-film deposition techniques [49] or thin piezoelectric plate bonding techniques [50] can be used for ultrasonic transducers operating at frequency up to 100 MHz. To achieve a higher frequency response, one has to minimize the insertion loss [51]. In this regard, Peuzin and Tasson have suggested that, by using a periodic domain reversal structure, one can obtain a high frequency, narrow-band hypersounic transducer on LiNbO$_3$ [52].

Referring to Figure 3, the operational principle of utilization of a periodic domain inverted LiNbO$_3$ structure as an acoustic resonator/transducer is as follows [53]. Under excitation by an alternating external electric field, the discontinuity of the piezoelectric coefficient at the interface of the domain walls can effectively result in a periodic distribution of sound δ-sources [54]. The ultrasonic waves emitted by such sound δ-sources will interfere with each other. The conditions for constructive interference are

$$k(a + b) = 2n\pi \qquad (3)$$

and the resonance frequency is determined by

$$f_n = nv/(a + b) \qquad (4)$$

where $(a + b)$ is the domain period, $k = \omega/v$ is the wave vector, $v = \sqrt{C_{33}^D/\rho}$ is the sound velocity, and C_{33} is the elastic constant, respectively. With a periodic modulation in the sign of the piezoelectric coefficient h$_{33}$ along the Z-axis, excitation of a pure longitudinal ultrasound wave can be realized. In this configuration, the spontaneous polarization vector P_s in the periodical domain exhibits a "head-to-head" or "tail-to-tail" configuration. Note that the resonant frequency is now solely determined by the period of the domain reversal structure instead of the plate thickness as in the conventional method. Moreover, if the periodical domain structure has the spontaneous polarization vector aligned in an "anti-parallel" configuration, the piezoelectric constant (h$_{22}$ or h$_{15}$) will be periodically reversed along the X-axis, and a quasi-longitudinal (h$_{22}$) or quasi-shear (h$_{15}$) wave can be excited.

Compared with the conventional thin-film or thin-plate techniques, ultrasonic resonators and transducers based on periodically poled domain reversed structures have the following advantages. First of all, the resonant frequency can be precisely determined by the *periodicity* alone. For example, an acoustic resonator of operation frequency near 1 GHz can be realized with a domain periodicity of 7.4 μm. Second, the load impedance of the resonator can be adjusted by the *number* of periods. As a result, very low insertion loss can be achieved. For example, transducers with an insertion loss close to 0 dB at 555 MHz has been reported [45].

In the polarization switching methods that will be discussed as follows, examples will be given on, but not limited to, lithium niobate (LiNbO$_3$) and its isomorphism lithium tantalate (LiTaO$_3$). These polarization switching techniques, however, can be applied to other ferroelectric crystals and materials having inversion symmetry as well. At the present time, LiNbO$_3$ and LiTaO$_3$ are the two widely used ferroelectric crystals due to the excellent properties of relatively wide optical transparency range, large electro-optics, nonlinear-optics, pyroelectric, acoustic-optics [55], and piezoelectric constants [56, 57]. To date, most of the LiNbO$_3$ or LiTaO$_3$ single crystals are produced commercially by the Czochralski (CZ) technique at a growth temperature in the vicinity of its melting point

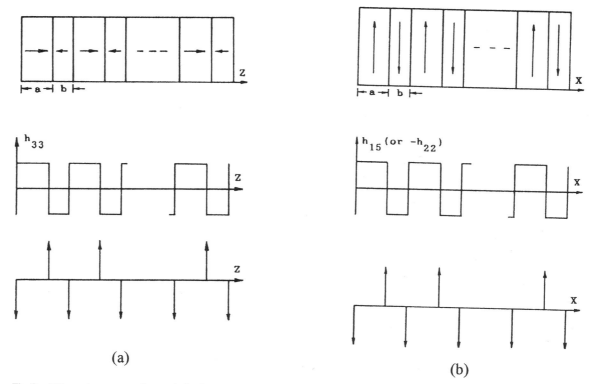

Fig. 3. Ultrasonic resonance in a periodically poled domain reversed structure on LiNbO$_3$. In case (a), the periodical domain lies in a "head-to-head" configuration. The periodic modulation in the piezoelectric constant h_{33} and its discontinuity at the domain boundary can generate a δ sound source and propagate longitudinally. In case (b), the periodical domains are anti-parallel to each other to generate a quasi-longitudinal (h_{22}) or a quasi-shear (h_{15}) wave. Reprinted with permission from [53], copyright 1992, American Institute of Physics.

of 1250 or 1650°C, respectively. However, the relative large coercive field, ~21 kV/mm, makes the process of room temperature polarization switching on this material system a great challenge. In comparison, polarization switching on other ferroelectric material systems such as barium titanate (BaTiO$_3$), triglycine sulfate (TGS), PZT, and KTP becomes relatively easier due to a much lower coercive field of 1–3 kV/mm involved in the domain reversal process.

Note that much of the current understanding on the kinetics of the polarization switching process, however, is based on that of previous experience with BaTiO$_3$ and TGS. They offer valuable information on the kinetics of domain reversal in the *low*-field regime. On the other hand, by using the-state-of-the-art pulsed poling techniques, one can further explore the domain switching process in the *high*-field regime. Discussion of the polarization switching process in this chapter will be organized as follows. We will first review the material processing issues that lead to the realization of single domain ferroelectric crystal. The imaging methods to examine domain nucleation, domain wall motion, and defect structure will be discussed next. Regarding the process of polarization switching of a layered structure, we will first discuss the thermal procedures that can induce a thin layer of surface domain inversion on single domain crystals and the related switching mechanism. Discussion on the *in situ* growth techniques for making periodically

poled domain reversal structures will follow. We then focus on the *ex situ* periodic poling techniques involving the use of the e-beam writing, scanning force microscopy, and pulsed electric field. The state-of-the-art nanoscale domain engineering that is attractive to the remnant storage application on the PZT and BaTiO$_3$ material system will also be addressed. Discussion of the stoichiometry control and optical damage resistant impurity effects on the polarization switching process will also be included. Future applications of the periodical domain reversal structures on nonlinear- and electro-optics will be concluded in the summary section.

2. PROCESSING OF SINGLE DOMAIN CRYSTALS

2.1. Czochralski Growth Method

2.1.1. General Remarks

Lithium niobate (LiNbO$_3$) and lithium tantalate (LiTaO$_3$) are the man-made crystals first flux-grown in 1949. They are found to be ferroelectric based on the observation of hysteresis loops reported by Matthias and Remeika [58]. Large size LiNbO$_3$ and LiTaO$_3$ single crystals were first grown by the CZ technique by Ballman in 1965 [59]. To date most of the commercially available 3 inch LiNbO$_3$ and LiTaO$_3$ single crystal wafers are grown

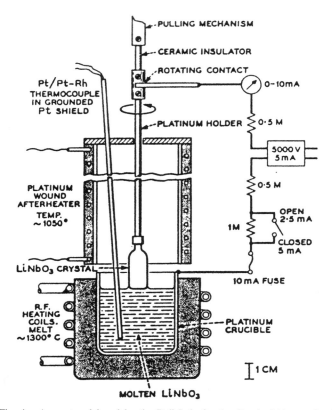

Fig. 4. Apparatus deigned by the Bell Lab. for the Czochralski growth of LiNbO₃. Note the *in situ* electric poling setup attached to the rotation contact and the Pt after-heater to reduce the axial temperature gradient. Reprinted with permission from [61], copyright 1966, Elsevier Science.

by the standard CZ method and are of congruent composition, but a modified double-crucible CZ method can be applied to grow crystals of stoichiometric composition [60]. Illustrated in Figure 4 is one such apparatus for the CZ growth of LiNbO₃ single crystals reported by the Bell Lab. Group in the early 1960s [61]. In preparing the charge for the crystal growth, for example on LiNbO₃, the Li_2CO_3 and Nb_2O_5 powders with purity in excess of 4N are used as the raw materials. The powders are mixed at the appropriate proportions in a ball mill and calcined at temperature of 1000°C for 10–12 hr [62, 63]. The material is then loaded into a platinum (Pt) crucible and RF heated by an induction coil.

It is noted that the RF heater can normally produce large axial temperature gradients unless an active after-heater, a Pt cylinder placed upon the crucible as shown in Figure 4, is used. Recent study, however, indicates the use of a SiC resistive heater instead can yield less temperature gradients [64]. By adjusting the height of the induction coil, a temperature gradient of 40–80°C/cm can be realized along the axial direction. This temperature gradient will later be shown to be a crucial parameter in making a single domain crystal due to the pyroelectric-induced self-poling effects. The crystal growth temperature can be controlled by a Pt/Pt10%–Rh thermocouple via a computer controlled, proportional integral differential algorithm. To maintain the same thermal conditions at various

part of the crystallization front, and to keep the melt homogeneous, the crystal should be rotated as it grows. The optimum growth conditions vary with the different system designs. But many investigators believe an initial axial gradient of 50 to 100°C/cm and a rotation rate of 5 to 40 rpm, in addition to a slow pulling rate of 0.1–3 mm/hr, are desirable to ensure a growth of flat interface [65–67]. After growth the boules are equilibrated in the after-heater and are then slowly cooled, *in situ*, to room temperature. Normally in the furnace, there is a supply of a slow oxygen purge to prevent the crystal from oxygen loss and to avoid brown discoloration [68]. By doing so, the resultant boules appear clear and can be free from cracks. Using this method, mass production of LiNbO₃, LiTaO₃, and related crystals with congruent composition has become commercially possible in recent years. We pass by noting that other crystal growth or processing techniques such as top seeded solution growth or vapor transport equilibration can be used to obtain stoichiometric LiNbO₃, LiTaO₃ single crystals and will be discussed in Section 3.5.

2.1.2. Congruent Issue

The phase equilibrium diagram of the Li_2O–Nb_2O_5 system was first proposed by Reisman and Holtzberg in 1958 [69]. Although commonly referred to as "LiNbO₃," the crystal composition in the phase diagram can range from a near-stoichiometric value to a lithium-poor composition as low as ~45 mol% Li_2O [70]. However, for a congruent-grown crystal, the composition of the melt and the crystal are identical. The crystal can solidify at the melt composition, and both the melt and the crystal compositions are independent of the fraction of the melt that has been solidified. By growing crystals from a congruent melt, one would expect the crystal composition be less susceptible to the temperature fluctuation or growth rate change. From the production point of view, congruent crystallization can thus allow the maximization of the crystal growth and the total melt fraction that has been crystallized [71].

We note the compositional variation of a ferroelectric crystal can be conveniently determined by measuring the difference in the ferroelectric transition temperature (i.e., the Curie point, T_c) for samples sliced from the top and bottom of each boule. Traditionally this can be done by dielectric or differential thermal analyzing (DTA) techniques. By definition the difference in T_c would be zero if the samples are of congruent composition. In the dielectric measurement, on cycling the temperature of a sample, a dielectric anomaly due to the ferroelectric phase transition can be detected. It appears as a sharp peak in the phase difference between the sample and the reference arms of an ac impedance bridge circuit [72]. In the DTA measurement, on temperature-cycling the sample, T_c is determined from the extrapolated departure from the baseline of the temperature hysteresis curve [73]. The congruent composition of LiNbO₃ has thus been determined to contain 48.38–48.45 mol% Li_2O with a measured $T_c \sim 1140$°C Near the congruent composition, the variation of the Curie point is as follows: $T_c = 9095.2 - 369.05C + 4.228C^2$, where C is the

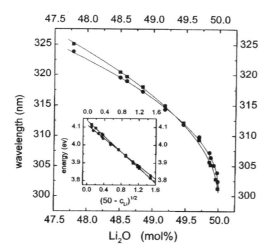

Fig. 5. The difference in the extraordinary (symbol ■) and ordinary (symbol ●) absorption edge of LiNbO₃ at $T = 22°C$ as a function of the crystal's composition. Reprinted with permission from [83], copyright 1997, American Institute of Physics.

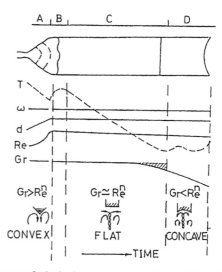

Fig. 6. Summary of a hydrodynamic theory analysis of the melt flow effects on the change of the interface shape at various stages of the Czochralski growth of LiNbO₃. Reprinted with permission from [87], copyright 1995, Elsevier Science.

concentration of Li_2O in units of mol%. We pass by noting that the increase of T_c with an improved stoichiometry control has also been noted on LiTaO₃ [74]. For example, it is noted that a congruent LiTaO₃ with 48.5 mol% Li_2O has a $T_c \sim 600°C$, whereas the stoichiometric mixture of Li_2CO_3 and Ta_2O_5 powders shows a $T_c \sim 690°C$.

Note that the recent development of optical methods has also enabled the characterization of a ferroelectric crystal over a wide range of temperatures [75], wavelengths [76], dopings [77, 78], and compositions [79]. For a review on this subject, the reader can refer to the work by Wöhlecke et al. [80]. One such efficient characterization method is the use of the tuning curve of SHG phase matching temperature to characterize the homogeneity and the composition of LiNbO₃ [81]. Note that the SHG technique can also be engaged as a nondestructive method to image the domain structures of LiNbO₃ and LiTaO₃ [82]. The optical measurement on the birefringence and the UV absorption edge can also be used to determine the composition of a ferroelectric crystal. For example, data shown in Figure 5 exhibit the difference in the ordinary and extraordinary polarized absorption edges of a LiNbO₃ single crystal at various crystal compositions [83].

2.1.3. Interface Issue

From the study of the crystal growth procedure, it is noted that the crystal quality is related to the shape of the liquid–solid interface. Conventional wisdom has recognized that a good crystal quality is obtained when the interface is flat, while a nonflat interface could result in dislocation, strain, and inhomogeneous distribution of the impurities. In particular, it is noted that the melt flow near the interface plays an important role in determining the interface shape. Here, we will focus on the kinetics of crystal growth and the relevant control parameters commonly encountered in the CZ growth method.

Our current knowledge of the kinetics of the CZ growth is based on a hydrodynamic theory developed by Kobayashi et al. [84]. It is noted that, in a pure melt, flows from free convection, surface tension driven flow, forced convection due to the crystal rotation, and crucible rotation are the important factors related to the crystal growth. The parameters affecting the melt flow are the axial temperature, rotation rate, melt depth, and the position of the crucible in the furnace. The pertinent dimensionless numbers used in the hydrodynamic analysis are the Grashof (Gr), Reynolds (Re), and Marangoni (Ma) numbers in the melt [85, 86]. In particular, the Gr number is a measure of the flow caused by the buoyancy force. It is affected by the melt depth and the temperature between the crystal and the crucible. The Re number of a forced flow is influenced by the rotational speed and the crystal radius. The Ma number is a measure of the surface tension-driven flow and is influenced by the temperature variation at the melt-free surface.

Noted that with a low Gr/Re^2 ratio, the process of forced convection dominates in the melt; whereas in the case of high Gr/Re^2 ratio, the process of free convection dominates. For an intermediate ratio, both free and forced convection coexist and they are separated by a stagnant surface. In summary, at a fixed Gr number, the heat transfer mechanism would change from a free convection dominant mode into a forced convection dominant mode as the Re number increases. This change of flow process can further bring up a change in the temperature distribution in the melt. As a result, the interface shape evolves according to the temperature distribution of the isotherm near the liquid–solid interface.

Shown in Figure 6 is one such example of the interface shape of a CZ-grown LiNbO₃ single crystal [87]. Besides the dimensionless numbers of Gr and Re, the relevant growth parameters illustrated in the figure are the crystal diameter d, rotation speed ω, and evolution of temperature T. We note at

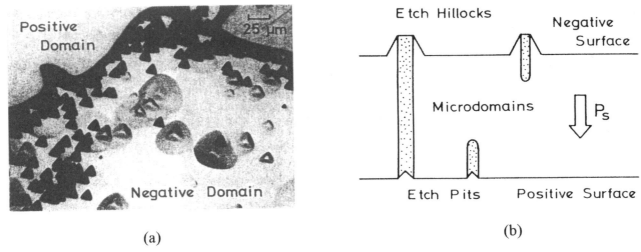

Fig. 7. Surface topography on the C faces of LiNbO$_3$ by an etchant of 1 : 2 volume ratio of HF and HNO$_3$. (a) Black triangular in the negative domain is the hillock etch pattern observed on the $-C$ face. (b) The correspondence between the etch pit on the $+C$, etch hillock on the $-C$, and the microdomain in LiNbO$_3$. Reprinted with permission from [95], copyright 1975, American Institute of Physics.

a small crystal diameter, the free conventional flow becomes the dominant mode in the melt. This process results in a small longitudinal temperature distribution in the central part of the interface and therefore forms a *convex* interface shape. Note that this procedure continues during the "necking" and "shoulder" process (in zones A and B of Figure 6) of the crystal growth. In the growth of the crystal "body" region (zone C), it is desirable to maintain a *flat* interface to ensure crystal quality. This condition can be fulfilled by maintaining a constant ratio of Gr $=$ (Re)^n where $1.5 \leq n \leq 2.5$. Under this circumstance, both the free and the forced convection coexist in the melt, resulting in a homogeneous temperature distribution lying underneath the crystal growth region and ensuring a *flat* interface. Near the end of the crystal growth (zone D), the reduction of the melt depth affects the Gr number and results in a dominant forced convection flow. This procedure then produces a large temperature gradient in the center part of the interface and changes its shape to *concave*. We note that similar flow analysis can be applied to explain the interface shape of other oxide crystals such as Gd$_3$Ga$_5$O$_{12}$ [88], (Sr,Ba)N$_2$O$_6$ [89], and Y$_3$Al$_5$O$_{12}$ [90].

2.2. Domain Characterization

2.2.1. Domain Observation by Etching Techniques

The as-grown single LiNbO$_3$ crystal normally exhibits a multidomain structure due to (i) segregation issues in the melt composition and (ii) minimization of the total free energy. Crystals containing multiple domains usually contain considerable strains and low-angle grain boundaries [91]. These side effects make a multidomain LiNbO$_3$ crystal susceptible to the processing environment and limit its useful application.

In many of the as-grown ferroelectric crystals, domains of opposite parity cannot be distinguished from each other under the conventional optical microscopy due to the lack of phase contrast. This is because the second-rank dielectric tensor does not have a sign change at the opposite domain boundaries. To reveal the domain structure, conventional practice suggests that one can apply the destructive methods to bring up the topographic contrast between the domains. Chemical etching with a differential etch rate on the opposite domains represents one such approach. After etching, details of the domain structures can be revealed under the optical or scanning electronic microscope. It is found that molten KOH at 400°C or solution consisting of 1 : 2 ratio of HF : HNO$_3$ at the boiling point ($\sim$110°C) can preferentially etch the polar $-C$ face over the $+C$ face of LiNbO$_3$ [18]. During the HF treatment, it is noted that the fluorine ions diffuse into the crystal but mainly onto the positive ferroelectric domain [92]. Note that the polarity assignment of a positive or negative face is based on a pyroelectric test. The conventional sign is that the spontaneous polarization P_s increases with the decrease of temperature and that the change can be detected as a current flow in the external circuit [93, 94].

The surface topography of an etched LiNbO$_3$ multidomain structure can be summarized as follows. As illustrated in Figure 7, the etch patterns consist of etch hillocks on the $-C$ surface, etch pits on the $+C$ surface, and needle-shape etch patterns on the Y surface [95]. One can further identify a one to one correspondence between the etch pit on the $+C$ surface, and the etch hillock on the $-C$ surface of LiNbO$_3$. These observations suggest that the pit and hillock originate from the same source. One possibility of such a source is due to the formation of a microdomain with a spontaneous polarization opposite that of the original crystal. Note that similar etch patterns have also been observed on the surfaces of LiTaO$_3$ crystal when etched in a 1 : 2 ratio of HF : HNO$_3$ solution [96].

Moreover, we note there are other modern noninvasive methods that take advantage of the piezoelectric or electro-optic response to reveal the kinetics of domain formation and wall motion on ferroelectrics. Details of these methods will be discussed in Section 2.4. Due to the lattice distortion [97, 98], or strain [99] induced with the domain poling process, the X-ray imaging techniques can also be applied to observe the domain structure.

2.2.2. Single Domain Poling

Using the selective etching method, one can further examine the details of domain structures on a single crystal. Along the C-axis of the as-grown $LiNbO_3$ single crystal, it is noted that the domain tends to consist of a concentric annual ring structure extending down to the crystal [100]. In a growth direction perpendicular to the C-axis, the multidomains tend to occur as flat sandwich structures parallel to the C-axis. The presence of impurities in the crystal can also affect the appearance of the domain structures. Shown in Figure 8 is one example of multidomain structures on a 1% $MgO:LiNbO_3$ single crystal grown along the C-axis of a congruent composition [101]. The appearance of a concentric ring structure consisting of alternative positive and negative domains can be clearly seen from the etched C-faces. In fact, the ring-shaped structure corresponds to the growth striations from the solid–liquid interface. As mentioned above, the strains and grain boundaries associated with the multidomains can cause stability issues on the device application and one generally requires a "poling" process to restore the crystal into a single domain state.

To retain a single domain structure, conventional wisdom has shown that the poling process can be proceeded by applying an electric field along the crystal's polar C-axis when it is cooled through the Curie point. One suggestion of such a poling mechanism is due to the coupling of the spontaneous polarization to a temperature gradient field, i.e., via the pyroelectric effect [102]. When the crystal is heated into a paraelectric state, a small external electric field can effectively displace the Li ions across the oxygen plane. This action can impress an ordered $LiNbO_3$ structure as shown in Figure 1 such that its polarization state can be "locked" or "frozen" in place when the crystal is cooled below the Curie point [18]. Note that this poling procedure can be applied *in situ* during the crystal growth or *ex situ* after the growth.

One such *in situ* poling apparatus is already included in Figure 4 of the CZ growth oven. Note that the rotating contact on the top of the Pt holder can be used to apply a voltage such that an electric field can be developed between the high resistant cooler end of the crystal and the low resistant hot end near the melt. It is found that an applied voltage as little as 0.5 V, corresponding to 2 mA/cm^2 across a 1000-ohm seed, is sufficient to pole a single domain $LiNbO_3$ crystal. As the crystal grows and the resistance rises, one has to increase the applied voltage to keep the poling current constant and to maintain a constant field near the crystal growth region [103]. However, caution must be exercised in deploying the electrical parameters to avoid the electrolysis of the melt and the decomposition of the seeding crystal [104].

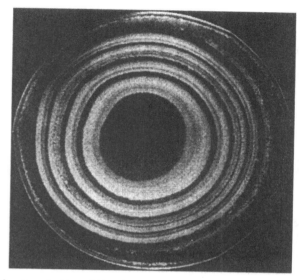

Fig. 8. Ferroelectric multidomain structures revealed on the etched surface of the 1 mol% MgO doped as-grown $LiNbO_3$ crystal of the congruent composition. Note the ring-shaped domain structure corresponds to the growth striation form the solid–liquid interface. Reprinted with permission from [101], copyright 1990, Elsevier Science.

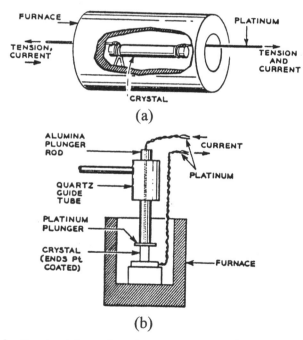

Fig. 9. Experimental setup for performing *ex situ* electrical poling at temperature near the Curie point (T_c) of $LiNbO_3$ for (a) long and (b) thin samples. Reprinted with permission from [61], copyright 1966, Journal of Physics and Chemistry of Solids.

For *ex situ* poling of a single domain crystal, the apparatus shown in Figure 9 is the one originally designed by the Bell Lab. Group [18]. Referring to Figure 9, a long ferroelectric crystal to be poled is hung by the Pt wires anchored in grooves cut into the crystal. A thin substrate can be coated with a Pt electrode on one face and the other upper face contacted by a Pt plunger rode.

A poling field can then be applied between the Pt electrodes. For *ex situ* poling of LiNbO$_3$, it is found that at a temperature range from 1100 to 1200°C, a smaller applied field on the order of 1 V/cm is sufficient to complete a single domain poling process within 15 min [103]. Note that if a poled single domain LiNbO$_3$ crystal is reheated at a temperature above its Curie point (~1140°C) but below its melting point of 1260°C, after a cooling procedure, a multidomain structure would reappear due to the occurrence of a para- to ferroelectric phase transition. On the other hand, an extended annealing of the poled LiNbO$_3$ bulk single crystal at a temperature below its Curie point will not produce any changes in the domain configuration. However, this will be not the case when there is a periodically poled domain reversed structure on the LiNbO$_3$ substrate. We will address the domain stability issue in Section 3.5.

For the *ex situ* poling of LiTaO$_3$ whose Curie point is around 600°C, a typical procedure is to anneal the crystal in air at 750°C and subsequently cool it at a rate of 2°C/min to room temperature with an electric field applied along the crystal *C*-axis. However, from common practice one can note that the poling field could vary from a few volts/cm at temperatures near the Curie point up to 5×10^3 V/mm at 200°C [105].

For *ex situ* poling of KTP, one notes that its Curie point T_c at about 940°C is very close to the material's decomposition temperature of 1050°C [106]. As a result, the poling procedure of bulk single domain KTP crystal is preferred to perform at a lower temperature of 500°C with a current density of 5 mA/cm^2 [107]. Because of the high ionic conductivity of KTP with the temperature, the voltage required to keep a constant poling current density varies from 700 V at 200°C to 25 V at 500°C.

Considering the temperature effect on the decrease of the coercive field and the spontaneous polarization, it is recognized that electric field poling of LiNbO$_3$ and LiTaO$_3$ can be much easier proceeding at temperature near T_c than at room temperature. However, there are processing concerns on the stability issues of the periodically poled domains upon which an *ex situ* poling process at room temperature becomes a preferred choice. We will address this issue by the state-of-the-art pulsed field poling techniques in Section 3.5.

2.3. Kinetics of Polarization Switching in the Low-Field Regime

Much of our current understanding on the kinetics of domain inversion and domain motion is credited to the early study of the polarization switching process on BaTiO$_3$ [108, 109], TGS [110, 111], and Rochelle Salt [112] (e.g., KH$_2$PO$_4$ [113]) in the 1950s and the 1960s. These crystals are genetically known to have relatively low coercive field and spontaneous polarization in the ferroelectric state as listed in Table I. Study of the poling process on these material systems therefore can provide us with valuable information on the kinetics of polarization switching in the low-field regime.

Characteristics of the polarization switching process in the low-field regime can be summarized in the following steps:

(i) nucleation of the new 180° reversed domains and their forward growth through the thickness of the crystal, (ii) sideways expansion of the domains, and (iii) coalescence of the domains [114]. Step (i) can also include the forward growth of the nuclei, which can exist *a priori* in the initial state. In the low-field regime, the domain nucleation rate exhibits an exponential dependence on the applied poling field. Moreover, the domain motion in the forward direction generally is found to be much faster than that in the lateral direction.

A theory suggests that upon the application of an external poling field, electrons tunnel into the region where the spontaneous polarization vector fluctuates. This region forms a "fluctuon" which is a bound state of electrons with fluctuation, and plays a role as a repolarization nucleus to initiate the polarization switching process [115]. Another theory suggests that clusters of reversed polarization nucleate under the applied field in the vicinity of a random field attached to the quenched defects. Polarization switching occurs due to the breakdown of the ferroelectric long-range order associated with the polar cluster state [116]. Yet another theory suggests that the growth front of the reversed domain exhibits an ellipsodial shape which is related to the release of electromagnetic energy. This specific electromagnetic energy is used to compensate the surface energy increase in the domain wall and the energy dissipation due to the finite velocity of domain motion [117].

In the following discussion, we will review the optical and electrical characterization methods that enable the observations of domain nucleation and domain wall motion in the polarization switching process. Among the optical characterization techniques, we will discuss those involving the use of polarized light, differential etching, and piezoelectric response upon which the spatial resolution of domain characterization increases accordingly from micro- to nanometer. In the electrical characterization, we will address the issue of transient current analysis and the use of the abstracted switching parameters to distinguish different mechanisms governing the process of domain nucleation and domain wall motion [118].

2.4. Optical Characterization of Domain Switching

2.4.1. Polarized Light Method

The polarized light method is an old technique used to characterize the optical properties of crystals and ceramics [119]. The cross-polarized microscope method is originally proposed by Merz in 1954 to examine the domain structure of BaTiO$_3$ under *stress* [120]. The idea is based upon the fact that the *C*-axis of BaTiO$_3$ in the tetragonal phase is the optic axis of the crystal. As a result, the *c* domain would look dark when viewed through a microscope put between two cross-polarizers, whereas the *a* domain would be bright.

This method is then adapted to study the kinetics of sidewise 180° domain motion in BaTiO$_3$ [121]. Illustrated in Figure 10 is one such experimental setup designed by Miller and Savage [122]. In this configuration, transparent electrodes are deposited on the ferroelectric axis of BaTiO$_3$ which in turn is inserted between a pair of cross-polarizers. As the applied poling field exceeds the small coercive field ~ 0.5 kV/cm of $BaTiO_3$, one can discern a change in the anti-parallel domain configuration continuously during the polarization reversal process. This procedure also enables a direct observation of the sidewise motion of the $180°$ reversed domain walls. When

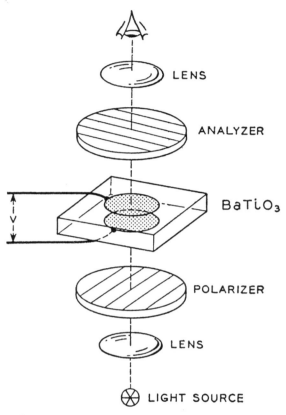

Fig. 10. The cross-polarized method used to observe the 180°C domain wall motion during the polarization reversal in BaTiO₃. Reprinted with permission from [122], copyright 1960, American Institute of Physics.

viewed through the cross-polarizers, the domain wall velocity can be calculated from the measurement of the time for the wall under observation to move through a known distance. A motion picture can be made of the poling process but the resolved domain wall velocity is limited to $\sim 10^{-2}$ cm/s. This method is particularly suitable for investigating the dynamics of polarization switching of $BaTiO_3$ in the *low*-field regime (<2 kV/cm).

A modern revised imaging technique of domain motion, however, takes advantage of the electro-optic (EO) effect to resolve a phase contrast between the opposite domains [123]. Schematically shown in Figure 11 is the setup of an electro-optic imaging microscopy, in which the imaging is taken in a transmission mode using a pair of cross-polarizers in conjunction with a Nomarski objective. The use of liquid electrode in the domain reversal procedure can assist a high-field poling without causing the dielectric breakdown and minimizing the fringing field effect. We will be discussing these issues in Section 3.5. The operational principle of the EO imaging technique is as follows. When an external field E is applied across a ferroelectric domain, an increase of refractive index occurs for domain with P_s opposite to E whereas the refractive index decreases when P_s is parallel to E.

For congruent grown $LiNbO_3$ and $LiTaO_3$, the first order EO coefficient of $r_{13} = r_{23} \sim 8.4$ pm/V. When the $LiNbO_3$ and $LiTaO_3$ is poled at a coercive field of 21 kV/mm, the induced in-plane refractive index difference, i.e., $\Delta n_2 = n^3 r_{23} E_3 \sim 9.1 \times 10^{-4}$ across a domain boundary, will result in light scattering when it transmits through the crystal. Using this method, video imaging of the $180°$ domain nucleation and its lateral motion is possible. The successive video frames in Figure 12 represent a typical evolution picture on the domain nucleation, lateral domain motion, and coalescence on $LiTaO_3$ at a poling field of 20.76 kV/mm [124]. The image resolution is 6×6 μm/per pixel across a 160×120 pixel image space, and the frame rate is $1/30$ s. Using this video imaging method, the sidewise velocity of the $180°$ domain wall of $LiTaO_3$ is calculated

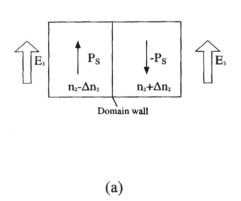

(a)

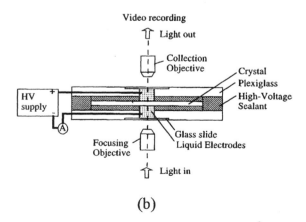

(b)

Fig. 11. (a) The principle of electro-optic imaging of a 180° domain is by the induced refraction index change $\Delta n_2 = -(1/2)n^3 r_{23} E_3$. The index decreases (increases) when E_3 is (anti)-parallel to P_s, thus resulting in a phase contrast when transmits a light. (b) Schematic drawing of the electro-optic imaging microscopy setup for the *in situ* observation of domain reversal and lateral motion. Reprinted with permission from [123], copyright 1999, American Institute of Physics.

Fig. 12.   The *in situ* video recording on the nucleation and growth of 180° domains in congruent LiTaO$_3$ under a DC poling field of 20.76 kV/mm. All successive frames are taken 15 s apart from each other. The merged domain fronts (V, VI, VII) are marked with arrows. Reprinted with permission from [124], copyright 1999, American Institute of Physics.

to be 2300 μm/s at the merged domain fronts when measured at a constant field of 21.2 kV/mm.

This method can also be used to investigate the nature of the poling field on the domain wall motion. For example, it

is noted that the wall velocity of LiTaO$_3$ under a steady-state field poling is larger by one order of magnitude than that in the pulsed-field poling case [125]. This result suggests that the latter poling process is beneficial in controlling the lateral domain

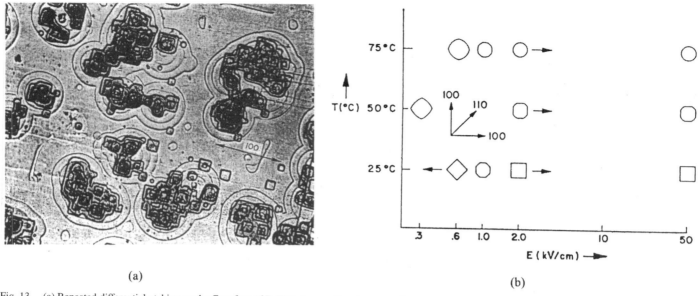

(a)

(b)

Fig. 13. (a) Repeated differential etching on the C surface of BaTiO$_3$ in revealing the successive positions of the 180° domain wall motion. (b) The phase diagram of domain shape schematically drawn as a function of the applied field and temperature. Note in (a) that the large, round shape corresponds to domain wall moving at 75°C and a poling field of 2.0 kV/cm, and the small square corresponds to that moving at 25°C and 2.7 kV/cm. Reprinted with permission from [128], copyright 1964, American Institute of Physics.

motion to make a fine-pitch of switched domain. By using this method, one can further examine the details of the domain nucleation structures. For example, it is noted that the domains nucleate and grow as a six-folded polygon on LiNbO$_3$ whereas it appears as a triangular shape on LiTaO$_3$ [126].

2.4.2. Repeated Differential Etching Method

We now discuss an old but valuable technique of repeated differential etching that can be used to examine the domain wall motion with better resolution at high velocity than that of the optical method mentioned just above [127]. This procedure involves the use of multiple partial forward-switching and full back-switching pulses to reveal the domain nucleation and domain wall growth. During each pulse of the partial forward switching, the domain walls move to the new positions to be revealed by the next back-switching process. When the crystal fully is back-switched, the positions of the domain walls in existence can be revealed by the diluted HF etch. Illustrated in Figure 13a is one such photograph of a BaTiO$_3$ surface revealing the successive positions of the 180° domain walls after the HF etching [128]. The large, round colored figures are positions of domain walls moving at 75°C at 2 kV/cm, whereas the small, square, dark figures are that moving at 25°C at a larger field of 2.7 kV/cm. Summarized in Figure 13b is the phase diagram of the BaTiO$_3$ domain shape as a function of the poling temperature and applied field. Note the square BaTiO$_3$ domain shape commonly observed at room temperature becomes rounded at 50 and 75°C and almost exactly circular at 75°C at 1 kV/cm. Note that this repeated differential etching method can also be applied to study the nucleation rate of the reversed domain on the surface of a BaTiO$_3$ crystal by applying successive poling

pulses [129]. Analysis of the domain switching rate and wall velocity of BaTiO$_3$ will be given in Section 2.5 of the electrical characterization methods.

2.4.3. Piezoelectric Imaging Method

A more refined nondestructive imaging method is to take advantage of the differential piezoelectric response of a ferroelectric crystal to reveal the topographic distribution of the domain structure. This method is sensitive to the polarization distribution at the ferroelectric surfaces and can offer a lateral resolution down to nanometer scale [130]. Shown in Figure 14 is one such experimental setup combined with a state-of-the-art scanning force microscope (SFM) [131]. The operation principle is based on the laser deflection from the SFM cantilever onto a four-quadrant photodetector to resolve the surface topography [132]. Note that electronic feedback is established by scanning the cantilever in a constant force (i.e., contact mode) such that the corresponding perpendicular and lateral oscillation of the cantilever can follow the interaction of the surface polarization with the applied modulation field.

A recent study suggests that the image contrast from SFM indeed reflects an interaction between the electrostatic force and the piezoelectric response [133]. To be more specific, the applied voltage $U = U_0 \sin \omega t$ between the SFM tip and the crystal can induce a converse piezoelectric effect which in turn can result in an expansion, contraction, or shear strain on the ferroelectric crystal. The corresponding change of the crystal thickness (Δ) varies according to $\Delta = -d_{ij} U_0 \sin \omega t$, where d_{ij} is the piezoelectric coefficient of the surface domain [134]. Since the piezoelectric constant d_{ij} is a second-rank tensor, there would be a sign change when the domain switches its polarization state. The converse piezoelectric response of d_{15} and d_{33} would then cause an in-plane torsion and z-axis vibration, respectively, of the cantilever to follow the surface distribution of the crystal's polarization state. When the laser beam is deflected from the oscillating cantilever, the corresponding phase change can be demodulated by the lock-in technique. As a result, one can reveal the in- and out-plane polarization contrast simultaneously.

Note that this polarization sensitive method is in particular valuable to characterize the dynamic response of the domain reversal process. For example, a domain width less than 8 nm has been measured on the 180° domain boundaries in TGS [135]. This method also allows a reconstruction of the surface crystallography of ferroelectric BaTiO$_3$ ceramics

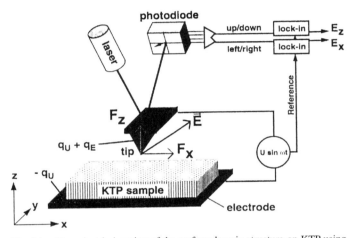

Fig. 14. Piezoelectric imaging of the surface domain structure on KTP using a modified SFM setup. Note the topographic image contrast arises from the surface polarization interaction via the converse piezoelectric effect of the biased SFM cantilever. The position dependence of the laser beam deflection from the distorted and vibrated SFM tip would then transfer into a local distribution of the surface domain structure. Reprinted with permission from [131], copyright 1998, American Institute of Physics.

by recording the corresponding three-dimensional polarization distribution [136]. The surface domain structures of ferroelectric TGS and BaTiO$_3$ crystals have also been observed [137]. Shown in Figure 15 are one such piezoelectric probing of the (a) the topographic image of the (a, c) domain in BaTiO$_3$, and the distribution of the (b) out-plane polarization P_z of the c domain, and (c) in-plane polarization P_x of the a domain, respectively. Note in Figure 15b and c that the bright areas correspond to polarization vectors pointing along the $+C$ axis and the upper right (i.e., [100]) direction, respectively; whereas the dark regions reflect an anti-parallel polarization vector.

We pass by noting that the idea of piezoelectric scanning can also be applied to other systems such as the electrostatic force microscope to study the domain structure, for example on LiNbO$_3$ [138] or TGS [139]. Similarly, the confocal scanning or near-field optical microscope can also be used to reveal the surface domain structures on LiNbO$_3$ [140], LiTaO$_3$ [141], BaSrTiO$_3$ [142], and TGS [143].

2.5. Electrical Characterization of Domain Switching in Low Coercive-Field Materials

In the early development of the polarization switching theory, the popular Fatuzzo–Merz three-stage model suggests that the dynamical response of domain reversal consists of a first step of domain nucleation on the ferroelectric surface [144]. In the second step the domains grow as thin spikelike cones and move forward through the thickness of the crystal. Finally, after the cones reach the opposite surface of the crystal, they gradually expand by a sidewise motion.

A typical electrode configuration for an electrical characterization of the polarization switching process is illustrated in Figure 16 [145]. Note that at a high-poling field, sidewise spreading of the domain exceeding the electrode area can take place. A guard electrode introduced in the setup of Figure 16 can prevent the effects of fringing field on the switching and

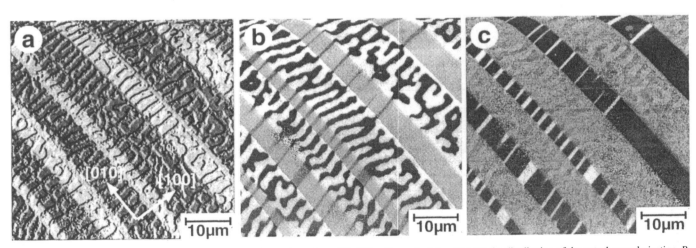

Fig. 15. (a) SFM topographic imaging of the (a, c) domain on the surface of BaTiO$_3$. Shown in (b) and (c) are the distribution of the out-plane polarization P_z of the c domain, and the in-plane polarization P_x of the a domain, respectively. Note the bright areas in (b) and (c) correspond to the polarization vector pointing to the $+C$ axis and the [100] direction, respectively. Reprinted with permission from [137], copyright 1998, Springer–Verlag.

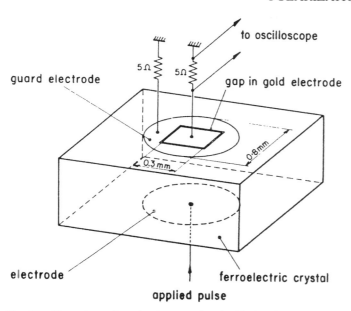

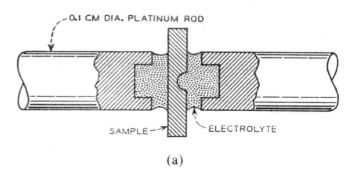

(a)

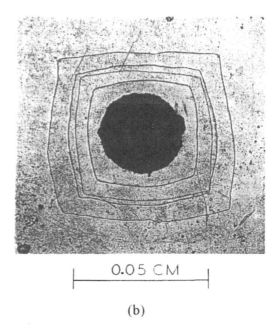

(b)

Fig. 16. Electrode configuration in preventing the fringing field effect from the electrical poling process of domain switching. Reprinted with permission from [145], copyright 1965, American Institute of Physics.

Fig. 17. (a) The dimpled BaTiO₃ sample in a liquid electrode configuration used to observe the sidewise 180° domain motion. (b) The partially switched and etched BaTiO₃ surface revealing the expansion of the 180° reversed domain out of the dimpled area. Note the square domain shape poled at 0.32 kV/cm and room temperature agrees with that analyzed in Figure 13b. Reprinted with permission from [146], copyright 1958, American Physical Society.

hence eliminate the signal distortion in the waveform analysis. Note that the change rate of the spontaneous polarization, $d(2P_s \cdot A)/dt$, where A is the area under polarization reversal, generates a polarization switching current. To study the dynamic response of the domain reversal process, conventional wisdom suggests that one can first apply a forward electrical pulse to align all the dipoles in one direction, followed by a second short pulse of opposite polarity to measure the switching current i_s as function of time.

Quantities of interest to the switching analysis are the maximum averaged switching current $i_{\max}$ and t_s, the switching time. The transient current response of i_s can be measured as a voltage drop across a series resistor by a fast storage oscilloscope. By convention, t_s is arbitrarily defined as the time at which the switching current (i_s) has fallen to 0.1 of its maximum value ($i_{\max}$). Note that the total amount of the switched charges on a fixed poling area remains constant, independent of whether the domain switching is performed in the *forward* or *reverse* poling direction. They can be expressed as $Q_s = 2P_s \cdot A = t_s \cdot i_{\max} \cdot f$, where f is a factor between 0.5 to 1 and depends on the shape of the transient current. If f remains a constant in the poling procedure, $i_{\max}$ and $1/t_s$ would show exactly the same dependence on the field. A general rule is that with an increase of the applied poling field, both of the $i_{\max}$ and $1/t_s$ would increase correspondingly.

To characterize the electrical poling process, ferroelectric crystals with low coercive field naturally become the materials of choice. By doing so, the switching parameters of $i_{\max}$ and t_s can be measured over a wide range of applied field, temperature, and impurity. One such example is on the electrical poling of the c domain BaTiO₃ crystals. In the following we will review the characteristics of polarization switching on BaTiO₃

and the governing mechanism responsible for the dynamics of the polarization switching process.

2.5.1. Sidewise 180° Domain Motion in BaTiO₃

In 1958 Miller and Savage made the first observation of the sidewise 180° domain motion on BaTiO₃ by mounting a dimpled sample between the liquid electrode as shown in Figure 17 [146]. This setup is suitable to perform electrical poling in the low-filed regime ($E < 0.5$ kV/cm). The contact liquid electrode consists of saturated LiCl aqueous electrolyte. During the polarization switching process, the dimpled area of BaTiO₃ contacted to the liquid electrode behaves as a region of high field compared with the surrounding area. Polarization reversal will therefore begin in the dimpled region and expand sidewise. Also illustrated in Figure 17 is the C-face photograph of a dimpled BaTiO$_3$ which has been partially switched at a poling field of 0.32 kV/cm at room temperature and etched in HF for four times. The four concentric, square-etched patterns indicate a single 180° domain grown out of the dimpled area as suggested above. Note that the long sides of the switched domain make a 45° angle with respect to the crystalline a axis of BaTiO$_3$. This observation agrees well with what has been shown in Figure 13b. Moreover, from the etch figure one can determine the sidewise domain velocity by calculating the distance (x) it travels during each time interval of switching, i.e., $v_s = x/t_s$.

In the low-field regime ($E < 0.5$ kV/cm), the sidewise velocity of the 180° domain motion in BaTiO$_3$ can vary from a low value of 10^{-6} cm/s at $E = 0.2$ kV/cm up to 10^{-2} cm/s at $E = 0.5$ kV/cm. When the data are plotted in a logarithmic scale, however, revealed is an exponential dependence on the poling field according to [147]

$$v_s = v_\infty \exp\left(\frac{-\delta}{E}\right) \tag{5}$$

The δ will be shown in a later discussion to be the activation field in the Miller–Weinreich theory. It is found that δ depends also on the impurity concentration in the crystal. For example, $\delta = 3.9$ kV/cm for a 0.02 mol% Fe$_2$O$_3$ doped BaTiO$_3$, while it drops to 1.9 kV/cm in a 0.1 mol% AgNO$_3$ doped crystal. As a short note, in the low-field poling regime, one set of the measured *forward* and *sidewise* domain velocity of a 0.5 mm thick BaTiO$_3$ crystal at room temperature is [148]

$$
\begin{aligned}
v_f &= (5500 \text{ cm/s}) \exp\left(\frac{(-1.8 \text{ kV/cm})}{E}\right) \\
v_s &= (500 \text{ cm/s}) \exp\left(\frac{(-4.0 \text{ kV/cm})}{E}\right)
\end{aligned}
\tag{6}
$$

Notice that in the low-field poling of BaTiO$_3$, the domain forward motion speed is at least on order of magnitude larger than that in the lateral direction. It is worth noting that the corresponding activation field for the nucleation of new 180° reverse domains of BaTiO$_3$ at room temperature is ∼10 kV/cm. This observation indicates that the activation field for new domain nucleation is higher than that for sidewise domain expansion and that both are greater than that for the forward domain growth. The domain nucleation therefore can be regarded as the rate-limiting mechanism in the polarization switching process.

To study the domain reversal process of BaTiO$_3$ in the high-field regime (e.g., at E higher than 1 kV/cm), caution should be exercised to avoid the electrical short and dielectric breakdown. Conventional practice has shown that this can be done by mounting the BaTiO$_3$ crystal on a Teflon plate and using paraffin or epoxy resin to seal the crystal to avoid the electrical short in a pulsed field. To characterize the fast lateral 180° domain motion in the high-field regime, one can apply the repeated differential etching technique as discussed in Section 2.4.1. The sidewise velocity can thus be evaluated by dividing the displacement of an average domain by the pulse duration.

Since one expects the switching time t_s to decrease with the poling field, it is necessary to make a distinction between the processes of the faster domain forward motion against the

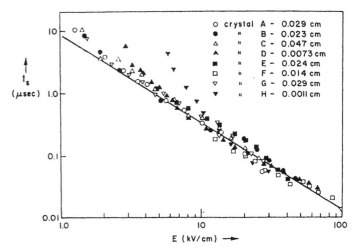

Fig. 18. The polarization switching time t_s of BaTiO$_3$ as a function of the poling field and crystal thickness. Note in the high-field regime ($E \gg 1$ kV/cm), there is essentially no systematic dependence of t_s on the thickness but reveals a $E^{-1.4}$ dependence. Reprinted with permission from [149], copyright 1962, American Institute of Physics.

slower lateral motion. The difference can be made clear by investigating a thickness dependence of the switching time as shown in Figure 18 [149]. Note that above a threshold field of 1 kV/cm, there is essentially no systematic dependence of t_s on the BaTiO$_3$ sample thickness over a wide thickness range from 10 to 300 μm. Instead, a sublinear dependence of $t_s = 9E^{-1.4}$ can be clearly resolved despite a three order of magnitude difference in the switching time. The dramatic change in the dependence of the sidewise velocity and switching time from an exponential dependence in the low-field poling regime to a sublinear dependence will be discussed as follows.

2.5.2. Nucleation Theory of Polarization Switching

The above observations of the switching parameter dependence in BaTiO$_3$ can be explained by a domain nucleation model developed by Miller and Weinreich [150]. This theory assumes that the domain wall motion results from a repeated nucleation of steps along existing parent 180° domain walls. It shows that the nucleation takes place along an existing 180° domain boundary and serves to propagate the boundary in a sidewise direction. As a result, the surface nucleation rate is the controlling factor in propagating the domain wall. The energy change due to the formation of a nucleus of volume V and domain wall area A can be written as

$$\Delta U = -2(P_s \cdot E)V + \sigma_w A + U_d \tag{7}$$

where σ_w is the wall energy per unit area and is ∼0.4 erg/cm^2 on BaTiO$_3$, P_s is the spontaneous polarization, and U_d is the depolarization energy. By assuming that the plane of the 180° domain wall is parallel to the ferroelectric axis and that the nucleation steps are triangular in shape, the wall velocity can be written as

$$v = v_\infty \sum_{n=1}^{\infty} e^{-(\delta/E)n^{3/2}} \qquad (8)$$

where δ is the activation field and v_∞ is the extrapolated wall velocity at an applied field $E = \infty$. To confirm the validity of this nucleation model analysis, let us take a first-order expansion of Eq. (7) to examine the behavior of the switching parameters in the low-field regime. In doing so, the Miller–Weinreich theory predicts a wall velocity dependence as $v = v_\infty \exp(-\delta/E)$, which agrees with the empirical formulation in Eq. (5).

Moreover, at applied fields above a threshold of 2 kV/cm, the summation series changes over and begins to vary as E^x with x dropping from 1.45 to 1.4, and more slowly to 1.34 as the field goes from 3 to 450 kV/cm in the high-field regime. This result also agrees well the observation in Figure 18. Notice that from an algebraic summation over the series in Eq. (7), one obtains $x \to 4/3$ as $E \to \infty$ [127].

Alternatively, in a modified Fatuzzo theory, when the sidewise motion is a predominant mechanism of polarization switching, the power-law dependence of t_s can be expressed as [151]

$$t_s = E^{-1.33} exp(\alpha^*/E) \qquad (9)$$

Equation (9) represents the case of polarization switching on BaTiO$_3$ or thiourea [152]. In the low-field poling regime, the exponential term is predominant, whereas in the high-field regime, the exponential term saturates and the term of $E^{-1.33}$ takes over.

Notice that the activation field δ in the Miller–Weinreich theory depends on $(P_s)^3 T^{-1} \varepsilon^{-1/2}$, where ε is the small signal dielectric constant perpendicular to P_s. For example, the activation field of undoped BaTiO$_3$ at 370 K is only half of its value at room temperature [153]. By heating the sample from 3 to 75°C, the domain wall velocity can be increased by a factor of 3 at a fixed switching field. This property makes the high-temperature poling procedure attractive in lowering the required switching field and should be valuable for performing polarization switching on ferroelectrics of high coercive field.

We pass by noting that a linear dependence of $1/t_s$ on E^x has also been observed on other ferroelectrics having low coercive field. In particular, when the ferroelectric crystal is poled in a high-field regime, a linear dependence of the sidewise 180° domain motion can be observed in Rochelle salt [154] and gadolinium molybdate (Gd$_2$(MoO$_4$)$_3$, GMO) [155]. For an account of the linear dependence on the sidewise domain motion in GMO, the reader can refer to [156].

3. PERIODICAL POLARIZATION SWITCHING ON FERROELECTRICS WITH HIGH COERCIVE FIELD

So far, we have reviewed the basics on the polarization switching process of BaTiO$_3$ which has a low coercive field

~0.5 kV/cm. We also note the effects of temperature and impurity in reducing the activation field. We have also discussed the optical and electrical characterization techniques to examine the domain structures and to analyze the dynamic response during the polarization switching process.

In the following, we will apply these established techniques to study the domain reversal process in ferroelectric nonlinear optical crystals whose coercive field (E_c) is at least one order of magnitude larger than that previously discussed on BaTiO$_3$ and TGS. Ferroelectric crystals of this kind include LiNbO$_3$, its isomorphism LiTaO$_3$ (with $E_c \sim 21$ kV/mm), KTP, and its family (with $E_c \sim 2.5$ kV/mm). We will begin with single domain crystals and discuss the methods of making periodical domain reversal QPM structures on the surface and the bulk of these nonlinear crystals. The discussion will be organized as follows. We will first review the thermal process that leads to a thin layer of surface domain inversion on a single crystal. These techniques more or less invoke the diffusion procedure and often give rise to a change of the surface crystallographic structure. As a result, they will be associated with a change in the refractive index and optical nonlinearity. We then move to the growth-striation and field-poling techniques that not only can result in a periodical reversal of the spontaneous polarization and odd-rank tensors but also leave the crystal structure and other even-rank tensor properties of the ferroelectrics unchanged. Details on the laminar growth of periodical domain reversed QPM structure by the laser-enhanced pedestal growth and by the off-axis CZ crystal growth will be reviewed. The periodically poled QPM structures can also be made through the state-of-the-art e-beam or scanning force writing and by the pulsed-field poling techniques.

3.1. Heat-Induced Surface Domain Inversion

Nakamura et al. first observe that when an uncovered Z-cut LiNbO$_3$ crystal is heated at temperature close to its Curie point (~1140°C), domain reversal can take place near the $+C$ surface without the need of external fields [157]. Shown in Figure 19 is the cross section of an etched $+Y$ face of LiNbO$_3$ that has been subjected to a heat treatment in a quartz furnace at 1110°C, followed by a cooling process to room temperature at a rate of 50°C/min. The data of Figure 19 are obtained after (a) 4 hr treatment in a flowing Ar gas and (b) 5 hr heating in air, respectively. It is found that the reversed domain nucleates at the $+C$ surface of the crystal and propagates along the crystal C-axis, and the reverse domain boundary stops at the median plane of the crystal and becomes ragged as shown in Figure 19b.

It is further noted that the thickness of the inversion layer depends strongly on the conditions of heat treatment. The relevant parameters are the temperature, time, and atmosphere. At temperature below 1050°C, nearly no surface inversion can be observed. For extended heat treatment at 1110°C in air, the thickness of the inversion layer saturates at half thickness of the substrate, whereas it can be less than 50 μm when treated in an Ar flowing gas containing water vapor. Indeed by adding water

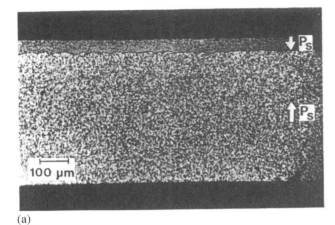

(a)

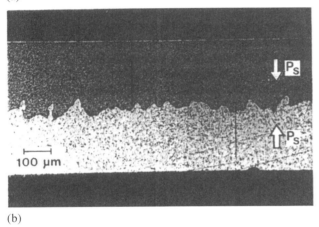

(b)

Fig. 19.   Etched y face of LiNbO$_3$ after heat treatment at 1110°C for (a) 4 h in a flowing Ar gas containing water vapor, and (b) 5 h in air. Reprinted with permission from [157], copyright 1987, American Institute of Physics.

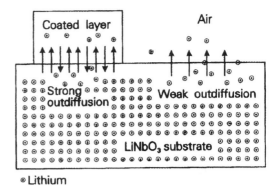

Fig. 20.   The principle of surface coverage induced domain inversion via the enhanced Li out-diffusion process on a MgO covered LiNbO$_3$. Reprinted with permission from [162], copyright 1992, IEEE.

experiments are sorted out by examining the surface coverage and vacancy effects as discussed below.

3.1.1. Surface Coverage Induced Piezoelectric Effects

The thermal induced domain inversion process is confirmed by Webjörn et al. by placing a silicon dioxide (SiO$_2$) patterned LiNbO$_3$ in a gas of flowing dry Ar heated at 1000°C [161]. It is noted that by depositing a dielectric thin film on the ferroelectric surface, one can further enhance the domain inversion process. Shown in Figure 20 is one such example illustrating the magnesium oxide (MgO) coverage induced lithium out-diffusion (MILO) process on Z-cut LiNbO$_3$ [162]. By heating the substrate at 1000°C for 0.3–1 h, enhanced out-diffusion of Li$_2$O can be observed on the MgO covered surface with respect to the uncovered area. Since the extraordinary refractive index of LiNbO$_3$ increases approximately with the decrease of the lithium concentration [163], this MILO process is useful for fabricating a transverse electric (TE) and transverse magnetic (TM) mode waveguide in the Y- and the Z-cut LiNbO$_3$ substrates, respectively.

A second example of enhanced surface domain inversion is on SiO$_2$ covered LiNbO$_3$ [164]. In this case, a Z-cut LiNbO$_3$ periodically covered with a thin layer of 100 nm SiO$_2$ is subjected a heat treatment at 1000°C for 2 h in a flowing dry O$_2$ environment. This process then leads to a surface domain inversion layer with a periodicity of 7 μm and a depth of 1.3 μm. It is noted the phenomenon of domain inversion can only take place under the SiO$_2$ covered region and the heating temperature has to be critically larger than 950°C. However, at temperatures above 1100°C, the SiO$_2$ layer shrinks and becomes cracked, and the domain inversion can occur even in the region without being covered by SiO$_2$ as in the case of Nakamura's experiment [157].

These two experiments suggest that at a lower curing temperature ($T < 1000$°C), the enhanced surface domain inversion on the dielectric covered LiNbO$_3$ is related to the stress developed by the difference in the thermal expansion coefficients between the dielectric layer and the substrate. It is thus suggested that the piezoelectric effect, among other possible

vapor to the diffusion gas, one can effectively suppress the out-diffusion of Li$_2$O from LiNbO$_3$ [158]. The rationale is due to the fact that the hydrogen can enter the crystal and result in a reduction of the Li mobility and suppression of its out-diffusion from the crystal.

We pass by noting that a similar spontaneous domain inversion process has also been observed on TGS when subjected to a thermal shock. The latter process could either be due to a cooling or warming shock [159]. Note that TGS has a relatively low coercive field ~0.5 kV/cm and a low Curie point around 50°C [160]. During the thermal treatment of TGS, it is found that the density of the reversed domain increases with the temperature difference between the baths that give rise to the thermal shock. It is suggested that the driving force of this type of domain inversion arises from the polarization discontinuities within the crystal when it is subjected to a thermal shock.

Combined with the information shown above, one might think that the Li out-diffusion mechanism and the induced internal field via the pyroelectric, piezoelectric, or space charge effects are responsible for the domain inversion process. In order to decide which factor plays the dominant role, supportive

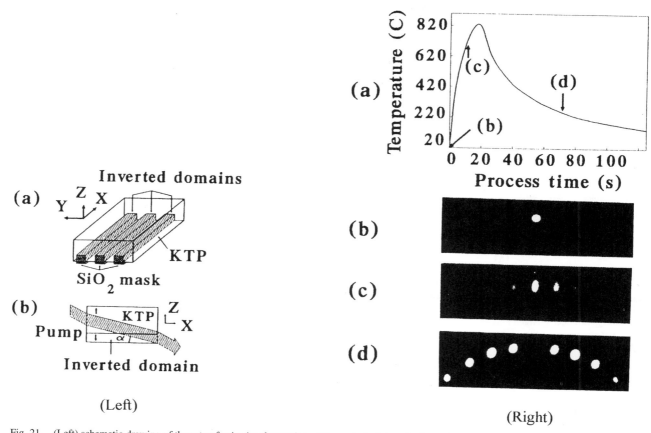

Fig. 21. (Left) schematic drawing of the setup for *in situ* observation of the periodical domain formation on the (a) SiO$_2$ covered KTP crystal using the (b) nonlinear Talbot effect to map the near- and far-field diffraction patterns of the converted SHG signal. (Right) Far-field SHG diffraction pattern at various stage of the heating process. Note in (c) the domain inversion process begins at temperature $T = 730°C$ and continues in (d) with the cooling procedure untill $T = 250°C$. Reprinted with permission from [165], copyright 1996, American Institute of Physics.

mechanisms, is responsible for the thermal-induced surface domain inversion on the dielectric covered ferroelectric crystals.

In order to support this analysis, a third example is taken on a SiO$_2$ covered flux-grown Z-cut KTP. Illustrated in Figure 21 is the experimental setup for making an *in situ* observation of the domain inversion process [165]. A thin, 10 nm SiO$_2$ covered KTP sample is heated in a H$_2$–O$_2$ burner up to 840°C at a rate of 70°C/s to introduce a strain in the KTP crystal. The surface domain inversion structure can be indirectly characterized from the far-field SHG pattern using the nonlinear Talbot effect. By doing so, the sequential formation of a periodical domain structure can be inferred from the far-field SHG pattern of Figure 21. The data of Figure 21 indicate that the domain inversion process starts at a temperature of 730°C and continues as the substrate temperature drops to 250°C during the cooling procedure. From a near-field measurement of the SHG pattern, one can further infer that the period and the depth of the domain reversed QPM structure are 4.5 and 5.5 μm, respectively. We note this strain-induced surface domain inversion temperature is far below its Curie point of 940°C but higher than that in obtaining single domain KTP crystal (~500°C) as discussed in Section 2.2.2.

3.1.2. Vacancy Induced Space Charge Effects

Several models including the oxygen vacancy and the Nb antisite defect have been proposed as the mechanism responsible for inducing the surface domain reversal in the heated LiNbO$_3$ crystals. However, a correct theory should answer why the domain inversion only takes place on the $+C$ surface instead of on the $-C$ face of the crystal. An equivalent electric field is generally assumed to arise from the heating process and enables the domain reversal process in the absence of external field. From the basics of electrostatics, such a self-generated field can result from a gradient distribution of impurities, defects, or nonuniform distribution of the spontaneous polarization caused by the heat treatment. It can also result from the pyroelectric effect if there is a temperature gradient in the crystal. It can also be due to the piezoelectric effect if there is a strain in the crystal. Among many of these possible sources, a common feature associated with the high temperature treatment is the occurrence of several types of lattice vacancies [166, 167]. In the following, two possible sources, i.e., the oxygen vacancy and the niobium antisite models, will be shown to give rise to a similar space charge field. This type of space charge field is found to have an

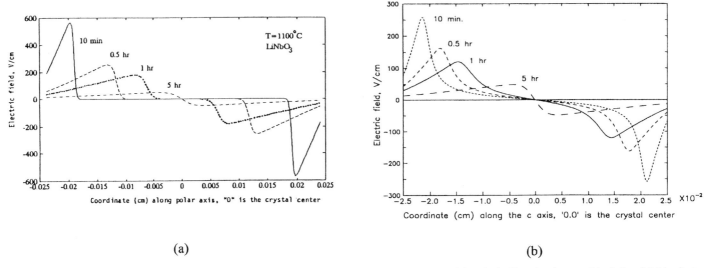

(a)   (b)

Fig. 22.   (a) Space charge field distribution in the heated Z-cut LiNbO$_3$ using the oxygen-vacancy model analysis. Note the change of the induced field polarity with respect to the C-axis of the crystal. The crystal thickness is 500 μm. Reprinted with permission from [168], copyright 1993 American Institute of Physics. (b) Space charge field distribution in the heated Z-cut LiNbO$_3$ using the niobium antisite model analysis. Reprinted with permission from [173], copyright 1994 American Institute of Physics. Note in (a) and (b) that the surface domain inversion takes place only on the $+C$ face of LiNbO$_3$ due to the anti-parallelism of the space charge field with respect to the original P_s in the crystal.

axial polarity such that the surface domain inversion can only be induced on the $+C$ face of LiNbO$_3$.

3.1.2.1. Oxygen Vacancy Model

In the oxygen vacancy model, it assumes that the oxygen out-diffusion process can result in a positively charged oxygen vacancy (V_O^+) accompanied by an electron to keep valid of the charge neutrality condition [168]. Assuming the distribution of the oxygen vacancy concentration follows a complementary error function in the crystal, calculation of the corresponding space charge field in Figure 22a at a heating temperature of 1100°C exhibits a change in the field's polarity with respect to the center of the crystal axis. Moreover, one notes that the resultant space charge field and its action range could vary from 550 V/cm, 50 μm for a 10 min heat treatment procedure to 45 V/cm, 250 μm at a longer treatment time of 5 h. In Section 2.2.2, we mention that the electrical poling of a single domain LiNbO$_3$ crystal at temperature above 1100°C only requires a small switching field less than 5 V/cm. Since the calculated space charge field exceeds 5 V/cm, one would expect to have a domain inversion process that takes place during the high-temperature heating process.

Now, one has to answer the following questions: why the heat-induced domain inversion only occurs on the $+C$ face and why it moves inward and saturates at the center of the crystal. The answers can be gathered from Figure 22a within the frame of the Miller–Weinreich theory. They are summarized as follows. First, the induced space charge field exhibits a nonuniform distribution and its peak value shifts from surface to bulk as the duration of heat treatment increases. Since the induced field in the $+C$ surface is anti-parallel to that of the original P_s in the crystal and its value exceeds the corresponding coercive

field, one would expect that nucleation of the reversed domain begins at the $+C$ surface and propagates into the bulk. Second, there is an axial dependence of the sign change in the induced field. This indicates a change of the field's polarity to parallel to that of P_s in the lower half part of the crystal. This explains why the domain inversion cannot take place on the $-C$ face of LiNbO$_3$ and why the thickness of the inversion layer saturates as the domain boundaries meet the median plane where there is no space charge field left.

3.1.2.2. Niobium Antisite Model

In parallel to the oxygen vacancy model, the niobium antisite model can also be used to explain the heat-induced domain inversion process on the $+C$ surface of LiNbO$_3$ as well. The rationale is based on the observation that the commercially available LiNbO$_3$ and LiTaO$_3$ single crystals are made of congruent composition. From the lattice structure analysis, one can notice that roughly 94.1% of the lithium (Li) sites are occupied by the Li ions (Li$_{Li}$) while the rest of the 5.9% Li missing sites are occupied by the Nb^{5+} ions (Nb$_{Li}$)$^{4+}$. In this case, there will exist compensating vacancies residing at the Nb sites $(V_{Nb})^{5-}$ to maintain the charge neutrality condition [169].

In the niobium antisite model, it is assumed that during the Li$_2$O out-diffusion process, there is an accumulation of the niobium antisite (Nb$_{Li}$)$^{4+}$ and the compensating vacancy defects $(V_{Nb})^{5-}$ in the bulk. During a high-temperature treatment, the reduction process takes place and renders the LiNbO$_3$ an n-type crystal [170]. In the reduction process, the vacancy defects $(V_{Nb})^{5-}$ are consumed, resulting a distribution of the (Nb$_{Li}$)$^{4+}$ antisite defects which are balanced by the nonuniform distribution of free electrons [171]. In the calculation, the distribution of the niobium antisite defects $(Nb_{Li})^{4+}$ is also assumed to follow a complementary error function as derived in the out-diffusion of Li_2O [172]. From a gradient distribution in the niobium antisite defect $(Nb_{Li})^{4+}$ and the compensated free electrons, one can calculate a space charge field distribution as shown in Figure 22b [173]. Comparing the results shown in Figure 22, one can notice a great similarity in the induced field obtained from the niobium antisite and the oxygen vacancy models. Therefore, according to our previous discussion, the space charge field acting in the niobium defect model can also induce a surface domain inversion on the $+C$ face of $LiNbO_3$ as well.

Since the out-diffusion of Li_2O is regarded as the source of domain inversion in heat-treated $LiNbO_3$ crystals, other methods capable of increasing the Li deficiency should also result in similar effects as well. In the following we will discuss two addition processes, i.e., the proton exchange and the Ti diffusion techniques, that are commonly used to induce surface domain inversion in $LiNbO_3$ and $LiTaO_3$.

3.2. Diffusion-Induced Surface Domain Inversion

3.2.1. Proton-Exchange Technique

3.2.1.1. LiNbO₃

A commonly used method to enhance the Li out-diffusion is to immerse the $LiNbO_3$ in hot acids or in certain hydrate melts so that the Li ions diffusing out of the crystal $+C$ face can be substituted by an equal number of protons (H^+) via indiffusion [174]. The bath temperature in the process typically ranges from 200 to 350°C, depending on the melt, but is far less than that in the high-temperature induced LiO_2 out-diffusion process near the Curie point. This process is genetically known as "proton-exchange" (PE). In strong acids, such as HNO_3 or H_2SO_4, when the substitution process is complete, the newly formed compound $HNbO_3$ has a cubic perovskite structure different from the original twisted perovskite $LiNbO_3$ structure. Moreover, in a less acidic environment, such as in benzoic acid (C_6H_5COOH) or pryophosphoric acid $(H_4P_2O_7)$, as much as 50% of the Li^+ ions near the surface of $LiNbO_3$ can be replaced by the H^+ ions, thereby forming a Li-deficient $H_xLi_{1-x}NbO_3$ surface layer. The resultant $H_xLi_{1-x}NbO_3$ structures exhibit several crystallographic phases according to the different PE sources [175–177]. The PE process can also result in a refractive index increase in the extraordinary direction that makes it useful in waveguide fabrication on $LiNbO_3$ [178]. A summary of the extraordinary refractive index change (Δn_e) and strain ε_{33} perpendicular to the crystal surface at different crystallographic phases of $H_xLi_{1-x}NbO_3$ has been reported [179].

However, the existence of these structure phases often causes severe propagation loss in the waveguide [180] and a decrease in the electro-optic and nonlinear-optic coefficients [181–184]. There is also an enhanced photorefractive effect associated with the PE waveguide [185]. These observations suggest a degradation of the ferroelectric properties due to the PE process. It is noted that after a specific annealing procedure, loss in the

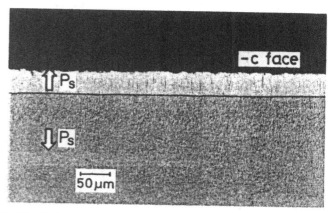

Fig. 23. Etched y face of $LiTaO_3$ after proton-exchange in benzoic acid at 220°C for 9 h and heat treated at 590°C for 3 h. Note that the domain inversion only takes place in the $-C$ face of $LiTaO_3$. Reprinted with permission from [189], copyright 1990, American Institute of Physics.

linear propagation and optical nonlinearity can be largely restored [186]. Taking advantage of these recovery effects on the annealed proton-exchange $LiNbO_3$ waveguide structure, a normalized SHG conversion efficiency of 85 %/W/cm² at 488 nm [187] and an IR parametric generation efficiency of 130%/W/cm² in the 1.5–2 μm spectral range [188], respectively, have been reported.

3.2.1.2. LiTaO₃

Lithium tantalate ($LiTaO_3$), an isomorphism of $LiNbO_3$, has a lower Curie temperature at ~600°C and a short wavelength transparency down to 280 nm. The PE process has also been developed on $LiTaO_3$ using acids and melts similar to those for $LiNbO_3$. However, a major difference between these two processes is that the PE-induced domain inversion takes place on the $-C$ face of $LiTaO_3$ whereas it is on the $+C$ face of $LiNbO_3$.

Illustrated in Figure 23 is one such example of the etched Y-face of $LiTaO_3$ revealing a ~45 μm thick inversion layer taken place on the $-C$ face [189]. The PE process consists of immersing a $LiTaO_3$ crystal in benzoic acid at 220°C for 9 h followed by heat treatment at 590°C for 3 h. Note that there exists a temperature threshold for the heat treatment, i.e., 460°C, below which the domain inversion cannot form no matter how long the proton exchange process takes [190]. Normally the thickness of a PE layer on $LiTaO_3$ is on the order of 1 μm, but the corresponding domain inversion layer after heat treatment can reach a thickness on the order of 50 μm. This observation suggests that the PE layer on the $-C$ face may serve as the domain nucleation layer whose polarity is opposite that of P_s in the original $LiTaO_3$ crystal. The domain reversal process may resume when the crystal is subjected to a heat treatment at temperature above a certain threshold. There are several sources of the temperature-driven poling field, among which a study on the dielectric and electric response near the Curie point has precluded the pyroelectric effect as the possible source [191].

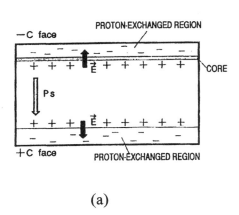

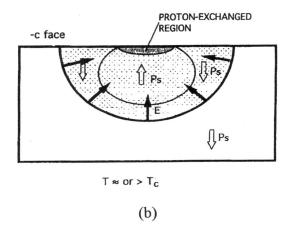

(a)   (b)

Fig. 24.   (a) Schematic representation of the domain inversion process taking place on the $-C$ face of LiTaO$_3$ due to an internal field caused by the diffusion of proton and lithium ions. Reprinted with permission from [192], copyright 1993, American Institute of Physics. Note that the space charge field is maximized at the PE region and decays sidewise. (b) Lateral and vertical expansion of the inverted domain on the $-C$ face of LiTaO$_3$ due to the heat treatment at temperature below or larger than the Curie point (T_c). Reprinted with permission from [197], copyright 1994, American Institute of Physics.

A plausible explanation to the PE induced domain inversion on LiTaO$_3$ is given by Mizuuchi et al. Referring to Figure 24a, this model suggests that an internal field caused by the diffusion of proton and lithium is responsible for the surface domain inversion [192]. It states that during a heat treatment procedure at a temperature near T_c, a diffusion process occurs such that H$^+$ diffuses from the PE region toward the substrate and Li$^+$ out-diffuses in the opposite direction. A space charge field therefore can arise from a faster distribution of the in-diffused H$^+$ with respect to the slower out-diffusion of Li$^+$ and the Li vacancy $(V_{Li})^{-1}$ lagged behind. The field distribution is such that it has a maximum value near the interface of the PE region but decays sidewise. Note that the direction of the induced field points toward the surface normal on both sides of the LiTaO$_3$ substrate. One therefore obtains from this model analysis an electric field on the $-C$ face side of LiTaO$_3$ opposite that of the original P_s of the crystal and another field on the $+C$ face side parallel to P_s. As a result, domain inversion is expected to take place only on the $-C$ face of the proton-exchanged LiTaO$_3$.

The dynamic response of the heat-driven domain inversion process on a proton-exchanged LiTaO$_3$ can be understood as follows. With the space charge field maximized at the interface of the PE area, it would bring up a nucleation layer of reversed domain. This "core region" as indicated in Figure 24a would serve as the nucleation layer and propagates the reversed domain during the subsequent heat treatment. Now recall from the Miller–Weinreich nucleation theory that the activation energy δ decreases with temperature. The expansion of the reversed domain in the lateral and vertical direction would benefit from a high-temperature treatment above a certain threshold, in this case 460°C. Note that this heat-induced surface domain inversion process on the proton-exchanged LiTaO$_3$ differs from that on LiNbO$_3$. In the latter case, the nucleation of the 180° reversed domain begins on the $+C$ face and propagates toward the $-C$ face of the proton-exchanged LiNbO$_3$.

Using this method, the Matsushita research group conducted a series of periodical domain reversal studies in making QPM structures on LiTaO$_3$ for blue SHG in the early 1990s [193–196]. The procedure to make a periodically domain reversed QPM structure is as follows. A 30 nm protective mask of Ta is first sputtered on the $-C$ face of LiTaO$_3$, followed by a patterned definition using CF$_4$ etching. With a PE processing condition of crystal immersion in pyrophosphoric acid at 260°C, it is found that the depth of the domain inversion layer exhibits a diffusion characteristic. Moreover, the domain inversion layer can reach a saturation value after a 10 min treatment in high-temperature soaking above 510°C. However, the sidewise domain expansion does depend on the soaking temperature and has a semicircular shape when it is heated over the Curie point. This observation can be understood from the space charge field distribution model near the proton-diffused region as shown in Figure 24b [197]. At heat treatment above T_c, the process of polarization switching requires less activation energy compared with that below T_c. As a result of the high-temperature soaking process, polarization switching can easily take place around the proton-diffused region which serves as the nucleation mechanism. The reversed domain, driven by the space charge at the interface of the PE region, can thereby grow laterally and vertically to exhibit a semicircular shape.

We pass by noting that the process of PE on LiTaO$_3$ can also result in a reduction of the optical nonlinearity [198] and a change of the surface crystallographic structure [199]. The elastic properties of LiTaO$_3$ such as the surface acoustic velocity also decreases with the PE process [200]. Depending on the processing condition, the recovery of the optical nonlinearity and the SHG efficiency may only have a narrow window, for example, at a annealing time of 7 min at 380°C [201]. Combined with the PE technique and postannealing process, a normalized SHG efficiency of 1500%/W/cm^2 has been reported on the LiTaO$_3$ QPM waveguide device emitting a blue light of 1 mW at 429 nm [202].

3.2.2. Ti In-Diffusion Process

In addition to the above thermal process involving the diffusion of Li_2O and H^+ to enable the domain inversion on $LiNbO_3$ and $LiTaO_3$, the Ti in-diffusion process offers an alternative method to achieve surface domain inversion on Z-cut $LiNbO_3$. Conventionally the process of Ti diffusion is used for waveguide fabrication by increasing the refractive index of a $LiNbO_3$ surface layer which is coated with Ti [203]. A typical waveguide process is to perform the Ti diffusion at ~1000°C in an oxygen environment. This would result in an approximately 1-μm-thick Ti-diffused waveguide after 1 h of annealing. Note that accompanying the high-temperature Ti waveguide-making process is the change of surface crystallography of $LiNbO_3$. It is recently reported that a solid solution of rutile and lithium triniobate $(Li_{0.25}Nb_{0.75}O_2)_{1-x}(TiO_2)_x$ can be formed [204]. This new rutilelike phase can disappear only after prolonged annealing (>6 h) at 1000°C.

To observe surface domain inversion on a Ti-coated $LiNbO_3$, the process, however, requires a higher temperature treatment. A typical diffusion temperature ranges from 1000 to 1100°C with a diffusion time of 5 to 10 h in air [205]. Note that the Ti diffusion process only results in a surface domain inversion near the $+C$ face of $LiNbO_3$ but not on the $-C$ face. This observation is very similar to that of surface domain inversion obtained by the heat treatment and the proton-exchange process on $LiNbO_3$. However, the high-temperature process is known to result in optical loss due to the replacement of Li or Nb cations by Ti ions [206] and is not a welcome factor in the design of optical integrated devices [207]. Also note that accompanying the Ti-induced domain inversion is a significant reduction in the electro-optic effects on the $+C$ face of $LiNbO_3$ [208].

The mechanism responsible for the Ti-induced surface domain inversion in $LiNbO_3$ still remains a controversial issue over the past decade. It is postulated that the low diffusivity of Ti in the $+C$ surface can increase the Ti surface concentration. This effect can reduce the Curie point in the Ti-doped region of $LiNbO_3$ and may play an important role in the polarization switching process. By doing so, Peuzin suggests that a gradient distribution in the Ti concentration may result in an internal field to overcome the coercive field near the Curie point and to form a surface domain inversion layer [209]. On the other hand, Miyazawa proposes a spontaneous polarization gradient model to explain the domain reversal process [210]. As illustrated in Figure 25, the Ti-concentration gradient in the polar C-faces of $LiNbO_3$ can result in a nonuniform distribution of the spontaneous polarization. This phenomenon in turn can create a surface electrical field pointing inward from the C-surfaces to the $LiNbO_3$ substrate but disappears in the bulk. At a diffusion temperature near T_c, the induced field can overcome the corresponding coercive field and initiate the domain reversal process. This result is similar to the case discussed in the oxygen vacancy or the niobium antisite model in Section 3.1.2. Note that with the direction of the induced field E on the $+C$ face opposite to that of the original P_s in the bulk, a polarization switching process can be initiated with the resultant reversed

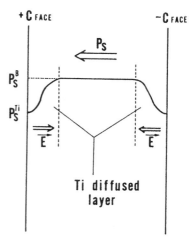

Fig. 25. Schematic drawing of the Ti concentration gradient effects in the space charge field and the surface domain inversion on the $+C$ face of Ti diffused $LiNbO_3$. Reprinted with permission from [210], copyright 1986, American Institute of Physics.

domain layer confined on the $+C$ face. No domain inversion will occur on the $-C$ surface due to the parallelism of the surface field E on P_s.

We pass by noting that there are several common features associated with the aforementioned diffusion-induced surface domain inversion process. First, the domain inversion layers are confined to *one* polar surface of the crystals, i.e., on the $+C$ face of Z-cut $LiNbO_3$ by the heat-treatment, proton-exchange, and Ti-diffusion process and on the $-C$ face of $LiTaO_3$ by a combined proton-exchange/heat-treatment process. Second, the heat treatment temperature has to be close to T_c in order to lower the activation energy of domain nucleation. In this way, the surface domain reversal can take place in conjunction with the heat-induced field. Third, these diffusion processes inevitably involve a modification of the surface crystallographic structure and result in a change of the refractive index and a decrease in the electro-optic, nonlinear optic, and elastic properties. The degradation of odd-rank tensor properties such as the electro-optic and optical nonlinear coefficients in these diffusion processes is disadvantageous to the applications of opto-electronic devices. A suitable postannealing process is needed to recover the optical loss but the processing window and the recovery may be very limited.

3.2.3. Ion-Exchange Process on the KTP Family

In parallel to the diffusion-induced surface domain inversion on $LiNbO_3$ and $LiTaO_3$, there is a counterprocess, called "ion-exchange," available to KTP crystal and its family. KTP crystal is known for its high optical nonlinearity and large electro-optical effect that make this material system promising for the integrated optics application. KTP is also known to be very chemically stable up to a temperature about 1000°C [211]. Moreover, the potassium (K) ions of KTP are relatively mobile and can be exchanged with other ions from molten salts. This is based upon the fact that a solid-state solution can exist in the MTiOPO$_4$ structure where M can be K, Rb, Tl, Cs, NH$_4$, or any combination of these ions and solid solutions between MTiOPO$_4$ and MTiOAsO$_4$ [212]. A typical ion-exchange process consists of a first metal-masking step on the Z-cut KTP, followed by immersing the sample in an ion-exchange bath containing molten nitrate salt of Rb, Cs, or Tl heated at a temperature between 300 to 400°C for 30 min to 4 h. This ion-exchange process appears to be diffusion-limited but the exchange rate depends on the substrate conductivity.

It is noted that by substituting the K ions in the KTP lattice with the alkaline-earth ions of appropriate ionic radii, the surface conductivity can be increased and so will be the ion-exchange rate. This observation can be ascribed to the potassium vacancy effect regarding the ionic conductivity of KTP [213]. By doing so, an increase of surface refractive index as high as 0.23 and diffusion depth of 15 μm have been achieved in an ion-exchange process of Tl and Ba/Rb on KTP. These results are very promising for opto-electronic applications where a periodic modulation of the refractive index and the electro-optic coefficient are needed. One such example is the development of ion-exchange periodically segmented KTP techniques to achieve visible and ultraviolet SHG laser sources in the 1990s [214–216]. A relevant issue on the optical damage resistance of KTP is that it depends on the impurities incorporated into the crystal. The impurity essentially plays a role in shifting the absorption edge of the crystal [217]. It is reported that in a Rb-exchanged periodically poled KTP SHG device, no optical damage is observed at intensity up to 80 kW/cm^2 in a spectral range from 700 to 900 nm [218].

We pass by noting that the increase of surface refractive index in an ion-exchanged KTP crystal is accompanied with the formation of a surface inversion layer. To observe the opposite domains in the ion-exchange KTP area, one can apply the differential etching technique by immersing the sample in a molten salt consisting of a 2 : 1 mole ratio of KOH and HNO$_3$ at 220°C [219]. The etchant only attacks the $-C$ face of the KTP crystal but leaves the $+C$ face nearly untouched. The etched domains can then be examined under a conventional optical or scanning microscope. The domain can also be observed using the electronic toning technique upon which the pyroelectricity developed in the cooled KTP faces, i.e., $V \propto (dP_s/dT)$ [220], would attract toner particles of opposite charge polarity.

3.3. High-Energy Beam Poling Process

We note there are concerns regarding the domain stability and the processing parameter control that desire a poling procedure taken at a much lower temperature than near T_c. One method capable of inducing polarization switching in LiNbO$_3$ and LiTaO$_3$ at a temperature significantly below T_c is to use an energetic beam to result in a transit vacancy to allow the movement of Li atoms. This mechanism takes advantage of the fact that an oxygen molecule has a size comparable to that of a separate oxygen ion [33]. Referring to Figure 1, if the oxygen ions in the triangular plane are excited by an energetic radiation, then a possible relaxation route would be the formation of a metastable oxygen

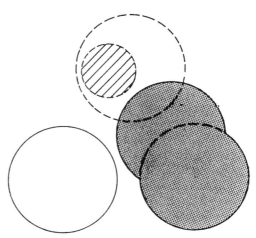

Fig. 26. Possible relaxation route of forming a metastable oxygen molecule (shaded area) when a LiNbO$_3$ or LiTaO$_3$ crystal is excited by an energetic radiation. The Li atom (hatched circle) can make a quick move through the transient opening (dotted circle) in the original oxygen triangular plane and to reverse the spontaneous polarization. Reprinted with permission from [33], copyright 1986, American Institute of Physics.

molecular ion as shown in Figure 26. In this way, a temporary opening of a vacancy in the oxygen triangle plane would allow the Li atom to move when subjected to an applied field along the C-axis. With the Li atom moving to the opposite side of the oxygen plane, it would alter the site ordering sequence and reverse the polarization of the crystal. For example, as irradiated with an electron beam of 1.8 MeV in energy, it is noted that domain inversion can take place on LiTaO$_3$ at a low poling field of 900 V/cm, and on LiNbO$_3$ at a much smaller field of 10 V/cm, respectively. Moreover, the poling temperature now can be as low as 400 and 600°C, separately, for the above two ferroelectric crystals.

Another example of high-energy beam poling involves the use of a laser beam to generate a thermal pulse and to facilitate the electrical field poling process. This technique takes advantage of a rapid rise in the surface temperature to near T_c when the ferroelectric crystal is illuminated with a focused laser beam. For LiNbO$_3$, successful domain inversion has been reported by the use of a focused 532 nm SHG from a pulsed Nd : YAG laser and from 488 nm irradiation from a CW argon laser [221].

The processing condition of using a focused 532 nm laser beam to facilitate the domain inversion on Z-cut LiNbO$_3$ is as follows. When the laser width and the repetition rate are fixed at 5 ns and 10 Hz, a laser energy of 40–90 mJ in conjunction with an applied electric field of 187 V/cm is found able to induce domain inversion on LiNbO$_3$ at a base temperature of 400°C [222]. Note a thin ~90 nm copper film deposited on the LiNbO$_3$ surface can enhance the absorption of the laser power and help the heat transfer process. At a pulse energy of 80 mJ, it is estimated that a transient temperature rise of 820°C above the base temperature can be reached on the irradiated LiNbO$_3$ surface but there exhibits an exponential decay depth ~0.25 μm of the transient temperature distribution. The corresponding figure for a 90 mJ pulse is 915°C. By adding up a 400°C base temperature difference, these calculations suggesting a transient temperature rise above the Curie point of $LiNbO_3$ can be easily reached on the sample surface by a high-energy laser pulse. In doing so, this process could result in a temporary opening of vacancy on the distorted oxygen triangular plane as in the case of e-beam induced surface domain inversion on $LiNbO_3$ just mentioned above.

So far we have discussed the processing techniques that can lead to the realization of a single domain or a surface domain structure on ferroelectric crystals. Polarization switching of ferroelectric crystals can be pursued *in situ* during the crystal growth or *ex situ* by high-temperature poling or the diffusion induced domain reversal process. The domain reversal structures are suitable for the fabrication of memory, electro-optic, and nonlinear optics devices. However, to improve the efficiency and performance of these ferroelectric opto-electronic devices, it is desirable to have a device structure with periodically reversed domain as discussed in Section 1. To enhance the storage density of ferroelectric memory devices, development of nanoscale domain engineering techniques becomes important. In the following two sections we will discuss methods that enable the growth of periodical domain reversal structures. Development of the state-of-the-art field poling and nanoscale domain engineering techniques will also be reviewed.

3.4. Bulk Periodical Domain Reversal: *In Situ* Growth Methods

3.4.1. Laser-Heated Pedestal Growth of Periodical Domain Reversal Structure

The laser-heated pedestal growth (LHPG) method, a variant of the float-zone process, is a convenient method for growing small diameter crystals such as Al_2O_3, YAG, and $LiNbO_3$ [223]. Schematically drawn in Figure 27 is a LHPG design that can be used to grow the $LiNbO_3$ single domain fiber crystal or the periodical domain reversal structure [224]. In the design, 10.6 μm radiation from a CO_2 laser is focused on the tip of a source rod of $LiNbO_3$ to make a small molten droplet. The source rod may be fabricated from single crystal [225], polycrystalline, sintered [226], or pressed powder [227] materials. A typical source rod diameter ranges from 500 μm to 2 mm. The seed rod, which determines the crystallographic orientation of the fiber, is then withdrawn from the droplet as the fresh source material is supplied. The crystals can be grown in air or in a 4 : 1 mixture of helium and oxygen. In doing so, crystal fibers of $LiNbO_3$ have been grown successfully along the c-axis and the a-axis.

The $LiNbO_3$ fiber grown along the c-axis is virtually a single domain crystal when the diameter is less than 800 μm. However, for the crystal fiber grown along the a-axis, it normally reveals a bi-domain configuration having the P_s aligned in a back-to-back configuration with the domain boundary along the a-axis [228]. These observations can be ascribed to the thermo-

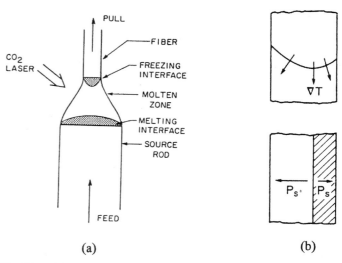

(a)   (b)

Fig. 27. (a) Schematic diagram of a LHPG apparatus for growing single domain crystal fibers. Reprinted with permission from [224], copyright 1984, American Institute of Physics. (b) Ferroelectric domain structure in a-axis crystal fibers based on an asymmetric thermoelectric field effect in a *radial* temperature distribution in the melt. Reprinted with permission from [228], copyright 1986, Elsevier Science.

electricity induced domain inversion effect during the crystal growth. The rationale is as follows.

A temperature gradient in the growth melt can create a spatially varying theromelectric field according to

$$E = Q\nabla T \qquad (10)$$

It is important to note that the theromoelectric power factor, Q, is a second-rank tensor and has a nonvanishing coefficient in both of the para- and ferroelectric states. The thermoelectric coefficient of Q_{33} is measured on the order of 0.8 mV/°C at a temperature close to the melting point of $LiNbO_3$. Estimated from a large *axial* temperature gradient ~1000°C/cm along the growth interface shown in Figure 27a, the resultant thermoelectric field will be on the order of 1 V/cm. This field strength is sufficient to drive a c-axis grown $LiNbO_3$ fiber crystal to make a para- to ferroelectric state phase transition at a temperature near T_c. Moreover, note that this thermoelectric poling field will also result in a P_s of the fiber crystal pointing toward the melting interface, i.e., along the direction of the temperature gradient field.

Similarly, the occurrence of a bi-domain structure in the a-axis grown $LiNbO_3$ fiber crystal can be understood by considering a *radial* temperature gradient distribution at the growth interface as shown in Figure 27b. Note that when the projected *radial* temperature gradient field changes its sign at the domain boundary, so will the induced thermoelectric field. The corresponding thermoelectric effect across the boundary would then set up a domain structure with P_s aligning in a back-to-back configuration.

The bi-domain feature in the a-axis grown LHPG fiber crystal technique can be further improved to obtain a periodic

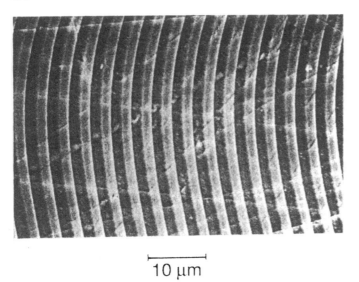

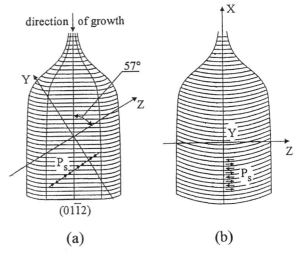

Fig. 29. Possible arrangement of the periodically poled domain reversed structures on LiNbO₃ grown by the off-axis Czochralski method, showing the (a) "head-to-head" and (b) anti-parallel domain configuration. Reprinted with permission from [231], copyright 1997, Elsevier Science.

Fig. 28. Micrograph of a periodically poled LiNbO₃ a-axis grown by the LHPG method. Note that the curvature of the domains corresponds to the convex shape of the growth interface shown as the freezing interface in Figure 27a. Reprinted with permission from [229], copyright 1991, American Institute of Physics.

3.4.2. Off-Axis Czochralski Growth of Periodical Domain Reversal

By taking advantage of the thermal gradient effects, it is now possible to grow bulk periodically domain reversed ferroelectric crystals by the commonly used CZ method. In the following, we will discuss the use of the "off-axis" CZ method to grow periodical domain reversal structures with P_s aligned in either a "head-to-head" or a "anti-parallel" configuration as shown in Figure 29 [231]. The "off-axis" CZ method is named after the work by Ming et al. when it was found that by shifting the rotation axis of crystal not coinciding with the symmetry axis of the temperature field, one can induce a temperature fluctuation related to the crystal rotation [232]. This temperature fluctuation will be shown in the following discussion to cause a spatial modulation in the distribution of the impurity or the composition in the crystal. This effect can in turn generalize a space charge field that plays the role of causing *in situ* periodical domain poling during the CZ growth of ferroelectric crystals.

domain reversed QPM structure. The rationale is based upon a periodic modulation of the laser heating power and therefore the corresponding freezing interface position of the domain boundary. Using a 5% MgO doped LiNbO₃ crystal as the source rod, a 0.8-mm-diameter fiber crystal with a domain period as small as 2 μm has been reported [229]. Combined with the small dimension of the molten zone and the larger thermal gradient in the LHPG process, one can obtain a freezing interface with a precisely controlled position [230]. During the laser off time, the molten zone volume shrinks to about half the initial size, the freezing speed is accelerated, and this process would result in a MgO-enriched crystal segment. The poling field in this case is thought to originate from a spatial gradient distribution in the composition or impurity due to the temperature modulation effect.

In the LHPG mode of growing periodical reversal domain structures, a typical averaged growth speed is 2 mm/min and the heating laser is periodically interrupted for a duration of 20–40 ms. The period of the domain structure in μm can be empirically determined as $\Lambda = v_{pull}/f_{mod}$, where v_{pull} is the crystal pull rate in μm/s and f_{mod} is the heating laser power modulation frequency in Hz. For example, at a crystal pull rate of 2 mm/min, and $f_{mod} = 12.35$ Hz, growth of a Mg : LiNbO₃ domain reversal structure of period $\Lambda = 2.7$ μm can be obtained as shown in Figure 28. However, note that the domain reversal structure exhibits a curvature following the convex shape of the growth interface. In addition, the domains are discontinuous along the ⟨0001⟩ plane bisecting the rod at the center. Caution therefore must be exercised to offset the pump fundamental beam from the rod center in performing the SHG experiment.

3.4.2.1. a-Axis Grown Periodically Poled LiNbO₃

Shown in Figure 30 is one such example illustrating a one-to-one correspondence between the temperature fluctuation and the induced growth striation for a periodical domain reversed structure grown on the a-axis pull LiNbO₃ crystal [232]. Here we note that the polarity variation of the periodical domain reversed structure domain corresponds to the fluctuation in the growth temperature. In particular, one can further note that the period of the reversed domain also depends on the crystal rotational speed. When the crystal stops rotation, the surface striations disappear; but when the crystal begins to rotate, the striations appear immediately. The spacing of the rotational striations therefore can be changed at will by means of adjusting

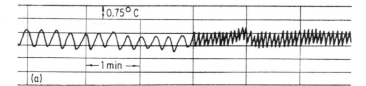

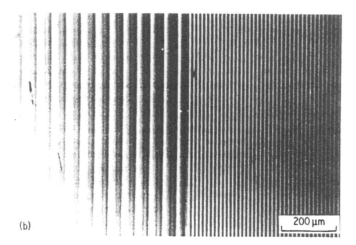

Fig. 30. The correspondence between the growth temperature fluctuation to that of the periodically poled domain reversed structure in the *a*-axis grown LiNbO₃, single crystal by the off-axis Czochralski method. Reprinted with permission from [232], copyright 1982, Kluwer Academic.

Fig. 31. Photograph of the periodically poled domain reversed structures in the *a*-axis grown LiNbO₃ by the off-axis Czochralski method showing a period of (a) 2.7 μm, (b) 5.2 μm, and (c) 15.0 μm. Reprinted with permission from [233], copyright 1996, American Institute of Physics.

the ratio of the crystal pulling rate to the rotation rate. In addition, by introducing a periodical modulation in the heating power, one can also obtain a growth of the periodical domain reversed crystal.

In the design of the asymmetric temperature field distribution, there are three important parameters, i.e., the axial temperature gradient above the melt surface, the axial temperature gradient below the melt surface, and the radial temperature gradient in the melt surface, that need to be taken into consideration [233]. Unlike the conventional CZ growth of single domain crystal as discussed in Section 2.1. it is recommended to have a convex solid–liquid interface for the growth of periodical domain reversed ferroelectric crystals. If the shape of the solid–liquid interface is exactly flat, which indicates a much smaller temperature gradient in this region, the periodic domain structure cannot occur. By carefully adjusting these growth parameters, a periodically poled domain reversed LiNbO₃ single crystal with a period ranging from 2 to 15 μm and a period number exceeding 250 has been successfully grown along the crystal *a*-axis as shown in Figure 31 [233].

In order to accentuate the growth striations, early study on the off-axis CZ growth of the periodically poled domain reversed structure has found it desirable to dope the melt with 1 wt% of yttrium [234]. Subsequent study has shown that this doping method can be applied to grow periodically poled domain reversed structures on LiNbO₃ crystal doped with 0.01–0.2 wt% Cr₂O₃ [235], or on Er:LiNbO₃ [236], Nd:MgO:LiNbO₃ [237–239], Yb:MgO:LiNbO₃ [240],

Yb:LiNbO₃ [241], and Nd:LiTaO₃ [242]. There are, however, certain difficulties associated with the growth of PPLN by the CZ method. They include the unequal thickness between the positive and the negative domains, and the occurrence of islandlike domains in the periodical structures [243]. There also appears to be a critical concentration gradient to activate the domain reversal process.

A theory has been developed to analyze the formation of CZ-grown periodically poled domain reversed structures on LiTaO₃ and LiNbO₃. It is based on the influence played by a nonuniform distribution of the impurities such as yttrium (Y) in the para- to ferroelectric phase transition [244]. It is found that the domain walls always occur at the place where the gradient of the concentration of Y changes its sign. It is suggested that the coupling between the order parameter and the self-generated external field, which can arise from the dopant diffusion or the strain-induced polarization effects, is responsible for the CZ growth of periodically poled domain reversed structures.

3.4.2.2. Temperature Effects

A recent study on the off-axis CZ growth of periodically poled ferroelectric crystals indicates that the cooling rate can greatly influence the formation and the structure of PPLN. It is noted that by introducing a nonaxial asymmetric temperature field, one can also create a temperature gradient along the solid–liquid interface. For a slow cooling rate of $40°C/h$ in bringing down the temperature of the grown PPLN crystal from the melting point, the periodically poled domain reversed structures appear preferentially at the edges of the crystal. The growth mechanism of PPLN in this case corresponds to the surface rotation striation already shown in Figure 30. In this case, a large periodical variation of the Nb composition can be observed along the convex solid–liquid interface [245]. However, at a quenching procedure with a fast cooling rate of $1000°C/h$ in lowering down the crystal temperature, the periodical domain reversal structure can also form at the center of the crystal where the interface is flat. In conjunction with this, one can observe a nearly constant distribution of the Nb and the Er impurity level at the interface.

These phenomena can be ascribed to a competition between the internal freezing force arisen from the cooling process when the crystal makes a ferroelectric transition from a paraelectric state [246], and the build-in internal field resultant from the thermal gradient effect mentioned in Section 3.4.1. If the freezing force is larger than the electric force, the crystal would develop into a polydomain structure to minimize the free energy. On the other hand, when the crystal is quenched to room temperature, a low freezing force occurs during this procedure. If the freezing force is low enough to be compensated by the electrical field, the PPLN structure can then appear on the whole radius of the crystal.

The issue of impurity distribution in the periodically poled off-axis growth of PPLN has also been re-examined recently [247]. For PPLN grown from Er, Nd, or Y doped $LiNbO_3$ in a slow cooling procedure, the periodically poled domain structure is found to preferentially appear at the borders of the convex solid–liquid interface. In all of the above doping cases, the distribution of the impurity concentration is nearly constant along the direction where the domain structure changes its sign of polarity periodically. Note that a much larger periodic variation in the Nb concentration can be observed on the rare earth doped (e.g., Er and Nd) PPLN than on the transition metal doped (e.g., Y) PPLN. These observations suggest a somewhat different growth mechanism of PPLN from that of the impurity fluctuation model as discussed earlier [232, 244]. As a result, a combined effect due to the nature of the dopant and its position in the $LiNbO_3$ lattice site [248], and the corresponding impurity segregation effect, may play an important role in the formation of the PPLN structure.

3.4.2.3. c-Axis Grown Opposite Domain LiNbO₃ (PPLN)

It is noted that the off-axis CZ method can further be applied to grow c-axis pull $LiNbO_3$ with the domain structure aligned in a "head-to-head" or "tail-to-tail" configuration as shown in Figure 29a. A structure of this type also has been known as an "acoustic superlattice" (ASL) owing to an efficient energy conversion and propagation of the electromagnetic energy into the elastic energy in the structure. After its first introduction by Peuzin and Tasson in 1976, followed by the extensive work at Nanjin University, China in the 1990s, research on acoustic superlattices has attracted a great deal of interest and exhibits promising applications such as acoustic transducers and filters [249].

There are two methods in growing the c-axis pull ASL. The first one is the off-axis CZ method by growing the crystal in an asymmetric temperature field. The period (Λ) of the domain structure can be determined by $\Lambda = (v_{pull} + v_{dec})/n_{rot}$, where v_{pull} is the pulling rate of the crystal, v_{dec} is the decreasing rate of the free surface of the melt during the crystal growth, and n_{rot} is the rotation rate of the pulling rod. This method is in particular suitable for growing a chirped structure upon which the period of the alternatively polarized domain varies along the growth direction to extend the useful bandwidth of ASL in device applications. For example, a wide frequency response of an ASL transducer in a frequency range from 250 to 850 MHz has been realized using this method [250].

In the second method, the crystal seed is put in the center of the temperature field with the rotation axis remaining static during the growth. How can a periodical modulation of the domain polarity come about? The solution resides on the observation that the distribution coefficient of the impurity such as Cr decreases linearly with the current density [251]. The resultant impurity gradient distribution is similar to that obtained by the off-axis rotation method except now the crystal seed is sited in the center of the temperature field. The space charge field associated with the gradient distribution of the impurity can then play the role of domain poling during the crystal growth and thus forms the periodically poled domain reversed structure. Therefore if one periodically modulates the applied current density, then a periodic distribution of dopant along the growth direction can be induced. A typical growth condition is to add 0.5 wt% yttrium in the melt, and the pulling rate is 3.5 mm/h. In this current modulation technique, a typical current pulse is 10 s with a 50% on–off duty cycle. The crystal seed is chosen as the positive pole with a current density of ~ 15 mA/cm^2 flown into the crucible which is chosen as the negative pole. By doing so, a periodically poled domain reversed PPLN structure with nearly equal positive and negative domain thickness ~ 10 μm and good uniformity over 400 periods has been reported [252].

We pass by noting that the tensor characteristics of the periodical poled ASL bear a great resemblance to those in the 1-D ionic-type photonic crystal [253]. A one-to-one correspondence can be found in the microwave absorption, dielectric abnormality, and polariton dispersion relations that originally exist in the ionic crystals. The periodic modulation in the impurity distribution, however, results in a side effect of periodic modulation in the dielectric constant. Note that the latter is detrimental to the phase-matching condition and electro-optic application where a fixed index relation is a desirable system parameter.

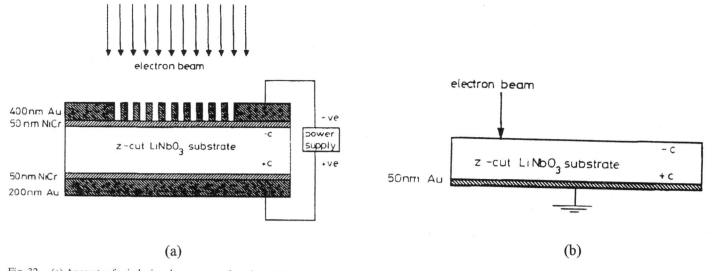

Fig. 32. (a) Apparatus for inducing doman reversal on the $-C$ face of LiNbO$_3$ by electron-beam bombardment. Reprinted with permission from [254], copyright 1990, IEEE. Note that the mechanism of domain reversal is similar to that involved in Figure 26. (b) Arrangement for domain reversal by a electron-beam lithograph. Reprinted with permission from [256], copyright 1991, IEEE. Note that in (b) there is no need for the external poling field.

3.5. Bulk Periodical Domain Reversal: *Ex Situ* Poling Methods

3.5.1. Electron-Beam Assisted Poling

We have previously discussed in Sections 3.1 and 3.2 the use of a heat-treatment or thermal-diffusion process at a temperature near T_c to induce surface domain inversion on the $+C$ surface of LiNbO$_3$. With the crystal's original spontaneous polarization P_s aligned parallel to the thermal induced poling field, these methods cannot result in domain inversion on the $-C$ surface of LiNbO$_3$ where the fabrication of the waveguide is preferred. Here we note that by exposing the $-C$ face of LiNbO$_3$ to an energetic electron beam, the process of polarization switching can be realized at temperature much lower than T_c with a poling field as low as 10 V/cm.

Illustrated in Figure 32a is one such arrangement to achieve domain inversion using an electron-beam bombardment technique [254]. Here a bilayer thin film of NiCr/Au is deposited on the $-C$ as a masking electrode to cover the region where domain reversal is not desired. A NiCr/Au layer is used to make voltage contact with the $+C$ face. By exposing the $-C$ face of the patterned LiNbO$_3$ to an electron beam of 10 kV energy, surface domain inversion can take place on the $-C$ face of LiNbO$_3$ at an applied poling field as low as 10 V/cm at a substrate heating temperature of 580°C. The corresponding mechanism for domain reversal is similar to that of the high-energy beam poling process discussed in Section 3.3.

Soon after the electron-beam bombardment effect was discovered, researchers in Japan quickly realized that by taking advantage of the electron-beam scanning or writing technique, domain inversion in LiNbO$_3$ can even occur at *room* temperature *without* the needs of external field. The rationale is as follows [255]. When the uncoated $-C$ surface of LiNbO$_3$ is exposed to the electron beam, it will also be susceptible to the charging effect. The accelerated electrons can then generate a transient vacancy in the oxygen triangle plane and open a pathway for the Li motion. This surface charging effect in turn acts as a local poling field to assist the move of the Li atom into an order structure in this domain inversion process.

The experimental setup is as simple as that shown in Figure 32b and the poling process contains the following procedures [256]. First, the area exposed to the electron-beam writing, i.e., the $-C$ face, must remain uncoated. But the $+C$ face must be grounded via a good, conducting path. The pattern is then written with an acceleration voltage typically around 25 kV. Care must be exercised not to deliver too much beam current (typically less than 10 nA) to avoid the surface cracking but the beam current should be high enough to keep good resolution for the e-beam writing procedure [257]. The e-beam writing can be performed in either a dotted-line or a continuous-line scanning mode. In the latter case the spot of the focused e-beam is scanned with a constant velocity without any dwelling time as in the former case. However, it is found that as the e-beam writing is performed in a continuous-line scanning mode, the domain inversion occurs only in a segmented region [258, 259]. This behavior seems related to the neutralization of the deposited charge by the inversion of the surface polarization.

Note that it is suggested that the excessive charge can accumulate on the scanned region due to the incomplete neutralization process. The subsequent e-beam position and the scanning electron microscope (SEM) focusing procedure will thus be affected by the lateral electric field induced by the surface charges accumulated in the previously scanned area. Therefore good condition of the domain inversion cannot be maintained over a wide area by the e-beam writing technique [260]. Moreover, a study on the e-beam writing of PPLN has shown that the high-energy irradiation can induce a varia-

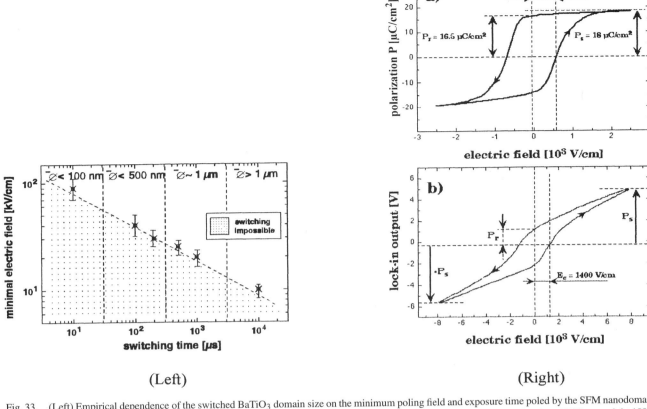

Fig. 33.   (Left) Empirical dependence of the switched BaTiO₃ domain size on the minimum poling field and exposure time poled by the SFM nanodomain technique. (Right) Hysteresis loop recorded during (a) bulk and (b) nanodomain poling of BaTiO₃. Reprinted with permission from [265], copyright 1999, Institute of Physics. Note the increase of coercive field from a bulk value of 0.5 to 1.4 kV/cm in the nanodomain case revealing a dominant 180° domain nucleation process.

tion on the optical refractive index in Ti : LiNbO₃ but not in pure LiNbO₃ [261]. Using this e-beam writing/poling technique, a periodically poled domain structure has also been realized on other materials such as LiTaO₃ [262] and KTP [263].

3.5.2. Nanodomain Poling

Just as the high-energy electron beam in the SEM can be used to facilitate the examination as well as the "writing" of a reversed domain structure, so will be the field distribution near the nanotip of a SFM as mentioned in Section 2.4.3. With the tip radius of a SFM cantilever typically around 20–40 nm as used in Figure 14, a well focused electric field distribution can limit the size of domain nucleation down to a nanometer scale. By choosing an adequate polarity for the voltage applied to the conductive tip of SFM, one can operate the SFM in a writing mode to manipulate a nanodomain switching process. To reveal this procedure on BaTiO₃, a working area ~20 × 20 μm² is first defined by scanning the SFM tip across the above area with the counterelectrode biased to a high negative voltage (U_2, $E > 800$ V/cm). By doing so, one can set up in the working area a predefined polarization status [264]. Next, the tip is moved to a desired spot with a positive bias applied to the counterelectrode for a specific exposure (switching) time τ.

An empirical dependence of the domain size on the minimal electric field and the required exposure (switching) time is illustrated in Figure 33 [265]. Referring to this figure, to perform a nanodomain switching with a feature size less than 100 nm, for example, the minimum switching electric field $E > 50$ kV/cm and an exposure (switching) time $\tau < 30$ μs are required. Note that a domain size down to 25 nm has thus been realized on BaTiO₃ using the SFM technique.

In addition, in the nanodomain switching process of BaTiO₃, an increase in the coercive field with the decrease of the switched domain size appears. This can be seen from the hysteresis loop recorded during the polarization switching process also shown in Figure 33. Note that an increase of coercive field from (a) the bulk value of 550 V/cm to (b) 1.4 kV/cm in the nanodomain can be clearly resolved. This observation suggests that the process of nanodomain switching is limited by the nucleation energy. Moreover, the lack of saturation in the nanodomain hysteresis loop as compared with the bulk hyeteresis curve of polarization switching can be ascribed to a high-field enhanced domain growth in the former poling situation.

We pass by noting that other scanning force microscope techniques, such as atomic force microscopy (AFM) [266] which takes advantage of the piezoelectric response, can also be applied for the nondestructive imaging and switch-

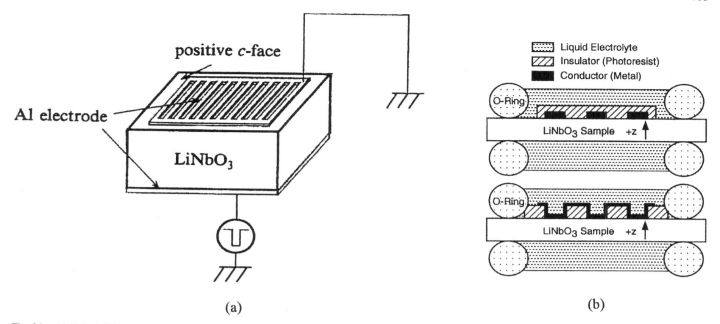

Fig. 34. (a) Pulsed-field poling apparatus originally designedi by the Sony Corp. Reprinted with permission from [274], copyright 1993, American Institute of Physics. (b) Liquid electrode configuration used to form an isopotential surface to control the fringing field around the metallic pattern. Reprinted with permission from [294], copyright 1995, Optical Society of America.

ing of the ferroelectric domain [267]. Domain switching by the AFM technique has been reported on the ferroelectric $Sr_xBa_{1-x}Nb_2O_6$ crystal at a coercive field of 0.2 kV/mm [268], on guanidinium aluminum sulfate hexahydrate crystal at a coercive field of 0.15 kV/mm [269], on PZT thin film with a threshold field of 3.2 kV/mm [270], and on PZT/SrRuO$_3$ heterostructure at a coercive field of 7–10 kV/mm [271].

3.5.3. Electrical Field Poling Method

The widely used ferroelectric LiNbO$_3$ and LiTaO$_3$ nonlinear optical crystals have long been regarded as "frozen ferroelectrics" due to on the inability to perform the polarization switching process at room temperature [272]. The major difficulty indeed arises from a high coercive field ~21 kV/mm near the crystal's avalanche breakdown [273]. However, in 1993, scientists at Sony Corp. successfully demonstrated a pulse-field poling technique to achieve periodical domain inversion in bulk LiNbO$_3$ crystal. This technique has ever since greatly revived the research activity on the polarization switching of ferroelectric crystals characterized by large coercive field.

The original setup of Sony's pulsed-field poling apparatus is schematically drawn in Figure 34a [274]. A typical pulsed field duration is 100 μs with a magnitude of 24 kV/mm to overcome the coercive field encountered in the polarization switching procedure. It is found that by applying the pulsed field across a thin, Z-cut LiNbO$_3$ substrate periodically patterned with a Al-film electrode, the nucleation of the 180° inverted domain can first appear on the +C face of LiNbO$_3$ under the electrode and then grow along the C-axis. To obtain a periodically poled domain

reversed QPM structure, the external field must be terminated before the reversed domain grown out of the region defined by the patterned electrode. By doing so, the reversed domain can propagate all the way down to the opposite side of the substrate and appears as a mirror image of its counterpart on the +C face. To make a distinction from its antecedent of surface domain inversion, this process will be called as bulk domain inversion.

When combined with the proton-exchange technique in making a waveguide structure as discussed in Section 3.2, a record high 426 nm blue SHG efficiency of 600%/W/cm^2 was observed on a 2.8 μm PPLN in 1993 by the Sony Corp. research group. Soon after this report, periodical domain reversal by pulsed field poling evolved as a core technology to make fine-pitch QPM structures on LiNbO$_3$ [275, 276] and on other ferroelectric crystals such as LiTaO$_3$ [277–279], KTiOPO$_4$ (KTP) [280–282], KTiOAsO$_4$ (KTA) [283, 284], RbTiOAsO$_4$ (RTA) [285, 286], RbTiOPO$_4$ (RTP) [287], potassium lithium tantalate niobate (KLTN) [288], and strontium barium niobate (SBN) [289, 290]. To date, this field-poling technology has reached a state such that multiperiod QPM structures with periods as low as 6 μm are now commercially available for LiNbO$_3$, LiTaO$_3$, KTP, and RTA.

There are, however, several important issues associated with the pulsed-field poling techniques on the domain's uniformity, stability, and optical damage resistance that deserve further discussion. By doing so, one can ensure the improvement of the device performance of the periodically poled domain reversed QPM structures for the nonlinear- and electro-optic applications.

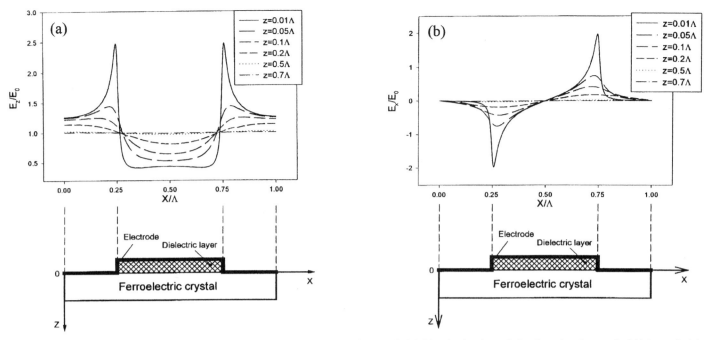

Fig. 35.   Distribution of the (a) normal (E_z) and (b) tangential (E_x) component of the applied field under the electrode in a ferroelectric crystal of thickness d. Λ is the period of the designed QPM structure. Reprinted with permission from [293], copyright 1998, American Institute of Physics.

3.5.3.1. Fringing Field Effect

One consideration in applying the short-pulse poling technique is the design of the pulse shape, i.e., the magnitude and duration of the poling pulse. Let us take the periodical poling of a 250-μm-thick LiNbO$_3$ substrate with 500 μs pulses for example [291]. It is noted that the process of domain reversal cannot take place at poling voltages less than 5.6 kV, but as the applied voltage exceeds 6.0 kV and above, catastrophic breakdown occurs with damages visibly seen on both sides of the sample. To further make a fine-pitch domain reversal QPM structure, caution must be exercised in restraining the fringing field effect [292]. The appearance of a field singularity in the latter arises from the discontinuity of dielectric distribution in the patterned ferroelectric surface where an external poling voltage would be applied.

Illustrated in Figure 35 are the calculated normal (E_z) and tangential (E_x) field distribution under the electrode in a ferroelectric crystal of thickness d [293]. At a depth of z larger than 0.5Λ in the crystal, where Λ is the period of the designed QPM structure, a homogeneous field distribution can be maintained. However, near the surface edge of the electrode, E_z is substantially higher than the corresponding bulk poling field ($E_0 = V/d$) by a factor of 2.5 due to the fringing field effect. By the same token, a large, nonvanishing E_x of similar strength can occur near the edge of the metal electrode. According to the Miller–Weinreich theory discussed in Section 2.5, one would expect the polarization switching to start at the edge of the electrode. It is suggested the injection of the compensation charge on to the metal electrode during the polarization switching pro-

cess will be swept into the dielectric-coated region and cause domain broadening to occur in the undesirable region.

3.5.3.2. Liquid Electrode Technique

Improved control in the fringing field effect and release of the domain broadening effect can be achieved through the use of liquid electrode technique introduced by the Stanford University Group [294]. The liquid electrode fixture, as shown in Figure 34b, consists of two electrolyte-containing chambers in which the LiNbO$_3$ sample is pressed against the viton O-rings. The electrolyte, made of saturated LiCl in deionized water, connects the sample to the circuit, forms an isopotential surface, and helps to control the fringing fields around the pattern electrode. This arrangement can permit application of fields exceeding 25 kV/mm without causing breakdown on the sample and without the need for surrounding oil or vacuum. Using this liquid electrode technique in conjunction with the pulsed-field poling, multigrating domain reversed QPM structures with a period varying from 20 to 35 μm suitable for optical parametric generation can be fabricated on a 3″ LiNbO$_3$ wafer scale [295].

3.5.3.3. Back-Switching Technique

In making fine-pitch QPM structures with a domain reversed period less than 6 μm, the introduction of the back-switching procedure by the Stanford University group in 1999 can be regarded as the turning point in the development of the pulsed-field poling technique. In this method, a deliberate design in the poling waveform including the stages of domain forward-

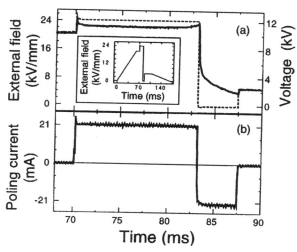

Fig. 36. (a) Design and (b) measurement of the back-switch poling waveform on a 500-μm-thick LiNbO$_3$ substrate. Reprinted with permission from [296], copyright 1999, American Institute of Physics.

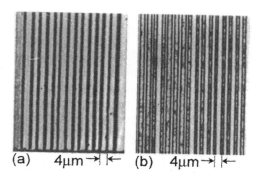

Fig. 37. Photograph of (a) y-face and (b) $-Z$-face of 4-μm-period domains in back-switched poled 500-μm-thick LiNbO$_3$ substrate. Reprinted with permission from [296], copyright 1999, American Institute of Physics.

growth, back-switching, and domain stabilization is carefully exercised as illustrated in Figure 36a [296].

Unlike the conventional poling procedure, the initial forward-growth procedure is sustained until 100% of domain reversal is completed under the patterned electrode area [297]. Then the applied voltage is rapidly decreased to allow a spontaneous back-switching of the inverted domain as indicated by the negative current also shown in Figure 36b. This current back flow arises from an unscreened internal field effect to be discussed in the next section. In the last stage, the termination of charge back flow and the procedure of domain stabilization can be realized by applying an external voltage larger than the value of the instantaneous decaying internal field. Using this method, periodically poled domain reversal QPM structures with a period as short as 4 μm can be fabricated over a 3″ LiNbO$_3$ wafer as shown in Figure 37. It is interesting to note that multiplication of the domain spatial frequency to a doubling or tripling of the original QPM designed period can also be realized using this novel back-switching technique [298].

3.5.3.4. Domain Inhomogeneity

The nonlinear frequency conversion efficiency of the QPM device is hinged upon the homogeneity of the electrically poled periodical domain structure [299, 300]. The inhomogeneity can reside in the absence of domain inversion in the periodically poled QPM structure, and the stochastic disturbance in the domain boundary and period. Departure from the ideal QPM condition in the periodicity can also lead to the demise of the conversion efficiency [301, 302].

The loss of conversion efficiency can also be attributed to the photorefractive effect commonly observed in bulk LiNbO$_3$ and LiTaO$_3$ crystals. In this case, one would observe a decollimation and scattering of the laser beam as it propagates along the crystal [303]. It can also be observed as a beam distortion on the SHG signal as the light propagates in the periodically poled LiNbO$_3$ [304] and LiTaO$_3$ [305]. The photorefractive effect, also known as optical damage, is due to the optically induced change in the refractive index. The latter can arise from a combined action of the optical induced space-charge field via the electro-optic effects. The photorefractive effect is detrimental to applications in nonlinear frequency generation, electro-optic modulation, and acoustic-optic deflection where a stabilized refractive index is the key issue of the system design [306]. In the congruent-grown LiNbO$_3$ and LiTaO$_3$ single crystals, index changes on the order of 10^{-3} can occur. When a waveguide is excited with a 488 nm laser at a density greater than 200 W/cm^2, it is found the refractive index change reaches a rate of 2.8×10^{-3} and 4.9×10^{-4} cm^2/μW on Ti : LiNbO$_3$ and Zn : LiTaO$_3$, respectively [307].

Regarding the optical damage resistance in the periodically poled QPM device, theoretical analysis indicates that it could be enhanced by three orders of magnitude due to the reduced photogalvanic effects in the longitudinal current flown on the periodically poled structure [308, 309]. Experimentally, however, these QPM waveguides are known to degrade when used to generate high intensity short wavelength light [310]. Recent investigation has further confirmed that the photorefractive effect can broaden the QPM curve and reduce the conversion efficiency even at an irradiation power less than 20 mW [311].

Part of the reason for the reduced optical damage resistance on QPM devices is the deviation from a 50% duty cycle in the designed grating period [312]. Another possible cause is the inhomogeneous distribution in the refractive index on the domain walls [313]. It is suggested that the inhomogeneity in the refractive index may be attributed to the electric charge accumulated at the domain boundaries during the electrical-poling process [314]. The electric field produced by the charged domain walls can thus lead to variation in the refractive index via the electro-optic effect [315]. Note that the formation and evolution of the charged domain walls in LiNbO$_3$ has recently been observed by combining the pulsed-poling technique with a cross-polarizer microscope [316].

In order to eliminate the photorefractive effect on the electrically poled QPM devices, an annealing procedure to cure the substrate is necessary. Conventional experience suggests

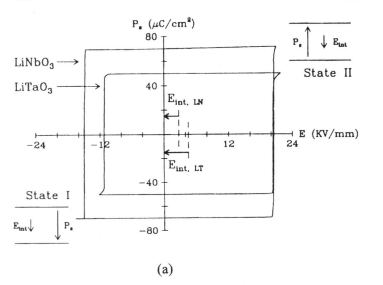

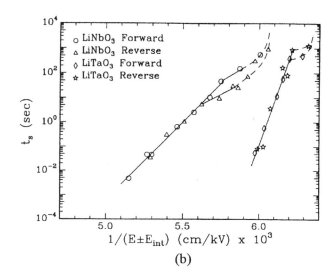

(a)   (b)

Fig. 38.   (a) Asymmetric hysteresis loops recording the internal field effect on the polarization switching process of congruent-grown Z-cut LiNbO$_3$ and LiTaO$_3$. The loop offsets $E_{int,LN} = 2.75$ kV/mm and $E_{int,LT} = 4.56$ kV/mm are the measured internal fields in LiNbO$_3$ and LiTaO$_3$, respectively. (b) Measurement of the switching time in the *low*-field forward and reverse poling, data are fitted by using Eq. (10) with $E_{int,LN} = 2.22$ kV/mm and $E_{int,LT} = 4.12$ kV/mm. Reprinted with permission from [324], copyright 1999, Elsevier Science.

that by annealing the QPM devices at temperature above 80°C for LiTaO$_3$ and 150°C for LiNbO$_3$, respectively, the poling-induced photorefractive effect can be eliminated. At higher annealing temperatures above 500°C, however, the periodically poled domains of LiNbO$_3$ start to merge, indicating degradation in the domain stabilization issue [314].

We pass by noting that the decrease of P_s with temperature can be used to reduce the coercive field in the polarization switching process. It is found that by increasing the LiTaO$_3$ poling temperature from 22 to 250°C, a fourfold reduction in the coercive field can be achieved [317]. However, the high-temperature process is accompanied by a dramatic increase in the transient conductivity. For a flux grown KTP which normally has a high conductivity around 10^{-7} W^{-1} cm^{-1} at room temperature, the increase of conductivity makes it difficult to maintain a sufficient voltage over the crystal to facilitate the electrical poling [318]. In this case, a low-temperature poling process would be more beneficial to make polarization switching on KTP and its isomorphous crystals [319]. At a low temperature of 170 K, it is found that the K$^+$ ionic conductivity can dramatically decrease to a low value of 10^{-12} W^{-1} cm^{-1}. This low-temperature process has found great success in making larger area ($30 \times 30 \times 0.5$ mm^3) periodically poled QPM structures on KTP with the grating period ranging from 4 to 39 μm, albeit the coercive field is increased by a factor of 5 to 12 kV/mm.

3.5.3.5.   Internal Field Effects

Here we address the issue of internal field on the dynamics of polarization switching in ferroelectric crystals. We note that the domain switching process is controlled by the local field in the crystal. The local field arises from a sum of (a) the ex-

ternal field produced by the voltage applied to the electrodes, (b) the depolarization field due to the bound charges associated with the instantaneous domain configuration, and (c) the internal field associated with the crystal defects. There exist two types of screening mechanisms to decrease the depolarization field effect [320]. The first is the external screening due to the fast redistribution of charges flowing into the electrode from the external circuit. The second is the bulk screening due to the surface layers proposed by Drougard and Landauer in explaining the domain motion of BaTiO$_3$ [321, 322]. For ferroelectrics with low conductivity such as LiNbO$_3$ and LiTaO$_3$, bulk screening due to the residual bulk charges is an inefficient process and can be neglected in the discussion.

However, it is noted even with a complete external screening or surface-layer screening of the depolarization field, there still exists an internal field in the crystal. The internal field, first observed in LiTaO$_3$ in 1996, is attributed to the non-stoichiometric point defects in the crystal grown from the congruent melt [323]. Referring to Figure 38a, the presence of an internal field can manifest itself in the asymmetric form of the polarization hysteresis loop [324]. The figure exhibits the relative orientation of the spontaneous polarization P_s and internal field E_{int} in LiNbO$_3$ and LiTaO$_3$, respectively. From there one can identify an internal field of 2.75 and 4.56 kV/mm, respectively, and points along the $+Z$ direction on the congruent-grown LiTaO$_3$ and LiNbO$_3$. The presence of an internal field can further cause a process of polarization back-switching at the termination of the external poling field. As a result, it is desirable to exert a domain stabilization time during the forward poling procedure to ensure the switched domain does not flip back. A more recent study indicates that due to the internal field induced back-switching, the reversed domain reveals fast relaxation with a time constant on the order of a sub-

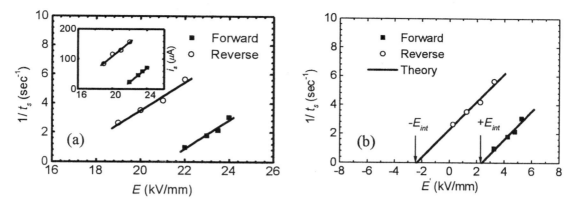

Fig. 39. (a) Dependence of the switching time (t_s) and peak averaged switching current (i_s) of congruent-grown Z-cut LiNbO$_3$ in the *high*-field forward and reverse poling. (b) Kinetic analysis of the lateral 180° reversed domain motion of (a) in the *high*-field regime using Eq. (12). Note the interaction with the abscissa at $\pm E_{\text{int}} = 2.37$ kV/mm. Reprinted with permission from [327], copyright 1999, American Institute of Physics.

second [325]. However, once the domain reversal is complete and remains stabilized, the internal field tends to align along the new polarization direction of the reversed domain. But this process turns out to be temperature dependent. At room temperature, the realignment process is inefficient and the internal field remains in its original direction of the unpoled state. At temperatures higher than 200°C, the realignment of the internal field can be completed within 30 s [323].

The internal field effects on the 180° domain switching time are further explored in Figure 38b. Referring to the figure, one further notes that the presence of an internal field can cause a switching time difference in the *forward* and *reverse* poling directions. When the electrical poling is performed in a *low*-field regime, according to the Miller–Weinreich nucleation theory, the domain switching time should exhibit an exponential dependence on the applied field as discussed in Section 2.5 and Eq. (5). However, in the presence of an internal field E_{int}, the forward and reverse switching time can be formulated in a modified form of

$$t_f = t_{f0} \exp\left[\delta_f/(E - E_{\text{int}})\right]$$
$$t_r = t_{r0} \exp\left[\delta_r/(E + E_{\text{int}})\right] \tag{11}$$

respectively. Here the condition of *low*-field poling is approximately defined as having the *forward* poling field ($E - E_{\text{int}}$) less than 20 kV/mm and *reverse* poling field ($E + E_{\text{int}}$) less than 16 kV/mm. One set of the fitting parameters to Figure 38b is $\ln(t_{f0}(\text{sec}))_{\text{LN}} = \ln(t_{r0}(\text{sec}))_{\text{LN}} = -81 \pm 10$ and $\delta_{f,\text{LN}} = \delta_{r,\text{LN}} = 1470 \pm 135$ kV/mm for LiNbO$_3$, and $\ln(t_{f0}(\text{sec}))_{\text{LT}} = \ln(t_{r0}(\text{sec}))_{\text{LT}} = -241 \pm 15$ and $\delta_{f,\text{LT}} = \delta_{r,\text{LT}} = 3980 \pm 240$ kV/mm for LiTaO$_3$. Note here the fitting parameter δ of the activation field on LiNbO$_3$ and LiTaO$_3$ is by a factor of 70 and 190, respectively, larger than the corresponding coercive field ~21 kV/cm in the crystal. In comparison, for BaTiO$_3$ which has a low coercive field of 0.5 kV/mm, the theory suggests a proportional factor of 8 for the activation field on the coercive field [326]. The latter seems to agree very well with the experimental observation in Eq. (6).

Moreover, the presence of an internal field in the ferroelectric crystal can also cause an axial anisotropy in the polarization switching rate and averaged peak switching current. In the *reverse* poling direction as shown in Figure 39a, polarization switching of LiNbO$_3$ can take place at a smaller field, exhibit a larger switching current, and substantiate a faster switching time compared with the *forward* poling case [327]. Note when the domain polarization of LiNbO$_3$ is switched in the *high*-field regime, i.e., at a *forward* field ($E - E_{\text{int}}$) larger than 22 kV/mm and *reverse* field ($E + E_{\text{int}}$) larger than 17 kV/mm, a *linear* dependence of these switching parameters on the applied field can be clearly resolved. The apparent sidewise velocity (v_s) of the 180° domain motion in this condition can further be formulated by

$$v_s = \mu_s\left[E - (E_{\text{th}} \pm E_{\text{int}})\right] \tag{12}$$

Here we use the internal field E_{int} to signify the axial anisotropy, and μ_s the lateral mobility of the 180° domain motion, similar to the case in gadolinium molybdate (GMO). ($E_{\text{th}} \pm E_{\text{int}}$) in Eq. (12) corresponds to the effective switching field in the *forward* ($+$) and *reverse* ($-$) poling, respectively. By using such a linear fitting procedure in Figure 39b with a reduced field unit of $E' = E - E_{\text{th}}$, we can determine an internal field from the intersection with the abscissa at $\pm E_{\text{int}} = 2.37$ kV/mm and a E_{th} of 18.74 kV/mm. The internal field of LiNbO$_3$ determined in this linear fitting procedure agrees well with that obtained from the hysteresis loop analysis of Figure 38a. Moreover, from the slope of $d(1/t_s)/d(E')$ of Figure 39a, one can further deduce a lateral mobility μ_s of 1.6 ± 0.1 mm^2/kV/s.

By taking advantage of the pulsed-field poling technique, now it is possible to re-examine the kinetics of the polarization switching process on ferroelectrics characterized with a larger coercive field. At this moment we recall from Table I that the KTP crystal is known to have a coercive field of 2.5 kV/mm at room temperature which is well within the amenable range of most of the poling apparatus. Illustrated in Figure 40 is one such example taken on the polarization switching parameters (t_s, i_s) of KTP in a field ranging from the *low*- (1.5 kV/mm) to the

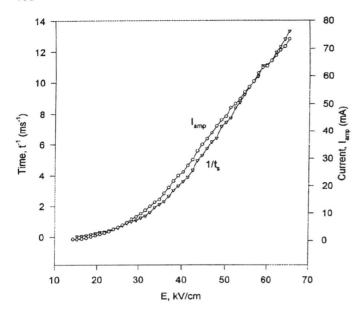

Fig. 40. Examination of the reciprocal switching time ($1/t_s$) and maximum switching current (I_m) of KTP crystals as a function of the switching field. Note the characteristics of the switching curves can be categorized into three regions of low-, intermediate-, and high-field poling regime, respectively; see text for discussion. Reprinted with permission from [328], copyright 1997, Institute of Physics.

high-field (7.0 kV/mm) poling regime [328]. One first observes from the figure that the overall behavior of the peak switching current (i) is proportional to the switching rate ($1/t_s$) in the wide poling-field regime. Further analysis of the data shown in Figure 40 suggests that the characteristics of the switching rate and the averaged peak switching current can be categorized into three parts. In the first part of the *low*-field regime from 1.5 to 2.5 kV/mm, the switching time follows an *exponential* dependence by

$$t_s = t_\infty \exp(\delta_s/E) \qquad (13)$$

Here the activation field is fitted with $\delta_s = 11.2$ kV/mm. Note that the proportionality factor of the activation field to the coercive field is 4.5 in KTP which is on the same order of that in BaTiO$_3$ but is much smaller than that obtained from Figure 38b on LiTaO$_3$ and LiNbO$_3$. According to the Miller–Weinreich theory, in the *low*-field regime, the polarization switching process is dominated by the domain nucleation mechanism as suggested by Eq. (13).

In the second part of the curve, i.e., in the intermediate field regime from 2.5 to 4.2 kV/mm, the switching rate $1/t_s$ reveals a power dependence on E by $1/t_s = kE^n$, where the power coefficient n is 3.7. In the third part of the *high*-field regime at which $E > 4.2$ kV/mm, there a *linear* dependence of the switching rate on the applied field occurs, i.e., $1/t_s = aE$, where the constant a is proportional to the lateral mobility. In this *high*-field poling regime, the switching rate dependence agrees with the analysis shown in Eq. (12). Note in this regime, the sidewise expansion of the inverted domain plays a dominant role in determining the switching time.

3.5.3.6. Stoichiometric Control

VTE Technique. The presence of larger internal field and high coercive field in LiNbO$_3$ and LiTaO$_3$ has made the fabrication of fine-pitch periodically poled QPM devices on congruent-grown crystals a difficult task. Motivated by the fact that the internal field originated from the nonstoichiometric defects, rich activity has arisen in the growth of stoichiometric crystals. Among them, the vapor transport equilibration (VTE) technique contains a simple annealing process. The stoichiometric control in the VTE process is via the mechanism of vapor transport and solid state diffusion [329]. In the case of LiNbO$_3$, the procedure consists of annealing the substrate in close proximity to a large mass of lithium niobate powder of a desirable composition. For a typical 500-μm-thick LiNbO$_3$ substrate, a minimum processing time of 60 h is needed for a powder composition at the Li-rich phase boundary, whereas a maximum of 400 h is needed on the Li-poor boundary.

CZ Method. On the other hand, using a modified double-crucible Czochralski method, the National Institute for Research in Inorganic Materials (NIRIM) group in Japan has recently reported a series of successful growth on the near-stoichiometric LiNbO$_3$ and LiTaO$_3$ single crystal [330–332]. In this method, the compositions of the inner and outer melt are kept as Li-rich and stoichiometric, separately. The Li-rich melt contains 58.5 or 60 mol% of Li$_2$O in the growth of LiNbO$_3$ or LiTaO$_3$, respectively. The feeding and supply of the stoichiometric powder into the outer melt can be automatically adjusted from the weight increase of the growing crystal. A typical crystal pulling rate in the "double crucible" CZ growth of LiNbO$_3$ is 0.5 mm/h and the seed rotation rate is 5 rpm. The composition of LiNbO$_3$ and LiTaO$_3$ crystal can be estimated from the Curie point measurement by the DTA or the UV absorption method as described in Section 1. Note that in the conventional "single crucible" CZ growth method, the crystal's stoichiometry can also be controlled by adding K$_2$O in the melt. For example, by adding 7–10 wt% of K$_2$O in the congruent melt, growth of near-stoichiometric LiNbO$_3$ single crystal has recently been reported [333].

TSSG Method. Alternatively, stoichiometric LiNbO$_3$ single crystals can also be grown by the high-temperature top seeded solution growth method (HT-TSSG) in a potassium containing flux [334]. This method is based upon a recent observation that the K$^+$ ions in the growth melt can improve the incorporation of Li ions into the solid phase [335]. With a K$_2$O/LiNbO$_3$ composition ratio ranging from 0.16 to 0.195, growth of stoichiometric LiNbO$_3$ crystals has been reported [336]. A typical crystal pulling rate in the TSSG growth of LiNbO$_3$ is 0.2 mm/h and the seed rotation rate is 6 rpm. In the seeding procedure, it is preferred to keep the crystallization temperature below the melting temperature by more than 150°C and it has to be further decreased as the crystal grows. For a successful HT-TSSG growth of stoichiometric LiNbO$_3$ single crystal, the system requires a steep axial temperature gradient of 200–700°C/cm.

Despite the crystal growth beginning at a temperature below the Curie point, the steep thermal gradient can result in growth of single domain crystal without the need of any poling procedure as mentioned in Section 2.2.2.

3.5.3.7. Field Reduction

With the improvement of stoichiometric control on LiNbO$_3$ and LiTaO$_3$, one can expect a reduction in the internal field and the coercive field as well. For LiTaO$_3$, a dramatic drop of the internal field from 4 kV/mm in the congruent-grown crystal down to 0.1 kV/mm in the stoichiometric case has recently been reported [337]. One the other hand, the polarization switching field further exhibits a great deal of reduction from a high value of 21 kV/mm in the congruent crystal down to 1.7 kV/mm in the stoichiometric case, reaching a reduction factor of 13. Here we note that the coercive field reduction can also be activated by impurities in the crystals. One such example is by the optical damage resistant impurities that will be discussed in the next section. Another example is on the Nd doped LiTaO$_3$ congruent crystal upon which the coercive field is found to decrease with the Nd concentration and reaches 16.8 kV/mm at a Nd doping level of 0.6 wt% [338]. However, note that the latter type of doped crystal still suffers from photorefraction effects.

For a near-stoichiometri LiNbO$_3$ (i.e., of a composition ratio of 49.8 mol% in Li$_2$O) grown by the double-CZ method, it is found that the internal field almost disappears compared with a large residual value ~2.5 kV/mm in the congruent-grown case [339]. The switching field reduction is also very promising. A small forward switching field of 4–5 kV/mm is observed on the near-stoichiometric LiNbO$_3$ and reaches a reduction factor of 4 when compared with the congruent case of 21 kV/mm. Moreover, further coercive field reduction can be credited to the stoichiometric LiNbO$_3$ crystal grown by the HT-TSSG method. The forward switching field can be lowered down to a vanishing small value of 0.2 kV/mm [340]. Here we summarize in Figure 41 the reported forward switching field of LiNbO$_3$ and illustrate the principle of stoichiometric control for the switching field reduction.

Furthermore, we note that the improvement of stoichiometry control in the ferroelectric crystal can change the Curie point but leave the spontaneous polarization of the crystal nearly unchanged. The minimization of the nonstoichiometric defects can also lead to great improvement on the electro- and mechanic-optical properties of the crystals. Regarding the commonly used EO coefficients of r_{33} and r_{22} on LiNbO$_3$, enhancement factors larger than 20% [341] and 60% [342], respectively, have been noted on the stoichiometric crystals. The EO coefficient of r_{33} increases from 31.5 to 38.3 pm/V, whereas the increase of r_{22} is from 6.07 to 9.89 pm/V in the near-sotichiometric case. For an account of the stoichiometry effect on the elastic properties of LiNbO$_3$, the reader can refer to the work by Bernabe et al. [343].

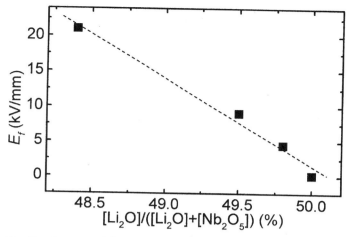

Fig. 41. Summary of the reported forward switching fields on LiNbO$_3$ single crystals of congruent and stoichiometric compositions. Data collected from [339, 340]. Reprinted with permission, copyright 2000, IEEE.

3.5.3.8. Optical Damage Resistance

One unexpected side effect associated with the stoichiometric crystal is on the increase of the photorefractive (optical damage) effect. From the two-beam coupling experiments, it is found that the undoped and Fe doped stoichiometric LiNbO$_3$ crystals exhibit stronger photorefractive effects than the Fe-doped congruent crystals. Recent investigation has further shown that strong beam distortion, or the so-called beam fanning phenomenon, can occur on a stoichiometric LiNbO$_3$ when transmitting a green light at an intensity as low as 64 W/cm^2 [344]. The increase of photorefractive effects in stoichiometric LiNbO$_3$ can cause serious problems for these materials to be used in the electro-optic and nonlinear-optic applications. Here we address the methods of restoring the optical damage resistance by selectively doping the ferroelectric nonlinear crystals with suitable impurities.

Mg Dopant. In the early development of optical damage-resistant congruent LiNbO$_3$ crystal, it is noted by doping the crystal with MgO above a threshold concentration of 4.6 at%, one can greatly reduce the holographic grating erasure time [345]. From the study of light-induced charge transport properties, it is concluded that the damage resistance is ascribed to the increase of the crystal's photoconductivity by the Mg dopant which in turn leads to a demise of the light-induced refractive index change [346]. One such advantage on the optical damage resistant application is to maintain a high-gain amplification process over a wide range of laser intensities by doping the Fe : LiNbO$_3$ photorefractive crystal with the optical damage resistant impurities [347].

To resolve the issue of photorefractive effects in the stoichiometric LiNbO$_3$ crystal, one of the promising solutions is to employ MgO doping during the crystal growth [348]. Here the photorefractive damage effects measured as a function of the MgO doping in the LiNbO$_3$ crystals are illustrated

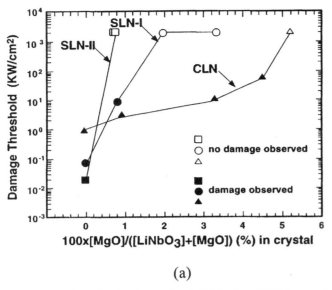

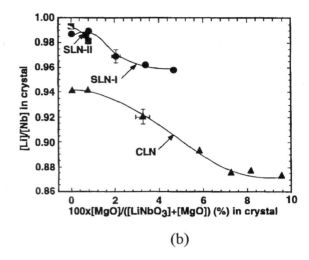

(a)   (b)

Fig. 42. (a) Photorefractive damage threshold behavior of LiNbO$_3$ crystals as a function MgO concentration. SLN-I (II) sample has a [Li]/[Nb] ratio of 0.96–0.98 (0.98–0.993), whereas the CLN sample has a ratio of 0.87–0.94. Reprinted with permission from [349], copyright 2000, American Institute of Physics. (b) [Li]/[Nb] ratio as a function of the MgO concentration in CLN, SLN-I, and SLN-II LiNbO$_3$ single crystals. Reprinted with permission from [351], copyright 2000, Elsevier Science.

in Figure 42a [349]. Note that both of the congruent and the stoichiometric samples exhibit a MgO doping threshold above which the optical damage resistance can be dramatically increased. The damage threshold is defined as the intensity of a CW Nd:YAG SHG laser upon which the transmitted laser beam exhibits a distorted and elongated pattern along the crystal c-axis due to the photorefractive effect at an exposure time of 10 min.

As inferred from the figure, for the stoichiometric crystal having a higher [Li]/[Nb] ratio of 0.98–0.993 (sample SLN-II), a smaller MgO doping threshold of 0.8 mol% is sufficient to eliminate the photorefractive effect at a light intensity of 2 MW/cm^2. In comparison, a higher doping of 5 and 1.8 mol%, respectively, is required for the congruent and the near stoichiometric LiNbO$_3$ sample (SLN-I with a [Li]/[Nb] ratio of 0.96–0.99) to reach a similar damage threshold. The dramatic increase of the optical damage resistance in the doped stoichiometric LiNbO$_3$ crystals is attributed to the decrease of the space charge field arising from a combined effect of increased photoconductivity and decreased photogalvanic current by the MgO dopant [350].

For a further correlation of the doping effect to the crystal's stoichiometry, one can infer from Figure 42b a nonlinear decrease in the [Li]/[Nb] ratio with an increase of the MgO doping in the congruent-grown and the stoichiometric LiNbO$_3$ crystal [351]. This behavior can be ascribed to a two-threshold phenomenon first observed in the MgO doped congruent grown LiNbO$_3$ [352]. To explain the observation in Figure 42b, one can resort to a Li-site vacancy model analysis as described in [353–356].

The theory is based on a charge compensation mechanism of the Mg^{2+} ions to replace the Li vacancy $(V_{Li})^-$ and the

Nb antisite defects $(Nb_{Li})^{4+}$ [357]. During this doping procedure, the Mg replacement mechanism changes twice at the threshold concentration. At first the Mg^{2+} ions will initially replace the $(Nb_{Li})^{4+}$ defects and decrease the $(V_{Li})^-$ vacancy in LiNbO$_3$. In this case, the [Li]/[Nb] ratio of the crystal is kept constant at $\sim$0.94. A complete replacement process would take place at a Mg concentration of 3 mol% in the congruent-grown LiNbO$_3$ crystal. Upon increase of the Mg doping concentration, the $(V_{Li})^-$ vacancy density begins to increase at a compensation of reducing the [Li]/[Nb] ratio down to 0.84. A further increase of the Mg doping above 8 mol% will allow the dopant to enter the Nb and Li sites simultaneously, maintain the [Li]/[Nb] ratio in the crystal unchanged, and result in a decrease in the $(V_{Li})^-$ vacancy again.

The dependence of the [Li]/[Nb] ratio in the MgO doped stoichiometric LiNbO$_3$ samples (SLN-I and SLN-II) can also be explained by using the same Li vacancy model as discussed above. As seen from Figure 42b, the Mg doping dependence of the [Li]/[Nb] ratio on the SLN-I sample is consistent with the two-threshold phenomena. At the Mg doping concentration up to 0.5 mol%, the [Li]/[Nb] ratio is nearly kept constant at 0.99 due to the replacement of the residual $(Nb_{Li})^{4+}$ antisite defects by the Mg^{2+} ions in the near-stoichiometric crystal. The [Li]/[Nb] ratio then exhibits a sharp decrease down to $\sim$0.96 at a further increase of the Mg doping concentration to a threshold of 1.8 mol%. In this case, the Mg^{2+} ions will be incorporated into the Li sites and this process will result in the creation of Li vacancy $(V_{Li})^-$. A further increase of the Mg doping passing this threshold would cause the Mg dopant to enter the Nb site.

With the availability of the optical damage resistant LiNbO$_3$ crystals, the next question one naturally would raise is how the doping affects the polarization switching process. At the first

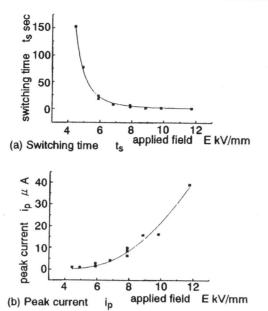

Fig. 43. Switching field dependence of the (a) switching time (t_s) and (b) averaged peak switching current (i_p) in a Z-cut 5 mol% MgO : LiNbO$_3$ crystal. The fitting procedure follows an exponential form of Miller–Weinreich theory of Eq. (5); see text for discussion. Reprinted with permission from [362], copyright 1996, American Institute of Physics.

glance, due to the extremely small diffusion coefficient of Ti in MgO : LiNbO$_3$ [358] and an increase of the Curie point with the MgO doping [359], one would expect the use of a thermal process as discussed in Sections 3.1 and 3.2 be less efficient in making a surface domain inversion in doped LiNbO$_3$. Moreover, due to the fact that the spontaneous polarization P_s in LiNbO$_3$ is insensitive to the stoichiometry and impurity in the crystal, one would expect a coercive field of the same order of magnitude as that of the undoped case in the doped LiNbO$_3$ crystal.

A recent corona discharge study, however, has shown that the polarization switching field in bulk periodical poling of a 5 mol% MgO doped LiNbO$_3$ crystal can be less than 6 kV/mm, reaching a quarter of that in the undoped congruent LiNbO$_3$ crystal [360]. From a direct electric poling study on 4.8 mol% MgO : LiNbO$_3$, the reported coercive field can be as low as 2.5 kV/mm but the dielectric breakdown can easily take place in the crystal [361]. In the *low*-field poling regime, the polarization switching parameters of a 5.0 mol% MgO : LiNbO$_3$ crystal are found to have the characteristics resembling those of the BaTiO$_3$. As illustrated in Figure 43, both the switching time and the averaged peak switching current in the electric poling of a 5 mol% MgO : LiNbO$_3$ crystal exhibit an exponential dependence on the applied field [362]. This observation indicates that in the *low*-field regime, the kinetics of polarization switching in MgO doped LiNbO$_3$ is dominated by the 180° domain nucleation mechanism as in the case of polarization switching of BaTiO$_3$ discussed in Section 2.5.

When the polarization switching time on the 5 mol% MgO : LiNbO$_3$ is replotted with an exponential form of $t_s =$

$C \exp(\delta/E)$, one can obtain a fitting parameter of the activation field δ as 34.4 kV/mm. Here we summarize that in the *low*-field poling regime, the activation field (δ) to the coercive field (E_c) ratio is commonly kept at 4–8 in several ferroelectric crystals examined so far in this chapter. For example, a pair value of $(\delta, E_c) = $ (4 kV/mm, 0.5 kV/mm) can be found in BaTiO$_3$ (Eq. (6)), (11.2 kV/mm, 2.5 kV/mm) in KTP (Eq. (13)), and (34.4 kV/mm, 5 kV/mm) in 5 mol% MgO : LiNbO$_3$. However, in Eq. (11), a relatively large ratio is observed on the undoped LiNbO$_3$ with $(\delta, E_c) = $ (1470 kV/mm, 21 kV/mm) and (3980 kV/mm, 21 kV/mm) in LiTaO$_3$, respectively.

We pass by noting that the reverse domain on the MgO : LiNbO$_3$ is found to have a needlelike shape and nucleate preferentially on the existing domains which agrees with the Miller–Weinreich theory discussed in Section 2.5. However, it is found that at a low-field poling of 4.45 kV/mm, the polarization switching occurs with the domain walls moving faster in the sidewise direction than in the forward direction [356]. The latter observation would cause difficulty in making fine-pitch periodically poled QPM structure upon which the lateral control of the reverse domain motion is critical. In this regard, it is desirable not only to reduce the coercive field in the electric poling procedure but also in the same time to achieve precise control of the lateral domain motion.

Other Optical Damage Resistant Dopants. Mg has been widely used as an efficient optical damage resistant dopant for LiNbO$_3$ in the near IR and visible range; however, it can result in an enhanced photorefractive effect in the UV regime. One such example is that a short wavelength of 351 nm, the two-beam coupling efficiency and photorefractive recording ability of the MgO : LiNbO$_3$ can be largely increased [363]. On the other hand, we note that there are other kinds of optical damage resistant dopants that have been developed on the congruent-grown LiNbO$_3$ crystal in the 1990s. They include the divalent dopant of ZnO [364] and the trivalent dopant of Sc$_2$O$_3$ [365] and In$_2$O$_3$ [366]. In these doped LiNbO$_3$ crystals, the enhancement of optical damage resistance is related to the doping level which reaches a threshold. The latter observation is now understood as arising from the increase of the crystal's photoconductvity by the dopant-induced reduction of the Nb$_{Li}$ antisite defects [367].

We have recently reported that decrease of the coercive field and precise control of the 180° domain lateral motion can be simultaneously maintained during the bulk periodical poling procedure on ZnO : LiNbO$_3$ in the *high*-field regime [368]. Moreover, the switching field is found to decrease with the ZnO doping and a threshold (E_{th}) and internal (E_{int}) fields, corresponding to those analyzed in Eq. (12), as low as 2.5 kV/mm and 0.5 kV/mm, respectively, are observed on a 8 mol% ZnO doped congruent LiNbO$_3$ crystal. The substantial decreases of E_{th} and E_{int} are ascribed to the suppression of the non-stoichiometric point defects by the substitution of Zn^{2+} ions in the lattice site.

Illustrated in Figure 44 are the evolution of the polarization switching current on an 8 mol% ZnO : LiNbO$_3$ crystal in

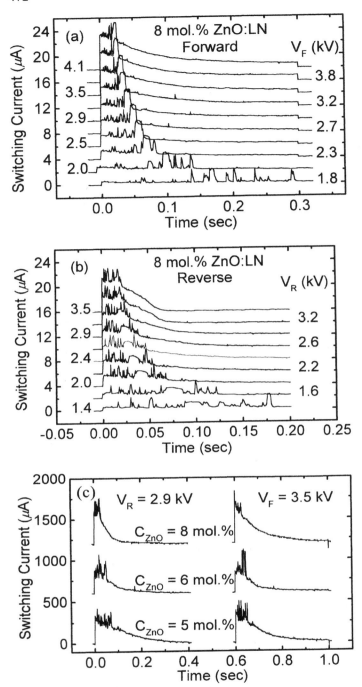

the 180° domain reversal process on the ZnO doped LiNbO$_3$, whereas in the undoped case it has to be larger than 22 kV/mm. A close examination of Figure 44 further reveals the occurrence of Barkhausen spikes and nonvanishing ohmic current in the current waveforms. The reappearance of the compressed Barkhausen spikes in the *successive* forward and reverse poling procedure signifies a reversible process of fusion [369] and pinning [370] of the lateral 180° domain motion by the localized defects. These phenomena have been consistently observed, for example, in the domain reversal of Rochelle salt [371] and MgO : LiNbO$_3$ [362].

The details of the ZnO doping effects on the polarization switching current are further explored in Figure 44c and have been vertically shifted for comparison. We first note an increase of the averaged peak switching current with the ZnO doping. Since the switching current measures the exchange rate of the spontaneous polarization (P_s), an increase in the switching current would suggest a faster switching rate due to the fact that P_s is known insensitive to the impurity level in the crystal. In support of this analysis, we illustrate in Figure 45 the switching rate ($1/t_s$) dependence on the ZnO doping at a concentration of (a) 8 and (b) 5 mol%, respectively. As inferred from the figure, an increase of $1/t_s$ with the ZnO doping can be clearly resolved.

More importantly, we note from Figure 45 a *linear* dependence of $1/t_s$ on E. As previously discussed in Section 2.5, this phenomenon has been known to characterize a sidewise motion of the 180° domain wall of Rochelle salt, gadolinium molybdate, and BaTiO$_3$, and KTP in the *high*-field poling regime. In comparison, the switching rates of undoped LiNbO$_3$, MgO : LiNbO$_3$, BaTiO$_3$, and KTP in the *low*-field poling regime all exhibit an *exponential* dependence. This unique *linear* dependence thus enables us to take advantage of Eq. (12) to analyze the lateral domain motion of ZnO : LiNbO$_3$ in the *high*-field regime. This fitting procedure can also lead us to make a decisive measure of E_{int} and E_{th} in ZnO : LiNbO$_3$ from the intersection with the abscissa of Figure 45a and b. As illustrated in Figure 45c, here we note that a significant decrease of E_{th} and E_{int} down to 2.5 and 0.5 kV/mm, respectively, can be realized on 8 mol% ZnO : LiNbO$_3$.

The dramatic decrease of the switching field indeed bears close relation to the ZnO doping effects on the crystal structure. Recent investigation on the electro-optical properties of ZnO : LiNbO$_3$ has suggested a compensation mechanism of the Zn^{2+} ions to the Li vacancy (V_{Li})$^-$ and Nb antisite (Nb$_{Li}$)$^{4+}$ defects [372]. It is concluded that up to a doping concentration of 6.4 mol%, the Zn^{2+} ions are localized at the Li site, whereas in a higher concentration (>7.6 mol%), the Zn^{2+} are partially incorporated at the Nb site. The action of high Zn doping in the congruent grown LiNbO$_3$ crystal thus results in (i) a substantial reduction of the nonstoichiometric point defects of (V_{Li})$^-$ and (Nb$_{Li}$)$^{4+}$ and (ii) an increase of the lattice constant (a, c) due to the substitution of larger Zn^{2-} ions in the lattice site. The significant decrease of E_{th} and E_{int} is therefore ascribed to the effective suppression of the nonstoichiometric point defects by the substitution of Zn into the lattice site. Moreover, when the ZnO doping effects on the dilation of the lattice constant a are

Fig. 44. Evolution in the waveforms of switching currents on a Z-cut, congruent-grown 8 mol% ZnO : LiNbO$_3$ single crystal in the (a) *forward* and (b) *reverse* poling. (c) is the switching current waveforms on ZnO : LiNbO$_3$ at a doping level of 5, 6, and 8 mol% at a *forward* and *reverse* poling voltage of 3.5 and 2.9 kV. The substrate thickness is 500 μm. Reprinted with permission from [368], copyright 2001, American Institute of Physics.

(a) *forward* and (b) *reverse* poling, respectively. The experiment is conducted at room temperature using the LiCl liquid electrode technique in a pulsed-field poling configuration as schematically shown in Figure 34. Here we note that an applied field (E) as low as 3 kV/mm is sufficient to initiate

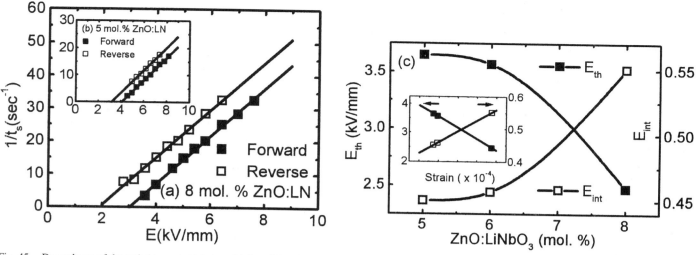

Fig. 45. Dependence of the switching rate ($1/t_s$) on (a) 8 mol% and (b) 5 mol% ZnO : LiNbO$_3$ single crystals in the *forward* and *reverse* poling direction. The linear dependence of $1/t_s$ on E can be formulated by Eq. (11) and the resultant internal (E_{int}) and threshold (E_{th}) field are shown in (c). Reprinted with permission from [368], copyright 2001, American Institute of Physics.

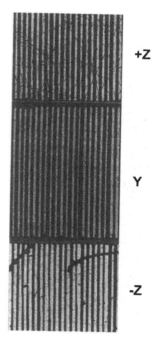

Fig. 46. Micrograph of ZnO : PPLN forward poled at 5 kV/mm. The substrate thickness is 500 μm and the QPM period is 20 μm. Reprinted with permission from [368], copyright 2001, American Institute of Physics.

transformed into the lattice tensile strain ε, a linear dependence of E_{int} and E_{th} on ε can be seen in the inset of Figure 45c. This observation indicates that the residual small E_{int} in the highly doped LiNbO$_3$ is related the strain induced piezoelectric effects in the crystal [373].

By taking advantage of the high ZnO doping effect in reducing the switching and internal field, we have successfully fabricated a periodically poled domain reversed QPM structure on a 5 mol% ZnO : LiNbO$_3$ at a low poling field of 5 kV/mm.

As inferred from Figure 46, good control of the QPM domain structure with a 50% duty cycle over the 20 μm period, and high aspect ratio over the 3 mm sample length and 500-μm-thick substrate, has thus been realized on the ZnO : LiNbO$_3$.

4. CONCLUSION

The emergent applications of using ferroelectric thin films and crystals for the (i) high-density memory and storage and (ii) highly efficient linear and nonlinear optoelectroinc devices have had a great impact on the research and technical development of ferroelectric domain reversal techniques over the past decades. The most renowned example, as discussed in Section 2.5, is the Miller–Weinreich theory developed in the 1960s that suggests that the 180° polarization switching process is dominated by the surface domain nucleation process. Progress on the crystal's stoichiometry improvement and domain motion control, however, has enabled the fabrication of nanodomain structures on ferroelectrics and achieved periodical domain reversed QPM structure with period as short as 2 μm. This achievement has further enriched the device application reaching the physical limit. One such example is the current development of the ferroelectric capacitor suitable for the deep submicron memory applications and capable of handling the gigabyte level data storage integration [374]. Another example is all-optical beam deflection and switching using the ferroelectric wavegudie structure [375].

In the following, we summarize the technical development and outlook of the ferroelectric polarization switching techniques and applications discussed in this chapter. While most of the single domain bulk crystals are now commercially produced by the conventional CZ or by the flux-grown methods, the need for periodical domain reversal structures to ensure better device performance has resulted in several renovations in the polarization switching techniques. A unique *in situ* growth technique of the periodically poled QPM domain reversal structure, as discussed in Section 3.4.2, is to introduce a temperature gradient along the crystal growth axis and the solid–liquid interface, respectively. In this way, a modulation in the distribution of impurity and composition can result in a space charge field to affect the domain polarity in the crystal growth procedure. During the crystal growth of a periodically poled structure, one can also co-dope the crystal with the laser active ions such as Er, Yb, and Nd. By doing so, the optical process of SHG due to the QPM nonlinear effect and laser action due to frequency up-conversion can simultaneously take place and result in multicolor laser emission. For example, a 980 nm InGaAs laser diode can be QPM converted into a blue 490 nm SHG and up-converted to a green 547 nm in a Er : PPLN with a period of 5.3 μm [376]. Another example is to use the *in situ* growth of chirped PPLN doped with Nd to widen the QPM curve and to achieve a combined process of SHG, sum frequency generation, and up-conversion. Using this method, simultaneous generation of red, green, blue, and ultraviolet laser emission has been reported [377, 378].

Progress on the postgrowth *ex situ* poling techniques now enables the polarization switching process to be taken at temperatures much lower than the corresponding Curie point (T_c). By doing so, one avoids the conventional process involving the para- to ferroelectric phase transition. Use of the scanning force microscope with an improved piezoresponsibility can further enable the fabrication and observation of ferroelectric structure on a nanometer scale. A reversed domain with feature sizes less than 50 nm can be routinely made on BaTiO$_3$, TGS, PZT, and KTP ferroelectric crystals as discussed in Section 3.5.2. This nanodomain fabrication technique can also impact the device applications invoking the electromechanic properties of ferroelectrics such as on the sensors [379] and the actuators [380, 381]. Needless to say, it will bring a challenge to memory device design for applications on gigabyte data storage.

Moreover, it is found that by taking advantage of the stoichiometric control over the LiNbO$_3$ and LiTaO$_3$ crystals, one can achieve a two order of magnitude reduction in the coercive field ($E_c \sim 21$ kV/mm) and yield improved electro-optic and nonlinear-optical tensor properties. As discussed in Section 3.5.3, the use of K$_2$O additive in the flux-grown TSSG and the excess Li$_2$O in the CZ melt-grown methods now succeed in the commercialization of LiNbO$_3$ and LiTaO$_3$ single domain crystals with a composition near the stoichiometry. For ferroelectric crystal with a smaller coercive field, the domain reversed QPM structure can be made with a larger aperture size of several mm in thickness to allow the handling with high optical power applications. Conventionally this only can be done with KTP [382] and its family crystals such as RTA [383], which has a low coercive field of $\sim$2.5 kV/mm. To make a large aperture size on ferroelectric crystals with large coercive fields such as on LiNbO$_3$ and LiTaO$_3$, one has to apply a sophisticated diffusion-bonded technique that uses Van der Waals forces to bring two flat surfaces together [384]. By doing so, PPLN with a 3-mm thickness can be diffusion-bonded by three 1-mm-thick crystals [385]. In this method, it is not easy to align each peri-

odically poled domain to make a perfect stack. However, with the coercive field greatly reduced in the stoichiomteric LiTaO$_3$, a 2-mm-thick PPLT has been tested and revealed to have superior parametric oscillation performance than that of the undoped PPLT [386].

The side effect of photorefractive damage in the stoichiometric LiNbO$_3$ crystal can be eliminated by adding a small amount $\sim$1 mol% of MgO impurity in the crystal. For the widely used congruent-grown LiNbO$_3$ crystal, coercive field reduction by a factor of 4 and an increase of the optical damage resistance can be simultaneously achieved by adding 5 mol% MgO in the crystal. Experimental observation has shown that the optical parametric oscillation (OPO) [387] performance on PPMgLN at room temperature is comparable to that of PPLN heated to 95°C to prevent photorefractive damage [388]. This indicates that QPM-OPO generation of mid-IR lasers by a PPMgLN crystal can be operated in a steady state at room temperature. Moreover, we note that a far greater 10-fold reduction in the coercive and internal field can be resolved in the ZnO doped LiNbO$_3$. The mechanism is ascribed to the effective increase of the crystal's photoconductivity, and suppression of the nonstoichiometric (V_{Li})$^-$ and (Nb$_{Li}$)$^{4+}$ antisite defects by the substitution of Zn^{2+} ions into the lattice site.

The more recent inventions of the pulse-field poling and the polarization back-switching procedure have several advantages over the conventional thermal poling process of diffusion and annealing in controlling the reverse domain motion as discussed in Section 3.1. First of all, the domain inversion can be proceeded at room temperature and the domain shape and growth rate will not be diffusion-limited. Second, the inverted domain is no longer restricted to the ferroelectric surface and can extend through the thickness of the substrate. This result can also enlarge the usable aperture of the ferroelectric crystal and facilitate high optical power application [389]. The recent commercialization of periodically poled QPM ferroelectric crystals of PPLN, PPLT, PPKTP, and PPRTA represents one such exciting achievement. We pass by noting that the electrical poling techniques can have also practical applications on the fabrication of integrated optics components such as lenses [390], directional coupler switches [391], modulators [392], and beam deflection devices [393]. A deflection sensitivity of 12 mrad/kV for a SHG blue laser operation has been reported [394]. By engineering the reversed domain with a chirped period, the QPM structure can further be used to stretch or compress the width of the converted laser pulses [395]. In addition, the domain reversal structure can also be periodically poled on the X-cut [396] or on the inclined substrate [397]. The deep domain inversion formed in the electrically poled X-cut LiNbO$_3$ has found special application in the high-speed bandpass intergrated-optics modulator with a large 15 GHz bandwidth center at the operation frequency of 21 GHz [398].

We conclude by noting that the availability of these periodically poled QPM devices has opened up promising applications on the ferroelectric integrated optics [399, 400]. The periodically poled crystals also lay down a test ground of many novel ideas such as cascade second-order optical switching [401], phase-shifting [402], wavelength-shifting [403], and optical rectification for terahertz generation [404]. One current major interest is on the efficient generation of laser sources to bridge the spectral range not amenable by the conventional solid or gas laser technology. The difference frequency generation (DFG) of multiple-channel wavelength [405] or cascade wavelength shifting [406, 407] around 1.5 μm on QPM-PPLN is one such example that can be fit into the design of the WDM optical communication system. The QPM laser system of this kind is known to have excellent characteristics of low noise, transparent to bit-rate and data format, and multichannel conversion capabilities [408].

On the other hand, efficient laser operations in the CW, Q-switched, and mode-lock modes have also been realized on the periodically poled QPM devices and span a wide spectral range from short wavelength generation by the optical process of SHG [409] to mid-IR OPO by DFG [410]. For example, UV lasers at wavelength shorter than 400 nm have been reported on femtosecond SHG of QPM-LiNbO$_3$ [411, 412] on CW SHG of QPM-LiTaO$_3$ [413], and on SFG of QPM-LiNbO$_3$ [414]. The QPM-SHG of blue laser with peak power of 17 mW and modulation bandwidth exceeding 20 MHz that is sufficient for the rewritable optical disk systems with a 10 gigabyte storage capacity has already been demonstrated [415]. Widely tunable near- to mid-IR laser sources based on the femtosecond [416, 417], picosecond [418], and CW [419] QPM-OPO of PPLN, PPRTA, and PPKTA have also been reported. The generation of near- to mid-IR pulsed lasers is of interest to the time-resolved spectroscopic measurement of molecules, condensed matter, and analysis of the carrier dynamics [420]. Another practical consideration of these mid-IR QPM-OPO lasers is on gas sensing, for example, of the methane (P branch) around 3.4 μm [421], and CO$_2$ (vibrational transition) around 1.5 μm [422].

We further note that the poling techniques developed for polarization switching of the ferroelectric crystals can also be applied to other materials system. For materials originally characterized with inversion symmetry and lack of optical nonlinearity, the use of electrical poling in conjunction with thermal treatment or UV illumination can result in a great deal of enhancement on the electro-optic and nonlinear-optic response [423]. One such exciting achievement can be found in the review articles on field poling on glass [424] and polymer [425, 426]. For example, electric poling on silica can be performed with a field $\sim$1–5 kV/mm at temperature around 300°C in the ambient of air or vacuum [427], or with the assistance of high-energy UV lasers [428]. The pulsed-field poling method can also be applied to induce domain inversion in the polymeric thin films at room temperature [429, 430]. We finally note that the concept of a periodically poled QPM structure and its application on electro-optics and nonlinear optics is far-reaching and can be applied to materials systems besides the ferroelectric crystals. One such example is QPM-SHG in the *ex situ* periodically poled optical fiber [431], and polymer waveguide structures in the surface [432] and edge [433] emitting configurations. There is also a one-to-one correspondence in the *in situ* growth of QPM structures on GaAs layers [434], AlGaAs waveguides [435], and AlGaAs quantum wells [436] where a periodical modulation of the nonlinear susceptibility can be introduced to enhance the nonlinear conversion efficiency.

REFERENCES

1. W. Käzik, in "Solid State Physics" (F. Seitz and D. Turnbull, Eds.), Vol. 4. Academic, New York, 1957.
2. R. Bruchhaus, H. Huber, D. Pitzer, and W. Wersing, *Ferroelectrics* 127, 137 (1992).
3. O. Kolosov, A. Gruverman, J. Hatano, K. Takahashi, and H. Tokumoto, *Phys. Rev. Lett.* 74, 4309 (1995).
4. O. Auciello, J. F. Scott, and R. Ramesh, *Phys. Today* 51, 22 (1998).
5. A. Gruverman, H. Tokumoto, A. S. Prakash, S. Aggarwal, B. Yang, M. Wuttig, R. Ramesh, O. Auciello, and T. Venkatesan, *Appl. Phys. Lett.* 71, 3492 (1997).
6. W. Jo, D. C. Kim, and J. W. Hong, *Appl. Phys. Lett.* 76, 390 (2000).
7. A. Yariva and P. Yeh, "Optical Waves in Crystals." Wiley, New York, 1984.
8. C. Kittel, "Introduction to Solid State Physics," 6th ed., Chapt. 13. Wiley, New York, 1986.
9. C. Kittel, *Rev. Mod. Phys.* 21, 541 (1949).
10. T. Mitsui and J. Furuichi, *Phys. Rev.* 159, 193 (1953).
11. T. Mitsui, I. Tatsuzaki, and E. Nakamura, "An Introduction to the Physics of Ferroelectrics." Gordon and Breach, New York, 1976.
12. V. L. Ginzburg, *Zh. Eksp. Teor. Fiz.* 15, 759 (1945).
13. A. F. Devonshire, *Philos. Mag.* 40, 1040 (1949).
14. O. G. Vendik, S. P. Zubko, and L. T. Ter-Martirosayn, *Appl. Phys. Lett.* 73, 37 (1998).
15. F. Jona and G. Shirane, "Ferroelectric Crystals." Pergamon, New York, 1962.
16. E. Fatuzzo and W. J. Merz, "Ferroelectricity." North-Holland, Amsterdam, 1967.
17. J. C. Burfoot, "Ferroelectrics." Van Nostrand, London, 1967.
18. K. Nassau, H. J. Levinstein, and G. M. Loiacono, *J. Phys. Chem. Solids* 27, 983 (1966).
19. S. C. Abrahams, J. M. Reddy, and J. L. Berstein, *J. Phys. Chem. Solids* 27, 997 (1966).
20. A. M. Prokhorov and Yu. S. Kuz'minov, "Physics and Chemistry of Crystalline Lithium Niobate." Hilger, New York, 1990.
21. R. S. Weis and T. K. Gaylord, *Appl. Phys. A* 37, 191 (1985).
22. S. C. Abrahams and J. L. Bernstein, *J. Phys. Chem. Solids* 28, 1685 (1967).
23. S. C. Abrahams, W. C. Hamilton, and A. Sequeira, *J. Phys. Chem. Solids* 28, 1693 (1967).
24. S. C. Abrahams, E. Buehler, W. C. Hamilton, and S. J. Laplaca, *J. Phys. Chem. Solids* 34, 521 (1973).
25. D. C. Sinclair and A. R. West, *Phys. Rev. B* 39, 13586 (1989).
26. V. K. Yanovskii and V. I. Voronkova, *Phys. Status Solidi A* 93, 665 (1980).
27. G. M. Loiacono and R. A. Stolzenberger, *Appl. Phys. Lett.* 53, 1498 (1988).
28. J. D. Bierlein and H. Vanherzeele, *J. Opt. Soc. Am. B* 6, 622 (1989).
29. F. Laurell, M. G. Roelofs, W. Bindloss, H. Hsiung, A. Suna, and J. D. Bierlein, *J. Appl. Phys.* 71, 4664 (1992).
30. Landolt-Börnstein, "Numerical Data and Functional Relationship in Sciences and Technology," New series III-28 a.
31. "Ferroelectric Thin Film: Synthesis and Basic Properties" (C. A. Paz de Araujo, J. F. Scott, and G. W. Taylor, Eds.). Gordon & Breach, New York, 1996.
32. J. F. Nye, "Physical Properties of Crystal." Oxford Univ. Press, London, 1985.
33. P. W. Haycock and P. D. Townsend, *Appl. Phys. Lett.* 48, 698 (1986).

34. S. Kielich, *Ferroelectrics* 4, 257 (1972).
35. M. E. Lines and A. M. Glass, "Principles and Applications of Ferroelectrics and Related Materials." Clarendon, Oxford, 1977.
36. G. A. Smolenskii, V. A. Bokov, V. A. Isupov, N. N. Krainik, R. E. Pasynkov, and A. I. Sokolov, "Ferroelectrics and Related Materials." Gordon & Breach, New York, 1984.
37. "Piezoelectricity" (G. W. Taylor, J. J. Gagnepain, T. R. Meeker, T. Nakamura, and L. A. Shuvalov, Eds.). Gordon and Breach, Switzerland, 1992.
38. G. I. Stegeman and C. T. Seaton, *J. Appl. Phys.* 58, R57 (1985).
39. L. E. Myers, R. C. Eckardt, M. M. Fejer, R. L. Byer, and W. R. Bosenberg, *Opt. Lett.* 21, 591 (1996).
40. K. Shinozaki, Y. Miyamoto, H. Okayama, T. Kamijoh, and T. Nonake, *Appl. Phys. Lett.* 58, 1934 (1991).
41. T. Suhara, H. Ishizuki, M. Fujimura, and H. Nishihara, *IEEE Photon. Technol. Lett.* 11, 1027 (1999).
42. M. Arbore, A. Galvanauskas, D. Harter, M. Chou, and M. Fejer, *Opt. Lett.* 22, 1341 (1997).
43. Y.-S. Lee, T. Meade, V. Perlin, H. Winful, T. B. Norris, and A. Galvanauskas, *Appl. Phys. Lett.* 76, 2505 (2000).
44. M. Yamada, M. Saitoh, and H. Ooki, *Appl. Phys. Lett.* 69, 3659 (1996).
45. Y.-Y. Zhu, N.-B. Ming, W.-H. Jiang, and Y.-A. Shui, *Appl. Phys. Lett.* 53, 1381 (1988).
46. R. W. Boyd, "Nonlinear Optics." Academic Press, San Diego, 1992.
47. J. A. Armstrong, N. Bloembergen, J. Ducuing, and P. S. Pershan, *Phys. Rev.* 127, 128 (1962); N. Bloembergen, U. S. Patent 3384433, 1968.
48. M. M. Fejer, G. A. Magel, F. H. Jundt, and R. L. Byer, *IEEE J. Quantum Electron.* 28, 2631 (1992).
49. N. F. Foster, G. A. Coquin, G. A. Rozgonyi, and F. A. Vannatta, *IEEE Trans. Sonic Ultrason.* SU-15, 28 (1968).
50. E. A. Gerber, T. Lukaszek, and A. Ballato, *IEEE Trans. Microwave Theory Tech.* MTT-34, 1002 (1986).
51. A. H. Meitzler and E. K. Sittig, *J. Appl. Phys.* 40, 4341 (1969).
52. J. C. Peuzin and M. Tasson, *Phys. Stat. Sol. A* 37, 119 (1976).
53. Y.-Y. Zhu and N.-B. Ming, *J. Appl. Phys.* 72, 904 (1992).
54. H. E. Bommel and K. Dransfeld, *Phys. Rev.* 117, 1245 (1960).
55. M. Rotter, A. Wixforth, W. Ruile, D. Bernklau, and H. Riechert, *Appl. Phys. Lett.* 73, 2128 (1998).
56. H. Tanaka and N. Wakatsuki, *Japan. J. Appl. Phys.* 37, 2868 (1998).
57. N. Wakatsuki, H. Yokoyama, and S. Kudo, *Japan. J. Appl. Phys.* 37, 2970 (1998).
58. B. T. Matthias and J. P. Remeika, *Phys. Rev.* 76, 1886 (1949).
59. A. A. Ballman, *J. Am. Ceram. Soc.* 48, 112 (1965).
60. K. Kitamura, J. K. Yamamoto, N. Iyi, S. Kimura, and T. Hayashi, *J. Cryst. Growth* 116, 327 (1992).
61. K. Nassau, H. J. Levinstein, and G. M. Loiacono, *J. Phys. Chem. Solids* 27, 989 (1966).
62. K. Nassau, H. J. Levinstein, and G. M. Loiacono, *J. Phys. Chem. Solids* 27, 983 (1966).
63. N. Niizeki, T. Yamada, and H. Toyoda, *Japan. J. Appl. Phys.* 6, 318 (1967).
64. J. M. Lopez, M. A. Caballero, M. T. Santos, L. Arizmendi, and E. Dieguez, *J. Cryst. Growth* 128, 852 (1993).
65. M. T. Santos, J. C. Rojo, A. Cintas, L. Arizmendi, and E. Dieguez, *J. Cryst. Growth* 156, 413 (1995).
66. K. Polgar, A. Peter, L. Kovacs, G. Corradi, and Z. Szaller, *J. Cryst. Growth* 177, 211 (1997).
67. Y. Furukawa, K. Kitamura, S. Takekawa, K. Niwa, Y. Yajima, N. Iyi, I. Mnushkina, P. Guggenheim, and J. M. Martin, *J. Cryst. Growth* 211, 230 (2000).
68. R. L. Byer, J. F. Young, and R. S. Feigelson, *J. Appl. Phys.* 41, 2320 (1970).
69. A. Reisman and F. Holtzberg, *J. Am. Ceram. Soc.* 80, 6503 (1958).
70. L. O. Svaasand, M. Eriksrud, G. Nakken, and A. P. Grande, *J. Cryst. Growth* 22, 230 (1974).
71. P. F. Bordui, R. G. Norwood, C. D. Bird, and G. D. Calvert, *J. Cryst. Growth* 113, 61 (1991).
72. J. R. Carruthers, G. E. Petertson, M. Grasso, and P. M. Bridenbaugh, *J. Appl. Phys.* 42, 1846 (1971).
73. H. M. O'Bryan, P. K. Gallagher, and C. D. Brandle, *J. Am. Ceram. Soc.* 68, 493 (1985).
74. K. Kitamura, Y. Furukawa, K. Niwa, V. Gopalan, and T. E. Mitchell, *Appl. Phys. Lett.* 73, 3073 (1998).
75. D. H. Jundt, *Opt. Lett.* 22, 1553 (1997).
76. V. G. Dmitriev, G. G. Gurzadyan, and D. N. Nikogosyan, "Handbook of Nonlinear Optical Crystals." Springer-Verlag, Berlin, 1990.
77. K. Kasemir, K. Betzler, B. Matzas, B. Tiegel, T. Wahlbrink, M. Wöhlecke, B. Gather, N. Rubinina, and T. Volk, *J. Appl. Phys.* 84, 5191 (1998).
78. D. E. Zelmon, D. L. Small, and D. Jundt, *J. Opt. Soc. Am. B* 14, 3319 (1997).
79. U. Schlarb and K. Betzler, *Phys. Rev. B* 48, 15613 (1993).
80. M. Wöhlecke, G. Corradi, and K. Betzler, *Appl. Phys. B* 63, 323 (1996).
81. C. a. d. Horst, K.-U. Kasemir, and K. Betzler, *J. Appl. Phys.* 84, 5158 (1998).
82. S. Kurimura and Y. Uesu, *J. Appl. Phys.* 81, 369 (1997).
83. L. Kovacs, G. Ruschhaupt, K. Polgar, G. Corradi, and M. Wöhlecke, *Appl. Phys. Lett.* 70, 2801 (1997).
84. N. Kobayashi and T. Arizumi, *Jpn. J. Appl. Phys.* 9, 361 (1970).
85. N. Kobayashi and T. Arizumi, *J. Cryst. Growth* 30, 177 (1975).
86. N. Kobayashi, *J. Cryst. Growth* 52, 425 (1981).
87. M. T. Santos, J. C. Rojo, A. Cintas, L. Arizmendi, and E. Dieguez, *J. Cryst. Growth* 156, 413 (1995).
88. K. Takago, T. T. Fukazawa, and M. Ishii, *J. Cryst. Growth* 32, 86 (1976).
89. Y. Furuhata, *J. Japan. Assoc. Cryst. Growth* 3, 1 (1976).
90. Y. Miyazawa, Y. Mori, S. Homma, and K. Kitamura, *Mater. Res. Bull.* 13, 675 (1978).
91. H. Cerva, P. Pongratz, and P. Skalicky, *Philos. Mag. A* 54, 199 (1986).
92. V. Bermudez, F. Caccavale, C. Sada, F. Segato, and E. Dieguez, *J. Cryst. Growth* 191, 589 (1998).
93. A. G. Chynoweth, *J. Appl. Phys.* 27, 78 (1956).
94. G. D. Boyd, R. C. Miller, K. Nassau, W. L. Bond, and A. Savage, *Appl. Phys. Lett.* 5, 234 (1964).
95. N. Ohnishi and T. Iizuka, *J. Appl. Phys.* 46, 1063 (1975).
96. T. Ueda, Y. Takai, R. Shimizu, H. Yagyu, T. Matsushima, and M. Souma, *Jpn. J. Appl. Phys.* 39, 1200 (2000).
97. Z. H. Hu, P. A. Thomas, A. Snigirev, I. Snigireva, A. Souvorov, P. G. Smith, G. W. Ross, and S. Teat, *Nature* 392, 690 (1998).
98. Z. W. Hu, P. A. Thomas, and W. P. Risk, *Phys. Rev. B* 59, 14259 (1999).
99. S. Kim, V. Gopalan, and B. Steiner, *Appl. Phys. Lett.* 77, 2051 (2000).
100. K. Nassau, H. J. Levinstein, and G. M. Loiacono, *Appl. Phys. Lett.* 6, 228 (1965).
101. Y. Furukawa, M. Sato, F. Nitanda, and K. Ito, *J. Cryst. Growth* 99, 832 (1990).
102. M. Tasson, H. Legal, J. C. Peuzin, and F. C. Lissalde, *Phys. Stat. Sol. A* 31, 729 (1975).
103. K. Nassau and H. J. Levinstein, *Appl. Phys. Lett.* 7, 69 (1965).
104. V. Bermudez, P. S. Dutta, M. D. Serrano, and E. Dieguez, *J. Appl. Phys.* 81, 862 (1997).
105. A. A. Ballman and H. Brown, *Ferroelectrics* 4, 189 (1972).
106. V. K. Yanovskii, V. I. Voronkova, A. P. Lenov, and S. Y. Stefanovich, *Sov. Phys. Solid State* 27, 1508 (1985).
107. J. D. Bierlein and F. Ahmed, *Appl. Phys. Lett.* 51, 1322 (1987).
108. W. J. Merz, *J. Appl. Phys.* 27, 938 (1956).
109. R. C. Miller and A. Savage, *J. Appl. Phys.* 32, 714 (1961).
110. C. F. Pulvari and W. Kuebler, *J. Appl. Phys.* 29, 1742 (1958).
111. G. W. Taylor, *J. Appl. Phys.* 47, 593 (1966).
112. T. Mitsui and J. Furuichi, *Phys. Rev.* 90, 193 (1953).
113. J. Bornarel and J. Lajzerowicz, *J. Appl. Phys.* 39, 4339 (1968).
114. E. Fatuzzo and W. J. Merz, "Ferroelectricity," Chaps. 7 and 8. North-Holland, Amsterdam, 1967.
115. M. Molotskii, R. Kris, and G. Rosenman, *J. Appl. Phys.* 88, 5318 (2000).
116. D. Viehland and Y.-H. Chen, *J. Appl. Phys.* 88, 6696 (2000).
117. B. Wang and Z. Xiao, *J. Appl. Phys.* 88, 1464 (2000).
118. E. Fatuzzo, *Phys. Rev.* 127, 1999 (1962).

119. P. W. Forsbergh, *Phys. Rev.* 76, 1187 (1949).
120. W. J. Merz, *Phys. Rev.* 95, 690 (1954).
121. R. C. Milller and A. Savage, *Phys. Rev. Lett.* 2, 294 (1959).
122. R. C. Milller and A. Savage, *J. Appl. Phys.* 31, 662 (1960).
123. V. Gopalan and T. Mitchell, *J. Appl. Phys.* 85, 2304 (1999).
124. V. Gopalan, S. A. A. Gerstl, A. Itagi, T. E. Mitchell, Q. X. Jia, T. E. Schlesinger, and D. D. Stancil, *J. Appl. Phys.* 86, 1638 (1999).
125. V. Gopalan and T. E. Mitchell, *J. Appl. Phys.* 83, 941 (1998).
126. V. Gopalan, Q. X. Jia, and T. E. Mitchell, *Appl. Phys. Lett.* 75, 2482 (1999).
127. H. L. Stadler and P. J. Zachmanidis, *J. Appl. Phys.* 34, 3255 (1963).
128. H. L. Stadler and P. J. Zachmanidis, *J. Appl. Phys.* 35, 2895 (1964).
129. H. L. Stadler and P. J. Zachmanidis, *J. Appl. Phys.* 35, 2625 (1964).
130. R. Lüth, H. Haefke, K.-P. Meyer, E. Meyer, L. Howald, and H.-J. Güntherodt, *J. Appl. Phys.* 74, 7461 (1993).
131. L. M. Eng, H.-J. Güntherodt, G. Rosenman, A. Skliar, M. Oron, M. Katz, and D. Eger, *J. Appl. Phys.* 83, 5973 (1998).
132. L. M. Eng, M. Friedrich, J. Fousek, and P. Günter, *J. Vac. Sci. Technol. B* 14, 1191 (1996).
133. V. Likodimos, X. K. Orlik, L. Pardi, M. Labardi, and M. Allegrini, *J. Appl. Phys.* 87, 443 (2000).
134. K. Franke, *Ferroelectrics Lett.* 19, 35 (1995).
135. L. M. Eng, M. Friedrich, J. Foousek, and P. Günter, *Ferroelectrics* 191, 211 (1997).
136. L. M. Eng, H.-J. Güntherodt, G. A. Schneider, U. Köpke, and J. Muñoz Saldaña, *Appl. Phys. Lett.* 74, 233 (1999).
137. M. Abplanalp, L. M. Eng, and P. Günter, *Appl. Phys. A* 66, S231 (1998).
138. H. Bluhm, A. Wadas, R. Wiesendanger, A. Roshko, J. A. Aust, and D. Nam, *Appl. Phys. Lett.* 71, 146 (1997).
139. J. W. Hong, K. H. Noh, S.-I. Park, S. I. Kwun, and Z. G. Khim, *Phys. Rev. B* 58, 5078 (1998).
140. O. Tikhomirov, B. Red'kin, A. Trivelli, and J. Levy, *J. Appl. Phys.* 87, 1932 (2000).
141. T. J. Yang and U. Mohideen, *Phys. Lett. A* 250, 205 (1998).
142. C. Hubert, J. Levy, A. C. Carter, W. Chang, S. W. Kiechoefer, J. S. Horwitz, and D. B. Chrisey, *Appl. Phys. Lett.* 71, 3353 (1997).
143. X. K. Orlik, M. Labardi, and M. Allegrini, *Appl. Phys. Lett.* 77, 2042 (2000).
144. E. Fatuzzo and W. J. Merz, *Phys. Rev.* 116, 61 (1959).
145. B. Binggeli and E. Fatuzzo, *J. Appl. Phys.* 36, 1431 (1965).
146. R. C. Miller and A. Savage, *Phys. Rev.* 112, 755 (1958).
147. R. C. Miller and A. Savage, *Phys. Rev.* 115, 1176 (1959).
148. H. L. Stadler, *J. Appl. Phys.* 37, 1947 (1966).
149. H. L. Stadler, *J. Appl. Phys.* 33, 3487 (1962).
150. R. C. Miller and G. Weinreich, *Phys. Rev.* 117, 1460 (1960).
151. E. Fatuzzo, *Phys. Rev.* 127, 1999 (1962).
152. G. J. Goldsmith and J. G. White, *J. Chem. Phys.* 31, 1175 (1959).
153. A. Savage and R. C. Miller, *J. Appl. Phys.* 31, 1546 (1960).
154. T. Mitsui and J. Furuichi, *Phys. Rev.* 95, 558 (1954).
155. R. B. Flippen, *J. Appl. Phys.* 46, 1068 (1975).
156. B. Y. Shur, E. L. Rumyantsev, V. P. Kuminov, A. L. Subbotin, and E. V. Nikolaeva, *Phys. Solid State* 41, 112 (1999).
157. K. Nakamura, H. Ando, and H. Shimizu, *Appl. Phys. Lett.* 50, 1413 (1987).
158. J. L. Jackel, V. Ramaswamy, and S. P. Lyman, *Appl. Phys. Lett.* 38, 509 (1981).
159. A. G. Chynoweth and W. L. Feldmann, *J. Phys. Chem. Solids* 15, 225 (1960).
160. M. Hayashi, *J. Phys. Soc. Japan* 33, 739 (1972).
161. J. Webjörn, F. Laurell, and G. Arvidsson, *IEEE Photon. Technol. Lett.* 1, 316 (1989).
162. C.-S. Lau, P.-K. Wei, C.-W. Wu, and W.-S. Wang, *IEEE Photon. Technol. Lett.* 4, 872 (1992).
163. M. N. Armenise, *IEE Proc. J.* 135, 85 (1982).
164. M. Fujimura, T. Suhara, and H. Nishihara, *Electron. Lett.* 27, 1209 (1991).
165. K. S. Buritskii, E. M. Dianov, V. A. Maslov, V. A. Chernykh, and E. A. Shcherbakov, *J. Appl. Phys.* 79, 3345 (1996).
166. P. J. Jorgensen and R. W. Bartlett, *J. Phys. Chem. Solid* 30, 2639 (1969).
167. O. F. Schirmer, O. Thiemann, and M. Wöhlecke, *J. Phys. Chem. Solids* 52, 185 (1991).
168. V. D. Kugel and G. Rosenman, *Appl. Phys. Lett.* 62, 2902 (1993).
169. S. C. Abrahams and P. Marsh, *Acta Cryst. B* 42, 61 (1986).
170. P. J. Jorgensen and R. W. Bartlett, *J. Phys. Chem. Solids* 30, 2639 (1969).
171. D. M. Smyth, *Ferroelectrics* 50, 93 (1983).
172. J. R. Carruthers, I. P. Kaminow, and L. W. Stulz, *Appl. Opt.* 13, 2333 (1974).
173. L. Huang and N. A. F. Jaeger, *Appl. Phys. Lett.* 65, 1763 (1994).
174. C. E. Rice and J. L. Jackel, *J. Solid State Chem.* 41, 308 (1982).
175. C. E. Rice, *J. Solid State Chem.* 64, 188 (1986).
176. Y. N. Korkishko and V. A. Fedorov, "Ion Exchange in Single Crystals for Integrated Optics and Optoelectronics." Cambridge International Science, Cambridge, MA, 1999.
177. J. Rams and J. M. Cabrera, *J. Opt. Soc. Am. B* 16,401 (1999).
178. J. L. Jackel, C. E. Rice, and J. J. Veselka, *Appl. Phys. Lett.* 41, 607 (1982).
179. Y. N. Korkishko and V. A. Fedorov, *IEEE J. Select. Topics Quantum Electron.* 2, 187 (1996).
180. K. Ito and K. Kawamoto, *Japan. J. Appl. Phys.* 37, 3977 (1998).
181. F. Laurell, M. G. Roelofs, and H. Hsiung, *Appl. Phys. Lett.* 60, 301 (1992).
182. M. L. Bortz, L. A. Eyres, and M. M. Fejer, *Appl. Phys. Lett.* 61, 2012 (1993).
183. T. Veng, T. Skettrup, and K. Pedersen, *Appl. Phys. Lett.* 69, 2333 (1996).
184. J. Rams, J. Olivares, and J. M. Cabrera, *Elec. Lett.* 33, 322 (1997).
185. E. E. Robertson, R. E. Eason, Y. Yokoo, and P. J. Chandler, *Appl. Phys. Lett.* 70, 2094 (1997).
186. Y. N. Korkishko, V. A. Fedorov, and F. Laurell, *IEEE J. Select. Topics Quantum Electron.* 6, 132 (2000).
187. M. L. Bortz, S. J. Field, M. M. Fejer, D. W. Nam, R. G. Waarts, and D. F. Welch, *IEEE J. Quantum Electron.* 30, 2953 (1994).
188. L. Chanvillard, P. Aschiéri, D. B. Ostrowsky, M. De Micheli, L. Huang, and D. J. Bamford, *Appl. Phys. Lett.* 76, 1089 (2000).
189. K. Nakamura and H. Shimizu, *Appl. Phys. Lett.* 56, 1535 (1990).
190. K. Mizuuchi and K. Yamamoto, *Appl. Phys. Lett.* 72, 5061 (1992).
191. K. Nakamura, M. Hosoya, and A. Tourlog, *J. Appl. Phys.* 73, 1390 (1993).
192. K. Mizuuchi, K. Yamamoto, and H. Sato, *Appl. Phys. Lett.* 62, 1860 (1993).
193. K. Mizuuchi, K. Yamamoto, and T. Taniuchi, *Appl. Phys. Lett.* 58, 2732 (1991).
194. K. Mizuuchi, K. Yamamoto, and T. Taniuchi, *Appl. Phys. Lett.* 59, 1538 (1991).
195. K. Mizuuchi and K. Yamamoto, *Appl. Phys. Lett.* 60, 1283 (1992).
196. K. Yamamoto, K. Mizuuchi, and T. Taniuchi, *IEEE J. Quantum Electron.* 28, 1909 (1992).
197. K. Mizuuchi, K. Yamamoto, and H. Sato, *J. Appl. Phys.* 75, 1311 (1994).
198. W.-Y. Hsu, C. S. Willand, V. Gopalan, and M. C. Gupta, *Appl. Phys. Lett.* 61, 2263 (1992).
199. H. Åhlfeldt, *J. Appl. Phys.* 76, 3255 (1994).
200. J. Kushibiki, M. Miyashita, and N. Chubachi, *IEEE Photon. Technol. Lett.* 8, 1516 (1996).
201. V. Wong and M. C. Gupta, *J. Appl. Phys.* 84, 6513 (1998).
202. S.-Y. Yi, S.-Y. Shih, Y.-S. Jin, and Y.-S. Son, *Appl. Phys. Lett.* 68, 2493 (1996).
203. P. Kaminov and J. R. Carruthers, *Appl. Phys. Lett.* 22, 326 (1973).
204. E. Zolotoyabko, Y. Avrahami, W. Sauer, T. H. Metzger, and J. Peisl, *Appl. Phys. Lett.* 73, 1352 (1998).
205. S. Miyazawa, *J. Appl. Phys.* 50, 4599 (1979).
206. E. Zolotoyabko and Y. Avrahami, *Mater. Lett.* 24, 215 (1995).
207. H. Nishihara, "Optical Integrated Circuits." McGraw–Hill, New York, 1989.
208. S. Thaniyavarn, T. Findakly, D. Booher, and J. Moen, *Appl. Phys. Lett.* 46, 933 (1985).
209. J. C. Peuzin, *Appl. Phys. Lett.* 48, 1104 (1986).

210. S. Miyazawa, *J. Appl. Phys.* 48, 1105 (1986).

211. V. A. Kalesensakas, N. I. Pavlova, I. J. Rez, and J. P. Grigas, *Sov. Phys. Collect.* 22, 68 (1982).

212. J. D. Bierlein, A. Ferretti, L. H. Brixner, and W. Y. Hsu, *Appl. Phys. Lett.* 50, 1216 (1987).

213. M. G. Roelofs, P. A. Morris, and J. D. Bierlein, *J. Appl. Phys.* 70, 720 (1991).

214. C. J. van der Poel, J. D. Bierlein, J. B. Brown, and S. Colak, *Appl. Phys. Lett.* 57, 2074 (1990).

215. F. Laurell, J. B. Brown, and J. D. Bierlein, *Appl. Phys. Lett.* 62, 1872 (1993).

216. D. Eger, M. Oron, M. Katz, and A. Zussman, *J. Appl. Phys.* 77, 2205 (1995).

217. C. Zaldo, J. Carvajal, R. Solé, D. Diaz, D. Bravo, and A. Kling, *J. Appl. Phys.* 88, 3242 (2000).

218. D. Eger, M. A. Arbore, M. M. Fejer, and M. L. Bortz, *J. Appl. Phys.* 82, 998 (1997).

219. F. Laurell, M. G. Roelofs, W. Bindloss, H. Hsiung, A. Suna, and J. D. Bierlein, *J. Appl. Phys.* 71, 4664 (1992).

220. A. M. Glass, *J. Appl. Phys.* 40, 4699 (1969).

221. K. Daneshvar and D.-H. Kang, *IEEE Quantum Electron.* 36, 85 (2000).

222. M. Houe and P. D. Townsend, *Appl. Phys. Lett.* 66, 2667 (1995).

223. R. S. Feigelson, W. L. Kway, and R. K. Route, in *Proc. SPIE* 484, 133 (1984).

224. M. M. Fejer, J. L. Nightingale, G. A. Magel, and R. L. Byer, *Rev. Sci. Instrum.* 55, 1791 (1984).

225. C. A. Burrus and J. Stone, *Appl. Phys. Lett.* 26, 318 (1975).

226. C. A. Burrus and L. A. Coldren, *Appl. Phys. Lett.* 31, 383 (1977).

227. J. Stone and C. A. Burrus, *J. Appl. Phys.* 49, 2281 (1978).

228. Y. S. Luh, R. S. Feigelson, M. M. Fejer, and R. L. Byer, *J. Cryst. Growth* 78, 135 (1986).

229. G. A. Magel, M. M. Fejer, and R. L. Byer, *Appl. Phys. Lett.* 56, 108 (1990).

230. D. H. Jundt, G. A. Magel, M. M. Fejer, and R. L. Byer, *Appl. Phys. Lett.* 59, 2657 (1991).

231. I. I. Naumova, N. F. Evlanova, O. A. Gilko, and S. V. Lavrishchev, *J. Cryst. Growth* 181, 160 (1997).

232. N.-B. Ming, J.-F. Hong, and D. Feng, *J. Mater. Sci.* 17, 1663 (1982).

233. Y.-L. Lu, Y.-Q. Lu, X.-F. Cheng, C.-C. Xue, and N.-B. Ming, *Appl. Phys. Lett.* 68, 2781 (1996).

234. D. Feng, N.-B. Ming, J.-F. Hong, Y.-S. Yang, J.-S. Zhu, Z. Yang, and Y.-N. Wang, *Appl. Phys. Lett.* 37, 607 (1980).

235. A. Feisst and P. Koidl, *Appl. Phys. Lett.* 47, 1125 (1985).

236. V. Bermudez, J. Capmany, J. G. Sole, and E. Dieguez, *Appl. Phys. Lett.* 73, 593 (1998).

237. Y.-L. Lu, Y.-Q. Lu, C.-C. Xue, and N.-B. Ming, *Appl. Phys. Lett.* 68, 1467 (1996).

238. Y.-Q. Lu, J.-J. Zheng, and N.-B. Ming, *Appl. Phys. B* 67, 29 (1998).

239. I. I. Naumova, N. F. Evlanova, S. A. Blokhin, and S. V. Lavrishchev, *J. Cryst. Growth* 187, 102 (1998).

240. J. Capmany, E. Montoya, V. Bermudez, D. Callejo, E. Dieguez, and L. E. Bausa, *Appl. Phys. Lett.* 76, 1374 (2000).

241. J. Capmany, V. Bermudez, and E. Dieguez, *Appl. Phys. Lett.* 74, 1534 (1999).

242. K. S. Abedin, T. Tsuritani, M. Sato, and M. Ito, *Appl. Phys. Lett.* 70, 10 (1997).

243. Y.-L. Lu, Y.-Q. Lu, X.-F. Cheng, G.-P. Luo, C.-C. Xue, and N.-B. Ming, *Appl. Phys. Lett.* 68, 2642 (1996).

244. J. Chen, Q. Zhou, J.-F. Hong, W.-S. Wang, N.-B. Ming, and C.-G. Fang, *J. Appl. Phys.* 66, 336 (1989).

245. V. Bermudez, D. Callejo, and E. Dieguez, *J. Cryst. Growth* 207, 303 (1999).

246. F. Parlinski, *Ferroelectrics* 172, 1 (1995).

247. V. Bermudez, D. Callejo, F. Caccavale, F. Segato, F. Agullo-Rueda, and E. Dieguez, *Solid State Commun.* 114, 555 (2000).

248. A. Lorenzo, H. Jaffrezic, B. Roux, G. Boulon, and J. Garcia-Sole, *Appl. Phys. Lett.* 67, 3735 (1995).

249. H. Gnewuch, N. K. Zayer, C. N. Pannell, G. W. Ross, and P. G. R. Smith, *Opt. Lett.* 25, 305 (2000).

250. S.-D. Cheng, Y.-Y. Zhu, Y.-L. Lu, and N.-B. Ming, *Appl. Phys. Lett.* 66, 291 (1995).

251. A. Feisst and A. Rauber, *J. Cryst. Growth* 63, 337 (1983).

252. Z.-L. Wan, Q. Wang, Y.-X. Xi, Y.-Q. Lu, Y.-Y. Zhu, and N.-B. Ming, *Appl. Phys. Lett.* 77, 1891 (2000).

253. Y.-Q. Lu, Y.-Y. Zhu, Y.-F. Chen, S.-N. Zhu, N.-B. Ming, and Y.-J. Feng, *Science* 284, 1822 (1999).

254. R. W. Keys, A. Loni, R. M. De La Lu, C. N. Ironside, J. H. Marsh, B. J. Luff, and P. D. Townsend, *Electron. Lett.* 26, 188 (1990).

255. H. Ito, C. Takyu, and H. Inaba, *Electron. Lett.* 27, 1221 (1991).

256. M. Yamada and K. Kishima, *Electron. Lett.* 27, 828 (1991).

257. A. C. G. Nutt, V. Gopalan, and M. C. Gupta, *Appl. Phys. Lett.* 60, 2828 (1992).

258. M. Fujimura, T. Suhara, and H. Nishihara, *Electron. Lett.* 28, 721 (1992).

259. M. Fujimura, T. Suhara, and H. Nishihara, *Electron. Lett.* 28, 1868 (1992).

260. M. Fujimura, K. Kintaka, T. Suhara, and H. Nishihara, *J. Lightwave Technol.* 11, 1360 (1993).

261. C. Restoin, C. Darraud-Taupiac, J. L. Decossas, J. C. Vareille, J. Hauden, and A. Martinez, *J. Appl. Phys.* 88, 6665 (2000).

262. M. C. Gupta, W. Kozlovsky, and A. C. G. Nutt, *Appl. Phys. Lett.* 64, 3210 (1994).

263. M. C. Gupta, W. P. Risk, A. C. G. Nutt, and S. D. Lau, *Appl. Phys. Lett.* 63, 1167 (1993).

264. L. M. Eng, M. Abplanalp, and P. Günter, *Appl. Phys. A* 66, S679 (1998).

265. L. M. Eng, *Nanotechnology* 10, 405 (1999).

266. F. Saurenbach and B. D. Terris, *Appl. Phys. Lett.* 56, 1703 (1990).

267. O. Kolosov, A. Gruverman, J. Hatano, K. Takahashi, and H. Tokumoto, *Phys. Rev. Lett.* 74, 4309 (1995).

268. Y.-G. Wang, W. Kleemann, T. Woike, and R. Pankrath, *Phys. Rev. B* 61, 3333 (2000).

269. A. Gruverman, O. Kolosov, J. Hatano, K. Takahashi, and H. Tokumoto, *J. Vac. Sci. Technol. B* 13, 1095 (1995).

270. S. Hong, E. L. Colla, E. Kim, D. V. Taylor, A. K. Tagantsev, P. Muralt, K. No, and N. Setter, *J. Appl. Phys.* 86, 607 (1999).

271. C. H. Ahn, T. Tybell, L. Antognazza, K. Char, R. H. Hammond, M. R. Beasley, Ø. Fisher, and J.-M. Triscone, *Science* 276, 1100 (1997).

272. L. Camlibel, *J. Appl. Phys.* 40, 1690 (1969).

273. H. D. Megaw, *Acta Cryst.* 7, 187 (1954).

274. M. Yamada, N. Nada, M. Saitoh, and K. Watanabe, *Appl. Phys. Lett.* 62, 435 (1993).

275. G. M. Ross, M. Pollnau, P. G. R. Smith, W. A. Clarkson, P. E. Britton, and D. V. Hanna, *Opt. Lett.* 23, 171 (1998).

276. G. D. Miller, R. G. Batchko, W. M. Tulloch, D. R. Weise, M. M. Fejer, and R. L. Byer, *Opt. Lett.* 22, 1834 (1997).

277. K. Mizzucchi and L. Yamamoto, *Appl. Phys. Lett.* 66, 2943 (1995).

278. S.-N. Zhu, Y.-Y. Zhu, Z.-Y. Zhang, H. Shu, H.-F. Wang, J.-F. Hong, C.-Z. Ge, and N.-B. Ming, *J. Appl. Phys.* 77, 5481 (1995).

279. J.-P. Meyn and M. M. Fejer, *Opt. Lett.* 22, 1214 (1997).

280. Q. Chen and W. P. Risk, *Electron. Lett.* 30, 1516 (1994).

281. S. Wang, V. Pasiskevicius, F. Laurell, and H. Karlsson, *Opt. Lett.* 23, 1883 (1998).

282. D. Eger, M. B. Oron, A. Bruner, M. Katz, Y. Tzuk, and A. Englander, *Appl. Phys. Lett.* 76, 406 (2000).

283. G. Rosenman and A. Skliar, *Cryst. Rep.* 44, 112 (1999).

284. G. Rosenman, A. Skliar, Y. Findling, P. Urenski, A. Englander, P. A. Thomas, and Z. W. Hu, *J. Phys. D* 32, L49 (1999).

285. H. Karlsson, F. Laurell, P. Henriksson, and G. Arvidsson, *Electron. Lett.* 32, 556 (1996).

286. W. P. Risk and G. M. Loiacono, *Appl. Phys. Lett.* 69, 311 (1996).

287. H. Karlsson, F. Laurell, and L. K. Cheng, *Appl. Phys. Lett.* 74, 1519 (1999).

288. X. Tong, A. Yariv, M. Zhang, A. J. Agranat, R. Hofmeister, and V. Leyva, *Appl. Phys. Lett.* 70, 224 (1997).

289. A. S. Kewitsch, T. W. Towe, G. J. Salamo, A. Yariv, M. Zhang, M. Segev, E. J. Sharp, and R. R. Neurgaonkar, *Appl. Phys. Lett.* 66, 1865 (1995).

290. Y. Y. Zhu, J. S. Fu, R. F. Xiao, and G. K. L. Wong, *Appl. Phys. Lett.* 70, 1793 (1997).

291. W. K. Burns, W. McElhanon, and L. Goldberg, *IEEE Photon. Lett.* 6, 252 (1994).

292. K. Kintaka, M. Fujimura, T. Suhara, and H. Nishihara, *J. Lightwave Technol.* 14, 462 (1996).

293. G. Rosenman, Kh. Garb, A. Skliar, M. Oron, D. Eger, and M. Katz, *Appl. Phys. Lett.* 73, 865 (1998).

294. L. E. Myers, R. C. Eckardt, M. M. Fejer, R. L. Byer, W. R. Bosenberg, and J. W. Pierce, *J. Opt. Soc. Am. B* 12, 2102 (1995).

295. L. E. Myers, R. C. Eckardt, M. M. Fejer, R. L. Byer, and W. R. Bosenberg, *Opt. Lett.* 21, 591 (1996).

296. R. G. Batchko, V. Y. Shur, M. M. Fejer, and R. L. Byer, *Appl. Phys. Lett.* 75, 1673 (1999).

297. R. G. Batchko, M. M. Fejer, R. L. Byer, D. Woll, R. Wallenstein, V. Y. Shur, and L. Erman, *Opt. Lett.* 24, 1293 (1999).

298. V. Ya. Shur, E. L. Rumyantsev, E. V. Nikolaeva, E. I. Shishkin, D. V. Fursov, R. G. Batchko, L. A. Eyres, M. M. Fejer, and R. L. Byer, *Appl. Phys. Lett.* 76, 143 (2000).

299. S. Helmfrid and G. Arvidsson, *J. Opt. Soc. Am. B* 8, 797 (1991).

300. S. Helmfrid, G. Arvidsson, and J. Webjörn, *J. Opt. Soc. Am. B* 10, 222 (1992).

301. P. Baldi, P. Aschieri, S. Nouth, M. De Micheli, D. B. Ostrowsky, D. Delacourt, and M. Papuchon, *IEEE J. Quantum Electron.* 31, 997 (1995).

302. Y. Dikmelik, G. Akgün, and O. Aytür, *IEEE J. Quantum Electron.* 35, 897 (1999).

303. E. Krätiz and R. Orlowski, *Opt. Quantum Electron.* 12, 495 (1980).

304. V. Pruneri, P. G. Kazansky, J. Webjörn, P. S. J. Russel, and D. C. Hanna, *Appl. Phys. Lett.* 67, 1957 (1995).

305. K. S. Abedin, T. Tsuritani, M. Sato, and H. Ito, *Appl. Phys. Lett.* 70, 10 (1997).

306. F. S. Chen, *J. Appl. Phys.* 40, 3389 (1969).

307. O. Eknoyan, H. F. Taylor, W. Matous, T. Ottinger, and R. R. Neurgaonkar, *Appl. Phys. Lett.* 71, 3051 (1997).

308. M. Taya, M. C. Bashaw, and M. M. Fejer, *Opt. Lett.* 21, 857 (1996).

309. B. Sturman, M. Aguilar, F. Agullo-Lopez, V. Pruneri, and P. G. Kazansky, *J. Opt. Soc. Am. B* 14, 2641 (1997).

310. D. Eger, M. A. Arbore, M. M. Fejer, and M. L. Bortz, *J. Appl. Phys.* 82, 998 (1997).

311. C. Q. Xu, H. Okayama, and Y. Ogawa, *J. Appl. Phys.* 87. 3203 (2000).

312. L. Goldberg, R. W. McElhanon, and W. K. Burns, *Electron. Lett.* 31, 1576 (1995).

313. V. Gopalan and M. C. Gupta, *J. Appl. Phys.* 80, 6099 (1996).

314. M. Yamada and M. Saitoh, *J. Appl. Phys.* 84, 2199 (1998).

315. Y.-Q. Lu, J.-J. Zheng, Y.-L. Lu, N.-B. Ming, and Z.-Y. Xu, *Appl. Phys. Lett.* 74, 123 (1999).

316. V. Y. Shur, E. L. Rumyantsev, E. V. Nikolaeva, and E. I. Shishkin, *Appl. Phys. Lett.* 77, 3636 (2000).

317. C. C. Battle, S. Kim, V. Gopalan, K. Barkocy, M. C. Gupta, Q. X. Jia, and T. E. Mitchell, *Appl. Phys. Lett.* 76, 2436 (2000).

318. H. Karlsson and F. Laurell, *Appl. Phys. Lett.* 71, 3474 (1997).

319. G. Rosenman, A. Skliar, D. Eger, M. Oron, and M. Katz, *Appl. Phys. Lett.* 73, 3650 (1998).

320. V. Y. Shur, E. L. Rumyantsev, R. G. Batchko, G. D. Miller, M. M. Fejer, and R. L. Byer, *Phys. Solid State* 41, 1681 (1999).

321. M. E. Drougard and D. Landauer, *J. Appl. Phys.* 30, 1663 (1959).

322. R. C. Miller and A. Savage, *J. Appl. Phys.* 32, 714 (1961).

323. V. Gopalan and M. C. Gupta, *Appl. Phys. Lett.* 68, 888 (1996).

324. V. Gopalan, T. E. Mitchell, and K. E. Sicakfus, *Solid State Commun.* 109, 111 (1999).

325. J. H. Ro and M. Cha, *Appl. Phys. Lett.* 77, 2391 (2000).

326. C. F. Pulvari and W. Kuebler, *J. Appl. Phys.* 29, 1315 (1958).

327. L.-H. Peng, Y.-C. Fang, and Y.-C. Lin, *Appl. Phys. Lett.* 74, 2070 (1999).

328. G. Rosenman, A. Skliar, M. Oron, and M. Katz, *J. Phys. D* 30, 277 (1997).

329. P. F. Bordui, R. G. Norwood, D. H. Jundt, and M. M. Fejer, *J. Appl. Phys.* 71, 875 (1992).

330. Y. Furukawa, M. Sata, K. Kitamura, and F. Nitanda, *J. Cryst. Growth* 128, 909 (1993).

331. K. Kitamura, Y. Furukawa, and N. Iye, *Ferroelectrics* 202, 21 (1997).

332. K. Kitamura, *Ceram. Trans.* 60, 37 (1995).

333. M. D. Serrano, V. Bermudez, L. Arizmendi, and E. Dieguez, *J. Cryst. Growth* 210, 670 (2000).

334. K. Polgar, A. Peter, L. Kovacs, G. Corradi, and Zs. Szaller, *J. Cryst. Growth* 177, 211 (1997).

335. G. I. Malovichko, V. G. Grachev, L. P. Yurchenko, V. Y. Proshko, E. P. Kokanyan, and V. T. Gabrielyan, *Phys. Stat. Sol.* 133, K29 (1992).

336. K. Polgar, A. Peter, I. Foldvari, Zs. Szaller, *J. Cryst. Growth* 218, 327 (2000).

337. K. Kitamura, Y. Furukawa, K. Niwa, V. Gopalan, and T. E. Mitchell, *Appl. Phys. Lett.* 73, 3073 (1998).

338. K. S. Abedin, T. Tsuritani, M. Sato, H. Ito, K. Shimamura, and T. Fukuda, *Opt. Lett.* 20, 1985 (1995).

339. V. Gopalan, T. E. Mitchell, Y. Furukawa, and K. Kitamura, *Appl. Phys. Lett.* 72, 1981 (1998).

340. A. Grisard, E. Lallier, K. Polgar, and A. Peter, *Electron. Lett.* 36, 1043 (2000).

341. T. Fujimwra, M. Takahashi, M. Ohama, A. J. Ikushima, Y. Furukawa, and K. Kitamura, *Electron. Lett.* 35, 499 (1999).

342. F. Abdi, M. Aillerie, P. Bourson, M. D. Fontana, and K. Polgar, *J. Appl. Phys.* 84, 2251 (1998).

343. A. de Bernabe, C. Prieto, and A. de Andres, *J. Appl. Phys.* 79, 143 (1996).

344. K. Niwa, Y. Furukawa, S. Takekawa, and K. Kitamura, *J. Cryst. Growth* 208, 492 (2000).

345. D. A. Bryan, R. Gerson, and H. E. Tomaschke, *Appl. Phys. Lett.* 44, 847 (1984).

346. R. Sommerfeldt, L. Holtmann, E. Krätzig, and B. C. Grabmaier, *Phys. Stat. Sol. A* 106, 89 (1988).

347. N. Y. Kamber, J. Xu, S. M. Mikha, G. Zhang, X. Zhang, S. Liu, and G. Zhang, *J. Appl. Phys.* 87, 2684 (2000).

348. K. Niwa, Y. Furukawa, S. Takekawa, and K. Kitamura, *J. Cryst. Growth* 208, 493 (2000).

349. Y. Furukawa, K. Kitamura, S. Takekawa, A. Miyamoto, M. Terao, and N. Suda, *Appl. Phys. Lett.* 77, 2494 (2000).

350. Y. Furukawa, K. Kitamura, S. Takekawa, K. Niwa, and H. Hatano, *Opt. Lett.* 23, 1892 (1998).

351. Y. Furukawa, K. Kitamura, S. Takekawa, K. Niwa, Y. Yajima, N. Iyi, I. Mnushkina, P. Guggenheim, and J. M. Martin, *J. Cryst. Growth* 211, 230 (2000).

352. B. C. Grabmaier and F. Otto, *J. Cryst. Growth* 79, 682 (1986).

353. S. C. Abrahams and P. Marsh, *Acta Cryst. B* 42, 61 (1986).

354. O. F. Schirmer, O. Thiemann, and M. Wöhlecke, *J. Phys. Chem. Solid* 52, 185 (1991).

355. F. P. Safaryan, R. S. Feigelson, and A. M. Petrosyan, *J. Appl. Phys.* 85, 8079 (1999).

356. Y. Kong, J. Xu, X. Chen, C. Zhang, W. Zhang, and G. Zhang, *J. Appl. Phys.* 87, 4410 (2000).

357. N. Iyi, K. Kitamura, Y. Yajima, and S. Kimura, *J. Solid State Chem.* 118, 148 (1995).

358. C. Q. Xu, H. Okayama, and M. Kawahara, *Appl. Phys. Lett.* 64, 2504 (1994).

359. B. C. Grabmaier and F. Otto, *J. Cryst. Growth* 79, 682 (1986).

360. A. Harada and Y. Nihei, *Appl. Phys. Lett.* 69, 2629 (1996).

361. K. Mizuuchi, K. Yamamoto, and M. Kato, *Electron. Lett.* 22, 2091 (1996).

362. A. Kurodo, S. Kurimura, and Y. Uesu, *Appl. Phys. Lett.* 69, 1565 (1996).

363. J. Xu, G. Zhang, F. Li, X. Zhang, Q. Sun, S. Liu, F. Song, Y. Kong, X. Chen, H. Qiao, J. Yao, and L. Zhao, *Opt. Lett.* 25, 129 (2000).

364. T. R. Volk, V. I. Pryalkin, and N. M. Rubinina, *Opt. Lett.* 15, 996 (1990).

365. J. K. Yamamoto, K. Kitamura, N. Iyi, S. Kimura, Y. Furukawa, and M. Sato, *J. Cryst. Growth* 128, 920 (1993).

366. T. Volk, M. Wöhlecke, N. Rubinina, N. V. Razumovski, F. Jermann, C. Fischer, and E. Böwer, *Appl. Phys. A* 60, 217 (1995).

367. T. Volk, N. Rubinina, and M. Wöhlecke, *J. Opt. Soc. Am. B* 11, 1681 (1994).
368. L.-H. Peng, Y.-C. Zhang, and Y.-C. Lin, *Appl. Phys. Lett.* 78, 4 (2001).
369. A. G. Chynoweth, *Phys. Rev.* 110, 1316 (1958).
370. T. J. Yang, V. Gopalan, P. J. Swart, and U. Mohideen, *Phys. Rev. Lett.* 82, 4106 (1999).
371. T. Mitsui and J. Furuichi, *Phys. Rev.* 95, 558 (1954).
372. F. Abdi, M. Aillerie, M. Fontana, P. Bourson, T. Volk, B. Maximov, S. Sulyanov, N. Rubinina, and M. Wöhlecke, *Appl. Phys. B* 68, 795 (1999).
373. H. Cerva, P. Pongratz, and P. Skalicky, *Philos. Mag. A* 54, 199 (1986).
374. T. Kawakubo, K. Abe, S. Komatsu, K. Sano, N. Yanase, and H. Mochizuki, *IEEE Elec. Dev.* 18, 529 (1997).
375. D. Kip, M. Wesner, E. Krätzig, V. Shandarov, and P. Moretti, *Appl. Phys. Lett.* 72, 1960 (1998).
376. J.-J. Zheng, Y.-Q. Lu, G.-P. Luo, J. Ma, Y.-L. Lu, N.-B. Ming, J.-L. He, and Z.-Y. Xu, *Appl. Phys. Lett.* 72, 1808 (1998).
377. J. Capmany, B. Bermúdez, D. Callejo, J. G. Solé, and E. Dieguéz, *Appl. Phys. Lett.* 76, 1225 (2000).
378. J. Capmany, *Appl. Phys. Lett.* 78, 144 (2001).
379. "Sensor Technology and Devices" (L. Ristic, Ed.). Artech House, Boston, 1994.
380. I. J. Busch-Vishniac, *Phys. Today*, July, 1998, p. 28.
381. S. H. Chang and S. S. Li, *Rev. Sci. Instrum.* 70, 2776 (1999).
382. J. Hellström, V. Pasiskevicius, F. Laurell, and H. Karlsson, *Opt. Lett.* 24, 1233 (1999).
383. H. Karlsson, M. Olson, R. Wallenstein, G. Arvidsson, F. Laurell, U. Bäder, A. Borsutzky, R. Wallenstein, S. Wicksröm, and M. Gustafsson, *Opt. Lett.* 24, 330 (1999).
384. J. Haisma, B. A. C. M. Spierings, U. K. Biermann, and A. A. van Gorkum, *Appl. Opt.* 33, 1154 (1994).
385. M. J. Missey, V. Dominic, L. E. Myers, and R. C. Eckardt, *Opt. Lett.* 23, 664 (1998).
386. T. Hatanaka, K. Nakamura, T. Taniuchi, H. Ito, Y. Furukawa, and K. Kitamura, *Opt. Lett.* 25, 651 (2000).
387. G. D. Boyd and D. A. Kleinman, *J. Appl. Phys.* 39, 3597 (1968).
388. M. Nakamura, M. Sugihara, M. Kotoh, H. Taniguchi, and K. Tadatomo, *Jpn. J. Appl. Phys.* 38, L1234 (1999).
389. J. Hellström, V. Pasiskevicius, H. Karlsson, and F. Laurell, *Opt. Lett.* 25, 174 (2000).
390. M. Yamada, M. Saitoh, and Ooki, *Appl. Phys. Lett.* 69. 3659 (1996).
391. S. A. Samson, R. F. Tavlykaev, and R. V. Ramaswamy, *IEEE Photon. Technol. Lett.* 9, 197 (1997).
392. W. Wang, R. Tavlykaev, and R. Ramaswamy, *IEEE Photon. Technol. Lett.* 9, 610 (1997).
393. J. Li, H. C. Cheng, M. L. Kawas, D. N. Lambeth, T. E. Schlesingerm, and D. D. Stancil, *IEEE Photon. Technol. Lett.* 8, 1486 (1996).
394. V. Gopalan, M. J. Kawas, M. C. Gupta, T. E. Schlesinger, and D. D. Stancil, *IEEE Photon. Technol. Lett.* 8, 1704 (1996).
395. M. A. Arbore, A. Galvanauskas, D. Harter, M. H. Chou, and M. M. Fejer, *Opt. Lett.* 22, 1341 (1997).
396. S. Sonoda, I. Tsuruma, and M. Hatori, *Appl. Phys. Lett.* 70, 3078 (1997).
397. S. Sonoda, I. Tsuruma, and M. Hatori, *Appl. Phys. Lett.* 71, 3048 (1997).
398. T. Kishino, R. F. Tavlykaev, and R. V. Ramaswamy, *Appl. Phys. Lett.* 76, 3852 (2000).
399. A. Grisard, E. Lallier, G. Garry, and P. Aubert, *IEEE J. Quantum. Electron.* 33, 1627 (1997).
400. M. Houe and P. D. Townsend, *J. Phys. D* 28, 1747 (1995).
401. M. Asobe, I. Yokohama, H. Itoh, and T. Kaino, *Opt. Lett.* 22, 274 (1997).
402. P. Vidaković, D. J. Lovering, J. A. Levenson, J. Webjörn, and P. St. J. Russell, *Opt. Lett.* 22, 277 (1997).
403. I. Cristiani, G. P. Banfi, V. Degiorgio, and L. Tartara, *Appl. Phys. Lett.* 75, 1198 (1999).
404. Y.-S. Lee, T. Meade, M. DeCamp, T. B. Norris, and A. Galvanauskas, *Appl. Phys. Lett.* 77, 1244 (2000).
405. M. H. Chou, K. R. Parameswaran, and M. M. Fejer, *Opt. Lett.* 24, 1157 (1999).
406. C. G. Treviño-Palacios, G. I. Stegeman, P. Baldi, and M. P. De Micheli, *Electron. Lett.* 34, 2157 (1998).
407. M. H. Chou, I. Brener, M. M. Fejer, E. E. Chaban, and S. B. Christman, *IEEE Photon. Technol. Lett.* 11, 653 (1999).
408. C. Q. Xu, K. Fujita, Y. Ogawa, and T. Kamjioh, *Appl. Phys. Lett.* 74, 1933 (1999).
409. L. E. Myers and W. R. Bosenberg, *IEEE J. Quantum Electron.* 33, 1663 (1997).
410. J. Webjörn, S. Siala, D. W. Nam, R. G. Waarts, and R. J. Lang, *IEEE J. Quantum Electron.* 33, 1673 (1997).
411. Y.-Q. Lu, Y.-L. Lu, C.-C. Xue, J.-J. Zheng, X.-F. Chen, G.-P. Luo, N.-B. Ming, B.-H. Feng, and X.-L. Zhang, *Appl. Phys. Lett.* 69, 3155 (1996).
412. M. A. Arbore, M. M. Fejer, M. E. Fermann, A. Hariharan, A. Galvanauskas, and D. Harter, *Opt. Lett.* 22, 13 (1997).
413. K. Mizuuchi and K. Yamamoto, *Opt. Lett.* 21, 107 (1996).
414. K. Kintaka, M. Fujimura, T. Suhara, and H. Nishihara, *Electron. Lett.* 33, 1459 (1997).
415. Y. Kitaoka, T. Yokoyama, K. Mizuuchi, K. Yamamoto, and M. Kato, *Electron. Lett.* 33, 1638 (1997).
416. K. C. Burr, C. L. Tang, M. A. Arbore, and M. M. Fejer, *Opt. Lett.* 22, 1458 (1997).
417. D. T. Reid, G. T. Kennedy, A. Miller, W. Sibbett, and M. Ebrahimzadeh, *IEEE J. Selected Topics Quantum Electron.* 4, 238 (1998).
418. S. D. Butterworth, P. G. R. Smith, and D. C. Hanna, *Opt. Lett.* 22, 618 (1997).
419. W. R. Bosenberg, A. Drobshoff, J. J. Alexander, L. E. Myers, and R. L. Byer, *Opt. Lett.* 21, 713 (1997).
420. A. Galvanauskas, M. A. Arbore, M. M. Fejer, M. E. Fermann, and D. Harter, *Opt. Lett.* 22, 105 (1997).
421. K. Fradkin-Kashi, A. Arie, P. Urenski, and G. Rosenman, *Opt. Lett.* 25, 743 (2000).
422. P. E. Powers, T. J. Kulp, and S. E. Bisson, *Opt. Lett.* 23, 159 (1998).
423. R. A. Myers, N. Mukherjee, and S. R. J. Brueck, *Opt. Lett.* 16, 1732 (1991).
424. A. J. Ikushima, T. Fujiwara, and K. Saito, *J. Appl. Phys.* 88, 1201 (2000).
425. D. M. Burland, R. D. Miller, and C. A. Walsh, *Chem. Rev.* 94, 31 (1994).
426. R. Blum, M. Sprave, J. Sablotny, and M. Eich, *J. Opt. Soc. Am. B* 15, 318 (1998).
427. V. Pruneri, F. Samoggia, G. Bonfrate, P. G. Kazansky, and G. M. Yang, *Appl. Phys. Lett.* 74, 2423 (1999).
428. T. Fujiwara, M. Takahashi, and A. J. Ikushima, *Appl. Phys. Lett.* 71, 1032 (1997).
429. V. Taggi, F. Michelotti, M. Bertolotti, G. Petrocco, V. Foglietti, A. Donval, E. Toussaere, and J. Zyss, *Appl. Phys. Lett.* 72, 2794 (1998).
430. Z. Z. Yue, D. An, R. T. Chen, and S. Tang, *Appl. Phys. Lett.* 72, 3420 (1998).
431. V. Pruneri and P. G. Kazansky, *IEEE Photon. Technol. Lett.* 9, 185 (1997).
432. A. Otomo, G. I. Stegeman, W. H. G. Horsthuis, and G. R. Möhlmann, *Appl. Phys. Lett.* 68, 3683 (1996).
433. S. Tomaru, T. Watanabe, M. Hikita, M. Amano, Y. Shuto, I. Yokohoma, T. Kaino, and M. Asobe, *Appl. Phys. Lett.* 68, 1760 (1996).
434. L. Becouarn, B. Gerard, M. Brévignon, J. Lehoux, Y. Gourdel, and E. Lallier, *Electron. Lett.* 34, 2409 (1998).
435. S. J. B. Yoo, R. Bhat, C. Caneau, and M. A. Koza, *Appl. Phys. Lett.* 66, 3410 (1995).
436. J. P. Bouchard, M. Têtu, S. Janz, D.-X. Xu, Z. R. Wasilewski, P. Piva, U. G. Akano, and I. V. Mitchell, *Appl. Phys. Lett.* 77, 4247 (2000).

Chapter 9

HIGH-TEMPERATURE SUPERCONDUCTOR AND FERROELECTRIC THIN FILMS FOR MICROWAVE APPLICATIONS

Félix A. Miranda, Joseph D. Warner

NASA Glenn Research Center, Communication Technology Division, Cleveland, Ohio, USA

Guru Subramanyam

Department of Electrical and Computer Engineering, University of Dayton, Dayton, Ohio, USA

Contents

1. INTRODUCTION

This chapter deals with the materials and electrical properties of high-temperature superconductor (HTS) and ferroelectric thin films, and their applications in microwave communications components, devices, and systems. From the microwave applications point of view, neither HTS nor ferroelectric thin films can be considered a mature and well-established technology. This is so, because none of them commanded substantial attention before a couple of decades ago. In the case of HTS, the reason is obvious since these materials were not discovered until 1986. For ferroelectrics, the lack of effective processing techniques, particularly those necessary to grow high-quality thin films, hindered their insertion into practical microwave circuits. In fact, their thin-film form was seriously considered for applications only after materials processing and deposition techniques were applied successfully in growing high-quality HTS films. Thus, for the purpose of clarity, and to maintain a fair chronological sense regarding the evolution of HTS and ferroelectric thin-film-based technology, we will introduce HTS materials first and then we will proceed to discuss ferroelectrics.

In this chapter, we will discuss the design, fabrication, and performance of HTS-based and thin-film ferroelectric-based microwave components and devices, including hybrid (i.e., HTS/ferroelectric) microwave devices and circuits. We certainly do not pretend to cover in this short chapter all the advances and information that is actually known regarding these types of components. However, we have attempted to provide a general view of several practical and versatile microwave components rather than discuss in intricate details the complexities and advantages of a particular one. Finally, it is worth mentioning that the experimental results discussed here very often represent proof-of-concept (POC). Therefore, in view of the continuous optimization of the material properties of HTS and fer-

Handbook of Thin Film Materials, edited by H.S. Nalwa
Volume 3: Ferroelectric and Dielectric Thin Films

ISBN 0-12-512911-4/$35.00

roelectric thin films, and the advances in processing techniques, the results presented here could be subjected to modification (hopefully improvements) as the fields gain strength and become more mature.

2. HIGH-TEMPERATURE SUPERCONDUCTING MATERIALS

The phenomenon of superconductivity was observed for the first time in 1911 while the Dutch physicist Heike Kammerling Onnes was cooling mercury (Hg) with the goal of studying the behavior of metals at low temperatures [1]. In general terms, a material can be classified as a superconductor if it exhibits the following two fundamental properties: first, the complete loss of electrical resistivity when cooled below a transition temperature known as the *critical temperature* T_c, and second, the expulsion of magnetic flux from the bulk of the sample (i.e., perfect diamagnetism) in the superconducting state. The former was the phenomenon observed by Onnes in the aforementioned experiment, while the latter phenomenon was discovered by the German physicist Meissner and his graduate student Ochsenfeld in 1933 [2]. Since 1911, many materials have been found to be superconductors (e.g., lead (Pb), $T_c = 7.2$ K and niobium (Nb), $T_c = 9.5$ K). However, even 75 years after the initial discovery of superconductivity, the highest T_c reported was $T_c \cong 23$ K, corresponding to the niobium–germanium (Nb$_3$Ge) compound. Therefore, before 1986, practical applications of superconductor materials were limited to superconducting magnets and superconducting quantum interference devices (SQUIDS) that operate at temperature of $T_c/2$ or lower. Very sensitive sensor systems have been made with SQUIDS and ultra-low-noise amplifiers that operate at 4 K. These systems could only be utilized in very specialized applications because they required liquid helium (LHe) or special cryocoolers to keep the superconducting materials at operating temperature. These stringent requirements limited the practical use of such systems to those applications which were enabled by these magnetic devices and sensors.

At the end of 1986 and beginning of 1987, Bednorz and Muller [3], and Wu et al. [4] observed the onset of superconductivity at 35 K on the La-Ba-Cu-O (LBCO) compound and at 90 K on the YBa$_2$Cu$_3$O$_{7-\delta}$ (YBCO) compound, respectively. These events marked the discovery of a new type of superconductors known today as high-temperature superconductors (HTS) or ceramic cuprate superconductors. Immediately afterwards, other HTS compounds, such as Bi$_2$Sr$_2$Ca$_{n-1}$Cu$_n$O$_x$ ($n = 1$–3) (BSCCO), Tl$_2$Ba$_2$Ca$_{n-1}$Cu$_n$O$_x$ ($n = 1$–3) (TBCCO), and HgBa$_2$CaCu$_2$O$_{6+\delta}$ (HBCCO) [5–7], were discovered. The properties of some of these compounds nearly matched or exceeded those of Nb$_3$Ge or Nb at temperatures near liquid nitrogen (77.3 K). Consequently, there has been enormous activity aimed not only at achieving a better understanding of the mechanism of superconductivity in these compounds, but also at investigating potential applications in thin-film-based and bulk-based technology. The advent of HTS com-

pounds offers the user the following implementation advantages over previously known superconductors. First, with a T_c above 77 K, either liquid nitrogen (LN$_2$) or less expensive cryocoolers can be used for practical applications. Second, they have a critical current density of over 1×10^6 A/cm^2 and surface resistance at 10 GHz of 4×10^{-4} Ω or less at 77 K. Third, there are several deposition techniques available for growing HTS films at relatively low cost.

Progress in the deposition of HTS thin films on low-loss microwave substrates such as lanthanum aluminate (LaAlO$_3$), magnesium oxide (MgO), and sapphire (Al$_2$O$_3$), has resulted in thin films with microwave surface resistance (R_s) values orders of magnitude lower than those of conventional conductors such as gold (Au) and copper (Cu) at 77 K and below 20 GHz [8–10]. These developments have led to demonstrations of HTS-based passive microwave circuits such as resonators and filters [11, 12]. In the semiconductor arena, interesting results at cryogenic temperatures have been observed in the last decade. For example, carrier mobility approaching 10^6 V/cm^2·sec has been measured in quantum well devices when the temperature is lowered from room temperature to 100 K or below, depending on the material and the doping level. This is particularly true for the case of silicon–germanium (SiGe) and III–V compound semiconductor based devices [13, 14]. For example, for SiGe heterojunction bipolar transistors (HBTs), the current gain, cut-off frequency (f_T), and maximum oscillation frequency (f_{max}) all rise significantly as the temperature drops. Thus, realization of hybrid circuits consisting of passive HTS-based microwave components and active semiconductor devices for integration into microwave subsystems could offer additional advantages such as reduced loss and noise [15]. Further, it has also been demonstrated that HTS and ferroelectric thin-film technology could be used to fabricate tunable HTS/ferroelectric microwave components [16–18], as we will discuss in more detail in Section 3 of this chapter.

2.1. Properties of High-Temperature Superconductors

In this section, we define and briefly review the important parameters of high-temperature superconductors. The important parameters of HTS defined include the critical temperature, critical current density, and the critical fields. Keep in mind that the useful temperature range of operation of a superconductor is typically below $0.6T_c$, as other important superconducting properties are near saturation at this temperature. Some of the characteristic structural properties are also discussed briefly.

The critical temperature (T_c) is defined as the temperature below which a superconductor possesses no dc electrical resistivity or the temperature below which a superconductor exhibits perfect diamagnetic behavior at zero applied magnetic field. The critical current density (J_c) is one of the most important superconducting properties, for engineering applications. It is the maximum current density (i.e., current/cross-sectional area of the conductor) a superconductor can support before becoming a normal conductor. The third characteristic is the magnetic

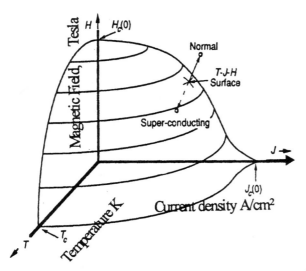

Fig. 1. Three-dimensional space defined by T_c, J_c, and H_c for a typical type-I superconductor.

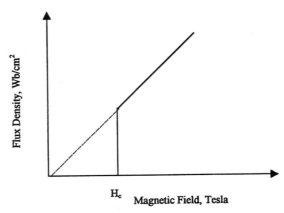

Fig. 2. Typical magnetic properties of a type-I superconductor. The flux density in the superconductor is zero when the magnetic field level is below H_c.

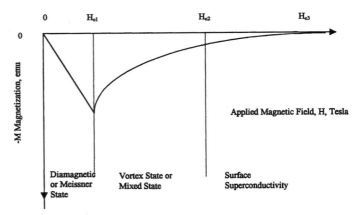

Fig. 3. Magnetic properties of a type-II superconductor.

critical field (H_c). It is defined as the field value above which the superconductor material will become normal.

A further classification of superconductors has been found to be needed because of their behavior in a magnetic field. The superconductors are classified as type-I and type-II materials depending upon their magnetic properties. The critical fields for a superconductor are H_c for type-I and H_{c1} and H_{c2} for type-II superconductors. For type-I superconductor as H_c is the maximum magnetic field below which a superconductor exhibits diamagnetic behavior, and above which the material is normal. The three-dimensional phase diagram (T_c, H_c, and J_c) of type-I superconductors is shown in Figure 1. The material is a superconductor below the given surface and normal above it. In type-I superconductors, the eddy current responsible for the shielding of the magnetic field below H_c is restricted to a thin layer near the surface. The eddy current elements decrease exponentially from the surface. The depth at which the current density is $1/e$ of that at the surface is called the penetration depth (λ). The penetration depth is very small near 0 K, and increases dramatically as the temperature approaches T_c. The penetration depth at 0 K ranges from 100–1500 Å for type-I materials [19]. Above the critical field, the magnetic field completely penetrates a type I superconductor, quenching superconductivity, as shown in Figure 2. Figure 2 also illustrates the T-dependence of resistivity in a superconductor at $T > T_c$.

In type-II superconductors, H_{c1} represents the lower critical field, the maximum field below which the material completely excludes magnetic flux from the interior (similar to H_c in a type-I material). When the field exceeds H_{c1}, the magnetic flux partially penetrates a superconductor to form a mixed state until the field reaches the upper critical field H_{c2}. When $H > H_{c2}$, the upper critical field, the magnetic flux completely penetrates the superconductor, and superconductivity is largely confined to the surface of a material. In the mixed or vortex state, magnetic flux penetrates through small tubular regions of the order of the coherence length (ξ) (a length scale that char-

acterizes superconducting electron pair coupling), called vortices (or flux tubes), with each vortex containing one quantum of flux, Φ_0 [20]. Abrikosov, in his study of type-II superconductors, determined that $\Phi_0 = h/2e$, where h is Planck's constant and e is the electronic charge [20]. The vortices form a periodic lattice called the Abrikosov vortex lattice. The resistivity of a superconductor may still be zero in the mixed state, provided the vortices are pinned or trapped. As the applied magnetic field (H_a) approaches H_{c2}, the number of vortices increases until there can no longer be any more addition of vortices, at which point the material becomes a normal conductor. Figure 3 shows the magnetic properties of type-II superconductors. In the mixed state, each vortex resides in a normal region, which is separated by superconducting regions.

Since T_c, J_c, and H_c values are relatively low in type-I superconductors compared to type-II, type-II superconductors are generally more suitable for electrical and electronic applications. Due to the complex nature of the cuprate ceramic superconductors, and operation at higher temperatures, the ac losses in HTS materials (type-II superconductors) are generally higher compared to type-I low temperature superconductors (LTS) materials [21, 22].

Table I. T_c and Crystalline Properties of Rare-Earth-Based Materials

Element	T_c (K)	a (Å)	b (Å)	c (Å)
YBa$_2$Cu$_3$Q$_x$	94[a]	3.8237(8)[b]	3.8874(8)[b]	11.657(2)[b]
	>90[c]	3.827(1)[c]	3.877(1)[c]	11.708(6)[c]
LaBa$_2$Cu$_3$O$_x$	75[d]	3.8562(8)[b]	3.9057(16)[b]	11.783(3)[b]
PrBa$_2$Cu$_3$O$_x$		3.922(2)[b]		
		3.905(2)[c]	3.905(2)[c]	11.660(10)[c]
NdBa$_2$Cu$_3$O$_x$	92.0[b]	3.8546(8)[b]	3.9142(12)[b]	11.736(2)[b]
SmBa$_2$Cu$_3$O$_x$	88.3[b]	3.855(2)[b]	3.899(2)[b]	11.736(2)[b]
	>90[c]	3.891(1)[c]	3.894(1)[c]	11.6601(1)[c]
EuBa$_2$Cu$_3$O$_x$	90 ± 1,[e] 54.6[f]	3.8448(8)[b]	3.9007(10)[b]	11.704(3)[b]
	>90[c]	3.869(2)[c]	3.879(3)[c]	11.693(6)[c]
GdBa$_2$Cu$_3$O$_x$	92.2[b]	3.8397(12)[b]	3.8987(18)[b]	11.703(3)[b]
	>90[c]	3.854(2)[c]	3.896(2)[c]	11.702(7)[c]
TbBa$_2$Cu$_3$O$_x$	35 ± 1[e]			
DyBa$_2$Cu$_3$O$_x$	92 ± 1,[e] 91.2[b]	3.8284(8)[b]	3.8888(8)[b]	11.668(2)[b]
	>90[c]	3.830(3)[c]	3.885(3)[c]	11.709(3)[c]
HoBa$_2$Cu$_3$O$_x$	93 ± 2,[e] 92.2[b]	3.8221(8)[b]	3.8879(8)[b]	11.670(2)[b]
	88,[d] >90[c]	3.846(1)[c]	3.881(1)[c]	11.640(2)[c]
ErBa$_2$Cu$_3$O$_x$	92 ± 2,[e] 91.5[b]	3.8153(8)[b]	3.8847(12)[b]	11.659(2)[b]
	>90[c]	3.812(3)[c]	3.851(4)[c]	11.626(2)[c]
TmBa$_2$Cu$_3$O$_x$	>90[c]	3.8101(8)[b]	3.8821(17)[b]	11.656(2)[b]
	>90[c]	3.829(3)[c]	3.860(3)[c]	11.715(2)[c]
YbBa$_2$Cu$_3$O$_x$	91 ± 1,[e] 85.6[b]	3.7989(8)[b]	3.8727(10)[b]	11.650(2)[b]
	87[d]			
LuBa$_2$Cu$_3$O$_x$	85[d]			

[a] From [113].

[b] From [114].

[c] From [115].

[d] From [116].

[e] From [117].

[f] From [118].

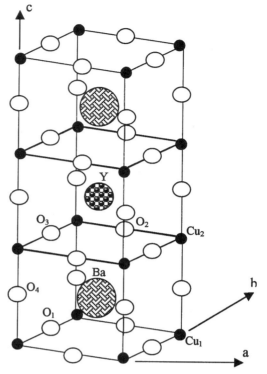

Fig. 4.   Crystal structure of the YBCO high-temperature superconductor.

2.1.1. Physical Properties of High-Temperature Superconductors

The presence of one or more copper oxide (CuO_2) planes in the unit cell is a common feature of all HTS materials. The most popular cuprate materials are $YBa_2Cu_3O_{7-\delta}$ (henceforth referred to as YBCO), $Bi_2Sr_2Ca_{n-1}Cu_nO_{2n+4}$ (where $n = 2, 3$) (henceforth referred to as Bi2212 and Bi2223 for $n = 2$, and $n = 3$, respectively), $Tl_2Ba_2Ca_{m-1}Cu_mO_{2m+4}$ (henceforth referred to as Tl2201, Tl2212, and Tl2223), and $HgBa_2Ca_{m-1}Cu_mO_{2m+2}$ (where $m = 1, 2, 3$). Table I lists the mature cuprate superconductors, their superconducting properties, and lattice parameters. In YBCO, there are two square planar CuO_2 planes stacked in the c-direction, separated by an intercalating layer of barium and copper atoms and a variable number of oxygen atoms. The conventional wisdom is that the CuO_2 planes are the conduction channels of superconductivity, while the intercalating layers provide carriers or act as charge reservoirs necessary for superconductivity, although this view is not shared universally [23]. The charge density, the number of superconducting charge carriers per unit volume, is determined by the overall chemistry of the system and by the charge transfer between the CuO_2 planes and the CuO chains. The charge density in a HTS material (10^{19}/cm^3) is two orders of magnitude lower than in conventional LTS (10^{21}/cm^3). Remarkably, the oxygen content in the system changes the oxidation states of the copper chain atoms which, in turn, affects their ability for charge transfer, charge density, and superconducting properties. Depending upon the oxygen content, the YBCO material could have a non-superconducting tetragonal ($a = b \neq c$) phase ($6 < \delta < 6.5$) to a 92 K superconducting orthorhombic ($a \neq b \neq c$) phase ($6.5 < \delta < 7$). When fully oxygenated, YBCO possesses an orthorhombic unit cell with typical dimensions of $a = 3.85$ Å, $b = 3.88$ Å, and $c = 12.0$ Å, and a $T_c = 92$ K. Figure 4 shows the crystal structure of the YBCO 123 superconductor, showing the aforementioned conduction and binding layers.

All of the HTS materials possess a fundamental limitation. That limitation is crystalline anisotropy; that is, they possess different structural and electrical properties in different directions. Superconducting properties such as critical current density (J_c) within the a–b plane and critical magnetic field (H_c) perpendicular to the a–b planes (xy) are superior to those perpendicular to these directions. A major challenge for researchers has been to grow materials with a highly preferred crystal orientation to take advantage of high J_c and low R_s in the a–b planes. The HTS materials also exhibit higher penetration depths compared to LTS materials. The penetration depth is an important parameter for high-frequency applications of superconductors. As mentioned before, the magnetic field decays in the form of

$$H = H_0 e^{-x/\lambda} \qquad (1)$$

where H_0 is the field at the surface, x is the depth through the sample, and λ is the penetration depth of the superconductor, analogous to the skin depth in conventional electrical conductors. The penetration depth is a function of temperature, denoted by $\lambda(T)$. The penetration depth can be modeled through the use of the Gorter–Casimir two-fluid model [24], the London theory [21, 25], and the BCS theory in the Pippard limit ($\xi_0 \ll \lambda$) near and below T_c [26], and is given by the equation

$$\lambda(T) = \frac{\lambda(0)}{\sqrt{1 - (T/T_c)^4}} \qquad (2)$$

The penetration depth is a frequency-independent parameter, in contrast to frequency-dependent skin depth of normal conductors. This means that at a given temperature, little or no dispersion is introduced in superconducting components up to frequencies as high as tens of gigahertz, in contrast to dispersion present in normal metals. Furthermore, lower losses in superconductors allows the use of substrates with higher dielectric constants, leading to a reduction in physical size of the devices, without sacrificing performance. This feature represents another advantage for HTS thin-film-based circuits. Compact delay lines, filters, and resonators are possible with high quality factor (Q) due to low conductor losses with respect to their normal-conductor counterparts. The challenge in achieving these properties is in developing processes to make HTS materials with smooth surface morphology to minimize high-frequency ac conductor losses.

While T_c and H_c values of a superconductor are generally intrinsic properties of a specific material, J_c, R_s, and λ values are a function of sample microstructure, and can vary by several orders of magnitude due to various deposition and processing techniques. Although high J_cs in HTS materials are achieved in bulk, thin films, and wires, several processing-related problems need to be addressed. These problems relate to grain boundaries, metallurgical defects, flux pinning mechanisms, and flux creep. Grain boundaries are interfacial regions between adjacent grains, and could contain impurity phases, normally conducting or insulating, extending beyond the coherence length of these superconductors, which is typically less than 30 Å along the a–b-plane. These grain boundaries can create superconductor-normal conductor junctions, or insulator-superconductor Josephson junctions, the critical currents of which are much lower than a homogeneous superconductor, and hence they are called weak links. The grain boundaries are particularly a problem in polycrystalline bulk superconductors, as they affect the bulk critical current density. Grain boundaries also contribute to surface resistance of a superconductor at high frequencies, as they increase the residual ac losses. Processing of HTS materials to improve J_c and R_s requires ways to minimize the grain-boundary effects, and to enhance the flux pinning sites in the superconductors.

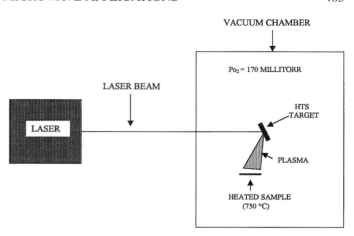

Fig. 5. Schematic representation of the laser ablation technique.

2.2. High-Temperature Superconducting Thin Films

Several techniques are used for the deposition of HTS thin films. Among these are laser ablation, off-axis magnetron sputtering, co-evaporation, sequential evaporation, and others. All of these methods strive to produce epitaxial or textured films with the right stoichiometry and composition. A brief description of these deposition techniques is given below.

2.2.1. Laser Ablation

A technique that has proven to be very effective for the deposition of HTS thin films is pulsed laser deposition (PLD) [27, 28]. In this technique, laser pulses (KrF, ArF excimer lasers, or a Nd-YAG laser are among the most commonly used) are fired onto a stoichiometric target of the HTS material to be deposited (YBCO, BSCCO, etc.). This produces a plasma "plume" of ejected material from the target which condenses onto a substrate mounted on a heated holder and kept at temperatures between 650 and 850°C in a partial atmosphere of oxygen. A schematic of this configuration is shown in Figure 5. The main advantage of this technique is that it can preserve stoichiometry of complex composition, and therefore the resulting films have the target's stoichiometry. Since the films are grown in an oxygen atmosphere (e.g., 100 mtorr), the superconducting phase is attained *in situ* during the laser ablation process, eliminating the need of a post deposition or "*ex situ*" annealing treatment. Thus, the major advantages of this technique are that a stiochiometric target can be used as the source material, there is a high deposition rate (e.g., 100 Å/min), and there is easy optimization of the film composition and crystallinity by the adjustment of the main deposition parameters. Those deposition parameters are the energy density of the laser pulses, the wavelength, the distance between the target and substrate, and the deposition temperature. For example, a useful set of deposition parameters for YBCO films on LaAlO$_3$ or r-plane sapphire with cerium oxide (CeO$_2$) buffer layer is the following: a laser wavelength of 193 nm (ArF) or 248 nm (KrF), laser beam energy density at the target of 2 J/cm^2/pulse, pulse rate of 5 pulses per second, oxygen pressure of 120 mT, laser beam 60°

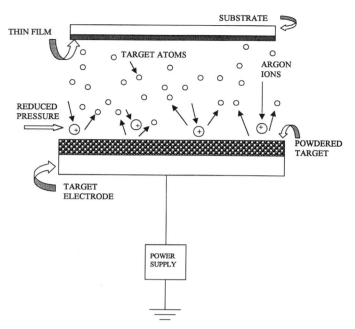

Fig. 6.   Schematic representation of the rf magnetron sputtering technique.

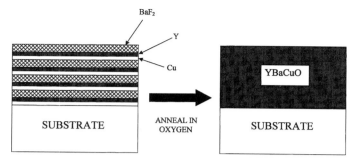

Fig. 7.   Schematic representation of the sequential evaporation technique showing the as-deposited multilayer structure of the film (left), and the superconducting film after annealing (right).

the growing film due to negative oxygen ions, because the ions lose a considerable amount of energy due to inelastic collisions. A small oxygen pressure ($\sim$20 mtorr) is also used to keep negative ion effects small. During the deposition, the substrate is placed off-axis from the ion source and is attached with a thermally conductive adhesive (e.g., silver paint) to a block (e.g., stainless steel) held at temperatures near 650°C. Typical deposition rates and dc power are of order 200 Å per hour and 45 W, respectively. The most important aspect in designing an off-axis magnetron sputtering system is to minimize the magnetic flux crossing the substrate.

2.2.3. Sequential Evaporation

The deposition of YBCO superconducting thin films using sequential evaporation was first reported by Tsaur et al. [33]. In this technique, the multilayer film is made by electron beam evaporation of alternating thin layers of either Cu, Ba, and Y, or Cu, BaF$_2$, and Y. The basic three-layer stack is repeated to give the desired thickness. This technique allows for the deposition of films with little spatial variation of stoichiometry across the substrate, as all components of the film are evaporated from the same point in space. The stoichiometry of the films is easily adjusted by controlling the thickness of the individually deposited layers. However, this deposition technique requires postdeposition annealing of the film in order to attain the superconducting phase. The annealing time is typically 0.5 hr at 850°C. During the annealing, the sample is exposed to ultra-high-purity oxygen and must be bubbled at room-temperature water if BaF$_2$ was used as the Ba source. The water vapor hydrolyzes the BaF$_2$ to form barium oxide (BaO) and hydrogen fluoride gas (HF). Dry oxygen is then used for the remainder of the annealing process. The temperature is then ramped down to 450°C at a rate of -2°C per minute. The samples are held at this temperature for 6 hr and then the temperature is ramped to room temperature, also at a rate of 2°C per minute. A schematic representation of the as-deposited multilayer structure of the film and the superconducting film after the annealing is shown in Figure 7.

from normal of the target, target-to-substrate distance of 6 cm, and a deposition temperature of 725°C. These parameters have resulted in YBCO films with T_c of 90 K or better as obtained from dc resistivity versus temperature measurements. The major disadvantage of PLD is the difficulty of depositing uniform films on wafers larger than 7.5 cm in diameter.

2.2.2. Magnetron Sputtering

In this technique, a composite oxide sintered target is spread on a metal plate, usually copper, which acts as the cathode of the sputtering chamber. A power supply generates an rf input power ($\sim$100 to 300 W) which drives the sputtering gas ions (generally any of the following combinations of argon (Ar) and oxygen (O$_2$); Ar(80%) + O$_2$(20%), Ar(70%) + O$_2$(30%), Ar (90%) + O$_2$(10%) or even pure Ar [29–31]) toward the sintered oxide target. When the sputtering gas ions reach the target, atoms are sputtered out of the target and driven toward the substrate. All the deposition process is carried out at sputtering gas pressures of approximately 3×10^{-2} to 8×10^{-2} torr. During the deposition, the substrates are kept within the 650–700°C temperature range. The growth rate of the films depends upon the rf power used for a particular deposition, but it is commonly within 24 to 70 Å per minute. An "*in situ*" annealing in an O$_2$ atmosphere is usually required in order to improve the value of T_c for the films. A schematic representation of this deposition technique is shown in Figure 6.

A similar technique, called off-axis dc magnetron sputtering, which gives a superior material, is also used to grow HTS films [32]. This technique uses high Ar gas pressures ($\sim$150 mtorr) in order to enhance the collision frequency of sputtered atoms, thereby making the deposition process more efficient. This also minimizes or eliminates the resputtering of

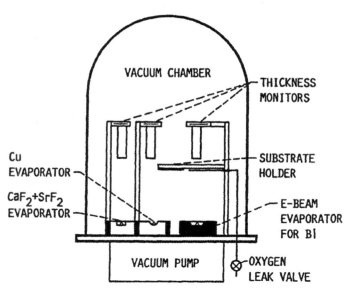

Fig. 8. Co-evaporation setup for growing Bi-Sr-Ca-Cu-O thin films.

2.2.4. Co-evaporation

In this deposition process, the evaporation of the primary components is performed at the same time, with a separate source for each element (metal or fluoride sources). Since the evaporated amount of each of the primary components is monitored by three independent thickness monitors, this method offers the advantage of very good control and flexibility in the final composition of the sample.

In general, the oxygen pressure during the deposition is approximately 10^{-5} to 10^{-6} torr, the deposition rate is 1 to 10 Å per second, and the substrate is maintained at room temperature. Since the high-T_c superconducting phase is formed at temperatures well above room temperature, a postdeposition annealing, similar to sequential evaporation, is required when films are deposited using this technique. Co-evaporation from Y, BaF$_2$, and Cu sources, followed by oxygen annealing, has become one of the most common methods of making YBCO HTS thin films [34–36]. Y, Cu, and BaF$_2$ are evaporated from three separate sources, forming an amorphous film of approximately correct metal stoichiometry. A schematic of the deposition setup put together by Kalkur et al. [37] at the University of Colorado, Colorado Springs, for growing Bi-Sr-Ca-Cu-O superconducting thin films is shown in Figure 8. In this process, bismuth (Bi) is evaporated using an electron beam and copper is evaporated from a tungsten (W) boat. Calcium (CaF$_2$) and strontium fluorides (SrF$_2$) can be evaporated together in one tungsten boat since their melting and evaporation temperatures are very close (the melting and evaporation temperatures are 1473°C and 2489°C for SrF$_2$, and 1423°C and 2500°C for CaF$_2$) [38], which implies that their vapor pressures at a given temperature are close. The mixing ratio in the boat is made to equal the composition of Sr and Ca in the final film. The composition of the deposited film is monitored by three independent quartz thickness monitors. Before the film deposition, the vacuum system is pumped to a pressure of less than 10^{-6} torr. Then

oxygen is leaked into the system through a nozzle near the substrate holder. The chamber pressure in the system during the deposition process is maintained at around 5×10^{-5} torr. Evaporation rates of the components during the process range from 0.8 to 3.2 Å per second. Finally, the as-deposited film is annealed in a furnace following a two-step procedure. The first step of the annealing is performed at 750°C in wet oxygen for about 30 to 60 minutes. This step is performed to decompose the fluorides so that the fluorine will react with the H$_2$O molecules, forming volatile hydrogen fluoride gas (HF). The second annealing step is performed to form the superconducting phase. This step is performed at 850°C for about 5 to 15 minutes. Afterwards, the sample is allowed to cool down slowly (~ 2°C per minute) to room temperature before it is removed from the deposition system.

2.2.5. Deposition of Thallium-Based High-Temperature Superconductors

Unlike the yttrium- and bismuth-based HTS compounds, the Tl-based HTS compound cannot be made simply by sintering or evaporation of the metal oxides because of the volatility of the thallium. The two most interesting phases of this group are Tl-2212 and Tl-2223. Tl-2212 can be made as single-phase, c-axis oriented thin films with transition temperatures 100 to 105 K. While Tl-2223 cannot be made easily as a single-phase pure compound, its T_c of 125 K is the highest of any of the thallium-based compounds. The method of making Tl-based superconductors can be dangerous, and appropriate safety procedures must be followed to prevent poisoning from the thallium. The basic way to prepare the bulk or thin-film compounds is the same as the methods mentioned earlier except that the thallium is left out. After the bulk material is sintered or the thin is film deposited, the material is then heated in a closed container filled with oxygen and thallium oxide at temperatures between 600 and 800°C. This method provides an overpressure of thallium oxide, which then forms the thallium HTS compound.

The roughness and the multiphase nature of the Tl-based films will limit their use to operating temperatures above 75 K, where they have an advantage of having higher T_c and J_c and lower R_s values than the YBCO and BSCCO films. But this advantage is large for applications where the power need to cool the sample and the size and weight of the cooler are important. Some of those applications are for space communications and some cellular base stations.

2.2.6. Microwave Substrates

The use of high-temperature superconducting thin films for microwave applications has been subjected to the availability of substrates with low and thermally stable dielectric constant, low losses, and good lattice match with the HTS films. Soon after 1986, the deposition of HTS thin films onto SrTiO$_3$ substrates produced high-quality films (i.e., with T_cs near 90 K) mainly due to the excellent lattice match between the SrTiO$_3$ and the

Table II.   Microwave Substrates for HTS Thin Films[a]

Material	Structure (298 K)	Dielectric constant	Loss tangent (298 K)	Lattice size (Å)	Lattice mismatch	Remarks
MgO	cubic	9.8	3.0×10^{-4}	4.178	$a = 11.0\%$ $c = 12.8\%$	Small areas, good for *in situ* film growth, reacts with O_2
LaAlO$_3$	pseudocubic	22–24	5.8×10^{-3}	3.792	$a = 0.7\%$ $c = 2.6\%$	Large area, twinning very high-quality films
LaGaO$_3$	orthorhombic	25	1.8×10^{-3}	3.902	$a = 2.1\%$ $c = 0.2\%$	Large area, phase transitions at 140 and 400°C may cause surface roughness (steps)
SrTiO$_3$	cubic	~300 @ 300 K ~1900 @ 80 K ~18000 @ 4.2 K	~0.03 @ 300 K ~0.06 @ 80 K	3.905	$a = 2.2\%$ $c = 0.3\%$	Small area, high-quality films
YSZ	cubic	27	5.4×10^{-3}	3.648	$a = 4.6\%$ $c = 6.3\%$	Large area substrates

[a]From [119].

superconducting copper oxides. This excellent lattice match resulted in J_c values greater than 1×10^6 A/cm^2 at 77 K and zero magnetic field for currents along the a–b plane [39], and values an order of magnitude lower for currents along the c-axis [40]. Nevertheless, the microwave properties of the SrTiO$_3$ substrate are limited because of its large and strongly temperature-dependent dielectric constant of approximately 300 at room temperature and over 1000 at 77 K [41, 42], and its large loss tangent at microwave frequencies [43], which result in degradation of the microwave transmission properties.

On the other hand, magnesium oxide (MgO), with a dielectric constant of 9.8 and a low loss tangent of approximately 10^{-4} at room temperature [44], is a convenient substrate for microwave applications. However, it has a large lattice mismatch with the HTS oxides, which makes epitaxial film growth more difficult. Also, it is hygroscopic, requiring careful handling and storage conditions. The lack of good epitaxial match between film and substrate has proven to be detrimental to the overall superconducting transport properties of the HTS films. The same applies for the yttria-stabilized zirconia (YSZ), with a dielectric constant of 27 and a loss tangent of 10^{-3} at room temperature, but with a considerable lattice mismatch with the HTS films [45].

The lanthanum aluminate (LaAlO$_3$) substrate overcomes the limitations of the SrTiO$_3$ and the MgO substrates. This substrate has a relative dielectric constant of 23–25 at room temperature, changing less than 10% when cooled to cryogenic temperatures [46]. It also has loss tangents of 10^{-4} and 10^{-5} at room temperature and 77 K, respectively, and an excellent lattice match with the high-T_c superconductors [47]. These properties make LaAlO$_3$ very suitable for operation at microwave frequencies. However, this substrate is twinned and gives a slight variation of properties across the wafer.

The properties of the LaGaO$_3$ substrate are very similar to those of LaAlO$_3$, and it is not twinned. It has a dielectric constant of 25 at room temperature and a good lattice match with

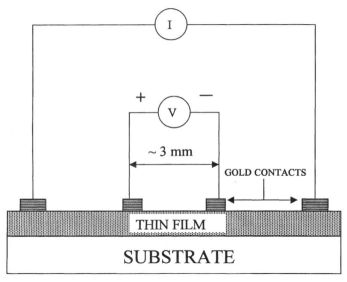

Fig. 9.   Four-point probe setup for measurements on HTS thin films.

the high-T_c superconductors [45]. Table II summarizes some of the most relevant properties of the most commonly used substrates for HTS thin-film deposition.

2.3.   Characterization of High-Temperature Superconducting Thin Films

Sample characterization techniques for determining the critical parameters of the HTS thin films, such as the T_c, J_c, and R_s, are discussed in this section. Accurate measurement techniques are needed for effective characterization of thin films.

2.3.1.   Transition Temperature (T_c)

One of the most commonly used characterization techniques for HTS materials is measurement of T_c. A rather simple way to do this is to use the so-called four-point probe measurement technique. Figure 9 shows a schematic of this configuration implemented at the NASA Glenn Research Center. In order to cancel the contribution of any thermal voltage, two measurements are performed with the current reversed between the measurements. The resistance is then determined from the average of these measurements. The criterion for the determination of T_c is generally fixed at the point where the resistivity of the sample being measured falls below the noise level of the instrument (e.g., 10^{-8} Ω cm). During the actual measurement, the current density is typically between 2 and 10 A/cm^2.

Another method of measuring T_c is to measure the magnetization (M) of the sample in a constant magnetic field. As the sample is cooled through T_c, M of the sample goes from nearly zero to -4π. The highest temperature at which $M = -4\pi$ is T_c. In practice, one measures the change in M, and the temperature at which that change reaches a maximum is T_c. In the literature, other useful temperatures are given to characterize HTS materials. Those are T_{onset}, $T_{mid-point}$, and ΔT_{90-10}. The T_{onset} is the maximum temperature at which some of the sample has turned superconducting. $T_{mid-point}$ is where resistance (R) or the magnetization (M) is one-half of their values at T_{onset} and T_c, respectively. ΔT_{90-10} is the width of the transition defined as $\Delta T_{90-10} = T_{90\%} - T_{10\%}$, where $T_{90\%}$ is the temperature at which $R = 0.9R_N$ or $M = 0.1M_M$ and $T_{10\%}$ is the temperature at which $R = 0.1R_N$ or $M = 0.9M_M$, where R_N and M_M are the dc resistance and the magnetization at T_{onset} and T_c, respectively.

2.3.2. Critical Current Density Measurements

Figure 10 shows a schematic of a geometry typically used for the electrical transport measurements. In this geometry, a current is connected between pads 1 and 2, and corresponding voltage measurements are taken across either pads 3 and 4, or pads 5 and 6. A plot of current density versus temperature for a Tl-Ba-Ca-Cu-O (TBCCO: 2212) thin film measured using this setup is shown in Figure 11. Note that this particular sample exhibits a J_c of the order of 3×10^5 A/cm^2 at 77 K. Currently, values above 10^6 A/cm^2 are typical of good-quality YBCO and TBCCO HTS samples.

2.3.3. Microwave Measurements

Microwave measurements of the HTS thin films provide a convenient probe to be used in attempting to identify the conduction mechanisms and the nature of the superconducting state of these compounds [48]. Whereas the dc resistance measurements mentioned earlier provide information about the normal state above T_c, and other techniques, such as magnetization measurements, give information on the superconducting state below T_c, microwave measurements can give useful information on both the superconducting and the normal states [49]. Another main objective of the microwave studies of the HTS thin films is to evaluate the potential of these materials for microwave devices applications [50]. In an attempt to uncover the intrinsic properties and the ultimate performance of these

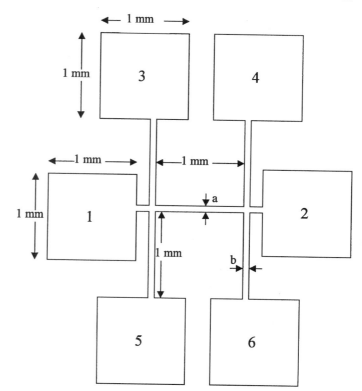

Fig. 10.   Schematic of the test structure for current density measurements.

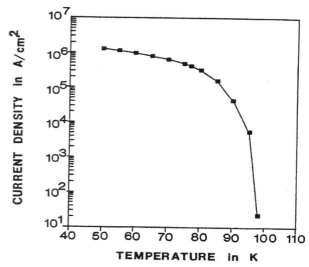

Fig. 11.   Current density versus temperature performed on a Tl-2212 thin film.

oxides at microwave frequencies, surface resistance (R_s) measurements of very-high-quality thin films have been carried out [39, 51]. Another parameter of fundamental importance in the characterization of these materials is the microwave conductivity ($\sigma^* = \sigma_1 - j\sigma_2$). Contrary to the normal conductivity, the microwave or ac conductivity is a complex quantity, and more painstaking techniques are necessary to measure it directly. Measurements of σ^* in the normal and the superconducting states can be used to determine R_s and the magnetic penetration depth (λ).

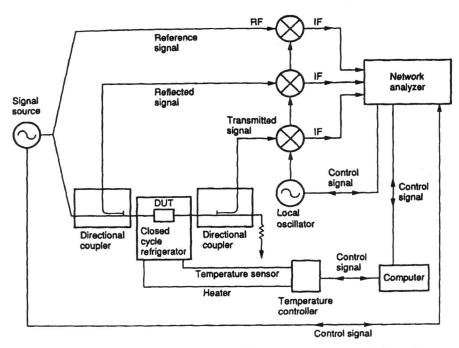

Fig. 12.   Schematic of the experimental setup for the microwave power transmission method.

As an example of a technique used for the microwave characterization of HTS thin films, we are going to discuss the microwave power transmission and phase measurement method. The power transmission and phase measurement technique is implemented using a Hewlett–Packard 8510 automatic network analyzer connected to a helium gas closed-cycle refrigerator by Ka-band (26.5–40.0 GHz) waveguides. A schematic and pictures of the experimental setup are shown in Figures 12 and 13, respectively. These measurements are performed under vacuum ($<10^{-3}$ Torr), and inside the vacuum chamber the sample is clamped between two waveguide flanges mounted on top of the cold head of the refrigerator. These waveguides are typically made of thin-wall stainless steel to minimize heat conduction, with their inner surfaces gold-plated to reduce microwave energy losses. In our particular case, the flanges are made of brass for ease of machining, and their inner surfaces are also gold-plated. Inside the waveguides, there are vacuum-sealed mica windows. The temperature of the sample is monitored with cryogenic sensors (e.g., silicon diode sensors) mounted on the waveguide flanges that supported the sample. The method could also be implemented using coaxial waveguides up to 50 GHz.

To show how this technique works, we are going to consider the following typical example. Figure 14 shows the dc resistance versus temperature for a YBCO thin film deposited onto a LaAlO$_3$ substrate by laser ablation. A $T_c = 86.3$ K was measured for this film. The X-ray diffraction pattern revealed that this film is single phased with a predominantly c-axis orientation. Also shown in Figure 14 is the measured temperature dependence of the power transmission coefficient (T) (ratio of the transmitted power to incident power) corresponding to the same sample. In the normal state, the behavior of the transmitted power with decreasing temperature is similar to that of

the dc resistance. However, at temperatures just below the onset temperature for the transition from the normal to the superconducting state, the transmitted power drops abruptly, falling monotonically with decreasing temperature, until a lower limit is reached. This behavior is typical of high-quality HTS films.

One can calculate the surface resistance and the magnetic penetration depth of the film from the microwave complex conductivity. The real and imaginary parts of the complex conductivity are given in terms of the power transmission coefficient T and the phase shift ϕ by the following mathematical expressions:

$$R = \left\{ \frac{2n}{T^{1/2}} \left[n \cos(k_0 n t) \sin(k_0 t + \phi) \right.\right.$$
$$\left. - \sin(k_0 n t) \cos(k_0 t + \phi) \right]$$
$$\left. - n(n^2 - 1) \sin(k_0 n t) \cos(k_0 n t) \right\}$$
$$\times \left\{ k_0 d \left[n^2 \cos^2(k_0 n t) + \sin^2(k_0 n t) \right] \right\}^{-1} \quad (3)$$

and

$$I = \left\{ \frac{2n}{T^{1/2}} \left[n \cos(k_0 n t) \cos(k_0 t + \phi) \right.\right.$$
$$\left. + \sin(k_0 n t) \sin(k_0 t + \phi) \right]$$
$$\left. - 2n^2 \cos^2(k_0 n t) - (n^2 + 1) \sin^2(n k_0 t) \right\}$$
$$\times \left\{ k_0 d \left[n^2 \cos^2(k_0 n t) + \sin^2(k_0 n t) \right] \right\}^{-1} \quad (4)$$

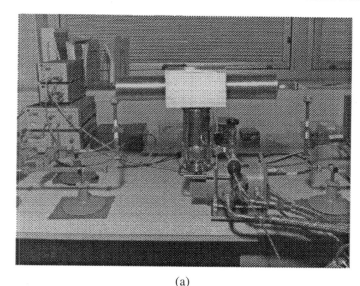

(a)

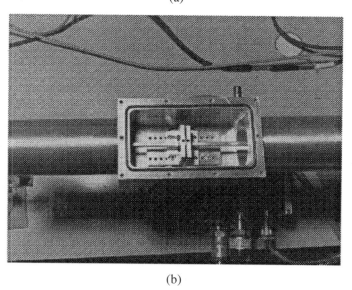

(b)

Fig. 13. Experimental setup for the microwave power transmission method: (a) external view, (b) internal view.

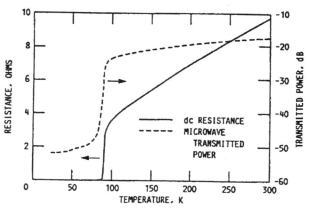

Fig. 14. Dc resistance and microwave transmitted power at 35 GHz versus temperature for a $YBa_2Cu_3O_{7-\delta}$ thin film (2665 Å thick) on a $LaAlO_3$ substrate.

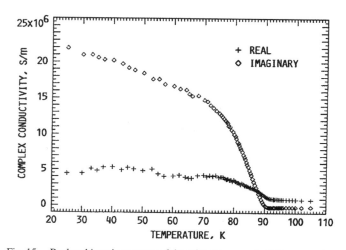

Fig. 15. Real and imaginary parts of the microwave conductivity versus temperature at 35 GHz for a YBCO thin film (2655 Å thick) on a $LaAlO_3$ substrate.

where k_0 is the wave number of the normal incident transverse electric wave propagating in the rectangular waveguide, d is the thickness of the film, t is the thickness of the substrate with refraction index n, $R = 1 + 4\pi\sigma_2/\omega\varepsilon$, $I = 4\pi\sigma_1/\omega\varepsilon$, $\omega/2\pi = f$ is the frequency of the wave, and ε is the relative dielectric constant of the material.

Figure 15 shows the temperature dependence of σ_1 and σ_2 for the YBCO/LAO sample at 35 GHz. The conductivity at room temperature ($\sim 3.9 \times 10^5$ S/m) compared reasonably well with reported values of dc conductivities in this type of films [52]. The change in σ_1 with increasing temperature exhibits a metallic behavior down to the onset temperature, at which $\sigma_1 \sim 1.3 \times 10^6$ S/m. In the normal state, σ_2 was close to zero, as expected for a good conductor. Note that both σ_1 and σ_2 increased upon going through the onset temperature, with σ_1 reaching values of 4×10^6 and 4.8×10^6 S/m at 76 and 50 K,

respectively, and σ_2 reaching values of approximately 1.3×10^7 and 1.8×10^7 S/m at these same temperatures.

The magnetic penetration can be calculated using σ_2 and London's expression:

$$\lambda = (\mu_0\omega\sigma_2)^{-1/2} \qquad (5)$$

In this case, penetration depth values of $\lambda = 0.57$ and 0.40 μm were obtained at 76 and 25 K, respectively, as shown in Figure 16. However, for very-high-quality HTS films, the best values reported so far are $\lambda = 0.20$–0.25 μm at 77 K, and $\lambda_0 = 0.14$ μm, where λ_0 is the value of the penetration depth at absolute zero [53]. The larger values of λ measured for the sample used as an example here can be explained in terms of the existence of residual inhomogeneities, which can produce grain-boundary Josephson junctions, resulting in a larger effective penetration depth. Furthermore, even for films whose X-ray diffraction pattern reveals predominantly c-axis orientation, there could still be a-axis-oriented grains in the film. The presence of a-axis-oriented grains increases λ, since the penetration depth for shielding currents along the c-axis is greater than that for shielding currents in the a–b plane [54].

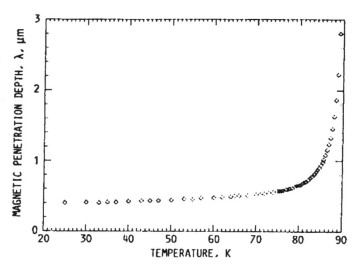

Fig. 16.   Magnetic penetration depth versus temperature for a YBCO thin film (2655 Å thick) on a LaAlO₃ substrate.

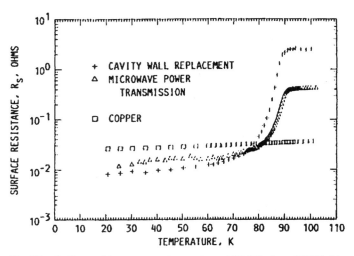

Fig. 17.   Surface resistance versus temperature at 36 GHz for a YBCO thin film (2665 Å) on LaAlO₃ as measured by the cavity wall replacement method and by the microwave power transmission method. The R_s for copper is also plotted for comparison.

The surface resistance for films in the superconducting state can be obtained by using the expression [55, 56]

$$R_s = R_N \left\{ \left(\left[\left(\frac{\sigma_1}{\sigma_N} \right)^2 + \left(\frac{\sigma_2}{\sigma_N} \right)^2 \right]^{1/2} - \frac{\sigma_2}{\sigma_N} \right) \times \left[\left(\frac{\sigma_1}{\sigma_N} \right)^2 + \left(\frac{\sigma_2}{\sigma_N} \right)^2 \right]^{-1} \right\}^{1/2} \quad (6)$$

where

$$R_N = \left(\frac{\omega \mu_0}{2\sigma_N} \right)^{1/2} \quad (7)$$

and σ_N are the surface resistance and the conductivity, respectively, at the onset temperature as determined from microwave power transmission measurements. The R_s values at 36 GHz were obtained assuming an f^2 dependence for R_s, where f is the frequency. The R_s versus temperature for the sample under discussion is shown in Figure 17. $R_s = 24$ mΩ and 12 mΩ were measured for this sample at 76 K and 25 K, respectively. The surface resistance of the HTS sample could also be measured by looking at the change in Q of a TE₀₁₁-mode (OFHC) copper cavity resonant at 36 GHz when one of its end walls is replaced with the superconducting sample. An R_s of 25 mΩ at 76 K was measured. The R_s values for the YBCO film measured using the resonant cavity are plotted in Figure 17. Also plotted is the R_s of copper for comparison.

Note that both techniques give an R_s that decreases rapidly when the sample is cooled through the transition temperature and then levels off at lower temperatures, showing a residual surface resistance that changes very slowly with decreasing temperature. Although there is a considerable discrepancy between the R_s values obtained with the two techniques at temperatures not far below T_c, the agreement is better at lower temperatures.

The normal skin depth,

$$\delta_N = \frac{2R_s}{\omega \mu_0} \quad (8)$$

for this sample calculated from the value of R_s at 87 K, as measured by the cavity technique, is approximately 5.4 μm. The largeness of this value relative to the thickness of the film suggests that a great deal of energy is leaking through the substrate, an effect that would result in an overestimation of R_s. Because of the inhomogeneous nature of the HTS films, it is probable that leakage can persist at temperatures lower than T_c, but not at temperatures far below T_c since at these temperatures, most of the film is superconducting. Thus, there is better correlation at low temperature between the measurements using the two techniques.

We must mention that the results obtained using the microwave power transmission technique are more susceptible to being influenced by the intrinsic behavior of the superconducting intragranular material as well as by nonintrinsic losses due to normal inclusions and grain-boundary effects in the interior of the film. Therefore, the R_s values obtained with this technique may be affected more by the nonintrinsic properties of the films than those measured using the cavity technique, which is only sensitive to the surface properties of the film. Nevertheless, the microwave power transmission technique provides an alternative way for determining R_s.

2.3.4. Comparison of TBCCO and YBCO High-Temperature Superconducting Thin Films for Microwave Applications

Of the two compounds, Tl-2212 is the more useful for microwave communication applications and is usually preferred over YBCO for applications that need or desire a working temperature between 75 and 80 K. Tl-2212 has the lowest R_s value of any of the HTS compounds within this range of temperatures. But for applications where the operating temperature is below 75 K, YBCO is preferred because then it has the lowest R_s value.

The higher operating temperatures of Tl-2212 directly impact system design and performance in an advantageous way. Some advantages are lower power needs for the cooler, compressor, and controller, lower temperature dependence of the reactance, which allows a less sophisticated temperature controller, and lower cost of the cooling and control systems. For satellite communication purposes, all three advantages can add up to tens to hundreds of thousand of dollars. For ground applications, these advantages can result in thousands to ten thousand dollars saving per terminal.

YBCO does have its advantages: it is safer and cheaper to produce; it has a smoother surface; it has lower R_s values below 75 K; it has higher critical currents; and it has a higher H_{c1}. Its smoothness allows smaller feature sizes to be patterned, and its higher critical currents and H_{c1} allow higher microwave power without going to Bessel function-type filters.

2.4. High-Temperature Superconductor Circuits

The advent of high-temperature superconductivity did open the doors to the development of HTS-based microwave components for communication applications. By the early 1990s, discrete components such as resonators, filters, phase shifters, and local oscillators were already demonstrated [57–70]. These early demonstrations strengthened the confidence of the communication engineers and satellite circuit designers in the potential of this technology to address issues such as low power loss, low cost, reduced size and mass, and better circuit performance when integrated into wireless and satellite communication systems. In this section, we are going to provide some examples of circuits demonstrated by our group in the past. Other important microwave HTS circuits demonstrated in the recent past are summarized in Table III.

Table III.   Summary of Recently Demonstrated HTS Microwave Circuits

No.	Authors	Topic	Reference	Comments
1	Talisa et al.	Dynamic range considerations for high-temperature superconducting filter applications to receiver front-ends	[57]	HTS filters for preselection in microwave receiver front-ends
2	Mansour et al.	Design of superconductive multiplexers using single-mode and dual-mode filters	[58]	Demonstrates the feasibility of building C-band compact superconductive multiplexers
3	Swanson and Forse	An HTS end coupled CPW filter at 35 GHz	[59]	
4	Barner et al.	Design and performance of low noise hybrid superconductor/semiconductor 7.4 GHz received down converter	[60]	When cooled to 77 K, the down converter plus cables inside the cryogenic refrigerator had a noise figure of approximately 0.7 dB with conversion gain of 18 dB
5	Lee et al.	Fabrication of YBCO superconducting dual-mode resonator for satellite communications	[61]	C-band; unloaded Q measured to be 1312 at 77 K
6	Shen et al.	High-power HTS planar filters with novel back-side coupling	[62]	2.88 GHz, 0.7% equal ripple bandwidth, 2-pole TE_{01} mode filters
7	Lancaster et al.	Miniature superconducting filters	[63]	
8	Herd et al.	Twenty-GHz broadband microstrip array with electromagnetically coupled high-T_c superconducting feed network	[64]	A novel antenna architecture is described that provides a broadband radiating aperture to be used as a scanning array with compatible HTS phase shifters
9	Fiedziuszko et al.	Low loss multiplexers with planar dual-mode HTS resonators	[65]	The basic resonator/filter structures (including HTSSE-I resonator/filter) suitable for these applications are described
10	Feng et al.	High temperature superconducting resonators and switches: design, fabrication and characterization	[66]	
11	Berkowitz et al.	Demonstration of a 20 GHz phase shifter using high temperature superconducting SNS junctions	[67]	
12	Aminov et al.	High Q-tunable YBCO disk resonator filters for transmitter combiners in radio base stations	[68]	Unloaded $Q \sim 29,000$ below 70 K
13	Yoshida et al.	Design and performance of miniature superconducting coplanar waveguide filters	[69]	
14	Hong et al.	An HTS microstrip filter for mobile communications	[70]	

2.4.1. Ring, Linear, and Coplanar Waveguide Resonators

Superconducting thin-film based ring resonators have been fabricated for operation at frequencies from the X-band (8 to 12 GHz) up to the Ka-band (26.5 to 40.0 GHz). Because of its geometry with no open ends, this structure is practically free of radiative losses. For example, a YBCO on LAO microstrip ring resonator designed for operation at 35 GHz became one of the first Ka-band microwave circuits fabricated with HTS thin films. A schematic showing the general configuration of this circuit is shown in Figure 18. The resonator consists of superconducting strips over normal metal ground planes. This resonator showed significant improvement in Q (six to seven times higher) over identical gold resonators at 20 K, as shown in Figure 19. The ring resonator is not only relevant as a versatile microwave component, it is also used as a characterization tool. Using a microstrip loss model, the surface resistance of the YBCO was determined to be 9 mΩ at 77 K and 35 GHz. This corresponds to a R_s of 735 $\mu\Omega$ at 10 GHz. Also, from the change in resonant frequency with temperature, a value for the magnetic penetration depth (λ_0) of 0.3 μm was obtained. This value is nearly within a factor of 2 of the theoretical value ($\sim$0.14 μm).

Microstrip ring resonators have been fabricated also using Tl-Ba-Ca-Cu-O thin films. A TBCCO on LAO microstrip ring resonator was designed for a fundamental resonance at 8.4 GHz. A schematic of this resonator is shown in Figure 20. This resonator consists of a ring circumference $\sim$1λ_g (where λ_g is the guide wavelength) with a mean radius (R) of 1489 μm, and its line width (W), as well as that of the feed line, is 160 μm. Coupling to the resonator is attained across a coupling gap (G) of 45 μm. Also shown in Figure 20 is a microstrip linear resonator designed for resonance at the same frequency as the ring resonator. This linear resonator was patterned simultaneously with the ring resonator with the purpose of comparing its unloaded Q values with those of the ring resonator, given that both circuits had been fabricated on the same material and under the same patterning conditions. For this resonator, the W, G, and feed line length ($L1$) were the same as for the ring resonator, while its length (L) was 4674 μm. A plot of the unloaded Q versus temperature for the linear and ring resonators is shown in Figure 21. Also shown is the unloaded Q for their gold counterpart for comparison purposes. Observe that in this case, the two HTS resonators have approximately the same Q at the same temperature, suggesting a high degree of homogeneity and uniformity of the superconducting properties across the film.

Ease of fabrication and performance reliability are two requirements that HTS thin films should meet for practical microwave applications. Because of their geometrical attributes of having the ground planes on the same surface as the signal transmission line, coplanar waveguide (CPW) structures are advantageous for HTS-based microwave integrated circuits. Figure 22 shows a schematic representation of a CPW resonator. The unloaded quality factor for TBCCO and gold versions of this resonator are shown in Figure 23. The unloaded Q val-

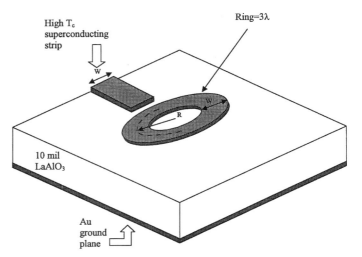

Fig. 18.   A schematic drawing of the ring resonator circuit. For this case, the ring was three wavelengths (3λ) in circumference at 35 GHz. The line width (W) was 143 μm and the substrate thickness was 254 μm (10 mil). The calculated impedance was 38 Ω.

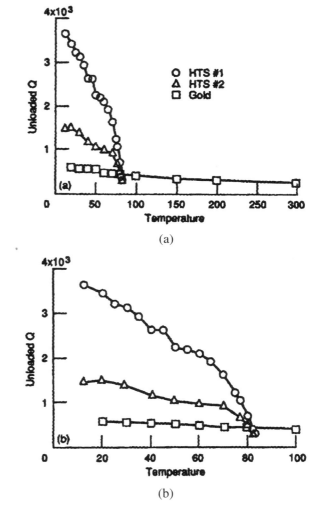

Fig. 19.   Measured Q values for gold and superconducting resonators over the range of (a) 0 to 300 K and (b) 0 to 100 K, at 35 GHz. The Q of the gold resonator increases by only a factor of 2 in cooling from 300 to 20 K.

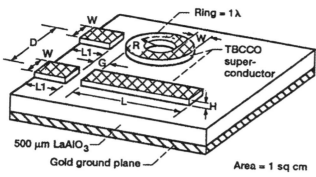

Fig. 20. A schematic representation of the microstrip ring and linear resonators. The microstrip ring resonator was one wavelength in circumference at 8.4 GHz. $W = 160\ \mu$m, $G = 45\ \mu$m, $R = 1489\ \mu$m, $L = 4674\ \mu$m, $D = 4000\ \mu$m, $L1 = 2000\ \mu$m, and $H = 0.8\ \mu$m. The calculated line impedance was 53 Ω.

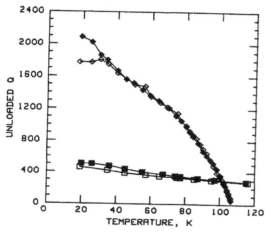

Fig. 21. Unloaded quality factor versus temperature for TBCCO-based ring ($\diamond$) and linear ($\blacklozenge$) 8.4-GHz resonators on LaAlO$_3$. The unloaded quality factors for all-gold ring ($\square$) and linear ($\blacksquare$) resonators are also shown for comparison.

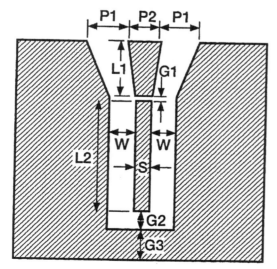

Fig. 22. Top view of a conductor-backed coplanar waveguide resonator (9.230×9.230 mm^2). $P1 = 0.533$ mm, $P2 = 0.559$ mm, $L1 = 1.000$ mm, $L2 = 7.020$ mm. $W = 0.530$ mm, $S = 0.200$ mm, $G1 = 0.050$ mm, $G2 = 0.530$ mm, and $G3 = 0.630$ mm. The cross-hatched sections represent the metallized portion of the circuit (HTS or gold).

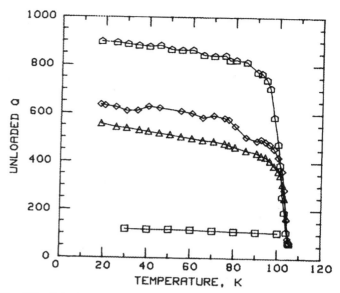

Fig. 23. Unloaded quality factor versus temperature for TBCCO-based 10.8-GHz conductor-backed CPW resonators on LaAlO$_3$. Sample 1 ($\triangle$), sample 2 ($\diamond$), and sample 3 ($\bigcirc$) are TBCCO. A gold version of the resonator ($\square$) is shown for comparison.

ues measured using these resonators can be used to evaluate the surface resistance (R_s) of the superconductor film. Since these circuits are tested with appropriate shielding, radiation losses are typically neglected in the calculation of R_s. Likewise, the dielectric losses are neglected since these circuits are fabricated in very-low-loss dielectric substrates and the overall circuit losses are dominated by the conductor losses. Therefore, the total quality factor (Q_T) becomes

$$\frac{1}{Q_T} = \frac{1}{Q_c} + \frac{1}{Q_d} + \frac{1}{Q_r} \sim \frac{1}{Q_c} \qquad (9)$$

where Q_c, Q_d, and Q_r correspond to the quality factor associated with the conductor, dielectric, and radiation losses of the circuit, respectively. For the CPW configuration, the Q_c is defined in terms of the attenuation and phase propagation constants (α_c and β, respectively) of the signal along the conducting microstrip, as shown in the following equation:

$$Q_c = \frac{\beta}{2\alpha_c} = \left(\frac{240\pi^2 f_0}{c R_s}\right)\left[\left(\frac{1}{Z_0}\right)\left(\frac{\partial Z_0}{\partial n}\right)\right]^{-1} \qquad (10)$$

where f_0 is the resonant frequency, c is the speed of light in a vacuum, Z_0 is the characteristic impedance of the transmission line, $\partial/\partial n = 2[\partial/\partial W - \partial/\partial S - \partial/\partial t]$ with t being the HTS film thickness, and W and S are the dimensions shown in Figure 22. For example, for this specific case, $(1/Z_0)(\partial Z_0/\partial n) = 0.0154\ \mu$m^{-1}. Since LaAlO$_3$ has a very low tan δ of $\sim 8 \times 10^{-5}$ at 77 K [47], Eq. (9) becomes $Q_T \sim Q_c$, and therefore,

$$Q_c = 5.127 \times 10^{-10} f_0/R_s \qquad (11)$$

or

$$R_s(\Omega) = 5.127 \times 10^{-10} f_0/Q_c \qquad (12)$$

In the case of the ring resonator, since its configuration is different from that of the CPW, the R_s of the conductor is related to the attenuation constant α_c as follows:

$$4\pi\alpha_c Z_0 = B\big[(C + D)R_{s1} + CR_{s2}\big] \quad (13)$$

where $B = 1 - (\omega'/4h)^2$, $C = (1/h)[1 - t/(\pi\omega')]$, $D = [2/(\pi\omega')][\pi + \ln(2h/t)]$, and $\omega' = \omega + (t/\pi)[\ln(2h/t) + 1]$. R_{s1} and R_{s2} are the surface resistance of the microstrip and the ground plane, respectively, and ω', ω, h, and t are the effective electrical microstrip width, the actual ring width, the substrate height, and the film thickness, respectively. For the gold resonator, $R_{s1} = R_{s2}$, and thus the surface resistance of gold (R_{sAu}) can be expressed in terms of the Q_0 for the gold resonator (Q_{0Au}) as

$$R_{sAu} = \frac{4\pi^2 Z_0}{[\lambda_g Q_{0Au}]B[2C + D]} \quad (14)$$

For the case of the example under discussion, $\lambda_g = 1.05$ cm, $Z_0 = 53\ \Omega$, $\omega = 160\ \mu$m, $h = 508\ \mu$m, and $t = 1.5\ \mu$m. Using these values in Eq. (14) gives

$$R_{sAu}(\Omega) = \frac{4.8799}{Q_{0Au}} \quad (15)$$

For the case when the ring resonator is an HTS one with gold ground plane, $R_{s1} \neq R_{s2}$. Thus, combining Eqs. (10) and (13) gives [71]

$$R_{ss}(\Omega) = R_{sAu} - \frac{4\pi^2 Z_0}{\lambda_g B(C + D)}\left\{\frac{1}{Q_{0Au}} - \frac{1}{Q_{0s}}\right\} \quad (16)$$

where Q_{0s} and R_{ss} are the unloaded quality factor of the HTS resonator with gold ground plane and the surface resistance of the HTS film, respectively. As before, the geometrical factor for the superconducting part of Eq. (16) would change slightly as λ changes with temperature. Therefore, once the surface resistance for the gold resonator is found, one can obtain the surface resistance of the superconductor in a superconducting microstrip with a gold ground plane configuration. Using the aforementioned geometrical parameters in Eq. (16) and $t = 0.7\ \mu$m gives

$$R_{ss} = R_{sAu} - 4.6634\left(\frac{1}{Q_{0Au}} - \frac{1}{Q_{0s}}\right) \quad (17)$$

Figure 24 shows a plot of R_s versus temperature for Tl-based CBCPW and ring resonators. Also shown is the R_s for gold as measured using the all-gold CBCPW and the ring resonators. In this example, the R_s for the CBCPW was found to be $\sim$6.6 mΩ at 10.680 GHz and near 77 K. This value is approximately one-eighth of its gold counterpart ($\sim$52.7 mΩ at 77 K and 10.803 GHz). For the TBCCO ring resonator, an $R_s \sim 4.5$ mΩ was obtained at 77 K and 8.301 GHz; this value is approximately 30% that of its gold counterpart (14.9 mΩ, 77 K, and 8.404 GHz); this ratio was the same for all temperatures below 77 K. When the gold ground plane is replaced by a superconducting ground plane, the obtained values for the ring resonator R_s become smaller, as reported by Subra-

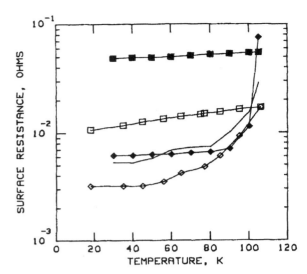

Fig. 24. Surface resistance (R_s) versus temperature at 8.4 GHz of a TBCCO-based ring resonator ($\diamond$) and at 10.4 GHz of a TBCCO-based CBCPW resonator ($\blacklozenge$). The R_s for all-gold ring ($\square$) and CBCPW ($\blacksquare$) resonators are also shown for comparison. The full curve represents the R_s expected at 10.7 GHz obtained from the ring resonators assuming $R_s \propto f^2$ dependence.

manyam et al. [72]. For rf sputtered Tl-based ring resonators, $R_s \sim 2.75$ mΩ at 12 GHz; $R_s \sim 1.3$ mΩ at 8.3 GHz assuming $R_s \propto f^2$ dependence [72]. The full curve in Figure 24 represents the R_s values expected at 10.7 GHz, the resonant frequency of the CBCPW resonator, obtained from those of the ring resonators by assuming an $R_s \propto f^2$ dependence. Note that the agreement between the two sets of values improves for low temperatures. This is expected since the losses for the superconducting transmission line are expected to diminish at low temperatures.

For superconducting transmission lines, the shift in the resonant frequency with decreasing temperature can be used to determine the penetration depth (λ) of the material. For a superconducting microstrip line with a normal ground plane, the inductance (L) and the capacitance (C) per unit length of the structure are [73, 74]

$$L = L_s + L_n = \frac{\mu_0 h}{w}\left[1 + \frac{\lambda}{h}\coth\left(\frac{t}{\lambda}\right)\right] + \frac{R_n}{w}\coth\left(\frac{t'}{\delta}\right)(2\pi h)^{-1}\left[1 - \left(\frac{w'}{4h}\right)^2\right] \quad (18)$$

and

$$C = \varepsilon_0 \varepsilon_{eff} w/h \quad (19)$$

where t and t' are the thickness of the HTS and normal conductor microstrip, respectively, w is the microstrip width, w' is the conductor width corrected for thickness (i.e., the effective electrical microstrip width), $R_n = (\mu_0 \omega \rho)^{1/2}$ is the surface resistance of the gold ground plane, δ is the normal skin depth, and $\omega = 2\pi f$, with f the frequency. The phase velocity propagating on a loss less transmission line is

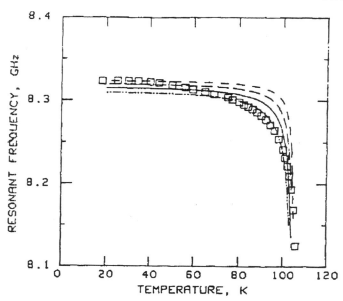

Fig. 25. Measured ($\square$) and calculated values for the resonant frequency of the fundamental mode of a TBCCO-based ring resonator versus temperature. The calculated values are generated using the two-fluid model [$\lambda_0 = 800$ nm (- - - -), $\lambda_0 = 1000$ nm (– – –), $\lambda_0 = 1200$ nm (——), $\lambda_0 = 1400$ nm (·· – ··)]. The best fit is obtained with $\lambda_0 = 1200$ nm.

$$
\begin{aligned}
v_{\mathrm{ph}}(T) &= \frac{1}{(LC)^{1/2}} \\
&= \frac{c}{(\varepsilon_{\mathrm{eff}})^{1/2}} \left[1 + \left(\frac{\lambda}{h}\right) \coth\left(\frac{t}{\lambda}\right) \right. \\
&\quad \left. + \left(\frac{w R_{\mathrm{n}}}{2\pi h^2 \omega \mu_0}\right) \coth\left(\frac{t'}{\delta}\right) \left[1 - \left(\frac{w'}{4h}\right)^2 \right]^{-1/2} \quad (20)
\end{aligned}
$$

For ring resonators that satisfy $w/h < 1$ and where their mean circumference ($l = 2\pi R$) $\ll w$, we have that at resonance, $l = n\lambda_{\mathrm{g}} = n v_{\mathrm{ph}}/f$, where n is the order of the resonance. Thus one can rewrite f as a function of v_{ph} and l

$$
f = n v_{\mathrm{ph}}/l \quad (21)
$$

From the above equations and the Gorter–Casimir temperature dependence of the penetration depth, $\lambda(T) = \lambda_0[1 - (T/T_{\mathrm{c}})^4]^{-1/2}$ an estimate of the zero temperature penetration depth (λ_0) can be obtained. Figure 25 shows a plot of the resonant frequency versus temperature for the TBCCO ring resonator discussed above. Also shown are different fits to these data generated by assuming different values of λ_0. Note that the best fit to the experimental data is obtained with $\lambda_0 \sim 1200$ nm. This value is approximately twice as large as that obtained for the unpatterned film using the power transmission method. Several factors may account for this discrepancy. The larger λ_0 obtained from the ring resonator may be related to material defects such as rough edges along the microstrip line caused by the patterning process. The frequency shift approach to determine λ_0 is also limited by the uncertainties in physical dimensions of the film and substrates in the patterned devices, the uncertainties in dielectric properties of the LaAlO$_3$ substrate, and the assumption of the temperature dependence for λ.

2.4.2. Filters and Phase Shifters

One of the most important applications of HTS in communication systems is in preselect filters. Planar HTS preselect filters offer the possibility of replacing heavy and bulky waveguide or dielectric resonator filters. The development of HTS planar filters and multiplexers (input and output) are expected to save mass and volume, with equivalent or better performance compared to the conventional filters. With the maturity level of current HTS thin-film technology, enhanced performance is expected in planar HTS filters and multiplexers with respect to passband loss, noise figure, and spectral efficiency. For the transmit side of the communication circuits, another important requirement is the high power capability. HTS circuits with high power handling capability have been demonstrated already, with power levels as high as 100 W [62]. In this section we provide several examples of filter circuits demonstrated in the recent past.

Superconducting microstrip bandpass filters are typically implemented using Chebyshev response due to the ease of design. Recently, quasi-elliptic function response filters have been the choice for HTS microstrip bandpass filters [70]. This is primarily due to their advantage of needing fewer resonators to achieve the same selectivity. HTS preselect filters with low insertion loss and very steep roll-off at the band edges have been demonstrated [70]. One such example is a bandpass filter covering a sub-band of 15 MHz of a receive band which ranges from 1710 MHz to 1785 MHz, demonstrated by Hong et al. [70]. Figure 26a shows an eight-pole HTS microstrip filter on MgO substrate of $0.3 \times 39 \times 22.5$ mm^3. The filter is comprised of eight microstrip meander open loop resonators. The attractive feature of this filter is the capability of cross coupling between nonadjacent resonators, essential for realizing quasi-elliptic function response. The filters were fabricated using double-sided YBCO thin films on both MgO and LAO substrates. The filter was packaged inside a test housing and cooled down to 55 K. Figure 26b shows the performance of the filter at 55 K. The filter exhibited an excellent performance with two diminishing transmission zeros near the passband edges, as shown in the figure. It is worth mentioning that such a performance was obtained after tuning.

Multiplexer filter banks used for channel separation in communication satellites require components that are as small and lightweight as possible. The filters used in multiplexers have been successfully made using waveguide or coaxial techniques, yet they are large and heavy. Microstrip filters using conventional conductors, which potentially can be used in filter banks, are small, but are often too lossy for multiplexer applications. Conventional superconductors are less lossy, but have to be cooled with liquid helium or expensive cryostats to 4–5 K. Once again, the availability of HTS thin films has brought a solution to these problems. Microstrip filters using HTS thin films are just as small and lightweight, but can display even less power loss than waveguide or coaxial counterparts [75]. Considering that filter banks are composed of many filters, the gain achieved by using a lower-loss filter is often multiplied. For example, for

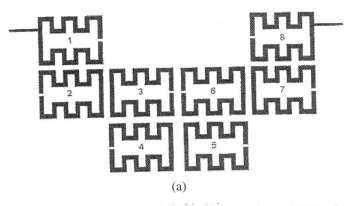

(a)

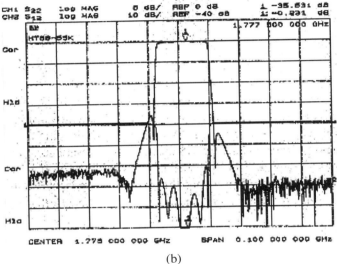

(b)

Fig. 26. (a) Schematic diagram of an eight-pole quasi-elliptic function HTS filter on a 0.3-mm MgO substrate [70]. (b) Measured performance of the HTS filter in Figure 26 at 55 K [70]. Reproduced with permission from [70], copyright 2000, IEEE.

satellite multiplexer applications, the most frequently used filter configurations are quasi-elliptical [76] dual-mode cavity and dielectric resonator filters. Although these filters provide superior performance, they are highly labor intensive to produce and possess the drawbacks of large size, large mass, and high cost. Dual-mode HTS planar filter structures have therefore been proposed as the next-generation filter solution [77]. With recently developed advanced cryogenic refrigerators [78], a number of filters and other rf components, for example a 24-channel multiplexing payload, will have less weight, smaller size, and improved performance compared to current systems.

A recent example of a C-band HTS multiplexer was given by Mansour et al. [58]. In that work, input channel filters were designed using a dual-mode resonator as shown in Figure 27a, where coupling between two modes is controlled by the spacing G and the offset L. The advantage of this resonator configuration is the ease of realizing cross coupling, which is essential for quasi-elliptic and self-equalization functions. An eight-pole dual-mode resonator filter is shown in Figure 27b. This filter consists of four dual-mode resonators. The filter layout permits easy realization of the cross-coupling coefficients M14

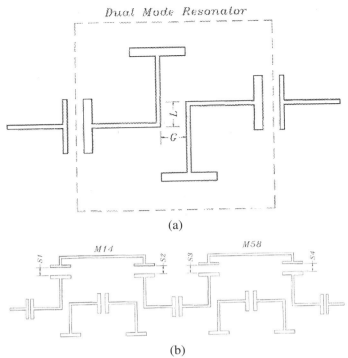

(a)

(b)

Fig. 27. Schematic diagrams of (a) dual-mode resonator, (b) eight-pole dual mode planar filter.

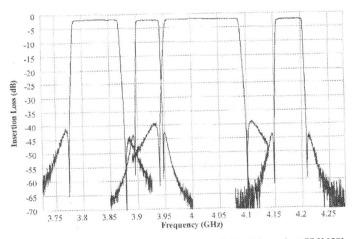

Fig. 28. Measured rf performance of the four individual channels at 77 K [58]. Reproduced with permission from [58], copyright 2000, IEEE.

and M58. The sign of the coupling (positive or negative) can be controlled by the length of the coupling elements as well as the gaps S1, S2, S3, and S4. The HTS multiplexer was designed to duplicate the requirements of the INTELSAT 8 program using 10-pole dual-mode resonator filters. Four operational channels were realized with bandwidths of 34, 41, 72, and 112 MHz using single- and double-sided TBCCO thin films on LAO substrates. Figure 28 illustrates the rf performance of the four-channel multiplexer at 77 K, with the filters packaged as part of a 60-channel multiplexer [58]. Note that the overall loss is the sum of the filter insertion loss, loss of the input/output cables, loss of the output isolator, and loss of the channel dropping

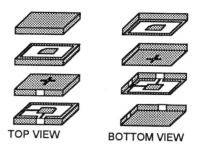

Fig. 29. Physical layout of a four-pole dual-mode multilayer filter.

Fig. 30. A stand-alone multilayer filter chip.

circulators. Comparing the 10-pole self-equalized filter with a 10-pole self-equalized dielectric resonator filter, more than 50% reduction in mass and size can be achieved using the HTS technology [58].

Among the dual-mode planar structures being considered, one advantageous configuration is the multilayer-stacked filter, since it does not require a metal enclosure [79]. More importantly, this configuration can be integrated without much difficulty onto the same substrate with other superconducting and semiconducting components. For example, C-band, four-pole bandpass filters fabricated with YBCO on MgO and LAO substrates have been reported [79, 80]. These filters are compact, with typical volumes of 0.46 cm^3. This represents less than 1% of dual-mode cavities or dielectric loaded resonators ($\sim$400 to 2000 cm^3). A schematic representation of one of these filters is shown in Figure 29. The filter consists of two dual-mode patch resonators staked together with coupling between the layered resonators achieved through a slot iris. Coupling between the orthogonal mode of each resonator is achieved through a 45° cut on the corner of the patch resonators. (There are other ways to introduce such a perturbation, as for example with a stub or a screw. However, it is up to the filter designer to select the approach that works better for a given design.) The rf signal is fed to the bottom layer through feed lines directly coupled to the resonator. To improve electrical contact between the layers, the circuit patterns are mirrored on both sides of the substrate, with dimensions adjusted to allow for slight misalignment in stacking the layers. The finalized substrates are then glued together using a thermally matched, nonconductive epoxy [81]. By doing so, the deliverable filter is a stand-alone device, as shown in Figure 30. The input and output interfaces can also be modified to a number of configurations for various integration schemes.

The experimental performance of the YBCO version of the filter measured at 77 K in liquid nitrogen (LN$_2$) are shown in Figure 31. The two finite transmission zeros are clearly shown as predicted by the prototype circuit. However, with no tuning capability on this miniaturized design, full optimization of this type of filter could be challenging.

Another application of HTS thin films is in phase shifters. Romanofsky and Bhasin [82] designed an HTS phase shifter where the phase shifter can be controlled either optically or by applying a dc current to an HTS-based switch. As shown in Figure 32, this phase shifter consists of two transmission lines of different length (i.e., the additional length for phase shift)

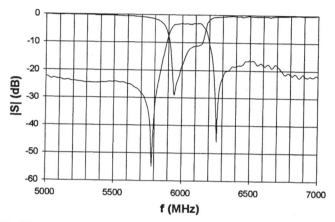

Fig. 31. Measured return loss and rejection of the YBCO/MgO multilayer filter at 77 K.

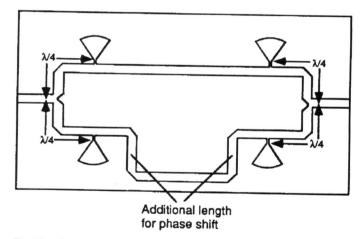

Additional length
for phase shift

Fig. 32. Schematic of an optically controlled HTS phase shifter. The λ/4 radial chokes provide virtual ground to short circuit at the main line to turn it off. A laser spot illuminating (or a dc current applied to) at the thin branch line isolates the main line from the choke to turn it on.

with a common input and output via T-junctions. The phase shifter contains HTS switches joining the transmission line with four radial stubs, located a distance of λ/4 from the T-junctions at the input and output. These switches provide the switching function for the phase shift. When the thin branch line is not illuminated (or in the case of the dc-current operated version, when no dc current is passed through it), the radial chokes provide virtual ground to short circuit at the transmission line to turn it off. On the other hand, the transmission line is "on" when the thin branch line is illuminated; this results in an increase of the resistance of the branch line isolating the line from the radial stub. In Figure 32, when the upper branch is on and the lower one is off, the signal goes through the shorter path (i.e., no phase shift). In the opposite case, the signal goes through the long path with an additional phase shift determined by the difference in electrical length between the two paths. A 360° phase shifter can be fabricated by combining several of the aforementioned sections in series. A phase shifter such as this one is characterized by low insertion loss, and its operating speed is determined by the switching mechanism employed.

3. INTRODUCTION TO FERROELECTRIC MATERIALS

Ferroelectric materials belong to a class of crystals which exhibit spontaneous polarization along one or more crystal axes. "Ferrroelectric" is a misnomer, though an understandable one. Theoretically, ferroelectric and ferromagnetic materials can be described in a similar fashion. The hysteresis loops of polarization against electric field in the case of ferroelectrics are similar to magnetization against magnetic field for ferromagnetics. Few ferroelectric materials contain any iron at all.

Ferroelectric crystals are characterized by polarization vectors that are oriented in two diametrically opposite directions (denoted by convention as $+$ and $-$) in an applied external electric field. The $+$ and $-$ polarization states in a ferroelectric crystal are due to displacements of positive metallic and negative oxygen ions in opposite directions. This displacement automatically reduces the symmetry of the crystal from cubic to tetragonal. Thermodynamically stable, these states can be switched by applying an external electric field known as the coercive field E_c. The polarization versus electric field characteristics for a ferroelectric bulk material are shown in Figure 33, identifying the important parameters of the material. P_r is the remanent polarization, the spontaneous polarization that remains aligned with the previous applied field. E_c is the magnitude of the reverse electric field required to decrease the net polarization to zero. The net polarization in the ferroelectric material is due to (1) spontaneous polarization, (2) electronic polarization, and (3) ionic polarization [83]. Such hysteretic behavior of the ferroelectric material is useful, for example, in nonvolatile memories, as one can use the positive polarized state for storing a "logic 1" and the negative polarized state for storing a "logic zero." Also, the slope of the polarization versus the electric field curve gives the relative dielectric constant of the material, indicating the nonlinear relationship of the relative dielectric constant with the dc electric field.

The temperature dependence of the polarization characteristics of a typical ferroelectric material is an interesting one. Some ferroelectric materials, for example, lead zirconium titanate ($Pb(Zr_xTi_{1-x})O_3$ or PZT), are transformed from a fer-

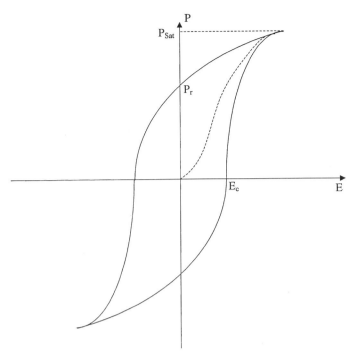

Fig. 33. Polarization versus electric field for a ferroelectric material showing the hysteretic nature.

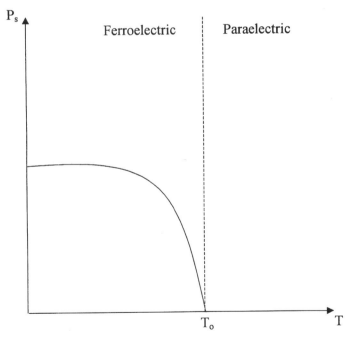

Fig. 34. Spontaneous polarization versus temperature showing the two distinct phases of a ferroelectric material.

roelectric (low-temperature) phase, in which there exists spontaneous polarization, to a non-ferroelectric (high-temperature) phase called the paraelectric phase, in which there is no spontaneous polarization, at the "Curie temperature" (about 670 K for PZT). Figure 34 shows the typical temperature dependence of polarization for ferroelectric materials. The temperature depen-

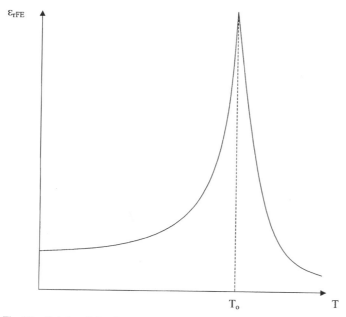

Fig. 35. Relative dielectric constant versus temperature for a ferroelectric material showing the anomalous behavior near T_c.

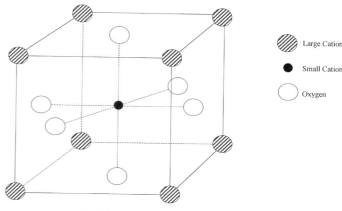

Fig. 36. Unit cell of a perovskite structure.

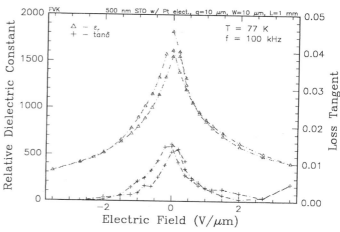

Fig. 37. Relative dielectric constant versus electric field for a thin-film STO at 77 K.

dence of polarization also leads to temperature dependence of the relative dielectric constant of the material. The relative dielectric constant of a ferroelectric material (ε_{rFE}), when plotted against temperature, exhibits Curie–Weiss behavior as shown in Figure 35. Some ferroelectric materials, for example, barium magnesium fluoride ($BaMgF_4$ or BMF), do not exhibit such a phase transition, even up to their melting points. The temperature dependence of the ε_{rFE} exhibits Curie–Weiss behavior, proportional to $1/(T - T_c)$. When one plots the ε_{rFE} versus T, ε_{rFE} rises anomalously near the T_c, as shown in Figure 35.

3.1. Structural Compatibility of Ferroelectrics with High-Temperature Superconducting Materials

Some of the ferroelectrics and the HTS materials fall under the broad category of "perovskites," whose unit cell is in the general form of ABO_3, where A is a divalent ion, B is a trivalent ion, and O is the oxygen atom, as shown in Figure 36. Ferroelectric materials can have unit cell structures with different degrees of complexity. In PZT, when a negative or positive voltage is applied on opposite faces of a crystal, the small Ti^{4+} or Zr^{4+} ions in the center of the cubic lattice are displaced up or down, while the O^{2-} ions move down or up. The displacement of the positive and negative ions results in the polarization that characterizes ferroelectric materials [83].

3.2. Electrical and Microwave Properties of Ferroelectric Materials

Since the polarization is nonlinearly dependent upon electric field, the ε_{rFE} is also nonlinearly dependent upon electric field. Currently, ferroelectric materials are used in bulk, thick-film,

and thin-film forms. The properties of a ferroelectric material in these various forms are quite different, depending on several parameters. For low-temperature electronic applications, strontium titanate ($SrTiO_3$, henceforth referred to as STO) is the most useful ferroelectric, especially for integration with HTS in the thin-film form. STO is actually thought of as an incipient ferroelectric because in its bulk form it does not exhibit a T_c. STO typically has an order of magnitude higher dielectric constant value as well as an order of magnitude lower loss tangent in the bulk form compared to the thin films of the same material. The T_c and the dielectric properties of the thin films depend upon the processing methods, processing conditions, thickness of the films, morphology, and residual stresses in the films due to lattice mismatches with the substrates. Annealed films have been shown to reduce residual stress on STO thin films and to improve the ε_r and the loss tangent of the films on substrates such as LAO and MgO. Figure 37 shows the electric field dependence of the ε_r characteristics for a 500-nm-thick STO ferroelectric thin-film material under the influence of an external field measured using capacitance versus voltage (C–V) technique on a patterned interdigitated capacitor at low rf frequen-

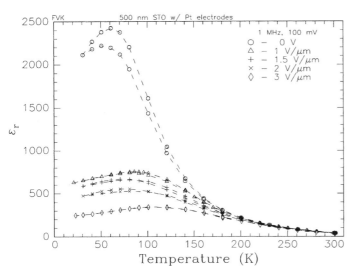

Fig. 38.   Relative dielectric constant versus temperature for a thin-film STO on LaAlO$_3$ for various electric fields.

cies. As one can see, the ε_{rFE} is reduced from approximately 2000 at zero field to approximately 300 at a high dc electric field in a nonlinear fashion. As one can see, there is a large change in ε_{rFE} below the electric field of 2 V/μm. Also, the loss tangent varies with the applied field, as shown in the same figure. The loss tangent is below 0.01 at moderate electric fields.

One of the important criteria for the use of ferroelectric thin films in tunable circuits is the large dielectric tunability of the ferroelectric thin film with low additional microwave dielectric losses due to the insertion of the ferroelectric thin films. Dielectric tunability is defined as the ($\varepsilon_{r\text{ at zero bias}} - \varepsilon_{r\text{ at large bias}})/\varepsilon_{r\text{ at zero bias}}$. Dielectric tunability as high as 90% is attainable in STO thin films at moderate loss-tangent values (typical values between 0.005–0.01 at GHz frequencies) [84, 85]. Figure 38 shows the ε_r versus T for increasing electric fields for a 500-nm thin-film STO characterized using an interdigitated capacitor structure. This figure is an important one, as it illustrates the main principle behind the use of STO ferroelectrics for HTS-based tunable circuits. If one maintains the sample at a constant temperature of 77 K, the ε_{rSTO} can be changed from a high value of approximately 2700 at zero bias to a lower value of 350 at a high bias field. If one can maintain the loss tangent reasonably low in the STO thin films, then the combination of HTS and the STO ferroelectric thin films could result in low-loss electrically tunable components. Recent progress in processing of high-quality STO thin films has resulted in large dielectric tunability with low loss tangents up to GHz frequencies [85].

3.3. High-Temperature Superconductor Ferroelectric Tunable Components

In Section 2 of the chapter, we have shown several examples of HTS-based planar microwave components (e.g., filters, resonators) demonstrated to date. The use of HTS thin films in

place of normal conductors (e.g., gold, copper) has reduced conductor losses and consequently improved the performance of the circuits. Therefore, design, fabrication, and optimization of the HTS/ferroelectric hybrid circuits can be of great interest for the development of low-loss and tunable microwave components and systems for wireless and satellite communications. Recent developments in ferroelectric tunable components are summarized in Table IV [86–93]. In this section, we summarize important results from our investigations in recent years.

Ferroelectric tunable microwave components are currently being investigated by incorporating a thin-film ferroelectric in microstrip as well as CPW transmission lines. In this section, we introduce the tunable microstrip and CPW structures, and present some theoretical and experimental results, which will allow for easier selection of geometric parameters for ferroelectric tunable microstrip and CPW components.

3.3.1. Tunable Microstrip Structure

Ferroelectric thin films can be easily incorporated into a microstrip structure. The cross section of the modified two-layered microstrip structure used in various ferroelectric tunable components is shown in Figure 39. The modified microstrip structure consists of a dielectric substrate (e.g., LAO or MgO, typically 254 to 500 μm thick), a ferroelectric thin-film layer (thickness "t" varying between 300 and 2000 nm for various applications), a gold or YBCO thin film (2 μm thick or 300–600 nm thick, respectively) for the top conductor, and a 2-μm-thick gold ground plane. The dielectric properties of the ferroelectric thin film, and the thickness of the ferroelectric film, play a fundamental role in the frequency or phase tunability and the overall insertion loss of the circuit. The geometry of the multilayered microstrip can be optimized by performing electromagnetic analysis with simulation packages such as Sonnet em$^{\circledR}$ [94] and the Zeland's IE3D software [95], to obtain correlations between parameters such as the characteristic impedance (Z_0), the effective dielectric constant (ε_{eff}), and attenuation (α), as a function of frequency, temperature, linewidth-to-substrate height ratio (W/H), ferroelectric film thickness (t), and the relative dielectric constant and loss tangent of the ferroelectric thin film (ε_r, and tan δ, respectively) [96]. As an example of this process, let us consider a microstrip line on a LaAlO$_3$ (LAO) substrate 254 μm thick. The first step in a microstrip circuit design is to determine the W/H ratio required for a chosen characteristic impedance. Figure 40 shows Z_0 versus W/H characteristics applicable to K-band microstrip circuit design on LAO substrates, in the absence of a ferroelectric tuning layer. For $Z_0 = 25, 50, 75$ Ω, the required W/H values are 1.63, 0.345, and 0.08, respectively. When a ferroelectric thin-film layer is inserted, the effective dielectric constant (ε_{eff}) and the characteristic impedance (Z_0) of the microstrip are tunable between a minimum and a maximum value due to the bias-dependent change in the relative dielectric constant of the ferroelectric thin film (ε_{rFE}). The tunability in Z_0 and ε_{eff} is higher for the smaller W/H, that is, higher characteristic impedance. Percentage change in Z_0 and ε_{eff} is higher

Table IV. Summary of Various Works in Ferroelectric Tunable Components

No.	Authors	Topic	Reference	Comments
1	Sengupta et al.	BSTO and non-ferroelectric oxide composites for use in phased array antennas and other electronic devices	[86]	Bulk ferroelectrics for phase-shifting applications
2	Galt et al.	Ferroelectric thin-film characterization using HTS resonators	[87]	Microstrip resonator incorporating a ferroelectric capacitor
3	Findikoglu et al.	CPW tunable filter	[88]	YBCO/STO/LAO CPW bandpass filter; $f_c = 2.5$ GHz; >15% tunability
4	Abbas et al.	A distributed ferroelectric superconducting transmission line phase shifter	[89]	Analysis of ferroelectric tunable multilayer structures
5	De Flaviis and Alexopoulus	Low-loss ferroelectric based for high power antenna scan beam system	[90]	A phase shift larger than 160° with insertion loss below 3 dB and power consumption below 1 mW is achieved
6	Kozyrev et al.	Ferroelectric films: nonlinear properties and applications in microwave devices	[91]	The speed of resonator tuning the microwave loss variations are measured as 1.9 GHz phase-shifter and 20 GHz tunable filter operating at $T = 300$ K
7	Nagra and York	Distributed analog phase shifters with low insertion loss	[92]	Optimally designed circuits exhibit 0–360° phase shift at 20 GHz with a maximum insertion loss of 4.2 dB
8	Oates and Dionne	Magnetically tunable superconducting resonators	[93]	3-pole 1% bandwidth filter with 10-GHz center frequency 1 dB insertion loss; tuning range 13% obtained

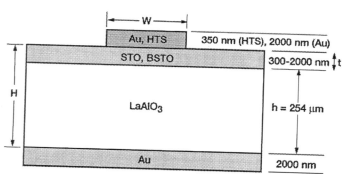

Fig. 39. Cross section of a microstrip transmission line (Au or HTS) fabricated on a ferroelectric thin film coated (e.g., STO or BSTO) on a dielectric substrate (e.g., LaAlO₃).

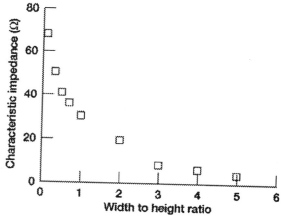

Fig. 40. Characteristic impedance (Z_0) versus the geometric factor (W/H) for a microstrip line on 10 mil LAO substrate.

for the 50-Ω compared to the 25-Ω microstrip line; however, it is the highest for the 75-Ω microstrip line. Table V summarizes these results for 20 GHz. This trend prevails also at frequencies of 10 and 15 GHz. This is an important result, as one can choose higher characteristic impedance to obtain larger frequency tunability since it is directly related to the percentage change in Z_0 and ε_{eff}.

To study the effect of ferroelectric film thickness on the characteristic impedance of the microstrip line, the microstrip was modeled for several values of ε_{rFE} for three different thicknesses. Figure 41 shows the variation in Z_0 for the three different thicknesses. Higher thickness of the ferroelectric film leads to higher tunability in Z_0, as shown in the figure. For a given

value of tan δ, the attenuation becomes larger at higher frequencies, and for a given frequency, it becomes larger as the value of tan δ increases. Note also that for a given frequency and tan δ value, the attenuation increases with film thickness, as shown in Figure 42. At higher frequencies, because of the skin depth effect, more rf field is concentrated in the ferroelectric film and less is concentrated in the dielectric substrate, resulting in larger insertion loss. As the value of the ε_{rFE} increases, the attenuation also increases. This is a consequence of mismatches resulting from the decrease in Z_0 and the increase in ε_{eff} with increasing ε_{rFE}. Note also that as the dielectric losses of the ferroelectric film worsen (i.e., tan $\delta = 0.1$), the attenuation increases dra-

Table V. Modeled Percentage of Change in Z_0 and ε_{eff} for 25, 50, and 75-Ω Microstrip Lines versus ε_{rFE} at 20 GHz for Three Different Thicknesses of Ferroelectric Films; $\tan\delta = 0.01$

Thickness of FE film, nm	25 Ω		50 Ω		75 Ω	
	Z_0	ε_{eff}	Z_0	ε_{eff}	Z_0	ε_{eff}
300	5	10	11	26	17	46
900	10	23	19	54	26	82
2000	18	40	27	87	33	124

% change $Z_0 = [((Z_0 \text{ for } \varepsilon_{\text{rFE}} = 3000) - (Z_0 \text{ for } \varepsilon_{\text{rFE}} = 300))/ (Z_0 \text{ for } \varepsilon_{\text{rFE}} = 300)] \times 100$.

% change $\varepsilon_{\text{eff}} = [((\varepsilon_{\text{eff}} \text{ for } \varepsilon_{\text{rFE}} = 3000) - (\varepsilon_{\text{eff}} \text{ for } \varepsilon_{\text{rFE}} = 300))/ (\varepsilon_{\text{eff}} \text{ for } \varepsilon_{\text{rFE}} = 300)] \times 100$.

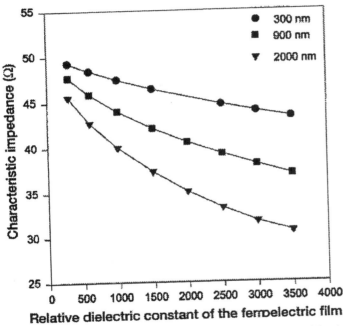

Fig. 41. Modeled characteristic impedance (Z_0) versus the relative dielectric constant of the ferroelectric thin film (ε_{rFE}) on LaAlO$_3$ for various thicknesses of the ferroelectric thin film.

matically, with respect to those corresponding to $\tan\delta = 0.01$ and 0.005.

These modeled correlations could be used to analyze the experimental performance of the ferroelectric tunable microstrip circuits. For example, let us consider data from ring resonators designed for operation at the third-order resonance and fabricated with Au/STO/LAO thin-film multilayered structures [97]. These resonances satisfy the condition $\pi d = 3\lambda_g$ (d is the mean diameter of the ring and λ_g is the guided wavelength). Knowing the resonant frequency (f_3) of the resonators, one can determine the effective dielectric constant (ε_{eff}) of the resonant structure; that is,

$$\varepsilon_{\text{eff}} = \left(\frac{c}{f_3 d}\frac{3}{\pi}\right)^2 = 0.91189\left[\frac{c}{f_3 d}\right]^2 \qquad (22)$$

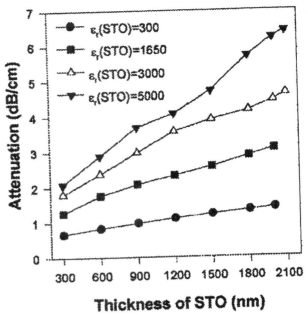

Fig. 42. Attenuation versus thickness of the ferroelectric thin film for different values of the ε_{rFE}.

We fabricated two types of ring resonators with characteristic impedance (Z_0) of 25 and 50 Ω. The mean diameters (d) of these rings are $d = 3388$ μm and 3670 μm for the 25 and 50 Ω, respectively. Therefore, the ε_{eff} for the two types of rings are given by

$$\varepsilon_{\text{eff}}(25\ \Omega) = 7.15 \times 10^{21}/f_3^2 \qquad \text{and}$$
$$\varepsilon_{\text{eff}}(50\ \Omega) = 6.09 \times 10^{21}/f_3^2 \qquad (23)$$

Table VI shows experimental data for several ring resonators, including the corresponding range of variation for ε_{eff}.

Consider first the 25-Ω ring resonators at room temperature. The experimental and simulated values of ε_{eff} for the circuits with no ferroelectric film agree within 0.4%. For rings 3, 5, and 6, the experimental room temperature ε_{eff} values correspond to simulated values using $\varepsilon_{\text{rSTO}}$ values of 275 to 325. Ring 4 has a resonant frequency corresponding to an $\varepsilon_{\text{rSTO}} = 150$. As the rings are cooled to 77 K and below, the dielectric constant of the LAO drops slightly, causing a drop of 0.33 in ε_{eff} for the bare LAO ring. After adjusting for this LAO change, a comparison of experimental data and simulations for the 2-μm-thick STO rings indicates a range of $\varepsilon_{\text{rSTO}}$ from 2900 to 1540 for biases of 10 V to 458 V for sample 5. For sample 6, the measured range corresponds to $\varepsilon_{\text{rSTOs}}$ of 3300 to 1700 for biases from 10 to 491 V. The initial 10 V applied to the sample causes the ring to go from overcoupled to undercoupled, greatly sharpening and shifting the resonance. The resonant frequencies obtained above 10 V are believed to be more accurate for this purpose. The upper dielectric constant values are close to those measured at 1 MHz using interdigital capacitors on similar samples. These rings are not expected to be fully tuned to $\varepsilon_{\text{rSTO}} = 300$ under the voltages given here. The relevant electric

Table VI. Frequency and Effective Dielectric Constant Tuning Range of
Au/SrTiO$_3$/LaAlO$_3$ Ring Resonators versus Temperature and Bias

Sample No.	Design and STO thickness	f_3 at room temperature (GHz)	T (K)	Maximum V_R (V)	f_3 range (GHz)	ε_{eff} range
1	Au 50-Ω ring, 300 nm	19.152	40	300	14.6–16.2	28.57–23.21
1	Au 50-Ω ring, 300 nm	19.152	77	350	15.1–16.8	26.71–21.57
2	Au 25-Ω ring, no STO	20.275	40	N/A	20.48	17.05
2	An 25-Ω ring, no STO	20.275	77	N/A	20.47	17.06
3	Au 25-Ω ring, 300 nm	19.350	40	450	19.0–20.0	19.81–17.88
3	An 25-Ω ring, 300 nm	19.350	77	350	19.3–19.9	19.20–18.06
4	Au 25-Ω ring, 1 μm	19.762	52	250	17.0–18.1	24.74–21.8
5	Au 25-Ω ring, 2 μm	19.435	77	458	15.75–17.64	28.82–22.98
6	Au 25-Ω ring, 2 μm, partially etched	19.420	77	491	15.27–17.26	30.63–24.00

field is that between the ring and the ground plane of the ring resonator. For samples 5 and 6, the maximum electric fields are 1.93 and 1.80×10^4 V/cm. Using Au/STO/LAO interdigital capacitors at 1 MHz, it has been found that, upon applying such a field, $\varepsilon_{\text{rSTO}}$ reach values near 1300 at 77 K.

For sample 3 of Table VI, which has a thinner STO film, ε_{eff} was computed to be 17.88 under a ring bias of 450 V and at 40 K. Also, the experimental value obtained for the effective dielectric constant of the same ring measured at 77 K under a bias of 350 V is 18.06, which is within 5.7% of the modeled value for $\varepsilon_{\text{rSTO}} = 1000$, which is 19.092 at 20 GHz. Consider now sample 4 of Table VI, which corresponds to a 1-μm STO thin film; it shows $\varepsilon_{\text{eff}} = 21.8$ under a 250-V dc bias and at 52 K. For this structure, the total electric field between the ring and the ground plane for the above bias is approximately 10 kV/cm. Using the same capacitor technique, the authors have found that, upon applying such a field, $\varepsilon_{\text{rSTO}}$ reach values near 1700 at 50 K. Thus, if one compares the modeled data for ε_{eff} corresponding to $\varepsilon_{\text{rSTO}} = 1500$ at 20 GHz (which is the closest modeled value available) with the experimental data, it can be seen that the modeled value of $\varepsilon_{\text{eff}} = 21.327$ differs from 21.8 by 2.2%, which represents an acceptable correspondence between the modeled and the experimental data. Note that in general, the modeled data agree very well with the experimental data. The major reasons for discrepancies arise from the fact that experimentally only the STO almost directly under the ring is influenced by the electric field, in contrast to the simulations where the STO is tuned everywhere across the sample. Also, there is some variation in $\varepsilon_{\text{rSTO}}$ between films, without any simple way to measure it at these frequencies. Discrepancies between the actual and expected thickness of the films, due primarily to fabrication tolerances, could also exist. Nevertheless, it is evident from Table VI that the films presented here are very tunable and of high quality. The good agreement between the modeled and experimental data discussed above demonstrates the usefulness of the modeled plot in selecting geometrical and film thickness parameters for a targeted application.

Considering the 50-Ω ring (sample 1 of Table VI), it shows an effective dielectric constant of 23.21 at 40 K and 16 GHz under a dc bias of 300 V. This bias translates into a field of 12 kV/cm. At 40 K, this bias results in an experimental $\varepsilon_{\text{rSTO}} = 1700$. Thus, the modeled ε_{eff} corresponding to this dielectric constant $\varepsilon_{\text{rSTO}} = 1500$ (closest modeled data available to 1700) at 15 GHz is 18.031, which represents a 22% discrepancy. However, by using the part of Eq. (24) corresponding to 50 Ω and the $3\lambda_g$ frequency of resonance for this ring resonator at room temperature (19.152 GHz), one obtains $\varepsilon_{\text{eff}} = 16.6$. Reported values for $\varepsilon_{\text{rSTO}}$ at room temperature fall between 275 and 325. Thus, when comparing the experimental value with the modeled effective dielectric constant of $\varepsilon_{\text{eff}} = 16.356$ corresponding to $\varepsilon_{\text{rSTO}} = 300$, they agree within 1.5%. It is possible that the narrower lines have a greater discrepancy between experiment and simulation because of the larger extent of the microwave fringing fields away from the lines in regions where the dc bias is not affecting the film as strongly in actuality as it does in the model.

Figure 43 shows the magnitude of the transmission scattering parameter (S_{21}) for 1-cm-long, 50-Ω Au/STO/LAO microstrip line, with a 2-μm STO thin film. The data were obtained at 77 K in the 10 to 20 GHz frequency range. Observe that at no bias and at 16.5 GHz, the insertion loss $I_L = 4.68$ dB. These experimental data are not de-embedded, meaning that any contribution from the SMA launchers used for the measurement to the overall insertion loss has not been subtracted from the data. Typically, the launcher contribution is not higher than 0.2 dB. Since no bias is applied and we are operating at 77 K, it is reasonable to consider $\varepsilon_{\text{rSTO}}$ to be somewhere in the range of 3000 to 5000. Figure 44 shows modeled data at 15 GHz for a Au/STO/LAO line with a 2-μm STO film for $\varepsilon_{\text{rSTO}} = 5000$ and tan $\delta = 0.01$, resulting in an attenuation of 2.67 dB, which is within a factor of 2 of the measured result. When modeling for this line is done for tan $\delta = 0.05$ and 0.1, the attenuation becomes 4.21 and 6.22 dB, respectively. The modeled data corresponding to tan $\delta = 0.05$ is within 11% of the values measured experimentally. Other sources, such as conductor loss

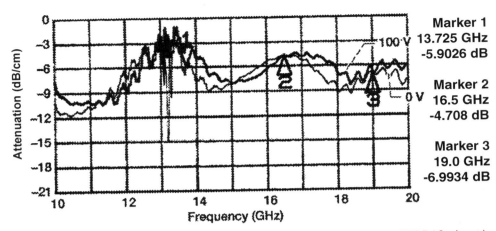

Marker 1
13.725 GHz
-5.9026 dB

Marker 2
16.5 GHz
-4.708 dB

Marker 3
19.0 GHz
-6.9934 dB

Fig. 43. Experimental data for the attenuation versus frequency for a 1-cm-long, 50-Ω Au/STO/LAO microstrip line with a 2-μm-thick STO film at 77 K.

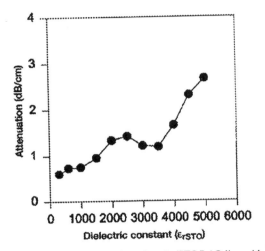

Fig. 44. Modeled data for attenuation of an Au/STO/LAO line with a 2-μm-thick STO for $\varepsilon_{rFE} = 5000$ and $\tan \delta = 0.01$, at a frequency of 15 GHz.

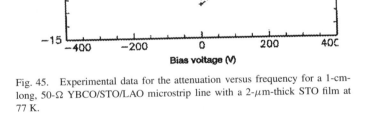

Fig. 45. Experimental data for the attenuation versus frequency for a 1-cm-long, 50-Ω YBCO/STO/LAO microstrip line with a 2-μm-thick STO film at 77 K.

and mismatches, may also contribute to the observed discrepancy between experimental and modeled data. Figure 45 shows a YBCO/STO (2 μm thick)/LAO microstrip line at 77 K. Note that at 15 GHz and 400 V dc, the attenuation is approximately 4 dB, which is also consistent with $\tan \delta = 0.05$. These examples also show that the modeled data are helpful in assessing the expected losses of a thin-film ferroelectric-based tunable circuit.

3.3.2. Tunable Coplanar Waveguide Structure

Although microstrip circuits are the most common transmission line component for microwave frequencies, the ground plane is difficult to access for shunt connections necessary for active devices, when used in MMICs. Coplanar waveguide (CPW) is an attractive alternative, especially due to the ease of monolithic integration, as the ground conductor runs adjacent to the conductor strip. As an example, let us consider the modeling of tunable components such as a CPW transmission line and a

CPW linear resonator performed using a modified CPW structure shown in Figure 46. The structure consists of a dielectric substrate (typically lanthanum aluminate (LAO) 254 μm thick), a ferroelectric thin-film layer (thickness t modeled for $t = 300$ to 2000 nm), a 2-μm-thick gold thin film for the conductor (center line), and ground lines (adjacent parallel lines). The important geometrical dimensions are indicated in the figure. Assuming that the ferroelectric thin film is epitaxially grown on the dielectric substrate, the critical design parameters for the tunable circuits, that is, the characteristic impedance Z_0 and the effective dielectric constant ε_{eff} will be a function of the electric field-dependent ε_{rFE} and $\tan \delta$ of the ferroelectric thin film. Important geometrical features controlling the above parameters are the width of the center conductor (W), spacing between the ground lines (S), thickness of the ferroelectric thin film (t), and the thickness of the substrate (h). The dielectric properties of the ferroelectric thin film, and the thickness of the ferroelectric film, are expected to determine the overall insertion loss of

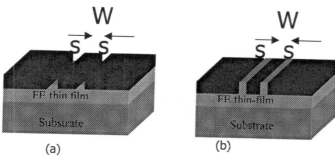

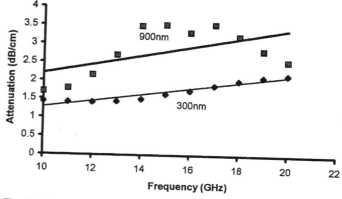

Fig. 46. The cross-sectional view of the ferroelectric tunable coplanar waveguide (CPW) structures. (a) CPW with no ferroelectric thin film between the ground line and center conductor. (b) CPW with ferroelectric thin film between the ground line and center conductor. The conductor-backed CPW will add a gold metallic ground plane to the structure.

Fig. 48. Attenuation versus frequency for two different thicknesses of a BSTO ferroelectric thin film on a LaAlO$_3$ for a CPW line; $\varepsilon_{rFE} = 300$.

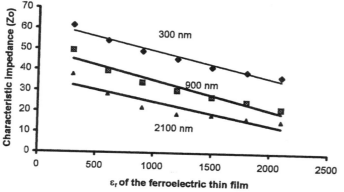

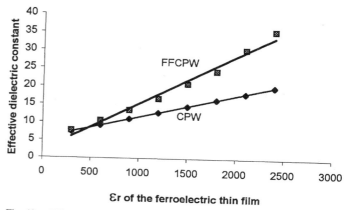

Fig. 47. Modeled characteristic impedance (Z_0) of the CPW line versus the relative dielectric constant of the ferroelectric thin film (ε_{rFE}) for ferroelectric thin films of various thicknesses on a LaAlO$_3$ substrate.

Fig. 49. Effective dielectric constant versus ε_{rFE} for CPW and FFCPW structures.

the circuit as well. The CPW and conductor backed coplanar waveguide (CBCPW) transmission lines of 1 cm length were simulated using Sonnet em® electromagnetic simulation software, to obtain the relationships between the critical design parameters and the dielectric properties of the ferroelectric thin film [98]. The transmission lines were designed for characteristic impedances of 50 and 75 Ω in the absence of the ferroelectric layer.

Figure 47 shows the characteristic impedance versus ε_{rBSTO} characteristics for a 75-Ω CPW on LaAlO$_3$ substrate, for various thicknesses of the ferroelectric thin film. For a 300-nm thickness of the ferroelectric film, one can achieve a Z_0 from 40 Ω to 60 Ω when one reduces the ε_{rFE} from 2100 down to 300. This thickness of the ferroelectric film is ideal for designs centered around 50 Ω. As in microstrip structure, higher ferroelectric thickness leads to improved tunability in the ε_{eff}, at the expense of increasing attenuation, as one would expect. Figure 48 shows the effect of thickness for 300-nm and 900-nm ferroelectric thin film on the attenuation of the CPW line for one fixed value of ε_{rBSTO} of 300. The attenuation in the frequency range of 10 to 20 GHz is almost double for the 900-nm thickness compared to the 300-nm case. In general, the CPW and CBCPW yield very similar attenuation and dispersion charac-

teristics for ε_{rFE} values below 1500. At higher ε_{rFE} values, the dispersion is higher in the CBCPW lines.

Another possibility for improved tunability is to use the ferroelectric filled CPW (FFCPW) structure, shown in Figure 46b. For the same ferroelectric thickness, the dielectric tunability of the FFCPW is predicted to be higher, due to the larger filling factors possible with this structure, as compared to CPW structure with an airgap between the center conductor and the ground line. The ε_{eff} versus ε_{rBSTO} for a CPW and a FFCPW are plotted for a 50-Ω design, as shown in Figure 49. The FFCPW is highly tunable, as one could compare the two in Figure 49. The large tunability in ε_{eff} in this structure is primarily due to the improved filling factor, resulting in higher ε_{eff}. This result is significant, as one could apply a smaller electric field for a required tunability if one uses the FFCPW line instead of CPW.

3.4. Tunable Ferroelectric Circuit Prototypes

The fabrication and testing of the tunable microwave devices have been discussed elsewhere [99, 100]. The HTS and ferroelectric films considered here have been deposited by laser ablation on LAO and MgO substrates and have thicknesses within the 0.3- to 2.0-μm range. However, other techniques such as magnetron sputtering and metallorganic chemical vapor deposition have been used to grow ferroelectric thin films [101, 102]. The microwave testing of the devices could be performed in air at low bias, or under vacuum to avoid dielectric breakdown of the air if high dc bias is required to induce tunability. Since some ferroelectrics are more tunable at low temperatures (e.g., STO), measurements are also performed with the samples mounted on the cold finger of a helium gas closed-cycle refrigerator via a custom-made test fixture. The microwave characterization is generally done using a network analyzer (e.g., an HP-8510-C Automatic Network Analyzer) to measure the reflection and transmission scattering parameters (S_{11} and S_{21}, respectively). When dealing with high dc bias (e.g., 300 to 500 V), it is necessary to develop experimental approaches to protect the instrumentation. Application of large dc voltages to the microstrip line-based tunable thin-film ferroelectric components has been achieved using custom-made bias tees designed to withstand the application of up to 700 volts dc at K-band frequencies [103].

3.4.1. Tunable Ring Resonators

Ring resonators are very useful microwave components for material characterization, as well as components in filters, and local oscillators. The versatility of these components has increased largely by integrating ferroelectric thin films for tunability. For example, tunable Au/STO/LAO and YBCO/STO/LAO conductor/ferroelectric/dielectric (CFD) ring resonators have been developed for Ku- and K-band frequencies [97]. Figure 50 shows the schematic diagram of these structures. The side-coupled resonators are used for evaluating the tunability. Under appropriate biasing schemes, these "band-stop" resonators exhibit sharp resonances with unloaded $Q (f_0/\Delta f_{3\,dB})$ as high as 15,000. The resonant frequency can be tuned beyond 1 GHz (e.g., 16.6 GHz at zero bias to nearly 17.8 GHz at a dc voltage of 400 V) without degrading the sharpness of the resonance (see Fig. 51). The effect of the applied dc electric field on the parameters such as the insertion loss, center frequency, the return loss, and the unloaded Q of the microstrip resonators has been studied experimentally [97]. The various biasing schemes studied are also shown in Figure 50. Changing the dc voltage on the ring, V_R, mainly changes the ε_{eff} of the ring and its resonant frequency. The voltage difference, $V_R - V_L$, controls the coupling of the ring to the line, allowing one to tune the resonator from over-, through critical, to undercoupled. Maximum coupling to the ring is obtained using the biasing scheme A ($V_R - V_L = 0$), which results in higher frequency tunability, but overcoupled resonances having low Q values. Biasing scheme B allows one to tune for a critical coupling and a sharp resonance, but only in a narrow range of frequency. By adjusting both V_R and V_L simultaneously (bias scheme C), one can optimize the coupling and sharpen the third-order resonance while maintaining large frequency tunability. The magnitude of S_{21} for this differential biasing scheme is shown in Figure 51. The data shown in Figure 51 exhibit sharp band-stop characteristics with unloaded Q values as high as 15,000. Figure 52

shows data from a ring resonator of the same dimensions using a YBCO/STO/LAO structure. The resonators of Figures 51 and 52 also differ in that the Au resonator had the STO etched in the region beyond 2 mm from the ring (Fig. 50b), while the YBCO resonator was not etched.

These results must be discussed in the context of the performance of dielectric resonator oscillators (DROs), which represent current state of the art at high frequencies. Reported values of Q_0 for DROs have been as high as 50,000 at 10 GHz [104]. However, their manufacturing cost, lack of electronic tunability, and nonplanar geometry limit their versatility for insertion in frequency-agile communication systems such as tunable local oscillators, broadband band-stop filters, and "notch" filters for wireless communications [105].

3.4.2. Tunable Local Oscillators

A key parameter of modern digital communication systems is the phase noise. Phase noise is introduced into a system primarily by the local oscillator (LO) used in a receiver. Currently, there are several ways to achieve stabilization of a LO. However, all of these approaches have intrinsic limitations. Crystal stabilized oscillators represent the state of the art at low frequencies. However, they are generally restricted to frequencies below several gigahertz. DROs are commercially available up to at least 20 GHz, are not appropriate for to electronic tuning or frequency locking. Further, their fabrication must be done independently of the oscillator circuit, and their three-dimensional geometry is not conducive to the fast and high-volume production of the optimized circuit. Thin-film ferroelectric technology can be used as a viable alternative to the aforementioned shortcomings. A tunable local oscillator using a 0.25-μm gate length pseudomorphic high electron mobility transistor (PHEMT) as the active component and a novel ferroelectric tunable microwave ring resonator with a center frequency of 16.7 GHz at zero bias as the passive component has been demonstrated by Romanofsky et al. (see Fig. 53) [106].

The circuit consists of a 25-Ω ring resonator side-coupled using a microstrip feed line, and the PHEMT (Avantek/HP 0.25 mm ATF-35076) portion of the voltage-controlled oscillator (VCO) with appropriate impedance matching networks constructed on a 0.25-mm-thick alumina. The ferroelectric tunable ring resonator is similar to the ones discussed in the previous section. At 77 K, the PHEMT exhibits a gain of approximately 7 dB at 16.5 GHz. An inductor is inserted between the source and ground to make the device unstable, by using a section of a 50-Ω microstrip line 1.08 mm long. An iterative computer routine is used for choosing the source impedance (Z_s) to maximize the negative resistance of the PHEMT while preserving the loop gain. With a Z_s of $j35\,\Omega$, a stability factor (K) of -0.499 is obtained. During operation, the ring is dc biased at a bias in the range between 0 and 250 V with the bias on the microstrip kept at 0 V. With this biasing scheme, the gate and the ring could be dc coupled. Because the tunability of STO increases at low temperatures, an operating temperature of 43 K was chosen to maximize the electric field-induced tunability of

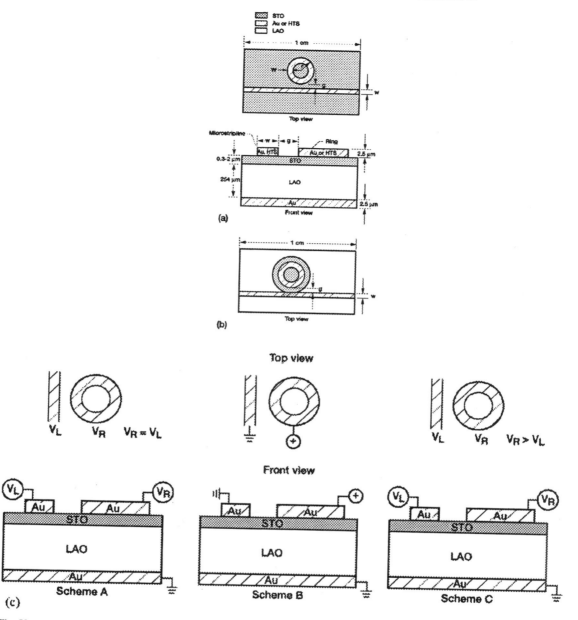

Fig. 50. (a) Microstrip side coupled 25-Ω ring resonator. $W = 406$ μm, $w = 89$ μm, $r = 1694$ μm, and $g = 25$ μm. (b) Ferroelectric in the circuit is limited only to the areas underneath the ring. (c) Various biasing schemes.

STO. Both the ring resonator and the PHEMT circuit are packaged on a brass fixture with conductive epoxy for testing inside a He gas closed-cycle cryogenic system. By applying 38 V dc to the ring, the oscillator frequency could be tuned by about 100 MHz around the center frequency (see Fig. 54). The figure shows the broadband and the narrowband measurements done at 43 K with the drain bias at 2.1 V and the gate bias of −0.2 V. By increasing the ring voltage to 250 V, the tuning range could be extended over 500 MHz. The merits of this VCO are its high performance potential, small size, simplicity of implementation, and its potential for low-cost and high-volume production.

3.4.3. Tunable Filters

Another important component that can benefit from the ferroelectric thin-film technology is the preselect filter in the receiver front-end. YBa$_2$Cu$_3$O$_{7-\delta}$ (YBCO)/STO HTS/ferroelectric thin-film-based K-band tunable microstrip bandpass filters on LaAlO$_3$ (LAO) dielectric substrates have been demonstrated [107]. The layout of the filter is shown in Figure 55. The two-pole filter was designed for a center frequency of 19 GHz and a 4% bandwidth, using the parallel coupled half-wavelength resonators. The two-pole bandpass filter was designed using identical quarter-wavelength coupled microstrip resonator sections designed for 50 Ω characteristic impedance.

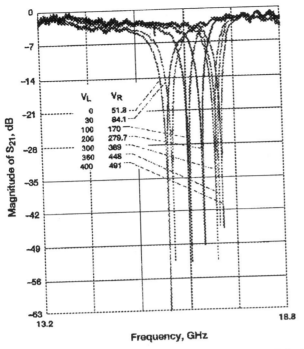

Fig. 51. Effect of dc bias on the 3λ ring resonant frequency and Q of the Au (2 μm)/STO (2 μm)/LAO (254 μm) ring resonator at 77 K and for the ring and line dc voltage values (V_R and V_L, respectively) shown in the figure. Voltage units are volts.

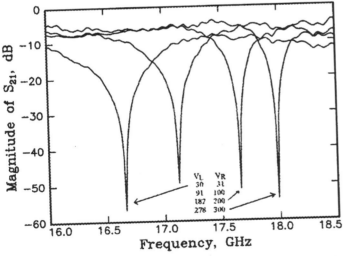

Fig. 52. Data from a second tunable ring resonator of the same design as in Figure 51 using a YBCO (0.35 μm)/STO (1 μm)/LAO (254 μm) circuit at 77 K. The ring and line dc voltage values (V_R and V_L, respectively) are shown in the figure. Voltage units are volts.

The coupling is achieved through the fringing fields of adjacent resonator sections. The coupled sections were designed using the 0.1-dB ripple Chebyshev filter design procedure [76, 107, 108]. To reduce the bandwidth, the spacing between the input and output sections is increased by the use of 45° angular coupled sections.

Figure 56 shows the bias dependence of S_{21} and S_{11} for one of the YBCO/STO/LAO bandpass filters (BPF). The cen-

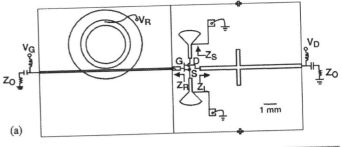

(a)

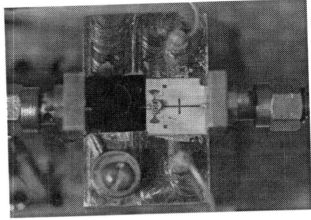

(b)

Fig. 53. (a) Layout of the complete voltage controlled oscillator (VCO) showing the ring resonator portion as well as the PHEMT portion. The PHEMT portion is fabricated on an alumina substrate as discussed in the text. (b) Actual circuit mounted for testing.

ter frequency shifts from 17.4 GHz at zero bias to 19.1 GHz at ±500 V bias, a tunability of 9% at 77 K. Note that the passband insertion loss, return loss, and the bandwidth improve with applied bias.

In these ferroelectric tunable filters, one can reduce the insertion loss of the passband, or maintain the passband relatively unchanged over a tunable frequency range depending on the biasing scheme employed:

1. Unipolar bias (UPB) where alternate nodes were biased positive, and ground.
2. Partial bipolar bias (PBB) where input and output lines are grounded, and the resonator sections biased positive and negative alternatively.
3. Full bipolar bias (FBB) where alternate sections (including the input and output lines) are biased positive and negative [109].

It is important to note that the effective dielectric constant of the microstrip structure depends upon the electric field between the coupled microstrip lines as well as the perpendicular field between the top conductor and the ground plane. In general, the FBB configuration gives the largest frequency tunability due to higher electric fields that can be applied in this configuration, and the PBB gives the lowest insertion loss in the passband in the ferroelectric tunable microstrip filters. Large tunability

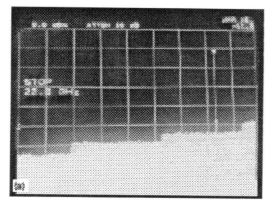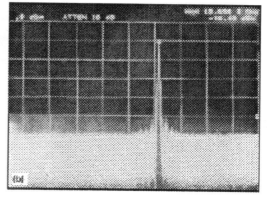

Fig. 54. VCO signals as measured on an HP 8566B Spectrum Analyzer at 43 K with $V_d = 2.1$ V, $V_g = -0.2$ V, $I_d = 13.9$ mA, and $V_{ring} = 38$ V. The scale on (a) is 2 to 22 GHz, and (b) shows a 500-MHz span with oscillation frequency at 16.696 MHz. In both parts, the vertical scale represents the insertion loss in dB.

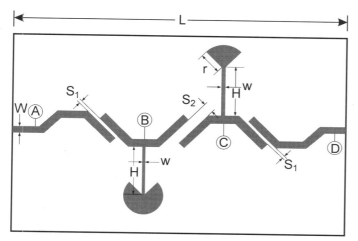

Fig. 55. The mask layout of a two-pole bandpass filter. The dimensions are $W = 86.25$ μm, $L = 6.8$ mm, $S_1 = 100$ μm, $S_2 = 300$ μm, $H = 1.33$ mm, $w = 12.5$ μm, and $r = 200$ μm.

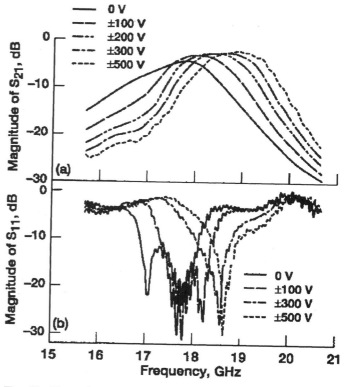

Fig. 56. The reflection S_{11} and transmission S_{21} characteristics of a YBCO/STO/LAO bandpass filter for different bipolar bias voltages at 77 K.

does not necessarily give the lowest insertion loss for the filters. A large frequency tunability of greater than 10% has been obtained in YBCO/STO/LAO microstrip bandpass filters operating below 77 K using the FBB biasing scheme described above. Remarkably, all of the filters tested to date under the FBB scheme have shown large tunability factors ($\geq 9\%$) at and below 77 K.

For comparison, Au/STO/LAO microstrip filter circuits have been tested for electrical tunability at temperatures below 77 K. Figure 57 shows the electrical tunability of a Au/STO/LAO biased using the bipolar biasing scheme. A tunability of approximately 11% was obtained at 40 K and at a dc bias of ± 200 V. This tunability is comparable to that exhibited by the YBCO/STO/LAO filters. However, the insertion loss exhibited by this filter was approximately 6 dB, compared to nearly 2 dB for HTS counterparts. The unloaded Q of the HTS resonator sections in the filters was estimated to be approximately 200 based on the model given in [76]. The major limiting factor for the unloaded Q is the dielectric losses in the STO layer. Mea-

surements of $\tan\delta$ values for laser ablated STO thin films at cryogenic temperatures and GHz frequencies range from 0.005 to 0.05 [84, 85]. An important finding has been that the percentage tunability remained essentially the same for a specific applied electric field, irrespective of the biasing scheme employed [109]. This finding created two new parameters called the sensitivity parameter (S) and the loss parameter (L). The sensitivity parameter is defined as the slope of the frequency shift versus peak dc electric field. Since the electric field is different across the filter, a peak electric field (E_{peak}) is used for the sensitivity and loss parameter calculations.

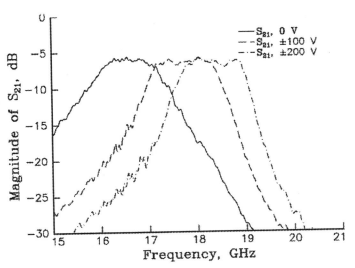

Fig. 57.   The transmission characteristics of an Au/STO/LAO filter for different bipolar bias voltages at 77 K.

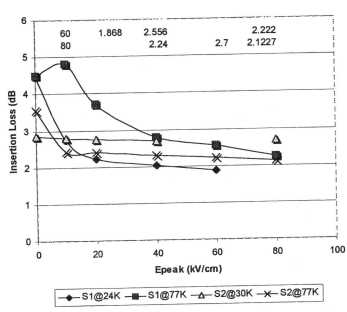

Fig. 59.   Insertion loss versus E_{peak} for the same filters as in Figure 58.

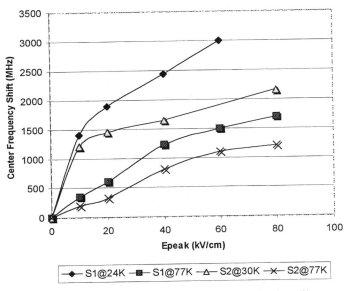

Fig. 58.   The center frequency shift versus E_{peak} for two bandpass filter samples at different temperatures.

The center frequency shift versus E_{peak} for two geometrically identical YBCO/STO/LAO bandpass filters is shown in Figure 58. This figure shows that the center frequency is tunable by several GHz with applied dc bias, at low temperatures. In sample 2, the center frequency shifted from 17.765 GHz at zero bias to 18.985 GHz at ±400 V bias, exhibiting a tunability of more than 6% at 77 K. Sample 1's frequency tunability improved to more than 17% at 24 K, as compared to sample 2's frequency tunability of 13% at 30 K. One can obtain the sensitivity parameter as the slope of the center frequency shift versus E_{peak} characteristics [109]. As shown in Figure 58, the

sensitivity parameter is the highest below 20 kV/cm, for both STO-based filters. The sensitivity parameter varies from 31 to 15 MHz cm/kV for HTS/STO/LAO-based filters at 77 K and is greater than 100 MHz cm/kV at 24 K for fields below 20 kV/cm.

Figure 59 shows the insertion loss versus E_{peak} characteristics. We define the loss parameter as the slope of the insertion loss versus E_{peak} characteristics [109]. As can be seen in the figure, the two samples have loss parameters that are different as well. What is evident from Figure 59 is that there are two regions, one below 20 kV/cm, where there is a large change in insertion loss, and the other above 20 kV/cm, in which the insertion loss varies less dramatically. Depending upon the design requirements, one could choose to operate in a region where the variation in insertion loss is low (requiring a higher applied field) or in a region where the insertion loss could be reduced to the lowest level possible. Note that the loss parameter is comparatively lower in sample 2, in which the tunability and sensitivity parameters are lower. As can be seen from Figures 58 and 59, large tunability at low fields below 20 kV/cm does not necessarily guarantee the lowest insertion loss for the filters. The loss parameter varies from 0.25 dB cm/kV at fields below 20 kV/cm to 0.005 dB cm/kV at higher fields. Higher sensitivity parameter and lower loss parameter indicates larger frequency tunability combined with lower insertion loss for the filters. These parameters are very useful for evaluating the material quality, as one could compare geometrically different tunable components for both sensitivity and loss parameters.

The impact of these filters can be evaluated at the component level as well as the subsystem level. At the component level, the filter's frequency agility allows for adjusting for Doppler effects, frequency hopping, and other communication applica-

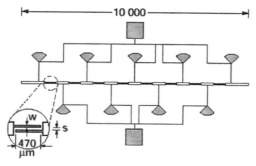

Fig. 60. Schematic of an eight-element coupled microstrip line phase shifter (CMPS) on a 254-μm-thick LaAlO$_3$ substrate. $S = 7.5$ μm and $W = 25$ μm. The microstrip lines are 84 μm wide. All units are in microns.

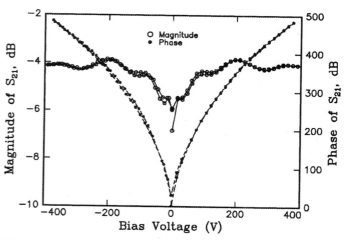

Fig. 61. Insertion loss and phase data for an eight-element YBCO (350 nm)/ STO (1.0 μm)/LAO coupled microstrip line phase shifter (CMPS). Data were taken at 77 K and at 16 GHz.

tions requiring the filter's passband reconfiguration. In addition, using a single tunable filter instead of fixed-frequency filter banks can add system flexibility. Also, low cost, ease of fabrication, and planar geometry make this filter technology very appealing for insertion into satellite receiver front-ends. The added flexibility may warrant the slightly increased insertion loss for some applications.

3.4.4. Thin-Film Ferroelectric Phase Shifters

Another area of application of the ferroelectric thin-film technology is the fabrication of compact, low-loss phase shifters. In general, phase-shifting elements can be realized through the use of ferrite materials, MMICs, and diodes (e.g., switched line, reflection, and loaded line). Typically, diode or MESFET-based phase shifters are digital with phase bits of 11.25, 22.5, 45, 90, and 180°. Losses increase with the number of bits (~2 dB/bit), and the discrete phase shift steps sometimes result in scanning granularity. Unfortunately, MMIC technology has not yet met the expectations of lower cost for phased array applications. Despite their high cost and the poor efficiency of power amplifiers (~15 %), MMICs remain the technology of choice for K- and Ka-band phased arrays. As an alternative, a new class of low-cost and reliable phase shifters based on ferroelectric thin-film technology has been developed. Because of the configuration of its main phase-shifting element, this type of phase shifter is known as the coupled microstrip line phase shifter (CMPS) [110, 111]. Figure 60 shows the schematic of an eight-element K-band CMPS fabricated with an YBCO (0.35 μm thick)/STO (1 μm thick)/LAO (254 μm thick) two-layered microstrip structure. A coupled microstrip line, which can be excited in the odd mode of the rf electric field, is the fundamental building block for the CMPS. Eight such sections are cascaded to obtain the desired phase shift. A biasing network is also shown in the figure to apply bias voltages to each section. By applying a differential dc bias between the coupled lines, one can reach fairly large electric fields between the lines (40 V/μm in Fig. 60), to effectively tune the relative dielectric constant of the ferroelectric thin film. Thus, the coupled microstrips provide more phase shift per unit length (at a specific dc bias) compared to a simple microstrip. However, a CMPS circuit requires careful design, as each coupled section is basically a one-pole bandpass filter. The phase shifters were optimized for low insertion loss and maximum relative phase shift. Further information on the phase-shifting and passband performance of CMPS circuits has been published previously [110, 111]. Figure 61 shows the magnitude and phase shift of S_{21} for this circuit using a 1-μm-thick STO film, tested at 40 K and 77 K. The phase shift at 40 K is greater because of the large dielectric tunability of STO at this temperature. Between the bias ranges of 75 and 375 V, a differential phase shift of 290° was observed at 16 GHz, while maintaining the insertion loss below 4.5 dB. This early YBCO prototype device achieved 500° insertion phase shift with 6 dB of loss (worst case) or a figure of merit of about 80°/dB. After much effort, this figure of merit has been nearly matched in room-temperature devices employing Au/Ba$_x$Sr$_{1-x}$TiO$_3$/MgO CMPS devices [111]. It is worth noting that the relative phase shifts increase in direct proportion to the thickness of the ferroelectric thin film. Also, the use of YBCO electrode instead of gold resulted in higher phase shift for the same thickness of the ferroelectric. This is possibly due to a residual low ε layer formed by the oxidized titanium adhesion layer used in the gold electrode-based circuit. The CMPS discussed here could enable the development of a low-cost, easy-to-fabricate, phase shifter technology with continuous phase-shifting capabilities from zero to over 360°. Phased array antennas, particularly reflect-arrays, will benefit from this phase-shifter technology.

4. SUMMARY AND CONCLUDING REMARKS

In this chapter, we have attempted to review the essential properties of high-temperature superconductors (HTS) and ferroelectric materials for applications in microwave communications. Planar HTS microstrip circuits are attractive for communication applications due to their lower conductor losses compared to normal conducting circuits at cryogenic temperatures.

A variety of HTS circuits have been demonstrated in the past decade, such as resonators, filters, and phase-shifting circuits aimed at high-performance (i.e., improved bandwidth, selectivity, low noise, and high Q) communication systems. As the demand for high bandwidth continues due to the latest developments in Internet and wireless communication technology, more applications are evolving at frequencies above the traditionally used C-band ($\sim$4 to 7 GHz) in which the integration of HTS is attractive due to the low-loss HTS thin films currently available. Among the HTS components that are useful for the communication industry are high-Q planar microstrip resonators, filters, phase shifters, and highly directive antennas. Initial applications of the HTS thin-film microstrip filters have been successful, exhibiting superior performance (e.g., lower insertion loss and steeper out-of-band rejection) than their gold-based counterparts. Preselect filters in the PCS and cellular base station receivers are currently being marketed by commercial companies such as Conductus Inc. (Sunnyvale, CA), Superconductor Technologies, Inc. (Santa Barbara, CA), and others. The growth of cellular and PCS systems indicates that the HTS filter industry will be growing at a steady pace. It has been predicted that the data traffic using wireless communication system will be more than 45% of all wireless traffic, compared to just 2% in 1999 [112]. This indicates the possibility of the HTS filter and receiver front-end being at the heart of the wireless infrastructure development.

However, it is known that for most filters and other resonant structures, attainment of optimal performance requires some level of additional tuning through either mechanical means (i.e., with tuning screws) or other coupling mechanisms. Therefore, incorporation of electronic tuning into HTS components without degradation of performance is very attractive. The result will be low-loss microwave components that could be fine tuned for optimal performance, with the additional attribute of being tunable over a broadband frequency range. This approach will greatly improve the practical applicability of HTS circuits in years to come.

Ferroelectric tunable microstrip as well as coplanar waveguide transmission line structures are potentially attractive for frequency-agile microwave communication systems. The most common ferroelectric tunable structure is based on conductor/ferroelectric/dielectric two-layered microstrip or CPW configuration. These tunable components enable a new class of frequency-agile components with large tunabilities and negligible additional losses when combined with HTS conductor for low-temperature applications. SrTiO$_3$ (STO) thin films with large dielectric tunability and low loss tangents at microwave frequencies have been the most promising ferroelectrics for integration with HTS circuits. Several tunable microwave components such as ring resonators, preselect filters, and phase shifters have been demonstrated using the HTS/STO/LAO tunable microstrip structure. Tunable Au/STO/LAO ring resonators with unloaded Q as high as 15,000 and frequency tunability factor of more than 12% have been demonstrated at 77 K. A Au/STO/LAO-based tunable ring resonator was successfully used as a stabilizing element in a PHEMT-based

voltage-controlled oscillator circuit. The VCO exhibited frequency tunability of more than 100 MHz with the ring resonator dc biased at 38 V. Several tunable bandpass filters were demonstrated with tunability factors greater than 9% at or below 77 K, with non-deembedded insertion loss as low as 1.5 dB at K-band frequencies using the YBCO/STO/LAO two-layered microstrip configuration. Also, continuous differential phase shifts of more than 360° at or below 77 K and Ku-band frequencies using coupled microstrip phase shifters have been demonstrated. These prototype demonstrations have equaled or exceeded the performance of typical HTS-based microstrip circuits. The attributes of these components, such as small size, light weight, and low loss, as well as their demonstrated performance, suggest that they can be used advantageously in satellite communication systems for Ku- and K-band applications.

REFERENCES

1. H. K. Onnes, *Commun. Phys. Lab., Univ. of Leiden* 120b, 3 (1911).
2. W. Meissner and R. Oschenfeld, *Naturwiss.* 21, 787 (1933).
3. J. G. Bednorz and K. A. Muller, *Z. Phys. B* 64, 189 (1986).
4. M. K. Wu, J. R. Ashburn, C. J. Torng, P. H. Hor, R. L. Meng, L. Gao, Z. J. Huang, Y. Q. Wang, and C. W. Chu, *Phys. Rev. Lett.* 58, 908 (1987).
5. H. Maeda, Y. Tanaka, M. Fukitomi, and T. Asano, *Japan. J. Appl. Phys.* 27, L209 (1988).
6. Z. Z. Sheng and A. M. Hermann, *Nature* 332, 138 (1988).
7. S. N. Putilin, E. V. Antipov, O. Chmaissem, and M. Marezio, *Nature, London* 362, 226 (1993).
8. D. S. Ginley, in "Thallium Based Superconducting Compounds" (A. M. Hermann and Y. Yakhmi, Eds.). World Scientific Publ., Singapore, 1994.
9. W. Holstein, L. A. Parisi, D. W. Face, X. D. Wu, S. R. Foltyn, and R. E. Muenchausen, *Appl. Phys. Lett.* 61(8), 982 (1992).
10. D. E. Oates, A. C. Anderson, and P. M. Mankiewich, *J. Supercond.* 3, 251 (1990).
11. Z. Y. Shen, C. Wilker, P. Pang, W. L. Holstein, D. Face, and D. J. Kountz, *IEEE Trans. Microwave Theory Tech.* 40, 2424 (1992).
12. R. R. Mansour, S. Ye, S. Peik, B. Jolley, V. Dokas, T. Romano, and G. Thomson, *IEEE MTT-S Int. Microwave Symp. Dig.* 2367 (1998).
13. H. Kroger, C. Hilhert, D. A. Gibson, U. Ghoshal, and L. N. Smith, *IEEE Proc.* 77, 1287 (1989).
14. J. D. Cressler, *IEEE Spectrum* 32, 49 (1995).
15. T. Van Duzer, *Cryogenics* 28, 527 (1988).
16. A. M. Hermann, R. M. Yandrofski, J. F. Scott, A. Naziripour, D. Galt, J. C. Price, J. Cuchario, and R. K. Ahrenkiel, *J. Supercond.* 7, 463 (1994).
17. F. A. Miranda, F. W. Van Keuls, R. R. Romanofsky, and G. Subramanyam, *Integrated Ferroelectrics* 22, 269 (1998).
18. F. A. Miranda, G. Subramanyam, F. W. Van Keuls, R. R. Romanofsky, J. D. Warner, and C. H. Mueller, *IEEE Microwave Theory Tech.* 41, 1181 (2000).
19. J. D. Doss, "Engineer's Guide to High Temperature Superconductivity." Wiley, New York, 1989.
20. A. Abrikosov, *Sov. Phys. JETP* 5, 1174 (1957).
21. F. London, "Superfluids, Macroscopic Theory of Superconductivity," Vol. 1, pp. 3–4. Wiley, New York, 1950.
22. J. Bardeen, L. N. Cooper, and J. Schreiffer, *Phys. Rev.* 108, 1175 (1957).
23. R. Beyers and T. M. Shaw, *Solid State Phys.* 42, 135 (1989).
24. C. J. Gorter and H. B. G. Casimir, *Phys. Z.* 35, 963 (1934).
25. B. D.Josephson, *Phys. Lett.* 1, 251 (1962).
26. A. B. Pippard, *Proc. Roy. Soc. London Ser. A* 191, 370 (1947).
27. D. Dijikkamp, T. Venkatesan, X. D. Wu, S. A. Shaheen, N. Jisrawi, Y. H. Min-Lee, W. L. McLean, and M. Croft, *Appl. Phys. Lett.* 51, 619 (1987).
28. D. C. Payne and J. C. Bravman, Eds., "Laser Ablation for Material Synthesis," Materials Research Society Symposium Proc. 191, Boston, 1990.

29. H. Adachi, K. Hurochi, K. Setsune, M. Kitabake, and K. Wasa, *Appl. Phys. Lett.* 51, 2263 (1987).

30. K. Mizuno, K. Higashino, K. Setsune, and K. Wasa, *Appl. Phys. Lett.* 56, 1469 (1990).

31. S. Takano, N. Hayashi, S. Okuda, and H. Tsuyanagi, *Phys. C* 162-164, 1535 (1990).

32. J. Talvacchio, J. Gavaler, M. Forrester, and G. Braggins, "Science and Technology of Thin Film Superconductors II" (R. D. McConnell and S. A. Wolf, Eds.). Plenum, New York, 1990.

33. B. Y. Tsaur, M. S. Diorio, and A. J. Strauss, *Appl. Phys. Lett.* 51, 858 (1987).

34. P. M. Mankiewich, J. H. Scoefield, W. J. Skocpol, R. E. Howard, A. H. Dayen, and E. Good, *Appl. Phys. Lett.* 51, 1753 (1987).

35. S. W. Chan, B. G. Bagley, L. H. Greene, M. Giroud, W. L. Feldman, K. R. Jenken, and B. J. Wilkins, *Appl. Phys. Lett.* 53, 1443 (1988).

36. J. R. Phillips, J. W. Mayer, J. A. Martin, and M. Natasi, *Appl. Phys. Lett.* 56, 1374 (1990).

37. T. S. Kalkur, R. Kwor, S. Jernigan, and R. Smith, "Conference of Science and Technology of Thin Film Superconductors," 1988.

38. R. C. West, Ed., "CRC Handbook of Chemistry and Physics," 69th ed. CRC Press, Boca Raton, FL, 1988.

39. N. Klein, G. Muller, H. Piel, B. Roas, L. Shultz, U. Klein, and M. Peiniger, *Appl. Phys. Lett.* 54, 757 (1989).

40. T. K. Worthington, W. J. Gallager, and T. R. Dinger, *Phys. Rev. Lett.* 59, 1160 (1987).

41. G. A. Samara and A. A. Giandini, *Phys. Rev.* 140, A954 (1965).

42. H. E. Weaver, *Phys. Chem. Solids II* 274 (1959).

43. A. F. Harvey, *Microwave Eng.* 253 (1963).

44. A. R. Von Hippel, "Dielectric Materials and Applications," MIT Press, Cambridge MA, 1954.

45. R. L. Sandstrom, E. A. Giess, W. J. Gallager, A. Segmuller, E. I. Cooper, M. F. Chisholm, A. Gupta, S. Shinde, and R. B. Laibowitz, *Appl. Phys. Lett.* 53, 1874 (1988).

46. F. A. Miranda, W. L. Gordon, K. B. Bhasin, B. T. Ebihara, V. O. Heinen, and C. M. Chorey, *Microwave Opt. Tech. Lett.* 3, 11 (1990).

47. R. W. Simon, C. E. Platt, A. E. Lee, G. S. Lee, K. P. Daly, M. S. Wire, J. A. Luine, and M. Urbanik, *Appl. Phys. Lett.* 53, 2677 (1988).

48. K. Khachaturyan, E. R. Weber, P. Tejedor, A. M. Stacey, and A. M. Portis, *Phys. Rev. B* 36, 8309 (1987).

49. S. Tyagi, *Phys. C* 156, 73 (1988).

50. S. Sridhar and W. L. Kennedy, *Rev. Sci. Instrum.* 59, 531 (1988).

51. J. P. Carini, A. M. Awasthi, W. Beyermann, G. Gruner, T. Hylton, K. Char, M. R. Beasley, and A. Kapitulnik, *Phys. Rev. B* 37, 9726 (1988).

52. Q. Hu and P. L. Richards, *Appl. Phys. Lett.* 55, 2444 (1989).

53. A. T. Firoy, A. F. Hebard, P. M. Mankiewich, and R. E. Howard, *Phys. Rev. Lett.* 61, 1419 (1988).

54. S. M. Anlage, H. Sze, H. J. Snortland, S. Tahara, B. Langley, C. B. Eom, M. R. Beasley, and R. Taber, *Appl. Phys. Lett.* 54, 2710 (1989).

55. J. I. Gittleman and B. Rosenblum, *IEEE Proc.* 52, 1138 (1964).

56. J. I. Gittleman and J. R. Matey, *J. Appl. Phys.* 65, 688 (1989).

57. S. H. Talisa, M. A. Robertson, B. J. Meier, and J. E. Sluz, *IEEE MTT-S Int. Microwave Symp. Dig.* 2, 997 (1994).

58. R. R. Mansour, S. Ye, B. Jolley, G. Thomson, S. F. Peik, T. Romano, W.-C. Tang, C. M. Kudsia, T. Nast, B. Williams, D. Frank, D. Enlow, G. Silverman, J. Soroga, C. Wilker, J. Warner, S. Khanna, G. Seguin, and G. Brassard, *IEEE Trans. Microwave Theory Tech.* 48, 1171 (2000).

59. D. G. Swanson, Jr. and R. J. Forse, *IEEE MTT-S Int. Microwave Symp. Dig.* 1, 199 (1994).

60. J. B. Barner, J. J. Bautista, J. G. Bowen, W. Chew, M. C. Foote, B. H. Fujiwara, A. J. Guern, B. J. Hunt, H. H. S. Javati, G. G. Ortiz, D. L. Rascoe, R. P. Vasquez, P. D. Wamhof, K. B. Bhasin, R. F. Leonard, R. R. Romanofsky, and C. M. Chop, *IEEE Trans. Appl. Supercond.* 5, 2075 (1995).

61. S. Y. Lee, K. Y. Kang, and D. Ahn, *IEEE Trans. Appl. Supercond.* 5, 2563 (1995).

62. Z. Y. Shen, C. Wilker, P. Pang, and C. Carter, *IEEE Trans. Microwave Theory Tech.* 44, 981 (1996).

63. M. J. Lancaster, F. Huang, A. Porch, B. Avenhaus, J. S. Hong, and D. Hung, *IEEE Trans. Microwave Theory Tech.* 44, 1339 (1996).

64. J. S. Herd, L. D. Poles, J. P. Kenney, J. S. Deroy, M. H.Champion, J. H. Silva, M. Davidovitz, K. G. Herd, W. J. Bocchi, and D. T. Hayes, *IEEE Trans. Microwave Theory Tech.* 44, 1383 (1996).

65. S. J. Fiedziuszko, J. A. Curtis, S. C. Holme, and R. S. Kwok, *IEEE Trans. Microwave Theory Tech.* 44, 1248 (1996).

66. M. Feng, F. Gao, Z. Zhou, J. Kruse, M. Heins, J. Wang, S. Remillard, R. Lithgow, M. Scharen, A. Cardona, and R. Forse, *IEEE Trans. Microwave Theory Tech.* 44, 1347 (1996).

67. S. J. Berkowitz, C. F. Shih, W. H. Mallison, D. Zhang, and A. S. Hirahara, *IEEE Trans. Appl. Supercond.* 7, 3056 (1997).

68. B. A. Aminov, A. Baumfalk, H. J. Chaloupka, M. Hein, T. Kasier, S. Kolesov, H. Piel, H. Medelius, and E. Wikborg, *IEEE MTT-S Int. Microwave Symp. Dig.* 1, 363 (1998).

69. K. Yoshida, K. Sashiyama, S. Nishioka, H. Shimakage, and Z. Wang, *IEEE Trans. Appl. Supercond.* 9, 3905 (1999).

70. J.-S. Hong, M. J. Lancaster, D. Jedamzik, R. B. Greed, and J. C. Mage, *IEEE Trans. Microwave Theory Tech.* 48, 1240 (2000).

71. J. H. Takemoto, F. K. Oshita, H. R. Fetterman, P. Kobrin, and E. Sovero, *IEEE Trans. Microwave Theory Tech.* 37, 1650 (1989).

72. G. Subramanyam, V. J. Kapoor, C. M. Chorey, and K. B.Bhasin, *J. Appl. Phys.* 72, 2396 (1992).

73. J. C. Swihart, *J. Appl. Phys.* 32, 461 (1961).

74. C. M. Chorey, K. S. Kong, K. B. Bhasin, J. D. Warner, and T. Itoh, *IEEE Trans. Microwave Theory Tech.* 39, 1480 (1991).

75. J. C. Sabatis, K. B. Bhasin, and F. A. Miranda, *Adv. Cryogenic Eng.* 41, 1996.

76. G. Matthaei, L. Young, and E. M. T. Jones, "Microwave Filters, Impedance-Matching Networks, and Coupling Structures," Artech House, Dedham, MA, 1980.

77. J. A. Curtis and S. J. Fiedziuszko, *IEEE MTT-S Dig.* 443 (1991).

78. See, for example, "Inframetrics Miniaturized Cryocoolers," Inframetrics, Inc., 16 Esquire Road, N. Billerica, MA 10862. Also DRS Infrared Technologies Inc., 13532 North Central Expressway, Dallas, TX 75243.

79. S. J. Fiedziusko and J. A. Curtis, *IEEE MTT-S Int. Microwave Symp. Dig.* 1203 (1992).

80. R. S. Kwok, S. J. Fiedziusko, F. A. Miranda, G. V. Leon, M. S. Demo, and D. Y. Bohman, *IEEE Trans. Appl. Supercond.* 7, 3706 (1997).

81. R. S. Kwok, S. J. Fiedziuszko, T. Schnabel, F. A. Miranda, N. C. Varaljay, and C. H. Mueller, *IEEE MTT-S Int. Microwave Symp. Dig.* 1377 (1999).

82. R. R. Romanofsky and K. B. Bhasin, NASA Tech. Memorandum 105184, 1991.

83. R. Ramesh, O. Auciello, and J. F. Scott, "The Physics of Ferroelectric Memories," Physics Today, p. 22 (July, 1998).

84. D. M. Dalbert, R. E. Stauber, J. C. Price, C. T. Rogers, and D. Galt, *Appl. Phys. Lett.* 72, 507 (1998).

85. W. Chang, J. S. Horwitz, W.-J. Kim, C. M. Gilmore, J. M. Pond, S. W. Kirchoefer, and D. B. Chrisey, "Materials Issues for Tunable RF and Microwave Devices," Vol. 603, pp. 181–186, 2000.

86. L. C. Sengupta, E. Ngo, M. E. O'Day, S. Stowell, and R. Lancto, "Proceedings of the 9th IEEE Symposium on Applications of Ferroelectrics," 1994.

87. D. Galt, J. C. Price, J. A. Beall, and T. E. Harvey, *IEEE Trans. Appl. Supercond.* 5, 2563 (1995).

88. A. T. Findikoglu, Q. X. Jia, X. D. Wu, G. J. Chen, and T. Venkatesan, *Appl. Phys. Lett.* 68, 1651 (1996).

89. F. Abbas, L. E. Davis, and J. C. Gallop, *IEEE MTT-S Int. Microwave Symp. Dig.* 3, 1671 (1996).

90. F. De Flaviis and N. G. Alexopoulos, *IEEE AP-S Int. Symp. Dig.* 316 (1997).

91. A. Kozyrev, A. Ivanov, V. Kies, M. Khazov, V. Osadchy, T. Samoliova, A. Pavlov, G. Koepf, D. Galt, and T. Rivkin, *IEEE MTT-S Int. Microwave Symp. Dig.* 2, 985 (1998).

92. A. S. Nagra and R. A. York, *IEEE Trans. Microwave Theory Tech.* 47, 1711 (1999).

93. D. E. Oates and G. F. Dionne, *IEEE Trans. Appl. Supercond.* 9, 2575 (1999).

94. Sonnet Software, Inc., Liverpool, New York, California.

95. Zeland Software, Inc., Fremont, CA.

96. F. A. Miranda, F. W. Van Keuls, G. Subramanyam, C. H. Mueller, R. R. Romanofsky, and G. Rosado, *Integrated Ferroelectrics* 24, 195 (1999).

97. F. W. Van Keuls, R. R. Romanofsky, D. Y. Bohman, and F. A. Miranda, *Integrated Ferroelectrics* 22, 883 (1998).

98. G. Subramanyam, Al. Zaman, N. Mohsina, F. A. Miranda, F. W. Van Keuls, R. Romanofsky, and P. Boolchand, *Integrated Ferroelectrics* 34, 197 (2001).

99. J. Synowczynski, L. C. Sengupta, and L. H. Chu, *Integrated Ferroelectrics* 22, 861 (1998).

100. F. A. Miranda, C. H. Mueller, C. D. Cubbage, K. B. Bhasin, R. K. Singh, and S. D. Harkness, *IEEE Trans. Appl. Supercond.* 5, 3191 (1995).

101. A. T. Findikoglu, Q. X. Jia, I. H. Campbell, X. D. Wu, D. Reagor, C. B. Mombourquette, and D. McMurry, *Appl. Phys. Lett.* 66, 3674 (1995).

102. K. L. Kaiser, M. D. Vaudin, L. D. Rotter, Z. L. Wang, J. P. Cline, C. S. Hwang, R. B. Marinenko, and J. G. Gillen, *Mater. Res. Soc. Symp. Proc.* 361, 355 (1995).

103. R. Romanofsky, unpublished.

104. A. P. S. Khanna, *Microwave & RF*, p. 120 (1992).

105. L. Zhu, B. Taylor, and J. Jarmuszewski, *RF Design*, Nov. (1996).

106. R. R. Romanofsky, F. W. Van Keuls, and F. A. Miranda, *J. Phys. IV France* 8, 171 (1998).

107. G. Subramanyam, F. W. Van Keuls, and F. A. Miranda, *IEEE MTT-S Int. Microwave Symp. Dig.* 1011 (1998).

108. G. Subramanyam, F. W. Van Keuls, and F. A. Miranda, *IEEE Microwave Guided Wave Lett.* 8, 78 (1998).

109. G. Subramanyam, F. Van Keuls, F. Miranda, C. Canedy, S. Aggarwal, T. Venkatesan, and R. Ramesh, *Integrated Ferroelectrics* 24, 273 (1999).

110. R. R. Romanofsky, F. W. Van Keuls, J. D. Warner, C. H. Mueller, S. A. Alterovitz, F. A. Miranda, and A. H. Qureshi, in "Materials Issues for Tunable RF and Microwave Devices" (Q. Jia, F. A. Miranda, D. E. Oates, and X. Xi, Eds.), Vol. 603. Materials Research Society Press, Boston, 2000.

111. F. W. Van Keuls, C. H. Mueller, F. A. Miranda, R. R. Romanofsky, J. S. Horwitz, W. Chang, and W. J. Kim, *Integrated Ferroelectrics* 28, 49 (2000).

112. R. Simon, *Superconductor Sci. Tech.* Spring, 9 (2000).

113. M. K. Wu, J. R. Ashurn, C. I. Tang, P. H. Hoi, R. L. Meng, L. Gao, Z. J. Huang, Y. Q. Wang, and C. W. Chu, *Phys. Rev. Lett.* 58, 908 (1987).

114. J. M. Tarascon, W. R. M. Kinnon, L. H. Greene, G. W. Hull, and E. M. Vagel, *Phys. Rev. B.* 36, 226 (1987).

115. Y. Le Page, T. Siegrist, S. A. Sunshine, L. F. Schneemeyer, D. W. Murphy, S. M. Zahurak, J. V. Waszczak, W. R. McKinnon, J. M. Tarascon, G. W. Hull, and L. H. Greene, *Phys. Rev. B.* 36, 3617 (1987).

116. P. H. Hoi, R. L. Meng, Y. Q. Wang, L. Gao, Z. Huang, J. Bechtold, F. Foretei, and C. W. Chu, *Phys. Rev. Lett.* 58, 1891 (1987).

117. S. E. Brown, J. D. Thompson, J. O. Willis, R. M. Aikin, E. Zringiebl, J. L. Smith, Z. Fish, and R. B. Schwarz, *Phys. Rev. B* 36, 2258 (1987).

118. M. Hikita, Y. Tajimo, A. Katsui, and Y. Hidaka, *Phys. Rev. B* 36, 7199 (1987).

119. E. A. Gutierrez, M. J. Deen, and C. L. Claeys, "Low Temperature Electronics: Physics, Devices, Circuits and Applications." Academic Press, San Diego, CA, 2000.

Chapter 10

Twinning in Ferroelectric Thin Films: Theory and Structural Analysis

S. Pamir Alpay

Department of Metallurgy and Materials Engineering, University of Connecticut, Storrs, Connecticut, USA

Contents

1. INTRODUCTION

During the past decade, perovskite ferroelectric thin films such as of $BaTiO_3$, $PbTiO_3$, and solid solutions of $PbTiO_3$-$PbZrO_3$ [$Pb(Ti_{1-x}Zr_x)O_3$ compounds] have received considerable interest because of their numerous potential device applications in the microelectronics industry as elements of non-volatile random access memories (NVRAM) because of their ability to reorient the direction of the spontaneous polarization, static random access memories (SRAM), high dielectric constant capacitors, optical waveguides, and pyroelectric detectors. They are also used as active components of sensors, actuators, and micro-electromechanical systems (MEMS). These applications require either a high electrical field response to an external mechanical field (direct piezoeffect) for sensors or a high strain response to an applied electrical field (converse piezoeffect) for actuators and micro-electromechanical systems (MEMS). The current interest is driven by the unique bulk dielectric, piezoelectric, pyroelectric, electro-optic, and acousto-optic properties of these ferroelectric ceramics [1, 2].

Progress in existing film deposition techniques, such as sol-gel processing and rf sputtering and the advent of new techniques such as pulsed laser deposition (PLD) and metal-organic chemical vapor deposition (MOCVD) made it possible to grow these ferroelectric ceramics as thin films with good compositional control on various substrates. However, if the film is

Handbook of Thin Film Materials, edited by H.S. Nalwa
Volume 3: Ferroelectric and Dielectric Thin Films

ISBN 0-12-512911-4/$35.00

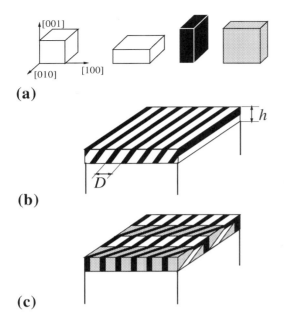

(a)

(b)

(c)

Fig. 1. Three kinds of domains and polydomain structures due to a cubic-orthorhombic phase transformation in an epitaxial layer [6].

Fig. 2. Polydomain plates as a result of structural transformation in AuMn alloy [12].

polycrystalline (i.e., consists of randomly oriented grains), then it must be poled to achieve maximum piezoelectric, pyroelectric, and electro-optic response, which obviously is less than the response of a single crystalline film. Furthermore, the presence of high-angle grain boundaries is detrimental to device performance and electrical properties [3]. Fortunately, if the substrate is chosen suitably such that the lattice mismatch between the substrate and the film is relatively small, then epitaxial growth (i.e., with the film and the substrate having the same crystallographic orientation) can be promoted in many cases. Epitaxy in the film provides a means to achieve superior physical properties when compared to polycrystalline films, because epitaxial films have a lower density of scattering centers such as grain boundaries [4].

Experimental studies indicate that epitaxial ferroelectric films usually exhibit twinning. The films are generally deposited at temperatures at which they are paraelectric. On cooling, the cubic paraelectric phase transforms to a noncentrosymmetric phase that is ferroelectric, i.e., has a spontaneous polarization. Twinning (or polydomain formation) is a typical strain relaxation mechanism in all constrained structural phase transformations [5, 6]. The product phase, after cooling, has a lower symmetry than the parent phase, giving rise to several energetically equivalent, crystallographically different variants or domains [7]. If this transformation occurs in a constrained media, as it is the case in an epitaxial couple, then formation of polydomain structures may reduce the total energy of the system. Figure 1, first published in 1976 [6], shows domain structures that have been predicted theoretically as a result of a structural phase transformation in an epitaxial film. As shown, transformation of a cubic phase to an orthorhombic phase should result in six domains, three of which are illustrated in Figure 1a. After the transformation, a twinned (polydomain)

film can fit better in average to a substrate than a single domain film. Polydomain formation decreases the elastic energy of an epitaxial system and thus is a thermodynamically driven phenomenon. Taking into account different contributions to the free energy of a polydomain film, it has been shown that there are some equilibrium polydomain structures, as illustrated in Figures 1b and c. The scale of the polydomain structures depends on the film thickness. For example, the domain period of a simple structure in Figure 1b is proportional to the square root of the film thickness [5, 6].

Three features—the existence of several equivalent variants, the possibility of lowering energy of the long-range field, and a square root dependence of domain period on film thickness—are usual attributes of magnetic domains in ferromagnetics and electric domains in ferroelectrics [8–10]. That is why the structural domains in epitaxial films are called elastic domains. The idea of elastic domains came from the theory of martensitic transformations, where this concept was used in the explanation of a twinned internal structure of plate-like crystals of a product phase (martensite) embedded in a matrix of a parent phase [11]. An example of a well-organized twin structure inside a martensite plate is presented in Figure 2 [12]. Elastic domains are in a sense structural domains that reduce the intensity of the elastic field and elastic energy in a manner similar to that in which electric (or magnetic) domains in ferroelectric (or ferromagnetic) materials relax the depolarizing (or demagnetizing) fields. The engineering of similar domain microstructures in epitaxial layers was the goal of earlier theoretical work [6]. However, the first experimental evidence on such kind of polydomain structure in epitaxial films was obtained many years later. A quantitative study of $YBa_2Cu_3O_7$ films deposited by magnetron sputtering onto a MgO single-crystal substrate found that the dependence of twin spacing on film thickness follows a square-root dependence [13]. Later, a modulated pattern in an epitaxial layer of ferroelectric (001) $PbTiO_3$ on (001) $KTaO_3$ was observed directly by transmission electron

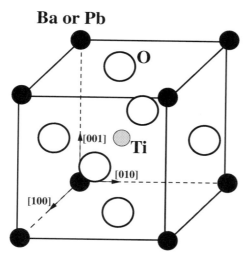

Ba or Pb

Fig. 3. Atomic arrangements in the unit cell of $BaTiO_3$ or $PbTiO_3$ above the Curie temperature (space group m3m).

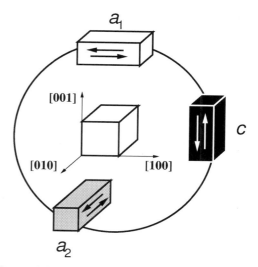

Fig. 4. Structural domains of the tetragonal lattice. The direction of the spontaneous polarization is also shown. The paraelectric cubic phase is in the center for reference.

microscopy (TEM) and by high-resolution X-ray diffraction (XRD) [14, 15].

The fundamental aim of this chapter is to provide a review of the theoretical work on polydomain formation in ferroelectric films. It also examines experimental studies to date on the domain structure and its effect on electrical and electromechanical properties. The goal is to determine theoretically the parameters of polydomain structures and the conditions for their formation, i.e., their dependence on the characteristics of the phase transformation, lattice misfits, and film thickness, as well as external fields. We formulate a simple but an effective quantitative theoretical model to explain observed experimental results and to predict and control the microstructure in epitaxial ferroelectric films. It should be noted that strain-controlled polydomain structures can be obtained as a result of not only the ferroelectric transformation, but also essentially all solid-solid phase transformations, including ordering, decomposition into isomorphic or polymorphic solid solutions, and polymorphic martensitic, ferroelastic, and ferromagnetic transformations. Therefore, the concepts developed in this chapter are not limited to ferroelectric thin films, but are equally applicable to any structural phase transformation.

For the sake of simplicity, we concentrate on a particular case: the cubic (paraelectric)-tetragonal (ferroelectric) phase transformation in epitaxial ferroelectric films. The cubic-tetragonal phase transformation observed in ferroelectric perovskites such as $BaTiO_3$ and $PbTiO_3$ is accompanied with the emergence of a permanent spontaneous polarization in the tetragonal lattice. The high-temperature cubic phase (space group m3m) is paraelectric and the atomic arrangements are shown in Figure 3. The low-temperature tetragonal phase (space group 4mm) is ferroelectric, because in $BaTiO_3$ the Ti atoms shift to off-center positions along the [001] axis, creating a permanent dipole moment in the lattice [1, 2]. In $PbTiO_3$, both Ti and O atoms are displaced from their original positions by 0.4 nm and 1.1 nm along [001] in the same direction, respectively, producing the same effect [2]. This structural transformation occurs at the Curie temperature (T_C), which is 120°C for $BaTiO_3$ and 490°C for $PbTiO_3$.

In epitaxial ferroelectric films, the sources of internal stresses are obvious: the structural phase transformation at T_C, lattice mismatch between the layers, and the difference in thermal expansion coefficients of the film and the substrate. Therefore, under certain conditions, it is possible to reduce internal stresses by twinning. The polydomain structure generally consists of two of the three possible orientational variants (or ferroelastic domains) of the tetragonal phase (Fig. 4) separated from each other by charge-neutral 90° domain walls. The interdomain interfaces are elastically compatible and thus stress-free [6]. The most common structure is the $c/a/c/a$ polydomain pattern, which consists of transversely modulated thin platelets of c-domains with the tetragonal axis perpendicular to the film-substrate interface and a-domains with the c-axis of the tetragonal film along either [100] or [010] directions of the substrate. These structures have been observed in epitaxial films of $BaTiO_3$ [16–21], $PbTiO_3$ [14, 15, 22–31], $Pb(Zr_xTi_{1-x})O_3$ [32–40], $Pb_{1-x}La_xTiO_3$ [24, 41–43], and $KNbO_3$ [44] on various cubic and pseudo-cubic substrates. TEM studies show that the interdomain interface is usually along {101} type planes [15, 26].

In recent years, along with experimental data have come significant developments in the theoretical description of polydomain formation in constrained ferroelectric films [15, 45–52] and multilayer heterostructures containing ferroelectric layers as active components [53–55]. The Landau–Ginzburg–Devonshire approximation for the free energy of ferroelectrics was used to take into account the temperature dependence of a structural transformation [15]. The stress state of domains in epitaxial films, the effects of thermal expansion, and misfit dislocations are explored in a series of articles by Speck and co-workers [46–48]. Considerable progress was achieved in recent studies, which include accurate calculations of microstresses and their energy for the interface between a thin polydomain film and a substrate and for the free surface of the film. Different methods of continuum elasticity theory, methods of continuum dislocations and disclinations [48], and Green's tensor function method [15, 45, 49, 51] give similar results. Accurate calculations of the stress, strain, and elastic energy for different domain arrangements, including individual domains, were performed by Pertsev and Zembilgotov [49, 50] and Sridhar et al. [51, 52]. Theoretical and experimental work was also performed on epitaxial rhombohedral ferroelectric films [56, 57]. Most recent works include accurate calculations of the nonuniform microstresses near the film–substrate interface or a free film surface for different polytwin and individual domains. The contribution of microstresses to elastic energy is important if the thickness of domain or the period of domain modulation is larger than the film thickness. However, it is difficult to estimate how closly these calculations describe reality, because they are performed in the frame of linear isotropic elastic theory. Meanwhile, for thick domains due to strong fields inside of them, the deviation from linearity is substantial. The theoretical approach presented here avoids this difficulty, because its task is to describe periodic polydomain structures with a period less than the film thickness. Assemblies of domains that determine average material properties of polydomain films and, consequently, material properties of multilayer heterostructures and composites are under consideration, rather than nonregular domain configurations or individual domains. The theory can quantitatively take into account the anisotropy when the primary effects of long-range elastic fields and their relaxation are analyzed. Besides, the theory gives a satisfactory description of nonlinear properties that are a result of secondary effects of the short-range field of microstresses. Furthermore, the present models either are not quantitative or are too complex to be used as practical tools to determine the equilibrium structure and its fundamental parameters. In addition, the possibility of formation of more complex hierarchical structures has not been taken into account. There is also the possibility of "domain engineering," i.e., manipulation of the polydomain structures, via external fields.

The theoretical part of this chapter, Section 2, starts with the crystallographic relations of domains in a polydomain structure, i.e., the compatibility of the domains. The thermodynamic analysis of the polydomain formation leads to the domain stability map, which is essentially a phase diagram that shows not only the regions of stability for possible domain configurations, but also the fraction of domains in a specific domain structure as functions of misfit strain, lattice parameters, and film thickness. Subsequently, we discuss the effect of external fields (particularly applied uniaxial stresses) on the domain stability map. We shown that with the application of such fields during film deposition and cooling, desired polydomain structures can be engineered. We demonstrate that in some cases, the polydomain formation can be completely suppressed, allowing one to obtain perfect single domain structures. We also present

theoretical background on the possibility of complete strain relaxation by the formation of so-called second-order polytwins [6, 58–60] and construct a domain stability map. Despite its simple form, the theory is rigorous and allows one to obtain quantitative results by modeling a real system on the basis of accurate information on concrete values of the theoretical parameters.

In the latter part of this section, we discuss one other strain relaxation mechanism—the formation of misfit dislocations at the film–substrate interface—and analyze their effect on the equilibrium domain structure. The role of misfit dislocations in relaxating elastic fields in an epitaxial film is similar to the role of free charges in decreasing electrostatic fields. But if the time of relaxation of charge is small in comparison with the time of domain formation, then the opposite should be true for misfit dislocation. The formation of a polydomain structure is a result of a decay of the initial phase with simultaneous relaxation of elastic fields. Thus, it is very likely that plastic deformation (i.e., generation, multiplication and motion) of misfit dislocations is not competitive with the domain formation as a mechanism of stress relaxation. This assumption does not exclude the possibility of relaxation through misfit dislocations when the film is in its paraphase. This relaxation can be easily included in the theory by renormalizing the initial misfit. Relaxation by cracking (or delamination) is not included, because it significantly deteriorates the mechanical and electrical properties of epitaxial film [61] and should be avoided in these films.

Section 3 reviews experimental methods of deposition and crystallographic and microstructural characterization of polydomain ferroelectric films. First, determination of the structural characteristics of polydomain assemblies by XRD is discussed. From these experiments, the fractions of the domains, their tilt, and internal stresses in the film can be determined. Furthermore, direct visualization techniques, such as TEM and the recently implemented atomic force microscopy (AFM) in the piezoresponse mode, give information on domain architecture, domain fractions, and domain period.

Section 4 of this chapter provides experimental verification of theoretical predictions. One of the main features of the developed theory is that it does not take into account the electrostatic effect from spontaneous polarization within the individual domains that make up the polydomain structure. We show that depolarizing fields do not affect the domain structure. For this reason, we discuss a specific experimental study [38] in detail. Next, we analyze the temperature dependence of the domain fraction of a $PbTiO_3$ film grown on (001) MgO obtained via XRD in conjunction with theoretical expectations. We also review the dependence film thickness of the domain structure. We compare theoretical predictions with experimental results and results from the literature. Finally, in the last part of this section, we provide experimental evidence on the formation of the three-domain architecture in relatively thicker films and also discuss twinning in $BaTiO_3$ films on various substrates.

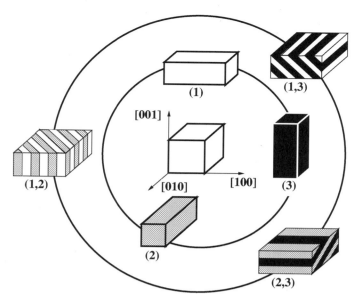

Fig. 5. Possible domain structures of a constrained layer undergoing a cubic-tetragonal transformation. The inner circle shows single-domain structures, and the outer circle shows two-domain polytwin structures.

2. THEORY

2.1. Crystallography of Polydomain Formation

When ferroelectric perovskite crystals undergo a cubic-tetragonal phase transformation at T_C, three different orientational variants (or domains) may form as shown in the inner circle of Figure 5, where the cubic phase is shown in the center for reference. The domains are spontaneously polarized along their longer axis (i.e., the c-axis), and the spontaneous polarization can be in two opposite but equivalent directions within the orientational variants. Therefore, there are six distinct electrical domains, as illustrated in Figure 5.

All phase transformations in solids are accompanied by a self-strain, and different structural (or orientational) domains correspond to different self-strains. The self-strains for the cubic-tetragonal transformation for a_1-, a_2-, and c-domains, respectively, are given by

$$\widehat{\varepsilon}_1 = \begin{pmatrix} \varepsilon_0' & 0 & 0 \\ 0 & \varepsilon_0 & 0 \\ 0 & 0 & \varepsilon_0 \end{pmatrix} \quad \widehat{\varepsilon}_2 = \begin{pmatrix} \varepsilon_0 & 0 & 0 \\ 0 & \varepsilon_0' & 0 \\ 0 & 0 & \varepsilon_0 \end{pmatrix}$$

$$\widehat{\varepsilon}_3 = \begin{pmatrix} \varepsilon_0 & 0 & 0 \\ 0 & \varepsilon_0 & 0 \\ 0 & 0 & \varepsilon_0' \end{pmatrix} \tag{1}$$

where $\varepsilon_0' = (c - a_0)/a_0$, $\varepsilon_0 = (a - a_0)/a_0$, a and c are the lattice parameters of the film in the ferroelectric state, and a_0 is the lattice parameter of the film in the paraelectric state. The variant with the tetragonal axis perpendicular to the film–substrate interface is known as the c-domain, whereas the two variants with their tetragonal axes in the plane of film–substrate interface are known as the a_1-domain and the a_2-domain in the ferroelectric thin film literature.

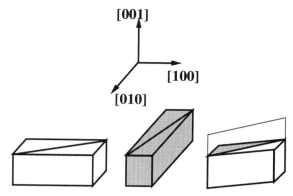

Fig. 6. The twin relation between a_1 and a_2 domains.

If a ferroelectric film is grown epitaxially on a cubic substrate such that $(001)_{film} \| (001)_{substrate}$, then the misfit due to the difference in lattice parameters of the film and the substrate may be described by the following misfit strain tensors for a_1-, a_2-, and c-domains, respectively:

$$\widehat{\varepsilon}_1^M = \begin{pmatrix} \varepsilon_T' & 0 & 0 \\ 0 & \varepsilon_M & 0 \\ 0 & 0 & \varepsilon_M \end{pmatrix} \quad \widehat{\varepsilon}_2^M = \begin{pmatrix} \varepsilon_M & 0 & 0 \\ 0 & \varepsilon_T' & 0 \\ 0 & 0 & \varepsilon_M \end{pmatrix}$$

$$\widehat{\varepsilon}_3^M = \begin{pmatrix} \varepsilon_M & 0 & 0 \\ 0 & \varepsilon_M & 0 \\ 0 & 0 & \varepsilon_T' \end{pmatrix} \tag{2}$$

Here $\varepsilon_T' = \varepsilon_M + \varepsilon_T$, $\varepsilon_M = (a - a_S)/a_S$ is the misfit strain between the substrate and one side of the base of the tetragonal substrate, $\varepsilon_T = (c - a)/a$ is the tetragonality of the film, and a_S is the lattice parameter of the substrate. The foregoing tensors are obtained from the linear relation between the self-strain and the misfit strain for each variant, since the strains are small [53].

The elastic energy of heterostructures may be reduced by the formation of polydomain structures (shown in the outer circle of Fig. 5) consisting of a uniform mixture of two of the three variants. The domains in the polydomain structures are related to each other as twins; i.e., their relative strains,

$$\Delta\widehat{\varepsilon}_{ij} = \widehat{\varepsilon}_i^M - \widehat{\varepsilon}_j^M = \widehat{\varepsilon}_i - \widehat{\varepsilon}_j \quad (i, j = 1, 2, 3) \tag{3}$$

are twinning shear. The twinning shear plane is the common plane for both domains, and thus the twin domains are compatible and stress-free. It means that the strain difference $\Delta\widehat{\varepsilon}_{ij}$ satisfies the equation of compatibility [5],

$$\mathbf{m} \times \Delta\widehat{\varepsilon}_{ij} \times \mathbf{m} = 0 \tag{4}$$

where $\mathbf{m}$ is a vector normal to the interdomain interface, which is a twinning plane. There are two solutions of Eq. (4), and thus each pair of domains can form two polytwins (or polydomains). For example, the tetragonal domains (1) and (2) with

$$\Delta\widehat{\varepsilon}_{12} = \widehat{\varepsilon}_1^M - \widehat{\varepsilon}_2^M = \begin{pmatrix} \varepsilon_T & 0 & 0 \\ 0 & -\varepsilon_T & 0 \\ 0 & 0 & 0 \end{pmatrix} \tag{5}$$

may form a polytwin with interfaces along (110), as shown in Figure 6, or a polytwin with interfaces along $(\bar{1}10)$. At the same domain fraction, both polytwins have the same average self-strain but different rotations. However, the rotations are not essential if the strains are small and should be neglected in the linear theory. The same holds for difference between $(\widehat{\varepsilon}_i - \widehat{\varepsilon}_j)$ and $(\widehat{\varepsilon}_i^M - \widehat{\varepsilon}_j^M)$ in Eq. (3) and the difference between ε_T' and $\varepsilon_M + \varepsilon_T$ in Eq. (2). Since both polytwins of the same domain composition considered identical, only three different polytwins of a tetragonal phase are shown in Figure 5. The (1, 2) structure is denoted in the literature as the $a_1/a_2/a_1/a_2$ configuration; the (1, 3) polytwin and the (2, 3) polytwin are referred to as simply the $c/a/c/a$ pattern [46], because, as we later show, their free energy densities are identical in the absence of external fields. We use both notations interchangeably.

2.2. Thermodynamics of Polydomain Formation

The free energy density of a polydomain film (i, j) composed of any of the two variants i and j, with self-strains $\widehat{\varepsilon}_i$ and $\widehat{\varepsilon}_j$, $i, j = 1, 2, 3$ and $i \neq j$, is a function of the fraction of the domain j, α_j, and consists of three components:

$$f_{i/j}(\alpha_j) = f_E + f_{\text{Micro}} + f_\gamma \tag{6}$$

The first term of Eq. (6), f_E, is the elastic energy density due to the misfit between the polydomain film and the substrate, $\widehat{\varepsilon}_i^M$ and $\widehat{\varepsilon}_j^M$. The misfit results in internal macrostresses that are uniformly distributed in the film [53]. If a fine and uniform mixture of platelets of compatible phases i and j forms [i.e., the interdomain interface is coherent and thus stress-free, as defined by Eq. (4)], then the average misfit strain of the two-phase mixture is

$$\widehat{\varepsilon}^M(\alpha_j) = (1 - \alpha_j)\widehat{\varepsilon}_i^M + \alpha_j\widehat{\varepsilon}_j^M \tag{7}$$

and f_E is given by

$$f_E = \frac{1}{2}\widehat{\varepsilon}^M \cdot \mathbf{G} \cdot \widehat{\varepsilon}^M = \left[(1-\alpha_j)e_i^M + \alpha_j e_j^M - \alpha_j(1-\alpha_j)e_{ij}^I\right] \tag{8}$$

In this expression,

$$e_i^M = \frac{1}{2}\widehat{\varepsilon}_i^M \cdot \mathbf{G} \cdot \widehat{\varepsilon}_i^M \qquad e_j^M = \frac{1}{2}\widehat{\varepsilon}_j^M \cdot \mathbf{G} \cdot \widehat{\varepsilon}_j^M \tag{9}$$

$$e_{ij}^I = \frac{1}{2}\Delta\widehat{\varepsilon}_{ij} \cdot \mathbf{G} \cdot \Delta\widehat{\varepsilon}_{ij} \tag{10}$$

and $\Delta\widehat{\varepsilon}_{ij} = \widehat{\varepsilon}_i^M - \widehat{\varepsilon}_j^M = \widehat{\varepsilon}_i - \widehat{\varepsilon}_j$. The energy e_{ij}^I is the part of the misfit energy that depends only on the difference between the phases, because $\Delta\widehat{\varepsilon}_{ij}$ is a transformational self-strain. e_{ij}^I can be interpreted as the energy of indirect interaction between the phases throughout a passive matrix. As follows from Eq. (8), the misfit energy of a polydomain film is always lower than the energies of single-domain states, and the energy defect grows with the indirect interaction energy e_{ij}^I. In the foregoing equations, $\mathbf{G}$ is the planar elastic modulus given by [7]

$$\mathbf{G}(\mathbf{n}) = \mathbf{C} - \mathbf{Cn}(\mathbf{nCn})^{-1}\mathbf{nC} \tag{11}$$

which depends on the orientation of the interfaces between layers ($\mathbf{n}$ is a vector normal to the interface) and the elastic modulus tensors $\mathbf{C}$ of single crystalline ferroelectric film and

Table I. Misfit Energies, e_i^M, of the Three Tetragonal Variants

Variant	Misfit energy, e_i^M
(1)	$\dfrac{Y}{2(1 - \nu^2)}\left[2\varepsilon_M^2(1 + \nu) + 2\varepsilon_M\varepsilon_T(1 + \nu) + \varepsilon_T^2\right]$
(2)	$\dfrac{Y}{2(1 - \nu^2)}\left[2\varepsilon_M^2(1 + \nu) + 2\varepsilon_M\varepsilon_T(1 + \nu) + \varepsilon_T^2\right]$
(3)	$\dfrac{Y}{(1 - \nu)}\varepsilon_M^2$

Table II. Energy of Indirect Interaction, e_{ij}^I, of Possible Polydomain Structures

Domain structure	Energy of indirect interaction, e_{ij}^I
(1, 2)	$\dfrac{Y}{(1 + \nu)}\varepsilon_T^2$
(1, 3)	$\dfrac{Y}{2(1 - \nu^2)}\varepsilon_T^2$
(2, 3)	$\dfrac{Y}{2(1 - \nu^2)}\varepsilon_T^2$

substrate. To make the results less complicated, the difference between elastic properties of the phases is neglected hereafter; i.e., the film/substrate system is considered elastically homogeneous. For a film with thickness h grown on substrate with thickness h_S such that $h_S \gg h$, this simplification is not important, because the stress due to layer misfit is concentrated in the film and $\mathbf{C}$ can be taken as the elastic modulus tensor of the film. Furthermore, the elastic inhomogeneity is small and can be neglected if the film/substrate couple is made up of similar materials.

For elastically isotropic media, $\mathbf{G}$ can be expressed in terms of the Young modulus Y and the Poisson ratio ν of the film only [55], and Eqs. (9) and (10) can be evaluated with the well-known formula for the plane stress state,

$$e = \frac{Y}{2(1 - \nu^2)}\left[\varepsilon_{xx}^2 + \varepsilon_{yy}^2 + 2\nu\varepsilon_{xx}\varepsilon_{yy}\right] \tag{12}$$

where x and y are the principal axes of the two-dimensional misfit strain and ε_{xx} and ε_{yy} are principal strains. Therefore, the misfit energies e_i^M and e_j^M and the energy of indirect interaction e_{ij}^I can be determined in terms of the misfit strain, tetragonality, and the elastic constants of the film, as illustrated in Tables I and II.

The second term of Eq. (6), f_{Micro}, is the energy of the microstresses [54]. Microstresses arise from periodic deviation of the actual misfit from the average misfit on the film–substrate interface. Micromisfit creates sign-alternating sources of stress with zero average power. According to the St. Venant principle, the stress field disappears in a distance about the period of source distribution. Therefore, microstresses are localized in a preinterface layer with a thickness approximately equal to the domain period D (Fig. 7). The problem of microstresses is very similar to the classical problem of stray fields of polydomain

Structure   Stress Distribution

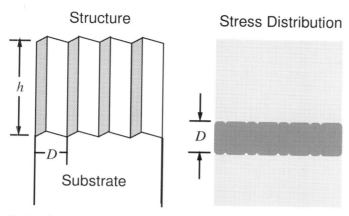

Fig. 7. Structure and model of microstress distribution at the interface between a polytwin structure and the substrate.

ferromagnetic and ferroelectric materials [9, 62]. For elastic domains, the first results were obtained in the 1970s [7, 11]. More accurate mathematical descriptions of the microstress energy have been recently developed by either Fourier analysis of the interfacial strains using Green's tensor function method [15, 45, 51, 52] or the methods of continuum dislocations and disclinations [47, 49]. These have similar forms to the energy of the stray fields of a plane-parallel arrangement of 180° domains in a ferromagnetic material developed by Kooy and Enz [62]. The exact expression for the microstress energy of a polydomain layer of thickness h embedded in an infinite elastic matrix as shown in Figure 7 has been developed by Sridhar et al. [51] and is given by

$$f_{\text{Micro}}\left(\alpha_j, \frac{D}{h}\right) = \frac{D}{2\pi^3 h} e_{ij}^{\text{I}} \sum_{n=1}^{\infty} \frac{1}{n^3} \sin^2(n\pi\alpha_j)$$
$$\times \left[1 - \exp\left(-2n\pi\frac{h}{D}\right)\right] \tag{13}$$

Recently, it was shown that the foregoing expression for the energy of microstresses of polydomain structures with $D/h \leq 1$ can be effectively approximated as [54]

$$f_{\text{Micro}} = \xi\alpha_j^2(1-\alpha_j)^2 e_{ij}^{\text{I}} \frac{D}{h} \tag{14}$$

where ξ is a numerical coefficient equal to 0.27.

The last term in Eq. (6) is the energy of the interdomain interfaces and is given by

$$f_\gamma = 2\sqrt{2}\frac{\gamma}{D} \tag{15}$$

where γ is the specific surface energy of the interdomain interface per unit area.

Therefore, by combining the individual components [i.e., Eqs. (8), (14), and (15)], the free energy density of a polydomain structure can be obtained. Minimization with respect to the domain period D, i.e., $\partial f_{i/j}/\partial D = 0$, yields the equilib-

rium domain period as

$$D^0 = \frac{\sqrt{lh}}{\alpha_j(1-\alpha_j)} \tag{16}$$

where $l = (2\sqrt{2}\gamma)/(\xi e_{ij}^{\text{I}})$ is a characteristic length in the range of 1–10 nm. For example, it has been experimentally shown [13] that $l \cong 2$ nm for polytwin films of a $YBa_2Cu_3O_{7-\delta}$ (YBCO) superconductor on (001) MgO substrates. Substitution of the equilibrium domain period back into Eq. (6) results in

$$f_{i/j}(\alpha_j) = f_{\text{E}} + e_{ij}^{\text{I}}\alpha_j(1-\alpha_j)\sqrt{h_{\text{cr}}/h} \tag{17}$$

and minimization with respect to α_j yields the equilibrium domain fraction

$$\alpha_j^0 = \frac{1}{2}\left[1 - \frac{\Delta e_{ij}^{\text{M}}}{(1-\eta)e_{ij}^{\text{I}}}\right] \tag{18}$$

where $\Delta e_{ij}^{\text{M}} = e_j^{\text{M}} - e_i^{\text{M}}$, $\eta = \sqrt{h_{\text{cr}}/h}$, and $h_{\text{cr}} = (8\sqrt{2}\xi\gamma)/e_{ij}^{\text{I}}$ is the critical thickness for the formation of a (i, j) polydomain structure below which this structure is not favored energetically and the film is in a single-domain state. The critical thickness is obtained by comparing the free energy densities of a single-domain state and a polydomain structure, i.e., $f_{i/j} = f_i = e_i^{\text{M}}$. Recent experimental studies indicate that the critical thickness may range from 10–150 nm [15, 25, 26, 39], depending on the film-substrate system and deposition methods and conditions.

The equilibrium free energy density of the polydomain structure can then be found by combining Eq. (17) with Eq. (18) and substituting for f_{E} [Eq. (8)] as

$$f_{i/j}(\alpha_j^0) = e_i^{\text{M}} - \frac{1}{4}e_{ij}^{\text{I}}(\alpha_j^0)^2 \qquad i, j = 1, 2, 3 \quad i \neq j \tag{19}$$

Note that there is also an additional but necessary condition imposed by Eq. (19) that must be met for a polydomain structure to be stable. The equilibrium free energy density, $f_{i/j}(\alpha_j^0)$, should have a minimum for $0 < \alpha_j^0 < 1$ if there exists a stable polydomain structure with the domain fraction α_j^0, i.e.

$$0 < \frac{1}{2}\left[1 - \frac{\Delta e_{ij}^{\text{M}}}{(1-\eta)e_{ij}^{\text{I}}}\right] < 1 \tag{20}$$

This equation ensures that the polydomain structure has a lower free energy density than the single-domain states with energies e_i^{M} or e_j^{M}. It also forces another condition on the critical thickness for polydomain formation given by

$$\frac{h_{\text{cr}}}{h} \leq \left[1 - \frac{\Delta e_{ij}^{\text{M}}}{e_{ij}^{\text{I}}}\right]^2 \tag{21}$$

which is almost always met for films with thicknesses slightly greater than the critical thickness. Equation (21) for a $(1, 3)$ or a $(2, 3)$ polydomain structure can be expressed in terms of the misfit strain and tetragonality as

$$\frac{h_{\text{cr}}}{h} \leq \frac{4[\varepsilon_{\text{M}}(1+\nu) + \varepsilon_{\text{T}}]^2}{\varepsilon_{\text{T}}^2} \tag{22}$$

2.3. Domain Stability Map

The next task is to find the equilibrium microstructures of the film after its transformation to the tetragonal phase state. All possible microstructures, both single-domain and polytwins, are presented in Figure 5. It is necessary to find the conditions of their stability and the equilibrium domain fractions in polytwins.

The equalities $\alpha_j^0 = 0$ and $\alpha_j^0 = 1$ determine the "second-order type" transitions between polytwin and single-domain states

$$\begin{aligned}
\alpha_j^0 = 0, \quad i \leftrightarrow (i, j): & \quad (1 - \eta)e_{ij}^I - \Delta e_{ij}^M = 0 \\
\alpha_j^0 = 1, \quad j \leftrightarrow (i, j): & \quad -(1 - \eta)e_{ij}^I - \Delta e_{ij}^M = 0
\end{aligned} \tag{23}$$

and the "first-order type" transitions between polytwins correspond to

$$f_{1/2} = f_{1/3}: \quad (\alpha_2^0)^2 e_{12}^I = (\alpha_3^0)^2 e_{13}^I \tag{24a}$$

$$f_{1/2} = f_{2/3}: \quad (\alpha_2^0)^2 e_{12}^I = (\alpha_3^0)^2 e_{23}^I \tag{24b}$$

$$f_{1/3} = f_{2/3}: \quad (\alpha_3^0)^2 e_{13}^I = (\alpha_3^0)^2 e_{23}^I \tag{24c}$$

The foregoing equations determine the areas of stability of polydomain and single-domain structures and lead to the construction of a domain stability map because the thermodynamics of an epitaxial film of a certain thickness is dictated by only two parameters, ε_M and ε_T, and can be graphically presented as a stability diagram on the ε_M–ε_T plane.

To develop this phase diagram, we first calculate the equilibrium domain fractions with the aid of Eq. (18) and Tables I and II in all possible polydomain structures in terms of the misfit strain ε_M and tetragonality ε_T. The equilibrium fraction of variant (3) in a (1, 3) structure is equal to its fraction in a (2, 3) structure and is given by

$$\alpha_3^0 = \frac{\varepsilon_T(2 - \eta) + 2\varepsilon_M(1 + \nu)}{2\varepsilon_T(1 - \eta)} \tag{25}$$

The fraction of variant (2) in a (1, 2) structure is $\alpha_2^0 = 1/2$ and is independent of the misfit strain, tetragonality, and film thickness.

For the sake of simplicity, let us assume that the film thickness is much greater than the critical thickness for polydomain formation of structures (1, 2), (1, 3), and (2, 3) such that $\eta = 0$. Therefore, the free energy density is reduced to the energy of macrostresses, f_E [see Eq. (17)]. The equilibrium domain fraction of variant (3) in a (1, 3)/(2, 3) structure can be simplified as

$$\alpha_3^0 = \frac{(1 + \nu)\varepsilon_M}{\varepsilon_T} + 1 \tag{26}$$

whereas $\alpha_2^0 = 1/2$. Table III lists the equilibrium elastic energies, $f_E(\alpha_j^0) = f_E^0$, for polydomain structures together with the elastic energies of single-domain states, which are equivalent to the misfit energies in Table I. The elastic energies of single-domain (1) and (2) structures are equal. It means that these structures are elastically equivalent and the same argument is also true for polydomain (1, 3) and (2, 3) structures

Table III. Equilibrium Elastic Energy Densities of All Possible Domain Structures

Domain structure	Equilibrium elastic energy density, f_E^0
(1) or (2)	$\dfrac{Y}{2(1 - \nu^2)}\left[2\varepsilon_M^2(1 + \nu) + 2\varepsilon_M\varepsilon_T(1 + \nu) + \varepsilon_T^2\right]$
(3)	$\dfrac{Y}{(1 - \nu)}\varepsilon_M^2$
(1, 2)	$\dfrac{Y}{4(1 - \nu)}(2\varepsilon_M + \varepsilon_T)^2$
(1, 3) or (2, 3)	$\dfrac{Y\varepsilon_M^2}{2}$

since their elastic energies are identical. Therefore, their formation is equally probable, and, we hereafter refer to them as the (1, 3)/(2, 3) or the $c/a/c/a$ structure whenever a distinction is not necessary.

In Table III, it can be seen that for any given pair of ε_M and ε_T ($\varepsilon_T \geq 0$), energies of single-domain states (1) or (2) are higher than those of either single-domain (3), two-domain (1, 2), or (1, 3)/(2, 3). Therefore, single-domain structures with their tetragonal axes in the plane of the film–substrate interface are not stable. When equilibrium between (1, 2) and (1, 3)/(2, 3) is considered [Eq. (21a)], the following relation defining the equilibrium between those domain structures can be obtained:

$$\varepsilon_T = \left[-2 \pm \sqrt{2(1 - \nu)}\right]\varepsilon_M \tag{27}$$

If the positive root of this relation is used to calculate the equilibrium domain fraction of variant of type (3) in a (1, 3)/(2, 3) structure, Eq. (26), it turns out to be negative (with $\nu = 0.3$ hereafter), which is physically impossible. The negative root is thus the boundary between (1, 2) and (1, 3)/(2, 3) on an ε_M–ε_T plane on which $\alpha_3^0 \cong 0.59$.

The boundary between (3) and (1, 3)/(2, 3) is given by the relation $\alpha_3^0 = 1$, i.e., $\varepsilon_M = 0$ for $\eta = 0$. Consequently, whenever the film is in compression ($a > a_s$, i.e., $\varepsilon_M > 0$), it stabilizes in a single-domain pattern consisting of the variant that has its tetragonal axis perpendicular to the film–substrate interface. The line

$$\varepsilon_T = -(1 + \nu)\varepsilon_M \tag{28}$$

determined from the condition $\alpha_3^0 = 0$, is the limit of stability of (1, 3)/(2, 3).

The foregoing discussion can be visualized in terms of a domain stability map illustrated in Figure 8 in the coordinates of misfit strain and tetragonality. The solid line given by the negative root of Eq. (24) separates the (1, 3)/(2, 3) polydomain structure from the (1, 2) structure. Along this line, the fraction of the variant (3) is 0.59 and increases up to 1 when the vertical axis is approached. The vertical axis is the phase stability limit of the (1, 3)/(2, 3) structure, which eventually transforms to single-domain (3) state when $\varepsilon_M > 0$.

Within the stability region of the (1, 3)/(2, 3) structure, the fraction of variant (3) (i.e., the c-domain fraction) is determined

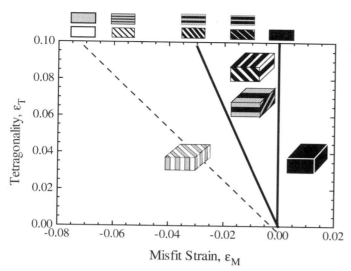

Fig. 8. Domain stability map for $\eta = 0$ showing stability regions of possible domain structures [55].

by Eq. (26). The $(1, 3)/(2, 3)$ structure abruptly transforms to the $(1, 2)$ structure with $\alpha_2^0 = 0.5$ when α_3^0 reaches 0.59 as misfit strain decreases. The $(1, 3)/(2, 3)$ structures with $\alpha_3^0 < 0.59$ become metastable and eventually unstable when α_3^0 reaches 0. The dashed line in Figure 8 given by Eq. (28) corresponds to the loss of metastability. The fraction of variant (3) in the $(1, 3)$ or the $(2, 3)$ structure approaches zero, and since neither the single-domain structures (1) nor (2) can be stable, a switch to a $(1, 2)$ domain mixture occurs.

The map shown in Figure 8 is applicable to any tetragonal film on a cubic substrate. Depending on the values of the misfit strain and the tetragonality, not only the stable domain structure but also the fractions of variants in a polydomain mixture can be deduced from the location of the ε_M–ε_T pair on this map. The three lines in Figure 8 are really "equifraction" lines on which α_3^0 is constant and additional equifraction lines can be introduced into the stability region of the $(1, 3)/(2, 3)$ structure by solving Eq. (26) in the range $0.59 < \alpha_3^0 < 1$. Since both ε_M and ε_T are temperature-dependent parameters, the map allows one to estimate the temperature dependence of domain fractions and conveniently combines the effects of lattice mismatch between the film and the substrate, the structural phase transformation, and the differences in the thermal expansion coefficients of the film and the substrate.

The domain stability maps for $0 < \eta \leq 1$ (i.e., for films with thickness h close or comparable to h_{cr}) are discussed in detail later in conjunction with the interpretation of experimental results. However, it is worth noting that for films with $\eta \neq 0$, the domain stability map is essentially the same in terms of the stability regions and the difference is only in the slopes of the stability lines and thus the areas of the stability regions (see the Appendix).

The domain stability map shows domain stability regions only for films with positive tetragonality. For films with a negative tetragonality, the lines that divide phase stabil-

ity regions should be extrapolated to negative values of ε_T. This can be particularly useful for epitaxial films of ferroelectric KNbO$_3$, which can be considered to have a negative tetragonality at room temperature. KNbO$_3$ undergoes a tetragonal–orthorhombic transformation at 225°C, and since the orthorhombic lattice parameters a_{or} and b_{or} are close, it can be assumed to have a negative tetragonality at room temperature. With some modifications, the domain stability map can be used to explain recent experimental observations of polydomain architectures of KNbO$_3$ films deposited on various cubic and pseudocubic substrates [44].

The theory developed herein is based on the concepts of linear elasticity. Its application to the ferroelectric transformation becomes doubtful near the critical temperature. For the transformations that are close to seconf-order type, the elastic properties near the transformation temperature are strongly temperature dependent and nonlinear. However, as shown in this section, the positions of the boundaries on the stability map do not depend on elastic moduli. Therefore, it is likely that this diagram is applicable at least qualitatively even near critical temperature. This assumption should of course be checked through nonlinear elastic theory calculations.

2.4. Effect of External Fields on the Domain Stability Map

The equilibrium between domain states and thus the domain stability maps are governed by the Gibbs free energy, rather than the free energy, when external electrical or stress fields are applied during cooling down from the deposition temperature. The Gibbs free energy density of a polytwin structure (i, j) is given by

$$\varphi_{i/j}(\alpha_j) = f_{i/j} - \alpha_j \widehat{\sigma} \cdot \Delta \widehat{\varepsilon}_{ij} - \alpha_j \mathbf{P}_s \cdot \mathbf{E} \qquad (29)$$

where $f_{i/j}$ is defined by Eq. (6), $\widehat{\sigma}$ is the external applied stress tensor, the product $\widehat{\sigma} \cdot \Delta \widehat{\varepsilon}_{ij}$ is the energy of the applied stress field, $\mathbf{P}_s$ is the vector of saturation polarization, $\mathbf{E}$ is the external electric field vector, and the product $\mathbf{P}_s \cdot \mathbf{E}$ is the electrostatic energy of the applied electrical field. A similar analysis as in Section 2.2 gives the equilibrium domain fraction as

$$\alpha_j^0 = \frac{1}{2}\left[1 - \frac{\widehat{\sigma} \cdot \Delta \widehat{\varepsilon}_{ij} + \mathbf{P}_s \cdot \mathbf{E} + \Delta e_{ij}^M}{(1 - \eta)e_{ij}^I}\right] \qquad (30)$$

To simplify the problem, we now investigate only the effect of an external uniaxial stress field in different directions on the domain stability map for a polydomain film with $\eta = 0$. The discussion can be easily expanded to an external uniaxial electric field or to the combination of an uniaxial external mechanical and an electrical field. However, special attention should be paid to external electrical fields, since these fields stabilize the same orientational variant with an opposite sense of polarization when the electrical field is reversed.

For $\widehat{\sigma} \| [001]$, the stress tensor contains only one component, σ_{zz}. Therefore, the energy of the external stress in Eq. (29), $\widehat{\sigma} \cdot \Delta \widehat{\varepsilon}_{ij}$, reduces to $\sigma_{zz}\varepsilon_T$ for the $(1, 3)/(2, 3)$ structure and to 0 for the $(1, 2)$ structure. The latter means that a stress applied parallel to [001] does not alter the domain fractions of a

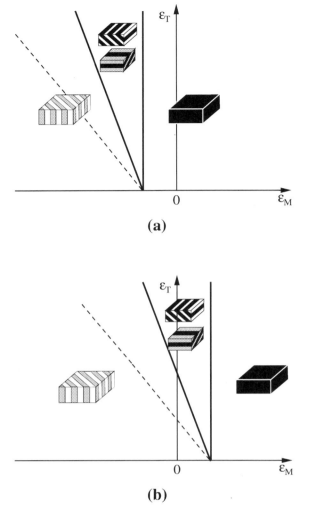

(a)

(b)

Fig. 9. Domain stability maps for a (a) tensile and (b) compressive uniaxial stress field parallel to the [001] direction applied during cooling down from the growth temperature [55].

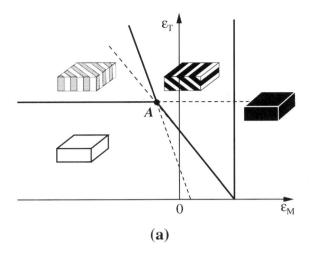

(a)

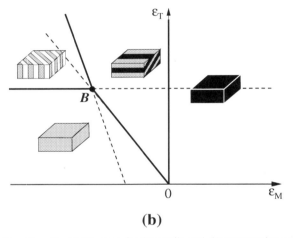

(b)

Fig. 10. Domain stability map for (a) tensile and (b) compressive uniaxial stress field parallel to the [100] direction applied during cooling down from the growth temperature [55].

(1, 2) structure, i.e., $\alpha_2^0 = 1/2$. However, the fraction of variant (3) in a (1, 3) or (2, 3) structure changes to

$$\alpha_3^0 = \frac{(1 + v)\varepsilon_M}{\varepsilon_T} + 1 + \frac{\overline{\sigma}}{\varepsilon_T} \quad (31)$$

where $\overline{\sigma} = \sigma_{zz}(1 - v^2)/Y$ is the normalized stress. Therefore, the boundary between single-domain (3) and polydomain (1, 3) or (2, 3), i.e., $\alpha_3^0 = 1$, shifts from $\varepsilon_M = 0$ to $\varepsilon_M = -\overline{\sigma}/(1+v)$. The limit of metastability ($\alpha_3^0 = 0$) is now given by

$$\varepsilon_T = -(1 + v)\varepsilon_M - \overline{\sigma} \quad (32)$$

and the boundary between the (1, 2) and the (1, 3)/(2, 3) structures is

$$\varepsilon_T = \left[-2 - \sqrt{2(1 - v)}\right]\left[\varepsilon_M + \overline{\sigma}/(1 + v)\right] \quad (33)$$

since at the boundary, $\varphi_{1/2} = \varphi_{2/3}$. The effect of the stress parallel to the [001] direction on the domain stability map is illustrated in Figure 9a for a tensile stress and in Figure 9b for

a compressive stress. An important conclusion is that a certain film-substrate system that normally would form a (1, 3)/(2, 3) structure (i.e., a $c/a/c/a$ pattern) or even a (1, 2) structure (i.e., $a_1/a_2/a_1/a_2$ pattern) in the absence of an external stress can be crystallized in the single-domain structure (3) (i.e., only c-domains) by applying a strong enough tensile stress along [001] during deposition and cooling down.

For a tensile stress $\widehat{\sigma} \| [100]$, the stress tensor contains only one component, σ_{xx}. Therefore, the energy of the external stress, $\widehat{\sigma} \cdot \Delta\widehat{\varepsilon}_{ij}$, becomes $-\sigma_{xx}\varepsilon_T$ for both the (1, 3) and the (1, 2) structures. The (2, 3) structure cannot form, because the tensile stress along [100] favors domain structures containing variant (1) or even the single-domain (1) structure. From Eq. (27), the fraction of variant (2) in a (1, 2) structure is

$$\alpha_2^0 = \frac{1}{2} - \frac{\overline{\sigma}}{2(1 - v)\varepsilon_T} \quad (34)$$

where $\overline{\sigma} = \sigma_{xx}(1 - v^2)/Y$ is the normalized stress. Since the formation of single-domain state (2) is not favored for positive tetragonalities, only the condition $\alpha_2^0 = 0$ is important. This yields a horizontal line on the domain stability map (Fig. 10a) given by $\varepsilon_T = \overline{\sigma}/(1 - \nu)$. This line defines the boundary below which the single-domain (1) is stable and above which the poly-domain (1, 2) is stable. The fraction of variant (3) in the (1, 3) structure is

$$\alpha_3^0 = \frac{(1 + \nu)\varepsilon_M}{\varepsilon_T} + 1 - \frac{\overline{\sigma}}{\varepsilon_T} \quad (35)$$

In the region where the single-domain state (1) is preferred over the polydomain (1, 2), i.e., $\varepsilon_T < \overline{\sigma}/(1 - \nu)$, the equilibrium between the single-domain structure (1) and the polydomain (1, 3) is given by $\alpha_3^0 = 0$ and is defined by $\varepsilon_T = -(1 + \nu)\varepsilon_M + \overline{\sigma}$. There is a triple point at which the single-domain state (1) and polydomains (1, 2) and (1, 3) coexist. This triple point is marked as A in Figure 10 and is defined by

$$\varepsilon_M = -\nu\overline{\sigma}/(1 - \nu^2) \qquad \varepsilon_T = \overline{\sigma}/(1 - \nu) \quad (36)$$

The boundary between single domain (3) and polydomain (1, 3) can be determined from the condition $\alpha_3^0 = 1$ as $\varepsilon_M = -\overline{\sigma}/(1 + \nu)$. When $\varepsilon_T > \overline{\sigma}/(1 - \nu)$, the boundary between the (1, 2) and the (1, 3) structure is obtained from the condition $\varphi_{1/2} = \varphi_{2/3}$ as

$$\varepsilon_T = \left[-2 - \sqrt{2(1 - \nu)} \right]\varepsilon_M + \frac{\overline{\sigma}}{(1 + \nu)} \left[1 - \frac{\nu\sqrt{2(1 - \nu)}}{(1 - \nu)} \right] \quad (37)$$

For a compressive stress $\widehat{\sigma} \| [100]$, as in the previous case, the stress tensor contains only one component, σ_{xx}. This time, however, single-domain (2) and polydomain (2, 3) are preferred over single-domain (1) and polydomain (1, 3). The energy of the external stress, $\widehat{\sigma} \cdot \Delta\widehat{\varepsilon}_{ij}$, becomes $-\sigma_{xx}\varepsilon_T$ and 0 for the (1, 2) and (2, 3) structures, respectively. The fraction of variants of type (2) is again given by Eq. (34), but the condition $\alpha_2^0 = 1$ [i.e., $\varepsilon_T = -\overline{\sigma}/(1 - \nu)$] defines the boundary between the single-domain (2) and polydomain (1, 2) structures. The fraction of variant (3) in the (2, 3) structure can be determined by Eq. (26), since the energy of the external stress is zero for this structure. Therefore, the boundary between the (2, 3) structure and the single-domain state (3) is given by $\alpha_3^0 = 1$ or $\varepsilon_M = 0$. When $\varepsilon_T < -\overline{\sigma}/(1 - \nu)$, there exists an equilibrium between structures (2) and (2, 3) defined by $\alpha_3^0 = 1$ or $\varepsilon_T = -(1 + \nu)\varepsilon_M$. If $\varepsilon_T > -\overline{\sigma}/(1 - \nu)$, then the boundary between structures (1, 2) and (2, 3) is

$$\varepsilon_T = \left[-2 - \sqrt{2(1 - \nu)} \right]\varepsilon_M - \frac{\overline{\sigma}}{(1 + \nu)} \left[1 + \frac{\sqrt{2(1 - \nu)}}{(1 - \nu)} \right] \quad (38)$$

which is obtained from the condition $\varphi_{1/2} = \varphi_{2/3}$. The domain stability map in the case of a compressive stress is shown in Figure 10b.

The triple point (marked B) at which the single-domain state (2) and polydomains (1, 2) and (2, 3) coexist has the coordinates

$$\varepsilon_M = \overline{\sigma}/(1 - \nu^2) \qquad \varepsilon_T = -\overline{\sigma}/(1 - \nu) \quad (39)$$

A tensile stress along the [010] direction results in a domain stability map which is identical to Figure 10a, whereas a com-

pressive stress along the same direction gives a map identical to Figure 10b. Therefore, it is theoretically possible to stabilize single-domain structures with their tetragonal axes in the plane of the film–substrate interface by applying a uniaxial stress along either the [100] or the [010] direction.

2.5. Cellular Polytwin Architecture

It is clear that the $c/a/c/a$ polytwin structure relaxes the internal stresses only partially by reducing the biaxial stress state to a uniaxial stress state. The internal stresses from lattice misfit can be completely eliminated if all of the three variants of the tetragonal phase are arranged such that the film has the same in-plane size as the substrate [6, 58–60, 63]. This can be achieved by a second-order polydomain structures (Fig. 11), which is constructed by twinning of the simple twins shown in Figure 5. For a homogenous mixture of domains, the condition for complete relaxation of the average stress is $\langle \varepsilon_{xx} \rangle = \langle \varepsilon_{yy} \rangle = 0$, where $\langle \varepsilon_{xx} \rangle$ and $\langle \varepsilon_{yy} \rangle$ are the average strains along the [100] and [010] directions, i.e.,

$$\alpha_1\widehat{\varepsilon}_1^M + \alpha_2\widehat{\varepsilon}_2^M + \alpha_3\widehat{\varepsilon}_3^M = \begin{pmatrix} 0 & 0 & 0 \\ 0 & 0 & 0 \\ 0 & 0 & (\alpha_1 + \alpha_2)\varepsilon_M + \alpha_3\varepsilon_T' \end{pmatrix} \quad (40)$$

Here α_1, α_2, and α_3 are the volume fractions of a_1-, a_2-, and c-domains, respectively ($\alpha_1 + \alpha_2 + \alpha_3 = 1$). Full relaxation is attained is attained for $\alpha_1 = \alpha_2$, and

$$\alpha_3^0 = \frac{2\varepsilon_M}{\varepsilon_T} + 1 \quad (41)$$

if $-(\varepsilon_T/2) < \overline{\varepsilon}_M < 0$. The architecture of the three-domain structures is dictated by the energy of the interdomain interfaces, the energy of the junctions where the three domains

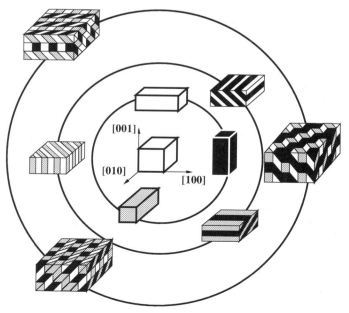

Fig. 11. Second-order polytwin structures consisting of a uniform mixture of all the three domains of the tetragonal phase [58].

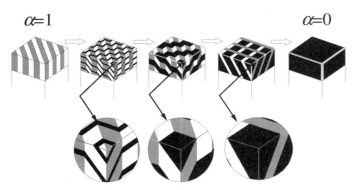

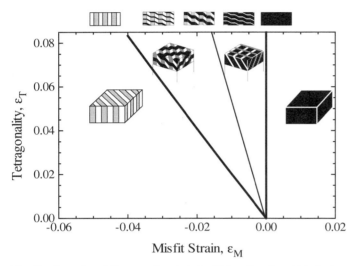

Fig. 12. Schematic evolution of the domain architecture as a function of the a-domain fraction, α.

Fig. 13. Domain stability map for relatively thicker films for which the formation of three-domain structures is possible.

come together, and the microstresses at the film–substrate interface. These factors increase the total energy of the system even though the elastic energy of macrostresses is effectively reduced to zero. Their interplay results in a critical thickness below which the three-domain state is not stable [58, 63]. This critical thickness is of one order of magnitude greater than the critical thickness for the simple twin structures. It can be shown that the second-order polydomain structure is favored for small c-domain fractions, whereas a cellular arrangement is more stable for relatively larger α_3 (Fig. 12).

The domain stability shown in Figure 13 is constructed by comparing the total elastic energies of all possible single-domain and polydomain structures following the formalism of Sections 2.2 and 2.3. It can be shown that a_1 and a_2 single domains are not stable for any given pair of ε_M and ε_T. Furthermore, the $c/a/c/a$ polytwins are also unstable, because for any pair of ε_M and ε_T, their energy is always larger than either the single domain c, the $a_1/a_2/a_1/a_2$ polydomain, or the three-domain structure. The lower and upper limits of the c-domain fraction α_3 [see Eq. (3)] for the three-domain structure define the phase stability boundaries between $a_1/a_2/a_1/a_2$ and the

three-domain state and the three-domain state and the single domain c, respectively ($\alpha_3 = 0$, i.e., $\varepsilon_T = -2\varepsilon_M$ and $\alpha_3 = 1$, i.e., $\varepsilon_M = 0$). Within the stability area of the three-domain structure, α_3 gradually changes from 0 to 1 as the misfit approaches 0.

2.6. Relaxation by Misfit Dislocations

Another important strain relaxation mechanism that usually takes place at the deposition temperature where the film is cubic is the formation of interface dislocations. These interface dislocations are called misfit dislocations because their presence reduces the internal stresses due to lattice mismatch. The equilibrium thermodynamic theory of misfit dislocations was developed by van der Merwe [64] and Matthews and Blakeslee [65, 66] and is now well established [61, 67, 68]. The elastic energy density of a pseudomorphic cubic film with thickness h at the growth temperature T_G on a cubic substrate with thickness h_s, such that $h_s \gg h$, is given by

$$f_G = \frac{Y}{1 - \nu} \left[\varepsilon_0(T_G) \right]^2 \tag{42}$$

where $\varepsilon_0(T_G) = [a_S(T_G) - a_0(T_G)/a_S(T_G)]$, $a_0(T_G)$, and $a_S(T_G)$ are the misfit strain, the lattice parameter of the substrate, and the lattice parameter of the film at T_G, respectively. The foregoing equation assumes that the film is elastically isotropic. The lattice misfit produces a biaxial strain state at the film–substrate interface with equal components.

If the film has a thickness exceeding a critical value, h_ρ, then the total free energy density of the film at the growth temperature can be partially reduced by the formation of a square grid of edge misfit dislocations to [46],

$$f_G = \frac{Y}{1 - \nu} \left[\varepsilon_0(T_G) - \rho |\mathbf{b}| \cos \lambda \right]^2 + \frac{\rho Y |\mathbf{b}|^2}{4\pi(1 + \nu)h} \ln \left(\frac{\alpha h}{|\mathbf{b}|} \right) \tag{43}$$

where ρ is the linear dislocation density, $|\mathbf{b}|$ is the magnitude of the Burgers vector, λ is the angle between the Burgers vector and a vector in the plane of the film–substrate interface, and α is the dislocation cutoff parameter [69].

The term $\rho |\mathbf{b}| \cos \lambda$ in Eq. (43) represents the amount of reduction in the misfit strain at the growth temperature, and the last term is the energy required to "create" the dislocations (i.e., the energy due to elastic strains close to the core of the dislocation). This term is similar to microstresses, since it disappears at a distance approximately 1.5–2 $|\mathbf{b}|$ from the dislocation line according to the St. Venant principle. This energy is infinitely large at the dislocation line, where the continuum approach becomes invalid [69, 70]. The critical thickness is the result of the interplay between these two opposing components, similar to the existence of a critical thickness for polydomain formation, which is due to the interplay between the elastic energy because of lattice misfit and microstresses.

The critical thickness for generation of the first misfit dislocations can be obtained by

$$\left(\frac{\partial f_G}{\partial \rho} \right)_{\rho=0} = 0 \tag{44}$$

from which the classical Matthews–Blakeslee (MB) criteria [65, 66] for a cubic perovskite material (the slip system is $\{101\}\langle 10\bar{1}\rangle$ based on TEM studies [71] and thus $\lambda = 45°$) can be derived as

$$h_\rho = \frac{L(h_\rho)}{\varepsilon_0(T_\mathrm{G})(1 - \nu)} \tag{45}$$

where

$$L(h_\rho) = \frac{a_0(1 + \nu)}{4\pi} \ln\left(\frac{2\sqrt{2}h_\rho}{a_0}\right) \tag{46}$$

with the substitutions of $\alpha = 4$, $\lambda = 45°$, and $\mathbf{b} = [10\bar{1}]$ (and thus $|\mathbf{b}| = \sqrt{2}a_0$). With the simplification [46]

$$\frac{L(h)}{\varepsilon_0(T_\mathrm{G})(1 - \nu)} \cong \frac{L(h_\rho)}{\varepsilon_0(T_\mathrm{G})(1 - \nu)} = h_\rho \tag{47}$$

the equilibrium dislocation density can be obtained as [55]

$$\rho(T_\mathrm{G}) \cong \frac{\varepsilon_0(T_\mathrm{G})}{a_0(T_\mathrm{G})}\left(1 - \frac{h_\rho}{h}\right) \tag{48}$$

Among the limitations of this theory are that it is based on the elastic continuum concept only and neglects the interaction of misfit dislocation and the dislocations present in the film. Furthermore, the real critical thickness for dislocation formation and the linear equilibrium dislocation density may differ significantly from the actual observed values because of kinetic reasons [61, 67, 68]. The thermodynamic theory assumes that the dislocations are somehow "created" at the film–substrate interface. Although the actual mechanisms of the misfit dislocation formation process are still not very well understood, there has been some speculation on the process of misfit dislocation generation [72]. If dislocations are present in the substrate on which the film is growing, then they grow naturally into the epitaxial film and reach the free surface of the film. These dislocations are sometimes called "threading" dislocations. Due to the epitaxial strain, these dislocations bend in the epitaxial film. Above a critical thickness, the strain energy stored in the film is sufficient to move the threading dislocations and to create misfit dislocations as it moves. However, misfit dislocations are also formed on dislocation-free substrates, where threading dislocations are not available for the generation of misfit dislocations. It is believed that in this case dislocations form by nucleation, most probably at defects at the free surface of the growing film [61].

The relaxation by misfit dislocations can be incorporated into the domain stability map using an "effective" substrate lattice parameter, $\bar{a}_\mathrm{s}$, given by [38, 55]

$$\bar{a}_\mathrm{s}(T) = \frac{a_\mathrm{s}(T)}{\rho(T_\mathrm{G})a_\mathrm{s}(T) + 1} \tag{49}$$

One can use $\bar{a}_\mathrm{s}$ to calculate the misfit strain ε_M assuming that no additional dislocations form during cooling down and the thermodynamically predicted equilibrium dislocation density at T_G is attained. We show in the following sections that the former assumption is valid based on TEM evidence. The latter assumation can be verified only by the determination of the actual

dislocation density via high-resolution TEM. However, if the film is much thicker than the theoretical critical thickness, as is the case for most of the film–substrate systems in this study, Eq. (48) provides a reasonably good approximation for the dislocation density.

3. EXPERIMENTAL METHODS

3.1. Deposition Techniques

The discussion and comparison of the numerous deposition methods of electronic ceramics is far beyond the scope of this chapter. A comprehensive discussion and details of the techniques mentioned here can be found in this handbook. However, it is worthwhile to give a very brief overview not only for the sake of completeness but also because of different internal stress states due to the choice of the deposition technique, which may alter the polydomain structure in ferroelectric thin films. The reason for the growth stresses is probably the nonequilibrium point defect concentration within the film, although the exact nature is not completely understood [4].

During the past decade, ferroelectric films and multilayer heterostructures have been successfully fabricated by various vapor phase deposition techniques, including plasma and ion-beam sputter deposition, pulsed laser deposition (PLD), molecular beam epitaxy (MBE), and chemical vapor deposition (CVD) and its derivative metalorganic-CVD (MOCVD). To produce high-quality multicomponent oxide films and heterostructures that are fully compatible with existing semiconductor process technology and to be able to fully exploit the unique characteristics of ferroelectric films, the following qualities are necessary [73]: (a) good stoichiometric control, (b) high deposition rate, (c) an ability to produce conformal deposition, and (d) scalability to cover large-area substrates with uniform composition and thickness. Sputter deposition is a classical method that is relatively easy and gives good compositional control. The physics and technology of numerous sputter and plasma approaches have been extensively discussed [74]. PLD is a fairly new technique that was introduced in the late 1980s to fabricate high-T_c superconducting oxides in thin film form [75]. Since then, it has been successfully applied to produce hundreds of different oxide and even metallic thin films [76]. Although both methods produce films with good stoichiometric control and have high deposition rates, issues of conformal deposition and scalability have hampered their commercial use. However, they are very widely used in research laboratories and universities, because they offer rapid exploration of new and exciting materials and have low startup costs. MBE is usually thought of mainly in the context of semiconductor films in which simple Knudsen cell sources are used. The growth of ferroelectrics requires reactive deposition of metals such as Pb and Ti in an oxygen atmosphere on a heated substrate. The frontrunner for the large-scale industrial fabrication of ferroelectric films and heterostructures is MOCVD [77], which is basically a CVD process [74] that uses metal-organic precursors. MOCVD produces a film that is uniform in composition and thickness over

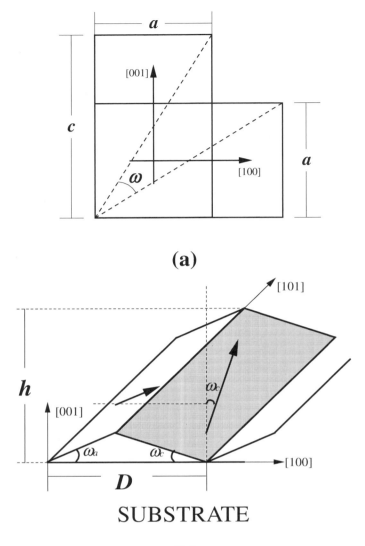

(a)

SUBSTRATE

(b)

Fig. 14. (a) The tilt in polydomain films consisting of a- and c-domains when brought together at the [101] interface due to the tetragonality of the lattice. (b) The accommodation of tilt in polydomain films consisting of a- and c-domains. The tetragonality is highly exaggerated.

large areas. Its fundamental limitation is the availability of suitable precursors, which reduces the technique's adaptability to new materials and even minor variations in the composition by doping.

3.2. Crystallographic Characterization via XRD: Determination of Domain Fractions, Tilts, and Internal Stresses

The basic parameters of the developed theory are the domain fractions and the domain period. The former can be obtained from standard θ–2θ XRD patterns, from θ-rocking curves; the latter, only by direct imaging via TEM or AFM.

The domain fractions in an epitaxial polydomain ferroelectric film consisting of a- and c-domains [i.e., the $(1, 3)/(2, 3)$ or the $c/a/c/a$ domain pattern], can be simply determined from the relative integrated intensities of $00l$- and $h00$-type reflections of the film [32, 41, 78, 79]. However, this method gives only qualitative and sometimes even inaccurate values because (a) if any one of the lattice parameters of the film and substrate are close, then the diffraction of the film may be hidden in the peak of the substrate, and (b) there is a tilt in the a- and c-domains away from the $(h00)$ or $(00l)$ planes of the substrate because of the tetragonality of the film. The former problem can be overcome by tilting the sample around the angle θ by as much as a few degrees. The latter problem, however, is more difficult. The theoretical tilt angle can be easily determined from the geometry of the polydomain structure (Fig. 14a) as

$$\omega = 2 \tan^{-1}\left(\frac{c}{a}\right) - \frac{\pi}{2} \tag{50}$$

The tetragonality of bulk PbTiO$_3$ at room temperature is ~1.06, corresponding to a tilt of ~3.5°. Experimental results show that this theoretically predicted tilt of a-domains is not observed [15, 26, 36] and is always less than ω. This is because the tilt is accommodated in both a- and c-domains, depending on their volume fractions. The theoretical value of ω can be detected only in films with $D/h \gg 1$, which should consist of large blocks of a- and c-domains rather than the thin lamellar domain configuration, and to our knowledge these types of polydomain microstructures have not been observed. The tilt accommodation is schematically illustrated in Figure 14b. From the geometry, the tilts in each domain can be calculated as

$$\frac{\tan \omega_a}{\tan \omega_c} = \frac{\alpha_3^0}{1 - \alpha_3^0} \tag{51}$$

where α_3^0 is the equilibrium domain fraction of c-domains, ω_a is the tilt in the a-domains, and ω_c is the tilt in the c-domains. Therefore, depending on the χ-angle resolution of a four-circle diffractometer (see Fig. 15 for the four different rotations), $h00$-type reflections of the film may disappear [26] and the relative integrated intensities of the $00l$ and $h00$ peaks of the film from the standard θ–2θ XRD pattern to calculate domain fractions usually gives an overestimation of c-domain abundance [25]. It should be noted that if the X-ray diffractometer is perfectly aligned with respect to the χ angle, even if the film has a significant amount of a-domains, then the $h00$-type peaks will be absent from the θ–2θ XRD pattern, because they will not satisfy Bragg's diffraction condition.

As discussed in Section 2.1, the domain boundaries may be along two different but elastically equivalent compatible planes. For the $(1, 3)$ structure with

$$\Delta\widehat{\varepsilon}_{13} = \widehat{\varepsilon}_1^M - \widehat{\varepsilon}_3^M = \begin{pmatrix} \varepsilon_T & 0 & 0 \\ 0 & 0 & 0 \\ 0 & 0 & -\varepsilon_T \end{pmatrix} \tag{52}$$

the interdomain interface may be along (101), as shown in Figure 5, or along $(\bar{1}01)$, whereas for the $(2, 3)$ structure, this interface lies on either (011) or $(0\bar{1}1)$. Figure 16 shows

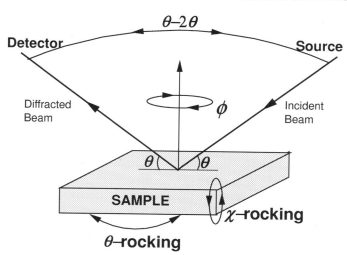

Fig. 15.   Simplified four-circle XRD geometry.

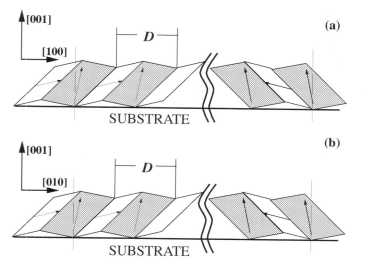

Fig. 16.   Schematic drawing showing a polydomain structure consisting of equivalent $c/a/c/a$ domain configurations on (a) the (100) plane and (b) (010) plane.

the schematic domain structure of two equivalent polydomain structures and their rotated versions. This domain structure results in a fourfold tilt (rather than the simple tilt as shown in Fig. 14b) of the $(h00)$ and $(00l)$ planes film away from the $[00l]$ direction of the substrate along $[h00]$, $[\bar{h}00]$, $[0k0]$, and $[0\bar{k}0]$. This fourfold splitting of the a- and c-domains is readily observed in θ-rocking curves [15, 42, 43] and θ–χ scans (area maps) [18, 25–28, 30, 36]. The fourfold symmetry of c-domains is less pronounced, but nonetheless existent, for films with a high amount of c-domains, since the angle ω_c is small. Therefore, to obtain the domain populations in the twinned film more accurately, the integrated intensities of the θ–χ scans of the $00l$ and $h00$ peaks of the film should be used. The integrated intensity of the area map of a $00l$ reflection corresponds to a lobe over the θ–χ plane. For a $h00$ reflection of the film, there are four such lobes arranged in a fourfold symmetry in the θ–χ plane [25–27]. The volume of a lobe can be estimated by as-

suming that the intensity has the form of a Gaussian distribution function in both θ and χ and that the standard deviation, s, of both θ and χ are equal as

$$V_c = \int_{-\infty}^{\infty} \int_{-\infty}^{\infty} I_c \, dA$$

$$\cong I_{\max} C(s) \int_{-\infty}^{\infty} \int_{-\infty}^{\infty} e^{-\theta^2/(2s^2)} e^{-\chi^2/(2s^2)} \, d\chi \, d\theta \quad (53)$$

where $C(s)$ is a function of the standard deviation, I_c is the intensity distribution function, and $I_{\max}$ is the maximum intensity of the c-domain lobe. For most cases, the standard deviation can be taken as the full width at half-maximum (FWHM). The integral can be evaluated by relating it to the error function as

$$V_c \cong 2\pi I_{\max} C(s) s^2 \quad (54)$$

Therefore, the fraction of c-domains, α_3^0, can be estimated as

$$\alpha_3^0 \cong \frac{V_c}{V_c + 4V_a} \quad (55)$$

where V_a is the volume of one of the a-domain lobes and is multiplied by 4 to take into account the fourfold symmetry. Usually, the domain fractions can be obtained by direct numerical integration of the θ–χ scans using computer software, making the foregoing approximations unnecessary. However, Eqs. (54) and (55) clearly illustrate that if the relative intensities of the θ-rocking curves of $00l$ and $h00$ peaks of the film are used (i.e., areas under the θ-rocking curves of $00l$ and $h00$ peaks) to calculate the c-domain fraction as

$$\alpha_3^0 \cong \frac{A_c}{A_c + 4A_a} \quad (56)$$

as given by Kang and Baik [43], then one must be sure that the FWHM of the c-domain lobe and the four a-domain lobes are equal, i.e., $s_a = s_c = s$, where s_a and s_c are the standard deviations of the intensity function of the a- and c-domains, respectively. Otherwise, significant errors might be introduced.

In addition to the domain fractions and tilts, the out-of-plane lattice parameters obtained from standard θ–2θ XRD patterns may also give information on the stress state of the film. If the structure consists of a- and c-domains, then the component of the strain tensor in a direction normal to the film–substrate interface can be obtained from these data as

$$\varepsilon_{zz}^a = \frac{a - a^0}{a^0} \qquad \varepsilon_{zz}^c = \frac{c - c^0}{c^0} \quad (57)$$

where a and c are the experimentally measured lattice parameters from $h00$ and $00l$ reflections of the film, respectively, and a^0 and c^0 are the lattice parameters of the free-standing (unstressed) film.

If a fine and uniform mixture of alternating platelets of a- and c-domains forms, then the strains ε_{zz}^a and ε_{zz}^c must be identical (i.e., $\varepsilon_{zz}^a = \varepsilon_{zz}^c = \varepsilon_{zz}$). If the film is assumed to be elastically isotropic, then this value is related to the in-plane components of the internal strain tensor via

$$\varepsilon_{zz} = -\frac{\nu}{1-\nu}(\varepsilon_{xx} + \varepsilon_{yy}) \qquad (58)$$

where ε_{xx} and ε_{yy} can be determined theoretically from the misfit strain tensors [Eq. (2)] for the $(1, 3)$ structure as

$$\varepsilon_{xx} = (1 - \alpha_3^0)(\varepsilon_T + \varepsilon_M) + \alpha_3^0 \varepsilon_M = (1 - \alpha_3^0)\varepsilon_T + \varepsilon_M$$

$$\varepsilon_{yy} = (1 - \alpha_3^0)\varepsilon_M + \alpha_3^0 \varepsilon_M = \varepsilon_M \qquad (59)$$

For the elastically equivalent $(2, 3)$ structure, the values of ε_{xx} and ε_{yy} should be interchanged. To determine the in-plane stresses, it should be noted that although there is a normal strain in the [001] direction, there is no constraint, and thus $\sigma_{zz} = 0$. Also, there is no shear strain such that $\tau_{xy} = \tau_{xz} = \tau_{yz} = 0$. Therefore, the in-plane internal stresses are determined by

$$\sigma_{xx} = \frac{Y}{1-\nu^2}(\varepsilon_{xx} + \nu\varepsilon_{yy})$$

$$\sigma_{yy} = \frac{Y}{1-\nu^2}(\nu\varepsilon_{xx} + \varepsilon_{yy}) \qquad (60)$$

For the $(2, 3)$ structure, the values of σ_{xx} and σ_{yy} should be interchanged. If the domain structure is the single-domain (3) state, then the foregoing equations become simpler, since $\varepsilon_{xx} = \varepsilon_{yy} = -\varepsilon_M$,

$$\sigma_{xx} = \sigma_{yy} = -\frac{Y}{1-\nu}\varepsilon_M \qquad (61)$$

and result in a biaxial stress state with equal orthogonal components. If there is relaxation by misfit dislocations, then the misfit strain in the foregoing equations should be replaced by the effective misfit strain. It should be noted that for the three-domain structures, as discussed in Section 2.5, the in-plane stresses are zero. Thus if this structure forms, then there will be no shifts of the a- and c-peaks from their unstressed (bulk) positions.

The determination of internal stresses by XRD methods in epitaxial semiconductor thin films is very common [80] and has been improved by X-ray double-crystal diffractometry [81] and high-resolution XRD [82, 83]. The state of the art in XRD methods for epitaxial semiconductor thin films now extends to the determination of film thickness [84], composition [85], and the effect of misfit dislocations [83, 86]. However, there is lack of research and publications on the measurement of internal stresses in epitaxial ferroelectric films, and the existing studies are limited to polycrystalline ferroelectric films [87–90]. There are multiple reasons for this. One reason is the obscurity of the lattice parameters of the stress-free lattice parameters of the thin film, i.e., a^0 and c^0, which represent the reference state for the calculation the out-of-plane strains [Eq. (54)]. Unstrained lattice parameters in thin films may differ from values for bulk materials because of thin film anomalies such as variations in stoichiometry, presence of a high concentration of point defects such as vacancies and interstitials due to nonequilibrium deposition techniques, and interface effects [91]. Furthermore, literature values for unstrained lattice parameters for some materials, or for materials that can be grown only in thin film form, such as PZT ($PbZr_{1-x}Ti_xO_3$) solid solutions, may not

be available. Another reason may be related to the tilt in the domains and relative domain fractions. As was shown, the domain fraction is a function of film thickness. If the film is somewhat nonuniform in thickness, then the resulting internal stress field is nonuniform within the film and cannot be determined using the simple equations for the internal stresses. For epitaxial single-domain PZT [92], relaxor ferroelectric [93, 94], and $(Ba,Sr)TiO_3$ films [95], the internal stresses were measured by XRD methods. The reference state for the strain was established by measuring the lattice parameter of the target used in the PLD process. The effects of internal stresses on electrical and electromechanical properties were theoretically determined and verified by dielectric and piezoelectric measurements.

3.3. Direct Imaging of Polydomain Structures

Due to the fine lateral length scale of the twins in ferroelectric films (10–500 nm), optical methods used for bulk ferroelectrics cannot be used to visualize the microstructure. The polydomain architecture in ferroelectric thin films can be readily observed by TEM, although sample preparation is somewhat tedious. One recent development in the imaging of ferroelectric films in the nanoscale is the implementation of AFM in the piezoresponse mode [96–106]. AFM techniques have also been used to investigate the scaling of ferroelectric properties [107–110]. Very briefly, piezoresponse microscopy is based on detection of the piezoelectric response of the ferroelectric thin film when probed using an alternating-current (ac) field. This response leads to a change in the film thickness, the sign of which depends on the polarization vector direction. Thus, regions of film with opposite polarization states will vibrate out of phase on the action of this ac field. The amplitude and phase of the film vibration give a measure of the magnitude and sign of the piezoelectric coefficient and hence that of the local polarization. Since the piezoresponse for the c-oriented domains is much larger than that for the a-domains, it is clearly possible to observe the a-domains as regions of low piezosignal. Another advantage of the AFM is that the average longitudinal piezoelectric coefficient as a function of the applied field can also be obtained in the piezoresponse detection mode with the "AFM tip/top electrode/ferroelectric/bottom electrode" configuration, which provides a virtually uniform electrical field in the thickness direction.

4. CORRELATION BETWEEN EXPERIMENT AND THEORY

In this section we compare experimental results with theoretical predictions. First, we review a special experiment designed to illustrate that the twins in epitaxial ferroelectric thin films are really elastic domains. Next, we discuss the temperature and film thickness dependence of the polydomain structures. In the subsequent subsection, we analyze experimentally the three-domain structure in relatively thicker films. Finally, we give domain stability maps for $BaTiO_3$ on some popular substrates.

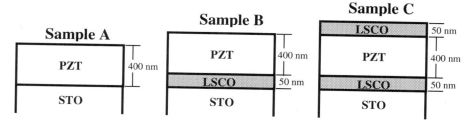

Fig. 17.   Three different types of samples used in the experiments [38].

This section is designed to illustrate the use of the theoretical model and the experimental techniques described in Sections 2 and 3, respectively, in conjunction with important characteristics of the twin structures in epitaxial ferroelectric films.

4.1. Effect of Electrode Layers on the Polydomain Structure

In capacitor or memory elements, ferroelectric films are sandwiched between metallic or conducting perovskite oxide electrode layers to form a multilayer heterostructure. Among others, $La_{0.5}Sr_{0.5}CoO_3$ (LSCO) electrodes provide better fatigue, retention, and imprint characteristics than conventional platinum electrodes [34]. In addition, the choice of LSCO as the electrode minimizes both epitaxial and thermal stresses, since its lattice parameter matches well with Pb-based perovskites and thermal expansion coefficients of perovskites are similar [$\sim(10$–$12) \times 10^{-6}\,°C^{-1}$] [111]. Recently, we reported that in samples of 400-nm-thick (001) $PbZr_{0.2}Ti_{0.8}O_3$ (PZT) films on (001) $SrTiO_3$ (STO) substrates with and without heteroepitaxially grown LSCO electrodes, identical domain structure and domain fractions were seen [38]. The important conclusion of this study was that the domain structure in epitaxial ferroelectric films is determined by mechanical factors only; i.e., the 90° domains in ferroelectric films are elastic domains. In this section we review the experimental results in conjunction with theoretical predictions. Three different types of samples, as illustrated in Figure 17, were deposited at 650°C (001) epitaxial-grade STO substrates by PLD. Epitaxial growth in the samples was established from ϕ-scans and the presence of only 00l-type reflections in the θ–2θ diffraction pattern. TEM was used to obtain plan-view and cross-sectional microstructures of samples A and B.

Figure 18a shows the θ–2θ XRD patterns from 40–50° (high χ-angle resolution) of all of the samples. The absence of the 200-PZT peak (the approximate location of which is marked in Fig. 18a) may lead to an incorrect conclusion that the PZT layer is completely c-axis oriented. As mentioned before, this is due to the tilting of the a-domains away from the (00l) planes of the substrate because of the tetragonality of the PZT film [$c/a \sim 1.05$ at room temperature, corresponding to a theoretical tilt angle of $\sim 2.9°$ via Eq. (50)]. The results are illustrated in Figures 18b and 18c for sample A. Samples B and C gave almost identical rocking curves and are not shown.

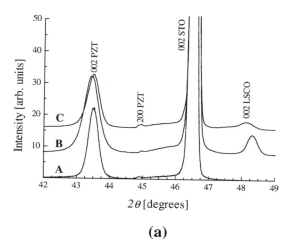

(a)

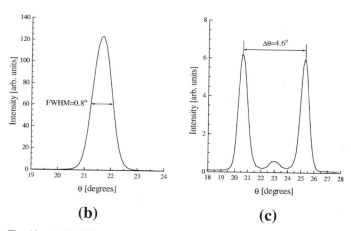

(b)   **(c)**

Fig. 18. θ–2θ XRD patterns (a) in the range 40–50° for samples A, B, and C, and θ-rocking curves of sample A around 002 PZT peak (b) and around 200 PZT peak (c) [38].

The rocking curves clearly indicate ·to the presence of a-domains and the domain structure is the $c/a/c/a$ pattern, i.e., the $(1,3)/(2,3)$ structure. The FWHM of the 002 rocking curves for all of the samples gives similar values (0.80° for A, 0.85° for B, and 0.81° for C). The FWHM of the 200 peaks are also close to 0.8°, which makes it possible to use the integrated intensities (i.e., the areas below the rocking curves) to estimate the domain populations in the PZT layer via Eq. (56). The c-domain fraction in samples A, B, and C are almost equal (0.83 for A, 0.85 for B and C).

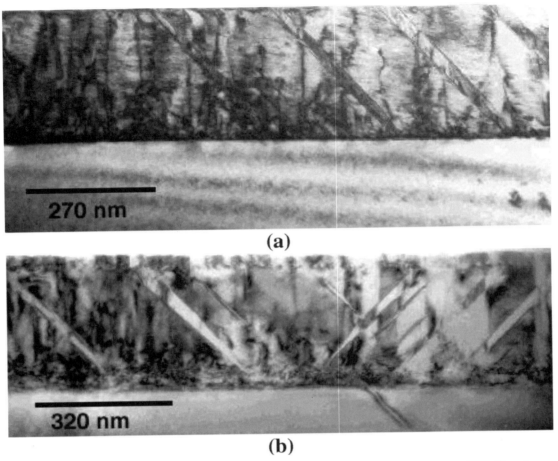

Fig. 19. Cross-sectional bright field images of (a) sample A and (b) sample C with orientation close to [001]. These images show that the *c*-domains are dominant in the PZT film. Vertical wavy lines across the film are dislocations formed before the transformation in PZT [38].

The tilt of the *a*-domains from the [002] direction of the substrate (the distance of the *a*-domain lobes divided by 2, i.e., $\Delta\theta/2$) is 2.30° for sample A, 2.35° for sample B, and 2.33° for sample C. These measured values of the tilt angles are less than the theoretically predicted value of 2.9° because, as discussed in Section 3.3, the tilt due to the tetragonality of the PZT layer is accommodated in both the *a*- and *c*-domains, causing also a fourfold splitting in the 002 PZT peak, which cannot be detected because of the resolution of the X-ray equipment. The tilt in the *c*-domains can be estimated using the relation (3.2) by taking $\omega_a \cong 2.32°$ and $\alpha_3^0 \cong 0.85$ as $\omega_c \cong 0.41°$.

Figures 19a and b show the cross-sectional TEM micrographs of samples A and C, respectively. The microstructure has a pattern consisting of alternating thin *a*- and *c*-domains with a (101) interdomain interface. The period of the alternation is approximately 200 nm (i.e., $D/h < 1$), and a very approximate estimate of the fraction of *c*-domains is ~0.85, which is consistent with XRD diffraction results. Numerous dislocations with the dislocation line normal to the film plane are present. The linear dislocation density in the film is estimated to be ~1×10^5–10^6 cm^{-1}. These threading dislocations emanate from the film–substrate interface of sample A and from

the bottom electrode–film interface of sample C and reach the free surface. The threading dislocations pass through the twins, which implies that the dislocation generation has preceded the domain formation process and that no additional dislocations were formed during the cooling-down from the deposition temperature. A similar microstructure was observed for sample B (not shown).

The formation of the microstructure of the 400-nm-thick PZT film on the STO substrate (sample A) is a result of a combination of two different mechanisms for strain relaxation: misfit dislocation generation at the deposition temperature and polydomain formation below T_C. One would expect that, since $a_s < a$ at room temperature (RT = 25°C), the PZT film should consist of only *c*-domains, because this configuration gives the minimum amount of strain energy density in the film when compared to a polydomain state consisting of a mixture of either *c*- and *a*-domains (*c/a/c/a* domain pattern) or two orientational variants of *a*-domains ($a_1/a_2/a_1/a_2$ domain pattern). However, the observed domain structure at RT is the *c/a/c/a* structure with the *c*-domain fraction equal to ~0.83 for sample A. At the deposition temperature ($T_G = 650°C$), the PZT film has a cubic structure with a lattice parameter of

$a_0 \cong 0.4001$ nm, resulting in $\sim$1.75% strain due to mismatch between the STO substrate and the film. The stresses in the PZT film are compressive and can be relaxed by misfit dislocation generation at the film–substrate interface. This relaxation can be taken into account using an "effective" substrate lattice parameter $\bar{a}_s$ defined in Eq. (47). The critical thickness h_ρ is around 8–10 nm for PZT on STO, calculated from the MB criteria [Eq. (43)]. If the equilibrium dislocation density is attained at T_G [$\rho \cong 4.2 \times 10^6$ cm^{-1}, from Eq. (46)], then the strain due to lattice mismatch is reduced by $\sim$98% ($h_\rho/h = 0.02$–0.025). Thus at T_G, the in-plane (or lateral) dimensions of the film and the substrate become almost identical, and the film is in an unstressed state.

When the deposition process is complete, the film is cooled down from T_G; when T_C is reached, the structural phase transformation occurs. In the interval $T_G > T > T_C$, we assume that the thermal strain due to the difference in the thermal expansion coefficients of STO and PZT (both $\sim$11 $\times 10^{-6}$ °C^{-1}) is negligible. Because of the cubic-tetragonal transformation, the in-plane dimensions of the film decreases, and the film can relax resultant tensile stresses by forming the $c/a/c/a$ polydomain structure.

The domain stability map for the 400-nm PZT film on (001) STO substrate is shown in Figure 20. To take into account the misfit dislocation relaxation at the deposition temperature, the map is constructed in terms of the "effective" misfit strain, $\bar{\varepsilon}_M = (a - \bar{a}_s)/\bar{a}_s$, rather than the misfit strain. The effect of film thickness is accounted for by taking $\eta = 0.25$. Experimental observations show that $h_{cr} = 25$ nm for epitaxial PbTiO$_3$ on (001) STO [25], and we assume that this thickness is approximately the same for the PZT film. The boundary

between the $c/a/c/a$ or $(1,3)/(2,3)$ structure and the single-domain c or (3) structure is given by $\varepsilon_T \cong -10.4\bar{\varepsilon}_M$, whereas its boundary to the $a_1/a_2/a_1/a_2$ or $(1,2)$ structure is defined as $\varepsilon_T \cong -3.0\bar{\varepsilon}_M$. (For the determination of the slopes, see the Appendix.) The ε_T–$\bar{\varepsilon}_M$ pairs are evaluated as a function of temperature and plotted in the map at 50°C intervals up to T_C, which are the solid circles in Figure 20. It can be seen that the stable domain structure is the $c/a/c/a$ pattern at temperatures below T_C. Equifraction lines that are the solutions for $0.6 < \alpha_3^0 < 1$ of Eq. (25) with $\eta = 0.25$ and $\nu = 0.3$ are shown as thin lines in the stability region of the $c/a/c/a$ structure in Figure 19.

The effective misfit strain is given by $\bar{\varepsilon}_M = (a - \bar{a}_s)/\bar{a}_s$ and its value can be calculated to be -0.0093 at RT using Eq. (47) with $\rho \cong 4.2 \times 10^6$ cm^{-1}. The tetragonality is $\sim$0.051, and Eq. (25) estimates the c-domain fraction as $\sim$0.85 [i.e., the equifraction line with $\alpha_3^0 \cong 0.85$ passes through the point $\{\bar{\varepsilon}_M(RT), \varepsilon_T(RT)\}$ in Fig. 20]. This theoretical prediction is in good agreement not only with the experimentally observed value of α_3^0 of sample A ($\sim$0.83), but also with the c-domain fractions of samples B and C ($\sim$0.85), which contain an intermediate 50-nm-thick LSCO layer as the bottom electrode.

4.2. Temperature Dependence of the Polydomain Structures

The temperature dependence of the domain structure has been addressed in several studies [4, 15, 28, 31, 43, 55]. As an example, the domain stability map for a 210-nm-thick (001) PbTiO$_3$ epitaxial film on (001) MgO is shown in Figure 21 [31]. The parameter η was determined from a similar study [15] as $\eta \cong 0.3$, where a c-domain fraction of 0.91 was observed for a 40-nm PbTiO$_3$ film on MgO deposited by MOCVD. Therefore, we assume that $h_{cr} \cong 20$ nm such that $\eta \cong 0.3$. The stabil-

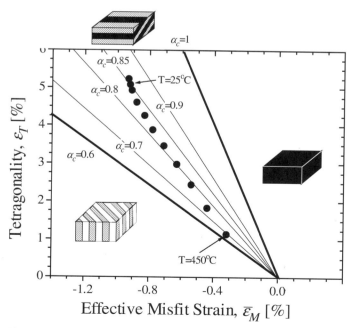

Fig. 20. Domain stability map for PZT on (001) STO. Filled circles represent the expected theoretical effective misfit strain and tetragonality pairs at different temperatures for a 400-nm-thick PZT film [38].

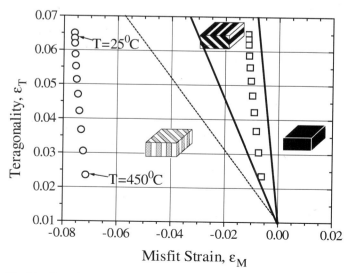

Fig. 21. Domain stability map for PbTiO$_3$ on (001) MgO for $\eta = 0.3$. Circles show the data for unrelaxed film, and squares are for a completely relaxed film in a temperature range of 0–450°C [31].

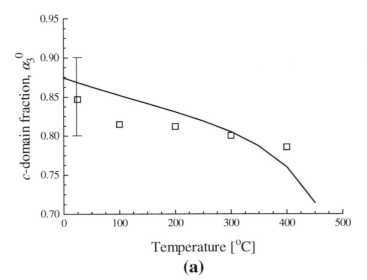

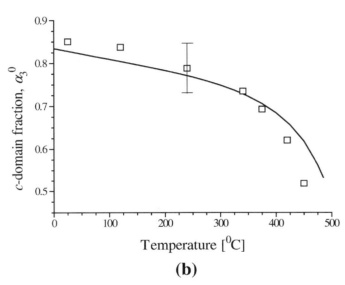

Fig. 22. (a) Fraction of c-domains of a 210-nm-thick PbTiO$_3$ film on (001) MgO as a function of temperature. Solid line shows theoretical data and open squares are experimental results [31]. (b) Plot of theoretical (solid line) and experimental (solid squares) fractions of c-domains 280-nm-thick PbTiO$_3$ film on (001) MgO as a function of temperature [15].

ity and metastability lines were determined using the approach summarized in the Appendix. If the relaxation by misfit dislocations is not taken into account, then the stable domain structure is an $a_1/a_2/a_1/a_2$ domain pattern (open circles). However, as discussed in the previous section, the experimentally observed structure is the $c/a/c/a$ pattern due to misfit dislocation formation at the deposition temperature, and the c-domain fraction gradually decreases as the transition temperature is approached. The theoretical c-domain fraction as a function of temperature, together with the experimental observations obtained from the ratio of 200 and 002 peaks [31], are shown in Figure 22a. It can be seen that the theoretical predictions provide a good fitting to the experimental results. As illustrated in Figure 22b,

the data of Kwak et al. [15] on the temperature dependence of c-domains show similar behavior (open squares) for the same film–substrate system. The same theoretical methodology of this section is equally applicable to this data using the cited 280-nm film thickness and $T_G = 550°C$ [15]. The solid line is the theoretical prediction for which the details were reported elsewhere [55]. Similar temperature dependence was also observed for epitaxial (Pb,La)TiO$_3$ films [43].

4.3. Thickness Dependence of Polydomain Structures

The thickness dependence of the domain structure has been investigated in detail [15, 25, 39, 112, 113]. It has been shown experimentally that there exists a critical thickness below which relaxation of internal stresses by polydomain formation is not feasible. For illustrative purposes, we summarize results of our own work in conjunction with theoretical predictions. Using PLD, (001) PbTi$_{0.8}$Zr$_{0.2}$O$_3$ films (PZT) of thicknesses varying from 60–400 nm were deposited on (001) LaAlO$_3$ substrates with top and bottom 50-nm-thick LSCO electrode layers at 650°C [39]. Results indicate that the 60-nm-thick sample has a single-domain c structure, whereas all of the other samples have a polydomain $c/a/c/a$ structure, and the c-domain fraction gradually decreases as the film thickness increases.

The critical thickness for misfit dislocation formation h_ρ is around 5–6 nm for PZT on LAO, calculated from the MB criteria [Eq. (43)]. If the theoretical equilibrium dislocation density is attained at T_G [from Eq. (46), $\rho \cong 1.09 \times 10^6$ cm^{-1} for the sample with PZT layer thickness of 60 nm and 1.2×10^6 cm^{-1} for 400 nm thickness], then the strain due to lattice mismatch is reduced by ~90% and ~98.5% for the samples with PZT layer thicknesses of 60 nm and 400 nm, respectively.

The critical thickness for domain formation, h_{cr}, is required for the construction of the domain stability maps and the theoretical estimation of the domain fractions. Experimental results show that h_{cr} is close to 60 nm. The domain stability maps for PZT films with thicknesses of 150 nm ($\eta = 0.63$) and 400 nm ($\eta = 0.39$) on (001) LAO substrate are shown in Figures 23a and b, respectively. These thicknesses define the experimental range for which polydomain formation was observed. The stability regions are determined by the thermodynamic equilibrium between possible domain structures and are a function of the film thickness, as illustrated in Figure 23. The boundary between the $c/a/c/a$ structure and the single-domain c structure is given by $\varepsilon_T \cong -4.04\bar{\varepsilon}_M$ and $\varepsilon_T \cong -6.60\bar{\varepsilon}_M$ for the 150-nm-thick and 400-nm-thick PZT layers, respectively. The boundary to the $a_1/a_2/a_1/a_2$ structure is defined for both thicknesses as $\varepsilon_T \cong -3.0\bar{\varepsilon}_M$. As before, the slopes were calculated numerically. The ε_T–$\bar{\varepsilon}_M$ pairs are evaluated as a function of temperature and plotted in the maps at 50°C intervals up to T_C, which are the solid squares in Figure 23. It can be seen that the stable domain structure is the $c/a/c/a$ pattern at room temperature for both thicknesses.

The c-domain fraction in the $c/a/c/a$ structure, α_3^0, can be evaluated theoretically using Eq. (25) with the substitution of

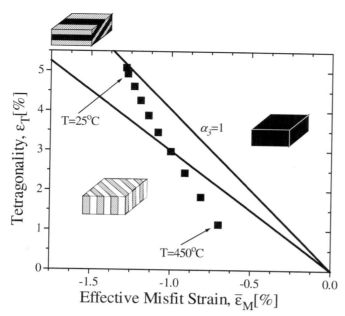

(a)

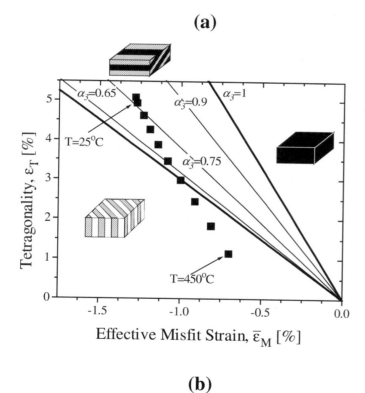

(b)

Fig. 23. Domain stability maps for PbTi$_{0.8}$Zr$_{0.2}$O$_3$ films deposited on (001) LaAlO$_3$ with thickness of (a) 150 nm and (b) 400 nm. The parameter η is 0.63 and 0.39, respectively.

the effective misfit strain. Equifraction lines that are the solutions of the above equation for α_3^0 are shown as thin lines in the stability region of the $c/a/c/a$ structure in Figure 23b. The effective misfit strain is given by $\overline{\varepsilon}_M = (a - \overline{a}_s)/\overline{a}_s$ where $\overline{a}_s$ is defined in Eq. (47). It should be noted that $\overline{\varepsilon}_M$ is also a function

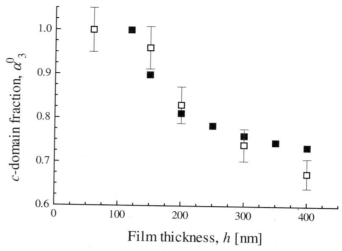

Fig. 24. Thickness dependence of the c-domain fraction of PbTi$_{0.8}$Zr$_{0.2}$O$_3$ films deposited on (001) LaAlO$_3$. Open squares represent experimental results, and solid squares are theoretical predictions.

of film thickness through $\overline{a}_s$ which depends on the dislocation density and thus the film thickness. This indicates that the thickness dependence of the domain fraction is due to an interplay between two relaxation processes; misfit dislocation formation at T_C and the formation of the polydomain structure below T_G. The theoretical results are illustrated in Figure 24 along with experimental observations. It can be seen that the theoretical prediction is in good agreement with experimental results.

The thickness dependence of the polydomain structure has also been experimentally studied by Kwak et al. [15] and Hsu and Raj [25] for PbTiO$_3$ films of varying thickness on (001) MgO and (001) SrTiO$_3$ substrates, respectively. Experimental results indicate that the polydomain structure is the $c/a/c/a$ structure and the fraction of c-domains is decreasing as the film thickness increases, a behavior similar to our results. Using the cited experimental conditions and the critical thickness for polydomain formation, i.e., $T_G = 550°C$ and $h_{cr} \cong 25$ nm for the data of Kwak et al. [15] and $T_G = 645°C$ and $h_{cr} = 30$ nm for the data of Hsu and Raj [25], a similar theoretical analysis can be carried out. The results show that the theoretical approach is in accordance with experimental results, as illustrated in Figures 25 and 26.

4.4. Three-Domain Architecture

Recently, the theoretically predicted three-domain architecture (see Section 2.4) has been observed experimentally by plan-view TEM [63] and by AFM [105, 106] in relatively thick samples (~450–500 nm) of (001) PbZr$_{0.2}$Ti$_{0.8}$O$_3$ films on (001) STO and LAO. The plan-view TEM micrograph in Figure 27 displays the evolution of the microstructure in epitaxial PbZr$_{0.2}$Ti$_{0.8}$O$_3$ films from the simple polytwin $c/a/c/a$ to the cellular polydomain mixture as film thickness is increased from 200 nm to 450 nm. The cellular polytwin consists of all three possible domains of the tetragonal ferroelectric phase. The film is divided into cells of c-domains in the (001) plane bound by a_1- and a_2-domains along the [100] and [010] directions, respectively. It should be noted that since the interdomain interfaces may be inclined by 45° or 135° with respect to the [100] and [010] directions, these cells are not rectangular boxes. Their exact three-dimensional geometry depends on the specific crystallographic planes of the interdomain interfaces. The

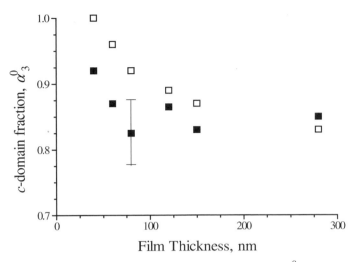

Fig. 25. Thickness dependence of the c-domain fraction α_3^0 for (001) PbTiO$_3$/MgO. Solid squares are experimental values [15], and open squares are calculated values with $h_{cr} \cong 25$ nm.

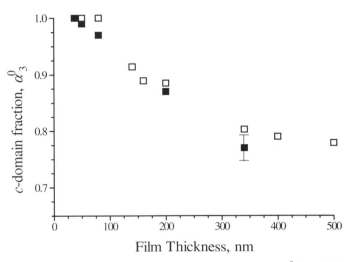

Fig. 26. Thickness dependence of the c-domain fraction α_3^0 for (001) PbTiO$_3$/SrTiO$_3$. Solid squares are experimental values [25], and open squares are calculated values with $h_{cr} \cong 30$ nm.

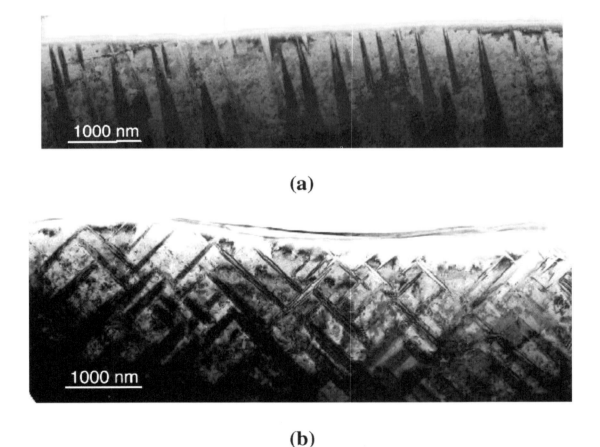

(a)

(b)

Fig. 27. Bright-field TEM micrographs showing the structural evolution in (001) PbZr$_{0.2}$Ti$_{0.8}$O$_3$ films on (001) SrTiO$_3$ as a function of film thickness: (a) simple $c/a/c/a$ polytwin in 200-nm-thick films and (b) cellular polytwin in 450-nm-thick films.

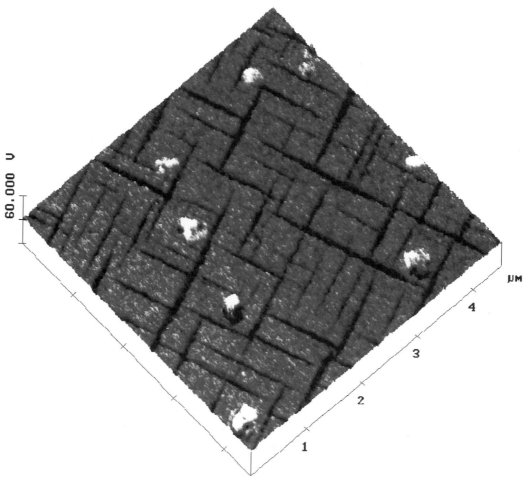

Fig. 28. AFM image showing the cellular domain architecture in 400-nm-thick (001) PbZr$_{0.2}$Ti$_{0.8}$O$_3$ films on (001) SrTiO$_3$ film [105].

fraction of c-domains is approximately 80%. As predicted, the fractions of the a_1 and a_2 domains are equal. Furthermore, the high-resolution TEM reveals the presence a pseudo-orthogonal network of misfit dislocations with the Burgers vectors $\mathbf{b} = [100]$ and $\mathbf{b} = [010]$. XRD results indicate that the epitaxial stresses indeed are completely relieved. Theoretical analysis based on the domain stability map shown in Figure 13 successfully explains experimental observations. The cellular domain structure has also been observed by AFM in similar ferroelectric films AFM [105, 106]. Figure 28 shows the AFM image of the three-domain architecture of a 400-nm-thick (001) PbZr$_{0.2}$Ti$_{0.8}$O$_3$ films on (001) STO.

4.5. Domain Stability Maps for BaTiO$_3$ on (001) MgO, Si, and SrTiO$_3$

The effects of substrate selection and various deposition techniques and conditions on the domain selection of BaTiO$_3$ films ($a = 0.399$ nm, and $c = 0.403$ nm for bulk BaTiO$_3$ at room temperature) have been experimentally investigated [16–21]. BaTiO$_3$ is not commonly used as a nonvolatile memory el-

ement, mainly because of its smaller saturation polarization when compared to PbTiO$_3$ or PZT solid solutions. However, films of BaTiO$_3$ and its solid solutions with Sr (Ba$_{1-x}$Sr$_x$TiO$_3$, BST) have received considerable interest as capacitor elements for the next generation dynamic random access memory (DRAM) technology due to their large dielectric constant [114, 115]. BST solid solutions are also potential candidates for applications in frequency-agile microwave electronic components, including phase shifters, varactors, tunable filters, and antennas because of their good tunability (large variation in the dielectric constant with the applied field). To provide some theoretical insight, Figure 29 presents domain stability maps for a 500-nm-thick BaTiO$_3$ film grown at 600°C on (001) MgO, (001) Si, and (001) SrTiO$_3$ substrates. The map was constructed by assuming that the critical thickness for domain formation is small compared to the film thickness, i.e., $h \gg h_{cr}$ and thus $\eta \cong 0$.

The data shown in Figure 29 correspond to films completely relaxed by misfit dislocations at the growth temperature. It should be noted that were the films not relaxed at all, the expected equilibrium domain structure for (001)

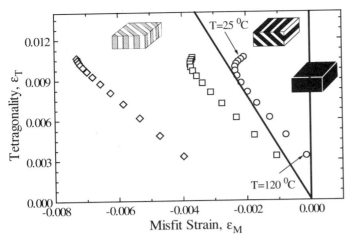

Fig. 29.   Domain stability maps for completely relaxed BaTiO₃ films on (001) MgO (circles), (001) Si (diamonds) and (001) SrTiO₃ (squares) in the temperature range of 0–120°C [55].

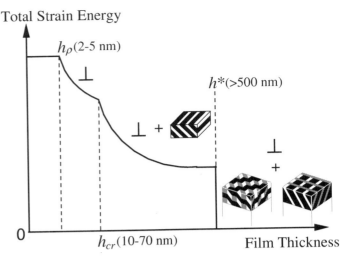

Fig. 30.   Summary of relaxation mechanisms in epitaxial ferroelectric films as a function of film thickness. The thickness axis is not to scale.

BaTiO₃/MgO and (001) BaTiO₃/Si would be the $a_1/a_2/a_1/a_2$ pattern whereas the unrelaxed (001) BaTiO₃/SrTiO₃ system would have a single domain c-structure, since the lattice misfit between SrTiO₃ and BaTiO₃ is compressive. BaTiO₃ on (001) Si is still expected to form the $a_1/a_2/a_1/a_2$ polydomain structure even if it is completely relaxed by misfit dislocations, since the misfit strain between BaTiO₃ and Si ($a = 0.543$ nm) is relatively large as compared to the other substrates. It has been experimentally verified that BaTiO₃ films grown on (001) MgO by PLD are oriented along (00l) and that the film is completely relaxed [18]. This is in accordance with our map shown in Figure 27 (open circles).

5. SUMMARY AND CONCLUDING REMARKS

The spontaneous polarization within the individual domains that make up the polydomain structure produces depolarizing fields at the film–substrate and film–free surface interfaces. However, the depolarization does not affect the domain structure and is compensated for by free charges in the electrode layers of a multilayer heterostructure consisting of a substrate and a ferroelectric film sandwiched between top and bottom electrodes. This indicates that the 90° domain structure in epitaxial ferroelectric films is determined by mechanical factors only, which is the basis for the thermodynamic approach of this chapter.

The microstructure of epitaxial ferroelectric films is determined by the interplay of two different strain relaxation mechanisms: relaxation by misfit dislocations above the transition temperature, and formation of polytwin structures below it. Figure 30 summarizes internal stress relaxation mechanisms as a function of film thickness in these films. Up to a critical thickness h_ρ, the film is fully stressed due to the lattice mismatch and thermal expansion coefficient difference between the film and the substrate. At h_ρ, the formation of misfit dislocations becomes feasible, which reduces the internal stresses gradually

as a function of film thickness in the region $h_\rho < h < h_{cr}$. After the critical thickness for polydomain formation has been reached (h_{cr}), twinning can further relieve the internal stresses. In the thickness range $h_{cr} < h < h^*$, the microstructure consists of misfit dislocations that have formed at the deposition temperature and the simple polytwins. However, the simple domain structure can only reduce the biaxial stress state to a uniaxial stress state. Therefore, the total strain energy levels off. If the film thickness exceeds a third critical thickness h^*, then there is the possibility of formation of the three-domain architectures discussed in Section 2.4. Whether it is the second-order polytwin or the cellular arrangement, the three-domain structures provide complete relaxation of internal stresses.

The microstructure obviously effects electrical and electromechanical response of ferroelectric films. Two limiting cases can be considered: immobile and mobile domain walls. If the domain wall motion is restricted, then the dielectric and piezoelectric properties are well below their bulk values. The reason for the lowering of these properties in single-domain or simple polydomain structures is the two-dimensional clamping of the film by the substrate, which is much thicker than the film [116–119], and/or the pinning of the polydomain boundaries by such defects as vacancies and dislocations. For the more complex three-domain structures, further constraint or clamping is imposed by the intricate domain structure, which further reduces the electrical and electromechanical properties [120]. In the case of mobile domain walls, elastic domain movement is shown to give a large extrinsic contribution to the electric and electromechanical compliances [59, 121]. The variation in the domain structure as a function of an applied uniaxial stress field or electrical field along [001] is shown in Figure 31. If the domain walls can move, then the structure adapts itself to the external conditions by varying the domain fractions. In this case, both fields alter the fraction of c domains, preserving the equality of the fractions of a_1 and a_2 domains. Thus, the three-domain heterostructure may change between a single domain structure and the equidomain $a_1/a_2/a_1/a_2$ polytwin. This structural variation is accompanied by a self-strain, as shown in Figure 31, which results in a large piezoelectric response (depending on the tetragonality of the ferroelectric film).

The three-domain structure shown in Figures 27 and 28 may also play an important role in the switching characteristics of ferroelectric films. It was shown experimentally that the switching in (001) $PbZr_{0.2}Ti_{0.8}O_3$ films on (001) $LaAlO_3$ substrates proceeds faster in thicker films (~400 nm) compared to thinner films (of thickness $h \leq 100$ nm) under pulsed testing conditions [39]. The activation field, E_α, a measure of the impedance to switching, fell substantially as the film thickness increased. The microstructure of the films observed via TEM changed from a single-domain state ($h < 60$ nm) to a polydomain structure ($h \geq 60$ nm) with gradual decline in the c-domain fraction coupled with increase in the density of 90° domain walls. The polydomain formation was accompanied by a drop in the saturation polarization and relaxation of internal stresses [92]. Although it is difficult to single out the formation of polydomain structure as the only reason for faster switching in these films with increasing film thickness, together with AFM studies that show heterogeneous nucleation of reversed domains at 90° domain walls in 400-nm $PbZr_{0.2}Ti_{0.8}O_3$ films on STO substrates [105, 106] (see Fig. 32), it is reasonable to hypothesize that they make a distinct contribution to the process of polarization reversal. However, this may be due to a pure size effect or may be related to the differing internal stress levels in the films.

Acknowledgments

I express my sincere gratitude to Dr. A. L. Roytburd and Dr. R. Ramesh for introducing me to the filed of phase transformations in constrained media and ferroelectric films. I also thank my collaborators and co-workers over the past years: Dr. S. Aggarwal, Dr. L. Salamanca-Riba, V. Nagarajan, C. S. Ganpule, and Hao Li at the University of Maryland, and Dr. L. A. Bendersky and Dr. M. D. Vaudin at NIST. Special thanks to Taylan Güyer for his help in preparing the illustrations in this article.

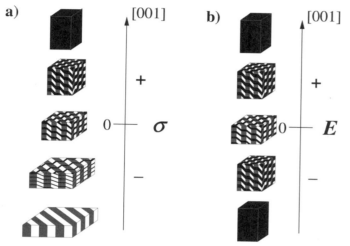

Fig. 31. Effect of (a) stress and (b) electric field along the [001] direction on the polydomain structure of an epitaxial film.

APPENDIX

The slopes of the limits of the stability regions can be determined using Eqs. (23) and (24) for cases $0 < \eta \leq 1$. The

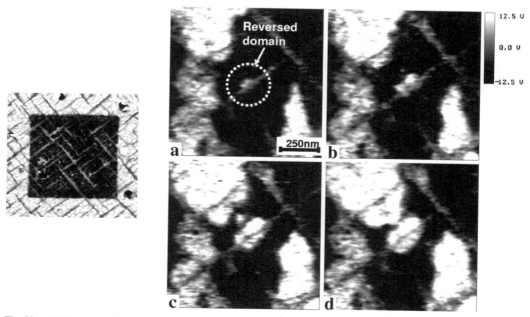

Fig. 32. AFM micrograph of 400-nm-thick $PbZr_{0.2}Ti_{0.8}O_3$ film on a $SrTiO_3$ substrate. The reversed polarization state is "written" inside the black square with the tip of the AFM. The time evolution of the polarization reversal is shown in (a)–(d). Nucleation starts at the 90°-domain wall boundaries [106].

limit of the stability of the $(1, 3)/(2, 3)$ region is defined as $0 \leq \alpha_j^0 \leq 1$. The upper limit is the boundary between this structure and the single-domain (3) state and is given by

$$\alpha_3^0 = 1 \qquad - (1 - \eta)e_{13}^{\mathrm{I}} - \Delta e_{13}^{\mathrm{M}} = 0 \qquad (A.1)$$

where

$$e_{13}^{\mathrm{I}} = e_{23}^{\mathrm{I}} = \frac{Y}{2(1 - \nu^2)}\varepsilon_{\mathrm{T}}^2 \qquad (A.2)$$

and

$$\Delta e_{13}^{\mathrm{M}} = \Delta e_{23}^{\mathrm{M}} = e_3^{\mathrm{M}} - e_1^{\mathrm{M}} = e_3^{\mathrm{M}} - e_2^{\mathrm{M}}$$
$$\Delta e_{13}^{\mathrm{M}} = \frac{Y}{(1 - \nu)}\left[\varepsilon_{\mathrm{M}}^2 - \frac{1}{2(1 + \nu)}\left[2\varepsilon_{\mathrm{M}}^2(1 + \nu)\right.\right.$$
$$\left.\left. + 2\varepsilon_{\mathrm{M}}\varepsilon_{\mathrm{T}}(1 + \nu) + \varepsilon_{\mathrm{T}}^2\right]\right] \qquad (A.3)$$

The metastability line is defined as $\alpha_j^0 = 0$ and is given by

$$\alpha_3^0 = 0 \qquad (1 - \eta)e_{13}^{\mathrm{I}} - \Delta e_{13}^{\mathrm{M}} = 0 \qquad (A.4)$$

The boundary between the $(1, 2)$ and the $(1, 3)/(2, 3)$ structures is defined as

$$f_{1/2} = f_{1/3} = f_{1/3}$$
$$(\alpha_2^0)^2 e_{12}^{\mathrm{I}} = (\alpha_3^0)^2 e_{13}^{\mathrm{I}} = (\alpha_3^0)^2 e_{23}^{\mathrm{I}} \qquad (A.5)$$

where

$$\alpha_2^0 = \frac{1}{2}$$
$$\alpha_3^0 = \frac{\varepsilon_{\mathrm{T}}(2 - \eta) + 2\varepsilon_{\mathrm{M}}(1 + \nu)}{2\varepsilon_{\mathrm{T}}(1 - \eta)} \qquad (A.6)$$

and

$$e_{12}^{\mathrm{I}} = \frac{Y}{(1 + \nu)}\varepsilon_{\mathrm{T}}^2 \qquad (A.7)$$

The simple solutions for the case $\eta = 0$ were given in Section 2.3. For cases when $0 < \eta \leq 1$, the Eqs. (A.1), (A.4), and (A.5) can be solved for ε_{T} by numerical methods if the critical thickness for the formation of $(1, 3)/(2, 3)$ structure, and thus the parameter η, are known and ν is assumed to be 0.3.

REFERENCES

1. F. Jona and G. Shirane, "Ferroelectric Crystals." Pergamon Press, New York, 1962.
2. M. E. Lines and A. M. Glass, "Principles and Applications of Ferroelectrics and Related Materials." Oxford University Press, Oxford, U.K., 1977.
3. R. Ramesh, W. K. Chan, B. Wilkens, H. Gilchrist, T. Sands, J. M. Tarascon, V. G. Keramidas, D. K. Fork, J. Lee, and A. Safari, *Appl. Phys. Lett.* 61, 1537 (1992).
4. J. S. Speck, A. Seifert, W. Pompe, and R. Ramesh, *J. Appl. Phys.* 76, 477 (1994).
5. A. L. Roytburd, *Phase Trans.* 45, 1 (1993).
6. A. L. Roitburd, *Phys. Stat. Sol. A* 37, 329 (1976).
7. A. G. Khachaturyan, "Theory of Structural Transformations in Solids." Wiley, New York, 1983.
8. L. D. Landau and E. M. Lifshitz, *Phys. Zs. Sowjet.* 8, 153 (1935).
9. C. Kittel, *Phys. Rev.* 70, 965 (1948).
10. L. D. Landau and E. M. Lifshitz, "Electrodynamics of Continuous Media." Pergamon Press, New York, 1984.
11. A. L. Roitburd, in "Solid-State Physics," Vol. 33, p. 317 (H. Ehrenreich, F. Seitz, and D. Turnbull, Eds.). Academic Press, New York, 1978.
12. H. Warlimont and L. Delaey, "Martensitic Transformations in Copper, Silver- and Gold-Based Alloys." Pergamon Press, Oxford, U.K., 1974.
13. S. K. Streiffer, E. M. Zielinski, B. M. Lairson, and J. C. Bravman, *Appl. Phys. Lett.* 58, 2171 (1991).
14. B. S. Kwak, A. Erbil, B. J. Wilkens, J. D. Budai, M. F. Chrisholm, and L. A. Boatner, *Phys. Rev. Lett.* 68, 3733 (1992).
15. B. S. Kwak, A. Erbil, J. D. Budai, M. F. Chrisholm, L. A. Boatner, and B. J. Wilkens, *Phys. Rev. B* 49, 14865 (1994).
16. L. A. Wills, B. W. Wessels, D. S. Richeson, and T. J. Marks, *Appl. Phys. Lett.* 60, 41 (1992).
17. D. L. Kaiser, M. D. Vaudin, L. D. Rotter, Z. L. Wang, J. P. Cline, C. S. Hwang, R. B. Marinenko, and J. G. Gillen, *Appl. Phys. Lett.* 66, 2801 (1995).
18. V. Srikant, E. J. Tarsa, D. R. Clarke, and J. S. Speck, *J. Appl. Phys.* 77, 1517 (1995).
19. S. Kim, S. Hishita, Y. M. Kang, and S. Baik, *J. Appl. Phys.* 78, 5604 (1995).
20. S. Kim, T. Manabe, I. Yamaguchi, T. Kumagai, and S. Mizuta, *J. Mater. Res.* 12, 1141 (1997).
21. C. S. Hwang, M. D. Vaudin, and G. T. Stauf, *J. Mater. Res.* 12, 1625 (1997).
22. Y. Gao, G. Bai, K. L. Merkle, Y. Shi, H. L. M. Chang, Z. Shen, and D. J. Lam, *J. Mater. Res.* 8, 145 (1993).
23. Y. Gao, G. Bai, K. L. Merkle, H. L. M. Chang, and D. J. Lam, *Thin Solid Films* 235, 86 (1993).
24. S. G. Ghonge, E. Goo, R. Ramesh, T. Sands, and V. G. Keramidas, *Appl. Phys. Lett.* 63, 1628 (1993).
25. W.-Y. Hsu and R. Raj, *Appl. Phys. Lett.* 67, 792 (1995).
26. C. M. Foster, Z. Li, M. Buckett, D. Miller, P. M. Baldo, L. E. Rehn, G. R. Bai, D. Guo, H. You, and K. L. Merkle, *J. Appl. Phys.* 78, 2607 (1995).
27. C. M. Foster, W. Pompe, A. C. Daykin, and J. S. Speck, *J. Appl. Phys.* 79, 1405 (1996).
28. R. S. Batzer, B. M. Yen, D. Liu, H. Chen, H. Kubo, and G. R. Bai, *J. Appl. Phys.* 80, 6235 (1996).
29. C. D. Theis and D. G. Schlom, *J. Mater. Res.* 12, 1297 (1997).
30. B. M. Yen and H. Chen, *J. Appl. Phys.* 85, 853 (1999).
31. S. P. Alpay, A. S. Prakash, S. Aggarwal, P. Shuk, M. Greenblatt, R. Ramesh, and A. L. Roytburd, *Scripta Mater.* 39, 1435 (1998).
32. R. Takayama and Y. Tomita, *J. Appl. Phys.* 65, 1666 (1989).
33. R. Ramesh, T. Sands, and V. G. Keramidas, *Appl. Phys. Lett.* 63, 731 (1993).
34. R. Ramesh, H. Gilchrist, T. Sands, V. G. Keramidas, R. Haakenaasen, and D. K. Fork, *Appl. Phys. Lett.* 63, 3592 (1993).
35. A. E. M. De Veirman, J. F. M. Cillessen, M. De Keijser, R. M. Wolf, D. J. Taylor, A. A. Staals, and G. J. M. Dormans, *Mater. Res. Soc. Symp. Proc.* 341, 329 (1994).
36. C. M. Foster, G. R. Bai, R. Csencsits, J. Vetrone, R. Jammy, L. A. Wills, E. Carr, and J. Amano, *J. Appl. Phys.* 81, 2349 (1997).
37. B. A. Tuttle, T. J. Headley, B. C. Bunker, R. W. Schwartz, T. J. Zender, C. L. Hernandez, D. C. Goodnow, R. J. Tissot, and J. Michael, *J. Mater. Res.* 7, 1876 (1992).
38. S. P. Alpay, V. Nagarajan, L. A. Bendersky, M. D. Vaudin, S. Aggarwal, R. Ramesh, and A. L. Roytburd, *J. Appl. Phys.* 85, 3271 (1999).
39. V. Nagarajan, I. G. Jenkins, S. P. Alpay, H. Li, S. Aggarwal, L. Salamanca-Riba, A. L. Roytburd, and R. Ramesh, *J. Appl. Phys.* 86, 595 (1999).
40. K. S. Lee, Y. M. Kang, and S. Baik, *J. Mater. Res.* 14, 132 (1999).
41. K. Iijima, R. Takayama, Y. Tomita, and I. Ueda, *J. Appl. Phys.* 60, 2914 (1986).
42. Y. M. Kang, J. K. Ku, and S. Baik, *J. Appl. Phys.* 78, 2601 (1995).
43. Y. M. Kang and S. Baik, *J. Appl. Phys.* 82, 2532 (1997).

44. M. J. Nystrom, B. W. Wessels, J. Chen, and T. J. Marks, *Appl. Phys. Lett.* 68, 761 (1996).

45. S. Little and A. Zangwill, *Phys. Rev. B* 49, 16659 (1994).

46. J. S. Speck and W. Pompe, *J. Appl. Phys.* 76, 466 (1994).

47. J. S. Speck, A. C. Daykin, A. Seifert, A. E. Romanov, and W. Pompe, *J. Appl. Phys.* 78, 1696 (1995).

48. A. E. Romanov, W. Pompe, and J. S. Speck, *J. Appl. Phys.* 79, 4037 (1996).

49. N. A. Pertsev and A. G. Zembilgotov, *J. Appl. Phys.* 78, 6170 (1995).

50. N. A. Pertsev and A. G. Zembilgotov, *J. Appl. Phys.* 80, 6401 (1996).

51. N. Sridhar, J. M. Rickman, and D. J. Srolovitz, *Acta Mater.* 44, 4085 (1996).

52. N. Sridhar, J. M. Rickman, and D. J. Srolovitz, *Acta Mater.* 44, 4097 (1996).

53. A. L. Roytburd, *J. Appl. Phys.* 83, 228 (1998).

54. A. L. Roytburd, *J. Appl. Phys.* 83, 239 (1998).

55. S. P. Alpay and A. L. Roytburd, *J. Appl. Phys.* 83, 4714 (1998).

56. S. K. Streiffer, C. B. Parker, A. E. Romanov, M. J. Lefevre, L. Zhao, J. S. Speck, W. Pompe, C. M. Foster, and G. R. Bai, *J. Appl. Phys.* 83, 2742 (1998).

57. A. E. Romanov, M. J. Lefevre, J. S. Speck, W. Pompe, S. K. Streiffer, and C. M. Foster, *J. Appl. Phys.* 83, 2754 (1998).

58. A. L. Roytburd and Y. Yu, *Ferroelectrics* 144, 137 (1993).

59. A. L. Roytburd, in "Thin Film Ferroelectric Materials and Devices," p. 71 (R. Ramesh, Ed.). Kluwer, Norvell, MA, 1997.

60. S. P. Alpay and A. L. Roytburd, *Mater. Res. Soc. Symp. Proc.* 474, 407 (1997).

61. W. D. Nix, *Metall. Trans. A* 20, 2217 (1989).

62. C. Kooy and U. Enz, *Phillips Res. Reprints* 15, 7 (1960).

63. A. L. Roytburd, S. P. Alpay, L. A. Bendersky, V. Nagarajan, and R. Ramesh, *J. Appl. Phys.* 89, 553 (2001).

64. J. H. van der Merve, *J. Appl. Phys.* 34, 123 (1963).

65. J. W. Matthews and A. E. Blakeslee, *J. Crystal Growth* 27, 118 (1974).

66. J. W. Matthews and A. E. Blakeslee, *J. Crystal Growth* 29, 273 (1975).

67. L. B. Freund, *MRS Bull.* 17, 52 (1992).

68. L. B. Freund and W. D. Nix, *Appl. Phys. Lett.* 69, 173 (1996).

69. J. P. Hirth and J. Lothe, "Theory of Dislocations," 2nd ed. Wiley, New York, 1982.

70. J. W. Christian, "The Theory of Transformations in Metals and Alloys, Part I: Equilibrium and General Kinetic Theory," 2nd ed. Pergamon Press, Oxford, U.K., 1975.

71. M. Tanaka and Y. Himiyama, *Acta Crystallogr. A* 21, S264 (1975).

72. L. B. Freund, *J. Appl. Mech.* 54, 553 (1987).

73. O. Auciello and R. Ramesh, *MRS Bull.* 21, 31 (1996).

74. J. L. Vossen and W. Kern, "Thin Film Processes." Academic Press, New York, 1978.

75. D. Dijkkamp, T. Venkatesan, X. D. Wu, S. A. Shaheen, N. Jisrawi, Y. H. Min-Lee, W. L. McLean, and M. Croft, *Appl. Phys. Lett.* 51, 619, 861 (1987).

76. D. B. Chrisey and G. K. Hubler, "Pulsed Laser Deposition of Thin Films." Wiley, New York, 1994.

77. M. de Keijser and G. J. M. Dormans, *MRS Bull.* 21, 37 (1996).

78. E. V. Sviridov, V. A. Alyoshin, Y. I. Golovko, I. N. Zakharchenko, V. M. Mukhortov, and V. P. Dudkevich, *Phys. Stat. Sol. A* 121, 157 (1990).

79. T. Ogawa, A. Senda, and T. Kasanami, *Japan. J. Appl. Phys.* 30, 2145 (1991).

80. J. Hornstra and W. J. Bartels, *J. Crystal Growth* 44, 513 (1978).

81. W. J. Bartels and W. Nijman, *J. Crystal Growth* 44, 518 (1978).

82. C. Bocchi, C. Ferrari, P. Franzosi, A. Bosacchi, and S. Franchi, *J. Crystal Growth* 132, 427 (1993).

83. P. van der Sluis, *J. Phys. D: Appl. Phys.* 26, A188 (1993).

84. J. Chaudhuri, S. Shah, and J. P. Harbison, *J. Appl. Phys.* 66, 5373 (1989).

85. M. S. Goorsky, T. F. Kuech, M. A. Tischler, and R. M. Potemski, *Appl. Phys. Lett.* 59, 2269 (1991).

86. B. Heying, X. H. Wu, S. Keller, Y. Li, D. Kapolnek, B. P. Keller, S. P. DenBaars, and J. S. Speck, *Appl. Phys. Lett.* 68, 643 (1996).

87. B. D. Desu, *J. Electrochem. Soc.* 140, 2981 (1993).

88. C. H. Peng and S. B. Desu, *J. Am. Ceramic Soc.* 77, 1486 (1994).

89. B. A. Tuttle, T. J. Headley, H. N. Al-Shareef, J. A. Voigt, M. Rodriguez, J. Michael, and W. L. Warren, *J. Mater. Res.* 11, 2309 (1996).

90. D. Fu, T. Ogawa, H. Suzuki, and K. Ishikawa, *Appl. Phys. Lett.* 77, 1532 (2000).

91. G. Cornella, S.-H. Lee, W. D. Nix, and J. C. Bravman, *Appl. Phys. Lett.* 71, 2949 (1997).

92. A. L. Roytburd, S. P. Alpay, V. Nagarajan, C. S. Ganpule, S. Aggarwal, E. D. Williams, and R. Ramesh, *Phys. Rev. Lett.* 85, 190 (2000).

93. V. Nagarajan, C. S. Ganpule, B. Nagraj, S. Aggarwal, S. P. Alpay, A. L. Roytburd, E. D. Williams, and R. Ramesh, *Appl. Phys. Lett.* 75, 4183 (1999).

94. V. Nagarajan, S. P. Alpay, C. S. Ganpule, B. Nagraj, S. Aggarwal, E. D. Williams, A. L. Roytburd, and R. Ramesh, *Appl. Phys. Lett.* 77, 438 (2000).

95. C. L. Canedy, H. Li, S. P. Alpay, L. Salamanca-Riba, A. L. Roytburd, and R. Ramesh, *Appl. Phys. Lett.* 77, 1695 (2000).

96. O. Kolosov, A. Gruverman, J. Hatano, K. Takahashi, and H. Tokumoto, *Phys. Rev. Lett.* 74, 4309 (1995).

97. G. Zavala, J. H. Fendler, and S. Trollier-McKinstry, *J. Appl. Phys.* 81, 7480 (1997).

98. J. A. Christman, R. R. Woolcott Jr., A. I. Kingon, and R. J. Nemanich, *Appl. Phys. Lett.* 73, 3851 (1998).

99. O. Auciello, A. Gruverman, H. Tokumoto, S. A. Prakash, S. Aggarwal, and R. Ramesh, *MRS Bull.* 23, 33 (1998).

100. E. L. Colla, D. V. Taylor, A. K. Tagantsev, and N. Setter, *Appl. Phys. Lett.* 72, 2478 (1998).

101. E. L. Colla, S. Hong, D. V. Taylor, A. K. Tagantsev, N. Setter, and K. No, *Appl. Phys. Lett.* 72, 2763 (1998).

102. T. Tybell, C. H. Ahn, and J.-M. Triscone, *Appl. Phys. Lett.* 72, 1454 (1998).

103. T. Tybell, C. H. Ahn, and J.-M. Triscone, *Appl. Phys. Lett.* 75, 856 (1999).

104. C. Durkan, D. P. Chu, P. Migliorato, and M. E. Welland, *Appl. Phys. Lett.* 76, 366 (2000).

105. C. S. Ganpule, V. Nagarajan, H. Li, A. S. Ogale, D. E. Steinhauer, S. Aggarwal, E. Williams, R. Ramesh, and P. De Wolf, *Appl. Phys. Lett.* 77, 292 (2000).

106. C. S. Ganpule, V. Nagarajan, S. B. Ogale, A. L. Roytburd, E. D. Williams, and R. Ramesh, *Appl. Phys. Lett.* 77, 3275 (2000).

107. M. Alexe, A. Gruverman, C. Harnagea, N. D. Zakharov, A. Pignolet, D. Hesse, and J. F. Scott, *Appl. Phys. Lett.* 75, 1158 (1999).

108. A. Gruverman, *Appl. Phys. Lett.* 75, 1452 (1999).

109. C. S. Ganpule, A. Stanishevsky, Q. Su, S. Aggarwal, J. Melngailis, E. Williams, and R. Ramesh, *Appl. Phys. Lett.* 75, 409 (1999).

110. C. S. Ganpule, A. Stanishevsky, S. Aggarwal, J. Melngailis, E. Williams, R. Ramesh, V. Joshi, and C. P. de Araujo, *Appl. Phys. Lett.* 75, 3874 (1999).

111. Landolt-Börnstein, "Numerical Data and Functional Relationships in Science and Technology," Vol. 16 (K.-H. Hellwege and A. M. Hellwege, Eds.). Springer-Verlag, Berlin, 1981.

112. K. S. Lee and S. Baik, *J. Appl. Phys.* 87, 8035 (2000).

113. W. K. Choi, S. K. Choi, and H. M. Lee, *J. Mater. Res.* 14, 4677 (1999).

114. S. Summerfelt, in "Thin Film Ferroelectric Materials and Devices," p. 1 (R. Ramesh, Ed.). Kluwer, Norvell, MA, 1997.

115. N. Mikami, in "Thin Film Ferroelectric Materials and Devices," p. 43 (R. Ramesh, Ed.). Kluwer, Norvell, MA, 1997.

116. K. Lefki and G. J. M. Dormans, *J. Appl. Phys.* 76, 1764 (1994).

117. A. Kholkin, A. Seifert, and N. Setter, *Appl. Phys. Lett.* 72, 3374 (1998).

118. N. A. Pertsev, A. G. Zembilgotov, and A. K. Tagantsev, *Phys. Rev. Lett.* 80, 1988 (1998).

119. N. A. Pertsev, A. G. Zembilgotov, S. Hoffman, R. Waser, and A. K. Tagantsev, *J. Appl. Phys.* 85, 1698 (1999).

120. S. P. Alpay and A. L. Roytburd, unpublished.

121. N. A. Pertsev, G. Arlt, and A. G. Zembilgotov, *Phys. Rev. Lett.* 76, 1364 (1996).

Chapter 11

FERROELECTRIC POLYMER LANGMUIR–BLODGETT FILMS

Stephen Ducharme

Department of Physics and Astronomy, Center for Materials Research and Analysis, University of Nebraska, Lincoln, Nebraska, USA

S. P. Palto and V. M. Fridkin

Shubnikov Institute of Crystallography, Russian Academy of Sciences, Moscow 117333, Russia

Contents

Handbook of Thin Film Materials, edited by H.S. Nalwa
Volume 3: Ferroelectric and Dielectric Thin Films

ISBN 0-12-512911-4/$35.00

1. INTRODUCTION

Interest in ferroelectric thin films has focused on developing technologies that exploit unique characteristics of ferroelectricity, technologies such as ultrasonic transduction, infrared imaging, compact capacitors, and nonvolatile memories. As with many technologies, this focus tends to enhance, rather than impede, progress in the understanding of the underlying physics by stimulating intense study of material properties.

Unique insight into the nature of ferroelectricity is emerging from the study of two-dimensional ferroelectric films, which are made by Langmuir–Blodgett deposition of vinylidene fluoride copolymers. This chapter reviews recent results from the study of ferroelectric Langmuir–Blodgett (LB) films as thin as one monolayer. The films are highly crystalline and exhibit excellent ferroelectric properties. Foremost among the results are the discovery of *two-dimensional ferroelectricity* with a thickness-independent phase transition temperature; demonstration of double hysteresis and the *ferroelectric critical point* for the first time in a polymer; the first direct measurement of the *intrinsic ferroelectric coercive field* in any ferroelectric; discovery of a new *surface phase transition* at 20°C; and discovery of a *lattice-stiffening phase transition* at 160 K. In addition, we have been able to study in unprecedented detail the relationship between *piezoelectric and pyroelectric response* and the spontaneous polarization; the *nonlinear dielectric response* near the phase transition; the *scaling of extrinsic properties* with thickness; and images of the polarization across the film in real time during switching by *pyroelectric scanning microscopy*.

This section continues with a review of ferroelectricity and ferroelectric polymers. Section 2 describes the fabrication techniques based on Langmuir–Blodgett deposition used to fabricate the ultrathin ferroelectric polymer films, and Section 3 summarizes film structure and morphology. The ferroelectric and related properties of the films are reviewed in Section 4. Section 5 covers several applications proposed to exploit unique features of the ferroelectric polymer LB films.

1.1. Ferroelectricity

From the time of the first studies of Rochelle salt by Valasek in the early 1920s [1, 2], ferroelectric materials have attracted the attention of researchers both for their unique properties and for their utility. A ferroelectric material exhibits a stable macroscopic polarization that can be repeatedly switched by an external electric field between equal-energy states of opposite polarization. This definition excludes materials which exhibit metastable polarization due to, for example, trapped charge in electrets [3] or nonequilibrium orientation of dipoles in poled electro-optic polymers [4]. It also excludes other pyroelectric materials with an equilibrium macroscopic polarization that cannot be reversibly switched, usually because the constituent dipoles have insufficient rotational freedom within the elastic limit of the material, as in many organic nonlinear optical crystals [5]. A ferroelectric usually, but not necessarily, undergoes a continuous and reversible phase transition as it is cooled from a nonpolar paraelectric state to a polar ferroelectric state. Ferroelectric phase transitions can be either first-order or second-order depending on whether or not there is an enthalpy difference between the phases at the transition temperature. There are also improper ferroelectric phase transitions, where polarization is not the order parameter, as in ferroelectric liquid crystals where the director axis is the order parameter and polarization is generally in a different direction [6, 7]. It is also possible to have a discontinuous first-order phase transition, which cannot be described with a unique order parameter. For the remainder of this chapter, we will focus on the properties of ferroelectrics with a proper continuous first- or second-order phase transition, for which the order parameter is some component of the polarization.

Ferroelectrics with a continuous proper phase transition are generally of the displacive or order–disorder types. Displacive ferroelectrics, like barium titanate and other perovskites, are characterized by a phonon soft-mode transition from a nonpolar paraelectric phase to a polar ferroelectric phase, with the simultaneous appearance of a dipole moment in the unit cell and macroscopic ordering of the dipole moments. Order–disorder ferroelectrics, like potassium dihydrogen phosphate, generally have built-in dipoles that individually have two or more equivalent orientations and which condense by a diffusive soft mode into a macroscopically ordered polarization at the phase transition. Displacive ferroelectrics are often modeled as a mean-field formalism, while the Ising formalism seems more appropriate for order–disorder ferroelectrics.

Many excellent general books on ferroelectricity have been published, including the classic text by Jona and Shirane [8], the more comprehensive treatise of Lines and Glass [9], and other useful but more specialized texts [10–16].

1.1.1. Phenomenology of Ferroelectric Materials

Mean-field models of the continuous ferroelectric phase transition treat the polarization as an order parameter. Ginzburg developed such a model of ferroelectricity in the 1940s [17–19], based on the Landau theory of continuous phase transitions [20–22]. A similar treatment was developed by Devonshire shortly thereafter [23–25], and many aspects of the problem were further developed by Lifshitz, Levanyuk, Sannikov, Indenbom, and others. Here we give a brief review of the main features of the Landau–Ginzburg–Devonshire (LGD) formalism for the uniaxial ferroelectric.

The Landau theory of continuous phase transitions [20–22] is based on a thermodynamic free energy that can be expressed as a power series in a small quantity called the order parameter, the polarization P in the case of ferroelectricity. The LGD formalism requires that the ferroelectric phase have a point group symmetry that is a subgroup of the paraelectric phase symmetry. Since polarization is a vector, symmetry requires that only even powers of polarization appear in the expansion unless there is an external field to extrinsically break the symmetry. We shall discuss only the uniaxial case, where the electric polarization and any applied field are parallel to a unique crystalline axis, the axis of broken inversion symmetry in the polar phase. The LGD form of the Gibbs free-energy density can be written [17, 18]

$$G = G_0 + \frac{\alpha}{2}P^2 + \frac{\beta}{4}P^4 + \frac{\gamma}{6}P^6 - PE \quad (1)$$

where E is the electric field, G_0 is the free-energy density of the paraelectric phase at zero field, and the expansion coefficients α, β, and γ are in general dependent on temperature and pressure. (The effects of pressure can be explicitly included with additional terms in the free-energy density [25].) The equilibrium condition corresponds to the absolute minimum of the free-energy density.

The Landau theory requires that the first coefficient α vanish at the so-called Curie temperature T_0, so that the simplest form for the first coefficient is

$$\alpha = \frac{1}{\varepsilon_0 C}(T - T_0) \quad (2)$$

where T is the temperature, $C > 0$ is the Curie–Weiss constant, and ε_0 is the permittivity of free space. The coefficients C, β, and γ are generally assumed independent of temperature, but it is often best to confine the LGD analysis to temperatures near the Curie temperature to ensure this. The sign of the second coefficient β determines whether the transition is first-order ($\beta < 0$) or second-order ($\beta > 0$).

The constitutive relation between polarization P and electric field E is obtained from the minimum of the free energy (Eq. (1)), yielding the expression

$$E = \alpha P + \beta P^3 + \gamma P^5 \quad (3)$$

The zero-field relative polarizability χ can be calculated from the constitutive relation Eq. (3). The inverse polarizability in the paraelectric phase is

$$\frac{1}{\chi} = \varepsilon_0 \frac{\partial E}{\partial P} = \frac{(T - T_0)}{C} \quad (T > T_C) \quad (4)$$

This is the so-called Curie–Weiss law, which is built into the definition of α. The total dielectric constant of the medium $\varepsilon = \varepsilon_\infty + \chi$ contains contributions ε_∞ from the background electronic polarizability.

The second-order LGD ferroelectric is described by the free-energy density in Eq. (1) with expansion coefficient values $C > 0$, $\beta > 0$, and $\gamma = 0$. The second-order LGD ferroelectric has two distinct phases, a nonpolar paraelectric phase and a polar ferroelectric phase. The free-energy plots in Figure 1a illustrate the distinction between these phases. At temperatures above the equilibrium phase transition temperature $T_C = T_0$, the free energy has a single minimum at $P = 0$. This is the paraelectric phase, which has no spontaneous polarization at zero applied electric field. At temperatures below T_C in zero field, there are two equivalent minima, where the magnitude of the equilibrium spontaneous polarization is

$$P_S = \sqrt{\frac{|\alpha|}{\beta}} = \sqrt{\frac{(T_0 - T)}{\varepsilon_0 C \beta}} \quad (T < T_C) \quad (5)$$

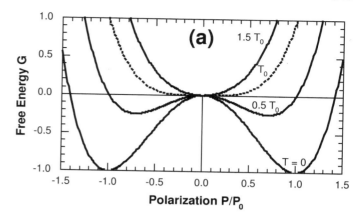

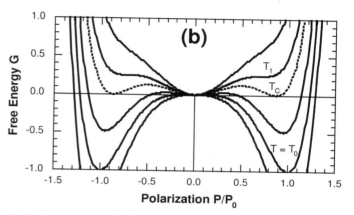

Fig. 1. Free-energy density (Eq. (1)) from the LGD phenomenology at various temperatures and zero electric field. The temperature increases from bottom to top. (a) Second-order transition; (b) first-order transition.

The zero-field equilibrium polarization $P(E = 0) = \pm P_S$ can point in either direction along the symmetry axis (recall that this is the uniaxial case), corresponding to the two energetically equivalent states of the ferroelectric crystal at zero electric field. The spontaneous polarization in the ferroelectric phase decreases steadily with increased temperature and vanishes at the transition temperature, as shown in Figure 2a. At a second-order phase transition, the order parameter (spontaneous polarization) decreases smoothly to zero; there is no jump in the order parameter, nor is there an enthalpy difference between phases. The relative polarizability is given by Eq. (4) in the paraelectric phase and, in the ferroelectric phase, by

$$\frac{1}{\chi} = \frac{2(T_0 - T)}{C} \quad (T < T_C) \quad (6)$$

The first-order LGD ferroelectric, corresponding to expansion coefficients $\beta < 0$ and $\gamma > 0$ in Eq. (1), has two distinct phases as is evident from the free-energy plots shown in Figure 1b. The phase transition occurs at $T_C = T_0 + (3/16)\varepsilon_0 C \beta^2/\gamma$, above the Curie temperature T_0, where the polarization at the absolute minimum energy switches abruptly from $P = 0$ to $\pm P_S$. Minimizing the free-energy density (Eq. (1)) at zero electric field, we obtain the magnitude of the spontaneous polarization in the ferroelectric phase as a function

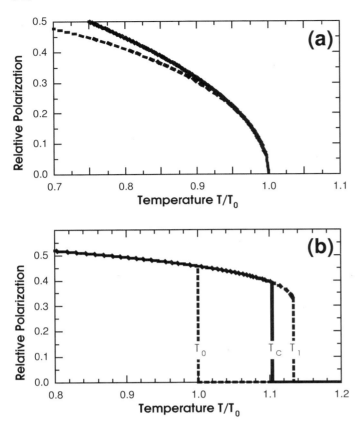

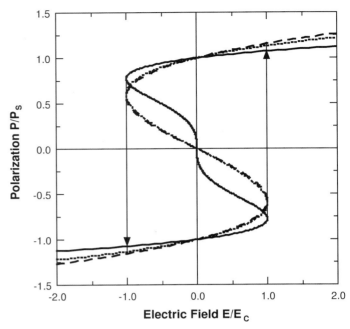

Fig. 2. Spontaneous polarization from the mean-field ferroelectric models: (a) the second-order LGD ferroelectric from Eq. (5) (solid line) and the second-order mean-field Ising model (dashed line); (b) the first-order LGD ferroelectric from Eq. (7), where the solid lines denote equilibrium solutions and the dashed lines denote metastable solutions.

Fig. 3. Theoretical hysteresis loops in the ferroelectric state $(T < T_C)$ from the uniaxial ferroelectric mean-field models: LGD first-order (solid line), LGD second-order (dashed line), and mean-field Ising (dotted line).

of temperature (see Fig. 2b),

$$P_S = \sqrt{\frac{-\beta}{2\gamma}\left(1 + \sqrt{1-t}\right)} \quad (T < T_C) \qquad (7)$$

where the reduced temperature is $t = 4\alpha\gamma/\beta^2 = 4\gamma(T - T_0)/(\varepsilon_0 C\beta^2)$. There is a jump in the magnitude of the spontaneous polarization by $\Delta P_S = \sqrt{-3\beta/4\gamma}$ at the phase transition temperature T_C. As is usual for a first-order phase transition, there is a transition enthalpy and a region of metastable phase coexistence from T_0 to $T_1 = T_0 + \frac{1}{4}\varepsilon_0 C\beta^2/\gamma$. The coexistence region is manifest in thermal hysteresis because if the crystal is not in equilibrium, the paraelectric phase can persist on cooling through T_C as far as T_0 due to the local minimum at $P = 0$, and the ferroelectric phase can persist on heating through T_C as far as T_1 due to the local minima at $\pm P_S$ (compare Figs. 1b and 2b). Section 3.2.2 describes the results of studies of phase coexistence and thermal hysteresis in the ferroelectric copolymer LB films. Another special feature of the first-order phase transition is the increase of the transition temperature T_C with an applied electric field up to a critical point T_{Cr}, as detailed in Section 4.3. The relative polarizability is given by Eq. (4) in the

paraelectric phase and, in the ferroelectric phase, by

$$\frac{1}{\chi} = \frac{4(T_0 - T)}{C} + \frac{\varepsilon_0\beta^2}{\gamma}\left(1 + \sqrt{1-t}\right) \quad (T < T_C) \qquad (8)$$

The Ising model may be more appropriate for a ferroelectric with an order–disorder phase transition. The application of the Ising model to electric dipoles coupled by a mean field leads to the following expression for the order parameter $\eta = P_S/P_0 = \tanh[(\eta J_0 + Ep_0)/(kT)]$ [9, 16, 25, 26], where J_0 is the pseudo-spin interaction constant, p_0 is the dipole moment, $P_0 = p_0 N$ is the saturated polarization, N is the number density of dipoles, and k is the Boltzmann constant. The spontaneous polarization at zero electric field decreases smoothly from P_0 at absolute zero temperature to 0 at the transition temperature $T_C = T_0 \equiv J_0/k$, as shown by the dashed line in Figure 2a. The phase transition is of the second order as there is no polarization jump at T_C and no enthalpy. Like the LGD models, the mean-field Ising model is more properly confined to temperatures near T_0, where the equation for the order parameter η can be expanded to yield the second-order LGD form (Eq. (5)) [25, 26].

The nonlinear dependence of the polarization on the electric field $P(E)$ in Eq. (3) leads to a polarization hysteresis loop, the defining characteristic of a ferroelectric material. The hysteresis loops shown in Figure 3 for the ferroelectric state $(T < T_C)$ illustrate the bistable nature of the polarization over a range of electric fields. The positive and negative values of the zero-field polarization $P(0) = \pm P_S$ correspond to the equivalent minima in the free-energy density. The value of the polarization measured at zero applied field after saturation at high field is called the remanent (remaining) polarization. Ideally, the remanent polarization is equal to the spontaneous polarization P_S (Eq. (5) or Eq. (7)), though the measured remanent polarization is often lower because of incomplete saturation, sample inhomogeneity, or the reappearance of domains. The intrinsic ferroelectric coercive field E_C is the magnitude of the opposing electric field at which the polarization becomes single-valued and must reverse direction, as shown by the arrows in Figure 3. The values of the coercive field measured in real ferroelectric materials are orders of magnitude smaller than this intrinsic value because polarization reversal is invariably initiated locally by nucleation and spread by domain growth. The first experimental measurements of the intrinsic coercive field are detailed in Section 4.1.

The behavior of real crystals is complicated by imperfections such as inhomogeneities, impurities, and grain boundaries. Imperfections can strongly influence the ferroelectric properties, giving a diffuse character to the phase transition by permitting phase coexistence or smearing out the hysteresis loop so the edges are not so sharp. Imperfections also nucleate new domains so that the measured coercive field is almost always much less than the intrinsic value expected in a perfect crystal. Domains with opposite polarization can be nucleated by imperfections, but they also arise naturally in a finite sample to reduce the energy of the external electric field. The macroscopic equilibrium state corresponds to a balance between the reduction in the external field energy and the added energy of the domain walls. An effective method to produce and maintain the monodomain state in a crystal is to coat the polar faces with conducting electrodes and "pole" it by saturating the polarization with an applied field well above the coercive field. Even without an external potential, the electrodes will maintain uniform potential on the polar crystal faces and inhibit the formation of opposing domains.

1.1.2. Ferroelectric Thin Films

Ferroelectric materials are key to many technologies currently employed or under development, but thin films in particular are increasingly important, especially in integrated hybrid devices such as the smart card ferroelectric memories introduced recently [27]. Thin films are convenient for studying a wide range of dielectric and electronic properties, as summarized in the following paragraphs, and often exhibit unique thickness-dependent effects.

From the point of view of electrostatics, the spontaneous polarization of a ferroelectric crystal is equivalent to a surface charge on the polar faces. The bound surface charge density is equal to the normal component of the polarization discontinuity, with positive surface charge on the "top" surface at the head of the polarization vector and negative surface charge on the "bottom" surface at the tail. Consider a freshly polarized ferroelectric crystal with faces cut normal to the spontaneous polarization and thickness much less than the other dimensions, as shown in Figure 4a. The single-domain state is energetically unstable and will tend to minimize the external electrostatic "depolarization" energy by breaking up into domains with op-

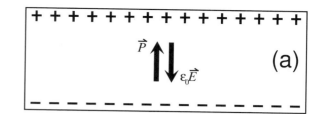

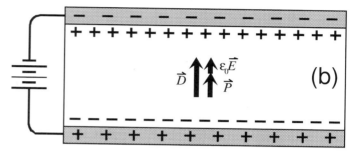

Fig. 4. Cross section of a ferroelectric crystal with surfaces normal to the spontaneous polarization axis. The arrows represent the internal field vectors. (a) The open circuit case with no external field or free charge. (b) The uniform potential case with conducting electrodes on the polar faces connected to an external potential difference.

posing polarization or by accumulating sufficient free charge from internal conduction or external sources.

Conducting electrodes are a convenient means to control the potential on the polar surfaces and inhibit formation of opposing domains. This is illustrated in Figure 4b, where the electrodes are shown connected to an external voltage source supplying a constant potential difference V. The voltage source supplies sufficient surface charge to fix the electric field $E = V/L$ in the sample (neglecting fringing fields at the edges and free charge in the crystal). This configuration is useful for measuring any *changes* in the spontaneous polarization, because a change in polarization at constant potential will be accompanied by a charge $\Delta Q = A \Delta P$ delivered to the external circuit, where A is the area of the electrodes on the polar faces. For example, polarization hysteresis loops can be recorded by the Sawyer–Tower method [28] of placing a reference capacitor in series with the sample and voltage source to measure the charge produced while cycling the voltage. Complete polarization reversal produces twice the polarization charge. Another method for recording polarization reversal is the Merz method of placing a resistor in series with the sample and voltage source to measure the switching current while cycling the voltage [29]. Since the spontaneous polarization depends on temperature, a change in the temperature of the sample at constant potential will also produce a surface charge by the pyroelectric effect. Strain will also result in a change in polarization and produce a charge by the converse piezoelectric effect. The pyroelectric and piezoelectric effects are covered in Section 4.4.

1.1.3. Finite-Size Effects

Ultrathin crystalline films offer the possibility of exploring phase transitions in the crossover region between two and three

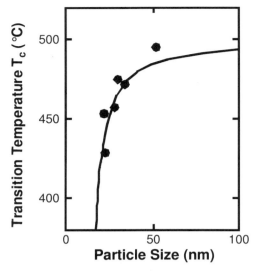

Fig. 5. Finite-size effect on the transition temperature in sintered PTiO₃ crystals. Reprinted with permission from K. Ishikawa et al., *Phys. Rev. B* 37, 5852 (1988). Copyright 1988 American Physical Society.

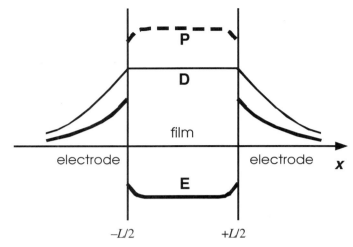

Fig. 6. Schematic of the electrostatic quantities for a ferroelectric thin film enclosed between conducting electrodes. Reprinted with permission from Gordon and Breach Publishers from D. R. Tilley, in "Ferroelectric Thin Films: Synthesis and Basic Properties" (C. Paz de Araujo, J. F. Scott, and G. F. Taylor, Eds.), p. 11. Copyright 1996 Overseas Publishers Association, N.V.

dimensions. Second-order ferromagnetic phase transitions have been observed in monolayer magnetic films [30, 31], where the surface anisotropy energy stabilizes the two-dimensional ferromagnetic state at finite temperature [32]. But ferroelectrics more often exhibit finite-size effects suppressing T_C, observed in nanocrystals as small as 20 nm in diameter, as shown in Figure 5 [33–35], and in vinylidene fluoride copolymer films as thin as 60 nm [36], and possibly in lead titanate films as thin as 10 nm [37]. Recent switching studies in lead zirconate-titanate (PZT) films only 4 nm thick indicate that they may be the thinnest ferroelectric perovskite films made so far [38]. These results also can be interpreted as bulk ferroelectricity suppressed by surface energies and the depolarization field, and imply that the bulk ferroelectric state has a minimum critical size [39–42].

A number of magnetic materials have magnetic surface layers that show a second-order ferromagnetic–paramagnetic phase transition with higher Curie temperature than the bulk [43]. Similarly, a distinct ferroelectric transition in surface layers with a transition temperature higher than in the bulk has been reported by Scott in studies of KNO₃ thin films [40, 41]. These surface layer phases have been modeled by a bulk three-dimensional free energy modified by surface terms.

The free-energy analysis of the ferroelectric state in thin-film samples of bulk ferroelectric materials should also account for the depolarization energy mentioned in Section 1.1.1 and the surface energy connected with the truncation of the spontaneous polarization. Consider a uniaxial film such as the one depicted in Figure 4. The depolarization energy can suppress the phase transition or cause the film to break up into antiparallel domains and may be important even in crystals with conducting electrodes on the polar surfaces because of charge screening in the ferroelectric or in the electrodes. The total LGD free energy should include a polarization discontinuity

term proportional to $|\Delta P|^2$ at the surface and another term proportional to $|\nabla P|^2$ for any polarization gradients in the ferroelectric crystal. In ferroelectric thin films, these two factors are important and lead to a finite-size effect, the effect of thickness on ferroelectric and related properties of the film.

The following analysis of the surface energy contributions is from the review by Tilley [39], which summarizes the results of several reports [10, 44–48]. Consider a uniaxial ferroelectric film, where the spontaneous polarization P is perpendicular to the film surfaces located at positions $z = \pm L/2$ (see Fig. 6). If the potential across the film is zero, the Gibbs free energy per unit area has the form

$$G' = G'_0 + \int_{-L/2}^{L/2} \left[\frac{\alpha}{2} P^2 + \frac{\beta}{4} P^4 + \frac{\gamma}{6} P^6 + \frac{g}{2} \left(\frac{\partial P}{\partial z} \right)^2 \right] dz$$
$$+ \frac{g}{2\delta} (P_+^2 + P_-^2) \qquad (9)$$

where $P_\pm$ are the values of the spontaneous polarization on the surfaces at $z = \pm L/2$, $g > 0$ is the polarization gradient coupling constant, and δ is the surface energy decay length. (The material is assumed embedded in a nonpolar medium so that the surface discontinuities are given by $\Delta P_\pm = P_\pm$. But it is also possible for the ferroelectric material to induce polarization into an adjacent dielectric layer, with interesting consequences considered in recent papers [49–51].) The equation of state derived from the generalized free energy (Eq. (9)) has the form of the Euler–Lagrange equation, which can be solved numerically subject to the boundary conditions $\partial P/\partial z \pm P/\delta = 0$, at $z = \pm L/2$ [39].

If coupling to the surface polarization discontinuity is positive ($\delta > 0$), there is a decrease in the magnitude of the polarization P near the polar surfaces (see Fig. 6) and a decrease in the film's phase transition temperature T_C with decreased thickness. The opposite case with negative coupling

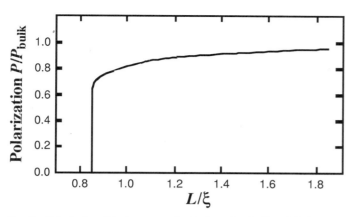

Fig. 7. Polarization P in the center of a thin ferroelectric film of thickness L, from the mean-field theory for positive surface energy, $\delta > 0$. Reprinted with permission from Gordon and Breach Publishers from B. D. Qu et al., *Ferroelectrics* 152, 219 (1994). Copyright 1994 Overseas Publishers Association, N.V.

($\delta < 0$) leads to an increase in the magnitude of the polarization near the polar surfaces and an increase in T_C with decreased thickness. In either case, the ferroelectric state of the film does not break into separate bulk and surface states with different transition temperatures, but retains a single state with a single transition temperature T_C. Figure 7 shows the finite-size effect for positive coupling ($\delta > 0$) on the equilibrium polarization in the center of the film (where the polarization has maximum value) as a function of the film thickness L. Here P_{Bulk} is the spontaneous polarization in the bulk material ($L = \infty$) and the thickness L is normalized to the ferroelectric correlation length $\xi = \sqrt{g|\alpha|}$. The ferroelectric state is completely suppressed in films below the critical thickness $L_C \approx 0.85\xi$.

The depolarization energy is another mechanism that tends to suppress ferroelectricity, decreasing the polarization P and the transition temperature T_C, but it is necessary to account for screening by free charge in the ferroelectric and the electrodes. Consider a ferroelectric film with thickness L covered on both sides by conducting electrodes (Fig. 4b). Charge accumulates within the electrodes and within the film to screen the polarization discontinuity at the surface of the ferroelectric film (see Fig. 6). The charge does not completely screen the polarization because the charge must be spread out over a finite Thomas-Fermi screening length L_S. The analysis of the screening effect on the depolarization field reveals the spatial dependence of the electric field (see Fig. 6) and amounts to a renormalization of the LGD coefficients in Eq. (1). There is finite-size suppression of ferroelectricity, with critical thickness $L_C = 2\chi C L_S / T_{C\text{-Bulk}}$, similar to the suppression caused by a positive surface coupling coefficient δ. The results from sintered PbTiO$_3$ crystals [33] are in good qualitative agreement with the screened depolarization mechanism, represented by the solid line shown in Figure 5 [39].

The finite-size effect in the three-dimensional Ising model was analyzed by de Gennes [52] and applied to ferroelectrics by others [11, 44, 48, 53]. The Ising model, useful for order–disorder ferroelectrics such as KDP, also describes the de-

pendence of the spontaneous polarization P_S and transition temperature T_C on the film thickness, predicting a critical thickness L_C. At the surface of the ferroelectric, the interaction between the neighbor dipoles should be much different than in the bulk, reducing correlation among the dipoles and suppressing ferroelectricity.

Though the ferroelectric finite-size effect can arise from several distinct mechanisms, such as the depolarization, screening, and surface energy mechanisms outlined above, or possibly extrinsic effects due to surface pinning or inhomogeneity, all depend on the ratio of surface area to volume and are therefore fundamentally connected with three-dimensional (bulk) ferroelectricity. The absence of significant finite-size effects in ferroelectric vinylidene fluoride copolymer films as thin as 1 nm is evidence that they are two-dimensional ferroelectrics (Section 4.5).

1.2. Ferroelectric Polymers: Vinylidene Fluoride Copolymers

The study and application of ferroelectric polymers dates back to approximately 1969 with the discovery of piezoelectricity in polyvinylidene fluoride (PVDF) by Kawai [54]. The ferroelectric polymers—mainly PVDF and its copolymers [55], and the odd nylons [56–58]—are crystalline and exhibit clear ferroelectric phase transitions. There are also a number of ferroelectric liquid crystals, some polymers, with application to display technology [7], but these are improper ferroelectrics, where the order parameter is obtained from the molecule's director axis, not its dipole moment, which generally points in a different direction. The properties of ferroelectric polymers are reviewed by Furukawa [55, 59], by Lovinger [60, 61], and in the books edited by Wang, Herbert, and Glass [14] and by Nalwa [62].

Ferroelectric polymers exhibit the interesting and useful properties common to ferroelectric and other noncentrosymmetric materials, including pyroelectric and piezoelectric effects [55, 63], bulk photovoltaic (photogalvanic) currents [64], electro-optic effects [65], second-harmonic generation [66], and possible photorefractive effects [67]. The rapid commercialization of piezoelectric transducers made from PVDF is a model of technology transfer [14, 62, 68, 69].

1.2.1. Molecular and Crystal Structure

Polyvinylidene fluoride (PVDF), a nonconjugated linear fluorinated hydrocarbon, is the prototypical ferroelectric polymer. The vinylidene fluoride C$_2$H$_2$F$_2$ monomers form a linear carbon–carbon chain with structure $-(\text{CH}_2-\text{CF}_2)-$. The monomer has a dipole moment pointing roughly from the fluorines to the hydrogens. The all-trans $TTTT$ conformation (Fig. 8a) has a net dipole moment of about 7×10^{-30} essentially perpendicular to the chain axis [70], while the alternating trans-gauche $TG T\overline{G}$ conformation (Fig. 8b) has a net dipole moment with components both perpendicular and parallel to the chain. The polymer chains can have different conformations defined by the sequence of dihedral bond angles. The

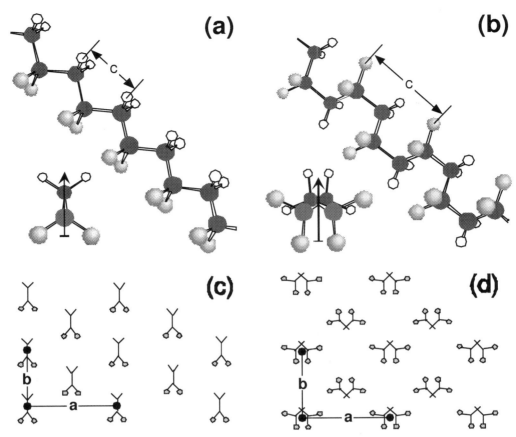

Fig. 8. Conformations and crystalline forms of PVDF: (a) in the all-trans conformation (inset, end view of a chain); (b) in the alternating trans-gauche conformation (inset, end view of a chain); (c) end-on view of the crystal structure of the ferroelectric β phase, composed of close-packed all-trans chains; (d) end-on view of the crystal structure of the paraelectric α phase, composed of close-packed trans-gauche chains. Reprinted with permission from L. M. Blinov et al., *Physics Uspekhi* 43, 243 (2000). Copyright 2000 Turpion Limited.

trans bond (T) has a dihedral angle of approximately ~180° and the left and right *gauche* bonds (G and $\overline{G}$) have dihedral angles of approximately ±60°. The most common conformations are all-trans $TTTT$ (Fig. 8a), alternating left-right trans-gauche $TGT\overline{G}$ (Fig. 8b), and helical $TGTG$ or $T\overline{G}T\overline{G}$. All three conformations form straight cylindrical chains that crystallize in a quasi-hexagonal packing.

The all-trans ($TTTT$) conformation crystallizes in an orthorhombic m2m structure with the chains along the crystal c-axis and the dipoles aligned approximately along the crystal b-axis as shown in Figure 8c [71–75]. This so-called "β phase" (also called "phase I") is polar and a uniaxial ferroelectric, as the polarization can be repeatedly switched between opposite but energetically equivalent directions along the 2-fold b-axis. The β-phase unit cell nominally consists of two $-(CH_2-CF_2)-$ formula units, one along the c-axis parallel to the chains (see Fig. 8a) times two in the plane perpendicular to the c-axis (see Fig. 8c). The unit cell dimensions are approximately [75]: $c = 0.256$ nm along the chain axis; $b = 0.491$ along the polarization direction, the 2-fold axis; and $a = 0.858$ nm perpendicular to the chain axis and to the polarization. (It is possible that the unit cell is twice as big, containing two monomers along the chain, because a ±7° dihe-

dral tilt-ordering would make the c-axis period two monomers long, or $c \approx 0.512$ nm [72].)

The paraelectric phase is composed of chains with the alternating trans-gauche ($TGT\overline{G}$) conformation (Fig. 8b), packing with no macroscopic polarization (Fig. 8d). The precise structure of the paraelectric phase is not clear. Published reports propose a variety of structures including single-crystal structures and multiphase composites. The fully crystalline "α phase" or "phase-II" structure shown in Figure 8d consists of opposing polar sublattices of the trans-gauche chains, producing zero net polarization, both parallel and perpendicular to the chains. The α-phase unit cell nominally consists of four $-(CH_2-CF_2)-$ formula units, two along the c-axis parallel to the chains and two in the plane perpendicular to the c-axis. The unit cell dimensions are approximately [76]: $c = 0.462$ nm; $b = 0.496$ nm; and $a = 0.964$ nm. It is possible that the α-phase structure is antiferroelectric, but published reports reveal only direct conversion to the all-trans conformation in the β phase on application of an electric field [77–81], not rotation of the trans-gauche dipoles into the polar alternating trans-gauche α_P, or "phase II$_P$," which may be only metastable [72, 75, 82].

The crystal structure of the paraelectric phase has been variously described as orthorhombic mmm [76, 83], monoclinic 2/m [72, 84, 85], and hexagonal [86, 87]. Some have proposed that the macroscopic paraelectric phase is composed of a random packing of the trans-gauche chains or a mixture of microcrystalline regions, each with a different packing [82]. Other proposals for the paraelectric phase structure include a helical conformation in hexagonal packing [88], though infrared and Raman spectroscopies indicate a preponderance of the trans-gauche conformation [89, 90]. The question is further complicated by the common observation that PVDF and its copolymers exhibit a number of metastable crystalline phases that are difficult to separate from the true equilibrium phases [75, 91, 92].

Standard synthetic methods unavoidably introduce 3–6% monomer reversal defects with structure $-(CH_2-CF_2)-(CF_2-CH_2)-(CH_2-CF_2)-$ [14, 93, 94]. PVDF can also be copolymerized, apparently in a random sequence, with trifluoroethylene (TrFE) $-(CHF-CF_2)-$ and tetrafluoroethylene (TeFE) $-(CF_2-CF_2)-$. The reversed monomers and the intentionally incorporated TrFE or TeFE units function as defects, tending to lower the melting and ferroelectric phase transition temperatures, as can chain ends, bends, and folds. The copolymers also have a slightly larger unit cell and smaller average dipole moment than pure PVDF, owing to the replacement of some of the hydrogen atoms by the larger fluorine atoms. The differences between PVDF and these copolymers are mostly quantitative, so it is useful to begin with a review of the main properties of PVDF.

Molecular modeling of PVDF and its copolymers is part of the more general study of crystalline polymers, especially the linear hydrofluorocarbon systems, from fully hydrogenated polyethylene to fully fluorinated polytetrafluoroethylene (Teflon™). These studies have determined stable conformations, crystal structures, band structures, and other physical properties [95, 96]. Molecular modeling of the structure and properties of PVDF and its copolymers dates back to the early 1970s. The early work of Farmer et al. [72] stands out because of the clarity of the analysis, though the authors had at their disposal relatively little computing power compared to later work. They investigated the stability of crystal structures of the two main conformations, all-trans and alternating trans-gauche, finding that at least 14% defects (e.g., TrFE monomers, head-head defects, chain ends) are needed to stabilize the α phase. This was in reasonable agreement with experimental results, which showed that a minimum of 20% TrFE was required to produce a stable paraelectric α phase below the melting point [14, 97].

Increased attention to the study and application of PVDF in the late 1970s and early 1980s, coupled with significant advances in computing power, encouraged further molecular modeling focused on dynamical features that might affect the phase transition and switching [98–102]. These calculations yielded the propagation speed of reversed-polarization kinks, or solitons, along the chains, and therefore the upper limit of domain-wall velocity [98, 99, 102]. Calculations considering

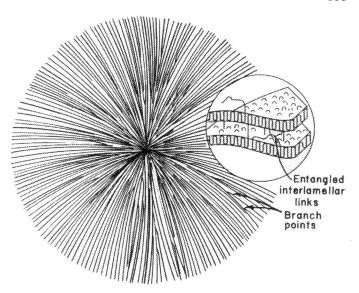

Fig. 9. Spherulite structure showing radial arrangement of lamellar crystals. Reprinted with permission from Kluwer Academic Publishers from "The Applications of Ferroelectric Polymers" (T. T. Wang, J. M. Herbert, A. M. Glass, Eds.), 1988. Copyright 1988 Chapman and Hall.

the full crystal were in reasonable agreement with switching measurements [102, 103]. More recent modeling has focused again on refining the crystal structure, in hopes of deciding the structure of the paraelectric α phase [82, 104] and to better model the phase transition [105–107]. Karasawa and Goddard [82] found that the trans-gauche conformation could crystallize in four different packing arrangements with nearly identical energies, and proposed that the actual paraelectric α phase is a statistical mix of these four packing arrangements. The microscopic models have so far proven inadequate for studying the ferroelectric–paraelectric phase transition because they neglect dipole–dipole interactions and treat interchain interactions only in the mean-field approximation. Modern computers and codes are much more powerful and will likely produce greatly improved models of the structure and dynamics of ferroelectric polymers (see progress reported in [108]).

Though piezoelectricity was discovered in PVDF over 30 years ago [54], progress in determining the structure and understanding structure–property relationships has been considerably impeded by imperfect samples. Samples made by solvent-casting and spin-coating methods are polymorphous, containing amorphous material and lamellar crystallites, often with more than one crystalline phase. The needle-shaped crystallites tend to form radial spherulite balls as shown in Figure 9 [14, 109–111], where the crystalline regions are strained by the dangling and looped chain segments and the embedding amorphous material. The relative proportion of amorphous and crystalline phases depends strongly on the thermal, mechanical, and electrical treatments of the films, making it even more difficult to identify equilibrium features associated with ferroelectricity. Despite incomplete understanding of its fundamental ferroelectric behavior, the piezoelectric properties of PVDF were employed early on in transducers [14, 68, 69].

PVDF essentially does not crystallize from the melt and crystallizes only poorly from solution. Films must be treated mechanically by stretching to align the polymer chains, and electrically to align the polarization axes of the crystallites. Mechanical and electrical alignment results in macroscopic piezoelectric behavior, though the films still contain about 50% amorphous material and the crystallites are incompletely oriented [14, 55, 60]. But copolymers of PVDF with trifluoroethylene (TrFE) and tetrafluoroethylene (TeFE) crystallize readily from the melt and can be stretched and electrically polarized to over 90% crystallinity [14, 60, 61, 73]. Films formed by spinning on textured Teflon have shown particularly good crystallinity and orientation, permitting detailed studies of intrinsic elastic anisotropy [112]. Recently, refined sample treatments significantly improved the crystallinity in copolymer films, apparently eliminating amorphous material and lamellae, and producing a single crystalline phase, with highly oriented crystallites [86, 87, 113]. Though much has been learned about the fundamental nature of ferroelectricity and related properties of PVDF copolymers, the polymorphous and polycrystalline samples made by solvent crystallization limit the detail and accuracy of measurements and leave many questions concerning the fundamental properties unanswered.

1.2.2. Ferroelectric and Related Properties

For many years following the discovery of piezoelectricity in PVDF [54], there was some doubt that it was a true ferroelectric material, even though there was evidence of reversible polarization, because molecular films, such as electrets [3], can exhibit these properties due to nonequilibrium effects connected with charge injection. Also, for reasons explained below, PVDF does not undergo a paraelectric–ferroelectric phase transition. The polymorphous structure of PVDF samples further muddied the issue. This uncertainty did not prevent application of PVDF to electromechanical transducers, as the piezoelectric properties were strong and stable enough for practical use.

Convincing new evidence for ferroelectricity came from the copolymers of PVDF with trifluoroethylene P(VDF-TrFE) [109] and tetrafluoroethylene P(VDF-TeFE) [114], which do exhibit a clear ferroelectric–paraelectric phase transition. Addition of the larger and less polar TrFE and TeFE units suppresses the transition temperature by reducing the average dipole moment of the chains, expanding the lattice, and introducing defects. This suppression of ferroelectricity on addition of TrFE is illustrated in Figure 10, which shows the resulting decrease in transition temperature [97], and in Figure 11, which shows the decrease in spontaneous polarization. (The relatively low remanent polarization measured in PVDF films is due to the low crystallinity of about 50%.)

Extrapolation of the phase transition and melting temperatures to higher VDF content (see Fig. 10) also explains why pure PVDF does not exhibit the ferroelectric–paraelectric phase

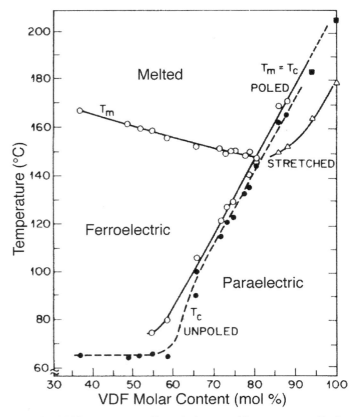

Fig. 10. Melting temperature T_m and phase transition temperature T_C for various VDF-TrFE copolymers. Reprinted with permission from K. Koga and H. Ohigashi, *J. Appl. Phys.* 59, 2142 (1986). Copyright 1986 American Institute of Physics.

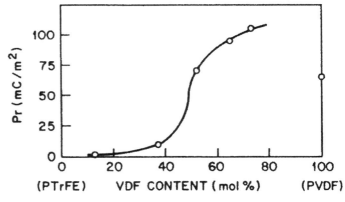

Fig. 11. Remanent polarization P_r for various VDF-TrFE copolymers. Reprinted with permission from Kluwer Academic Publishers from "The Applications of Ferroelectric Polymers" (T. T. Wang, J. M. Herbert, A. M. Glass, Eds.), 1988. Copyright 1988 Chapman and Hall.

transition, because the transition temperature is above the melting point, even for copolymers with up to 20% TrFE. Later studies showed that the ferroelectric–paraelectric phase transition in PVDF could be observed at high pressure [115, 116].

Most copolymer studies use a composition near 70% VDF and 30% TrFE, P(VDF-TrFE 70 : 30), because they have the most distinct ferroelectric properties and can be made nearly

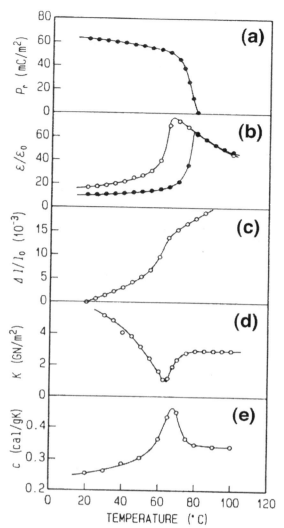

Fig. 12. Properties of P(VDF-TrFE 52 : 48) near the ferroelectric–paraelectric phase transition on heating: (a) remanent polarization on heating; (b) dielectric constant for an unpoled sample (solid circles) and after poling (open circles); (c) strain parallel to the polarization axis; (d) bulk modulus of elasticity; (e) specific heat capacity. Reprinted with permission from Gordon and Breach Publishers from T. Furukawa, *Phase Transitions* 18, 143 (1989). Copyright 1989 Overseas Publishers Association, N.V.

100% crystalline without stretching treatments. The 70 : 30 copolymer has the highest spontaneous polarization of about 0.1 C/m², a first-order ferroelectric–paraelectric phase transition at approximately 100°C, and large thermal hysteresis. At the phase transition, the structure changes from the ferroelectric-all-trans ($TTTT$) conformation arranged in a dipole-aligned structure (Fig. 8a, c) to the paraelectric alternating trans-gauche ($TGT\overline{G}$) conformation arranged in a nonpolar structure (Fig. 8b, d), though the true packing and crystal symmetry of the paraelectric phase are not yet identified, as discussed in Section 1.2.1.

The ferroelectric–paraelectric phase transition is accompanied by significant changes in thermodynamic properties, as displayed in Figure 12 [55]. The remanent polarization P_r

(Fig. 12a) decreases to zero at the transition because the order parameter must vanish in the paraelectric phase. Measurements of the remanent polarization in the copolymers typically decrease smoothly to zero indicative of a second-order transition (Fig. 2a), not with a sharp drop expected for a first-order transition (Fig. 2b). Furukawa has succeeded in obtaining much charper polarization drops and sharp dielectric peaks at the transition in carefully prepared samples [59]. The smearing of the phase transition may have several causes. The first cause of smearing is sample inhomogeneity. The second cause is phase coexistence; the finite enthalpy of the transition makes it difficult to maintain equilibrium during real measurements. Third, it has been suggested that the phase transition in the ferroelectric polymers is inherently diffusive because of the relatively short-range and weak interactions that dominate the transition, compared to the more familiar perovskites [117, 118]. The nature of the interactions in the copolymers is also evident in the relatively large spontaneous strain between the paraelectric and ferroelectric phases [74, 81, 119, 120] and the large difference between the equilibrium transition temperature and the critical point (Section 4.3) [121].

More convincing evidence for the first-order nature of the transition comes from thermal hysteresis, like the hysteresis shown in Figure 2b. Because of the finite enthalpy difference between the phases, and the resulting phase coexistence, the transition occurs at a higher temperature on heating and a lower temperature on cooling than the equilibrium transition temperature T_C. This thermal hysteresis and phase coexistence are also evident in structure measurements, which can quantify the amount of each phase (Section 3.2.2) [74, 81, 119]. Since thermal hysteresis is a nonequilibrium effect, it is important to distinguish the strong nonequilibrium effects accompanying a first-order transition and the weak nonequilibrium effects possible with both first-order and second-order transitions. In the weak nonequilibrium case, the thermal hysteresis will narrow and tend to vanish as the rate of temperature change is decreased (toward true equilibrium). On the other hand, the strong nonequilibrium of the first-order transition is due to the enthalpy difference between the phases. Thermal hysteresis and phase coexistence will typically persist without significant narrowing, even in quasi-equilibrium, as the heating or cooling rate is decreased by orders of magnitude. Other clear indicators of the first-order phase transition include the spontaneous strain (Fig. 12c), the softening evident in the bulk modulus (Fig. 12d), and a peak in the specific heat (Fig. 12e). These quantities, like the remanent polarization and the capacitance, all exhibit thermal hysteresis (not shown in Fig. 12a–e). The ferroelectric copolymer LB films exhibit these and other features of the first-order ferroelectric phase transition, as described in the following sections.

2. LANGMUIR–BLODGETT FILM FABRICATION

Langmuir–Blodgett (LB) deposition is a long-established means of obtaining uniform molecular monolayers. An amphiphilic molecule is dispersed on the surface of a liquid subphase, usually water, and compressed laterally into a dense film one molecule thick—a monolayer. This film can then be transferred to a substrate by dipping. Thicker films are built up by repeated dipping. One of the impressive possibilities of the LB method is an opportunity to vary the multilayer structure of LB films with an accuracy of one monolayer, when the kind of molecules and their orientation within a single monolayer can be controlled.

In 1995, we began making crystalline films of PVDF copolymers by LB deposition. Though these polymers are not good amphiphiles, we were able to produce excellent ferroelectric films with polarization hysteresis and a clear ferroelectric–paraelectric phase transition. The LB technique for the first time permitted us to prepare ultrathin ferroelectric films as thin as one monolayer (ML), about 5 Å, thus allowing study of ferroelectricity at films thicknesses never before possible. This section describes the methods used to produce these ultrathin ferroelectric films.

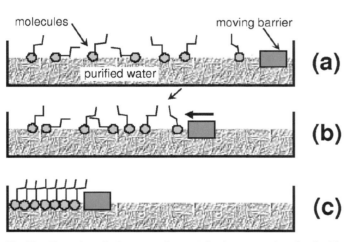

Fig. 13. Formation of a dense monolayer confined to a water (or other liquid) surface by: (a) dispersal of molecules at low area density; (b) two-dimensional compression by sweeping a movable barrier across the liquid surface; (c) holding the monolayer at constant pressure with the barrier.

2.1. Langmuir–Blodgett Films

The fact that oil forms thin films on water is known from ancient times. The Japanese art of "Suminagashi," or deposition of thin films from a water surface, has been known for more than a millennium and is still in use [122]. The modern study of molecular monolayers begins with the work of Agnes Pockels [123] and Lord Rayleigh [124], who realized more than a century ago that soap films on water lead to a change in surface pressure and that the films were very thin, about 1 nm, leading to the hypothesis that these were *monomolecular* films.

Irving Langmuir performed systematic investigations of the behavior of fatty acid molecules on a water surface, confirming the monomolecular nature of the films [125]. The fatty acids and other amphiphilic molecules were dispersed at low area density on the surface of a water trough and compressed to form dense monolayers, as shown in Figure 13. Together with Katharine Blodgett, Langmuir also developed a method to transfer multilayer films onto solid substrates by pushing or drawing the substrates through the floating Langmuir monolayer, as shown in Figure 14, so that multilayer structures could be formed as a result of successive transfer of individual monolayers [126, 127]. Now, molecular monolayers on the surface of a liquid substrate are commonly called "Langmuir" films, and films consisting of one or more Langmuir monolayers transferred to a solid substrate are called "Langmuir–Blodgett" (LB) films. Langmuir and Schaefer developed a variation of the LB technique, where the monolayers were transferred by touching the substrate horizontally to the monolayer (see Fig. 15), which proved useful for making some kinds of films [128]. Other early investigations of Langmuir and Langmuir–Blodgett films carried out by Trapeznikov [129] and others are summarized in Gaines' book [130]. In the second half of the twentieth century, due to the possibilities for creating artificial molecular systems with detailed control at a molecular level, the investigations of Langmuir and Langmuir–Blodgett films were expanded beyond basic studies of their physical chemistry to the study and control of potentially useful properties. Several scientists, most notably Kuhn [131], demonstrated many possibilities for application of tailored LB films.

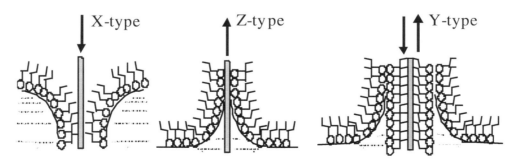

Fig. 14. Transfer of molecular monolayers on a solid substrate from a liquid subphase by the three variations of the Langmuir–Blodgett method: X-type pushing down through the monolayer into the subphase to produce films with the hydrophobic side of the monolayers closest to the substrate; Z-type drawing the substrate up from the subphase through the monolayer to produce films with the hydrophilic side of the monolayers closest to the substrate; Y-type alternate pushing and drawing to produce films with alternating monolayer orientations.

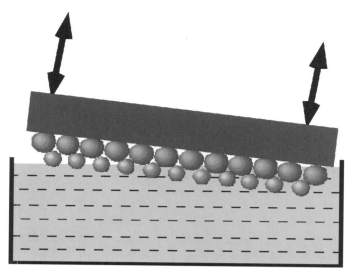

Fig. 15. Diagram of the horizontal Langmuir–Schaefer deposition method for transferring monolayers to a solid substrate. Reprinted with permission from L. M. Blinov et al., *Physics Uspekhi* 43, 243 (2000). Copyright 2000 Turpion Limited.

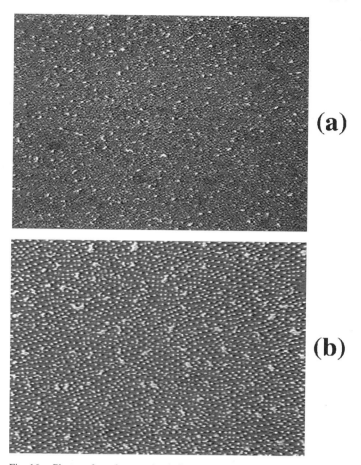

Fig. 16. Photos of a polymer microball monolayer: (a) on the water surface; and (b) on the substrate after it was transferred by LB deposition. The diameter of the microballs is 4.5 μm. Reprinted with permission from Gordon and Breach Publishers from S. P. Palto et al., *Mol. Cryst. Liq. Cryst.* 326, 259 (1999). Copyright 1999 Overseas Publishers Association, N.V.

Polar materials exhibit useful properties, such as piezoelectricity and nonlinear optical response [132], that are forbidden by symmetry in centrosymmetric materials. However, most molecular dipoles naturally crystallize in centrosymmetric structures, thus negating macroscopic effects. Noncentrosymmetric films of many polar molecules can be made using the X-type or Z-type LB deposition (Fig. 14) or by horizontal LB dipping (Fig. 15). There has been much work, particularly in the 1980s [68, 132–135], on polar LB films demonstrating strong piezoelectricity [136, 137], pyroelectricity [78, 132, 138, 139], photovoltaic generation [140], electro-optic response [132], and second-harmonic generation [141–143]. The Langmuir–Blodgett method can be used to build ultrathin oriented crystalline films, without resorting to complex methods such as molecular-beam or chemical deposition, that often require epitaxy on crystalline substrates. Such ultrathin molecular crystals exhibit unique properties and are useful for studying finite-size effects and the physics of two-dimensional crystals [144–146].

At first glance, PVDF and its copolymers seem to be unacceptable for LB deposition because the molecules (see Fig. 8) are not amphiphilic. A typical amphiphile is a fatty acid molecule, which has a hydrophilic polar head and a hydrophobic alkyl tail. An interesting confirmation that amphiphilic properties are not necessary is the formation of Langmuir monolayers of micron-size plastic spheres (Fig. 16a) and LB transfer to a solid substrate (Fig. 16b) [147]. The recent review by Arslanov [148] covers the influence of chemical structure and external conditions in the organization of planar ensembles.

Even though PVDF molecules are not good amphiphiles, good monolayer and multilayer films can be prepared using the standard LB technique (Fig. 14) or the horizontal variation (Fig. 15) [149–151]. It turns out that PVDF and its copolymers, consisting of macromolecules with low solubility in water,

readily form metastable monolayers on a water subphase, as was demonstrated in 1967 by Ferroni et al. [152]. It is remarkable that the fabrication of PVDF LB films was demonstrated about two years before the discovery of piezoelectricity in polymorphous PVDF [54], but the study of ferroelectric LB films of PVDF copolymers began nearly three decades later [149–151, 153–155].

2.2. Langmuir–Blodgett Fabrication of Copolymer Films

The discovery of the first two-dimensional ferroelectrics was enabled by LB deposition of PVDF and its copolymers. For film fabrication, we used an LB trough of a special design shown in Figure 17 [156]. The trough consists of two sections, a buffer zone that has a clean water surface (free of the material molecules), and a deposition zone confining the Langmuir monolayer of the molecules to be deposited. The Langmuir films are prepared from solution of the copolymer of 0.01–0.10% weight concentration in dimethyl-sulfoxide (DMSO) or in acetone. A small amount of the solution is spread on the

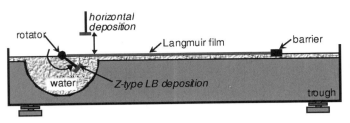

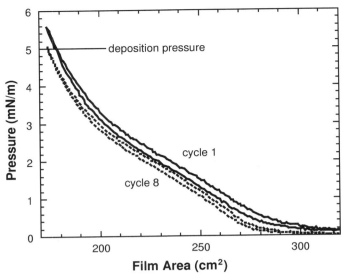

Fig. 17. Side view of the multipurpose LB trough. A clean water surface is maintained to the left of the rotator and the Langmuir film is compressed between the moving barrier and the rotator. For Z-type LB deposition, the substrate is attached to the rotator arm and passed first down through the clean water surface and then up through the Langmuir film to pick up a monolayer. For horizontal (Langmuir–Schaefer) deposition, the substrate is lowered from above into contact with the Langmuir film and then pulled back up with a new monolayer attached.

Fig. 19. Pressure–area isotherms of a PVDF Langmuir film at 23°C for the first and eighth cycles.

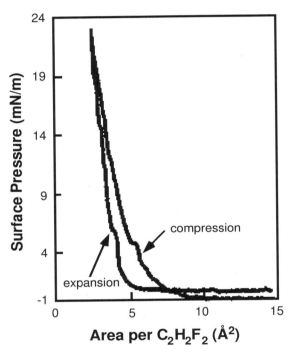

Fig. 18. Pressure–area isotherm of a P(VDF-TrFE 70:30) Langmuir monolayer. Reprinted with permission from Gordon and Breach Publishers from S. Palto et al., *Ferro. Lett.* 19, 65 (1995). Copyright 1995 Overseas Publishers Association, N.V.

water surface of the main section, where the molecules can be compressed by the moving barrier to form a dense monolayer.

Important information about the monolayer state on a water surface can be obtained from the pressure–area isotherm, the dependence of the surface pressure on film area. The pressure–area isotherm of P(VDF-TrFE 70:30) shown in Figure 18 exhibits the typical behavior of a Langmuir film, with a low-pressure 'gas' phase at large area followed by rapidly increasing pressure as the free area is eliminated and the film passes through liquid and solid phases [149]. If the pressure gets too high, the film collapses and generally does not completely re-disperse, as shown by the hysteresis in the pressure–area cycle in Figure 18.

At a surface pressure of about 5 mN/m, the measured area per $-(CH_2-CF_2)-$ repeating unit of the polymer molecule is 5.7 $Å^2$, close to the estimated value of about 6 $Å^2$ from the molecular model [149]. We believe that at this pressure we have close packing of the polymer molecules with their polymer chains in the plane of the water surface, so that the layer thickness is about 5 Å, the thickness of the individual polymer chains. Since both acetone and DMSO are soluble in water, they can carry a significant quantity of the material into the bulk of the water, resulting in apparently low values for the area per monomer. This problem is worse with DMSO, which has a higher density than water. Precautions must be taken to minimize the loss of material by depositing the solution in small drops and allowing time between drops for the solution to spread out on the surface. On the isotherm (Fig. 18), there is a small region just above 5 mN/m where a decrease in area does not lead to an increase of the pressure, perhaps due to some local collapse. In fact, AFM and SEM measurements detailed in Section 3.1 show that collapsed regions occupying ~5% of the total area exist even after deposition at only 5 mN/m pressure [157, 158]. Films formed at significantly higher surface pressures buckle and fold, producing poor samples [159]. The Langmuir films for LB deposition are prepared by slow compression to no more than 5 mN/m pressure, well below the collapse pressure. These monolayers exhibit highly reversible pressure–area isotherms, even after multiple compression cycles, as demonstrated in Figure 19.

The Langmuir monolayer can be transferred to a substrate when the substrate passes through the compressed monolayer. Division of the trough into two sections allows the three main types of monolayer deposition—X-, Y-, and Z-type—which transfer the film at right angles to the substrate (see Fig. 14). The traditional Y-type deposition consists of dipping the substrate into the trough through the Langmuir film and then removing it, producing a bilayer with each stroke. In Z-type deposition, the substrate passes into the water through the clean surface and out through the surface with the Langmuir film, while X-type deposition is done in the reverse order. We found that Z-type deposition frequently made good ferroelectric films of PVDF and its copolymers. We also obtained consistently good films using the horizontal Langmuir–Schaefer method [128], where monolayer transfer proceeds by picking up the film with the substrate nearly parallel to the water surface, as shown in Figure 15.

2.3. Sample Preparation

The generic sample structure shown in Figure 20a consists of a flat substrate, a thin bottom electrode, the multilayer LB film, and a thin top electrode. The actual structure of each sample was chosen to suit particular measurements. Films formed well on a variety of surfaces, including glass, silicon, aluminum, chromium, indium tin oxide, and with some success, on gold. Films for STM and AFM measurements were deposited directly on freshly cleaved pyrolytic graphite or on highly polished electronic-grade silicon. Films deposited directly on silicon substrates were suitable for X-ray, neutron, and electron diffraction, spectroscopic ellipsometry, reflectometry, and photoelectron spectroscopy.

Samples for dielectric or other electronic measurements required conducting electrodes above and below the LB film (Fig. 20a). The electrodes consisted of aluminum films approximately 25–50 nm thick deposited by vacuum evaporation on the substrate through a striped mask, covered with the LB film and then the top electrodes evaporated through a striped mask at right angles to the bottom electrodes, as shown in Figure 20b, so that the sample had several spots that could be probed independently. Some samples, prepared for *in situ* X-ray diffraction studies of structure changes in an electric field, had larger aluminum electrodes to allow *in situ* application of a uniform electric field over the X-ray beam area.

Films were prepared with careful attention to purity and cleanliness of both substrate and trough surfaces. Substrates were precleaned with acetone and alcohol. The LB trough (Fig. 17) was thoroughly cleaned and filled with triple-distilled or ultrapurified (>18 MΩ cm) water. The water surface was repeatably compressed and suctioned until there was no measurable pressure change with area, indicating a clean water surface. Substrates were further cleaned immediately before deposition by passing them through the clean water surface, and the surface was again compressed and suctioned to remove any residue left behind by the substrate. Langmuir monolayers of the polymer were prepared as described in Section 2.2, typically by spreading a solution of 0.01% by weight of P(VDF-TrFE 70 : 30) copolymer in dimethyl-sulfoxide and compressing to 5 mN/m or less. Even in this low pressure range, the effective thickness of the samples depends slightly on deposition pressure, as the measured capacitance is highest at nominally zero deposition pressure, as shown in Figure 21. The polymer presumably aggregates on the water surface, held together by its own perimeter tension.

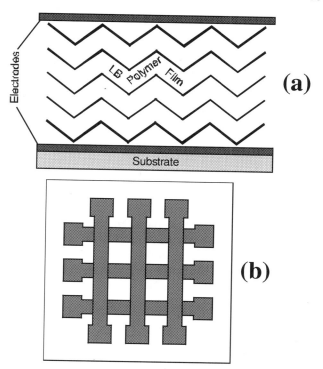

Fig. 20. Schematic diagram of the typical sample structure: (a) cross section showing a flat substrate, metal bottom electrode, multilayer LB polymer film; (b) top view showing the crossed stripe electrodes sandwiching the LB film.

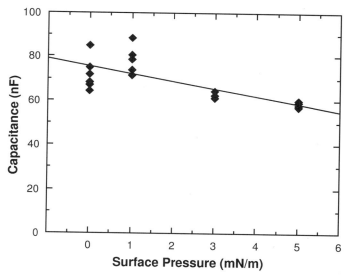

Fig. 21. Capacitance of various spots on four 12-ML P(VDF-TrFE 70 : 30) LB films made at low deposition pressures.

Most films were deposited by the horizontal Langmuir–Schaefer method (Fig. 15), taking care to let the film dry and the trough pressure stabilize between layers and obtaining each layer from a different spot on the trough surface. Completed samples were annealed at 120°C for at least 1 hr. The results summarized below note when samples were prepared by significant variations in this procedure. Further details of deposition methods and conditions are reported in [151, 160, 161].

3. FILM STRUCTURE AND MORPHOLOGY

This section reviews studies of the structure and morphology of the ferroelectric polymer LB films. The nominal film structure depicted in Figure 20a consists of tightly packed layers with straight chains lying parallel to the substrate. The polymer monolayers were well aligned over areas of several square millimeters or more, as confirmed by scanning tunneling microscopy (STM), low-energy electron diffraction (LEED), angle-resolved inverse photoemission spectroscopy, second-harmonic generation, and electron energy-loss spectroscopy (EELS). The multilayer films are crystalline, as confirmed by X-ray diffraction and neutron diffraction. There is no sign of the lamellar crystallites, amorphous material, or other phases often found in solvent-formed films, though there is some reconstruction evident in the surface layers. Scanning electron microscopy (SEM) and atomic force microscopy (AFM) show excellent planar morphology of the films.

3.1. Monolayer Structure

Atomic-resolution STM images like those in Figure 22 show that the monolayers have excellent short-range order. The polymer chains are straight, tightly packed, and mutually parallel [118, 154, 162]. Though the STM images show excellent ordering along the chains (this is expected as the covalent bonds rigidly hold the 2.6-Å spacing between monomers), some of the images, such as the one of a 2-ML film on graphite shown in Figure 23a, exhibit some disorder in the interchain spacing and registration. The Fourier transform of this STM image (Fig. 23b) and the LEED image of a 5-ML film on silicon (Fig. 23c) both have a series of evenly spaced lines perpendicular to the chains [119]. These regularly spaced lines confirm the excellent order along the chains and reveal the disorder perpendicular to the chains. The LEED patterns from other films showed distinct spots in both directions and only faint lines from slight interchain disorder. Though individual monolayers can have poor interchain order (i.e., Fig. 23a as compared to Fig. 22), annealing significantly improves the crystallinity of the multilayer LB films. Unlike the polymorphous films, the LB films do not contain amorphous material or lamellae.

The twofold in-plane rotational symmetry of the LEED data in Figure 23 indicates that chains are generally aligned in the same direction over areas of several square millimeters or more, though the films may in fact be polycrystalline, perhaps lacking full translational symmetry [119, 162]. The twofold rotational symmetry is also evident in EELS [163], second-harmonic generation (see Fig. 24) [164, 165], and angle-resolved inverse photoemission results [119, 162, 163].

There is evidence for surface reconstruction, as shown schematically in Figure 25 [119]. The STM images in Figures 22 and 23a show a highly flattened surface, though this is to be expected of LB layers deposited directly on the atomically flat graphite substrate. The results from LEED [119, 162], EELS [163], and angle-resolved inverse photoemission [119,

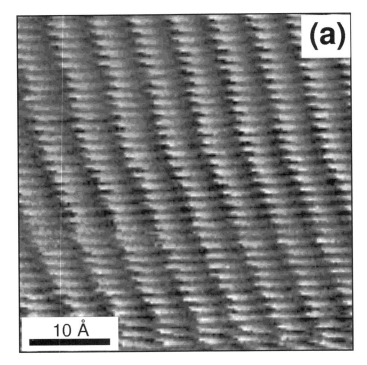

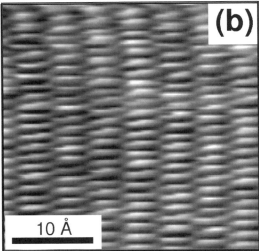

Fig. 22. Scanning tunneling microscopy (STM) images of samples consisting of one LB monolayer of the copolymer P(VDF-TrFE 70 : 30) on a graphite substrate. (a) STM image with the chains running diagonally up and to the left. (b) another 1-ML film on graphite with particularly good short-range order. Reprinted with permission from A. V. Bune et al., *Nature* 391, 874 (1998). Copyright 1998 Macmillan Magazines Ltd. (b) STM image with chains running vertically.

162], are also consistent with significant reconstruction in the top layer or two in the LB films as thick as 5 ML, deposited on silicon. The reconstructed surface understandably exhibits a separate phase transition, resembling the bulk ferroelectric phase transition [118, 166], but at a lower temperature, with the same conformation change from all-trans to alternating trans-gauche forms [119, 162]. This surface phase transition, described in Section 4.6, does not appear to be ferroelectric; that is, polarization is not the order parameter [163].

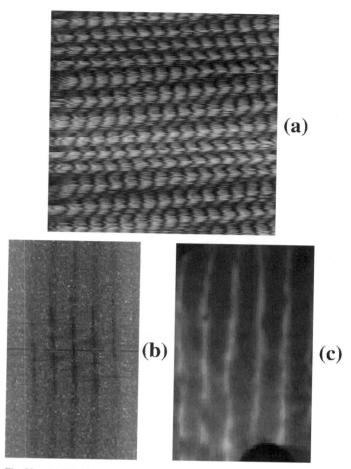

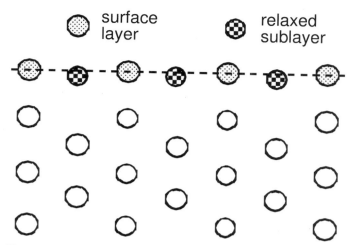

Fig. 25. Schematic representation of surface reconstruction, viewing the polymer chains end-on. Reprinted with permission from J. Choi et al., *Phys. Rev. B* 61, 5760 (2000). Copyright 2000 American Physical Society.

Fig. 23. (a) STM image of a 2-ML LB film on graphite. (b) The Fourier transform of the STM image in (a). (c) The LEED image of a 5-ML LB film on Si (100). Reprinted with permission from J. Choi et al., *Phys. Lett. A* 249, 505 (1999). Copyright 1999 Elsevier Science.

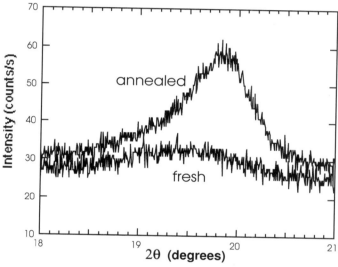

Fig. 26. Effect of annealing on the crystallinity of a 30-ML P(VDF-TrFE 70:30) LB film, revealed by X-ray diffraction. The lower curve was recorded at room temperature with the fresh film as grown and the upper curve after annealing for one hour at 120°C and cooling at 1°C/min to room temperature.

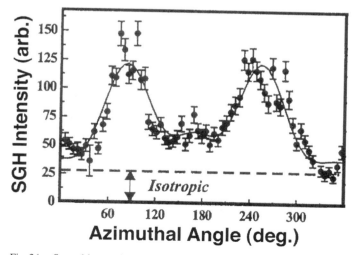

Fig. 24. Second-harmonic generation from a 15-ML P(VDF-TrFE 70:30) LB film on a fused quartz substrate. The azimuthal angle is the angle the plane of incidence makes with an arbitrary direction on the substrate. Reprinted with permission from O. A. Aktsipetrov et al., *Opt. Lett.* 25, 411 (2000). Copyright 2000 Optical Society of America.

3.2. Crystal Structure

The planar structure of the LB films permits us to monitor the crystalline phase using theta-two-theta X-ray diffraction, which measures spacing normal to the film. The ferroelectric and paraelectric phases are readily distinguished by their different layer spacings, about 4.5 Å in the all-trans ferroelectric β phase and about 4.8 Å in the alternating trans-gauche paraelectric α phase. Freshly deposited films generally have poor crystallinity, as indicated by the broad weak diffraction peak in Figure 26, but annealing for 1 h at 120°C is sufficient to achieve a stable crystal structure, as confirmed by the stronger and narrower peak [81].

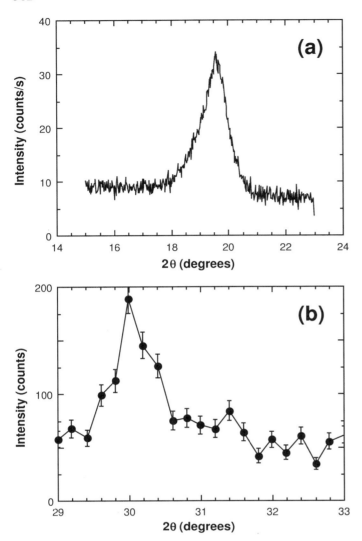

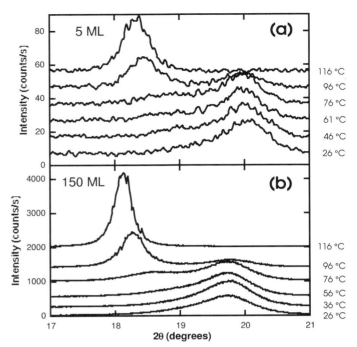

Fig. 28. Phase coexistence revealed by X-ray diffraction during heating of P(VDF-TrFE 70 : 30) LB films: (a) a 5-ML film; (b) a 150-ML film. Reprinted with permission from J. Choi et al., *Phys. Rev. B* 61, 5760 (2000). Copyright 2000 American Physical Society.

Fig. 27. Diffraction measurements of the layer spacing in the theta-two-theta geometry from P(VDF-TrFE 70 : 30) LB films on silicon (100) substrates. (a) X-ray diffraction from a 100-ML film, yielding a layer thickness of 4.5 Å. (b) Neutron diffraction measurement from a 50-ML film, yielding a layer thickness of 4.4 Å.

3.2.1. Measurement of the Layer Spacing

Both X-ray [119] and neutron [167] diffraction studies of the ferroelectric polymer LB films show excellent order perpendicular to the film, with a β-phase layer spacing of 4.5 Å in good agreement with studies of bulk films where this peak was associated with (100) and (200) reflections [71–75]. X-ray diffraction measurements, such as the one shown in Figures 26 and 27a, were carried out using a Rigaku theta-two-theta system with K_α radiation from a fixed copper anode (1.54 Å wavelength). This system produced a clear peak at $2\theta = 19.5°$ ($d = 4.5$ Å) at 25°C from films 4 ML to 150 ML thick [81, 119]. Analysis of the 19.5° peaks shows that the structural coherence length was comparable to the film thickness. Neutron diffraction measurements, such as the one shown in Figure 27b, were carried with 14.718-eV neutrons from the High-Flux Isotope Reactor at the Oak Ridge National Laboratory [167, 168].

3.2.2. Thermal Hysteresis and Phase Coexistence

The first-order ferroelectric phase transition exhibits phase coexistence, even under quasi-equilibrium conditions, because of the local minima in the free energy (see Fig. 1b). The phase coexistence region is the range of temperatures near T_C with a metastable phase in addition to the stable one. The ferroelectric phase may persist as the crystal is heated past the transition temperature T_C through its metastable region, to a maximum temperature T_1, as shown in Figure 2b. Similarly, the paraelectric phase may persist as the crystal is cooled through the transition temperature T_C through its metastable region, to a minimum temperature T_0, also shown in Figure 2b. The crystal can maintain a mixture of the two phases within the phase coexistence range between T_0 and T_1 (see Fig. 2b).

X-ray diffraction measurements are the simplest and most direct means for quantifying phase coexistence, because the out-of-plane layer spacings of the ferroelectric phase (4.5 Å) and the paraelectric phase (4.8 Å) are easily resolved, even in LB films only a few monolayers thick [81, 119]. The coexistence of the two phases on heating is evident in the X-ray diffraction data of Figure 28. At low temperatures there is only the ferroelectric peak near $2\theta = 19.5°$, while the paraelectric peak near 18.5° appears at 76°C (near T_C) and is the only peak evident at 116°C, well above the $T_1 \approx 100°C$ upper limit of the coexistence region. The reverse occurs on cooling; there is only the paraelectric peak above T_C, and both paraelectric and ferroelectric peaks from T_C down to $T_0 \approx 34°C$, which is the lower limit of the coexistence region. The coexistence region is

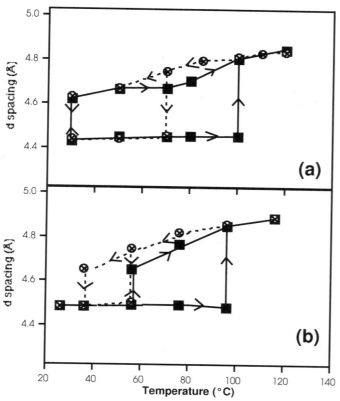

Fig. 29. Thermal hysteresis of the layer spacing of P(VDF-TrFE 70 : 30) LB films on heating (squares) and cooling (circles): (a) a 5-ML film; (b) a 150-ML film. Reprinted with permission from J. Choi et al., *Phys. Rev. B* 61, 5760 (2000). Copyright 2000 American Physical Society.

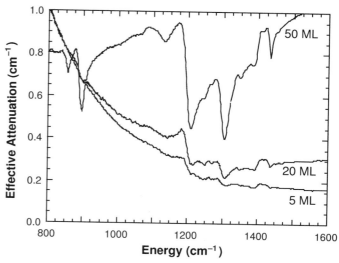

Fig. 30. Infrared ellipsometry spectra of P(VDF-TrFE 70 : 30) LB films on silicon substrates with thickness 5, 20, and 50 ML recorded at 25°C.

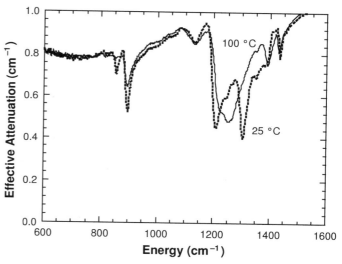

Fig. 31. Infrared ellipsometry spectra from a 50-ML P(VDF-TrFE 70 : 30) LB film on a silicon substrate, in the ferroelectric phase at 25°C (dotted line) and in the paraelectric phase at 100°C (solid line).

shown in Figure 29, plots of the layer spacings obtained from the data in Figure 28 and similar data obtained during cooling [81, 119]. Notice that the phase coexistence is evident in approximately the same temperature range in LB films 5 ML (2.5 nm) and 150 ML (75 nm) thick, and was observed in much thicker polymorphous films studied in detail by Legrand [74]. The similarities in the coexistence data from the ultrathin LB films and the much thicker polymorphous films [74] indicate that the nature of the phase transition is the same, independent of sample thickness or the presence of lamellae and amorphous material.

The vibrational modes of the polymer chains can also be used to distinguish the two phases. Certain vibrational modes exist only in one conformation, either the all-trans conformation (Fig. 8a) of the ferroelectric phase, or the alternating trans-gauche conformation (Fig. 8b) of the paraelectric phase. These modes can be probed by infrared (IR) spectroscopy [89, 90, 169–171], Raman spectroscopy [89, 90, 170], or electron energy-loss spectroscopy (EELS).

Because the ferroelectric LB films are so thin, we used IR ellipsometry to monitor the vibrational modes, because of the superior sensitivity of ellipsometry over transmission and reflection spectroscopies. Infrared spectroscopic ellipsometry reveals many modes characteristic of the polymer chains, even in samples as thin as 5 ML (2.5 nm), as shown in Figure 30.

The strong peaks at 846 cm^{-1}, 1186 cm^{-1}, and 1294 cm^{-1} are unique to the all-trans structure of the ferroelectric phase. The IR ellipsometry data from the 50-ML film (Fig. 31) show that these three peaks are evident only in the ferroelectric phase at 25°C, but not in the paraelectric phase at 100°C, where there is a different peak at about 1238 cm^{-1} from a mode unique to the alternating trans-gauche conformation. We have also followed some of these modes across the surface phase transition using EELS (Fig. 32), where the modes at 500 cm^{-1} and 840 cm^{-1} are evident only in the all-trans ferroelectric phase at low temperature, not in the alternating trans-gauche paraelectric phase at high temperature [163].

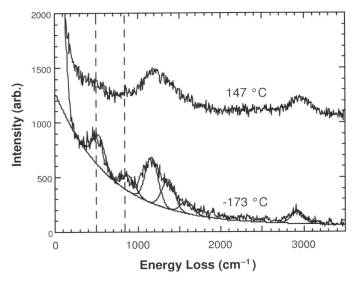

Fig. 32. Electron energy-loss spectra from a 5-ML P(VDF-TrFE 70:30) LB film on a silicon substrate, showing vibrational modes in the paraelectric α phase at 147°C (top) and the ferroelectric β phase at −173°C (bottom). Reprinted with permission from J. Choi et al., *J. Phys. Cond. Matter* 12, 4735 (2000). Copyright 2000 American Physical Society.

3.2.3. Field-Induced Phase Change

Another example of phase coexistence is the observation of continuous conversion of the paraelectric phase to the ferroelectric phase by an external electric field. Application of an electric field E raises the ferroelectric transition temperature $T_C(E)$ up to a critical point (T_{Cr}, E_{Cr}), as was demonstrated by Merz with barium titanate in 1953 [172], and for the first time with a ferroelectric polymer by us (see Section 4.3) [121]. This electric-field induced phase conversion is also evident in the structure [81]. A 50-ML LB film of P(VDF-TrFE 70:30) was annealed at 120°C, to improve crystal structure and eliminate all traces of the ferroelectric phase, and then cooled to 100°C, well above the zero-field transition temperature $T_C(0) \approx 80°C$. Application of an electric field to the sample at 100°C converts the paraelectric phase to the ferroelectric phase, as shown by the X-ray diffraction data in Figure 33a. At zero field, there is only the peak at 18.2° from the paraelectric phase. The peak at 20° from the ferroelectric phase appears at a threshold field of about 5 V and rises steadily until it saturates at about 15 V, as the paraelectric peak decreases in compensation. This steady conversion of the paraelectric phase to the ferroelectric phase is summarized in Figure 33b. On reducing the electric field, the ferroelectric phase converts back to the paraelectric phase, but at a lower field than on heating, another example of hysteresis connected with phase coexistence.

3.2.4. Stiffening Transition

The neutron diffraction studies revealed a previously unknown phase transition marked by a stiffening of the lattice [168]. The stiffening is evident in a fivefold increase in the Debye temperature below about 160 K, as shown in Figure 34. The Debye

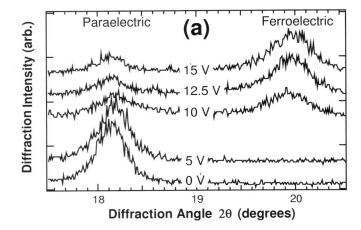

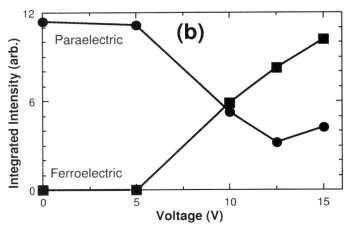

Fig. 33. Phase coexistence from θ–2θ X-ray diffraction of a 50-ML P(VDF-TrFE 70:30) LB film recorded at 100°C as a function of increasing electric field: (a) diffraction intensity showing the α-phase peak near 18.2° and the β-phase peak near 19°; (b) integrated diffraction peak intensities showing conversion from the α phase to the β phase as the field is increased from 0 to 600 MV/m.

temperature is obtained by fitting the integrated intensity of the out-of-plane neutron diffraction peaks (like Fig. 27b) to an exponential Debye–Waller dependence on temperature. The stiffening transition is not accompanied by a dielectric anomaly or a change in layer spacing, indicating that it is not a ferroelectric transition [168]. It may be connected with a subtle in-chain ordering of the ±7° CF$_2$ tilt proposed from steric considerations and molecular modeling [72].

3.3. Morphology

Scanning electron microscopy (SEM) images of the ferroelectric polymer LB films exhibit good planar morphology over large areas, as shown in Figure 35. The film is highly planar except for some ridges, or "mountain ranges," occupying about 5% of the total area, probably due to buckling at collisions between solid regions on the trough before monolayer deposition. The close-up of a Y-shaped intersection in Figure 35b illustrates the "tectonic" nature of the creation of these ridges. The excel-

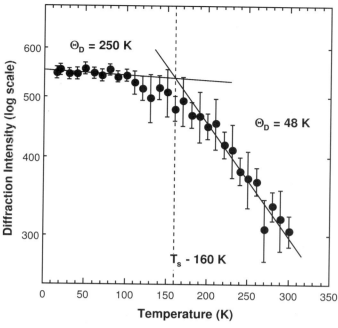

Fig. 34. Neutron diffraction (010) peak intensities showing a lattice stiffening transition near 160 K in a 100-ML P(VDF-TrFE 70 : 30) LB film. Reprinted with permission from C. N. Borca et al., *Phys. Rev. Lett.* 83, 4562 (1999). Copyright 1999 American Physical Society.

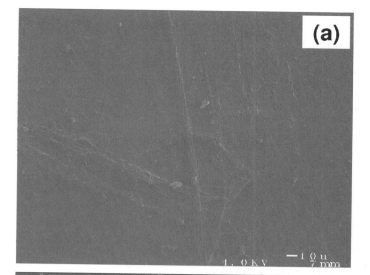

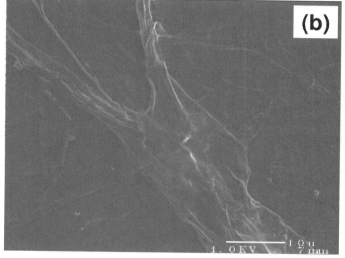

Fig. 35. SEM images of a 50-ML copolymer film recorded with 4-keV electrons. The white bars mark a length of 10 μm. (a) Low-magnification image showing large flat areas bounded by "mountain ranges." (b) High-magnification close-up of a Y-shaped ridge intersection.

lent results of the dielectric and structural studies are in part due to the relatively large proportion of the planar region.

The films can be switched on the nanometer scale by electrically biasing the tip of an atomic force microscope (AFM). Individual bits of information were thus written and then read out using electric field microscopy, which measures the electric potential above the film as the tip passes over [157]. Other recent AFM studies of polymorphous copolymer films reveal their fibrous lamellar nature [173], in contrast to the highly planar crystalline LB films.

A new technique of pyroelectric scanning microscopy (PSM) detailed in Section 4.4.2 allows us to image the film polarization in real time by scanning a modulated laser beam across the electrodes and measuring the pyroelectric current with a lock-in amplifier [158]. The PSM technique is useful for monitoring domains and domain dynamics and can be used to test films and devices.

4. FERROELECTRIC PROPERTIES

This section reviews the ferroelectric properties of copolymer LB films as thin as 2 ML (1 nm). The films exhibit all the main features of a ferroelectric with a first-order phase transition, including reversible polarization hysteresis; dielectric anomalies with thermal hysteresis; structure change at the phase transition; phase coexistence; piezoelectric and pyroelectric responses; double hysteresis; and a critical point in the field–temperature phase diagram. In fact, the results from the copolymer LB films

are remarkably consistent with the results from polymorphous samples that are thicker by up to six orders of magnitude.

The highly oriented crystalline structure and nanometer thickness of the LB copolymer films have allowed detailed study of their ferroelectric and related properties. The films exhibit a clear first-order bulk ferroelectric phase transition near $T_C = 80°C$ even in films as thin as 1 nm, showing that P(VDF-TrFE) is essentially a two-dimensional ferroelectric wholly consistent with the mean-field LGD theory. The LB films show clear and reversible polarization hysteresis, like the loops shown in Figure 36 with total switched charge up to 0.1 C/m², close to the maximum value obtained from solvent-formed films and to the expected saturation value considering the density of dipoles. The remarkably high value of the coercive field, about 500 MV/m in films 15 nm and thinner, is the first appearance of the elusive nucleation-independent intrinsic

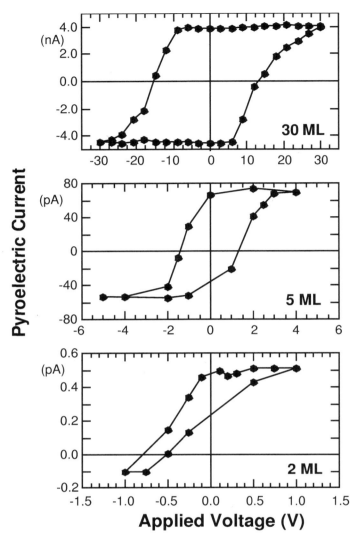

Pyroelectric Current

Fig. 36. Polarization hysteresis loops, measured at 25°C by the pyroelectric technique in LB films of P(VDF-TrFE 70 : 30). Reprinted with permission from A. V. Bune et al., *Nature* 391, 874 (1998). Copyright 1998 Macmillan Magazines Ltd.

coercive field. The dielectric response is a useful indicator of the phase transition and also of polarization hysteresis through the so-called "butterfly" capacitance hysteresis loops. The films exhibit single hysteresis below $T_C \approx 80°C$ and double hysteresis above T_C up to the critical point at $T_{Cr} = 145°C$. The switching in many films is slow, about 10 s, but some switch in a few microseconds. In addition, we observed a separate surface phase transition, which seems to have a nonferroelectric nature, at a lower temperature in the thinnest films.

4.1. The Ferroelectric Coercive Field

The defining characteristic of a ferroelectric is the ability to repeatably reverse, or switch, the polarization by application of a sufficiently large external electric field, as illustrated by the theoretical hysteresis loops in Figure 3. Yet since the discovery of ferroelectricity more than 80 years ago by Valasek [1],

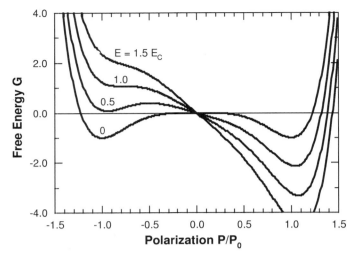

Fig. 37. Free energy of the first-order LGD model at various electric field values at temperature T_0.

there has been little mention in the literature of the intrinsic coercive field connected with collective polarization reversal, which is a consequence of the instability of the macroscopic polarization state in an opposing electric field. This seeming inattention to a fundamental property is because the measured value of the ferroelectric coercive field has invariably been much smaller than the intrinsic value predicted by theory. The low coercive field observed in real ferroelectric crystals and films is a result of extrinsic effects. Extrinsic switching is generally initiated by nucleation of domains with reversed polarization, which then grow and coalesce by domain-wall motion [9, 29, 174]. Nucleation can occur at fixed defects in the crystal or at grain boundaries. Nucleation mechanisms and domain-wall dynamics must be inserted into theoretical models in order to explain the experimental observations of switching kinetics with low extrinsic coercive fields in real ferroelectric materials [9]. In fact, the study and modeling of extrinsic switching seems to dominate the field and is central to the application of ferroelectric films to nonvolatile random-access memories [27, 175, 176].

Intrinsic polarization reversal is illustrated by plotting the free-energy density (Eq. (1)) in an electric field E, as shown in Figure 37. Consider a crystal in zero electric field in the negative polarization state represented by the well on the left of Figure 37. Application of an opposing electric field will raise the energy of that well and lower the energy of its counterpart at positive polarization on the right. The polarization should then tunnel from the now metastable left-hand well into the lowest-energy right-hand well at positive polarization, even at very small electric field. However, the polarization state is a collective property of the entire crystal, and macroscopic tunneling is highly improbable unless there is a local nucleation site where a finite volume of crystal can reverse polarization and a domain wall can propagate this reversal through the crystal. If there is no source of nucleation, or if nucleation or domain-wall motion are somehow inhibited, then the crystal will not reverse polarization until the barrier between the opposite polarization states is entirely eliminated, at the intrinsic coercive field E_C or above.

4.1.1. General Features of the Intrinsic Coercive Field

We can estimate an upper limit on the intrinsic ferroelectric coercive field E_C in a uniaxial ferroelectric thin film by calculating the electrostatic energy due to an external electric field necessary to balance the energy stored in the depolarization field $E_d = P_S/(\varepsilon\varepsilon_0)$, where P_S is the spontaneous polarization, ε is the ferroelectric contribution to the dielectric constant along P_S, and ε_0 is the permittivity of free space. The energy density associated with the spontaneous polarization is $u_P = (1/2)P_S^2/(\varepsilon\varepsilon_0)$. Application of an external field $\mathbf{E}$ in a direction opposite to the spontaneous polarization would contribute an energy density $u_E = \mathbf{E}\cdot\mathbf{P_S} = -EP_S$. (This contribution applies to all kinds of ferroelectrics, whether displacive, order–disorder, or other. The energy density of an external field interacting with dipoles of moment $\mathbf{m}$ and number density N is $u_E = N\mathbf{E}\cdot\langle\mathbf{m}\rangle$, leading to the same result.) The intrinsic coercive field E_C is then approximately the value of the applied field that produces an energy comparable to the polarization energy, $|u_E| \approx |u_P|$, so that the intrinsic coercive field is approximately half the depolarization field E_d,

$$E_C \approx \frac{1}{2}\frac{P_S}{\varepsilon\varepsilon_0} \qquad (10)$$

The intrinsic coercive field in the LGD phenomenology is obtained from the relationship $P(E)$ in Eq. (3), corresponding to the arrows in Figure 3 [9, 19, 25, 177]. The coercive field in the first-order LGD model can be written [177]

$$E_C \approx \frac{3\sqrt{3}}{25\sqrt{5}}\frac{P_S}{\varepsilon\varepsilon_0}\left(1 - \frac{t}{6}\right) \qquad (11)$$

The value of the intrinsic coercive field near the Curie temperature T_0 ($t = 0$) is about $1/10$ of the depolarization field. Similarly, the coercive field in the second-order LGD model is

$$E_C = \frac{1}{3\sqrt{3}}\frac{P_S}{\varepsilon\varepsilon_0} \qquad (12)$$

or about $1/5$ of the depolarization field at all temperatures. The coercive field in the second-order Ising–Devonshire model is approximately equal to the LGD second-order result (Eq. (12)) near $T_C = T_0$ [26].

The value of the intrinsic coercive field near the Curie temperature T_0 obtained from the mean-field models (Eq. (11), Eq. (12)) ranges from 10 to 20% of the depolarization field, a little less than the value 50% estimated from basic energy considerations (Eq. (10)), even as the spontaneous polarization and dielectric constant vary strongly with temperature. The approximations hold well near T_0 for all models, as shown in Figure 38. The essential existence and approximate value of the intrinsic coercive field do not depend on the nature of the ferroelectric transition, whether it is first-order or second-order, displacive or order–disorder, from permanent dipoles or induced dipoles.

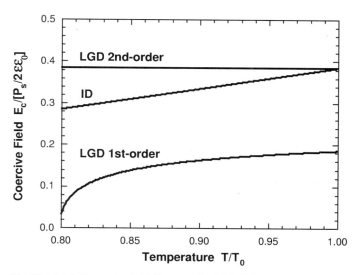

Fig. 38. Intrinsic coercive field E_C, normalized to half the depolarization field calculated from the first-order Landau–Ginzburg model (LGD-1), the second-order Landau–Ginzburg model (LGD-2), and the mean-field Ising–Devonshire model (ID).

4.1.2. Measurement of the Intrinsic Coercive Field

Measurements of the coercive field in bulk ferroelectric crystals have invariably yielded values much smaller than the intrinsic value, as summarized in Table I. While most traditional ferroelectrics have extremely low extrinsic coercive fields, the ferroelectric polymer polyvinylidene fluoride (PVDF) and its copolymers showed considerably higher coercive fields, typically 50 MV/m in polymorphous films. In 1986, Kimura and Ohigashi [36] reported that copolymer films thinner than 1 μm showed increasing coercive fields with decreasing thickness, up to a then-record coercive field of 125 MV/m in a P(VDF-TrFE 65 : 35) film 60 nm thick. But the solvent-spinning techniques used to make the films would not yield thinner films of adequate quality, so direct measurement of the intrinsic coercive field seemed just beyond reach. As Figure 39 shows, the rising coercive field with decreasing film thickness L reported by Kimura and Ohigashi continues with the thinner P(VDF-TrFE 70 : 30) LB films to follow a $L^{-0.7}$ power law scaling down to a thickness of 15 nm [160], consistent with a suppression of nucleation or domain-wall motion.

The LB copolymer films of thickness 15 nm (30 monolayers) or less all had coercive fields of about 500 MV/m at 25°C, comparable to the theoretical intrinsic value [177] calculated from the first-order LGD free-energy (Eq. (1)) coefficients listed in Table II that were determined in prior studies [63, 121]. The measured hysteresis loop shown in Figure 40 is in good agreement with the intrinsic hysteresis loop, though the pyroelectric response was measured instead of polarization (see Section 4.4), so the vertical scale is not a proper quantitative comparison with theory [177].

Because of uncertainties in the LGD parameters used to calculate the theoretical intrinsic coercive field, the good coincidence between the measured and expected values of the

Table I.   Typical Experimental Values of the Coercive Field E_C Compared to the Depolarization Field E_d for Several Ferroelectric Materials

	ε	P_S (C/m^2)	E_C^{exp} (MV/m)	$E_d = P_S/(\varepsilon\varepsilon_0)$ (MV/m)	$E_C^{exp}/\frac{1}{2}E_d$
BaTiO$_3$[8]	150	0.26	0.20	196	0.0020
Triglycine sulfate (TGS) [228]	43	0.028	0.011	74	0.00030
KD$_2$PO$_4$ [229]	43	0.062	0.34	163	0.0042
PZr$_{0.25}$Ti$_{0.75}$O$_3$ (PZT) 100-nm-thin film [183]	200	0.38	10	215	0.093
Polyvinylidene fluoride (PVDF) [68]	11	0.065	55	667	0.16
P(VDF-TrFE 75:25) 60-nm-thin film [36]	10	0.10	125	1129	0.22
P(VDF-TrFE 70:30) ≤15-nm-thin films [118, 190]	8	0.10	480	1412	0.68
				LGD 1st order	0.18
				LGD 2nd order	0.38
				ID	0.38

All values were obtained near T_0.

Table II.   Values of the LGD Coefficients (see Eq. (1)) for P(VDF-TrFE) Copolymers in SI Units

		LB films (70:30 copolymer) [63, 118, 121, 177, 190]	NL spectroscopy (70:30 copolymer) [189]	Spun film (65:35 copolymer) [59]
Curie constant	C	1500 ± 300 K	2000 ± 1000 K	3227 K
Curie temperature	T_0	30 ± 7°C		40°C
Transition temperature	T_C	78 ± 2°C		102°C
Coercive field slope	dE_C/dT	7.7 ± 0.5 × 10^6 V/Km		
Spontaneous polarization	$P_S(T_0)$	0.10 ± 0.02 C/m^2		0.09 C/m^2
LGD coefficient	β	−1.9 ± 0.2 × 10^{12}	−2 × 10^{12}	−1.5 × 10^{12}
LGD coefficient	γ	1.9 ± 0.2 × 10^{14}	5 ± 3 × 10^{14}	1.9 × 10^{14}

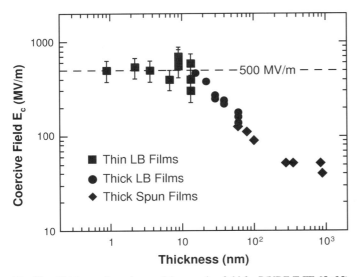

Fig. 39.   Thickness dependence of the coercive field for P(VDF-TrFE 65:35) spun and P(VDF-TrFE 70:30) LB films of the ferroelectric copolymer P(VDF-TrFE). Reprinted with permission from S. Ducharme et al., *Phys. Rev. Lett.* 84, 175 (2000). Copyright 2000 American Physical Society.

coercive field did not ensure that intrinsic switching had been achieved. Necessary additional evidence came from the absence of finite-size scaling below 15 nm (Fig. 39), which implies that nucleation and domain-wall motion no longer limit the coercive field. The most convincing evidence is the dependence on temperature (Fig. 41), showing the excellent agreement between the theoretical intrinsic coercive field (solid line, from Eq. (11)) and the measured coercive field (squares) [177].

Though nucleation and domain-wall motion do not limit the value of the intrinsic coercive field, they may play a key role in switching dynamics. Pertsev and Zembil'gotov [102] analyzed switching transients from PVDF films in terms of the motion of reversed-polarization kinks, or solitons, in the all-trans chains, and estimated an "activation" field of 900–1500 MV/m, which is somewhat larger than the measured value of 500 MV/m.

What is the reason for the transition from the nucleation-limited extrinsic coercive field to the nucleation-independent extrinsic coercive field at 15 nm thickness? This behavior is consistent with the dominance of nucleation; as the films are made thinner, nucleation volume is reduced and becomes

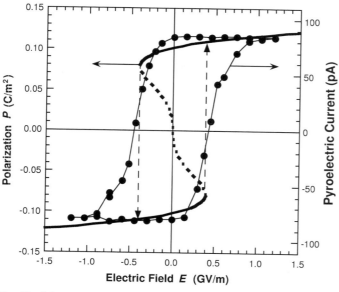

Fig. 40. Measured hysteresis loop (points) and the theoretical relation $P(E)$ for the first-order LGD model, from Eq. (3) (solid and dotted lines). Reprinted with permission from S. Ducharme et al., *Phys. Rev. Lett.* 84, 175 (2000). Copyright 2000 American Physical Society.

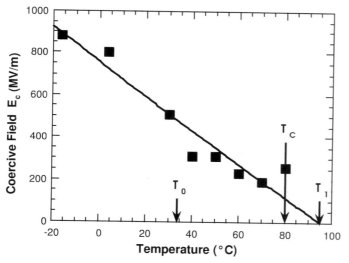

Fig. 41. Temperature dependence of the intrinsic coercive field E_C (solid line) and experimental values (squares), measured in a 30-ML P(VDF-TrFE 70 : 30) LB film. Reprinted with permission from S. Ducharme et al., *Phys. Rev. Lett.* 84, 175 (2000). Copyright 2000 American Physical Society.

energetically less favorable [178, 179] until it is completely inhibited, at 15 nm in the LB films. Recent improvements in the quality of solvent-crystallized vinylidene fluoride copolymers should help elucidate the effects of nucleation and domain-wall motion in thicker crystals [86, 87, 180]. There may soon be more results bearing on this question, as several groups have reported making good films with thickness less than 50 nm from several other ferroelectric materials [37, 38, 181–183].

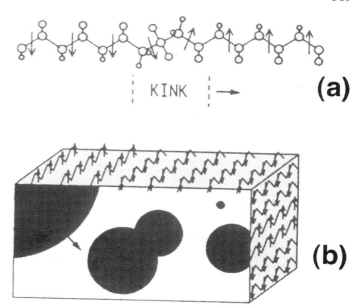

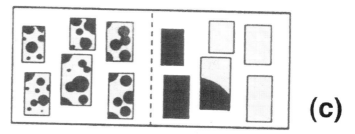

Fig. 42. Proposed mechanism of domain growth in PVDF. (a) Intrachain propagation by a kink in an all-trans chain bounding lengths of opposite polarization; (b) interchain propagation in a direction normal to the chains in a crystallite; (c) intercrystal propagation coalescing the mosaic of domains in a polycrystal or lamellar sample. Reprinted with permission from Gordon and Breach Publishers from T. Furukawa, *Phase Transitions* 18, 143 (1989). Copyright 1989 Overseas Publishers Association, N.V.

4.1.3. The Extrinsic Coercive Field

The common picture of extrinsic switching consists of nucleation followed by domain-wall motion and coalescence. A defect in the crystal or at a grain boundary can nucleate local polarization reversal in the presence of an opposing applied field. This local reversal requires relatively little energy compared to the collective reversal of an entire crystal in intrinsic switching, so the appearance and stability of nucleation regions generally follow an exponential activation law depending on temperature, local strain energies, and the applied electric field. Under the influence of the opposing field, a small nucleation region can spread, usually with a well-defined domain wall, until it reaches an impediment, such as a grain boundary. It has proven difficult to completely prevent extrinsic switching by reducing the number of defects or by pinning domain walls.

The nucleation and domain-wall motion processes for ferroelectric polymers are depicted in Figure 42 [55]. Nucleation might consist of a short segment of reversed polarization at a

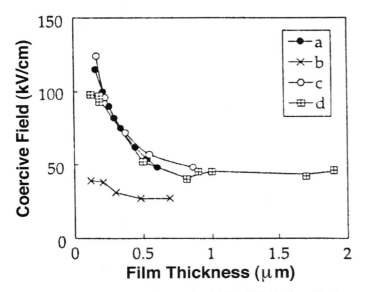

Fig. 43. Finite-size effect on the coercive field in PZT thin films. The data are from: (a) [231]; (b) [232]; (c) [188]; (d) [187]. Reprinted with permission from Gordon and Breach Publishers from A. K. Tagantsev, *Ferroelectrics* 184, 79 (1996). Copyright 1996 Overseas Publishers Association, N.V.

defect or at the end of a polymer chain. This reversal, or "kink," can propagate rapidly along the chain of the PVDF molecule (Fig. 42a), covering typical straight sections of the polymer chain in nanoseconds [99, 102]. So reversed-polarization regions can grow quickly along the chains, but much more slowly perpendicular to the chains, which interact by much weaker van der Waals forces. In the polymorphous films, this highly anisotropic behavior is usually not evident, as extrinsic switching progresses among the lamellar crystals much more slowly on average, whether within crystallites (Fig. 42b) or between them in thin polymorphous films (Fig. 42c). In the multilayer LB films with all the chains lying parallel to each other in the film plane, the switching speed may be limited only by the interaction between chains within the crystals (Fig. 42b).

The extrinsic coercive field measured in thin films and small particles often exhibits a finite-size effect, an increase of the coercive field with a decrease in the film thickness, as is the case for copolymer films from 15 to 200 nm thick (see Fig. 39). The coercive fields measured in PZT films (Fig. 43) show a similar finite-size effect below 500 nm, independent of the method used to prepare the films [178, 179]. Several mechanisms have been proposed for this finite-size effect in thin films—the dependence of the coercive field E_C on the thickness L [29, 40, 51, 174, 178, 184, 185]. One mechanism is connected with the minimum nucleation radius. If this radius is comparable to (or larger) than the film thickness, there is insufficient available nucleation volume to stabilize the initial reversed-polarization nucleus. This minimum-volume coercive field increases as an inverse power $L^{-2/3}$ in the thickness, quite close to the $L^{-0.7}$ dependence observed in the copolymer thin films [36, 160]. Another mechanism is based on the domain-wall pinning by grain boundaries or defects. This pinning-limited coercive field depends on the ratio between bulk and surface pinning en-

ergies and therefore is proportional to L^{-1}, a dependence observed in some ferroelectric thin films [186]. Another possible mechanism for reduced extrinsic coercive field is connected with charge screening and the Debye screening length L_D. If $L_D > L/2$, then screening charge creates an electric field $E_D = eNL/(2\varepsilon\varepsilon_0)$, where N is the volume density of available charge. But L_D is typically much smaller than the actual film thicknesses and so this mechanism has not been verified to our knowledge.

An extrinsic finite-size effect can arise from the presence of surface dielectric layers, nonferroelectric layers with low dielectric constant. These layers are common in perovskite films for a variety of reasons, such as nonstoichiometry and impurity effects near the electrodes [29, 187, 188]. A composite thin film consisting of a ferroelectric layer of thickness L and two surface dielectric layers with total thickness l will have a reduced electric field in the ferroelectric layer, so that the internal field is decreased by the factor $[1 + C_F/C_D]^{-1} = [1 + (l/L)(\varepsilon/\varepsilon_D)]^{-1}$, where ε_D is the dielectric constant of the dielectric layer. For PZT films, Tagantsev evaluated this expression, taking into account the nonlinear dielectric response $\varepsilon(E)$ extracted from the measured hysteresis loops. According to this mechanism, the apparent coercive field should decrease as L/l decreases, or as $\varepsilon/\varepsilon_D$ increases, but measurements reveal the opposite behavior [36, 160, 177–179, 187, 188]. It seems that the surface layer mechanism alone cannot account for the coercive field finite-size effect.

4.2. Dielectric Properties

The dielectric constant of the ferroelectric film includes a background value ε_∞ plus a contribution from the ferroelectric polarizability χ (see Eqs. (4), (6), and (8)). The dielectric constant of the second-order ferroelectric diverges at the transition temperature $T_C = T_0$, though this peak, or "dielectric anomaly," is usually broadened by sample or temperature inhomogeneity. The first-order ferroelectric similarly has a peak in the dielectric constant at the phase transition temperature T_C, though the dielectric constant does not formally diverge, and inhomogeneity can broaden this peak, too. An additional complication with the first-order ferroelectric is the existence of thermal hysteresis and phase coexistence. In practice, the dielectric peak falls at a higher temperature on heating than on cooling, as the current phase overshoots the equilibrium transition temperature T_C in either direction. The thermal hysteresis in the dielectric response of several LB films of the P(VDF-TrFE 70 : 30) copolymer is evident in Figure 44. By convention, we assign the temperature of the dielectric peak on cooling (78°C in the 30-ML film) to the transition temperature T_C, even though this is probably below the equilibrium value. Comparison of the X-ray coexistence data (Fig. 29) with the dielectric data (Fig. 44) indicates that the dielectric peaks occur at the first appearance of the new phase (the ferroelectric phase on cooling or the paraelectric phase on heating). The magnitude of the dielectric constant at 25°C is about 8 in the LB films [118], a little lower than the typical values of 10–12 measured in the polymorphous films [36, 55, 68].

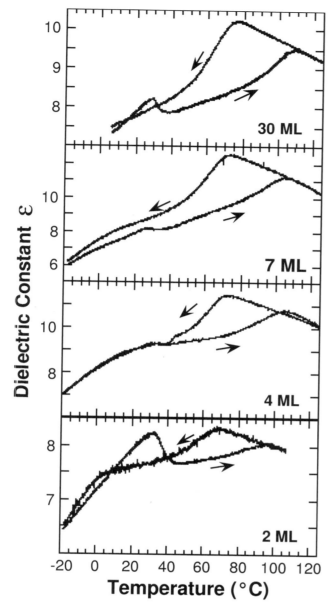

Fig. 44. Temperature dependence of the dielectric constant of P(VDF-TrFE 70:30) LB films with different thicknesses. Reprinted by permission from A. V. Bune et al., *Nature* 391, 874 (1998). Copyright 1998 Macmillan Magazines Ltd.

4.2.1. Nonlinear Dielectric Response and Determination of the LGD Coefficients

Nonlinear dielectric spectroscopy is useful for determining additional features in the dielectric response. Several LGD coefficients can be extracted by measuring the linear and nonlinear dielectric response. When a small-amplitude voltage $U(t) = U_0 \sin(\omega t)$ is applied to a capacitor containing a nonlinear dielectric, then the resulting total current is [189]

$$I(t) \approx \frac{U_0}{R} \sin(\omega t) + \omega U_0 C(0) \cos(\omega t)$$
$$+ \omega U_0^2 \sin(2\omega t) \frac{dC(U)}{dU} \qquad (13)$$

where $C(U)$ is the nonlinear capacitance of the film and R is the (constant) resistance. The nonlinearity from the third term in Eq. (13) generates higher-frequency components that can be extracted by Fourier analysis of the current. The dielectric properties are contained in the coefficients A_k proportional to the amplitudes of the first five Fourier coefficients oscillating at frequencies $k\omega$ ($k = 1$ to 5). If the experimental conditions are chosen so that intermodulation contributions into the second and third harmonics are small (see [189] for the details), then the coefficients A_k from a nonlinear capacitor made from a first-order LGD ferroelectric film are

$$A_2 = P_S(3\beta + 10\gamma P_S^2) \qquad A_3 = \beta + 10\gamma P_S^2$$
$$A_4 = \gamma P_S \qquad\qquad A_5 = \gamma \qquad (14)$$

where the equilibrium spontaneous polarization P_S is given by Eq. (7).

The expressions in Eq. (14) are valid for homogeneously polarized ferroelectrics. If the sample has no net polarization, only the odd harmonics exist in the resultant response. The sample polarization should first be saturated by a sufficiently large voltage. A useful check of the method is to measure Fourier components with a sample polarized first with one polarity and then with the opposite polarity. According to Eq. (14), the coefficients A_3 and A_5 should be the same for both polarities, as they are symmetric in the spontaneous polarization P_S, but the coefficients A_2 and A_4 should change sign (or their phases Φ_2 and Φ_4 shift by 180°). Close to the ferroelectric–paraelectric phase transition, one can have large changes in both the amplitudes and phases of the Fourier coefficients A_k. The temperature dependence of the measured values A_k can be a good test of the LGD approach, which assumes that the parameters T_0, C, β, and γ in the free-energy density (Eq. (1)) are independent of temperature.

In the case of the first-order LGD phase transition, where $\beta < 0$ and $\gamma > 0$, and according to Eq. (14), there must be temperature points T_2 and T_3 where the values A_2, A_3 change sign and therefore $A_2(T_2) = 0$ and $A_3(T_3) = 0$. Some of the LGD parameters can be calculated from the A_k measurements at these special temperatures T_2 and T_3 as follows [189]:

$$\beta = -\frac{1}{2}A_3(T_2) \qquad \gamma = \frac{1}{20}A_3(T_2)\left(\frac{A_3(T_2)}{A_2(T_3)}\right)^2 \qquad (15)$$

The value of the Curie constant C is obtained from these values of β and γ and temperature hysteresis interval $\Delta T = T_1 - T_0 = (1/4)\varepsilon_0 C\beta^2/\gamma$. The values of the LGD coefficients so determined can then be used to calculate the expected values of the A_k coefficients at other temperatures and thus test the assumption of constant LGD coefficients.

We implemented this technique by making simultaneous measurements of the five harmonic terms A_k and their relative phases Φ_k using a computer virtual instrument diagrammed in Figure 45 [189]. The samples consisted of 20-ML P(VDF-TrFE 70:30) LB films prepared as described in Section 2.3. The film polarization was saturated with a brief 15-V pulse. The virtual instrument (Fig. 45) supplied a voltage with amplitude

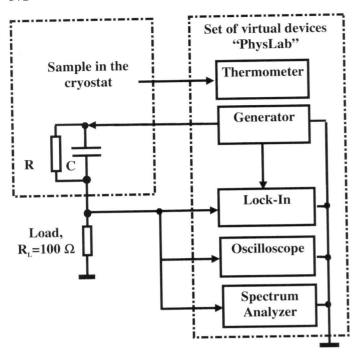

Fig. 45. Electronic components of the nonlinear dielectric spectroscopy apparatus. Reprinted with permission from S. P. Palto et al., *J. Exp. Theor. Phys.* 90, 872 (2000). Copyright 2000 Journal of Experimental and Theoretical Physics.

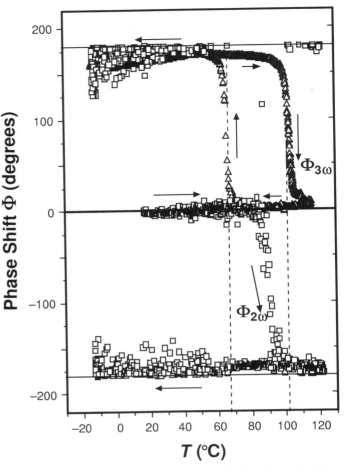

Fig. 47. Temperature dependencies of the phase shifts $\Phi_{2\omega}$ (squares) and $\Phi_{3\omega}$ (triangles). Reprinted with permission from S. P. Palto et al., *J. Exp. Theor. Phys.* 90, 872 (2000). Copyright 2000 Journal of Experimental and Theoretical Physics.

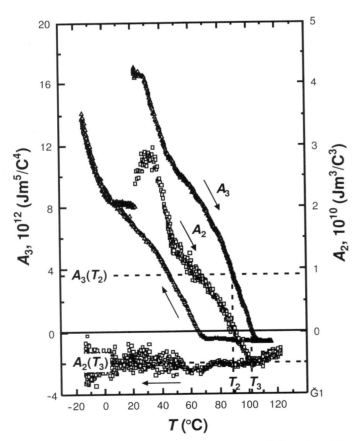

Fig. 46. Temperature dependencies of the modulation coefficients A_3 and A_2. Reprinted with permission from S. P. Palto et al., *J. Exp. Theor. Phys.* 90, 872 (2000). Copyright 2000 Journal of Experimental and Theoretical Physics.

$U_0 = 5$ V and frequency 1000 Hz, recorded the resulting current, and extracted the coefficients A_k and their phases Φ_k.

The measured values of the coefficients A_2 and A_3 are shown in Figure 46 and the corresponding phases Φ_k are shown in Figure 47. These data allow a clear interpretation in terms of the LGD approach. Under heating, at temperatures higher than 110°C the coefficient A_3 becomes constant, so that P_S is constant (Eq. (14)) and likely zero. On cooling, the coefficient A_3 remains constant, and P_S is zero down to 70°C. The jumps in the signal phases in Figure 47 further support the interpretation that the phase changes occur at 110°C on heating and 70°C on cooling. According to Eq. (14), there are temperature points $T_2 = 89$°C and $T_3 = 102$°C where A_2 and A_3 are equal to zero and change their sign, respectively. The value of the coefficient $A_5 = \gamma$ is nearly independent of temperature, except near T_C, where phase coexistence complicates the situation [189].

The values of the LGD parameters extracted from these data are $\beta = -2 \times 10^{12}$ J C^{-4} m^5, $\gamma = (5 \pm 3) \times 10^{14}$ J C^{-6} m^9, and $C = (2 \pm 1) \times 10^3$ K, in good agreement with values obtained by more traditional measurements on LB films of the same composition [63, 118, 121, 177, 190] and the values from polymorphous films of similar composition [59], as summa-

Table III. Comparison of the Piezoelectric and the Pyroelectric Coefficients of 30-ML P(VDF-TrFE 70 : 30) LB Films and Polymorphous Films

Material	d_{33} (eff) (pm/V)	p_3 (eff) (μC/m^2 K)	P_S (C/m^2)
LB film	-20 ± 2 [63]	-20 ± 4 [63]	0.1 [154]
Polymorphous film	-40 [198]	-35 [230]	0.1 [55]

rized in Table II. The calculation of the pyroelectric coefficient from the data using analysis described in [189] gives a value $p = -70 \ \mu$C/(m^2 K), higher than the value of $-20 \ \mu$C/(m^2 K) obtained from more direct measurements with similar LB films (see Section 4.4.1 and Table III) [63].

Close examination of the A_2 and A_3 data in Figure 46 reveals features deviating from the basic LGD forms in Eq. (14), possibly due to sample inhomogeneity (see [191] for details). For instance, above 110°C, which is above the coexistence limit and well into the paraelectric phase where $P_S = 0$, the value of A_2 does not drop to zero as it should, according to Eq. (14). This suggests the existence of a small quantity of polar phase up to rather high temperatures of about 140°C. Such a mixed-phase sample cannot be so clearly interpreted in terms of the LGD approach.

4.2.2. Dielectric-Layer Model

Samples consisting of a ferroelectric crystal bounded by conducting electrodes can have dielectric boundary layers that act as series capacitors and can drop some of the applied voltage. In this section, we consider how such layers might affect the measured capacitance and apply the analysis to ferroelectric polymer LB films as an illustration [191]. Model the composite sample with surface area S as two capacitors in series, $C_1(T) = \varepsilon'(T)\varepsilon_0 S/d_1$ for the ferroelectric film of thickness d_1 and $C_2 = \varepsilon_2\varepsilon_0 S/d_2$ for the capacitance of the dielectric layers of total thickness d_2. We will assume that the dielectric constant ε_2 of the dielectric layers is independent of temperature and that the dielectric constant of the ferroelectric film $\varepsilon' = \varepsilon_\infty + \chi$ includes a temperature-independent background value ε_∞ plus the contribution from the ferroelectric polarizability χ (Eq. (4) or Eq. (8)). We also assume that the thickness values d_1 and d_2 are independent of temperature (though the ferroelectric film expands by about 5% from the ferroelectric phase to the paraelectric phase [119]). The sample capacitance $C(T)$ is

$$\frac{1}{C(T)} = \left[\frac{1}{C_1(T)} + \frac{1}{C_2} \right] \tag{16}$$

Since the dielectric layer is assumed to have constant capacitance, it is possible to extract the capacitance of the ferroelectric layer $C_1(T)$ through a series of measurements at different temperatures. It is convenient to introduce the parameter [191]

$$\eta(T_x, T) = \frac{\varepsilon_0 S}{d} \left[\frac{1}{C(T_x)} - \frac{1}{C(T)} \right] \tag{17}$$

where $C(T_x)$ is the value of the sample capacitance at some reference temperature T_x. The permittivity of the ferroelectric

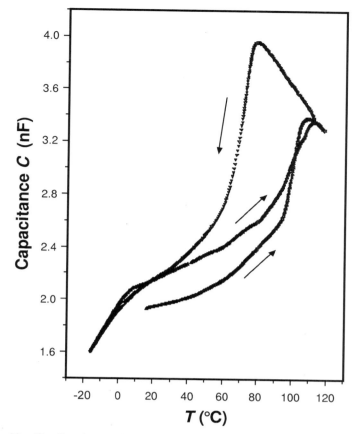

Fig. 48. Capacitance of a 20-ML P(VDF-TrFE 70 : 30) LB film. Arrows show the direction of the temperature change. Reprinted with permission from S. P. Palto et al., *J. Exp. Theor. Phys.* 90, 301 (2000). Copyright 2000 Journal of Experimental and Theoretical Physics.

part becomes

$$\varepsilon'(T) = \left[\frac{1}{\varepsilon'(T_x)} - \eta(T, T_x) \right]^{-1} \tag{18}$$

To reconstruct the temperature dependence of the dielectric constant $\varepsilon'(T)$ of the ferroelectric layer, we need an accurate value $\varepsilon'(T_x)$ at the reference temperature T_x. Because $\varepsilon'(T) \geq \varepsilon_\infty$, then $1/\eta_{max} \geq \varepsilon'(T_x) \geq \varepsilon_\infty$, where η_{max} is the maximum value of $\eta(T_x, T)$ measured at the ferroelectric phase transition temperature, which may differ on heating and cooling due to thermal hysteresis. Below the phase transition $\varepsilon'(T)$ decreases, so by choosing $T_x < T_0$, the allowed interval for $\varepsilon'(T_x)$ can be made narrow enough to allow reasonable reconstruction of the dependence $\varepsilon'(T)$ according to Eq. (18). However, this procedure must be applied with care, as a small error in the value of $\varepsilon'(T_x)$ can lead to large errors in the values of $\varepsilon'(T)$ near the dielectric anomalies.

The P(VDF-TrFE 70 : 30) LB films were prepared using acetone as a solvent and the standard Z-type LB deposition as described in Section 2. The films of 20 monolayers were deposited onto glass substrates with evaporated aluminum electrodes (see Fig. 20b) and the overlap area $S = 1.00 \pm 0.03$ mm^2. The dependence of the capacitance on temperature is shown in Figure 48. If we calculate the dielectric constant directly from

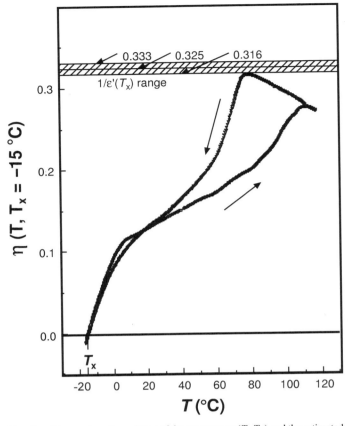

Fig. 49.   Temperature dependence of the parameter $\eta(T, T_x)$ and the estimated range of the parameter $1/\varepsilon'(T_x)$. Reprinted with permission from S. P. Palto et al., *J. Exp. Theor. Phys.* 90, 301 (2000). Copyright 2000 Journal of Experimental and Theoretical Physics.

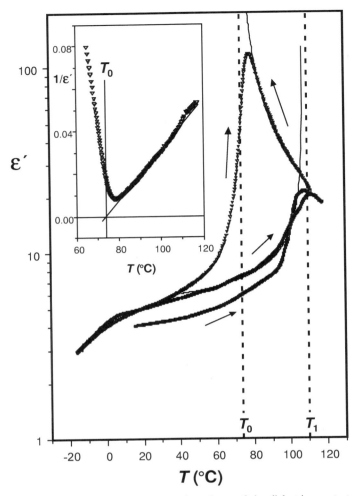

Fig. 50.   Reconstructed temperature dependence of the dielectric constant $\varepsilon'(T)$ for the ferroelectric part of the LB sample. The solid curves in the hysteresis region are the theoretical curves calculated according to the LGD phenomenology. The inset shows the Curie–Weiss behavior of the inverse dielectric constant $1/\varepsilon'(T)$. Reprinted with permission from S. P. Palto et al., *J. Exp. Theor. Phys.* 90, 301 (2000). Copyright 2000 Journal of Experimental and Theoretical Physics.

the data in Figure 48, ignoring any contribution from dielectric layers, then we find the value of $\varepsilon' = 2.5$ at room temperature, which is approximately the value measured in the bulk samples and less than the background value $\varepsilon_\infty \approx 4$ obtained from thicker polymorphous films [192]. The low sample capacitance may be due to parasitic capacitance from a thin aluminum-oxide dielectric layer in the electrodes. Any non-ferroelectric layer in the sample can play the role of the parasitic capacitance. We chose the reference temperature $T_x = -15°C$ at which there is little observed thermal hysteresis.

The experimental dependence of $\eta(T, T_x)$ and the allowed range of values for $1/\varepsilon'(T_x) = \{0.316 \text{ to } 0.333\}$ are shown in Figure 49, where for the upper limit we have taken the value $1/\varepsilon' = 0.33$ measured at $T = -150°C$ and 1000 Hz [192]. Since the estimated measurement uncertainty is ~3%, any value of $\varepsilon'(T_x)$ from this interval can be taken for the reconstruction of the $\varepsilon'(T)$. In the case of a second-order phase transition, the value $\varepsilon'(T_x) = 1/\eta_{max}$ can be taken, which will lead to an infinite value of the permittivity at the Curie point. For the first-order phase transition, the infinite value is not allowed and therefore we chose the midpoint of the interval, or $\varepsilon'(T_x) = 0.325$, for use in further analysis. This choice serves well for illustration, but it is not unique and may exaggerate the effect of any surface layer.

The reconstructed values of $\varepsilon'(T)$ obtained using the correction parameter $\varepsilon'(T_x) = 0.325$ are shown in Figure 50. The dielectric anomalies are much more pronounced than in the original data in Figure 48, so it is necessary to emphasize that the maximum value of ε' strongly depends on the choice of the parameter $\varepsilon'(T_x)$. The basic characteristics of the phase transition, such as the Curie–Weiss behavior, and other parameters of the LGD phenomenology are retained. The inset in Figure 50 shows the modified Curie–Weiss law that would result from persistence of the paraelectric phase extending into the coexistence region, as the sample is cooled from well above T_1 (see Fig. 2b). This amounts to assuming maximum survival of the metastable states allowed by LGD in the coexistence region $\Delta T = T_1 - T_0 = \beta^2/4\alpha_0\gamma$ and will result in maximum thermal hysteresis in the dielectric data. On heating from below T_0 in the ferroelectric phase up to T_1, the LGD contribution to the dielectric constant comes only from the polar β phase and is given by Eq. (8) for temperatures $T < T_1$. Similarly, on cooling from above T_1 in the paraelectric phase down to T_0, the dielectric constant arises only from the nonpolar α phase and is given by Eq. (4) for temperatures $T > T_0$. The values for the LGD coefficients extracted from the corrected $\varepsilon'(T)$ data in Figure 50 are $C = 807$ K, $\beta = -3.9 \times 10^{12}$ J m^5 C^{-4}, $\gamma = 7.9 \times 10^{14}$ J m^9 C^{-6}, $\varepsilon_\infty = 4.3$, and $T_0 = 74°$C [191]. These values were used with Eq. (4) and Eq. (8) to calculate the expected form of $\varepsilon'(T)$, which is shown by the lines in Figure 50 [191]. The discrepancy between theory and the data at temperatures close to T_C and T_1 suggests that, close to these temperatures, the assumptions of negligible phase coexistence and a homogeneous sample are not valid. If there is significant phase coexistence, additional information can be obtained from nonlinear dielectric spectroscopy measurements.

4.3. Double Hysteresis and the Critical Point

The first-order ferroelectric phase transition temperature can be raised above the zero-field transition temperature T_C by application of an external electric field, up to a critical point (T_{Cr}, E_{Cr}) [121, 172]. The shift in T_C is demonstrated in the ferroelectric copolymer by the shift in the dielectric anomalies to higher temperature, as shown in Figure 51. Application of the field raises the transition point on both heating and cooling. X-ray diffraction measurements confirm that the applied field can convert the paraelectric α-phase structure to the ferroelectric β-phase structure (see Section 3.2.2). Above the critical temperature T_{Cr}, the ferroelectric phase can no longer be sustained even at very high applied field. The existence of this ferroelectric critical point (E_{Cr}, T_{Cr}) in the field–temperature (E–T) phase diagram was predicted by Ginzburg [18, 19] and Devonshire [25] from the Landau mean-field theory of continuous phase transitions.

A simple way to measure the phase boundary $T_C(E)$ is to monitor the hysteresis loops as a function of temperature. Several representative hysteresis loops plotted from Eq. (3) are shown in Figure 52. At a temperature below T_C, there is only a single hysteresis loop centered at zero electric field. Above T_C, there are two hysteresis loops, each centered at a nonzero field, up to the critical temperature T_{Cr}, beyond which there is no hysteresis, only the dielectric nonlinearity. The shift of T_C with electric field, double hysteresis, and the presence of a critical point have been clearly demonstrated in BaTiO$_3$ [172] and KDP (KH$_2$PO$_4$) [193, 194], both proper ferroelectrics with first-order phase transitions. Double hysteresis, the increase of T_C with applied field, and a critical point were also observed in a ferroelectric liquid crystal [195], a first-order improper ferroelectric in which the order parameter is molecular tilt, not electric polarization. Furukawa et al. reported studies of ferroelectric P(VDF-TrFE 52:48) films that exhibited a clear increase in the transition temperature with applied field, and nonlinear dielectric hysteresis above T_C likely due to double hysteresis, but found no indication of the critical point [79].

Polarization hysteresis is manifest in the dielectric "butterfly" capacitance curves $C(E)$ shown in Figure 53. The butterfly curves peak near the coercive field because the capacitance

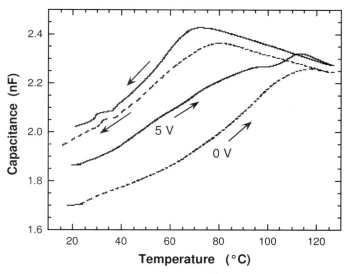

Fig. 51. Temperature dependence of the capacitance at zero bias voltage (dashed line) and under bias voltage of 5 V (solid line) in a 30-ML P(VDF-TrFE 70:30) LB film. Reprinted with permission from S. Ducharme et al., *Phys. Rev. B* 57, 25 (1998). Copyright 1998 American Physical Society.

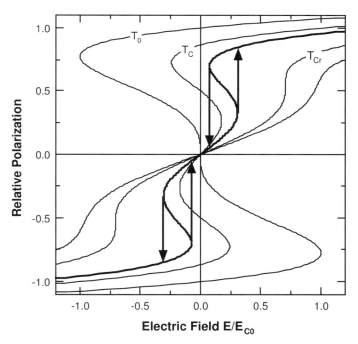

Fig. 52. Polarization plots from Eq. (3) for a first-order LGD ferroelectric. Double hysteresis loops (heavy line) exist in the intermediate range of temperatures $T_C < T < T_{Cr}$.

$C(E) \propto \varepsilon(E) \propto dD/dE$ is maximum at the point where the polarization is unstable. The dielectric butterfly curves from a 30-ML copolymer LB film shown in Figure 53 reveal all three stages, single hysteresis below T_C, double hysteresis up to T_{Cr}, and no hysteresis above T_{Cr}. Below the zero-field ferroelectric phase transition temperature of $T_C(0) = 80 \pm 10°$C, there is one hysteresis loop centered at zero field, giving two peaks in capacitance at the coercive field $E_C(T)$, one on each

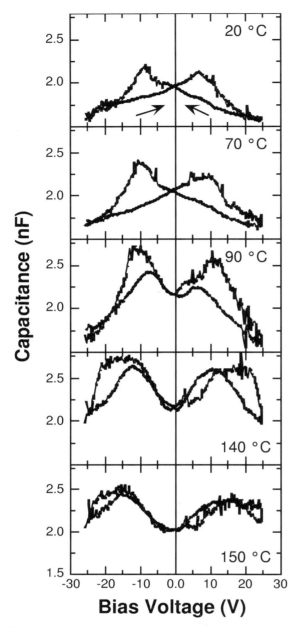

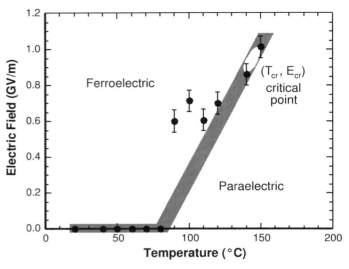

Fig. 54. Ferroelectric E–T phase diagram of a 30-ML LB film at atmospheric pressure showing the zero-field phase transition temperature $T_C = 80 \pm 10°C$ and the ferroelectric critical point $T_{Cr} = 145 \pm 5°C$. The data points were derived from the "butterfly" hysteresis curves like those in Figure 53. Reprinted with permission from S. Ducharme et al., *Phys. Rev. B* 57, 25 (1998). Copyright 1998 American Physical Society.

The butterfly curves like those in Figure 53 permit us to construct the E–T phase diagram shown in Figure 54, where the electric field at the boundary between the ferroelectric and paraelectric phases is obtained from the midpoint between the pairs of peaks at either positive or negative bias. Since we observed double hysteresis at 140°C and not at 150°C, the critical point is contained within the ellipse centered at $T_{Cr} = 145 \pm 5°C$ and $E_{Cr} = 930 \pm 100$ MV/m [121]. Table IV gives values of dT_C/dE, E_{Cr}, T_{Cr} for the copolymer P(VDF-TrFE 70 : 30), $BaTiO_3$, KDP, and a ferroelectric liquid crystal. Ferroelectric polymers have a relatively large span of the field-induced phase transition temperatures $T_{Cr} - T_C(0) = 65 \pm 11°C$ and a relatively small slope dT_C/dE, as compared to other ferroelectric materials such as $BaTiO_3$ [9, 172], KDP [193, 194], and ferroelectric liquid crystals [195], because of the relatively weak van der Waals interaction in the polymers [55].

4.4. Pyroelectric and Piezoelectric Properties

The piezoelectric and pyroelectric effects are well established in PVDF and its copolymers. However, there is still uncertainty about the contributions from the amorphous phase and the polycrystalline nature of the typical bulk films formed by solvent techniques [54, 66, 196–198]. Detailed measurements of the pyroelectric response (charge generated on heating or cooling) and piezoelectric response (strain due to an electric field) in the ferroelectric copolymer LB films confirm with unprecedented precision that both properties are directly proportional to the spontaneous polarization, a basic result predicted by LGD theory. We have also imaged the spatial distribution of polarization in the films by pyroelectric scanning microscopy (PSM), measurement of the pyroelectric current as a focused laser beam

Fig. 53. Butterfly capacitance hysteresis curves of a 30-ML P(VDF-TrFE 70 : 30) LB film at temperatures spanning both the Curie point $T_C = 80 \pm 10°C$ and the critical point $T_{Cr} = 145 \pm 5°C$. The capacitance was measured at 1 kHz while the field was ramped at 0.02 V/s during measurement. Reprinted with permission from S. Ducharme et al., *Phys. Rev. B* 57, 25 (1998). Copyright 1998 American Physical Society.

side of zero bias. Above T_C, but below the critical temperature T_{Cr}, there are four peaks in the capacitance due to the two hysteresis loops arranged antisymmetrically about zero bias, as shown in Figure 52. The dual hysteresis loops occur because the sample is not ferroelectric at zero field but the ferroelectric state is induced at sufficiently high field. There is no hysteresis, only dielectric nonlinearity, observed at 150°C, and because the sample is above the ferroelectric critical temperature $T_{Cr} \approx 145°C$.

Table IV. Ferroelectric Phase Transition Parameters for P(VDF-TrFE 70 : 30) Films made by Langmuir–Blodgett Deposition and Solvent Spinning, Potassium Dihydrogen Phosphate (KDP), Barium Titanate (BaTiO$_3$), and for a Ferroelectric Liquid Crystal C7[a]

	T_{C0} (°C)	T_{Cr} (°C)	E_{Cr} (V/m)	dT_C/dE (°Cm/V)
LB copolymer	80 ± 10	145 ± 5	$0.93 \pm 0.1 \times 10^9$	$7.0 \pm 2 \times 10^{-8}$
Polymorphous film [55]	102	no data	no data	12×10^{-8}
BaTiO$_3$[172]	108	116	0.6×10^6	6.5×10^{-4}
KDP [193, 194]	-61	-60	0.83×10^6	1.25×10^{-6}
C7[a] [195]	55.0	55.8	5.0×10^6	1.6×10^{-7}

[a] 4-(3-methyl-2-chloropentanoyloxy)4′ -heptyloxybiphenyl.

is scanned across the film. The new PSM technique allows us to probe crystallite and domain structures with fine spatial and temporal resolution.

4.4.1. Pyroelectric and Piezoelectric Measurements

Heating a ferroelectric crystal changes its polarization (see Fig. 2), a phenomenon called the pyroelectric effect. For small changes in temperature, the polarization change is described by pyroelectric coefficients $p_i = \lfloor \partial P_i / \partial T \rfloor_{(\sigma, \mathbf{E})}$, where the designation $(\sigma, \mathbf{E})$ specifies that the measurements are made at constant stress σ and constant electric field $\mathbf{E}$ [199]. Pyroelectric measurements are usually made by heating or cooling the crystal at a rate $\dot{T} = dT/dt$ and measuring the current from electrodes applied to the polar faces (see Fig. 4b). For a uniaxial constrained film, the pyroelectric current is given by

$$J = A p_{\text{eff}} \dot{T} \qquad (19)$$

where the effective pyroelectric coefficient p_{eff} depends on the true pyroelectric coefficients p_i and also on the piezoelectric and elasticity tensors, because the thin film constrained on a rigid substrate is stress-free normal to the film, but strain-free in the film plane [63, 200].

The pyroelectric response can be measured by the Chynoweth method [201], where a light beam modulated with an optical chopper illuminated the sample. Some of the light is absorbed by the sample, causing a small temperature modulation. The pyroelectric current due to the temperature modulation (Eq. (19)) can be measured synchronously with a lock-in amplifier referenced to the chopper frequency. The Chynoweth method is convenient for monitoring the sample polarization state. For example, the pyroelectric response shows the ferroelectric–paraelectric phase transition (Fig. 55) as a peak just below the transition temperature, followed by a decrease to zero at T_C, as the polarization goes to zero in the paraelectric phase. Integration of the pyroelectric response with temperature yields the spontaneous polarization, as demonstrated in Figure 55. This method is especially useful for tracing out the hysteresis loop when the dynamic Sawyer–Tower [28] method is inadequate because switching is too slow or if the sample is conducting and its RC time constant is less than the ferroelectric switching time [63, 149, 151, 153–155]. The polarization hysteresis loop

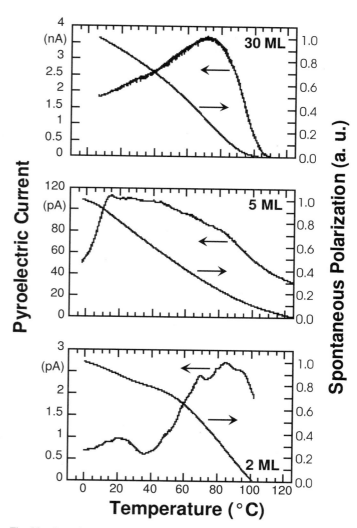

Fig. 55. Pyroelectric current from films of P(VDF-TrFE 70 : 30) measured by the Chynoweth method and the calculated spontaneous polarization obtained by integrating the pyroelectric response over temperature. Reprinted by permission from A. V. Bune et al., *Nature* 391, 874 (1998). Copyright 1998 Macmillan Magazines Ltd.

is evident in the pyroelectric current measurements shown in Figure 36 and Figure 40, which were recorded by biasing the sample at each voltage for several minutes, then removing the voltage and measuring the pyroelectric response at zero bias. In this way, the true equilibrium remanent hysteresis loop can be recorded. The Chynoweth method is usually not suitable for making absolute measurements of the pyroelectric coefficient because of the difficulty in calibrating the temperature change. The pyroelectric coefficient can be measured more accurately by slow heating of the sample in a temperature-controlled oven and measuring the current or evolved charge. The results from the ferroelectric polymer LB films are in reasonable agreement with measurements on the polymorphous films (see Table III), and demonstrate the effects of substrate constraint on the effective pyroelectric coefficient [63, 155].

When stress σ is applied to a ferroelectric crystal, there will be a change in the polarization due to the piezoelectric effect. The converse effect is a strain induced by an applied electric field. For small perturbations at constant temperature, both the direct and converse piezoelectric effects are described by the piezoelectric tensor coefficients $d_{kij} = [\partial D_k / \partial \sigma_{ij}]_{(\mathbf{E}, T)} = [\partial S_{ij} / \partial E_k]_{(\sigma, T)}$, where σ_{ij} are the components of the stress tensor, S_{ij} are the components of the strain tensor, D_k are the components of the electric displacement vector, and E_k are the components of the electric field vector [199]. The designations $(\mathbf{E}, T)$ and (σ, T) specify that the measurements are made at constant temperature T and either constant stress σ (i.e., at low frequency in an unconstrained sample) or constant electric field $\mathbf{E}$ (e.g., $V = 0$). Consider a uniaxial thin film with spontaneous polarization normal to the film and conducting electrodes applied to the polar faces (Fig. 4b). Application of an electric field $E = V/L$ to the film will cause a change in thickness given by

$$\Delta L = L d_{\text{eff}} E = d_{\text{eff}} V \qquad (20)$$

where the effective piezoelectric coefficient d_{eff} depends on the true piezoelectric coefficients d_{ijk} and also on the elasticity tensors, again because the film is constrained [63, 200].

Even in the nanometer-thick LB films, changes in sample thickness of order 10^{-3} nm are readily measured using a Mach–Zehnder interferometer like the one diagrammed in Figure 56 [202]. An ac voltage applied to the film modulates the sample thickness (Eq. (20)) so that the laser beam reflected from the top electrode is phase shifted by an amount proportional to the thickness change and therefore to the voltage modulation amplitude. The phase-modulated beam from the signal arm was mixed with the beam from the reference arm to produce an intensity-modulated signal in the optical detector, which signal was measured with a lock-in amplifier referenced to the function generator frequency. By this interferometric method, it is possible to measure modulation amplitudes with a sensitivity of about a millionth of a wavelength, or less than one picometer [202].

Both the piezoelectric and pyroelectric responses in a ferroelectric are proportional to the net spontaneous polarization. In the case of the LGD first-order ferroelectric, the pyroelectric and piezoelectric coefficients in the uniaxial thin-film geometry

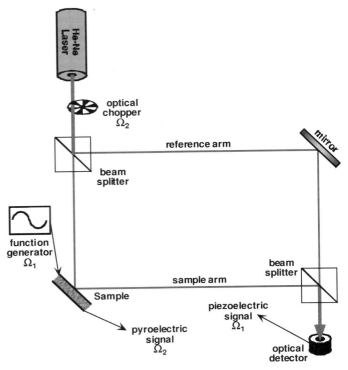

Fig. 56.   Diagram of the apparatus used to simultaneously measure the pyroelectric response and the piezoelectric response of the ferroelectric polymer LB films.

are [19, 25, 63, 198]

$$p_3 = -\chi_{33} P_{\text{S}}/C \qquad d_{333} = 2\chi_{33}\varepsilon_0 k_{3333} P_{\text{S}} \qquad (21)$$

where k_{3333} is the quadratic electrostriction coefficient in the paraelectric phase, χ_{33} is the ferroelectric contribution to the dielectric constant (Eq. (8)), and C is the Curie constant. If the sample region probed by the light beam has opposing domains, the pyroelectric and piezoelectric responses will be given by the spatial average of the spontaneous polarization.

To more precisely test the predictions of Eq. (21), the pyroelectric current (Eq. (19)) and piezoelectric thickness change (Eq. (20)) were measured simultaneously in the ferroelectric copolymer LB films by combined interferometric and thermal modulation methods using the apparatus diagrammed in Figure 56 [63]. (The chopper and function generator frequencies were both well below 2 kHz, where the sample dielectric response is independent of frequency, and so both measurements are effectively in the dc limit [63].) The proportional relationships (Eq. (21)) are clearly demonstrated by the data shown in Figure 57, as both the piezoelectric signal coefficient and the pyroelectric current follow the familiar polarization hysteresis loop. The same values of the effective piezoelectric coefficient and the pyroelectric current are plotted against each other in Figure 58, producing a straight line, even though the data are from points all around the hysteresis loop. This mutual proportionality between the piezoelectric and pyroelectric responses is further confirmation of their identical dependence on the spontaneous polarization (Eq. (21)) [63].

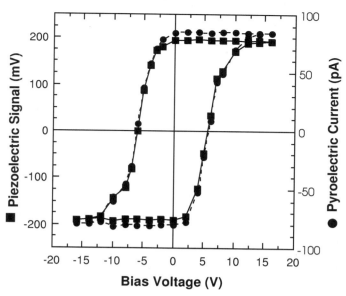

Fig. 57. Hysteresis loop of the effective piezoelectric signal and pyroelectric current from a 30-ML P(VDF-TrFE 70 : 30) LB film recorded with the apparatus diagrammed in Figure 56. Reprinted with permission from A. V. Bune et al., *J. Appl. Phys.* 85, 7869 (1999). Copyright 1999 American Institute of Physics.

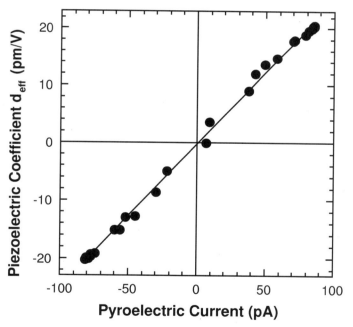

Fig. 58. Plot of the effective piezoelectric coefficient versus the pyroelectric current for all polarization states, from the data in Figure 57. Reprinted with permission from A. V. Bune et al., *J. Appl. Phys.* 85, 7869 (1999). Copyright 1999 American Institute of Physics.

4.4.2. Pyroelectric Scanning Microscopy

We have developed a new technique for imaging the polarization of ferroelectric films and devices—pyroelectric scanning microscopy (PSM). PSM consists of scanning a focused, modulated, laser beam across a capacitor made of ferroelectric material and recording the pyroelectric current from the elec-

trodes, as in the Chynoweth method described above. The scanned pyroelectric data is assembled into an image, like the one shown in Figure 59 of a 12-ML copolymer LB film recorded at room temperature with 20-μm separation between points and 2-kHz modulation frequency. The steep "cliffs" are actually the edges of the electrodes and demonstrate the high spatial resolution. Independent measurements show the resolution was about 17 μm, limited by the diameter of the beam focused on the sample. It is also possible to make simultaneous piezoelectric and pyroelectric images by employing the interferometer diagrammed in Figure 56.

The PSM technique can also produce time-resolved images of dynamic processes such as switching. This is achieved by recording switching transients at each image spot and combining the time–space data into image frames, provided that the switching transients are repeatable. This method allows us to monitor inhomogeneity in dynamics and should allow us to record domain motion. Figure 60 shows one frame of a movie we made of the polarization profile of a ferroelectric LB film during switching [203]. The film was initially saturated with positive "up" polarization, as represented by the values +1.0 (relative to P_S) in Figure 60. Then a large opposing voltage was applied to the film, causing it to reverse polarization, but different parts of the film switched at different rates, so that the film polarization is nonuniform at 4 s, about halfway through switching. (See the complete *Polarization Switching Movie*, uncensored and uncut, at physics.unl.edu/directory/ducharme/ducharme.html.)

4.5. Two-Dimensional Ferroelectricity

It was expected that ferroelectricity would be suppressed as crystal size or film thickness was reduced, and studies of small particles and thin films of ferroelectric oxides were consistent with these predictions. As we measured the ferroelectric properties of the copolymer LB films of diminishing thickness, down to 1 nm, we found that the ferroelectric properties hardly changed. The phase transition temperature decreased by only 10°C (Fig. 61), and other key properties such as pyroelectric and dielectric responses were also little changed. This indicated that the ferroelectric state in the copolymers is inherently two-dimensional, as the ferroelectric properties were essentially independent of thickness.

Measurements of the dielectric constant, pyroelectric response, and polarization hysteresis all demonstrated the essentially two-dimensional character of the ferroelectric state in these films. Figure 44 shows that multilayer ferroelectric LB films have the dielectric anomalies and thermal hysteresis that mark the usual bulk first-order ferroelectric phase transition found at 77°C on cooling (108°C on heating) in the 30-ML film and decreasing to 68°C (98°C) in the 2-ML film (Fig. 61). The vanishing of the pyroelectric response shown in Figure 55 also locates the phase transition temperature, though with less precision than the dielectric measurements. Therefore, the "bulk" first-order phase transition at $T_C \approx 70$–80°C remains even in the 2-ML films—a result that challenges the theoretical

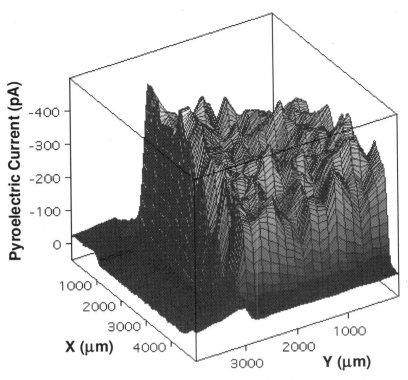

Fig. 59.   Pyroelectric scanning microscopy (PSM) image of a polarized 12-ML LB film of P(VDF-TrFE 70 : 30).

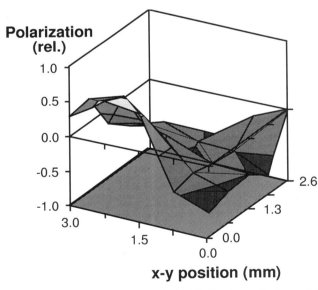

Fig. 60.   Time-resolved PSM image of a 30-ML film 4 seconds after applying a switching voltage, about halfway through switching.

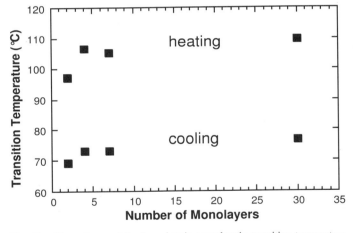

Fig. 61.   Dependence of the ferroelectric–paraelectric transition temperature on film thickness for P(VDF-TrFE 70 : 30) LB films. Reprinted with permission from L. M. Blinov et al., *Physics Uspekhi* 43, 243 (2000). Copyright 2000 Turpion Limited.

predictions that three-dimensional ferroelectricity must vanish in films below a minimum critical thickness. The hysteresis loops like those shown in Figure 36 exhibit good saturation for 4-ML and thicker films and incomplete saturation for the 2-ML film. The hysteresis loops for the 5-ML and 2-ML films (Fig. 36) have considerable vertical bias, probably due to interactions with the substrate or the top electrode, consistent

with the earlier observations on the dynamics of switching LB films [153, 160].

The results from the ferroelectric copolymer LB films lead to the following conclusions. First, there is no apparent critical thickness in LB ferroelectric films as thin as 1 nm, much thinner than any previous ferroelectric films. Second, there is no finite-size effect in LB ferroelectric films with thickness in the interval 1–15 nm, contrary to the predictions reviewed in Section 1.1.3. Third, in the region of two-dimensionality, these films reveal the intrinsic coercive field, meaning that ferroelectric switching is not dependent on nucleation and domain propagation processes. There is no sign of a minimum critical thickness or suppression of the ferroelectric phase that might result from surface and depolarization energies expected for three-dimensional ferroelectricity [39], and observed in other ferroelectrics [33, 47, 184]. Therefore, these LB ferroelectric films must be considered essentially two-dimensional ferroelectrics.

The two-dimensional nature of the LB films means that the ferroelectric state may be generated by coupling only within the plane of the film. Any coupling between planes is weak and possibly responsible for the slight reduction of transition temperature in the thinnest films [118]. The generalized Ising model for ultrathin films [48] is a more appealing approach to modeling ferroelectricity in two-dimensional polymer films because the dipole moments have restricted freedom—they can rotate only about the chain axis and are further inhibited from rotations about the axis by both interchain steric interactions and intrachain dihedral stiffness. We expect that an appropriate Ising model could be constructed in either of two ways. One way is to consider anisotropic ferroelectric coupling between chains with strong ferroelectric coupling in the plane and a weak coupling perpendicular to the plane. This would still resemble a three-dimensional state, but could remain stable even in one monolayer. Another approach is to construct an inherently two-dimensional ferroelectric state in one plane and add weak interplane coupling through a mean field shared by all layers. Both approaches should achieve ferroelectricity at finite temperature in nanometer-thick films. Both approaches also suggest that the surface layers, boundaries between the ferroelectric film and the electrodes or other outside material, may have a Curie point different from the interior "bulk" layers because they couple with only one other ferroelectric layer.

4.6. Surface Phase Transition

The dielectric constant data in Figure 44 and the pyroelectric current data in Figure 55 also reveal a phase transition at a lower temperature of approximately 20°C in films of 30 ML or less. The STM, LEED, and inverse photoemission studies of the surface phase transition near 20°C demonstrate that the surface is reconstructed such that some polymer chains are pushed up toward the surface as shown in Figure 25. Since the ferroelectric state is essentially two-dimensional, the reconstructed surface layer(s) should have a significantly different transition, or perhaps no transition at all. Surface phase transitions are known in many systems. For example, there are surface ferromagnetic transitions [43], surface structural transitions in Si (001) [204–206] and Ge(001) [207], and surface nonmetal-to-metal transitions [208, 209].

There is indeed clear evidence for a distinct surface transition in the copolymer LB films. The transition is evident in the dielectric response (Fig. 44) near 20°C on heating and 0°C on cooling, well below the 80°C bulk transition, and is strongest in the thinnest films [118]. The surface phase transition is accompanied by a doubling of the Brillouin zone, as verified by

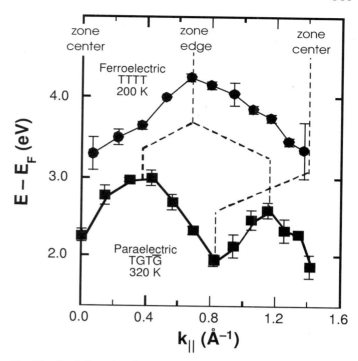

Fig. 62. Band dispersion, from angle-resolved inverse photoemission spectroscopy, where the LUMO states for the in-plane component of the incident electron wave vector are along the polymer chains, with a 5-ML P(VDF-TrFE 70 : 30) LB film on Si. Reprinted with permission from J. Choi et al., *Phys. Lett. A* 249, 505 (1999). Copyright 1999 Elsevier Science.

angle-resolved inverse photoemission spectroscopy [119, 162, 166, 210]. This doubling, shown in Figure 62, is consistent with the conversion of the all-trans chain conformation (Fig. 8a) to the alternating trans-gauche conformation (Fig. 8b) that has been shown to occur at the bulk phase transition near 80°C [71, 73, 74]. There is also the appearance of a dynamic Jahn–Teller distortion, perhaps due to a soft mode or modes connected with the conformation change [210]. Inverse photoemission spectroscopy reveals a shift of the Fermi level from 1 eV below the conduction band to degenerate with the conduction band, consistent with a major change in the conduction band wave functions [166]. An increase in the work function [119] also suggests a structural change leading to a reorientation of the dipoles toward the surface [166].

Note that subsequent tests showed that the dielectric anomaly is only observable in the presence of water vapor, and therefore it is probably not a true ferroelectric transition. The transition is still evident from other measurements, all made in vacuum with no water present in the film, showing a transition in the structure [119, 162, 163], the electronic properties [162, 166, 210], or the nonlinear optical response (see Fig. 63) [164, 165]. The absence of a true dielectric anomaly and measurements of the symmetry of the surface phases above and below the surface transition temperature [163] are consistent with a non-ferroelectric character.

Though the surface phase transition is probably not a proper ferroelectric transition, it is accompanied by the same conformational change, all-trans to trans-gauche, occurring at the bulk

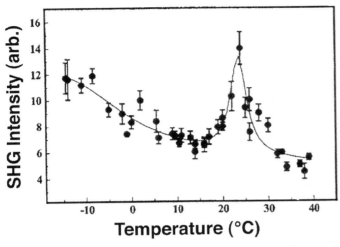

Fig. 63. Surface phase transition (on heating) is evident in the second-harmonic generation signal from a 15-ML LB film of the P(VDF-TrFE 70:30) copolymer film on a fused quartz substrate. Reprinted with permission from O. A. Aktsipetrov et al., *Opt. Lett.* 25, 411 (2000). Copyright 2000 Optical Society of America.

ferroelectric transition, and therefore the intrachain interactions probably play a similar role in both transitions.

4.7. Switching Dynamics

As discussed in Section 4.1.2, the ferroelectric copolymer LB films 1–15 nm thick generally exhibit intrinsic switching, but switching in these films proved to be confoundingly slow, about 10 *seconds* (see Fig. 64a). This is in stark contrast to the polymorphous solvent-formed films, which switch as fast as 1/10 microsecond [55, 211]. Therefore one might expect the LB films to switch much more quickly, given that the switching fields are 5–10 times larger, though the suppression of nucleation in these films does seem to negate this extrapolation from the fast-switching results. The mechanism of switching must be much different in the intrinsic case, and may be slowed by fluctuations. Examine carefully the free-energy curves shown in Figure 37. For switching fields of $E > E_C$, the original (negative) polarization state is completely destabilized; there is no more local minimum and the barrier has vanished. The polarization will transit smoothly down the free-energy slope to the new absolute minimum at positive polarization.

4.7.1. Fast Switching

We have found that the ferroelectric copolymer LB films, even films thinner than 15 nm, can be made to switch much faster [212]. The contrast between the switching times is illustrated in Figure 64. If switching is initiated by a reverse-potential pulse (which should exceed the coercive voltage $E_C L$), the switching time is given by the timing of the delayed peak in the switching current transient. (The initial peak is just the RC charging transient of the film's capacitance through the load resistance.) The films with high intrinsic coer-

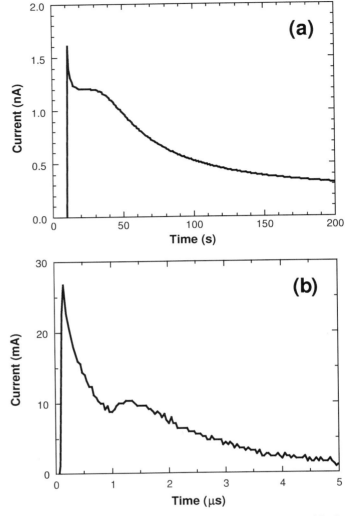

Fig. 64. Switching transients of P(VDF-TrFE 70:30) LB films with aluminum electrodes: (a) a 30-ML film on a glass substrate after applying a 14-V switching voltage; (b) an 8-ML film on a silicon substrate after applying a 5-V switching voltage.

cive field (~500 MV/m) have switching times of 10 s or more (Fig. 64a), while films with much lower extrinsic coercive fields (~50 MV/m) switch as fast as 1.2 μs (Fig. 64b), 10 million times faster! The switching mechanism can often be inferred by measuring the dependence of switching speed on the switching field and the temperature. The field dependence of the switching rates for fast and slow samples is shown in Figure 65. The functional dependence of switching rate on reversing voltage from the slow-switching samples (Fig. 65a) is not so clear, but the fast-switching films seem to exhibit a quadratic dependence (Fig. 65b).

Ideally, the switched charge per unit area should be twice the spontaneous polarization, which is about 100 mC/m² for the copolymer films [154]. The approximate switched charge can be measured by the current hysteresis loops of the Merz method [29]. In the Merz method, we apply a triangular voltage with frequency below the inverse switching time and record

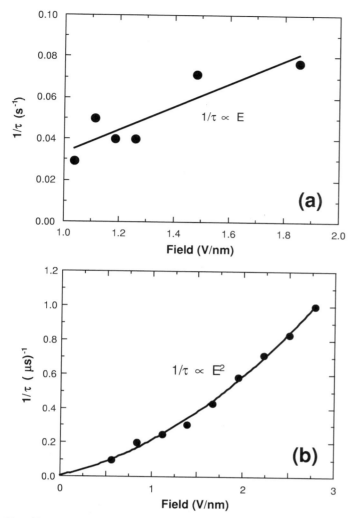

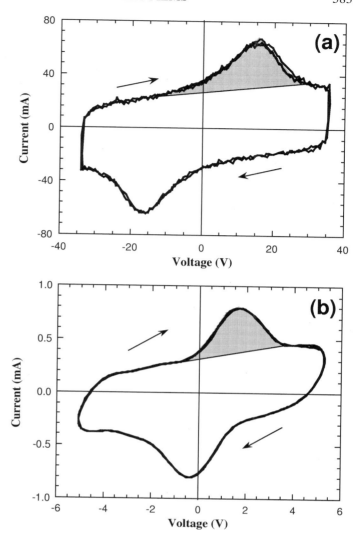

Fig. 65. Dependence of switching rate on the reversing electric field in P(VDF-TrFE 70 : 30) copolymer LB films: (a) a slow-switching film 30 ML thick; (b) a fast-switching film 8 ML thick.

Fig. 66. Merz current loops from P(VDF-TrFE 70 : 30) LB films (a) from a 10-ML film on a glass substrate, recorded at 200 Hz. The shaded area corresponds to a switched polarization of 8 mC/m²; (b) from a 4-ML film on a silicon (100) substrate, recorded at 2000 Hz. The shaded area corresponds to a switched polarization of 25 mC/m².

the resulting current, as shown in Figure 66. The current peaks in excess of the usual capacitive current are due to polarization reversal, and their area yields the switched polarization. A 10-ML P(VDF-TrFE 70 : 30) LB film deposited on a glass substrate had a switching time of about 50 μs and a switched polarization of 8 mC/m² calculated from the Merz data shown in Figure 66a. A 5-ML ferroelectric P(VDF-TrFE 70 : 30) LB film on a silicon (100) substrate exhibited slightly faster switching, about 15 μs, and larger switched polarization, 25 mC/m² (see Fig. 66b). Both films were deposited at a surface pressure of 3 mN/m and had evaporated aluminum electrodes.

The essential difference between the slow- and fast-switching samples remains unclear, though the correlation between switching speed and coercive field offers one possibility. The slow samples have a high intrinsic coercive field and therefore ordinary nucleation is suppressed. The fast samples have lower coercive fields comparable to the polymorphous films and therefore probably switch by ordinary nucleation and domain-growth processes.

The cause, and control, of fast switching may prove crucial to the application of the ferroelectric polymer LB films to nonvolatile random-access memories. These results highlight three desirable material properties achieved in the ferroelectric copolymer LB films: high speed, low switching voltage, and large switchable polarization.

4.7.2. Conductance Switching

One of the surprises from the study of the first ferroelectric copolymer LB films was a conductance switching phenomenon not observed before in ferroelectric materials [153]. Figure 67 shows that the conductance of a 30-ML copolymer LB film switches from low to high and back to low in coincidence with polarization reversal. The conductance change is three orders of magnitude, a remarkable dynamic range for a completely non-

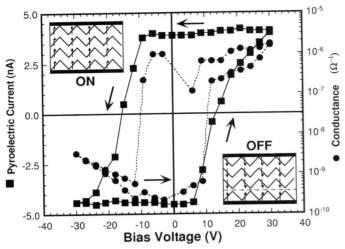

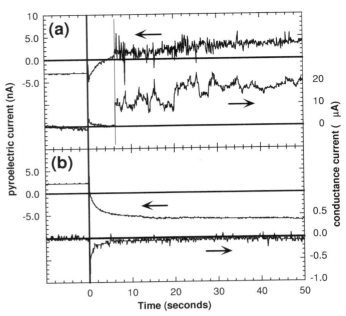

Fig. 67. Hysteresis in the pyroelectric current (squares) and in the dc conductance (circles) of a 30-ML P(VDF-TrFE 70 : 30) LB film. Insets: proposed polarization of the film layers in the two saturated states, the ON state of the fully ordered ferroelectric LB film (upper left) and the OFF state of the partially ordered LB film (lower right). Reprinted with permission from A. Bune et al., *Appl. Phys. Lett.* 67, 3975 (1995). Copyright 1995 American Institute of Physics.

Fig. 68. Kinetics of polarization and conductance switching in a 30-ML P(VDF-TrFE 70 : 30) LB film. (a) Switching ON. The sample was first polarized in the OFF state with a −25 V bias voltage for 15 minutes. The negative bias voltage was removed and a positive +25 V bias voltage applied at Time = 0 seconds. (b) Switching OFF. The sample was first polarized in the ON state with a +25 V bias voltage for 15 minutes. The positive bias voltage was removed and a negative −25 V bias voltage applied at Time = 0 seconds. Reprinted with permission from A. Bune et al., *Appl. Phys. Lett.* 67, 3975 (1995). Copyright 1995 American Institute of Physics.

volatile and reversible switch. It appears that the conductance switching is controlled by the polarization state, a hypothesis also consistent with the switching dynamics. Similar conductance switching has been recently observed in barium titanate thin films [213].

The conductance switching mechanism seems to be connected with the barrier presented by a single polar layer pinned by the substrate or top electrode. Additional evidence is available in the dynamics of the conductance and polarization during switching, as shown in Figure 68. When the film is initially saturated in the ON state and the applied voltage reversed, the switching-OFF process is much faster than the bulk polarization reversal (Fig. 68b), because the reversal of the first layer presents a barrier and switches the conductance OFF early in the polarization reversal process. When the film is initially saturated in the OFF state and the applied voltage reversed, there is a delay in the appearance of the high-conductance ON state until nearly all the bulk polarization of the film is switched, because the transition from low conductance (OFF) to high conductance (ON) occurs only as the polarization of the *last* monolayer switches. The OFF state is maintained because one or a few layers are pinned to the substrate or to the top electrode. We find that the films are partially polarized in the direction of the ON state as grown. This hypothesis is supported by the observation of asymmetric (biased) hysteresis loops in films thinner than 20 layers [118, 160]. The cause of layer pinning is unknown.

The large conductance in the ON state is consistent with a tunneling current as in the Fowler–Nordheim mechanism [214], a mechanism that has been proposed for other thin organic films [215]. It is also possible that the ON state is dominated by thermally activated hopping over the same barrier, with an exponential (Arrhenius) temperature dependence. In the OFF state, the negative charge sheet at the polarization discontinuity presents a barrier to electron tunneling (or thermally activated hopping). The electrodes may also play an important role (besides the physical pinning of the adjacent layers) in the conductance switching due to carrier injection. However, we observe ferroelectric and conductance switching with several different materials composing the top and bottom electrodes, leading to the conclusion that the conductance switching is probably controlled by an interlayer barrier.

4.8. Other Results from Polar Langmuir–Blodgett Films

4.8.1. Copper Phthalocyanine Films

Recently we prepared and investigated LB films of Cu-phthalocyanine that exhibit ferroelectric behavior [216]. It seems likely that the LB method can be used to obtain ferroelectric films from some other molecular systems and permit broader study of finite-size effects in ultrathin ferroelectric films.

4.8.2. Internal Field Measurement with Stark Spectroscopy

An interesting method to probe the internal electric field and charge distribution in thin films is based on Stark spectroscopy. The concept is illustrated in Figure 69. A thin dye LB film is inserted into the subject film, a ferroelectric copolymer LB film in this case. This is possible because of a unique property of

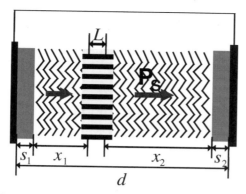

Fig. 69. Schematic diagram of a sample of the ferroelectric film (wavy lines) with built-in anthraquinone dye probe multilayer. The anthraquinone dye forms nonpolar LB bilayer films.

the Langmuir–Blodgett method to make heterogeneous molecular structures with monolayer precision. If the dye film shows the dc Stark effect, that is, the dye optical spectrum is changed in an electric field, then it can be used as a probe of this electric field from the subject film [217–221]. The dye will respond to the local field instead of the external field, though there is some unavoidable distortion of the local field by the dye layer itself.

Let us make a model for the electric field at the position of the dye layer. We will assume that any charge injection is confined to a narrow region near the electrodes, so that there is negligible space charge near the dye layer. Using the continuity of electric displacement D in the dielectric part of the sample and assuming zero dc potential difference between electrodes, the following set of equations can be written for the case shown in Figure 69 [219, 220]:

$$\varepsilon_0 E_f + P_f = \varepsilon \varepsilon_0 E_L$$
$$\int_{s_1} E_1(x)\,dx + E_f(x_2 + x_1) + E_L L + \int_{s_2} E_2(x)\,dx = 0 \quad (22)$$

The fields E_f, and E_L are the static electric fields in the ferroelectric and dye films, respectively, P_f is the total polarization of the ferroelectric film ($P_f = \chi \varepsilon_0 E_f + P_S$), and ε is the dielectric constant of the dye film. We consider the ferroelectric and dye films to be dielectrics with total thickness $x_1 + x_2$ and L, respectively, while space charge can exist in the electrodes over distances s_1 and s_2. The first expression in Eq. (22) ensures the continuity of the electric displacement at the border of the ferroelectric and dye films. The terms in the second equation give the work produced by the electric field in different parts of the sample when we virtually bring the probe charge from one electrode into the other. The total work is zero if no dc voltage is applied to the sample (i.e., the electrodes are shorted together). It is evident that the integrals actually can be considered as work functions for the charge from electrodes into the dielectric part of the ferroelectric film. If we define these functions as

$$\int_{s_1} E_1(x)\,dx = \Delta\varphi_1 \qquad \int_{s_2} E_2(x)\,dx = -\Delta\varphi_2 \quad (23)$$

then the static electric field at the dye film is

$$\varepsilon_0 E_L = \frac{P_f(x_2 + x_1) + \varepsilon_0(\Delta\varphi_2 - \Delta\varphi_1)}{L + \varepsilon(x_2 + x_1)} \quad (24)$$

An important property of Eq. (24) is that the sign of the electric field E_L depends not only on the sign of the spontaneous polarization P_S, but also on the difference of work functions and the sign of the field E_f, which influences the total polarization. In general, the work function difference is not zero, because it can depend on the variation of the spontaneous polarization vector $\mathbf{P_S}$ close to the given electrode, which is evident in the results described below. In the case of thick ferroelectric films ($x_1 + x_2 \gg L + s_1 + s_2$), the role of work functions and E_f becomes negligible and the field is determined only by spontaneous polarization, as was reported in [219, 220]. In this limit, the field in the dye films, is dominated by the field in the ferroelectric film, so that $E_L \approx P_S/(\varepsilon \varepsilon_0)$.

The anthraquinone dye films are not polar, so they exhibit a Stark effect quadratic (or higher even power) in the total electric field [217]. If an ac voltage $U(t) = U_0 \sin(\omega t)$ is applied to the sample, then the resulting change in the relative absorption due to the Stark effect

$$\Delta A = C_b\big(E_0 \sin(\omega t) + E_L\big)^2$$
$$= C_b\left[2E_0 E_L \sin(\omega t) - \frac{1}{2}E_0^2 \cos(2\omega t) + \frac{1}{2}E_0^2 + E_L^2\right] \quad (25)$$

has a component at the second harmonic 2ω due solely to the applied field $E_0 = U/d$ and at the fundamental frequency 1ω due to the cross term proportional to $E_0 E_L$ between the applied field E_0 and the internal field E_L from the ferroelectric film. Independent measurement of LB films consisting only of anthraquinone bilayers determined the Stark modulation constant $C_b = 6 \times 10^{-22}$ (m/V)^2 per bilayer at frequency 15,700 cm^{-1}. Assuming that the Stark modulation constant C_b was not changed when the film was built into the ferroelectric matrix, the absolute values of the fields can be determined from the harmonic coefficients of the Stark modulation signal as follows:

$$E_0 = \sqrt{\frac{\Delta A_2}{n C_b}} \qquad E_L = \frac{\Delta A_1}{2\Delta A_2} E_0 \quad (26)$$

where ΔA_1 and ΔA_2 are the modulation amplitudes measured at the fundamental and second harmonics (the first and second terms in Eq. (25)), and n is the number of anthraquinone dye bilayers embedded in the ferroelectric LB films.

Measurements were made on a sample fabricated as depicted in Figure 69 with six bilayers ($L \approx 20$ nm) of anthraquinone dye in the middle of a 70-ML ($x_1 + x_2 = 35$ nm) ferroelectric LB film. The experimental Stark modulation spectra corresponding to the fundamental (1ω) and second harmonic (2ω) are shown in Figure 70. The weak response at the fundamental frequency in Figure 70a indicates that even the fresh sample is slightly polar at zero applied field, possibly due to the layer pinning mentioned in Section 4.7.2. If an external dc field

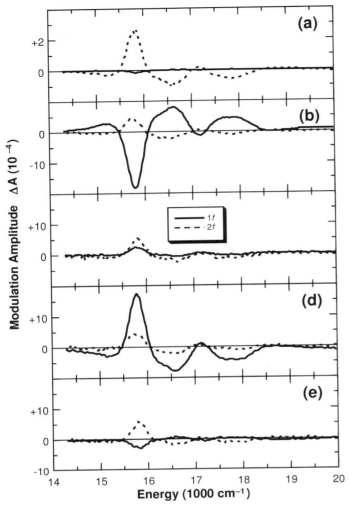

Fig. 70. Stark effect spectra at the fundamental frequency ($1f$) and the second harmonic ($2f$) showing bistable switching of the internal static electric field. The spectra were measured for the applied voltage of amplitude $U_0 = 5$ V at frequency 1000 Hz: (a) for a fresh (nonpolarized) sample; (b) while applying a bias potential of $+10$ V across the film; (c) relaxed state after the positive bias voltage was turned off; (d) while applying a bias potential of -10 V across the film; (e) relaxed state after the negative bias voltage was turned off.

is applied (10 V to polarize the sample), then the response at the fundamental frequency is dramatically increased (Fig. 70b). Note that after the polarizing external field is removed, the response at the first harmonic remains (indicating that the film is still partially polarized), but its sign becomes opposite to the sign of the polarizing external field used, Figure 70c. Thus the field probed by the dye is opposite to the spontaneous polarization, as it should be for the thick samples. This result can be explained in terms of Eq. (24), which takes into account the work function difference. The subsequent polarization by the external voltage of opposite sign (Fig. 70d) results in reversal of the sign of the Stark modulation after the voltage was removed (Fig. 70e). The symmetric sign reversal (compare Fig. 70c and e) can be explained in terms of symmetric reversal of the work functions in Eq. (24). Thus, the sponta-

neous polarization vector orientation influences the electrode work function, perhaps by charge injection. The absolute value of the static field calculated using Eq. (26) is 190 MV/m, about 1/5 the value of the internal depolarization field $P_S/(\varepsilon\varepsilon_0)$.

5. APPLICATIONS OF FERROELECTRIC LANGMUIR–BLODGETT FILMS

The parent polymer poly(vinylidene fluoride) has been in wide use for nearly 20 years in piezoelectric transducers for electromechanical actuators, soft-touch switches, strain gauges, and in sonar and ultrasound transducers. The LB polymer films have large pyroelectric response over a useful range $-50°C$ to $+100°C$, making them candidate infrared sensors for low-cost uncooled infrared imaging systems. The films exhibit a novel $1000:1$ conductance switching when the film polarization is reversed, and this might be exploited in nonvolatile random-access data storage with nondestructive readout.

Important advantages of using polymers in electronic devices are the low production costs, ease and flexibility (literally and figuratively) of fabrication in a variety of thin-film forms, and resistance to degradation caused by strain. Crystalline polymer films need not be grown epitaxially, a cumbersome requirement with many other high-quality crystalline films, yet polymers adhere well to a wide variety of substrates. The main disadvantages of polymer ferroelectric materials concern material stability due to chemical and thermal vulnerability, such as low melting and thermal decomposition temperatures as well as susceptibility to oxidation and photodecomposition. While chemical and photochemical degradation can be prevented by proper encapsulation, the relatively low melting and decomposition temperatures do limit the uses of polymers.

5.1. Nonvolatile Memory

Switching dynamics are often controlled by nucleation rates. In the thinnest high-quality LB ferroelectric films, where nucleation is suppressed, we measured switching times over 10 seconds and coercive fields of about 500 MV/m. But by changing film fabrication and post-treatment, we were able to decrease the coercive field to about 50 MV/m and switching time to a microsecond. This is presumably due to the introduction of nucleation centers not inhibited by the nanometer thickness of the films. We are currently investigating the source(s) of nucleation and developing methods that best control nucleation and fast switching. The fast-switching behavior in the ultrathin LB films is very encouraging for the production of nonvolatile memories because the operating voltage is low, less than 5 V, and the switched charge is large enough, about 25 mC/m^2. The hybrid ferroelectric field-effect device is the current choice for commercial ferroelectric memories made with PZT and similar perovskite ferroelectrics [27, 175, 176]. The field-effect transistor (FET) sensing scheme is sensitive, simple, and can be made with standard semiconductor fabrication techniques. This

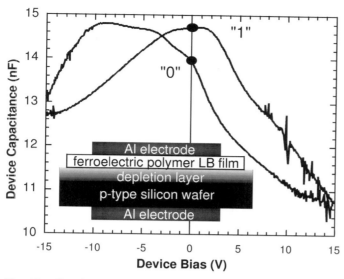

Fig. 71. Capacitance hysteresis loop from a hybrid ferroelectric–semiconductor memory element, showing the bistable state at zero bias due to the two opposite polarization states of the ferroelectric films. Inset: diagram of the device structure.

configuration also seems an excellent choice for the LB ferroelectric polymers for many of the same reasons.

A first test of the feasibility of a field-effect memory element based on the ferroelectric copolymer LB films was made using a simple capacitor structure consisting of a ferroelectric polymer LB film and electrodes on top of a *p*-type silicon wafer with an ohmic back electrode (inset to Fig. 71b). We demonstrated bistability (Fig. 71b) of the device capacitance. By applying a large enough bias, we could polarize the ferroelectric film either up or down, and thus grow or shrink the depletion layer in the silicon wafer, changing the overall capacitance of the device at zero bias.

5.2. Wide-Band Imaging

Pyroelectric detectors and imaging systems for use at infrared wavelengths have two distinct advantages over quantum detectors. The pyroelectric detectors are inherently broad spectrum, relying only on the incorporation of a suitable radiation-absorbing element, and they do not need to be cooled. Current pyroelectric imaging technology relies on expensive cut-and-polished single crystals. The LB ferroelectric polymer films should be inexpensive to make with large area and have an added advantage that they have very low thermal mass, for high sensitivity and quick response. The PSM results show that the pyroelectric signal can be well localized for high pixel density. The ferroelectric LB films can be deposited directly on an integrated charge-sensing array and covered with counter electrodes and an absorbing coating. With proper coatings, it should be possible to make an array sensitive to wavelengths ranging from the microwave to the X-ray region.

5.3. Piezoelectric Transducers

PVDF is now widely used in electromechanical actuators, switches, strain gauges, accelerometers, and acoustic transducers for ultrasound and sonar. A recent issue of the *IEEE Transactions on Ultrasonics, Ferroelectrics* (Vol. 47, No. 6, 1 November 2000, pp. 1275–1454) was devoted to PVDF-based transducers. The piezoelectric response of the untreated ferroelectric polymer LB films is comparable to that of the polymorphous films [63], but we do not yet have a direct measurement of the acoustic coupling efficiency, a key property in ultrasonic applications. Recent electron irradiation studies demonstrate conversion of the P(VDF-TrFE) copolymer film from ferroelectrics into relaxors, producing strains of several percentage [222–224]. We expect a similar improvement in piezoelectric response in irradiated LB copolymer capacitors.

5.4. Capacitors and Energy Storage

The dielectric strength of insulators is generally limited by physical defects, pinholes and particles, or by avalanche breakdown. The record dielectric strength of about 200 MV/m was previously held by solvent-spun PVDF films of about 60 nm thickness [36]. But the high quality of the LB copolymer films, coupled with the low operating voltages (so avalanche breakdown is not possible), resulted in much higher dielectric strength. The films hold off electric fields over 3 GV/m (3 *billion* volts per meter), more than ten times higher than that obtained from thin polymorphous films of the copolymers. This is possible because the high-quality films at least 2 ML thick lack significant electrical shorts, and since this field strength is achieved with only a few volts bias, avalanche breakdown is avoided. Since the energy density of a capacitor is proportional to the dielectric constant divided by the square of its thickness for a given operation voltage, decreasing the thickness is more effective than increasing the dielectric constant. These nanometer-thick films, even with a low dielectric constant of only $\kappa \approx 8$ [118], have achieved record high energy density of over 400 Joules per cubic centimeter of dielectric, far more than any other capacitor and even competitive with advanced batteries. Recent electron irradiation studies demonstrated conversion of the P(VDF-TrFE) copolymer film from ferroelectrics into relaxors, increasing the dielectric constant fivefold, and stabilized this value over a wide temperature range [222–224]. We expect a similar improvement in energy storage capacity in properly treated LB copolymer capacitors; initial trials have demonstrated a threefold increase in the capacitance of irradiated LB films. A multilayer capacitor made from these films could revolutionize portable power, lightening many mechanical and electronic devices that use heavy capacitors or batteries. The main challenge is to find a suitable electrode with sufficient conductivity and compatible fabrication at comparable thickness (about 10 nm). As an added bonus, these capacitors, unlike batteries, should work well in the frigid Nebraska and Moscow winters.

5.5. Fabrication Issues

Langmuir–Blodgett deposition produces nanometer-thick films with excellent uniformity and coverage, critical features for both the capacitor and memory applications. The low anticipated cost of the devices stems from the use of standard silicon integration technology to make the charge-sensing and ancillary electronics and the low inherent cost of coating these ICs with the LB polymer films. The ferroelectric copolymer LB films have some advantages over similar inorganic ferroelectric active layers (like PZT), which often must be processed at high temperatures that are damaging to silicon integrated circuits.

Though there is at present no mass-production LB manufacturing to our knowledge, there is no fundamental impediment to developing this fabrication technology, and small-scale demonstrations of LB manufacturing systems have provided encouraging results [134]. Batch systems rely on automated control, dipping, and periodic monolayer replenishment [225], while continuous-flow systems, which use the flow for monolayer compression [226, 227], are more likely to satisfy the demands of continuous production and would work well with tape and spool substrate handling. LB deposition can be scaled to very large areas, enabling applications and efficiencies not practical with other deposition technologies.

6. CONCLUSIONS

The fabrication of ultrathin ferroelectric polymer films by Langmuir–Blodgett deposition has provided an exciting system for detailed study, revealing valuable insights into the nature of the vinylidene-fluoride polymers specifically and ferroelectricity in general. The LB films prove superior to the polymorphous films for many purposes, both scientific and technological, because of the ability to control thickness with monolayer precision, the natural orientational alignment of straight polymer chains without amorphous material or lamellae, and the convenient crystallographic orientation with the polarization axis perpendicular to the film.

The ability to make such thin ferroelectric films led to two key discoveries. First, there is no minimum critical thickness for the ferroelectric state, contrary to many predictions based on three-dimensional mean-field theory, demonstrating the essentially two-dimensional nature of the ferroelectric state, and closing the dimensionality gap between ferromagnetism and ferroelectricity. Second, the finite-size effect is evident in the coercive field and other switching properties for films thicker than 15 nm, but not in thinner films, demonstrating intrinsic switching not significantly influenced by nucleation and domain dynamics. These discoveries may well be repeated with other ferroelectrics, as the LB method is applied to other molecular materials, and techniques improve for making good nanometer-thick films from some of the oxide ferroelectrics.

The studies completed so far leave several other fundamental questions unanswered. The connection between structure and ferroelectric properties is still poorly understood. We still do not understand how the intrachain, interchain, and long-range interactions contribute to ferroelectricity in the polymer system. Progress in this area will require concerted effort in both experiment and modeling. The nature of the surface transition and the depth of surface reconstruction remain unclear. We are only beginning to probe the nature of the stiffening transition at 160 K, but it appears connected to subtle intrachain ordering mediated by interchain interactions, features that may also shed light on the nature of the ferroelectric–state.

To date, we have studied LB films of the P(VDF-TrFE 70 : 30) composition almost exclusively, though LB films of PVDF and other copolymers exhibit generally the same properties as the 70 : 30 films. Further studies with other copolymer compositions should also be fruitful, particularly in probing the relative contributions of the interactions, as should studies of the copper-phthalocyanine and other potentially ferroelectric molecular films. In addition, LB deposition permits the fabrication of more complex composites by providing precise control of the composition of each monolayer. These composites may exhibit enhanced properties, such as the expected dielectric resonance effect, or even unexpected behavior, much as the P(VDF-TrFE 70 : 30) LB films provided their own surprises.

The ultrathin ferroelectric LB films have unique potential for use in many applications. Some applications, such as ultrasonic transducers, pyroelectric imaging arrays, and nonvolatile memories, follow from the expected advantages of making such thin films with the same basic properties as their bulk counterparts. Other applications unique to the LB films may arise from unexpected opportunities, such as the record dielectric strength attained with the first ferroelectric polymer LB films.

Acknowledgments

We thank our co-workers and students for assistance in preparing and reviewing this chapter, particularly Shireen Adenwalla, Mengjun Bai, Candace Bacon, Peter Dowben, Brad Petersen, Shawn Pebley, Matt Poulsen, Timothy Reece, and Alexander Sorokin. Work at the University of Nebraska was supported by the USA National Science Foundation, the USA Office of Naval Research, the Nebraska Research Initiative, and the J. A. Woollam Company. Work at the Institute of Crystallography was supported by the Russian Foundation for Basic Research (#99-02-16484) and the Inco-Copernicus Programme (#IC15-CT96-0744).

REFERENCES

1. J. Valasek, *Phys. Rev.* 15, 537 (1920).
2. J. Valasek, *Phys. Rev.* 17, 475 (1921).
3. G. M. Sessler, Ed., "Electrets," Vol. 33, Springer-Verlag, Berlin, 1987.
4. P. N. Prasad and D. J. Williams, "Introduction to Nonlinear Optical Effects in Molecules and Polymers." Wiley, New York, 1991.
5. D. S. Chemla and J. Zyss, Eds., "Nonlinear Optical Properties of Organic Molecules and Crystals," Vol. 1. Academic Press, Orlando, 1987; H. S. Nalwa and S. Miyta, Eds., "Nonlinear Optics of Organic Molecules and Polymers," CRC Press, Boca Raton, 1997.

6. P. G. de Gennes and J. Prost, "The Physics of Liquid Crystals." Clarendon Press, Oxford, 1993.

7. V. G. Chigrinov, "Liquid Crystal Devices: Physics and Applications." Artech House, Boston, 1999; H. S. Nalwa, Ed., "Handbook of Advanced Electronic and Photonic Materials," Vol. 1–10, Academic Press, Boston, 2001.

8. F. Jona and G. Shirane, "Ferroelectric Crystals." Macmillan, New York, 1962.

9. M. E. Lines and A. M. Glass, "Principles and Applications of Ferroelectrics and Related Materials." Clarendon, Oxford, 1977.

10. D. R. Tilley, "Phase Transitions in Thin Films in Ferroelectric Ceramics." Birkhäuser, Basel, 1993.

11. R. Blinc and B. Zeks, "Soft Modes in Ferroelectrics and Antiferroelectrics." North-Holland, Amsterdam, 1974.

12. T. Mitzui, I. Tatsuzaki, and E. Nakamura, "An Introduction to the Physics of Ferroelectrics." Gordon and Breach, New York, 1976.

13. V. M. Fridkin, "Ferroelectric Semiconductors." Consultants Bureau, New York, 1980.

14. T. T. Wang, J. M. Herbert, and A. M. Glass, Eds., "The Applications of Ferroelectric Polymers." Chapman and Hall, New York, 1988.

15. Y. Xu, "Ferroelectric Materials and their Applications." North-Holland, Amsterdam, 1991.

16. B. A. Strukov and A. P. Levanuk, "Ferroelectric Phenomena in Crystals." Springer-Verlag, Berlin, 1998.

17. V. Ginzburg, Zh. Eksp. Teor. Fiz. 15, 739 (1945).

18. V. Ginzburg, J. Phys. USSR 10, 107 (1946).

19. V. Ginzburg, Zh. Eksp. Teor. Fiz. 19, 39 (1949).

20. L. D. Landau, Zh. Eksp. Teor. Fiz. 7, 627 (1937).

21. L. D. Landau, Phys. Z. Sowjun. 11, 545 (1937).

22. L. D. Landau and E. M. Lifshitz, "Statistical Physics: Part I." Pergamon, Oxford, 1980.

23. A. F. Devonshire, Phil. Mag. 40, 1040 (1949).

24. A. F. Devonshire, Phil. Mag. 42, 1065 (1951).

25. A. F. Devonshire, Advances in Physics 3, 85 (1954).

26. V. M. Fridkin and S. Ducharme, Phys. Solid State 43, 1320 (2000).

27. O. Auciello, J. F. Scott, and R. Ramesh, Phys. Today 51, 22 (1998).

28. C. B. Sawyer and C. H. Tower, Phys. Rev. 269 (1930).

29. W. J. Merz, J. Appl. Phys. 27, 938 (1956).

30. W. Dürr, M. Taborelli, O. Paul, R. Germar, W. Gudat, D. Pescia, and M. Landolt, Phys. Rev. Lett. 62, 206 (1989).

31. M. Farle and K. Baberschke, Phys. Rev. Lett. 58, 511 (1987).

32. N. D. Mermin and H. Wagner, Phys. Rev. Lett. 17, 1133 (1966).

33. K. Ishikawa, K. Yoshikawa, and N. Okada, Phys. Rev. B 37, 5852 (1988).

34. S. Schlag and H.-F. Eicke, Solid State Commun. 91, 993 (1994).

35. M. Tanaka and Y. Makino, Ferro. Lett. 24, 13 (1998).

36. K. Kimura and H. Ohigashi, Japan. J. Appl. Phys. 25, 383 (1986).

37. J. Karasawa, M. Sugiura, M. Wada, M. Hafid, and T. Fukami, Integrated Ferroelectrics 12, 105 (1996).

38. T. Tybell, C. H. Ahn, and J.-M. Triscone, Appl. Phys. Lett. 75, 856 (1999).

39. D. R. Tilley, in "Ferroelectric Thin Films: Synthesis and Basic Properties" (C. Paz de Araujo, J. F. Scott, and G. F. Taylor, Eds.), p. 11. Gordon and Breach, Amsterdam, 1996.

40. J. F. Scott, Phase Transitions 30, 107 (1991).

41. J. F. Scott, Physica B 150, 160 (1988).

42. C. L. Wang and S. R. P. Smith, J. Phys. Cond. Matter 7, 7163 (1995).

43. P. A. Dowben, D. N. McIlroy, and D. Li, in "Handbook on the Physics and Chemistry of Rare Earths" (J. K. A. Gschneidner and L. Eyring, Eds.). Elsevier, Amsterdam, 1997.

44. M. G. Cottam, D. R. Tilley, and B. Zeks, J. Phys. C: Solid State Phys. 17, 1793 (1984).

45. H. M. Duiker, Static and Dynamic Properties of Ferroelectric Thin Film Memories, Ph.D. Dissertation, University of Colorado, 1989.

46. B. D. Qu, P. L. Zhang, Y. G. Wang, C. L. Wang, and W. L. Zhong, Ferroelectrics 152, 219 (1994).

47. J. F. Scott, M. S. Zheng, R. B. Godfrey, C. A. Paz de Araujo, and L. D. McMillan, Phys. Rev. B 35, 4044 (1987).

48. C. L. Wang, W. L. Zhong, and P. L. Zhang, J. Phys. Cond. Matter 3, 4743 (1992).

49. S. Li, J. A. Eastman, Z. Li, C. M. Foster, R. E. Newnham, and L. E. Cross, Phys. Lett. A 212, 341 (1996).

50. S. Li, J. A. Eastman, J. M. Ventroner, R. E. Newnham, and L. E. Cross, Philos. Mag. B 76, 47 (1997).

51. A. M. Bratkovsky and A. P. Levanyuk, Phys. Rev. Lett. 84, 3177 (2000).

52. P. G. de Gennes, Solid State Commun. 1, 132 (1963).

53. C. L. Wang and S. R. P. Smith, J. Phys. Cond. Matter 8, 3075 (1996).

54. H. Kawai, Japan. J. Appl. Phys. 8, 975 (1969).

55. T. Furukawa, Phase Transitions 18, 143 (1989).

56. B. A. Newman, J. I. Scheinbeim, J. W. Lee, and Y. Takase, Ferroelectrics 127, 229 (1992).

57. B. Z. Mei, J. I. Scheinbeim, and B. A. Newman, Ferroelectrics 144, 51 (1993).

58. S. Esayan, J. I. Scheinbeim, and B. A. Newman, Appl. Phys. Lett. 67, 623 (1995).

59. T. Furukawa, Ferroelectrics 57, 63 (1984).

60. A. J. Lovinger, Science 220, 1115 (1983).

61. A. J. Lovinger, in "Developments in Crystalline Polymers—1" (D. C. Basset, Ed.). Applied Science, London, 1982.

62. H. S. Nalwa, Ed., "Ferroelectric Polymers." Marcel Dekker, New York, 1995.

63. A. V. Bune, C. Zhu, S. Ducharme, L. M. Blinov, V. M. Fridkin, S. P. Palto, N. N. Petukhova, and S. G. Yudin, J. Appl. Phys. 85, 7869 (1999).

64. K. A. Verkhovskaya, R. Danz, and V. M. Fridkin, Sov. Phys. Solid State 29, 1268 (1987).

65. M. H. Berry and D. N. Gookin, "Proceedings of the Conference on Nonlinear Optical Properties of Organic Materials," 1988, Vol. 971, p. 154. SPIE, San Francisco, 1988.

66. J. G. Bergman, J. H. McFee, and G. R. Crane, Appl. Phys. Lett. 18, 203 (1971).

67. K. A. Verkhovskaya, V. M. Fridkin, A. V. Bune, and J. F. Legrand, Ferroelectrics 134, 7 (1992).

68. M. A. Marcus, Ferroelectrics 40, 29 (1982).

69. R. G. Kepler and R. A. Anderson, Advances in Physics 41, 1 (1992).

70. R. G. Kepler, in "Ferroelectric Polymers" (H. S. Nalwa, Ed.), p. 183. Marcel Dekker, New York, 1995.

71. J. B. Lando and W. W. Doll, J. Macromolecular Science—Physics B2, 205 (1968).

72. B. L. Farmer, A. J. Hopfinger, and J. B. Lando, J. Appl. Phys. 43, 4293 (1972).

73. A. J. Lovinger, G. T. Davis, T. Furukawa, and M. G. Broadhurst, Macromol. 15, 323 (1982).

74. J. F. Legrand, Ferroelectrics 91, 303 (1989).

75. K. Tashiro, in "Ferroelectric Polymers" (H. S. Nalwa, Ed.), p. 63. Marcel Dekker, New York, 1995.

76. M. A. Bachmann and J. B. Lando, Macromol. 14, 440 (1981).

77. T. Furukawa, M. Date, and E. Fukada, Ferroelectrics 57, 63 (1980).

78. K. Kimura and H. Ohigashi, Appl. Phys. Lett. 43, 834 (1983).

79. T. Furukawa, A. J. Lovinger, G. T. Davis, and M. G. Broadhurst, Macromol. 16, 1885 (1983).

80. T. Kajiyama, N. Khuwattanasil, and A. Takahara, J. Vac. Sci. Tech. B 16, 121 (1998).

81. M. Poulsen, Use of an External Electric Field to Convert the Paraelectric Phase to the Ferroelectric Phase in Ultra-Thin Copolymer Films of P(VDF-TrFE), B.S. Thesis, University of Nebraska, Lincoln, 2000.

82. N. Karasawa and W. A. Goddard, III, Macromol. 25, 7268 (1992).

83. V. V. Kochervinskii, Russian Chemical Reviews 65, 865 (1996).

84. Y. Takahashi, M. Kohyama, Y. Matsubara, H. Iwane, and H. Tadokoro, Macromol. 14, 1841 (1981).

85. Y. Takahashi, Y. Matsubara, and H. Tadokoro, Macromol. 15, 334 (1982).

86. H. Ohigashi, K. Omote, and T. Gomyo, Appl. Phys. Lett. 66, 3281 (1995).

87. H. Ohigashi, K. Omote, H. Abe, and K. Koga, Japan. J. Appl. Phys. 68, 1824 (1999).

88. G. T. Davis, T. Furukawa, A. J. Lovinger, and M. G. Broadhurst, Macromol. 15, 329 (1982).

89. K. Tashiro and M. Kobayashi, *Phase Transitions* 18, 213 (1989).
90. K. Tashiro, R. Tanaka, K. Ushitora, and M. Kobayashi, *Ferroelectrics* 171, 145 (1995).
91. A. J. Lovinger, *Japan. J. Appl. Phys.* 24-2, 18 (1985).
92. A. J. Lovinger, G. E. Johnson, H. E. Bair, and E. W. Anderson, *J. Appl. Phys.* 56, 2142 (1984).
93. R. E. Cais and H. M. Kometani, *Macromol.* 18, 1354 (1985).
94. R. E. Cais and H. M. Kometani, *Macromol.* 17, 1887 (1984).
95. J. J. Ladik, "Quantum Theory of Polymers as Solids." Plenum Press, New York, 1988.
96. B. Wunderlich, "Crystal Structure, Morphology, Defects." Academic Press, New York, 1973.
97. K. Koga and H. Ohigashi, *J. Appl. Phys.* 59, 2142 (1986).
98. A. J. Hopfinger, A. J. Lewanski, T. J. Slukin, and P. L. Taylor, in "Solitons and Condensed-Matter Physics" (A. R. Bishop and T. Schneider, Eds.), p. 330. Springer-Verlag, New York, 1979.
99. H. Dvey-Aharon, T. J. Sluckin, and P. L. Taylor, *Phys. Rev. B* 21, 3700 (1980).
100. N. C. Banik, F. P. Boyle, T. J. Sluckin, P. L. Taylor, S. K. Tripathy, and A. J. Hopfinger, *J. Chem. Phys.* 72, 3191 (1980).
101. R. Zhang and P. L. Taylor, *J. Appl. Phys.* 73, 1395 (1993).
102. N. A. Pertsev and A. G. Zembil'gotov, *Sov. Phys. Solid State* 33, 165 (1991).
103. J. D. Clark and P. L. Taylor, *Phys. Rev. Lett.* 49, 1532 (1982).
104. K. Tashiro, Y. Abe, and M. Kobayashi, *Ferroelectrics* 171, 281 (1995).
105. S. Ikeda and H. Suda, *Phys. Rev. E* 56, 3231 (1997).
106. G. J. Kavarnos and R. W. Holman, *Polymer* 35, 5586 (1994).
107. R. W. Holman and G. J. Kavarnos, *Polymer* 37, 1697 (1996).
108. R. E. Cohen, Ed., "Fundamental Physics of Ferroelectrics-2000," Vol. 535. American Institute of Physics, Melville, New York, 2000.
109. T. Yagi, M. Tatemoto, and J. Sako, *Polymer J.* 12, 209 (1980).
110. M. G. Broadhurst and G. T. Davis, *Ferroelectrics* 32, 177 (1981).
111. M. V. Fernandez, A. Suzuki, and A. Chiba, *Macromol.* 20, 1806 (1987).
112. J. K. Krüger, B. Heydt, C. Fischer, J. Baller, R. Jiménez, K.-P. Bohn, B. Servet, P. Galtier, M. Pavel, B. Ploss, M. Beghi, and C. Bottani, *Phys. Rev. B* 55, 3497 (1997).
113. K. Omote, H. Ohigashi, and K. Koga, *J. Appl. Phys.* 81, 2760 (1997).
114. A. J. Lovinger, *Macromol.* 16, 1529 (1983).
115. K. Matsushige, *Phase Transitions* 18, 247 (1989).
116. E. Bellet-Amalric, J. F. Legrand, M. Stock-Schweyer, and B. Meurer, *Polymer* 35, 34 (1994).
117. R. L. Moreira, R. P. S. M. Lobo, G. Medeiros-Ribeiro, and W. N. Rodrigues, *J. Polymer Sci. B* 32, 953 (1984).
118. A. V. Bune, V. M. Fridkin, S. Ducharme, L. M. Blinov, S. P. Palto, A. V. Sorokin, S. G. Yudin, and A. Zlatkin, *Nature* 391, 874 (1998).
119. J. Choi, C. N. Borca, P. A. Dowben, A. Bune, M. Poulsen, S. Pebley, S. Adenwalla, S. Ducharme, L. Robertson, V. M. Fridkin, S. P. Palto, N. N. Petukhova, and S. G. Yudin, *Phys. Rev. B* 61, 5760 (2000).
120. Z.-Y. Cheng, V. Bharti, T.-B. Xu, S. Wang, Q. M. Zhang, T. Ramotowski, F. Tito, and R. Ting, *J. Appl. Phys.* 86, 2208 (1999).
121. S. Ducharme, A. V. Bune, V. M. Fridkin, L. M. Blinov, S. P. Palto, A. V. Sorokin, and S. Yudin, *Phys. Rev. B* 57, 25 (1998).
122. T. Ishii and J. Muru, *Thin Solid Films* 179, 109 (1989).
123. A. Pockels, *Nature* 43, 437 (1891).
124. L. Rayleigh, *Proc. R. Soc. London* 47, 364 (1890).
125. I. Langmuir, *J. Am. Chem. Soc.* 39, 1848 (1917).
126. I. Langmuir, *Trans. Faraday Soc.* 15, 62 (1920).
127. K. B. Blodgett, *J. Am. Chem. Soc.* 57, 1007 (1935).
128. I. V. Langmuir and V. J. Schaefer, *J. Am. Chem. Soc.* 60, 1351 (1938).
129. V. A. Trapeznikov, *J. Phys. Chem. (USSR)* 19, 228 (1945).
130. G. L. Gaines, "Insoluble Monolayers at Liquid–Gas Interfaces." Interscience, New York, 1966.
131. H. Kuhn, D. Möbius, and H. Bücher, in "Physical Methods of Chemistry" (A. Weissberger and B. Rossiter, Eds.), p. 577. Wiley, New York, 1972.
132. G. G. Roberts, *Ferroelectrics* 91, 21 (1989).
133. L. M. Blinov, *Sov. Phys. Usp.* 31, 623 (1988).
134. G. G. Roberts, Ed. "Langmuir–Blodgett Films." Plenum, New York, 1990.
135. M. C. Petty, "Langmuir–Blodgett Films: An Introduction." Cambridge University Press, Cambridge, 1996.
136. Y. Higashihata, J. Sake, and T. Yagi, *Ferro.* 32, 85 (1981).
137. K. Tashiro and H. Tadokoro, *Macromol.* 19, 961 (1983).
138. L. M. Blinov, N. N. Davydova, V. V. Lazarev, and S. G. Yudin, *Soviet Physics-Solid State* 24, 1523 (1982).
139. V. M. Fridkin, A. L. Shlensky, K. A. Verkhovskaya, V. D. Bilke, N. N. Markevich, H. Pietch, and M. Sudow, *Electrophotography* 23, 193 (1984).
140. H. Sasabe, T. Furuno, and K. Takimoto, *Synthetic Metals* 28, C787 (1989).
141. O. A. Aktsipetrov, N. N. Akhmediev, E. D. Mishina, and V. R. Novak, *JETP Lett.* 37, 207 (1983).
142. O. A. Aktsipetrov, N. N. Akhmediev, I. M. Baranova, E. D. Mishina, and V. R. Novak, *J. Exp. Theor. Phys.* 11, 249 (1985).
143. R. H. Selfridge, T. K. Moon, P. Stroeve, J. Y. S. Lam, S. T. Kowel, and A. Knoesen, *Proc. SPIE—Int. Soc. Opt. Eng.* 971, 197 (1988).
144. C. Duschl, W. Frey, and W. Knoll, *Thin Solid Films* 160, 251 (1988).
145. M. Sastry, S. Pal, D. V. Paranjape, A. Rajagopal, S. Adhi, and S. K. Kulkarni, *J. Chem. Phys.* 99, 4799 (1993).
146. S. S. Shiratori, K. Tachi, and K. Ikezaki, *Synthetic Metals* 84, 833 (1997).
147. S. P. Palto, L. M. Blinov, V. F. Petrov, S. V. Jablonsky, S. G. Yudin, H. Okamoto, and S. Takenaka, *Mol. Cryst. Liq. Cryst.* 326, 259 (1999).
148. V. V. Arslanov, *Russian Chemical Reviews* 63, 3 (1994).
149. S. Palto, L. Blinov, A. Bune, E. Dubovik, V. Fridkin, N. Petukhova, K. Verkhovskaya, and S. Yudin, *Ferro. Lett.* 19, 65 (1995).
150. L. M. Blinov, V. M. Fridkin, S. P. Palto, A. V. Sorokin, and S. G. Yudin, *Thin Solid Films* 284-285, 469 (1996).
151. A. Sorokin, S. Palto, L. Blinov, V. Fridkin, and S. Yudin, *Molecular Materials* 6, 61 (1996).
152. E. Ferroni, G. Gabrielli, and M. Puggelli, *La Chimica e L'Industria (Milan)* 49, 147 (1967).
153. A. Bune, S. Ducharme, V. M. Fridkin, L. Blinov, S. Palto, N. Petukhova, and S. Yudin, *Appl. Phys. Lett.* 67, 3975 (1995).
154. S. Palto, L. Blinov, E. Dubovik, V. Fridkin, N. Petukhova, A. Sorokin, K. Verkhovskaya, S. Yudin, and A. Zlatkin, *Europhys. Lett.* 34, 465 (1996).
155. S. Palto, L. Blinov, A. Bune, E. Dubovik, V. Fridkin, N. Petukhova, K. Verkhovskaya, and S. Yudin, *Ferroelectrics* 184, 127 (1996).
156. S. G. Yudin, S. P. Palto, V. A. Khavrichev, S. V. Mironenko, and M. I. Barnik, *Thin Solid Films* 210, 46 (1992).
157. L. M. Blinov, R. Barberi, S. P. Palto, M. P. De Santo, and S. G. Yudin, *J. Appl. Phys.* 89, 3960 (2001).
158. S. Ducharme, M. Bai, M. Poulsen, S. Adenwalla, S. P. Palto, L. M. Blinov, and V. M. Fridkin, *Ferroelectrics* 252, 191 (2001).
159. R. C. Advincula, W. Knoll, L. Blinov, and C. W. Frank, in "Organic Thin Films, Structure and Applications" (C. W. Frank, Ed.), p. 192. American Chemical Society, Washington, 1998.
160. L. M. Blinov, V. M. Fridkin, S. P. Palto, A. V. Sorokin, and S. G. Yudin, *Thin Solid Films* 284-285, 474 (1996).
161. A. V. Sorokin, Preparation and Investigation of Monolayers and Thin Films of a Ferroelectric Polymer, Ph.D. Dissertation, Institute of Crystallography, Moscow, 1997 (in Russian).
162. J. Choi, P. A. Dowben, S. Ducharme, V. M. Fridkin, S. P. Palto, N. Petukhova, and S. G. Yudin, *Phys. Lett. A* 249, 505 (1999).
163. J. Choi, S.-J. Tang, P. T. Sprunger, P. A. Dowben, J. Braun, E. W. Plummer, V. M. Fridkin, A. V. Sorokin, S. P. Palto, N. Petukhova, and S. G. Yudin, *J. Phys. Cond. Matter* 12, 4735 (2000).
164. O. A. Aktsipetrov, T. V. Misuryaev, T. V. Murzina, L. M. Blinov, V. M. Fridkin, and S. P. Palto, *Opt. Lett.* 25, 411 (2000).
165. O. A. Aktsipetrov, T. V. Misuryaev, T. V. Murzina, S. P. Palto, N. N. Petukhova, V. M. Fridkin, Y. G. Fokin, and S. G. Yudin, *Integrated Ferroelectrics* 35, 23 (2001).
166. J. Choi, P. A. Dowben, S. Pebley, A. V. Bune, S. Ducharme, V. M. Fridkin, S. P. Palto, and N. Petukhova, *Phys. Rev. Lett.* 80, 1328 (1998).

167. C. N. Borca, J. Choi, S. Adenwalla, S. Ducharme, P. A. Dowben, L. Robertson, V. M. Fridkin, S. P. Palto, and N. Petukhova, *Appl. Phys. Lett.* 74, 347 (1999).

168. C. N. Borca, S. Adenwalla, J. Choi, P. T. Sprunger, S. Ducharme, L. Robertson, S. P. Palto, J. Liu, M. Poulsen, V. M. Fridkin, H. You, and P. A. Dowben, *Phys. Rev. Lett.* 83, 4562 (1999).

169. K. Tashiro, K. Takano, M. Kobayashi, Y. Chatani, and H. Tadokoro, *Polymer* 22, 1312 (1981).

170. K. Tashiro, K. Takano, and M. Kobayashi, *Ferroelectrics* 57, 297 (1984).

171. K. J. Kim, G. B. Kim, C. L. Vanlencia, and J. F. Rabolt, *J. Polymer Sci. B: Polymer Phys.* 32, 2435 (1994).

172. W. J. Merz, *Phys. Rev.* 91, 513 (1953).

173. T. Fukuma, K. Kobayashi, T. Horiuchi, P. Yamada, and K. Matsushige, *Japan. J. Appl. Phys.* 39, 3830 (2000).

174. W. J. Merz, *Phys. Rev.* 95, 690 (1954).

175. D. Bondurant and F. Gnadinger, *IEEE Spectrum* 89, 30 (1989).

176. J. F. Scott, *Ferroelectrics Review* 1, 1 (1998).

177. S. Ducharme, V. M. Fridkin, A. Bune, S. P. Palto, L. M. Blinov, N. N. Petukhova, and S. G. Yudin, *Phys. Rev. Lett.* 84, 175 (2000).

178. A. K. Tagantsev, *Integrated Ferroelectrics* 16, 237 (1997).

179. A. K. Tagantsev, *Ferroelectrics* 184, 79 (1996).

180. M. Hikosaka, K. Sakurai, H. Ohigashi, and T. Koizumi, *Japan. J. Appl. Phys.* 32, 2029 (1993).

181. E. D. Specht, H.-M. Christen, D. P. Norton, and L. A. Boatner, *Phys. Rev. Lett.* 80, 4317 (1998).

182. J. F. M. Cillessen, M. W. J. Prins, and R. M. Wolf, *J. Appl. Phys.* 81, 2777 (1997).

183. T. Maruyama, M. Saitoh, I. Sakai, T. Hidaka, Y. Yano, and T. Noguchi, *Appl. Phys. Lett.* 73, 3524 (1998).

184. K. Dimmler, M. Parris, D. Butler, S. Eaton, B. Pouligny, J. F. Scott, and Y. Ishibashi, *J. Appl. Phys.* 61, 5467 (1987).

185. Y. Ishibashi, in "Ferroelectric Thin Films: Synthesis and Basic Properties" (C. Paz de Araujo, J. F. Scott, and G. W. Taylor, Eds.), p. 135. Gordon and Breach, Amsterdam, 1996.

186. N. I. Lebedev and A. S. Sigov, *Integrated Ferroelectrics* 4, 21 (1994).

187. Y. Sakashita, H. Segawa, K. Tominaga, and M. Okada, *J. Appl. Phys.* 13, 785 (1993).

188. C. D. E. Lakeman and D. A. Payne, *Ferroelectrics* 152, 145 (1994).

189. S. P. Palto, G. N. Andreev, N. N. Petukhova, S. G. Yudin, and L. M. Blinov, *J. Exp. Theor. Phys.* 90, 872 (2000).

190. L. M. Blinov, V. M. Fridkin, S. P. Palto, A. V. Bune, P. A. Dowben, and S. Ducharme, *Physics Uspekhi* 43, 243 (2000).

191. S. P. Palto, A. M. Lotonov, K. A. Verkhovskaya, G. N. Andreev, and N. D. Gavrilova, *J. Exp. Theor. Phys.* 90, 301 (2000).

192. T. Furukawa and G. E. Johnson, *Appl. Phys. Lett.* 38, 1027 (1981).

193. K. Okada and H. Sugie, *Phys. Lett. A* 37, 337 (1971).

194. V. V. Gladkii and E. V. Sidnenko, *Soviet Physics-Solid State* 13, 2592 (1972).

195. C. Bahr and G. Heppke, *Phys. Rev. A* 39, 5459 (1989).

196. K. Tashiro, M. Kobayashi, H. Tadokori, and E. Fukada, *Macromol.* 13, 691 (1980).

197. E. Fukada, *Phase Transitions* 18, 135 (1989).

198. T. Furukawa and N. Seo, *Japan. J. Appl. Phys.* 24, 675 (1990).

199. P. Nye, "Physical Properties of Crystals." Oxford Press, London, 1967.

200. K. Lefki and G. J. M. Dormans, *J. Appl. Phys.* 76, 1764 (1994).

201. A. G. Chynoweth, *J. Appl. Phys.* 27, 78 (1956).

202. S. Ducharme and J. Feinberg, *IEEE J. Quantum Electron.* 23, 2116 (1987).

203. C. Bacon, "Profile of a Ferroelectric Polymer." University of Nebraska, Lincoln, 1998.

204. B. Gubanka, M. Donath, and F. Passek, *Phys. Rev. B* 54, R11153 (1996).

205. B. Gubanka, M. Donath, and F. Passek, *J. Mag. Mater.* 161, L11 (1996).

206. D. Li, M. Freitag, J. Pearson, A. Q. Qui, and S. D. Bader, *Phys. Rev. Lett.* 72, 3112 (1994).

207. L. Gavioli, M. G. Betti, and C. Mariani, *Phys. Rev. Lett.* 77, 3869 (1996).

208. S. D. Kevan and N. G. Stoffel, *Phys. Rev. Lett.* 53, 702 (1984).

209. S. D. Kevan, *Phys. Rev. B* 32, 2344 (1985).

210. J. Choi, P. A. Dowben, C. N. Borca, S. Adenwalla, A. V. Bune, S. Ducharme, V. M. Fridkin, S. P. Palto, and N. Petukhova, *Phys. Rev. B* 59, 1819 (1999).

211. T. Furukawa, H. Matsuzaki, M. Shiina, and Y. Tajitsu, *Japan. J. Appl. Phys.* 24, L661 (1985).

212. S. Ducharme, S. P. Palto, L. M. Blinov, and V. M. Fridkin, in "Proceedings of the Fundamental Physics of Ferroelectrics-2000" (R. E. Cohen, Ed.), Vol. 535, p. 354. American Institute of Physics, Aspen, CO, 2000.

213. K. Abe, S. Komatsu, N. Yanese, K. Sano, and T. Kawakubo, *Japan. J. Appl. Phys.* 36, 5846 (1997).

214. R. H. Fowler and L. Nordheim, *Proc. R. Soc. London. Ser. A* 119, 173 (1928).

215. Y. Isono and H. Nakano, *J. Appl. Phys.* 75, 4557 (1994).

216. S. G. Yudin, L. M. Blinov, N. N. Petukhova, and S. P. Palto, *JETP Lett.* 70, 633 (1999).

217. L. M. Blinov, A. V. Ivaschenko, S. P. Palto, and S. G. Yudin, *Thin Solid Films* 160, 271 (1988).

218. L. M. Blinov, S. P. Palto, A. A. Udalyev, and S. G. Yudin, *Thin Solid Films* 179, 351 (1989).

219. L. M. Blinov, K. A. Verkhovskaya, S. P. Palto, and A. V. Sorokin, *Appl. Phys. Lett.* 68, 2369 (1996).

220. L. M. Blinov, K. A. Verkhovskaya, S. P. Palto, A. V. Sorokin, and A. A. Tevosov, *Crystallography Reports* 41, 310 (1996).

221. L. M. Blinov, L. M. Palto, and S. G. Yudin, *J. Mol. Electron.* 5, 45 (1989).

222. Q. Zhang, V. Bharti, and X. Zhao, *Science* 280, 2101 (1998).

223. Q. M. Zhang, Z. Y. Cheng, and V. Bharti, *Appl. Phys. A* 70, 307 (2000).

224. V. Bharti, H. S. Xu, G. Shanthi, Q. M. Zhang, and K. Liang, *J. Appl. Phys.* 87, 452 (2000).

225. I. R. Peterson, *Thin Solid Films* 134, 135 (1985).

226. A. Barraud and M. Vandevyver, *Thin Solid Films* 99, 221 (1983).

227. Z. Lu, F. Qian, Y. Zhu, X. Yang, and Y. Wei, *Thin Solid Films* 243, 371 (1994).

228. P. J. Lock, *Appl. Phys. Lett.* 19, 390 (1971).

229. G. A. Samara, *Ferroelectrics* 5, 25 (1973).

230. R. Kohler, G. Gerlach, P. Padmini, G. Hofmann, and J. Bruchhaus, *J. Korean Physical Society* 32, 1744 (1998).

231. A. K. Tagantsev, C. Pawlaczyk, K. Brooks, and N. Setter, *Integrated Ferroelectrics* 4, 1 (1994).

232. P. K. Larsen, G. J. M. Dormans, D. J. Taylor, and P. J. van Veldhoven, *J. Appl. Phys.* 76, 2405 (1994).

Chapter 12

OPTICAL PROPERTIES OF DIELECTRIC AND SEMICONDUCTOR THIN FILMS

I. Chambouleyron

Institute of Physics "Gleb Wataghin," State University of Campinas—UNICAMP, 13073-970 Campinas, São Paulo, Brazil

J. M. Martínez

Institute of Mathematics and Computer Science, State University of Campinas—UNICAMP, 13083-970 Campinas, São Paulo, Brazil

Contents

1. THEORY

1.1. Historical Note

Interference phenomena of light waves occur frequently in nature, as in the colored reflection produced by feathers and wings of birds and many insects. The interference pattern given by oily substances onto water has been certainly noticed by man since ancient times. The origin of such effects, however, remained unclear until relatively recent times. The first systematic study on interference was undertaken by Newton [1], the well-known Newton rings, obtained from multiple reflections of light in the tiny air layer left between a flat and a little convex piece of glass. The effect, however, was not properly understood at that moment due to Newton's belief that light beams were the propagation of a stream of particles, Newton rejecting the idea of light being a wavelike perturbation of the ether. In 1690, Huygens [2] published his *Treatise on Light*, in which he proposed that light was a disturbance propagating through the ether as sound moves through air. There is no reference to wavelengths, though, or a connection of these and color. In 1768, Euler [3] advanced the hypothesis that the colors we see depends on the wavelength of wave light, in the same way as the pitch of the sound we hear depends on the wavelength of sound. In 1801, in the lecture *On the theory of light and colors*, delivered before the Royal Society, Young sketched the current theory of color vision in terms of three different (pri-

Handbook of Thin Film Materials, edited by H.S. Nalwa
Volume 3: Ferroelectric and Dielectric Thin Films
Copyright © 2002 by Academic Press

ISBN 0-12-512911-4/$35.00

mary) colors. Young's experiments on the interference of two light beams [4] were made around 1800 and published in 1807. They gave the ultimate proof of the wave nature of light and allowed the determination of the wavelength of visible light. In the words of Herschel [5], "... Dr. Young has established a principle in optics, which, regarded as a physical law, has hardly its equal for beauty, simplicity, and extent of applications, in the whole circle of science.... The experimental means by which Dr. Young confirmed this principle, which is known in optics by the name of interference of the rays of light, were as simple and satisfactory as the principle itself is beautiful.... Newton's colors of thin films were the first phenomena to which the author applied it with full success." Based on the experiments of Malus on the polarization of light by ordinary reflection at the surface of a transparent body, Young proposed a few years later that light waves were transverse perturbations of the ether and not longitudinal, as sound waves. The consequences of the finding kept physicists busy for the next hundred years. It was Fresnel that put Young's results on a firm mathematical basis. He established the relations between the amplitudes of the reflected and refracted waves and the incident waves, known as the Fresnel formulae.

Optical coatings remained a subject of laboratory curiosity in the sense that no industrial applications occurred until the twentieth century, with the notable exception of mirror production. Until the late Middle Ages, mirrors were simply polished metals. At a certain unknown moment, the Venetian glass workers introduced glass silvering using an amalgam of tin and mercury [6]. The method lasted until the nineteenth century, when the backing of glass was made with silver. The development of vacuum apparatus techniques allowed the evaporation of a large class of materials, with the ensuing application to optical systems, among others. The methods to deposit thin films in a controllable way increased at a rapid pace, as did the applications to different fields of industry. Today, advanced electronic and optical devices are manufactured involving single or multilayered structures of materials of very different nature: electro-optic and optical materials, semiconductors, insulators, superconductors, and all kinds of oxide and nitride alloys. This rapid development demands accurate deposition methods, sometimes with *in situ* characterization, as well as powerful algorithms for fast corrections to predicted performance.

1.2. The General Problem

Figure 1 sketches the general problem. A beam of mono- or polychromatic light falls onto a *p*-layered optical structure. Within the structure, or at the interfaces, light may be absorbed and/or scattered. In general, the thickness of the different optical coatings composing the structure varies between 0.05 and a few times the wavelength of the incident light beam. An abrupt interface between materials of different optical properties (or a discontinuity taking place in a distance small compared with the light wavelength) provokes a partial reflection of the light

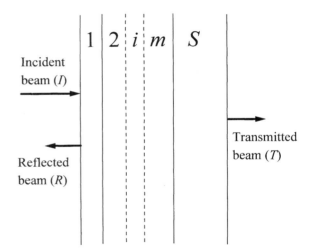

Fig. 1. The figure sketches the general optical coating problem. A light beam impinges onto a stack of dielectric thin layers supported by thick substrate. The radiation is partly transmitted and partly reflected, both depending on photon energy and layer thickness. Part of the radiation may be internally absorbed.

beam. Sometimes, a change of phase also occurs. As the distances between layer boundaries are comparable to the photon wavelength, multiplyreflected or transmitted beams are coherent with one another. Consequently, the total amount of light reflected and transmitted by the structure results from the algebraic sum of the amplitudes of these partially reflected and partially transmitted beams. This property is at the base of a large number of applications. Depending on the specific case, the substrate (*S*) supporting the structure may be transparent or absorbing. Self-supported structures may be built also but, in what follows, the existence of a thick supporting substrate will be assumed. Hence, in the present context, a thin film is equivalent to an optical coating.

The structure depicted in Figure 1 suggests three different problems:

(a) *Optical response.* Calculation of the spectral response of a *p*-layered structure deposited onto a substrate of known properties. The optical constants of the coating materials as a function of photon wavelength are given, as well as the thickness of each layer.

(b) *Structure design.* The problem here is to find the appropriate materials, the number of layers, and their respective thicknesses in order to meet a desired spectral performance.

(c) *Reverse optical engineering.* Retrieval of the optical parameters and the thicknesses of the layers composing the structure from measured spectral responses.

In mathematical terms, the *optical response* problem is a *direct* problem. The phenomenon is governed by a partial differential wave equation (coming from Maxwell equations) where we assume that the parameters and boundary conditions are given. The direct problem consists of computing the state of the wave within each layer, using the information mentioned above.

Analytic solutions can be given if we assume that the incident wave is a *pure* wave and that the boundary layers are regular. In more complicated situations, numerical solutions are necessary. However, even in the case where analytic calculations are possible, the practical computation of the response can be very costly. This is because pure waves do not exist in most optical experiments, and the true response corresponds to an average among many waves of different wavelength. Although it is possible to give the transmitted and reflected energies of pure waves in a closed analytic form, their analytic integration cannot be done and the numerical integration process is, computationally, very expensive.

Design and *reverse engineering* problems are inverse problems in the sense that, in both cases, the response of the system (or a part of it, such as transmitted or reflected energy for a set of wavelengths) is known, but (some of) the parameters producing this response must be estimated. In *design* problems, we have a "target response" but there is no *a priori* knowledge of whether or not it is possible to obtain that response for some set of boundary conditions. In *reverse optical* problems, the response of the structure has already been measured but the specific conditions under which it has been generated are not known. The key tool for an efficient solution of *design* and *reverse engineering* problems is an adequate code that solves the optical response problem. All computational devices used to solve those problems obey the following general scheme: (i) try a set of parameters; (ii) solve the optical response problems using them; (iii) accept the parameters if the response is satisfactory and change them in a suitable way if it is not. This is also the scheme of most optimization algorithms, which try to minimize (or maximize) some objective function subject to a set of constraints. The choice of the objective (or merit) function to be used depends on the nature of the problem. Smoothness considerations usually lead to the use of some kind of sum of squares of experimental data.

Some features complicating the optimization of *design* and *reverse engineering* problems are:

(1) *Local nonglobal minimizers*. In some problems, the solution is well defined but, out of this solution, the merit function presents a highly oscillatory behavior with similar functional values. In other words, the function has many local (nonglobal) minimizers which are useless for design or estimation purposes. Most efficient optimization algorithms have guaranteed convergence to local minimizers (in fact, something less than that) but not to global minimizers, so they tend to get stuck in the attraction basin of those undesirable local minimizers.

(2) *Ill-posedness*. Some mathematical problems can be highly underdetermined, the solution not being unique. Adding to this feature the fact that additional errors due to small inadequacies of the model can be present, we have situations where an infinite set of mathematical approximate solutions can be found, but where the *true physical solution* may not correspond to the point with smallest merit function value. This feature is typical in the so-called *inverse problems* in the estimation literature. The ill-posedness problem can be partially surmounted

by introducing, in the model, some prior information on the behavior of the parameters to be estimated. When our knowledge of this behavior is poor, we can use regularization processes [7]. Finally, instead of the *true* expensive models, simplified direct algorithms can be used which, sometimes, are sufficient to find suitable estimates of the parameters, or a good enough approximation to them.

(3) *Expensive models*. Problems (1) and (2) become more serious if the solution of the associated direct problem is computationally expensive. Optimization algorithms need, essentially, trial-and-error evaluations of the optical response and, obviously, if this evaluation demands a great deal of computer time, the possibility of obtaining good practical solutions in a reasonable period of time is severely diminished. Partial solutions to all the difficulties mentioned above have been found. The multiple-local-minimizers problem can be attacked by the use of *global optimization algorithms* that, in principle, are able to jump over undesirable local nonglobal solutions. However, all these procedures are computationally expensive, and their effectiveness is linked to the possibility of overcoming this limitation.

For the sake of completeness, let us note that all these remedies may fail. No global optimization method guarantees the finding of global minimizers of any function in a reasonable time. The information that must be introduced into the problem to enhance the probability of finding a good estimate of the parameters is not always evident. Moreover, it is not always clear in which way the information must be introduced in the model. Finally, it is not known how much an optical response problem can be simplified without destroying its essential characteristics. As in many inverse problems exhaustively analyzed in the literature, each particular situation must be studied from scratch (although, of course, experience in similar situations is quite useful) and can demand original solutions. In this review, we consider the reverse solution of some simple optical structures.

The structure shown in Figure 1 calls for some additional considerations.

(1) In principle, the angle of incidence of the impinging beam may be any angle, the simplest case being that of normal incidence. The mathematical formulation of the problem for light arriving at angles other than normal to the surface of the coatings becomes more involved. The general case is treated in a coming section.

(2) The coatings and the substrate are optically homogeneous and isotropic. The surfaces and the interfaces between layers are perfectly flat and parallel to each other. In other words, we will not consider the case of rough surfaces producing some scattering of the light, nor the existence of thickness gradients or inhomogeneities. The effects produced by these deviations to the ideal case will not be discussed. See [8, 9].

(3) In this chapter, we review the properties and applications of semiconductor and dielectric thin films; that is, metal coatings are not included.

1.3. Light–Matter Interaction

Dielectric and semiconductor coatings are films of materials having strong ionic or directed covalent bonds. In most cases, they are transparent to visible and/or infrared light. The interaction of the electromagnetic radiation with these films is treated by applying boundary conditions to the solutions of Maxwell equations at the boundary between different media. In the field of optical coatings, the wavelength of the light is always very much larger than interatomic dimensions. Thus the interaction of light and matter is averaged over many unit cells. As a consequence, the optical properties within each layer can be described macroscopically in terms of phenomenological parameters, the so-called optical constants or optical parameters. As shown below, these are the real and the imaginary parts of a complex index of refraction $\widetilde{n}$. The real part, $n(\lambda)$, is the ratio of the velocity of light in vacuum to the velocity of light of wavelength (λ) in the material. The imaginary part, $-\kappa(\lambda)$, is an attenuation coefficient measuring the absorption of light with distance. Using Maxwell equations, it is possible to relate these frequency-dependent "constants" to other optical parameters such as the dielectric constant and the conductivity.

The coating materials are composed of charged particles: bound and conduction electrons, ionic cores, impurities, etc. These particles move differently with oscillating electric fields, giving rise to polarization effects. At visible and infrared light frequencies, the only contribution to polarization comes from the displacement of the electron cloud, which produces an induced dipole moment. At frequencies smaller than these, other contributions may appear, but they are of no interest to the purpose of the present work. The parameters describing these optical effects, that is, the dielectric constant ε, the dielectric susceptibility χ, and the conductivity σ, can be treated as scalars for isotropic materials. In what follows, dielectric and semiconductor coating materials are considered nonmagnetic and have no extra charges other than those bound in atoms.

To find out what kind of electromagnetic waves exist inside dielectric films, we take $\rho = -\nabla \cdot \mathbf{P}$ and $\mathbf{j} = \partial\mathbf{P}/\partial t$, where ρ is an effective charge, $\mathbf{P}$ is the polarization induced by the electromagnetic wave, assumed to be proportional to the electric field, and $\mathbf{j}$ is the corresponding current density averaged over a small volume. Under these conditions, average field Maxwell equations in MKS units read:

$$\nabla \cdot \mathbf{E} = -\frac{\nabla \cdot \mathbf{P}}{\varepsilon_0}$$

$$\nabla \times \mathbf{E} = -\frac{\partial \mathbf{B}}{\partial t}$$

$$\nabla \cdot \mathbf{B} = 0$$

$$c^2 \nabla \times \mathbf{B} = \frac{\partial}{\partial t}\left(\frac{\mathbf{P}}{\varepsilon_0} + \mathbf{E}\right)$$

where the symbols have their usual meaning. Note that the normal component of the electric field $\mathbf{E}$ is not conserved at the interface between materials of different polarizability. Instead, $\mathbf{D} = \varepsilon_0 + \mathbf{P}$, called electrical displacement, is conserved across such interfaces. The solutions to these equations have the form of harmonic plane waves with wave vector k:

$$\mathbf{E} = \mathbf{E}_0 \exp\left[i(\omega t - k \cdot r)\right]$$

$$\mathbf{H} = \mathbf{H}_0 \exp\left[i(\omega t - k \cdot r)\right]$$

and represent a wave traveling with a phase velocity $\omega/k = c/n$, where c is the speed of light in vacuum and n is the index of refraction. When optical absorption is present, the wave vector and the index are complex quantities. From the Maxwell equations, a dispersion relation $k^2 = \varepsilon(\omega/c)^2$ is obtained relating the time variation with the spatial variation of the perturbation.

In general, then, the wave vector k and the dielectric constant ε are complex quantities, that is, $\widetilde{k} = k_1 - ik_2$ and $\widetilde{\varepsilon} = \varepsilon_1 - i\varepsilon_2$. It is useful to define a complex index of refraction:

$$\widetilde{n} \equiv \widetilde{k}\left(\frac{c}{\omega}\right) = n - i\kappa$$

For isotropic materials, k_1 and k_2 are parallel and

$$\varepsilon_1 = n^2 - \kappa^2 \qquad \varepsilon_2 = 2n\kappa$$

with the converse equations,

$$n^2 = \frac{1}{2}\left[\varepsilon_1 + \left(\varepsilon_1^2 + \varepsilon_2^2\right)^{1/2}\right] \qquad \kappa^2 = \frac{1}{2}\left[-\varepsilon_1 + \left(\varepsilon_1^2 + \varepsilon_2^2\right)^{1/2}\right]$$

In the photon energy region where ε is real, $n = \varepsilon^{1/2}$ is also real and the phase (ω/k) and group ($\partial\omega/\partial k$) velocities are equal to c/n. In general, the velocity is reduced to $v(\lambda) = 1/\sqrt{\varepsilon_c(\lambda)}$ in a medium of a complex dielectric constant ε_c. The real part of n determines the phase velocity of the light wave, the imaginary part determining the spatial decay of its amplitude. The absorption coefficient α measures the intensity loss of the wave. For a beam traveling in the z direction, $I(x) = I(0)\exp(-\alpha z)$, which means $\alpha = 2\omega\kappa/c = 4\kappa\kappa/\lambda$. See [8].

Optical Properties of Dielectric Films

The main absorption process in semiconductors and dielectrics originates from the interaction of light with electrons [10–12]. If the photon has a frequency such that its energy matches the energy needed to excite an electron to a higher allowed state, then the photon may be absorbed. The electron may be an ion-core electron or a free electron in the solid. If the energy of the incoming photon does not match the required excitation energy, no excitation occurs and the material is transparent to such radiation. In nonmetal solids, there is a minimum energy separating the highest filled electron states (valence band) and the lowest empty ones (conduction band), known as the energy band-gap. Electron transitions from band to band constitute the strongest source of absorption. In dielectrics, such as glass, quartz, some salts, diamond, many metal oxides, and most plastic materials, no excitation resonances exist in the visible spectrum, because the valence electrons are so tightly bound that photons with energy in the ultraviolet are necessary to free them. Ideally, photons having an energy smaller than the band-gap are not absorbed. On the contrary, metals are good reflectors and are opaque to visible and infrared radiation. Both effects derive from the existence of a large density of free electrons which are able to move so freely that no electric field may propagate within the solid. The reflecting properties of metals find numerous optical applications, as in mirrors.

The optical properties of semiconductors lie between those of metals and those of insulating dielectrics. Semiconductors are normally transparent in the near infrared and absorbing in the visible spectrum, whereas the absorption in dielectrics is strong in the ultraviolet. Thus, the fundamental absorption edge of semiconductors lies, approximately, between 0.5 eV ($\lambda \sim$ 2500 nm) and 2.5 eV ($\lambda \sim 500$ nm). Within a small energy range around the fundamental absorption edge, semiconductors go, ideally, from high transparency to complete opacity. However, the presence of impurities, free conduction electrons or holes, and/or other defect states may affect the transparency of semiconductor and dielectric materials at photon energies smaller than the band-gap.

The situation just depicted is ideal in the sense that we have considered *perfect* dielectric and semiconductor crystalline materials. In most cases, however, the deposited optical coatings are either microcrystalline or amorphous. Microcrystalline layers are formed by the aggregation of randomly oriented small crystals having a typical size of the order of the film thickness. The regions between crystallites, or grain boundaries, are highly imperfect regions having a large density of localized electron states located at energies between the valence and conduction bands. These surface or interface defect states constitute a source of absorption; that is, electrons are excited from the valence band to localized states and/or from them to the conduction band. In all cases, however, the absorption due to defects is smaller than that of the fundamental edge, but it may affect in a nonnegligible way the optical properties of the coating.

The fundamental characteristic of an amorphous material is the topological disorder. The atoms of an amorphous network do not form a perfect periodic array as in crystalline materials. The absence of periodicity, or of long-range order, has important consequences on the optical properties of the films. Disorder induces the appearance of localized electron states within the band-gap. These localized states are of two types: coordination defects, also called dangling bonds or deep defects, and valence and conduction band tail states (shallow states). Both contribute to the absorption of light. Most important, due to the lack of periodicity, the electron wave vector $\vec{k}$ is no longer a good quantum number and the selection rules, valid for optical transitions in crystals, break down. As a consequence, more transitions are allowed between electron states with an energy separation equal to the photon energy. The absorption coefficient due to-band-to band transitions is much larger in amorphous dielectrics and semiconductors than in their crystalline parents. The density of defect states and the corresponding optical absorption strength depend strongly on deposition methods and conditions.

The Absorption Edge of Semiconductors and Dielectrics

In general, the process of electron excitation by photon absorption is very selective. The sharpness of absorption lines is clearly noted in the dark lines of atomic spectra. The requirement of energy matching, called resonance, must also be met in solids, but solids have such an abundant range of possible excitations that absorption occurs in a broad energy range. Even so, many of the optical properties of materials around resonances can be understood in terms of the properties of a damped harmonic oscillator, often called the Lorentz oscillator. In the model, it is assumed that the electrons are bound to their cores by harmonic forces. In the presence of an external field $E_0 \exp(-i\omega t)$, their motion is given by

$$m\frac{\partial^2 r}{\partial t^2} + m\gamma\frac{\partial r}{\partial t} + m\omega_0^2 r = -eE\exp(-i\omega t)$$

where m and e are the mass and the charge of the electron, ω_0 is the natural frequency of the oscillator, and γ is a damping term. Solving for the electron displacement $r = r_0\exp(-i\omega t)$, one gets

$$r = \frac{-eE/m}{(\omega_0^2 - \omega^2) - i\gamma\omega}$$

The response of a system having a density N/V of such oscillators can be obtained in terms of the induced polarization:

$$P = -erN/V = \frac{(e^2 N/mV)E}{(\omega_0^2 - \omega^2) - i\gamma\omega}$$

The dielectric constant corresponding to this polarization is

$$\widetilde{\varepsilon} = \varepsilon_1 - i\varepsilon_2 = 1 - \text{const} \cdot \frac{1}{(\omega_0^2 - \omega^2 - i\gamma\omega)}$$

with

$$\varepsilon_1 = n_r^2 - \kappa^2 = 1 - K\frac{(\omega_0^2 - \omega^2)}{(\omega_0^2 - \omega^2)^2 + \gamma^2\omega^2}$$
$$\varepsilon_2 = 2n_r\kappa = K\frac{\gamma\omega}{(\omega_0^2 - \omega^2)^2 + \gamma^2\omega^2}$$

where K is a constant. Figure 2 shows the shape of the real and imaginary parts of the dielectric constant of a Lorentz oscillator. This model is extremely powerful because, in real systems, it describes quite accurately not only the absorption between bands, but also that corresponding to free electrons and phonon systems. The reason for the general applicability of the Lorentz oscillator is that the corresponding quantum-mechanical equation for ε, not discussed here, has the same form as the classical equation for the damped oscillator.

In a solid dielectric, resonances at different frequencies having different loss values contribute to the dielectric response with terms having a shape similar to that shown in Figure 2. The overall dielectric response results from the addition of all these contributions, as shown in Figure 3 for a hypothetical semiconductor. For photon energies smaller than the band-gap ($\hbar\omega < E_G$), the material is transparent or partially absorbing. This is the *normal dispersion* frequency range in which both

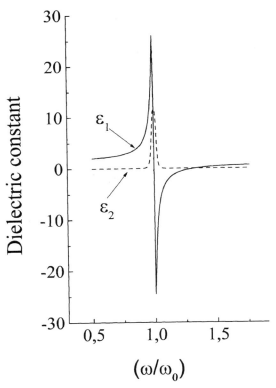

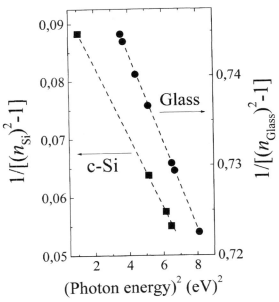

Fig. 4. Single-effective-oscillator representation of the index of refraction of crystalline silicon and of an optical glass as a function of the square of photon energy.

Fig. 2. Characteristic behavior of the real and the imaginary part of the dielectric constant of a classical damped electron oscillator as a function of the relative excitation frequency (Lorentz oscillator).

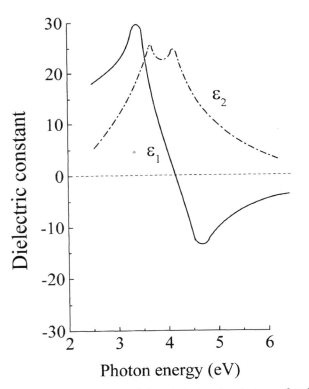

Fig. 3. Real and imaginary parts of a hypothetical semiconductor as a function of photon energy.

the index of refraction and the absorption increase with photon energy. Moreover, the slope of $\partial n/\partial(\hbar\omega)$ also increases with increasing photon energy. It is important to remark at this point that the refractive index dispersion data below the interband absorption edge of a large quantity of solids can be plotted using a single-effective-oscillator fit of the form $n^2 - 1 = F_d F_0/(F_0^2 - \hbar^2\omega^2)$, where F_0 is the single-oscillator energy, and F_d is the dispersion energy [13]. These parameters obey simple empirical rules in large groups of covalent and ionic materials. Figure 4 shows the single-effective-oscillator representation of the refractive index of an optical glass and of crystalline silicon. The fit is very good for a wide photon energy range.

Let us end the section with a few words on the shape of the absorption edge in crystalline and amorphous layers. The absorption coefficient has an energy dependence near the absorption edge expressed as $\alpha \approx (\hbar\omega - E_G)^s$, where s is a constant. In the one-electron approximation, the values taken by s depend on the selection rules for optical transitions and on the material's band structure. For allowed transitions, $s = 1/2$ for direct transitions (i.e., GaAs) and $s = 2$ for indirect transitions (i.e., Si or Ge). In amorphous materials, the absorption edge always displays an exponential dependence on photon energy. The characteristic energy of the exponential edge, called Urbach energy, depends on the material and on deposition conditions [14, 15].

For reasons that will become clear in a forthcoming section, knowledge of the physics behind the dependence of the optical constants on photon energy may be of great help in solving optical reverse engineering problems.

1.4. Basic Formulae for Transmitted and Reflected Waves

1.4.1. Single Interface

Consider the complex wave function

$$\widetilde{u}_I(z,t) = \widetilde{E}_I \exp\left[i\left(\omega t - \widetilde{k}_0 z\right)\right] \tag{1}$$

where ω is real and nonnegative, whereas $\widetilde{E}_I$ and $\widetilde{k}_0$ are complex. By (1), the wavelength λ is $2\pi/|\Re(\widetilde{k}_0)|$. ($\Re(z)$ will denote always the real part of the complex number z and $\Im(z)$ denotes its imaginary part.) So,

$$\left|\Re\left(\widetilde{k}_0\right)\right| = \frac{2\pi}{\lambda} \tag{2}$$

Let the complex number

$$\widetilde{n}_0 = n_0 - i\kappa_0$$

denote the generalized index of refraction in the $x-$ domain of u, and

$$\widetilde{k}_0 = \pm\frac{\omega\widetilde{n}_0}{c} \tag{3}$$

If $\Re(k) > 0$, the wave travels in the forward direction, whereas if $\Re(k) < 0$, the wave goes backwards.

Assume now that (1) is an "incident" wave, defined for $z \leq L$, arriving at an abrupt interface (xy plane) between two different media at $z = L$. As sketched in Figure 5, this wave generates a "transmitted" wave $\widetilde{u}_T$, defined for $z \geq L$, and a "reflected" wave $\widetilde{u}_R$, defined for $z \leq L$. Let us write

$$\widetilde{u}_T(z,t) = \widetilde{E}_T \exp\left[i\left(\omega_T t - \widetilde{k}_T z\right)\right] \tag{4}$$

and

$$\widetilde{u}_R(z,t) = \widetilde{E}_R \exp\left[i\left(\omega_R t - \widetilde{k}_R z\right)\right] \tag{5}$$

Assume that the complex number $\widetilde{n}_1$ is the generalized index of refraction of the layer $z > L$.

Continuity in $z = L$ imposes:

$$\widetilde{E}_I \exp\left[i\left(\omega t - \widetilde{k}_0 L\right)\right] + \widetilde{E}_R \exp\left[i\left(\omega_R t - \widetilde{k}_R L\right)\right]$$
$$= \widetilde{E}_T \exp\left[i\left(\omega_T t - \widetilde{k}_T L\right)\right]$$

for all real t. So,

$$\widetilde{E}_I \exp\left(-i\widetilde{k}_0 L\right)\exp[i\omega t] + \widetilde{E}_R \exp\left(-i\widetilde{k}_R L\right)\exp[i\omega_R t]$$
$$= \widetilde{E}_T \exp\left(-i\widetilde{k}_T L\right)\exp[i\omega_T t]$$

for all $t \in \mathbb{R}$. Since this identity is valid for all t, we must have

$$\omega_R = \omega_T = \omega$$

Now, the direction of the reflected wave must be opposite to that of the incident wave, and the direction of the transmitted wave must be the same. So, by (3),

$$\widetilde{k}_R = -\widetilde{k}_0 \qquad \widetilde{k}_T = \frac{\widetilde{n}_1}{\widetilde{n}_0}\widetilde{k}_0$$

Therefore,

$$\widetilde{E}_I \exp\left(-i\widetilde{k}_0 L\right) + \widetilde{E}_R \exp\left(i\widetilde{k}_0 L\right) = \widetilde{E}_T \exp\left(-i\frac{\widetilde{n}_1}{\widetilde{n}}\widetilde{k}_0 L\right) \tag{6}$$

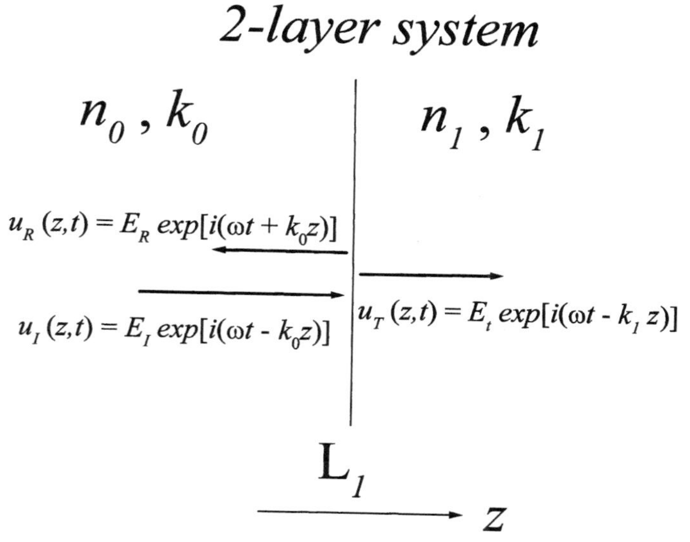

2-layer system

$$n_0, k_0 \qquad n_1, k_1$$

$$u_R(z,t) = E_R \exp[i(\omega t + k_0 z)]$$

$$u_I(z,t) = E_I \exp[i(\omega t - k_0 z)] \quad | \quad u_T(z,t) = E_t \exp[i(\omega t - k_1 z)]$$

$$L_1$$

$$\longrightarrow z$$

Fig. 5. Incident, reflected, and transmitted waves at an abrupt interface between two different media.

Now, using continuity of the derivative with respect to z in $z = L$, and the above expressions of $\omega_R, \omega_T, \widetilde{k}_R, \widetilde{k}_T$, we have

$$\left(-i\widetilde{k}_0\right)\widetilde{E}_I \exp\left[i\left(\omega t - \widetilde{k}_0 z\right)\right] + \left(i\widetilde{k}_0\right)\widetilde{E}_R \exp\left[i\left(\omega t + \widetilde{k}_0 z\right)\right]$$
$$= -i\widetilde{k}\frac{\widetilde{n}_1}{\widetilde{n}_0}\widetilde{E}_T \exp\left[i\left(\omega t - \frac{\widetilde{n}_1}{\widetilde{n}_0}\widetilde{k}_0 z\right)\right]$$

for $z = L$ and for all $t \in \mathbb{R}$. Therefore,

$$\widetilde{E}_I \exp\left[i\left(\omega t - \widetilde{k}_0 L\right)\right] - \widetilde{E}_R \exp\left[i\left(\omega t + \widetilde{k}_0 L\right)\right]$$
$$= \frac{\widetilde{n}_1}{\widetilde{n}_0}\widetilde{E}_T \exp\left[i\left(\omega t - \frac{\widetilde{n}_1}{\widetilde{n}_0}\widetilde{k}_0 L\right)\right]$$

Thus,

$$\widetilde{E}_I \exp\left(-i\widetilde{k}_0 L\right)\exp[i\omega t] - \widetilde{E}_R \exp\left(i\widetilde{k}_0 L\right)\exp[i\omega t]$$
$$= \frac{\widetilde{n}_1}{\widetilde{n}_0}\widetilde{E}_T \exp\left[-i\frac{\widetilde{n}_1}{\widetilde{n}_0}\widetilde{k}_0 L\right]\exp[i\omega t]$$

which implies

$$\widetilde{E}_I \exp\left(-i\widetilde{k}_0 L\right) - \widetilde{E}_R \exp\left(i\widetilde{k}_0 L\right) = \frac{\widetilde{n}_1}{\widetilde{n}_0}\widetilde{E}_T \exp\left[-i\frac{\widetilde{n}_1}{\widetilde{n}_0}\widetilde{k}_0 L\right] \tag{7}$$

Adding (6) and (7), we get

$$2\widetilde{E}_I \exp\left(-i\widetilde{k}_0 L\right) = \left[1 + \frac{\widetilde{n}_1}{\widetilde{n}_0}\right]\widetilde{E}_T \exp\left[-i\frac{\widetilde{n}_1}{\widetilde{n}_0}\widetilde{k}_0 L\right]$$

Therefore,

$$\left[1 + \frac{\widetilde{n}_1}{\widetilde{n}_0}\right]\widetilde{E}_T = 2\widetilde{E}_I \exp\left[-i\widetilde{k}_0 L\left(1 - \frac{\widetilde{n}_1}{\widetilde{n}_0}\right)\right]$$

so,

$$\widetilde{E}_T = \frac{2\widetilde{n}}{\widetilde{n} + \widetilde{n}_1}\widetilde{E}_I \exp\left[i\widetilde{k}_0 L\left(\frac{\widetilde{n}_1}{\widetilde{n}_0} - 1\right)\right] \tag{8}$$

Subtracting (7) from (6), we obtain

$$2\widetilde{E}_R \exp(i\widetilde{k}_0 L) = \left(1 - \frac{\widetilde{n}_1}{\widetilde{n}_0}\right)\widetilde{E}_T \exp\left[-i\frac{\widetilde{n}_1}{\widetilde{n}_0}\widetilde{k}_0 L\right]$$
$$= \left(1 - \frac{\widetilde{n}_1}{\widetilde{n}_0}\right)\frac{2\widetilde{n}_0}{\widetilde{n}_0 + \widetilde{n}_1}\widetilde{E}_I \exp(-i\widetilde{k}_0 L)$$

So,

$$\widetilde{E}_R = \frac{\widetilde{n}_0 - \widetilde{n}_1}{\widetilde{n}_0 + \widetilde{n}_1}\widetilde{E}_I \exp(-2i\widetilde{k}_0 L) \qquad (9)$$

Summing up:

If the incident wave acting on $z < L$ is

$$\widetilde{u}_I(z,t) = \widetilde{E}_I \exp\left[i\left(\omega t - \widetilde{k}_0 z\right)\right]$$

then the reflected wave ($z < L$) is

$$\widetilde{u}_R(z,t) = \frac{\widetilde{n}_0 - \widetilde{n}_1}{\widetilde{n}_0 + \widetilde{n}_1}\widetilde{E}_I \exp(-2i\widetilde{k}_0 L)\exp\left[i\left(\omega t + \widetilde{k}_0 z\right)\right] \quad (10)$$

and the transmitted wave ($z > L$) is

$$\widetilde{u}_T(z,t) = \frac{2\widetilde{n}_0}{\widetilde{n}_0 + \widetilde{n}_1}\widetilde{E}_I \exp\left[i\widetilde{k}_0 L\left(\frac{\widetilde{n}_1}{\widetilde{n}_0} - 1\right)\right]$$
$$\times \exp\left[i\left(\omega t - \frac{\widetilde{n}_1}{\widetilde{n}_0}\widetilde{k}_0 z\right)\right] \qquad (11)$$

1.4.2. Recursive Formulae for m Layers

Assume now, as shown in Figure 6, a system of m layers with generalized index of refraction $\widetilde{n}_0, \widetilde{n}_1, \ldots, \widetilde{n}_{m-1}$, respectively, separated by abrupt interfaces at $z = L_1, z = L_2, \ldots,$ $z = L_{m-1}$, where $L_1 < L_2 < \cdots < L_{m-1}$. We identify each layer with its refractive index. The incident wave (defined for $z < L_1$) in the layer $\widetilde{n}_0$ is given by (1). The incident wave generates multiple reflected and transmitted waves in all the layers, as sketched in Figure 7. It is easy to verify that the ω-coefficient of all these generated waves is ω. Moreover, in the layer $\widetilde{n}_{m-1}$ ($z > L_{m-1}$), the k-coefficient of the transmitted wave will be $\widetilde{k}_0\widetilde{n}_{m-1}/\widetilde{n}_0$. The incident wave will also generate a reflected wave in $z < L_1$ whose k-coefficient will be $-\widetilde{k}_0$.

Mutatis mutandi, if the incident wave is defined in $z > L_{m-1}$ and given by $E_I \exp[i(\omega t + \widetilde{k}_{m-1} z)]$, the transmitted wave in $z < L_1$ has a k-coefficient equal to $\widetilde{k}_{m-1}\widetilde{n}_0/\widetilde{n}_{m-1}$ and the reflected wave in $z > L_{m-1}$ will have $-\widetilde{k}_{m-1}$ as k-coefficient. The ω-coefficient is always the same.

In this section, we deduce how the E-coefficient of the transmitted and reflected waves depends on the parameters of the incident wave. We define the coefficients $t_{0,m-1}, r_{0,m-1}, t_{m-1,0}, r_{m-1,0}$ in the following way:

(a) $t_{0,m-1}\widetilde{E}$ is the E-coefficient of the transmitted wave in the $\widetilde{n}_{m-1}$ layer generated by the wave (1) in layer $\widetilde{n}_0$.

(b) $r_{0,m-1}\widetilde{E}$ is the E-coefficient of the reflected wave in the $\widetilde{n}_0$ layer generated by the wave (1) in layer $\widetilde{n}_0$.

(c) $t_{m-1,0}\widetilde{E}$ is the E-coefficient of the transmitted wave in the layer $\widetilde{n}_0$, generated by a wave (1) defined in layer $\widetilde{n}_{m-1}$.

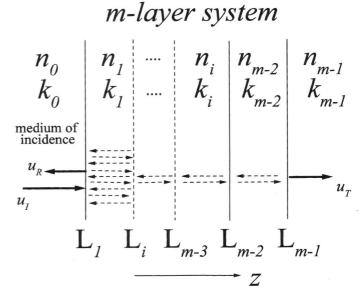

Fig. 6. Structure of an m-layered optical film system. The figure illustrates multiple reflections inside each layer characterized by a refractive index and an extinction coefficient.

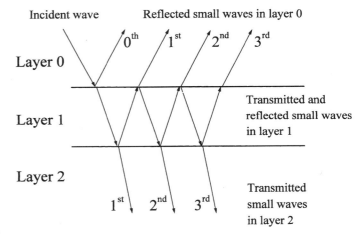

Fig. 7. Sketch of the multiple transmitted and reflected waves in a multilayer system.

(d) $r_{m-1,0}\widetilde{E}$ is the E-coefficient of the reflected wave in the $\widetilde{n}_{m-1}$ layer generated by a wave (1) defined in layer $\widetilde{n}_{m-1}$.

For completeness, we will write, when necessary:

$$t_{0,m-1} = t_{0,m-1}(\widetilde{k})$$
$$= t_{0,m-1}(\widetilde{k}_0, \widetilde{n}_0, \widetilde{n}_1, \ldots, \widetilde{n}_{m-1}, L_1, L_2, \ldots, L_{m-1})$$
$$r_{0,m-1} = r_{0,m-1}(\widetilde{k})$$
$$= r_{1,m}(\widetilde{k}_0, \widetilde{n}_0, \widetilde{n}_1, \ldots, \widetilde{n}_{m-1}, L_1, L_2, \ldots, L_{m-1})$$
$$t_{m-1,0} = t_{m-1,0}(\widetilde{k})$$
$$= t_{m,1}(\widetilde{k}_{m-1}, \widetilde{n}_0, \widetilde{n}_1, \ldots, \widetilde{n}_{m-1}, L_1, L_2, \ldots, L_{m-1})$$
$$r_{m-1,0} = r_{m-1,0}(\widetilde{k})$$
$$= r_{m-1,0}(\widetilde{k}_{m-1}, \widetilde{n}_0, \widetilde{n}_1, \ldots, \widetilde{n}_{m-1}, L_1, L_2, \ldots, L_{m-1})$$

From the previous section, we know that

$$t_{0,1}(\widetilde{k}) = \frac{2\widetilde{n}_0}{\widetilde{n}_0 + \widetilde{n}_1} \exp\left[\frac{i\widetilde{k}_0 L_1(\widetilde{n}_1 - \widetilde{n}_0)}{\widetilde{n}_0}\right] \qquad (12)$$

$$r_{0,1}(\widetilde{k}) = \frac{\widetilde{n}_0 - \widetilde{n}_1}{\widetilde{n}_0 + \widetilde{n}_1} \exp\left[-2i\widetilde{k}_0 L_1\right] \qquad (13)$$

$$t_{1,0}(\widetilde{k}) = \frac{2\widetilde{n}_1}{\widetilde{n}_0 + \widetilde{n}_1} \exp\left[\frac{i\widetilde{k}_1 L_1(\widetilde{n}_0 - \widetilde{n}_1)}{\widetilde{n}_1}\right] \qquad (14)$$

$$r_{1,0}(\widetilde{k}) = \frac{\widetilde{n}_1 - \widetilde{n}_0}{\widetilde{n}_1 + \widetilde{n}_0} \exp\left[-2i\widetilde{k}_1 L_1\right] \qquad (15)$$

Computing $t_{0,m-1}$

Let us compute $t_{0,m-1}$ for $m \geq 3$ assuming that $t_{0,m-2}$, $r_{0,m-2}$, $t_{m-2,0}$, $r_{m-2,0}$ have already been computed.

Consider an incident wave $\widetilde{E}_0 \exp[i(\omega t - \widetilde{k}_0 z)]$ in layer $\widetilde{n}_0$, defined for $z < L_1$. The transmitted wave in $z > L_{m-1}$ will be the sum of an infinite number of "small-waves" resulting from partial transmission and partial reflection at the interfaces. The k-coefficient of all these transmitted small-waves will be given by

$$k \text{ of the "small-waves"} = \frac{\widetilde{k}_0 \widetilde{n}_{m-1}}{\widetilde{n}_0}$$

The E-coefficient of the first small-wave comes from a direct transmission from layer $\widetilde{n}_0$ to layer $\widetilde{n}_{m-2}$ (through L_1, L_2, L_{m-2}) and, finally transmission through $z = L_{m-1}$. Therefore,

$$E \text{ of the 1st small-wave}$$
$$= t_{0,m-2}\widetilde{E}_0 \frac{2\widetilde{n}_{m-2}}{\widetilde{n}_{m-2} + \widetilde{n}_{m-1}}$$
$$\times \exp\left[i\widetilde{k}_0 L_{m-1} \frac{\widetilde{n}_{m-2}}{\widetilde{n}_0}\left(\frac{\widetilde{n}_{m-1}}{\widetilde{n}_{m-2}} - 1\right)\right]$$

The E-coefficient of the second small-wave comes from calculating the transmission from layer $\widetilde{n}_0$ to layer $\widetilde{n}_{m-2}$ (through L_1, L_2, L_{m-2}) [reflection in $z = L_{m-1}$, reflection in $[L_{m-2}, \ldots, L_1]$] and, finally, transmission through $z = L_{m-1}$.

Therefore,

$$E \text{ of the 2nd small-wave}$$
$$= t_{0,m-2}\widetilde{E}_0\left\{\frac{\widetilde{n}_{m-2} - \widetilde{n}_{m-1}}{\widetilde{n}_{m-2} + \widetilde{n}_{m-1}} \exp\left[-2i\frac{\widetilde{k}_0 \widetilde{n}_{m-2}}{\widetilde{n}_0} L_{m-1}\right]\right.$$
$$\left. \times r_{m-2,0}\left(-\frac{\widetilde{k}_0 \widetilde{n}_{m-2}}{\widetilde{n}_0}\right)\right\}$$
$$\times \frac{2\widetilde{n}_{m-2}}{\widetilde{n}_{m-2} + \widetilde{n}_{m-1}} \exp\left[\frac{i\widetilde{k}_0 L_{m-1}(\widetilde{n}_{m-1} - \widetilde{n}_{m-2})}{\widetilde{n}_0}\right]$$

Proceeding in the same way for all $\nu > 2$, we have

$$E \text{ of the } \nu\text{th small-wave}$$
$$= t_{0,m-2}\widetilde{E}_0\left\{\frac{\widetilde{n}_{m-2} - \widetilde{n}_{m-1}}{\widetilde{n}_{m-2} + \widetilde{n}_{m-1}} \exp\left[-2i\frac{\widetilde{k}_0 \widetilde{n}_{m-2}}{\widetilde{n}_0} L_{m-1}\right]\right.$$
$$\left. \times r_{m-2,0}\left(-\frac{\widetilde{k}_0 \widetilde{n}_{m-2}}{\widetilde{n}_0}\right)\right\}^{\nu-1}$$

$$\times \frac{2\widetilde{n}_{m-2}}{\widetilde{n}_{m-2} + \widetilde{n}_{m-1}} \exp\left[\frac{i\widetilde{k}_0 L_{m-1}(\widetilde{n}_{m-1} - \widetilde{n}_{m-2})}{\widetilde{n}_0}\right]$$

Adding all the above small-waves, we obtain

$$t_{0,m-2}E_0 \frac{2\widetilde{n}_{m-2}}{\widetilde{n}_{m-2} + \widetilde{n}_{m-1}} \exp\left[\frac{i\widetilde{k}_0 L_{m-1}(\widetilde{n}_{m-1} - \widetilde{n}_{m-2})}{\widetilde{n}_0}\right]$$
$$\times \left\{1 - \frac{\widetilde{n}_{m-2} - \widetilde{n}_{m-1}}{\widetilde{n}_{m-2} + \widetilde{n}_{m-1}} \exp\left[-2i\frac{\widetilde{k}_0 \widetilde{n}_{m-2}}{\widetilde{n}_0} L_{m-1}\right]\right.$$
$$\left. \times r_{m-1,0}\left(-\frac{\widetilde{k}_0 \widetilde{n}_{m-2}}{\widetilde{n}_0}\right)\right\}^{-1}$$

Therefore,

$$t_{0,m-1} = t_{0,m-1}(\widetilde{k})$$
$$= t_{0,m-2} \frac{2\widetilde{n}_{m-2}}{\widetilde{n}_{m-2} + \widetilde{n}_{m-1}} \exp\left[\frac{i\widetilde{k}_0 L_{m-1}(\widetilde{n}_{m-1} - \widetilde{n}_{m-2})}{\widetilde{n}_0}\right]$$
$$\times \left\{1 - \frac{\widetilde{n}_{m-2} - \widetilde{n}_{m-1}}{\widetilde{n}_{m-2} + \widetilde{n}_{m-1}} \exp\left[-2i\frac{\widetilde{k}_0 \widetilde{n}_{m-2}}{\widetilde{n}_0} L_{m-1}\right]\right.$$
$$\left. \times r_{m-2,0}\left(-\frac{\widetilde{k}_0 \widetilde{n}_{m-2}}{\widetilde{n}_0}\right)\right\}^{-1} \qquad (16)$$

A brief explanation on notation becomes necessary. When we write, for example, $r_{m-2,0}(-(\widetilde{k}_0 \widetilde{n}_{m-2}/\widetilde{n}_0))$, this means "the function $r_{m-2,0}$ applied on $-(\widetilde{k}_0 \widetilde{n}_{m-2}/\widetilde{n}_0)$." When we write $t_{0,m-2}$, this means $t_{0,m-2}(\widetilde{k})$. When, say, t_{ij} is used without argument, it means $t_{ij}(\widetilde{k})$. Otherwise the argument is explicitly stated, as in the case of $r_{m-2,0}$ in the formula above.

Computing $r_{0,m-1}$

Assume, again, that the incident wave $\widetilde{E}_0 \exp[i(\omega t - \widetilde{k}_0 z)]$ is defined for $z < L_1$. The reflection in $[L_1, L_2, \ldots, L_{m-1}]$ generates a reflected wave, also defined in $z < L_1$. This wave is the sum of an infinite number of "small-waves" and an additional wave called here the zeroth-wave. The k-coefficient of all these waves is $-\widetilde{k}_0$.

The zeroth-wave is the first reflection of the incident wave at $[L_1, \ldots, L_{m-1}]$, so

$$E \text{ of the 0th wave} = r_{0,m-2}(\widetilde{k}_0)\widetilde{E}_0$$

The E-coefficient of the first small-wave comes from transmission through $L_1, \ldots, L_{m-2}$, followed by reflection at $z = L_{m-1}$ and, finally, transmission through $L_{m-2}, \ldots, L_1$. So,

$$E \text{ of the 1st small-wave}$$
$$= t_{0,m-2}\widetilde{E}_0 \frac{\widetilde{n}_{m-2} - \widetilde{n}_{m-1}}{\widetilde{n}_{m-2} + \widetilde{n}_{m-1}} \exp\left[-2i\frac{\widetilde{k}_0 \widetilde{n}_{m-2}}{\widetilde{n}_0} L_{m-1}\right]$$
$$\times t_{m-2,0}\left(-\frac{\widetilde{k}_0 \widetilde{n}_{m-2}}{\widetilde{n}_0}\right)$$

The E-coefficient of the second small-wave comes from transmission through $L_1, \ldots, L_{m-2}$, followed by reflection at $z = L_{m-1}$ [then reflection at $[L_{m-2}, \ldots, L_1]$, then reflection at $z = L_{m-1}$] and, finally, transmission through $L_{m-2}, \ldots, L_1$. So,

E of the 2nd small-wave

$$= t_{0,m-2}\widetilde{E}_0 \frac{\widetilde{n}_{m-2} - \widetilde{n}_{m-1}}{\widetilde{n}_{m-2} + \widetilde{n}_{m-1}} \exp\left[-2i\frac{\widetilde{k}_0\widetilde{n}_{m-2}}{\widetilde{n}_0}L_{m-1}\right]$$

$$\times \left\{ r_{m-2,0}\left(-\frac{\widetilde{k}_0\widetilde{n}_{m-2}}{\widetilde{n}_0}\right)\frac{\widetilde{n}_{m-2} - \widetilde{n}_{m-1}}{\widetilde{n}_{m-2} + \widetilde{n}_{m-1}} \right.$$

$$\left. \times \exp\left[-2i\frac{\widetilde{k}_0\widetilde{n}_{m-2}}{\widetilde{n}_0}L_{m-1}\right]\right\} t_{m-2,0}\left(-\frac{\widetilde{k}_0\widetilde{n}_{m-2}}{\widetilde{n}_0}\right)$$

So far, the E-coefficient of the νth small-wave comes from transmission through $L_1, \ldots, L_{m-2}$, then reflection at $z = L_{m-1}$ [then reflection at $[L_{m-2}, \ldots, L_1]$, then reflection at $z = L_{m-1}$] $\nu - 1$ times and, finally, transmission through $L_{m-2}, \ldots, L_1$. So,

E of the νth small-wave

$$= t_{0,m-2}\widetilde{E}_0 \frac{\widetilde{n}_{m-2} - \widetilde{n}_{m-1}}{\widetilde{n}_{m-2} + \widetilde{n}_{m-1}} \exp\left[-2i\frac{\widetilde{k}_0\widetilde{n}_{m-2}}{\widetilde{n}_0}L_{m-1}\right]$$

$$\times \left\{ r_{m-2,0}\left(-\frac{\widetilde{k}_0\widetilde{n}_{m-2}}{\widetilde{n}_0}\right)\frac{\widetilde{n}_{m-2} - \widetilde{n}_{m-1}}{\widetilde{n}_{m-2} + \widetilde{n}_{m-1}} \right.$$

$$\left. \times \exp\left[-2i\frac{\widetilde{k}_0\widetilde{n}_{m-2}}{\widetilde{n}_0}L_{m-1}\right]\right\}^{\nu-1} t_{m-2,0}\left(-\frac{\widetilde{k}_0\widetilde{n}_{m-2}}{\widetilde{n}_0}\right)$$

Therefore, the E-coefficient of the reflected wave defined in $z < L_1$ is

$$r_{0,m-2}(\widetilde{k})\widetilde{E}_0 + t_{0,m-2}\widetilde{E}_0\frac{\widetilde{n}_{m-2} - \widetilde{n}_{m-1}}{\widetilde{n}_{m-2} + \widetilde{n}_{m-1}}$$

$$\times \exp\left[-2i\frac{\widetilde{k}_0\widetilde{n}_{m-2}}{\widetilde{n}_0}L_{m-1}\right] t_{m-2,0}\left(-\frac{\widetilde{k}_0\widetilde{n}_{m-2}}{\widetilde{n}_0}\right)$$

$$\times \left\{ 1 - r_{m-2,0}\left(-\frac{\widetilde{k}_0\widetilde{n}_{m-2}}{\widetilde{n}_0}\right)\frac{\widetilde{n}_{m-2} - \widetilde{n}_{m-1}}{\widetilde{n}_{m-2} + \widetilde{n}_{m-1}} \right.$$

$$\left. \times \exp\left[-2i\frac{\widetilde{k}_0\widetilde{n}_{m-2}}{\widetilde{n}_0}L_{m-1}\right]\right\}^{-1}$$

So,

$$r_{0,m-1}(\widetilde{k}) = r_{0,m-2}(\widetilde{k}) + t_{0,m-2}\frac{\widetilde{n}_{m-2} - \widetilde{n}_{m-1}}{\widetilde{n}_{m-2} + \widetilde{n}_{m-1}}$$

$$\times \exp\left[-2i\frac{\widetilde{k}_0\widetilde{n}_{m-2}}{\widetilde{n}_0}L_{m-1}\right] t_{m-2,0}\left(-\frac{\widetilde{k}\widetilde{n}_{m-2}}{\widetilde{n}_0}\right)$$

$$\times \left\{ 1 - r_{m-2,0}\left(-\frac{\widetilde{k}_0\widetilde{n}_{m-2}}{\widetilde{n}_0}\right)\frac{\widetilde{n}_{m-2} - \widetilde{n}_{m-1}}{\widetilde{n}_{m-2} + \widetilde{n}_{m-1}} \right.$$

$$\left. \times \exp\left[-2i\frac{\widetilde{k}_0\widetilde{n}_{m-2}}{\widetilde{n}_0}L_{m-1}\right]\right\}^{-1} \qquad (17)$$

Computing $t_{m-1,0}$

Assume that the wave $\widetilde{E}_{m-1} \exp[i(\omega t + \widetilde{k}_{m-1}z)]$ is defined for $z > L_{m-1}$, where the refraction index is $\widetilde{n}_{m-1}$ (see Fig. 6). The k-coefficient of the transmitted wave defined for $z < L_1$, as

well as the k-coefficient of all the small-waves whose addition is the transmitted wave, will be $\widetilde{k}_{m-1}\widetilde{n}_0/\widetilde{n}_{m-1}$.

As in the previous cases, the transmitted wave is the sum of infinitely many small-waves.

The E-coefficient of the first small-wave is obtained by transmission through $z = L_{m-1}$ (from "right" to "left") followed by transmission through $[L_{m-2}, \ldots, L_1]$. So,

E of the 1st small-wave

$$= \widetilde{E}_{m-1}\frac{2\widetilde{n}_{m-1}}{\widetilde{n}_{m-1} + \widetilde{n}_{m-2}} \exp\left[\frac{i\widetilde{k}_{m-1}L_{m-1}(\widetilde{n}_{m-2} - \widetilde{n}_{m-1})}{\widetilde{n}_{m-1}}\right]$$

$$\times t_{m-2,0}\left(\frac{\widetilde{k}_{m-1}\widetilde{n}_{m-2}}{\widetilde{n}_{m-1}}\right)$$

The E-coefficient of the second small-wave is obtained by transmission through L_{m-1} (from right to left), followed by [reflection at $[L_{m-2}, \ldots, L_1]$, reflection at $z = L_{m-1}$] and, finally, transmission through $[L_{m-2}, \ldots, L_1]$. So,

E of the 2nd small-wave

$$= \widetilde{E}_{m-1}\frac{2\widetilde{n}_{m-1}}{\widetilde{n}_{m-1} + \widetilde{n}_{m-2}} \exp\left[\frac{i\widetilde{k}_{m-1}L_{m-1}(\widetilde{n}_{m-2} - \widetilde{n}_{m-1})}{\widetilde{n}_{m-1}}\right]$$

$$\times r_{m-2,0}\left(\frac{\widetilde{n}_{m-2}\widetilde{k}_{m-1}}{\widetilde{n}_{m-1}}\right)\frac{\widetilde{n}_{m-2} - \widetilde{n}_{m-1}}{\widetilde{n}_{m-2} + \widetilde{n}_{m-1}}$$

$$\times \exp\left(\frac{2i\widetilde{k}_{m-1}\widetilde{n}_{m-2}L_{m-1}}{\widetilde{n}_{m-1}}\right) t_{m-2,0}\left(\frac{\widetilde{k}_{m-1}\widetilde{n}_{m-2}}{\widetilde{n}_{m-1}}\right)$$

The E-coefficient of the νth small-wave is obtained by transmission through L_{m-1} (from right to left), followed by [reflection at $[L_{m-2}, \ldots, L_1]$, reflection at $z = L_{m-1}$] $\nu - 1$ times, and finally, transmission through $[L_{m-2}, \ldots, L_1]$. So,

E of the νth small-wave

$$= \widetilde{E}_{m-1}\frac{2\widetilde{n}_{m-1}}{\widetilde{n}_{m-1} + \widetilde{n}_{m-2}} \exp\left[\frac{i\widetilde{k}_{m-1}L_{m-1}(\widetilde{n}_{m-2} - \widetilde{n}_{m-1})}{\widetilde{n}_{m-1}}\right]$$

$$\times \left\{ r_{m-2,0}\left(\frac{\widetilde{n}_{m-2}\widetilde{k}_{m-1}}{\widetilde{n}_{m-1}}\right)\frac{\widetilde{n}_{m-2} - \widetilde{n}_{m-1}}{\widetilde{n}_{m-2} + \widetilde{n}_{m-1}} \right.$$

$$\left. \times \exp\left(\frac{2i\widetilde{k}_{m-1}\widetilde{n}_{m-2}L_{m-1}}{\widetilde{n}_{m-1}}\right)\right\}^{\nu-1} t_{m-1,1}\left(\frac{\widetilde{k}_{m-1}\widetilde{n}_{m-2}}{\widetilde{n}_{m-1}}\right)$$

Adding all the small-waves, we obtain

E of the transmitted wave in $z < L_1$

$$= \widetilde{E}_{m-1}\frac{2\widetilde{n}_{m-1}}{\widetilde{n}_{m-1} + \widetilde{n}_{m-2}} \exp\left[\frac{i\widetilde{k}_{m-1}L_{m-1}(\widetilde{n}_{m-2} - \widetilde{n}_{m-1})}{\widetilde{n}_{m-1}}\right]$$

$$\times t_{m-2,0}\left(\frac{\widetilde{k}_{m-1}\widetilde{n}_{m-2}}{\widetilde{n}_{m-1}}\right)$$

$$\times \left\{ 1 - r_{m-2,0}\left(\frac{\widetilde{n}_{m-2}\widetilde{k}_{m-1}}{\widetilde{n}_{m-1}}\right)\frac{\widetilde{n}_{m-2} - \widetilde{n}_{m-1}}{\widetilde{n}_{m-2} + \widetilde{n}_{m-1}} \right.$$

$$\left. \times \exp\left(\frac{2i\widetilde{k}_{m-1}\widetilde{n}_{m-2}L_{m-1}}{\widetilde{n}_{m-1}}\right)\right\}^{-1}$$

Therefore,

$$
t_{m-1,0}(\widetilde{k}) = \frac{2\widetilde{n}_{m-1}}{\widetilde{n}_{m-1} + \widetilde{n}_{m-2}} \exp\left[\frac{i\widetilde{k}_{m-1}L_{m-1}(\widetilde{n}_{m-2} - \widetilde{n}_{m-1})}{\widetilde{n}_{m-1}}\right]
$$

$$
\times\, t_{m-2,0}\left(\frac{\widetilde{k}_{m-1}\widetilde{n}_{m-2}}{\widetilde{n}_{m-1}}\right)
$$

$$
\times\, \left\{ 1 - r_{m-2,0}\left(\frac{\widetilde{n}_{m-2}\widetilde{k}_{m-1}}{\widetilde{n}_{m-1}}\right)\frac{\widetilde{n}_{m-2} - \widetilde{n}_{m-1}}{\widetilde{n}_{m-2} + \widetilde{n}_{m-1}} \right.
$$

$$
\left. \times \exp\left(\frac{2i\widetilde{k}_{m-1}\widetilde{n}_{m-2}L_{m-1}}{\widetilde{n}_{m-1}}\right) \right\}^{-1} \tag{18}
$$

Computing $r_{m-1,0}(\widetilde{k})$

Assume, again, that the incident wave $\widetilde{E}_{m-1}\exp[i(\omega t + \widetilde{k}_{m-1}z)]$ is defined for $z > L_{m-1}$. We wish to compute the reflected wave defined, also, in the layer $\widetilde{n}_{m-1}$. The k-coefficient, as always, will be $-\widetilde{k}_{m-1}$. The reflected wave we wish to compute is the sum of infinitely many small-waves plus a zeroth-wave.

The zeroth-wave is the reflection of the incident wave at $z = L_{m-1}$. So,

E of the 0th wave

$$
= \widetilde{E}_{m-1}\frac{\widetilde{n}_{m-1} - \widetilde{n}_{m-2}}{\widetilde{n}_{m-1} + \widetilde{n}_{m-2}}\exp(-2i\widetilde{k}_{m-1}L_{m-1})
$$

The first small-wave comes from transmission from right to left through $z = L_{m-1}$, followed by reflection at $[L_{m-2}, \ldots, L_1]$, followed by transmission from left to right through $z = L_{m-1}$. So,

E of the 1st small-wave

$$
= \widetilde{E}_{m-1}\frac{2\widetilde{n}_{m-1}}{\widetilde{n}_{m-1} + \widetilde{n}_{m-2}}\exp\left[\frac{i\widetilde{k}_{m-1}L_{m-1}(\widetilde{n}_{m-2} - \widetilde{n}_{m-1})}{\widetilde{n}_{m-1}}\right]
$$

$$
\times\, r_{m-2,0}\left(\frac{\widetilde{k}_{m-1}\widetilde{n}_{m-2}}{\widetilde{n}_{m-1}}\right)
$$

$$
\times\, \frac{2\widetilde{n}_{m-2}}{\widetilde{n}_{m-2} + \widetilde{n}_{m-1}}\exp\left[-i\widetilde{k}_{m-1}L_{m-1}\frac{\widetilde{n}_{m-2}}{\widetilde{n}_{m-1}}\frac{\widetilde{n}_{m-1} - \widetilde{n}_{m-2}}{\widetilde{n}_{m-2}}\right]
$$

The second small-wave comes from transmission from right to left through $z = L_{m-1}$, then reflection at $[L_{m-2}, \ldots, L_1]$, then [reflection at $z = L_{m-1}$, reflection at $[L_{m-2}, \ldots, L_1]$] and, finally, transmission (left–right) through $z = L_{m-1}$. So,

E of the 2nd small-wave

$$
= \widetilde{E}_{m-1}\frac{2\widetilde{n}_{m-1}}{\widetilde{n}_{m-1} + \widetilde{n}_{m-2}}\exp\left[\frac{i\widetilde{k}_{m-1}L_{m-1}(\widetilde{n}_{m-2} - \widetilde{n}_{m-1})}{\widetilde{n}_{m-1}}\right]
$$

$$
\times\, r_{m-2,0}\left(\frac{\widetilde{k}_{m-1}\widetilde{n}_{m-2}}{\widetilde{n}_{m-1}}\right)
$$

$$
\times\, \frac{\widetilde{n}_{m-2} - \widetilde{n}_{m-1}}{\widetilde{n}_{m-2} + \widetilde{n}_{m-1}}\exp\left(\frac{2i\widetilde{k}_{m-1}\widetilde{n}_{m-2}L_{m-1}}{\widetilde{n}_{m-1}}\right)
$$

$$
\times\, r_{m-1,1}\left(\frac{\widetilde{k}_{m-1}\widetilde{n}_{m-2}}{\widetilde{n}_{m-1}}\right)
$$

$$
\times\, \frac{2\widetilde{n}_{m-2}}{\widetilde{n}_{m-2} + \widetilde{n}_{m-1}}\exp\left[-i\widetilde{k}_{m-1}L_{m-1}\frac{\widetilde{n}_{m-1} - \widetilde{n}_{m-2}}{\widetilde{n}_{m-1}}\right]
$$

The νth small-wave comes from transmission (right–left) through $z = L_{m-1}$, then reflection at $[L_{m-2}, \ldots, L_1]$, then [reflection at $z = L_{m-1}$, reflection at $[L_{m-2}, \ldots, L_1]$] $\nu - 1$ times, and, finally, transmission left–right through $z = L_{m-1}$. So,

E of the νth small-wave

$$
= \widetilde{E}_{m-1}\frac{2\widetilde{n}_{m-1}}{\widetilde{n}_{m-1} + \widetilde{n}_{m-2}}\exp\left[\frac{i\widetilde{k}_{m-1}L_{m-1}(\widetilde{n}_{m-2} - \widetilde{n}_{m-1})}{\widetilde{n}_{m-1}}\right]
$$

$$
\times\, r_{m-2,0}\left(\frac{\widetilde{k}_{m-1}\widetilde{n}_{m-2}}{\widetilde{n}_{m-1}}\right)
$$

$$
\times\, \left\{ \frac{\widetilde{n}_{m-2} - \widetilde{n}_{m-1}}{\widetilde{n}_{m-2} + \widetilde{n}_{m-1}}\exp\left(\frac{2i\widetilde{k}_{m-1}L_{m-1}\widetilde{n}_{m-2}}{\widetilde{n}_{m-1}}\right) \right.
$$

$$
\left. \times\, r_{m-2,0}\left(\frac{\widetilde{k}_{m-1}\widetilde{n}_{m-2}}{\widetilde{n}_{m-1}}\right) \right\}^{\nu-1}
$$

$$
\times\, \frac{2\widetilde{n}_{m-2}}{\widetilde{n}_{m-2} + \widetilde{n}_{m-1}}\exp\left[-i\widetilde{k}_{m-1}L_{m-1}\frac{\widetilde{n}_{m-1} - \widetilde{n}_{m-2}}{\widetilde{n}_{m-1}}\right]
$$

Adding all these waves, we obtain

E of the reflected wave in $z > L_{m-1}$

$$
= \widetilde{E}_{m-1}\frac{\widetilde{n}_{m-1} - \widetilde{n}_{m-2}}{\widetilde{n}_{m-1} + \widetilde{n}_{m-2}}\exp(-2i\widetilde{k}_{m-1}L_{m-1})
$$

$$
+ \widetilde{E}_{m-1}\frac{2\widetilde{n}_{m-1}}{\widetilde{n}_{m-1} + \widetilde{n}_{m-2}}\exp\left[\frac{i\widetilde{k}_{m-1}L_{m-1}(\widetilde{n}_{m-2} - \widetilde{n}_{m-1})}{\widetilde{n}_{m-1}}\right]
$$

$$
\times\, \left\{ 1 - \frac{\widetilde{n}_{m-2} - \widetilde{n}_{m-1}}{\widetilde{n}_{m-2} + \widetilde{n}_{m-1}}\exp\left(\frac{2i\widetilde{k}_{m-1}L_{m-1}\widetilde{n}_{m-2}}{\widetilde{n}_{m-1}}\right) \right.
$$

$$
\left. \times\, r_{m-2,0}\left(\frac{\widetilde{k}_{m-1}\widetilde{n}_{m-2}}{\widetilde{n}_{m-1}}\right) \right\}^{-1}
$$

$$
\times\, r_{m-2,0}\left(\frac{\widetilde{k}_{m-1}\widetilde{n}_{m-2}}{\widetilde{n}_{m-1}}\right)\frac{2\widetilde{n}_{m-2}}{\widetilde{n}_{m-2} + \widetilde{n}_{m-1}}
$$

$$
\times\, \exp\left[-i\widetilde{k}_{m-1}L_{m-1}\frac{\widetilde{n}_{m-1} - \widetilde{n}_{m-2}}{\widetilde{n}_{m-1}}\right]
$$

So,

$$
r_{m-1,0}(\widetilde{k}_{m-1})
$$

$$
= \frac{\widetilde{n}_{m-1} - \widetilde{n}_{m-2}}{\widetilde{n}_{m-1} + \widetilde{n}_{m-2}}\exp(-2i\widetilde{k}_{m-1}L_{m-1})
$$

$$
+ \frac{2\widetilde{n}_{m-1}}{\widetilde{n}_{m-1} + \widetilde{n}_{m-2}}\exp\left[\frac{i\widetilde{k}_{m-1}L_{m-1}(\widetilde{n}_{m-2} - \widetilde{n}_{m-1})}{\widetilde{n}_{m-1}}\right]
$$

$$
\times\, \left\{ 1 - \frac{\widetilde{n}_{m-2} - \widetilde{n}_{m-1}}{\widetilde{n}_{m-2} + \widetilde{n}_{m-1}}\exp\left(\frac{2i\widetilde{k}_{m-1}L_{m-1}\widetilde{n}_{m-2}}{\widetilde{n}_{m-1}}\right) \right.
$$

$$
\left. \times\, r_{m-2,0}\left(\frac{\widetilde{k}_{m-1}\widetilde{n}_{m-2}}{\widetilde{n}_{m-1}}\right) \right\}^{-1}
$$

$$
\times\, r_{m-2,0}\left(\frac{\widetilde{k}_{m-1}\widetilde{n}_{m-2}}{\widetilde{n}_{m-1}}\right)\frac{2\widetilde{n}_{m-2}}{\widetilde{n}_{m-2} + \widetilde{n}_{m-1}}
$$

$$
\times\, \exp\left[-i\widetilde{k}_{m-1}L_{m-1}\frac{\widetilde{n}_{m-1} - \widetilde{n}_{m-2}}{\widetilde{n}_{m-1}}\right]
$$

Thus,

$$
\begin{aligned}
&r_{m-1,0}(\widetilde{k}_{m-1}) \\
&= \frac{\widetilde{n}_{m-1} - \widetilde{n}_{m-2}}{\widetilde{n}_{m-1} + \widetilde{n}_{m-2}} \exp\left(-2i\widetilde{k}_{m-1}L_{m-1}\right) \\
&\quad + \frac{4\widetilde{n}_{m-1}\widetilde{n}_{m-2}}{(\widetilde{n}_{m-1} + \widetilde{n}_{m-2})^2} \exp\left[\frac{2i\widetilde{k}_{m-1}L_{m-1}(\widetilde{n}_{m-2} - \widetilde{n}_{m-1})}{\widetilde{n}_{m-1}}\right] \\
&\quad \times r_{m-2,0}\left(\frac{\widetilde{k}_{m-1}\widetilde{n}_{m-2}}{\widetilde{n}_{m-1}}\right) \\
&\quad \times \left\{1 - \frac{\widetilde{n}_{m-2} - \widetilde{n}_{m-1}}{\widetilde{n}_{m-2} + \widetilde{n}_{m-1}} \exp\left(\frac{2i\widetilde{k}_{m-1}L_{m-1}\widetilde{n}_{m-2}}{\widetilde{n}_{m-1}}\right)\right. \\
&\quad \left. \times r_{m-2,0}\left(\frac{\widetilde{k}_{m-1}\widetilde{n}_{m-2}}{\widetilde{n}_{m-1}}\right)\right\}^{-1} \qquad (19)
\end{aligned}
$$

1.4.3. Organizing the Computations

The analysis of Eqs. (16)–(19) reveals that for computing, say, $t_{0,m-1}(\widetilde{k})$, we need to compute $t_{0,\nu}(\widetilde{k})$ for $\nu < m-1$ and $r_{\nu,0}(-\widetilde{k}\widetilde{n}_{\nu-1}/\widetilde{n}_0)$ for $\nu < m-1$.

Analogously, for computing $r_{0,m-1}(\widetilde{k})$, we need to compute $t_{0,\nu}(\widetilde{k})$ for $\nu < m-1$, $t_{\nu,0}(-\widetilde{k}\widetilde{n}_{\nu-1}/\widetilde{n}_0)$ for $\nu < m-1$, $r_{\nu,0}(-\widetilde{k}\widetilde{n}_{\nu-1}/\widetilde{n}_0)$ for $\nu < m-2$, and $r_{0,\nu}(\widetilde{k})$ for $\nu = m-2$.

Let us see this in the following tables, for $m = 4$. Under the $(j + 1)$th column of the table, we write the quantities that are needed to compute the quantities that appear under column j. For example, we read that for computing $t_{0,4}(\widetilde{k})$, we need $t_{0,3}(\widetilde{k})$ and $r_{3,0}(-\widetilde{k}\widetilde{n}_3/\widetilde{n}_0)$, and so on.

Computing Tree of $t_{0,4}(\widetilde{k})$

$t_{0,4}(\widetilde{k})$	$t_{0,3}(\widetilde{k})$	$t_{0,2}(\widetilde{k})$	$t_{0,1}(\widetilde{k})$
			$r_{1,0}(-\widetilde{k}\widetilde{n}_1/\widetilde{n}_0)$
		$r_{2,0}(-\widetilde{k}\widetilde{n}_2/\widetilde{n}_0)$	$r_{1,0}(-\widetilde{k}\widetilde{n}_1/\widetilde{n}_0)$
	$r_{3,0}(-\widetilde{k}\widetilde{n}_3/\widetilde{n}_0)$	$r_{2,0}(-\widetilde{k}\widetilde{n}_2/\widetilde{n}_0)$	$r_{1,0}(-\widetilde{k}\widetilde{n}_1/\widetilde{n}_0)$

Computing Tree of $r_{0,4}(\widetilde{k})$

$r_{0,4}(\widetilde{k})$	$t_{0,3}(\widetilde{k})$	$t_{0,2}(\widetilde{k})$	$t_{0,1}(\widetilde{k})$
			$r_{1,0}(-\widetilde{k}\widetilde{n}_1/\widetilde{n}_0)$
		$r_{2,0}(-\widetilde{k}\widetilde{n}_2/\widetilde{n}_0)$	$r_{1,0}(-\widetilde{k}\widetilde{n}_1/\widetilde{n}_0)$
	$t_{3,0}(-\widetilde{k}\widetilde{n}_3/\widetilde{n}_0)$	$t_{2,0}(-\widetilde{k}\widetilde{n}_2/\widetilde{n}_0)$	$t_{1,0}(-\widetilde{k}\widetilde{n}_1/\widetilde{n}_0)$
			$r_{1,0}(-\widetilde{k}\widetilde{n}_1/\widetilde{n}_0)$
		$r_{2,0}(-\widetilde{k}\widetilde{n}_2/\widetilde{n}_0)$	$r_{1,0}(-\widetilde{k}\widetilde{n}_1/\widetilde{n}_0)$
	$r_{0,3}(\widetilde{k})$	$t_{0,2}(\widetilde{k})$	$t_{0,1}(\widetilde{k})$
			$r_{1,0}(-\widetilde{k}\widetilde{n}_1/\widetilde{n}_0)$
		$t_{2,0}(-\widetilde{k}\widetilde{n}_2/\widetilde{n}_0)$	$t_{1,0}(-\widetilde{k}\widetilde{n}_1/\widetilde{n}_0)$
			$r_{1,0}(-\widetilde{k}\widetilde{n}_1/\widetilde{n}_0)$
		$r_{0,2}(\widetilde{k})$	$t_{0,1}(\widetilde{k})$
			$t_{1,0}(-\widetilde{k}\widetilde{n}_1/\widetilde{n}_0)$
			$r_{0,1}(\widetilde{k})$

Therefore, all the computation can be performed using the following conceptual algorithm.

Algorithm

Step 1. Compute $t_{0,1}(\widetilde{k})$, $r_{0,1}(\widetilde{k})$, $t_{1,0}(-\widetilde{k}\widetilde{n}_1/\widetilde{n}_0)$, $r_{0,1}(-\widetilde{k}\widetilde{n}_1/\widetilde{n}_0)$.

Step 2. For $\nu = 3, \dots, m-1$, Compute $t_{0,\nu}(\widetilde{k})$, $r_{0,\nu}(\widetilde{k})$, $t_{\nu,0}(-\widetilde{k}\widetilde{n}_{\nu-1}/\widetilde{n}_0)$, $r_{0,\nu}(-\widetilde{k}\widetilde{n}_{\nu-1}/\widetilde{n}_0)$.

1.4.4. Computation by Matricial Methods

The previous sections provide a practical way of computing transmissions and reflections and understanding how complex waves are generated as summations of simpler ones. In this section, we compute the same parameters using a more compact calculation procedure, although some insight is lost [16]. The assumptions are exactly the ones of the previous sections. In layer $\widetilde{n}_\nu$, for $\nu = 0, 1, \dots, m-1$, we have two waves given by

$$E_T^\nu \exp\left[i\left(wt - \widetilde{k}_\nu z\right)\right]$$

and

$$E_R^\nu \exp\left[i\left(wt + \widetilde{k}_\nu z\right)\right]$$

The first is a summation of "transmitted small-waves" and the second is a summation of "reflected small-waves." As before, $k = 2\pi/\lambda$,

$$\widetilde{k}_0 = \widetilde{k} \quad \text{and} \quad \widetilde{k}_\nu = \frac{\widetilde{k}\widetilde{n}_\nu}{\widetilde{n}_0}$$

for all $\nu = 0, 1, \dots, m-1$. We can interpret that $E_T^0 \exp[i(wt - \widetilde{k}z)]$ is the "incident wave." Since there are no reflected waves in the last semi-infinite layer, we have that

$$E_R^{m-1} = 0$$

Using the continuity of the waves and their derivatives with respect to x at the interfaces $L_1, \dots, L_{m-1}$, we get, for $\nu = 1, 2, \dots, m-1$,

$$E_T^{\nu-1} \exp\left(-i\widetilde{k}_{\nu-1}L_\nu\right) + E_R^{\nu-1} \exp\left(i\widetilde{k}_{\nu-1}L_\nu\right) = E_T^\nu \exp\left(-i\widetilde{k}_\nu L_\nu\right) + E_R^\nu \exp\left(i\widetilde{k}_\nu L_\nu\right)$$

and

$$-\widetilde{k}_{\nu-1}E_T^{\nu-1} \exp\left(-i\widetilde{k}_{\nu-1}L_\nu\right) + \widetilde{k}_{\nu-1}E_R^{\nu-1} \exp\left(i\widetilde{k}_{\nu-1}L_\nu\right) = -\widetilde{k}_\nu E_T^\nu \exp\left(-i\widetilde{k}_\nu L_\nu\right) + \widetilde{k}_\nu E_R^\nu \exp\left(i\widetilde{k}_\nu L_\nu\right)$$

In matricial form, and using $\widetilde{k}_\nu = \widetilde{k}\widetilde{n}_\nu/\widetilde{n}_0$, we obtain

$$
\begin{pmatrix} 1 & 1 \\ -\widetilde{n}_{\nu-1} & \widetilde{n}_{\nu-1} \end{pmatrix}
\begin{pmatrix} \exp(-i\widetilde{k}_{\nu-1}L_\nu) & 0 \\ 0 & \exp(i\widetilde{k}_{\nu-1}L_\nu) \end{pmatrix}
\begin{pmatrix} E_T^{\nu-1} \\ E_R^{\nu-1} \end{pmatrix}
$$
$$
= \begin{pmatrix} 1 & 1 \\ -\widetilde{n}_\nu & \widetilde{n}_\nu \end{pmatrix}
\begin{pmatrix} \exp(-i\widetilde{k}_\nu L_\nu) & 0 \\ 0 & \exp(i\widetilde{k}_\nu L_\nu) \end{pmatrix}
\begin{pmatrix} E_T^\nu \\ E_R^\nu \end{pmatrix}
$$

Therefore,

$$\begin{pmatrix} E_T^\nu \\ E_R^\nu \end{pmatrix} = \begin{pmatrix} \exp(-i\widetilde{k}_\nu L_\nu) & 0 \\ 0 & \exp(i\widetilde{k}_\nu L_\nu) \end{pmatrix}$$

$$\times \frac{1}{2\widetilde{n}_\nu} \begin{pmatrix} \widetilde{n}_\nu + \widetilde{n}_{\nu-1} & \widetilde{n}_\nu - \widetilde{n}_{\nu-1} \\ \widetilde{n}_\nu - \widetilde{n}_{\nu-1} & \widetilde{n}_\nu + \widetilde{n}_{\nu-1} \end{pmatrix}$$

$$\times \begin{pmatrix} \exp(-i\widetilde{k}_{\nu-1}L_\nu) & 0 \\ 0 & \exp(i\widetilde{k}_{\nu-1}L_\nu) \end{pmatrix} \begin{pmatrix} E_T^{\nu-1} \\ E_R^{\nu-1} \end{pmatrix}$$

Let us write, for $\nu = 1, \ldots, m-1$,

$$A_\nu = \frac{1}{2\widetilde{n}_\nu} \begin{pmatrix} \widetilde{n}_\nu + \widetilde{n}_{\nu-1} & \widetilde{n}_\nu - \widetilde{n}_{\nu-1} \\ \widetilde{n}_\nu - \widetilde{n}_{\nu-1} & \widetilde{n}_\nu + \widetilde{n}_{\nu-1} \end{pmatrix}$$

Then,

$$\begin{pmatrix} E_T^{\nu+1} \\ E_R^{\nu+1} \end{pmatrix}$$

$$= \begin{pmatrix} \exp(-i\widetilde{k}_{\nu+1}L_{\nu+1}) & 0 \\ 0 & \exp(i\widetilde{k}_{\nu+1}L_{\nu+1}) \end{pmatrix}$$

$$\times A_{\nu+1} \begin{pmatrix} \exp(-i\widetilde{k}_\nu[L_{\nu+1}-L_\nu]) & 0 \\ 0 & \exp(i\widetilde{k}_\nu[L_{\nu+1}-L_\nu]) \end{pmatrix}$$

$$\times A_\nu \begin{pmatrix} \exp(-i\widetilde{k}_{\nu-1}L_\nu) & 0 \\ 0 & \exp(i\widetilde{k}_{\nu-1}L_\nu) \end{pmatrix} \begin{pmatrix} E_T^{\nu-1} \\ E_R^{\nu-1} \end{pmatrix}$$

Let $d_\nu \equiv L_{\nu+1} - L_\nu$ ($\nu = 1, \ldots, m-2$) be the thickness of layer ν. We define, for $\nu = 1, \ldots, m-2$,

$$M_\nu = A_{\nu+1} \begin{pmatrix} \exp(-i\widetilde{k}_\nu d_\nu) & 0 \\ 0 & \exp(i\widetilde{k}_\nu d_\nu) \end{pmatrix}$$

Then, setting for simplicity and without loss of generality, $L_1 = 0$,

$$\begin{pmatrix} E_T^{m-1} \\ E_R^{m-1} \end{pmatrix} = \begin{pmatrix} \exp(i\widetilde{k}_{m-1}L_{m-1}) & 0 \\ 0 & \exp(i\widetilde{k}_{m-1}L_{m-1}) \end{pmatrix} M_{m-2}$$

$$\times \cdots \times M_1 A_1 \begin{pmatrix} E_T^0 \\ E_R^0 \end{pmatrix}$$

Define

$$M = M_{m-2} \times \cdots \times M_1 A_1 = \begin{pmatrix} M_{11} & M_{12} \\ M_{21} & M_{22} \end{pmatrix}$$

and

$$M' = \begin{pmatrix} \exp(i\widetilde{k}_{m-1}L_{m-1}) & 0 \\ 0 & \exp(i\widetilde{k}_{m-1}L_{m-1}) \end{pmatrix} M$$

$$= \begin{pmatrix} M'_{11} & M'_{12} \\ M'_{21} & M'_{22} \end{pmatrix}$$

Using $E_R^{m-1} = 0$, we obtain that

$$E_R^0 = -\frac{M'_{21}}{M'_{22}} E_T^0 = -\frac{M_{21}}{M_{22}} E_T^0$$

and

$$E_T^{m-1} = \left(M'_{11} - \frac{M'_{12}M'_{21}}{M'_{22}} \right) E_T^0$$

$$= \exp(i\widetilde{k}_{m-1}L_{m-1}) \left(M_{11} - \frac{M_{12}M_{21}}{M_{22}} \right) E_T^0$$

So,

$$r_{0,m-1} = -\frac{M_{21}}{M_{22}}$$

and

$$t_{0,m-1} = \exp(i\widetilde{k}_{m-1}L_{m-1}) \left(M_{11} - \frac{M_{12}M_{21}}{M_{22}} \right) \qquad (20)$$

In this deduction, we assumed that the transmitted wave in the final layer n_{m-1} is $E_T^{m-1} \exp[i(wt - \widetilde{k}_{m-1}z)]$. For this reason, the factor $\exp(i\widetilde{k}_{m-1}L_{m-1})$ appeared in the final computation of $t_{1,m-1}$. In other words, according to (20), the transmitted wave in the final layer is

$$\left(M_{11} - \frac{M_{12}M_{21}}{M_{22}} \right) E_T^0 \exp[i(wt - \widetilde{k}_{m-1}(z - L_{m-1}))]$$

For energy computations, since $|\exp(i\widetilde{k}_{m-1}L_{m-1})| = 1$, the presence of this factor in the computation of $t_{0,m-1}$ is irrelevant.

1.4.5. Transmitted and Reflected Energy

Assume that the layers $z < L_1$ and $z > L_{m-1}$ are transparent. In this case, the coefficients n_0 and n_{m-1} are real, so we have

$$\text{incident energy} = n_0|\widetilde{E}|^2$$
$$\text{transmitted energy} = n_{m-1}|E\text{-coefficient of the transmitted wave}|^2$$

and

$$\text{reflected energy} = n_0|E\text{-coefficient of the reflected wave}|^2$$

Accordingly, we define

$$\text{transmittance} = \frac{\text{trasmitted energy}}{\text{incident energy}}$$

and

$$\text{reflectance} = \frac{\text{reflected energy}}{\text{incident energy}}$$

In other words,

$$\text{transmittance} = \frac{n_{m-1}}{n_0}|t_{0,m-1}|^2 \qquad (21)$$

and

$$\text{reflectance} = |r_{0,m-1}|^2 \qquad (22)$$

If *all* the layers were transparent, we would necessarily have that

$$\text{incident energy} = \text{transmitted energy} + \text{reflected energy}$$

and, in that case:

$$\text{transmittance} + \text{reflectance} = 1$$

Assume now that the layer $L_{m-2} < z < L_{m-1}$ is also transparent, so that $\tilde{n}_{m-2}$ is also real. Consider the transmission coefficient $t_{0,m-1}$ as a function of L_{m-1}, keeping fixed all the other arguments. By (16), we see that $|t_{0,m-1}|^2$ is periodic and that its period is $\lambda n_0/(2n_{m-2})$.

Assume now that all the layers are transparent and, without loss of generality, $L_1 = 0$. We consider $|t_{0,m-1}|^2$ as a function of the thicknesses $d_1 \equiv L_2 - L_1, \ldots, d_{m-2} \equiv L_{m-1} - L_{m-2}$. As above, it can be seen that $|t_{0,m-1}|^2$ is periodic with respect to each of the above variables, and that its period with respect to d_i $(i = 1, 2, \ldots, m-2)$ is $\lambda n_0/(2n_i)$.

1.5. Nonnormal Incidence and Linear-System Computations

In the previous section, we deduced explicit formulae for transmitted and reflected energies when light impinges normally to the surface of the layers. Explicit formulae are important because they help to understand how waves effectively behave. However, it is perhaps simpler to perform computations using an implicit representation of transmitted and reflected waves inside each layer. As we are going to see, in this way the electric and magnetic vectors arise as solutions of a single linear system of equations. This approach allows us to deal in a rather simple way with a more involved situation: the case in which the incidence is not normal, so that the plane of propagation is not parallel to the interfaces. The case considered in the previous section is a particular case of this.

Assume that, in the three-dimensional space xyz, we have m layers divided by the interface planes $z = L_1, \ldots, z = L_{m-1}$ and characterized by the complex refraction indices $\tilde{n}_0, \ldots, \tilde{n}_{m-1}$. Therefore, we have, for $\nu = 0, 1, \ldots, m-1$,

$$\tilde{n}_\nu = n_\nu - i\kappa_\nu$$

where, as always, n_ν represents the real refraction index and κ_ν is the attenuation coefficient. We assume that $\kappa_0 = \kappa_{m-1} = 0$, so the first and last layer are transparent.

Suppose that light arrives at the first surface $z = L_1$ with an angle θ_0 with respect to the normal to the surface. This means that the vector of propagation of the incident light in the first layer is

$$\mathbf{s}_0 = (\sin(\theta_0), 0, \cos(\theta_0))$$

Accordingly (for example, invoking Snell's law), the angle with the normal of the transmitted wave in layer ν is

$$\theta_\nu = \frac{n_\nu}{n_0} \sin(\theta_0)$$

and, consequently, the vector of propagation of the transmitted wave in the layer ν will be

$$\mathbf{s}_\nu = (\sin(\theta_\nu), 0, \cos(\theta_\nu))$$

Reflected waves are generated in the layers $0, 1, \ldots, m-2$. By Snell's laws, their vectors of propagation are $(\sin(\theta_\nu), 0, -\cos(\theta_\nu))$ for $\nu = 0, 1, \ldots, m-2$.

Incident light in layer n_0 is represented by the electric vector $\mathbf{E}$ and the magnetic vector $\mathbf{H}$. Electromagnetic theory [8] tells that both $\mathbf{E}$ and $\mathbf{H}$ are orthogonal to $\mathbf{s}_0$. Moreover, the relation between these vectors is

$$\mathbf{H} = n_0 \mathbf{s}_0 \times \mathbf{E} \tag{23}$$

where $\times$ denotes the vectorial product. We consider that the electromagnetic vectors are linearly polarized, so the considerations above lead to the following expression for $\mathbf{E}$:

$$\mathbf{E} = (-E_p \cos(\theta_0), E_y, E_p \sin(\theta_0)) \times \exp\{i[\omega t - k(x \sin(\theta_0) + z \cos(\theta_0))]\} \tag{24}$$

where $k = 2\pi/\lambda$ and λ is the wavelength.

By (23), we also have

$$\mathbf{H} = n_0(-E_y \cos(\theta_0), -E_p, E_y \sin(\theta_0)) \times \exp\{i[\omega t - k(x \sin(\theta_0) + z \cos(\theta_0))]\}$$

The incidence of light on the plane $z = L_1$ produces transmitted vectors $\mathbf{E}_T^1, \ldots, \mathbf{E}_T^{m-1}$ in the layers $1, \ldots, m-1$ and the corresponding magnetic vectors $\mathbf{H}_T^1, \ldots, \mathbf{H}_T^{m-1}$. Simultaneously, reflected waves are produced, represented by the electric vectors $\mathbf{E}_R^0, \ldots, \mathbf{E}_R^{m-2}$ in the layers $0, 1, \ldots, m-2$ and the corresponding magnetic vectors $\mathbf{H}_R^0, \ldots, \mathbf{H}_R^{m-2}$.

As in (24), for $\nu = 1, \ldots, m-1$, taking into account the velocity of light in layer ν and the attenuation factor, we can write

$$\mathbf{E}_T^\nu = (-E_{T,p}^\nu \cos(\theta_\nu), E_{T,y}^\nu, E_{T,p}^\nu \sin(\theta_\nu)) \times \exp\left\{i\left[\omega t - k\frac{n_\nu}{n_0}(x \sin(\theta_\nu) + z \cos(\theta_\nu))\right]\right\} \times \exp\left(-\frac{k\kappa_\nu z}{n_0 \cos(\theta_\nu)}\right) \tag{25}$$

So, by a relation similar to (23) in layer ν,

$$\mathbf{H}_T^\nu = n_\nu(-E_{T,y}^\nu \cos(\theta_\nu), -E_{T,p}^\nu, E_{T,y}^\nu \sin(\theta_\nu)) \times \exp\left\{i\left[\omega t - k\frac{n_\nu}{n_0}(x \sin(\theta_\nu) + z \cos(\theta_\nu))\right]\right\} \times \exp\left(-\frac{k\kappa_\nu z}{n_0 \cos(\theta_\nu)}\right) \tag{26}$$

Similarly, for the reflected fields, with $\nu = 0, 1, \ldots, m-2$, we have

$$\mathbf{E}_R^\nu = (E_{R,p}^\nu \cos(\theta_\nu), E_{R,y}^\nu, E_{R,p}^\nu \sin(\theta_\nu)) \times \exp\left\{i\left[\omega t - k\frac{n_\nu}{n_0}(x \sin(\theta_\nu) - z \cos(\theta_\nu))\right]\right\} \times \exp\left(\frac{k\kappa_\nu z}{n_0 \cos(\theta_\nu)}\right) \tag{27}$$

and

$$\mathbf{H}_R^\nu = n_\nu(E_{R,y}^\nu \cos(\theta_\nu), -E_{R,p}^\nu, E_{R,y}^\nu \sin(\theta_\nu)) \times \exp\left\{i\left[\omega t - k\frac{n_\nu}{n_0}(x \sin(\theta_\nu) - z \cos(\theta_\nu))\right]\right\} \times \exp\left(\frac{k\kappa_\nu z}{n_0 \cos(\theta_\nu)}\right) \tag{28}$$

From electromagnetic theory, we know that the tangential component of the electric and the magnetic fields must be continuous at the interfaces. This means that

(i) The x and y components of $\mathbf{E} + \mathbf{E}_R^0$ must coincide with the x and y components of $\mathbf{E}_T^1 + \mathbf{E}_R^1$ for $z = L_1$.

(ii) The x and y components of $\mathbf{H} + \mathbf{H}_R^0$ must coincide with the x and y components of $\mathbf{H}_T^1 + \mathbf{H}_R^1$ for $z = L_1$.

(iii) The x and y components of $\mathbf{E}_T^\nu + \mathbf{E}_R^\nu$ must coincide with the x and y components of $\mathbf{E}_T^{\nu+1} + \mathbf{E}_R^{\nu+1}$ for $z = L_\nu$ for $\nu = 1, \ldots, m - 3$.

(iv) The x and y components of $\mathbf{H}_T^\nu + \mathbf{H}_R^\nu$ must coincide with the x and y components of $\mathbf{H}_T^{\nu+1} + \mathbf{H}_R^{\nu+1}$ for $z = L_\nu$ for $\nu = 1, \ldots, m - 3$.

(v) The x and y components of $\mathbf{E}_T^{m-2} + \mathbf{E}_R^{m-2}$ must coincide with the x and y components of $\mathbf{E}_T^{m-1}$ for $z = L_{m-1}$.

(vi) The x and y components of $\mathbf{H}_T^{m-2} + \mathbf{H}_R^{m-2}$ must coincide with the x and y components of $\mathbf{H}_T^{m-1}$ for $z = L_{m-1}$.

Let us define, for $\nu = 0, 1, 2, \ldots, m - 2$,

$$\beta_\nu = \exp\left(\frac{k\kappa_\nu L_\nu}{n_0 \cos(\theta_\nu)} + \frac{ikn_\nu L_\nu \cos(\theta_\nu)}{n_0}\right) \quad (29)$$

$$\gamma_\nu = \exp\left(\frac{k\kappa_{\nu-1} L_\nu}{n_0 \cos(\theta_{\nu-1})} + \frac{ikn_{\nu-1} L_\nu \cos(\theta_{\nu-1})}{n_{r0}}\right) \quad (30)$$

Then, by (24)–(30), the condition (iii) takes the form

$$E_{T,p}^{\nu-1} \frac{\cos(\theta_{\nu-1})}{\gamma_\nu} - E_{R,p}^{\nu-1} \cos(\theta_{\nu-1})\gamma_\nu$$
$$- E_{T,p}^\nu \frac{\cos(\theta_\nu)}{\beta_\nu} + E_{R,p}^\nu \cos(\theta_\nu)\beta_\nu = 0 \quad (31)$$

$$E_{T,y}^{\nu-1} \frac{1}{\gamma_\nu} + E_{R,y}^{\nu-1} \gamma_\nu - E_{T,y}^\nu \frac{1}{\beta_\nu} - E_{R,y}^\nu \beta_\nu = 0 \quad (32)$$

for $\nu = 1, \ldots, m - 3$. Analogously, condition (iv) takes the form

$$E_{T,y}^{\nu-1} \frac{n_{\nu-1} \cos(\theta_{\nu-1})}{\gamma_\nu} - E_{R,y}^{\nu-1} n_{\nu-1} \cos(\theta_{\nu-1})\gamma_\nu$$
$$- E_{T,y}^\nu \frac{n_\nu \cos(\theta_\nu)}{\beta_\nu} + E_{R,y}^\nu n_\nu \cos(\theta_\nu)\beta_\nu = 0 \quad (33)$$

$$E_{T,p}^{\nu-1} \frac{n_{\nu-1}}{\gamma_\nu} + E_{R,p}^{\nu-1} n_{\nu-1}\gamma_\nu - E_{T,p}^\nu \frac{n_\nu}{\beta_\nu} - E_{R,p}^\nu n_\nu \beta_\nu = 0 \quad (34)$$

Moreover, adopting the definition $\mathbf{E}_T^0 = \mathbf{E}$ and writing $\mathbf{E}_R^{m-1} = \mathbf{0}$ (since there is no reflection in the last layer), we see that (31)–(34) also represent the continuity conditions (i), (ii), (v), and (vi). So, far, (31) and (34) form a system of $2(m-1)$ linear equations with $2(m-1)$ complex unknowns ($E_{T,p}^\nu, \nu = 1, \ldots, m-1$ and $E_{R,p}^\nu, \nu = 0, \ldots, m-2$). On the other hand, (32)–(33) form a similar system whose equations are $E_{T,y}^\nu, \nu = 1, \ldots, m-1$ and $E_{R,y}^\nu, \nu = 0, \ldots, m-2$. Both systems have a very special band-structure (each row of the matrix of coefficients has only four nonnull entries) and can

be efficiently solved by Gaussian elimination. As a result, we can compute the transmitted energy and the reflected energy in the usual way:

$$\boxed{\text{Transmitted energy} = n_{m-1}\left|\mathbf{E}_T^{m-1}\right|^2}$$

and

$$\boxed{\text{Reflected energy} = n_0\left|\mathbf{E}_R^0\right|^2}$$

1.6. The Effect of a Thick Substrate

Assume that we have a stack of p films deposited on a thick substrate. By thick substrate we mean that the substrate thickness is much larger than the wavelength of light, whereas the film thicknesses are of the same order of magnitude as that wavelength. Consider first the case of an infinitely thick substrate. The transmittance is then defined in the substrate and reflections at the back substrate surface are neglected. Suppose, for example, that the incident medium is air ($n_0 = 1$) and one has an incident radiation of wavelength 995 nm. $n_1 = 2.1$ and $\kappa_1 = 0.1$ are the index of refraction and the extinction coefficient, respectively, of a unique $d = 127$-nm-thick film ($p = 1$) deposited onto a semi-infinite transparent substrate of refractive index $n_S = 1.57$. In this case, the transmittance is

$$T_{\text{semi-inf}} = 0.673819 \quad (35)$$

The above optical configuration is not realistic because, in general, the substrate supporting the films has a finite thickness. Reflections at the back surface of the transparent substrate will occur and the transmittance is now measured in an additional layer (usually air) which, in turn, is considered to be semi-infinite. Assuming a substrate thickness of exactly 10^6 nm = 1 mm, and performing the corresponding four-layer calculation (($p + 3$)-layer in the general case), we obtain

$$T_{\text{thick}=10^6 \text{ nm}} = 0.590441 \quad (36)$$

The difference between $T_{\text{semi-inf}}$ and $T_{\text{thick}=10^6}$ is significant. Performing the same calculations with a substrate of thickness $10^6 + 150$ nm, we get

$$T_{\text{thick}=(10^6+150) \text{ nm}} = 0.664837 \quad (37)$$

Again, an important difference appears between $T_{\text{thick}=10^6 \text{ nm}}$ and $T_{\text{thick}=(10^6+150) \text{ nm}}$. Note that, in practice, the two substrates are indistinguishable, their respective thicknesses differing by only 0.15 μm. This leads to the question of what we actually measure when we perform such experiments in the laboratory. Remember that the transmittance is a periodic function of the substrate thickness with a period of one-half wavelength. So, assuming random substrate thickness variations ($d_S \approx d_S \pm \Delta d_S$), with $\Delta d_S \approx \lambda_s/2$, we are led to conjecture that the measured transmission is better approximated by an average of transmissions over the period. This involves the integration of the transmittance as a function of the substrate thickness. In the above-mentioned example, this operation gives

$$T_{\text{average}} = 0.646609 \quad (38)$$

We arrive at the same value when we consider other indeterminations or perturbations to "exact" situations which normally happen in experimental setups, for example, slight deviations from normal incidence, or the fact that the measured wave (λ) is not a "pure" wave but contains wavelengths between $\lambda - \Delta\lambda$ and $\lambda + \Delta\lambda$, where $\Delta\lambda$ depends on the physical configuration of the measuring apparatus. It is worth mentioning that this $\pm\Delta\lambda$ "slit effect" alone does not explain the behavior of the transmission in real experiments. Slit widths commonly used in spectrophotometers are not large enough to eliminate completely interference effects that appear when one consider pure waves.

Fortunately, the integral leading to T_{average} in the general case can be solved analytically. The analytical integral has been done in [17]. The final formula is simple and admits a nice interpretation, which we give below. (See [17].)

Assume that the refractive index of the (transparent) substrate is n_S and the refractive index of the (transparent) final layer is n_F. By the arguments of previous sections on the two-layers case, when a wave defined in the substrate arrives at the back surface, it is refracted to the last layer with its energy multiplied by

$$\mathcal{T} = \frac{n_F}{n_S} \left[\frac{2n_S}{n_S + n_F} \right]^2 \tag{39}$$

and is reflected with its energy multiplied by

$$\mathcal{R} = \left[\frac{n_S - n_F}{n_S + n_F} \right]^2 \tag{40}$$

Consider, for a moment, that the substrate is semi-infinite and define

$T = $ transmittance from layer n_0 to the semi-infinite layer n_S of a wave with length λ;

$\overline{T} = $ transmittance from layer n_S to layer n_0 of a wave with length $\lambda(n_0/n_S)$;

$\overline{R} = $ reflectance in the same situation considered for $\overline{T}$ ($\overline{T} + \overline{R} \leq 1$).

In the above example, we have $\overline{T} = 0.673819$ and $\overline{R} = 0.186631$.

Assume a unit energy incident wave. Reasoning in energy terms (that is, not taking into account phase changes), we detect the total transmitted energy as the sum of infinitely many "small energies," as follows.

The first "small energy" comes from transmission through the films (T) followed by transmission through the substrate ($\mathcal{T}$), so it is equal to $T\,\mathcal{T}$.

The second "small energy" comes from transmission through the films (T), followed by reflection at the back-substrate ($\mathcal{R}$), followed by reflection at the back surface of the stack of films ($\overline{R}$), followed by transmission through the substrate ($\mathcal{T}$). Therefore, this "small energy" is equal to $R\,\mathcal{R}\,\overline{R}\,\mathcal{T}$.

The νth "small energy" comes from transmission through the films (T), followed by [reflection at the back-substrate ($\mathcal{R}$) and reflection at the back surface of the stack of films ($\overline{R}$)] $\nu - 1$ times, followed by transmission through the substrate ($\mathcal{T}$). Therefore, this "small energy" is equal to $T\,[\mathcal{R}\,\overline{R}]^{\nu-1}\,\mathcal{T}$.

Summing all the "small energies," we obtain

$$\text{Average transmittance} = \frac{T\mathcal{T}}{1 - \mathcal{R}\overline{R}}$$

Analogously, it can be obtained that

$$\text{Average reflectance} = \frac{T\overline{T}\mathcal{R}}{1 - \mathcal{R}\overline{R}}$$

The proof that these formulae correspond to the average integration of the transmittance and reflectance along the period can be found in [17].

1.6.1. Substrate on Top or Between Thin Films

Suppose now that a stack of thin films is covered by a thick transparent material. All the considerations above can be repeated, and the formulae for average transmittance and average reflectance turn out to be

$$\text{Average transmittance} = \frac{T\mathcal{T}}{1 - \mathcal{R}\overline{R}}$$

and

$$\text{Average reflectance} = R + \frac{T^2\mathcal{R}}{1 - \mathcal{R}\overline{R}}$$

where now

$$T = \frac{4n_0 n_S}{(n_0 + n_S)^2} \quad \text{and} \quad R = \left(\frac{n_0 - n_S}{n_0 + n_S} \right)^2$$

$\mathcal{T}$ is the transmittance from the layer n_S (first layer now) to the last semi-infinite layer of a wave of length $\lambda n_0/n$, and $\mathcal{R}$ is the corresponding reflectance.

In some applications, a transparent thick substrate is covered with stacks of films on both sides with the purpose of getting some desired optical properties. As before, since the thickness of the substrate is much larger than the wavelength and, thus, than the period of transmittance and reflectance as functions of this thickness, a reasonable model for the measured transmittance and reflectance considers integrating these functions with respect to the period of the substrate. As shown in previous sections, if n_S and d_S are the refractive index and the thickness of the substrate, respectively, the period of the transmittance and of the reflectance as a function of d_S is $\lambda n_0/(2n_S)$, for an incident wave of length λ. At first sight, it is hard to integrate analytically T and R over the period. However, reasoning in energy terms, as done in a previous subsection, leads to the desired result. Let us define:

$T_1 = $ transmittance at λ through the upper stack of films, from layer n_0 to layer n_S;

$R_1 = $ reflectance of the previous situation;

$\overline{T}_1 = $ transmittance, for wavelength $\lambda n_0/n_S$, through the upper stack of films from layer n_S to layer n_0;

$\overline{R}_1 = $ reflectance corresponding to the $\overline{T}_1$ situation;

T_2 = transmittance, for wavelength $\lambda n_0/n_S$, through the upper stack of films, from layer n_0 to layer n_S;

R_2 = reflectance of the $\overline{T}_2$ situation.

Then, we get

$$\text{Transmittance from top stack to the substrate} = \frac{T_1 T_2}{1 - R_2 \overline{R}_1}$$

and

$$\text{Reflectance from top stack to the substrate} = R_1 + \frac{T_1 \overline{T}_1 R_2}{1 - R_2 \overline{R}_1}$$

This leads us to the previous case, a top substrate, with an effective transmittance and reflectance given by the above formulae.

1.7. Computational Filter Design

In this section, we consider that all the layers are transparent and, moreover, $n_0 = n_{m-1} = 1$. Therefore, the transmitted energy is $|t_{0,m-1}|^2$ and the reflected energy is $|r_{0,m-1}|^2$. For given wavelength λ and refraction coefficients $n_1, \ldots, n_{m-2}$, these energies are periodic functions of $d_1, \ldots, d_{m-2}$, the thicknesses of the intermediate layers. For some applications (lens design), it is necessary to determine the thicknesses of these layers in such a way that some objectives in terms of transmission and reflection are fulfilled [18]. For example, suppose that we want to design a system by the alternated superposition of four films of two different materials with refraction indices $n_L = 1.8$ and $n_H = 3.4$, in such a way that the transmission for a wavelength of 1000 nm is maximized, whereas the transmission for wavelengths λ such that $|\lambda - 1000| \leq 10$ is minimal. We can do this solving the following optimization problem:

$$\text{Maximize } f(d_1, \ldots, d_{m-2}) \quad (41)$$

where

$$f(d_1, \ldots, d_{m-2})$$
$$= \left| t_{5,0}\left(\lambda = 1000, (d_1, \ldots, d_{m-2})\right)\right|^2$$
$$- \text{Average}_{|\lambda - 1000| \geq 10} \left| t_{6,1}\left(\lambda, (d_1, \ldots, d_{m-2})\right)\right|^2$$

Note that, although the function $|t_{5,0}|^2$ is periodic with respect to each thickness, this is not the case of f. As a result, f has many local maximizers that are not true solutions of (41).

The gradient of f can be computed recursively using the scheme presented in the previous section, so that a first-order algorithm for optimization can be used to solve the problem. We used the algorithm [19] (see also [20]) for solving the problem, with 500 nm as an initial estimate of all the thicknesses. For these thicknesses, the transmission for $\lambda = 1000$ is 0.181 and the average of the "side-transmissions" is 0.515. The algorithm has no difficulties in finding a (perhaps local) minimizer using eight iterations and less than five seconds of computer time in a Sun workstation. The resulting transmission for $\lambda = 1000$ is 0.9999, whereas the average of the side-transmissions is 0.358. The thicknesses found are 556.3, 508.7, 565.6, and 509.9 nm, respectively.

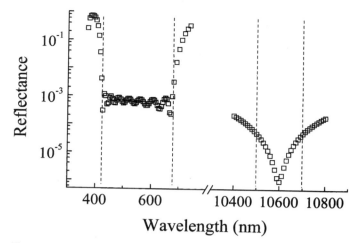

Fig. 8. Reflectance in the visible and infrared spectral regions of the optimized dual-band antireflection coating.

Dual-band antireflection filter. Consider, as a final example, the following design problem. Given air as the incident medium, a nonabsorbing sibstrate (sub), and two nonabsorbing coating materials (L and H), produce an antireflection that reduces simultaneously the reflectance to the lowest possible values in two spectral regions: (1) 420 nm to 680 nm; and (2) 10500 nm to 10700 nm.

The foregoing requirements are to be for dielectric and substrate layers having the following index of refraction:

$$n_{\text{sub}}(1) = 2.60 \qquad n_{\text{sub}}(2) = 2.40$$
$$n_L(1) = 1.52 \qquad n_L(2) = 1.43$$
$$n_H(1) = 2.32 \qquad n_H(2) = 2.22$$

There is restriction neither to the thickness of the individual layers nor to the number of layers. The dual-band problem has been solved as an optimization problem. The objective function was the sum of the reflectances in the two bands. This function was minimized with respect to the thicknesses using the algorithm described in [19]. This algorithm can be freely obtained in [20]. The minimization was performed in several steps, increasing the number of films that are present in the stack. Given the solution for a number of films, an additional film was added with zero thickness, in an optimal position determined by the needle technique [21]. According to this technique, the position of the new zero-thickness film is the best possible one in the sense that the partial derivative that corresponds to its position is negative and minimal. In other words, the potentiality of decreasing locally the objective function for the needle-position of the zero-film is maximized.

Figure 8 shows the results of the optimization process for a total of 20 layers. The total thickness of the stack is 1560 nm. Note that the maximum reflection in the two bands is approximately 0.1%, the reflectance being smaller in the infrared than in the visible spectral region. It is clear that this is not the only solution to the problem [22]. In principle, an increasing number of layers should improve the performance of the dual-band antireflection filter. The manufacturing of filters always entails small random variations of the individual layer thickness. When a random thickness variation of, say, 1% is applied to the stack, a tenfold increase of the mean reflectance is normally found. The effect becomes more severe as the number of layers increases.

1.8. Optimization Algorithm for Thin Films

The method used for solving the optimization problems derived from thin-film calculations is a trust-region box-constraint optimization software developed at the Applied Mathematics Department of the State University of Campinas. During the last five years, it has been used for solving many practical and academic problems with box constraints, and it has been incorporated as a subalgorithm of Augmented Lagrangian methods for minimization with equality constraints and bounds. This software is called BOX-QUACAN here. Its current version is publicly available as the code EASY! in `www.ime.unicamp.br/~martinez`. The problems solved in the present context are of the type

$$\text{Minimize } f(x)$$

subject to $x \in \Omega \subset \mathbb{R}^n$. Ω is a set defined by simple constraints. In our case, the thicknesses must be nonnegative or restricted to some lower bounds. The art of minimization with simple constraints is known as box-constrained optimization. Actually, this is a well-developed area of numerical analysis.

BOX-QUACAN is a box-constraint solver whose basic principles are described in [19]. It is an iterative method which, at each iteration, approximates the objective function by a quadratic and minimizes this quadratic model in the box determined by the natural constraints and an auxiliary box that represents the region where the quadratic approximation is reliable (trust region). If the objective function is sufficiently reduced at the (approximate) minimizer of the quadratic, the corresponding trial point is accepted as the new iterate. Otherwise, the trust region is reduced. The subroutine that minimizes the quadratic is called QUACAN and the main algorithm, which handles trust-region modifications, is BOX.

Assume that Ω is an n-dimensional box, given by

$$\Omega = \left\{ x \in \mathbb{R}^n \mid \ell \leq x \leq u \right\}$$

So, the problem consists of finding an approximate solution of

$$\text{Minimize}_x \, f(x) \quad \text{subject to } \ell \leq x \leq u \qquad (42)$$

BOX-QUACAN uses sequential (approximate) minimization of second-order quadratic approximations with simple bounds. The bounds for the quadratic model come from the intersection of the original box with a trust region defined by the ∞-norm. No factorization of matrices are used at any stage. The quadratic solver used to deal with the subproblems of the box-constrained algorithm visits the different faces of its domain using conjugate gradients on the interior of each face and "chopped gradients" as search directions to leave the faces. See [19, 23, 24] for a description of the 1998 implementation of

QUACAN. At each iteration of this quadratic solver, a matrix-vector product of the Hessian approximation and a vector is needed. Since Hessian approximations are cumbersome to compute, we use the "Truncated Newton" approach, so that each *Hessian × vector* product is replaced by an incremental quotient of ∇f along the direction given by the vector.

The box-constraint solver BOX is a trust-region method whose convergence results have been given in [19]. Roughly speaking, if the objective function has continuous partial derivatives and the Hessian approximations are bounded, every limit point of a sequence generated by BOX is stationary. When one uses true Hessians or secant approximations (see [25]) and the quadratic subproblems are solved with increasing accuracy, quadratic or superlinear convergence can be expected. In our implementation, we used a fixed tolerance as stopping criterion for the quadratic solver QUACAN, because the benefits of high-order convergence of BOX would not compensate increasing the work of the quadratic solver. It is interesting to note that, although one usually talks about "Hessian approximations," second derivatives of the objective function are not assumed to exist at all. Moreover, global convergence results hold even without the Hessian-existence assumption.

2. APPLICATIONS OF THIN FILMS

In this section, we discuss a few situations where the theoretical considerations of the previous section are put to work. The subject of thin-film applications is very vast and impossible to be detailed in a book chapter. The selection to follow considers, as usual, the most classical structures and those with which the authors have been particularly involved. For a discussion of remaining subjects we refer the readers to the specialized literature [8, 9, 16, 18, 27–30].

2.1. Antireflection Coatings

The reflectance between two different (thick) dielectric materials for a light beam falling normally to the interface is given approximately by $R = (n_1 - n_2)^2 / (n_1 + n_2)^2$ and it may be quite important in dense materials. In an air/glass interface, around 4% of the incident light is reflected. Hence, for a normal lens, the loss amounts to $\approx 8\%$. For an optical system including a large number of lenses, the optical performance may be severely affected. For a silicon solar cell ($n_{Si} \approx 4$), the optical losses, of the order of 35%, reduce considerably the conversion efficiency.

The amount of reflected light may be diminished significantly by depositing thin intermediate antireflection (AR) layers. The subject is addressed in most textbooks on elementary optics as well as in the specialized literature on the physics of thin films. An exact theoretical derivation of reflection-reducing optical coatings from Maxwell equations and boundary conditions between layer and substrate has been given by Mooney [31]. The analysis considers normal incidence

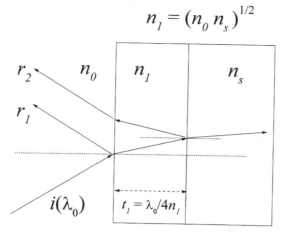

$$n_1 = (n_0 \, n_s)^{1/2}$$

Fig. 9. Single antireflection layer onto a thick substrate. The figure shows the conditions of zero reflectance at wavelength λ_0.

Table I.   Refractive Indices of Some Materials

Material	Index	Material	Index
MgF$_2$	1.3–1.4	Si$_3$N$_4$	1.9–1.95
SiO$_2$	1.4–1.5	TiO$_2$	2.3–2.4
Al$_2$O$_3$	1.7–1.8	Ta$_2$O$_5$	2.1–2.3
SiO	1.8–1.9	ZnS	2.5
SnO$_2$	1.8–1.9	Diamond	2.4

MgF$_2$ ($n \approx 1.38$) can be used, allowing a larger than 2% reduction of the reflectance. Table I lists the refractive indices of some materials.

Note that the values given in Table I correspond to refractive indices measured in the transparent regions of the spectrum, in other words, far from the absorption edge, where the index of refraction increases with photon energy. Variations of the refractive index with photon energy can be included in the design of antireflection coatings. The calculations, however, become more involved and numerical methods are required.

A double-layer coating using two different dielectrics, 1 and 2, allows zero reflectance at two different wavelengths. The condition for an equal intensity of reflection at each of the three interfaces is now expressed as $n_0/n_1 = n_1/n_2 = n_2/n_s$. Quarter-wavelength-thick layers of such materials, $\lambda_0/4n_1$ and $\lambda_0/4n_2$, produce destructive interference at wavelengths $3/4\lambda_0$ and $3/2\lambda_0$, at which the phase difference Θ between the beams reflected at the three interfaces equals 120° and 240°, respectively, as shown in Figure 10. The theory can be extended to a larger number of layers giving an achromatic antireflection coating. Three appropriate AR layers offers the possibility of zero reflectance at three different wavelengths, etc. As the number of layers increases, however, the benefits may not compensate the additional technical difficulties associated with the deposition of extra films.

The refractive index match of a large number of layers needed to obtain zero reflection may become difficult to meet. However, as the resulting reflectance is approximately given by the vector sum of the Fresnel coefficients at the interfaces, a zero reflectance at a given wavelength can be met using the film thickness as a design variable. The required condition is that the Fresnel coefficients $r_1, \ldots, r_n$ form a closed polygon at the selected wavelength [17].

The same kind of analysis may be applied to the circumstance where a high reflectance is desired at a particular wavelength. Besides metals, dielectric and semiconductor materials can be used for this purpose. Quarter-wavelength layers are still used, but the refractive index of the coating should be greater than that of the substrate; in fact, the difference should be as large as possible. ZnS, TiO$_2$, and Ta$_2$O$_5$ are used onto glass for applications in the visible spectrum. For infrared radiation, a collection of very dense materials exists, such as Si ($n \sim 3.5$), Ge ($n \sim 4$), and the lead salts (PbS, PbSe, and PbTe), to cite a few. With these materials, it is possible to have reflected intensities larger than 80%.

and isotropic flat parallel dielectric materials. The analytical expressions for the reflectivity, defined as the ratio of the intensity of the reflected beam to that of the incident beam, is

$$R = \left\{ \left[(\varepsilon_1)^{1/2} + (\varepsilon_S)^{1/2} \right]^2 \left[1 - (\varepsilon_1)^{1/2} \right]^2 \right.$$
$$+ \left[(\varepsilon_1)^{1/2} - (\varepsilon_S)^{1/2} \right]^2 \left[1 + (\varepsilon_1)^{1/2} \right]^2$$
$$\left. + 2(\varepsilon_1 - \varepsilon_S)(1 - \varepsilon_1) \cos(2k_1 t) \right\}$$
$$\times \left\{ \left[(\varepsilon_1)^{1/2} + (\varepsilon_S)^{1/2} \right]^2 \left[1 + (\varepsilon_1)^{1/2} \right]^2 \right.$$
$$+ \left[(\varepsilon_1)^{1/2} - (\varepsilon_S)^{1/2} \right]^2 \left[1 - (\varepsilon_1)^{1/2} \right]^2$$
$$\left. + 2(\varepsilon_1 - \varepsilon_S)(1 - \varepsilon_1) \cos(2k_1 t) \right\}^{-1}$$

where medium 0 is air, medium 1 is the optical coating, and S is the substrate. The minimum R occurs for $t = \pi/2k_1 = \lambda_0/4\sqrt{\varepsilon_1}$ and for all odd integral multiples of this thickness. Under these circumstances, the beams reflected at the air/layer and at the layer/substrate interfaces interfere destructively, as shown in Figure 9. The minimum reflectance reads

$$R_{\min}(\lambda_1) = \left[\frac{(\varepsilon_S)^{1/2} - \varepsilon_1}{(\varepsilon_S)^{1/2} + \varepsilon_1} \right]^2$$

The condition for zero reflection of a single coating can be obtained at a wavelength λ_0 if $\varepsilon_1(\lambda_0) = \sqrt{\varepsilon_s(\lambda_0)}$, or $n_1 = \sqrt{n_s}$ if the material is transparent at λ_0. The relation guarantees that the intensity reflected at each of the two interfaces is the same. If air is not the medium where the light beam comes from ($\varepsilon_0 > 1$), the condition for zero reflection for odd multiples of a quarter-wavelength thickness becomes $\varepsilon_1 = (\varepsilon_0\varepsilon_s)^{1/2}$. For the case of a typical glass $n_G \approx 1.5$/air interface, the antireflection layer should possess an index $n_1 = 1.225$ for a perfect match. Stable materials having this low index value do not exist. Instead,

$$n_0/n_1 = n_1/n_2 = n_2/n_s$$

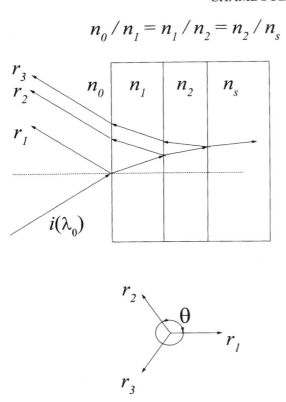

Fig. 10. Quarter-wavelength double antireflection coating. The figure illustrates the conditions for index match to produce zero reflectance at wavelengths $3/4\lambda_0$ and $3/2\lambda_0$.

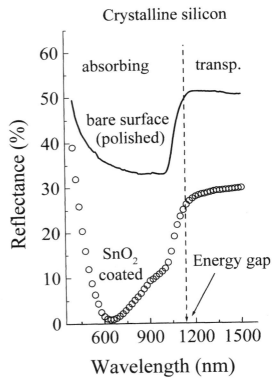

Fig. 11. Reflectance (%) of a polished crystalline silicon wafer as a function of wavelength. The open circles indicate the reflectance measured after spraying a 77-nm-thick SnO_2 layer.

Multiple-layer reflector structures employ an odd large number of stacked dielectric films of quarter-wavelength thickness. The stacking alternates layers of a large and a small index of refraction. The reflected rays of all the interfaces of the structure emerge in phase for the selected wavelength, at which a reflectance of nearly 100% can be obtained. The high reflectance, however, exists for a rather narrow frequency range because the phase change at the interfaces depends on wavelength. The expressions giving the reflectance and the transmittance of such structures as a function of wavelength are rather lengthy but can be optimized without difficulty using the formulation given in the previous section.

2.1.1. Antireflection Coatings for Solar Cells

Solar cells are devices that directly convert sunlight into electricity. We will not review here the basic mechanisms of solar electricity generation, but simply note that antireflection coatings are of utmost importance to improve the conversion efficiency of the device. As a consequence, the subject has received much attention in the last decades, when solar electricity generation became an interesting alternative for many terrestrial applications. Silicon, either single-crystal or polycrystalline, is the base semiconductor material dominating the industrial manufacturing of solar cells. Other semiconductor alloys, such as hydrogenated amorphous silicon (a-Si:H), CdTe, and $CuInSe_2$, are emerging materials for photovoltaic applica-

tions. Let us consider the reflectance of a single-crystal silicon solar cell. Figure 11 illustrates the situation. The continuous line is the measured reflectance of a polished, 400-μm-thick, Si wafer ($n_{Si} \geq 3.4$). The vertical dashed line roughly indicates the boundary between complete absorption and transparency. In the transparent region, the reflectance is high due to the large index of silicon and to the contribution of the two faces of the wafer to the overall reflectance. For wavelengths smaller than $\sim$1130 nm, the reflectance decreases because of light absorption in the wafer, which diminishes the contribution of the reflection at the back Si–air interface. At $\lambda \lesssim 1050$ nm, the back interface does not contribute any more and the measured reflectance corresponds to the front interface only. The increased reflectance at $\lambda \lesssim 1050$ nm derives from an increasing n_{Si}. The open symbols indicate the reflectance measured of the same Si wafer after spraying an SnO_2 AR coating. The situation depicted in Figure 11 deserves some additional comments.

(a) First, in order to work as an electric generator, photons must be absorbed in the silicon substrate; that is, the photons of interest possess an energy larger than the Si band-gap [$\hbar\omega > E_g(Si) \approx 1.1$ eV, or $\lambda < \lambda_{gap} \approx 1130$ nm].

(b) Second, the solar spectrum is very broad [32]. The continuous line in Figure 12 shows the irradiance of sunlight, or its spectral intensity distribution, in the wavelength range of interest for the present discussion. Note the radiance units on the left ordinate. The area below the complete curve gives

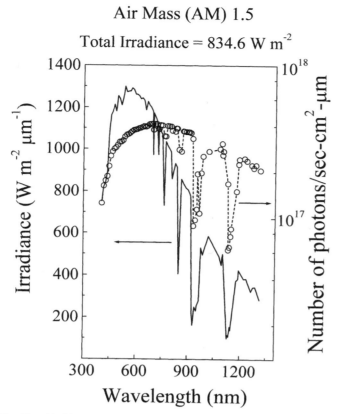

Fig. 12. Air Mass 1.5 solar spectrum. Full line: irradiance (power density) per unit wavelength interval. Open circles: photon flux density per unit energy interval. Note that for solar cells (quantum detectors), it is the reflected photon flux that has to be minimized, not the irradiance.

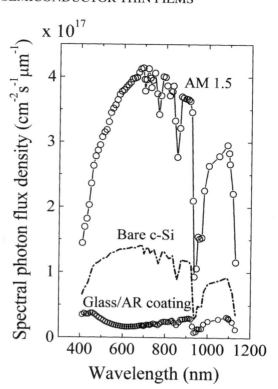

Fig. 13. Spectral photon flux density under AM 1.5 irradiation conditions. The dotted-dashed curve indicates the photon flux density reflected by a polished silicon surface (solar cell). The open circles at the bottom of the figure show the photon reflection in a system consisting of a cover glass, an optimized AR coating, and a crystalline silicon solar cell. The optimization finds the index of refraction and the thickness of the AR coating that minimizes the sum of reflected useful photons of the AM 1.5 solar spectrum. See text.

$834.6 \, \mathrm{W \cdot m^{-2}}$, and this value corresponds to a clear day with the sun inclined 48° (AM 1.5) with respect to the normal [32]. The spectrum shows several absorption bands originating mainly from water vapor and carbon oxides.

(c) Third, absorbed photons means working in a photon energy region where the optical constants of the substrate silicon strongly depend on wavelength.

(d) Fourth, solar cells are quantum detectors; that is, electrons are excited through photon absorption. In other words, in order to maximize the photocurrent of the solar cell, the total amount of reflected photons has to be minimized in the energy range of interest, and not the reflectance (intensity). Remember that photon energy is inversely proportional to wavelength. The open circles in Figure 12 show the photon density per unit wavelength interval versus wavelength for the same radiance spectrum given by the continuous line. The area below this curve for energies smaller than the energy gap gives the total number of absorbed photons. This is the spectrum to be considered for photovoltaic AR coatings.

(e) Fifth, the optimization of the photocurrent in a solar cell must also consider: (1) the internal collection efficiency, which measures the amount of collected electrons at the electrical contacts per absorbed photon, and (2) the variations of the refractive index with wavelength. In a good device, the collection efficiency $\eta \simeq 1$ for most absorbed photons. In general, it decreases at short wavelengths due to surface effects. At long wavelengths, photon absorption decreases because of a decrease of the absorption coefficient, and the photocurrent drops.

Summarizing, the optimization of an antireflection coating on a solar cell should include the variations of the refractive index of the substrate material in the range of high absorption and the quantum nature of the conversion process (photon spectral distribution of the solar spectrum).

As an illustrative case, we present below the results of minimizing the reflected photon flux in a solar converter system consisting of a 4-mm protecting glass ($n = 1.52$), an AR coating (to be optimized), and a crystalline silicon solar cell. The input data are the index of refraction of the glass cover, the optical properties of crystalline silicon, and the photon flux density of the AM 1.5 solar spectrum (see Fig. 12). We consider normal incidence for the photon flux and the 400- to 1130-nm photon wavelength range. The minimization algorithm finds $n = 2.41$ and $d = 66$ nm for the AR coating that will maximize the energy output of the system. The results on reflectance are shown in Figure 13.

The above optimization will just consider the photon distribution of a particular sunlight spectrum. In fact, the important parameter to be maximized is the electric energy delivered by the device. In other words, a complete optimization should consider the position of the sun in the sky along the day and along the year (angle of incidence), as well as the variations of the solar spectrum with prevailing weather conditions.

2.2. A Reverse Engineering Problem: The Retrieval of the Optical Constants and the Thickness of Thin Films from Transmission Data

Thin-film electronics includes thin-film transistors and amorphous semiconductor solar cells. In many circumstances, the thickness d and the optical properties $n(\lambda) - i\kappa(\lambda)$ must be known with some degree of accuracy. As shown in a previous section, measurements of the complex amplitudes of the light transmitted and reflected at normal or oblique incidence at the film and substrate sides, or different combinations of them, enable the explicit evaluation of the thickness and the optical constants in a broad spectral range. For electronic applications, however, the interesting photon energy range covers the neighborhood of the fundamental absorption edge in which the material goes from complete opacity to some degree of transparency. As optical transmittance provides accurate and quick information in this spectral range, efforts have been made to develop methods allowing the retrieval of the thickness and the optical constants of thin films from transmittance data only.

Figure 14 shows three transmission spectra typical of thin isotropic and homogeneous dielectric films, bounded by plane-parallel surfaces and deposited onto transparent thick substrates. They correspond to: (a) a film, thick enough to display interference fringes coming from multiple coherent reflections at the interfaces and being highly transparent below the fundamental absorption edge; (b) a thick film, as above, but strongly absorbing in all the measured spectral range; and (c) a very thin film not showing any fringe pattern in the spectral region of interest. The interference fringes of films having a transparent spectral region, such as the ones shown by film 14a, can be used to estimate the film thickness and the real part of the index of refraction in this region [8, 33–35]. The approach is known as the envelope method, first proposed by Manifacier et al. [34]. The calculation of the optical constants in the absorbing region, however, cannot be estimated in a simple way. Usually, a functional dependence for $n(\lambda)$ is assumed, in most cases a function of the type $n = A\lambda^{-2} + B\lambda + C$ (where A, B, and C are constants), to extrapolate the index values estimated in the transparent region of the spectrum to higher photon energies [35]. Knowing d and $n(\lambda)$, the absorption coefficient can be calculated from the general expression of the transmission. In general, these approximations are of limited accuracy and result in a poor estimate of the absorption coefficient. Moreover, films not possessing a transparent spectral region (see film 14b), or too thin to display optical fringes in the measured range (see film 14c), cannot be solved by these approximate methods, which use transmission data only.

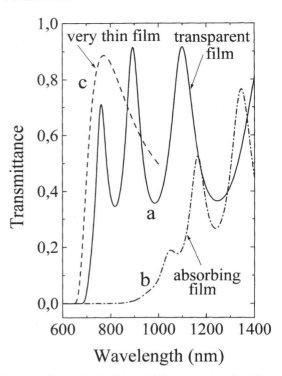

Fig. 14. Transmittance in the 600- to 1400-nm wavelength of three typical semiconductor thin films. Spectrum a corresponds to a relatively thick film displaying a fringe pattern in a region of very weak absorption. Spectrum b sketches the transmittance of a thick absorbing semiconductor film. There is no useful fringe pattern in the measured wavelength range. Spectrum c illustrates the transmittance of a rather thin film not displaying a fringe pattern.

Before discussing new retrieval algorithms, let us first consider in some detail the optical transmission measurements through thick transparent substrates (or layers).

2.2.1. Measured Transmission and the Four-Layer Case

In real cases, what is measured is not the transmittance of pure waves with a single λ. The detectors always indicate a signal corresponding to a wave packet of a finite bandwidth ($\pm \Delta\lambda$) centered around λ. Hence, the measured transmittance data of an *air/film/thick substrate/air* system ($n_0/n_1/n_S/n_0$) represent an average transmittance, or a signal integrated in a very narrow frequency band, the width of which depends on a chosen slit or other experimental conditions. Therefore, what is measured is not $|t_{m-1,0}(\lambda)|^2$ but:

$$\text{Measured transmission} = \frac{1}{2\Delta\lambda} \int_{\lambda-\Delta\lambda}^{\lambda+\Delta\lambda} |t_{m-1,0}(\lambda)|^2 \, d\lambda$$

where $\pm\Delta\lambda$ corresponds to the mentioned experimental bandpass centered around λ.

As mentioned in a previous section, $|t_{m-1,0}(\lambda)|^2$ is a periodic function of film thickness, with period $\lambda/(2n_1)$. Usually, the length of this period is smaller than the substrate thickness, say $[\lambda/(2n_1)]/d_S \sim 10^{-3}$–$10^{-4}$. Such values are, in turn, of the order of, or smaller than, the error associated to the measured substrate thickness. As a consequence, a reasonable approximation to the measured transmittance is an average over the entire period:

Measured transmission (2)

$$= \frac{2n_1}{\lambda} \int_{d_S}^{d_S+\lambda/(2n_1)} |t_{m,1}(\lambda, d_S)|^2 \, d(d_S) \qquad (43)$$

The integral (43) can be calculated analytically and, clearly, does not depend on the substrate thickness d_S. The discussion can be easily extended to the multilayer case.

Assume now, as above, that we are in the mentioned four-layer situation: *air* (n_0)/*film* (n_1)/*thick substrate* (n_S)/*air* (n_0). For the sake of simplicity of notation, let us call, from now on, d, n, and κ the thickness, the refractive index, and the attenuation coefficient of the film, and s the refractive index of the substrate. In this case, (43) reduces to

$$T = \text{Measured transmission (2)}$$
$$= \frac{Ax}{B - Cx + Dx^2} \qquad (44)$$

where

$$A = 16s(n^2 + \kappa^2) \qquad (45)$$

$$B = [(n+1)^2 + \kappa^2][(n+1)(n+s^2) + \kappa^2] \qquad (46)$$

$$C = [(n^2 - 1 + \kappa^2)(n^2 - s^2 + \kappa^2) - 2\kappa^2(s^2 + 1)]2\cos\varphi$$
$$- \kappa[2(n^2 - s^2 + \kappa^2)$$
$$+ (s^2 + 1)(n^2 - 1 + \kappa^2)]2\sin\varphi \qquad (47)$$

$$D = [(n-1)^2 + \kappa^2][(n-1)(n-s^2) + \kappa^2] \qquad (48)$$

$$\varphi = 4\pi nd/\lambda \qquad x = \exp(-\alpha d) \qquad \alpha = 4\pi\kappa/\lambda \qquad (49)$$

2.2.2. Pointwise Optimization Approach

Consider now the problem of estimating the absorption coefficient, the refractive index, and the thickness of a thin film, as in the above example, using only transmittance data for (many) different wavelengths. The theoretical transmission is given by (44), where the refractive index of the substrate s is known and $n(\lambda)$, $\alpha(\lambda)$, and d are the unknowns. At first glance, this problem is highly underdetermined since, for each wavelength, the single equation

| Theoretical transmission = Measured transmission | (50) |

has three unknowns d, $n(\lambda)$, $\alpha(\lambda)$, and only d is repeated for all values of λ. One way to decrease the degrees of freedom of (50), at a point that only physically meaningful estimated parameters are admissible, is to consider a prior knowledge of the functions $n(\lambda)$, $\alpha(\lambda)$. The idea of assuming a closed formula for n and α depending on a few coefficients has been explored in envelope methods [33–35]. These methods are efficient when the transmission curve exhibits a fringe pattern, representing a rather large zone of the spectrum, where $\alpha(\lambda)$ is almost null. In other cases, the fulfillment of (50) can be very crude or the curves $n(\lambda)$, $\alpha(\lambda)$ are physically unacceptable.

In [36, 37], instead of imposing a functional form for $n(\lambda)$ and $\kappa(\lambda)$, the phenomenological constraints that restrict the variability of these functions were stated explicitly, so that the estimation problem takes the form:

Minimize $\sum_{\lambda}$[Theoretical transmission(λ)

$$- \text{Measured transmission}(\lambda)]^2$$

subject to Physical Constraints $\qquad (51)$

In this way, well-behaved functions $n(\lambda)$ and $\kappa(\lambda)$ can be obtained without severe restrictions that could damage the quality of the fitting (50).

At this point, it is convenient to say that the main inconvenience of the pointwise constrained optimization approach [36, 37] is that (51) is a rather complex large-scale linearly constrained nonlinear programming problem whose solution can be obtained only by means of rather sophisticated and not always available computer codes that can deal effectively with the sparsity of the matrix of constraints. See [38, 39].

A second approach was established in [40] for solving the estimation problem. In the new method, (51) is replaced by an unconstrained optimization problem. We solved this problem using a very simple algorithm introduced recently by Raydan [41], which realizes a very effective idea for potentially large-scale unconstrained minimization. It consists of using only gradient directions with steplengths that ensure rapid convergence. The reduction of (51) to an unconstrained minimization problem needs the calculation of very complicated derivatives of functions, which requires the use of automatic differentiation techniques. The present authors used the procedures for automatic differentiation described in [42]. Before addressing the unconstrained optimization approach, let us consider the main ideas behind the pointwise optimization, as well as its effectiveness in retrieving the optical constants and the thickness of *gedanken* and real films.

A set of experimental transmittance data $(\lambda_i, T^{\text{meas}}(\lambda_i)$, $i = 1, \ldots, N$, $\lambda_{\min} \leq \lambda_i < \lambda_{i+1} \leq \lambda_{\max}$ for all $i = 1, \ldots, N)$ is given, $s(\lambda)$ is known, and we want to estimate d, $n(\lambda)$, and $\alpha(\lambda)$. As said, the problem is highly underdetermined because, for a given d and any λ, the following equation must hold:

$$T^{\text{calc}}(\lambda, s(\lambda), d, n(\lambda), \kappa(\lambda)) = T^{\text{meas}}(\lambda)$$

Equation (44) has two unknowns $n(\lambda)$ and $\alpha(\lambda)$ and, therefore, its set of solutions is, in general, a curve in the two-dimensional $(n(\lambda), \alpha(\lambda))$ space. As a consequence, the set of functions (n, α) that satisfy (44) for a given d is infinite and, roughly speaking, is represented by a nonlinear manifold of dimension N in $\mathbb{R}^{2N}$.

However, physical constraints reduce drastically the range of variability of the unknowns $n(\lambda)$, $\alpha(\lambda)$. The type of constraints which relate n and α at different wavelengths derive from a prior knowledge of the physical solution. To illustrate the approach, let us consider, for example, the behavior of the optical constants of an amorphous (a-)semiconductor thin film deposited onto a thick glass substrate. In the neighborhood of the fundamental absorption edge of a-semiconductors, it is known that:

PC1: $n(\lambda) \geq 1$, $\alpha(\lambda) \geq 0$ for all $\lambda \in [\lambda_{\min}, \lambda_{\max}]$;
PC2: $n(\lambda)$ and $\alpha(\lambda)$ are decreasing functions of λ;
PC3: $n(\lambda)$ is convex;
PC4: in a $\log(\alpha)$ vs. $\hbar\omega$ (photon energy) plot, the absorption coefficient α of an a-semiconductor has an elongated $\int$-like shape. Thus, for the spectral region corresponding to the exponential absorption edge, and smaller photon energies, there exists $\lambda_{\mathrm{infl}} \in [\lambda_{\min}, \lambda_{\max}]$ such that $\alpha(\lambda)$ is convex if $\lambda \geq \lambda_{\mathrm{infl}}$ and concave if $\lambda < \lambda_{\mathrm{infl}}$.

The constraints **PC1**, **PC2**, **PC3**, and **PC4** relate unknowns of (50) for different indices i. So, (50) cannot be considered independent anymore and its degrees of freedom are restricted. From these considerations, the problem of determining d, n, and α in a selected spectral region can be modeled minimizing

$$\sum_{i=1}^{N} \left[T_i^{\mathrm{calc}}\left(\lambda, s(\lambda), d, n(\lambda), \alpha(\lambda)\right) - T^{\mathrm{meas}}(\lambda) \right]^2$$

subject to constraints **PC1**, **PC2**, **PC3**, and **PC4**. For obvious reasons, this is called a *pointwise constrained optimization approach* for the resolution of the estimation problem.

Observe that, assuming **PC2**, **PC1** is satisfied under the sole assumption $n(\lambda_{\max}) \geq 1$ and $\kappa(\lambda_{\max}) \geq 0$.

The constraints **PC2**, **PC3**, and **PC4** can be written, respectively, as

$$n'(\lambda) \leq 0 \qquad \kappa'(\lambda) \leq 0 \quad \text{for all } \lambda \in [\lambda_{\min}, \lambda_{\max}] \quad (52)$$
$$n''(\lambda) \geq 0 \quad \text{for all } \lambda \in [\lambda_{\min}, \lambda_{\max}] \quad (53)$$
$$\kappa''(\lambda) \geq 0 \quad \text{for } \lambda \in [\lambda_{\mathrm{infl}}, \lambda_{\max}] \quad \text{and}$$
$$\kappa''(\lambda) \leq 0 \quad \text{for } \lambda \in [\lambda_{\min}, \lambda_{\mathrm{infl}}] \quad (54)$$

Clearly, the constraints

$$n''(\lambda) \geq 0 \quad \text{for all } \lambda \in [\lambda_{\min}, \lambda_{\max}] \quad \text{and} \quad n'(\lambda_{\max}) \leq 0$$

imply that

$$n'(\lambda) \leq 0 \quad \text{for all } \lambda \in [\lambda_{\min}, \lambda_{\max}]$$

Moreover,

$$\kappa''(\lambda) \geq 0 \quad \text{for all } \lambda \in [\lambda_{\mathrm{infl}}, \lambda_{\max}] \quad \text{and} \quad \kappa'(\lambda_{\max}) \leq 0$$

imply that

$$\kappa'(\lambda) \leq 0 \quad \text{for all } \lambda \in [\lambda_{\mathrm{infl}}, \lambda_{\max}]$$

Finally,

$$\kappa'(\lambda_{\min}) \leq 0 \quad \text{and} \quad \kappa''(\lambda) \leq 0 \quad \text{for all } \lambda \in [\lambda_{\min}, \lambda_{\mathrm{infl}}]$$

imply that

$$\kappa'(\lambda) \leq 0 \quad \text{for all } \lambda \in [\lambda_{\min}, \lambda_{\mathrm{infl}}]$$

Therefore, **PC2** can be replaced by

$$n'(\lambda_{\max}) \leq 0 \qquad \kappa'(\lambda_{\max}) \leq 0 \qquad \kappa'(\lambda_{\min}) \leq 0$$

Summing up, the assumptions **PC1–PC4** will be satisfied if, and only if,

$$n(\lambda_{\max}) \geq 1 \qquad \kappa(\lambda_{\max}) \geq 0 \qquad (55)$$
$$n'(\lambda_{\max}) \leq 0 \qquad \kappa'(\lambda_{\max}) \leq 0 \qquad (56)$$
$$n''(\lambda) \geq 0 \quad \text{for all } \lambda \in [\lambda_{\min}, \lambda_{\max}] \qquad (57)$$
$$\kappa''(\lambda) \geq 0 \quad \text{for all } \lambda \in [\lambda_{\mathrm{infl}}, \lambda_{\max}] \qquad (58)$$
$$\kappa''(\lambda) \leq 0 \quad \text{for all } \lambda \in [\lambda_{\min}, \lambda_{\mathrm{infl}}] \qquad (59)$$
$$\kappa'(\lambda_{\min}) \leq 0 \qquad (60)$$

So, the continuous least-squares solution of the estimation problem is the solution $(d, n(\lambda), \kappa(\lambda))$ of

Minimize
$$\int_{\lambda_{\min}}^{\lambda_{\max}} \left| T^{\mathrm{calc}}\left(\lambda, s(\lambda), d, n(\lambda), \kappa(\lambda)\right) - T^{\mathrm{meas}}(\lambda) \right|^2 d\lambda \quad (61)$$

subject to the constraints (55)–(60).

For a given value of the thickness d, the constraints **PC1**, **PC2**, **PC3**, and **PC4** define a polyhedron P with nonempty interior in $\mathbb{R}^{2N}$. (The discretization of any pair of nonnegative, strictly convex, and decreasing functions n and α is an interior point of $\mathbb{P}$.) Therefore, the intersection of $\mathbb{P}$ with the set of points that satisfy (50) is nonempty, and defines a set of $\mathbb{R}^{2N}$ that, in general, contains more than one point. In principle, a global solution of (51), (55)–(60) could not be a unique solution of the estimation problem (which clearly exists in actual real-world situations because of physical reasons and, in numerical *gedanken* experiments, by construction). However, if the intersection of $\mathbb{P}$ with the set defined by (50) is small, we can expect that any (approximate) solution of (51), (55)–(60) is the closest to the true solution of the estimation problem. Theoretical verification of this fact is very hard, or even impossible, except for very small values of N, but an experimental study was designed in order to establish the reliability of the pointwise constrained optimization approach.

To illustrate the application of the above procedure, consider the optical transmission data of computer-generated films of known optical properties and thickness deposited onto a transparent substrate of known index of refraction. These *gedanken* experiments intend to simulate representative films for electronic applications. The goodness of the retrieval process is evaluated by comparing the retrieved values with the known solution. The transmission data, generated in the 500- to 1000-nm wavelength range, are rounded off to three digits to mimic the precision of the information normally available with experimental data. The simulation includes a glass substrate, assumed to be transparent in the spectral range under consideration. The numerical examples consider an absorption coefficient of the film having an exponential dependence on photon energy. The idea behind this particular choice of a rather steep $\alpha(\hbar\omega)$ is to investigate the powerfulness of the method for the retrieval of very small and very large α's.

The calculated transmittance of two films (1 and 2) of thickness 100 nm and 600 nm, respectively, is shown in Figure 15. As expected, film 1 does not show any fringe pattern in the

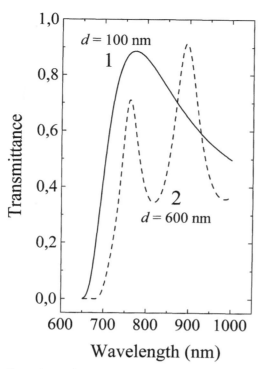

Fig. 15. Transmittance in the 650- to 1000-nm wavelength range of two computer-generated thin films supported by a transparent substrate of known properties. The thickness of the films 1 and 2 are 100 and 600 nm, respectively. Reproduced with permission from [36], copyright 1997, Optical Society of America.

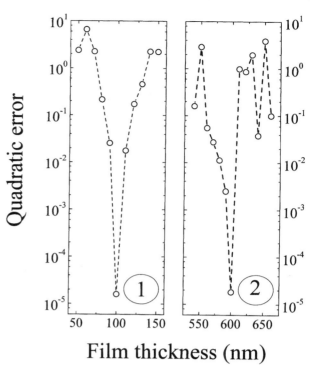

Fig. 16. Quadratic error of the minimization process as a function of trial thickness for films 1 and 2 of Figure 13. Note the excellent retrieval of the true thickness and the presence of local nonglobal minimizers. Reproduced with permission from [37], copyright 1998, Elsevier Science.

spectral range under consideration. Due to the photon energy dependence of the adopted α, and the film thickness, no transmission is measured for $\lambda < 650$ nm. Note that, using only the transmission data in the $650 \leq \lambda \leq 1000$ nm range, none of the films under consideration can be solved by approximate, or envelope, methods.

The results of the minimization process **as a function of the film thickness** d, which is the optimization variable, are shown in Figure 16. The least value of the quadratic error corresponds to the true film thickness. In the tests, the initial guesses for n and α are far from the values used to generate the transmission data. Figure 17 shows the retrieved optical constants of these *gedanken* films as a function of photon energy as given by the minimization process at the optimum thickness. They are compared in Figure 17 with the "true" values used to generate the transmission data. The precision of the retrieval is outstanding.

2.2.3. Unconstrained Optimization

The idea in the second approach [40] is to eliminate, as far as possible, the constraints of the problem, by means of suitable changes of variables. Roughly speaking, we are going to put the objective function of (61) as depending on the second derivatives of $n(\lambda)$ and $\kappa(\lambda)$ plus functional values and first derivatives at λ_{max}. Moreover, positivity will be guaranteed expressing the variables as squares of auxiliary unknowns. In fact,

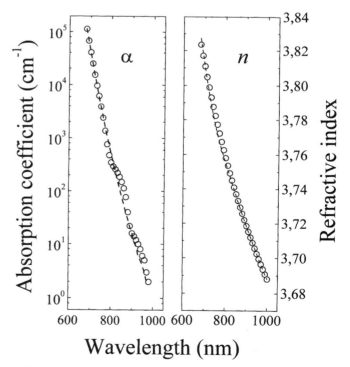

Fig. 17. Retrieved absorption coefficient and index of refraction of film 2 of Figure 13. Dashed line: α and n values used to calculate the transmittance shown in Figure 13. Open circles: retrieved α and n values. The agreement between original and retrieved values is very good.

we write

$$n(\lambda_{\max}) = 1 + u^2 \qquad \kappa(\lambda_{\max}) = v^2 \tag{62}$$

$$n'(\lambda_{\max}) = -u_1^2 \qquad \kappa'(\lambda_{\max}) = -v_1^2 \tag{63}$$

$$n''(\lambda) = \omega(\lambda)^2 \quad \text{for all } \lambda \in [\lambda_{\min}, \lambda_{\max}] \tag{64}$$

$$\kappa''(\lambda) = z(\lambda)^2 \quad \text{for all } \lambda \in [\lambda_{\mathrm{infl}}, \lambda_{\max}] \tag{65}$$

$$\kappa''(\lambda) = -z(\lambda)^2 \quad \text{for all } \lambda \in [\lambda_{\min}, \lambda_{\mathrm{infl}}] \tag{66}$$

At this point, in order to avoid a rather pedantic continuous formulation of the problem, we consider the real-life situation, in which data are given for a set of N equally spaced points on the interval $[\lambda_{\min}, \lambda_{\max}]$. So, we define

$$h = \frac{\lambda_{\max} - \lambda_{\min}}{N - 1}$$

and

$$\lambda_i = \lambda_{\min} + (i - 1)h \quad i = 1, \dots, N$$

Consequently, the measured value of the transmission at λ_i will be called T_i^{meas}. Moreover, we will use the notation n_i, κ_i, ω_i, z_i in the obvious way:

$$n_i = n(\lambda_i) \qquad \kappa_i = \kappa(\lambda_i)$$

$$\omega_i = \omega(\lambda_{i+1}) \qquad z_i = z(\lambda_{i+1})$$

for all $i = 1, \dots, N$. Discretization of the differential relations (62)–(66) gives

$$n_N = 1 + u^2 \qquad v_N = v^2 \tag{67}$$

$$n_{N-1} = n_N + u_1^2 h \qquad \kappa_{N-1} = \kappa_N + v_1^2 h \tag{68}$$

$$n_i = \omega_i^2 h^2 + 2n_{i+1} - n_{i+2} \quad i = 1, \dots, N-2 \tag{69}$$

$$\kappa_i = z_i^2 h^2 + 2\kappa_{i+1} - \kappa_{i+2} \quad \text{if } \lambda_{i+1} \geq \lambda_{\mathrm{infl}} \tag{70}$$

$$\kappa_i = -z_i^2 h^2 + 2\kappa_{i+1} - \kappa_{i+2} \quad \text{if } \lambda_{i+1} < \lambda_{\mathrm{infl}} \tag{71}$$

Finally, the objective function of (61) is approximated by a sum of squares, giving the optimization problem

$$\text{Minimize} \quad \sum_{i=1}^{N} \left[T^{\mathrm{calc}}(\lambda_i, s(\lambda_i), d, n_i, \kappa_i) - T_i^{\mathrm{meas}}\right]^2 \tag{72}$$

subject to

$$\kappa_1 \geq \kappa_2 \tag{73}$$

Since n_i and κ_i depend on u, u_1, v, v_1, ω, z, and λ_{infl} through (67)–(71), the problem (72) takes the form

$$\boxed{\text{Minimize} f(d, \lambda_{\mathrm{infl}}, u, u_1, v, v_1, \omega_1, \dots, \omega_{N-2}, z_1, \dots, z_{N-2})}$$
$$\tag{74}$$

subject to (73). We expect that the constraint (73) will be inactive at a solution of (74), so we consider the unconstrained problem (74).

The unknowns that appear in (74) have different natures. The thickness d is a dimensional variable (measured in nanometers in our real-life problems) that can be determined using the data $T^{\mathrm{meas}}(\lambda_i)$ for (say) $\lambda_i \geq \lambda_{\mathrm{bound}}$, where λ_{bound}, an upper bound for λ_{infl}, reflects our prior knowledge of the physical problem.

For this reason, our first step in the estimation procedure will be to estimate d using data that correspond to $\lambda_i \geq \lambda_{\mathrm{bound}}$. To accomplish this objective, we solve the problem

$$\text{Minimize} \quad \overline{f}(u, u_1, v, v_1, \omega, z)$$
$$\equiv \sum_{\lambda_i \geq \lambda_{\mathrm{bound}}} \left[T^{\mathrm{calc}}(\lambda_i, s(\lambda_i), d, n_i, \kappa_i) - T_i^{\mathrm{meas}}\right]^2 \tag{75}$$

for different values of d, and we take as estimated thickness the one that gives the lowest functional value. In this case, the constraint (73) is irrelevant since it is automatically satisfied by the convexity of κ and the fact that the derivative of κ at $\lambda_{\min}$ is nonpositive. From now on, we consider that d is fixed, coming from the above procedure.

The second step consists of determining λ_{infl}, together with the unknowns u, u_1, v, v_1, ω, z. For this purpose, note that, given d and λ_{infl}, the problem

$$\text{Minimize} \quad \sum_{i=1}^{N} \left[T^{\mathrm{calc}}(\lambda_i, s(\lambda_i), d, n_i, \kappa_i) - T_i^{\mathrm{meas}}\right]^2 \tag{76}$$

is (neglecting (73)) an unconstrained minimization problem whose variables are u, u_1, v, v_1, ω, z ($2N$ variables). We solve this problem for several trial values of λ_{infl}, and we take as estimates of n and κ the combination of variables giving the lowest value. To minimize this function and to solve (75) for different trial thicknesses, we use the unconstrained minimization solver introduced in [41]. Let us emphasize at this point that the unconstrained optimization method has proven to be as reliable as the pointwise method. The comparison has been performed on numerically generated films and on real films deposited onto glass and onto crystalline substrates. We illustrate the point in the following subsection.

Testing the Algorithm with a-Si:H Films Deposited onto Glass Substrates. The unconstrained formulation applied to real thin films is considered now [43]. A series of a-Si:H samples with thicknesses varying from 100 nm to 1.2 μm, were deposited by the plasma decomposition of SiH_4 in a chemical vapor deposition (CVD) system under identical nominal conditions at an excitation frequency of 13.56 MHz. The deposition time was varied in order to produce films of different thickness but, otherwise, with identical properties. Figure 18 shows the optical transmission spectra of some samples of the series. The deposition time diminishes from (a) to (c), resulting in thinner films, as confirmed by a decreasing number of interference fringes. Note that the spectrum of sample 18a displays a reasonable fringe pattern in the region of weak absorption ($\lambda > 1000$ nm). The approximate properties of this film can be extracted from the position and the magnitude of the transmission maxima and minima appearing as an interference pattern [35]. However, the transmittance spectra of samples 18b and 18c, which possess a reduced (or absent) fringe pattern, can not be analyzed with envelope methods.

As already stated, the optimization process looks for a thickness that, subject to the physical constraints of the problem,

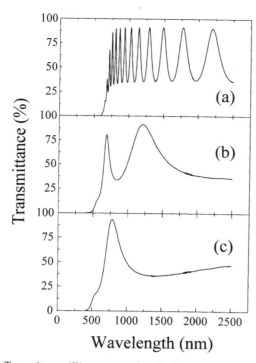

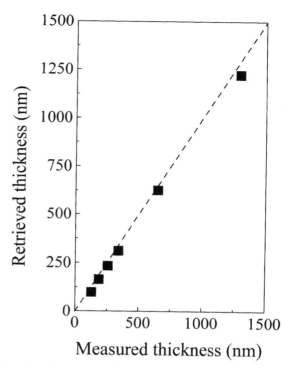

Fig. 18. Transmittance (%) versus wavelength of three a-Si:H films deposited onto glass of thickness: (a) 1220 nm, (b) 160 nm, and (c) 100 nm. Reproduced with permission from [43], copyright 2000, American Institute of Physics.

Fig. 19. Optically retrieved thickness versus mechanical measurement of a series of a-Si:H thin films. Reproduced with permission from [43], copyright 2000, American Institute of Physics.

minimizes the difference between the measured and the calculated spectra. The minimization starts sweeping a thickness range Δd_r divided into thickness steps Δd_s and proceeds decreasing Δd_r and Δd_s until the optimized thickness d_{opt} is found. In the present case, the starting Δd_r and Δd_s are 5 μm and 100 nm, respectively.

The thickness of the films can be measured by independent methods and compared with the ones obtained from the minimization process. To this aim, part of the deposited films was etched away and the height of the step (film thickness) measured with a Dektak profilometer. Figure 19 shows that the agreement between retrieved and measured thickness is very good for the whole series of samples. The small differences between optically and mechanically determined thickness may have several origins: (i) the precision and accuracy of the transmitted data may not be sufficient for a perfect thickness retrieval; (ii) there may be a relative error of the mechanical measurement; and (iii) there may exist a real thickness difference between the etched region and the region used for the transmittance measurements.

Figure 20 shows the retrieved optical constants of the films. The top part of Figure 20 displays the index of refraction n as a function of photon energy, different symbols being used for each film. Note that, for energies $\hbar\omega < 2.2$ eV, the indexes of refraction of all samples agree to a good extent. Although the films have been deposited under (nominally) identical conditions, and large differences in their properties are not expected, it is well known that the optical constants of a-Si:H films depend, to a certain extent, on film thickness. In that sense, we do

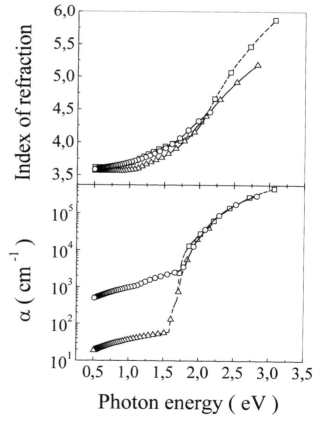

Fig. 20. Retrieved index of refraction and absorption coefficient of the films of Figure 16. Open circles: $d = 100$ nm; open squares: $d = 160$ nm; open triangles: $d = 1220$ nm. See text. Reproduced with permission from [43], copyright 2000, American Institute of Physics.

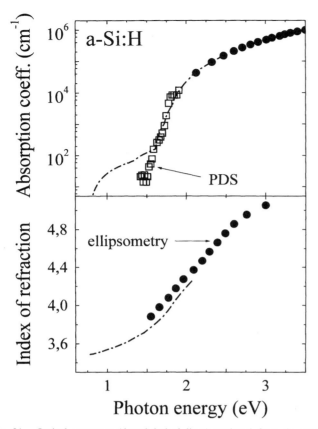

Fig. 21. Optical constants (dotted-dashed lines) retrieved from the transmittance data only of a 625-nm-thick a-Si:H film using the unconstrained optimization method. The retrieved values are compared with values measured on the same film with independent methods of characterization: ellipsometry and photothermal deflection spectroscopy. Reproduced with permission from [36], copyright 1997, Optical Society of America.

not expect that films having a thickness difference of more than one order of magnitude do display identical optical constants. The results shown in Figure 20, however, are a clear indication that the method works satisfactorily.

The bottom of Figure 20 shows the retrieved absorption coefficient α as a function of photon energy. The absorption coefficient at photon energies $\hbar\omega > 1.7$ eV is perfectly retrieved in all cases, even for the $d = 98$-nm-thick film. The retrieval of α at $\hbar\omega < 1.7$ eV depends, as expected, on film thickness. For a typical $d \approx 600$ nm a-Si:H film, values of α are correctly retrieved down to 100 cm^{-1}. The correctness of the retrieved optical constants has been confirmed in some samples by independent measuring techniques, that is, photothermal deflection spectroscopy and ellipsometry, as shown in Figure 21. Note that the absorption coefficient for the 625-nm-thick a-Si:H film shown in Figure 21 has been retrieved in a three-orders-of-magnitude span and down to 100 cm^{-1}. The small differences appearing between the retrieved refractive index and the ellipsometric determination of n may originate from the fact that, although the films have been grown simultaneously in the same run, they are deposited onto different substrates, glass for the optically retrieved n and crystalline silicon for the ellipsomet-

ric measurements. It is known that the nature of the substrate affects the properties of the films [36].

Summarizing, the unconstrained optimization method has been successful in retrieving the optical constants and the thickness of thin a-Si:H films (100 nm $< d < 1.2$ μm) from transmittance data only. The method, being applicable to absorbing films and to films not displaying any fringe pattern, constitutes a significant improvement over other retrieval methods.

3. CONCLUSIONS

The simulation and optimization of the behavior of dielectric layered devices can be improved by means of the massive incorporation of well-known and novel techniques of numerical mathematics. We have given some examples throughout this work, but many open problems remain and many working programs can be traced taking into account the real needs of the industry. The following topics are the subject of current research in our group at the University of Campinas.

1. Although the solution of the basic direct problem is well known, the optimization of the calculations that lead to computing electric and magnetic fields in a multilayer structure is far from being complete. Fast calculations are important if we want to obtain practical and useful responses in reasonable time. The organization of computations in an optimal (or nearly optimal) way without loss of numerical stability is a problem area of most practical importance.

2. Related to the previous topic is the problem of computing the derivatives of the parameters to be estimated (for example, the thicknesses) in an optimal way. Techniques of automatic (or computational) differentiation are presently well known. However, in the presence of particular classes of problems, calculations can be much improved taking into account the specific characteristics of them. Roughly speaking, automatic differentiation involves two essential techniques: direct differentiation and reverse differentiation. The first is "easier," uses less memory, but is more time-consuming than the second. We are using direct differentiation in our present codes but we feel that combining this with the reverse mode and, especially, introducing the special features of the problems in a clever way, will improve both accuracy and efficiency of this computation. Needless to say, fast and accurate computation of derivatives is an essential ingredient for the efficacy of optimization algorithms.

3. We have worked in the reverse engineering problem related to the estimation of parameters of thin films for wavelengths in the infrared zone. We used very strongly empirical knowledge about the behavior of the refraction index and absorption coefficient for those wavelengths. Analogous techniques can be used for the estimation of other types of materials using data in different spectral zones, if we employ in a suitable way phenomenological and empirical knowledge of the parameters to be estimated on those zones.

4. Current models for thin-film estimation are based on transparent substrates. This is an oversimplification since, in nature, such types of materials do not exist and the ones available in commerce (glasses, sapphire, c-Si, etc.) have nonnegligible absorption coefficients. Moreover, both models and practical experiments show that the substrate thickness (associated, of course, to its absorption) has a meaningful effect on the transmission measures. It seems, thus, that more comprehensive models that take into account the substrate thickness and its absorption could represent more accurately the reality and, in the estimation problem, could provide better guesses of the optical parameters.

5. More complex models for transmission and reflection can involve averaging of the responses over ranges of wavelengths (or thicknesses, or incidence angles, or polarization angles). The oscillatory behavior of light causes large variations of the functions to be integrated over small ranges of the integration parameter. Therefore, numerical integration procedures are very expensive (they demand thousands of integration points) and analytical methods are generally impossible. The practical consequence is that optimization procedures based on merit functions that involve those models could be very slow. On one hand, this means that the development of fast and accurate problem-oriented quadrature methods is necessary. On the other hand, optimization algorithms oriented to "expensive functions" must be used, and many methods exploiting the special features of our problems are yet to be discovered.

7. All the inverse problems considered in this work are "global optimization problems" in the sense that we are interested in the absolute minimizer of the merit functions and not, merely, in a local minimizer. However, these problems are not convex and, generally, possess many local minimizers and stationary points. Algorithms for global minimization have been devised in the last 30 years, since this is, probably, the most important problem in the broad area of optimization. However, no algorithm can guarantee to find global minima in reasonable time for a reasonably large scope of functions. Moreover, there is not a sufficiently comprehensive theory that explains the behavior of global optimization methods and serves as a guide for the invention of new ones. It is not surprising, thus, that we generally use "local" optimization methods for solving global optimization problems, with the hope that they will work, perhaps with the help of judicious or repetitive choices of initial points. It is worth mentioning that some successful optimization algorithms consist of strategies for changing initial points, after finding probable local minimizers using local optimization. In any case, when one is in the presence of a practical problem, useful global optimization procedures are the ones that exploit the problem characteristics. To a large extent, this has yet to be done in many of our estimation problems. There is still a paradox in the consideration of this item. In inverse problems, it is not always true that the global optimizer of the "obvious" merit function is the best estimate. If the problem is very ill-conditioned and the data are subject to errors, the "true solution" can have no meaning at all. The art of finding a good merit function (by means of, perhaps, regularization techniques) is an independent problem that has been exhaustively studied in many particular engineering problems.

8. In filter design problems, we require a certain behavior of the reflectance or the transmittance over a relative large range of wavelengths. Usually, we deal with that requirement by means of an ad hoc objective (merit) function. For example, if we want small reflectance over the range $[\lambda_{min}, \lambda_{max}]$, we minimize the sum of reflectance over that range. However, this approximation does not reflect exactly our requirement: a reflectance smaller than (say) 0.1% can be perfectly satisfactory, and it is not worthwhile to waste "optimization effort" in improving it if such a goal has been attained. A more adequate model could be to express the requirements as a set of inequalities such as (say) "Reflectance <0.1% for "all" λ in $[\lambda_{min}, \lambda_{max}]$." In this case, the design procedure does not require unconstrained or simple-constrained optimization as before, but rather nonlinearly constrained optimization (with many constraints!). Moreover, the constraints are not absolutely independent since, obviously, the response at some wavelengths is related to the response at their neighbors. Finally, the variations of the response with respect to the wavelength make it uncertain how many (and which) wavelengths must be involved in the optimization. All these considerations lead to more complex models than the ones considered up to now, at least, under the optimization framework. They represent in a closer way the practical requirements of calculations, but it is not clear whether their practical implementation is invalidated by their complexity, with the present state of computational architectures.

10. Practical designs of filters usually include "sensitivity requirements." We wish not only a given response for given wavelengths but also some stability with respect to thickness variations, since it is not realistic to assume that films with arbitrary mathematical thickness precision could be synthesized. It is relatively easy to analyze the sensitivity of a given solution, but the challenging question is to incorporate the sensitivity requirement in the original problem, in a correct mathematical-modelling framework and in a realistic way from the practical point of view.

11. A broad interface field between numerical mathematics and the engineering electronic devising field that deserves a great deal of further exploiting is "solar-cell optimization." The direct-problem mathematics related to modelling the incidence of solar rays on this type of devices seems to be sufficiently understood, but there are many optimization problems that need to be solved in practice. Solar rays do not fall vertically and with uniform intensity during all day, all year, and in different regions of the world. Incorporating these real features into models aiming to optimize materials, dimensions, and angles could be of maximal practical and economical significance.

Acknowledgment

This work was partially supported by the Brazilian agencies: Fundação de Amparo à Pesquisa do Estado de São Paulo (FAPESP) and Conselho Nacional de Desenvolvimento Científico e Tecnológico (CNPq).

REFERENCES

1. I. Newton, in "Optics." Great Books, Vol. 34. Encyclopedia Britannica, Chicago, 1952.

2. C. Huygens, in "Treatise on Light." Great Books, Vol. 34. Encyclopedia Britannica, Chicago, 1952.

3. L. Euler, in "Lettres a une Princesse d'Allemagne," pp. 86–100. Courcier Editeurs, Paris, 1812.

4. T. Young, "A Course of Lectures on Natural Philosophy and the Mechanical Arts." Johnson, London, 1807; Johnson, New York, 1971.

5. J. F. W. Herschel, "Natural Philosophy," p. 258. A. & R. Spottiswoode, London, 1831.

6. C. Coulston Gillespie, Ed., "A Diderot Pictorial Encyclopaedia of Trades and Industry," Plates 255 and 256. Dover Publications, New York, 1959.

7. A. N. Tikhonov and V. Ya. Arsenin, "Solution of Ill-Posed Problems." Winston-Wiley, New York, 1977.

8. M. Born and E. Wolf, "Principles of Optics," 6th ed. Pergamon Press, United Kingdom, 1993.

9. O. S. Heavens, "Optical Properties of Thin Solid Films." Dover Publications, New York, 1991.

10. R. A. Smith, "Semiconductors," 2nd ed. Cambridge University Press, London, 1979.

11. M. L. Cohen and J. Chelikowsky, "Electronic Structure and Optical Properties of Semiconductors," 2nd ed. Springer Series on Solid State Science, Vol. 75. Springer-Verlag, Berlin, Heidelberg, 1989.

12. P. T. Yu and M. Cardona, "Fundamentals of Semiconductors." Springer-Verlag, Berlin, Heidelberg, 1996.

13. S. H. Wemple and M. DiDomenico, Jr., *Phys. Rev. B* 3, 1338 (1971).

14. G. D. Cody, in "Semiconductors and Semimetals," Vol. 21, Part B, (J. Pankove, Ed.). Academic Press, New York, 1984.

15. A. R. Zanatta, M. Mulato, and I. Chambouleyron, *J. Appl. Phys.* 84, 5184 (1998).

16. F. Abeles, *Rev. Opt.* 32, 257 (1953); also *J. Opt. Soc. Am.* 47, 473 (1957).

17. H. M. Liddell, "Computer-Aided Techniques for the Design of Multilayer Filters." Adam Hilger, Bristol, UK, 1980.

18. H. A. Macleod, "Thin Film Optical Filters." Adam Hilger, Bristol, UK, 1969.

19. A. Friedlander, J. M. Martínez, and S. A. Santos, *Appl. Math. Optim.* 30, 235 (1994).

20. www.ime.unicamp.br/~martinez

21. A. V. Tikhonravov, M. K. Trubetskov, and G. W. DeBell, *Appl. Optics* 35, 5493 (1996).

22. See *Tech. Dig. Ser.-Opt. Soc. Am.* 9, 236 (1998).

23. A. Friedlander and J. M. Martínez, *SIAM J. Optim.* 4, 177 (1994).

24. R. H. Bielschowsky, A. Friedlander, F. M. Gomes, J. M. Martínez, and M. Raydan, *Investigación Operativa* 7, 67 (1998).

25. J. E. Dennis and R. B. Schnabel, "Numerical Methods for Unconstrained Optimization and Nonlinear Equations." Prentice-Hall, Englewood Cliffs, NJ, 1983.

26. O. S. Heavens, "Thin Film Physics." Methuen, London, 1970.

27. G. Hass, Ed., "Physics of Thin Films." Academic Press, New York, 1960.

28. Z. Knittl, "Optics of Thin Films." Wiley, Czechoslovakia, 1976.

29. L. Eckertova, "Physics of Thin Films," 2nd ed. Plenum, New York, 1986.

30. M. Ward, "The Optical Constants of Bulk Materials and Films." Institute of Physics, UK, 1994.

31. R. L. Mooney, *J. Opt. Soc. Am.* 35, 574 (1945).

32. Commission of the European Communities, Document EUR 6423 (1979).

33. A. M. Goodman, *Appl. Opt.* 17, 2779 (1978).

34. J. C. Manifacier, J. Gasiot, and J. P. Fillard, *J. Phys. E: Sci. Instrum.* 9, 1002 (1976).

35. R. Swanepoel, *J. Phys. E: Sci. Instrum.* 16, 1214 (1983); 17, 896 (1984).

36. I. Chambouleyron, J. M. Martinez, A. C. Moretti, and M. Mulato, *Appl. Opt.* 36, 8238 (1997).

37. I. Chambouleyron, J. M. Martinez, A. C. Moretti, and M. Mulato, *Thin Solid Films* 317, 133 (1998).

38. R. B. Murtagh and M. A. Saunders, MINOS User's Guide, Report SOL 77-9, Department of Operations Research, Stanford University, CA, 1977.

39. R. B. Murtagh and M. A. Saunders, *Math. Program.* 14, 41 (1978).

40. E. G. Birgin, I. Chambouleyron, and J. M. Martínez, *J. Comp. Phys.* 151, 862 (1999).

41. M. Raydan, *SIAM J. Optim.* 7, 26 (1997).

42. E. G. Birgin, Ph.D. Thesis, IMECC-UNICAMP, Brazil, 1998. [In Portuguese.]

43. M. Mulato, I. Chambouleyron, E. G. Birgin, and M. Martínez, *Appl. Phys. Lett.* 77, 2133 (2000).

Index